Handbook of Whale Optimization Algorithm

Handbook of Whale Optimization Algorithm

Variants, Hybrids, Improvements, and Applications

Edited by

Seyedali Mirjalili

Academic Press is an imprint of Elsevier
125 London Wall, London EC2Y 5AS, United Kingdom
525 B Street, Suite 1650, San Diego, CA 92101, United States
50 Hampshire Street, 5th Floor, Cambridge, MA 02139, United States

ISBN: 978-0-323-95365-8

For information on all Academic Press publications
visit our website at https://www.elsevier.com/books-and-journals

Publisher: Mara E. Conner
Acquisitions Editor: Chris Katsaropoulos
Editorial Project Manager: Tom Mearns
Production Project Manager: Fahmida Sultana
Cover Designer: Mark Rogers

Typeset by VTeX

Contents

14. U-WOA: an unsupervised whale optimization algorithm based deep feature selection method for cancer detection in breast ultrasound images

Payel Pramanik, Rishav Pramanik, Anurup Naskar, Seyedali Mirjalili, and Ram Sarkar

15. Constraint optimization: solving engineering design problems using Whale Optimization Algorithm (WOA)

Syed Kumayl Raza Moosavi, Malik Naveed Akhter, Muhammad Hamza Zafar, and Majad Mansoor

16. F-WOA: an improved whale optimization algorithm based on Fibonacci search principle for global optimization

Saroj Kumar Sahoo, Salpa Reang, Apu Kumar Saha, and Sanjoy Chakraborty

List of contributors

Mohmad Hussein Abdalla, Department of Computer Science, University of Raparin, Sulaimani, Iraq

Abdelaziz A. Abdelhamid, Department of Computer Science, Faculty of Computer and Information Sciences, Ain Shams University, Cairo, Egypt

Laith Abualigah, Computer Science Department, Al al-Bayt University, Mafraq, Jordan
Hourani Center for Applied Scientific Research, Al-Ahliyya Amman University, Amman, Jordan
MEU Research Unit, Middle East University, Amman, Jordan
Department of Electrical and Computer Engineering, Lebanese American University, Byblos, Lebanon
School of Computer Sciences, Universiti Sains Malaysia, Pulau Pinang, Malaysia
School of Engineering and Technology, Sunway University Malaysia, Petaling Jaya, Malaysia

Mahmood Ahmadi, Department of Computer Engineering and Information Technology, Razi University, Kermanshah, Iran

Aram M. Ahmed, Department of Information Technology, College of Science and Technology, University of Human Development, Sulaimani, Iraq
Department of Information Technology, Kurdistan Institution for Strategic Studies and Scientific Research, Sulaimani, Iraq

Adedotun T. Ajibare, Faculty of Information and Communications Technology, Rosebank College, Cape Town, South Africa

Malik Naveed Akhter, School of Mechanical and Manufacturing Engineering, National University of Sciences and Technology, Islamabad, Pakistan

Lateef A. Akinyemi, Department of Electronic and Computer Engineering, Lagos State University, Epe, Nigeria

Ali Ala, Industrial Engineering & Management, Shanghai Jiao Tong University, Shanghai, China

Mohammed Azmi Al-Betar, Artificial Intelligence Research Center (AIRC), College of Engineering and Information Technology, Ajman University, Ajman, United Arab Emirates
Department of Information Technology, Al-Huson University College, Al-Balqa Applied University, Irbid, Jordan

Nabeel Salih Ali, Information Technology Research and Development Center (ITRDC), University of Kufa, Najaf, Iraq

Zaid Abdi Alkareem Alyasseri, Information Technology Research and Development Center (ITRDC), University of Kufa, Najaf, Iraq
Institute of Informatics and Computing in Energy, College of Computing and Informatics, Universiti Tenaga Nasional, Kajang, Selangor, Malaysia
Department of Business Administration, College of Administrative and Financial Sciences, Imam Ja'afar Al-Sadiq University, Baghdad, Iraq

Erfan Amini, Department of Civil, Environmental and Ocean Engineering, Stevens Institute of Technology, Hoboken, NJ, United States

Bahman Arasteh, Department of Computer Engineering, Faculty of Engineering and Natural Science, Istinye University, Istanbul, Turkey

K. Ashwini, Department of Electronics and Communication Engineering, National Institute of Technology, Warangal, India

Mehmet Enes Avcu, Unmanned Systems Engineering Department, Titra Technology, Ankara, Turkey

Mohammed A. Awadallah, Department of Computer Science, Al-Aqsa University, Gaza, Palestine
Artificial Intelligence Research Center (AIRC), Ajman University, Ajman, United Arab Emirates

Mahdi Azizi, Department of Civil Engineering, University of Tabriz, Tabriz, Iran

Ziad Azzi, Department of Civil and Environmental Engineering, Florida International University, Miami, FL, United States

E. Balasubramanian, Centre for Autonomous System Research, Department of Mechanical Engineering, Vel Tech Rangarajan Dr Sagunthala R&D Institute of Science and Technology, Chennai, India

Mahla Basiri, Department of Civil Engineering, University of Tabriz, Tabriz, Iran

Abdelbaki Benayad, LIMOSE Laboratory, University of M'Hamed Bougara Boumerdes, Boumerdes, Algeria

Diptendu Bhattacharya, National Institute of Technology, Agartala, Jirania, Tripura, India

Amel Boustil, LIMOSE Laboratory, University of M'Hamed Bougara Boumerdes, Boumerdes, Algeria

Seyed Mostafa Bozorgi, Department of Computer Engineering, North Tehran Branch, Islamic Azad University, Tehran, Iran

Malik Braik, Department of Computer Science, Al-Balqa Applied University, As-Salt, Jordan

Sanjoy Chakraborty, Department of Computer Science and Engineering, National Institute of Technology Agartala, Agartala, Tripura, India
Department of Computer Science and Engineering, Iswar Chandra Vidyasagar College, Belonia, Tripura, India

Amit Chhabra, Department of Computer Engineering and Technology, Guru Nanak Dev University, Amritsar, India

Yashar Dadrasajirlou, Civil Engineering, Semnan University, Semnan, Iran

Lamees Mohammad Dalbah, Artificial Intelligence Research Center (AIRC), College of Engineering and Information Technology, Ajman University, Ajman, United Arab Emirates

Leandro S.P. da Silva, School of Mechanical Engineering, University of Adelaide, Adelaide, SA, Australia

Marwa M. Eid, Faculty of Artificial Intelligence, Delta University for Science and Technology, Mansoura, Egypt

Serdar Ekinci, Department of Computer Engineering, Batman University, Batman, Turkey

Stephen O. Ekwe, Department of Electrical, Electronic and Computer Engineering, Cape Peninsula University of Technology, Cape Town, South Africa

Hossien B. Eldeeb, Department of Electrical and Electronics Engineering, Özyeğin University, Istanbul, Turkey

El-Sayed M. El-kenawy, Department of Communications and Electronics, Delta Higher Institute of Engineering and Technology, Mansoura, Egypt

Nava Eslami, Department of Computer Engineering, North Tehran Branch, Islamic Azad University, Tehran, Iran

Amir Faraji, Construction Management Department, Faculty of Architecture, KHATAM University, Tehran, Iran
School of Engineering, Design and Built Environment, Western Sydney University, Sydney, NSW, Australia

Sepehr Faridmarandi, Department of Civil and Environmental Engineering, Florida International University, Miami, FL, United States

Mansoureh Shahabi Ghahfarokhi, Department of Civil and Environmental Engineering, Florida International University, Miami, FL, United States

Farhad Soleimanian Gharehchopogh, Department of Computer Engineering, Urmia Branch, Islamic Azad University, Urmia, Iran

Mojtaba Ghasemi, Department of Electronics and Electrical Engineering, Shiraz University of Technology, Shiraz, Iran

Tripti Goel, Department of Electronics and Communication Engineering, National Institute of Technology Silchar, Assam, India

Harun Gökçe, Faculty of Technology, Department of Industrial Design Engineering, Gazi University, Ankara, Turkey

Shahzaib Farooq Hadi, National University of Sciences and Technology, Islamabad, Pakistan

Hozan K. Hamarashid, Department of Information Technology, Kurdistan Institution for Strategic Studies and Scientific Research, Sulaimani, Iraq

Saad Harous, Department of Computer Science, College of Computing and Informatics, University of Sharjah, Sharjah, United Arab Emirates

Bryar A. Hassan, Department of Computer Science, College of Science, Charmo University, Sulaimani, Iraq

Abdelhameed Ibrahim, Computer Engineering and Control Systems Department, Faculty of Engineering, Mansoura University, Mansoura, Egypt

Gültekin Işık, Department of Computer Engineering, Igdir University, Igdir, Turkey

Saadat Izadi, Department of Computer Engineering and Information Technology, Razi University, Kermanshah, Iran

Davut Izci, Department of Computer Engineering, Batman University, Batman, Turkey

Jafar Jafari-Asl, Department of Civil Engineering, Faculty of Engineering, University of Sistan and Baluchestan, Zahedan, Iran

Norziana Jamil, Institute of Informatics and Computing in Energy, College of Computing and Informatics, Universiti Tenaga Nasional, Kajang, Selangor, Malaysia

Ravi Kumar Jatoth, Department of Electronics and Communication Engineering, National Institute of Technology, Warangal, India

Soleiman Kadkhoda Mohammadi, Department of Electrical and Electronic Engineering, Boukan Branch, Islamic Azad University, Boukan, Iran

Hojat Karami, Civil Engineering, Semnan University, Semnan, Iran

Okba Kazar, University of Kalba, Sharjah, United Arab Emirates

Muhammad Najeeb Khan, School of Mechanical Engineering, Shri Mata Vaishno Devi University, Katra, Jammu & Kashmir, India

Mohammad Reza Khayyambashi, Faculty of Computer Engineering, University of Isfahan, Isfahan, Iran

Mohamed Akram Khelili, Department of Computer Science, University of Biskra, Biskra, Algeria
Numidia Institute of Technology, Algies, Algeria

Nima Khodadadi, Department of Civil and Architectural Engineering, University of Miami, Coral Gables, FL, United States

Ouajdi Korbaa, Laboratory MARS, LR17ES05, ISITCom, University of Sousse, Sousse, Tunisia
ISITCom, University of Sousse, Sousse, Tunisia

Imene Latreche, Department of Computer Science, University of Biskra, Biskra, Algeria

Jaffer Majidpour, Department of Computer Science, University of Raparin, Sulaimani, Iraq

Parijata Majumdar, National Institute of Technology, Agartala, Jirania, Tripura, India

Sharif Naser Makhadmeh, Artificial Intelligence Research Center (AIRC), College of Engineering and Information Technology, Ajman University, Ajman, United Arab Emirates

Ghaith Manita, Laboratory MARS, LR17ES05, ISITCom, University of Sousse, Sousse, Tunisia
ESEN, University of Manouba, Manouba, Tunisia

Majad Mansoor, Dept. of Automation, University of Science and Technology of China, Hefei, China

Yassine Meraihi, LIST Laboratory, University of M'Hamed Bougara Boumerdes, Boumerdes, Algeria
Systems Engineering and Telecommunications Laboratory, University of Boumerdes, Boumerdes, Algeria

Farid MiarNaeimi, Faculty of Engineering, Department of Civil Engineering, University of Sistan and Baluchestan, Zahedan, Iran

Seyedali Mirjalili, Centre for Artificial Intelligence Research and Optimisation, Torrens University Australia, Brisbane, QLD, Australia
University Research and Innovation Center, Obuda University, Budapest, Hungary
Yonsei Frontier Lab, Yonsei University, Seoul, South Korea

Seyedeh Zahra Mirjalili, Centre for Artificial Intelligence Research and Optimisation, Torrens University Australia, Brisbane, QLD, Australia

Seyed Mohammad Mirjalili, Department of Engineering Physics, Polytechnique Montréal, Montreal, QC, Canada

Sanjoy Mitra, Tripura Institute of Technology, Agartala, Narsingarh, Tripura, India

Ali Mohammadzadeh, Department of Computer Engineering, Shahindezh Branch, Islamic Azad University, Shahindezh, Iran

Naufel B. Mohammed, Department of Information Technology, Kurdistan Institution for Strategic Studies and Scientific Research, Sulaimani, Iraq

Diab Mokeddem, Department of Electrical Engineering, Faculty of Technology, University of Ferhat Abbas Setif-1, Setif, Algeria

Ahmadreza Montazerolghaem, Faculty of Computer Engineering, University of Isfahan, Isfahan, Iran

Syed Kumayl Raza Moosavi, School of Electrical Engineering and Computer Science, National University of Sciences and Technology, Islamabad, Pakistan
National University of Sciences and Technology, Islamabad, Pakistan

R. Murugan, Department of Electronics and Communication Engineering, National Institute of Technology Silchar, Assam, India

Z. Mustaffa, Universiti Malaysia Pahang, Pekan, Pahang, Malaysia

Mohammad H. Nadim-Shahraki, Faculty of Computer Engineering, Najafabad Branch, Islamic Azad University, Najafabad, Iran

Mahdieh Nasiri, Department of Mechanical Engineering, Stevens Institute of Technology, Hoboken, NJ, United States

Anurup Naskar, Department of Computer Science and Engineering, Jadavpur University, Kolkata, India

Dallel Nasri, Department of Electrical Engineering, Faculty of Technology, University of Ferhat Abbas Setif-1, Setif, Algeria

Hathiram Nenavath, Department of Electronics and Communication Engineering, National Institute of Technology, Jamshedpur, India

Mehdi Neshat, Centre for Artificial Intelligence Research and Optimisation, Torrens University Australia, Brisbane, QLD, Australia

Kaniaw A. Noori, Database Department, Sulaimani Polytechnic University, Sulaimani, Iraq
Department of Database Technology, Technical College of Informatics, Sulaimani Polytechnic University, Sulaimani, Iraq

Maha Nssibi, Laboratory MARS, LR17ES05, ISITCom, University of Sousse, Sousse, Tunisia
ENSI, University of Manouba, Manouba, Tunisia

Sima Ohadi, Department of Civil Engineering, Faculty of Engineering, University of Sistan and Baluchestan, Zahedan, Iran

Sunday O. Oladejo, School for Data Science and Computational Thinking, Stellenbosch University, Stellenbosch, South Africa

D.N. Kiran Pandiri, Department of Electronics and Communication Engineering, National Institute of Technology Silchar, Assam, India

Payel Pramanik, Department of Computer Science and Engineering, Jadavpur University, Kolkata, India

Rishav Pramanik, Department of Computer Science and Engineering, Jadavpur University, Kolkata, India

Shko M. Qader, Database Department, Sulaimani Polytechnic University, Sulaimani, Iraq
Department of Computer Science, Kurdistan Technical Institute, Sulaimani, Iraq
Department of Information Technology, Computer Science Institute, Sulaimani Polytechnic University, Sulaimani, Iraq
Department of Information Technology, University College of Goizha, Sulaimani, Iraq

Mahdi Rahbar, Department of Computer Science, Saint Louis University, St. Louis, MO, United States

Chnoor Maheadeen Rahman, Department of Computer Science, College of Science, Charmo University, Sulaimani, Iraq

R.R. Rajalaxmi, Department of CSE, Kongu Engineering College, Perundurai, India

D. Rajamani, Centre for Autonomous System Research, Department of Mechanical Engineering, Vel Tech Rangarajan Dr Sagunthala R&D Institute of Science and Technology, Chennai, India

Awf Abdulrahmam Ramadhan, Public Health Department, College of Health and Medical Techniques - Shekhan, Duhok Polytechnic University, Duhok, Iraq

Amar Ramdane-Cherif, LISV Laboratory, University of Versailles St-Quentin-en-Yvelines, Velizy, France
Systems Engineering Laboratory of Versailles (LISV), University of Paris-Saclay, Velizy, France

Tarik A. Rashid, Department of Computer Science and Engineering, School of Science and Engineering, University of Kurdistan Hewler, Erbil, KRI, Iraq

Salpa Reang, Department of Mathematics, National Institute of Technology Agartala, Agartala, Tripura, India

Souad Refas, LIST Laboratory, University of M'Hamed Bougara Boumerdes, Boumerdes, Algeria

C. Roopa, Department of CSE, Kongu Engineering College, Perundurai, India

Apu Kumar Saha, Department of Mathematics, National Institute of Technology Agartala, Agartala, Tripura, India

Ashim Saha, Department of Computer Science and Engineering, National Institute of Technology Agartala, Agartala, Tripura, India

İsmail Şahin, Faculty of Technology, Department of Industrial Design Engineering, Gazi University, Ankara, Turkey

Saroj Kumar Sahoo, Department of Mathematics, National Institute of Technology Agartala, Agartala, Tripura, India

Taybeh Salehnia, Department of Computer Engineering and Information Technology, Razi University, Kermanshah, Iran

Ram Sarkar, Department of Computer Science and Engineering, Jadavpur University, Kolkata, India

Nataliia Y. Sergiienko, School of Mechanical Engineering, University of Adelaide, Adelaide, SA, Australia

Alaa Sheta, Computer Science Department, Southern Connecticut State University, New Haven, CT, United States

Milad Baghalzadeh Shishehgarkhaneh, Department of Civil Engineering, Islamic Azad University of Tabriz, Tabriz, Iran
Department of Construction Management, Islamic Azad University of Tabriz, Tabriz, Iran
Department of Civil Engineering, Faculty of Engineering, Monash University, Clayton, VIC, Australia

Haval Sidqi, Department of Database Technology, Technical College of Informatics, Sulaimani Polytechnic University, Sulaimani, Iraq

Amit Kumar Sinha, School of Mechanical Engineering, Shri Mata Vaishno Devi University, Katra, Jammu & Kashmir, India

M. Siva Kumar, Centre for Autonomous System Research, Department of Mechanical Engineering, Vel Tech Rangarajan Dr Sagunthala R&D Institute of Science and Technology, Chennai, India

Sihem Slatnia, Department of Computer Science, University of Biskra, Biskra, Algeria

K. Sruthi, Department of IT, Kongu Engineering College, Perundurai, India

M.H. Sulaiman, Universiti Malaysia Pahang, Pekan, Pahang, Malaysia

Sylia Mekhmoukh Taleb, LIST Laboratory, University of M'Hamed Bougara Boumerdes, Boumerdes, Algeria
Systems Engineering and Telecommunications Laboratory, University of Boumerdes, Boumerdes, Algeria

Noor Tayfor, Department of Computer Science, Kurdistan Technical Institute, Sulaimani, Iraq

R. Thangarajan, Department of IT, Kongu Engineering College, Perundurai, India

Satılmış Ürgün, Faculty of Aeronautics and Astronautics, Kocaeli University, Kocaeli, Türkiye

Selma Yahia, LIST Laboratory, University of M'Hamed Bougara Boumerdes, Boumerdes, Algeria
Systems Engineering and Telecommunications Laboratory, University of Boumerdes, Boumerdes, Algeria

Samaneh Yazdani, Department of Computer Engineering, North Tehran Branch, Islamic Azad University, Tehran, Iran

Halil Yiğit, Department of Information Systems Engineering, Kocaeli University, Kocaeli, Türkiye

Hassaan Bin Younis, National University of Sciences and Technology, Islamabad, Pakistan

Muhammad Hamza Zafar, Department of Engineering Sciences, University of Agder, Grimstad, Norway

Mohsen Zare, Department of Electrical Engineering, Faculty of Engineering, Jahrom University, Jahrom, Iran

Farouq Zitouni, Department of Computer Science, Kasdi Merbah University, Ouargla, Algeria

Preface

Welcome to "Whale Optimization Algorithm: Variants, Improvements, Hybrids, and Applications." In the ever-evolving world of Artificial Intelligence, optimization algorithms play a crucial role in solving complex problems across various domains. Among the plethora of meta-heuristic techniques, the Whale Optimization Algorithm (WOA) has emerged as one of the most well-regarded and widely-used approaches.

The WOA has garnered significant attention due to its effectiveness in addressing optimization problems, both in scientific research and industrial applications. However, harnessing the full potential of this algorithm requires tackling numerous challenges. These challenges range from dealing with multiple objectives and constraints to handling binary decision variables, large-scale search spaces, dynamic objective functions, and noisy parameters, to name just a few.

This handbook aims to provide you with an in-depth analysis of the Whale Optimization Algorithm and the existing methods in the literature that address these challenges. We delve into the fundamental concepts and principles behind the WOA, exploring its strengths and limitations. By reviewing the extensive body of literature surrounding this algorithm, we offer insights into its applications across various domains.

But this book goes beyond a comprehensive analysis of the WOA. It also presents a collection of improvements, variants, and hybrids that have been developed to enhance its performance and overcome specific challenges. These novel approaches push the boundaries of the algorithm, offering new possibilities for optimization in different contexts. By incorporating these advancements, you will be equipped with a broader toolkit to tackle complex optimization problems.

Furthermore, this handbook showcases a range of real-world applications that demonstrate the practical applicability of the methods presented within. From engineering to finance, from healthcare to logistics, the WOA finds its place in diverse domains, addressing critical challenges and delivering valuable solutions.

I would like to express my gratitude to the researchers, practitioners, and enthusiasts who have contributed to the development and understanding of the Whale Optimization Algorithm. Without their dedication and expertise, this book would not have been possible.

I hope that this handbook serves as a valuable resource for both researchers and practitioners in the field of optimization. Whether you are a seasoned expert or a newcomer to the domain, we believe that the insights and methodologies presented here will inspire new ideas, facilitate problem-solving, and foster innovation.

Enjoy your journey into the world of the Whale Optimization Algorithm, where the search for optimal solutions meets the vastness of possibility.

Seyedali Mirjalili
21/05/2023

Chapter 1

Presenting appointment scheduling with considering whale optimization algorithm in healthcare management

Ali Ala[a] and Seyedali Mirjalili[b,c]

[a]*Industrial Engineering & Management, Shanghai Jiao Tong University, Shanghai, China,* [b]*Centre for Artificial Intelligence Research and Optimisation, Torrens University Australia, Brisbane, QLD, Australia,* [c]*University Research and Innovation Center, Obuda University, Budapest, Hungary*

1.1 Introduction

Healthcare operations now attract much attention from healthcare optimization challenges to provide more convenient help at a reduced cost. A scheduling system may reduce patient wait times, make it simpler to provide healthcare services, and improve the efficiency of the healthcare framework. However, there may be a patient-life threat, overworked staff, patient disease incidence, and patient scheduling overload due to expanding medical care demands and its absence or unavailability. The objectives of such scheduling can be categorized into optimizing patient pleasure, lowering waiting times, maximizing equality policies, and reducing service costs in the healthcare industry. A strategy entitled health scheduling optimizes accessibility to medical facilities efficiently. Appointment scheduling is obviously focused on ensuring patient satisfaction. Also, appointment scheduling is addressed in different services for first consultations, personal visits, and elective surgeries in various aspects. Chen et al. [1] have concentrated on a few well metaheuristics whose binary variants can be effectively modified to solve the problems of feature selection and whose parts can also be enhanced. Finding the best solution to a particular problem while complying with extremely complex restrictions is a common requirement in optimization problems. Applying artificial intelligence and the principle of queues can also be used to resolve other scheduling issues in healthcare, such as waiting times. Ning and Cao [2] divided task allocations and task sequencing planning into two components while tackling the inter-combined operation challenge. Wang and Fung [3] addressed a problem model that considered the path cost and program execution capability before employing a WOA to solve it. A variety of applications for optimization and artificial networks in healthcare research were represented by Reuter and Kuhl [4]. They applied these optimization approaches to the variety of scheduling techniques for healthcare, including convolutional neural networks (CNN), artificial neural networks (ANN), particle swarm optimization (PSO), and whale optimization algorithms (WOA). Moreover, the fairness of healthcare scheduling is critical to ensure equitable forecasts for healthcare operations, and optimization models help explain, audit, and reduce this algorithmic bias. Researchers have highlighted a variety of worries about fairness and optimization in healthcare operations due to the high costs aspect of the health system. According to this concept of fairness, a patient who shares many features with another should not receive less favorable clinical treatment. An essential requirement is using some concept of similarity to apply comparable optimization to the same patients. Recovering individual scheduling interests, which means that every doctor has a different preferred schedule, was given by Zhan et al. [5] as a further crucial issue on fairness. Additionally, Güler and Geçici [6] discussed optimization models that have the capacity to justify regression on continuously dependent variable, such as in the case of heath care costs. Aladwani [7] suggested comparing estimates for healthcare costs for people with health issues and factors about substance usage to patients without these issues. The number of iterations from a learned model and the sensitive values group must be separate. Numerous researchers have examined the analysis of health services and optimization techniques with great success. Kang et al. [9] examined the most recent findings in research on relevant optimization problems and evolutionary algorithms and frameworks in the healthcare operation systems. Considering the probability distributions for the healthcare system and processing times, Ferreira and Vasconcelos [10] presented various evolutionary optimizations defined as main solution strategies to reduce the overall cost of patients waiting and physicians' idle time.

The Whale Optimization Algorithm (WOA) is a candidate of a series of swarm intelligence optimization techniques that were motivated by humpback whales' hunting strategies. It was designed and initiated by Mirjalili and Lewis [8],

Handbook of Whale Optimization Algorithm. https://doi.org/10.1016/B978-0-32-395365-8.00007-5

and it started to be used more frequently to solve various engineering optimization issues. The practical findings further demonstrated that WOA outperforms other approaches in terms of accuracy, high relative optima minimization, quick convergence time, and accurate balancing between exploitation and exploration. Mafarja and Mirjalili [11] implemented the Simulated Annealing (SA) algorithm as a searching technique to improve the WOA's ability to select the ideal subset of characteristics and increase classification performance. Kadam and Jadhav [12] presented a method for density minimization in neural networks that are further tuned with WOA to get the most significant possible space between the variables. For reducing the average delivery time and overall patient lengths of stay, the optimization model applies a heuristics approach to improve the scheduling of patient appointments. The patient waiting time and overall operation time have reduced by around 5%, according to the results. Sun et al. [13] submitted a proposal for more in-depth investigation in the research to build out strategies to optimize the number of patient appointments, reduce impacted patient waiting times, and boost patient satisfaction. Zhuang and Vincent [14] developed the response set applications to resolve the suggested combinatorial optimization issue that displayed a satisfactory evaluation employed in artificial intelligence. The chapter's primary goal is to disseminate a thorough summary of the relevant research papers on using WOA in healthcare operations to address challenging optimization issues. Additionally, the review highlights the study's findings regarding potential problems for the following investigators.

The remaining chapters are structured as follows. Section 1.2 provides a general description of the WOA's framework, while Section 1.3 classifies various healthcare operation methodologies. Section 1.4 gives a summary of WOA's application to various technical optimization issues. Several computations and methods using different models are examined in Section 1.5. In Section 1.6, the article's conclusion is presented.

1.2 Whale optimization algorithm

WOA is a computational intelligence method suggested for issues involving ongoing optimization. Aljarah et al. [15] demonstrated that this algorithm performs as well as or better than some existing algorithmic strategies. The WOA was motivated by the humpback whales' hunting habits. Each response in WOA is regarded as a whale. In this answer, a whale attempts to fill in a new location in the search area referenced as the group's greatest member. The whales utilize two different techniques to both attack and locate their prey. In the first, the prey is enclosed, while bubbles traps are made in the second. (See Fig. 1.1.)

WOA mimics the social behavior of humpback whales, beginning with a set of random solutions and using the three-phase encirclement and prey siege operation. The status of the search agents is updated in each iteration. In the prey siege phase, the whales assume that the ideal option, for now, is prey to surround the prey. The following equations are used to model this:

$$\vec{D} = \left|\vec{C}.\vec{X^*}(t) - X(t)\right| \tag{1.1}$$

$$\vec{X}(t+1)\vec{X^*}(t) - \vec{A}.\vec{D} \tag{1.2}$$

where t repeats the current, $\vec{D}$, $\vec{C}$ and coefficient vectors, $\vec{X^*}(t)$ is the best solution currently available, $\vec{X}$ is Place vector, $|\,|$ is the Absolute value point, (.) Multiplication is the element point in element, it may be mentioned that if a superior option exists, it needs to be updated in each $\vec{X^*}$ iteration. Vectors $\vec{A}$ and $\vec{C}$ are calculated with Eqs. (1.3) and (1.4):

$$\vec{A} = 2\vec{a}.\vec{r} - \vec{a} \tag{1.3}$$

$$\vec{C} = 2.\vec{r} \tag{1.4}$$

where $\vec{a}$ reduces linearly from two to zero during repetitions, and $\vec{r}$ is a random vector at intervals of zero to one. Humpback whales consider the bubbling net method of attack by swimming around the prey in a contractile ring and all at the same along the circular path depicted in Fig. 1.2.

For modeling this behavior, it anticipates that the whale makes a decision one of them with a 50% probability of contracting a siege mechanism or a spiral model. Therefore, the mathematical model of this step is defined as Eq. (1.5):

$$\vec{X}(t+1) = \begin{cases} \vec{X^*}(t+1) - \vec{A}.\vec{D} & \text{if } p < 0.5 \\ \vec{D}.e^{bt}.\cos(2\pi l) + \vec{X^*}(t) & \text{if } p \geq 0.5 \end{cases} \tag{1.5}$$

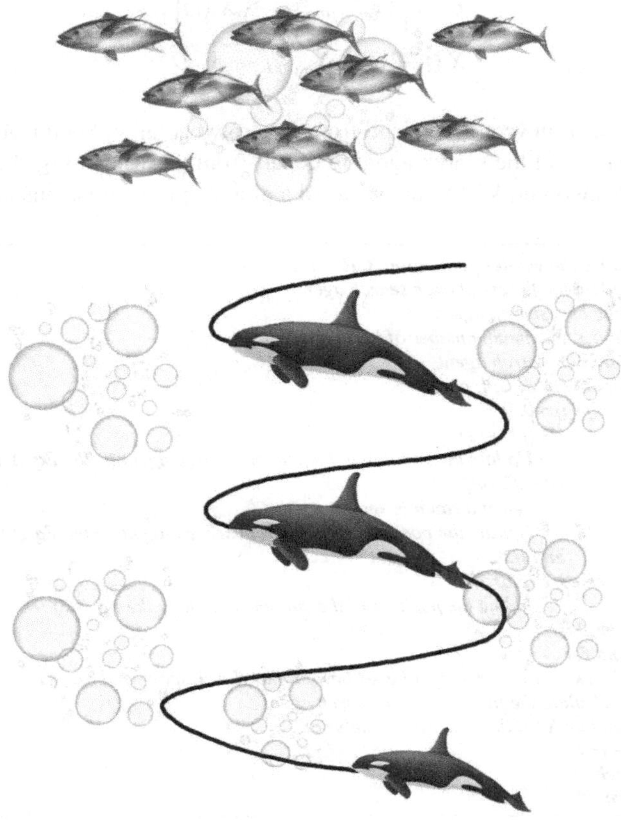

FIGURE 1.1 Creating bubble-net hunting manners by whales for the WOA approach.

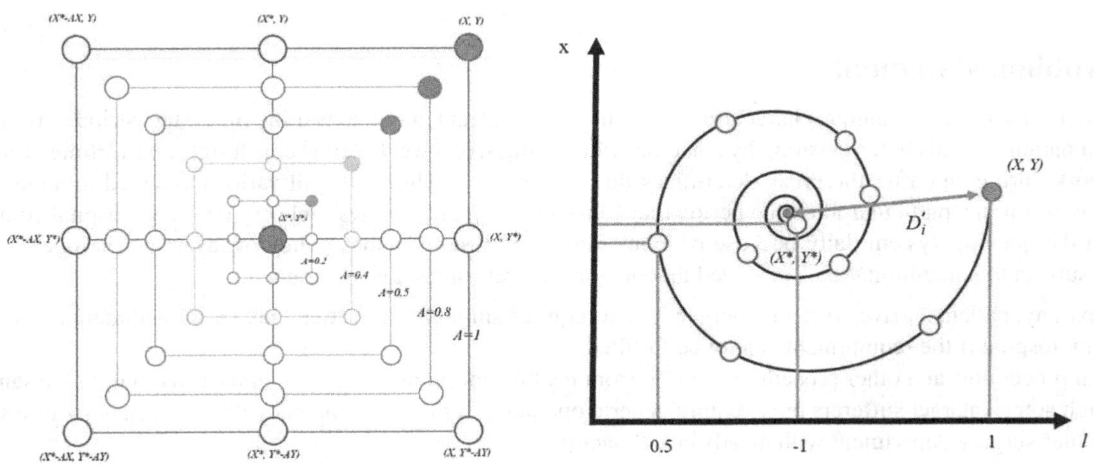

FIGURE 1.2 (a) Shrinking encircling mechanism and (b) spiral updating position of the mechanism of WOA [8].

where $\vec{D}$ is obtained from the relation $\left|\vec{C}.\vec{X}^*(t) - X(t)\right|$, and refers to the distance of the *i-th* whale to the prey, b is a constant coefficient for defining a logarithmic helical shape, l is a random number between -1 to +1 and the random number is P when P is between zero and one. Also, the random values $\vec{A}$ are between -1 to +1 that demonstrates how close the searching agent was to the target whale optimization algorithm. In the tracking for prey to update the position of the search agent, instead of operating the best tracking agent's data, the agent's random selection is considered so that its mathematical model is represented as Eqs. (1.6) and (1.7).

$$\vec{D} = \left| \vec{C} . \overrightarrow{X_{rand}}(t) - \vec{X}(t) \right| \quad (1.6)$$

$$\vec{X}(t+1) = \vec{X}_{rand} - \vec{A} . \vec{D} \quad (1.7)$$

where $\vec{X}_{rand}$ was a randomized location vector obtained from the crowd at large, Vector $\vec{A}$ is utilized with random values greater than +1 or less than -1 to compel the search agent to depart from the whale. Fig. 1.3 depicts the WOA algorithm's pseudo-code. From a theoretical viewpoint, WOA can be called a global optimizer because it contains discovery capability.

```
Initialize the whales population Xi (i = 1, 2, ..., n)
Calculate the fitness of each search agent
X*=the best search agent
while (t < maximum number of iterations)
   for each search agent
   Update a, A, C, l, and p
      if1 (p<0.5)
         if2 (|A| < 1)
            Update the position of the current search agent by the Eq. (1.1)
         else if2 (|A| ≥ 1)
            Select a random search agent (Xrand)
            Update the position of the current search agent by the Eq. (1.7)
         end if2
      else if1 (p≥0.5)
            Update the position of the current search by the Eq. (1.5)
   end if1
   end for
   Check if any search agent goes beyond the search space and amend it
   Calculate the fitness of each search agent
   Update X* if there is a better solution
   t=t+1
end while
return X*
```

FIGURE 1.3 Pseudo-code of the WOA algorithm based on various equations.

1.3 Problem statement

The outpatients have specific unique characteristics, such as minimal and maximal waiting times, the period of time between subsequent patient arrivals before visiting by a doctor, waiting times, and weekly time constraints on overtime. The minimal time on work during and after the break determines this timeframe. Additionally, all various assigned to none or several doctors may result in a particular idle time period that considers the patient or available options. The hospital admits happy patients in the queuing system daily because of many unoccupied rooms and a policy of fairness. The operations listed below are subject to some limitations, provided that doctors and resources are constrained.

1. The next day, patients arrive with an emergency that requires surgery. The patient can be immediately transferred to a different hospital if the requirement cannot be fulfilled.
2. Surgical procedures and other procedures—aside from health conditions, surgery not be carried out on the same day.
3. Although some cataract sufferers only require general operation in the clinic, others with emergency require surgery in a particular surgery department with an advanced machine.

1.4 Different method of WOA

The hybridizing algorithm is a general strategy for maximizing the benefits of two algorithms while minimizing their drawbacks. Combining these strategies has improved the outcomes using each technique separately and has worked effectively to solve the defined problem. We can finally improve the investigation and utilization of the technique by algorithm hybridization. Laskar et al. [16] suggested the Hybrid Whale-PSO (Particle Swarm Optimization) Method (HWPSO), a new neighborhood hybrid meta-heuristic algorithm to handle challenging optimization issues. The WOA algorithm has strong exploration capabilities; thus, the suggested approach makes an innovative effort to overcome the restrictions of a PSO evaluation stage that is successful when combined with a WOA.

1.5 Computational model

The objective function and constraints of the presented mixed integer linear programming approach are formulated as follows:

$$Min\ Z = \sum_{p=1}^{P}\sum_{o=1}^{O}\sum_{s=1}^{S}\sum_{t=1}^{T}\sum_{d=1}^{D} C_{pos} \times x_{postd} + \sum_{o=1}^{O}\sum_{d=1}^{D} EC_{od} \times \alpha_{od} + \sum_{o=1}^{O}\sum_{d=1}^{D} LC_{od} \times \beta_{od} + \sum_{d=1}^{D} U_d \times Z_d \quad (1.8)$$

$$\sum_{o=1}^{O}\sum_{s=1}^{S}\sum_{t=1}^{T}\sum_{d=1}^{D} x_{postd} = 1 \qquad \forall p \quad (1.9)$$

$$\overline{x}_{postd} = \sum_{t'=\max(t-du_p+1,1)}^{t} x_{post'd} \qquad \forall p, o, s, t, d \quad (1.10)$$

$$\sum_{\substack{p=1 \\ p\neq q}}^{P}\sum_{s=1}^{S}\sum_{t=t'}^{t'+du_q+cl_q-1} x_{postd} \leq 1 - \sum_{s=1}^{S} x_{qost'd} \qquad \forall q, o, t', d \quad (1.11)$$

$$\sum_{p=1}^{P}\sum_{s=1}^{S} \overline{x}_{postd} \leq 1 \qquad \forall o, t, d \quad (1.12)$$

$$\sum_{p=1}^{P}\sum_{o=1}^{O} \overline{x}_{postd} \leq 1 \qquad \forall s, t, d \quad (1.13)$$

$$\sum_{p=1}^{P}\sum_{o=1}^{O}\sum_{s=1}^{S}\sum_{t=1}^{T}\sum_{d=da_p+1}^{D} x_{postd} = 0 \quad (1.14)$$

The objective function of Eq. (1.8) is to decrease the cost of allocating patients to multiple sections and doctors, the costs of not using and overuse multiple rooms, and the overuse of an intensive care unit. Constraint (1.9) ensures that each patient (emergency and regular) is assigned to a specific department and doctor on a particular day and period. Limit (1.10) provides that each patient is allocated the number of time blocks required for time block health check. Constraint (1.11) ensures that at least cl_q of time is allocated for cleaning each department after visiting doctor for each patient. Constraint (1.12) specifies that only one operation is performed in each department in each time block in a day. Constraint (1.13) states that in each time block in a day, each surgeon can only perform one operation in one health department. Constraint (1.14) ensures that each patient is assigned to a specific department and surgeon before a particular time.

1.6 Solution approach

In order to test and analyze the WOA's performance thoroughly, experiment sets are undertaken in this section to demonstrate how well it performs while solving computational problems of various aspects. The testing findings were summarized in the last row of their report. The convergence approach is the best, and sensitivity analyses were performed on the investigation set to assess and contrast the convergence performance of the suggested WOA. By using MATLAB® 2019b programming language, all operations were fairly carried out in identical circumstances on a laptop running Windows 11 with a Core i7, 3.5 GHz CPU, and 16 GB of RAM.

1.7 Results analysis and discussion

In this section, we first verify the suggested MILP model using a numerical solution. The outcomes of a sensitivity analysis using the same scenario are then shared. We developed test scenarios on several scales to assess the effectiveness of our WOA. The WOA variables are set, and test scenario generation details are provided. To verify the suggested model, five physicians and 20 patients are given an instance lasting two days with 2 hours each. Table 1.1 contains the variables for this instance. When a patient has a scheduled appointment on a favorable day or hour, the punishment coefficient is set to 0. If not, it is equal to 100/10.

TABLE 1.1 Parameters used in the validation example for patients visiting based on hours and days.

		1	2	3	4	5	6	7	8	9	10
	j										
vp_j		0.8	0.8	0.5	0.8	0.8	0.6	0.5	0.6	0.7	0.8
	d										
PC_d^t	1	100	100	100	0	100	0	100	100	100	0
	5	100	100	0	100	100	100	0	0	0	100
	k										
PC_j^k	1	10	10	10	10	10	0	10	0	0	0
	5	10	10	0	0	10	10	0	10	10	10

TABLE 1.2 Comparison results of the WOA & Lingo in various problem instances.

size	instance number	days	doctors	urgent patients	regular patients	WOA		LINGO Solver	
						Best solution	process time (sec)	Best solution	process time (sec)
Small	1	1	5	10	20	10.41	1	10.41	14.13
	2	1	5	10	20	30.0	2	30.0	15.62
	3	1	5	10	20	33.33	3	33.33	15.88
	4	1	5	10	20	52.08	3	52.06	23.34
	5	1	5	10	20	55.83	16	55.83	31.72
Medium	6	2	5	10	20	158.41	318	180.41	325.86
	7	2	5	10	20	220.33	528.25	233.33	537.41
	8	2	5	10	20	280.2	418.61	294.66	427.55
	9	2	5	10	20	315.0	493.26	320.6	504.56
	10	2	5	10	20	330.20	386.47	426.50	390.95
Large	11	2	5	10	20	415.5	615.56	473.75	629.24
	12	2	5	10	20	427.50	432.15	482.00	463.41
	13	2	5	10	20	559.16	705.14	615.83	847.73
	14	2	5	10	20	410.32	632.55	429.54	784.22
	15	2	5	10	20	683.34	722.75	756.13	845.22

The suggested WOA was developed in MATLAB. The MILP framework with LINGO 8.0 and the WOA was applied to address provided test scenarios with various amounts of days, times, physicians, and patients. We performed the WOA five times on each instance because WOA is an optimization technique. We later compared two methods using various verifications.

Only two medium-scale examples and one small-scale scenario can have optimal solutions found using the MILP approach, according to Table 1.2. In similar problem situations, the WOA has also discovered these ideal solutions. However, when problem sizes grow, the MILP model cannot solve multiple test examples, such as all of the bigger ones, in under an hour. The WOA finds notably better solutions than those obtained by the MILP approach for medium-scale situations.

Fig. 1.4 illustrates these two strategies' effectiveness in producing the optimal outcome. The WOA is demonstrated to be more efficient than the MILP method. According to the data, we can say that the WOA performs better than MILP method. The WOA gains from swiftness and strong convergence. There is the problem of premature convergence and optimal local solutions for complicated optimization algorithms. The Whale Optimization Algorithm is a powerful optimization tool, but it has some limitations when applied to healthcare appointment scheduling:

1. The algorithm is unsuitable for very large-scale problems as it can take a long to converge.
2. The algorithm does not consider any external factors, such as patient preferences or doctor availability, which can impact the scheduling process.
3. The algorithm does not consider any constraints that healthcare regulations or policies may impose.
4. WOA is designed to optimize a single objective function, which may not be suitable for healthcare scheduling where multiple objectives need to be taken into account.

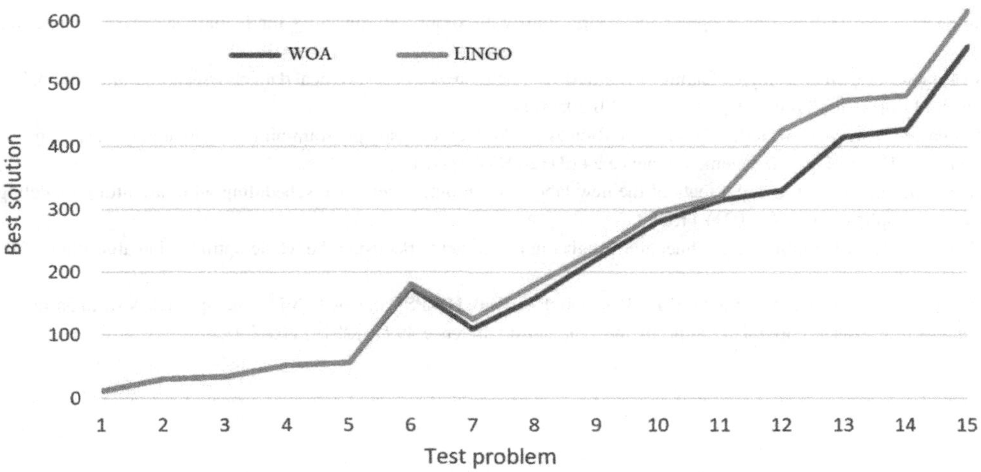

FIGURE 1.4 Comparison between MILP (LINGO) and the WOA in terms of the best solution.

Therefore, while the Whale optimization algorithm can be used to optimize healthcare appointment scheduling in some cases, it is important to determine these limitations when deciding whether or not to consider this approach.

1.8 Conclusion and future directions

Healthcare operations are now a necessary component of practically every healthcare service. We looked into issues with scheduling patients and doctors in the healthcare service, where one physician and one patient are assigned to each session. Although during the preparation phase, we also took patients' preferences as well as the availability of doctors into account. By assigning a shorter length of stay in the queue, the purpose was to boost the patient's satisfaction with their scheduled appointment. By evaluating the MILP approach for more complex examples, a WOA method was developed to address multiple instances of the issue. The proposed WOA could, according to computational findings, outperform the MILP in most situations in terms of processing time efficiency and perfect solution.

In future work, Authors can provide the suggested algorithm for handling practical feature selection uncontrolled challenges by utilizing heuristic operations. Also, exploring the use of the WOA for predicting patient no-show rates and optimizing appointment schedules accordingly. Moreover, analyzing the impact of different parameters, such as decision support systems, on the performance of the WOA in healthcare appointment scheduling scenarios.

References

[1] P.S. Chen, W.T. Huang, T.H. Chiang, G.Y.H. Chen, Applying heuristic algorithms to solve inter-hospital hierarchical allocation and scheduling problems of medical staff, International Journal of Computational Intelligence Systems 13 (1) (2020) 318–331.

[2] G.Y. Ning, D.Q. Cao, Improved whale optimization algorithm for solving constrained optimization problems, Discrete Dynamics in Nature and Society 2021 (2021) 8832251.

[3] J. Wang, R.Y. Fung, Adaptive dynamic programming algorithms for sequential appointment scheduling with patient preferences, Artificial Intelligence in Medicine 63 (1) (2015) 33–40.

[4] M. Reuter-Oppermann, N. Kühl, Artificial intelligence for healthcare logistics: an overview and research agenda, Artificial Intelligence and Data Mining in Healthcare (2021) 1–22.

[5] Y. Zhan, Z. Wang, G. Wan, Home service routing and appointment scheduling with stochastic service times, European Journal of Operational Research 288 (1) (2021) 98–110.

[6] M.G. Güler, E. Geçici, A decision support system for scheduling the shifts of physicians during COVID-19 pandemic, Computers & Industrial Engineering 150 (2020) 106874.

[7] T. Aladwani, Scheduling IoT healthcare tasks in fog computing based on their importance, Procedia Computer Science 163 (2019) 560–569.

[8] S. Mirjalili, A. Lewis, The whale optimization algorithm, Advances in Engineering Software 95 (2016) 51–67.

[9] C.W. Kang, M. Imran, M. Omair, W. Ahmed, M. Ullah, B. Sarkar, Stochastic-Petri net modeling and optimization for outdoor patients in building sustainable healthcare system considering staff absenteeism, Mathematics 7 (6) (2019) 499.

[10] I. Ferreira, A. Vasconcelos, A supervised learning model for medical appointments no-show management, International Journal of Medical Engineering and Informatics 14 (1) (2022) 90–104.

[11] M.M. Mafarja, S. Mirjalili, Hybrid whale optimization algorithm with simulated annealing for feature selection, Neurocomputing 260 (2017) 302–312.
[12] V.J. Kadam, S.M. Jadhav, Optimal weighted feature vector and deep belief network for medical data classification, International Journal of Wavelets, Multiresolution and Information Processing 18 (02) (2020) 2050006.
[13] Y. Sun, U.N. Raghavan, V. Vaze, C.S. Hall, P. Doyle, S.S. Richard, C. Wald, Stochastic programming for outpatient scheduling with flexible inpatient exam accommodation, Health Care Management Science 24 (3) (2021) 460–481.
[14] Z.Y. Zhuang, F.Y. Vincent, Analyzing the effects of the new labor law on outpatient nurse scheduling with law-fitting modeling and case studies, Expert Systems with Applications 180 (2021) 115103.
[15] I. Aljarah, H. Faris, S. Mirjalili, Optimizing connection weights in neural networks using the whale optimization algorithm, Soft Computing 22 (1) (2018) 1–15.
[16] N.M. Laskar, K. Guha, I. Chatterjee, S. Chanda, K.L. Baishnab, P.K. Paul, HWPSO: a new hybrid whale-particle swarm optimization algorithm and its application in electronic design optimization problems, Applied Intelligence 49 (1) (2019) 265–291.

Chapter 2

Recent advances of whale optimization algorithm, its versions and applications

Zaid Abdi Alkareem Alyasseri[a,b], Nabeel Salih Ali[a], Mohammed Azmi Al-Betar[c], Sharif Naser Makhadmeh[c], Norziana Jamil[b], Mohammed A. Awadallah[d,e], Malik Braik[f], and Seyedali Mirjalili[g]

[a]Information Technology Research and Development Center (ITRDC), University of Kufa, Najaf, Iraq, [b]Institute of Informatics and Computing in Energy, College of Computing and Informatics, Universiti Tenaga Nasional, Kajang, Selangor, Malaysia, [c]Artificial Intelligence Research Center (AIRC), College of Engineering and Information Technology, Ajman University, Ajman, United Arab Emirates, [d]Department of Computer Science, Al-Aqsa University, Gaza, Palestine, [e]Artificial Intelligence Research Center (AIRC), Ajman University, Ajman, United Arab Emirates, [f]Department of Computer Science, Al-Balqa Applied University, As-Salt, Jordan, [g]Centre for Artificial Intelligence Research and Optimisation, Torrens University Australia, Brisbane, QLD, Australia

2.1 Introduction

Over the past decades, there has been a tremendous need for optimization solutions through a wide variety of research fields such as scheduling and planning, cloud computing and IOT, engineering, network security, machine and deep learning, etc. The real-world optimization problems are non-convex, non-linear, constrained, multimodal, and non-continuous with a huge search space. Therefore, the traditional methods cannot easily tackle this type of problem [1]. Therefore, the emergence of the metaheuristic (MH) algorithms has been attracting the attention of the optimization algorithmic designer to cope with the complexity of real-world optimization problems [2].

In the optimization domain, the MH algorithms are known as a general optimization framework that initiates with one or more provisional solutions. Also, they have an iterative improvement loop whereby a set of intelligent operations controlled by control parameters are invoked utilizing the survival-of-the-fittest principle and navigate the problem search space carefully through exploring several regions at the same time and exploit the accumulated knowledge acquired in previous iterations to end up with promising solutions. The majority of MH algorithms have been established based on natural phenomena. The MH algorithms are classified into two main types: local search-based and population-based algorithms [3]. The local search-based algorithm initiated with one solution and that solution is iteratively improved. Population-based algorithms initiated with a set of random solutions where these solutions are evolved using recombination, mutation, and selection operators [1]. Later in the optimization domain, the MH algorithms have been categorized into five different types according to the emulated phenomena: evolutionary-based, swarm-based, physical-based, chemical-based, and human-based algorithms [4].

In particular, the swarm-based algorithms imitate the survivable behavior of animals living in a group. They are formulated as MH algorithms where their operators considered the leader-follower principle in searching for food or hunting a prey [5]. The initial idea is established in Particle Swarm Optimization (PSO) algorithm [6] and Ant Colony Optimization (ACO) Algorithm [7]. Nowadays, a large number of swarm-based algorithms are introduced such as rat swarm optimizer [8], Artificial hummingbird algorithm [9], sparrow search algorithm [10], snake optimizer [11], Dwarf mongoose optimization algorithm [12], White Shark Optimizer [13], Chimp optimization algorithm [14], Horse herd optimization algorithm [15], Grey wolf optimizer [16], Moth-flame optimization algorithm [17], and many more.

One successful swarm-based algorithm stemmed by humpback whales in hunting fish in the oceans is called Whale Optimization Algorithm (WOA) [18]. It has several advantages such as it has very strong operators to explore several search space regions and digging into each search space region to discover the local optima. It is simple in adaptation, easy-to-use, parameter-free, derivative-free, and sound and complete. Therefore, it has been used to a wide range of real-world optimization problems from different domains such as electrical and power system, wireless and network system, environment and materials engineering, classification and clustering, structural and mechanical engineering, feature selection, image processing, robotics, medical and healthcare, scheduling domain, and many others.

Handbook of Whale Optimization Algorithm. https://doi.org/10.1016/B978-0-32-395365-8.00008-7

In this comprehensive review, initially, the growth and the importance of the WOA in the different research field have been analyzed using different measurements such as number of WOA-related articles, number of citations, research topics, authors, institutions, etc as shown in Section 2.2. The theory and principles of WOA is then discussed and presented in Section 2.3. Due to the fact that the search space structure of the most real-world optimization problem is highly dimensional, continuous, non-linear, non-convex, discrete, and constrained. The WOA have been modified and hybridized to be more convenient to the shape of optimization problem search space. Therefore, different versions of WOA have been introduced in Section 2.4. The set applications and real-world optimization problems tackled by WOA in different domains is summarized and discussed in Section 2.5. The available open source code is also highlighted in Section 2.6. Finally, this review paper is end up with a solid conclusion about WOA to the interested audiences as well as the possible research directions can be addressed in the future for WOA as shown in Section 2.7.

2.2 The growth of whale optimizer algorithm

This section represents a detailed analysis of the growth of the Whale Optimizer Algorithm (WOA) for the period from 2021 to August 2022. In general, this analysis included the number of research published for the WOA per year, the number of citations obtained per year, and the selection of top authors working in WOA as well as the most important research centers or institutions which are interested in applying WOA in their works. In addition to many other statistics of WOA works, as shown in the following subsections.

To determine these results, we used the following query within the Scopus database.

```
TITLE ( "whale optimization algorithm" ) AND ( LIMIT-TO ( PUBYEAR , 2022 ) OR LIMIT-TO ( PUBYEAR , 2021 ) OR LIMIT-TO ( PUBYEAR , 2020 ) ) AND ( LIMIT-TO ( DOCTYPE , "ar" ) ) AND ( LIMIT-TO ( PUBSTAGE , "final" ) ) AND ( LIMIT-TO ( LANGUAGE , "English" ) )
```

2.2.1 No. publications per year

The number of research published in the application of any algorithm, especially in high-reputation journals, is one of the most important criteria for the success of that algorithm. Therefore, Fig. 2.1 shows the total number of WOA works published per year. The figure shows that there is a gradually increased between 2021 and 2022, as the number of research papers in 2021 was approximately 60, while in 2022 there were 90 research papers.

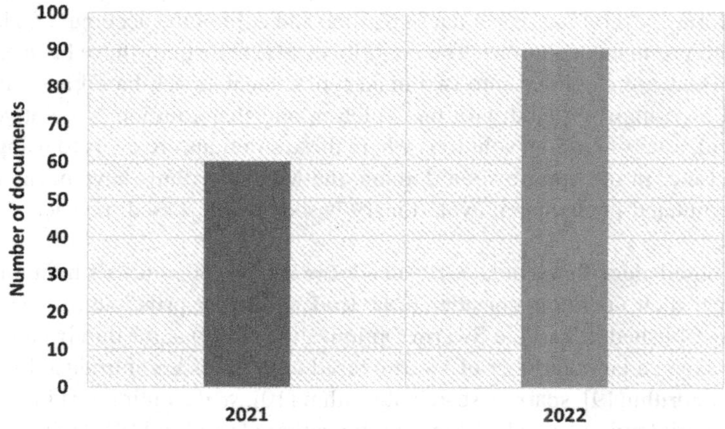

FIGURE 2.1 No. publications per year.

2.2.2 No. publications per publisher

Fig. 2.2 shows the top publishers who accept to hold the WOA works under their journals. It is clear that Elsevier, with 48 articles, achieves the first rank, MDPI with 27 research papers second rank, Springer Nature with 25 papers, and the rest publishers as presented in Fig. 2.2.

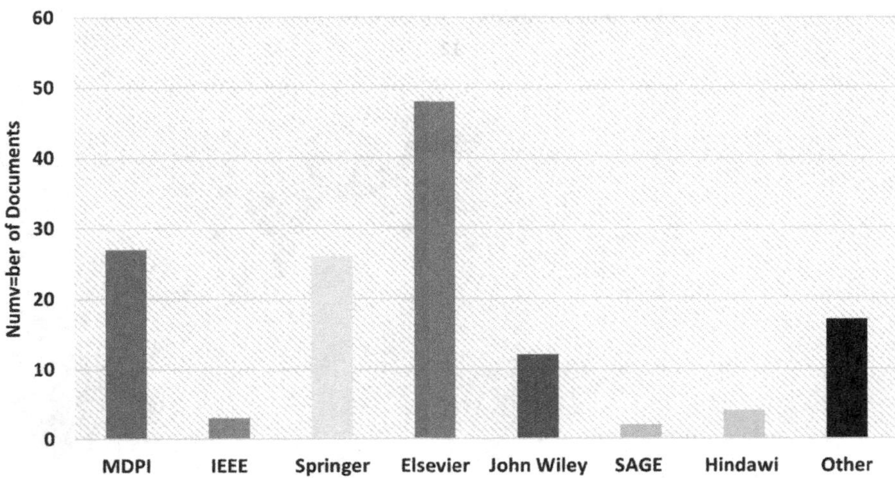

FIGURE 2.2 No. Publications Per Publisher.

2.2.3 No. publications per affiliation

Fig. 2.3 shows the top institutions' published works in WOA. The researchers team from Zagazig University-Egypt focus on WOA as main part of their research work, they published more than 22 articles. In the second place is Minia University with 14 WOA articles. In the third place is Fayoum University with 13 WOA works. The rest of institutions can be found in the figure.

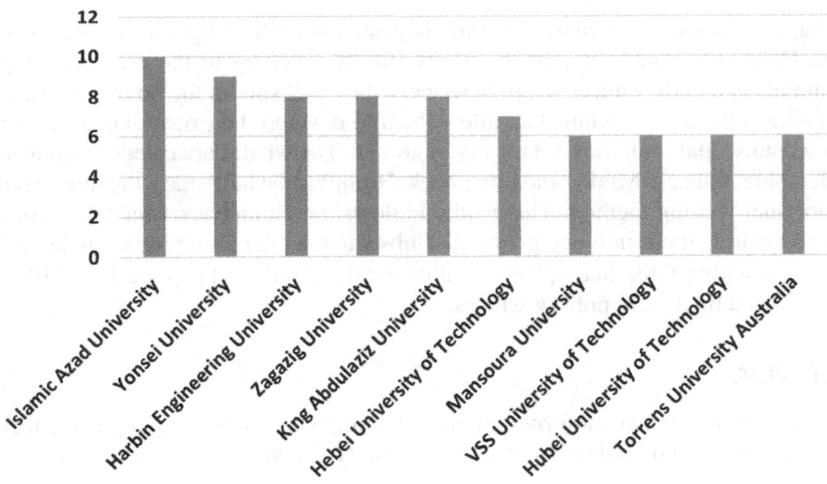

FIGURE 2.3 No. Publications Per Affiliation.

2.2.4 No. publications per country

Next Fig. 2.4 shows the ranking of WOA works by countries. The researchers from Egypt, China, and Saudi Arabia have achieved the top three places with WOA works, where the Egypt has ranked first with 65 papers, then China and Saudi Arabia with 30 and 27, respectively. The rest of WOA countries ranking presented in Fig. 2.4.

2.3 Fundamentals to whale optimizer algorithm

The whale optimization algorithm (WOA) is a robust metaheuristic that mimics the social behavior of humpback whales in nature. The WOA was proposed by Mirjalili in 2016 [18]. This section illustrated and discussed the inspiration for the whale and the mathematical model of WOA.

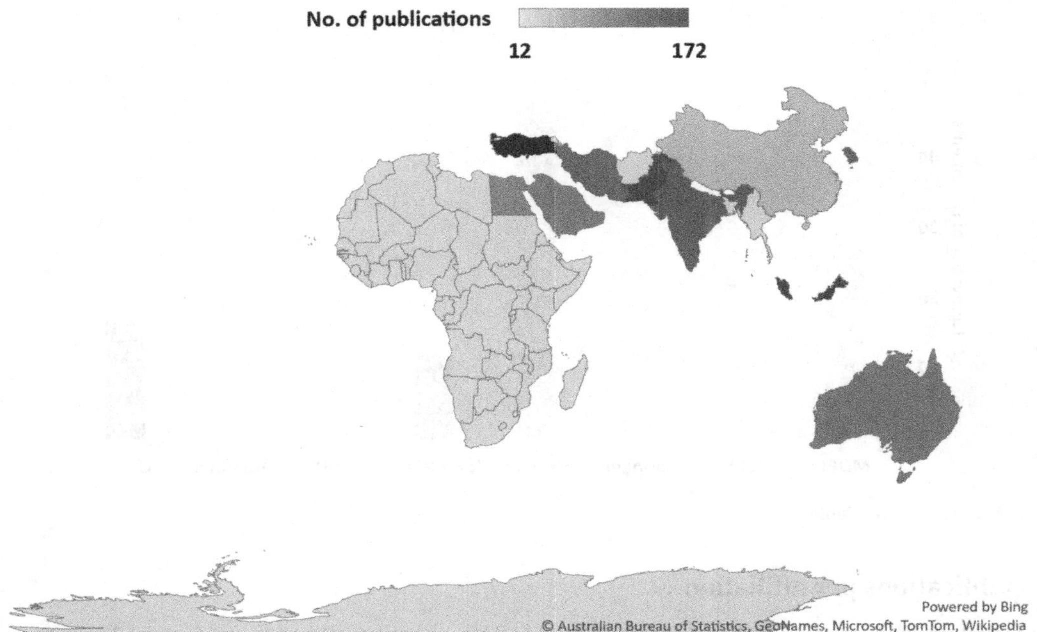

FIGURE 2.4 No. Publications Per Country.

2.3.1 Inspiration of WOA

Whales are one of the large mammals in oceans, if not the largest, where the length of the adults is up to 30 meters with weights up to 180 tons. Despite the innocence of their outward, they belong to the predators' top hierarchy. The most interesting of these mammals is that they never sleep, otherwise, they will sink to the bottom of the oceans.

The whales have spindle cells in their brains that allow them to develop their own dialect, emotions, and social living way either in a group or as individuals, but they are mostly in groups. The whales are categories into seven different species, including right, Sei, killer, blue, finback, Minke, and humpback. Humpback whales have the most exciting hunting behavior, where they use the bubble-net feeding method. This method allows the Humpback whales to produce a large number of bubble around the prey to encircle them into one prey ball. Subsequently, the Humpback whales will try to move toward, hunt and attack the prey ball within three main phases, called coral, lobtail, and capture loop [19,20]. Fig. 2.5 shows the bubble-net feeding method used by the Humpback whales.

2.3.2 Procedure of WOA

This section provides, firstly, the mathematical formulation of all humpback whales hunting and attacking phases, including the search for prey, encircling prey, and bubble-net feeding. Secondly, the WOA optimization stages are deeply illustrated.

2.3.2.1 Encircling prey

Humpback whales have the ability to determine the prey position in an ocean and start hunting behavior once they reach it. In WOA, the prey position is assumed as the fittest candidate solution, and the whale with the best position near the prey position is considered the best search agent, and the other Humpback whales will update their positions in accordance with the best search agent. Such following behavior is mathematically modeled as follows:

$$D = |C \times X^*(t) - X(t)| \tag{2.1}$$

$$X(t+1) = X^*(t) - A \times D \tag{2.2}$$

where t is the current iteration, X^* is the best solution's position that should be updated iterative, X is the current solution's position, $C\&A$ are two coefficients, which can be calculated as follows:

$$A = 2 \times a \times r - a \tag{2.3}$$

$$C = 2 \times r \tag{2.4}$$

FIGURE 2.5 Bubble-net feeding behavior of humpback whales.

where a is a value that linearly and iterative decreased from two to zero and r is a random number in the range zero and one.

2.3.2.2 *Bubble-net attacking method (exploitation phase)*

As mentioned previously, the bubble-net method used by Humpback whales is unique among all sea creatures. This section formulates this method mathematically in two stages: The shrinking encircling mechanism and the Spiral updating position.

1. **Shrinking Encircling Mechanism.** This stage is performed utilizing A that decreases in range a and $-a$ in changeable according to the value of a that decreases linearly, as shown in Eq. (2.3). If the value of A is set to be in the range $[-1, 1]$, the current position ($X(t)$) of a search agent will be updated to be between $X(t)$ and the best agent's position ($X^*(t)$), as shown in Fig. 2.6.
2. **Spiral Updating Position.** In this stage, the distance between the current position of a search agent and the best position, subsequently, a spiral is created using the spiral equation between the current position and the best position of the search agents, as follows:

$$X(t+1) = D' \times e^{b \times l} \times \cos(2 \times \pi \times l) + X^*(t) \tag{2.5}$$

where b denotes a constant variable used to define the logarithmic spiral' shape, l is a value generated randomly in range $[-1, 1]$, and D' is the distance between the current search agent and the best one, which can be calculated as follows:

$$D' = |X^*(t) - X(t)| \tag{2.6}$$

Notably, these whales are moving around the prey ball or best positions in a shrinking spiral-shaped circle. Accordingly, whales can choose whether using shrinking circles or spiral-shaped paths based on a 50% probability. Such movements and probability are mathematically formulated as follows:

$$X(t+1) = \begin{cases} X^*(t) - A \times D & \text{if } p \leq 0.5 \\ D' \times e^{b \times l} \times \cos(2 \times \pi \times l) + X^*(t) & \text{if } p \geq 0.5 \end{cases} \tag{2.7}$$

where p is a random value between zero and one.

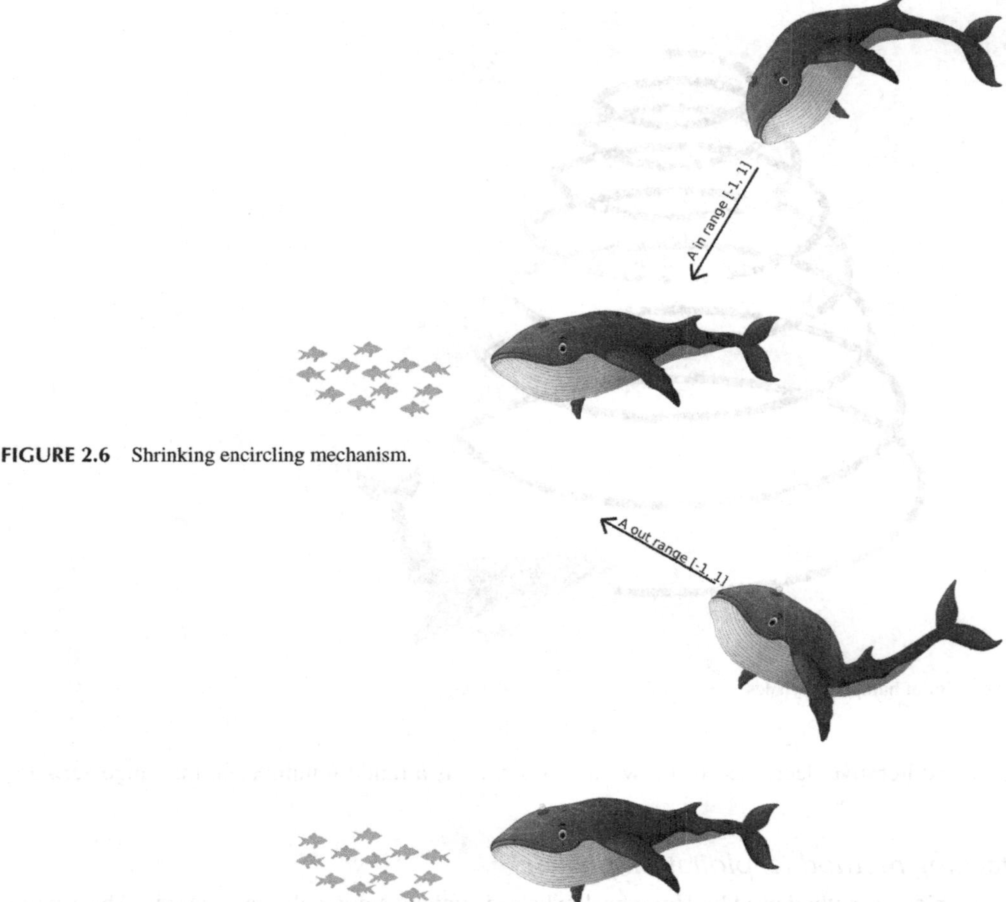

FIGURE 2.6 Shrinking encircling mechanism.

FIGURE 2.7 Search for prey.

2.3.2.3 Search for prey (exploration phase)

Another searching behavior, called exploration, can be utilized in WOA using the values of A. As mentioned previously, the exploitation phase can be performed when the value of A is between -1 and 1. Accordingly, the exploration phase will be performed when A value is less than -1 or greater than 1, in other words, when its value is not in the range $[-1, 1]$. Once this phase is performed, the current search agent will move away from the best search agent and will update its position on the bases of a search agent chosen randomly, as shown in Fig. 2.7. This phase is mathematically formulated as follows:

$$D = |C \times X_r - X| \tag{2.8}$$

$$X(t+1) = X_r - A \times D \tag{2.9}$$

where X_r denotes a random search agent.

2.3.2.4 WOA optimization steps

The WOA begins by generating a set of random solutions to create the population. Subsequently, the search agents (solutions) will change their positions iteratively in the search space according to Eq. (2.9). At the same time, the parameter a will be decreased linearly from 2 to 0 to perform the exploitation and exploration phases. In the exploitation phase, the current search agent will move toward the best one, whereas in the exploration phase will move based on a random solution. Finally, the WOA is terminated when the termination criterion is achieved.

2.4 Variants of WOA algorithm

This section will review the variants of WOA and recent research papers where the original version research of the WOA is explained in Section 2.4.1. The modified version works of the WOA reviewed in Section 2.4.2. Hybridizing version of the WOA reviewed in Section 2.4.4.

2.4.1 Original versions of WOA

In the wireless and network system (WNS) domain the original WOA is implemented in several works such as in [21] WSN with an efficient IoT services placement plan. The simulation experiment results showed that the WOA achieved better results compared with the other metaheuristic algorithms using several measures such as energy consumption, service acceptance ratio, resource usage, and eliminating service delay. Also, WOA proposed for WSN in [22]. The proposed method called (WOA_BiLSTM_Attention) is used to predict urban traffic flow. The proposed model performs better than other models, such as conventional neural networks and neural network models optimized by the same WOA algorithm according to various metrics, namely MAPE, RMSE, MAE, and R2. Another work of the original WOA with WSN is proposed in [23]. The proposed system significantly reduces total power losses and costs, besides improving the voltage profile and voltage stability index. The percentage of the loss reduction in both IEEE 33-bus and Dada 46-bus networks were 33.74% and 22.24%, with 27.60% and 25.60% annual net savings, respectively.

In the Electrical and power system (EPS) field the original WOA is proposed for several works such as in [24]. The proposed method suggested FinFET model resulted in its superior performance parameters when simulated according to the optimal value of WFin and HFin. Another work of the WOA in [25] proposed for electrical and power systems called (WOANFIS). The proposed method achieves better RMSE, MAE, and correlation coefficient (R2) values, which were 0.00113, 0.0047, and 0.98, respectively. Also, the WOA proposed for EPS in [26]. The proposed method is called Polymer Electrolyte Membrane Fuel Cells (PEMFC). The WOA algorithm highly improves the precision of degradation prediction. In [27] WOA is proposed for renewable energy systems. The simulation results of the conducted system based on accuracy, stability, and robustness metrics have shown that the proposed system exhibits high accuracy and validity in solving the optimization problem of a hybrid renewable energy system. In [28] the hybrid FOA-WOA gained 11.6% and 1.8% better than ACO and GA regarding packet delivery ratio. Besides, it achieved 57.6% and 27.3% better than ACO and GA concerning the delay. Also, it attains 15.3% and 36.4% better than ACO and GA according to energy consumption.

In forecasting wind power forecasting [29], the original WOA has proposed for prediction model that has a higher accuracy of ultra-short-term wind power among different prediction models in the literature.

For the environment and materials engineering [30], the WOA is proposed for isotoping separation cascades. The results of the proposed algorithm have shown a good agreement after evaluation against other well-known schemes. In [31], the original WOA estimates the soil moisture of maize. The results of the proposed algorithm (SVM-SWOA) evaluating with different SVM versions, the proposed algorithm showed that the SVM-SWOA improved 14%, 13%, 41.5%, and 14% over SVM-WOA at 60 cm depth for MAE, RMSE, MAPE, and MBE, respectively, and 20%, 29.5%, 44.5%, and 38% over SVM, respectively. Consequently, the SVM-SWOA can support adequate smart agriculture and precision irrigation guidance. In [32], the WOA is proposed for accurate prediction of rock squeezing. The results of the optimized WOA-SVM model have shown that the WOA-SVM attained higher accuracy (approximately 0.9565) than other un-optimized individual classifiers (SVM, ANN, and GP). Moreover, it has a high sensitivity for the percentage strain to the model. Therefore, x, H and K are the best combinations of parameters for the presented model.

The original WOA is proposed in [33] for robotics safety and reliability of autonomous navigation systems. The proposed algorithm has shown optimized results for effective and efficient path planning according to planning efficiency, shorter execution time, faster convergence speed, and higher solution precision. Besides, the WOA has better results by minimizing the fitness value than other algorithms.

For the Image and Signal Processing domain, the original WOA proposed in [34]. The proposed algorithm chooses the optimal codebook in image compression. Based on the results, the WOA algorithm performs better than PSO, bat, and firefly algorithms concerning compression efficiency and signal-to-noise ratio. Besides, the proposed method has a higher PSNR index of about 17% than the Linde-Buzo-Gray method in compression. Also, It has a compression execution time of 60.48%, and 10.21, 4.79%, 5.09%, and 3.94% decreased, respectively, among the FireFly, Bat, and Differential evolution, Improved Particle Swarm Optimization, and Improved Differential Evolution methods.

For the global optimization problems in [35], the WOA is proposed for premature convergence and quickly falling into local optimum. The evaluation of the improved algorithm has shown that the WOA provides fast speed and high accuracy for convergence and has the potential to jump out of the local optimum effectively.

2.4.2 Modified versions of WOA

In this work [36], a modified version of WOA called (REM-WOA) is proposed. The proposed algorithm has achieved high performance with better convergence behavior and robust global exploration efficiency among 9 different algorithms using a global function benchmark. Besides, it has the best efficiency-based exploration adopted with three real design case studies.

In [37] a modified version of WOA is proposed for the Electrical and power system problem. The proposed algorithm called IWOA is implemented to improve poor accuracy and stability problems in electric vehicle charging stations. The results of the IWOA have shown the ability to effectively apply it to location and sizing issues to eliminate costs for the whole society. Another work of a modified version of WOA is proposed for electrical and power systems [38]. From the experimental results, the MWOA technique has the reliability to obtain global or near-global optimal settings of control variables accurately. Besides, it can solve real-world optimization problems among other competitive, robust methods.

Also, a modified WOA version is proposed for clustering to improve the QoS performance of integrated energy system wireless sensor networks (IESWSNs) in [39]. The proposed algorithm called HPCP-QCWOA and evaluated with various scenarios, and the results of the simulation experiments have shown increasing the proposed HPCP-QCWOA scheme concerning lifetime by 28.78%, 25.50%, and 11.22% among O-LEACH, LDIWPSO, and ARSH-FATI-CHS based clustering algorithms respectively. Another work of the modified WOA version is proposed for clustering changing topology rapidly in VANETs for Intelligent Transportation Systems (ITS) on highways [40]. Based on the results, the proposed i-WOA method has an optimal number of cluster heads (CHs) in various scenarios compared with related methods. The regression analysis has shown the improvement in cluster optimization for VANETs with application in ITS, eliminating the cost of communication and routing overhead, consequently increasing the network lifetime.

For the environment and materials engineering sewer pipeline inspection another modified WOA version is proposed which is called (WOAPCD) [41]. The performance of the WOAPCD method is evaluated with a natural sewerage system. The results have shown that the WOAPCD has the potential to accurately and effectively reconstruct the 3D model of the sewer besides supporting valuable information for quantifying siltation conditions. Finally, the WOAPCD performs better than PSO and GA concerning the fitting error and modeling speed. Also, in [42], a modified version of the WOA is proposed for the electrical and power system optimal dispatching problem. The results have shown that the presented regional interconnection operation of the microgrid under the established dispatching strategy effectively reduces the operating system's cost, besides the improved WOA to solve the optimal dispatching problem.

For the global optimization problem the modified WOA version is proposed in [43] for enhancing the ability of cooperative coevolution among populations. The performance of the proposed method (MCCWOA) has outperformed regarding efficiency and significance among several peers in the literature after applying various statistical, diversity, and convergence analyses. Another work of global optimization development search space is proposed using a modified version of WOA which is called NWOA algorithm [44]. The proposed algorithm has enhanced the optimization accuracy and the convergence speed significantly. Also, it provides a powerful role for the algorithm in multimodal functions. Finally, it has better optimization performance compared to other inspired optimization algorithms. For the large-scale global optimization and converging slowly problems, a new version of the WOA is proposed in [45]. The proposed MWOA-CS produced better convergence speed and accuracy. Moreover, it has shown more effective and powerful performances than WOA and other optimization algorithms for solving large-scale global optimization problems.

In the field of image and signal processing a modified version of WOA is proposed in [46]. The proposed method is used for the multilevel thresholding image segmentation problem. The convergence speed and the RAV-WOA accuracy are significantly better than other algorithms based on the experimental results with image segmentation experiments in high and low thresholds on a set of benchmark images. Besides, it has better quality and stability in multi-threshold image segmentation than other algorithms. Also, in the medical and healthcare managing health and chronic illnesses, a modified version of the WOA has proposed [47]. The results of the WOA-LSTM model have shown effective performance after applying a series of experiments for performance evaluation and its efficient throughput concerning user recommendation.

In the wind power field the original WOA is used to generat a AI model to provide a higher accuracy of ultra-short-term wind power among different prediction models [29].

2.4.2.1 Binary WOA

Feature selection [48] in the work a binary version of WOA for online product sentiment analysis is proposed. The main findings of the proposed algorithm called REWOA-DBN model improved the system complexity and running time of the classifier and achieved better classification accuracy (96.86%) than other optimizers and classifiers. Also, in [49] a binary version of WOA is proposed for the high-dimensional microarray data problem. The presented AltWOA method has shown its superiority after being evaluated with standard eight high-dimensional microarray datasets, among other techniques regarding features selected and accuracy. In [50] a combining Improved Aquila Optimizer (IAO) results when using the WOA operators have shown that the system significantly increases its impact on the AO performance. Besides, the outcomes of the IAO achieved better results than other optimizers that used feature selection techniques such as particle swarm optimization (PSO), differential evaluation (DE), mouth flame optimizer (MFO), firefly algorithm, and genetic algorithm (GA). In [51] proposed composite framework achieves remarkable results concerning classification accuracy and detection rate and outperforms all other human recognition models in the literature. In [52] selecting significant features from a high-dimensional microarray dataset. The results of the proposed iWOA algorithm obtained its identification and stable feature selection technique based on the strength of the stability index agreement. In [53] Search for the optimal feature combinations. According to the two most datasets, such as TOX_171, Colon, and Prostate_GE, the produced SBWOA obtained the highest accuracy compared to competitors' other methods. In [54] students' performance prediction (SPP) problem. The suggested EWOA algorithms obtained significant results as wrapper feature selection with selected transfer functions compared with other existing methods. Moreover, the LDA classifier was the multiple reliable classifier with both datasets based on the acquired results.

For medical and healthcare, in [55] a binary version WOA is proposed for time-averaged serum albumin (TSA) in HD patients. The results have shown the superior performance of the WOFS model with the lowest Akaike information criterion (AIC) value in the multifactor analysis of TSA for HD patients. Consequently, the TSA-associated model has the ability to nutritional status monitoring in HD patients. In [56] medical and healthcare early detection of lung tumor. The proposed technique gains a remarkable accuracy classification such as Deep Convolutional Neural Network (DCNN), Convolutional Neural Network (CNN), UNet, and ASPP-UNet-WOA are 93.45%, 91.67%, 95.75%, and 98.68% respectively among other techniques in state of the art. Another work of binary WOA was proposed in [57] for medical and healthcare diagnosis and prediction of coronary artery disease (CAD). The proposed method obtained actual results in selecting the optimal feature subsets among 17 features for each primary artery diagnosis with accuracy is 89.68%, 88.71%, and 85.81% for LAD, LCX, and RCA test sets, respectively. Moreover, the classification performance of the stacking model attained by the KNN-based WOA method is better than related recent ML algorithms. Finally, the suggested feature selection methods outdo its performance on various wrappers compared to other metaheuristics algorithms.

For the classification of early disease diagnostics, in [58] a binary of the WOA is proposed. The classification accuracy of the modified MEWOA method among three datasets, namely INbreast, MIAS, and CBIS-DDSM, is 99.7%, 99.8%, and 93.8%, respectively. Moreover, the MEWOA algorithm outperforms recent related approaches. Another work of binary WOA is presented in [59] for classification plant identification. The experimental results have shown that the produced method obtained improves average accuracy is 5% than other different algorithms under consideration.

In [60] a binary WOA is proposed for the global optimization classification performance. The proposed MSWOA-SSELM was implemented with three wells and acquired vital results compared with other classification models based on several criteria such as Accuracy (ACC), Root Mean Square Error (RMSE), and Mean Absolute Error (MAE) that are 96.2567%, 0.0749%, and 0.3870% respectively. Furthermore, the presented MSWOA method registered its superiority and effectiveness for solving global optimization problems. In [61], a binary WOA is proposed for optimizing neural network hyperparameters. The proposed 3D-WOA method achieved 89.85% and 80.60% accuracy for Fashion MNIST and Reuters datasets, respectively, and it can be used for hyperparameters optimization successfully.

For the robotics field, in [62] Accuracy for the dynamic fatigue classification. The produced methods gained competitor results among other existing methods demonstrating their feasibility and effectiveness in predicting dynamic muscle fatigue. The introduced method has an average accuracy of 85.50% and 84.75% in both ankle dorsiflexion (DF) and ankle plantarflexion (PF) sequentially.

2.4.3 Multi-objective WOA

In [63] a multi-objective of the WOA is proposed for electrical and power systems. The performance of the proposed CamWOA is evaluated based on benchmark functions and efficiency. The results have shown that the CamWOA has the most benchmark functions among other state-of-the-art inspired algorithms. Besides, the efficiency of the CamWOA was also evaluated by solving a multiobjective engineering problem about the control of switched reluctance motors. Also, in [64] a multiobjective Combined Heat and Power Economic Emission Dispatch (MO-CHPEED) problem is proposed. The results have shown the effectiveness and robustness of the proposed method for getting better average and STD values. Regarding the MO-CHPEED problem, the DCWOA model gained better fitness, the best compromise solution obtained, and the convergence traits. Another work of multiobjective of WOA is proposed in [65]. The produced mWOASA technique can significantly reduce both torque ripple coefficient, integral square error of speed (ISE(speed)), and integral square error of current (ISE(current)) compared with hybrid WOASA and WOA, respectively.

In the feature selection domain, a multiobjective WOA version for finding the minimum number of features that result in high classification accuracy is proposed [66]. The produced algorithm evaluated twelve benchmark datasets and compared them with seven standard algorithms. The results demonstrated that the presented method obtained higher classification accuracy for several subsets that included a smaller number of features, revealing its efficacy and ability to gain efficient and optimal feature selection as a multiobjective problem.

2.4.4 Hybridized versions of WOA

For the global optimization enhancement a hybrid version of the WOA is proposed in [67]. The proposed ESSAWOA has shown superior performance among other WOA, SSA, and related optimization algorithms. Moreover, the hybrid ESSAWOA algorithm can acquire a suitable solution in the search space to solve optimization problems. Also, in [68] proposed a hybrid algorithm has better performance concerning convergence speed with a range of 2% improvement is obtained that approximately 64% improvement is achieved compared to FA and 39% improvement compared to WOA. Besides, the hybrid method can be used to find the optimal prediction horizon and control horizon as well as the Q and R matrices such that the system responses satisfy desired settling time and maximum overshoot criteria by the selected objective function for the hybrid firefly–whale optimization algorithm. In [69] proposed hybrid algorithm addresses reusable launch vehicle (RLV) reentry trajectory optimization problems. In [70] proposed a hybrid WOA algorithm for compressed sensing image reconstruction. The results show that the hybrid algorithm achieved better performance, which is significant for producing a global search with faster convergence than traditional reconstruction algorithms. Another hybrid version of the WOA is proposed in [71]. The proposed method used for nonlinear systems and unconstrained optimization problems. The results of the hybrid WOFPA appeared to outperform the WOFPA compared to other existing algorithms regarding attaining the optimum solutions for most nonlinear systems and optimization problems. Further, it proves its efficiency and superiority over other competitor methods. In [72] proposed a method for High-dimensional problems. The proposed HWOAG obtained a strong ability and higher efficiency for searching compared to the algorithms revealed in the literature and the WOA algorithm. These experimental results are applied to high-dimensional (i.e., 1000-, 2000-, 4000-, and 8000-dimensional) benchmark functions and clustering datasets for Fuzzy C-Means (FCM) optimization.

In [73] a hybrid version of the WOA is proposed for optimizing the controller parameters of an islanded microgrid. The hybrid method NSWOA has the best optimum reaching solution by an average of 4 iterations less than other existing algorithms. Further, the NSWOA obtained results much faster because the required computational time is 2.9201 s among other existing NSGA-II and SPEA algorithms.

In [74] the WOA algorithm is proposed for path planning and control over multiple mobile robots in static and dynamic environments. The results of the conducted technique show the efficiency and attained 20.63% in path lengths as a significant improvement compared to related competitor methods.

Finally, in [75] the WOA algorithm is applied for breast cancer diagnosis. The hybrid HAW-RP method was evaluated according to several metrics such as accuracy, complexity, and computational time using various breast cancer datasets. The HAW-RP variant obtained higher accuracy of 99.2%, 98.5%, 96.3%, 98.8%, 98.7%, and 99.1% with the low-complexity ANN model compared to HAW-LM and HAW-GD for WBCD, WDBC, WPBC, DDSM, MIAS, and INbreast, respectively.

2.4.4.1 WOA with other algorithms

In [28] a hybrid version of the WOA algorithm is applied for energy efficiency and delay issues in MANET. The hybrid FOA-WOA gained 11.6% and 1.8% better than ACO and GA regarding packet delivery ratio. Besides, it achieved 57.6% and 27.3% better than ACO and GA concerning the delay. Also, it attains 15.3% and 36.4% better than ACO and GA according to energy consumption.

Another hybrid version of the WOA is proposed in [76] for flow shop scheduling problems. The results of the numerical experiments for the proposed improved IMOWOA algorithm that were implemented with real-world cases in a Chinese company's digital hot-rolling workshop have shown that the presented IMOWOA proves it is superior to SPEA2 and NSGA-II. Moreover, the actual case implementation achieves the ability of the PSO algorithm to successfully tackle the presented MOHFSP-DRP very well and applied to a real-world hot-rolling shop.

2.5 Applications of whale optimizer algorithm

The WOA is powerfully used to solve diverse optimization issues that regard various domains such as Global optimization, electrical and energy system, network and wireless system, materials and environmental engineering, classification and clustering, mechanical and structural engineering, selection of features, image processing, robotics, healthcare and

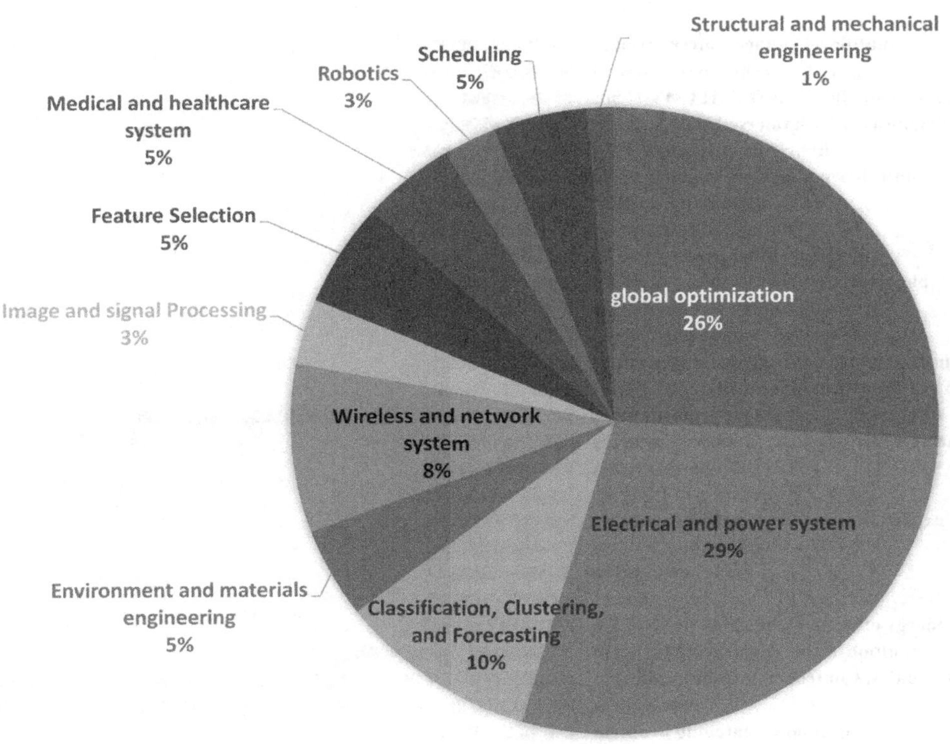

FIGURE 2.8 Overall WOA applications.

medicine, scheduling domain, and numerous others. The optimization problems that are resolved by the WOA in these purposes are different in their nature to encompass multiobjective, highly dimensional, continuous, non-linear, non-convex, discrete, real-world, large-scale global, and NP-hard problems. This section presents WOA applications, diverse concerns involving variants of WOA and its significant findings. This stimulates researchers to investigate further efforts in demonstrating WOA in varied other applications. Fig. 2.8 shows the WOA applications domains.

According to the findings that have been seen from Fig. 2.8, high numbers of WOA applications are applied to solve the issues belonging to the electrical and power system research domain. In this domain, the optimization problems addressed by WOA are economic/emission dispatch [64,77–79], Power system [38,67,80,81], electronics, control, and communication [23,82–84], Electric Vehicle (EV) charging [37,85,86], forecasting the electricity consumption [87–89], multiobjective engineering problem [63,65], optimal dispatching problem [42], energy efficiency and delay in MANET [28], wind power [22], coal consumption [90], fractional chaotic systems [32], exploration and exploitation capability [91], Forecasting of PEMFC [26], renewable energy system [27], Parameter identification in battery systems [92], Power Quality in photovoltaic (PV) wind [93], scheduling loads of residential consumers [94], web service composition (WSC) [95], Photovoltaic Power Prediction [96], photovoltaic-biowaste energy system [97], Optimal chiller loading (OCL) [98], Winding faults detection [99], optimal power flow [100], controller parameters in a microgrid [73], hybrid electric ship (HES) [101], fault diagnosis of oil-immersed transformers [102], proton exchange membrane fuel cells (PEMFCs) systems [103], gravity energy storage system [104], and multilevel voltage source inverter [105]. These applications are presented in Fig. 2.9. In the global optimization domain, the WOA can provide a high-efficacy solution for different application in large-scale global optimization problems namely convergence and exploitability challenges [36,43–45,106–114], exploration issues of the search space [36,44,70,113–116], complicated and multidimensional problems [72,117–119], Multilevel Thresholding Image [86,120,121], optimizing engineering applications [122], classification performance [60], optimal control problem (OCP) [68], Multiobjective optimization problems [123], optimize neural network hyperparameters [61], NP-hard combinatorial optimization problem [124], Seismic Inversion Problem [111], large-sized waste products [125], unconstrained optimization problems [71], shift-invariance [126], Search traveling salesman problem (TSP) [127], curve design issues [128], bearing failure diagnosis [129]. Fig. 2.10 presented WOA applications in the global optimization domain.

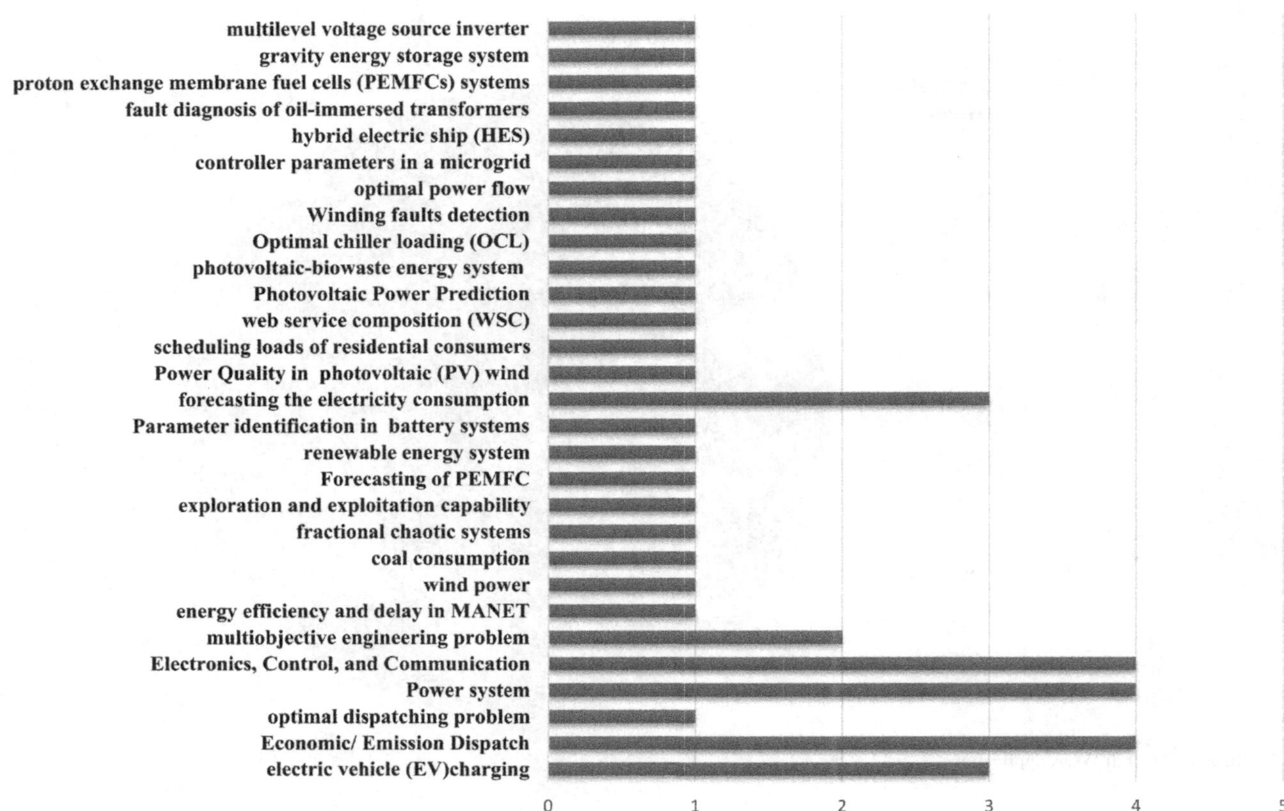

FIGURE 2.9 WOA Publication in different applications.

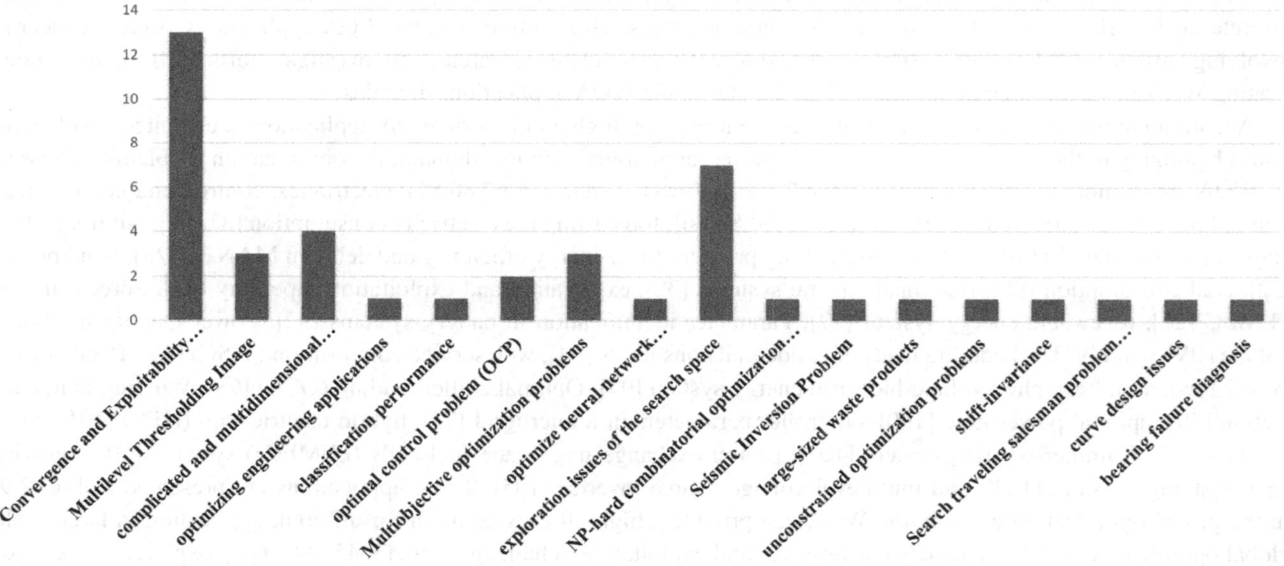

FIGURE 2.10 WOA Publication in Global Optimization.

In diverse domains such as classification, clustering, and forecasting, the WOA is applied successfully to various issues namely Prediction and classification [58,59,83,130–134], Data clustering [40,83,135,136], data Forecasting [29,137], and QoS performance in WSNs [39]. These applications are shown in Fig. 2.11.

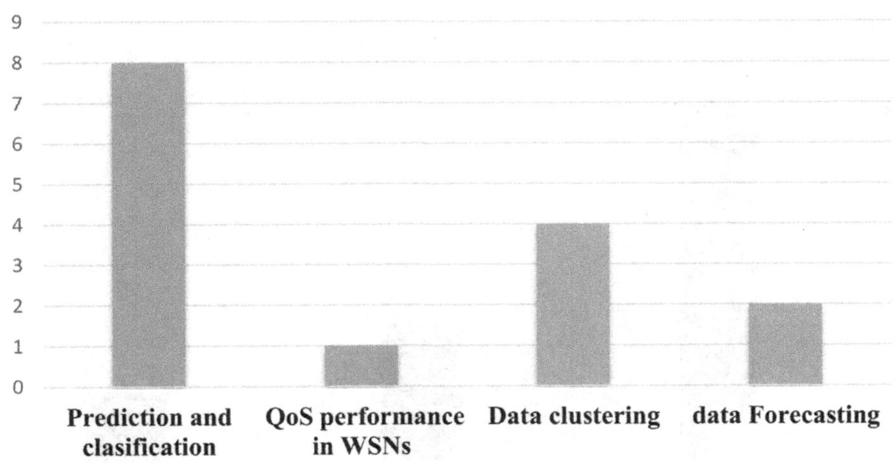

FIGURE 2.11 WOA Publication in Clustering Classification and Forecasting.

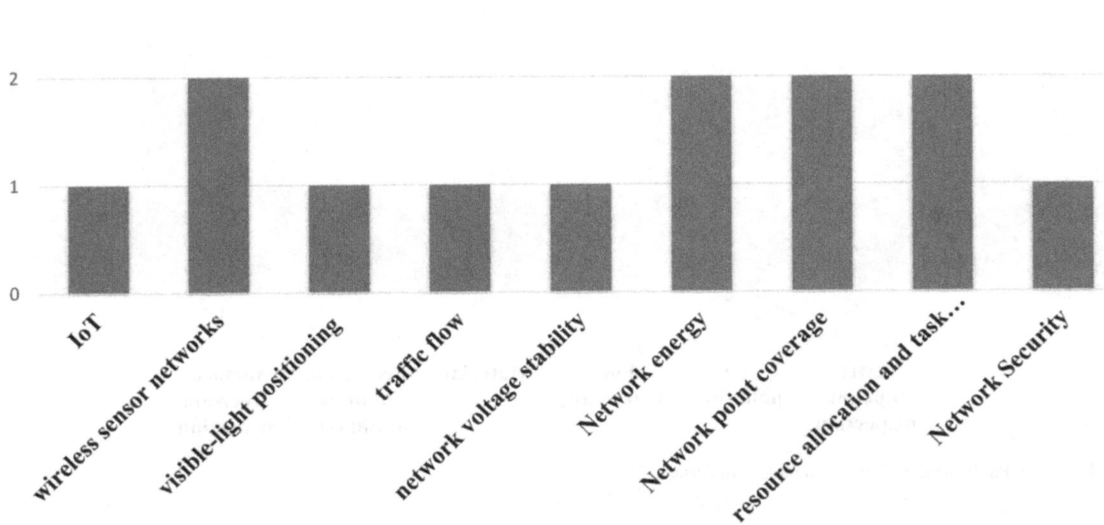

FIGURE 2.12 WOA Publication in Wireless networking.

In wireless and network system domain, the WOA algorithm is achieving the ability of providing a vital solution to different application problems in wireless sensor networks [138,139], Network energy [140,141], Network point coverage [142,143], resource allocation and task scheduling [144,145], traffic flow [24], visible-light positioning [146], network voltage stability [33], IoT [21], and Network Security [147]. Fig. 2.12 presented WOA applications in the wireless and network system domain.

In the feature selection domain, the WOA has demonstrated superior results and confirmed its scalability in solving high-dimensional and complex problems such as online feature selection [35,48,53,54,66], gene selection [49,52], and Search Optimization [25]. Fig. 2.13 presents the WOA applications in the feature selection domain.

Fig. 2.14 presents the WOA applications in the domain of environment and materials engineering. Different problems were resolved by WOA including Materials [51,148], sewer pipeline inspection [41], Air pollution [30], Food Processing [31], prediction of ionic liquid (IL) [149], surface precision inspection [150].

The WOA algorithm in the domain of medical and healthcare has shown promising solutions for Medical Application [56,57,75,119], Patient Monitoring [47,55], General Healthcare [151]. The medical applications are displayed in Fig. 2.15.

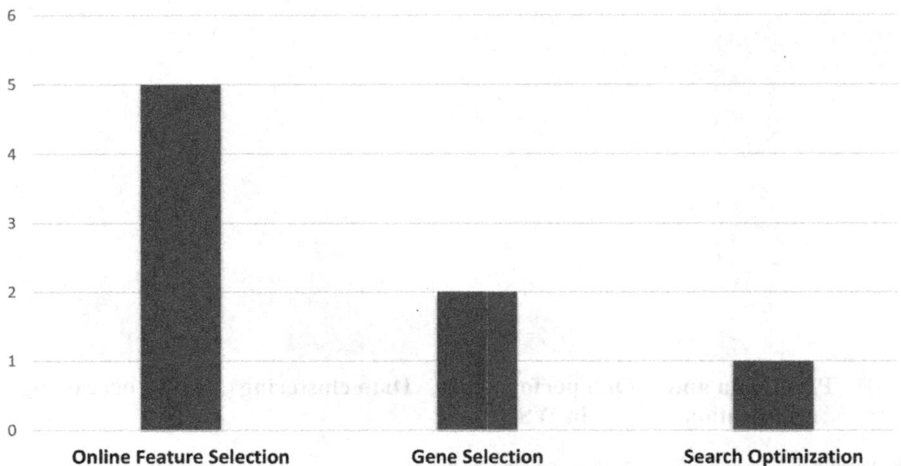

FIGURE 2.13 WOA Publication in Feature Selection.

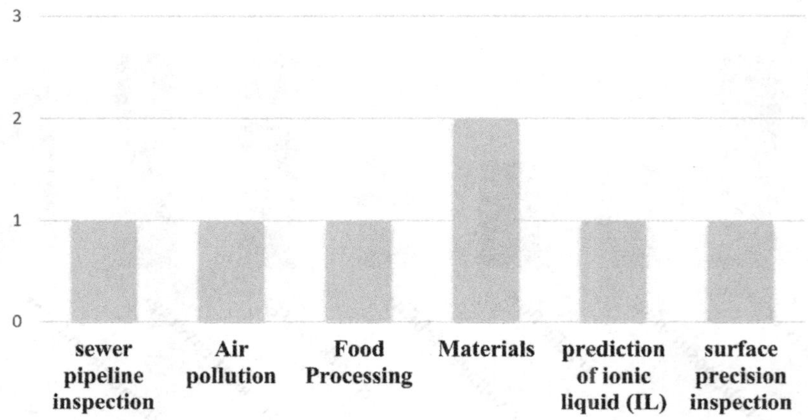

FIGURE 2.14 WOA Publication in Environment and materials.

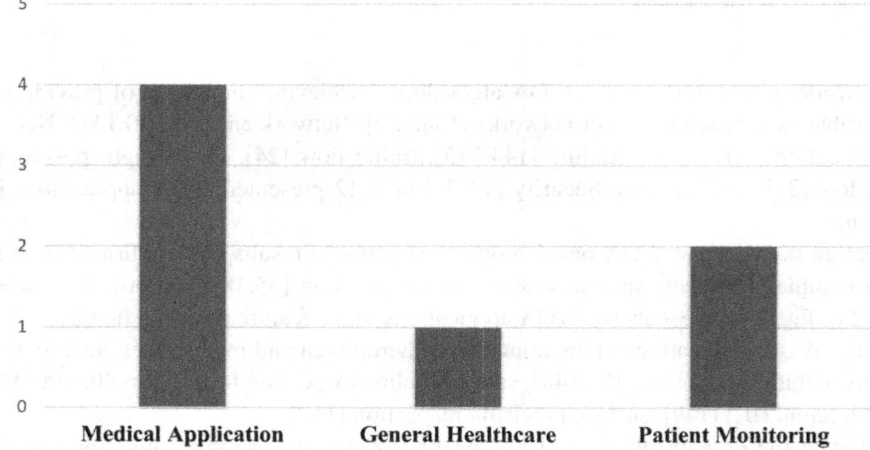

FIGURE 2.15 WOA Publication in Medical Healthcare.

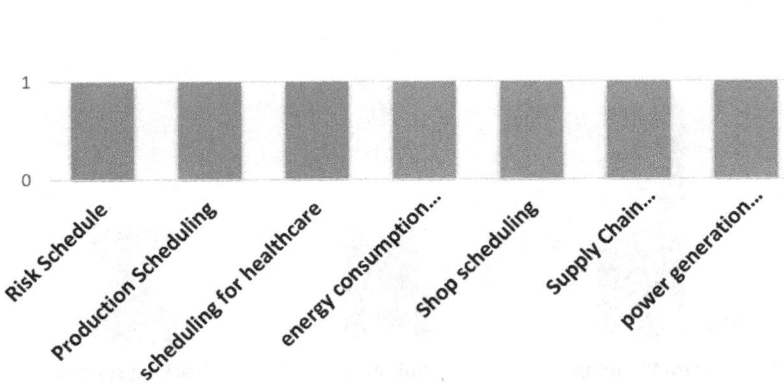

FIGURE 2.16 WOA Publication in Scheduling.

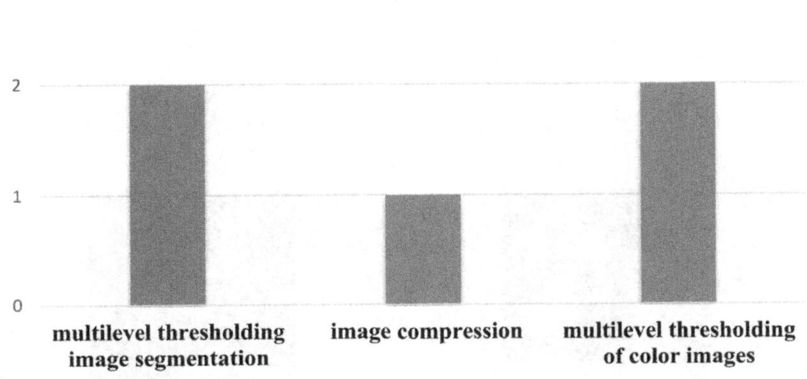

FIGURE 2.17 WOA Publication in Image and Signal Processing.

The WOA achieved advanced promising and robust solutions for the most challenging problems in the scheduling domain such as Risk Schedule [152], Production Scheduling [153] scheduling for healthcare [154], energy consumption scheduling [155], Shop scheduling [76], and Supply Chain Management [156], power generation scheduling [157]. Fig. 2.16 presents the applications of WOA in the scheduling domain.

And the WOA algorithm proves its superiority in the image-processing domain by solving multilevel thresholding image segmentation [46,158], multilevel thresholding of color images [159,160], and image compression [34]. The applications of WOA in the image-processing domain have shown in Fig. 2.17.

Various problems have been solved successfully by WOA in the robotics research domain including Path Planning [29,50,74], Robotics [161], and Tracking systems [62]. The mentioned applications are presented in Fig. 2.18.

Finally, the WOA algorithm has demonstrated robust results for solving challenging problems in structural and mechanical engineering domains. These issues such as the Identification of damage in plate structures [162], and ground response in short buildings [163]. The WOA applications in the domain of structural and mechanical engineering are depicted in Fig. 2.19.

2.6 Open source software of whale optimizer algorithm

The WOA algorithm has drawn a lot of scientific interest, as this review demonstrates. We have included the primary links for all conceivable open-source codes in this section to make it simple for next interested researchers to utilize, use, or alter the WOA method.

The source code for Mirjalili and Lewis [18] original Matlab® implementation of the WOA algorithm can be found on the MathWorks File Exchange repository.[1]

1. https://www.mathworks.com/matlabcentral/fileexchange/55667-the-whale-optimization-algorithm.

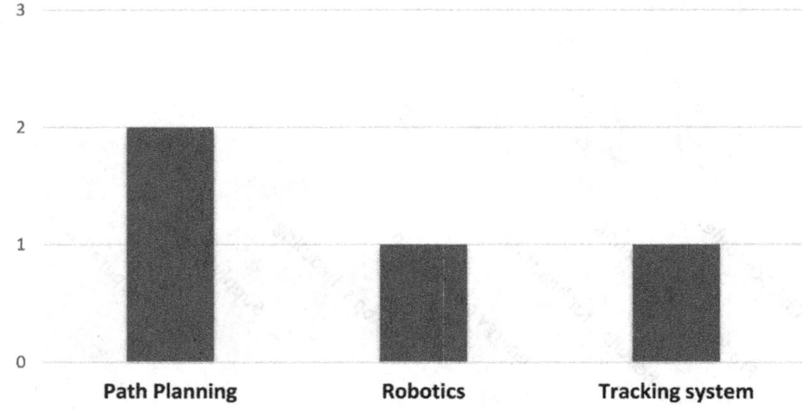

FIGURE 2.18 WOA Publication in Robotics.

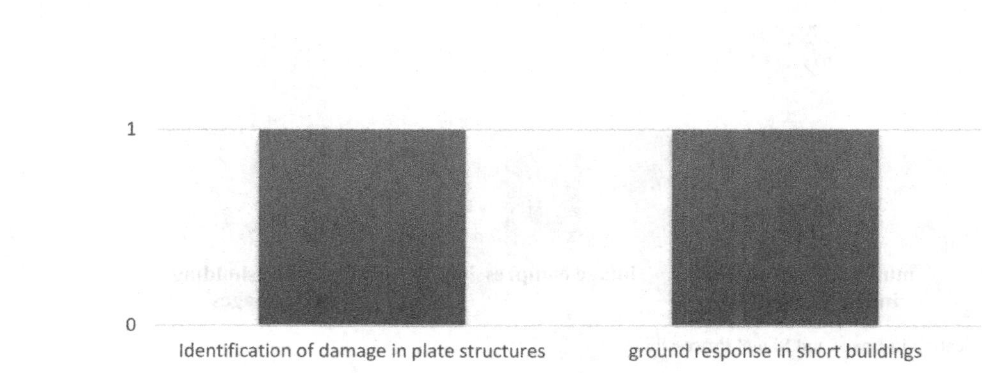

FIGURE 2.19 WOA Publication in structural and mechanical engineering.

Researchers can also find the Matlab code, motivation, and mechanism on the author's webpage.[2] The code was additionally posted on Github.[3]

Another open-source code for a Enhanced whale optimization algorithm (E-WOA) version can be found on the following website.[4]

2.7 Conclusions

The WOA is a robust metaheuristic that mimics the social behavior of humpback whales in nature. The WOA was proposed by Mirjalili in 2016. In this chapter, a whale optimization algorithm (WOA) is reviewed and a result of the numerous optimization challenges that WOA has been used to solve in the short time since its development.

The fundamentals of WOA and its variations that rely on its initial development were covered in this review. Furthermore, all of its variations were discussed, including Binary, Adaptive, Opposition-based learning, Multi-swarm WOA, and others. The performance of the WOA algorithm was improved by hybridizing it with/into other optimization approaches, such as fractional-order, Evolutionary algorithms, Swarm Intelligence, and others.

The applications of WOA have reviewed and summarized and the open sources of WOA variants has been provides in this chapter as well.

2. https://seyedalimirjalili.com/woa.
3. https://github.com/Diptiranjan1/PSO-vs-WOA.
4. https://seyedalimirjalili.com/woa.

Conflict of interest

The authors declare that they have no conflict of interest.

Acknowledgment

This research is supported by UNITEN Postdoc Fellowship 2023.

References

[1] A.E. Eiben, J.E. Smith, et al., Introduction to Evolutionary Computing, vol. 53, Springer, 2003.

[2] R.A. Zitar, M.A. Al-Betar, M.A. Awadallah, I.A. Doush, K. Assaleh, An intensive and comprehensive overview of JAYA algorithm, its versions and applications, Archives of Computational Methods in Engineering (2021) 1–30.

[3] S.N. Makhadmeh, M.A. Al-Betar, A.K. Abasi, M.A. Awadallah, I.A. Doush, Z.A.A. Alyasseri, O.A. Alomari, Recent advances in butterfly optimization algorithm, its versions and applications, Archives of Computational Methods in Engineering (2022) 1–22.

[4] Z.A.A. Alyasseri, O.A. Alomari, M.A. Al-Betar, S.N. Makhadmeh, I.A. Doush, M.A. Awadallah, A.K. Abasi, A. Elnagar, Recent advances of bat-inspired algorithm, its versions and applications, Neural Computing & Applications (2022) 1–36.

[5] F. Fausto, A. Reyna-Orta, E. Cuevas, Á.G. Andrade, M. Perez-Cisneros, From ants to whales: metaheuristics for all tastes, Artificial Intelligence Review 53 (2020) 753–810.

[6] J. Kennedy, R. Eberhart, Particle swarm optimization, in: Proceedings of ICNN95-International Conference on Neural Networks, vol. 4, IEEE, 1995, pp. 1942–1948.

[7] M. Dorigo, G. Di Caro, Ant colony optimization: a new meta-heuristic, in: Proceedings of the 1999 Congress on Evolutionary Computation-CEC99 (Cat. No. 99TH8406), vol. 2, IEEE, 1999, pp. 1470–1477.

[8] G. Dhiman, M. Garg, A. Nagar, V. Kumar, M. Dehghani, A novel algorithm for global optimization: rat swarm optimizer, Journal of Ambient Intelligence and Humanized Computing 12 (2021) 8457–8482.

[9] W. Zhao, L. Wang, S. Mirjalili, Artificial hummingbird algorithm: a new bio-inspired optimizer with its engineering applications, Computer Methods in Applied Mechanics and Engineering 388 (2022) 114194.

[10] J. Xue, B. Shen, A novel swarm intelligence optimization approach: sparrow search algorithm, Systems Science & Control Engineering 8 (2020) 22–34.

[11] F.A. Hashim, A.G. Hussien, Snake optimizer: a novel meta-heuristic optimization algorithm, Knowledge-Based Systems 242 (2022) 108320.

[12] J.O. Agushaka, A.E. Ezugwu, L. Abualigah, Dwarf mongoose optimization algorithm, Computer Methods in Applied Mechanics and Engineering 391 (2022) 114570.

[13] M. Braik, A. Hammouri, J. Atwan, M.A. Al-Betar, M.A. Awadallah, White shark optimizer: a novel bio-inspired meta-heuristic algorithm for global optimization problems, Knowledge-Based Systems 243 (2022) 108457.

[14] M. Khishe, M.R. Mosavi, Chimp optimization algorithm, Expert Systems with Applications 149 (2020) 113338.

[15] F. MiarNaeimi, G. Azizyan, M. Rashki, Horse herd optimization algorithm: a nature-inspired algorithm for high-dimensional optimization problems, Knowledge-Based Systems 213 (2021) 106711.

[16] S. Mirjalili, S.M. Mirjalili, A. Lewis, Grey wolf optimizer, Advances in Engineering Software 69 (2014) 46–61.

[17] S. Mirjalili, Moth-flame optimization algorithm: a novel nature-inspired heuristic paradigm, Knowledge-Based Systems 89 (2015) 228–249.

[18] S. Mirjalili, A. Lewis, The whale optimization algorithm, Advances in Engineering Software 95 (2016) 51–67.

[19] W.A. Watkins, W.E. Schevill, Aerial observation of feeding behavior in four baleen whales: Eubalaena glacialis, Balaenoptera borealis, Megaptera novaeangliae, and Balaenoptera physalus, Journal of Mammalogy 60 (1979) 155–163.

[20] J.A. Goldbogen, A.S. Friedlaender, J. Calambokidis, M.F. McKenna, M. Simon, D.P. Nowacek, Integrative approaches to the study of baleen whale diving behavior, feeding performance, and foraging ecology, BioScience 63 (2013) 90–100.

[21] M. Ghobaei-Arani, A. Shahidinejad, A cost-efficient IoT service placement approach using whale optimization algorithm in fog computing environment, Expert Systems with Applications 200 (2022), https://doi.org/10.1016/j.eswa.2022.117012.

[22] D. Zhang, Z. Chen, Y. Zhou, Wind power interval prediction based on improved whale optimization algorithm and fast learning network, Journal of Electrical Engineering and Technology 17 (2022) 1785–1802, https://doi.org/10.1007/s42835-022-01014-5.

[23] G. Kaur, S. Gill, M. Rattan, Whale optimization algorithm approach for performance optimization of novel Xmas tree-shaped FinFET, Silicon 14 (2022) 3371–3382, https://doi.org/10.1007/s12633-021-01077-5.

[24] X. Xu, C. Liu, Y. Zhao, X. Lv, Short-term traffic flow prediction based on whale optimization algorithm optimized BiLSTM_Attention, Concurrency and Computation: Practice and Experience 34 (2022), https://doi.org/10.1002/cpe.6782.

[25] A. Ewees, Z. Algamal, L. Abualigah, M. Al-Qaness, D. Yousri, R. Ghoniem, M. Elaziz, A Cox proportional-hazards model based on an improved Aquila optimizer with whale optimization algorithm operators, Mathematics 10 (2022), https://doi.org/10.3390/math10081273.

[26] K. Chen, A. Badji, S. Laghrouche, A. Djerdir, Polymer electrolyte membrane fuel cells degradation prediction using multi-kernel relevance vector regression and whale optimization algorithm, Applied Energy 318 (2022), https://doi.org/10.1016/j.apenergy.2022.119099.

[27] A. Yahiaoui, A. Tlemçani, Enhanced whale optimization algorithm for sizing of hybrid wind/photovoltaic/diesel with battery storage in Algeria desert, Wind Engineering 46 (2022) 844–865, https://doi.org/10.1177/0309524X211056529.

[28] K. Saminathan, R. Thangavel, Energy efficient and delay aware clustering in mobile adhoc network: a hybrid fruit fly optimization algorithm and whale optimization algorithm approach, Concurrency and Computation: Practice and Experience 34 (2022), https://doi.org/10.1002/cpe.6867.

[29] A. Zhu, Q. Zhao, X. Wang, L. Zhou, Ultra-short-term wind power combined prediction based on complementary ensemble empirical mode decomposition, whale optimisation algorithm, and Elman network, Energies 15 (2022), https://doi.org/10.3390/en15093055.

[30] S. Dadashzadeh, M. Aghaie, A. Zolfaghari, Optimal design of separation cascades using the whale optimization algorithm, Annals of Nuclear Energy 172 (2022), https://doi.org/10.1016/j.anucene.2022.109020.

[31] B. He, B. Jia, Y. Zhao, X. Wang, M. Wei, R. Dietzel, Estimate soil moisture of maize by combining support vector machine and chaotic whale optimization algorithm, Agricultural Water Management 267 (2022), https://doi.org/10.1016/j.agwat.2022.107618.

[32] S. Wang, W. Hu, I. Riego, Y. Yu, Improved surrogate-assisted whale optimization algorithm for fractional chaotic systems ' parameters identification, Engineering Applications of Artificial Intelligence 110 (2022), https://doi.org/10.1016/j.engappai.2022.104685.

[33] M. Okelola, O. Adebiyi, S. Salimon, S. Ayanlade, A. Amoo, Optimal sizing and placement of shunt capacitors on the distribution system using whale optimization algorithm, Nigerian Journal of Technological Development 19 (2022) 39–47, https://doi.org/10.4314/njtd.v19i1.5.

[34] J. Rahebi, Vector quantization using whale optimization algorithm for digital image compression, Multimedia Tools and Applications 81 (2022) 20077–20103, https://doi.org/10.1007/s11042-022-11952-x.

[35] R. Rajappan, T. Kondampatti Kandaswamy, A composite framework of deep multiple view human joints feature extraction and selection strategy with hybrid adaptive sunflower optimization-whale optimization algorithm for human action recognition in video sequences, Computational Intelligence 38 (2022) 366–396, https://doi.org/10.1111/coin.12499.

[36] J. Liu, J. Shi, F. Hao, M. Dai, A reinforced exploration mechanism whale optimization algorithm for continuous optimization problems, Mathematics and Computers in Simulation 201 (2022) 23–48.

[37] J. Cheng, J. Xu, W. Chen, B. Song, Locating and sizing method of electric vehicle charging station based on improved whale optimization algorithm, Energy Reports 8 (2022) 4386–4400, https://doi.org/10.1016/j.egyr.2022.03.077.

[38] M. Suhail Shaikh, C. Hua, S. Raj, S. Kumar, M. Hassan, M. Mohsin Ansari, M. Ali Jatoi, Optimal parameter estimation of 1-phase and 3-phase transmission line for various bundle conductor's using modified whale optimization algorithm, International Journal of Electrical Power & Energy Systems 138 (2022), https://doi.org/10.1016/j.ijepes.2021.107893.

[39] Y. Liu, C. Li, Y. Zhang, M. Xu, J. Xiao, J. Zhou, HPCP-QCWOA: high performance clustering protocol based on quantum clone whale optimization algorithm in integrated energy system, Future Generation Computer Systems 135 (2022) 315–332, https://doi.org/10.1016/j.future.2022.05.001.

[40] G. Husnain, S. Anwar, An intelligent probabilistic whale optimization algorithm (i-WOA) for clustering in vehicular ad hoc networks, International Journal of Wireless Information Networks 29 (2022) 143–156, https://doi.org/10.1007/s10776-022-00555-w.

[41] W. Liu, Y. Shao, K. Chen, C. Li, H. Luo, Whale optimization algorithm-based point cloud data processing method for sewer pipeline inspection, Automation in Construction 141 (2022), https://doi.org/10.1016/j.autcon.2022.104423.

[42] Z. Wang, Z. Dou, J. Dong, S. Si, C. Wang, L. Liu, Optimal dispatching of regional interconnection multi-microgrids based on multi-strategy improved whale optimization algorithm, IEEJ Transactions on Electrical and Electronic Engineering 17 (2022) 766–779, https://doi.org/10.1002/tee.23566.

[43] F. Zhao, H. Bao, L. Wang, J. Cao, J. Tang, Jonrinaldi, A multipopulation cooperative coevolutionary whale optimization algorithm with a two-stage orthogonal learning mechanism, Knowledge-Based Systems 246 (2022), https://doi.org/10.1016/j.knosys.2022.108664.

[44] J. Xi, Y. Chen, X. Liu, X. Chen, Whale optimization algorithm based on nonlinear adjustment and random walk strategy, Journal of Network Intelligence 7 (2022) 306–318.

[45] G. Sun, Y. Shang, R. Zhang, An efficient and robust improved whale optimization algorithm for large scale global optimization problems, Electronics (Switzerland) 11 (2022), https://doi.org/10.3390/electronics11091475.

[46] G. Ma, X. Yue, An improved whale optimization algorithm based on multilevel threshold image segmentation using the Otsu method, Engineering Applications of Artificial Intelligence 113 (2022), https://doi.org/10.1016/j.engappai.2022.104960.

[47] P. Sumathi, N. Malarvizhi, Whale optimization algorithm with deep learning-based usability recommendation model for medical mobile, International Journal of Engineering Trends and Technology 70 (2022) 251–257, https://doi.org/10.14445/22315381/IJETT-V70I5P227.

[48] A. Mehbodniya, M. Rao, L. David, K. Joe Nige, P. Vennam, Online product sentiment analysis using random evolutionary whale optimization algorithm and deep belief network, Pattern Recognition Letters 159 (2022) 1–8, https://doi.org/10.1016/j.patrec.2022.04.024.

[49] R. Kundu, S. Chattopadhyay, E. Cuevas, R. Sarkar, AltWOA: altruistic whale optimization algorithm for feature selection on microarray datasets, Computers in Biology and Medicine 144 (2022), https://doi.org/10.1016/j.compbiomed.2022.105349.

[50] Z. Yan, J. Zhang, J. Zeng, J. Tang, Three-dimensional path planning for autonomous underwater vehicles based on a whale optimization algorithm, Ocean Engineering 250 (2022), https://doi.org/10.1016/j.oceaneng.2022.111070.

[51] J. Zhou, S. Zhu, Y. Qiu, D. Armaghani, A. Zhou, W. Yong, Predicting tunnel squeezing using support vector machine optimized by whale optimization algorithm, Acta Geotechnica 17 (2022) 1343–1366, https://doi.org/10.1007/s11440-022-01450-7.

[52] U. Khaire, R. Dhanalakshmi, Stability investigation of improved whale optimization algorithm in the process of feature selection, IETE Technical Review (Institution of Electronics and Telecommunication Engineers, India) 39 (2022) 286–300, https://doi.org/10.1080/02564602.2020.1843554.

[53] J. Too, M. Mafarja, S. Mirjalili, Spatial bound whale optimization algorithm: an efficient high-dimensional feature selection approach, Neural Computing & Applications 33 (2021) 16229–16250, https://doi.org/10.1007/s00521-021-06224-y.

[54] T. Thaher, A. Zaguia, S. Al Azwari, M. Mafarja, H. Chantar, A. Abuhamdah, H. Turabieh, S. Mirjalili, A. Sheta, An enhanced evolutionary student performance prediction model using whale optimization algorithm boosted with sine-cosine mechanism, Applied Sciences (Switzerland) 11 (2021), https://doi.org/10.3390/app112110237.

[55] C.-H. Yang, Y.-S. Chen, S.-H. Moi, J.-B. Chen, L.-Y. Chuang, Identifying the association of time-averaged serum albumin levels with clinical factors among patients on hemodialysis using whale optimization algorithm, Mathematics 10 (2022), https://doi.org/10.3390/math10071030.

[56] M. Alkhonaini, S. Haj Hassine, M. Obayya, F. Al-Wesabi, A. Hilal, M. Hamza, A. Motwakel, M. Duhayyim, Detection of lung tumor using ASPP-Unet with whale optimization algorithm, Computers, Materials & Continua 72 (2022) 3511–3527, https://doi.org/10.32604/cmc.2022.024583.

[57] Z. Jin, N. Li, Diagnosis of each main coronary artery stenosis based on whale optimization algorithm and stacking model, Mathematical Biosciences and Engineering 19 (2022) 4568–4591, https://doi.org/10.3934/mbe.2022211.

[58] S. Zahoor, U. Shoaib, I. Lali, Breast cancer mammograms classification using deep neural network and entropy-controlled whale optimization algorithm, Diagnostics 12 (2022), https://doi.org/10.3390/diagnostics12020557.

[59] A. Altameem, S. Kumar, R. Poonia, A. Saudagar, Plant identification using fitness-based position update in whale optimization algorithm, Computers, Materials & Continua 71 (2022) 4719–4736, https://doi.org/10.32604/cmc.2022.022177.

[60] W. Yang, K. Xia, S. Fan, L. Wang, T. Li, J. Zhang, Y. Feng, A multi-strategy whale optimization algorithm and its application, Engineering Applications of Artificial Intelligence 108 (2022), https://doi.org/10.1016/j.engappai.2021.104558.

[61] A. Brodzicki, M. Piekarski, J. Jaworek-Korjakowska, The whale optimization algorithm approach for deep neural networks, Sensors 21 (2021), https://doi.org/10.3390/s21238003.

[62] Q. Liu, Y. Liu, C. Zhang, Z. Ruan, W. Meng, Y. Cai, Q. Ai, sEMG-based dynamic muscle fatigue classification using SVM with improved whale optimization algorithm, IEEE Internet of Things Journal 8 (2021) 16835–16844, https://doi.org/10.1109/JIOT.2021.3056126.

[63] N. Saha, S. Panda, Cosine adapted modified whale optimization algorithm for control of switched reluctance motor, Computational Intelligence 38 (2022) 978–1017, https://doi.org/10.1111/coin.12310.

[64] V. Kumar Jadoun, G. Rahul Prashanth, S. Suhas Joshi, K. Narayanan, H. Malik, F. García Márquez, Optimal fuzzy based economic emission dispatch of combined heat and power units using dynamically controlled whale optimization algorithm, Applied Energy 315 (2022), https://doi.org/10.1016/j.apenergy.2022.119033.

[65] N. Saha, S. Panda, Hybrid modified whale optimisation algorithm simulated annealing technique for control of SRM, International Journal of Applied Metaheuristic Computing 12 (2021) 123–147, https://doi.org/10.4018/IJAMC.2021070105.

[66] A. Got, A. Moussaoui, D. Zouache, Hybrid filter-wrapper feature selection using whale optimization algorithm: a multi-objective approach, Expert Systems with Applications 183 (2021), https://doi.org/10.1016/j.eswa.2021.115312.

[67] A. Kumar, S. Suhag, Whale optimization algorithm optimized fuzzy-PID plus PID hybrid controller for frequency regulation in hybrid power system, Journal of the Institution of Engineers (India): Series B 103 (2022) 633–648, https://doi.org/10.1007/s40031-021-00656-9.

[68] M. Cimen, Y. Yalçın, A novel hybrid firefly–whale optimization algorithm and its application to optimization of MPC parameters, Soft Computing 26 (2022) 1845–1872, https://doi.org/10.1007/s00500-021-06441-6.

[69] Y. Su, Y. Dai, Y. Liu, A hybrid hyper-heuristic whale optimization algorithm for reusable launch vehicle reentry trajectory optimization, Aerospace Science and Technology 119 (2021), https://doi.org/10.1016/j.ast.2021.107200.

[70] T. Kavitha, K. Prasad, Hybridizing ant lion with whale optimization algorithm for compressed sensing MR image reconstruction via l_1 minimization: an ALWOA strategy, Evolutionary Intelligence 14 (2021) 1985–1995, https://doi.org/10.1007/s12065-020-00475-9.

[71] M. Tawhid, A. Ibrahim, Solving nonlinear systems and unconstrained optimization problems by hybridizing whale optimization algorithm and flower pollination algorithm, Mathematics and Computers in Simulation 190 (2021) 1342–1369, https://doi.org/10.1016/j.matcom.2021.07.010.

[72] X. Zhang, S. Wen, Hybrid whale optimization algorithm with gathering strategies for high-dimensional problems, Expert Systems with Applications 179 (2021), https://doi.org/10.1016/j.eswa.2021.115032.

[73] Q. Islam, A. Ahmed, S. Abdullah, Optimized controller design for islanded microgrid using non-dominated sorting whale optimization algorithm (NSWOA), Ain Shams Engineering Journal 12 (2021) 3677–3689, https://doi.org/10.1016/j.asej.2021.01.035.

[74] S. Kumar, D. Parhi, A. Kashyap, M. Muni, Static and dynamic path optimization of multiple mobile robot using hybridized fuzzy logic-whale optimization algorithm, Proceedings of the Institution of Mechanical Engineers, Part C: Journal of Mechanical Engineering Science 235 (2021) 5718–5735, https://doi.org/10.1177/0954406220982641.

[75] P. Stephan, T. Stephan, R. Kannan, A. Abraham, A hybrid artificial bee colony with whale optimization algorithm for improved breast cancer diagnosis, Neural Computing & Applications 33 (2021) 13667–13691, https://doi.org/10.1007/s00521-021-05997-6.

[76] W. Yankai, W. Shilong, L. Dong, S. Chunfeng, Y. Bo, An improved multi-objective whale optimization algorithm for the hybrid flow shop scheduling problem considering device dynamic reconfiguration processes, Expert Systems with Applications 174 (2021), https://doi.org/10.1016/j.eswa.2021.114793.

[77] C. Paul, P. Roy, V. Mukherjee, Application of chaotic quasi-oppositional whale optimization algorithm on CHPED problem integrated with wind-solar-EVs, International Transactions on Electrical Energy Systems 31 (2021), https://doi.org/10.1002/2050-7038.13124.

[78] W. Yang, Z. Peng, Z. Yang, Y. Guo, X. Chen, An enhanced exploratory whale optimization algorithm for dynamic economic dispatch, Energy Reports 7 (2021) 7015–7029, https://doi.org/10.1016/j.egyr.2021.10.067.

[79] M. Tahmasebi, J. Pasupuleti, F. Mohamadian, M. Shakeri, J. Guerrero, M. Basir Khan, M. Nazir, A. Safari, N. Bazmohammadi, Optimal operation of stand-alone microgrid considering emission issues and demand response program using whale optimization algorithm, Sustainability (Switzerland) 13 (2021), https://doi.org/10.3390/su13147710.

[80] A. Rasool, N. Abbas, K. Sheikhyounis, Determination of optimal size and location of static synchronous compensator for power system bus voltage improvement and loss reduction using whale optimization algorithm, Eastern-European Journal of Enterprise Technologies 1 (2022) 26–34, https://doi.org/10.15587/1729-4061.2022.251760.

[81] B. Dasu, S. Mangipudi, S. Rayapudi, Small signal stability enhancement of a large scale power system using a bio-inspired whale optimization algorithm, Protection and Control of Modern Power Systems 6 (2021), https://doi.org/10.1186/s41601-021-00215-w.

[82] S. Natarajan, A. Loganathan, Analysis of energy management controller in grid-connected PV wind power system coupled with battery using whale optimisation algorithm, Iranian Journal of Science and Technology - Transactions of Electrical Engineering 46 (2022) 77–90, https://doi.org/10.1007/s40998-021-00469-y.

[83] K. Ghany, A. AbdelAziz, T. Soliman, A.-M. Sewisy, A hybrid modified step whale optimization algorithm with tabu search for data clustering, Journal of King Saud University: Computer and Information Sciences 34 (2022) 832–839, https://doi.org/10.1016/j.jksuci.2020.01.015.

[84] L. Abood, B. Oleiwi, Design of fractional order PID controller for AVR system using whale optimization algorithm, Indonesian Journal of Electrical Engineering and Computer Science 23 (2021) 1410–1418, https://doi.org/10.11591/ijeecs.v23.i3.pp1410-1418.

[85] Y. Gu, M. Liu, Fair and privacy-aware EV discharging strategy using decentralized whale optimization algorithm for minimizing cost of EVs and the EV aggregator, IEEE Systems Journal 15 (2021) 5571–5582, https://doi.org/10.1109/JSYST.2021.3050565.

[86] M. Abd Elaziz, S. Lu, S. He, A multi-leader whale optimization algorithm for global optimization and image segmentation, Expert Systems with Applications 175 (2021), https://doi.org/10.1016/j.eswa.2021.114841.

[87] M. Rostum, H. Moustafa, I. Ziedan, A. Zamel, A combined effective time series model based on clustering and whale optimization algorithm for forecasting smart meters electricity consumption, COMPEL - The International Journal for Computation and Mathematics in Electrical and Electronic Engineering 41 (2022) 209–237, https://doi.org/10.1108/COMPEL-04-2021-0150.

[88] B. Pu, F. Nan, N. Zhu, Y. Yuan, W. Xie, UFNGBM (1, 1): a novel unbiased fractional grey Bernoulli model with Whale Optimization Algorithm and its application to electricity consumption forecasting in China, Energy Reports 7 (2021) 7405–7423, https://doi.org/10.1016/j.egyr.2021.09.105.

[89] X. Xiong, X. Hu, H. Guo, A hybrid optimized grey seasonal variation index model improved by whale optimization algorithm for forecasting the residential electricity consumption, Energy 234 (2021), https://doi.org/10.1016/j.energy.2021.121127.

[90] M. Jalaee, A. Ghaseminejad, S. Jalaee, N. Zarin, R. Derakhshani, A novel hybrid artificial intelligence approach to the future of global coal consumption using whale optimization algorithm and adaptive neuro-fuzzy inference system, Energies 15 (2022), https://doi.org/10.3390/en15072578.

[91] A. Darvish Falehi, An optimal second-order sliding mode based inter-area oscillation suppressor using chaotic whale optimization algorithm for doubly fed induction generator, International Journal of Numerical Modelling: Electronic Networks, Devices and Fields 35 (2022), https://doi.org/10.1002/jnm.2963.

[92] T.-C. Pan, E.-J. Liu, H.-C. Ku, C.-W. Hong, Parameter identification and sensitivity analysis of lithium-ion battery via whale optimization algorithm, Electrochimica Acta 404 (2022), https://doi.org/10.1016/j.electacta.2021.139574.

[93] S. Suman, D. Chatterjee, M. Anand, R. Mohanty, A. Bhattacharya, An improved harmonic reduction technique for the PV-wind hybrid generation scheme using modified whale optimization algorithm (MWOA), UPB Scientific Bulletin, Series C: Electrical Engineering and Computer Science 84 (2022) 231–248.

[94] S. Nethravathi, V. Murali, A novel residential energy management system based on sequential whale optimization algorithm and fuzzy logic, Distributed Generation and Alternative Energy Journal 37 (2022) 557–586, https://doi.org/10.13052/dgaej2156-3306.3739.

[95] X. Teng, Y. Luo, T. Zheng, X. Zhang, An improved whale optimization algorithm based on aggregation potential energy for QoS-driven web service composition, Wireless Communications and Mobile Computing 2022 (2022), https://doi.org/10.1155/2022/9741278.

[96] B. Gao, H. Yang, H.-C. Lin, Z. Wang, W. Zhang, H. Li, A hybrid improved whale optimization algorithm with support vector machine for short-term photovoltaic power prediction, Applied Artificial Intelligence 36 (2022), https://doi.org/10.1080/08839514.2021.2014187.

[97] H. Sun, A. Ebadi, M. Toughani, S. Nowdeh, A. Naderipour, A. Abdullah, Designing framework of hybrid photovoltaic-biowaste energy system with hydrogen storage considering economic and technical indices using whale optimization algorithm, Energy 238 (2022), https://doi.org/10.1016/j.energy.2021.121555.

[98] A. Mohammadbeigi, A. Maroosi, M. Hemmati, Optimal chiller loading for energy conservation using a hybrid whale optimization algorithm based on population membrane systems, International Journal of Modelling and Simulation 42 (2022) 101–116, https://doi.org/10.1080/02286203.2020.1843935.

[99] K. Vanchinathan, K. Valluvan, C. Gnanavel, C. Gokul, R. Renold, An improved incipient whale optimization algorithm based robust fault detection and diagnosis for sensorless brushless DC motor drive under external disturbances, International Transactions on Electrical Energy Systems 31 (2021), https://doi.org/10.1002/2050-7038.13251.

[100] M. Nadimi-Shahraki, S. Taghian, S. Mirjalili, L. Abualigah, M. Elaziz, D. Oliva, EWOA-OPF: effective whale optimization algorithm to solve optimal power flow problem, Electronics (Switzerland) 10 (2021), https://doi.org/10.3390/electronics10232975.

[101] R. Yang, K. Li, K. Du, B. Shen, An ameliorative whale optimization algorithm (AWOA) for HES energy management strategy optimization, Regional Studies in Marine Science 48 (2021), https://doi.org/10.1016/j.rsma.2021.102033.

[102] Q. Fan, F. Yu, M. Xuan, Transformer fault diagnosis method based on improved whale optimization algorithm to optimize support vector machine, Energy Reports 7 (2021) 856–866, https://doi.org/10.1016/j.egyr.2021.09.188.

[103] M. Danoune, A. Djafour, Y. Wang, A. Gougui, The whale optimization algorithm for efficient PEM fuel cells modeling, International Journal of Hydrogen Energy 46 (2021) 37599–37611, https://doi.org/10.1016/j.ijhydene.2021.03.105.

[104] P. Dhar, N. Chakraborty, Optimal techno-economic analysis of a renewable based hybrid microgrid incorporating gravity energy storage system in Indian perspective using whale optimization algorithm, International Transactions on Electrical Energy Systems 31 (2021), https://doi.org/10.1002/2050-7038.13025.

[105] M. Alemi-Rostami, G. Rezazadeh, Selective harmonic elimination of a multilevel voltage source inverter using whale optimization algorithm, International Journal of Engineering, Transactions B: Applications 34 (2021) 1898–1904, https://doi.org/10.5829/ije.2021.34.08b.11.

[106] R. Kushwah, M. Kaushik, K. Chugh, A modified whale optimization algorithm to overcome delayed convergence in artificial neural networks, Soft Computing 25 (2021) 10275–10286, https://doi.org/10.1007/s00500-021-05983-z.

[107] M. Li, G. Xu, B. Fu, X. Zhao, Whale optimization algorithm based on dynamic pinhole imaging and adaptive strategy, Journal of Supercomputing 78 (2022) 6090–6120, https://doi.org/10.1007/s11227-021-04116-5.

[108] Q. Guo, L. Gao, X. Chu, H. Sun, Parameter identification for static var compensator model using sensitivity analysis and improved whale optimization algorithm, CSEE Journal of Power and Energy Systems 8 (2022) 535–547, https://doi.org/10.17775/CSEEJPES.2021.03540.

[109] C. Tang, W. Sun, M. Xue, X. Zhang, H. Tang, W. Wu, A hybrid whale optimization algorithm with artificial bee colony, Soft Computing 26 (2022) 2075–2097, https://doi.org/10.1007/s00500-021-06623-2.

[110] M. Li, G. Xu, Q. Lai, J. Chen, A chaotic strategy-based quadratic opposition-based learning adaptive variable-speed whale optimization algorithm, Mathematics and Computers in Simulation 193 (2022) 71–99, https://doi.org/10.1016/j.matcom.2021.10.003.

[111] R. Liang, Y. Chen, R. Zhu, A novel fault diagnosis method based on the KELM optimized by whale optimization algorithm, Machines 10 (2022), https://doi.org/10.3390/machines10020093.

[112] A. Ferrari, C. Silva, C. Osinski, D. Pelacini, G. Leandro, L. Coelho, Tuning of control parameters of the whale optimization algorithm using fuzzy inference system, Journal of Intelligent & Fuzzy Systems 42 (2022) 3051–3066, https://doi.org/10.3233/JIFS-210781.

[113] M. Li, G. Xu, Y. Fu, T. Zhang, L. Du, Improved whale optimization algorithm based on variable spiral position update strategy and adaptive inertia weight, Journal of Intelligent & Fuzzy Systems 42 (2022) 1501–1517, https://doi.org/10.3233/JIFS-210842.

[114] H. Jin, S. Lv, Z. Yang, Y. Liu, Eagle strategy using uniform mutation and modified whale optimization algorithm for QoS-aware cloud service composition, Applied Soft Computing 114 (2022), https://doi.org/10.1016/j.asoc.2021.108053.

[115] Q. Fan, Z. Chen, W. Zhang, X. Fang, ESSAWOA: enhanced whale optimization algorithm integrated with salp swarm algorithm for global optimization, Engineering With Computers 38 (2022) 797–814, https://doi.org/10.1007/s00366-020-01189-3.

[116] S. Chakraborty, A. Saha, R. Chakraborty, M. Saha, S. Nama Hswoa, An ensemble of hunger games search and whale optimization algorithm for global optimization, International Journal of Intelligent Systems 37 (2022) 52–104, https://doi.org/10.1002/int.22617.

[117] A. Ouladbrahim, I. Belaidi, S. Khatir, E. Magagnini, R. Capozucca, M. Abdel Wahab, Experimental crack identification of API X70 steel pipeline using improved artificial neural networks based on whale optimization algorithm, Mechanics of Materials 166 (2022), https://doi.org/10.1016/j.mechmat.2021.104200.

[118] Y. Sun, Y. Chen, Multi-population improved whale optimization algorithm for high dimensional optimization, Applied Soft Computing 112 (2021), https://doi.org/10.1016/j.asoc.2021.107854.

[119] S. Chakraborty, A. Saha, R. Chakraborty, M. Saha, An enhanced whale optimization algorithm for large scale optimization problems, Knowledge-Based Systems 233 (2021), https://doi.org/10.1016/j.knosys.2021.107543.

[120] B. Shivahare, S. Gupta, Multi-level image segmentation using randomized spiral-based whale optimization algorithm, Recent Patents on Engineering 15 (2021), https://doi.org/10.2174/1872212114999200730163151.

[121] J. Anitha, S. Immanuel Alex Pandian, S. Akila Agnes, An efficient multilevel color image thresholding based on modified whale optimization algorithm, Expert Systems with Applications 178 (2021), https://doi.org/10.1016/j.eswa.2021.115003.

[122] M. Alshayeji, B. Behbehani, I. Ahmad, Spark-based parallel processing whale optimization algorithm, Concurrency and Computation: Practice and Experience 34 (2022), https://doi.org/10.1002/cpe.6607.

[123] B. Aygün, H. Arslan, Block size optimization for pow consensus algorithm based blockchain applications by using whale optimization algorithm, Turkish Journal of Electrical Engineering & Computer Sciences 30 (2022) 406–419, https://doi.org/10.3906/elk-2105-217.

[124] H. Al-Khazraji, Comparative study of whale optimization algorithm and flower pollination algorithm to solve workers assignment problem, International Journal of Production Management and Engineering 10 (2022) 91–98, https://doi.org/10.4995/ijpme.2022.16736.

[125] Y. Zhang, Z. Zhang, C. Guan, P. Xu, Improved whale optimisation algorithm for two-sided disassembly line balancing problems considering part characteristic indexes, International Journal of Production Research 60 (2022) 2553–2571, https://doi.org/10.1080/00207543.2021.1897178.

[126] Q. Askari, I. Younas, M. Saeed, Emphasizing the importance of shift invariance in metaheuristics by using whale optimization algorithm as a test bed, Soft Computing 25 (2021) 14209–14225, https://doi.org/10.1007/s00500-021-06101-9.

[127] J. Zhang, L. Hong, Q. Liu, An improved whale optimization algorithm for the traveling salesman problem, Symmetry 13 (2021) 1–13, https://doi.org/10.3390/sym13010048.

[128] W. Guo, T. Liu, F. Dai, F. Zhao, P. Xu, Skewed normal cloud modified whale optimization algorithm for degree reduction of S-λ curves, Applied Intelligence 51 (2021) 8377–8398, https://doi.org/10.1007/s10489-021-02339-w.

[129] H. Wang, F. Wu, L. Zhang, Application of variational mode decomposition optimized with improved whale optimization algorithm in bearing failure diagnosis, Alexandria Engineering Journal 60 (2021) 4689–4699, https://doi.org/10.1016/j.aej.2021.03.034.

[130] A. Jaafari, M. Panahi, D. Mafi-Gholami, O. Rahmati, H. Shahabi, A. Shirzadi, S. Lee, D. Bui, B. Pradhan, Swarm intelligence optimization of the group method of data handling using the cuckoo search and whale optimization algorithms to model and predict landslides, Applied Soft Computing 116 (2022), https://doi.org/10.1016/j.asoc.2021.108254.

[131] M. Tair, N. Bacanin, M. Zivkovic, K. Venkatachalam, A chaotic oppositional whale optimisation algorithm with firefly search for medical diagnostics, Computers, Materials & Continua 72 (2022) 959–982, https://doi.org/10.32604/cmc.2022.024989.

[132] F. Xu, L. Hu, T. Jia, S. Du, Impact feature recognition method for non-stationary signals based on variational modal decomposition noise reduction and support vector machine optimized by whale optimization algorithm, Review of Scientific Instruments 92 (2021), https://doi.org/10.1063/5.0065197.

[133] X. Cui, E. Shaojun, D. Niu, B. Chen, J. Feng, Forecasting of carbon emission in China based on gradient boosting decision tree optimized by modified whale optimization algorithm, Sustainability (Switzerland) 13 (2021), https://doi.org/10.3390/su132112302.

[134] M. Wang, Z. Yan, J. Luo, Z. Ye, P. He, A band selection approach based on wavelet support vector machine ensemble model and membrane whale optimization algorithm for hyperspectral image, Applied Intelligence 51 (2021) 7766–7780, https://doi.org/10.1007/s10489-021-02270-0.

[135] D. Kotary, S. Nanda, R. Gupta, A many-objective whale optimization algorithm to perform robust distributed clustering in wireless sensor network, Applied Soft Computing 110 (2021), https://doi.org/10.1016/j.asoc.2021.107650.

[136] L. Wang, L. Gu, Y. Tang, Research on alarm reduction of intrusion detection system based on clustering and whale optimization algorithm, Applied Sciences (Switzerland) 11 (2021), https://doi.org/10.3390/app112311200.

[137] H. Peng, W.-S. Wen, M.-L. Tseng, L.-L. Li, A cloud load forecasting model with nonlinear changes using whale optimization algorithm hybrid strategy, Soft Computing 25 (2021) 10205–10220, https://doi.org/10.1007/s00500-021-05961-5.

[138] B. Rehman, M. Babar, A. Ahmad, M. Amir, W. Shahjehan, A. Sadiq, S. Mirjalili, A. Dehkordi, Joint user grouping and power control using whale optimization algorithm for NOMA uplink systems, PeerJ Computer Science 8 (2022), https://doi.org/10.7717/PEERJ-CS.882.

[139] M. Toloueiashtian, M. Golsorkhtabaramiri, S. Rad, Solving point coverage problem in wireless sensor networks using whale optimization algorithm, The International Arab Journal of Information Technology 18 (2021) 830–838, https://doi.org/10.34028/iajit/18/6/10.

[140] V. Godbole, Performance improvement in wireless sensor networks using whale optimisation algorithm (WOA) and butterfly optimisation algorithm (BOA), Defence S and T Technical Bulletin 15 (2022) 68–82.

[141] B. Priyanka, R. Jayaparvathy, D. DivyaBharathi, Efficient and dynamic cluster head selection for improving network lifetime in WSN using whale optimization algorithm, Wireless Personal Communications 123 (2022) 1467–1481, https://doi.org/10.1007/s11277-021-09192-7.

[142] R. Deepa, R. Venkataraman, Enhancing whale optimization algorithm with Levy flight for coverage optimization in wireless sensor networks, Computers & Electrical Engineering 94 (2021), https://doi.org/10.1016/j.compeleceng.2021.107359.

[143] M. Toloueiashtian, M. Golsorkhtabaramiri, S. Rad, An improved whale optimization algorithm solving the point coverage problem in wireless sensor networks, Telecommunication Systems 79 (2022) 417–436, https://doi.org/10.1007/s11235-021-00866-y.

[144] S. Hosseini, J. Vahidi, S. Tabbakh, A. Shojaei, Resource allocation optimization in cloud computing using the whale optimization algorithm, International Journal of Nonlinear Analysis and Applications 12 (2021) 343–360, https://doi.org/10.22075/ijnaa.2021.5188.

[145] S. Karthick, N. Gomathi, Galactic swarm-improved whale optimization algorithm-based resource management in internet of things, International Journal of Communication Systems 35 (2022), https://doi.org/10.1002/dac.5006.

[146] X. Meng, C. Jia, C. Cai, F. He, Q. Wang, Indoor high-precision 3D positioning system based on visible-light communication using improved whale optimization algorithm, Photonics 9 (2022), https://doi.org/10.3390/photonics9020093.

[147] K. SureshKumar, P. Vimala, Energy efficient routing protocol using exponentially-ant lion whale optimization algorithm in wireless sensor networks, Computer Networks 197 (2021), https://doi.org/10.1016/j.comnet.2021.108250.

[148] X. Lin, W. Lin, Whale optimization algorithm-based LQG-adaptive neuro-fuzzy control for seismic vibration mitigation with MR dampers, in: Shock and Vibration, vol. 2022, 2022.

[149] H. Bagheri, M. Sadegh Hosseini, H. Ghayoumi Zadeh, B. Notej, A. Fayazi, A novel modification of ionic liquid mixture density based on semi-empirical equations using Laplacian whale optimization algorithm, Arabian Journal of Chemistry 14 (2021), https://doi.org/10.1016/j.arabjc.2021.103368.

[150] Z. Wang, Y. Li, Y. Tang, Z. Shang, Enhanced precision inspection of free-form surface with an improved whale optimization algorithm, Optics Express 29 (2021) 26909–26924, https://doi.org/10.1364/OE.433975.

[151] P. Gupta, S. Bhagat, D. Saini, A. Kumar, M. Alahmadi, P. Sharma, Hybrid whale optimization algorithm for resource optimization in cloud e-healthcare applications, Computers, Materials & Continua 71 (2022) 5659–5676, https://doi.org/10.32604/cmc.2022.023056.

[152] F. Lu, T. Yan, H. Bi, M. Feng, S. Wang, M. Huang, A bilevel whale optimization algorithm for risk management scheduling of information technology projects considering outsourcing, Knowledge-Based Systems 235 (2022), https://doi.org/10.1016/j.knosys.2021.107600.

[153] W. Yang, Z. Yang, Y. Chen, Z. Peng, Modified whale optimization algorithm for multi-type combine harvesters scheduling, Machines 10 (2022), https://doi.org/10.3390/machines10010064.

[154] A. Ala, F. Alsaadi, M. Ahmadi, S. Mirjalili, Optimization of an appointment scheduling problem for healthcare systems based on the quality of fairness service using whale optimization algorithm and NSGA-II, Scientific Reports 11 (2021), https://doi.org/10.1038/s41598-021-98851-7.

[155] M. Abedini, M. Mostafai, Adaptive energy consumption scheduling of multi-microgrid using whale optimization algorithm, International Journal of Modeling, Simulation, and Scientific Computing 12 (2021), https://doi.org/10.1142/S1793962321500367.

[156] C. Govindasamy, A. Antonidoss, Enhanced inventory management using blockchain technology under cloud sector enabled by hybrid multi-verse with whale optimization algorithm, International Journal of Information Technology & Decision Making 21 (2022) 577–614, https://doi.org/10.1142/S021962202150067X.

[157] K. Yang, K. Yang, Short-term hydro generation scheduling of the Three Gorges Hydropower Station using improver binary-coded whale optimization algorithm, Water Resources Management 35 (2021) 3771–3790, https://doi.org/10.1007/s11269-021-02917-0.

[158] D.K. Patra, T. Si, S. Mondal, P. Mukherjee, Magnetic resonance image of breast segmentation by multi-level thresholding using moth-flame optimization and whale optimization algorithms, Pattern Recognition and Image Analysis 32 (2022) 174–186, https://doi.org/10.1134/S1054661822010060.

[159] S. Agrawal, R. Panda, P. Choudhury, A. Abraham, Dominant color component and adaptive whale optimization algorithm for multilevel thresholding of color images, Knowledge-Based Systems 240 (2022), https://doi.org/10.1016/j.knosys.2022.108172.

[160] M. Abdel-Basset, R. Mohamed, N. AbdelAziz, M. Abouhawwash, HWOA: a hybrid whale optimization algorithm with a novel local minima avoidance method for multi-level thresholding color image segmentation, Expert Systems with Applications 190 (2022), https://doi.org/10.1016/j.eswa.2021.116145.

[161] S. Dereli, A novel approach based on average swarm intelligence to improve the whale optimization algorithm, Arabian Journal for Science and Engineering 47 (2022) 1763–1776, https://doi.org/10.1007/s13369-021-06042-3.

[162] M. Huang, X. Cheng, Z. Zhu, J. Luo, J. Gu, A novel two-stage structural damage identification method based on superposition of modal flexibility curvature and whale optimization algorithm, International Journal of Structural Stability and Dynamics 21 (2021), https://doi.org/10.1142/S0219455421501698.

[163] Z. Liu, L. Zhang, J. Li, M. Mamluki, Predicting the seismic response of the short structures by considering the whale optimization algorithm, Energy Reports 7 (2021) 4071–4084, https://doi.org/10.1016/j.egyr.2021.06.095.

Chapter 3

A hybrid whale optimization algorithm with tabu search algorithm for resource allocation in indoor VLC systems

Selma Yahia[a], Yassine Meraihi[a], Seyedali Mirjalili[b,c], Sylia Mekhmoukh Taleb[a], Souad Refas[a], Amar Ramdane-Cherif[d], and Hossien B. Eldeeb[e]

[a]*LIST Laboratory, University of M'Hamed Bougara Boumerdes, Boumerdes, Algeria,* [b]*Centre for Artificial Intelligence Research and Optimisation, Torrens University Australia, Brisbane, QLD, Australia,* [c]*Yonsei Frontier Lab, Yonsei University, Seoul, South Korea,* [d]*LISV Laboratory, University of Versailles St-Quentin-en-Yvelines, Velizy, France,* [e]*Department of Electrical and Electronics Engineering, Özyeğin University, Istanbul, Turkey*

3.1 Introduction

Visible Light Communication (VLC) technology has generated considerable attention as a propitious solution for indoor wireless communication applications in 5G/6G networks [1]. This technology leverages the capabilities of white Light-Emitting Diodes (LEDs) to not only provide illumination, but also support high data rate communication. This makes VLC a complementary or alternative option to traditional Radio Frequency (RF) communication networks.

VLC systems have many advantages over traditional RF communications: they offer vast unregulated spectrum resources, do not generate electromagnetic interference, potentially offer greater security, and are environmentally friendly. Moreover, VLC technology is highly versatile and can be implemented in areas where RF transmission is restricted, such as hospitals and airplanes. Despite its various benefits, VLC systems face certain limitations, such as the dependency on Line-of-Sight (LoS). This issue can significantly reduce the coverage area of the LED and thus hinder the effective use of VLC in certain scenarios. Several works have been conducted to solve this problem using multi-user (MU) architectures [2–4]. In such scenarios, however, multiple transmitters are required to cover the entire communication area, which can cause severe interference between users. Therefore, an appropriate resource allocation scheme is needed to solve this problem.

Resource optimization in VLC systems is a vital activity. So far, some studies have focused on resource optimization in VLC systems [5–10]. For instance, the authors of [5], have formulated a resource allocation problem with the goal of maximizing the user satisfaction index under runtime constraints. This work investigates the optimal allocation of users to different subcarriers and LEDs using the simulated annealing (SA) algorithm. In [6], Gong et al. tackled the problems of sum power minimization and rate maximization by optimizing the power allocation of each LED (i.e., red, green, and blue colors) under lighting constraints. In [7], the power allocation issue in a downlink VLC system with multiple users is investigated, aiming To optimize the total throughput under several constraints such as chromaticity, bit error rate, amplitude, and luminance. In [8], a multi-cell VLC system is considered and the authors investigate the problem of user association, subcarrier and power allocation. The objective is to maximize the minimum throughput while meeting total power constraints. In [10], the authors present a proficient LED allocation plan that is designed to enhance the signal-to-interference-plus-noise ratio (SINR) of users by suppressing the interference caused by simultaneous LEDs data transmission. In this paper, we introduce a novel hybrid algorithm called WOATS, which combines the Whale Optimization Algorithm (WOA) with Tabu Search (TS) to solve the resource allocation problem in multi-user downlink VLC systems.

The main contributions of this paper are listed below

- We consider a multi-user downlink VLC system in which multiple LED transmit antennas and multiple users with multiple receive antennas are used. To reduce inter-symbol and co-channel interference while reducing users' battery power consumption, we consider enabling only the most appropriate PDs. The selection of these PDs is made to maximize both throughput and fairness for all users.
- To obtain the optimal PDs of each user, we involve a novel hybrid algorithm called WOATS. This algorithm combines the strengths of two well-known algorithms, WOA and TS.

Handbook of Whale Optimization Algorithm. https://doi.org/10.1016/B978-0-32-395365-8.00009-9

- Different user distribution scenarios are considered and the impact of the users number and PDs on the proposed algorithm efficiency is investigated for all considered scenarios.
- The proposed algorithm's performance is assessed by comparing it with other algorithms, such as WOA [11], TS [12], PSO [13], GA [14], AOA [15], MFO [16], GWO [17], SCA [18] algorithms.

Subsequent sections of this paper are laid out as: Section 3.2 describes the system model, whereas Section 3.3 deals with the formulation of the resource allocation problem. Section 3.4 proposes the hybrid WOATS algorithm to solve the resource allocation problem. In Section 3.5, we illustrate the results and analyze the performance of WOATS algorithm in different scenarios. Lastly, Section 3.6 provides the concluding remarks for this paper.

3.2 System model

We consider an indoor downlink VLC scenario in a room of dimension $l \times w \times h$ (see Fig. 3.1). This room is equipped with L light-emitting diodes (LEDs) with an optical power of P_{opt}, placed on the ceiling to serve as an access point (AP). We assume N receiving users move around inside this room, where each is equipped with M photodetectors (PDs) that are evenly distributed on the user's cell phone. Each of these PDs has a responsivity of R and an aperture diameter of D_r.

FIGURE 3.1 System Model.

For channel modeling, we adopt the non-sequential ray tracing approach of OpticStudio software, validated in [19], to model the indoor VLC channel, based on the IEEE VLC channel modeling reference [20]. This methodology accurately describes the interaction between rays emitted by a lighting source and the objects encountered Within a defined environment. The key phases of this methodology are presented in Fig. 3.2 and can be summarized as follows:

- A simulation environment with three dimensions has been established and imported into the OpticStudio, where the main characteristics of the system are included such as the wavelength reflectance of surfaces (floor, ceiling, and objects) and the transceiver location.

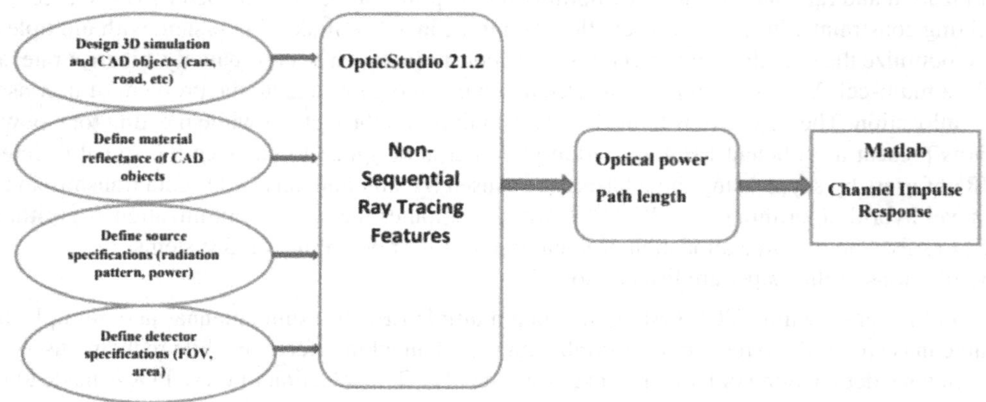

FIGURE 3.2 Main steps of ray tracing channel modeling.

- The different specifications of the transceiver including the light source (i.e. optical emission power, intensity profile, and orientations) and the detector (aperture, angles, and orientations) are defined.
- A non-sequential ray tracing is finally executed, and for each incoming ray, the path length and detected optical power are calculated. This output data is typically processed using Matlab® software to calculate the different channel statistics, such as channel impulse response (CIR).

The CIR can be formulated as [20]:

$$h(t) = \sum_{f=1}^{F} P_f \delta \left(t - \tau_f \right), \tag{3.1}$$

where P_f is the power and τ_f is the propagation time of the ith. F is the total number of rays reached at the receiver and δ is the Dirac delta function.

3.3 Problem formulation

As shown in Fig. 3.1, the multi-user downlink VLC system consists of multiple users, each with multiple PDs. However, to reduce the battery consumption of the phone, not all PDs are enabled. Thus, we select only the optimal PDs to be enabled, and the others are disabled. Consider M_a the number of activated PDs. Thus, for M_a PD activated at time t, the (M- M_a) other PDs are switched off. Assigning the operational PDs to various users is carried out to minimize the battery usage load and reduce inter-symbol interference, which in turn improves the overall network performance. The activation and deactivation states of all PDs linked to different users are summarized by the following assignment matrix A $(N \times M)$

$$A = \begin{pmatrix} a_{11} & a_{12} & \cdots & a_{1M} \\ a_{21} & a_{22} & \cdots & a_{2M} \\ \vdots & \ddots & \vdots & \vdots \\ a_{N1} & a_{N2} & \cdots & a_{NM} \end{pmatrix} \tag{3.2}$$

where a_{lk} represents a binary allocation variable related to the kth PD of the lth user. This variable is set to 1 in the case of an enabled PD and set to 0 in the case of a non-enabled PD. Since not all PDs are activated at a time t, the allocation variables related to the lth user sum up to M_a. In other words, $\left(\sum_{k=1}^{M} a_{lk} = M_a\right)$, $\forall l \in \{1, 2, \cdots, N\}$. We further define $\bar{a}$ as the vector representing the active PD index for each user, which is written as $\bar{a} = [\acute{a}_1, \acute{a}_2, ..., \acute{a}_M]$, where $\acute{a}_l \in a_1, \in a_2, ..., \in a_M$ is the index of the PD activated for the lth user at the $\bar{a}$ allocation vector. Thus, the values of the binary variables a_{lk} can be given by

$$a_{lk} = \begin{cases} 1, & k = \acute{a}_l \\ 0, & \text{otherwise} \end{cases} \tag{3.3}$$

Let denotes $H_{k,l}$ the channel gain at the kth PD of the lth, SNR at the lth user is given by

$$\gamma_l(\bar{a}) = \frac{R^2 \left(\sum_{k=1}^{M} P_t H_{k,l} a_{lk} \right)^2}{\sigma_t^2}, \tag{3.4}$$

where R is the PD responsivity and σ_t^2 is the total noise variance. Note that the bit error rate (BER) must be below the BER threshold, BER_{th}, to ensure reliable communication. In other words, $BER \leq BER_{th}$.

Therefore, the overall rate of data transfer can be represented by

$$C(\bar{a}) \approx \sum_{l=1}^{N} \underbrace{\left(\frac{B}{2\ln(2)} \ln \left(1 + \frac{\exp(1)\gamma_l(\bar{a})}{2\pi} \right) \right)}_{C_l(\bar{a})}, \tag{3.5}$$

where B is the bandwidth.

Based on (3.5), we further define the fairness between different users, which can be expressed as follows [21]

$$F(\bar{a}) = \frac{(C(\bar{a}))^2}{N\sum_{l-1}^{N} C_l(\bar{a})^2} \tag{3.6}$$

In our study, two main objectives are needed to be optimized, which are throughput and fairness. Thus, the problem under consideration Can be expressed as a problem of maximizing which can be formulated as follows:

$$\begin{aligned} &\max_{\bar{a}} (C(\bar{a})) \\ &\max_{\bar{a}} (F(\bar{a})) \end{aligned} \tag{3.7}$$

s.t.

$$\textbf{C1:} \quad a_{lk} = \begin{cases} 1, & k = \acute{a}_l \\ 0, & \text{otherwise} \end{cases} \quad \forall l, k \tag{3.8a}$$

$$\sum_{k-1}^{M} a_{lk} = M_a, \ \forall l$$

$$\begin{aligned} &\textbf{C2:} \quad BER \leq BER_{th} \\ &l \in \{1, 2, \cdots, N\},\ k \in \{1, 2, \cdots, M\}, \\ &\acute{a}_l \in \{1, 2, \ldots, M\} \end{aligned} \tag{3.8b}$$

3.4 Preliminaries

3.4.1 Whale Optimization Algorithm (WOA)

The Whale Optimization Algorithm, developed by Mirjalili and Lewis in 2016 [11], is a swarm-based optimization algorithm that models the foraging behavior of humpback whales, making it an effective tool for solving complex optimization problems. The humpback whale uses a particular hunting method called "bubble-net hunting" as part of its marine activities. The WOA can be divided into two search phases: exploitation and exploration. During the exploitation phase, the algorithm uses a bubble-net hunting strategy that involves prey encirclement and a spiral-shaped trajectory to refine the best solution found so far. In the exploration phase, the WOA algorithm updates the position of each whale based on a randomly selected search agent. The following parts present a detailed mathematical formulation of each stage of the WOA algorithm.

3.4.1.1 Exploitation

3.4.1.1.1 Encircling the prey

The WOA algorithm is inspired by the foraging behavior of humpback whales, which involves observing the location of prey and encircling them. In the algorithm, the best solution found so far is assumed to be near the optimal solution, and the other whales adjust their locations towards the best hunting whale. This behavior is mathematically defined by the following equations:

$$\boldsymbol{D} = \left|\boldsymbol{C} \cdot \boldsymbol{X}_p(t) - \boldsymbol{X}(t)\right| \tag{3.9}$$

$$\boldsymbol{X}(t+1) = \boldsymbol{X}_p(t) - \boldsymbol{A} \cdot \boldsymbol{D} \tag{3.10}$$

where

$$\boldsymbol{A} = 2\boldsymbol{a}\boldsymbol{r} - \boldsymbol{a} \tag{3.11}$$

$$\boldsymbol{C} = 2\boldsymbol{r} \tag{3.12}$$

where X_p denotes the position vector of the current best solution, X represents the position vector of an individual whale, and D is the distance between the whale and its prey. The algorithm also employs coefficient vectors A and C, and absolute value notation (| |) to describe the behavior of the whales. Furthermore, r is a random vector that ranges from 0 to 1, and a is a linearly decreasing parameter that starts at 2 and decreases to 0 during the iterations.

3.4.1.1.2 Spiral updating mechanism

The spiral updating mechanism is a key component of the WOA algorithm, and is used to update the position of a whale in the search space. To accomplish this, the algorithm first calculates the distance between the whale and its prey. Next, a spiral trajectory with a helix shape is generated to emulate the movement of humpback whales. This trajectory is mathematically defined using the following formula:

$$X(t+1) = \boldsymbol{D}' \cdot e^{bl} \cdot \cos(2\pi l) + \boldsymbol{X}_p(t) \tag{3.13}$$

$$\boldsymbol{D}' = |\boldsymbol{X}_p(t) - \boldsymbol{X}(t)| \tag{3.14}$$

where $X(t+1)$ is the updated position of the whales, l is a random number within the range of $[-1, 1]$, and b is a constant value used to control the shape of the logarithmic spiral. In the exploitation phase of the WOA algorithm, two search strategies are employed: the spiral trajectory and the shrinking encircling. The choice of which strategy to use is determined randomly, with a 50% probability of selecting either option. This decision is mathematically represented as follows:

$$\boldsymbol{X}(t+1) = \begin{cases} \boldsymbol{X}_p(t) - \boldsymbol{A} \cdot \boldsymbol{D} & \text{if } p' < 0.5 \\ \boldsymbol{D}' \cdot e^{bl} \cdot \cos(2\pi l) + \boldsymbol{X}_p(t) & \text{if } p' \geq 0.5 \end{cases} \tag{3.15}$$

where p' is a random number in $[0, 1]$.

3.4.1.2 Exploration phase

In the exploration phase of the WOA algorithm, humpback whales search randomly based on the position of other individuals. Each whale updates its position vector by selecting a random vector X_{rand} from the current population. The coefficient vector A controls the magnitude of the random search step and is selected from the range of $[-1, 1]$. If $|A| \leq 1$, the algorithm updates positions towards the current best solution, as described in Section 3.4.1.1.2, However, if $|A| > 1$, the algorithm performs an exploratory search to find the optimal solution. Mathematically, the exploration phase can be formulated as follows:

$$\boldsymbol{D} = |\boldsymbol{C} \cdot \boldsymbol{X}_{\text{rand}}(t) - \boldsymbol{X}(t)| \tag{3.16}$$

$$\boldsymbol{X}(t+1) = \boldsymbol{X}_{\text{rand}}(t) - \boldsymbol{A} \cdot \boldsymbol{D} \tag{3.17}$$

The main steps of WOA are given in Algorithm 3.1.

3.4.2 Tabu search algorithm (TS)

Tabu search is a powerful global optimization meta-heuristic algorithm inspired by the human memory process. It has a proven track record of successfully solving a wide range of complex combinatorial optimization problems [22]. This algorithm belongs to the category of single-base meta-heuristics, which means that it explores a candidate solution at each iteration. Initially, the TS algorithm generates a random solution as a starting point to search for improved solutions by examining a certain neighborhood. This can be achieved by exploring either the entire neighborhood or a subset of it. To ensure that the TS algorithm does not repeatedly explore the same solutions, it uses a tabu list to store the evaluated solutions. This list is updated if the solutions do not meet the evaluation criteria. Finally, the best solution that is not in the tabu list is chosen.

The pseudo-code of the Tabu Search algorithm is presented in Algorithm 3.2.

3.4.3 Hybrid algorithm (WOATS)

The resource allocation in VLC systems is proved to be an NP-hard problem. It is solved successfully using metaheuristics With an acceptable runtime. In this paper, we use the well known WOA algorithm to solve the considered problem. The

Algorithm 3.1 The WOA Algorithm.

```
Initialize WOA parameters
Randomly initialize the population of solutions
Evaluate the population and determine X_best
for t=1 to t_max do
    for i=1 to N do
        Update a, A, C, l and p'
        if p' < 0.5 then
            if |A| < 1 then
                Calculate X_i(t + 1) using Eq. (3.9)
            else
                Calculate X_t(t + 1) using Eq. (3.17)
            end if
        else
            Calculate X_t(t + 1) using Eq. (3.13)
        end if
    end for
    Evaluate the population and determine X_p
    t = t + 1
end for
Return the best solution
```

Algorithm 3.2 The pseudo-code of Tabu Search algorithm.

```
Initialize TS parameters: Maximum number of iterations T, Neighbors' number n, Tabu List TL.
Generate initial solution X
Set the current solution X as the best solution X_best
while (t < T) do
    for i = 1, 2, ..., n do
        Generate a random neighbor X' of X
        Evaluate neighbor' solution X'
        Update TL
        Update the best solution X_p
    end for
    t = t + 1
end while
return The best solution X_p
```

WOA algorithm has been employed to address a diverse range of problems, such as classification [23–25], path planning [26–30], clustering [31–33], placement problems [34–37]. The effectiveness of WOA has motivated us to investigate its potential application for solving the resource allocation problem in VLC systems. WOA may result in premature convergence leading the search space to converge prematurely on a suboptimal solution [38]. It is advised that researchers adapt it and integrate it with other strategies or meta-heuristics to improve the exploitation phase. In this sense, we propose a hybrid algorithm, WOATS, based on the incorporation of TS into the original WOA to improve the exploitation phase by searching for the WOA algorithm's most promising regions.

WOATS begins by generating initial solutions randomly in the search space. At each iteration, each solution is evaluated using a predefined fitness function. In a comparison phase, if it is the best whale X_p and it has an improvement in its fitness value during its last iteration, its position will be updated using the TS algorithm, else it will be updated using WOA algorithm. If the stopping criterion is achieved, WOATS will be terminated and the best solution X_p and its fitness value will be determined as outputs. Algorithm details for WOATS are provided in the following pseudo-code Algorithm 3.3.

Algorithm 3.3 The WOATS Algorithm.

```
Initialize parameters of WOA and TS
Randomly initialize the population of solutions
Evaluate the population and determine X_best
for t=1 to t_max do
  for i=1 to N do
    Update a, A, C, l and p'
    if X_i(t) = X_best and has an improvement in its fitness value in the last iteration then
      Calculate X_i(t + 1) using TS algorithm
    else
      if p' < 0.5 then
        if |A| < 1 then
          Calculate X_i(t + 1) using Eq. (3.9)
        else
          Calculate X_i(t + 1) using Eq. (3.17)
        end if
      else
        Calculate X_f(t + 1) using Eq. (3.13)
      end if
    end if
  end for
  Evaluate the population and determine X_best
  t = t + 1
end for
Return the best solution
```

3.4.4 Time complexity of the proposed WOATS algorithm

The WOATS complexity is depending on the problem dimension, the population size, and the maximum number of iterations. For every iteration, bubble-net hunting, random search, and local search strategies are implemented. Therefore, the total time complexity of the WOATS algorithm is given as follows:

$$O(WOATS) = O(O(initialization\ phase) + t_{max}(O(exploitation\ phase) + O(exploration\ phase) + O(local\ search)))$$
$$O(WOATS) = O(O(Nd) + t_{max}(O(Nd) + O(Nd) + O(nd)))$$
$$O(WOATS) = O(d(N + t_{max}(N + n)))$$
$$O(WOATS) = O(dt_{max}(N + n))$$

where t_{max} is the maximum number of iterations. N and d denote the population size and the problem dimension, respectively. n represents the number of neighborhoods defined in TS algorithm.

3.5 Numerical results

This section provides the numerical results for our proposed model. We consider two scenarios, where only one PD is activated in the first one and two DPs are activated in the second one. Performance analysis is then performed on this basis and the effect of different system parameters is studied. First, we focus on examining the effect of changing the number of users for the two scenarios considered. Then, a comparative study is conducted, in which we compare our algorithm with several established methods in terms of throughput, fairness, and fitness. In simulation analysis, we consider a room of dimension $10 \times 10 \times 3$, $M = 5$, $D_r = 1$ cm, $L = 9$, and $R = 0.54$ A/W. We further assume that $BER_{th} = 10^{-6}$, $B = 1$ MHz, $N_0 = 10^{-21}$ A^2/Hz, and different numbers of users, i.e., $N = 20, 30, 40, 50, 60, 70, 80, 90$, and 100. All simulations are

TABLE 3.1 Throughput, Fairness, and Fitness under various user numbers for the case of one activated PD.

	N	20	40	60	80	100
Throughput (Mbit/s)	WOATS	**295.1**	**582.2**	**878.4**	**1168**	**1472**
	WOA	294.8	578.5	877.5	1157	1455
	TS	288.4	578.5	864.3	1155	1444
	PSO	287.1	554.6	835.4	1107	1387
	GA	290.8	576.6	875.6	1165	1459
	AOA	287.3	563.8	842.8	1119	1400
	MFO	292.9	564.9	859.6	1145	1421
	GWO	294.2	577.5	873.4	1162	1457
	SCA	292	567.8	855.5	1143	1429
Fairness	WOATS	0.9971	**0.9977**	0.9976	**0.9976**	**0.9980**
	WOA	**0.9972**	0.9971	0.9979	0.9968	0.9975
	TS	0.9952	0.9969	0.9964	0.9963	0.9961
	PSO	0.9944	0.9960	0.9967	0.9952	0.9935
	GA	0.9967	0.9973	**0.9980**	0.9974	0.9972
	AOA	0.9954	0.9964	0.9966	0.9967	0.9957
	MFO	0.9971	0.9970	0.9969	0.9967	0.9961
	GWO	0.9968	0.9970	0.9975	0.9974	0.9973
	SCA	0.9962	0.9969	0.9964	0.9969	0.9961
Fitness	WOATS	**0.6543**	**0.6492**	**0.6522**	**0.6517**	**0.6532**
	WOA	0.6540	0.6475	0.6518	0.6483	0.6500
	TS	0.6456	0.6471	0.6456	0.6463	0.6464
	PSO	0.6438	0.6325	0.6341	0.6317	0.6319
	GA	0.6524	0.6459	0.6511	0.6498	0.6506
	AOA	0.6447	0.6372	0.6380	0.6355	0.6361
	MFO	0.6501	0.6427	0.6438	0.6433	0.6417
	GWO	0.6533	0.6472	0.6490	0.6484	0.6492
	SCA	0.6493	0.6414	0.6434	0.6426	0.6431

performed in Matlab software with a total number of iterations of 1000 and an average of 30 runs to mitigate the effects of randomness by obtaining average values of results.

3.5.1 Impact of changing the user count

In the following section, we evaluate the outcome of modifying the users number from 20 to 100 considering the case of only one activated PD. Fig. 3.3(a) and Table 3.1 show the effect of varying the number of users on the total throughput. It is observed that the proposed WOATS achieves higher throughput than other algorithms. For example, consider $N = 100$. The proposed WOATS can achieve a total throughput of 1472 Mb/s with an improvement of 17 Mb/s, 28 Mb/s, 85 Mb/s, 13 Mb/s, 72 Mb/s, 51 Mb/s, 15 Mb/s, and 43 Mb/s over WOA, TS, PSO, GA, AOA, MFO, GWO, and SCA, respectively.

In Fig. 3.3(b), we present the achieved fairness versus the number of users for all considered algorithms. It is observed that WOATS Can accomplish higher fairness in most of the considered cases. For example, let us consider $N = 40$ users. The fairness obtained is 0.9977 for WOATS algorithm. It reduces to 0.9971, 0.9969, 0.9960, 0.9973, 0.9973, 0.9964, 0.9970, 0.9970, and 0.9969 for WOA, TS, PSO, GA, AOA, MFO, GWO, and SCA, respectively.

In Fig. 3.3(c), we present the achieved fitness versus users number for all algorithms under consideration. It is clear that the WOATS algorithm achieves the best fitness in all scenarios under consideration. For example, consider $M = 20$ users. The fitnesses obtained are respectively 0.6492, 0.6475, 0.6471, 0.6325, 0.6459, 0.6372, 0.6427, 0.6472, and 0.6414 for WOATS, WOA, TS, PSO, GA, AOA, MFO, GWO, SCA algorithms.

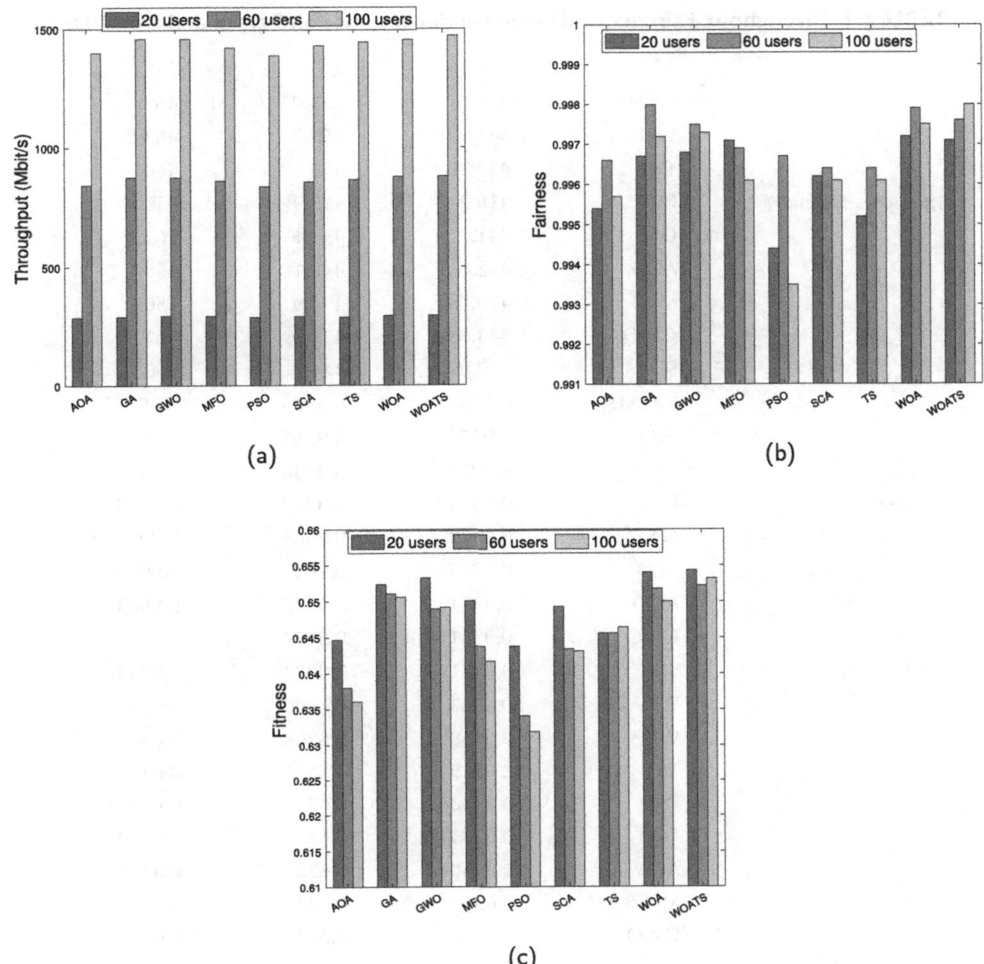

FIGURE 3.3 Throughput, fairness, and fitness under various numbers of users.

3.5.2 Impact of changing the number of activated PDs

This section evaluates the outcome of changing the number of activated PDs on the performance of the proposed approach (see Table 3.2).

Fig. 3.4(a) presents the achieved throughput for one, two, and three activated PDs, assuming different algorithms. It is noticed that an increase in the number of activated PDs resulted in higher throughput for most of the considered algorithms. The reason for this is that When the number of receivers increases, a greater amount of light emitted by the transmitters is captured, and as a result, the throughput increases. For instance, consider the WOATS algorithm. The throughput obtained is 444.8 Mb/s in the case of an activated PD. This increases to 455 Mb/s and 463.3 Mb/s in the case of two and three activated PDs, respectively.

Fig. 3.4(b) shows the fairness index obtained for one, two, and three switched-on PDs. Typically, it is found that it is better to use three PDs rather than one or two PDs. For example, consider PSO algorithm. The achieved fairness level is about 0.9983 for the case of three activated PD. This decreases to 0.9977 and 0.9943 when one and two PDs are activated, respectively.

Fig. 3.4(c) shows the fitness obtained for all the scenarios considered (i.e., one, two, and three activated PDs). It is observed that when only one PD is activated, the algorithms are much faster in finding the optimal solution than in the other cases. For example, consider the WOATS algorithm. The fitness obtained is 0.6567 for the case of a single-activated PD. This reduces to 0.6396 and 0.6437 for the cases of two and three activated PDs, respectively.

TABLE 3.2 Throughput, Fairness, and Fitness under various numbers of activated PDs.

	M	1	2	3
Throughput (Mbit/s)	WOATS	444.8	455.0	463.3
	WOA	441.9	452.7	463.0
	TS	439.2	451.5	460.5
	PSO	418.2	433.9	450.0
	GA	441.7	453.1	463.3
	AOA	432.0	442.1	455.1
	MFO	439.5	433.9	450.0
	GWO	443.6	433.9	462.5
	SCA	440.8	444.2	460.7
Fairness	WOATS	0.9981	0.9985	0.9987
	WOA	0.9977	0.9985	0.9979
	TS	0.9972	0.9986	0.9986
	PSO	0.9943	0.9977	0.9983
	GA	0.9972	0.9982	0.9981
	AOA	0.9969	0.9980	0.9983
	MFO	0.9971	0.9977	0.9983
	GWO	0.9981	0.9977	0.9985
	SCA	0.9981	0.9977	0.9984
Fitness	WOATS	0.6567	0.6396	0.6437
	WOA	0.6548	0.6365	0.6415
	TS	0.6515	0.6370	0.6424
	PSO	0.6338	0.6235	0.6249
	GA	0.6533	0.6382	0.6409
	AOA	0.6450	0.6285	0.6331
	MFO	0.6509	0.6235	0.6249
	GWO	0.6552	0.6235	0.6415
	SCA	0.6514	0.6312	0.6385

3.5.3 Convergence analysis

Herein, we analyze the convergence of the different algorithms considered. The convergence procedure is based on two parameters: the convergence speed and the convergence efficiency (fitness value). Fig. 3.5 shows the fitness values depending on the number of iterations. It is observed the superiority of the proposed WOATS in terms of convergence efficiency compared to WOA, TS, PSO, GA, AOA, MFO, GWO, SCA algorithms. In terms of convergence speed, Fig. 3.5 shows that the proposed WOATS has a moderate convergence speed where the optimal solutions can be obtained at iteration 180, 550, and 600 for $M = 20$ users, $M = 60$ users, $M = 100$ users, respectively.

3.6 Conclusion

In this paper, we have proposed a new hybrid approach named WOATS to solve a resource allocation problem in indoor VLC systems. We formulate a placement problem whose goal is to maximize the throughput and fairness of the users while satisfying the target network requirement. The performance of WOATS algorithm is assessed taking into account the user throughput, fairness, and fitness metrics in comparison with WOA, TS, PSO, GA, AOA, MFO, GWO, SCA algorithms. The Simulation results indicated the superiority of the proposed WOATS over WOA, TS, PSO, GA, AOA, MFO, GWO, SCA algorithms for all considered scenarios. As future work, we will focus on investigating the feasibility and performance of the proposed WOATS scheme in outdoor VLC systems, with the goal of further expanding its potential applications.

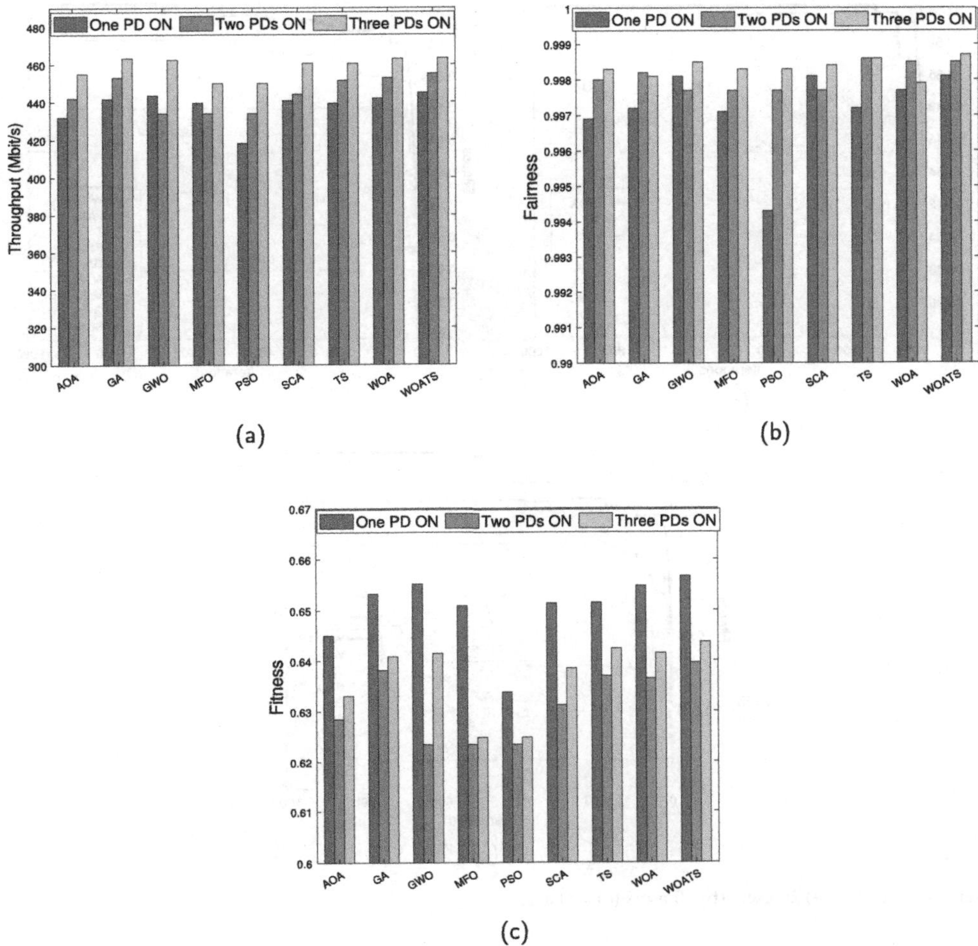

FIGURE 3.4 Throughput, fairness, and fitness under various numbers of activated PDs.

References

[1] M. Katz, I. Ahmed, Opportunities and challenges for visible light communications in 6G, in: 2020 2nd 6G Wireless Summit (6G SUMMIT), 2020, pp. 1–5.

[2] S. Al-Ahmadi, O. Maraqa, M. Uysal, S.M. Sait, Multi-user visible light communications: state-of-the-art and future directions, IEEE Access 6 (2018) 70555–70571.

[3] Q. Wang, Z. Wang, L. Dai, Multiuser MIMO-OFDM for visible light communications, IEEE Photonics Journal 7 (2015) 1–11.

[4] B. Li, J. Wang, R. Zhang, H. Shen, C. Zhao, L. Hanzo, Multiuser MISO transceiver design for indoor downlink visible light communication under per-led optical power constraints, IEEE Photonics Journal 7 (2015) 1–15.

[5] U.F. Siddiqi, S.M. Sait, M.S. Demir, M. Uysal, Resource allocation for visible light communication systems using simulated annealing based on a problem-specific neighbor function, IEEE Access 7 (2019) 64077–64091.

[6] C. Gong, S. Li, Q. Gao, Z. Xu, Power and rate optimization for visible light communication system with lighting constraints, IEEE Transactions on Signal Processing 63 (2015) 4245–4256.

[7] R. Jiang, Z. Wang, Q. Wang, L. Dai, Multi-user sum-rate optimization for visible light communications with lighting constraints, Journal of Lightwave Technology 34 (2016) 3943–3952.

[8] D. Bykhovsky, S. Arnon, Multiple access resource allocation in visible light communication systems, Journal of Lightwave Technology 32 (2014) 1594–1600.

[9] M.S. Demir, S.M. Sait, M. Uysal, Unified resource allocation and mobility management technique using particle swarm optimization for VLC networks, IEEE Photonics Journal 10 (2018) 1–9.

[10] Y.S. Eroğlu, I. Güvenç, A. Şahin, Y. Yapıcı, N. Pala, M. Yüksel, Multi-element VLC networks: led assignment, power control, and optimum combining, IEEE Journal on Selected Areas in Communications 36 (2017) 121–135.

[11] S. Mirjalili, A. Lewis, The whale optimization algorithm, Advances in Engineering Software 95 (2016) 51–67.

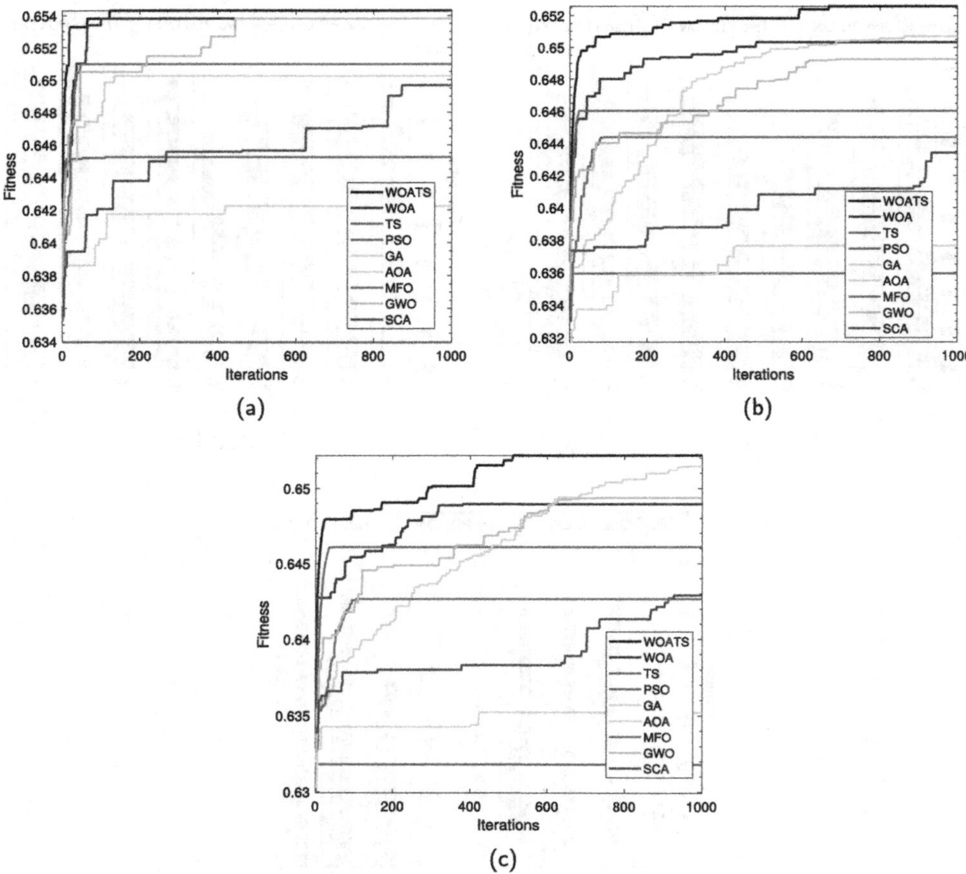

FIGURE 3.5 Convergence analysis (a) 20 users (b) 60 users (c) 100 users.

[12] R.A. Gallego, R. Romero, A.J. Monticelli, Tabu search algorithm for network synthesis, IEEE Transactions on Power Systems 15 (2000) 490–495.

[13] J. Kennedy, R. Eberhart, Particle swarm optimization, in: Proceedings of ICNN'95-International Conference on Neural Networks, vol. 4, IEEE, 1995, pp. 1942–1948.

[14] S. Forrest, Genetic algorithms, ACM Computing Surveys (CSUR) 28 (1996) 77–80.

[15] L. Abualigah, A. Diabat, S. Mirjalili, M. Abd Elaziz, A.H. Gandomi, The arithmetic optimization algorithm, Computer Methods in Applied Mechanics and Engineering 376 (2021) 113609.

[16] S. Mirjalili, Moth-flame optimization algorithm: a novel nature-inspired heuristic paradigm, Knowledge-Based Systems 89 (2015) 228–249.

[17] S. Mirjalili, S.M. Mirjalili, A. Lewis, Grey wolf optimizer, Advances in Engineering Software 69 (2014) 46–61.

[18] S. Mirjalili, SCA: a sine cosine algorithm for solving optimization problems, Knowledge-Based Systems 96 (2016) 120–133.

[19] H.B. Eldeeb, S.M. Mana, V. Jungnickel, P. Hellwig, J. Hilt, M. Uysal, Distributed MIMO for Li-Fi: channel measurements, ray tracing and throughput analysis, IEEE Photonics Technology Letters 33 (2021) 916–919.

[20] M. Uysal, F. Miramirkhani, T. Baykas, K. Qaraqe, IEEE 802.11 bb reference channel models for indoor environments, Technical Report, 2018.

[21] Y.S. Eroglu, A. Sahin, I. Guvenc, N. Pala, M. Yuksel, Multi-element transmitter design and performance evaluation for visible light communication, in: 2015 IEEE Globecom Workshops (GC Wkshps), IEEE, 2015, pp. 1–6.

[22] H.E. Kiziloz, T. Dokeroglu, A robust and cooperative parallel tabu search algorithm for the maximum vertex weight clique problem, Computers & Industrial Engineering 118 (2018) 54–66.

[23] M. Sharawi, H.M. Zawbaa, E. Emary, Feature selection approach based on whale optimization algorithm, in: 2017 Ninth International Conference on Advanced Computational Intelligence (ICACI), IEEE, 2017, pp. 163–168.

[24] M.M. Mafarja, S. Mirjalili, Hybrid whale optimization algorithm with simulated annealing for feature selection, Neurocomputing 260 (2017) 302–312.

[25] M. Tubishat, M.A. Abushariah, N. Idris, I. Aljarah, Improved whale optimization algorithm for feature selection in Arabic sentiment analysis, Applied Intelligence 49 (2019) 1688–1707.

[26] T.-K. Dao, T.-S. Pan, J.-S. Pan, A multi-objective optimal mobile robot path planning based on whale optimization algorithm, in: 2016 IEEE 13th International Conference on Signal Processing (ICSP), IEEE, 2016, pp. 337–342.

[27] J. Wu, H. Wang, N. Li, P. Yao, Y. Huang, H. Yang, Path planning for solar-powered UAV in urban environment, Neurocomputing 275 (2018) 2055–2065.

[28] S.V. Kumar, R. Jayaparvathy, B. Priyanka, Efficient path planning of AUVs for container ship oil spill detection in coastal areas, Ocean Engineering 217 (2020) 107932.

[29] Z. Yan, J. Zhang, Z. Yang, J. Tang, Two-dimensional optimal path planning for autonomous underwater vehicle using a whale optimization algorithm, Concurrency and Computation: Practice and Experience 33 (2021) e6140.

[30] A. Chhillar, A. Choudhary, Mobile robot path planning based upon updated whale optimization algorithm, in: 2020 10th International Conference on Cloud Computing, Data Science & Engineering (Confluence), IEEE, 2020, pp. 684–691.

[31] J. Nasiri, F.M. Khiyabani, A whale optimization algorithm (WOA) approach for clustering, Cogent Mathematics & Statistics 5 (2018) 1483565.

[32] B.M. Sahoo, H.M. Pandey, T. Amgoth, A whale optimization (WOA): meta-heuristic based energy improvement clustering in wireless sensor networks, in: 2021 11th International Conference on Cloud Computing, Data Science & Engineering (Confluence), IEEE, 2021, pp. 649–654.

[33] M. Kumar, A. Chaparala, OBC-WOA: opposition-based chaotic whale optimization algorithm for energy efficient clustering in wireless sensor network, Intelligence 250 (2019).

[34] S. Nasrollahzadeh, M. Maadani, M.A. Pourmina, Optimal motion sensor placement in smart homes and intelligent environments using a hybrid WOA-PSO algorithm, Journal of Reliable Intelligent Environments (2021) 1–13.

[35] P. Singh, S. Prakash, Optical network unit placement in Fiber-Wireless (FiWi) access network by whale optimization algorithm, Optical Fiber Technology 52 (2019) 101965.

[36] A. Al-Moalmi, J. Luo, A. Salah, K. Li, L. Yin, A whale optimization system for energy-efficient container placement in data centers, Expert Systems with Applications 164 (2021) 113719.

[37] K.M.S. Alzaidi, O. Bayat, O.N. Uçan, Multiple DGs for reducing total power losses in radial distribution systems using hybrid WOA-SSA algorithm, International Journal of Photoenergy 2019 (2019).

[38] H.M. Mohammed, S.U. Umar, T.A. Rashid, A systematic and meta-analysis survey of whale optimization algorithm, Computational Intelligence and Neuroscience 2019 (2019).

Chapter 4

Use of whale optimization algorithm and its variants for cloud task scheduling: a review

Ali Mohammadzadeh[a], Amit Chhabra[b], Seyedali Mirjalili[c,d], and Amir Faraji[e,f]

[a]*Department of Computer Engineering, Shahindezh Branch, Islamic Azad University, Shahindezh, Iran,* [b]*Department of Computer Engineering and Technology, Guru Nanak Dev University, Amritsar, India,* [c]*Centre for Artificial Intelligence Research and Optimisation, Torrens University Australia, Brisbane, QLD, Australia,* [d]*Yonsei Frontier Lab, Yonsei University, Seoul, South Korea,* [e]*Construction Management Department, Faculty of Architecture, KHATAM University, Tehran, Iran,* [f]*School of Engineering, Design and Built Environment, Western Sydney University, Sydney, NSW, Australia*

4.1 Introduction

Cloud computing is a web-based computing model that manages different cloud requests and offers clients speedy service [1]. Due to the cloud infrastructure's central management, users can utilize a variety of services from the cloud, such as managed services, adaptability, resource pooling, performance, expandability, throughput, high availability, etc. [2]. The following classifications can be used to categorize cloud computing: Private clouds that are administered internally by the corporation, free of bandwidth utilization, and constraints. Public clouds that everyone is allowed to join and use services on clouds. Private and public clouds are combined and called hybrid clouds. Users and providers of cloud services can profit from technology virtualization and dynamic resource scheduling methods [3].

Scheduling is a method to ensure ideal resource utilization across all tasks in the given time to obtain desired QoS. To achieve the best response for the objective function, tasks are scheduled on resources that are limited by restrictions. The purpose of the scheduling algorithm is to choose which resource will be used for which task. Efficient resource scheduling increases resource usage while simultaneously completing tasks in the shortest amount of time.

Scheduling problems are NP-hard problems that have been solved in the past by a variety of algorithms, including Round-Robin, FCFS, max-min, and min-min methods [4]. The goal of scheduling is to identify the optimal resource for a task's execution to enhance the scheduling algorithm's numerous qualities of service (QoS) criteria, such as execution cost, reliability, task rejection ratio, energy consumption, resource utilization, etc. Due to the continual growth in workload at cloud data centers, which may cause a shortage of cloud resources, task scheduling has become a real concern.

Conventional exhaustive, deterministic, and metaheuristic algorithms are frequently used to schedule tasks. Deterministic algorithms produce superior and quicker outcomes than conventional exhaustive techniques for scheduling issues, but deterministic algorithms do not apply to large-scale distributions of acquired data. A metaheuristic algorithm is highly efficient and effective at solving comparatively large and complicated issues. These algorithms use approximation techniques and iterative procedures to find optimal solutions quickly [5].

Several varieties of metaheuristic algorithms are offered for tasks or workflow scheduling, including Genetic Algorithm-based, Grey Wolf Optimization-based, Particle Swarm Optimization-based, Whale Optimization-based, and Hybrid Optimization-based in the cloud environment [6–9]. One of the interesting metaheuristic algorithms, WOA is employed by many different schemes to handle a wide range of issues, including issues with task and workflow scheduling in the cloud computing environment.

In this chapter, we investigate the cutting-edge WOA-based scheduling algorithms that have been suggested in the literature for the cloud environment. Additionally, we highlight the objectives and characteristics of various scheduling schemes and give a taxonomy of them according to the kind of WOA algorithm. We also explain how each scheme's WOA algorithm is enhanced or integrated with other metaheuristic algorithms to overcome scheduling issues.

The remainder of this chapter is organized as follows: A literature review of the scheduling objectives is introduced in Section 4.2. Section 4.3 is discussed the research methodology of the current study. The variants of metaheuristic scheduling algorithms in the cloud environment are stated in Section 4.4, and the standard WOA algorithm is discussed in

Handbook of Whale Optimization Algorithm. https://doi.org/10.1016/B978-0-32-395365-8.00010-5

Section 4.5. The suggested WOA-based scheduling methods are categorized in Section 4.6, and the objectives and simulator environments used in the scheduling techniques are covered in Section 4.7. In Section 4.8, the chapter concludes with conclusions.

4.2 Objective of scheduling

In this section, objectives are mentioned. We need to measure the work analysis in cloud computing on some metric. Service providers and cloud clients are the two main categories of entities in this environment. While achieving the aims, each part of the cloud environment has its objectives. The following list contains the basic scheduling criteria taken into consideration by the existing approaches:

Makespan: The maximum time it takes for the last task to be completed [10].
Budget: It is essentially a limit that is set by the user to decrease the overall process execution time, and tasks must be completed within the budget [11].
Quality of Service (QoS): QoS includes several client input restrictions, such as system cost, delivery dates, efficiency, and other factors [12].
Energy efficiency: A variety of scheduling algorithms were developed to improve performance while utilizing less energy, and making cloud computing services more environmentally friendly [13].
Deadline: Some of the decent scheduling algorithms attempt to keep the tasks executed inside by the distance between the task's presentation and the deadline for completion [14].
Performance: The overall efficacy of the scheduling algorithm in providing excellent services to consumers following their needs [15].
Efficiency: It indicates the best possible use of available resources by the user [16].
Cost: A client pays this amount to utilize a resource provided by the service provider.
Load balancing: The purpose is to distribute the whole load among the nodes in a cloud platform [17].
Security: Some features of the cloud environment make it more difficult than conventional networks to provide data privacy and security [18].
Resource utilization: To gain more profits by leasing finite resources to the user in a way that ensures full use of those resources [19].

4.3 Research methodology

We present a guide to finding appropriate papers on whale optimization algorithms and their variants for cloud task scheduling in this section. The main criteria of a rich knowledge survey are the discovery, collection, categorization, and evaluation of related articles. Researchers will be provided with a structured method to evaluate special topics by applying restrictions to the criteria of the topics. The goal of this paper is to describe how to discover crucial subjects in pertinent scopes.

This study desires for investigating critical objectives, simulation environments, and scheduling types utilized in various whale optimization papers with the main challenges and topics in the area of scheduling problems in the cloud environment. The search is done in September 2022, and Table 4.1 illustrated the results of our search for appropriate articles on WOA-based scheduling algorithms in several well-known academic online databases.

TABLE 4.1 Applied several well-known academic online databases.

Database	Related URL
Google Scholar	http://Scholar.google.com
Science Direct	http://www.sciencedirect.com
Springer	http://link.springer.com
IEEE explorer	http://ieeexplore.ieee.org
Hindawi	https://www.hindawi.com
Wiley	http://onlinelibrary.wiley.com

Fig. 4.1 illustrated the picking measure and evaluation framework of the offered taxonomy. The process of evaluating and choosing articles is outlined as follows:

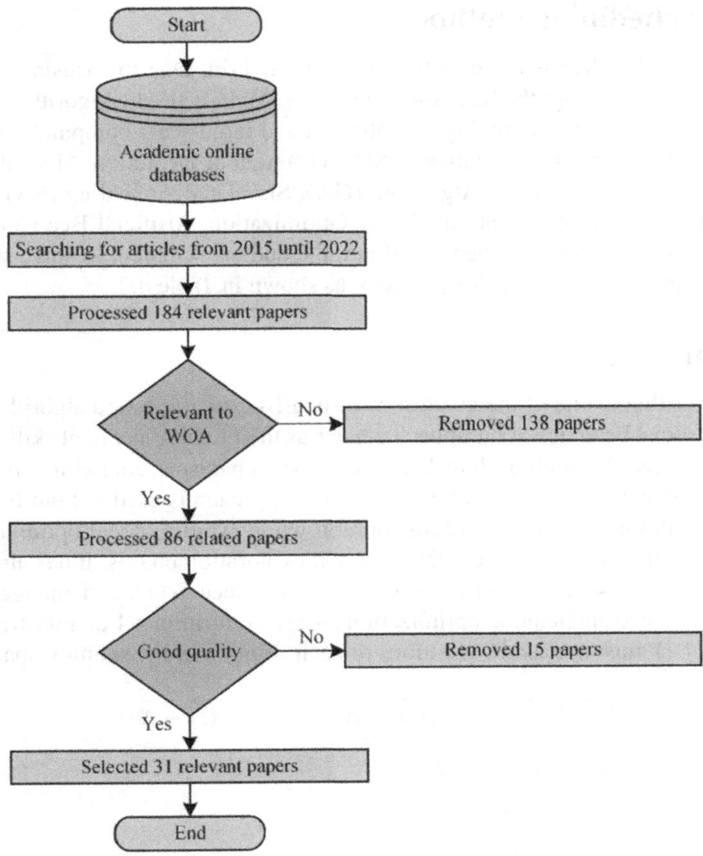

FIGURE 4.1 The flowchart of papers selection.

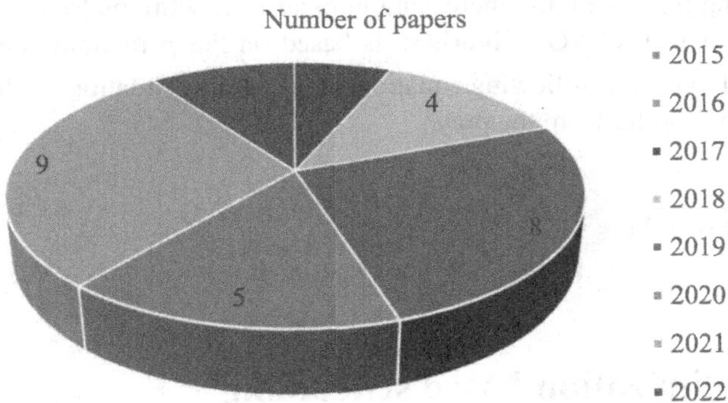

FIGURE 4.2 Distribution of published papers by year.

- Articles that were published at a certain time,
- Published papers concerning WOA-based scheduling algorithms to solve task scheduling problems,
- Quality has been used as a criterion to choose relevant articles.

In the process of searching the aforementioned online databases, suitable keywords like "meta-heuristic", "WOA algorithm", "whale optimization algorithm", "scheduling" and "task scheduling" were used. The findings show a total of 184 publications, which will be reduced to 31 papers to focus on WOA-based algorithms. Additionally, Fig. 4.2 illustrates the detailed distribution of these 31 chosen publications between 2015 and September 2022. According to the figure, the majority of papers have been published between 2019 and 2021 due to the topic's growing popularity with scientists.

4.4 Meta-heuristic scheduling methods

Scheduling methods are classified as heuristic, meta-heuristic, and hybrid. Due to focusing on the meta-heuristic whale optimization algorithm, we do not consider the heuristic algorithms. Meta-heuristic algorithms are an excellent procedure to generate or choose a superior solution for solving complicated and large-scale computational issues. Whale Optimization Algorithm (WOA), Particle Swarm Optimization (PSO), Differential Evolution Algorithm (DEA), Cuckoo Search (CS), Ant Colony Optimization (ACO), Genetic Algorithm (GA), Simulated Annealing (SA), BAT Optimization, Bacteria Foraging Optimization (BFO), Cat Optimization, Firefly Optimization, Artificial Bee Colony (ABC), and others are some of the meta-heuristic algorithms that have been used in the cloud environment to solve scheduling problem. Several meta-heuristic algorithms are utilized to resolve this problem, as shown in Table 4.2.

4.5 WOA algorithm

The Whale Optimization Algorithm is one of the greatest meta-heuristic swarm-based algorithms (WOA) [95]. The WOA algorithm is inspirited by the social behavior of humpback whales as their bubble-net hunt skill in nature. The following are the primary mechanisms used in the WOA algorithm: There are two mechanisms, encircling prey and Bubble-net attacking, that belong to the exploitation strategy. In contrast, the exploration approach is used to hunt for prey at random to prevent local optimum. WOA could explore the optimal overall solution while avoiding local optimums in a respectable amount of time because of the good balance between the exploitation and exploration phases. It has already been demonstrated by previous findings in the resolving of several practical issues [96–98]. Since WOA's advantages over other algorithms are clear, many academics use it to solve challenging optimization issues. Algorithm 4.1 demonstrates the WOA algorithm. To find the global solution, Eq. (4.1) must update the positions of each solution in the solution space.

$$\overrightarrow{X(t+1)} = \overrightarrow{X_{rand}(t)} - \vec{A}.\left|\vec{C}.\overrightarrow{X_{rand}(t)} - \overrightarrow{X(t)}\right| \tag{4.1}$$

$$\vec{A} = 2\vec{a}.\vec{r_1} - \vec{a} \tag{4.2}$$

$$\vec{C} = 2.\vec{r_2} \tag{4.3}$$

where t denotes the current iteration, $\overrightarrow{X(t)}$ represents the location vector of the solution acquired, $\overrightarrow{X_{rand}(t)}$ demonstrates the random possible position of the prey, A, and C denotes the coefficient vector, and $\bullet$ demonstrates element-by-element multiplication. a has a decreasing trend with the increasing number of repetitions from 2 to 0, and r_1, r_2, and p generate a random number between 0 and 1. The WOA algorithm is based on the p random value to perform the spiral-shaped movement and encircling of the target. The following equation simulates this situation. In this equation, $\overrightarrow{X^*(t)}$ demonstrates the best, possible location obtained so far from the prey.

$$\overrightarrow{X(t+1)} = \begin{cases} \overrightarrow{X^*(t)} - \vec{A}.\left|\vec{C}.\overrightarrow{X^*(t)} - \overrightarrow{X(t)}\right|, & r_3 < 0.5 \\ \vec{D'}e^{bl}\cos(2a\pi l) + \overrightarrow{X^*(t)}, & r_3 \geq 0.5 \end{cases} \tag{4.4}$$

4.6 Types of whale optimization-based scheduling

Numerous scheduling models, including Standard WOA, Multi-objective, Improved, and Hybrid WOA are presented in the literature as being based on WOA approaches.

4.6.1 Standard WOA

The scheduling issue in the cloud environment is solved using a discrete definition of the base Whale Optimization Algorithm (WOA), which enables efficient resource allocation and decreases the overall time of client requests for services. Each whale is constructed in the form of an array. A new idea of distance is developed for the distance function based on this array. This algorithm benefits from this function [99]. Shrinking and finding prey functions are made for straight motion, whereas spiral functions are intended for spiral movement. The appropriate modification of the value that the algorithm enters to the exploitation phase from the discovery phase is a crucial element in the efficient implementation of the method. The functions created for the whale algorithm's basic operators were able to thoroughly search the problem's solution space

TABLE 4.2 Comparison of various metaheuristic algorithms for task scheduling in cloud computing (2015-2022).

Algorithm	Simulation environment	Objectives																							
		Makespan	Accuracy	QoS	Execution time	Energy efficiency	Reliability	Deadline	Performance	Efficiency	Cost	Load balancing	Security	Resource utilization	Throughput	Schedule Length Ratio	Lateness	Coverage	Hypervolume	Task Failures	Out-of-bid Errors	Relative Errors	No. of Active Resource	Execution Overhead	Degree of Imbalance
TSPSO [20]	CloudSim	✓	-	-	✓	✓	-	-	-	-	-	-	-	-	-	-	-	-	-	-	-	-	-	-	-
MOPSO [21]	CloudSim	-	-	-	✓	-	-	-	-	-	-	-	-	-	✓	-	-	-	-	-	-	-	-	-	-
M-PSO [22]	CloudSim	-	-	-	✓	✓	-	-	-	-	-	-	-	-	-	-	-	-	-	-	-	-	-	-	-
PSO-COGENT [23]	CloudSim	✓	-	-	✓	✓	-	-	-	-	✓	-	-	✓	✓	-	-	-	-	-	-	-	-	-	-
PSO-hill climbing [24]	C# in Azure cloud	✓	-	-	-	-	-	-	-	-	-	-	-	-	-	-	-	-	-	-	-	-	-	-	-
MaOPSO [25]	CloudSim	-	-	-	✓	-	-	-	-	-	✓	-	-	✓	✓	-	-	-	-	-	-	-	-	-	-
CPSO [26]	CloudSim	✓	-	-	-	-	-	-	-	-	✓	-	-	-	-	-	-	-	-	-	-	-	-	-	-
FMPSO [27]	CloudSim	✓	-	-	-	-	-	-	-	-	✓	-	-	-	-	-	-	-	-	-	-	-	-	-	-
PSO Based [28]	Not stated	✓	-	-	✓	-	-	-	-	-	✓	-	-	-	-	-	-	-	-	-	-	-	-	-	-
PSO Based [29]	CloudSim	✓	-	-	-	-	-	-	-	-	-	-	-	-	-	-	-	-	-	-	-	-	-	-	-
QMPSO [30]	Not stated	✓	-	-	-	-	-	-	-	-	-	-	-	-	✓	-	-	-	-	-	-	-	-	-	-
CDCGA [31]	CloudSim	-	-	-	-	-	-	-	-	-	✓	-	-	-	-	-	-	-	-	-	-	-	-	-	-
BREA [32]	WorkflowSim	-	-	-	✓	-	-	-	-	-	✓	-	-	-	-	-	-	-	-	-	-	-	-	-	-
DSOS [33]	CloudSim	✓	-	-	✓	-	-	-	-	-	-	-	-	-	-	-	-	-	-	-	-	-	-	-	-
HEA-TaSDAP [34]	Amazon EC2 cloud	✓	-	-	-	-	-	-	-	-	-	-	-	-	-	-	-	-	-	-	-	-	-	-	-
BOGA [35]	Not stated	-	-	-	-	✓	✓	-	-	-	-	-	-	-	-	-	-	-	-	-	-	-	-	-	-
FOA-SA-LB [36]	Not stated	✓	-	-	-	✓	-	-	-	-	-	-	-	-	-	-	-	-	-	-	-	-	-	-	-
Genetic-based [37]	CloudSim	-	-	-	-	-	-	-	-	-	✓	-	-	-	-	-	-	-	-	-	-	-	-	-	-
GA-ETI [38]	Amazon EC2 cloud	✓	-	-	-	-	-	-	-	-	✓	-	-	-	-	-	-	-	-	-	-	-	-	-	-
E-DSOS [39]	Matlab®	✓	-	-	-	✓	-	-	-	-	-	-	-	-	-	-	-	-	-	-	-	-	-	-	-
HHSA [40]	Not stated	-	-	-	✓	-	-	-	-	-	✓	-	-	-	-	-	-	-	-	-	-	-	-	-	-
QGLBS [41]	Gridsim	-	-	-	✓	-	-	-	-	-	✓	-	-	-	-	-	-	-	-	-	-	-	-	-	-
Modify HEFT [42]	CloudSim	✓	-	-	-	-	-	-	-	-	-	-	-	-	-	-	-	-	-	-	-	-	-	-	-
MOBFOA [43]	Not stated	✓	-	-	✓	-	-	-	-	-	✓	-	-	-	-	-	-	-	-	-	-	-	-	-	-
DCOA [44]	Not stated	-	-	-	✓	✓	-	-	-	-	-	-	-	-	-	-	-	-	-	-	-	-	-	-	-
FDHEFT [45]	Not stated	✓	-	-	-	-	-	-	-	-	✓	-	-	-	-	-	-	-	-	-	-	-	-	-	-

continued on next page

TABLE 4.2 *(continued)*

Algorithm	Simulation environment	Objectives																							
		Makespan	Accuracy	QoS	Execution time	Energy efficiency	Reliability	Deadline	Performance	Efficiency	Cost	Load balancing	Security	Resource utilization	Throughput	Schedule Length Ratio	Lateness	Coverage	Hypervolume	Task Failures	Out-of-bid Errors	Relative Errors	No. of Active Resource	Execution Overhead	Degree of Imbalance
NNS [46]	Not stated	✓	-	-	-	-	-	-	-	-	-	-	-	-	-	-	-	-	-	-	-	-	-	-	-
Genetic-based [47]	WorkflowSim	-	-	-	-	-	-	-	-	-	✓	-	-	-	-	-	-	-	-	-	-	-	-	-	-
PCP-VM [48]	Not stated	✓	-	-	-	-	-	✓	-	-	✓	-	-	✓	-	✓	-	-	-	-	-	-	-	-	-
CMSOS [49]	CloudSim	✓	-	-	-	-	-	-	-	-	✓	-	-	-	-	-	-	-	-	-	-	-	-	-	-
OPTIC [50]	WorkflowSim	-	-	-	✓	-	-	-	-	-	-	-	-	-	-	-	-	-	-	-	-	-	-	-	-
GA-RTS [51]	Matlab	✓	-	-	✓	-	-	-	-	-	-	-	-	-	-	-	✓	-	-	-	-	-	-	-	-
LBBD with MILP [52]	Not stated	✓	-	-	-	-	-	-	-	-	-	-	-	-	-	-	-	-	-	-	-	-	-	-	-
MOWO [53]	iFogSim, EdgeCloudSim	-	-	-	✓	-	✓	-	-	-	✓	-	-	-	-	-	-	-	-	-	-	-	-	-	-
I_MaOPSO [54]	WorkflowSim	✓	-	-	-	✓	✓	-	-	-	✓	-	-	-	-	-	-	-	-	-	-	-	-	-	-
HIGA [55]	CloudSim, Matlab	✓	-	-	-	✓	-	-	-	-	-	-	-	-	-	-	-	-	-	-	-	-	-	-	-
MOPA [56]	CloudSim	-	-	-	-	-	-	-	-	-	-	-	-	-	-	-	-	✓	✓	-	-	-	-	-	-
CMI [57]	CloudSim	✓	-	-	-	-	-	-	-	-	✓	-	-	-	-	-	-	-	-	✓	✓	-	-	-	-
IWO-CA [58]	C#, Fog environment	-	-	-	-	✓	-	-	-	-	-	-	-	✓	-	-	-	-	-	-	-	-	-	-	-
SPSO [59]	Not stated	-	-	-	-	-	-	-	-	-	-	-	-	-	-	-	-	-	-	-	-	✓	-	-	-
ANN-based [60]	CloudSim, Matlab	✓	-	-	-	✓	-	-	-	-	-	-	-	-	-	-	-	-	-	-	-	-	✓	✓	-
BWM-VIKOR [61]	CloudSim	✓	-	-	-	-	-	✓	-	✓	✓	-	-	-	✓	-	-	-	-	-	-	-	-	-	-
MCoBRA [62]	Matlab	✓	-	-	-	✓	-	-	-	-	✓	-	-	-	-	-	-	-	-	-	-	-	-	-	-
MMHHO [63]	CloudSim	✓	-	-	✓	✓	-	-	-	-	-	-	-	✓	✓	-	-	-	-	-	-	-	-	-	-
LWMOGWO [64]	Matlab	-	-	-	-	✓	-	-	-	-	-	-	-	-	-	-	-	-	-	-	-	-	-	-	-
DMFO-DE [65]	iFogSim, EdgeCloudSim	-	-	-	-	✓	-	-	-	-	-	-	-	✓	-	-	-	-	-	-	-	-	-	-	-
CMSCW [66]	Java	-	-	-	-	-	-	-	-	-	✓	-	-	-	-	-	-	-	-	-	-	-	-	-	-
PICEA-g-based [67]	CloudSim	✓	-	-	-	✓	-	-	-	-	✓	-	-	-	-	-	-	-	-	-	-	-	-	-	-
PBMO-DALO [68]	WorkflowSim	✓	-	-	-	✓	-	-	-	-	-	-	-	-	-	-	-	-	-	-	-	-	-	-	-
FUPE [69]	Matlab, iFogSim	-	-	-	✓	-	-	-	-	-	-	-	-	✓	-	-	-	-	-	-	-	-	-	-	-
GA-TS [70]	Not stated	✓	-	-	-	-	-	-	-	-	-	-	-	-	-	-	-	-	-	-	-	✓	-	-	-
SPO [71]	CloudSim, Java	✓	-	-	-	-	-	-	-	-	-	-	✓	-	-	-	-	-	-	-	-	-	-	-	-

continued on next page

TABLE 4.2 *(continued)*

Algorithm	Simulation environment	Objectives																							
		Makespan	Accuracy	QoS	Execution time	Energy efficiency	Reliability	Deadline	Performance	Efficiency	Cost	Load balancing	Security	Resource utilization	Throughput	Schedule Length Ratio	Lateness	Coverage	Hypervolume	Task Failures	Out-of-bid Errors	Relative Errors	No. of Active Resource	Execution Overhead	Degree of Imbalance
DO-HHO [72]	iFogSim	✓	-	-	-	-	-	-	-	-	-	-	-	-	-	-	-	-	-	-	-	-	✓	-	-
PPR-RM [73]	Not stated	-	-	-	-	✓	✓	-	-	-	-	-	-	-	-	-	-	-	-	-	-	-	-	-	-
SOS-based [74]	Not stated	✓	✓	-	-	-	-	-	-	-	✓	-	-	-	-	-	-	-	-	-	-	-	-	-	-
HDECO [75]	CloudSim	-	-	-	✓	✓	-	-	-	-	✓	✓	-	-	-	-	-	-	-	-	-	-	-	-	-
LJFP & MCT-PSO [76]	CloudSim	✓	-	-	✓	✓	-	-	-	-	-	-	-	-	-	-	-	-	-	-	-	-	-	-	✓
HDD-PLB [77]	WorkflowSim	-	-	-	✓	-	-	-	-	-	✓	-	-	-	-	-	-	-	-	-	-	-	-	-	-
HWOA-based MBA [78]	CloudSim	✓	-	-	✓	-	-	-	-	-	✓	-	-	✓	-	-	-	-	-	-	-	-	-	-	-
HFSGA [79]	Not stated	✓	-	-	-	-	-	-	-	-	✓	-	-	-	-	-	-	-	-	-	-	-	-	-	-
mFA [80]	Not stated	-	-	-	-	✓	-	-	-	-	✓	-	-	-	-	-	-	-	-	-	-	-	-	-	-
TPVNS [81]	Matlab	✓	-	-	-	-	-	-	-	-	-	-	-	-	-	-	-	-	-	-	-	-	-	-	-
DEWTS [82]	CloudSim	-	-	-	✓	✓	-	-	-	-	-	-	-	✓	-	-	-	-	-	-	-	-	-	-	-
EEWS [83]	Java	✓	-	-	-	✓	-	-	-	-	-	-	-	-	-	-	-	-	-	-	-	-	-	-	-
MBHO [84]	WorkflowSim	✓	-	-	-	-	-	-	-	-	✓	-	-	✓	-	-	-	-	-	-	-	-	-	-	-
MALO [85]	CloudSim	✓	-	-	-	-	-	-	-	-	-	-	-	✓	-	-	-	-	-	-	-	-	-	-	-
DBOA [86]	iFogSim	✓	-	-	-	✓	-	-	-	-	-	-	-	-	-	-	-	-	-	-	-	-	-	-	-
HGALO-GOA [13]	WorkflowSim	✓	-	-	-	✓	-	-	-	-	✓	-	-	-	✓	-	-	-	-	-	-	-	-	-	-
IABC-MBOA-LB [87]	CloudSim	✓	-	-	-	✓	-	-	-	-	-	✓	-	-	-	-	✓	-	-	-	-	-	-	-	✓
HGALO-SCA [6]	WorkflowSim	✓	-	-	-	✓	-	-	-	-	✓	-	-	-	✓	-	-	-	-	-	-	-	-	-	-
MHO-S & MHO-D [88]	CloudSim	✓	-	-	✓	-	-	-	-	-	-	-	-	-	-	-	-	-	-	-	-	-	✓	-	-
HPC2N [89]	CloudSim	✓	-	-	-	✓	-	-	-	-	-	-	-	-	-	-	-	-	-	-	-	-	-	-	-
ALOPSO [90]	CloudSim	✓	-	-	-	✓	-	-	-	-	✓	✓	-	-	-	-	-	-	-	-	-	-	-	-	-
HGSOA-GOA [10]	WorkflowSim	✓	-	-	-	✓	-	-	-	-	✓	-	-	-	✓	-	-	-	-	-	-	-	-	-	-
IGWO [10]	CloudSim, Matlab	✓	-	-	-	✓	-	-	-	-	✓	-	-	-	-	-	-	-	-	-	-	-	-	-	-
PSO–GWO [91]	CloudSim	-	-	-	✓	-	-	-	-	-	✓	-	-	-	-	-	-	-	-	-	-	-	-	-	-
EM_WOA [92]	Real workflow app	-	-	-	-	✓	-	-	-	-	✓	-	-	-	-	-	-	-	-	-	-	-	-	-	-
F-NSPSO [93]	WorkflowSim	✓	-	-	-	✓	-	-	-	-	-	-	-	-	-	-	-	-	-	-	-	-	-	-	-
(ACO)-List [94]	Not stated	✓	-	-	-	-	-	-	-	-	✓	-	-	-	-	-	-	-	-	-	-	-	-	-	-

Algorithm 4.1 The Whale Optimization Algorithm.

```
Initialize a set of a random population
Evaluate all of the Whales and determine the best solution based on the fitness value
while the last condition is not satisfied
  for each whale
    calculate the parameters (a, A, C, and l)
    if (p < 0.5)
      if (|A| < 1)
        Update X(t + 1) using Eq. (4.4) (first line)
      else if (|A| ≥ 1)
        Update X(t + 1) using Eq. (4.1)
      end if
    else if (p ≥ 0.5)
      Update X(t + 1) using Eq. (4.4) (second line)
    end if
    update X* if there is a better answer
  end for
end while
```

and then find justifiable answers by optimizing those that were identified. The suggested method is contrasted with various algorithms already in use. Results show that the recommended method performs more effectively than others.

The WOA is utilized in [100] to choose the best task based on the score value. The optimization process will be split into three stages: surrounding the prey, using a bubble net assault, and looking for prey. (See Table 4.3.)

TABLE 4.3 Objective and scheduling scheme in Standard WOA.

Algorithm	Author	Sim. environment	Objectives	Scheduling Type
WOA [99]	Hosseini et al., 2021	Matlab	Makespan	Resource scheduling
WOA [100]	MS et al., 2020	CloudSim	Efficiency, Profit, Time, Memory, and CPU utilization	Task scheduling

4.6.2 Multi-objective WOA

To fulfill user needs and enhance cloud computing efficiency, effective workflow scheduling algorithms are used to increase resource usage. The whale optimizer algorithm [101] (WOA) is a novel framework that tries to optimize work completion for achieving QoS requirements like budget, and deadline, and manages the load on resources through effective task allocation in cloud environments. The achievements of the WOA approach were analyzed by deadline hit, makespan, and resource usage criteria as a Multi-objective Optimization Problem for the cloud systems. In comparison to other approaches (GA, PSO, ACO, and GWO), it demonstrates that WOA works well and can be implemented for practical systems. The WOA method is described in Algorithm 4.2.

The system receives workflow (B) as input data at random. The path that deviates from the sequential manner is then eliminated from the existing paths using the cycle elimination (BO) method. The workflow is then produced as input to the WOA (I). This algorithm optimizes the existing path to arrange the workflow. The fitness value of every agent will be determined by the WOA scheduler. Before a value is taken into consideration for the following generation, it is scored and either approved or refused. Following the completion of the total number of iterations, WOA changes the location of the current search agent. Calculate the fitness of all search agents to update it. The workflow is then submitted, and the cycle continues until the subsequent scheduling event. The $Nodes_{Migr}$ is utilized where it contains all the tasks that require relocation due to performance agreement violations or breakdowns. B_t is the running workflow in t^{th} time instance.

In [102], the WOA based on the multi-objective model with a dynamic voltage and frequency scaling approach was used to create an ideal task workflow scheduling strategy for mobile devices, considering several significant aspects. This study makes a balance between quality and power utilization by resolving the simultaneous optimization for makespan and energy consumption. In this work, a task scheduling workflow framework is created for the mobile environment and provides collaborative task scheduling on cloud and local servers based on the task execution order. Also, the DVFS technology is

Algorithm 4.2 The WOA-based Scheduling method.

Input data, Workflow: B = (Nodes, Edges)
CycleElimination: B_0 = (Nodes, Edges - Edges Queued)
Schedule: I = Whale Optimization(B_0)
Submit workflow: Execute(B)
Repeat
t = sleep until the next scheduling event
Select tasks for migration
$Nodes_{Migr} = \{N \in \text{Nodes} \mid \text{state}(N, t) = \text{failed} \bigvee \text{State}(N, t) = \text{running} \bigwedge \text{PC}(N, t) > f_N\}$
B_t = generate static DAG($B, I, t, Nodes_{Migr}$)
Cancel(N), $\forall N \in Nodes_{Migr}$
reschedule: I = Whale optimizers(B_t)
Until state(N, t) = completed, $\forall N \in$ Nodes $\bigwedge$ succ(N) = φ

used to further reduce energy consumption by dynamically adjusting the core's operating frequency and voltage. According to experimental findings, the scheme has several benefits over different approaches, such as PSO.

In [103] to further improve the efficiency of a cloud service with available computational resources, developed the technique known as Improved WOA for Cloud Task Scheduling (IWC). The suggested IWC has a superior speed of convergence and efficiency in determining the best scheduling strategies than the current metaheuristic algorithms, including PSO and ACO, according to the simulation-based research and the comprehensive implementation. Additionally, it can perform better in terms of system resource usage while interacting with both small and big tasks.

Principal implementation of cloud task scheduling using IWC can be partitioned into four key phases:

Phase 1: The initialization is the primary goal of this stage, and also involves setting up the locations of each particle and mapping the tasks and whales. The max number of iterations, the search space size, and the boundaries of the particles are additional implementation parameters that are introduced.
Phase 2: Each whale's fitness value is initially calculated using the location data. The whale with the lowest fitness value will be highlighted. The value of the current A as well as the randomly generated number p will determine the whales' exploration and exploitation behaviors.
Phase 3: In this phase applied the decrease and increase operators to control the population of whales when their locations are updated. Based on the clustering of whales, whales will be removed or added, and the population will be calculated using Eqs. (4.5) and (4.6).

$$n_{increment} = ps \times (PS_{max} - ps)^2 \times PS_{max}^{-2} \tag{4.5}$$

$$n_{decrement} = ps^2 \times (PS_{max} - ps) \times PS_{max}^{-2} \tag{4.6}$$

Here, ps is the present population size, while PS_{max} is the population size's upper bound. Using a generic clustering technique, the population was separated into $n_{increment}$ and $n_{decrement}$ groups in this phase. The best people in each group were then chosen to create sets, while the worst members in each class were eliminated.
Phase 4: If the number of repetitions is achieved, the search process will end; else, it will move on to Phase 2 for a new search. The location of the head whale will be transmitted to the decision variables a_{ij} as the optimal scheduling option for the cloud environment activities after the designated number of repetitions has been attained. The decision variable a_{ij} is set to 1 if the ith task is executed out on the jth virtual machine, and 0 otherwise.

A novel model of the Multi-Objective Cluster-based WOA is used to justify and evaluate the constraints associated with the appliance scheduling problem, such as the energy usage of the interior comfort index factors, which include air quality, humidity, and lighting quality [104]. Various initialization strategies and cluster-based methodologies will be utilized to tune the primary parameter of WOA under various MapReduce operations, which helps to regulate exploitation and exploration, to reach the goals.

In [105], proposed a multi-layer WOA scheduling model in the cloud computing environment. The three primary levels of the cloud computing scheduling challenge are the scheduling of virtual machines for client programs (user task layer), the scheduling of virtual machines to physical assets (task scheduling layer), and the scheduling of physical assets (data center layer). A multi-objective WOA-based task scheduling on the Gaussian cloud model (GCWOAS2) is utilized in the second layer of cloud computing.

The population-initialization process utilizes an opposition-based learning mechanism to define the optimal scheduling strategy. Next, a dynamic response factor is proposed to expand the search step size depending on the dynamic changes to search virtual machines at a wider range. Based on the Gaussian cloud model, this WOA model aims to improve optimization by making the scheduling process more random, causing the schedule to deviate from the local optimal. According to experimental findings, GCWOAS2 performs better in terms of resource efficiency and can not only schedule tasks faster than other metaheuristic algorithms currently in use, but it can also balance the demand on VMs.

Reddy and Kumar in 2018 [106], presented a Multi-objective WOA-based algorithm to schedule tasks in the cloud environment. The tasks are scheduled by WOA based on a fitness criterion. Resource consumption, service quality, and energy are the three main limitations that affect the fitness parameter. The suggested WOA organizes the tasks based on the aforementioned factors such that the cost and time associated with task execution on VMs are kept to a minimum.

One year later Reddy and Kumar 2019 [107], introduced the Regressive Whale Optimization (RWO) method for scheduling workflows in cloud computing. This algorithm varies from the conventional WOA in the location update stage, where the improvement is done with the addition of a regression-based location update. The task is scheduled using the RWO algorithm based on a fitness function. Using resource usage, energy, and QoS constraints, the tasks are distributed across the resources. As a result, while allocating the virtual machines in response to multiple user requests, the scheduling method complies with all three restrictions. The RWO algorithm's advantages include minimal cost and location discovery based on a global search. Additionally, the convergence of the solutions happens at the global phase.

The original WOA was outperformed by an improved method dubbed the Opposition-based Chaotic Whale Optimization Algorithm (OppoCWOA) with chaos theory, and jumping rate to achieve time-energy efficiency for resolving the task scheduling issue in fog environments [108]. With this approach, partial opposition was used instead of full opposition, improving the variety of the population and enhancing task scheduling performance while accelerating convergence. There is also a discussion of how various chaotic maps affect different parameters of WOA, and four different non-linear functions are presented to replace the linear function in this study. Moreover, a jumping rate was used to prevent the quick convergence that might happen when utilizing opposition-based techniques. According to the simulation findings, the suggested method performed noticeably better than some other optimization techniques at attaining time-energy efficient task scheduling.

Energy-aware Whale-optimized Task Scheduler (EWTS) concentrated on reducing energy usage along with makespan for separate task scheduling in the cloud environment with a variant of a multi-objective model [109].

The Whale scheduler (W-Scheduler) method is based on the multi-objective WOA method proposed to solve the task scheduling issue [110]. This method maintains the minimal cost and minimum makespan while allocating the tasks to the virtual machines in the best possible way. The central processing unit (CPU) and memory cost functions of each virtual machine are first calculated using the multi-objective model, and the budget cost function is then calculated by summing the cost functions of the CPU and memory. The makespan and the budget cost function are combined to provide the fitness value. According to the fitness score, the tasks are assigned to the virtual machines. The algorithm starts with a set of random solutions. It performs the search process based on the present answer and presumes that it is the best option. The optimal solution is then found by repeating this procedure.

In [111], a multi-objective mathematical model for the hybrid flow shop scheduling problem (MOHFSP-DRP) was suggested that takes into account both the devices' programmable processing modes and dynamic reconfiguration procedures (DRP). Through non-linear predation, a bi-objective optimization technique, and a carefully considered dynamic coding approach, an improved multi-objective WOA (IMOWOA) is given to solve the MOHFSP-DRP. This model's dual goals are to reduce energy usage and makespan. (See Table 4.4.)

4.6.3 Improved WOA

The improved whale optimization method (IWC), in [112] suggested a better scheduling efficiency approach. The WOA method has fewer parameters and a simpler operation when compared to other complex algorithms, but it suffers from low convergence precision, delayed convergence, and ease of falling into the local optimum. As a result, it got better in two areas: global search and local search.

In this method, a practical strategy is to be found the highest-scoring whale. To better find the optimal solution, applying the inertial weight method for the WOA was to enhance the local search capability and efficiently intercept the algorithm from attaining convergence speed. After each cycle is finished to increase the quality of learning, IWC utilized the add and delete operator to evaluate solutions. This algorithm evaluated against the ACO, PSO, and basic WOA for various numbers of jobs. The results showed that the IWC algorithm overcomes other algorithms in terms of cost, time, and load on the virtual machine.

TABLE 4.4 Objective and scheduling scheme in Multi-objective WOA.

Algorithm	Author	Sim. environment	Objectives	Scheduling Type
WOA [101]	Thennarasu et al., 2021	Hadoop under Linux environment	Makespan, Time, Deadline hit, and resource utilization	Workflow scheduling
WOA [102]	Peng et al., 2019	Matlab	Task execution position and sequence, Operating voltage, mobile devices frequency	Workflow scheduling
IWC [103]	Chen et al., 2019	Matlab	Load, Time, and Price cost	Task scheduling
MOWOA [104]	Faizal Omar et al., 2020	C++	Energy consumption	Appliances scheduling
GCWOAS2 [105]	Ni et al., 2021	Matlab	Time cost, Load cost, and Price cost	Task scheduling
WO [106]	Reddy and Kumar, 2018	Java	Time, Cost, Resource utilization, and Energy	Task scheduling
RWO [107]	Reddy and Kumar, 2019	Java	Time, Cost, Resource utilization, and Energy	Workflow scheduling
OppoCWOA [108]	Movahedi et al., 2021	CloudSim	Time, Energy	Task scheduling
EWTS [109]	Sharma and Garg, 2017	Matlab	Energy, Makespan	Task scheduling
W-Scheduler [110]	Sreenu and Sreelatha, 2017	CloudSim	Makespan, Cost	Task scheduling
IMOWOA [111]	Yankai et al. 2021	Matlab	Makespan, Energy	Flow shop scheduling

An Improved Whale Algorithm (IWA) is proposed for allocating dependent tasks in multiprocessing systems while reducing energy consumption and processing time. The processing cores reduce energy-supported Dynamic Voltage and Frequency Scaling (DVFS). IWA uses two major discretization approaches to change continuous variables into discrete ones and an initialization strategy to create a population of alternative schedules that preserve task priority. To decrease the load on busy cores, IWA utilizes the Load Balancing Improvement method. Finally, two particular crossover procedures were used to improve the candidate schedules' quality while maintaining the dependencies between tasks. Experimental findings demonstrate IWA's superiority.

The modified task scheduling algorithm based WOA (MWOA) [113], was presented as a way to enhance the task scheduling methodology. The order of queue methods is used in this method to store user tasks. New tasks are looked into and retained in the on-demand priority queue. The MWOA receives the result of the scheduled queue. This method outperforms existing algorithms and can solve optimization issues.

Another paper of the Improved WOA research group presented an enriched approach to the chaotic quantum whale optimization algorithm (CQWOA), intending to increase the degree of imbalance and reduce costs, resource utilization, and energy consumption [114]. By making use of chaotic mapping capabilities and quantum mechanism-based optimal virtual machine selection, this method considerably increases the global optimum. (See Table 4.5.)

TABLE 4.5 Objective and scheduling scheme in Improved WOA.

Algorithm	Author	Sim. environment	Objectives	Scheduling Type
IWC [112]	Jia et al., 2021	CloudSim	Cost, Time consumption, Load	Task scheduling
IWA [115]	Abdel-Basset et al., 2020	Real-world applications	Makespan, Fitness value, Energy, Tardiness, Flow time	Real-time task scheduling
MWOA [113]	Saravanan et al., 2019	CloudSim	Makespan, Energy consumption	Task scheduling
CQWOA [114]	Kiruthiga and Vennila, 2019	CloudSim	Makespan, Energy Consumption, Cost, Resource Utilization, Degree of imbalance	Task scheduling

4.6.4 Hybrid WOA

The Hybrid Whale Optimization Algorithm was suggested by Stromberg et al. [116] to solve the resource scheduling issue in cloud platforms. The WOA ABC exploration firefly search (WOA-AEFS) method, produced the best results in simulations using simulated data sets against the simple heuristics of WOA, PBACO, FCFS, and Min-Min, and it proved to be a reliable and effective optimization approach for dealing with the scheduling issue.

The modified hybrid whale algorithm (MHWA) [117] is combined with several methods such as a local search approach, reversed-block insertion, and swap mutation procedures for solving the scheduling of the Multimedia Data Objects (MDO) and to get better solutions. The MDO is a challenging and exceptional problem that we face in the World Wide Web (WWW) to reduce the reaction time for consumers to complete their activities quickly. The MDO scheduling problem can be imitated by a Two-Machine Flow Shop Scheduling Problem (T-MFSSP) that reduces the makespan. The basic population's solutions are classified into three groups:

(a) The answer produced by the Johnson algorithm is included in the MHWA's initialization.
(b) The solution is produced by the Nawaz-Enscore-Ham (NEH) algorithm, which is added to MHWA to improve the method's execution.
(c) The solutions are produced at random using the LRV approach to convert the continuous search space into a discrete space.

The diversity of the population is increased by all of these kinds of approaches. Through several repetitions, the mechanisms of reversed block addition and swapping mutation increase the quality of the outcomes. Johnson heuristic and EDD algorithm are two other algorithms that are evaluated with MHWA. In this work, the following three performance indicators are used: Average lateness, makespan, and time.

Johnson's method is outperformed by MHWA because it may acquire lower average task submission lateness. The practical findings demonstrate that, in terms of reducing the makespan and the average lateness for all cases, MHWA comes above EDD. The MHWA method is useful for scheduling MDO via the WWW in a feasible amount of time.

To solve the Permutation Flow Shop Scheduling Issue, Abdel-Basset et al. have developed a new hybrid whale algorithm (HWA) [118] that combines WOA with a local search approach, swap mutation, and the addition of reversed-block functions (PFSSP). WOA has been suggested as a strategy for dealing with continuous issues, then it is combined with several approaches to handle PFSSP as a discrete challenge. An efficient technique for converting continuous data into job permutations is the Largest Rank Value (LRV). In LRV, the continuous values are from highest to lowest. The representation, analysis, swapping mutation, inserting reversed-block function, and local search method are the five basic phases of this algorithm.

A swap mutation operation is also used to increase the variety of candidate schedules. To escape from the local optima, an insert-reversed block operation is also used. To enhance the performance of the method, HWA is combined with Nawaz-Enscore-Ham (NEH). NEH is an effective algorithm for handling scheduling issues. Based on NEH's default settings, 10% of the basic population is used for the search agent. It has been shown that HWA produces results that are equivalent to those of the current algorithms.

The hybridization of the WOA and the bat algorithm (WOA-BAT) is suggested in the other study for load balancing in a cloud-fog computing system [119]. One of the certain WOA downsides is the slowdown convergence speed imposed by seeking the best solution. Therefore, the BAT algorithm has been used to progress the discovery of WOA. Two techniques are used for the hybridization of these two algorithms: (1) integrating the BAT technique within the WOA search phase, and (2) implying the condition method whenever the locations of any search agents are altered. The preceding location has been changed if the new location is superior to the previous one.

To achieve load balancing, requests are sent by the consumers to the controller of the fog data center. This final step sends requests to the load balancer, which then distributes them across the virtual machines using the load balancing algorithm. Virtual machines are built on actual computers and managed by a VM administrator. WOA-BAT load balancing algorithm evaluated by the quality measures (Response and Processing time) and contrasted with other algorithms such as round robin, and particle swarm optimization strategies. The outcomes prove that the WOA-BAT algorithm performs better than the other algorithms.

A hybrid multi-objective algorithm (HGWWO) [120] was created to combine the benefits of two algorithms, the whale optimization algorithm and the grey wolf optimizer, to reduce the energy usage, costs, and overall execution time required for job scheduling on virtual machines. This method begins with the creation of a random set of search agents, after which the Assignment of tasks to VMs is dependent on their fitness value when the search agents are in search of prey. Eq. (4.7) is used to prioritize tasks for assignment to an existing VM, based on several objectives, and the best answer is determined as the task with the greatest value.

$$Fitness\ function(F) = \frac{1}{4}(FT_w + R_u + (1 - E) + (1 - L)) \tag{4.7}$$

In Eq. (4.7), FT_w stands for general task weight, which is supported by requester type and execution task importance. In a virtual machine, the terms R_u and E, which stand for the resources used and necessary energy, respectively, and L, which stands for round-trip latency, are used. The $(1 - E)$, $(1 - L)$ addition to the equation highlights the high priority and latency. Compared to the initial algorithms GWO and WOA, this method is capable of working at a higher level.

For the permutation flow shop scheduling issue, a memetic whale optimization algorithm (MWA) was presented in [121] and used two Nawaz-Enscore-Ham-based heuristics to establish the population. The continuous number is converted to task permutations using the smallest position value (SPV) method. In this method to increase the accuracy of the result, a local variable neighborhood search based on simulated annealing is applied. According to computational statistics, the MWA can outperform the other algorithms in terms of makespan.

The other hybrid algorithm has two meta-heuristic calculations consolidated, the Whale Optimization Algorithm and Harmony Search Algorithm (WHOA) [122]. This article proposed a task scheduling method based on Whale Harmony enhancement computation. The steps of this algorithm are presented in Algorithm 4.3.

Algorithm 4.3 The WHOA Scheduling method.

1: Initialize the population of whales
2: The fitness values of each search agent are computed
3: Set up the parameters for the Harmony search algorithm and execute New Harmony
4: Evaluate New Harmony Memory has been refreshed, and New Harmony is superior to Old Harmony
5: Harmony algorithm termination conditions are assessed; else, the previous step is repeated
6: Next, the WOA update formula used on the results obtained from the harmony search algorithm to modify the primary search agent location
7: Recalculate each search agent's fitness if a better solution becomes available
8: Verify the total number of iterations completed; else, repeat steps 2–8
9: Return the best solution

The outcomes of the experiment confirm that the WHOA effectively balances the workload and schedules the tasks. This algorithm reduces the cost and makespan.

In another work [123], a new hybrid method was introduced to solve the unrelated parallel machine scheduling problem (UPMSP) with machine-dependent setup time. This method called WOAFA combined the characteristics of two meta-heuristic algorithms, the whale optimization algorithm (WOA) and the firefly algorithm (FA), To solve the scheduling problem. To improve WOA's search capabilities, the FA is used as a local search. Typically, this method starts by generating a population of random integers that serves as the starting point for solving the UPMSP. Then, the WOA is used to optimize the solutions. The makespan is used as an objective function to analyze the solutions. Following that, either the WOA or FA operators are used to update each solution. A criterion based on probability is used to perform this switching. The optimization phases continued until the stop condition is reached. The evaluation results validated the WOAFA's impressive performance.

A novel hybridized whale optimization method was developed in [116], to address the resource scheduling issue in cloud systems. It is attempted to enhance the initial WOA by addressing the trade-off modifications between exploitation-exploration, taking into consideration that the essential processing mechanisms of any swarm algorithm are exploitation and exploration, and their adjusted balance. In WOA-AEFS first, adapted the exploration mechanism from the Artificial Bee Colony algorithm. Then, a novel exploration method is controlled by extra dynamic control factors. Finally, this strategy utilizes the firefly algorithm's search equation.

A hybrid oppositional differential evolution-enabled WOA (h-DEWOA) [124] method was developed to address Cloud Bag-of-Tasks Scheduling issues and reduce workload times and energy use to overcome WOA restrictions. The proposed h-DEWOA expands the basic WOA method by containing opposition-based learning, chaotic maps, differential evolution, and a fitness-based balancing mechanism. As a result, the algorithm executes with enhanced exploration, quicker convergence, and a sufficient tradeoff between exploration-exploitation. Then, to enhance resource allocation, an adjusted heuristic is included in this technique.

The TOPSIS method is an efficient approach for dealing with numerous criterion decision-making complexions in the real world. When TOPSIS was expanded to a fuzzy environment in [125], the fuzzy TOPSIS (FTOPSIS) method was proposed for scheduling tasks effectively based on task size. This approach regulates the requests' admission by reaching a goal QoS in terms of response time. As a result, access is restricted to ensure that requests that are approved do not experience delays that are longer than those specified in the SLA. Data packet loss rate and transmission time may both be effectively reduced by reducing SLA infractions and enhancing QoS. The new feature of FTOPSIS-WOA is that it significantly reduces the makespan of the cloud scheduling process while boosting the throughput of the cloud environment.

The WGOA hybrid algorithm integrated two algorithms, WOA and Genetic Algorithm to create a novel hybridized algorithm [126]. This algorithm certifies the optimization in terms of three competing criteria, Enactment Amelioration Rate, cost, and makespan. In this study, WOA is combined with genetic algorithm crossover and mutation actions. This hybridization is utilized to generate the multi-objective optimization process's nearly ideal solution. In terms of enactment amelioration rate, the findings obtained reveal a considerable decrease in the execution time.

GA-WOA is a useful approach for effectively scheduling tasks in a specific cloud utilizing the MAP reduction architecture [127]. The client's task is first used to extract the task characteristics. The features are minimized by using the maximized Rayleigh quotients of the FLDA (MRQFLDA) algorithm. Then, a map-reduce architecture is used to divide the huge tasks into smaller ones. The GA-WOA algorithm is then used to effectively schedule the tasks.

A new hybrid whale-genetic optimization algorithm was offered to optimize and balance makespan and energy consumption [128]. The genetic algorithm assists the original whale optimization method in escaping the local optimum point, where it has the propensity to fall. In GA, the survival of the fittest among individuals is simulated and followed across several iterations. New parameters that integrate the exploitation and exploration capabilities of GA and WOA, respectively, were established to decrease the execution cost by removing trap time or the local optimum point. (See Table 4.6.)

TABLE 4.6 Objective and scheduling scheme in Hybrid WOA.

Algorithm	Author	Sim. environment	Objectives	Scheduling Type
WOA-AEFS [116]	Strumberger et al., 2019	CloudSim	Makespan, Cost	Resource scheduling
MHWA [117]	Abdel-Basset et al., 2018	Java	Makespan, Average lateness	MDO scheduling
HWA [118]	Abdel-Basset et al., 2018	Java	Makespan	Flow shop scheduling
WOA-BAT [119]	Saoud and Recioui, 2022	Java	Response and Processing time	Task scheduling
HGWWO [120]	Ababneh, 2021	CloudSim	Energy consumption, Makespan, Cost, and Degree of imbalance	Task scheduling
MWA [121]	Lin et al., 2018	Not stated	Makespan	Flow shop Scheduling
WHOA [122]	Albert and Nanjappan, 2021	CloudSim	Makespan, Cost, Energy, Resource utilization, and Degree of imbalance	Task scheduling
WOAFA [123]	Al-qaness et al., 2021	Matlab	Makespan	Unrelated parallel machine scheduling
WOA-AEFS [116]	Strumberger et al., 2019	CloudSim	Makespan, Cost, Violation rate	Resource Scheduling
h-DEWOA [124]	Chhabra et al., 2022	CloudSim	Makespan, Energy utilization	Bag-of-Tasks Scheduling
FTOPSIS-WOA [125]	Samriya and Kumar, 2020	CloudSim	Makespan, Cost, Resource consumption, Response time, Degree of imbalance, Efficiency	Task scheduling
WGOA [126]	Natesan and Chokkalingam, 2019	CloudSim	Makespan, Cost	Task scheduling
GA-WOA [127]	Sanaj and Prathap, 2020	CloudSim	Cost, Time	Task scheduling
WGOA [128]	Sharma and Garg, 2022	Matlab	Makespan, Energy	Task scheduling

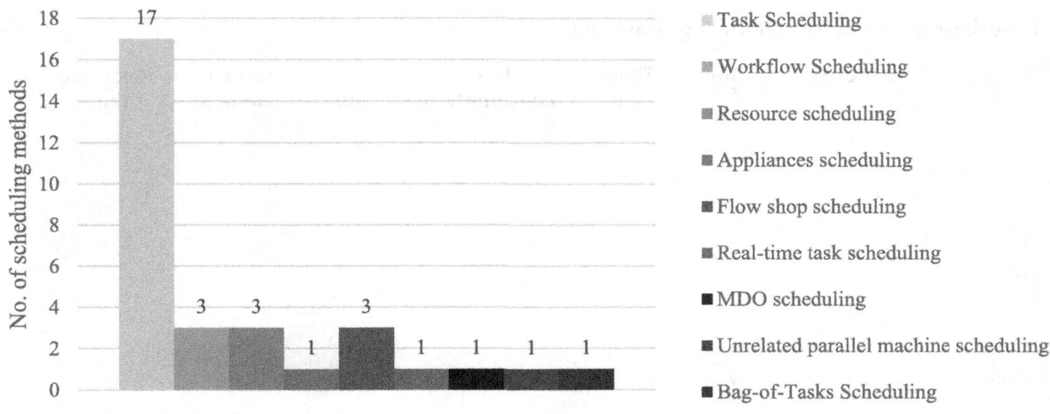

FIGURE 4.3 WOA Scheduling Schemes.

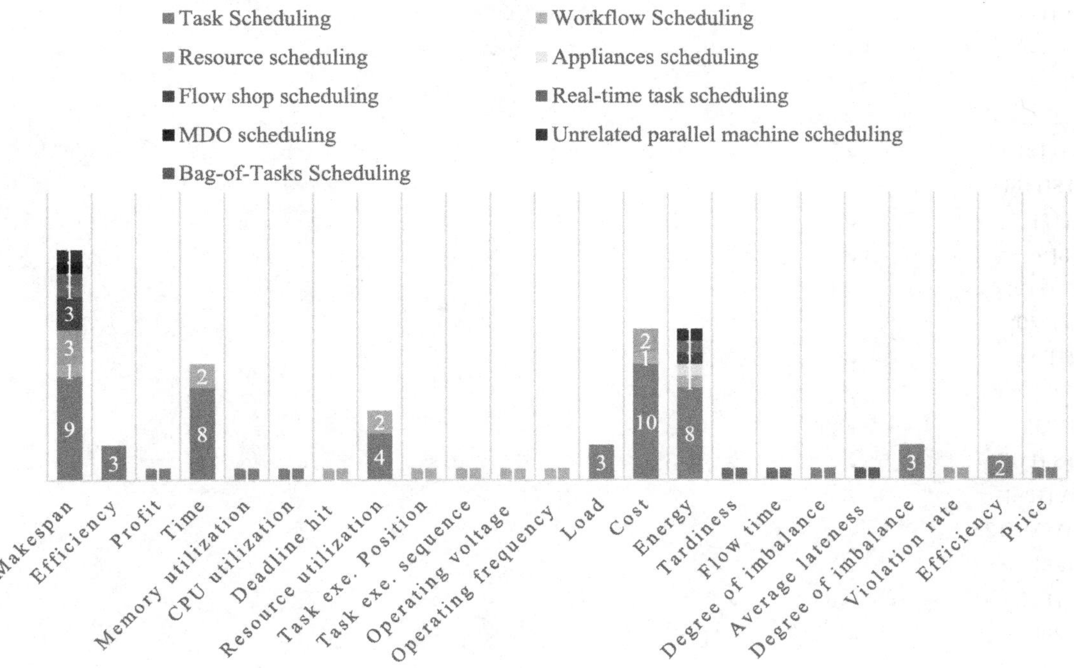

FIGURE 4.4 Objectives in WOA-based scheduling methods.

4.7 Discussions

This section presents the overall conclusion reached from a comparison of several WOA scheduling techniques used in the cloud environment. Different workflow and task scheduling approaches utilized in WOA-based scheduling are presented in Fig. 4.3. The majority of researchers placed a greater focus on task scheduling than other scheduling methods. The number of scheduling approaches that address each objective is demonstrated in Fig. 4.4. The makespan, time, cost, and energy objectives are given increased attention in task scheduling techniques. Also, in the WOA-based various scheduling schemes except task scheduling, the makespan, efficiency, load, and degree of imbalance objectives are approximated to be of equal importance.

As previously mentioned, every scheme under analysis takes various time-related aspects into account while scheduling. Kinds of time-based factors applied in the WOA-based scheduling strategies are shown in Table 4.7. In Fig. 4.5, we categorized the various scheduling strategies according to the kind of WOA algorithm that was applied. According to the obtained results, most of the studies have been multi-objective and hybrid.

By examining the obtained results, which are shown in Fig. 4.6, CloudSim, Matlab, and Java are the most used simulation tools in the implementation of the WOA-based Scheduling plans, respectively. The CloudSim toolkit, which is used in

TABLE 4.7 Time-based criteria for scheduling strategies.

Strategy	Makespan	Time	Time cost	Time consumption	Flow time	Average lateness	Response time	Processing time
WOA [99]	✓	-	-	-	-	-	-	-
WOA [100]	-	✓	-	-	-	-	-	-
WOA [101]	✓	✓	-	-	-	-	-	-
IWC [103]	-	✓	-	-	-	-	-	-
GCWOAS2 [105]	-	-	✓	-	-	-	-	-
WO [106]	-	✓	-	-	-	-	-	-
RWO [107]	-	✓	-	-	-	-	-	-
OppoCWOA [108]	-	✓	-	-	-	-	-	-
EWTS [109]	✓	-	-	-	-	-	-	-
W-Scheduler [110]	✓	-	-	-	-	-	-	-
IMOWOA [111]	✓	-	-	-	-	-	-	-
IWC [112]	-	-	-	✓	-	-	-	-
IWA [115]	✓	-	-	-	✓	-	-	-
MWOA [113]	✓	-	-	-	-	-	-	-
CQWOA [114]	✓	-	-	-	-	-	-	-
WOA-AEFS[116]	✓	-	-	-	-	-	-	-
MHWA [117]	✓	-	-	-	-	✓	-	-
HWA [118]	✓	-	-	-	-	-	-	-
WOA-BAT [119]	-	-	-	-	-	-	✓	✓
HGWWO [120]	✓	-	-	-	-	-	-	-
MWA [121]	✓	-	-	-	-	-	-	-
WHOA [122]	✓	-	-	-	-	-	-	-
WOAFA [123]	✓	-	-	-	-	-	-	-
WOA-AEFS [116]	✓	-	-	-	-	-	-	-
h-DEWOA [124]	✓	-	-	-	-	-	-	-
FTOPSIS-WOA [125]	✓	-	-	-	-	-	-	✓
WGOA [126]	✓	-	-	-	-	-	-	-
GA-WOA [127]	-	✓	-	-	-	-	-	-
WGOA [128]	✓	-	-	-	-	-	-	-

many scheduling algorithms including [100], [108], [110], [112], [113,114], [116], [120], [122], and [116,124–127], provides modeling and the construction of virtual machines (VMs) on virtual server resources in cloud services. Additionally, it provides jobs and their assignment to appropriate VMs using a variety of main scheduling methods.

We have discussed the difficulties and research hypotheses around cloud technology in this paragraph. Cloud technology is a field of innovation, that can be utilized by businesses and people, but like any new technology has disadvantages. We need efficient ways for cloud computing environments to satisfy cloud platform providers' standards and effectively use advantages to increase the life span of the cloud computing environment. Additionally, by examining the approaches, it has been shown that the optimization's intended objectives, including reducing makespan, energy efficiency, cost reduction, maximizing reliability, and throughput of cloud network are fulfilled. There are a few concerns that still need to be explored due to the dynamic nature of the cloud environment, such as measuring the use of renewable energy in the cloud, security, cost of service, greenhouse gas pollution, and scalability.

4.8 Conclusion and future work

Cloud computing provides a large number of different kinds of computing resources to users for executing their diverse applications. To efficiently utilize these resources and execute the applications promptly, an effective task scheduling ap-

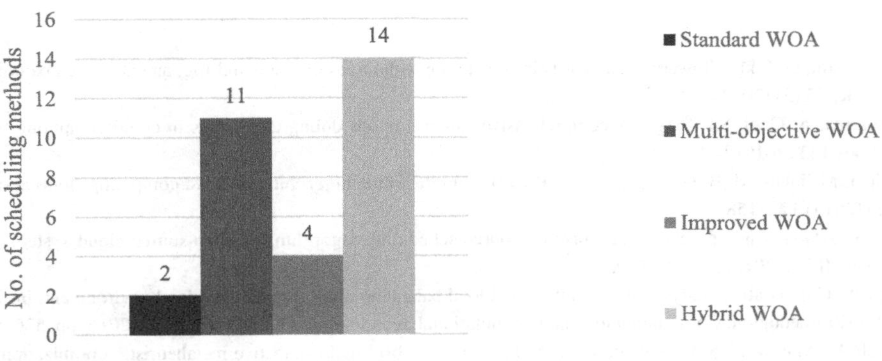

FIGURE 4.5 WOA algorithm types that are used to schedule plans.

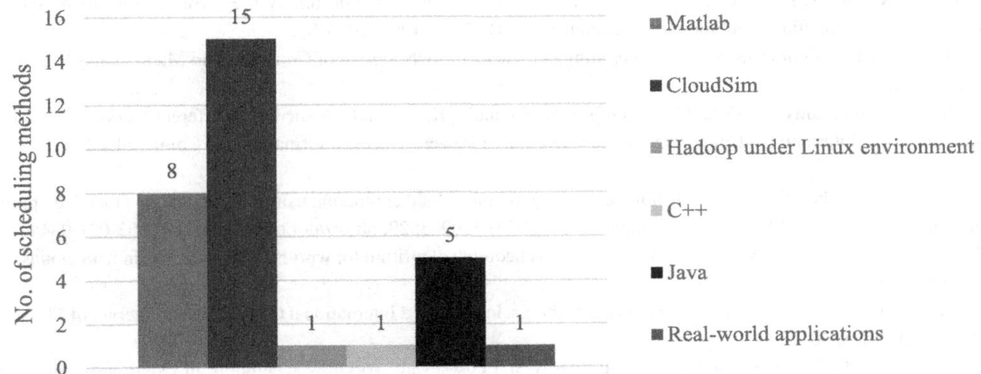

FIGURE 4.6 Simulation tools in the WOA based Scheduling plans.

proach is required. The task scheduling problem is an NP-complete issue even in its most basic form. Scheduling is one of the crucial processes that may lower the cost and duration of cloud task execution by properly assigning tasks to VMs. Additionally, it can save data center operating expenses while increasing resource scalability and availability for cloud service providers. The WOA can outperform well-known metaheuristic algorithms and provides an efficient searching approach for resolving optimization issues in a variety of applications, containing cloud computing scheduling.

In this study, we have made in-depth research and investigation of the existing WOA-based meta-heuristic optimization techniques for scheduling tasks in the cloud environment. We have also presented a categorization of these methods based on the kind of WOA algorithm that has been used for these solutions. Additionally, we have included a more extensive review of how each scheme's WOA algorithm has been enhanced and incorporated to address task/workflow scheduling issues. The dynamic and distributed characteristics of the environment of cloud computing led several researchers to concentrate on the modified and hybrid WOA algorithm to reach the ideal answer.

Future research should focus on scheduling workflows and tasks on federated and hybrid clouds as well as investigating and evaluating dynamic requests for diverse resources of the clouds that have varying degrees of dependability and reliance. Additionally, scheduling algorithms may be integrated with trust management systems. The reliance level of each cloud user that submits a task for processing can be regarded in addition to the other traditional scheduling criteria. The idea of secure scheduling is one of the topics that hasn't received as much attention in most scheduling techniques. While taking into consideration the rising number of securities cyberattacks on the cloud system, that attack the virtual machine and other cloud components like the cloud schedulers, the importance of this issue cannot be overstated. Consequently, future research should incorporate security and trust-related ideas into cloud scheduling so that hackers and trustworthy cloud consumers can be distinguished. Considering the security vulnerabilities of cloud environments, the cloud scheduler should include secure algorithms to stop diverse attacks.

Declaration of competing interest

The authors declare that they have no known competing financial interests or personal relationships that could have appeared to influence the work reported in this paper.

References

[1] R.M. Singh, L.K. Awasthi, G. Sikka, Towards metaheuristic scheduling techniques in cloud and fog: an extensive taxonomic review, ACM Computing Surveys (CSUR) 55 (3) (2022) 1–43.

[2] M. Kumar, S.C. Sharma, A. Goel, S.P. Singh, A comprehensive survey for scheduling techniques in cloud computing, Journal of Network and Computer Applications 143 (2019) 1–33.

[3] M. Masdari, F. Salehi, M. Jalali, M. Bidaki, A survey of PSO-based scheduling algorithms in cloud computing, Journal of Network and Systems Management 25 (1) (2017) 122–158.

[4] H. Zhong, K. Tao, X. Zhang, An approach to optimized resource scheduling algorithm for open-source cloud systems, in: 2010 Fifth Annual ChinaGrid Conference, IEEE, 2010, pp. 124–129.

[5] K. Pradeep, T.P. Jacob, Comparative analysis of scheduling and load balancing algorithms in the cloud environment, in: 2016 International Conference on Control, Instrumentation, Communication and Computational Technologies (ICCICCT), IEEE, 2016, pp. 526–531.

[6] A. Mohammadzadeh, M. Masdari, F.S. Gharehchopogh, A. Jafarian, A hybrid multi-objective metaheuristic optimization algorithm for scientific workflow scheduling, Cluster Computing 24 (2) (2021) 1479–1503.

[7] A. Mohammadzadeh, M. Masdari, F.S. Gharehchopogh, A. Jafarian, Improved chaotic binary grey wolf optimization algorithm for workflow scheduling in green cloud computing, Evolutionary Intelligence 14 (4) (2021) 1997–2025.

[8] I.M. Ibrahim, Task scheduling algorithms in cloud computing: a review, Turkish Journal of Computer and Mathematics Education (TURCOMAT) 12 (4) (2021) 1041–1053.

[9] V. Seethalakshmi, V. Govindasamy, V. Akila, G. Sivaranjini, K. Sindhuja, K. Prasanth, A survey of different workflow scheduling algorithms in cloud computing, in: 2019 International Conference on Computation of Power, Energy, Information and Communication (ICCPEIC), IEEE, 2019, pp. 293–297.

[10] A. Mohammadzadeh, M. Masdari, Scientific workflow scheduling in multi-cloud computing using a hybrid multi-objective optimization algorithm, Journal of Ambient Intelligence and Humanized Computing 14 (2023) 3509–3529, https://doi.org/10.1007/s12652-021-03482-5.

[11] K. Kalyan Chakravarthi, L. Shyamala, V. Vaidehi, Budget aware scheduling algorithm for workflow applications in IaaS clouds, Cluster Computing 23 (4) (2020) 3405–3419.

[12] M. Singh, G. Baranwal, Quality of service (QoS) in internet of things, in: 2018 3rd International Conference on Internet of Things: Smart Innovation and Usages (IoT-SIU), IEEE, 2018, pp. 1–6.

[13] A. Mohammadzadeh, M. Masdari, F.S. Gharehchopogh, Energy and cost-aware workflow scheduling in cloud computing data centers using a multi-objective optimization algorithm, Journal of Network and Systems Management 29 (3) (2021) 31.

[14] K. Dubey, S.C. Sharma, A novel multi-objective CR-PSO task scheduling algorithm with deadline constraint in cloud computing, Sustainable Computing: Informatics and Systems 32 (2021) 100605.

[15] J. Li, X. Zhang, L. Han, Z. Ji, X. Dong, C. Hu, OKCM: improving parallel task scheduling in high-performance computing systems using online learning, Journal of Supercomputing 77 (6) (2021) 5960–5983.

[16] M.U. Sana, Z. Li, Efficiency aware scheduling techniques in cloud computing: a descriptive literature review, PeerJ Computer Science 7 (2021) e509.

[17] Z. Tong, X. Deng, H. Chen, J. Mei, DDMTS: a novel dynamic load balancing scheduling scheme under SLA constraints in cloud computing, Journal of Parallel and Distributed Computing 149 (2021) 138–148.

[18] S. Meng, et al., Security-aware dynamic scheduling for real-time optimization in cloud-based industrial applications, IEEE Transactions on Industrial Informatics 17 (6) (2020) 4219–4228.

[19] M. Farid, R. Latip, M. Hussin, N.A.W.A. Hamid, Scheduling scientific workflow using multi-objective algorithm with fuzzy resource utilization in multi-cloud environment, IEEE Access 8 (2020) 24309–24322.

[20] R.K. Jena, Multi objective task scheduling in cloud environment using nested PSO framework, Procedia Computer Science 57 (2015) 1219–1227.

[21] E.S. Alkayal, N.R. Jennings, M.F. Abulkhair, Efficient task scheduling multi-objective particle swarm optimization in cloud computing, in: 2016 IEEE 41st Conference on Local Computer Networks Workshops (LCN Workshops), 2016, pp. 17–24.

[22] Z. Zhou, J. Chang, Z. Hu, J. Yu, F. Li, A modified PSO algorithm for task scheduling optimization in cloud computing, Concurrency and Computation: Practice and Experience 30 (24) (2018) e4970.

[23] M. Kumar, S.C. Sharma, PSO-COGENT: cost and energy efficient scheduling in cloud environment with deadline constraint, Sustainable Computing: Informatics and Systems 19 (2018) 147–164.

[24] N. Dordaie, N.J. Navimipour, A hybrid particle swarm optimization and hill climbing algorithm for task scheduling in the cloud environments, ICT Express 4 (4) (2018) 199–202.

[25] E. Alkayal, Optimizing Resource Allocation Using Multi-Objective Particle Swarm Optimization in Cloud Computing Systems, University of Southampton, 2018.

[26] T. Prem Jacob, K. Pradeep, A multi-objective optimal task scheduling in cloud environment using cuckoo particle swarm optimization, Wireless Personal Communications 109 (1) (2019) 315–331.

[27] N. Mansouri, B. Mohammad Hasani Zade, M.M. Javidi, Hybrid task scheduling strategy for cloud computing by modified particle swarm optimization and fuzzy theory, Computers & Industrial Engineering 130 (2019) 597–633.

[28] M. Sardaraz, M. Tahir, A hybrid algorithm for scheduling scientific workflows in cloud computing, IEEE Access 7 (2019) 186137–186146.

[29] N. Miglani, G. Sharma, Modified particle swarm optimization based upon task categorization in cloud environment, International Journal of Engineering and Advanced Technology (IJEAT) 8 (2019).

[30] U.K. Jena, P.K. Das, M.R. Kabat, Hybridization of meta-heuristic algorithm for load balancing in cloud computing environment, Journal of King Saud University: Computer and Information Sciences 34 (6, Part A) (2022) 2332–2342.

[31] A.A. Visheratin, M. Melnik, D. Nasonov, Workflow scheduling algorithms for hard-deadline constrained cloud environments, Procedia Computer Science 80 (2016) 2098–2106.

[32] N. Kaur, S. Singh, A budget-constrained time and reliability optimization BAT algorithm for scheduling workflow applications in clouds, Procedia Computer Science 98 (2016) 199–204.

[33] M. Abdullahi, M.A. Ngadi, S.i.M. Abdulhamid, Symbiotic Organism Search optimization based task scheduling in cloud computing environment, Future Generation Computer Systems 56 (2016) 640–650.

[34] L. Teylo, U. de Paula, Y. Frota, D. de Oliveira, Lúcia M.A. Drummond, A hybrid evolutionary algorithm for task scheduling and data assignment of data-intensive scientific workflows on clouds, Future Generation Computer Systems 76 (2017) 1–17.

[35] L. Zhang, K. Li, C. Li, K. Li, Bi-objective workflow scheduling of the energy consumption and reliability in heterogeneous computing systems, Information Sciences 379 (2017) 241–256.

[36] M. Lawanyashri, B. Balusamy, S. Subha, Energy-aware hybrid fruitfly optimization for load balancing in cloud environments for EHR applications, Informatics in Medicine Unlocked 8 (2017) 42–50.

[37] H.Y. Shishido, J.C. Estrella, C.F.M. Toledo, M.S. Arantes, Genetic-based algorithms applied to a workflow scheduling algorithm with security and deadline constraints in clouds, Computers & Electrical Engineering 69 (2018) 378–394.

[38] I. Casas, J. Taheri, R. Ranjan, L. Wang, A.Y. Zomaya, GA-ETI: an enhanced genetic algorithm for the scheduling of scientific workflows in cloud environments, Journal of Computational Science 26 (2018) 318–331.

[39] M. Sharma, A. Verma, A.K. Sangaiah, Chapter 8 - Energy-constrained workflow scheduling in cloud using E-DSOS algorithm, in: A.K. Sangaiah, M. Sheng, Z. Zhang (Eds.), Computational Intelligence for Multimedia Big Data on the Cloud with Engineering Applications, Academic Press, 2018, pp. 159–169.

[40] E.N. Alkhanak, S.P. Lee, A hyper-heuristic cost optimisation approach for Scientific Workflow Scheduling in cloud computing, Future Generation Computer Systems 86 (2018) 480–506.

[41] T. Alam, Z. Raza, Quantum genetic algorithm based scheduler for batch of precedence constrained jobs on heterogeneous computing systems, Journal of Systems and Software 135 (2018) 126–142.

[42] K. Dubey, M. Kumar, S.C. Sharma, Modified HEFT algorithm for task scheduling in cloud environment, Procedia Computer Science 125 (2018) 725–732.

[43] M. Kaur, S. Kadam, A novel multi-objective bacteria foraging optimization algorithm (MOBFOA) for multi-objective scheduling, Applied Soft Computing 66 (2018) 183–195.

[44] M. Tavana, S. Shahdi-Pashaki, E. Teymourian, F.J. Santos-Arteaga, M. Komaki, A discrete cuckoo optimization algorithm for consolidation in cloud computing, Computers & Industrial Engineering 115 (2018) 495–511.

[45] X. Zhou, G. Zhang, J. Sun, J. Zhou, T. Wei, S. Hu, Minimizing cost and makespan for workflow scheduling in cloud using fuzzy dominance sort based HEFT, Future Generation Computer Systems 93 (2019) 278–289.

[46] M. Melnik, D. Nasonov, Workflow scheduling using neural networks and reinforcement learning, Procedia Computer Science 156 (2019) 29–36.

[47] H. Hafsi, H. Gharsellaoui, S. Bouamama, Genetic-based multi-criteria workflow scheduling with dynamic resource provisioning in hybrid large scale distributed systems, Procedia Computer Science 159 (2019) 1063–1074.

[48] M. Adhikari, T. Amgoth, An intelligent water drops-based workflow scheduling for IaaS cloud, Applied Soft Computing 77 (2019) 547–566.

[49] M. Abdullahi, M.A. Ngadi, S.I. Dishing, S.i.M. Abdulhamid, B.I.e. Ahmad, An efficient symbiotic organisms search algorithm with chaotic optimization strategy for multi-objective task scheduling problems in cloud computing environment, Journal of Network and Computer Applications 133 (2019) 60–74.

[50] M. Guerine, et al., A provenance-based heuristic for preserving results confidentiality in cloud-based scientific workflows, Future Generation Computer Systems 97 (2019) 697–713.

[51] P.K. Muhuri, A. Rauniyar, R. Nath, On arrival scheduling of real-time precedence constrained tasks on multi-processor systems using genetic algorithm, Future Generation Computer Systems 93 (2019) 702–726.

[52] S. Li, W. Chen, Y. Chen, C. Chen, Z. Zheng, Makespan-minimized computation offloading for smart toys in edge-cloud computing, Electronic Commerce Research and Applications 37 (2019) 100884.

[53] V. De Maio, D. Kimovski, Multi-objective scheduling of extreme data scientific workflows in Fog, Future Generation Computer Systems 106 (2020) 171–184.

[54] S. Saeedi, R. Khorsand, S. Ghandi Bidgoli, M. Ramezanpour, Improved many-objective particle swarm optimization algorithm for scientific workflow scheduling in cloud computing, Computers & Industrial Engineering 147 (2020) 106649.

[55] M. Sharma, R. Garg, HIGA: harmony-inspired genetic algorithm for rack-aware energy-efficient task scheduling in cloud data centers, Engineering Science and Technology, an International Journal 23 (1) (2020) 211–224.

[56] Y. Wen, J. Liu, W. Dou, X. Xu, B. Cao, J. Chen, Scheduling workflows with privacy protection constraints for big data applications on cloud, Future Generation Computer Systems 108 (2020) 1084–1091.

[57] D.A. Monge, E. Pacini, C. Mateos, E. Alba, C. García Garino, CMI: an online multi-objective genetic autoscaler for scientific and engineering workflows in cloud infrastructures with unreliable virtual machines, Journal of Network and Computer Applications 149 (2020) 102464.

[58] P. Hosseinioun, M. Kheirabadi, S.R. Kamel Tabbakh, R. Ghaemi, A new energy-aware tasks scheduling approach in fog computing using hybrid meta-heuristic algorithm, Journal of Parallel and Distributed Computing 143 (2020) 88–96.

[59] Y. Zhang, Y. Liu, J. Zhou, J. Sun, K. Li, Slow-movement particle swarm optimization algorithms for scheduling security-critical tasks in resource-limited mobile edge computing, Future Generation Computer Systems 112 (2020) 148–161.

[60] M. Sharma, R. Garg, An artificial neural network based approach for energy efficient task scheduling in cloud data centers, Sustainable Computing: Informatics and Systems 26 (2020) 100373.

[61] E. Rafieyan, R. Khorsand, M. Ramezanpour, An adaptive scheduling approach based on integrated best-worst and VIKOR for cloud computing, Computers & Industrial Engineering 140 (2020) 106272.

[62] H. Habibi Tostani, H. Haleh, S.M. Hadji Molana, F.M. Sobhani, A Bi-Level Bi-Objective optimization model for the integrated storage classes and dual shuttle cranes scheduling in AS/RS with energy consumption, workload balance and time windows, Journal of Cleaner Production 257 (2020) 120409.

[63] M. Haris, S. Zubair, Mantaray modified multi-objective Harris hawk optimization algorithm expedites optimal load balancing in cloud computing, Journal of King Saud University: Computer and Information Sciences 34 (10) (2022) 9696–9709, https://doi.org/10.1016/j.jksuci.2021.12.003.

[64] B. Zhou, Y. Lei, Bi-objective grey wolf optimization algorithm combined Levy flight mechanism for the FMC green scheduling problem, Applied Soft Computing 111 (2021) 107717.

[65] O.H. Ahmed, J. Lu, Q. Xu, A.M. Ahmed, A.M. Rahmani, M. Hosseinzadeh, Using differential evolution and Moth–Flame optimization for scientific workflow scheduling in fog computing, Applied Soft Computing 112 (2021) 107744.

[66] P. Han, C. Du, J. Chen, F. Ling, X. Du, Cost and makespan scheduling of workflows in clouds using list multiobjective optimization technique, Journal of Systems Architecture 112 (2021) 101837.

[67] P. Paknejad, R. Khorsand, M. Ramezanpour, Chaotic improved PICEA-g-based multi-objective optimization for workflow scheduling in cloud environment, Future Generation Computer Systems 117 (2021) 12–28.

[68] R. Rani, R. Garg, Pareto based ant lion optimizer for energy efficient scheduling in cloud environment, Applied Soft Computing 113 (2021) 107943.

[69] S. Javanmardi, M. Shojafar, R. Mohammadi, A. Nazari, V. Persico, A. Pescapè, FUPE: a security driven task scheduling approach for SDN-based IoT–Fog networks, Journal of Information Security and Applications 60 (2021) 102853.

[70] M.S. Umam, M. Mustafid, S. Suryono, A hybrid genetic algorithm and tabu search for minimizing makespan in flow shop scheduling problem, Journal of King Saud University: Computer and Information Sciences 34 (9) (2022) 7459–7467.

[71] N. Maleki, A.M. Rahmani, M. Conti, SPO: a secure and performance-aware optimization for MapReduce scheduling, Journal of Network and Computer Applications 176 (2021) 102944.

[72] D. Javaheri, S. Gorgin, J.-A. Lee, M. Masdari, An improved discrete Harris hawk optimization algorithm for efficient workflow scheduling in multi-fog computing, Sustainable Computing: Informatics and Systems 36 (2022) 100787.

[73] M.I. Khaleel, PPR-RM: Performance-to-Power Ratio, Reliability and Makespan - aware scientific workflow scheduling based on a coalitional game in the cloud, Journal of Network and Computer Applications (2022) 103478.

[74] F. Li, T.W. Liao, W. Cai, Research on the collaboration of service selection and resource scheduling for IoT simulation workflows, Advanced Engineering Informatics 52 (2022) 101528.

[75] A.G. Delavar, R. Akraminejad, S. Mozafari, An energy-aware workflow scheduling method in heterogeneous green cloud environments based on a hybrid meta-heuristic algorithm, Computer Communications 195 (2022) 49–60, https://doi.org/10.1016/j.comcom.2022.08.006.

[76] S.A. Alsaidy, A.D. Abbood, M.A. Sahib, Heuristic initialization of PSO task scheduling algorithm in cloud computing, Journal of King Saud University: Computer and Information Sciences 34 (6, Part A) (2022) 2370–2382.

[77] A. Kaur, B. Kaur, Load balancing optimization based on hybrid Heuristic-Metaheuristic techniques in cloud environment, Journal of King Saud University: Computer and Information Sciences 34 (3) (2022) 813–824.

[78] N. Manikandan, N. Gobalakrishnan, K. Pradeep, Bee optimization based random double adaptive whale optimization model for task scheduling in cloud computing environment, Computer Communications 187 (2022) 35–44.

[79] S.M. Hussain, G.R. Begh, Hybrid heuristic algorithm for cost-efficient QoS aware task scheduling in fog–cloud environment, Journal of Computational Science (2022) 101828.

[80] A.C. Ammari, W. Labidi, F. Mnif, H. Yuan, M. Zhou, M. Sarrab, Firefly algorithm and learning-based geographical task scheduling for operational cost minimization in distributed green data centers, Neurocomputing 490 (2022) 146–162.

[81] S. Selvi, D. Manimegalai, Task scheduling using two-phase variable neighborhood search algorithm on heterogeneous computing and grid environments, Arabian Journal for Science and Engineering 40 (3) (2015) 817–844.

[82] Z. Tang, L. Qi, Z. Cheng, K. Li, S.U. Khan, K. Li, An energy-efficient task scheduling algorithm in DVFS-enabled cloud environment, Journal of Grid Computing 14 (1) (2016) 55–74.

[83] V. Singh, I. Gupta, P.K. Jana, An energy efficient algorithm for workflow scheduling in IaaS cloud, Journal of Grid Computing 18 (3) (2020) 357–376.

[84] F. Ebadifard, S.M. Babamir, Scheduling scientific workflows on virtual machines using a Pareto and hypervolume based black hole optimization algorithm, Journal of Supercomputing 76 (10) (2020) 7635–7688.

[85] L. Abualigah, A. Diabat, A novel hybrid antlion optimization algorithm for multi-objective task scheduling problems in cloud computing environments, Cluster Computing 24 (1) (2021) 205–223.

[86] M. Hosseinzadeh, et al., Improved butterfly optimization algorithm for data placement and scheduling in edge computing environments, Journal of Grid Computing 19 (2) (2021) 14.

[87] S. Janakiraman, M.D. Priya, Improved artificial bee colony using monarchy butterfly optimization algorithm for load balancing (IABC-MBOA-LB) in cloud environments, Journal of Network and Systems Management 29 (4) (2021) 39.

[88] S.P.M. Ziyath, S. Senthilkumar, RETRACTED ARTICLE: MHO: meta heuristic optimization applied task scheduling with load balancing technique for cloud infrastructure services, Journal of Ambient Intelligence and Humanized Computing 12 (6) (2021) 6629–6638.
[89] A. Chhabra, G. Singh, K.S. Kahlon, Multi-criteria HPC task scheduling on IaaS cloud infrastructures using meta-heuristics, Cluster Computing 24 (2) (2021) 885–918.
[90] J. Kakkottakath Valappil Thekkepuryil, D.P. Suseelan, P.M. Keerikkattil, An effective meta-heuristic based multi-objective hybrid optimization method for workflow scheduling in cloud computing environment, Cluster Computing 24 (3) (2021) 2367–2384.
[91] N. Arora, R.K. Banyal, A particle grey wolf hybrid algorithm for workflow scheduling in cloud computing, Wireless Personal Communications 122 (4) (2022) 3313–3345.
[92] L. Zhang, L. Wang, M. Xiao, Z. Wen, C. Peng, EM_WOA: a budget-constrained energy consumption optimization approach for workflow scheduling in clouds, Peer-to-Peer Networking and Applications 15 (2) (2022) 973–987.
[93] P. Soma, B. Latha, V. Vijaykumar, An improved multi-objective workflow scheduling using F-NSPSO with fuzzy rules, Wireless Personal Communications 124 (4) (2022) 3567–3589.
[94] Y. Wang, X. Zuo, Z. Wu, H. Wang, X. Zhao, Variable neighborhood search based multiobjective ACO-list scheduling for cloud workflows, Journal of Supercomputing 78 (2022) 18856–18886, https://doi.org/10.1007/s11227-022-04616-y.
[95] S. Mirjalili, A. Lewis, The whale optimization algorithm, Advances in Engineering Software 95 (2016) 51–67.
[96] N. Rana, M.S.A. Latiff, S.i.M. Abdulhamid, H. Chiroma, Whale optimization algorithm: a systematic review of contemporary applications, modifications and developments, Neural Computing & Applications 32 (20) (2020) 16245–16277.
[97] J. Nasiri, F.M. Khiyabani, A whale optimization algorithm (WOA) approach for clustering, Cogent Mathematics & Statistics 5 (1) (2018) 1483565.
[98] M.H. Nadimi-Shahraki, H. Zamani, S. Mirjalili, Enhanced whale optimization algorithm for medical feature selection: a COVID-19 case study, Computers in Biology and Medicine (2022) 105858.
[99] S.H. Hosseini, J. Vahidi, S.R. Kamel Tabbakh, A.A. Shojaei, Resource allocation optimization in cloud computing using the whale optimization algorithm, International Journal of Nonlinear Analysis and Applications 12 (Special Issue) (2021) 343–360.
[100] S. MS, J.P. P M, V. Alappatt, Profit maximization based task scheduling in hybrid clouds using whale optimization technique, Information Security Journal: A Global Perspective 29 (4) (2020) 155–168.
[101] S.R. Thennarasu, M. Selvam, K. Srihari, A new whale optimizer for workflow scheduling in cloud computing environment, Journal of Ambient Intelligence and Humanized Computing 12 (3) (2021) 3807–3814.
[102] H. Peng, W.-S. Wen, M.-L. Tseng, L.-L. Li, Joint optimization method for task scheduling time and energy consumption in mobile cloud computing environment, Applied Soft Computing 80 (2019) 534–545.
[103] X. Chen, et al., A WOA-based optimization approach for task scheduling in cloud computing systems, IEEE Systems Journal 14 (3) (2020) 3117–3128.
[104] M.F. Omar, N.M. Bakeri, M.N.M. Nawi, N. Hairani, K. Khalid, Methodology for modified whale optimization algorithm for solving appliances scheduling problem, Journal of Advanced Research in Fluid Mechanics and Thermal Sciences 76 (2) (2020) 132–143.
[105] L. Ni, X. Sun, X. Li, J. Zhang, GCWOAS2: multiobjective task scheduling strategy based on Gaussian cloud-whale optimization in cloud computing, Computational Intelligence and Neuroscience 2021 (2021).
[106] G. Narendrababu Reddy, S.P. Kumar, Multi objective task scheduling algorithm for cloud computing using whale optimization technique, in: International Conference on Next Generation Computing Technologies, Springer, 2017, pp. 286–297.
[107] G. Narendrababu Reddy, S. Phani Kumar, Regressive whale optimization for workflow scheduling in cloud computing, International Journal of Computational Intelligence and Applications 18 (04) (2019) 1950024.
[108] Z. Movahedi, B. Defude, An efficient population-based multi-objective task scheduling approach in fog computing systems, Journal of Cloud Computing 10 (1) (2021) 1–31.
[109] M. Sharma, R. Garg, Energy-aware whale-optmized task scheduler in cloud computing, in: 2017 International Conference on Intelligent Sustainable Systems (ICISS), IEEE, 2017, pp. 121–126.
[110] K. Sreenu, M. Sreelatha, W-Scheduler: whale optimization for task scheduling in cloud computing, Cluster Computing 22 (1) (2019) 1087–1098.
[111] W. Yankai, W. Shilong, L. Dong, S. Chunfeng, Y. Bo, An improved multi-objective whale optimization algorithm for the hybrid flow shop scheduling problem considering device dynamic reconfiguration processes, Expert Systems with Applications 174 (2021) 114793.
[112] L. Jia, K. Li, X. Shi, Cloud computing task scheduling model based on improved whale optimization algorithm, Wireless Communications and Mobile Computing 2021 (2021) 4888154.
[113] N.P. Saravanan, T. Kumaravel, An efficient Task Scheduling Algorithm using Modified Whale Optimization Algorithm in Cloud Computing, International Journal of Engineering and Advanced Technology (IJEAT) 9 (2019), https://doi.org/10.35940/ijeat.B3813.129219.
[114] G. Kiruthiga, S. Mary Vennila, An enriched chaotic quantum whale optimization algorithm based job scheduling in cloud computing environment, International Journal of Advanced Trends in Computer Science and Engineering 6 (4) (2019) 1753–1760.
[115] M. Abdel-Basset, D. El-Shahat, K. Deb, M. Abouhawwash, Energy-aware whale optimization algorithm for real-time task scheduling in multiprocessor systems, Applied Soft Computing 93 (2020) 106349.
[116] I. Strumberger, N. Bacanin, M. Tuba, E. Tuba, Resource scheduling in cloud computing based on a hybridized whale optimization algorithm, Applied Sciences 9 (22) (2019) 4893.
[117] M. Abdel-Basset, D. El-Shahat, I. El-henawy, A modified hybrid whale optimization algorithm for the scheduling problem in multimedia data objects, Concurrency and Computation: Practice and Experience 32 (4) (2020) e5137.
[118] M. Abdel-Basset, G. Manogaran, D. El-Shahat, S. Mirjalili, A hybrid whale optimization algorithm based on local search strategy for the permutation flow shop scheduling problem, Future Generation Computer Systems 85 (03) (2018).

[119] A. Saoud, A. Recioui, Hybrid algorithm for cloud-fog system based load balancing in smart grids, Bulletin of Electrical Engineering and Informatics 11 (1) (2022) 477–487.

[120] J. Ababneh, A hybrid approach based on grey wolf and whale optimization algorithms for solving cloud task scheduling problem, Mathematical Problems in Engineering 2021 (2021).

[121] C.-C. Lin, Z.-X. Wu, K.-W. Huang, Y.-M. Li, A hybrid whale optimization algorithm for flow shop scheduling problems, in: The Fourth International Conference on Electronics and Software Science (ICESS2018), Japan, 2018.

[122] P. Albert, M. Nanjappan, WHOA: hybrid based task scheduling in cloud computing environment, Wireless Personal Communications 121 (3) (2021) 2327–2345.

[123] M.A. Al-qaness, A.A. Ewees, M. Abd Elaziz, Modified whale optimization algorithm for solving unrelated parallel machine scheduling problems, Soft Computing 25 (14) (2021) 9545–9557.

[124] A. Chhabra, S.K. Sahana, N.S. Sani, A. Mohammadzadeh, H.A. Omar, Energy-aware bag-of-tasks scheduling in the cloud computing system using hybrid oppositional differential evolution-enabled whale optimization algorithm, Energies 15 (13) (2022) 4571.

[125] J. Samriya, N. Kumar, A QoS aware FTOPSIS-WOA based task scheduling algorithm with load balancing technique for the cloud computing environment, Indian Journal of Science and Technology 13 (35) (2020) 3675–3684.

[126] G. Natesan, A. Chokkalingam, Multi-objective task scheduling using hybrid whale genetic optimization algorithm in heterogeneous computing environment, Wireless Personal Communications 110 (4) (2020) 1887–1913.

[127] M. Sanaj, P.J. Prathap, An efficient approach to the map-reduce framework and genetic algorithm based whale optimization algorithm for task scheduling in cloud computing environment, Materials Today: Proceedings 37 (2021) 3199–3208.

[128] M. Sharma, R. Garg, WGOA: Whale-Genetic Optimization Algorithm for Energy Efficient Task Scheduling in Cloud Data Center, https://www.researchgate.net/profile/Mohan-Sharma-3/publication/327271124_WGOA_Whale-Genetic_Optimization_Algorithm_for_Energy_Efficient_Task_Scheduling_in_Cloud_Computing/links/5e5186c8299bf1cdb94000ba/WGOA-Whale-Genetic-Optimization-Algorithm-for-Energy-Efficient-Task-Scheduling-in-Cloud-Computing.pdf.

Chapter 5

Whale optimization algorithm and its application in machine learning

Nava Eslami[a], Mahdi Rahbar[b], Seyed Mostafa Bozorgi[a], and Samaneh Yazdani[a]
[a]*Department of Computer Engineering, North Tehran Branch, Islamic Azad University, Tehran, Iran,* [b]*Department of Computer Science, Saint Louis University, St. Louis, MO, United States*

5.1 Introduction

Machine learning is one of the fundamental and growing research areas in artificial intelligence. It aims to produce computer programs that learn from experience or data. Machine learning algorithms have been applied to many fields, such as natural language processing [1], signal processing [2], recommender systems [3] and etc.

To capture hidden patterns in data using a machine learning approach, raw data must first be collected. The data is then cleaned, and relevant features are extracted or selected using data preparation and processing algorithms. After that, a model is selected, and its parameters are adjusted according to a loss function. To remove expert intervention and design automated machine learning algorithms, optimization algorithms are employed in each stage of the machine learning cycle. For instance, many optimization algorithms are designed to tune hyperparameters to improve the performance of machine learning algorithms [4].

Swarm intelligence algorithms are very popular optimization algorithms that can search the solution space in an acceptable time without making assumptions about the characteristics of the search space. Their most important characteristics include searching the solution space simultaneously, utilizing randomness to escape from local optima, and sharing information among the population's individuals, which make them proper for solving complex problems [5]. In this chapter, the applications of a swarm intelligence algorithm, which is called Whale Optimization Algorithm (WOA) in machine learning, are considered [6]. Mirjalili and Lewis developed WOA by emulating the social life of humpback whales. In this chapter, various applications of WOA and its variants to solving machine learning problems such as feature selection or classification are presented.

This chapter is arranged as follows: Section 5.2 briefly describes WOA and its mathematical equations. Section 5.3 provides an overview of WOA's role in solving various machine learning tasks such as feature selection, parameter tuning for K-means clustering and SVM classifiers, artificial neural networks, and deep learning hyperparameter tuning. Section 5.4 provides a discussion on the limitations of WOA when applied to machine learning methods. Finally, the chapter conclusion and future research directions are given in Section 5.5.

5.2 Whale optimization algorithm

WOA is a population-based algorithm that is developed by Mirjalili and Lewis by emulating the hunting behavior of humpback whales [6]. Similar to other swarm intelligence algorithms, WOA starts with a population of randomly generated individuals called whales. Then, some steps are done iteratively, which are initialization, fitness evaluation, and updating the position of individuals. These steps are described as follows:

- **Initialization:** WOA starts with a predefined number of randomly generated individuals as a population. Each individual forms a solution in the search space.
- **Fitness evaluation:** the fitness of all individuals of population is calculated utilizing the objective function. The best individual with respect to the objective function is selected, which is shown by X^*.
- **Updating the position of individuals:** WOA utilizes three operators, which are encircling the prey, bubble-net attacking method, and searching for the prey, to update the position of individuals in the population. These explorative and exploitative operators are described as follows.

Handbook of Whale Optimization Algorithm. https://doi.org/10.1016/B978-0-32-395365-8.00011-7

- **Encircling the prey:** This operator is presented by simulating the behavior of humpback whales for recognizing and covering the location of the prey. Encircling prey changes the position of ith individual, according to Eq. (5.1).

$$X_i(t+1) = X^*(t) - A \cdot D \tag{5.1}$$

where $\boldsymbol{D}$ is

$$D = \left|C \cdot X^*(t) - X_i(t)\right| \tag{5.2}$$

where $X_i(t+1)$ and $X_i(t)$ are the position vector of the ith individual at iteration $t+1$ and in the current iteration t, respectively. $X^*(t)$ is the position vector of the best individual in the iteration t. $|\;|$ is the absolute value. In Eqs. (5.1) and (5.2), $\boldsymbol{A}$ and $\boldsymbol{C}$, which are coefficient vectors are calculated according to Eqs. (5.3) and (5.4).

$$A = 2 \cdot a \cdot r - a \tag{5.3}$$

$$C = 2 \cdot r \tag{5.4}$$

where a is linearly decreased from 2 to 0 over the course of iterations. $\boldsymbol{r}$ is a random vector in the range of [0, 1].
- **Bubble-net attacking method:** This explorative operator changes the position of ith individual according to Eq. (5.5).

$$X_i(t+1) = D.e^{bl}.cos\, cos\,(2\pi l) + X^*(t) \tag{5.5}$$

where $\boldsymbol{D}$ is given as follows:

$$D = \left|X^*(t) - X_i(t)\right| \tag{5.6}$$

In Eq. (5.5), l is a random number in the range of $[-1, 1]$ and b is a constant number.
- **Searching for the prey:** This explorative operator enables WOA to perform the global search according to Eq. (5.7).

$$X_i(t+1) = X_{rand}(t) - A \cdot D \tag{5.7}$$

where $\boldsymbol{D}$ is

$$D = |C \cdot X_{rand}(t) - X_i(t)| \tag{5.8}$$

In Eq. (5.7), X_{rand} is a randomly selected individual from the current population. $\boldsymbol{A}$ and $\boldsymbol{C}$ are defined in the previous subsection.

Algorithm 5.1 shows the pseudocode of the original WOA.

In Algorithm 5.1, *MaxIter* indicates the maximum number of iterations, and *rand* is a random number in the range of [0, 1].

5.3 WOA for various machine learning tasks

In the following subsections, a brief explanation on how WOA is utilized in different machine learning problems is given.

5.3.1 Feature selection

In recent years, the number of dimensions in datasets has grown dramatically, which has led to the "curse of dimensionality" [7] when using machine learning, especially classification algorithms. Dimension reduction algorithms, especially, feature selection techniques, are useful in solving this challenge [8]. Feature selection is categorized as a NP-hard optimization problem which can be classified as a pre-processing mechanism to select an optimal subset of features from all existing ones. Feature selection algorithms are basically categorized into three main categories, including filter, wrapper, and hybrid methods [9]. The filter-based methods evaluate features based on some data-dependent metrics namely, mutual information, correlation, information gain and etc. In contrast, wrapper-based methods usually utilize a classification algorithm to evaluate the quality of the selected features. Because most of the reviewed papers are based on wrapper-based feature selection algorithms, the wrapper-based methods are demonstrated in Fig. 5.1, schematically. As can be induced, hybrid methods incorporate the prominences of the filter and wrapper methods [10]. Recently, swarm intelligence algorithms have gained much attention for feature selection tasks due to their simplicity and their global search ability [11]. Among swarm

Algorithm 5.1 Pseudocode of WOA.

```
Randomly generate the initial individual X_i (i = 1, 2, ..., n)
Evaluate the fitness value of each individual in the population
Record the best individual so far (X*)
while (t < MaxIter)
  for each individual in the population
    Update A, C, l, a
    if (rand < 0.5)
      if (|A| < 1)
        Update the position of the ith individual as Eq. (5.1)
      else if (|A| ≥ 1)
        Randomly select an individual X_rand
        Change the position of ith individual as Eq. (5.5)
      end if
    else
      Generate the ith individual X_i(t) as Eq. (5.7)
    end
  end
  Evaluate each individual by the objective function
  Update the position of the best individual X*
  t = t + 1
end
```

intelligence algorithms, WOA has shown promising results in many optimization fields, especially feature selection problems [12]. The general form of feature selection using WOA is shown in Fig. 5.2 [6]. As can be seen in Fig. 5.2, each solution is called a whale consists of m different features. WOA-based feature selection algorithm is utilized to select the most discriminative features among all existing ones [12]. Moreover, we realized that to use a swarm intelligence algorithm for feature selection, there are two ways to represent the potential solutions. The first way is the standard representation (continuous representation) which is the primary representation when the algorithm is proposed for the first time [10]. In this representation style, generally, a transfer function is utilized to convert standard representation to binary one. Basically, the transfer functions have different general forms namely, S-shaped [13], V-shaped [13][14], U-shaped [15][16][17], Tapper-shaped [18], X-shaped [19][20], Z-shaped [21][22], time-varying [23] [24], and etc. S-shaped and V-shaped transfer functions are demonstrated in Fig. 5.3 [14]. The U-shaped, and X-shaped transfer functions are shown in Fig. 5.4 [16], and Fig. 5.5 [19], respectively. Moreover, S-shaped and V-shaped transfer functions' formulations are given in Table 5.1 [14]. The general forms of U-shaped and taper-shaped transfer functions are drowned in Eqs. (5.9) [16] and (5.10) [17], respectively. In Eq. (5.9), α and β are two control parameters that balance the slope and basin of the U-shaped transfer functions. Furthermore, in Eq. (5.10), A is a positive real-valued number and n adjusts the curvature of the taper-shaped transfer function.

$$U(x) = \alpha \left| x^{\beta} \right| \tag{5.9}$$

$$T(x) = (\left| \frac{x}{A} \right|)^{\frac{1}{n}}, \; x \in [-A, A], \; n \geq 1 \tag{5.10}$$

As a contrast, in the second method, which is called binary representation, each feature is represented by a binary value, where "1" means the corresponding feature is considered and "0" means the corresponding feature is not considered [25]. Based on these two representations, we classified WOA-based feature selection algorithms into two main categories. The rest of this subsection is organized as follows. In Section 5.3.1.1, WOA-based feature selection algorithms with standard representation are reviewed from different aspects. Finally, in Section 5.3.1.2, methods with binary representation are investigated, extensively.

5.3.1.1 Standard representation

As an instance for using WOA directly for feature selection, Shuaib et al. [27] introduced a spam filtering framework by utilizing WOA and rotation forest classifier. The proposed spam filtering model includes two main stages namely feature

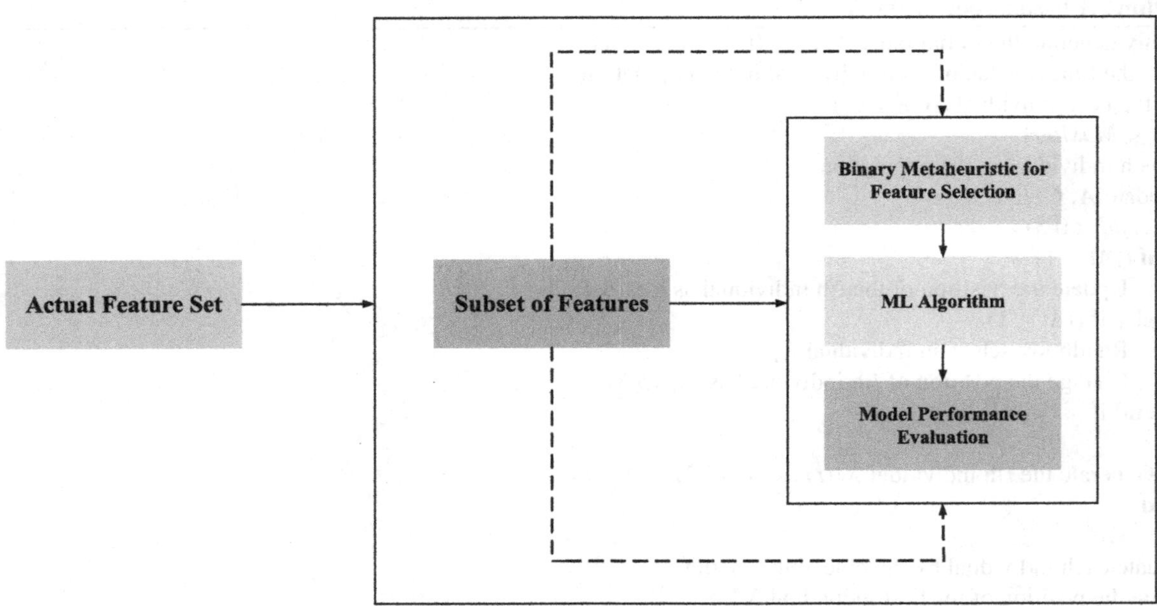

FIGURE 5.1 The schematic representation of wrapper-based methods [10].

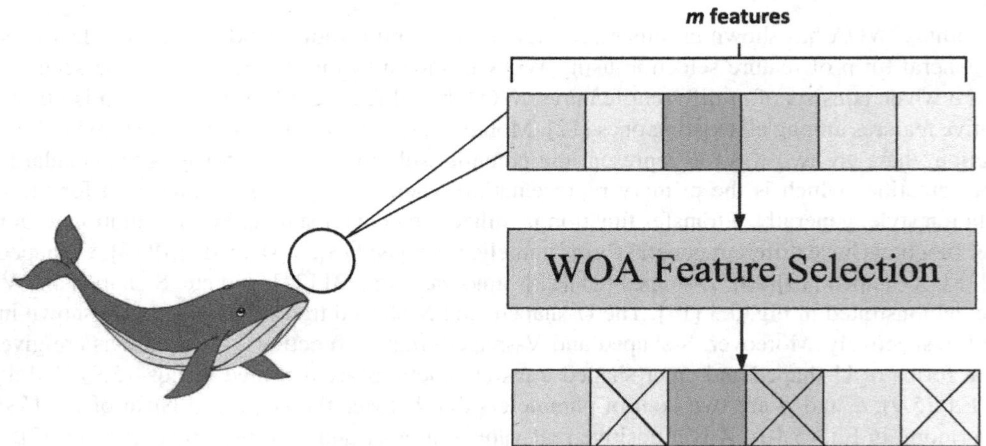

FIGURE 5.2 Feature selection using WOA.

TABLE 5.1 S-shaped and V-shaped families of transfer functions [26].

S-shaped family		V-shaped family	
Name	Transfer function	Name	Transfer function
S1	$T(x) = \frac{1}{1+e^{-2x}}$	V1	$T(x) = \left\|erf(\frac{\sqrt{\pi}}{2}x)\right\| = \left\|\frac{\sqrt{2}}{\pi}\int_0^{(\frac{\sqrt{\pi}}{2})x} e^{-t^2}dt\right\|$
S2	$T(x) = \frac{1}{1+e^{-x}}$	V2	$T(x) = \|tanh(x)\|$
S3	$T(x) = \frac{1}{1+e^{(\frac{-x}{2})}}$	V3	$T(x) = \left\|\frac{(x)}{\sqrt{1+x^2}}\right\|$
S4	$T(x) = \frac{1}{1+e^{(\frac{-x}{3})}}$	V4	$T(x) = \left\|\frac{2}{\pi}arctan(\frac{\pi}{2}x)\right\|$

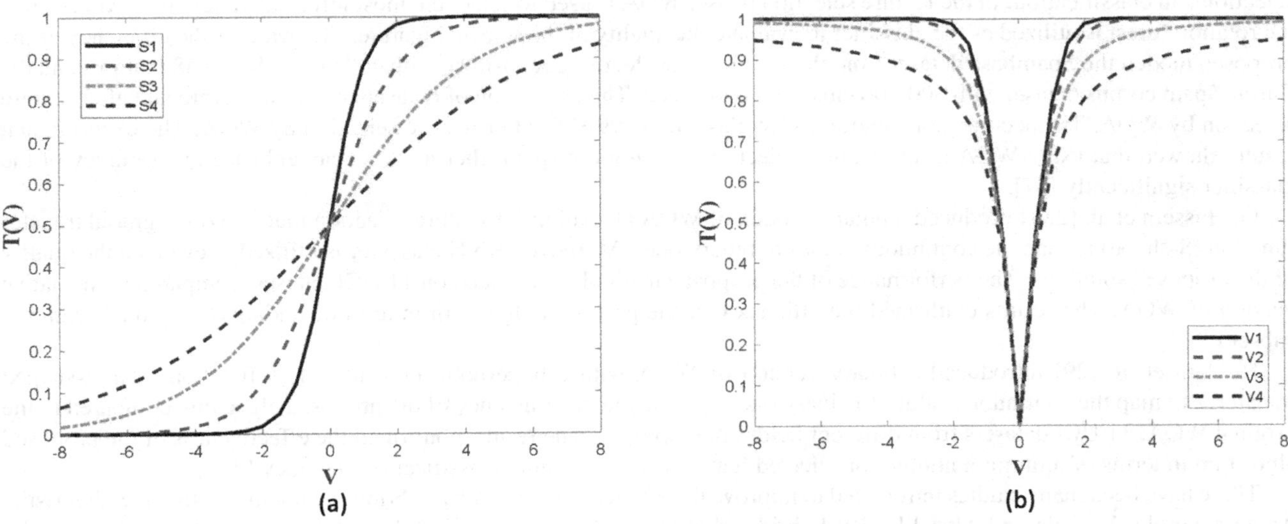

FIGURE 5.3 Transfer functions. (a) S-Shaped transfer functions. (b) V-shaped transfer functions [26].

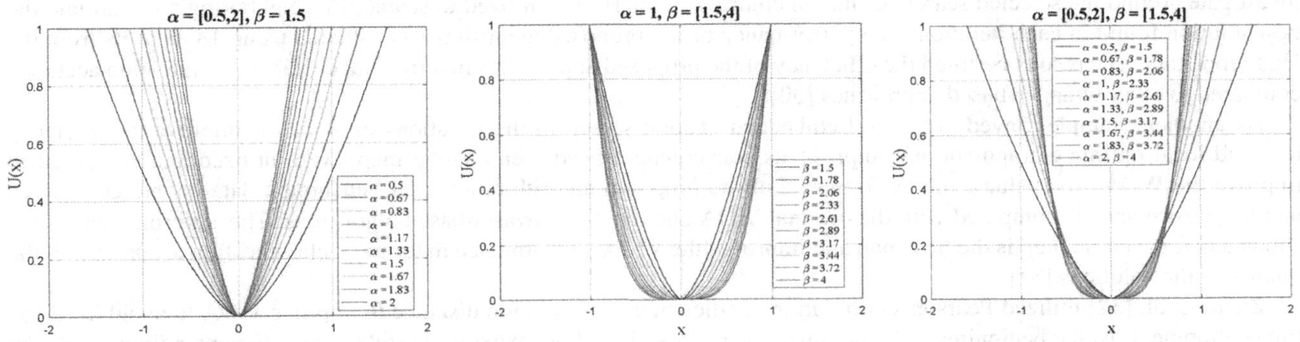

FIGURE 5.4 U-shaped transfer functions [15].

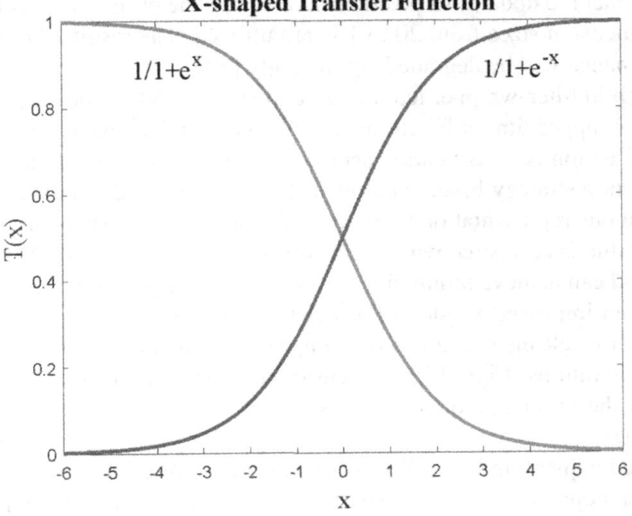

FIGURE 5.5 X-shaped transfer function [20].

selection and classification. In the feature selection phase, WOA is used to select the most informative features. Afterwards, the rotation forest is utilized as the classifier to evaluate the quality of the selected features. To evaluate the efficiency of the proposed model, the Spambase dataset from the UCI machine learning repository with 4601 records and 58 features and the Enron-Spam corpus dataset with 600,000 emails were utilized. The evaluation of datasets was done before and after feature selection by WOA. The accuracy of rotation forest classifier is 99.89% after feature selection by WOA. The experimental results showed that using WOA as the feature selection algorithm in spam filtering tasks can enhance the accuracy of the classifier significantly [27].

G. Hussein et al. [28] introduced a binary version of WOA algorithm for feature selection that utilizes Sigmoid transfer function (S-shape) to map the continuous values to binary ones. Moreover, KNN classifier is utilized to evaluate the quality of the proposed solutions. The performance of the proposed method is evaluated on 11 UCI datasets compared to the native version of WOA. The results confirmed the efficiency of the proposed algorithm in term of finding the optimal features' subset [28].

Hussein et al. [29] introduced a binary version of WOA with a hyperbolic tangent fitness function as a V-shaped transferor to map the continuous values to binary ones. To analyze the efficiency of the proposed algorithm compared to the original WOA, 11 UCI datasets from different fields are employed. The results confirmed the effectiveness of the proposed algorithm in terms of minimum number of selected features and maximum classification accuracy [29].

There have been many studies introduced to improve the efficiency of WOA by hybridizing it with another metaheuristic. As an example, Mafarja and Mirjalili [30] hybridized WOA with simulated annealing (SA) to enhance exploitation by searching the most promising areas found by WOA. SA was utilized following two hybrid models namely, Low-level teamwork hybrid model (LTH), and high-level hybrid model (HRH). In LTH, SA was used as an exploitative operator to investigate around the selected search agents. In contrast, in HRH, it is utilized to search the neighboring areas around the best solution found in each iteration. The performance of the proposed algorithms is evaluated using 18 datasets from the UCI repository. The results certified the efficiency of the proposed approaches in term of acceptable classification accuracy compared to other wrapper-based approaches [30].

As another example, Sayed et al. [31] embedded chaotic search in the iterations of WOA to improve its efficiency to avoid local optima entrapment and improve its convergence speed. Ten chaotic maps were utilized and compared to improve the WOA. To evaluate the performance of the proposed algorithm, ten UCI benchmark datasets are considered and its performance is compared with the original WOA and ten other swarm-based algorithms. The experimental results showed that the circle map is the best map and improves the WOA' performance in terms of classification accuracy and the quality of the solutions [31].

Zheng et al. [32] utilized Pearson's correlation coefficient and correlation distance to improve WOA to avoid local optima entrapment. At the beginning of the optimization process, based on Pearson's correlation coefficient, a filter algorithm is developed to rank features in each dataset. Two parameters are calculated using filter algorithm to adjust the weights of the relevance and redundancy. Then, an improved version of WOA is utilized as a wrapper method. In the proposed WOA, average of all dimensions' random numbers is used to calculate the probability that the humpback whale chooses a reduced surround or spiral movement to updates its position. To evaluate the proposed algorithm, ten different datasets from UCI repository with various dimension sizes from 20 to 100 are utilized. The results showed that in the high dimensional datasets, the algorithm's performance will be degraded significantly [32].

Got et al. [33] proposed a hybrid filter-wrapper feature selection using WOA. The proposed method is a multi-objective algorithm in which a filter and wrapper fitness functions are utilized simultaneously. Filter fitness function is mutual information and wrapper fitness function is classification accuracy. In this paper, an external archive update is introduced to ensure elitism and a leader selection strategy based on crowding distance is used to guide the population toward the pareto front. In order to convert continuous representation to binary one, the hyperbolic tangent transfer function is utilized. The efficiency of the proposed algorithm is compared with seven well-known algorithms on 12 benchmark datasets. The results showed that the proposed method can achieve minimal features' subset along with acceptable classification accuracy [33].

Kundu et al. [34] proposed an improved version of WOA by incorporating altruism concept for feature selection on microarray datasets. The concept of altruism is utilized to improve candidate solutions' scattering that are able to attain the global optimum during the iterations. Eight high-dimensional microarray datasets are used to evaluate the proposed algorithm. The results revealed the efficiency of the proposed method in terms of classification accuracy, and minimal number of selected features [34].

Agrawal et al. [35] introduced a quantum based WOA for wrapper feature selection. In this paper, a different quantum bit representation is utilized for population's individuals and the quantum gate operator is introduced as the variation operator. Moreover, modified mutation and crossover operators are developed for quantum-based exploration, shrinking and spiral movement of the whales in the proposed algorithm. The performance of the proposed algorithm is compared

TABLE 5.2 The summarized description of the reviewed papers.

Publication	Purpose	Algorithm	Dimension reduction?
Kundu et al. (2022) [34]	DNA microarray gene expression datasets Classification	WOA & concept of altruism	Yes
Got et al. (2021) [33]	Classification	WOA & mutual information	Yes
Agrawal et al. (2020) [35]	Classification	WOA & Quantum bit representation	Yes
Shuaib et al. (2019) [27]	Email spam filtering	WOA & Rotation Forest Algorithm	Yes
Zheng et al. (2019) [32]	Classification	WOA & Pearson's correlation coefficient and correlation distance	Yes
Tawhid & Ibrahim (2019) [38]	Classification	WOA & Rough Set Approach & logistic regression, Decision Tree (C4.5), and Naïve Bayes classifiers	Yes
G. Hussein et al. (2019) [28]	Classification	WOA with S-shaped transfer function & KNN	Yes
Sayed et al. (2018) [31]	Classification	WOA & Chaotic Search	Yes
Hussein et al. (2017) [29]	Classification	WOA with Hyperbolic tangent transfer function	Yes
Mafarja & Mirjalili (2017) [30]	Classification	WOA + SA & KNN	Yes
Sharawi et al. (2017) [37]	Classification	WOA & KNN	Yes

with eight well-known evolutionary, swarm and quantum-based algorithms on 14 datasets from different domains. The results demonstrated the improved performance in terms of the average fitness and classification accuracy compared to other comparative algorithms [35].

5.3.1.2 Binary representation

There are not many researches that have utilized binary representation of WOA for feature selection. Because, in the case of using binary representation, the original exploitation and exploration phases of WOA lose their applicability. Therefore, new procedures should be designed to overcome this obstacle [36].

Sharawi et al. [37] introduced a wrapper-based feature selection system based on WOA. In the proposed algorithm, the solution values are continuous and bounded to [0, 1] and KNN is used as the classifier. The quality of the proposed solutions is evaluated by a multi-objective fitness function that is correlated with classification accuracy and the number of selected features. The proposed feature selection algorithm was compared with particle swarm optimization (PSO) and genetic algorithm (GA) on 18 different datasets from UCI repository. The results showed that WOA can find the most informative features along with acceptable classification accuracy [37].

Tawhid and Ibrahim [38] developed a WOA algorithm for feature selection with binary representation. The updating procedures of the original WOA are changed to become appropriate for binary solutions. Then, rough set is used as the objective function to measure existence of redundant features in the solution. Moreover, a wrapper-based approach based on different well-known classifiers namely logistic regression, Decision Tree (C4.5), and Naïve Bayes is introduced. To verify the effectiveness of the proposed algorithm, 32 UCI datasets and seven state-of-the-art algorithms are utilized. The results showed that the proposed algorithm is able to find a minimal subset of features. In Table 5.2, the reviewed papers are summarized based on their purposes [38].

5.3.2 WOA for data clustering

Clustering is the most important unsupervised learning technique that successfully has been applied to many areas such as image segmentation [39], bioinformatic [40], intrusion detection [41] and so on.

Clustering partitions data instances into similar groups without any information about the labels of instances into disjoint clusters. Different clustering algorithms are presented which are typically can be divided into hierarchical clustering [42], density clustering and partitional clustering [43]. Clustering algorithms have some major limitations such as sensitivity to initialization. One way to alleviate these drawbacks is utilizing swarm intelligence algorithms. In this section, we focus on one of the most widely used partitional-based clustering which is Kmeans [43]. Kmeans is an iterative algorithm that starts with choosing randomly K data point as cluster centers. Each data instance assigns to the cluster centroid according

Euclidean distance. The center of each cluster is re-computed via calculating the mean of the current cluster instances. The process of re-assigning instances to the clusters and updating clusters' centroids are repeated until the value of clusters' centroids do not change.

Kmeans utilizes gradient descent to find out the optimal cluster centers. It suffers from stuck in the local optima and its performance depends on the initial value of centroids. To address these limitations, swarm intelligence algorithms such as WOA have been applied in Kmeans. Refs. [44][45][46] investigated the employing of WOA and its variations to address the Kmeans drawbacks. These proposed approaches follow similar iterative cycle. In this section, at first, this cycle is described and then a brief review of clustering algorithms based on WOA is investigated.

a. Basic steps for combining WOA and Kmeans clustering

This cycle consists of three steps which are described as follows:

- **Initialization:** WOA starts with a population of randomly generated N_{pop} individuals. Each individual of the population is a $k \times m$ vector where k and m are the number of clusters and the number of features, respectively. So, each individual represents a complete clustering solution and initialized by $k \times m$ data points that are selected randomly from dataset. Fig. 5.6 shows an individual.

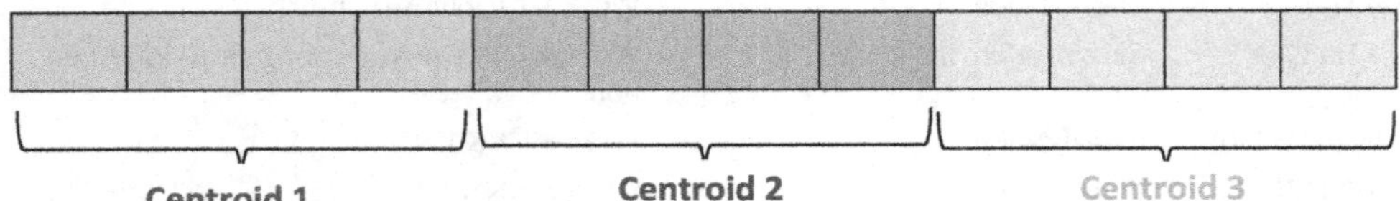

FIGURE 5.6 An individual when k is 3 and m is 4.

- **Fitness Evaluation:** The goodness of each individual is determined using the objective function. The sum of squared error (SSE) is utilized as the objective function to maximize similarity measure inside the clusters and minimize it between them. SSE calculates similarity based on the Euclidean distance according to Eq. (5.11).

$$SSE = \sum_{i=1}^{k} \sum_{j=1}^{|C_j|} \sqrt{\sum_{d=1}^{m} (centroid_{id} - r_{jd})} \tag{5.11}$$

where k is the number of clusters. *centroid* shows the dth dimension of ith cluster's centroid. r_{jd} indicates the dth dimension of ith instance that belongs to the ith cluster. $|C_j|$ shows the number of instances that belong to cluster C_j.

- **Update the Centroids:** the operators of WOA and its variants are applied to change the position of individuals (clusters' centroids).

b. Review of the proposed algorithms

Nasiri and Khyabani [46] utilized WOA for optimizing the Kmeans clustering's objective function. Their proposed algorithm was tested on eight datasets. Its performance was compared with five population-based algorithms such as ABC clustering [47] and PSO clustering. Simulation results verified the efficiency of the proposed algorithm. WOATS [44] is a hybrid swarm intelligence algorithm that combines WOA with Tabu search to improve WOA performance for optimizing the objective function of Kmeans clustering. Experiments on real-world datasets with different sizes illustrated the superiority of WOATS when it is compared with other original and hybrid swarm intelligence algorithms.

5.3.3 WOA for data classification

The performance of many classifiers highly depends on the values of their hyper parameters. For instance, SVM's learning and generalization ability is directly affected by selected values for its parameters such as penalty parameters [48]. Many parameter optimization algorithms have been proposed to select suitable values for classifiers' parameters in recent years. Among various parameter tuning algorithms, swarm intelligence algorithms have obtained comparative results as a global search. Many researches were presented to demonstrate the utilization of WOA in SVM algorithm to improve its performance. This sub-section provides an overview of utilizing WOA for optimizing the SVM's parameters.

WOA-SVM [49] utilized WOA to select suitable value for 8 parameters of SVM's kernel parameters. WOA-SVM which was applied for categorizing medical data utilized a multi-objective fitness function that includes sensitivity, specificity, and accuracy to evaluate population individuals. Simulation results indicated that WOA-SVM outperformed the other comparative algorithms in terms of sensitivity, specificity and accuracy.

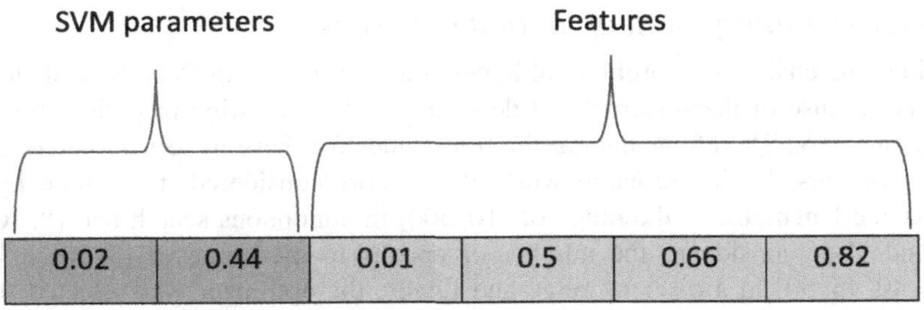

SVM parameters | Features

0.02	0.44	0.01	0.5	0.66	0.82

FIGURE 5.7 Individual encoding scheme used to represent an SVM-WOA for tuning the parameter of SVM and selecting relevant features.

SVM-WOA was proposed by Al-Zoubi et al. [50] in 2018 for detecting spam profiles on Twitter. SVM-WOA employed WOA for SVM's parameter tuning and feature selection, simultaneously. According to Fig. 5.7 each individual of WOA's population of SVM-WOA is a $1 \times (m+2)$ vector where m is the number of features and 2 corresponds to two SVM's parameters. As shown in Fig. 5.7, the real numbers in each individual are in the range of [0, 1]. SVM-OWA utilized an encoder and rounding mechanism to reveal the meaning of each real numbers. So, SVM-WOA utilized WOA for performing two optimization tasks, simultaneously. Experimental results on different lingual contexts indicated that SVM-WOA is able to select more informative features. As a result, its performance is better than the other classical comparative models.

In [51], a model which includes three phases, is proposed to predict the toxicity effects of Bio transfer Hepatic Drugs. The first and second phases consist of feature selection and sampling methods. In the third phase, SVM classifier was used to classify an unknown drug. In order to improve the classification accuracy of SVM, [51] utilized WOA to tune penalty and kernel parameters of SVM. The experimental results proved that applying WOA to optimize the parameter of SVM increases it accuracy.

5.3.4 WOA for neural network and deep neural network training

In the past few years, the significant success of Artificial Neural Networks (ANNs) and Deep Learning Models in different applications has strongly motivated scientists and researchers to take advantage of these methods in different areas. With all these advancements, there are a few challenging problems with these models that have been the center of attention for many research scientists. To have a highly accurate model, ANNs require two important considerations: 1. Precise weight values to have a faultless prediction. 2. Finding the best model architecture to avoid overfitting or underfitting.

5.3.4.1 Artificial neural network weights and parameters tuning

Haghnegahdar et al. [52] applied WOA on a neural network to detect the potential cyber-attacks on a new generation power system, namely, smart grids. WOA was utilized to initiate weights and find the best values for each neuron of the layers. The proposed model was tested with the Mississippi state university and Oak Ridge national laboratory databases of power systems attacks and showed the superiority of the trained neural network using WOA, among other methods and models.

Kushwah et al. [53] tackled the convergence time problem in neural networks using a modified WOA. Depending on the model architecture, sometimes the convergence of networks' weights can be time-consuming, or the weight values can get stuck in local optima and, therefore, can't get the best possible result. In this work, the authors combined the roulette wheel selection method with WOA to enhance its convergence speed and applied it to neural networks to get a faster convergence time and faultless weights. The results of the proposed model on 11 benchmark functions showed that it was successful in terms of convergence time.

Diop et al. [54] applied WOA to optimize a multi-layer perceptron for the accurate prediction of annual precipitation. In this study, the authors considered a population of 30 to update the parameters of a multi-layer perceptron. The reported results on three different lag times on 6 different architectures show that the best results are obtained on the optimized MLPs using WOA.

Dixit et al. [55] employed the WOA in texture detection. The proposed model uses WOA to optimize Convolutional Neural Networks on two levels. First, to optimize the values of Convolutional filters and second, to optimize the weights and biases of the following dense neural network. For evaluation of their performance, the proposed framework was tested on three well-known texture datasets, and in all cases, the trained CNN using WOA showed superiority.

5.3.4.2 Hyperparameters tuning for deep neural networks

Brodzicki et al. [56] tackled the challenging problem of hyperparameter tuning in Deep Neural Networks. The reason that this area is challenging is because of the sensitivity of deep models to overfitting or underfitting with slight changes in the values of the hyperparameters. Therefore, finding the best values for these hyperparameters can prevent the network from falling into these problems. In this research work, the authors considered three main hyperparameters to optimize. First, they consider the batch size in the range of $[10, 500]$ in continuous search and $[8, 16, 32, 64, 128, 256, 512]$ in a discrete one. Second, they considered the number of epochs in the range of [5–50] in continuous search and $[5, 10, 15, 20, 25, 30, 35, 40, 45, 50]$ in a discrete range and finally, the optimizer was selected from a pool of different optimizers including SGD, RMSProp, Adam, Adamax, Adagrad, Adadelta. The final results of the proposed framework on Fashion MNIST and Reuters data sets show the effectiveness of this approach among other trained models.

Murugan et al. [57] proposed WOANet which optimize the hyperparameters of deep neural network via WOA for the diagnosis of Covid-19 patients from the chest Computed Tomography (CT) images automatically. To prove the performance of WOANet, it was compared with the existing pre-trained network. The results indicated the performance of the proposed method in term of the accuracy of diagnosis.

5.4 Discussion

According the above-mentioned studies, WOA has been extensively applied for solving many real-world problems through machine learning methods. However, WOA has some limitations that needs further considerations. Firstly, the performance of WOA is highly depends to maintaining the balance between exploration and exploitation. The balance between exploration and exploitation is achieved through proper value selection of WOA's parameters which are too many. Secondly, existing researches evaluated the effectiveness of WOA in small and medium datasets. However, the performance of many population-based algorithms is degraded with high dimensional or large-scale datasets which is the real-world of data analysis task. Thus, the effectiveness of WOA for handling large-scale problems needs more researches. Thirdly, the extensive researches demonstrate that WOA suffers from local optima entrapment and premature convergence. This critical issue should be addressed to tackle with high-dimensional datasets.

5.5 Conclusion and future direction

This chapter investigates how WOA has been applied to improve the machine learning techniques. Example of utilizing WOA for solving the most common challenges in machine learning algorithms such as feature selection, parameter tuning of SVM classifier and Kmeans clustering, Artificial Neural Networks and Deep Learning hyperparameters tuning were discussed. Review of researches indicates that WOA is very efficient and widely used optimization algorithm to address challenges in machine learning algorithms. Since WOA and its variants are very new, they have a lot of chances for applying to different issues of machine learning approaches.

Future works may be exploring several aspects such as investigating the effect of applying new binary operators to WOA for feature selection on real-world problems. Furthermore, utilizing different search strategies by applying different search operators could make the WOA appropriate for solving different optimization problems with different characteristics. Applying different search strategies that enables WOA dynamically change its behavior during the search process is another research direction. Another worthwhile work is developing WOA for solving multitask optimization. Planning to extend WOA and its improvements to optimize the structure and weights/biases of different types of deep neural networks is another aspect for further studies. Moreover, searching behavior of other ocean creatures that have communication with humpback whales can be incorporated in WOA to have heterogeneous individuals with simple behaviors that can search the search space from different aspects and enhance the chance of finding the global optimum.

References

[1] T.P. Nagarhalli, V. Vaze, N.K. Rana, Impact of machine learning in natural language processing: a review, in: 2021 Third International Conference on Intelligent Communication Technologies and Virtual Mobile Networks (ICICV), Feb. 2021, pp. 1529–1534, https://doi.org/10.1109/ICICV50876.2021.9388380.

[2] M. Wasimuddin, K. Elleithy, A.-S. Abuzneid, M. Faezipour, O. Abuzaghleh, Stages-based ECG signal analysis from traditional signal processing to machine learning approaches: a survey, IEEE Access 8 (2020) 177782–177803, https://doi.org/10.1109/ACCESS.2020.3026968.

[3] I. Portugal, P. Alencar, D. Cowan, The use of machine learning algorithms in recommender systems: a systematic review, Expert Syst. Appl. 97 (May 2018) 205–227, https://doi.org/10.1016/j.eswa.2017.12.020.

[4] H. Song, I. Triguero, E. Özcan, A review on the self and dual interactions between machine learning and optimisation, Prog. Artif. Intell. 8 (2) (Jun. 2019) 143–165, https://doi.org/10.1007/s13748-019-00185-z.
[5] G. Wu, R. Mallipeddi, P.N. Suganthan, Ensemble strategies for population-based optimization algorithms – a survey, Swarm Evol. Comput. 44 (Feb. 2019) 695–711, https://doi.org/10.1016/j.swevo.2018.08.015.
[6] S. Mirjalili, A. Lewis, The whale optimization algorithm, Adv. Eng. Softw. 95 (May 2016) 51–67, https://doi.org/10.1016/j.advengsoft.2016.01.008.
[7] S. Solorio-Fernández, J.A. Carrasco-Ochoa, J.F. Martínez-Trinidad, A survey on feature selection methods for mixed data, Artif. Intell. Rev. 55 (4) (Apr. 2022) 2821–2846, https://doi.org/10.1007/s10462-021-10072-6.
[8] T. Dokeroglu, A. Deniz, H.E. Kiziloz, A comprehensive survey on recent metaheuristics for feature selection, Neurocomputing 494 (Jul. 2022) 269–296, https://doi.org/10.1016/j.neucom.2022.04.083.
[9] E.-S.M. El-Kenawy, et al., Novel meta-heuristic algorithm for feature selection, unconstrained functions and engineering problems, IEEE Access 10 (2022) 40536–40555, https://doi.org/10.1109/ACCESS.2022.3166901.
[10] P. Agrawal, H.F. Abutarboush, T. Ganesh, A.W. Mohamed, Metaheuristic algorithms on feature selection: a survey of one decade of research (2009-2019), IEEE Access 9 (2021) 26766–26791, https://doi.org/10.1109/ACCESS.2021.3056407.
[11] S. Yildirim, Y. Kaya, F. Kılıç, A modified feature selection method based on metaheuristic algorithms for speech emotion recognition, Appl. Acoust. 173 (Feb. 2021) 107721, https://doi.org/10.1016/j.apacoust.2020.107721.
[12] F.S. Gharehchopogh, H. Gholizadeh, A comprehensive survey: whale optimization algorithm and its applications, Swarm Evol. Comput. 48 (Aug. 2019) 1–24, https://doi.org/10.1016/j.swevo.2019.03.004.
[13] S. Mirjalili, A. Lewis, S-shaped versus V-shaped transfer functions for binary Particle Swarm Optimization, Swarm Evol. Comput. 9 (Apr. 2013) 1–14, https://doi.org/10.1016/j.swevo.2012.09.002.
[14] K.K. Ghosh, R. Guha, S.K. Bera, N. Kumar, R. Sarkar, S-shaped versus V-shaped transfer functions for binary Manta ray foraging optimization in feature selection problem, Neural Comput. Appl. 33 (17) (Sep. 2021) 11027–11041, https://doi.org/10.1007/s00521-020-05560-9.
[15] S. Mirjalili, H. Zhang, S. Mirjalili, S. Chalup, N. Noman, A novel U-shaped transfer function for binary particle swarm optimization, https://doi.org/10.1007/978-981-15-3290-0_19, 2020, pp. 241–259.
[16] S. Ahmed, K.K. Ghosh, S. Mirjalili, R. Sarkar, AIEOU: automata-based improved equilibrium optimizer with u-shaped transfer function for feature selection, Knowl.-Based Syst. 228 (Sep. 2021) 107283, https://doi.org/10.1016/j.knosys.2021.107283.
[17] Z. Beheshti, UTF: upgrade transfer function for binary meta-heuristic algorithms, Appl. Soft Comput. 106 (Jul. 2021) 107346, https://doi.org/10.1016/j.asoc.2021.107346.
[18] Y. He, F. Zhang, S. Mirjalili, T. Zhang, Novel binary differential evolution algorithm based on Taper-shaped transfer functions for binary optimization problems, Swarm Evol. Comput. 69 (Mar. 2022) 101022, https://doi.org/10.1016/j.swevo.2021.101022.
[19] K.K. Ghosh, P.K. Singh, J. Hong, Z.W. Geem, R. Sarkar, Binary social mimic optimization algorithm with X-shaped transfer function for feature selection, IEEE Access 8 (2020) 97890–97906, https://doi.org/10.1109/ACCESS.2020.2996611.
[20] Z. Beheshti, A novel x-shaped binary particle swarm optimization, Soft Comput. 25 (4) (Feb. 2021) 3013–3042, https://doi.org/10.1007/s00500-020-05360-2.
[21] S. Guo, J. Wang, M. Guo, Z-shaped transfer functions for binary particle swarm optimization algorithm, Comput. Intell. Neurosci. 2020 (Jun. 2020) 1–21, https://doi.org/10.1155/2020/6502807.
[22] M.H. Nadimi-Shahraki, A. Fatahi, H. Zamani, S. Mirjalili, Binary approaches of quantum-based avian navigation optimizer to select effective features from high-dimensional medical data, Mathematics 10 (15) (Aug. 2022) 2770, https://doi.org/10.3390/math10152770.
[23] Z. Beheshti, A time-varying mirrored S-shaped transfer function for binary particle swarm optimization, Inf. Sci. (NY) 512 (Feb. 2020) 1503–1542, https://doi.org/10.1016/j.ins.2019.10.029.
[24] Z. Beheshti, BMPA-TVSinV: a binary marine predators algorithm using time-varying sine and V-shaped transfer functions for wrapper-based feature selection, Knowl.-Based Syst. 252 (Sep. 2022) 109446, https://doi.org/10.1016/j.knosys.2022.109446.
[25] N. Lang, I. Zincir, N. Zincir-Heywood, Binary text representation for feature selection, https://doi.org/10.1007/978-3-030-63128-4_52, 2021, pp. 681–692.
[26] M. Mafarja, D. Eleyan, S. Abdullah, S. Mirjalili, S-shaped vs. V-shaped transfer functions for ant lion optimization algorithm in feature selection problem, in: Proceedings of the International Conference on Future Networks and Distributed Systems, Jul. 2017, pp. 1–7, https://doi.org/10.1145/3102304.3102325.
[27] M. Shuaib, et al., Whale optimization algorithm-based email spam feature selection method using rotation forest algorithm for classification, SN Appl. Sci. 1 (5) (May 2019) 390, https://doi.org/10.1007/s42452-019-0394-7.
[28] A.G. Hussien, A.E. Hassanien, E.H. Houssein, S. Bhattacharyya, M. Amin, S-shaped binary whale optimization algorithm for feature selection, https://doi.org/10.1007/978-981-10-8863-6_9, 2019, pp. 79–87.
[29] A.G. Hussien, E.H. Houssein, A.E. Hassanien, A binary whale optimization algorithm with hyperbolic tangent fitness function for feature selection, in: 2017 Eighth International Conference on Intelligent Computing and Information Systems (ICICIS), Dec. 2017, pp. 166–172, https://doi.org/10.1109/INTELCIS.2017.8260031.
[30] M.M. Mafarja, S. Mirjalili, Hybrid Whale Optimization Algorithm with simulated annealing for feature selection, Neurocomputing 260 (Oct. 2017) 302–312, https://doi.org/10.1016/j.neucom.2017.04.053.
[31] G.I. Sayed, A. Darwish, A.E. Hassanien, A new chaotic whale optimization algorithm for features selection, J. Classif. 35 (2) (Jul. 2018) 300–344, https://doi.org/10.1007/s00357-018-9261-2.
[32] Y. Zheng, et al., A novel hybrid algorithm for feature selection based on whale optimization algorithm, IEEE Access 7 (2019) 14908–14923, https://doi.org/10.1109/ACCESS.2018.2879848.

[33] A. Got, A. Moussaoui, D. Zouache, Hybrid filter-wrapper feature selection using whale optimization algorithm: a multi-objective approach, Expert Syst. Appl. 183 (Nov. 2021) 115312, https://doi.org/10.1016/j.eswa.2021.115312.

[34] R. Kundu, S. Chattopadhyay, E. Cuevas, R. Sarkar, AltWOA: Altruistic Whale Optimization Algorithm for feature selection on microarray datasets, Comput. Biol. Med. 144 (May 2022) 105349, https://doi.org/10.1016/j.compbiomed.2022.105349.

[35] R.K. Agrawal, B. Kaur, S. Sharma, Quantum based Whale Optimization Algorithm for wrapper feature selection, Appl. Soft Comput. 89 (Apr. 2020) 106092, https://doi.org/10.1016/j.asoc.2020.106092.

[36] M.H. Nadimi-Shahraki, et al., A systematic review of the whale optimization algorithm: theoretical foundation, improvements, and hybridizations, Arch. Comput. Methods Eng. (May 2023) 1–47, https://doi.org/10.1007/s11831-023-09928-7.

[37] M. Sharawi, H.M. Zawbaa, E. Emary, H.M. Zawbaa, E. Emary, Feature selection approach based on whale optimization algorithm, in: 2017 Ninth International Conference on Advanced Computational Intelligence (ICACI), Feb. 2017, pp. 163–168, https://doi.org/10.1109/ICACI.2017.7974502.

[38] M.A. Tawhid, A.M. Ibrahim, Feature selection based on rough set approach, wrapper approach, and binary whale optimization algorithm, Int. J. Mach. Learn. Cybern. 11 (3) (Mar. 2020) 573–602, https://doi.org/10.1007/s13042-019-00996-5.

[39] S. Chebbout, H.F. Merouani, Comparative study of clustering based colour image segmentation techniques, in: 2012 Eighth International Conference on Signal Image Technology and Internet Based Systems, Nov. 2012, pp. 839–844, https://doi.org/10.1109/SITIS.2012.126.

[40] M.R. Karim, et al., Deep learning-based clustering approaches for bioinformatics, Brief. Bioinform. 22 (1) (Jan. 2021) 393–415, https://doi.org/10.1093/bib/bbz170.

[41] G. Andresini, A. Appice, D. Malerba, Nearest cluster-based intrusion detection through convolutional neural networks, Knowl.-Based Syst. 216 (Mar. 2021) 106798, https://doi.org/10.1016/j.knosys.2021.106798.

[42] S.C. Johnson, Hierarchical clustering schemes, Psychometrika 32 (3) (Sep. 1967) 241–254, https://doi.org/10.1007/BF02289588.

[43] M.E. Celebi (Ed.), Partitional Clustering Algorithms, Springer International Publishing, Cham, 2015, https://doi.org/10.1007/978-3-319-09259-1.

[44] K.K.A. Ghany, A.M. AbdelAziz, T.H.A. Soliman, A.A.E.-M. Sewisy, A hybrid modified step Whale Optimization Algorithm with Tabu Search for data clustering, J. King Saud Univ, Comput. Inf. Sci. 34 (3) (Mar. 2022) 832–839, https://doi.org/10.1016/j.jksuci.2020.01.015.

[45] T. Singh, A novel data clustering approach based on whale optimization algorithm, Expert Syst. 38 (3) (May 2021), https://doi.org/10.1111/exsy.12657.

[46] J. Nasiri, F.M. Khiyabani, A whale optimization algorithm (WOA) approach for clustering, Cogent Math. Stat. 5 (1) (Jan. 2018) 1483565, https://doi.org/10.1080/25742558.2018.1483565.

[47] C. Zhang, D. Ouyang, J. Ning, An artificial bee colony approach for clustering, Expert Syst. Appl. 37 (7) (Jul. 2010) 4761–4767, https://doi.org/10.1016/j.eswa.2009.11.003.

[48] F.S. Gharehchopogh, H. Gholizadeh, A comprehensive survey: Whale Optimization Algorithm and its applications, Swarm Evol. Comput. 48 (Aug. 2019) 1–24, https://doi.org/10.1016/j.swevo.2019.03.004.

[49] N.P. Karlekar, N. Gomathi, OW-SVM: ontology and whale optimization-based support vector machine for privacy-preserved medical data classification in cloud, Int. J. Commun. Syst. 31 (12) (Aug. 2018) e3700, https://doi.org/10.1002/dac.3700.

[50] A.M. Al-Zoubi, H. Faris, J. Alqatawna, M.A. Hassonah, Evolving Support Vector Machines using Whale Optimization Algorithm for spam profiles detection on online social networks in different lingual contexts, Knowl.-Based Syst. 153 (Aug. 2018) 91–104, https://doi.org/10.1016/j.knosys.2018.04.025.

[51] A. Tharwat, Y.S. Moemen, A.E. Hassanien, Classification of toxicity effects of biotransformed hepatic drugs using whale optimized support vector machines, J. Biomed. Inform. 68 (Apr. 2017) 132–149, https://doi.org/10.1016/j.jbi.2017.03.002.

[52] L. Haghnegahdar, Y. Wang, A whale optimization algorithm-trained artificial neural network for smart grid cyber intrusion detection, Neural Comput. Appl. 32 (13) (Jul. 2020) 9427–9441, https://doi.org/10.1007/s00521-019-04453-w.

[53] R. Kushwah, M. Kaushik, K. Chugh, A modified whale optimization algorithm to overcome delayed convergence in artificial neural networks, Soft Comput. 25 (15) (Aug. 2021) 10275–10286, https://doi.org/10.1007/s00500-021-05983-z.

[54] L. Diop, S. Samadianfard, A. Bodian, Z.M. Yaseen, M.A. Ghorbani, H. Salimi, Annual rainfall forecasting using hybrid artificial intelligence model: integration of multilayer perceptron with whale optimization algorithm, Water Resour. Manag. 34 (2) (Jan. 2020) 733–746, https://doi.org/10.1007/s11269-019-02473-8.

[55] U. Dixit, A. Mishra, A. Shukla, R. Tiwari, Texture classification using convolutional neural network optimized with whale optimization algorithm, SN Appl. Sci. 1 (6) (Jun. 2019) 655, https://doi.org/10.1007/s42452-019-0678-y.

[56] A. Brodzicki, M. Piekarski, J. Jaworek-Korjakowska, The whale optimization algorithm approach for deep neural networks, Sensors 21 (23) (Nov. 2021) 8003, https://doi.org/10.3390/s21238003.

[57] R. Murugan, T. Goel, S. Mirjalili, D.K. Chakrabartty, WOANet: whale optimized deep neural network for the classification of COVID-19 from radiography images, Biocybern. Biomed. Eng. 41 (4) (Oct. 2021) 1702–1718, https://doi.org/10.1016/j.bbe.2021.10.004.

Chapter 6

Whale optimization algorithm - comprehensive meta analysis on hybridization, latest improvements, variants and applications for complex optimization problems

Parijata Majumdar[a], Sanjoy Mitra[b], Seyedali Mirjalili[c,d], and Diptendu Bhattacharya[a]

[a]National Institute of Technology, Agartala, Jirania, Tripura, India, [b]Tripura Institute of Technology, Agartala, Narsingarh, Tripura, India, [c]Centre for Artificial Intelligence Research and Optimisation, Torrens University Australia, Brisbane, QLD, Australia, [d]Yonsei Frontier Lab, Yonsei University, Seoul, South Korea

Highlights

- The WOA is reviewed with an emphasis on applications to optimization problems.
- The advantages, lacunas in the existing methods of variants of WOA are reviewed.
- This survey reviews hybridizations, latest variants, and modifications of WOA.

6.1 Introduction

Metaheuristic algorithms have drawn attention from the scientific community over the last 20 years. In general, single-solution and multiple-solution based are the two broader subdivision of metaheuristics [16]. Metaheuristics can be either single-solution or population-based [32] where the latter type has multiple candidate solutions that shares information about the search space resulting in dynamic exploration of the promising regions. Swarm Intelligence (SI) is the most promising field of population-based metaheuristics inspired by the social community of animals, birds, plants, and human beings which stores the information about the search space, saves the best solution, have few operators (i.e., selection, crossover, mutation), and has fewer number of tunable parameters in general.

Particle Swarm Optimization (PSO) [24], Ant Colony Optimization (ACO) [9], Cuckoo Search (CS) [50], Krill Herd (KH) algorithm [42], Artificial Bee Colony (ABC) [18], Bat inspired (BA) algorithm [51], Firefly Algorithm (FA) [52], Fruit Fly Optimization Algorithm (FOA) [49], League Championship Algorithm (LCA) [20], Grey Wolf Optimizer (GWO) [32], and Bird Mating Optimizer (BMO) [4] are popular SI algorithms that can be found in the existing literature.

The search space of metaheuristics is splitted into diversification and intensification phases [15], [14]. Diversification denotes the capability of exploring the promising regions of the search space extensively and intensification denotes the capability of local searching around the promising regions of search space obtained from intensification. Balancing between these phases has garnered huge attention in research field due to the stochastic nature of metaheuristics. In order to maintain equilibrium between the diversification and intensification phases and improving the global convergence ability of metaheuristics, a large number of metaheuristics have emerged so far. Due to the vast amount of research in metaheuristics, it is challenging to acknowledge each significant contribution due to the diversity of the bibliography.

Recently, a metaheuristic called Whale Optimization algorithm (WOA) has developed [31]. Whales are thought to be the most intellectual animals containing motion. The WOA was driven by the whale's hunting strategy. The whales prefer to hunt fish that swims close to the water surface such as small fish using the bubble net feeding distinctive hunting strategy used by the whales. With this technique, the whales swim around the prey and produce visible bubbles along a circular path. Recent research demonstrates the outstanding capability of WOA to resolve challenging engineering optimization issues [13]. Numerous fields of research including data mining, machine learning, robotics path planning, cloud computing,

Handbook of Whale Optimization Algorithm. https://doi.org/10.1016/B978-0-32-395365-8.00012-9

the Internet of Things (IoT), and training Artificial Neural Networks (ANN) have derived benefits from the simplicity, flexibility, stochastic nature, and fast convergence speed of WOA. The WOA's balanced local searching and global searching techniques of searching as well as its successful execution even with less number of parameters are some of the standout characteristics of WOA [13]. Additionally, it has the potential to inherit the effective functionality of an evaluation-based algorithm that incorporates crossover and mutation processes and has a higher convergence rate [13].

Contrary to other meta-heuristics, WOA suffers from some of the shortcomings like it may become stuck in local optima and fails to globally search the search space [10]. Due to these drawbacks, it has been modified and hybridized with other approaches or metaheuristics to address high-dimensional issues. Although the WOA is still in its early stages of deployment and was only recently created, its use in solving multidisciplinary optimization problems is expanding quickly. As a result, it is safe to assume that the theoretical and practical applications of WOA will continue to grow. The objective of this review is to provide a thorough overview of the relevant research works on WOA, the changes that have been made to the WOA over time to time, hybridizations made to the WOA, the WOA's multi-objective applications across several disciplines including different engineering fields to address challenging optimization tasks, etc.

In the subsequent section 6.2, the basic WOA is discussed. Section 6.3. WOA with respect to its modifications, hybridizations, and applications is reviewed in Section 6.4. The existing problems and future research avenues as perspectives are highlighted in Section 6.5. Finally, Section 6.6 draws conclusion.

6.2 Whale optimization algorithm

A swarm-based intelligent algorithm called Whale Optimization Algorithm (WOA) is developed to solve continuous optimization problems [31]. The distance between the food and humpback whales corresponds to the objective cost's value, while the positions of humpback whales are represented as decision variables. It consists of shrinking encircling prey, bubble-net attacking method which corresponds to exploitation phase and prey search which corresponds to exploration phase are used to determine a whale's time-dependent location. During hunting, whales can locate their prey and encircle them. The WOA considers the best candidate solution at the moment is either the target prey or approximates the optimum because the location of the optimal design is not known apriori in the search space. The best search agent will be found out and the other search agents will update their positions in close proximity to it. Encircling mechanism can be defined as:

$$\vec{D} = |\vec{C} \cdot \vec{G}^*(i) - \vec{G}(i)| \tag{6.1}$$

$$\vec{G}(i+1) = \vec{G}^*(i) - \vec{A} \cdot \vec{D} \tag{6.2}$$

$$\vec{A} = 2\vec{a} \cdot \vec{r} - \vec{a} \tag{6.3}$$

$$\vec{C} = 2 \cdot \vec{r} \tag{6.4}$$

where $\vec{G}^*$ is the general best position, $\vec{G}$ represents the whale position, i denotes the recent iteration, a denotes the linear reduction within the range of 2 to 0 with increasing number of iterations, and $r \in [0, 1]$.

During exploitation to simulate the bubble-net behavior of humpback whales, a spiral mathematical formulation is applied between the whale's and the prey's position to simulate the movement of whales in helix-shape [22]:

$$G(\vec{i+1}) = \vec{D}' \cdot e^{bl} \cdot cos(2\pi l) + \vec{G^*}(i) \tag{6.5}$$

$$G(\vec{i+1}) = \vec{G^*}(i) - \vec{A} \cdot \vec{D}, \text{ if } k < 0.5 \tag{6.6}$$

$$G(\vec{i+1}) = \vec{D}' \cdot e^{bl} \cdot cos(2\pi l) + \vec{G^*}(i), \text{ if } k \geq 0.5 \tag{6.7}$$

where constant k is the shape of logarithmic spiral and $a \in [-1, 1]$.

During exploration if $A > 1$ or $A < -1$, the search agent is updated by randomly choosing search agent at each iteration which can be stated as follows:

$$\vec{D} = |\vec{C} \cdot \vec{G_{rand}} - \vec{G}| \tag{6.8}$$

$$G(\vec{i+1}) = \vec{G_{rand}} - \vec{G} \cdot \vec{D} \tag{6.9}$$

where $\vec{G_{rand}}$ is arbitrarily nominated from whales in the current iteration.

Fig. 6.1 describes the flowchart of WOA.

Fig. 6.2 illustrates the spiral position updating of whales as per the location of the prey.

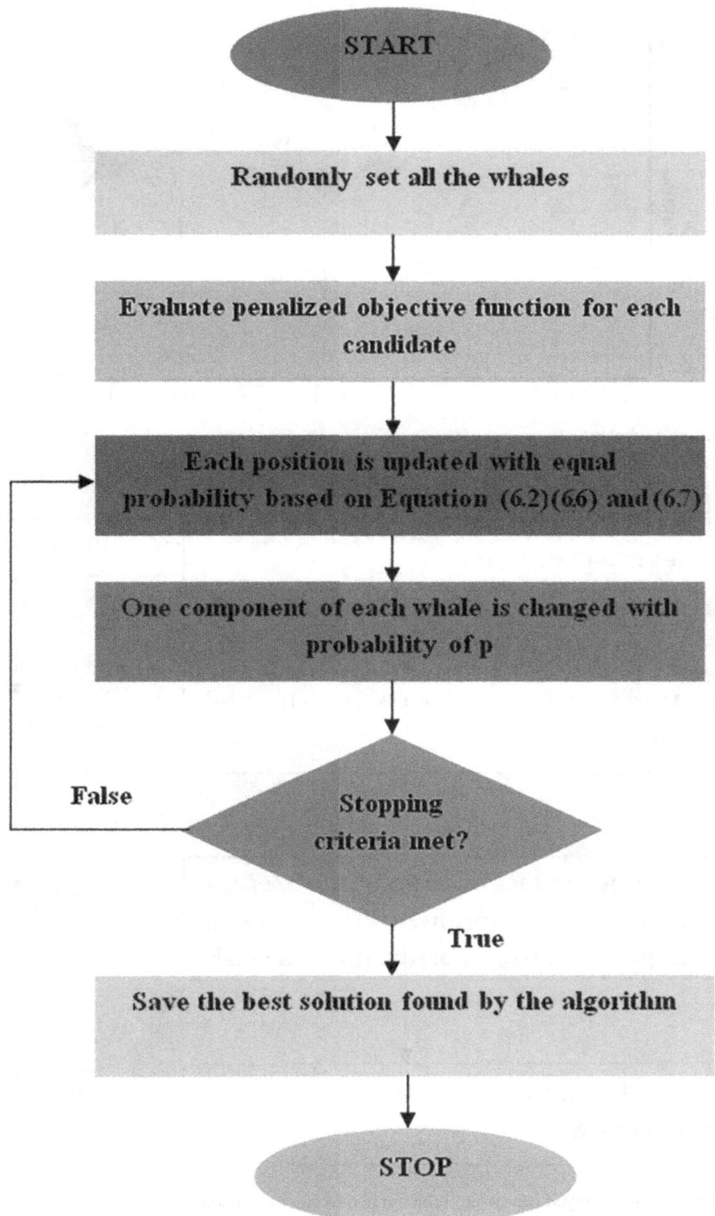

FIGURE 6.1 Flowchart of WOA [31].

6.3 Research methodology

The studies based on WOA that have been published in well-known scientific databases have been examined using exhaustive research. The extensive search is done depending on the guidelines given in [28] covering all the domains, where WOA is applied. Searching comprises all the authentic online databases, and then a rigorous verification of references and citations are done for ensuring the selection of best-quality peer-reviewed articles. Besides, the articles in the disciplines other than engineering and computer science was also considered for review. Finally, suitable publications are decided upon in accordance with inclusion and exclusion standards for review purposes as shown in Fig. 6.3. Two steps were followed in the keyword search procedure. With the help of the following keywords—"WOA", "latest developments on WOA", "Shrinking encircling prey", "modified WOA", "improved WOA", and "hybrid WOA", etc., most of the papers on WOA were retrieved from most of the pertinent databases. We continued searching till August 2022 in order to include the most recent publications in our review process. In total, 81 articles have been included in this survey that were found from exhaus-

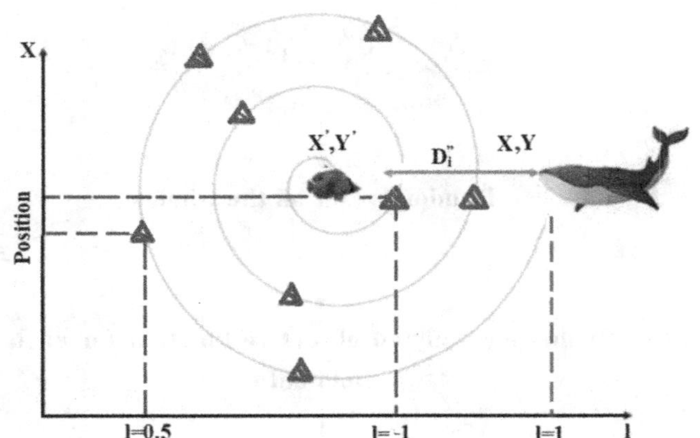

FIGURE 6.2 Spiral position updating of whales.

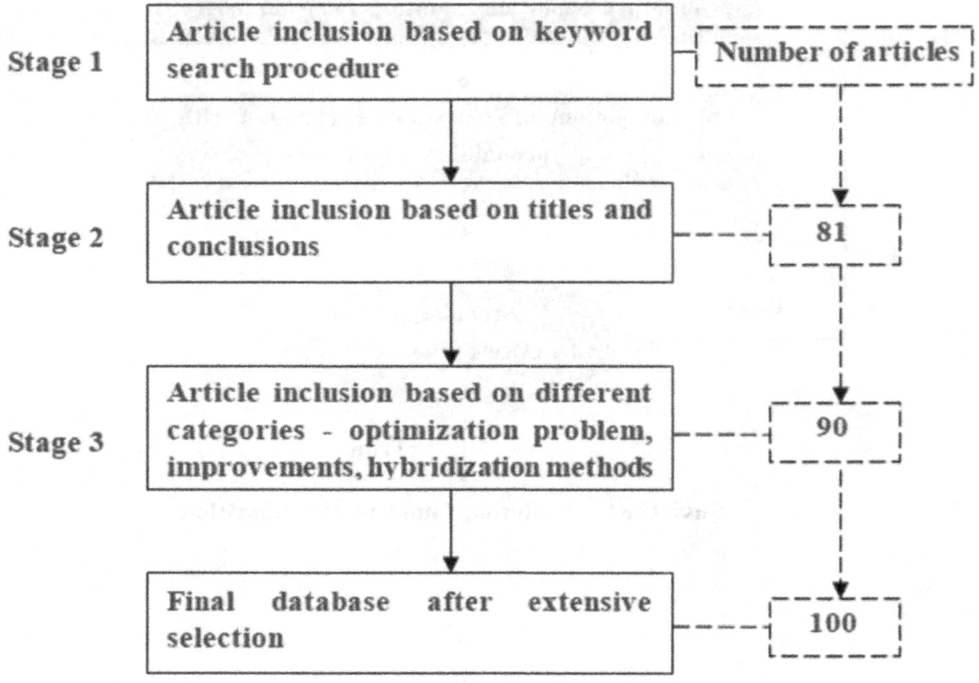

FIGURE 6.3 Number of WOA publications surveyed based on inclusion and exclusion standards.

tive searching of the keywords from the authentic publisher's databases available online such as Google Scholar, Scopus, ScienceDirect, World Scientific, IEEE Explorer, Springer Link, IGI Global, and Taylor & Francis, etc. The publications published each year by the relevant publishers are shown in Fig. 6.4 as concrete evidence of the rapidly expanding research on WOA. With a graphic depiction of the number of articles available, Fig. 6.5 supports the same. Fig. 6.6 shows different WOA articles published in different fields. Fig. 6.7 shows the different method wise distribution of WOA articles. Only the most pertinent research papers were kept and the unnecessary articles are deleted. The chosen papers were then marked for further processing after a comprehensive scrutiny was done on their titles, and conclusions. In total, 90 articles have been included in this survey that were found from comprehensive scrutiny done on their titles, and conclusions. Finally, the papers that were the most pertinent to this review were included. On the articles, a thorough data extraction procedure was carried out in order to prevent the information redundancy. Different categorization processes were done on data based on applications, method-wise categorization (improvements, hybridization and optimization problem solving, etc.), year and publisher-wise paper distribution. The extensive selection process finally returned 100 articles suitable for review [38].

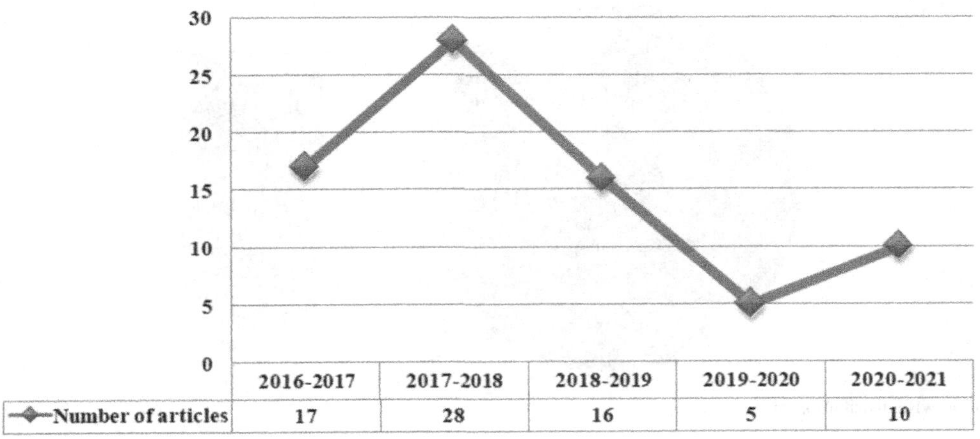

FIGURE 6.4 Number of publications per year.

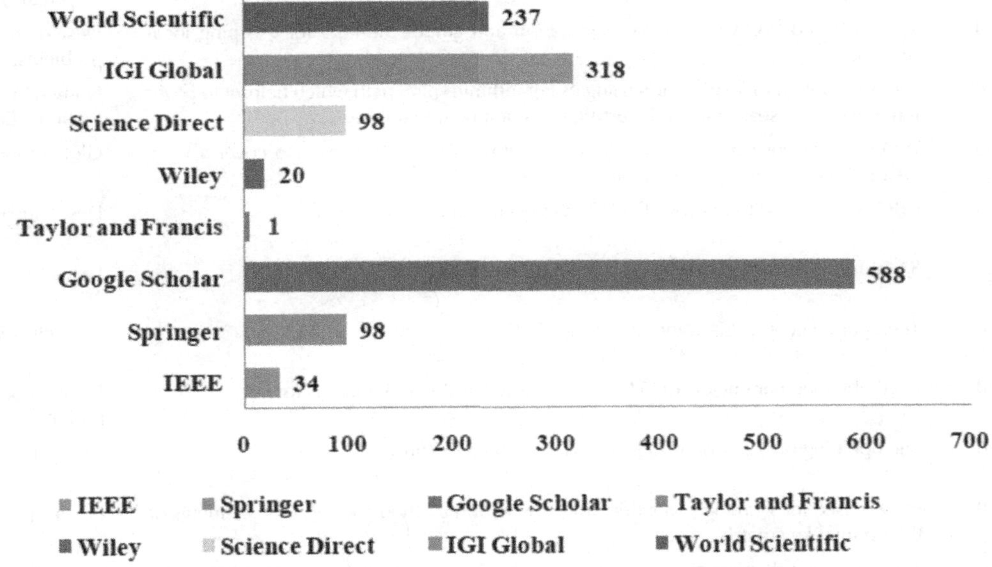

FIGURE 6.5 Number of articles by publishers.

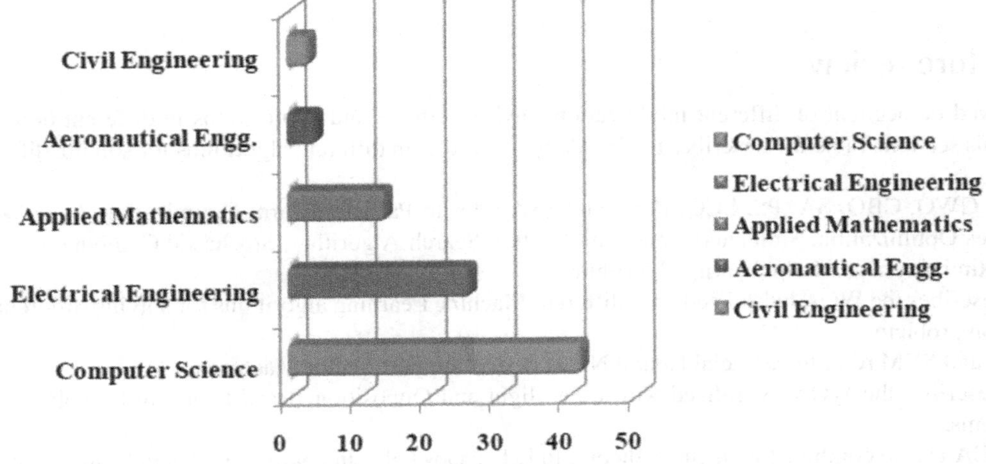

FIGURE 6.6 Number of WOA articles published in different fields.

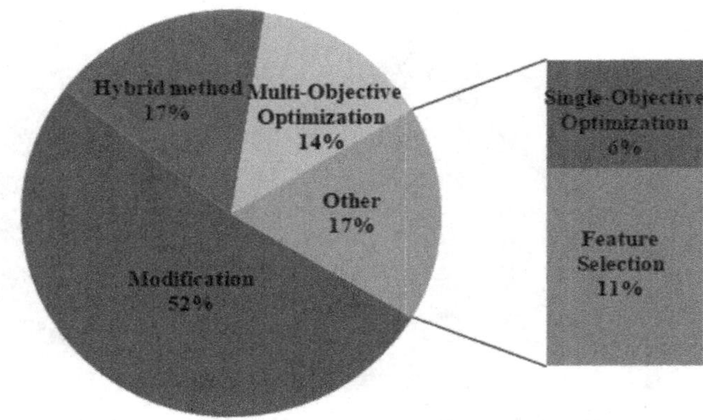

FIGURE 6.7 Method wise distribution of WOA articles.

TABLE 6.1 WOA hybridized with different metaheuristic algorithms.

Method	Contribution	Application
WOA hybridized with PSO [26]	The improved WOA is used for exploration and guides the PSO for escaping local optima.	Design optimization problems.
WOA hybridized with PSO [48]	WOA is applied for exploration using its logarithmic spiral path which helps it to cover large uncertain search space incurring less computational time.	Numerical Function Optimization.
WOA hybridized with GWO [43]	WOA is used for exploration using a spiral function which helps it to cover a wide area in an uncertain search region.	Global Optimization.
WOA hybridized with CBO [23]	CBO is integrated to improve the WOA's position updation.	Designing construction area.
WOA hybridized with SA [30]	SA is used to enhance the intensification of the WOA.	Feature selection.
WOA hybridized with GWO [17]	The optimal centroid is computed using GWO.	Data clustering.
WOA hybridized with PS [5]	Final obtained parameters of WOA are regarded as the initial points for PS.	Optimal power flow problem.
WOA hybridized with SCO [25]	The updating operator of SCO is used for catching the prey.	Parameter optimization.
WOA hybridized with BSO [39]	BSO is used for secret key identification. The hybrid method identifies the elements of the optimal key matrix.	Privacy preservation.
WOA hybridized with SSA [11]	The leader mechanism for exploitation of SSA is applied for WOA's position updation. Then the non-linear convergence is introduced to encircle prey and bubble-net attacking strategy. Lens OBL is used to compute opposite population.	Classical engineering design problems.

6.4 Literature review

WOA is reviewed on account of different modifications, hybridizations and applications in different domains which are elaborated in this section. Table 6.1 describes the WOA hybridized with different algorithms for solving different optimization problems.

Here, PSO, GWO, CBO, SA, PS, SCO, BSO, and SSA refer to Particle Swarm Optimization, Grey Wolf Optimizer, Colliding Bodies Optimization, Simulated Annealing, Pattern Search Algorithm, Stochastic Combinatorial Optimization, Bee Swarm Optimization, and Salp Swarm Algorithm.

Table 6.2 describes the WOA hybridized with different Machine Learning algorithms for solving different classification and optimization problems.

Here, ANN and SVM refer to Artificial Neural Network and Support Vector Machines.

Table 6.3 describes the WOA hybridized with Levy flight and Opposition Based Learning for solving different optimization problems.

Besides, WOA is also combined with chaos theory. In [21] chaos helps in controlling diversification and intensification of WOA to solve different optimization problems. In [54] chaos is introduced into the WOA to increase efficiency of a

TABLE 6.2 WOA hybridized with different Machine Learning algorithms.

Method	Contribution	Application
WOA hybridized with ANN [37]	The WOA develop new solutions in solution space and back propagation is used to find the global optima.	Classification.
WOA hybridized with ANN [3]	The best values for weights and biases of WOA are found to reduce error of ANN.	Optimization.
WOA hybridized with ANN [35]	The error performance of is measured using WOA and the updated weight with reduced error used to train the ANN.	Heterogeneous network.
WOA hybridized with SVM [19]	The WOA is used for feasible selection of appropriate kernel parameters.	Data classification.
WOA hybridized with SVM [46]	WOA optimizes the parameters of SVM so that the classification error can be reduced.	Classification
WOA hybridized with SVM [40]	WOA was used to obtain optimal features to train SVM.	e-mail classification.

TABLE 6.3 WOA hybridized with Levy flight and Opposition Based Learning.

Method	Contribution	Application
WOA hybridized with levy flight [29]	Levy flight trajectory increases the population diversity and enhances the capability of local optima avoidance.	Global optimization.
WOA hybridized with levy flight [45]	Levy flight strategy is allows the algorithm to jump out of local optima.	Large-scale global optimization problems.
WOA hybridized with levy flight [1]	Levy flight accelerates the convergence toward the global optimal solution and it can also overcome local optimum stagnation problems.	Knapsack problem.
WOA hybridized with Levy motion and Brownian motion [34]	The movement strategy of the whales is enhanced by using Levy motion for encircling the prey and uses Brownian motion to search for the prey that cooperates with canonical bubble-net attacking mechanism.	Controlling power flow of a power system.
WOA hybridized with Opposition Based Learning (OBL) [47]	Fitness of the opposite individuals is obtained using OBL. The individual having lower fitness is chosen for better population quality.	Constrained engineering problems.
WOA hybridized with Opposition Based Learning (OBL) [10]	OBL is used to increase the exploration capability of WOA.	Estimation of parameters of solar cells using diode models.
WOA hybridized with Opposition Based Learning (OBL) [2]	Search for the solution is done in opposite direction of the values obtained using WOA to evaluate whether the values of the opposite solution is better or not.	Optimization problems.
WOA hybridized with Opposition Based Learning (OBL) and GWO [55]	An individual updating way is used to reduce the computational complexity. A random opposition learning is used where WOA and GWO are combined for enhancing the global exploration of WOA. For balanced exploration and exploitation, switching parameter tuning, random differential disturbance and global-best spiral operator are used.	High dimensional problems like data clustering.
WOA hybridized with Opposition Based Learning (OBL) [27]	WOA is combined with the operators including OBL, non-linear parameter design, clustering and differential evolution to avoid local optima stagnation problem.	Seismic inversion problem.

MIMO radar. In [41] Chaos is used to manipulate the random parameters of WOA. Here, chaos with modification of exploration operators helps to achieve the optimal selected features. In [8] chaos is incorporated in WOA and Quasi-opposition is used to avoid the solution from getting trapped into local optima. In [12] five discrete chaotic maps were used for increasing the efficacy of the WOA for global optimization where the parameters were adjusted using the random number sequences from chaotic maps. Yin et al. [53] have developed a chaotic classification model for image extraction using WOA. Here, WOA is used to choose features by reducing the fitness value that overfits while eliminating the redundant features with better classification accuracy.

Qiao et al. [36] have developed a gas consumption prediction model which combines the adaptive filter and WOA for high prediction accuracy. In [7] to overcome the lack of exploration capability of WOA, the mutualism phase is combined with WOA to get rid of early convergence and balancing the exploration and exploitation of the search space simultaneously.

Combining mutualism with WOA also helps to avoid wastage of computational resources during excessive exploitation. In [6] WOA is combined with dynamic opposite learning and an adaptive encircling strategy for balancing the exploration and exploitation and premature convergence of WOA. In [44] a tent chaotic map initializes the initial population distribution. Then a new iteration-based update strategy controls the global and local search capabilities. Also, for further increasing the exploration, a feedback strategy is used for encircling prey. In [33] a WOA is used that utilizes a pooling mechanism along with migrating, preferential selecting and enriched encircling prey is proposed for medical feature selection.

6.5 Existing problems, applications, and future research avenues

The WOA algorithm still faces a difficult challenge for balancing the local and global search so that further research can be done on the best strategy to control diversification and intensification, which is a open research question. Many of the reviewed literatures on WOA suggest that in order to set the original WOA to fit into particular settings, some level of modifications must be made. This leads to a wide range of issues with numerous conditions and altered parameters. As a result, various versions of the WOA are required to be developed to resolve various issues in various contexts. So, comprehensive research of standardization in WOA is essential that will improve the applicability of WOA in different research domains. WOA is witnessing an emerging research trend in data mining, cloud computing, sensor networks, and robotics, etc., in computer science domain. In sensor networks, WOA is capable to handle resources inside the cluster area network and enhance longevity of sensor networks. Moreover, WOA is also witnessing an emerging research topic in cryptanalysis, applied mathematics to solve high-dimensional continuous function optimization problem. In applications of aeronautics engineering, construction engineering and solving multi-objective research problems also, WOA is widely applicable. In data science also, WOA must be tested to solve several real-time specific applications with a focus to large sized datasets because current literature does not consider the applications of WOA in big data applications. Although WOA is less sensitive to initial parameter settings, this issue still exists because the WOA must be executed using user-defined parameters. To get the best answer for each unique condition, the most optimized set of parameters must be found. This shows that choosing the best optimized parameter is still needed for attaining the overall performance improvement using WOA. Additionally, future studies should try to eliminate even the fewer number of parameters from WOA. Thus, parameter setup will not be necessary for the algorithm to function accordingly. More research needs to be done on how to integrate the WOA with other algorithms to increase its adaptability in solving real-life problems in different domains. In contrast, hybridizations can typically reach their boundary if problem instances with complicated search spaces can be effectively solved. Thus, extensive research should be carried out to hybridize different metaheuristics with WOA. The modifications of the WOA can be evaluated over the years in small datasets which is more prevalent in the existing literature reviewed above. Very limited or no literature is available that demonstrates the applicability of WOA on large sized datasets. This is true because big data has garnered a great deal of attention in both industry and scientific community. So, the efficiency of the advanced versions of WOA that can be found from the recent literature is only restricted to small sized datasets and thus it is recommended to extend the research on WOA for large size datasets [38].

6.6 Conclusion

This review is an attempt of exploring the potentials of WOA. Researchers are currently obsessed to WOA as it has simple structure, overcoming local optima problem and speedy convergence due to its efficient exploration and exploitation abilities. The aim is to summarize the background and use of WOA across different scientific domains through various modifications and hybridizations. Additionally, WOA can also might be used to resolve optimization problems like pattern recognition, unconstrained optimization and engineering design problems. It can be found from existing literature analysis that the main technique used for improvement of WOA is WOA combined with a chaotic map. In future, the WOA can be further improved and hybridized with other metaheuristics for solving continuous optimization, discrete optimization, and the NP-hard problems. Another intriguing area of development worth investigating is designing the multi-objective problem by tuning the single-objective parameters. Other challenging problems such as the chaotic sequence, the permutation flow shop, the assignment, the redundancy allocation strategy, and the multi-class support vector machine, etc., are some viable future research directions. This review also scrutinizes the WOA's applicability for resolving problems in different categories of multi-objective, combinatorial and real-world optimization problems with unknown search spaces for creating a way to investigate more multidisciplinary fields that the flexibility of WOA can unlock. There is also a major scope for future improvement in the WOA to solve different multi-dimensional and multimodel problems. Also in future, chaotic methods, discrete and binary methods, opposition and fuzzy methods needs to be incorporated in WOA which demands strenuous effort to solve complicated optimization tasks.

References

[1] M. Abdel-Basset, D. El-Shahat, A.K. Sangaiah, A modified nature inspired meta-heuristic whale optimization algorithm for solving 0–1 knapsack problem, International Journal of Machine Learning and Cybernetics (2017) 1–20.

[2] S.H. Alamri, Y.A. Alsariera, K.Z. Zamli, Opposition-based whale optimization algorithm, Advanced Science Letters 24 (2018) 7461–7464.

[3] I. Aljarah, H. Faris, S. Mirjalili, Optimizing connection weights in neural networks using the whale optimization algorithm, Soft Computing 22 (2016) 1–15.

[4] A. Askarzadeh, Bird mating optimizer: an optimization algorithm inspired by bird mating strategies, Communications in Nonlinear Science and Numerical Simulation 19 (2014) 1213–1228.

[5] B. Bentouati, L. Chaib, S. Chettih, A hybrid whale algorithm and pattern search technique for optimal power flow problem, in: 8th International Conference on Modelling, Identification and Control (ICMIC), IEEE, 2016, pp. 1048–1053.

[6] D. Cao, Y. Xu, Z. Yang, H. Dong, X. Li, An enhanced whale optimization algorithm with improved dynamic opposite learning and adaptive inertia weight strategy, Complex & Intelligent Systems (2022) 1–29.

[7] S. Chakraborty, A. Saha, S. Sharma, S. Mirjalili, R. Chakraborty, A novel enhanced whale optimization algorithm for global optimization, Computers & Industrial Engineering 153 (2021) 107086.

[8] H. Chen, W. Li, X. Yang, A whale optimization algorithm with chaos mechanism based on quasi-opposition for global optimization problems, Expert Systems with Applications 158 (2020) 113612.

[9] M. Dorigo, M. Birattari, T. Stutzle, Ant colony optimization, IEEE Computational Intelligence Magazine 1 (2006) 28–39.

[10] M. Elaziz, D. Oliva, Parameter estimation of solar cells diode models by an improved opposition-based whale optimization algorithm, Energy Conversion and Management 171 (2018) 1843–1859.

[11] Q. Fan, Z. Chen, W. Zhang, X. Fang, ESSAWOA: enhanced whale optimization algorithm integrated with salp swarm algorithm for global optimization, Engineering With Computers (2020) 1–18.

[12] Z. Garip, M. Cimen, D. Karayel, A. Boz, The chaos-based whale optimization algorithms global optimization, Chaos Theory and Applications 1 (2019) 51–63.

[13] F. Gharehchopogh, H. Gholizadeh, A comprehensive survey: whale optimization algorithm and its applications, Swarm and Evolutionary Computation 48 (2019) 1–24.

[14] D.E. Goldberg, Genetic Algorithms, Pearson Education India, 2013.

[15] J.H. Holland, Genetic algorithms, Scientific American 267 (1992) 66–73.

[16] W.G. Jackson, E. Ozcan, R.I. John, Move acceptance in local search metaheuristics for cross-domain search, Expert Systems with Applications 109 (2018) 131–151.

[17] A.N. Jadhav, N. Gomathi, WGC: hybridization of exponential grey wolf optimizer with whale optimization for data clustering, Alexandria Engineering Journal 57 (2018).

[18] D. Karaboga, An idea based on honeybee swarm for numerical optimization, Technical Report TR06, Erciyes University, Engineering Faculty, Computer Engineering Department, 2005.

[19] N.P. Karlekar, N. Gomathi, OW-SVM: ontology and whale optimization-based support vector machine for privacy-preserved medical data classification in cloud, International Journal of Communication Systems 31 (2018).

[20] A.H. Kashan, League championship algorithm: a new algorithm for numerical function optimization, in: International Conference of Soft Computing and Pattern Recognition, IEEE, 2009, pp. 43–48.

[21] G. Kaur, S. Arora, Chaotic whale optimization algorithm, Journal of Computational Design and Engineering 5 (2018) 275–284.

[22] A. Kaveh, A new optimization method: dolphin echolocation, Advances in Engineering Software 59 (2013) 53–70.

[23] A. Kaveha, M.R. Moghaddam, A hybrid WOA-CBO algorithm for construction site layout planning problem, Scientia Iranica, Sharif University of Technology 25 (2018) 1094–1104.

[24] J. Kennedy, R.C. Eberhart, Particle swarm optimization, in: IEEE International Conference on Neural Networks, 1995, pp. 1942–1948.

[25] S. Khalilpourazari, S. Khalilpourazary, SCWOA: an efficient hybrid algorithm for parameter optimization of multi-pass milling process, Journal of Industrial and Production Engineering, Taylor & Francis (2018) 1–14.

[26] N. Laskar, K. Guha, I. Chatterjee, S. Chanda, K. Baishnab, P. Paul, HWPSO: a new hybrid whale-particle swarm optimization algorithm and its application in electronic design optimization problems, Applied Intelligence (2018) 1–27.

[27] X. Liang, S. Xu, Y. Liu, L. Sun, A modified whale optimization algorithm and its application in seismic inversion problem, Mobile Information Systems 2022 (2022) 1–18.

[28] A. Liberati, D. Altman, J. Tetzlaff, C. Mulrow, P. Gotzsche, J. Ioannidis, M. Clarke, P. Devereaux, J. Kleijnen, D. Moher, The PRISMA statement for reporting systematic reviews and meta-analyses of studies that evaluate health care interventions: explanation and elaboration, PLoS Medicine 6 (2009) e1–34.

[29] Y. Ling, Y. Zhou, Q. Luo, Levy flight trajectory-based whale optimization algorithm for global optimization, IEEE Access 5 (2017) 6168–6186.

[30] M.M. Mafarja, S. Mirjalili, Hybrid whale optimization algorithm with simulated annealing for feature selection, Neurocomputing 260 (2017) 302–312.

[31] S. Mirjalili, A. Lewis, The whale optimization algorithm, Advances in Engineering Software 95 (2016) 51–67.

[32] S. Mirjalili, S.M. Mirjalili, A. Lewis, Grey wolf optimizer, Advances in Engineering Software 69 (2014) 46–61.

[33] M. Nadimi-Shahraki, H. Zamani, S. Mirjalili, Enhanced whale optimization algorithm for medical feature selection: a COVID-19 case study, Computers in Biology and Medicine 148 (2022) 105858.

[34] M.H. Nadimi-Shahraki, S. Taghian, S. Mirjalili, L. Abualigah, M. Abd Elaziz, D. Oliva, EWOA-OPF: effective whale optimization algorithm to solve optimal power flow problem, Electronics 10 (2021) 2975.

[35] D. Parambanchary, V.M. Rao, WOA-NN: a decision algorithm for vertical handover in heterogeneous networks, Wireless Networks (2018) 1–16.

[36] W. Qiao, Z. Yang, Z. Kang, Z. Pan, Short-term natural gas consumption prediction based on Volterra adaptive filter and improved whale optimization algorithm, Engineering Applications of Artificial Intelligence 87 (2020) 103323.

[37] J. Rajeshkumar, K. Kousalya, Diabetes data classification using whale optimization algorithm and backpropagation neural network, International Research Journal of Pharmacy 8 (2017) 219–222.

[38] N. Rana, M.S.A. Latiff, S.I.M. Abdulhamid, H. Chiroma, Whale optimization algorithm: a systematic review of contemporary applications, modifications and developments, Neural Computing & Applications 32 (2020) 16245–16277.

[39] S. Revathi, N. Ramaraj, S. Chithra, Brain storm-based whale optimization algorithm for privacy-protected data publishing in cloud computing, Cluster Computing (2018) 110.

[40] R.K. Saidala, N.R. Devarakonda, Bubble-net hunting strategy of whales based optimized feature selection for email classification, in: International Conference for Convergence in Technology (I2CT), 2017, pp. 626–631.

[41] G. Sayed, A new chaotic whale optimization algorithm for features selection, Journal of Classification (2018) 145.

[42] G.P. Singh, A. Singh, Comparative study of Krill Herd, firefly and cuckoo search algorithms for unimodal and multimodal optimization, International Journal of Intelligent Systems and Applications 6 (2014) 35.

[43] N. Singh, H. Hachimi, A new hybrid whale optimizer algorithm with mean strategy of grey wolf optimizer for global optimization, Mathematical and Computational Applications (2018) 1–32.

[44] G. Sun, Y. Shang, K. Yuan, H. Gao, An improved whale optimization algorithm based on nonlinear parameters and feedback mechanism, International Journal of Computational Intelligence Systems 15 (2022) 1–17.

[45] Y. Sun, X. Wang, Y. Chen, Z. Liu, A modified whale optimization algorithm for large-scale global optimization problems, in: Expert Systems with Applications, 2018.

[46] A. Tharwat, Y.S. Moemen, A.E. Hassanien, Classification of toxicity effects of biotransformed hepatic drugs using whale optimized support vector machines, Journal of Biomedical Informatics 68 (2017) 132–149.

[47] H.R. Tizhoosh, Opposition-based learning: a new scheme for machine intelligence, in: International Conference on Computational Intelligence for Modelling, Control and Automation and International Conference on Intelligent Agents, Web Technologies and Internet Commerce (CIMCA-IAWTIC'06), vol. 1, 2005, pp. 695–701.

[48] I. Trivedi, P. Jangir, A. Kumar, N. Jangir, R. Totlani, A novel hybrid PSO-WOA algorithm for global numerical functions optimization, Advances in Computer and Computational Sciences 554 (2018) 53–60.

[49] B. Xing, W.J. Gao, Fruit fly optimization algorithm, in: Innovative Computational Intelligence: A Rough Guide to 134 Clever Algorithms, Springer, Cham, 2014, pp. 167–170.

[50] X.S. Yang, S. Deb, Cuckoo search: recent advances and applications, Neural Computing & Applications 24 (2014) 169–174.

[51] X.S. Yang, A.H. Gandomi, Bat algorithm: a novel approach for global engineering optimization, Engineering Computations 29 (5) (2012) 464–483.

[52] X.S. Yang, A. Slowik, Firefly algorithm, in: Swarm Intelligence Algorithms, CRC Press, 2020, pp. 163–174.

[53] B. Yin, C. Wang, F. Abza, New brain tumor classification method based on an improved version of whale optimization algorithm, Biomedical Signal Processing and Control 56 (2020) 101728.

[54] P. Yuan, C. Guo, Q. Zheng, J. Ding, Sidelobe suppression with constraint for MIMO radar via chaotic whale optimisation, Electronics Letters 54 (2018) 311–313.

[55] X. Zhang, S. Wen, Hybrid whale optimization algorithm with gathering strategies for high-dimensional problems, Expert Systems with Applications 179 (2021) 115032.

Chapter 7

Near-fault ground motion attenuation of large-scale steel structure by upgraded whale optimization algorithm

Mahdi Azizi[a], Mahla Basiri[a], and Milad Baghalzadeh Shishehgarkhaneh[b]
[a]*Department of Civil Engineering, University of Tabriz, Tabriz, Iran,* [b]*Department of Civil Engineering, Islamic Azad University of Tabriz, Tabriz, Iran*

7.1 Introduction

In recent years, dynamic hazard mitigation in civil engineering has had a long and successful history, thanks to structural control. Control systems are active control systems based on energy absorption or dissipation. As a potential method to safeguard bridges, buildings, and other structures from loads caused by earthquakes, wind, etc., structural control had its roots in flexible space structures and quickly expanded to civil engineering structures. The Fuzzy Logic Controller (FLC), introduced by Mamdani and Assilian [1], is based on fuzzy logic [2] which is the most commonly used control methods in active systems is regarding a mathematical system that utilizes logical variables in a continuous range between 0 and 1, as opposed to classical systems that utilize discrete variables of either 0 or 1. Zadeh [3] introduced the fuzzy theory in his book, followed by proposing fuzzy algorithms [4], decision-making with fuzzy [5], and fuzzy ordering [6]. Lee [7] outlined a comprehensive procedure for developing an FLC and evaluating its performance. Han-Xiong and Gatland [8] proposed a novel approach to making FLC, in which the time-response and rule basis are connected via a phase plane.

Nonetheless, optimizing FLC designs using metaheuristic algorithms has become very important in recent years. García-Gutiérrez, Arcos-Aviles [9] used the cuckoo search (CS) algorithm to identify the optimal FLC parameters for a specific number of applications by optimizing FLC parameters. Azizi and Talatahari [10] optimized the fuzzy controllers implemented in steel structures with nonlinear behavior using the improved arithmetic optimization algorithm (IAOA). Castillo, Valdez [11] employed FLC of the Takagi–Sugeno (TS) type in conjunction with differential evolution (DE) and harmony search (HS) metaheuristic methods to identify the best design in the membership functions (MFs). Qais, Hasanien [12] produced optimum Sugeno FLCs to increase the fault ride-through (FRT) capability of grid-connected wind power plants (WPPs) employing the whale optimization algorithm (WOA). Furthermore, a version of the shark smell optimization (VSSO) method was used by Cuevas, Castillo [13] to optimize a Mamdani interval type-2 fuzzy controller's design (IT2-FLC). Soliman, Hasanien [14] suggested the chaotic-billiards optimizer (C-BO) method to create the FLC approach for grid-integrated WPPs with enhanced stability. Chrouta, Chakchouk [15] used the Cuckoo Search Algorithm (CSA) in conjunction with a FLC based on the TS model to achieve an irrigation station process's optimal control.

The optimization problem of FLC has been mainly explored in situations where linear structural materials and insufficient mathematical models of structures were used for structural analysis. The primary goal of this chapter is to introduce an improved metaheuristic algorithm for design optimization of FLC, while intelligently controlling the seismic vibration of structural buildings under severe earthquake records. The Whale Optimization Algorithm (WOA) [16] is utilized as the indispensable optimization method for the mentioned aim. At the same time, the Upgraded Whale Optimization Algorithm (UWOA) [17] is employed as an improved version of this approach to obtain more outstanding performance in solving the FLC optimal design problem. This approach uses the "continuous-time" concept instead of "discrete-time" concept of the standard WOA in which after updating each search agent's position, the algorithm assesses the fitness values for each agent and updates the global optimal solution throughout the optimization phase. While the structure's nonlinearity is considered and the mathematical model of the building structure is correct, the proposed UWOA is used for the FLC's optimum tuning in a three-story steel structure. Based on several devastating Near-Fault Earthquakes (NFE) data, the dynamic analysis's seismic inputs are deemed. The chosen NFE recordings are used as seismic input for the three-story steel structure, and the UWOA's capacity is assessed concerning the ground movements' destructive impacts.

Handbook of Whale Optimization Algorithm. https://doi.org/10.1016/B978-0-32-395365-8.00013-0

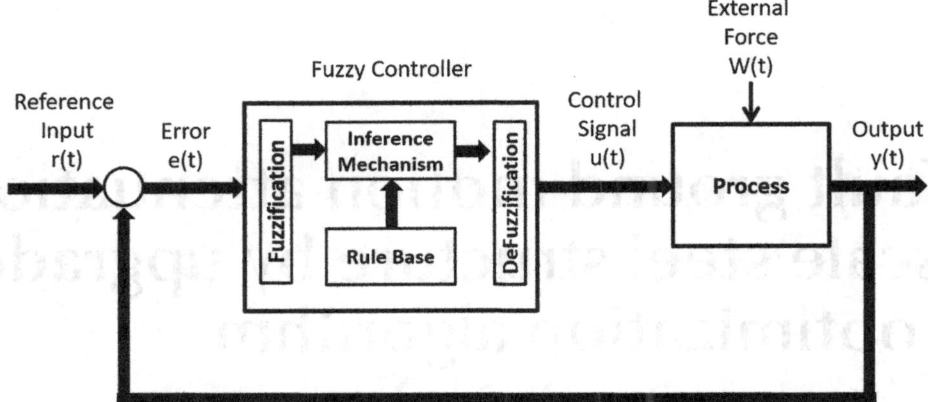

FIGURE 7.1 Fuzzy logic control system.

7.2 Fuzzy logic controller (FLC)

The goal of the fuzzy control was that computers learn to do it by witnessing the work of a human. The first experiment that was selected for the desired investigation for this purpose was the control of a steam engine sample. Qualitative information is communicated via verbal language (linguistic). Standard computers are incapable of analyzing qualitative data and human language. In order to integrate human knowledge with mathematical models, engineering systems require a framework that can organize human language or knowledge into a system. Fuzzy set theory is able to represent linguistic statements and approximate inferences, making it suitable for meeting these needs.

Many control methods commonly utilized are based on models, meaning that the design and adjustment of controllers rely on the mathematical representation of the system. Linear controllers are created concerning this model. Nonetheless, the mathematical model of the system is imprecise in some of situations, resulting in ineffective control techniques based on these models. In situations where the mathematical model is entirely unavailable, it's possible to design an effective control system using fuzzy logic based on expert knowledge. The FLC control algorithm is particularly valuable for managing the closed-loop of nonlinear systems where the mathematical models are not available. A closed-loop control system utilizing the FLC control algorithm is illustrated in Fig. 7.1 as an instance of such a system.

Fig. 7.1 illustrates the fuzzy control system, which comprises 3 key sections of a FLC as follows:

- Fuzzification;
- Fuzzy Rule base and Interfacing engine;
- Defuzzification.

The optimum membership functions and rule base cannot be derived from human knowledge and skill. One of the possible configurations for a controller is a fuzzy controller designed by a human. The current section introduces an FLC that relies on expert ability to develop an optimized FLC configuration for use in optimization issues. Eleven linguistic variables are employed to define a fuzzy space. The following variables are listed in Table 7.1.

The fuzzy controller is comprised of two input and one output variable. The output variable consists of eleven membership functions, while each input variable has eight. The input and output variables shown in Fig. 7.2 are anticipated that the membership functions will have triangle-shaped forms. In Table 7.2, the FLC rule is also included.

7.3 Optimization algorithms

7.3.1 Whale optimization algorithm (WOA)

There are 7 primary types of whales in nature, and they are all primarily considered predators. Since they must breathe ocean surface, they are unable to sleep. The other half of the brain is asleep. This fascinating capability makes them very intelligent and emotional creatures. The *humpback* whale could be deemed as one of the largest baleen whales in the world, and the most exciting feature of humpback whales is their unique hunting strategy. Their foraging style is known as bubble-net feeding behavior. As seen in Fig. 7.3, this foraging is accomplished by producing separate bubbles in a circular or "9"-shaped route. The formulation of WOA is regarding the mathematically modeled upward-spirals maneuver for use in the optimization process.

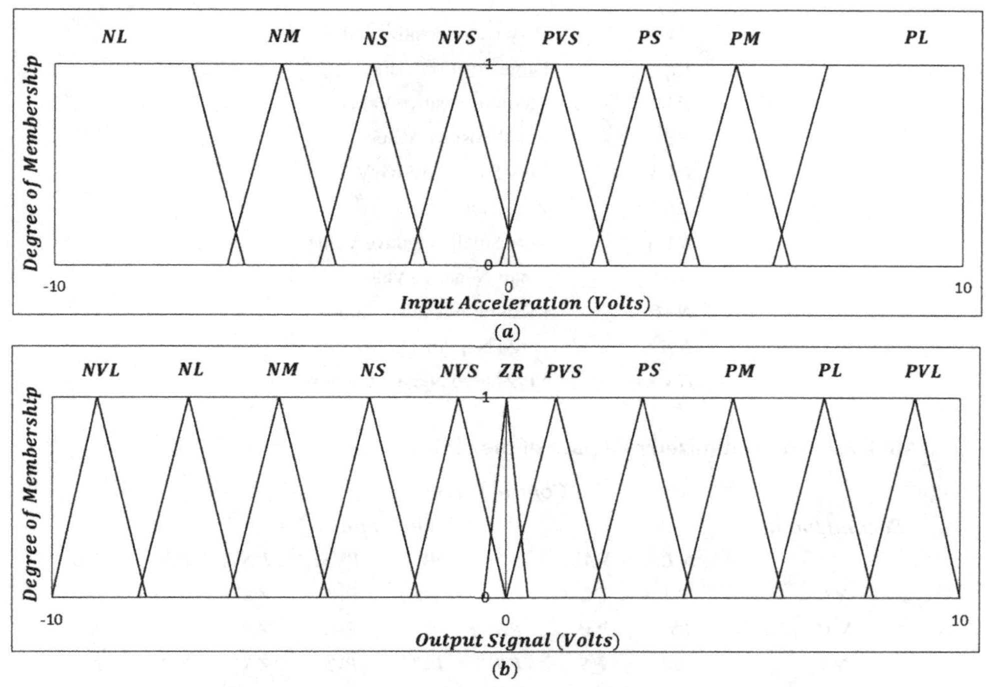

FIGURE 7.2 Various options for membership functions that can be used for the inputs (a) and the output (b) of a fuzzy logic controller (FLC).

FIGURE 7.3 Crating the spiral-shaped bubbles by humpback whales.

TABLE 7.1 Fuzzy linguistic variables.

PVL	Very Large Positive Value
PL	Large Positive Value
PM	Medium Positive Value
PS	Small Positive Value
PVS	Very Small Positive Value
ZR	Zero Value
NVS	Very Small Negative Value
NS	Small Negative Value
NM	Medium Negative Value
NL	Large Negative Value
NVL	Very Large Negative Value

TABLE 7.2 Non-optimized rule base of the FLC.

Control Force								
Second Input	*First Input*							
	NL	*NM*	*NS*	*NVS*	*PVS*	*PS*	*PM*	*PL*
NL	*PVL*	*PL*	*PM*	*PS*	*PVS*	*ZR*	*NVS*	*NS*
NM	*PL*	*PM*	*PS*	*PS*	*PVS*	*ZR*	*NVS*	*NS*
NS	*PM*	*PS*	*PS*	*PVS*	*PVS*	*ZR*	*NVS*	*NS*
NVS	*PM*	*PS*	*PVS*	*PVS*	*ZR*	*NVS*	*NS*	*NM*
PVS	*PM*	*PS*	*PVS*	*ZR*	*NVS*	*NVS*	*NS*	*NM*
PS	*PS*	*PVS*	*ZR*	*NVS*	*NVS*	*NS*	*NS*	*NM*
PM	*PS*	*PVS*	*ZR*	*NVS*	*NS*	*NS*	*NM*	*NL*
PL	*PS*	*PVS*	*ZR*	*NVS*	*NS*	*NM*	*NL*	*NVL*

Whales are known for their tendency to compass their prey and constantly update their location in order to find the most effective approach for hunting. In this sense, the target prey is deemed to be the best candidate solution at present. The following mathematical equations represent this approach:

$$\vec{D} = \left|\vec{C}.\vec{X}^*(t) - \vec{X}(t)\right| \tag{7.1}$$

$$\vec{X}(t+1) = \vec{X}^*(t) - \vec{A}.\vec{D} \tag{7.2}$$

$$\vec{A} = 2\vec{a}.\vec{r} - \vec{a} \tag{7.3}$$

$$\vec{C} = 2\vec{r} \tag{7.4}$$

where t shows the present iteration, $\vec{X}^*$ demonstrates the best solution's position vector, $\vec{X}$ shows the vector of position, $\vec{D}$ shows the current iteration's position vector, | | elucidates the absolute value, and . elucidates the element-by-element multiplication. In order to obtain an improved solution, the value of $\vec{X}^*$ needs to be updated at every iteration. $\vec{A}$ and $\vec{C}$ are vectors of coefficients, where $\vec{a}$ demonstrates a variable that linearly reduces from 2 to 0 throughout course of the iterations. Additionally, $\vec{R}$ shows a random vector with values between 0 and 1.

The bubble-net behavior of humpback whales is thought to be regulated by a mechanism for shrinking and circling, as well as a position-updating spiral. The process of encircling and shrinking is accomplished by decreasing the value of variable 'a' within the range of $\vec{A}$, as shown in Eq. (7.3). When the value of $\vec{A}$ decreases, it indicates that the updated positions of search agents may be located anywhere between the present best agent and the original position of the search agent. Fig. 7.4.a shows the various locations from (X, Y) to (X^*, Y^*) produced by $0 \leq A \leq 1$.

To create a helix-like movement similar to humpback whales, the Spiral updating position (illustrated in Fig. 7.4.b) calculates the distance between the prey at (X^*, Y^*) and the whales at (X, Y) using an equation in spiral form that establishes

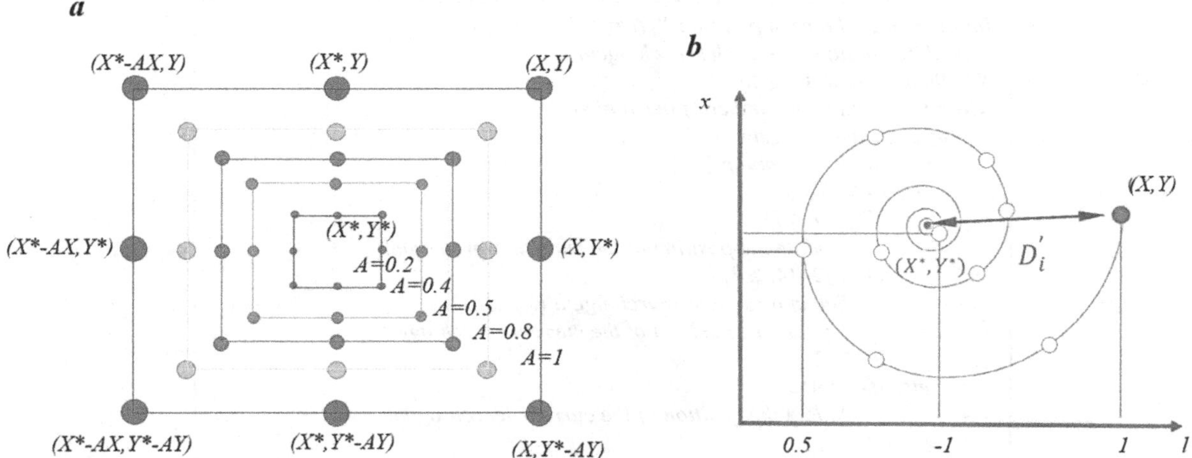

FIGURE 7.4 The WOA algorithm employs two strategies for the Bubble-net search mechanism: (a) shrinking encircling mechanism, and (b) spiral updating position.

their relationship.

$$\vec{D}' = \left|\vec{X}^*(t) - \vec{X}(t)\right| \tag{7.5}$$

$$\vec{X}(t+1) = \vec{D}'.e^{bl}.(\cos 2\pi l) + \vec{X}^*(t) \tag{7.6}$$

where $\vec{D}'$ shows the current iteration's position vector, the variable b represents a constant value that is used to model the logarithmic spiral shape, while l is a random vector that is uniformly distributed within the range of -1 to 1.

The probability of using Bubble-net searches for updating the location of the humpback whales while optimization is 50%. The mathematical model method is represented by the below equations, with p denoting the probability of each circling path:

$$\vec{X}(t+1) = \begin{cases} \vec{X}^*(t) - \vec{A}.\vec{D} & \text{if } p < 0.5 \\ \vec{D}'.e^{bl}.(\cos 2\pi l) + \vec{X}^*(t) & \text{if } p \geq 0.5 \end{cases} \tag{7.7}$$

The objective of a global search is to explore a wider search space, so the search agents update their positions regarding randomly chosen search agents rather than the current best search agent found. To achieve this, random values greater than one for $\vec{A}$ are considered. The mathematical model for this procedure is given by the following equations:

$$\vec{D} = \left|\vec{C}.\vec{X}_{rand} - \vec{X}(t)\right| \tag{7.8}$$

$$\vec{X}(t+1) = \vec{X}_{rand} - \vec{A}.\vec{D} \tag{7.9}$$

where $\vec{X}_{rand}$ shows a random whale or a random position vector selected from the present population. Fig. 7.5 demonstrates the pseudo-code of the WOA.

7.3.2 Upgraded WOA (UWOA)

In the WOA, the fitness value is evaluated after updating all search agents' position, and $\vec{X}^*$ is identified as the best solution up to that point. This evaluation is done at discrete time intervals called "iterations" in optimization issues. In the "continuous-time" approach, time changes constantly, and all updates are carried out once a single solution is generated. In this approach, the movement of each agent's new position affects the subsequent agents, whereas in the "discrete-time" approach, the novel positions are not used until an iteration is terminated.

The "continuous-time" notion is used to improve the performance of the conventional WOA while addressing FLC optimization problems. Following the update of each search agent's position, the UWOA evaluates each agent' fitness values and updates the optimal solution ($\vec{X}^*$). The pseudo-code for the UWOA can be found in Fig. 7.6.

```
Initialize the whales population Xi (i = 1, 2, ..., n)
Calculate the fitness of each search agent
X*=the best search agent
while (t < maximum number of iterations)
    for each search agent
    Update a, A, C, l, and p
        if1 (p<0.5)
            if2 (|A| < 1)
                Update the position of the current search agent
            else if2 (|A| ≥ 1)
                Select a random search agent (Xrand)
                Update the position of the current search agent
            end if2
        else if1 (p≥0.5)
                Update the position of the current search agent
        end if1
    end for
    Check if any search agent goes beyond the search space and amend it
    Calculate the fitness of each search agent
    Update X* if there is a better solution
    t=t+1
end while
return X*
```

FIGURE 7.5 Pseudo-code of the WOA.

```
Create random values for initial position of whales Xi (i = 1, 2, ..., n)
Evaluate the fitness value for each search agent
X*=the so far found best search agent
while (t < maximum number of iterations)
    for each search agent
    Update a, A, C, l, and p
        if1 (p<0.5)
            if2 (|A| < 1)
                Update the position of the current search agent
            else if2 (|A|≥1)
                Select a search agent randomly (Xrand)
                Update the position of the current search agent
            end if2
        else if1 (p≥0.5)
                Update the position of the current search
        end if1
    Control the position constraints for search agents and amend it
    Evaluate the fitness value for each search agent
    Update X* if a better solution is found
    end for
    t=t+1
end while
return X*
```

FIGURE 7.6 Pseudo-code of the UWOA.

The computational complexity of the UWOA is similar to WOA in the initialization stage. However, the complexity of the upgraded approach is different than the standard one in the primary loop of the method while the global best solution update is conducted after objective function evaluation process of each search agent. In other words, the big O notation declares a computational complexity of O(MaxIter×NSA) for the UWOA while NSA is the number of search agents and

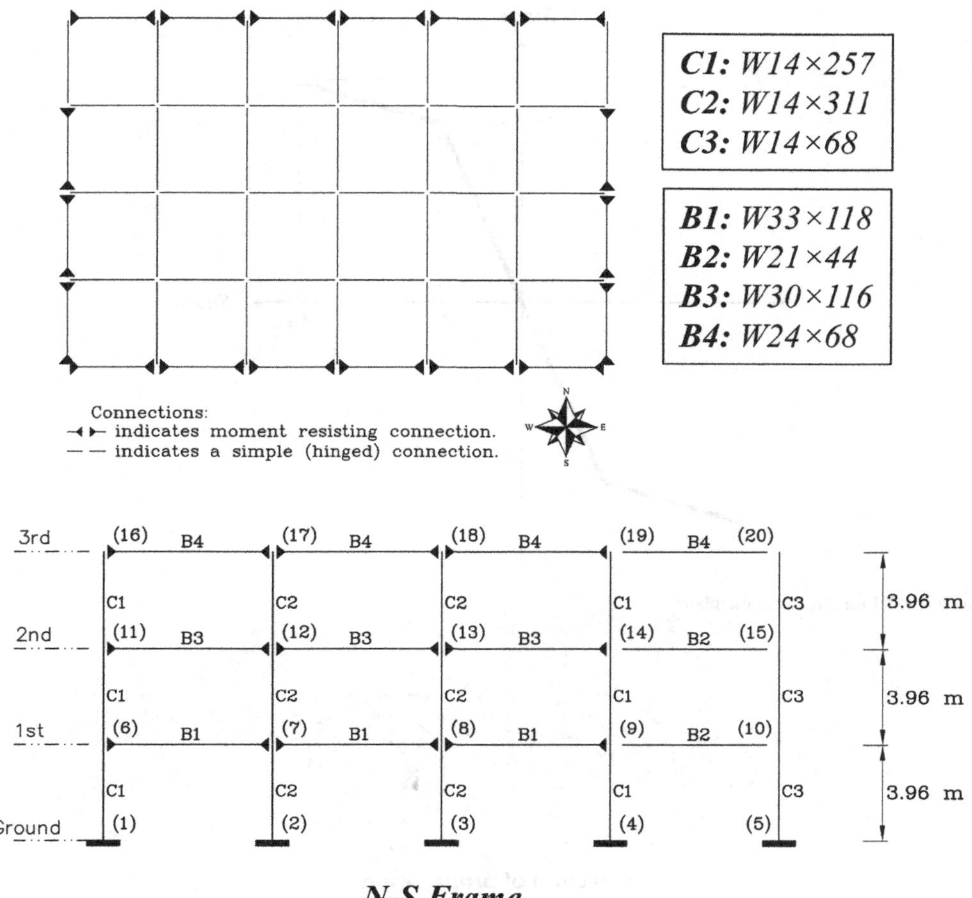

FIGURE 7.7 Benchmark design example with three stories.

MaxIter is the overall number of iterations. However, the complexity of the WAO is as O(MaxIter) so the computational complexity of the UWOA is somehow higher than the WOA.

7.4 Design example

The three-story benchmark building [18] is chosen as the numerical example in this chapter, as shown in Fig. 7.7. The steel perimeter moment-resisting frames make up this structure's lateral load-resisting system (MRFs). The building's inner bays have simple framing and composite flooring. The steel frame, floor slabs, roof, flooring, mechanical and electrical systems, as well as partitions and flooring, all contribute to the building's seismic mass. Because of the nonlinear nature of the materials and structural sections, structural members are susceptible to yielding during strong earthquakes. In this case, the structure's nonlinear behavior may vary significantly from a linear approximation. In the considered structure, the floor to floor heights are 3.96 m, the bay width is 9.15 m, the seismic masses in the 1st and 2nd floors are 9.57×105 while for the third level, the mass is considered as 2.95×105.

A bilinear hysteresis model is utilized to simulate the plastic hinges (yielding points) of benchmark buildings' structural components to depict the nonlinear behavior, as illustrated in Fig. 7.8.

7.4.1 Near-fault ground motion

The near-fault zone should be 20 to 60 kilometers away from a ruptured fault. The properties of ground motions in a given region are influenced by several factors such as the rupture mechanism, rupture direction, and continuous movement of the ground along the fault, known as "rupture-directivity" and "fling step" in this chapter. Multiple considerations are necessary when analyzing ground motions near an active fault. Following an earthquake, a shear rupture starts at a specific location

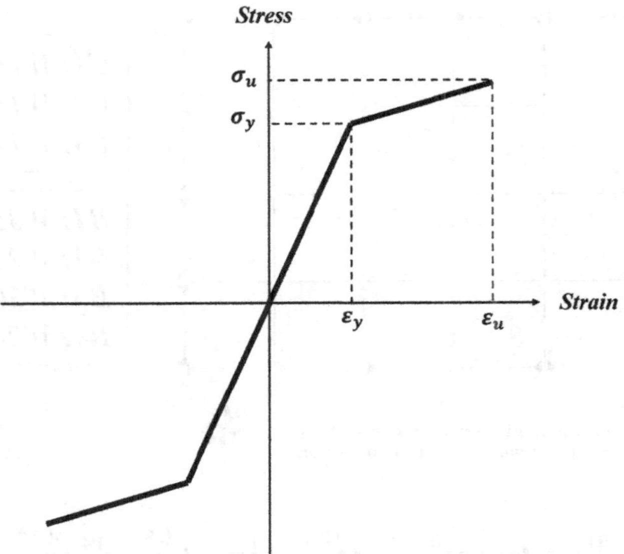

FIGURE 7.8 Bilinear model for structural members.

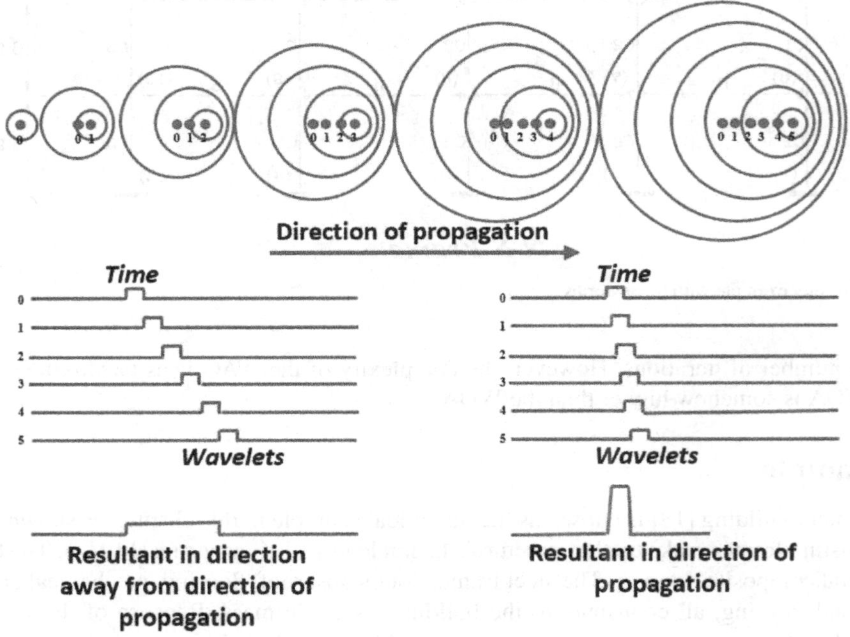

FIGURE 7.9 The direction in which fault rupture propagates at various locations.

on the fault and spreads at a higher speed than the shear wave velocity. Forward-directionality refers to the circumstance in which a rupture propagates toward one specific area. In this instance, the direction of the slide on the fault is likewise toward the site, and the majority of seismic energy comes in a single, enormous pulse of motion representing the cumulative consequences of the fault rupture. "Backwards-directivity" happens when a rupture propagates away from the site. In this circumstance, the duration and amplitude of ground motions are relatively lengthy and small [19]. These two features are shown in Fig. 7.9, with "site 5" displaying forward directivity and "site 0" demonstrating backward directivity [20].

Ground motion recordings near active faults often exhibit acceleration spikes, resulting in short-duration pulses with higher Peak Ground Acceleration (PGA) values. However, some pulses have longer durations and lower PGA values [19]. In addition, some near-fault earthquake (NFE) recordings exhibit impulsive ground motions that have lower frequencies

ID No.	Site Data		Source (Fault Type)	Site-Source Distance (km)			
	NEHRP Class	Vs_30 (m/sec)		Epicentral	Closest to Plane	Campbell	JoynerBoore
Pulse Records Subset							
1	D	203	Strike-slip	27.5	1.4	3.5	0.0
2	D	211	Strike-slip	27.6	0.6	3.6	0.6
3	B	1000	Normal	30.4	10.8	10.8	6.8
4	D	349	Strike-slip	16.0	1.0	3.5	1.0
5	C	371	Strike-slip	27.2	8.5	8.5	7.6
6	D	275	Strike-slip	9.0	4.4	4.4	0.0
7	C	713	Thrust	4.5	8.2	8.2	0.0
8	C	685	Strike-slip	44.0	2.2	3.7	2.2
9	D	282	Thrust	10.9	6.5	6.5	0.0
10	C	441	Thrust	16.8	5.3	5.3	1.7
11	B	811	Strike-slip	5.3	7.2	7.4	3.6
12	D	306	Thrust	26.7	0.6	6.7	0.6
13	C	714	Thrust	45.6	1.5	7.7	1.5
14	D	276	Strike-slip	1.6	6.6	6.6	0.0
No Pulse Records subset							
15	C	660	Thrust	12.8	5.5	5.5	3.9
16	D	223	Strike-slip	6.2	2.7	4.0	0.5
17	D	275	Strike-slip	18.9	7.3	8.4	7.3
18	C	660	Thrust	6.8	9.6	9.6	2.5
19	C	660	Thrust	6.5	4.9	4.9	0.0
20	C	376	Strike-slip	9.0	10.7	10.7	3.9
21	C	462	Strike-slip	7.2	3.9	3.9	0.2
22	C	514	Thrust	10.4	7.0	7.0	0.0
23	C	380	Thrust	8.5	8.4	8.4	0.0
24	D	281	Thrust	3.4	12.1	12.1	0.0
25	D	297	Strike-slip	19.3	4.8	5.3	1.4
26	C	434	Thrust	28.7	0.6	6.5	0.6
27	C	553	Thrust	8.9	11.2	11.2	0.0
28	C	553	Strike-slip	7.0	8.9	8.9	0.0

ID No.	Earthquake			Recording Station	
	M	Year	Name	Name	Owner
Pulse Records Subset					
1	6.5	1979	Imperial Valley-06	El Centro Array #6	CDMG
2	6.5	1979	Imperial Valley-06	El Centro Array #7	USGS
3	6.9	1980	Irpinia, Italy-01	Sturno	ENEL
4	6.5	1987	Superstition Hills-02	Parachute Test Site	USGS
5	6.9	1989	Loma Prieta	Saratoga - Aloha	CDMG
6	6.7	1992	Erzican, Turkey	Erzincan	--
7	7.0	1992	Cape Mendocino	Petrolia	CDMG
8	7.3	1992	Landers	Lucerne	SCE
9	6.7	1994	Northridge-01	Rinaldi Receiving Sta	DWP
10	6.7	1994	Northridge-01	Sylmar - Olive View	CDMG
11	7.5	1999	Kocaeli, Turkey	Izmit	ERD
12	7.6	1999	Chi-Chi, Taiwan	TCU065	CWB
13	7.6	1999	Chi-Chi, Taiwan	TCU102	CWB
14	7.1	1999	Duzce, Turkey	Duzce	ERD
No Pulse Records Subset					
15	6.8	6.8	Gazli, USSR	Karakyr	--
16	6.5	1979	Imperial Valley-06	Bonds Corner	USGS
17	6.5	1979	Imperial Valley-06	Chihuahua	UNAMUCSD
18	6.8	1985	Nahanni, Canada	Site 1	--
19	6.8	1985	Nahanni, Canada	Site 2	--
20	6.9	1989	Loma Prieta	BRAN	UCSC
21	6.9	1989	Loma Prieta	Corralitos	CDMG
22	7.0	1992	Cape Mendocino	Cape Mendocino	CDMG
23	6.7	1994	Northridge-01	LA - Sepulveda VA	USGS/VA
24	6.7	1994	Northridge-01	Northridge - Saticoy	USC
25	7.5	1999	Kocaeli, Turkey	Yarimca	KOERI
26	7.6	1999	Chi-Chi, Taiwan	TCU067	CWB
27	7.6	1999	Chi-Chi, Taiwan	TCU084	CWB
28	7.9	2002	Denali, Alaska	TAPS Pump Sta. #10	CWB

FIGURE 7.10 Characteristics of FEMA records of earthquakes near the fault.

in the velocity time history and a lower ratio of peak ground acceleration (PGA) to peak ground velocity (PGV) [21]. In this chapter, the Appendix A of the FEMA P695/2009 [22] directive is selected in this respect. In this code, 28 records of near-fault earthquakes are categorized and presented in two categories: records with pulse period (14 records) and records without pulse period (14 records). The specifications of these records are shown in Fig. 7.10, which are directly adapted from the above-mentioned code.

Under the title validation scenario, a new optimization scenario is developed in which we use the metaheuristic optimization method to optimize the fuzzy control system utilizing the 28 data stated. In this respect, a three-story building was chosen so that 28 seismic data could be included in the formulation of the objective function. 28 records with maximum accelerations mentioned in Fig. 7.11 will be used.

7.4.2 FLC implementation

To apply the FLC as a control method, control devices and sensors should always be connected to the structure regarding supplied specifics of fuzzy logic controllers. In the 3-story building, control activities are carried out using 3 sensors and 3 actuators. Specifically, each of the three stories is equipped with three accelerometers for sensing, and there are three fuzzy chips implemented across the three floors to control the actions of the actuators. The maximum control force that the structure's actuators can produce is 1000 kN. Fig. 7.12 depicts the main control scheme for the 3-story building and the fuzzy control scheme for the three-story structure.

ID No.	PEER-NGA Record Information				Recorded Motions	
	Record Seq. No.	Lowest Freq (Hz.)	File Names - Horizontal Records: FN Component	File Names - Horizontal Records: FP Component	PGA_{max} (g)	PGV_{max} (cm/s.)
Pulse Records Subset						
1	181	0.13	IMPVALL/H-E06_233	IMPVALL/H-E06_323	0.44	111.9
2	182	0.13	IMPVALL/H-E07_233	IMPVALL/H-E07_323	0.46	108.9
3	292	0.16	ITALY/A-STU_223	ITALY/A-STU_313	0.31	45.5
4	723	0.15	SUPERST/B-PTS_037	SUPERST/B-PTS_127	0.42	106.8
5	802	0.13	LOMAP/STG_038	LOMAP/STG_128	0.38	55.6
6	821	0.13	ERZIKAN/ERZ_032	ERZIKAN/ERZ_122	0.49	95.5
7	828	0.07	CAPEMEND/PET_260	CAPEMEND/PET_350	0.63	82.1
8	879	0.10	LANDERS/LCN_239	LANDERS/LCN_329	0.79	140.3
9	1063	0.11	NORTHR/RRS_032	NORTHR/RRS_122	0.87	167.3
10	1086	0.12	NORTHR/SYL_032	NORTHR/SYL_122	0.73	122.8
11	1165	0.13	KOCAELI/IZT_180	KOCAELI/IZT_270	0.22	29.8
12	1503	0.08	CHICHI/TCU065_272	CHICHI/TCU065_002	0.82	127.7
13	1529	0.06	CHICHI/TCU102_278	CHICHI/TCU102_008	0.29	106.6
14	1605	0.10	DUZCE/DZC_172	DUZCE/DZC_262	0.52	79.3
No Pulse Records Subset						
15	126	0.06	GAZLI/GAZ_177	GAZLI/GAZ_267	0.71	71.2
16	160	0.13	IMPVALL/H-BCR_233	IMPVALL/H-BCR_323	0.76	44.3
17	165	0.06	IMPVALL/H-CHI_233	IMPVALL/H-CHI_323	0.28	30.5
18	495	0.06	NAHANNI/S1_070	NAHANNI/S1_160	1.18	43.9
19	496	0.13	NAHANNI/S2_070	NAHANNI/S2_160	0.45	34.7
20	741	0.13	LOMAP/BRN_038	LOMAP/BRN_128	0.64	55.9
21	753	0.25	LOMAP/CLS_038	LOMAP/CLS_128	0.51	45.5
22	825	0.07	CAPEMEND/CPM_260	CAPEMEND/CPM_350	1.43	119.5
23	1004	0.12	NORTHR/0637_032	NORTHR/0637_122	0.73	70.1
24	1048	0.13	NORTHR/STC_032	NORTHR/STC_122	0.42	53.2
25	1176	0.09	KOCAELI/YPT_180	KOCAELI/YPT_270	0.31	73.0
26	1504	0.04	CHICHI/TCU067_285	CHICHI/TCU067_015	0.56	91.8
27	1517	0.25	CHICHI/TCU084_271	CHICHI/TCU084_001	1.16	115.1
28	2114	0.03	DENALI/ps10_199	DENALI/ps10_289	0.33	126.4

ID No.	As-Recorded Parameters			Normalization Factor	Normalized Motions	
	1-Sec.Spec. Acc. (g): FN Comp.	1-Sec.Spec. Acc. (g): FP Comp.	PGV_{PEER} (cm/s.)		PGA_{max} (g)	PGV_{max} (cm/s.)
Pulse Records Subset						
1	0.43	0.60	83.9	0.90	0.40	100.1
2	0.66	0.64	78.3	0.96	0.44	104.4
3	0.25	0.41	43.7	1.72	0.53	78.2
4	0.97	0.51	71.9	1.04	0.44	111.6
5	0.47	0.32	46.1	1.63	0.62	90.6
6	0.98	0.37	68.8	1.09	0.53	104.2
7	0.92	0.70	69.6	1.08	0.68	88.6
8	0.43	0.34	97.2	0.77	0.62	108.4
9	1.96	0.47	109.3	0.69	0.59	114.9
10	0.89	0.65	94.4	0.80	0.58	97.7
11	0.29	0.28	26.9	2.79	0.62	83.2
12	1.33	1.10	101.6	0.74	0.60	94.4
13	0.60	0.58	87.5	0.86	0.25	91.5
14	0.54	0.73	69.6	1.08	0.56	85.6
No Pulse Records Subset						
15	0.81	0.42	65.0	0.86	0.61	61.4
16	0.44	0.44	49.8	1.13	0.86	49.8
17	0.41	0.37	28.2	1.99	0.56	60.8
18	0.53	0.29	44.1	1.27	1.50	55.9
19	0.16	0.29	28.7	1.95	0.87	67.8
20	0.55	0.45	49.0	1.15	0.73	64.0
21	0.53	0.50	47.9	1.17	0.60	53.3
22	0.42	0.73	84.4	0.66	0.95	79.4
23	0.62	1.00	72.6	0.77	0.56	54.2
24	0.81	0.40	47.7	1.18	0.50	62.6
25	0.38	0.35	62.4	0.90	0.28	65.6
26	0.75	0.75	72.3	0.78	0.44	71.3
27	2.54	0.86	90.3	0.62	0.72	71.5
28	0.69	0.82	98.5	0.57	0.19	72.0

FIGURE 7.11 Other characteristics of the FEMA earthquakes near the fault.

7.4.3 Performance criteria

The fuzzy control method's performance used is evaluated based on the building's Performance Criteria (PC). To evaluate these criteria, the controlled responses' ratio by the FLC to the uncontrolled responses is calculated, which is then used as a measure of performance. The performance criteria are categorized into building responses, damage, and control device requirements, and an overview is provided in Table 7.3. The features are broken down to two groups. The first group includes the building responses and has 3 criteria: peak inter-story drift ratio (PC1), peak story acceleration (PC2), and peak base shear (PC3). The 2nd group relates to building damage and has 3 criteria as well. The ductility factor (PC4) represents the maximum curvature at the ends of the structural components during earthquakes. The second group of criteria includes two additional measures: the dissipated energy at the ends of the structural components (PC5) and the ratio of plastic hinges supported by the structure (PC6). The third group of criteria is related to the requirements of the control devices and comprises three criteria: the maximum control force (PC7), the stroke of the control device (PC8), and the control power (PC9). A detailed presentation of the considered PC can be found at Ref. [16].

7.5 Statement of the optimization problem

The optimization issue involves finding the best solution among a set of possible options. Typically, an optimization issue could be formulated as minimizing or maximizing an objective function, subject to a set of constraints. The general form of an optimization problem can be defined as follows:

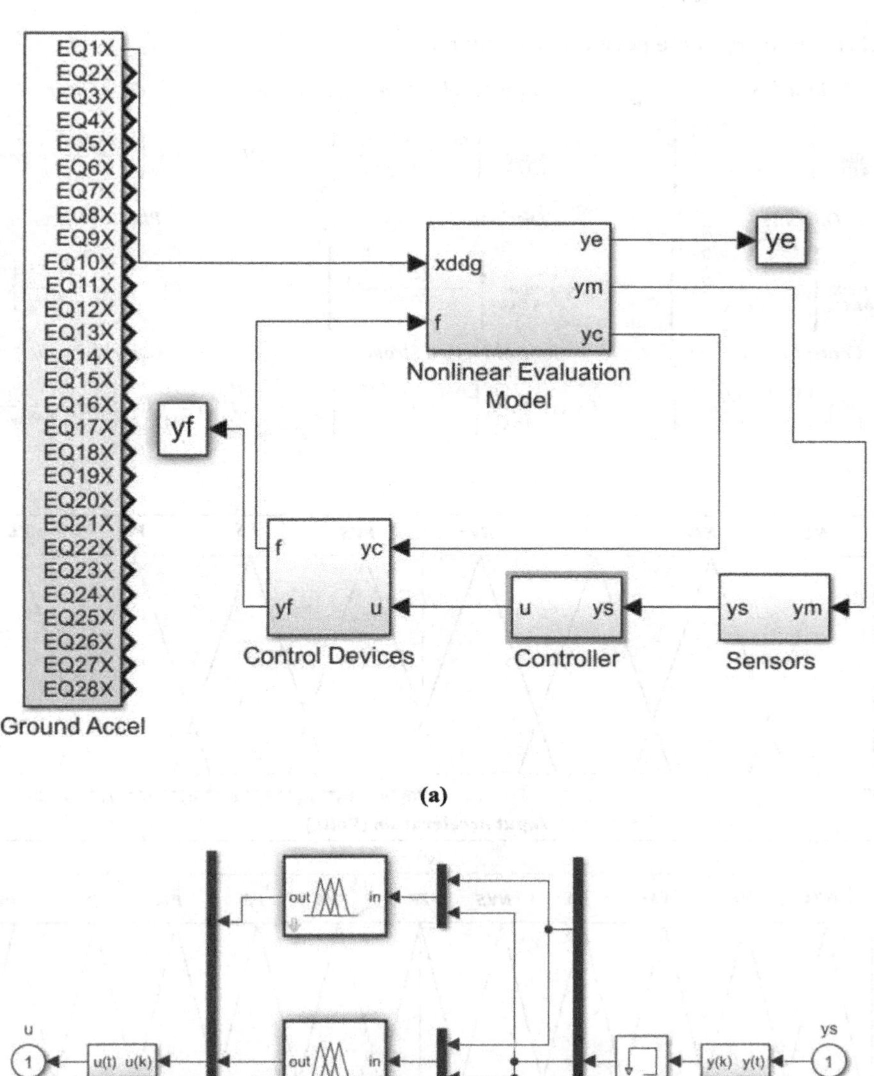

(a)

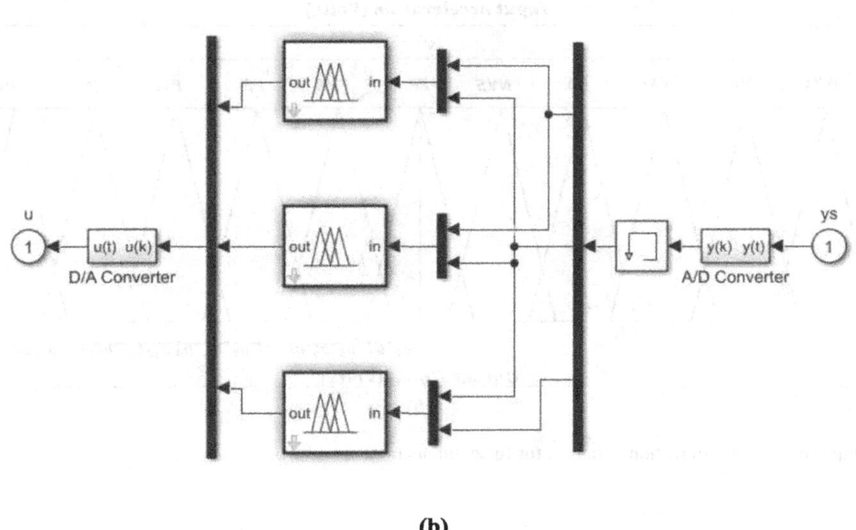

(b)

FIGURE 7.12 (a) Main control scheme for the 3-story building. (b) Fuzzy control scheme.

- A function $f : B \to R$ from some set B to the actual numbers.
- An element $x_0 \in B$ such that $f(x_0) \leq f(x)$ for every $x \in B$ (minimization problem) or such that $f(x_0) \geq f(x)$ for every $x \in B$ (maximization problem).

Generally speaking, B is a subset of the Euclidean space, often characterized by equality, inequality, or constraint requirements that B's members must meet. The domain B of f is referred to as the search space, whereas the parts of B are referred to as potential solutions or candidate solutions. f shows a function referred to as an "objective function". An

TABLE 7.3 Summary of the performance criteria.

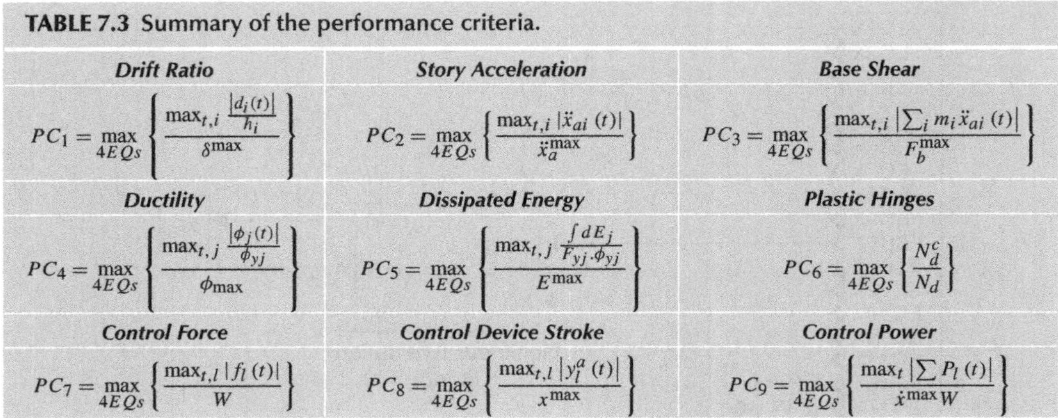

Drift Ratio	Story Acceleration	Base Shear
$PC_1 = \max\limits_{4EQs}\left\{\dfrac{\max_{t,i}\frac{\lvert d_i(t)\rvert}{h_i}}{\delta^{\max}}\right\}$	$PC_2 = \max\limits_{4EQs}\left\{\dfrac{\max_{t,i}\lvert \ddot{x}_{ai}(t)\rvert}{\ddot{x}_a^{\max}}\right\}$	$PC_3 = \max\limits_{4EQs}\left\{\dfrac{\max_{t,i}\lvert\sum_i m_i\ddot{x}_{ai}(t)\rvert}{F_b^{\max}}\right\}$
Ductility	**Dissipated Energy**	**Plastic Hinges**
$PC_4 = \max\limits_{4EQs}\left\{\dfrac{\max_{t,j}\frac{\lvert\phi_j(t)\rvert}{\phi_{yj}}}{\phi_{\max}}\right\}$	$PC_5 = \max\limits_{4EQs}\left\{\dfrac{\max_{t,j}\frac{\int dE_j}{F_{yj}\cdot\phi_{yj}}}{E^{\max}}\right\}$	$PC_6 = \max\limits_{4EQs}\left\{\dfrac{N_d^c}{N_d}\right\}$
Control Force	**Control Device Stroke**	**Control Power**
$PC_7 = \max\limits_{4EQs}\left\{\dfrac{\max_{t,l}\lvert f_l(t)\rvert}{W}\right\}$	$PC_8 = \max\limits_{4EQs}\left\{\dfrac{\max_{t,l}\lvert y_l^a(t)\rvert}{x^{\max}}\right\}$	$PC_9 = \max\limits_{4EQs}\left\{\dfrac{\max_t\lvert\sum P_l(t)\rvert}{\dot{x}^{\max}W}\right\}$

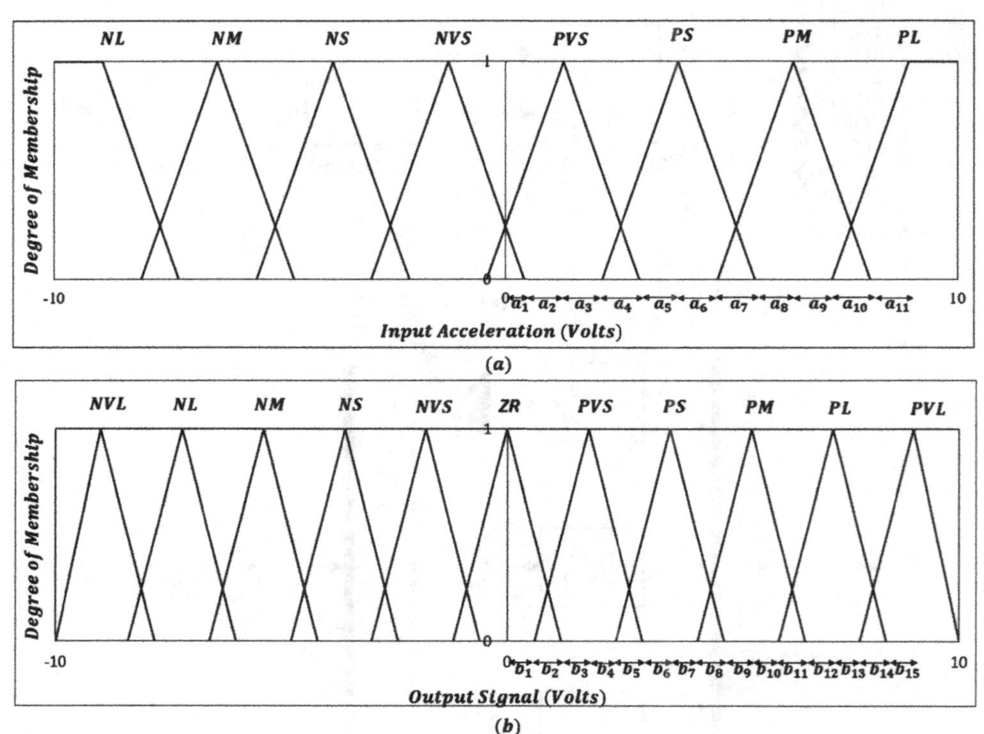

FIGURE 7.13 Membership functions' optimization variables for fuzzy inputs (a) and fuzzy output (b).

optimal solution is a potential solution that minimizes (or maximizes) the objective function. In structural control design, optimization problems are commonly formulated to minimize certain objectives.

This chapter utilizes the UWOA method to select variables regarding the membership functions' configuration and rule base for the optimal tuning of FLC parameters, which is considered the optimization problem. The optimization variables consist of a set of parameters, denoted as $a_1, a_2, \ldots, a_{11}$, which correspond to the first and second inputs of the FLC, as shown in Fig. 7.13a. In addition, the optimization variables for the FLC output are $b_1, b_2, \ldots, b_{15}$, as illustrated in Fig. 7.13b. Note that for the inputs and outputs of the FLC, it is assumed that the membership functions are symmetric. Table 7.4 presents the variables $c_1, c_2, \ldots, c_{64}$ for the FLC's rule basis. Note that the rule base fractures relate to the weight of each rule that the UWOA might optimize.

TABLE 7.4 The fuzzy rule base's variables.

Second Input	Control Force							
	First Input							
	NL	*NM*	*NS*	*NVS*	*PVS*	*PS*	*PM*	*PL*
NL	PVL/c1	PL/c9	PM/c17	PS/c25	PVS/c33	ZR/c41	NVS/c49	NS/c57
NM	PL/c2	PM/c10	PS/c18	PS/c26	PVS/c34	ZR/c42	NVS/c50	NS/c58
NS	PM/c3	PS/c11	PS/c19	PVS/c27	PVS/c35	ZR/c43	NVS/c51	NS/c59
NVS	PM/c4	PS/c12	PVS/c20	PVS/c28	ZR/c36	NVS/c44	NS/c52	NM/c60
PVS	PM/c5	PS/c13	PVS/c21	ZR/c29	NVS/c37	NVS/c45	NS/c53	NM/c61
PS	PS/c6	PVS/c14	ZR/c22	NVS/c30	NVS/c38	NS/c46	NS/c54	NM/c62
PM	PS/c7	PVS/c15	ZR/c23	NVS/c31	NS/c39	NS/c47	NM/c55	NL/c63
PL	PS/c8	PVS/c16	ZR/c24	NVS/c32	NS/c40	NM/c48	NL/c56	NVL/c64

The optimization issue is regarded as a single problem in this chapter, and the objective function is expressed using the structure's nonlinear responses. The objective function is in the following general state:

$$Obj = \frac{\sum_{i=1}^{n} P_i \frac{CR_i}{UR_i}}{\sum_{i=1}^{n} P_i} \tag{7.10}$$

In this equation, the objective function' weighting coefficient is denoted by P_i, which represents the absolute peak acceleration of the chosen earthquakes. The summation of P_i is calculated for all chosen earthquakes. The terms UR_i and CR_i represent the building's uncontrolled and controlled responses, respectively. The objective function is formulated using Eq. (7.11), where J_1 represents the maximum relative displacement (drift) of the floors. This criterion is employed to assess the controller's performance.

$$Obj = \frac{PGA_{EQ_1} * (J_1)_{EQ_1} + PGA_{EQ_2} * (J_1)_{EQ_2} + \cdots + PGA_{EQ_{28}} * (J_1)_{EQ_{28}}}{(PGA_{EQ_1} + PGA_{EQ_2} + \cdots + PGA_{EQ_{28}})} \tag{7.11}$$

The perpendicular component along the fault will be employed for the dynamic analysis of the three-story structure since the numerical investigation was only done in one direction of the structure plan.

7.6 Numerical results

In the current part, the progress of convergence for the UWOA and WOA is displayed concerning the selected objective function. Furthermore, using the UWOA approach, the performance criteria are identified. To provide an accurate assessment of the performance of the improved algorithm, it is necessary to compare the convergence history of UWOA and WOA. The convergence history regarding the objective function of Eq. (7.11) considering the three-story structure for both of the mentioned approaches is shown in Fig. 7.14. The calculated value of 0.89016 is the lowest value for the objective function found by the UWOA while the WOA calculated 0.89396.

The results of optimizing the fuzzy system in the three-story structure using the UWOA approach to minimize the maximum drift of the floors (J_1) are shown in Tables 7.5 and 7.6. The 28 distinct earthquakes evaluation criteria are listed individually for comparison reasons. It can be seen that the UWOA can reduce the drift of the 3-story building in 23 of the considered near-fault seismic records while in the other 5 records, the structural response is increased slightly. The maximum reduction in drift ratio of the structure is for the 27th record by reducing up to 55% of the structural response.

Fig. 7.15 depicts the displacement' time history diagram of the roof floor in a three-story building for the 28 records described in two states—uncontrolled and optimally controlled—so that a more thorough analysis of the optimal system could well be conducted. It can be seen that the optimum control scheme can reduce the roof displacement of the structure while it is also able to reduce the nonlinear responses of the structure in most of the considered records. In other words, Considering the fact that the range of vibrations in the 3-story structure has diminished for most of the selected records, by referring to the time history of the displacement of the roof floor of the above structure, it can be seen that the permanent displacement of the structural system in the case of using FLC has decreased significantly.

Table 7.7 provides the maximum displacement ratio between the roof floor in the controlled state and the uncontrolled state. It is clear that for 22 records, the maximum displacement of the roof of the three-story building has dramatically

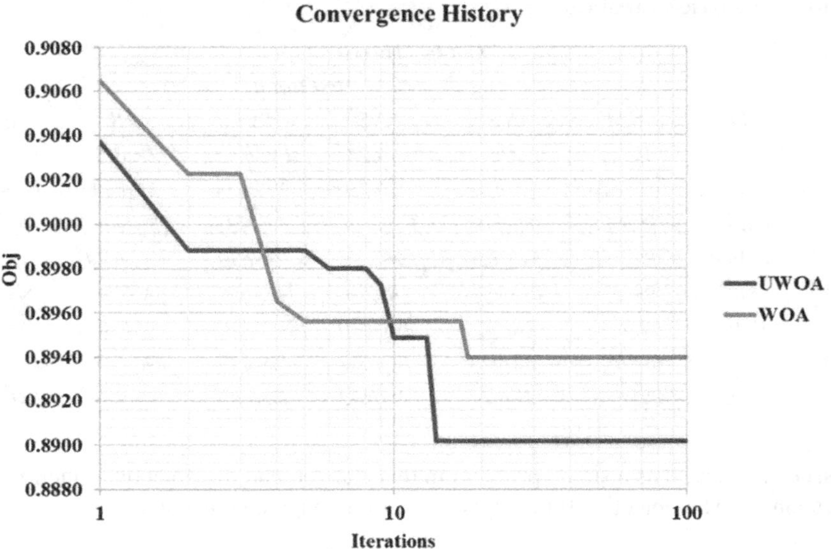

FIGURE 7.14 The convergence history for the 3-story building obtained by the UWOA and WOA.

TABLE 7.5 The results of the optimization of the fuzzy system in a 3-story structure using the UWOA algorithm.

	Earthquakes – FN Components													
EC	*1*	*2*	*3*	*4*	*5*	*6*	*7*	*8*	*9*	*10*	*11*	*12*	*13*	*14*
PC_1	***0.9455***	***0.9238***	***1.0819***	***0.7831***	***0.8882***	***0.9572***	***1.0086***	***0.6744***	***1.0009***	***0.8896***	***0.9097***	***0.7227***	***0.7468***	***0.9434***
PC_2	*1.1356*	*1.0974*	*1.2359*	*1.1553*	*0.9912*	*1.1611*	*1.0105*	*1.2451*	*1.1874*	*1.1069*	*0.9949*	*0.9899*	*1.0806*	*0.9959*
PC_3	*1.2100*	*1.0886*	*1.2566*	*1.1237*	*1.1148*	*1.2328*	*1.0951*	*1.1720*	*1.0902*	*1.1592*	*1.1815*	*1.1577*	*1.1318*	*1.1351*
PC_4	*0.9695*	*0.9087*	*1.0997*	*0.7922*	*0.9303*	*0.9968*	*0.8638*	*0.6405*	*1.0468*	*0.8812*	*0.8647*	*0.8352*	*0.7593*	*0.9820*
PC_5	*0.7934*	*0.7806*	*1.1561*	*0.7644*	*1.1088*	*0.9371*	*1.1023*	*0.2746*	*0.8862*	*0.9535*	*0.2942*	*0.8199*	*0.9471*	*1.0451*
PC_6	*1.0000*	*1.0000*	*1.0000*	*1.0000*	*0.8000*	*1.0909*	*0.9286*	*0.6364*	*1.0645*	*1.0000*	*0.7000*	*0.9310*	*1.0000*	*1.0000*
PC_7	*0.0199*	*0.0237*	*0.0258*	*0.0257*	*0.0129*	*0.0255*	*0.0258*	*0.0257*	*0.0257*	*0.0257*	*0.0154*	*0.0257*	*0.0155*	*0.0257*
PC_8	*0.3762*	*0.3383*	*0.4356*	*0.2923*	*0.4044*	*0.3572*	*0.3788*	*0.2691*	*0.3602*	*0.3335*	*0.3577*	*0.2582*	*0.2760*	*0.4040*
PC_9	*0.0373*	*0.0341*	*0.0359*	*0.0332*	*0.0302*	*0.0240*	*0.0449*	*0.0340*	*0.0535*	*0.0539*	*0.0191*	*0.0488*	*0.0185*	*0.0676*
PC_9	*0.9455*	*0.9238*	*1.0819*	*0.7831*	*0.8882*	*0.9572*	*1.0086*	*0.6744*	*1.0009*	*0.8896*	*0.9097*	*0.7227*	*0.7468*	*0.9434*

TABLE 7.6 The results of the optimization of the fuzzy system in a 3-story structure using the UWOA algorithm (continued).

	Earthquakes – FN Components													
EC	*15*	*16*	*17*	*18*	*19*	*20*	*21*	*22*	*23*	*24*	*25*	*26*	*27*	*28*
PC_1	***1.0051***	***0.9983***	***1.0046***	***0.9771***	***0.9978***	***0.9370***	***1.0083***	***0.8881***	***0.9667***	***0.8516***	***0.8558***	***0.8964***	***0.4592***	***0.9195***
PC_2	*1.4809*	*1.1557*	*1.1364*	*0.8760*	*1.0642*	*1.1800*	*1.1587*	*1.0592*	*0.9284*	*1.2036*	*1.0943*	*0.9562*	*1.0398*	*1.1444*
PC_3	*1.1963*	*1.1123*	*1.0828*	*1.2001*	*1.1687*	*1.2525*	*1.1435*	*1.0444*	*1.1623*	*1.1470*	*1.0457*	*1.1371*	*1.1629*	*1.2083*
PC_4	*1.0074*	*1.1246*	*0.9085*	*1.0160*	*0.9771*	*0.9617*	*1.0824*	*0.8455*	*1.0212*	*0.8316*	*0.7702*	*0.8896*	*0.5577*	*0.9167*
PC_5	*0.6965*	*0.7966*	*1.5178*	*1.7728*	—	*1.4148*	*1.4962*	*1.1125*	*1.2536*	*0.7705*	*1.0213*	*0.8982*	*0.9500*	*1.0359*
PC_6	*0.8462*	*0.9231*	*1.0000*	*1.0833*	—	*1.1250*	*1.0833*	*1.0323*	*1.0800*	*0.8889*	*0.9091*	*0.9600*	*1.0270*	*1.0690*
PC_7	*0.0257*	*0.0257*	*0.0155*	*0.0257*	*0.0083*	*0.0257*	*0.0246*	*0.0251*	*0.0255*	*0.0258*	*0.0155*	*0.0250*	*0.0257*	*0.0255*
PC_8	*0.4071*	*0.4828*	*0.3820*	*0.4221*	*0.4650*	*0.4358*	*0.4170*	*0.3628*	*0.3512*	*0.3291*	*0.3304*	*0.3513*	*0.2036*	*0.3680*
PC_9	*0.0373*	*0.0639*	*0.0274*	*0.0422*	*0.0156*	*0.0465*	*0.0292*	*0.0627*	*0.0591*	*0.0516*	*0.0336*	*0.0291*	*0.0515*	*0.0315*
PC_9	*1.0051*	*0.9983*	*1.0046*	*0.9771*	*0.9978*	*0.9370*	*1.0083*	*0.8881*	*0.9667*	*0.8516*	*0.8558*	*0.8964*	*0.4592*	*0.9195*

reduced, with the greatest amount of this decline reaching 53%. The maximum displacement of the structure rises by up to 16 percent for the 3rd record; however, for the other records a slight increase has been demonstrated.

7.7 Conclusion

This chapter examined the optimization of fuzzy logic controllers used in steel building structures employing the Upgraded Whale Optimization Algorithm (UWOA) as the primary optimization method. In this methodology, the "continuous-time" concept is utilized while the novel position of each agent influences the movement of the following agents, but in the "discrete-time" notion of standard WOA, the new positions are not employed until an iteration is accomplished. The primary results of the current research work are as follows:

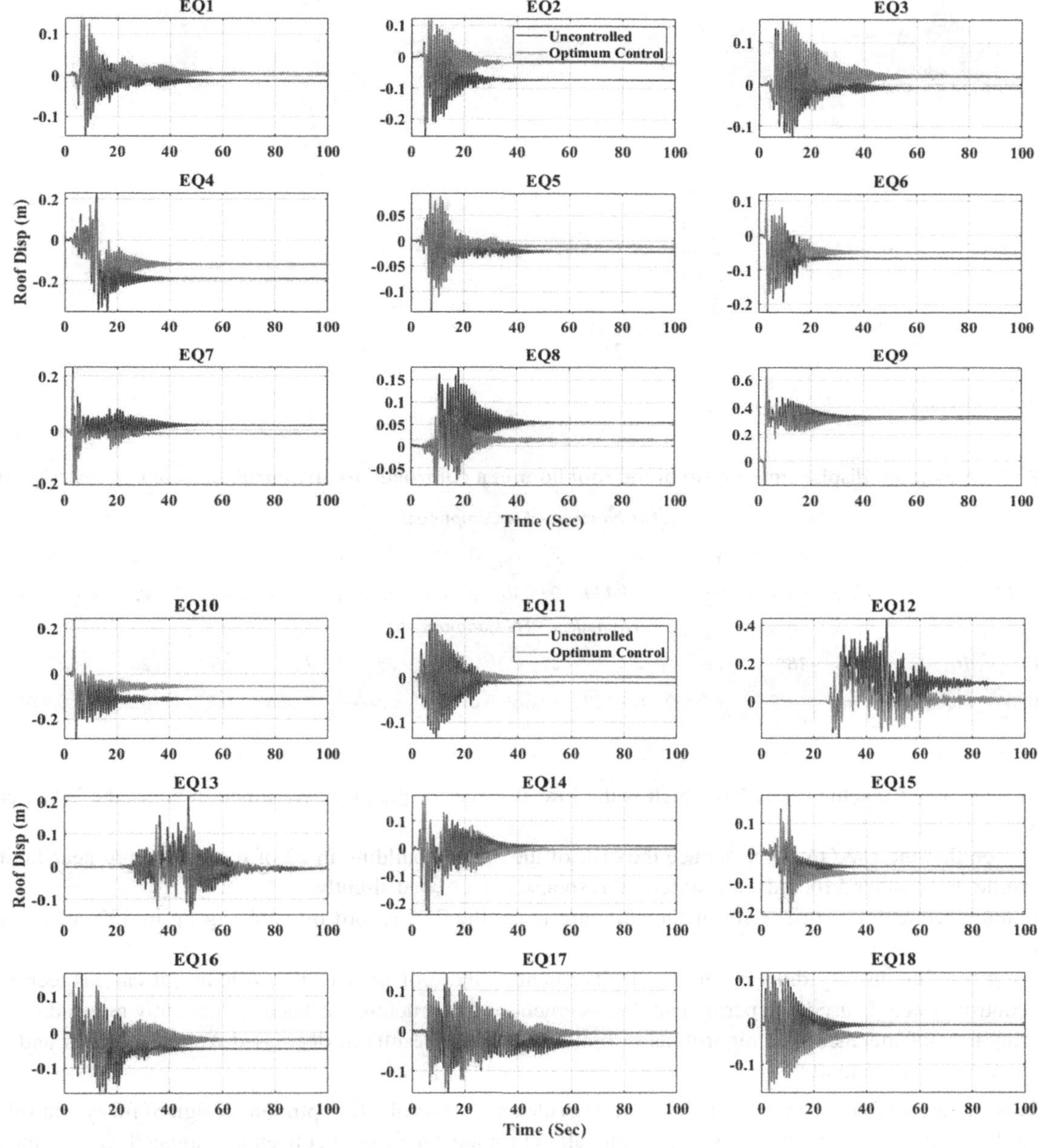

FIGURE 7.15 The time history of the displacement of the roof floor in the 3-story structure.

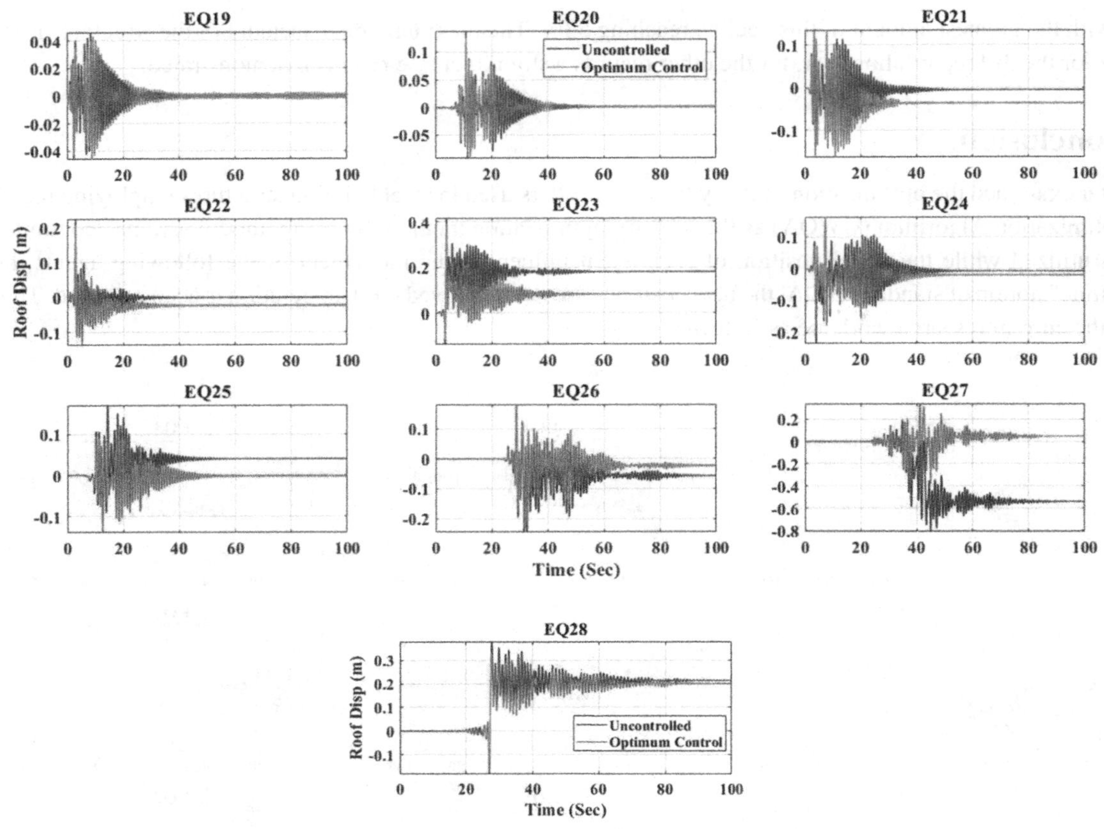

FIGURE 7.15 *(continued)*

TABLE 7.7 The maximum displacement ratio of the roof floor in a controlled to uncontrolled state in a 3-story structure.

	Earthquakes – FN Components													
	1	*2*	*3*	*4*	*5*	*6*	*7*	*8*	*9*	*10*	*11*	*12*	*13*	*14*
$\frac{Cnt}{Unc}$	*0.9313*	*0.9090*	*1.1653*	*0.7539*	*0.9445*	*0.9449*	*0.9050*	*0.6418*	*1.0022*	*0.8945*	*0.9090*	*0.7137*	*0.7346*	*0.9698*
	Earthquakes – FN Components													
	15	*16*	*17*	*18*	*19*	*20*	*21*	*22*	*23*	*24*	*25*	*26*	*27*	*28*
$\frac{Cnt}{Unc}$	*1.0205*	*0.8683*	*0.9366*	*1.0552*	*1.0000*	*0.9128*	*1.0366*	*0.9238*	*0.9440*	*0.8669*	*0.8221*	*0.9458*	*0.4742*	*0.8950*

- UWOA is capable of reaching 0.89016 which is the lowest value for the objective function while the WOA calculated 0.89396.
- It can be seen that the UWOA could reduce the drift of the 3-story building in 23 of the considered near-fault seismic records while in the other 5 records, the structural response is increased slightly.
- The maximum reduction in drift ratio of the structure is for the 27th record by reducing up to 55% of the structural response.
- By studying the time history diagram of the displacement of the roof floor of the structure, it can be seen that if the optimal control system is used, the permanent displacement of the structure has been significantly reduced.
- The employment of metaheuristic algorithms in optimizing FLCs results in decreased response values and structural damage in areas near the fault.

Regardless of the fact that the upgraded algorithm provided proper results for optimum design of fuzzy control systems, it is applicability in dealing with other types of optimization problems have not yet been investigated. One of the probable limitations of the utilized algorithm can be investigated in dealing with constraint optimization issues in which multiple types of equality and inequality constraint are utilized. Meanwhile, the capability of the upgraded algorithm in dealing with large search domains can also be another challenge.

For the future challenges, the feasibility of the UWOA can be assessed in dealing with constraint engineering optimization problems in which the capability of the upgraded method can be challenged by combination with different constraint handling schemes. However, the utilization of this upgraded algorithm in optimum design of other control systems including the passive tuned mass dampers by consideration of soil-structure interaction can be another challenge for the future.

References

[1] E.H. Mamdani, S. Assilian, An experiment in linguistic synthesis with a fuzzy logic controller, International Journal of Man-Machine Studies 7 (1) (1975) 1–13.

[2] M. Azizi, et al., Shape and size optimization of truss structures by Chaos game optimization considering frequency constraints, Journal of Advanced Research 41 (2022) 89–100.

[3] L.A. Zadeh, Fuzzy sets, in: Fuzzy Sets, Fuzzy Logic, and Fuzzy Systems: Selected Papers by Lotfi A Zadeh, World Scientific, 1996, pp. 394–432.

[4] L.A. Zadeh, On fuzzy algorithms, in: Fuzzy Sets, Fuzzy Logic, and Fuzzy Systems: Selected Papers by Lotfi A Zadeh, World Scientific, 1996, pp. 127–147.

[5] R.E. Bellman, L.A. Zadeh, Decision-making in a fuzzy environment, Management Science 17 (4) (1970) B-141-B-164.

[6] L.A. Zadeh, Similarity relations and fuzzy orderings, Information Sciences 3 (2) (1971) 177–200.

[7] C.C. Lee, Fuzzy logic in control systems: fuzzy logic controller. I, IEEE Transactions on Systems, Man and Cybernetics 20 (2) (1990) 404–418.

[8] L. Han-Xiong, H.B. Gatland, A new methodology for designing a fuzzy logic controller, IEEE Transactions on Systems, Man and Cybernetics 25 (3) (1995) 505–512.

[9] G. García-Gutiérrez, et al., Fuzzy logic controller parameter optimization using metaheuristic cuckoo search algorithm for a magnetic levitation system, Applied Sciences 9 (12) (2019) 2458.

[10] M. Azizi, S. Talatahari, Improved arithmetic optimization algorithm for design optimization of fuzzy controllers in steel building structures with nonlinear behavior considering near fault ground motion effects, Artificial Intelligence Review 55 (5) (2022) 4041–4075.

[11] O. Castillo, et al., Optimal design of fuzzy systems using differential evolution and harmony search algorithms with dynamic parameter adaptation, Applied Sciences 10 (18) (2020) 6146.

[12] M.H. Qais, H.M. Hasanien, S. Alghuwainem, Whale optimization algorithm-based Sugeno fuzzy logic controller for fault ride-through improvement of grid-connected variable speed wind generators, Engineering Applications of Artificial Intelligence 87 (2020) 103328.

[13] F. Cuevas, O. Castillo, P. Cortes, Optimal setting of membership functions for interval type-2 fuzzy tracking controllers using a shark smell metaheuristic algorithm, International Journal of Fuzzy Systems 24 (2) (2022) 799–822.

[14] M.A. Soliman, et al., Chaotic-billiards optimization algorithm-based optimal FLC approach for stability enhancement of grid-tied wind power plants, IEEE Transactions on Power Systems 37 (5) (2022) 3614–3629.

[15] J. Chrouta, et al., Modeling and control of an irrigation station process using heterogeneous cuckoo search algorithm and fuzzy logic controller, IEEE Transactions on Industry Applications 55 (1) (2019) 976–990.

[16] S. Mirjalili, A. Lewis, The whale optimization algorithm, Advances in Engineering Software 95 (May 2016) 51–67.

[17] M. Azizi, et al., Upgraded Whale Optimization Algorithm for fuzzy logic based vibration control of nonlinear steel structure, Engineering Structures 192 (2019) 53–70.

[18] Y. Ohtori, et al., Benchmark control problems for seismically excited nonlinear buildings, Journal of Engineering Mechanics 130 (4) (2004) 366–385.

[19] J.P. Stewart, et al., Ground motion evaluation procedures for performance-based design, Soil Dynamics and Earthquake Engineering 22 (9) (2002) 765–772.

[20] B.A. Bolt, Seismic input motions for nonlinear structural analysis, ISET Journal of Earthquake Technology 41 (2) (2004) 223–232.

[21] K. Galal, A. Ghobarah, Effect of near-fault earthquakes on North American nuclear design spectra, Nuclear Engineering and Design 236 (18) (2006) 1928–1936.

[22] Applied Technology Council, Quantification of building seismic performance factors, US Department of Homeland Security, FEMA, 2009.

Chapter 8

SDN-based optimal task scheduling method in Fog-IoT network using combination of AO and WOA

Taybeh Salehnia[a], Ahmadreza Montazerolghaem[b], Seyedali Mirjalili[c,d], Mohammad Reza Khayyambashi[b], and Laith Abualigah[e,f,g,h,i,j]

[a]*Department of Computer Engineering and Information Technology, Razi University, Kermanshah, Iran,* [b]*Faculty of Computer Engineering, University of Isfahan, Isfahan, Iran,* [c]*Centre for Artificial Intelligence Research and Optimisation, Torrens University Australia, Brisbane, QLD, Australia,* [d]*Yonsei Frontier Lab, Yonsei University, Seoul, South Korea,* [e]*Computer Science Department, Al al-Bayt University, Mafraq, Jordan,* [f]*Hourani Center for Applied Scientific Research, Al-Ahliyya Amman University, Amman, Jordan,* [g]*MEU Research Unit, Middle East University, Amman, Jordan,* [h]*Department of Electrical and Computer Engineering, Lebanese American University, Byblos, Lebanon,* [i]*School of Computer Sciences, Universiti Sains Malaysia, Pulau Pinang, Malaysia,* [j]*School of Engineering and Technology, Sunway University Malaysia, Petaling Jaya, Malaysia*

8.1 Introduction

The Internet of Things (IoT) is a collection of different devices that contain different software and hardware technologies to communicate with other devices using unique addressing methods [1,2]. The IoT devices collect data from their surroundings through various sensors and exchange them [2]. As a result, IoT system applied in various fields such as smart homes, smart cities, transportation, e-health care, agriculture, and industries. Cloud Computing (CC) [3] is an emerging computing technology that, due to its capabilities, can provide all the resources needed for the quality of IoT services for IoT. The CC system consists of a large number of Data Centers (DCs), each DC also consists of a large number of Virtual Machines (VMs). But due to the long geographical distance with IoT devices on the network Edge Computing (EC), the CC system is not suitable for delay-sensitive IoT devices such as emergency monitoring, and energy usage measurements from a smart grid, cause long delays that may not be acceptable for some applications in today's world [3,4]. Therefore, to solve this problem, the computing resources should be closer to the network EC devices, and the CC system is very suitable for this and can provide the resources needed to reduce the workload in cloud DC, facilitate task processing, facilitate networking, and facilitate the storage of data generated by IoT sensors, with the lowest amount of communication cost and delay [4–7]. Each server or Fog Computing (FC) node is a virtualized system equipped with a wireless communication unit, simpler processing and computing devices for data, and data storage cards. When FC nodes receive more task requests from IoT devices that exceed their capacity, they can offload some of their load to cloud layer DCs [8–10]. In other words, CC and FC are models of hosting services over the Internet for IoT devices. Fig. 8.1 shows the architecture of IoT-Fog-Cloud system, with CC in the top layer, FC in the middle layer, and IoT devices in the bottom layer.

Task Scheduling (TSch) is an effective method for efficient management of virtual resources of the FC and EC environment [11] based on specific constraints and deadlines by different users, which can be used to assign the set of requested tasks by users or existing IoT devices to FC and CC resources in order to execute them [12–16]. According to Fig. 8.1, in the proposed TSch model that is considered for scheduling the task requests of IoT devices in the FC system, Fog Broker (FB) is the main part and is located in the FC layer, which includes three main parts:

Task Administrator (TA), Resource Monitoring Service (RMS), and Task Scheduler (TSR).

The TA receives all task requests from various IoT devices, and then forwards them to the TSR, maintaining their required resources and attributes. Also, RMS is responsible for collecting information on FC resources and monitoring the status of FC resources. TSR unit is the main core of FB unit, and TSch algorithms are executed in it. According to the characteristics of the sent task requests as well as the capabilities of the available FC resources, the TSR schedules the tasks for execution and processing by assigning the appropriate FC nodes to the task requests. Finally, the processed task requests are sent back to the FB and from there to the respective users or IoT devices [12–16]. In order to allocate FC resources based on the demand of users or IoT devices, fully flexible infrastructure virtualization that uses IoT task

Handbook of Whale Optimization Algorithm. https://doi.org/10.1016/B978-0-32-395365-8.00014-2

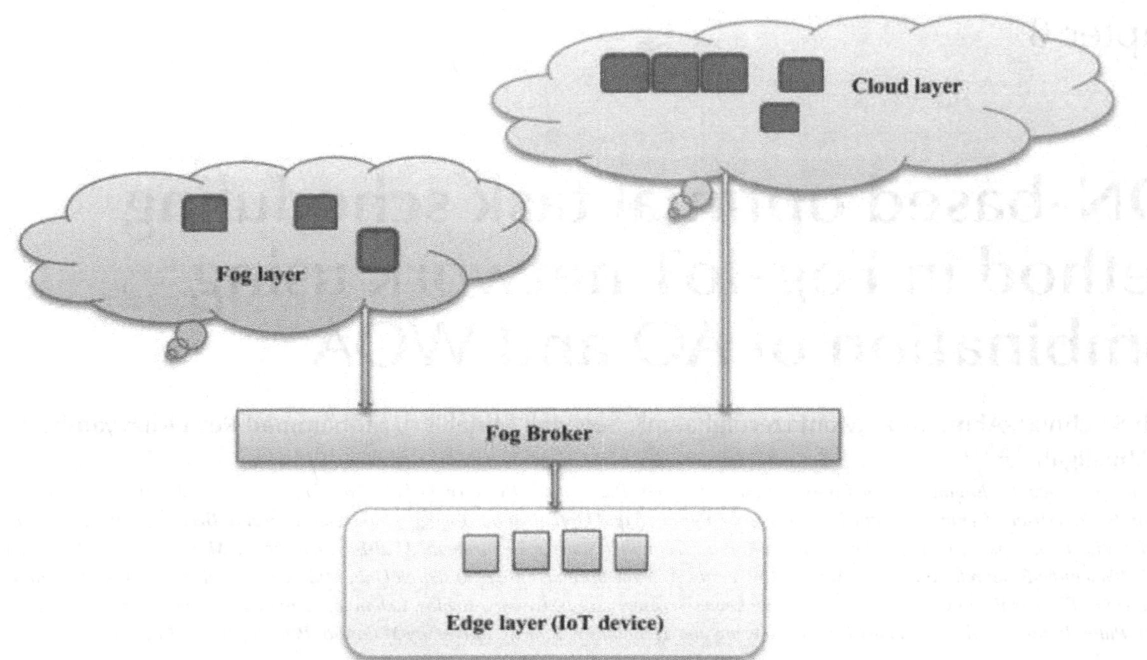

FIGURE 8.1 Three-layer system of IoT-Fog-Cloud.

request scheduling capabilities within the IoT-Fog System (IoTFS) to work on an active virtualization platform, and reduce latency for IoT devices, this chapter introduces a task requests scheduling method in Software Defined Networking (SDN)-based IoTFS. An SDN-IoT-Fog computing model is proposed, which reduces network latency by using a centralized IoTFS controller and coordinating network elements in the SDN controller layer. Therefore, it can be concluded that FC and CC environments allow different IoT users and applications to use different and many resources that exist in the respective computing environments virtually for each task request. This makes optimal scheduling methods more efficient than traditional methods.

Different optimization algorithms are used to solve different problems that are NP-HARD [17–21]. So, a hybrid Meta-Heuristic (MH) algorithm using the combination of Aquila Optimizer (AO) [22] and Whale Optimization Algorithm (WOA) [23], which is called AWOA, is proposed to schedule IoT task requests and allocate FC resources to IoT task requests to reduce the task completion time of IoT devices. The ability to ExploItation Phase (EIP) of WOA is stronger than the ability to ExploRation Phase (ERP) it during the search process. This weakness of WOA in the ERP causes a decrease in the quality of the final output. Also, in AO, the ERP is stronger than the EIP. Therefore, in this chapter, the high EIP ability of WOA is combined with the high ERP ability of AO, and it is used to solve the TSch problem in IoTFS. The purpose of the proposed SDN-based AWOA method is to optimize the task Execution Time (ET), Makespan Time (MT), and Throughput Time (TT), which are investigated to verify the QoS requirements of IoTFS. The TSch problem in FC, cloud, and Fog-Cloud environments is considered as an NP-hard problem. Therefore, optimization algorithms can be used to solve the TSch problem [18–21]. So, in this chapter, some different Meta Heuristic (MH) techniques are briefly described in the relevant works section. Also, some SDN-based TSch techniques are also described in the related works section. The continuation of the chapter is as follows: in Section 8.2, relevant works are described, problem formulation is described in Section 8.3, in Section 8.4 prerequisites are described, in Section 8.5 the proposed method is described, Section 8.6 includes Evaluation Metrics (EMs), evaluations, results and comparisons, and Section 8.7 contains conclusions.

8.2 Related works

In this section, a number of existing works in the field of scheduling in various multi-processor, FC, CC and Fog-Cloud computing environments that have been done using MH algorithms are briefly described. Some of these works are based on SDN.

8.2.1 Non-SDN based TSch algorithms

Boviri et al. [18] have proposed a TSch method based on an Improved Ant Colony Optimization (IACO) algorithm in multi-processor computing systems. The main purpose of the authors is to achieve an optimal order in order to perform the tasks. Qabaei Arani et al. [13] proposed a scheduling method in the FC environment using the Moth-Flame Optimization (MFO) algorithm. The authors purpose to reduce the ET of the entire tasks. Shukri et al. [19] proposed a scheduling method in the CC environment using the Multi-Verse Optimizer (MVO) algorithm. The purpose of the authors is to reduce the ET (MT) and execution cost, while increasing resources' utilization. Hasani Zade and Mansouri [24] proposed an optimal scheduling method in the CC environment using Fuzzy Improved Red Fox Optimization (FIRFO) algorithm and game theory named EGFIRFO, considering four objects (1) load balancing, 2) resource utilization, 3) MT, and 4) ET). Bahmanyar et al. [20] have proposed Multi-Objective Arithmetic Optimization Algorithm (MOAOA) for scheduling home appliances in energy management system (reduce the Peak to Average Ratio (PAR), reduce daily electricity costs, and increase User Comfort (UC)). Liu [21] has proposed a method in the CC layer based on the IACO algorithm. The purpose of IACO algorithm is reduce the ET, reduce the cost and load balancing rate based on tasks submitted by users.

Xiao et al. [25] have proposed a Chemical Reaction Optimization algorithm based on Adaptive Search Strategy (ASS-CRO) in order to schedule the tasks in the cloud. ASSCRO performs the scheduling using two steps. The first step is to search for the execution order of the tasks in the cloud, and the second step is to efficiently map the tasks on the VMs. Manikandan et al. [26] have proposed a Multi-Objective WOA-based Bee Algorithm (BA) for TSch in the cloud. The objects of this paper are minimize the MT and maximize the resource utilization. Dubey and Sharma [27] have used the combination of CRO and Partial Swarm Optimization (PSO) in order to schedule the tasks in the CC environment to improve the quality in terms of factors such as cost, energy, and MT and with deadline constraint. Elaziz et al. [28], have proposed a Single-Objective combined AEOSSA using the Artificial Ecosystem-Based Optimization (AEO) algorithm and Salp Swarm Algorithm (SSA) in order to schedule the IoT tasks requests in the Fog-Cloud environment to reduce MT. Salehnia et al. [29], have proposed a workflow scheduling method in the CC environment using the combination of MFO and SSA algorithms, which schedules different workflows as Multi-Objective.

8.2.2 SDN-based TSch algorithms

Vuppal and Swarnalatha [30] have proposed a load balancing scheduling method using SDN. In their work, a load balancing workflow scheduling method was made possible for the server cluster system, which could greatly improve the server load capacity. Javanmardi et al. [31] have proposed an SDN-based TSch method in the IoTFS using the Multi-Objective PSO (optimal response time and security) algorithm, which uses the concept of SDN to handle TCP SYN flood attacks. Shang et al. [32] proposed an SDN-based and bandwidth-based TSch method in the CC environment to accurately estimate the task ET, in which the SDN technique is used to provide bandwidth guarantee for the task. Sellami et al. [33] have proposed a low-latency and energy-aware TSch problem in Fog-IoT network that uses SDN-based Deep Q-Learning for dynamic TSch. Sellami et al. [34] proposed an SDN-based TSch using Deep Reinforcement Learning (DRL) for resource management and schedule of IoT traffic in EC networks, which attempts to reduce network latency by ensuring energy performance. Chalapathi et al. [35] have proposed an optimal delay-aware scheduling method for wireless SDN networked EC cloudlets in order to reduce latency and cost of computation.

Phan et al. [36] have proposed an SDN-based dynamic task offloading service in FC system. The main purpose is selecting an optimal task offloading and assisting the offloading path by providing an end-to-end bandwidth guarantee based on SDN technology.

Sellami et al. [37] proposed an energy-aware scheduling method using Blockchain-based DRL in a SDN-enabled IoT network. The authors' purpose was to increase reliability and reduce latency and increase energy performance in IoT networks with SDN. Liu et al. [38] have proposed an SDN scheme to load balance Edge-Cloud network traffic and improve response time. The MH algorithms mentioned in the literature that have been used to solve the scheduling problem in CC and FC environments include WOA [23], FA [39], PSO [40], and HHO [41], all of these algorithms are very powerful. They are not responsible for solving the scheduling problem. We discuss on the WOA in this chapter. WOA is more powerful than PSO, FA, and ACO algorithms. However, WOA has low power compared to SSA, AOA, and MFO, because, WOA is weaker than most algorithms including SSA, AOA, and MFO in the ERP, and this makes it unable to find global optimal solutions well. This leads to a decrease in the quality of the final output in solving various problems, including TSch. Therefore, in this chapter, the AO ERP is used to improve the WOA ERP.

AO is stronger than WOA in the ERP, so the ERP of AO is combined with the EIP of WOA to create a hybrid AWOA, that can achieve better results than WOA [23], AO [22], FA [39], PSO [40], and HHO [41]. Therefore, AWOA is used to solve the TSch problem in the FC system. The main factors to evaluate the proposed approach are MT, Objective Function

(OF), TT and Performance Improvement Rate (PIR), and the main purpose of AWOA and the comparative algorithms is to reduce the MT. Also, two different datasets are used to evaluate the effectiveness of the AWOA method. Then, the TSch output obtained from AWOA is given to SDN to finally obtain the best schedule for the FC system.

8.3 Problem formulation

The mathematical formulation of the TSch in the IoTFS is described below. It is assumed that $N1$ tasks (task requests of IoT devices) are independent as $T = \{T_1, T_2, \ldots, T_{N1}\}$ exists in the scheduling queue for scheduling purposes. A set of T tasks is sent to the FB to be processed and executed on the FC environment. The characteristics of each task from T set include task length (in terms of Millions of Instructions (MI)), size of files, memory requirement, and deadline (memory requirement, task length, deadline, and size of input and output files). Also, it is assumed that the FC environment includes a set of $N2$ sources or computing nodes in the form of $Node$, with $Node = \{Node_1, Node_2, \ldots, Node_{N2}\}$ is displayed. Each node of the set of FC system nodes has characteristics such as CPU processing rate (Millions of Instructions Per Second (MIPS)), network bandwidth, memory size, and storage capacity. The Expected Computation Time (ECT) of the task requests of IoT devices on each node of the FC system is represented by an ECT matrix of size $N1 \times N2$ [28]. TSR uses the ECT matrix to decide the scheduling of the relevant tasks, and each element of the ECT matrix (i.e. $ECT_{k,i}$) can be calculated as Eq. (8.1):

$$ECT_{k,i} = \frac{Task\_length_k}{Node.Pow_i} \tag{8.1}$$

In Eq. (8.1), $Task\_length_k$ is the length of task *k-th*, $Node.Pow_i$ is the processing speed of $Node_i$, $ECT_{k,i}$ represents the ECT of the *k-th* task on the *i-th* computing node [28]. The purpose of the AWOA is to find an optimal scheduling or to find the most optimal FC node index to execute IoT task requests in the scheduling queue so as to minimize the completion time or MT. Therefore, the OF for the optimizer algorithm is the same as MT, which is obtained using Eq. (8.2) [28]:

$$MT = \max_{i \in 1,2,\ldots,N2} \sum_{T_k \in T} Exe.Time(T_k) \tag{8.2}$$

where, $Exe.Time(T_k)$ is the ET of the *k-th* task on the *i-th* VM.

8.4 Prerequisites

In this section, at first SDN is described, and then, compared algorithms (FA, PSO, and HHO) are described, and then, AO and WOA, which are the main prerequisites for designing the proposed hybrid AWOA algorithm for SDN-based TSch in the IoTFS, are completely described. Then, in Section 8.5, the TSch problem based on SDN and AWOA in the IoTFS is described.

8.4.1 Software-defined networking

Due to the widespread use of protocols, as well as due to the distributed control that exists in the network layer elements (switches, routers, etc.), the combined management of these elements is difficult. Because these elements are implemented vertically in the network. The Control Plane (CP) (program plane) and Data Plane (DP) are implemented in network elements or devices. The CP is responsible for making traffic control decisions such as routing, scheduling, etc. The DP directs traffic based on relevant decisions made by the CP. This problem reduces flexibility in the network. Network layer operators must also configure each of the devices and network elements mentioned above using low-level commands individually.

SDN is a controller and an emerging network concept that removes all the mentioned limitations with the three purposes of 1) eliminating vertical structures, 2) separating the network logical plane from network devices, and 3) expanding centralized network control [42]. Using the concept of SDN, all network elements can be developed on a software platform, regardless of the brand name of the element or the corresponding product. In addition, the use of SDN provides an overview of the network status, a high flexibility of the network, and a simple and integrated management of the network. The DP in this architecture consists of network elements such as switching, forwarding, and others. These network elements are deprived of any control or software centers for automatic decision making. The CP consists of a set of programs, which include: firewall, routing, load balancing, etc. The communication protocol between the planes is a series of standard Open Application Programming Interfaces (APIs), including OpenFlow. The combination of SDN with CC/FC (IoTFS) and central management of scheduling tasks in the DP can be extremely effective in many aspects, including delay and performance [43].

8.4.2 Firefly algorithm

The FA [39] is an algorithm inspired by nature and based Swarm Intelligence (SI).

One of the important and fundamental features of FA is that it has a very good performance in searching for optimal solutions for multivariate problems and functions. This characteristic of FA makes FA different from existing optimization algorithms, and makes it an ideal choice for multivariate problems and functions.

Characteristic features related to the behavior of the Firefly and the flashing light pattern produced by them:

1. All Fireflies in FA are considered unisexual, and Fireflies in the problem space will be attracted to each other based on how much light or brightness they have.
2. In other words, in FA, for every two blinks a Firefly makes, a Firefly with less light is attracted to a Firefly with more light. Therefore, the attraction performance of Fireflies will be proportional to the brightness or light of each Firefly.

2.1. As the distance between two different Fireflies increases, their attractiveness and brightness also decrease. In other words, as two Fireflies move further apart, in addition to reducing their attraction to each other, their (visible) luminosity also decreases. If a Firefly is brighter than other Fireflies, this Firefly moves randomly through the environment (i.e., the Firefly is not attracted to any other Firefly).

3. The brightness of a Firefly is affected by the characteristic features of the OF or is determined by it. In maximization problems, brightness can be specified proportional to the value of OF. It is worth noting that it is possible to define the brightness of Fireflies in a similar way to the OF in Genetic Algorithms (GA).

8.4.3 Harris Hawks algorithm

The HHO [41] is a particle-based and nature-inspired MH algorithm that derives from the collaborative behavior and pursuit style of Harris's Hawks (HHs) in bait surprise. In this behavior and strategy which is intelligent, several falcons work together to surprise the bait, and attack to bait from different paths and directions. This behavior of HHs has been used to trap bait to solve optimization problems. In HHO, the most important and basic tactic of HHs to capture the bait is a joint, simultaneous and surprise attack on the bait from different directions by HHs, which is also known as the seven-kill strategy. In this case, HHs converge to the bait simultaneously. HHO performs the optimization process in two stages of ERP and EIP to solve the relevant problems.

8.4.3.1 In the ExploRation phase

In the HHO, the HHs are candidate solutions, and the best candidate solution at each step is considered as the desired or near-optimal bait. HHs sit and wait at random locations and bait is detected based on two strategies.

8.4.3.2 In the ExploItation phase

At this stage, HHs perform a surprise pounce by attacking the bait identified in the previous stage (the same famous seven-kills attack). Baits often try to escape from dangerous situations. According to the bait escape strategies and the pursuit strategies of HHs, there will be four strategies 1) soft besiege, 2) hard besiege, 3) soft siege with fast progressive dives, and 4) hard siege with fast progressive dives in the HHO algorithm to model the attack phase.

8.4.4 Partial swarm algorithm

In the early 1990s, various researches were conducted on the social behavior of animal groups. These studies indicated that some different animals that belong to a certain group, such as birds, fish, and others, are able to share information within their groups (herds) and such an ability to these animals. It conferred significant survival benefits. Inspired by these studies, Kennedy and Eberhart introduced the PSO [40] algorithm in 1995, which is a MH algorithm suitable for the optimization of non-linear continuous functions. The authors of the mentioned paper have inspired and created the PSO from the concept of SI, which usually exists in groups of animals such as herds.

The PSO algorithm has a lot in common with evolutionary computing techniques such as GA. The system starts by collecting random solutions and searching for optimization by updating solutions. Unlike GA, PSO algorithm does not have any evolutionary operators (CrossOver and mutation). In PSO, the potential solutions are called particles that fly through the problem space following optimally desired particles. The PSO algorithm starts with a group of random solutions (the particles) and then searches by updating solutions.

8.4.5 Aquila optimizer algorithm

The AO algorithm [22] is a SI algorithm in which there are four hunting strategies of Aquila for different types of bait. AO, such as other MH algorithms, consists of ERP and EIP, which performs the optimization process using these two phases to solve the desired problem. In fact, AO searches the Search Space (SS) to find the best solutions globally and locally by using ERP and EIP and finally converges to the final optimal solution. Brief description of AO algorithm:

Step 1: Extensive ERP as:

$$X(t+1) = X_{best}(t) \times (1 - \frac{t}{T}) + (X_{avg}(t) - X_{best}(t) \times rand) \tag{8.3}$$

$$X_{avg}(t) = \frac{1}{N1} \sum_{k=1}^{N1} X_k(t) \tag{8.4}$$

where $X_{best}(t)$ represents the best Location achieved so far, and $X_{avg}(t)$ represents the average Location of all Aquilas in the current iteration, t and T are the current iteration and the maximum number of iterations, respectively, $N1$ and $rand$ are the population size and random number between 0 and 1.

Step 2: The Location update in ERP as:

$$X(t+1) = X_{best}(t) \times lf(Dim) + X_R(t) + (y - x) \times rand \tag{8.5}$$

where $X_R(t)$ represents a random Location of the Aquila, and Dim is the dimension size, $lf(Dim)$ represents the levy flight (lf) function, which is given as:

$$lf(Dim) = s \times \frac{u \times \sigma}{|v|^{\frac{1}{\beta}}} \tag{8.6}$$

$$\sigma = \left[\frac{\Gamma(1+\beta) \times sin(\frac{\pi\beta}{2})}{\Gamma(\frac{1+\beta}{2}) \times \beta \times 2^{(\frac{\beta-1}{2})}} \right] \tag{8.7}$$

where $s = 0.01$ and $\beta = 1.5$ are constant values. u and v are random numbers between 0 and 1. y and x as:

$$\begin{cases} x = r \times sin(\theta) \\ y = r \times cos(\theta) \\ r = r_1 + 0.00565 \times Dim_1 \\ \theta = -w \times Dim_1 + \frac{3\times\pi}{2} \end{cases} \tag{8.8}$$

where, r_1 means the number of search cycles between 1 and 20, Dim_1 consists of integers from 1 to the dimension size (Dim) and is equal to 0.005.

Step 3: Extensive EIP as:

$$X(t+1) = (X_{best}(t) - X_{avg}(t)) \times \alpha - rand + ((upb - lob) \times rand + lob) \times \delta \tag{8.9}$$

where α and δ are the EIP adjustment parameters that fixed at 0.1, upb and lob are the upper and lower bounds of the problem.

Step 4: Limited EIP as:

$$X(t+1) = QF \times X_{best}(t) - (G_1 \times X(t)) \times rand) - G_2 \times lf(Dim) + rand \times G_1 \tag{8.10}$$

$$QF(t) = t^{(\frac{2\times rand-1}{(1-T)^2})} \tag{8.11}$$

$$G_1 = 2 \times rand - 1 \tag{8.12}$$

$$G_2 = 2 \times \left(1 - \frac{t}{T}\right) \tag{8.13}$$

where $A(t)$ is the current Location, $QF(t)$ represents the value of the quality function used to balance the search strategy, G_1 shows the movement parameter of the Aquila when tracking the bait, which is a random number between $[-1, 1]$, G_2 shows the flight slope when chasing bait, which decreases linearly from 2 to 0. The process of enhancing the solutions is done until the stopping condition is reached.

8.4.6 Whale optimization algorithm

The WOA [23] is a MH algorithm inspired by nature, which was designed by S. Mirjalili et al. and simulates the natural cooperative behavior of humpback whales. In WOA, the optimization process is performed using ERP and EIP. Two approaches of Encircling bait and Bubble-net attacking, are used to optimally update the Location of whales. In general, WOA is done in three stages or three phases, which are as follows:

a. Encircling bait

Once the best Search Agent (SA) or solution in SS is identified, other SAs try to update their location in the SS to the best SA. As:

$$\vec{D} = \left|\vec{C}.\vec{X}^*(t) - \vec{X}(t)\right| \tag{8.14}$$

$$\vec{A} = 2\vec{a}.\vec{r} - \vec{a} \tag{8.15}$$

$$\vec{C} = 2\vec{r} \tag{8.16}$$

where t denotes the current iteration, $\vec{A}$ and $\vec{C}$ are the coefficient vectors, $\vec{D}$ is the distance between the Location of $\vec{X}^*(t)$ and $\vec{X}(t)$, $X^*(t)$ the location vector is the best solution obtained at present, and $\vec{X}(t)$ is the location vector, $\vec{a}$ decreases linearly from 2 to 0 during iterations (in ERP and EIP), and $\vec{r}$ is considered a random vector between 0 and 1.

b. Explo1tation Phase

In this phase, a spiral equation is created between the whale's Location and the bait to mimic the spiral-shaped movement of the humpback whale:

$$\vec{X_i}(t+1) = \vec{D}'(t).e^{bl}.cos(2\pi l) + \vec{X}^*(t) \tag{8.17}$$

where $\vec{D}'$ refers to the distance from the $1-th$ SA to the bait (the best solution obtained so far), b is a constant for defining the shape of the logarithmic spiral, and $l \in [-1, 1]$. Each SA with a 50% probability ($p < 0.5$ or $p \geq 0.5$) chooses one of the contractile siege strategy or spiral models to update the SAs' Location during optimization. As:

$$\vec{X_i}(t+1) = \begin{cases} \vec{X}^*(t) - \vec{A}.\vec{D}, & p < 0.5 \\ \vec{D}'(t).e^{bl}.cos(2\pi l) + \vec{X}^*(t), & p \geq 0.5 \end{cases} \tag{8.18}$$

c. ExploRation Phase

The mathematical model of this behavior is as follows:

$$\vec{D} = \left|\vec{C}.\overrightarrow{X_{rand}} - \vec{X}\right| \tag{8.19}$$

$$\vec{X}(t+1) = \overrightarrow{X_{rand}} - \vec{A}.\vec{D} \tag{8.20}$$

where $\overrightarrow{X_{rand}}$ is the current population's randomly selected Location vector (random SA). A random SAs is selected in $|\vec{A}| > 1$ mode, while the best solution is selected when $|\vec{A}| < 1$ to update the Location of the SAs. The process of enhancing the solutions is done until the stopping condition is reached.

8.5 A proposed TSch method using SDN-based AWOA

In this section, the TSch problem in a IoTFS is described with the improvement of AO by WOA, which is called AWOA. In the proposed hybrid AWOA, solutions are updated in the EIP using WOA operators. Because in the ERP, AO is stronger than WOA, and in EIP, WOA is stronger than AO. The general framework of AWOA is given in Fig. 8.2. The proposed AWOA, in the initialization phase, starts to solve the TSch by defining, randomly generating, and randomly distributing solutions on the SS, which is the set of VMs in the IoTFS. This process is done by setting the initial value for a set of $N1$ solutions and converting them to integers. Then, the OF for each solution is calculated using the value of the OF, which is MT (Eq. (8.2)). Then, in the ERP and EIP, X solutions are expanded and updated. In the ERP, solutions are updated using AO operators, then, in the EIP, solutions are updated using WOA operators. The process of enhancing the solutions is done until the stopping condition is reached. The details of the AOWOA procedures are given in the subsections below.

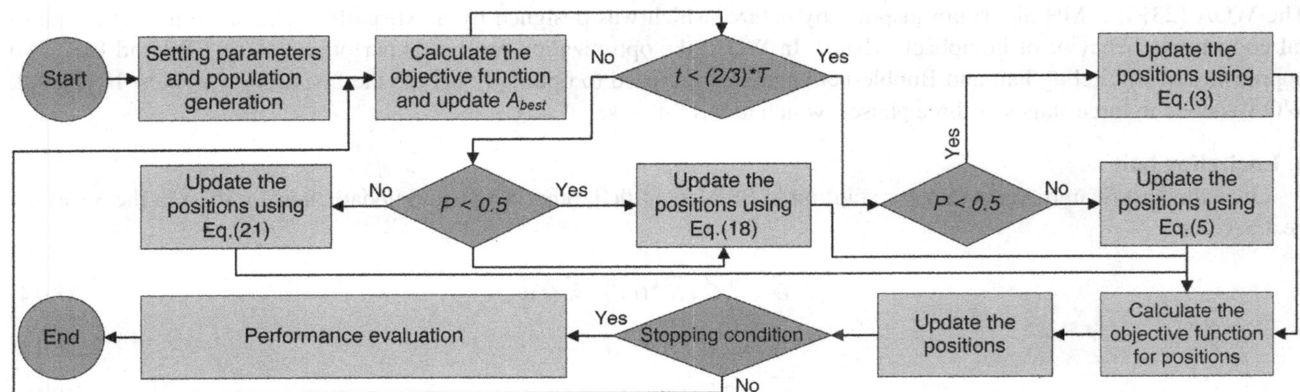

FIGURE 8.2 The general framework of AWOA.

8.5.1 Initialization phase

The first step in the proposed AWOA is the random generation and distribution of the population $\overrightarrow{X_i}$ to be suitable for the TSch in the IoTFS (that is, the random generation and distribution of solutions and converting them into integers). This conversion of integers is needed to modify the behavior of AO and its ability to solve discrete problems such as the TSch in the IoTFS, because the TSch in the IoTFS is an integer problem. Each solution is represented as a vector according to Eq. (8.21). The solutions produced using Eq. (8.22) and evaluated using Eq. (8.2).

$$\overrightarrow{X_i} = (x_1, x_2, \ldots, x_k) \quad \text{where} \quad 1 \le x_1, x_2, \ldots, x_k \le N2 \tag{8.21}$$

$$x_i = floor(lob + rand(0,1) \times (upb - lob)), \quad x_i \in \overrightarrow{X_i},\ i = 1, 2, \ldots, k \tag{8.22}$$

where, k represents the number of tasks in the priority queue, x_i represents the index or Location of the VMs, and $rand(0, 1)$ is a random number between 0 and 1. In Eq. (8.22), lob and upb are the limits of the search domain, upb is the number of VMs in the SS, and lob is equal to 1, $floor(.)$ is the same function used to convert real numbers into integers.

8.5.2 Updating the solutions phase

The TSch in IoTFS is considered as a minimization problem that uses the function used to calculate the MT. Therefore, at this stage, the proposed AWOA starts by finding the solution with the lowest MT and marks it as the best solution. Other solutions are then updated according to their behavior. These solutions use AO operators at this stage to explore the SS and update their Locations (Eq. (8.3) and Eq. (8.5)). However, the remaining solutions in the EIP are updated to explore the optimal Location using the WOA operators, which are defined as Eq. (8.23), which is obtained by changing Eq. (8.17), and is in the form of Eq. (8.23):

$$\overrightarrow{X_i}(t+1) = \overrightarrow{D}'(t).e^{bl}.cos(2\pi l) + \overrightarrow{X}^*(t) - G_2.lf(Dim) \tag{8.23}$$

In other words, $lf(Dim)$ is used in order to make the EIP as strong as possible in the WOA when combined with AO, and Eq. (8.18) is also changed to Eq. (8.24).

$$\overrightarrow{X_i}(t+1) = \begin{cases} \overrightarrow{X}^*(t) - \overrightarrow{A}.\overrightarrow{D}, & p < 0.5 \\ \overrightarrow{D}'(t).e^{bl}.cos\,(2\pi l) + \overrightarrow{X}^*(t) - G_2.lf(Dim), & p \geq 0.5 \end{cases} \tag{8.24}$$

The previous steps are repeated to update the obtained solutions until the stopping condition is met.

8.5.3 Computational complexity of AO, WOA, and AWOA

In MH algorithms, involved factors such as OF, number of SAs, number of problem dimensions, and maximum number of iterations of the algorithm in three phases of initialization, ERP, and EIP to calculate computational complexity. Computational complexity is one of the methods of evaluating algorithms. We have designed a combined algorithm by combining two AO and WOA. Therefore, we calculate the computational complexity for the AWOA and the AO and WOA. The WOA [23] for N SAs, T iteration and D dimension, has computational complexity as Eq. (8.25):

$$O(T \times N^2 + T \times N \times D) \tag{8.25}$$

The AO [22] has computational complexity as Eq. (8.26):

$$O(T \times N + T \times N \times D) \tag{8.26}$$

The AWOA is designed from a combination of the AO ERP and the WOA EIP, so, has computational complexity as Eq. (8.27):

$$O(T \times N) + O(T \times N \times D) + O(T \times N \times D) \tag{8.27}$$

which is equal to:

$$O(N \times (T \times D + T)) \tag{8.28}$$

Eq. (8.28) shows that the computational complexity of a combined algorithm is better than that of algorithms AO and WOA. AWOA, which uses a combination of ERP and EIP of AO and WOA, is faster than AO and WOA.

8.5.4 Proposed SDN based framework

In this section, we present the proposed framework based on SDN technology. The architecture of the proposed framework is shown in Fig. 8.3. This architecture has three planes: DP, CP and Application Plane (AP). The DP includes computing infrastructure and switching infrastructure. The computing infrastructure includes nodes in FC. The switching infrastructure includes OpenFlow switches. The CP includes the SDN controller. The AP also includes TA, RMS and TSR modules (algorithms). The connection between these planes is with the OpenFlow protocol and Open API. The SDN controller is logically centralized and is responsible for DP data collection. In other words, all DP equipment (including OpenFlow switches or IoT devices) are under the control of a centralized controller. After collecting the data by the controller, this data is provided to the AP so that appropriate decisions can be made for the scheduling of tasks. Network intelligence is at this plane. The mastermind of the network that makes the network intelligent are modules and algorithms based on artificial intelligence that are designed and implemented at the AP. In this case, there is no need to run the algorithm on individual equipment and devices, but it is executed in a centralized point and its output is sent to the devices through the controller. As a rule, the complexity is reduced and it is possible to get the right answer in a short time. This framework is a feasible plan for operationalizing intelligent algorithms for task management and planning, which are mainly faced with the problem of ET and implementation method.

Task management is challenging due to limitations such as real time and high number of tasks and nodes. Especially if it is implemented in a distributed manner and each IoT node is involved in decision making. Our proposed framework overcomes these challenges with a centralized approach using the new SDN technology. In this plan, the complexity of many DP equipment including switches or IoT devices is reduced and the logically centralized controller is responsible for collecting the necessary data and handling the planning of devices and tasks. Algorithms used in AP modules including TSR have been discussed in previous sections (AWOA). As a result, in order to achieve the maximum capacity of FC,

EC and CC networks, as well as scheduling tasks without delay, it is necessary to manage the network as effectively as possible, which can only be achieved with a central management and optimal TSch. In this regard, the framework based on SDN was presented in Fig. 8.3. The important advantage of this proposed framework is to have a global view of the entire FC, EC, and CC networks, which makes uniform, fast and optimal decisions. In addition, it should be noted that this framework has a software-oriented structure. This means that decisions are made algorithmically and software, making us completely independent of hardware. In a very diverse world with different brands of stylish equipment and IoT devices, it is considered an important advantage. This is due to the OpenFlow standard protocol, which can monitor and configure all network hardware and IoT equipment regardless of brand. Of course, according to the commands of the controller and the decisions of the modules. It should be noted that the modules are also successful and effective in achieving the optimal solution due to the implementation of intelligent algorithms (AWOA).

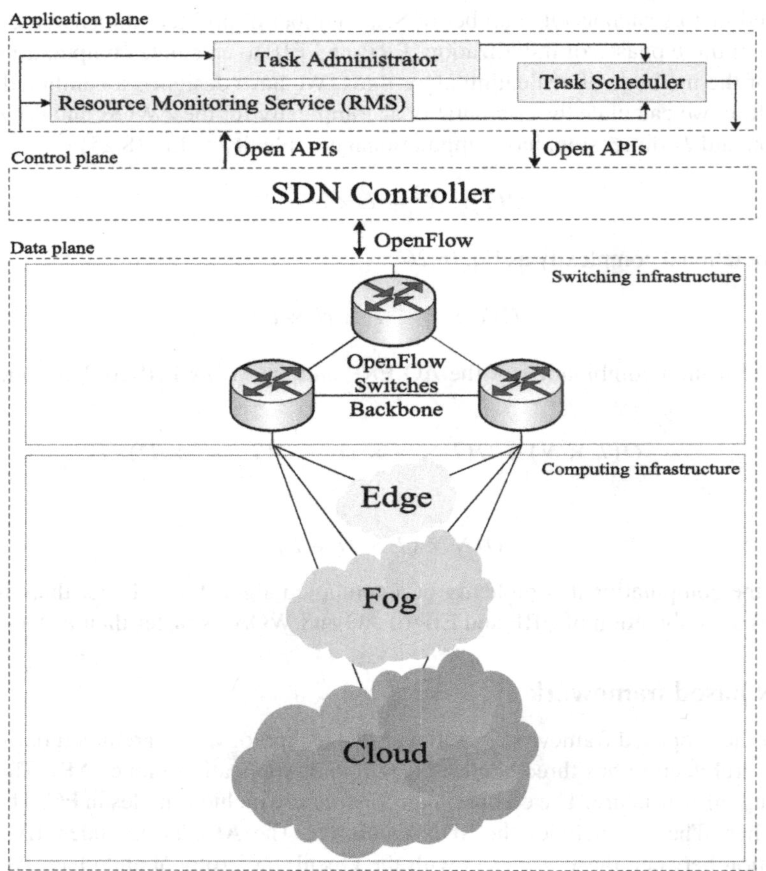

FIGURE 8.3 Proposed framework based on SDN.

8.5.5 The limitation of the proposed hybrid AO and WOA

In this chapter, the proposed hybrid AO and WOA or AWOA was designed to solve the TSch problem in the fog-IoT environment, in which the AO ERP is combined with the WOA EIP. In order to make AWOA work properly to solve the TSch problem in the fog-IoT system, as mentioned in the proposed method section, Eq. (8.17) and Eq. (8.18) in WOA were changed to Eq. (8.23) and Eq. (8.24), respectively. The first step in the AWOA is the random generation to be suitable for the TSch in the IoTFS (that is, the random generation and distribution of solutions and converting them into integers). This conversion of integers is needed to modify the behavior of AO and its ability to solve discrete problems such as the TSch in the IoTFS, because the TSch in the IoTFS is an integer problem. Each solution is represented as a vector according to Eq. (8.21). The solutions produced using Eq. (8.22) and evaluated using Eq. (8.2).

Considering that AO has a very high power, combining it with WOA has increased the ability of WOA to solve the TSch problem in the fog-IoT system. But this algorithm has only been evaluated and reviewed on the TSch problem in the fog-IoT system. This AWOA has not been evaluated for solving various other problems such as feature extraction, image processing, network routing and other engineering problems. Also, in this chapter, SDN-based AWOA is considered and used to improve TSch performance. Therefore, according to the problem we have considered in this chapter, our proposed method does not work with a specific protocol, the size of the problem space does not exceed a certain limit, it is not active in responsive environments, it does not have a security mechanism, and it is not able to handling requests is not real-time.

8.6 Evaluation metrics and experimental results

In this chapter, the simulation of the proposed TSch method and other compared algorithms has been done using MATLAB® R2018b software. All tests were performed on an ASUS laptop equipped with an Intel Core i5 processor with a frequency of 2.50 GHz and 6 GB of memory with a Windows 64-bit operating system. Floodlight controller based on OpenFlow protocol is used to implement the proposed framework based on SDN. This controller is based on Java and three modules of TA, RMS and TSR have been developed on it. Also, three OpenFlow virtual switches have been used, and we used Open vSwitch to implement them. Open vSwitch is virtual switche that run on common hardware. They also have the ability to collect data from network IoT devices using the OpenFlow protocol. They also apply and implement SDN controller decisions. The core of Floodlight and Open vSwitch is the Linux operating system, which makes it easy to develop designed modules that include the AWOA and monitor their performance and resource consumption (computational and memory complexity). In all experiments, the FC environment consists of 6 Physical Machines (PM) and 30 different VMs. Table 8.1 shows the characteristics of PMs and corresponding VMs. According to Table 8.1, the slowest VM has a speed of 100 MIPS and the fastest VM has a speed of 5000 MIPS.

TABLE 8.1 Experiment parameters.

	Specification	Amount
Client	Clients Count	[60, 120]
PM	Hosts Count	6
	Storage	1 TB
	Network Bandwidth	10 Gb/s
	RAM size	6 GB
VM	VMs count	30
	CPU capacity	[100, 5000]
	Storage	20 GB
	RAM size	1 GB
	Network Bandwidth	1 Gb/s
	Processor	Xen
	Processors' count	1

Two different datasets High Performance Computing Center North (HPC2N) and NASA Ames iPSC/860 [44] have been used, each with different number of 500, 1000, 1500, and 2000 tasks. The characteristics of these two datasets are given in Table 8.2. The tasks of each of these datasets are independent and non-preventive.

The proposed AWOA is compared with WOA [23], AO [22], FA [39], PSO [40], and HHO [41]. In order to achieve a more accurate and reliable estimation of the results of the algorithms, each algorithm has been executed 30 times for each experiment. Table 8.3 shows the fixed parameters used in the AWOA and the comparative algorithms.

In the following, first the EM used to evaluate the proposed TSch method and compare it with other algorithms are fully introduced, then the proposed method is compared with the relevant algorithms using the relevant EMs.

8.6.1 Evaluation metrics

In this section, the EMs used are introduced. The EMs used include computational complexity, MT, OF value, PIR, and TT. To evaluate the designed framework based on SDN, the metrics of resource consumption, transmission delay and convergence delay are also examined.

TABLE 8.2 Real workload characteristics of HPC2N and NASA iPSC.

	Specification	Amount
NASA iPSC	CPUs	240
	Users	257
	Tasks	202.871
	Utilization	60.1%
	Filename	NASA-iPSC-1993-3.1-cln.swf log
HPC2N	CPUs	128
	Users	69
	Tasks	18.239
	Utilization	46.6%
	Filename	HPC2N-2002-2.2-cln.swf log

TABLE 8.3 Parameters' settings in the proposed algorithm and compared algorithms.

	Specification	Amount
FA [39]	E_0	[−1, 1]
	α	0.5
PSO [40]	c_1	1.49
	c_2	1.49
AO [22]	A	0.1
	Δ	0.1
WOA [23]	A	[0, 2]
	b	1
	l	[−1, 1]
HHO [41]	B	0.2
	Γ	1
	W	0.9→0.4

8.6.1.1 Computational complexity

Computational complexity is a basic and efficient metric for evaluating and comparing MH algorithms. In this chapter, we have used the computational complexity metric for this comparison. Table 8.4 shows the computational complexity of WOA [23], AO [22], FA [39], PSO [40], and HHO [41], and AWOA algorithms. As can be seen from Table 8.4, the computational complexity of AWOA is lower than the compared algorithms.

TABLE 8.4 Computational complexity of algorithms.

Algorithms	Computational complexity
HHO [41]	O(N×(1+T×D+T))
FA [39]	O(N×(1+T×D+T))
PSO [40]	O(N×(1+T×D+T))
WOA [23]	O(N×(T×N+T×D))
AO [22]	O(N×(T+T×D))
AWOA	O(N×(T×D+T))

Table 8.4 shows that the computational complexity of a combined algorithm of AWOA is better than that of algorithms AO and WOA.

8.6.1.2 Objective function value

In problems that are solved using MH algorithms, the value of the OF is the main factor to determine the final optimal solution. Therefore, the value of the OF can be used as an evaluation metric. In minimization problems, the lower the value of the OF, it can be concluded that the solutions obtained by the corresponding optimizer algorithm are better. When it comes to comparing several MH algorithms, the MH algorithm and the value of the OF are effective in determining the near-optimal solution. Eq. (8.2) is used as the OF, which is the same as MT.

8.6.1.3 Makespan time

In TSch methods that discuss the ET and completion of tasks, the value of MT (in seconds) is a main metric for evaluating the relevant scheduling method and comparing it with other methods. The value of MT is equal to the end time of the last completed task in the corresponding set of tasks, which is calculated using Eq. (8.2) [28]. The lower the MT obtained by an algorithm, the better the corresponding algorithm performs.

8.6.1.4 Throughput time

This metric is used to measure the ET of the all tasks that have been successfully completed in a given period [28]. TT (in seconds) can be obtained using Eq. (8.29).

$$TT = \sum_{T_k \in T} Exe.Time(T_k) \tag{8.29}$$

A suitable scheduling algorithm is an algorithm that has a lower TT.

8.6.1.5 Performance improvement rate

The next metric is *PIR*, which obtains the percentage of performance improvement of an algorithm compared to other algorithms using the value of its OF [28]. *PIR* value is obtained using Eq. (8.30). The lower the *PIR* value of one algorithm compared to another algorithm, the closer the two algorithms are.

$$PIR(\%) = \frac{Fit_C - Fit_P}{Fit_P} \times 100 \tag{8.30}$$

where, Fit_C and Fit_P are the OF values obtained by each of the comparative algorithms and AWOA respectively.

8.6.1.6 Resource consumption of modules designed in SDN-based framework

The resource consumption of different floodlight controller modules has been monitored. Checking the consumption resources shows whether the designed modules bottleneck the controller or not? If the computational complexity of the designed modules is high, the controller will definitely have problems in consuming resources.

8.6.1.7 Transmission delay in SDN-based framework

One of the important evaluation metric in SDN-based frameworks is transmission delay. Because the control and data layers are separated and the main decisions are made in the controller and then communicated to the data layer for execution.

8.6.2 Results of comparison with existing works

In this section, the proposed TSch method has been compared with AO, WOA, PSO, HHO and FA using the mentioned evaluation metric. Table 8.5 shows the results of MT for NASA iPSC and HPC2N datasets. As can be seen from Table 8.5, for the NASA iPSC dataset, with 500, 1000, 1500, and 2000 tasks:

AWOA has 7.13, 6.09, 9.27 and 10.91 less MT than AO, respectively. AWOA has 30.22, 65.26, 133.18 and 109.30 less MT than WOA, respectively. AWOA has 41.17, 85.85, 176.05 and 188.67 lower MT than PSO, respectively. AWOA has 13.25, 21.93, 25.13 and 41.36 less MT than HHO, and, AWOA has 35.55, 71.56, 158.81 and 177.58 less MT than FA, respectively.

Also, for the HPC2N dataset, with 500, 1000, 1500, and 2000 tasks:

Compared to AO, AWOA has a lower MT value of 0.1274, 0.0902, 0.1120 and 0.0994 (*1.0e+04), respectively. Compared to WOA, AWOA has a lower MT value of 0.5986, 0.5283, 0.5111 and 0.5563 (*1.0e+04), respectively. Compared to PSO, AWOA has a lower MT value of 0.5384, 1.0331, 1.6316 and 2.2673 (*1.0e+04), respectively. AWOA has lower MT than

TABLE 8.5 Makespan Time for NASA iPSC and HPC2N datasets.

Dataset	NASA iPSC				HPC2N			
	Tasks				Tasks			
Algorithm	500	1000	1500	2000	500	1000	1500	2000
AWOA	41.39	85.19	132.56	160.45	3562.28	10462.65	17045.42	25462.32
AO	48.52	91.28	141.83	171.36	4836.52	11364.83	18165.35	26456.45
WOA	71.68	150.45	265.74	269.75	9548.49	15745.38	22156.72	31025.56
PSO	82.56	171.04	308.61	349.12	8946.52	20793.74	33361.63	48135.08
HHO	54.64	107.12	157.69	201.81	5572.43	12253.81	19732.12	28076.55
FA	76.94	156.75	291.37	338.03	8760.62	19682.11	31727.51	45749.89

HHO, 0.2010, 0.1791, 0.2687 and 0.2614 (*1.0e+04), respectively, and, AWOA has lower MT value than FA, 0.5198, 0.9219, 1.4682 and 2.0288 (*1.0e+04), respectively. The results of Table 8.5 show the superiority of AWOA over the comparative algorithms. Because AWOA has a lower MT than AO, WOA, PSO, HHO and FA for both datasets and has been able to reduce the time to complete the last task compared to other algorithms. Table 8.5 also shows the OF value obtained for all AWOA, AO, WOA, FA, PSO, and HHO algorithms. That is, the value of MT is the same as the value of OF. As can be seen from Table 8.5, AWOA has the lowest OF value compared to other algorithms. Fig. 8.4 shows the convergence diagram of AO, WOA, PSO, HHO, AWOA, and FA algorithms for the NASA iPSC dataset.

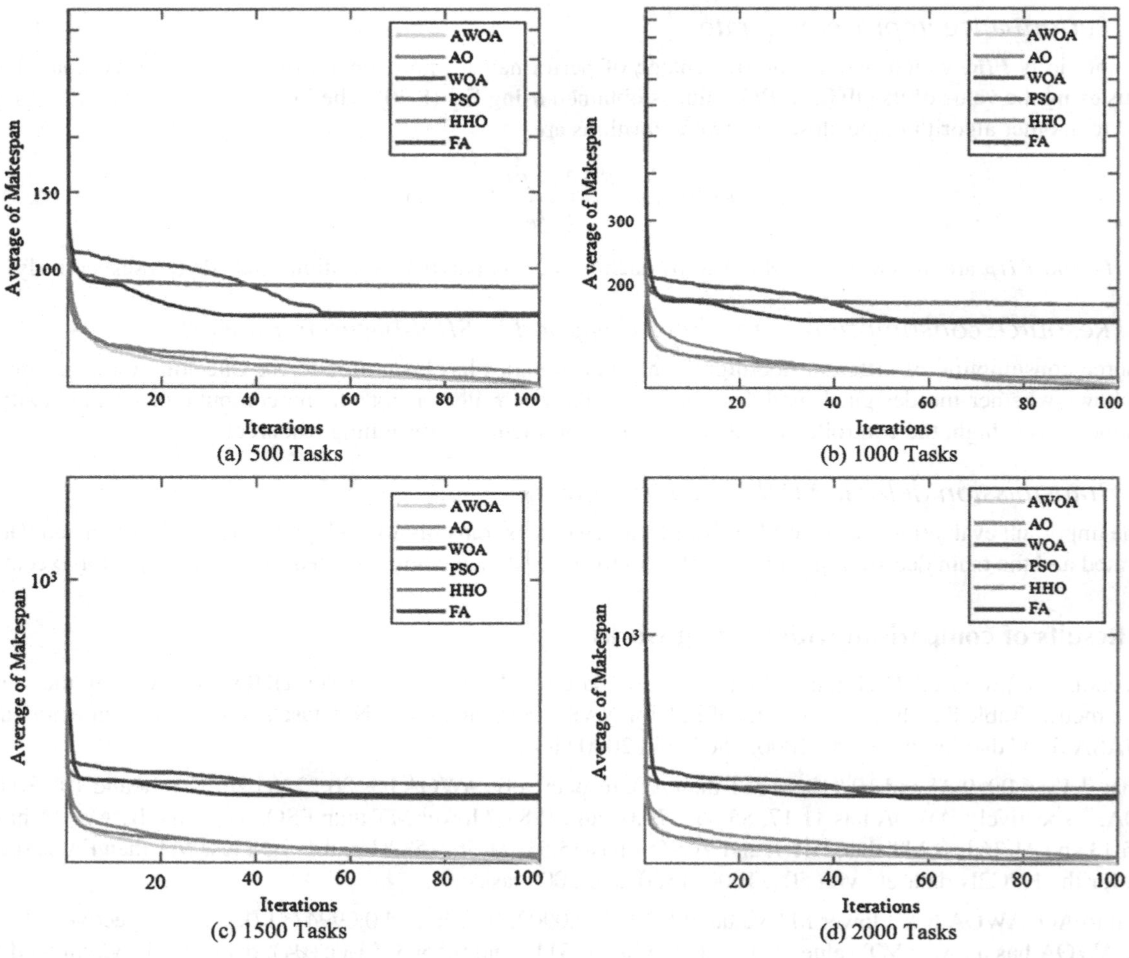

FIGURE 8.4 Convergence diagrams for NASA iPSC.

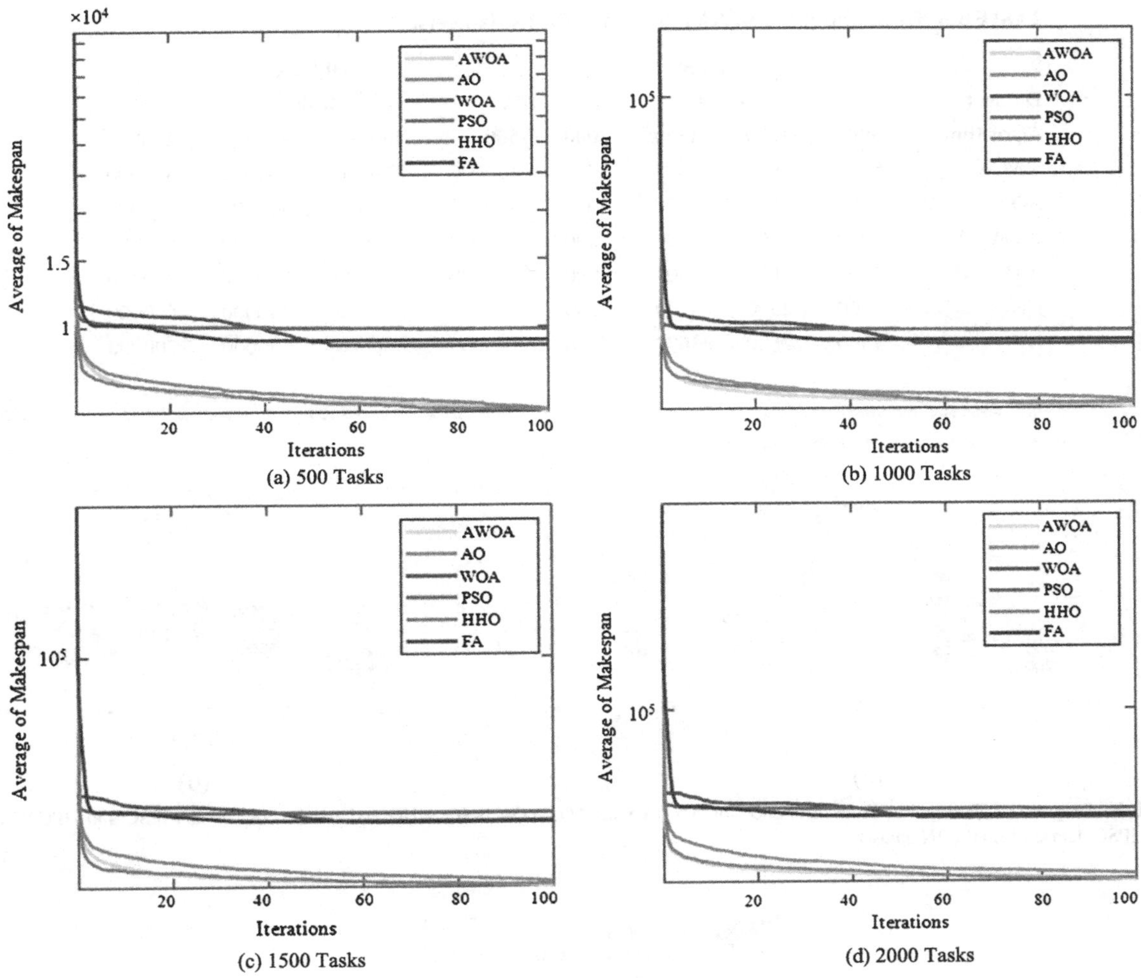

FIGURE 8.5 Convergence diagrams for HPC2N.

Fig. 8.5 shows the convergence diagram of AO, WOA, PSO, HHO, AWOA, and FA algorithms for the HPC2N dataset.

The curves of Fig. 8.4 and Fig. 8.5 show the convergence speed of the algorithms during different iterations where the algorithms finally find the optimal solution. Along the x-axis, it represents the number of iterations, and along the y-axis, it represents the OF value. Figs. 8.4 and 8.5 show that the AWOA converges to the optimal solution faster than other comparative algorithms. Therefore, the convergence curves of Figs. 8.4 and 8.5 have shown that the integration of the AO with the WOA can increase the convergence rate of the AWOA to the optimal solution for TSch and provide an opportunity to find optimal solutions. Table 8.5 shows the TT results for NASA iPSC and HPC2N datasets. According to Table 8.6, AWOA has a lower TT than AO, WOA, PSO, HHO, and FA and has been able to execute the set of tasks in less time than the algorithms reviewed.

Fig. 8.6 shows the difference diagram of AWOA with AO, WOA, PSO, HHO and FA for both NASA iPSC and HPC2N datasets. As can be seen from Fig. 8.6, AWOA has the lowest difference compared to AO and the highest difference compared to PSO.

In the following, we will examine the consumption resources of the modules designed in the controller. The results are shown in Figs. 8.7 and 8.8. CPU consumption is shown in Fig. 8.7 and memory consumption is shown in Fig. 8.8. As can be seen, the CPU consumption of the three designed modules (TA, RMS and TSR) is much lower than the others. As the number of tasks increases, the CPU consumption of the modules increases. This increase is much higher for other modules. For example, when we have 110 tasks, the TA module consumes less than 20% of the CPU, while the other modules (Others) occupy nearly 40% of the CPU. For RMS and TSR modules, CPU consumption is relatively low. As a result, these designed modules did not saturate the CPU of the controller.

TABLE 8.6 Throughput for NASA iPSC and HPC2N datasets.

Dataset	NASA iPSC				HPC2N			
	Tasks				Tasks			
Algorithm	500	1000	1500	2000	500	1000	1500	2000
AWOA	900	1900	2400	3700	130000	270000	360000	520000
AO	1000	2100	2800	3900	150000	310000	390000	570000
WOA	1200	2400	3400	4700	180000	350000	460000	660000
PSO	1400	2800	4300	5900	200000	380000	580000	780000
HHO	1100	2300	3200	4500	170000	340000	430000	620000
FA	1300	2500	3600	5000	190000	360000	480000	700000

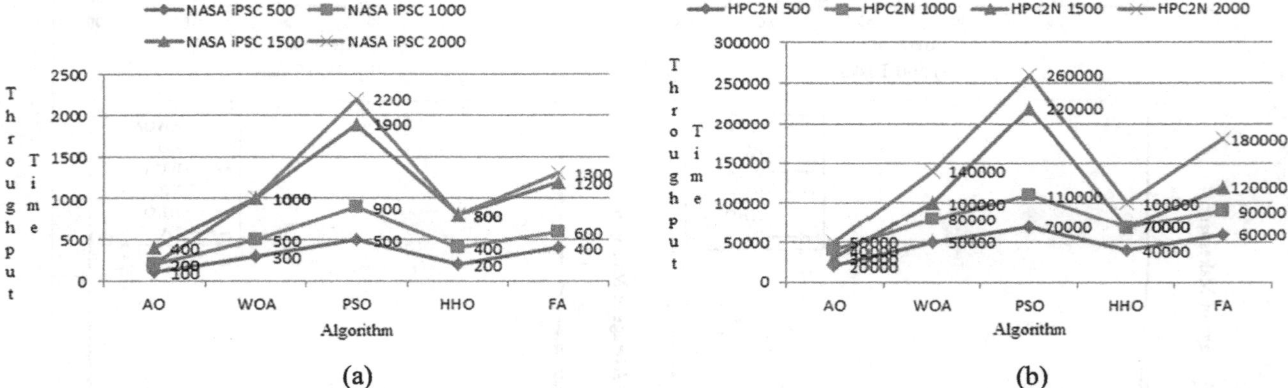

FIGURE 8.6 Diagram of Throughput Time difference for AWOA with AO, WOA, PSO, HHO and FA for both NASA iPSC and HPC2N datasets. (a) NASA iPSC dataset. (b) HPC2N dataset.

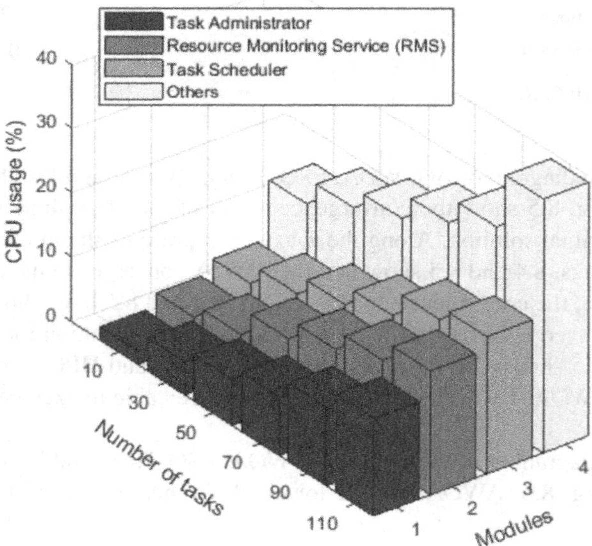

FIGURE 8.7 CPU consumption by modules designed in SDN controller.

The logic discussed about CPU consumption is also true for memory consumption. This is confirmed by Fig. 8.8. Of course, the comparison of Figs. 8.8 and 8.9 shows that the modules are more CPU-oriented than memory-oriented because they have a higher CPU consumption on average. In any case, the memory consumption of TA, RMS and TSR modules

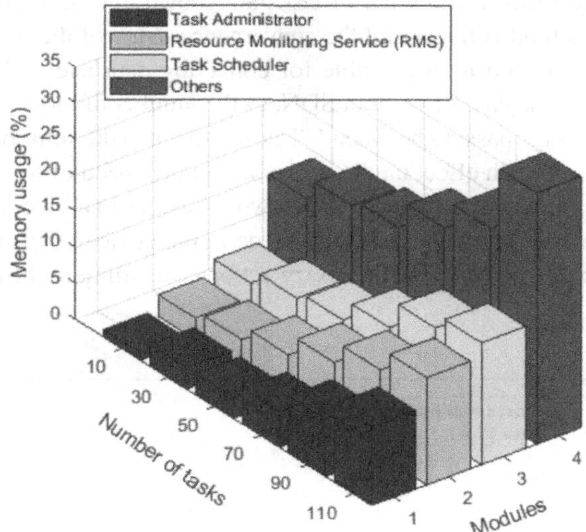

FIGURE 8.8 Memory consumption by modules designed in SDN controller.

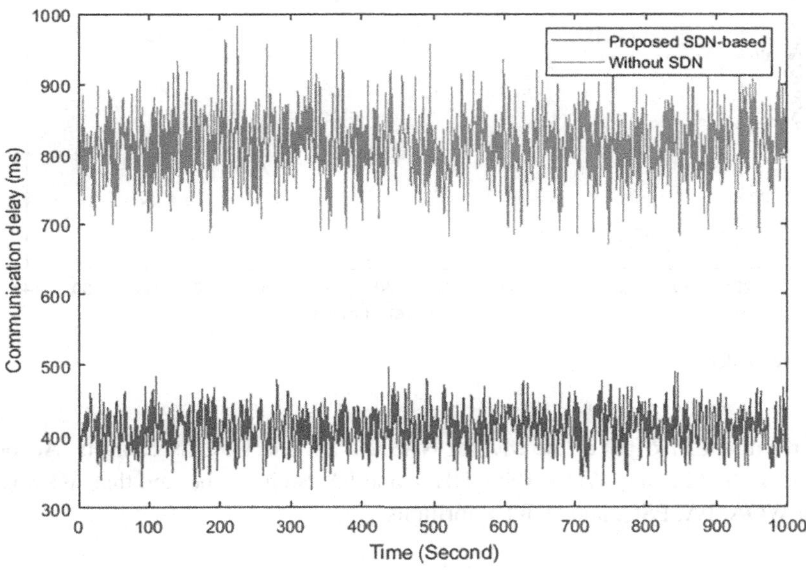

FIGURE 8.9 Transmission delay of commands issued by SDN controller modules to the infrastructure plan.

is less and almost half compared to others. So these designed modules do not make the controller bottleneck in terms of memory.

Fig. 8.9 shows the transmission delay for two modes proposed SDN-based and Without SDN. As can be seen, the transmission delay in the without SDN mode is almost twice that of the proposed SDN-based mode. The delay in the mode with SDN is approximately 400 milliseconds and in the mode without SDN is approximately 800 milliseconds. The lower this delay is, the faster the decisions made by the TSR module reach the IoT equipment at the DP. The opposite is also true. That is, the lower the transmission delay, the sooner the information collected from the DP will reach the RMS module and make decisions faster. The delay reduction that occurred in the case with SDN is due to centralization and that the very large network is logically centralized and directed and programmed by a centralized SDN controller. Imagine that the network was supposed to be distributed and at the DP and managed by IoT nodes or other network equipment. Certainly, management coordination between distributed nodes would impose a huge delay on the network. This issue can be seen to some extent in the Without SDN mode in Fig. 8.9.

The convergence delay results for different number of tasks are shown in Fig. 8.10. As can be seen, the convergence delay of the proposed SDN-based method is linear and the convergence delay of the without SDN method is exponential. Certainly, the presence of a controller who is responsible for collecting data and making decisions is very effective in convergence as quickly as possible. In the case of without SDN, as the number of tasks increases, the speed of convergence decreases because the number of message passing between IoT nodes to reach the solution increases. This is despite the fact that the number of tasks does not have much effect on the decision-making speed of the designed SDN controller. Because the SDN controller and its modules collect all the data of the network devices in a specific time period and centrally, and in the next period, execute the AWOA and reflect the decisions made to the devices. This issue helps to converge as quickly as possible to the answer and to quickly perform the necessary actions according to the decisions of the controller and the AWOA.

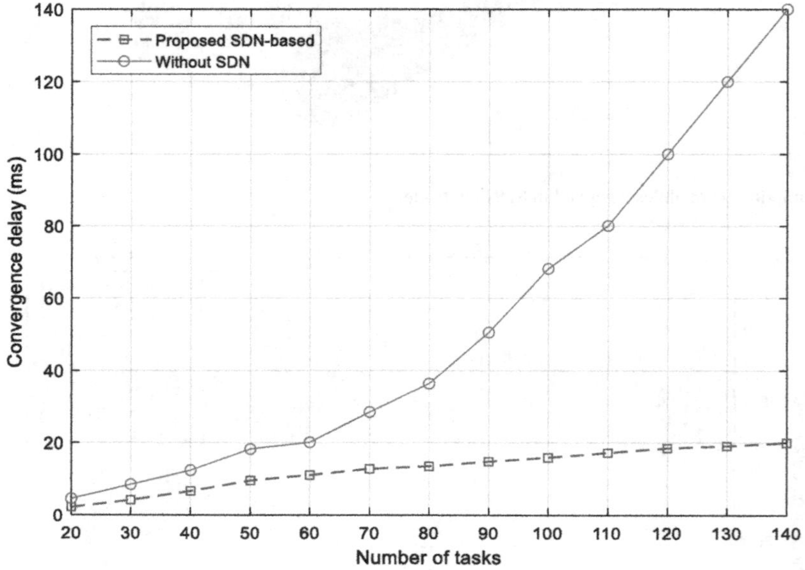

FIGURE 8.10 System convergence delay.

Table 8.7 shows the results of *PIR* (%) of the MT for NASA iPSC and HPC2N datasets. According to Table 8.7, AO has a lower *PIR* (%) value of the MT than WOA, PSO, HHO, and FA. So it can be said that AO was able to execute the set of tasks in less time than WOA, FA, PSO, and HHO algorithms.

TABLE 8.7 The PIR (%) of the Makespan Time for NASA iPSC and HPC2N datasets.

	NASA iPSC				HPC2N			
Dataset	Tasks				Tasks			
Algorithm	500	1000	1500	2000	500	1000	1500	2000
AO	4.28	1.34	1.85	1.42	1.18	1.59	1.27	1.39
WOA	6.43	2.93	3.76	2.05	1.32	2.48	1.76	1.54
FA	45.50	47.17	53.68	55.69	53.43	56.14	56.80	57.98
PSO	68.37	61.71	64.27	63.22	63.53	65.78	65.45	65.32
HHO	8.65	4.22	5.54	3.67	1.47	3.57	1.96	1.62

Fig. 8.11 shows the graph of *PIR* (%) of the MT resulting from AO, WOA, PSO, HHO and FA for both NASA iPSC and HPC2N datasets. As it is clear from Fig. 8.11, AWOA has the least difference compared to AO and the biggest difference compared to PSO.

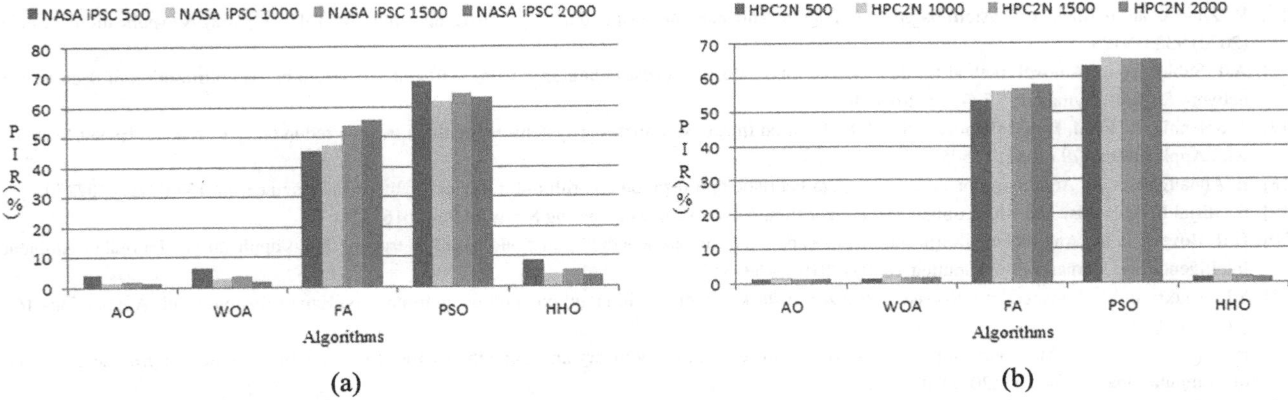

FIGURE 8.11 PIR (%) diagram of the Makespan Time for AO, WOA, PSO, HHO and FA for both NASA iPSC and HPC2N datasets. (a) NASA iPSC dataset. (b) HPC2N datasets.

8.7 Conclusion and future work

In this chapter, the scheduling problem of the IoT device requests in the FC was investigated as an optimization problem. The combination of WOA and AO algorithms was used to solve this problem, which is called AWOA. The WOA has a poor performance in the ERP, and this problem causes that the desired output for the WOA is not obtained in solving various problems, the ERP of the AO was used to improve the performance of WOA in the ERP. Because in AO, the ERP is stronger than the EIP. The performance of AWOA was evaluated using the metric of MT, OF value, PIR, and TT on two datasets and compared with HHO, FA, PSO, WOA, and AO algorithms, and it was proved that AWOA is better than algorithms it has better performance. Also, the proposed method was combined with SDN, and for the metrics of resource consumption, transmission delay, and convergence delay, the proposed method was evaluated in the state that is based on SDN and in the state that is without SDN, and it was proved that the scheduling algorithm performs better with SDN. Our future work in order to improve the WOA as much as possible is that it can be combined with other meta-heuristic algorithms such as MFO, Arithmetic Optimization Algorithm (AOA) [45] and Slime Mould Algorithm (SMA) [46] which are stronger than WOA in the ERP and to solve various problems, including feature extraction, image segmentation, network routing and TSch in CC and FC environments.

References

[1] A. Zanella, et al., Internet of things for smart cities, IEEE Internet of Things Journal 1 (2014) 22–32.
[2] L. Atzori, et al., The internet of things: a survey, Computer Networks 54 (2010) 2787–2805.
[3] M. Masdari, A. Khoshnevis, A survey and classification of the workload forecasting methods in cloud computing, Cluster Computing (2019) 1–26.
[4] F. Bonomi, et al., Fog computing and its role in the internet of things, in: Proceedings of the First Edition of the MCC Workshop on Mobile Cloud Computing, 2012, pp. 13–16.
[5] A. Shakarami, et al., Data replication schemes in cloud computing: a survey, Cluster Computing 24 (2021) 2545–2579.
[6] L.M. Vaquero, L. Rodero-Merino, Finding your way in the fog: towards a comprehensive definition of fog computing, ACM SIGCOMM Computer Communication Review 44 (2014) 27–32.
[7] M.R. Hossain, et al., A scheduling-based dynamic fog computing framework for augmenting resource utilization, Simulation Modelling Practice and Theory 111 (2021) 102336.
[8] M. Masdari, M. Zangakani, Green cloud computing using proactive virtual machine placement: challenges and issues, Journal of Grid Computing 18 (2020) 727–759.
[9] M. Yannuzzi, et al., Key ingredients in an IoT recipe: fog computing, cloud computing, and more fog computing, in: 2014 IEEE 19th International Workshop on Computer Aided Modeling and Design of Communication Links and Networks (CAMAD), 2014, pp. 325–329.
[10] A. Shahidinejad, et al., Context-aware multi-user offloading in mobile edge computing: a federated learning-based approach, Journal of Grid Computing 19 (2021) 1–23.
[11] Z. Tong, et al., A scheduling scheme in the cloud computing environment using deep Q-learning, Information Sciences 512 (2020) 1170–1191.
[12] M. Mtshali, et al., Multi-objective optimization approach for task scheduling in fog computing, in: 2019 International Conference on Advances in Big Data, Computing and Data Communication Systems, IcABCD, IEEE, 2019, pp. 1–6.
[13] M. Ghobaei-Arani, et al., An efficient task scheduling approach using moth-flame optimization algorithm for cyber-physical system applications in fog computing, Transactions on Emerging Telecommunications Technologies 31 (2) (2020) e3770.
[14] D. Zeng, et al., Joint optimization of task scheduling and image placement in fog computing supported software-defined embedded system, IEEE Transactions on Computers 65 (12) (2020) 3702–3712.

[15] W. Zhao, et al., Artificial ecosystem-based optimization: a novel nature-inspired meta-heuristic algorithm, Neural Computing & Applications 32 (13) (2020) 9383–9425.

[16] A.T. Sahlol, et al., A novel method for detection of tuberculosis in chest radiographs using artificial ecosystem-based optimisation of deep neural network features, Symmetry 12 (7) (2020) 1146.

[17] T. Salehnia, A. Fathi, Fault tolerance in LWT-SVD based image watermarking systems using three module redundancy technique, Expert Systems with Applications 179 (2021) 115058.

[18] L. Abualigah, et al., Aquila optimizer: a novel meta-heuristic optimization algorithm, Computers & Industrial Engineering 157 (2021) 107250.

[19] S. Mirjalili, A. Lewis, The whale optimization algorithm, Advances in Engineering Software 95 (2016) 51–67.

[20] H.R. Boveiri, et al., An efficient swarm-intelligence approach for task scheduling in cloud-based internet of things applications, Journal of Ambient Intelligence and Humanized Computing 10 (9) (2019) 3469–3479.

[21] S.E. Shukri, et al., Enhanced multi-verse optimizer for task scheduling in cloud computing environments, Expert Systems with Applications 168 (2021) 114230.

[22] B.M. Hasani Zade, N. Mansouri, Improved red fox optimizer with fuzzy theory and game theory for task scheduling in cloud environment, Journal of Computational Science 63 (2022) 101805.

[23] D. Bahmanyar, et al., Multi-objective scheduling of IoT-enabled smart homes for energy management based on Arithmetic Optimization Algorithm: a Node-RED and NodeMCU module-based technique, Knowledge-Based Systems 247 (2022) 108762.

[24] H. Liu, Research on cloud computing adaptive task scheduling based on ant colony algorithm, Optik 258 (2022) 168677.

[25] X. Xiao, et al., Adaptive search strategy based chemical reaction optimization scheme for task scheduling in discrete multiphysical coupling applications, Applied Soft Computing 121 (2022) 108748.

[26] N. Manikandan, et al., Bee optimization based random double adaptive whale optimization model for task scheduling in cloud computing environment, Computer Communications 187 (2022) 35–44.

[27] K. Dubey, S.C. Sharma, A novel multi-objective CR-PSO task scheduling algorithm with deadline constraint in cloud computing, Sustainable Computing: Informatics and Systems 32 (2021) 100605.

[28] M.A. Elaziz, et al., Advanced optimization technique for scheduling IoT tasks in cloud-fog computing environments, Future Generation Computer Systems 124 (2021) 142–154.

[29] T. Salehnia, et al., A workflow scheduling in cloud environment using a combination of Moth-Flame and Salp Swarm algorithms, Applied Soft Computing (2023).

[30] B. Vuppal, Swarnalatha, Software defined network using enhanced workflow scheduling in surveillance, Computer Communications 151 (2020) 196–201.

[31] S. Javanmardi, et al., FUPE: a security driven task scheduling approach for SDN-based IoT–Fog networks, Journal of Information Security and Applications 60 (2021) 102853.

[32] F. Shang, et al., The bandwidth-aware backup task scheduling strategy using SDN in Hadoop, Cluster Computing 22 (2019) 5975–5985.

[33] B. Sellami, et al., Energy-aware task scheduling and offloading using deep reinforcement learning in SDN-enabled IoT network, Computer Networks 210 (2022) 108957.

[34] B. Sellami, et al., Deep reinforcement learning for energy-efficient task scheduling in SDN-based IoT network, in: 2020 IEEE 19th International Symposium on Network Computing and Applications (NCA), 2020, p. 20340600.

[35] S.S. Chalapathi G., et al., An optimal delay aware task assignment scheme for wireless SDN networked edge cloudlets, Future Generation Computer Systems 102 (2020) 862–875.

[36] L.-A. Phan, et al., Dynamic fog-to-fog offloading in SDN-based fog computing systems, Future Generation Computer Systems 117 (2021) 486–497.

[37] B. Sellami, et al., Deep Reinforcement Learning for energy-aware task offloading in join SDN-Blockchain 5G massive IoT edge network, Future Generation Computer Systems 137 (2022) 363–379.

[38] Y. Liu, et al., A novel load balancing and low response delay framework for edge-cloud network based on SDN, IEEE Internet of Things Journal 7 (2020) 5922–5933.

[39] C.M. Wu, et al., A green energy-efficient scheduling algorithm using the DVFS technique for cloud datacenters, Future Generation Computer Systems 37 (2014) 141–147.

[40] J. Kennedy, R. Eberhart, Particle swarm optimization, in: Proceedings of ICNN'95-International Conference on Neural Networks, vol. 4, IEEE, 1995, pp. 1942–1948.

[41] A.A. Heidari, et al., Harris hawks optimization: algorithm and applications, Future Generation Computer Systems 97 (2019) 849–872.

[42] Montazerolghaem, et al., OpenSIP: toward software-defined SIP networking, IEEE Transactions on Network and Service Management 15 (1) (2018).

[43] M. Hamdan, et al., A comprehensive survey of load balancing techniques in software-defined network, Journal of Network and Computer Applications 174 (2021) 102856.

[44] Parallel workloads archive, http://www.cse.huji.ac.il/labs/parallel/workload/logs.html, 2020. (Accessed July 2020).

[45] L. Abualigah, et al., The arithmetic optimization algorithm, Computer Methods in Applied Mechanics and Engineering 376 (2021) 113609.

[46] S. Li, et al., Slime mould algorithm: a new method for stochastic optimization, Future Generation Computer Systems 111 (2020) 300–323.

Chapter 9

An enhanced whale optimization algorithm using the Nelder-Mead algorithm and logistic chaotic map

Farouq Zitouni[a] and Saad Harous[b]

[a]*Department of Computer Science, Kasdi Merbah University, Ouargla, Algeria,* [b]*Department of Computer Science, College of Computing and Informatics, University of Sharjah, Sharjah, United Arab Emirates*

9.1 Introduction

Optimization is the process of searching for optimal solutions to a given problem, usually modeled as a mathematical formulation, i.e., objective function and constraints [42]. There are three main optimization techniques: analytical, enumerative, and metaheuristics [5,6,34]. Analytical methods use gradient information to find the optimal solution. Enumerative schemes go over the search space and exhaustively search for the optimal solution. Metaheuristic algorithms manipulate some indices and frequently produce high-quality solutions. Several real-life optimization problems are hard to tackle by exact methods (i.e., analytical and enumerative techniques) because of their challenging nature, such as high-dimensionality, multimodality, and non-differentiability [10]. Hence, metaheuristic algorithms are the best approaches to solve them because they give near-optimal solutions in a reasonable amount of computational time.

Metaheuristic methods often provide satisfactory solutions with a few computing effort [16]. They search over a broad set of feasible solutions. They are generally nature-inspired: i.e., using various sources to propose new algorithms. For example, foraging behavior of ants [11], hunting behavior of whales [29], annealing of metals [43]. During the 1980s and 1990s, many researchers were interested in metaheuristic algorithms. Famous algorithms proposed during that period covered the simulated annealing [43], Tabu search [14,15], evolutionary computation [4], to name but a few. Before the No-Free-Lunch (NFL) theorem apparition in 1995 [46,47], researchers tried hard to design universal metaheuristic algorithm, i.e., algorithm capable of efficiently solving any optimization problem. The NFL theorem proved the non-existence of such an algorithm. The theorem states that no metaheuristic algorithm performs better than another when averaged over all functions. That is, all metaheuristic algorithms have similar performance when evaluated on all functions. Over the past three decades, many metaheuristic algorithms have been proposed and experimentally evaluated [57]. They could be broadly organized into four classes: evolutionary-based, swarm-based, physical-based, and human-based algorithms [56].

Metaheuristic algorithms can be individual-based or population-based algorithms [18]. The first family of algorithms use only one solution to investigate the search space. Simulated annealing [43], tabu search [14,15], iterated local search [26], guided local search [44], pattern search/random search [17], solis–wets algorithm [40], and variable neighborhood search [30] are some examples of individual-based metaheuristics. The second family of algorithms employ several solutions to cooperatively investigate the search space. Evolutionary [3] and swarm-based [51] algorithms are two typical examples of population-based metaheuristics.

Any metaheuristic algorithm must include two basic operations: exploration and exploitation [49]. In some references, exploration is called diversification, and exploitation is called intensification [50]. The diversification is used to investigate the search space globally (global optimization); so that the metaheuristic algorithm can find promising areas. The intensification is employed to scrutinize the search space locally (local optimization); so that the metaheuristic algorithm can inspect specific regions. Individual-based algorithms are slightly exploitive, whereas population-based algorithms are moderately explorative [12]. It is worth mentioning that the overuse of a specific operation should result in an unsatisfactory quality of final solutions [52]. If the search space is explored too much, the metaheuristic algorithm loses focus, and wastes effort. If the search space is exploited too much, the metaheuristic algorithm will likely get stuck in a local optimum. Hence, the proper balance between diversification and intensification is essential to find near optimal solutions in an efficient way.

Handbook of Whale Optimization Algorithm. https://doi.org/10.1016/B978-0-32-395365-8.00015-4

We introduce a novel hybrid metaheuristic algorithm for global optimization. The proposed optimizer combines the Nelder-Mead method [33] and the linear chaotic map [8] along with the basic whale optimization algorithm [29]. The proposed algorithm has the following properties:

- The Nelder-Mead algorithm is mainly used to shrink the search space range. This would help minimizing the search effort and avoiding to get stuck in local optimums.
- The linear chaotic map is employed to control the decreasing of the values of vector $\vec{a}$ in the basic whale optimization algorithm.

The algorithm's performance is evaluated on the CEC 2021 benchmark problem suite. It comprises ten challenging single objective test functions. For additional information on the CEC 2021 benchmark, please click here. The acquired results are compared to the original whale optimization and chaotic whale optimization metaheuristics. The proposed algorithm outperforms the other algorithms.

The remainder of the paper is structured as follows. Section 9.2 overviews recent related work regarding combining and hybridizing the whale optimization optimizer with some optimization algorithms. A summary of the whale optimization algorithm, Nelder-Mead method, and logistic chaotic map is given in Section 9.3. Section 9.4 outlines the proposed hybridization of the whale optimization algorithm with the Nelder-Mead method and logistic chaotic map. Section 9.5 discusses the experiment results and the comparison to other optimizer. Section 9.6 concludes the paper.

9.2 Related work

Hybrid metaheuristic algorithms are gaining increasing attention in the optimization community. During the last few years, the literature has witnessed a tremendous expansion concerning the hybridization and combination of various optimization methods to enhance and improve performance. In this section, we overview state-of-the-art metaheuristics regarding the hybridization of the whale optimization algorithm [29] with diverse optimization methods.

The work in [28] hybridizes the whale optimization and simulated annealing algorithms to solve the feature selection problem. In addition, the authors of [27] implant a modified differential evolution mechanism with a strong exploration capability in the whale optimization algorithm. Furthermore, the study in [7] inserts the mutualism and commensalism phases of the symbiotic organisms search into the whale optimization algorithm along with the mutation strategy of the differential evolution. Likewise, a whale optimization algorithm incorporated with an artificial bee colony is suggested in [41] to solve some classical drawbacks, such as slow convergence, low precision, and premature convergence. Finally, the proposal discussed in [38] hybridizes the whale optimization, differential evolution, and genetic algorithms to solve the unit commitment scheduling problem.

The work in [1] proposes a novel hybrid algorithm of slime mold and whale optimization algorithms to tackle the chest X-ray image segmentation problem. Likewise, the study in [37] suggests a new hybrid algorithm of the whale optimization and sine cosine algorithms with the Lévy flight for global optimization problems. Furthermore, the authors of [13] enhance the whale optimization algorithm by incorporating the mechanism of the salp swarm algorithm and lens opposition-based learning strategy. A new hybrid algorithm of the whale optimization algorithm and particle swarm optimization is presented in [22] to solve some design optimization problems in electronics. Finally, the paper [53] suggests a novel hybrid whale optimization algorithm with gathering strategies to deal with high-dimensional optimization problems.

The study in [20] proposes a novel hybrid algorithm of whale optimization and modified conjugate gradient algorithms. Also, the hybrid algorithm described in [25] introduces the niching technique to the whale optimization algorithm to enhance convergence speed and search coverage in solving global optimization problems. Likewise, the work presented in [54] describes a hybrid algorithm based on Pearson's correlation coefficient and correlation distance combined with the whale optimization algorithm to solve the feature selection problem. Furthermore, the paper [2] presents a hybrid of the whale optimization algorithm with a novel method called the local minima avoidance method to tackle multi-threshold color image segmentation problems. Finally, the work in [36] describes a hybrid machine-learning approach that utilizes rough neighborhood sets with a binary whale optimization algorithm for handwritten Arabic character recognition.

The solution proposed in [32] designs a new hybrid model of whale optimization and flower pollination algorithms to solve the feature selection problem. In addition, the study [39] presents a hybrid technique based on the mean grey wolf optimizer and whale optimization algorithm. Besides, the work proposed in [23] introduces a hybrid whale optimization algorithm using genetic and thermal exchange algorithms. Also, the algorithm presented in [21] proposes a novel hybrid using the whale optimization algorithm and grey wolf optimizer. Finally, the reference [9] suggests a hybrid algorithm based on the whale optimization algorithm, Tabu search, and local search procedures.

9.3 Overview of used algorithms

In this section, we give the building blocks of the algorithms used to propose the enhanced whale optimization algorithm. First, in Section 9.3.1, we overview the original whale optimization algorithm. Then, in Section 9.3.2, we describe the Nelder-Mead algorithm. Finally, in Section 9.3.3, we define the logistic chaotic map.

9.3.1 Whale optimization

The Whale Optimization Algorithm (WOA) is a swarm-based optimizer proposed to solve continuous optimization problems. It has shown outstanding performance compared to recent and classical state-of-the-art metaheuristic techniques [29]. In addition, it is a robust and easy to implement algorithm, which requires a few numbers of controlling parameters. The mechanism of food searching of humpback whales is depicted in Fig. 9.1.

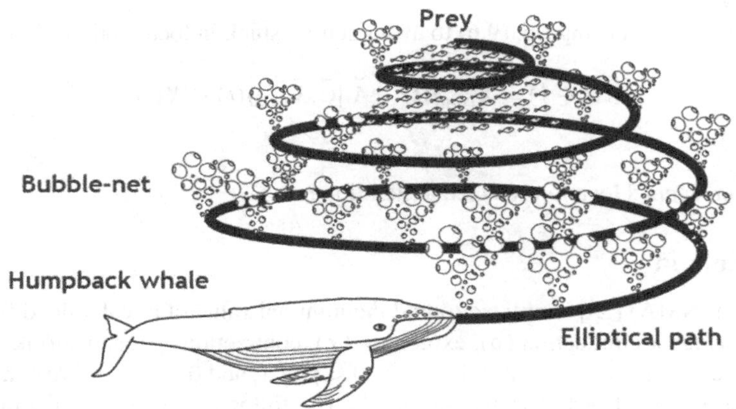

FIGURE 9.1 The mechanism of food searching of humpback whales.

Humpback whales possess a fascinating foraging strategy called the bubble-net hunting method [45]. They usually hunt schools of fish swimming near the ocean's surface. Scientists observed that whales dive profoundly (about 12 meters); then, they form many bubbles following a spiral pattern around the target; and finally swim up towards the surface - this maneuver is named the upward spiral. We describe the spiral bubble-net feeding maneuver and give its mathematical formulation. It comprises three stages: encircling prey, bubble-net attacking method (exploitation phase), and search for prey (exploration phase).

9.3.1.1 Encircling prey phase

Humpback whales can locate their prey's position and surround them. Provided that the location of the optimal solution is unknown in advance, the WOA initially considers that the existing best candidate solution is the target prey. Therefore, the best search individual is identified first, and the other search individuals will update their locations towards the former. Eqs. (9.1), (9.2), (9.3), and (9.4) describe the updating behavior.

$$\vec{D} = |\vec{C}.\vec{X}^*(t) - \vec{X}(t)| \tag{9.1}$$

$$\vec{X}(t+1) = \vec{X}^*(t) - \vec{A}.\vec{D} \tag{9.2}$$

$$\vec{A} = 2\vec{a}.\vec{r} - \vec{a} \tag{9.3}$$

$$\vec{C} = 2\vec{r} \tag{9.4}$$

where

$\vec{X}^*(t)$	:	The best solution at iteration t.
$\vec{X}(t)$ and $\vec{X}(t+1)$	:	The agent's location at iterations t and $t+1$.
$\vec{a}$	:	A linearly decreasing vector ranging from 2 to 0.
$\vec{r}$	:	A random uniformly distributed vector.
$\lvert . \rvert$	:	The absolute value.

9.3.1.2 Bubble-net attacking phase

To model the bubble-net behavior of humpback whales, we use the mathematical formulation of spirals. We employ Eq. (9.5) to mimic the helix-shaped movement of humpback whales.

$$\vec{X}(t+1) = e^{bl}\cos(2\pi l)|\vec{X}^*(t) - \vec{X}(t)| + \vec{X}^*(t) \tag{9.5}$$

where

b : A constant used to define the logarithmic spiral's shape.
l : A random number drawn within the range $[-1, 1]$.
p : A random number drawn within the range $[0, 1]$.

9.3.1.3 Search for prey phase

We update the search agents' location using Eq. (9.6) to avoid getting stuck in local optimums and have global optimizers.

$$\vec{X}(t+1) = \vec{X}_{rand}(t) - \vec{A}.|\vec{C}.\vec{X}_{rand}(t) - \vec{X}(t)| \tag{9.6}$$

where

$\vec{X}_{rand}(t)$: A random agent's location selected at iteration t.

9.3.2 Nelder-Mead method

The Nelder-Mead Algorithm (NMA) [33] is utilized to find the minimal value of a real-valued function $f(\mathbf{x})$, $\mathbf{x} \in \mathbb{R}^n$. The NMA has four parameters which are reflection (ρ), expansion (χ), contraction (γ), and shrinkage (σ). These coefficients should satisfy the following constraints: $\rho > 0$, $\chi > 1$, $\chi > \rho$, $0 < \gamma < 1$, and $0 < \sigma < 1$. We consider a simplex with $n+1$ vertices $\mathbf{x}_i$, where each one is a point of the search space $\mathbb{R}^n$. The following steps describe one iteration of the NMA's process.

1. **Sorting:** The set of vertices must be sorted in an ascending manner. Let $\mathbf{x}_1$ and $\mathbf{x}_{n+1}$ denote the locations with the smallest and largest values, respectively: i.e., $f(\mathbf{x}_1) \le f(\mathbf{x}_2) \le \ldots \le f(\mathbf{x}_n) \le f(\mathbf{x}_{n+1})$.
2. **Reflection:** Calculate the reflection point $\mathbf{x}_r$ using Eq. (9.7). The variable $\bar{\mathbf{x}}$ represents the centroid of the n first vertices. If the condition $f(\mathbf{x}_1) \le f(\mathbf{x}_r) < f(\mathbf{x}_n)$ is satisfied, then accept $\mathbf{x}_r$ and terminate the iteration.

$$\mathbf{x}_r = (1+\rho)\bar{\mathbf{x}} - \rho\mathbf{x}_{n+1} \tag{9.7}$$

3. **Expansion:** If $f(\mathbf{x}_r) < f(\mathbf{x}_1)$, then compute the expansion vertex $\mathbf{x}_e$ using Eq. (9.8). If $f(\mathbf{x}_e) < f(\mathbf{x}_r)$, then accept the vertex $\mathbf{x}_e$ and finish the iteration; otherwise, accept $\mathbf{x}_r$ and terminate the iteration.

$$\mathbf{x}_e = (1+\rho\chi)\bar{\mathbf{x}} - \rho\chi\mathbf{x}_{n+1} \tag{9.8}$$

4. **Contraction:** If $f(\mathbf{x}_r) \ge f(\mathbf{x}_n)$, then compute a contraction between $\bar{\mathbf{x}}$ and the better vertex of $\mathbf{x}_{n+1}$ and $\mathbf{x}_r$.
   a. **Outside:** If $f(\mathbf{x}_n) \le f(\mathbf{x}_r) < f(\mathbf{x}_{n+1})$, then calculate an outside contraction using Eq. (9.9). If $f(\mathbf{x}_{oc}) < f(\mathbf{x}_r)$, then accept the vertex $\mathbf{x}_{oc}$ and finish the iteration; otherwise, go to Step 5.

$$\mathbf{x}_{oc} = (1+\rho\gamma)\bar{\mathbf{x}} - \rho\gamma\mathbf{x}_{n+1} \tag{9.9}$$

   b. **Inside:** If $f(\mathbf{x}_r) \ge f(\mathbf{x}_{n+1})$, then perform an inside contraction using Eq. (9.10). If $f(\mathbf{x}_{ic}) < f(\mathbf{x}_{n+1})$, then accept the vertex $\mathbf{x}_{ic}$ and complete the iteration; otherwise, go to Step 5.

$$\mathbf{x}_{ic} = (1-\gamma)\bar{\mathbf{x}} + \gamma\mathbf{x}_{n+1} \tag{9.10}$$

5. **Shrinkage:** Compute the new set of vertices using Eq. (9.11). Then, the vertices used in the next iteration are $\mathbf{x}_1, \mathbf{v}_2, \ldots, \mathbf{v}_n, \mathbf{v}_{n+1}$

$$\mathbf{v}_i = \mathbf{x}_1 + \sigma(\mathbf{x}_i - \mathbf{x}_1),\ i = \{2, \ldots, n+1\} \tag{9.11}$$

9.3.3 Logistic chaotic map

This section introduces one of the most well-known dynamical systems: the logistic chaotic map. It is a one-dimensional and discrete-time map that, despite its simplicity, shows a surprising degree of complexity. Historically, it has been considered as one of the most essential and paradigmatic systems during the early days of study on deterministic chaos [8]. The logistic chaotic map is given by Eq. (9.12).

$$x_{n+1} = \lambda x_n (1 - x_n) \ , \ n = 0, 1, 2, \ldots \tag{9.12}$$

where

λ : A positive parameter ($0 < \lambda < 4$).

9.4 Proposed algorithm

We give the enhanced whale optimization algorithm and describe its steps. Algorithm 9.1 depicts the different instructions of the proposed optimizer.

- Lines 1 to 6: During the first stage of the enhanced WOA, we utilize the NMA to reduce the search space size. For this purpose, we randomly select $D+1$ vertices and iterate the NMA's process until the considered stopping criterion is satisfied. The set of vertices forms a simplex. The NMA's iterative procedure ceases when the volume of the corresponding simplex reaches a given value.
- Lines 7 to 9: In this phase, we define the new decision variables' intervals. We believe that if the search space size is appropriately reduced, this will allow a rapid convergence and avoid getting stuck in local optimums.
- Lines 10 to 12: We generate the initial population of the enhanced WOA.
- Lines 13 to 36: In this phase, we replicate the main body of the original WOA. In addition, the chaotic logistic map is used to update the components' values of vector $\vec{a}$.
- Line 37: Finally, we return the best candidate solution selected from the last population.

$$\vec{X} = (\alpha_1 (x_1^{max} - x_1^{min}) + x_1^{min}, \ldots, \alpha_D (x_D^{max} - x_D^{min}) + x_D^{min}) \tag{9.13}$$

$$V = \frac{1}{D!} |Det(M_E)| \ , \ M_E = [\mathbf{x}_2 - \mathbf{x}_1, \ldots, \mathbf{x}_{D+1} - \mathbf{x}_1] \tag{9.14}$$

9.5 Experimental results and discussion

The benchmark CEC 2021 on single objective bound constrained numerical optimization [31] is employed to examine the effectiveness of the enhanced WOA. The CEC 2021 test suite comprises four groups of test functions: unimodal (UM), basic (BC), hybrid (HD), and composition (CM). First, the UM functions possess one global optimum. They are utilized to evaluate the exploitation capacity of metaheuristics. Then, the BC functions have multiple global optimums. They are employed to assess, on the one hand, the exploration power of metaheuristics and, on the other hand, their ability to avoid getting stuck in local optimums. Finally, the HD and CM functions are derived from the hybridization and composition of several elementary test functions. They are used to observe the well-balancing between the exploration and exploitation of metaheuristics. The mathematical formulation and characteristics of the used UM, BC, HD, and CM functions are available in [31]. Table 9.1 summarizes the features of the used test functions.

All the experiments were run using the Java programming language on a workstation with a Windows 10 familial edition (64-bit). The processor is Intel(R) Core(TM) i7–9750H CPU @ 2.60 GHz 2.59 GHz, with 16 GB of RAM. The dimension of the search space (D) is set to 10 or 20. The population size (N) is set to $\lfloor 30 \times D^{1.5} \rfloor$, where the term $\lfloor x \rfloor$ denotes the truncation of the real number x. The maximum number of function evaluations for each dimension equals 200000 and 1000000, respectively. The range of allowed values for each decision variable is $[-100, 100]$. All the results are averaged over 50 independent runs.

- The reflection value ρ is set to 1; the expansion value χ is set to 2; the contraction value γ is set to 0.5; and the shrinkage value σ is set to 0.5 [33].
- The tolerance threshold θ is set to 10^5.
- The logistic chaotic map parameter λ is set to 2.
- The logarithmic spiral's shape is set to 1 [29].

The performance of the enhanced WOA (viz. NM-CWOA) is compared to the approaches described in [29] (viz. WAO) and [19] (viz. CWAO). The obtained statistical results are summarized in Tables 9.2 and 9.3 (i.e., we report the mean and

Algorithm 9.1: Pseudocode of the enhanced whale optimization algorithm.

Input: D: The search space dimension.
Input: $[x_1^{min}, x_1^{max}], \ldots, [x_D^{min}, x_D^{max}]$: The decision variables' domains.
Input: N: The population size.
Input: θ: The tolerance threshold.
Input: ρ, χ, γ, and σ: The reflection, expansion, contraction, and shrinkage parameters, respectively.
Input: λ: The logistic chaotic map parameter.
Input: b: The logarithmic spiral's shape.

/* Nelder-Mead algorithm */
Generate randomly $D+1$ vertices (viz. $\mathbf{x}_1, \ldots, \mathbf{x}_{D+1}$) using Eq. (9.13);
Calculate the geometric volume of the simplex $(\mathbf{x}_1, \ldots, \mathbf{x}_{D+1})$ using Eq. (9.14) (viz. V);
while $V \geq \theta$ **do**
    Iterate the NMA's process as described in Section 9.3.2;
    Calculate the new geometric volume of the simplex $(\mathbf{x}_1, \ldots, \mathbf{x}_{D+1})$ using Eq. (9.14);
end
/* Define the new boundaries of the search space */
for $i \leftarrow 1$ **to** D **do**
    $[x_i^{min}, x_i^{max}] \leftarrow [\min\{\mathbf{x}_1(i), \ldots, \mathbf{x}_{D+1}(i)\}, \max\{\mathbf{x}_1(i), \ldots, \mathbf{x}_{D+1}(i)\}]$;
end
/* Initialization of the initial population */
for $i \leftarrow 1$ **to** N **do**
    The location $\vec{X}_i(0)$ is generated using Eq. (9.13);
end
/* Optimization process */
for $t \leftarrow 1$ **to** $MaxIter$ **do**
    for $j \leftarrow 1$ **to** N **do**
        Update the components' values of the vector $\vec{a}$ using Eq. (9.12);
        Generate the random vector $\vec{r}$ using a uniform distribution;
        Update the components' values of vectors $\vec{A}$ and $\vec{C}$ using Eqs. (9.3) and (9.4), respectively;
        Generate the random number l within the range $[-1, 1]$;
        Generate the random number p within the range $[0, 1]$.;
        if $p < 0.5$ **then**
            if $|A| < 1$ **then**
                Update the location of the individual $\vec{X}_j(t)$ using Eq. (9.2);
            end
            else
                Update the location of the individual $\vec{X}_j(t)$ using Eq. (9.6);
            end
        end
        else
            Update the location of the individual $\vec{X}_j(t)$ using Eq. (9.5);
        end
    end
    for $i \leftarrow 1$ **to** N **do**
        Make sure that the position of the individual $\vec{X}_i(t)$ is within the decision variables' domains;
    end
end
/* Get the best solution */
Return the best solution $\vec{X}^*$;

TABLE 9.1 Summary of the benchmark CEC'21.

No.	Type	Function	Global optimum
F_1	UF	Shifted and Rotated Bent Cigar Function (CEC 2017 [48] $F1$)	100
F_2	BF	Shifted and Rotated Schwefel's Function (CEC 2014 [24] F_{11})	1100
F_3	BF	Shifted and Rotated Lunacek bi-Rastrigin Function (CEC 2017 [48] F_7)	700
F_4	BF	Expanded Rosenbrock's plus Griewangk's Function (CEC2017 [48] f_{19})	1900
F_5	HF	Hybrid Function 1 (N = 3) (CEC 2014 [24] F_{17})	1700
F_6	HF	Hybrid Function 2 (N = 4) (CEC 2017 [48] F_{16})	1600
F_7	HF	Hybrid Function 3 (N = 5) (CEC 2014 [24] F_{21})	2100
F_8	CF	Composition Function 1 (N = 3) (CEC 2017 [48] F_{22})	2200
F_9	CF	Composition Function 2 (N = 4) (CEC 2017 [48] F_{24})	2400
F_{10}	CF	Composition Function 3 (N = 5) (CEC 2017 [48] F_{25})	2500

TABLE 9.2 Statistical results of NM-CWOA in comparison to WAO and CWAO ($D = 10$).

	NM-CWOA		WAO			CWAO		
	Mean	STD	Mean	STD	P	Mean	STD	P
F_1	-1.94E-08	5.95E-06	1.38E-03	5.51E-03	+	2.28E-06	6.10E-04	+
F_2	-7.55E-07	5.78E-06	2.79E-04	5.35E-03	+	-4.35E-07	5.29E-04	+
F_3	-1.26E-06	5.51E-06	5.63E+00	5.58E-03	+	1.05E+01	4.81E-04	+
F_4	3.16E-02	5.56E-06	2.44E-01	5.20E-03	+	2.77E-01	5.65E-04	+
F_5	-1.64E-06	5.80E-06	-9.77E-05	5.72E-03	+	1.31E-04	5.92E-04	+
F_6	3.90E-02	5.14E-06	3.39E-02	5.47E-03	+	1.96E-02	5.78E-04	+
F_7	1.52E-02	5.52E-06	7.81E-03	5.58E-03	+	1.27E-03	5.97E-04	+
F_8	9.73E-07	4.94E-06	5.07E-04	4.97E-03	+	4.05E+01	6.11E-04	+
F_9	-1.34E-06	5.25E-06	2.89E-04	6.21E-03	+	9.33E+01	5.70E-04	+
F_{10}	4.29E-03	5.67E-06	4.64E+01	5.42E-03	+	3.90E+02	5.88E-04	+

TABLE 9.3 Statistical results of NM-CWOA in comparison to WAO and CWAO ($D = 20$).

	NM-CWOA		WAO			CWAO		
	Mean	STD	Mean	STD	P	Mean	STD	P
F_1	8.13E-09	6.00E-06	3.34E-04	6.02E-03	+	1.83E-04	5.06E-04	+
F_2	8.48E-07	5.60E-06	1.13E-03	5.11E-03	+	-6.19E-05	5.73E-04	+
F_3	2.02E+01	5.63E-06	1.28E+01	5.76E-03	+	-7.06E-05	5.67E-04	+
F_4	5.57E-01	6.32E-06	3.37E-01	5.55E-03	+	1.52E-02	5.99E-04	+
F_5	2.22E-07	5.63E-06	-1.87E-03	4.97E-03	+	1.32E-05	5.97E-04	+
F_6	8.02E-02	5.53E-06	9.01E-02	6.01E-03	+	1.78E-02	5.43E-04	+
F_7	3.48E-02	5.68E-06	3.65E-02	5.18E-03	+	1.03E-02	6.05E-04	+
F_8	9.67E+01	5.77E-06	2.13E-04	5.59E-03	+	1.55E-04	5.48E-04	+
F_9	1.80E+02	5.60E-06	-8.06E-04	5.95E-03	+	-9.57E-05	5.80E-04	+
F_{10}	4.00E+02	4.81E-06	4.88E+01	6.10E-03	+	1.47E-03	6.04E-04	+

standard deviation values). We perform the Wilcoxon signed-rank test [35] to check whether NM-CWOA is better/worst than WAO and CWAO in all dimensions. We consider the one-sided test with $\alpha = 0.05$.

The results in Tables 9.2 and 9.3 demonstrate that the performance of NM-CWOA is better than WOA and CWOA in 100% of the achieved experiments. Hence, NM-CWOA has outstanding exploitation and exploration of the search space. As a matter of fact, this is due to the use of the Nelder-Mead algorithm, which considerably reduced the size of the search region and consequently enhanced the quality of the obtained solutions.

Furthermore, we execute the Friedman statistical test [55] employing the IBM SPSS Statistics software to establish whether the performance of NM-CWOA, WOA, and CWOA is similar in dimensions 10 and 20 or not. We have three algorithms and ten standard deviation values for each algorithm. The value of α is set to 0.05, and the value of the degree of freedom is set to 2. The acquired ranks for NM-CWOA, WOA, and CWOA are 1, 3, and 2, respectively. We observe that NM-CWOA has the first rank value, which indicates that its performance is the best, followed by CWOA and WOA.

9.6 Conclusion and future scope

This work presented a new hybrid metaheuristic algorithm for global optimization. It combines the Nelder-Mead and the whale optimization optimizers. We used the Nelder-Mead algorithm in the first stage to shrink the search space size. It allows, on the one hand, minimizing the search effort and, on the other hand, avoiding getting stuck in local optimums and accordingly preventing the premature convergence issue. Moreover, we exploited the linear chaotic map equation to regulate the decreasing process of the values of vector $\vec{a}$ in the original whale optimization algorithm. In addition, we utilized the CEC 2021 benchmark test suite to assess the performance of the suggested hybrid optimizer by comparing it to the basic whale optimization and the chaotic whale optimization algorithms. Finally, we performed the Friedman and Wilcoxon statistical tests to examine the comparative study. The proposed optimizer outperforms the considered metaheuristic algorithms.

References

[1] M. Abdel-Basset, V. Chang, R. Mohamed, HSMA_WOA: a hybrid novel slime mould algorithm with whale optimization algorithm for tackling the image segmentation problem of chest X-ray images, Applied Soft Computing 95 (2020) 106642.

[2] M. Abdel-Basset, R. Mohamed, N.M. AbdelAziz, M. Abouhawwash, HWOA: a hybrid whale optimization algorithm with a novel local minima avoidance method for multi-level thresholding color image segmentation, Expert Systems with Applications 190 (2022) 116145.

[3] T. Back, Evolutionary Algorithms in Theory and Practice: Evolution Strategies, Evolutionary Programming, Genetic Algorithms, Oxford University Press, 1996.

[4] T. Bäck, D.B. Fogel, Z. Michalewicz, Evolutionary Computation 1: Basic Algorithms and Operators, CRC Press, 2018.

[5] R. Bellman, Mathematical Optimization Techniques, Univ of California Press, 1963.

[6] C. Blum, A. Roli, Metaheuristics in combinatorial optimization: overview and conceptual comparison, ACM Computing Surveys (CSUR) 35 (3) (2003) 268–308.

[7] S. Chakraborty, A.K. Saha, S. Sharma, R. Chakraborty, S. Debnath, A hybrid whale optimization algorithm for global optimization, Journal of Ambient Intelligence and Humanized Computing (2021) 1–37.

[8] M. Denker, W.A. Woyczynski, Introductory Statistics and Random Phenomena. Uncertainty, Complexity and Chaotic Behavior in Engineering and Science, with Mathematica Uncertain Virtual Worlds by Bernard Ycart, Birkhäuser, Boston, 1998.

[9] S.K. Dewi, D.M. Utama, A new hybrid whale optimization algorithm for green vehicle routing problem, Systems Science & Control Engineering 9 (1) (2021) 61–72.

[10] A. Díaz-Manríquez, G. Toscano-Pulido, W. Gómez-Flores, On the selection of surrogate models in evolutionary optimization algorithms, in: 2011 IEEE Congress of Evolutionary Computation (CEC), IEEE, 2011, pp. 2155–2162.

[11] M. Dorigo, C. Blum, Ant colony optimization theory: a survey, Theoretical Computer Science 344 (2–3) (2005) 243–278.

[12] K.-L. Du, M.N.S. Swamy, Introduction, in: Search and Optimization by Metaheuristics: Techniques and Algorithms Inspired by Nature, Springer International Publishing, Cham, 2006, pp. 1–28.

[13] Q. Fan, Z. Chen, W. Zhang, X. Fang, ESSAWOA: enhanced whale optimization algorithm integrated with salp swarm algorithm for global optimization, Engineering With Computers (2020) 1–18.

[14] F. Glover, Tabu search—part I, ORSA Journal on Computing 1 (3) (1989) 190–206.

[15] F. Glover, Tabu search—part II, ORSA Journal on Computing 2 (1) (1990) 4–32.

[16] T.F. Gonzalez, Handbook of Approximation Algorithms and Metaheuristics, CRC Press, 2007.

[17] R. Hooke, T.A. Jeeves, "Direct search" solution of numerical and statistical problems, Journal of the ACM (JACM) 8 (2) (1961) 212–229.

[18] N.S. Jaddi, S. Abdullah, Global search in single-solution-based metaheuristics, in: Data Technologies and Applications, 2020.

[19] G. Kaur, S. Arora, Chaotic whale optimization algorithm, Journal of Computational Design and Engineering 5 (3) (2018) 275–284.

[20] L.R. Khaleel, B.A. Mitras, Hybrid whale optimization algorithm with modified conjugate gradient method to solve global optimization problems, Open Access Library Journal 7 (6) (2020) 1–18.

[21] A. Korashy, S. Kamel, F. Jurado, A.-R. Youssef, Hybrid whale optimization algorithm and grey wolf optimizer algorithm for optimal coordination of direction overcurrent relays, Electric Power Components and Systems 47 (6–7) (2019) 644–658.

[22] N.M. Laskar, K. Guha, I. Chatterjee, S. Chanda, K.L. Baishnab, P.K. Paul, HWPSO: a new hybrid whale-particle swarm optimization algorithm and its application in electronic design optimization problems, Applied Intelligence 49 (1) (2019) 265–291.

[23] C.-Y. Lee, G.-L. Zhuo, A hybrid whale optimization algorithm for global optimization, Mathematics 9 (13) (2021) 1477.

[24] J.J. Liang, B.Y. Qu, P.N. Suganthan, Problem definitions and evaluation criteria for the CEC 2014 special session and competition on single objective real-parameter numerical optimization, Computational Intelligence Laboratory, Zhengzhou University, Zhengzhou China and Technical Report, Nanyang Technological University, Singapore, 2013, 635:490.
[25] X. Lin, X. Yu, W. Li, A heuristic whale optimization algorithm with niching strategy for global multi-dimensional engineering optimization, Computers & Industrial Engineering 171 (2022) 108361.
[26] H.R. Lourenço, O.C. Martin, T. Stützle, Iterated local search: framework and applications, in: Handbook of Metaheuristics, Springer, 2019, pp. 129–168.
[27] J. Luo, B. Shi, A hybrid whale optimization algorithm based on modified differential evolution for global optimization problems, Applied Intelligence 49 (5) (2019) 1982–2000.
[28] M.M. Mafarja, S. Mirjalili, Hybrid whale optimization algorithm with simulated annealing for feature selection, Neurocomputing 260 (2017) 302–312.
[29] S. Mirjalili, A. Lewis, The whale optimization algorithm, Advances in Engineering Software 95 (2016) 51–67.
[30] N. Mladenović, P. Hansen, Variable neighborhood search, Computers & Operations Research 24 (11) (1997) 1097–1100.
[31] A.W. Mohamed, A.A. Hadi, P. Agrawal, K.M. Sallam, A.K. Mohamed, Gaining-sharing knowledge based algorithm with adaptive parameters hybrid with IMODE algorithm for solving CEC 2021 benchmark problems, in: 2021 IEEE Congress on Evolutionary Computation (CEC), IEEE, 2021, pp. 841–848.
[32] H. Mohammadzadeh, F.S. Gharehchopogh, A novel hybrid whale optimization algorithm with flower pollination algorithm for feature selection: case study email spam detection, Computational Intelligence 37 (1) (2021) 176–209.
[33] J.A. Nelder, R. Mead, A simplex method for function minimization, The Computer Journal 7 (4) (1965) 308–313.
[34] G.C. Onwubolu, B. Babu, New Optimization Techniques in Engineering, vol. 141, Springer, 2013.
[35] D. Rey, M. Neuhäuser, Wilcoxon-signed-rank test, in: International Encyclopedia of Statistical Science, Springer, Berlin, Heidelberg, 2011, pp. 1658–1659.
[36] A.T. Sahlol, M. Abd Elaziz, M.A. Al-Qaness, S. Kim, Handwritten Arabic optical character recognition approach based on hybrid whale optimization algorithm with neighborhood rough set, IEEE Access 8 (2020) 23011–23021.
[37] A. Seyyedabbasi, WOASCALF: a new hybrid whale optimization algorithm based on sine cosine algorithm and Levy flight to solve global optimization problems, Advances in Engineering Software 173 (2022) 103272.
[38] A. Singh, A. Khamparia, A hybrid whale optimization-differential evolution and genetic algorithm based approach to solve unit commitment scheduling problem: WODEGA, Sustainable Computing: Informatics and Systems 28 (2020) 100442.
[39] N. Singh, H. Hachimi, A new hybrid whale optimizer algorithm with mean strategy of grey wolf optimizer for global optimization, Mathematical and Computational Applications 23 (1) (2018) 14.
[40] F.J. Solis, R.J.-B. Wets, Minimization by random search techniques, Mathematics of Operations Research 6 (1) (1981) 19–30.
[41] C. Tang, W. Sun, M. Xue, X. Zhang, H. Tang, W. Wu, A hybrid whale optimization algorithm with artificial bee colony, Soft Computing 26 (5) (2022) 2075–2097.
[42] A. Törn, A. Žilinskas, Global Optimization, vol. 350, Springer, 1989.
[43] P.J. Van Laarhoven, E.H. Aarts, Simulated annealing, in: Simulated Annealing: Theory and Applications, Springer, 1987, pp. 7–15.
[44] C. Voudouris, E.P. Tsang, Guided local search, in: Handbook of Metaheuristics, Springer, 2003, pp. 185–218.
[45] W.A. Watkins, W.E. Schevill, Aerial observation of feeding behavior in four baleen whales: Eubalaena glacialis, Balaenoptera borealis, Megaptera novaeangliae, and Balaenoptera physalus, Journal of Mammalogy 60 (1) (1979) 155–163.
[46] D.H. Wolpert, W.G. Macready, No free lunch theorems for optimization, IEEE Transactions on Evolutionary Computation 1 (1) (1997) 67–82.
[47] D.H. Wolpert, W.G. Macready, et al., No free lunch theorems for search, Technical report, Technical Report SFI-TR-95-02-010, Santa Fe Institute, 1995.
[48] G. Wu, R. Mallipeddi, P.N. Suganthan, Problem definitions and evaluation criteria for the CEC 2017 competition on constrained real-parameter optimization, National University of Defense Technology, Changsha, Hunan, PR China and Kyungpook National University, Daegu, South Korea and Nanyang Technological University, Singapore, Technical Report, 2017.
[49] J. Xu, J. Zhang, Exploration-exploitation tradeoffs in metaheuristics: survey and analysis, in: Proceedings of the 33rd Chinese Control Conference, IEEE, 2014, pp. 8633–8638.
[50] X.-S. Yang, Nature-Inspired Metaheuristic Algorithms, Luniver Press, 2010.
[51] X.-S. Yang, Swarm-based metaheuristic algorithms and no-free-lunch theorems, Theory and New Applications of Swarm Intelligence 9 (2012) 1–16.
[52] X.-S. Yang, S. Deb, S. Fong, Metaheuristic algorithms: optimal balance of intensification and diversification, Applied Mathematics & Information Sciences 8 (3) (2014) 977.
[53] X. Zhang, S. Wen, Hybrid whale optimization algorithm with gathering strategies for high-dimensional problems, Expert Systems with Applications 179 (2021) 115032.
[54] Y. Zheng, Y. Li, G. Wang, Y. Chen, Q. Xu, J. Fan, X. Cui, A novel hybrid algorithm for feature selection based on whale optimization algorithm, IEEE Access 7 (2018) 14908–14923.
[55] D.W. Zimmerman, B.D. Zumbo, Relative power of the Wilcoxon test, the Friedman test, and repeated-measures ANOVA on ranks, The Journal of Experimental Education 62 (1) (1993) 75–86.
[56] F. Zitouni, S. Harous, A. Belkeram, L.E.B. Hammou, The archerfish hunting optimizer: a novel metaheuristic algorithm for global optimization, Arabian Journal for Science and Engineering 47 (2) (2022) 2513–2553.
[57] F. Zitouni, S. Harous, R. Maamri, The solar system algorithm: a novel metaheuristic method for global optimization, IEEE Access 9 (2020) 4542–4565.

Chapter 10

Multi-criterion design optimization of contamination detection sensors in water distribution systems

Jafar Jafari-Asl[a], Sima Ohadi[a], and Seyedali Mirjalili[b,c,d]

[a]*Department of Civil Engineering, Faculty of Engineering, University of Sistan and Baluchestan, Zahedan, Iran,* [b]*Centre for Artificial Intelligence Research and Optimisation, Torrens University Australia, Brisbane, QLD, Australia,* [c]*Yonsei Frontier Lab, Yonsei University, Seoul, South Korea,* [d]*University Research and Innovation Center, Obuda University, Budapest, Hungary*

10.1 Introduction

Water distribution systems (WDSs) are considered as the most important and sensitive infrastructures of any country. WDSs have always been exposed to various intentional and accidental damages due to their spread and complexity [1]. These threats can reduce the resilience and efficiency of the networks and cause dissatisfaction and damage to consumers. One of the serious challenges in network operation is the possibility of contamination entrance. Due to the extent of WDSs, high number of joints and junctions, the existence of various components such as valves, tanks and pumps, the infiltration of contamination into these systems is inevitable. The contamination is transferred along the network by the flow as soon as entering the system. Ensuring the health of consumers against the vulnerability caused by contaminated water is one of the most expensive tasks in the operation and maintenance of WDSs. Detection of contamination entrance to distribution networks is done by monitoring stations. In this regard and in order to increase the reliability of the network, the installation of quality sensors can be considered as a high-efficiency strategy for contamination crisis managing of water distribution networks [2,3].

These sensors reduce the risk of possible dangers by detecting the contamination. Due to the high cost of installing and operating of sensors in all points of the network, choosing the best monitoring points to cover the entire network is a critical issue in the management of WDSs. Accordingly, most of the studies in the field of contamination detection during the last decades devoted to find the optimal location of monitoring stations [4–8].

Lee and Deininger (1992) focused on the optimal placement of contamination detection sensors in WDSs. They showed that the best set of sensor placements is the arrangement that maximizes the coverage of the network [9]. Bazargan Lari (2014) proposed a new model for the best placement of quality monitoring sensors in a WDS with a multi-objective (MO) approach. In this model, a MO version of Genetic Algorithm (GA) is used for optimization and Monte Carlo simulation (MCS) method used for the uncertainty quantification of pollution injection. Next, in order to choose the best point on the Pareto front, an evidential reasoning-based decision-making method is used [10].

Ref. [11] developed a game theory-based model for determining the best location of contamination detection sensors in a real case study from Iran. Minimizing the sensor installation costs and the time of contamination detection was considered as the objective functions of study. Ref. [12] introduced a MO-based approach to locate quality sensors in the WDSs using NSGA-II and EPANET simulation model on a real water network from Iran.

In the past decades, various metaheuristic-based models with different goals such as minimizing the time of contamination detection [6,11]; minimizing the volume of contaminated water consumed [12,13]; minimizing the number of monitoring stations [14], etc., have been proposed for optimally locating of quality sensors in WDSs, but complexity and sensitivity of this issue requires more comprehensive and extensive studies.

Moreover, most of these studies carried out as a single objective (SO) problem, but in the problem of optimally locating the quality sensors, different objectives exist which are usually in conflict with each other. Therefore, the aim of this study

Handbook of Whale Optimization Algorithm. https://doi.org/10.1016/B978-0-32-395365-8.00016-6

is to propose a MO framework to optimize the installation costs of contamination warning sensors in WDSs. To this end, the EPANET simulation model proposed in [15] is used in combination with a binary version of WOA.

10.2 Problem statement

The mathematical formulation MO optimization problem of contamination detection in WDSs through placement the quality sensors is explained in this section.

10.2.1 Objective functions

The worst-case impact-risk in WDS is the first objective function of the problem. Therefore, the impact-risk caused by the entry and consumption of contamination in the absence of a sensor is considered as the volume of contaminated water consumed. Therefore, assuming the contamination penetrates from node k in the time step s, the impact-risk caused in the node j and in the time step t is equal to [1]:

$$m_{t,j,s,k} = \frac{\delta}{\rho} C_{t,j,s,k} \times q_{t,j} \tag{10.1}$$

where $m_{t,j,s,k}$ indicates the impact-risk (volume) in the jth node and in the time step t due to the entrance of contamination from the kth node in the time step s, $C_{t,j,s,k}$ denotes contamination concentration in the jth node and in the time step t and $q_{t,j}$ is the volume of water consumed in the jth node and in the time step t. δ is the coefficient related to the duration of the simulation step in second. ρ also represents the specific weight of water.

It is noticeable that the concentration value that is less than the allowed concentration value is considered as zero. Also, the location of the sensors is not clear and should be determined in the optimization process. Therefore, the impact matrix ($M_{s,k}$) which represents the damage in all possible modes of contamination entrance, assuming that the pollution enters from the kth node in the time step s, is defined as below [1]:

$$M_{s,k} = \begin{bmatrix} m_{1,1,s,k} & \cdots & m_{1,nj,s,k} \\ \vdots & \ddots & \vdots \\ m_{nt,1,s,k} & \cdots & m_{nt,nj,s,k} \end{bmatrix}_{nt \times nj} \tag{10.2}$$

where nj is the number of demand nodes, nt is the number of quality simulation steps. The consumption of contaminated water will be continued until it is detected by the sensors, and after that the damage will be zero. So, the binary coefficient a has been used for this purpose. Therefore, if there is a sensor in the network, assuming that the flow is cut off in case of detection of pollution, the damage is modified according to the following equation as $m'_{t,j,s,k}$:

$$m'_{t,j,s,k} = a_{t,s,k} \times m_{t,j,s,k}, \quad a_{t,s,k} = \begin{cases} 1 & t \le \tau_{s,k} \\ 0 & t > \tau_{s,k} \end{cases} \tag{10.3}$$

where $\tau_{s,k}$ is the time step when the first sensor in the network detects the contamination that entered the network from the kth node and the time step s, and then the flow cut off along the network. In this research, the largest value of damages caused by the entry of contamination from various nodes is considered as the worst-case impact-risk (F_1) [1]:

$$F_1 = max_k \left(max_s \left(\sum_{t=1}^{nt} \sum_{j=1}^{nj} m'_{t,j,s,k} \right) \right) \tag{10.4}$$

Accordingly, the location of the sensors should be determined so that F_1 get minimize. That mean, the location of the warning sensors should be in such a way that the worst-case damage minimized assuming the uncertainty of the location and time of contamination entrance.

The second objective function of the problem is to minimize the sensors costs. Assuming the existence of N_s contamination warning sensors, the optimization model can be formulated as:

$$objective\ function\ 1: \text{Min}\, F_1 = max_k \left(max_s \left(\sum_{t=1}^{nt} \sum_{j=1}^{nj} m'_{t,j,s,k} \right) \right) \tag{10.5}$$

$$objective\ function\ 2: \text{Min}\, F_2 = N_s$$

$$X = [x_1 \ldots x_j \ldots x_{nj}]$$

$$x_j \in \{0, 1\}$$

where F_1 and F_2 represent the objective functions, include, the minimization of the worst-case impact-risk and minimization of total number of sensors, respectively. x_j denotes a binary decision variable that indicates the presence or absence of a sensor in node j.

10.2.1.1 Simulation model

To simulate the hydraulic and quality process of WDS, the EPANET simulation software was used. The EPANET software uses a dynamic implicit method for quality analysis. The derived flow from the hydraulic simulation process is used to solve a mass conservation equation for the substance within each pipe linking nodes i and j as follows [15]:

$$\frac{\partial c_{ij}}{\partial t} = -\left(\frac{Q_{ij}}{A_{ij}}\right)\left(\frac{\partial c_{ij}}{\partial L_{ij}}\right) + \theta(c_{ij}) \tag{10.6}$$

where, c_{ij} is the concentration of substance in pipe between nodes i, j $\left(\frac{mass}{m^3}\right)$, A_{ij} is the cross-sectional area of pipe between nodes i, j (m^2), and θ is the rate of constituent within pipe linking nodes i, j $\left(\frac{mass}{m^3/day}\right)$.

$$\frac{\partial c_{ij}}{\partial t} = -\left(\frac{Q_{ij}}{A_{ij}}\right)\left(\frac{\partial c_{ij}}{\partial L_{ij}}\right) + \theta(c_{ij}) \tag{10.7}$$

where:

$$c_{ij} = \frac{\sum_k qk_i c_{ki}\left(L_{k,i}, t\right) + M_i}{\sum_k qk_i + Q_{si}} \tag{10.8}$$

in which, $L_{k,i}$ is the length of pipe K connecting node i, M_i and Q_{si} are the substance mass injected by external source at node i and the source flow rate, respectively.

In this research, according to the nature of mentioned optimization problem, a BWOA is used to solve it. By changing the location of the sensors, this algorithm determines the best arrangement of sensors so that the maximum possible damage of the network minimized.

10.2.2 Whale optimization algorithm (WOA)

This algorithm is a population-based method that was first presented in 2016 by Ref. [16]. WOA has inspired from the intelligence behavior of humpback whales in the way of feeding and hunting prey. This hunting approach is called the bubble-net feeding method. The search process in WOA starts with creating a set of random answers which are the humpback whales' position [17]. Then, in each iteration of the algorithm, the whales update their position by using three main operators, including prey encirclement, Bubble-net attacking, and search for prey [18]. The mathematical model of the algorithm steps is given below.

10.2.2.1 Encircling prey

In the whale algorithm, the position of hunt is assumed as the best answer at the current iteration. Then other whales update their positions toward the best whale position as below [16,19]:

$$\vec{D} = \left|\vec{C} \cdot \vec{X^*}(t) - \vec{X}(t)\right| \tag{10.9}$$

$$\vec{X}(t+1) = \vec{X^*}(t) - \vec{A} \cdot \vec{D} \tag{10.10}$$

$$\vec{A} = 2\vec{a} \cdot \vec{r} - \vec{a} \tag{10.11}$$

$$\vec{C} = 2 \cdot \vec{r} \tag{10.12}$$

where $\vec{X^*}$ and $\vec{X}$ indicate the general best position and the positon vector, respectively. t is the current iteration. $\vec{a}$ represent linearly reduced within 2 to 0 over the iterations, and $\vec{r}$ is the random value in the range of [0, 1].

10.2.2.2 Exploitation phase (bubble-net attacking)

In this phase, the whale swims around the hunt along a shrinking circle and at t time in a spiral path. Eq. (10.10) simulates this simultaneous behavior in the exploitation phase [16,19]:

$$\vec{X}(t+1) = \begin{cases} \vec{X^*}(t) - \vec{A} \cdot \vec{D} & \text{if } p < 0.5 \\ \vec{D'} \cdot e^{bl} \cdot \cos(2\pi l) + \vec{X^*}(t) & \text{if } p \geq 0.5 \end{cases} \tag{10.13}$$

$$\vec{D'} = \left| \vec{X^*}(t) - \vec{X}(t) \right| \tag{10.14}$$

where b represents a constant number which define the logarithmic spiral shape, and l is a number in the range of [−1, 1]. p is a random value between [0, 1].

10.2.2.3 Exploration phase (search for prey)

In order to search for hunt in this step, the variety feature of vector $\vec{A}$ is used. Random values of $\vec{A}$ in the range greater than 1 or smaller than −1 cause the whales to move away from the reference whale, enabling the algorithm to search for the global optimum [20]:

$$\vec{D} = \left| \vec{C} \cdot \overrightarrow{X_{rand}} - \vec{X} \right| \tag{10.15}$$

$$\vec{X}(t+1) = \overrightarrow{X_{rand}} - \vec{A} \cdot \vec{D} \tag{10.16}$$

where $\overrightarrow{X_{rand}}$ is a random location of whale in iteration t.

To achieve the goals of the study, a multi objective version of WPA was used. This version was developed using the non-dominated sorting approach in [21]. As mentioned before, the nature of sensor placement problem needs to a metaheuristic algorithm with discrete search space. Therefore, the search space of MOWOA was converted into a binary space search using a transfer function as follows [22]:

$$TF\left(\vec{X}\right) = \frac{1}{1 + e^{(-10 \times (\vec{X} - 0.5))}} \tag{10.17}$$

$$\vec{X}_{binary}(t+1) = \begin{cases} 1, & TF \geq r \\ 0, & \text{otherwise} \end{cases} \tag{10.18}$$

where r is a random value between 0 and 1.

The performance of proposed binary MOWOA in solving the problem of contamination sensor placement in WDSs is compared with NSGA-II [23] and multi-objective version of Harris Hawk Optimizer (MOHHO) [24]. Note that the complexity of computational for algorithms is defined as $O(MN^2)$, where N and M are number of decision variables and objective functions. However, the complexity of computational of MOWOA is better than MOHHOA and NSGA-II.

10.3 Comparing metrics

To identify the most effective metaheuristic-based algorithms, the two well-known metrics, include inverted generation distance (IGD) and spacing metric (SP) are used. IGD is illustrated the distribution and convergence performance of algorithms. An algorithm in convergence and distribution is better than others, which has the smaller value for IGD. SP shows the uniformity of the options on the optimal Pareto. The smaller value of SP, shows the high efficiency of multi-objective algorithm [25].

$$IGD = \frac{\sum_{i=1}^{n} d_i}{N_s} \tag{10.19}$$

$$SP = \sqrt{\frac{\sum_{1}^{n}(\overline{d} - d_i)^2}{N_s - 1}} \tag{10.20}$$

where d_i denote the distance between the neighbor two options on the optimal Pareto, $\overline{d}$ is the average of these distances, and N_s illustrates the number of obtained solutions.

It is noteworthy that another criterion, which can be used to show the capabilities of multi-objective algorithms is the number of solutions on the Pareto front (N_s).

10.4 Case study

To show the resilience and the efficiency of proposed WOA-based model for optimizing the location of quality sensors in WDSs, a real network taken from Ref. [26] is investigated. The schematic of case study is presented in Fig. 10.1. This network has 126 nodes, 1 main reservoir with a constant head, 168 pipes, 2 pumps, 2 tanks, and 8 valves. Moreover, the demand of network subject to 4 different patterns.

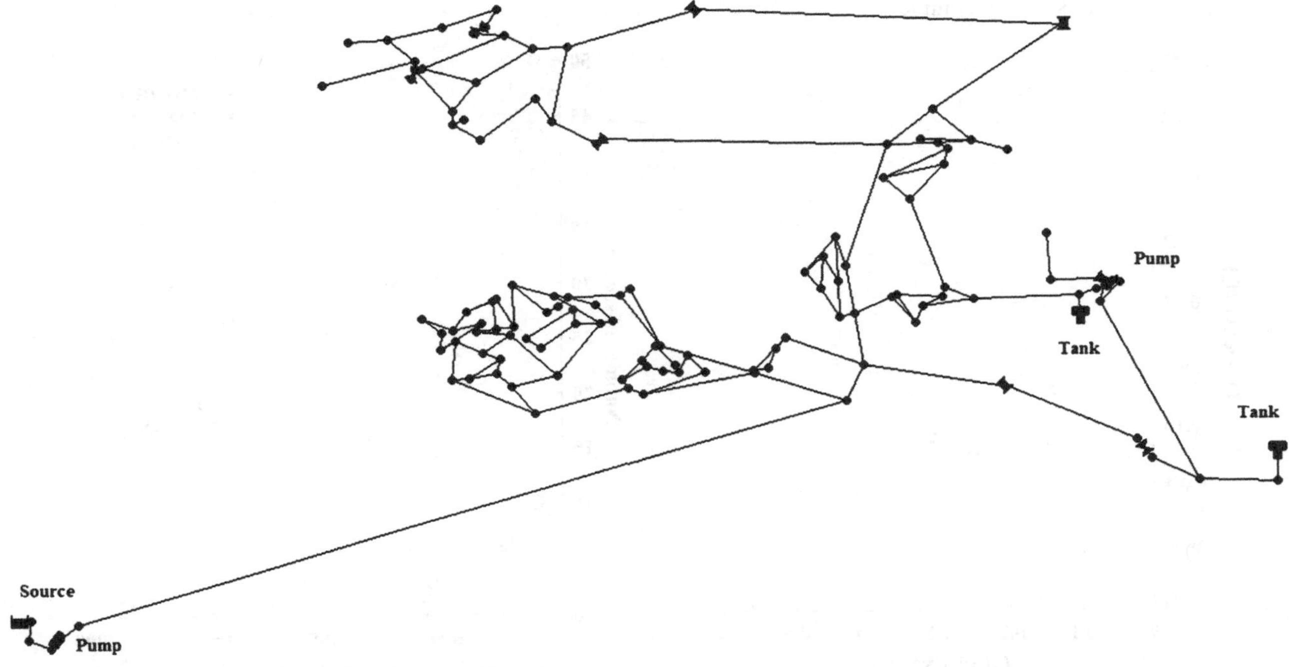

FIGURE 10.1 Layout of case study.

More details of network can be found in [26]. To simulate the contamination intrusion at network, the contamination scenarios were generated for each node with considering the contaminant concentration and injection duration equal to 10 mg/l and 2 hours, respectively. Also, a single injection location was considered for each contamination scenario, which may happen at any node and begin at any time. The hydraulic and quality characteristics of the network were reported with a time step of 1 hour and 5 minutes, respectively, for the full simulation length of 48 hours by using EPANET. Details of the developed hydraulic and quality simulation model are reported in Table 10.1.

TABLE 10.1 Detail of simulation analysis using EPANET model.

Parameter	Value
Duration of simulation (sec)	172800
Time step of hydraulic analysis (sec)	3600
Time step of water quality analysis (sec)	300
Head loss equation	Hazen–Williams

10.5 Results and discussion

The contamination scenarios and metaheuristic techniques were developed using MATLAB® R2019b and run on a laptop computer with an Intel Core i7-4510U 2.6 GHz processor and 4.00 GB RAM.

Given that the objective functions are not homogeneous, to increase the performance of algorithms, both of the objective functions were normalized using Eqs. (10.18) and (10.19).

$$f_1 = \frac{F_1(\mathrm{x}) - Min\ F_1(\mathrm{x})}{Max\ F_1(\mathrm{x}) - Min\ F_1(\mathrm{x})} \tag{10.21}$$

$$f_2 = \frac{N_s(\mathrm{x}) - Min\ F_2(\mathrm{x})}{Max\ F_2(\mathrm{x}) - Min\ F_2(\mathrm{x})} \tag{10.22}$$

where Max and Min indicate the highest and lowest values of objective functions, respectively. To create an impartial comparison, three algorithms were run 10 times with 100 initial population and 1000 iterations. Fig. 10.2 shows the best Pareto front of each algorithm. It is clear that the obtained options are closely ranked in relative merit and thus requirement to be compared based on several comparing criteria. The limited number of options on the Pareto front illustrates that the investigated problem is highly complex.

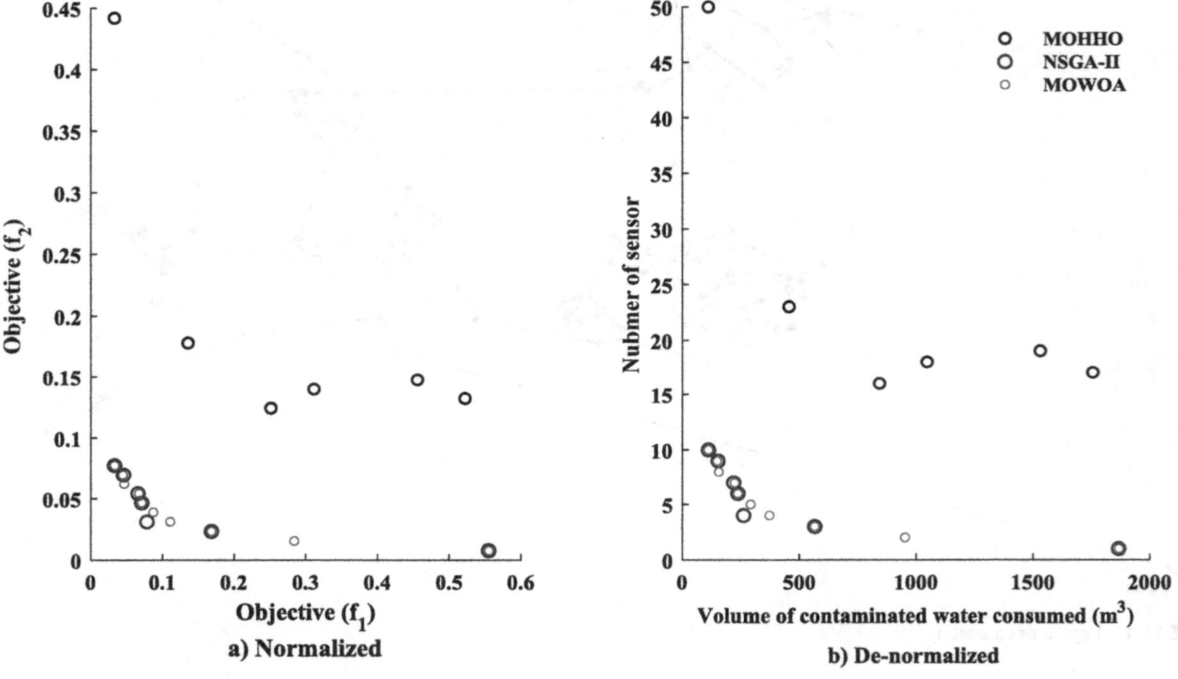

FIGURE 10.2 Trade of solution curves for algorithm; a) Normalized, b) De-normalized.

The values of GD and SP obtained by MOWOA, MOHHO, and NSGA-II over 10 simulations are shown in Table 10.2. Moreover, the statistical values of comparing metrics, include, the average and standard deviation of 10 simulations are reported in Table 10.3.

According to the results, it is clear that the number of options on the Pareto front obtained by the developed models is 10 which was attained by MOWOA. Therefore, the performance of proposed MOWOA in terms of solution numbers was better than MOHHO and NSGA-II. Furthermore, it can be observed that the mean values of IGD and SP obtained by MOWOA are lower than those obtained by the other algorithms. The overall observation is that MOWOA outperformed the NSGA-II, and MOHHO: Ns (MOWOA=10, NSGA-II=7, MOHHO=6), IGD (MOWOA=0.003, NSGA-II=0.008, MOHHO=0.015), and SP (MOWOA=0.033, NSGA-II=0.080, MOHHO=0.118). Fig. 10.3 illustrates the result of MOWOA in a Pareto front where each point represents one of the available options for the optimal location of contamination warning sensors on the investigated WDS.

In multi-objective problems with a convex Pareto front, an option from the knee point area is selected as the optimal option. However, choosing one option from the set of solutions can be confusing. Because, the all options on the Pareto

TABLE 10.2 Detail of SP and GD metrics obtained by algorithms.

NO.	NSGA-II		MOWOA		MOHHO	
	SP	IGD	SP	IGD	SP	IGD
1	0.042	0.0035	0.021	0.0023	0.122	0.0154
2	0.064	0.0053	0.023	0.0022	0.119	0.0140
3	0.056	0.0061	0.018	0.0018	0.123	0.0148
4	0.068	0.0065	0.025	0.0022	0.158	0.0199
5	0.164	0.0173	0.022	0.0021	0.098	0.0128
6	0.062	0.0063	0.017	0.0022	0.091	0.0116
7	0.160	0.0189	0.037	0.0033	0.141	0.0196
8	0.068	0.0065	0.049	0.0044	0.108	0.0136
9	0.063	0.0064	0.092	0.0083	0.107	0.0136
10	0.057	0.0047	0.022	0.0021	0.112	0.0138

TABLE 10.3 Statistical results of MOWOA, NSGA-II, and MOHHO.

NO.	NSGA-II		MOWOA		MOHHO	
	SP	IGD	SP	IGD	SP	IGD
Average	0.080	0.008	0.033	0.003	0.118	0.015
Std	0.041	0.005	0.021	0.002	0.019	0.003
Max	0.042	0.004	0.017	0.002	0.091	0.012
Min	0.164	0.019	0.092	0.008	0.158	0.020

front are optimal and acceptable solutions. Therefore, a fuzzy decision technique is used here to find the best compromise option. The base of fuzzy decision method is calculating the normalized membership value of objective functions as follows [25]:

$$\mu_i\,(Normalized) = \frac{\sum_{j=1}^{N_{obj}} \mu_{ij}}{\sum_{i=1}^{M} \sum_{j=1}^{N_{obj}} \mu_{ij}} \tag{10.23}$$

where:

$$\mu_i^j = \begin{cases} 1, & \text{if } f_i^j < f_{min}^j \\ \frac{f_{max}^j - f_i^j}{f_{max}^j - f_{min}^j}, & \text{if } f_{min}^j \le f_i^j \le f_{max}^j \\ 0, & \text{if } f_i^j > f_{max}^j \end{cases} \tag{10.24}$$

where M and N_{obj} are the number of options and objective functions, respectively. f_{min}^j is the minimum and f_{max}^j is the maximum value of objective function j. The best compromise solution is shown in Fig. 10.3. It is clear that the solution with the three installing sensor is identified as a fair solution. Choosing the best compromise solution for installing the three sensors at nodes 46, 117, and 122 led to decrease the second objective function up to 80%. Fig. 10.4 displays the distribution of sensors on the investigated network. It is clear that the proposed MOWOA-based model presented an optimal layout for locating of contamination warning sensors with good coverage of network nodes.

Details about the location of sensors and the worst-case impact-risk in the WDS of the case study, for all the solutions available on the trade-off diagram, are presented in Table 10.4.

If there is no contamination warning sensor in the investigated WDS, the worst-case impact-risk is equal to 3360.08 m^3 due to the intrusion of contamination. It is observed that with the optimal placement of one sensor (Option B), the volume of contaminated water consumed has decreased from 3360.08 to 1866.6 m^3 (44%). Also, by increasing the number of sensors from one to five, the potential damage has decreases up to 90%, which indicates the efficiency of placing the contamination warning sensors in WDSs. It is observed that by choosing about 10% of the network nodes for the optimal placement of sensors (Option A), worst-case impact-risk in the network has been reduced up to 96%. It should be noted that the meaning

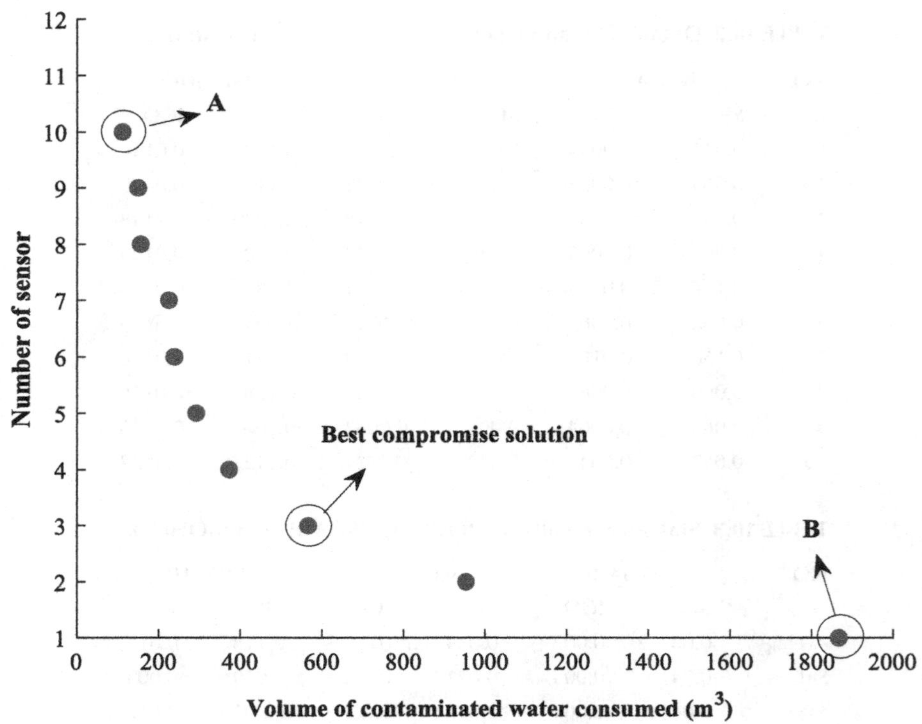

FIGURE 10.3 A trade-off between two objective functions.

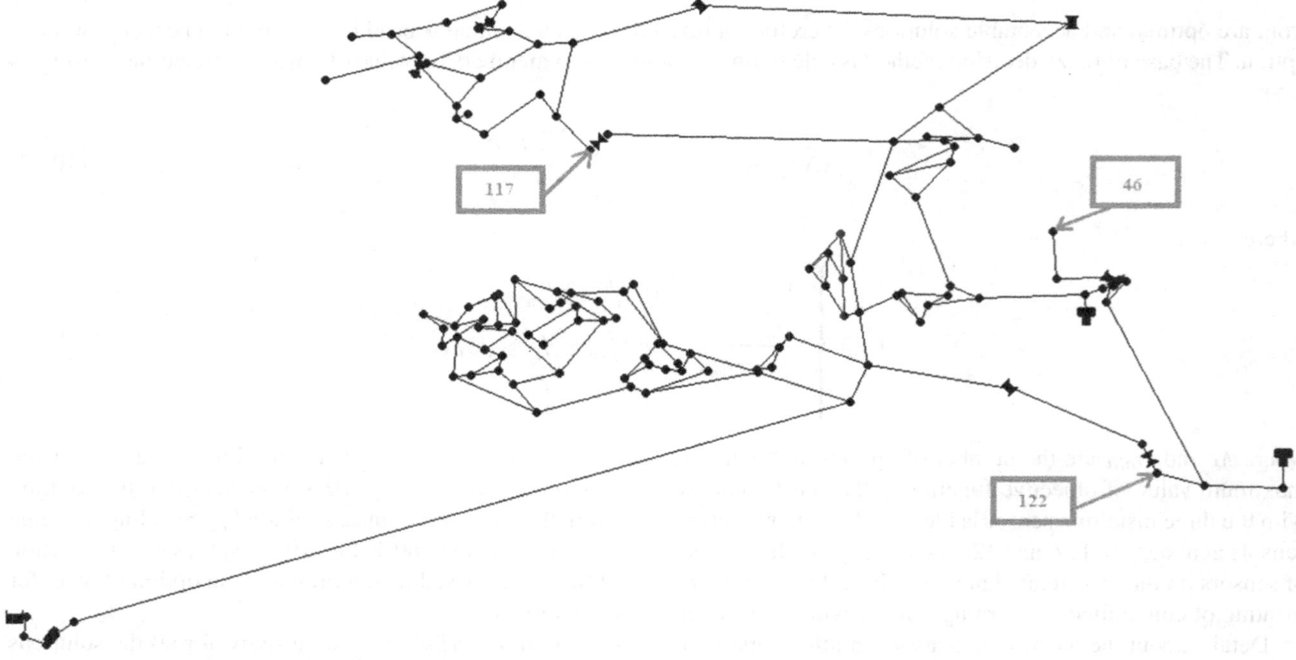

FIGURE 10.4 Best compromise solution selected.

of worst-case impact-risk is the most probable damage that calculates based on Eq. (10.2). Another very important matter is to identify some nodes for locating of sensors in all the obtained solutions by MOWOA. For example, the nodes 117 and 122 can be introduced as important and sensitive points of the network, which play a significant role in timely detection of contamination.

TABLE 10.4 Details of trade-off obtained by MOWOA.

Number of sensors	Sensor location (nodes)	Volume of contaminated water consumed (m^3)
1	117	1866.259
2	117, 122	952.936
3	**46, 117, 122**	**565.805**
4	46, 60, 117, 122	372.830
5	46, 92, 103, 117, 122	292.153
6	31, 46, 95, 104, 117, 122	237.974
7	31, 46, 99, 104, 117, 122, 125	224.508
8	46, 47, 59, 91, 103, 117, 122, 125	155.801
9	31, 46, 47, 59, 91, 103, 117, 122, 125	149.434
10	31, 46, 47, 63, 91, 103, 115, 117, 122, 125	110.924

Conclusion

In this study, a MO-based model was developed for installing contamination detection sensors in WDSs. EPANET software simulation model was utilized for generating the damage matrices with pollution injection scenarios. Binary MOWOA including the objective functions, include, the worst-case impact-risk and cost (number of sensors) was coupled to damage matrices created by EPANET software. Efficiency and applicability of the developed model was evaluated on a large scale WDS. Comparison of the investigated WDS results revealed that MOWOA outperformed NSGA-II and MOHHO based on the comparing metrics of spacing, number of obtained solutions, and inverted generation distance. The best run of the MOWOA was selected by fuzzy decision method. The location of contamination warning sensors of the best compromise option was presented. Installing three sensors according to the best compromise option, reduces the worst-case impact-risk up to 80%. Overall, it has been observed that the worst-case impact-risk decreases when the number of sensors increases. Despite the favorable results of the MOWOA compared to the other two algorithms, it is suggested to improve the performance of the algorithm to solve the complex MO-based quality sensor installation problem in WDSs by using chaos theory in future studies.

References

[1] H. Jafari, S. Nazif, T. Rajaee, A multi-objective optimization method based on NSGA-III for water quality sensor placement with the aim of reducing potential contamination of important nodes, Water Supply 22 (1) (2022) 928–944.

[2] R.G. Allen, C.E. Brockway, J.L. Wright, Weather station siting and consumptive use estimates, Journal of Water Resources Planning and Management 109 (2) (1983) 134–146.

[3] S.G. Vrachimis, R. Lifshitz, D.G. Eliades, M.M. Polycarpou, A. Ostfeld, Active contamination detection in water-distribution systems, Journal of Water Resources Planning and Management 146 (2020) 04020014.

[4] A. Ostfeld, J.G. Uber, E. Salomons, J.W. Berry, W.E. Hart, C.A. Phillips, et al., The battle of the water sensor networks (BWSN): a design challenge for engineers and algorithms, Journal of Water Resources Planning and Management 134 (6) (2008) 556–568.

[5] J.W. Berry, L. Fleischer, W.E. Hart, C.A. Phillips, J.P. Watson, Sensor placement in municipal water networks, Journal of Water Resources Planning and Management 131 (3) (2005) 237–243.

[6] C. Hu, G. Ren, C. Liu, M. Li, W. Jie, A Spark-based genetic algorithm for sensor placement in large scale drinking water distribution systems, Cluster Computing 20 (2) (2017) 1089–1099.

[7] S.S. Naserizade, M.R. Nikoo, H. Montaseri, A risk-based multi-objective model for optimal placement of sensors in water distribution system, Journal of Hydrology 557 (2018) 147–159.

[8] M.S. Khorshidi, M.R. Nikoo, M. Sadegh, Optimal and objective placement of sensors in water distribution systems using information theory, Water Research 143 (2018) 218–228.

[9] B.H. Lee, R.A. Deininger, Optimal locations of monitoring stations in water distribution system, Journal of Environmental Engineering 118 (1) (1992) 4–16.

[10] M.R. Bazargan-Lari, An evidential reasoning approach to optimal monitoring of drinking water distribution systems for detecting deliberate contamination events, Journal of Cleaner Production 78 (2014) 1–14.

[11] S. Liu, P. Auckenthaler, Optimal sensor placement for event detection and source identification in water distribution networks, Journal of Water Supply: Research and Technology. AQUA 63 (1) (2014) 51–57.

[12] A. Afshar, S.M. Miri Khombi, Multiobjective optimization of sensor placement in water distribution networks dual use benefit approach, International Journal of Optimization in Civil Engineering 5 (3) (2015) 315–331.

[13] A. Ostfeld, E. Salomons, Optimal layout of early warning detection stations for water distribution systems security, Journal of Water Resources Planning and Management 130 (5) (2004) 377–385.

[14] H. Jafari, T. Rajaee, S. Nazif, An investigation of the possible scenarios for the optimal locating of quality sensors in the water distribution networks with uncertain contamination, Journal of Water and Health 18 (5) (2020) 704–721.

[15] L.A. Rossman, EPANET 2 User's Manual, U.S. Environmental Protection Agency, Cincinnati, OH, 2000.

[16] S. Mirjalili, A. Lewis, The whale optimization algorithm, Advances in Engineering Software 95 (2016) 51–67.

[17] I. Aljarah, H. Faris, S. Mirjalili, Optimizing connection weights in neural networks using the whale optimization algorithm, Soft Computing 22 (1) (2018) 1–15.

[18] J. Jafari-Asl, M.E.A.B. Seghier, S. Ohadi, P. van Gelder, Efficient method using Whale Optimization Algorithm for reliability-based design optimization of labyrinth spillway, Applied Soft Computing 101 (2021) 107036.

[19] N. Rana, M.S.A. Latiff, S.I.M. Abdulhamid, H. Chiroma, Whale optimization algorithm: a systematic review of contemporary applications, modifications and developments, Neural Computing & Applications 32 (20) (2020) 16245–16277.

[20] G. Kaur, S. Arora, Chaotic whale optimization algorithm, Journal of Computational Design and Engineering 5 (3) (2018) 275–284.

[21] P. Jangir, N. Jangir, Non-dominated sorting whale optimization algorithm (NSWOA): a multi-objective optimization algorithm for solving engineering design problems, Global Journals of Research in Engineering 17 (F4) (2017) 15–42, https://engineeringresearch.org/index.php/GJRE/article/view/1643.

[22] J. Jafari-Asl, G. Azizyan, S.A.H. Monfared, M. Rashki, A.G. Andrade-Campos, An enhanced binary dragonfly algorithm based on a V-shaped transfer function for optimization of pump scheduling program in water supply systems (case study of Iran), Engineering Failure Analysis 123 (2021) 105323.

[23] K. Deb, S. Agrawal, A. Pratap, T. Meyarivan, A fast elitist non-dominated sorting genetic algorithm for multi-objective optimization: NSGA-II, in: Parallel Problem Solving from Nature PPSN VI: 6th International Conference Proceedings 6, Paris, France, September 18–20, 2000, Springer, Berlin, Heidelberg, 2000, pp. 849–858.

[24] U. Yüzgeç, M. Kusoglu, Multi-objective Harris Hawks optimizer for multiobjective optimization problems, BSEU Journal of Engineering Research and Technology 1 (1) (2020) 31–41.

[25] M.A. Mellal, M. Pecht, A multi-objective design optimization framework for wind turbines under altitude consideration, Energy Conversion and Management 222 (2020) 113212.

[26] A. Ostfeld, A. Kessler, I. Goldberg, A contaminant detection system for early warning in water distribution networks, Engineering Optimization 36 (5) (2004) 525–538.

Chapter 11

Balancing exploration and exploitation phases in whale optimization algorithm: an insightful and empirical analysis

Aram M. Ahmed[a,b], Tarik A. Rashid[c], Bryar A. Hassan[d], Jaffer Majidpour[e], Kaniaw A. Noori[f], Chnoor Maheadeen Rahman[d], Mohmad Hussein Abdalla[e], Shko M. Qader[f,g], Noor Tayfor[g], and Naufel B. Mohammed[b]

[a]*Department of Information Technology, College of Science and Technology, University of Human Development, Sulaimani, Iraq,* [b]*Department of Information Technology, Kurdistan Institution for Strategic Studies and Scientific Research, Sulaimani, Iraq,* [c]*Department of Computer Science and Engineering, School of Science and Engineering, University of Kurdistan Hewler, Erbil, KRI, Iraq,* [d]*Department of Computer Science, College of Science, Charmo University, Sulaimani, Iraq,* [e]*Department of Computer Science, University of Raparin, Sulaimani, Iraq,* [f]*Database Department, Sulaimani Polytechnic University, Sulaimani, Iraq,* [g]*Department of Computer Science, Kurdistan Technical Institute, Sulaimani, Iraq*

11.1 Introduction

The method of choosing the best answer among many possible ones for a given problem is called optimization. The size of the search space for numerous real-world issues is a major challenge with this procedure because it makes it impossible to check every answer in a fair amount of time. Stochastic techniques that are created to address these kinds of optimization issues include algorithms that draw inspiration from nature. They often fuse deterministic and randomized procedures to produce a variety of options and then, these options are iteratively compared and analyzed until an acceptable answer is found. Stochastic algorithms can be divided into types that are single solution-based and population-based [1]. In the first one, only one agent searches the domain to find the best solution. Simulated annealing algorithm [2] is an example of this type; in contrast, population-based types like particle swarm optimization (PSO) [3], involve many agents working and cooperating with one another in a decentralized way while seeking for the optimum solution. This later type is also known as swarm intelligence. Generally, these individuals conduct their searches in two ways: global search and local search, which are also known as diversification and intensification or exploration and exploitation phases respectively. Exploration entails venturing abroad to look for fresh territories on a global scale, whereas exploitation entails concentrating on previously explored areas in order to find superior answers. Too much of global search (Over-exploration) of an algorithm will result in agent diversification and broadening, and it is rare that near-optimal solutions will be found. Too much of local search (Over-exploitation) of an algorithm, on the other hand, increases the likelihood of trapping into local optima and failing to locate the near-global optimum. As a result, there is a trade-off issue with how the two phases are balanced and maintaining a good balance is crucial in any metaheuristic algorithm [1]. Therefore, it is essential to monitor the algorithm in terms of diversification and intensification in order to actually examine the search techniques that affect these two abilities.

There are enormous tactics and methods of metaheuristic algorithms to enforce diversification and intensification in the literature [4]. They can be classified according to selection based, fitness based, subpopulation based, replacement based, hybrid based, etc. Selection techniques that encourage selecting notable agent from the existing swarm are frequently used in population-based search methods this notable agent, which is also known as global best will either incorporate on the following cycle of the search process or it will be used in some solution updating method [5]. The agents who have the best-found answers among all candidate solutions, for instance, are guaranteed to survive for the following generation in greedy selection techniques, which is known to hasten convergence toward promising solutions. Additionally, when a single greedy selection mechanism is utilized in algorithms like Deferential Evolution [6], new solutions are only chosen if they enhance the original answer or solutions. Because it pushes an initial (diverse) population to improve independently from its beginning point, this selection technique has the potential to increase the exploration-exploitation ratio of solutions [7]. Furthermore, in some algorithms such as CSO [8] the roulette wheel method is employed, and it occasionally provides the

Handbook of Whale Optimization Algorithm. https://doi.org/10.1016/B978-0-32-395365-8.00017-8

opportunity for choosing undesirable solutions as well. This will help enhance the global search ability of the algorithm. Alternatively, some search algorithms accept any new solution regardless of its quality and do not use any sort of selection technique at all. While alternative search mechanisms must be used to balance the diversification and intensification rate, these approaches for conducting searches do not necessitate the use of well-known agents. In order to improve the exploitation capabilities of algorithms like PSO [9] or DCSO [1], search techniques that make use of "attraction operators" are taken into consideration. These methods aim to enhance a swarm by either directing them in the direction of the location of agents within the swarm who appear to be "excellent" or in the direction of the location of the finest solutions currently discovered by the algorithm. The architecture of the optimization process and search strategy determines how individuals are selected as attractors and how other individuals are drawn to these attractors. Individuals are predisposed to be attracted not just to the global best solution at a given iteration of the search process, but also to the best solution(s) recorded by each particle as the search process progresses, in the case of the Particle Swarm Optimization algorithm, for example (personal best solution) [10]. In addition, Changes in swarm number, replicate, removal, infusion methods, access to outer stores, or moving from a subpopulation into another can all help achieve the balance [4]. For example, in [11] and [12] two algorithms are hybridized and each algorithm has its own sup-population; whatever method performs better, its sub-population will increase at the expense of the other one. Then, once every while they exchange their best solutions to achieve a better balance.

AS a robust metaheuristic algorithm in the literature, the whale optimization algorithm [13], has suggested a creative plan to attain this equilibrium between exploration and exploitations. Additionally, it has produced better outcomes in a variety of applications [14]. It is crucial to examine both the behavior of each agent of WOA and the behavior of the swarm all together during the optimization process. Therefore, an insightful and empirical analysis of WOA is the main motivation behind this work.

The rest of this chapter is structured like: Section 11.2 discusses exploration-exploitation Tradeoffs in WOA. Section 11.3 describes the mathematical method, which is called dimension-wise diversity measurement, to compute the ratio of exploration and exploitation phases of WOA. Section 11.4 presents the results and analysis. Section 11.5 presents the conclusion of this study.

11.2 Exploration-exploitation tradeoffs in WOA

The exceptional hunting strategy of the humpback whales, which is called bubble-net feeding method is source inspiration for the WOA algorithm. During the optimization process, the best agent of each iteration is defined as the prey and the other agents in the swarm update their locations accordingly [13]. Therefore, the algorithm uses the encircling prey, Bubble-net attacking and searching for prey methods for the foraging of humpback whales, which are mathematically formulated in Eqs. (11.1) to (11.8).

$$\vec{D} = \left|\vec{C} \cdot \vec{X}^{*}(t) - \vec{X}(t)\right| \tag{11.1}$$

$$\vec{X}(t+1) = \vec{X}^{*}(t) - \vec{A} \cdot \vec{D} \tag{11.2}$$

$$\vec{A} = 2\vec{a} \cdot \vec{r_1} - \vec{a} \tag{11.3}$$

$$\vec{C} = 2 \cdot \vec{r_2} \tag{11.4}$$

$$\vec{a} = 2\left(1 - \frac{t}{T}\right) \tag{11.5}$$

$$\vec{X}(t+1) = \begin{cases} \vec{X}^{*}(t) - \vec{A} \cdot \vec{D} & \text{if } p < 0.5 \\ D' \cdot e^{bl} \cdot \cos(2\pi l) + \vec{X}^{*}(t) & \text{if } p \geq 0.5 \end{cases} \tag{11.6}$$

$$\vec{D} = \left|\vec{C} \cdot \vec{X}_{\text{rand}} - \vec{X}\right| \tag{11.7}$$

$$\vec{X}(t+1) = \vec{X}_{\text{rand}} - \vec{A} \cdot \vec{D} \tag{11.8}$$

Eqs. (11.1) and (11.2) formulate the position update for the search agents according to the prey, where $\vec{X}^{*}$ is the global best or prey and $\vec{X}$ is the current position of an agent. Eqs. (11.3) to (11.5) define $\vec{A}$, $\vec{C}$, a parameters, which are used to tune and control the three above-mentioned methods.

The algorithm uses the Bubble-net attacking method as an exploitation mechanism, which is composed of two attacking approaches, namely Shrinking encircling and Spiral updating. Eq. (11.6) represents their mathematical formulation. The first approach is accomplished by linearly lessening the value of $\vec{a}$ from 2 to 0 during the optimization process. Furthermore, this value lowers the fluctuation ranges of $\vec{A}$ as well. In addition, r_1 and r_2 are random numbers between 0 to 1, which gives an oscillation sense to $\vec{A}$ and $\vec{C}$ parameters. The second approach is accomplished by a spiral formula, which, replicates the humpback whales' helix-shaped motion. This equation considers the Euclidean distance between the individuals and the prey, which are specified through uses D', b, l parameters. The algorithm makes the assumption that there is a 50% chance that the spiral model or the declining encircling model will be chosen to update the location of whales. However, in either case the algorithm attempts to gradually push the agents to move towards the global best during the optimization process. Hence the algorithm avoids the premature convergence and escapes the possible local optima.

As to the exploration capabilities, the algorithm uses same encircling mechanism, which is controlled by $\vec{A}$ parameter. In this manner, if the value of $\vec{A}$ parameter was not in the range of -1 to 1, the algorithm pushes the agents to migrate towards a random selected agent rather than the global best. Eqs. (11.7) and (11.8) provide mathematical formulas for the exploring phase [14].

11.3 Dimension-wise diversity measurement

As mentioned earlier, Exploration and exploitation are the two primary search behaviors that swarm members often engage in. In the first, the agents are separating from one another and the spaces between them are growing wider. This stage is used to explore fresh territory and avoid any potential local optima traps. On the contrary, the agents are intensifying and getting closer together throughout the exploitation period. During this phase, they typically conduct local searches in their immediate area and congregate near the global optimum. Since, simply looking at the convergence graph and end results does not explain the exploration and exploitation ratio of WOA, dimension-wise diversity assessment [15] can be employed to quantitatively measure these phases and be able to do in-depth analysis. Additionally, this study replaced the mean in (11.9) with the median since it depicts the population's center more precisely [16].

$$\begin{aligned} \mathrm{Div_j} &= \frac{1}{\mathrm{n}} \sum_{\mathrm{i=1}}^{\mathrm{n}} \mathrm{meadian}\left(\mathrm{x^j}\right) - \mathrm{x_i^j}; \\ \mathrm{Div} &= \frac{1}{\mathrm{D}} \sum_{\mathrm{j=1}}^{\mathrm{D}} \mathrm{Div_j} \end{aligned} \tag{11.9}$$

where median $(\mathrm{x^j})$ is the population-wide median of dimension j. n is the swarm size and D is the dimension. $\mathrm{x_i^j}$ is the dimension j of individual i.

The average distance between each search agent's dimension j and that dimension's median is used to calculate the diversity in each dimension, or Div_j. The mean diversity for all dimensions is then calculated in Div. By averaging the following, it is possible to determine an algorithm's percentage of exploration and exploitation:

$$\begin{aligned} \mathrm{XPL\%} &= \left(\frac{\mathrm{Div}}{\mathrm{Div_{max}}}\right) * 100 \\ \mathrm{XPT\%} &= \left(\frac{|\mathrm{Div} - \mathrm{Div_{max}}|}{\mathrm{Div_{max}}}\right) * 100 \end{aligned} \tag{11.10}$$

where $\mathrm{Div_{max}}$ is the highest diversity value discovered over the entire optimization procedure. Furthermore, the levels of exploration and exploitation, $XPL\%$ and $XPT\%$, respectively, are complementary.

11.4 Results and analysis

Two sets of benchmarking functions were utilized to investigate the two very relevant factors (diversification and intensification) of the WOA algorithm. The first set includes of 23 traditional test functions which are of unimodal and multimodal types. Unimodal test functions often have a single global optimum in contrast to Multimodal test functions, which typically have several local optima. F1 to F7 are unimodal benchmarking functions that are used to benchmark the algorithms' capacity to perform global searches. Additionally, F8 through F23 are multimodal test functions that are used to assess the algorithms' capacity to do local searches. For a full explanation of unimodal and multimodal functions, see [13].

TABLE 11.1 Exploration and exploitation ratio of WOA for the classical test functions.

Classical Test Functions	Percentage of exploration and exploitation
F1	Exploration: 45.7196, Exploitation: 54.2804
F2	Exploration: 51.7746, Exploitation: 48.2254
F3	Exploration: 58.7007, Exploitation: 41.2993
F4	Exploration: 56.2336, Exploitation: 43.7664
F5	Exploration: 46.647, Exploitation: 53.353
F6	Exploration: 49.7779, Exploitation: 50.2221
F7	Exploration: 58.3987, Exploitation: 41.6013
F8	Exploration: 41.6489, Exploitation: 58.3511
F9	Exploration: 49.8063, Exploitation: 50.1937
F10	Exploration: 56.1418, Exploitation: 43.8582
F11	Exploration: 52.3485, Exploitation: 47.6515
F12	Exploration: 56.5455, Exploitation: 43.4545
F13	Exploration: 44.8052, Exploitation: 55.1948
F14	Exploration: 60.9302, Exploitation: 39.0698
F15	Exploration: 46.94, Exploitation: 53.06
F16	Exploration: 50.239, Exploitation: 49.761
F17	Exploration: 62.4855, Exploitation: 37.5145
F18	Exploration: 54.5377, Exploitation: 45.4623
F19	Exploration: 45.8668, Exploitation: 54.1332
F20	Exploration: 47.7914, Exploitation: 52.2086
F21	Exploration: 46.624, Exploitation: 53.376
F22	Exploration: 44.968, Exploitation: 55.032
F23	Exploration: 45.5886, Exploitation: 54.4114

The other set is composed of 10 modern benchmarking functions, namely CEC 2019. These benchmark functions, also called composite benchmark functions, are challenging solve. They are extended, rotated, shifted, and merged forms of typical test functions. The comprehensive description of these benchmark functions can be found in [17].

The WOA algorithm is tested 30 times independently for each benchmark function. 30 search agents searched during the period of 500 iterations for each run. The algorithm's parameter settings are set to default, and nothing was altered. The algorithm is widely used in the literature and has been applied on enormous applications. Therefore, this chapter does not include the numerical results of the mentioned test functions. It rather sufficed with calculating the ratio or percentage of exploration and exploitation for each benchmark function merely. Tables 11.1 and 11.2 present the results for the dimension-wise diversity measurement. As it can be seen, the exploration and exploitation percentages of the algorithm are very close to each other. For the traditional benchmark functions, the average of exploration and exploitation are 51.06606522% and 48.93393% respectively. Similarly, for the CEC 2019 test functions they are 50.8883% and 49.11215%. Therefore, it can be concluded that the algorithm performs the global and local search in a balanced manner.

In addition, Figs. 11.1 and 11.2 show the convergence curve as well as the graphical representation of exploration and exploitation capabilities of the WOA algorithm for some of the classical and modern benchmark functions. As it can be clearly seen from the figures, the algorithm usually performs high global search at the beginning, which generally ranges from 60% to 80%. Then, around iteration 300, this phenomenon reverses and the algorithm lean to decrease the diversification and increase the intensification until it reaches end of the iteration, where the exploration and exploitation phases are around 20% and 80% respectively. However, this is not the case for all the test functions. As shown in test function 12 of Fig. 11.1 and test function cec10 of Fig. 11.2, the algorithm does not tend to increase the exploitation phase and it remains around 40%.

There are many factors and elements that play important roles in an algorithm to produce robust results such as the formation of the algorithm, parameter tuning, the connection and communication between the agents, etc. The degree of

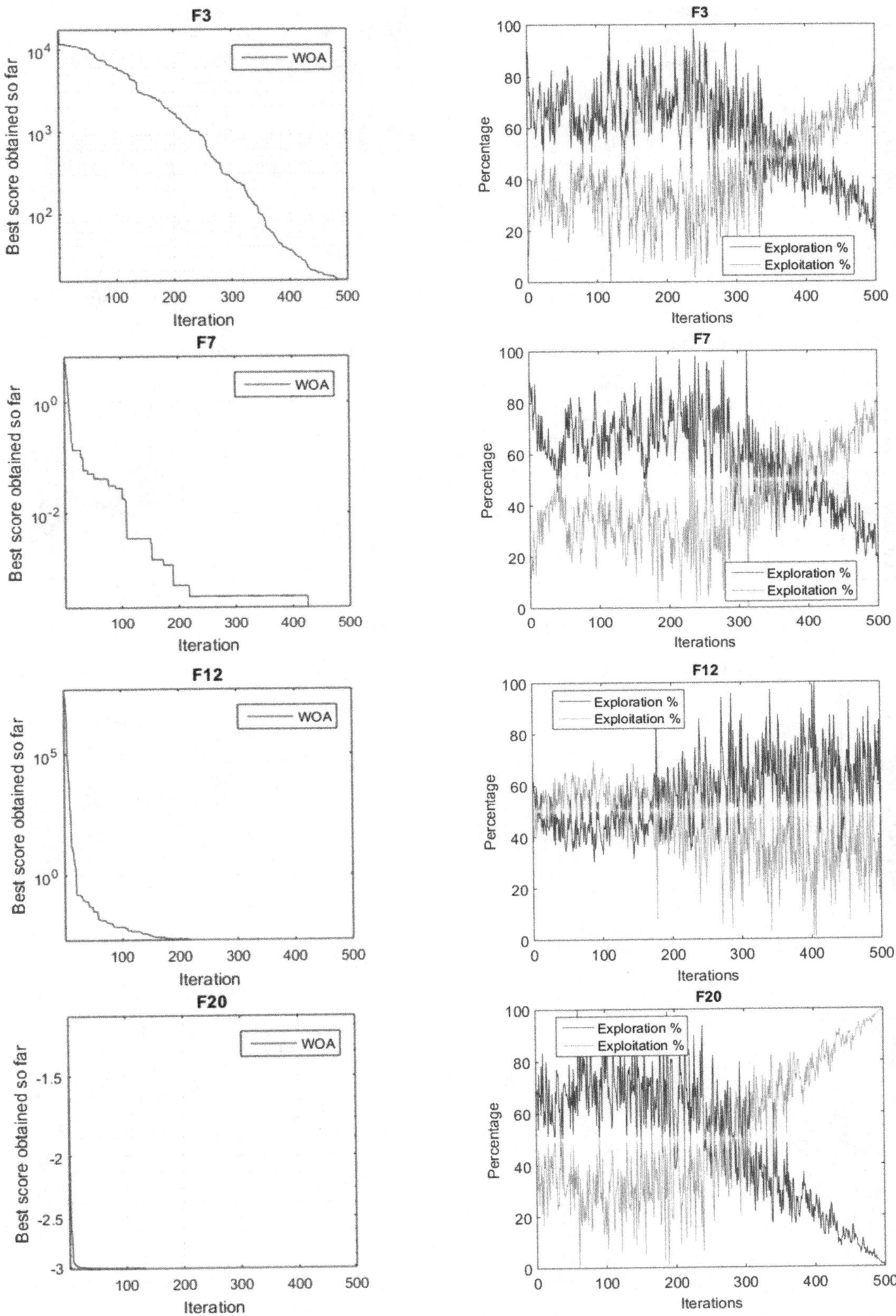

FIGURE 11.1 Convergence curve and visual representation of exploration and exploitation phases of the WOA algorithm for some of the classical test functions.

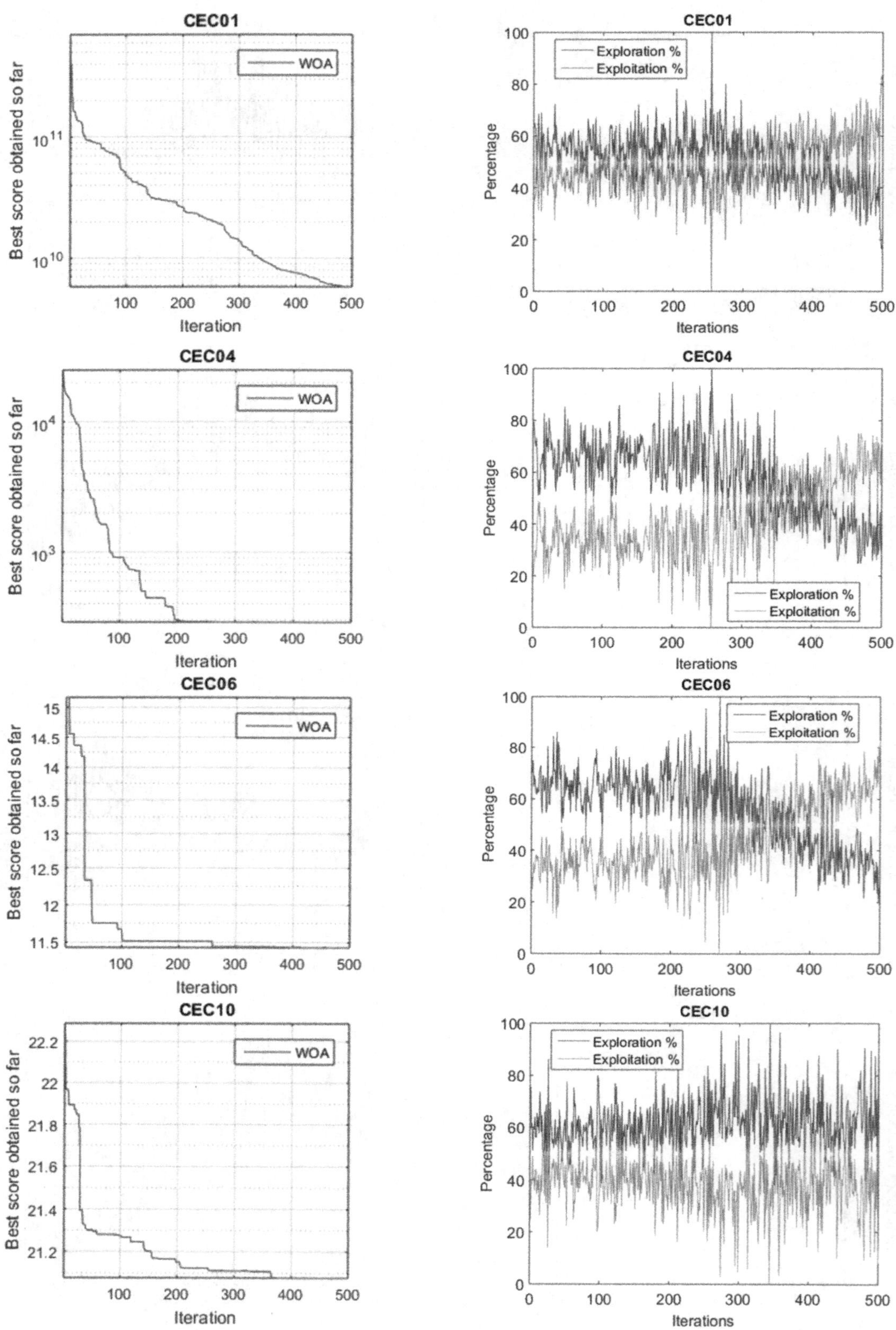

FIGURE 11.2 Convergence curve and visual representation of exploration and exploitation phases of the WOA algorithm for some of the CEC 2019 test functions.

TABLE 11.2 Exploration and exploitation ratio of WOA for CEC 2019 test functions.

CEC2019 Test Functions	Percentage of exploration and exploitation
Cec01	Exploration: 50.8883, Exploitation: 49.1117
Cec02	Exploration: 53.8929, Exploitation: 46.1071
Cec03	Exploration: 61.2, Exploitation: 38.8
Cec04	Exploration: 57.9113, Exploitation: 42.0887
Cec05	Exploration: 59.2973, Exploitation: 40.7027
Cec06	Exploration: 56.2303, Exploitation: 43.7697
Cec07	Exploration: 59.7024, Exploitation: 40.2976
Cec08	Exploration: 59.7618, Exploitation: 40.2382
Cec09	Exploration: 55.4782, Exploitation: 44.5218
Cec10	Exploration: 61.5279, Exploitation: 38.4721

diversification and intensification can also be an effective factor. As mentioned earlier, this degree in the case of WOA algorithm is close to 50% to 50%, which might reveal some of the mystery behind its success.

11.5 Summary

This chapter attempted to empirically investigate the WOA algorithm. To achieve this, the dimension-wise diversity measurement was used to quantitatively assess the convergence and diversity population of the algorithm in different periods of the optimization process. In the experiment, two sets of benchmark functions, which were composed of 23 traditional and 10 modern benchmark functions (CEC2019), were employed. The optimization results show that the algorithm's ratio of exploration to exploitation were quite similar to one another, i.e., it was close to 50% by 50%. This means that the algorithm performs the global and local searches in an even distribution and balanced manner. Moreover, the visual representation of the achieved results reveal that the algorithm possesses a great exploration rate in the beginning of the optimization process but then It gradually shifts in the direction of the local search.

References

[1] A.M. Ahmed, T.A. Rashid, S.A.M. Saeed, Dynamic Cat Swarm Optimization algorithm for backboard wiring problem, Neural Computing & Applications 33 (20) (2021) 13981–13997.

[2] D. Delahaye, S. Chaimatanan, M. Mongeau, Simulated annealing: from basics to applications, in: Handbook of Metaheuristics, Springer, Cham, 2019, pp. 1–35.

[3] J.C. Bansal, Particle swarm optimization, in: Evolutionary and Swarm Intelligence Algorithms, Springer, Cham, 2019, pp. 11–23.

[4] M. Črepinšek, S.H. Liu, M. Mernik, Exploration and exploitation in evolutionary algorithms: a survey, ACM Computing Surveys (CSUR) 45 (3) (2013) 1–33.

[5] H. Du, Z. Wang, W.E.I. Zhan, J. Guo, Elitism and distance strategy for selection of evolutionary algorithms, IEEE Access 6 (2018) 44531–44541.

[6] M. Pant, H. Zaheer, L. Garcia-Hernandez, A. Abraham, Differential evolution: a review of more than two decades of research, Engineering Applications of Artificial Intelligence 90 (2020) 103479.

[7] T. Huang, Z.H. Zhan, X.D. Jia, H.Q. Yuan, J.Q. Jiang, J. Zhang, Niching community based differential evolution for multimodal optimization problems, in: 2017 IEEE Symposium Series on Computational Intelligence (SSCI), IEEE, November 2017, pp. 1–8.

[8] A.M. Ahmed, T.A. Rashid, S.A.M. Saeed, Cat swarm optimization algorithm: a survey and performance evaluation, Computational Intelligence and Neuroscience 2020 (January 2020) 4854895, https://doi.org/10.1155/2020/4854895.

[9] S. Sengupta, S. Basak, R.A. Peters, Particle swarm optimization: a survey of historical and recent developments with hybridization perspectives, Machine Learning and Knowledge Extraction 1 (1) (2018) 157–191.

[10] B. Morales-Castañeda, D. Zaldivar, E. Cuevas, F. Fausto, A. Rodríguez, A better balance in metaheuristic algorithms: does it exist?, Swarm and Evolutionary Computation 54 (2020) 100671.

[11] O.H. Ahmed, J. Lu, A.M. Ahmed, A.M. Rahmani, M. Hosseinzadeh, M. Masdari, Scheduling of scientific workflows in multi-fog environments using Markov models and a hybrid salp swarm algorithm, IEEE Access 8 (2020) 189404–189422.

[12] O.H. Ahmed, J. Lu, Q. Xu, A.M. Ahmed, A.M. Rahmani, M. Hosseinzadeh, Using differential evolution and Moth–Flame optimization for scientific workflow scheduling in fog computing, Applied Soft Computing 112 (2021) 107744.

[13] S. Mirjalili, A. Lewis, The whale optimization algorithm, Advances in Engineering Software 95 (2016) 51–67.

[14] H.M. Mohammed, S.U. Umar, T.A. Rashid, A systematic and meta-analysis survey of whale optimization algorithm, Computational Intelligence and Neuroscience 2019 (2019) 8718571, https://doi.org/10.1155/2019/8718571.
[15] J. Xu, J. Zhang, Exploration-exploitation tradeoffs in metaheuristics: survey and analysis, in: Proceedings of the 33rd Chinese Control Conference, IEEE, July 2014, pp. 8633–8638.
[16] K. Hussain, M.N.M. Salleh, S. Cheng, Y. Shi, On the exploration and exploitation in popular swarm-based metaheuristic algorithms, Neural Computing & Applications 31 (11) (2019) 7665–7683.
[17] K.V. Price, N.H. Awad, M.Z. Ali, P.N. Suganthan, The 100-digit challenge: Problem definitions and evaluation criteria for the 100-digit challenge special session and competition on single objective numerical optimization, Nanyang Technological University, 2018.

Chapter 12

Equitable and fair performance evaluation of whale optimization algorithm

Bryar A. Hassan[a], Tarik A. Rashid[b], Aram M. Ahmed[c,d], Shko M. Qader[e,f], Jaffer Majidpour[g], Mohmad Hussein Abdalla[g], Noor Tayfor[h], Hozan K. Hamarashid[c], Haval Sidqi[i], Kaniaw A. Noori[i], and Awf Abdulrahmam Ramadhan[j]

[a]Department of Computer Science, College of Science, Charmo University, Sulaimani, Iraq, [b]Department of Computer Science and Engineering, University of Kurdistan Hewler, Erbil, KRI, Iraq, [c]Department of Information Technology, Kurdistan Institution for Strategic Studies and Scientific Research, Sulaimani, Iraq, [d]Department of Information Technology, College of Science and Technology, University of Human Development, Sulaimani, Iraq, [e]Department of Information Technology, Computer Science Institute, Sulaimani Polytechnic University, Sulaimani, Iraq, [f]Department of Information Technology, University College of Goizha, Sulaimani, Iraq, [g]Department of Computer Science, University of Raparin, Sulaimani, Iraq, [h]Department of Computer Science, Kurdistan Technical Institute, Sulaimani, Iraq, [i]Department of Database Technology, Technical College of Informatics, Sulaimani Polytechnic University, Sulaimani, Iraq, [j]Public Health Department, College of Health and Medical Techniques - Shekhan, Duhok Polytechnic University, Duhok, Iraq

12.1 Introduction

In a rapidly growing field such as optimization algorithms, all algorithms must be assessed accurately, equitably, and intelligently. This endeavor is not simple for various reasons. Due to the random nature of the processes involved, it is necessary to do several test runs of the algorithms to average out the effects of this unpredictability. Numerous published tests of algorithms provide a mean result, standard deviation, and the best and worst outcomes to illustrate the performance characteristics of the algorithm. When comparing several algorithms' performance, we face several problems. Among these are the utilized computer code's efficacy, the algorithms' unpredictability, and setting parameters at an equitable and similar level. This final issue deserves clarification. Consider that we wish to compare the firefly method to the simulated annealing technique. The firefly algorithm must provide the number of fireflies, the parameters, and 0. How can we choose these characteristics to compare the two approaches appropriately? A wrong choice of quenching factor or starting temperature for the particular test issue would result in poor performance of the simulated annealing methodology, preventing a fair comparison of the techniques. Comparing methods is not difficult, but prudence is required. In addition, specific algorithms may perform very well on one problem but poorly on another or minor problems but poorly on large ones. Therefore, a comprehensive collection of standard test problems that reflect the evolving difficulties of nonlinear optimization must be used. Statistical significance tests like Friedman's may be helpful in comparative research. See Daniel for further information on Friedman's experiment (1990). In addition, the tests must be done with identical settings, and efficiency assessments can be determined from time measurements or function evaluations. Function assessments provide a more precise comparison of testing in diverse computer settings. The evaluation test tasks for nature-inspired algorithms are vital. These should test the algorithm's fundamental properties. The essential property is that algorithms can calculate the global optimum of a function. Therefore, problems with several local optima must be included in the test set. Moreover, the distribution of these optimums is a crucial feature. Because both types of functions must be optimized and provide separate problems, specific test questions should yield optimum, tightly-packed answers. In contrast, others should generate ideal solutions that are widely spread and isolated. A sharp, practically discontinuous approach to the ideal valley or peak, or a steady descent to an undefined minimum, must also be considered. Fortunately, a rigorous and well-respected set of test problems reflecting these and other properties has been compiled, and new algorithms are routinely assessed on this set. The reader may be surprised to learn that accurate results may not be achieved during these tests. Some problems are intended to be pathologically tricky; hence, only highly approximate values for the optimum may be established. The search will be affected by the size or number of parameters of the problem. In such cases, we seek the algorithm that produces the most outstanding results, and academics in the field normalize the data to reflect this. Since this is not a collection of research papers, we refrained

Handbook of Whale Optimization Algorithm. https://doi.org/10.1016/B978-0-32-395365-8.00018-X

from conducting tests on monumental topics. Smaller jobs allow us to demonstrate the algorithm's essential properties so the reader can duplicate reasonably, even on a simple computer.

12.2 Background

Evaluation of the effectiveness of optimization algorithms is an intriguing area of study. However, several obstacles exist to conducting a fair and objective performance analysis of these algorithms. These aspects may include selecting initial settings for competing algorithms, system performance during the evaluation, algorithmic programming style, the proportion of randomization, and the cohorts and difficulty ratings of the chosen benchmark problems. In this chapter, we will choose the most up-to-date and well-known optimization strategies for equal evaluation versus WOA because the literature lacks equal experimental evaluation research on WOA. The selected algorithms are the Backtracking Search Optimization Method (BSA), Fitness Dependent Optimizer (FDO), Particle Swarm Optimization Algorithm (PSO), and Firefly Algorithm (FF). In this part, the prior algorithms involving WOA are briefly described.

12.2.1 WOA

WOA is a suggested swarm intelligence method for continuous optimization issues. This method has been demonstrated to perform as well as or better than specific existing algorithmic strategies [1]. WOA has drawn inspiration from the hunting behavior of humpback whales. In WOA, each solution is referred to as a whale. Using the best member of the group as a reference, a whale seeks to populate a new spot in the search space. Whales have two methods for locating prey and pursuing it. In the first, prey is encircled, and bubble nets are produced in the second. Exploration of the search space occurs during the whales' hunt for prey, whereas exploitation happens during their attack behavior. The complete description and source codes of the WOA are available in [1].

12.2.2 BSA

BSA is an iterative population-based meta-heuristic algorithm. BSA creates test populations to regulate the amplitude of the search-direction matrix, providing a robust capacity for global exploration [2]. During the crossover phase, BSA utilizes two random crossover procedures to interchange the corresponding elements of people in populations and test populations. In addition, the BSA has two selection methods. One is used to pick populations from the present and historical populations, while the other determines the best population. BSA may be broken down into five distinct processes: (a) Initiation, (b) Selection I, (c) Mutation, (d) Crossover, and (e) Selection II. The BSA's complete description and source codes are available in [2].

12.2.3 FDO

FDO is comparable to a swarm of bees reproducing. This method is based on how scout bees navigate several prospective colonies, searching for a new, appropriate hive. Each scout bee that looks for additional colonies offers a potential solution to this technique. The best hive among numerous good hives must be chosen to achieve convergence to optimality [3]. Starting with a random initiation of a synthetic scout population in the search space Xi (i=1,2,...,n), each scout bee position represents a newly discovered hive. The process then repeatedly looks for bee hives (solution). Scout bees randomly investigate various locations to locate superior hives; once a superior hive is located, the previously located hive is disregarded. In the same way, the algorithm discards the previously recognized alternative when it discovers a new, superior one. The artificial scout bee might not find a better option if the current activity continues (hive). In such an instance, it will carry on along its introductory course to arrive at a better answer. Nonetheless, the algorithm will go on to the present solution, which is the best choice so far, if the prior route does not result in a better solution. The FDO's complete description and source codes may be found in [3].

12.2.4 PSO

PSO is a self-adaptive, stochastic, population-based optimization method. The PSO initially creates the first particles and gives them starting velocities [4]. The ideal function value and position are calculated by assessing the objective function at each particle size. It selects new velocities based on the present velocity, the particles' ideal positions, and their neighbors' ideal positions. The particle positions, velocities, and neighbors are then regularly updated (the new location is the previous one, plus the velocity, tweaked to keep particles inside boundaries). Up until a halting condition is encountered, the algorithm iterates. PSO's complete source code and description are available in [4].

12.2.5 FF

FF is one of the optimization algorithms inspired by nature that Xin-She Yang devised at Cambridge in 2008 and made public by Yang (2009) [5]. A technique for optimizing functions with multiple optima was created using the behavior of firefly swarms. To encourage exploration of the solution space, it specifically used the notion that individual fireflies' brilliance attracted them together as well as a randomization component.

The release of the firefly method has led to the publication of several publications on its study, modification, and application to numerous real-world issues, as well as numerous accounts of its successful uses. The essential characteristics of this algorithm are based on the following principles created by Xin-She Yang:

1. Since fireflies are unisexual, they all have a strong attraction to one another.
2. The brightness of them affects how beautiful they are. Though perceived brightness reduces as two fireflies are separated, the less brilliant one will always be drawn to (and move towards) the brighter one.
3. It will move arbitrarily if no other fireflies are brighter than a particular firefly.

These three criteria may be used to create an optimization process where the brightness is inversely correlated with the value of the goal function. The firefly rules may be turned into algorithmic steps by creating the locations of an initial population of fireflies, calculating the objective function value for each firefly, and then putting these rules into practice over several generations. Public access to FF's source code may be found in [6].

12.3 Evaluation

The mean, standard deviation, and worst and best outcomes were calculated based on earlier studies [1], [7] that compared the performance characteristics of meta-heuristic algorithms to those of their rivals. For various reasons, comparing the effectiveness of various metaheuristic algorithms fairly and similarly is challenging. The choice and initialization of crucial variables, such as the size of the issue, the search space, and the number of iterations needed to solve the problem for each strategy, pose particular difficulties. Another reason to be concerned is that every algorithm depends on a randomization mechanism. It is required to test the algorithms several times to remove the impacts of this randomization and provide an average result. The programming language and technology used to construct the approach may also impact how well an algorithm performs in classification. All algorithms on the same system can utilize the same coding approach to prevent this impact. Selecting the issues or benchmark test functions for the algorithm evaluation is another challenging undertaking. For instance, an algorithm could work well for some categories of issues but not others.

Also, the size of a task may impact an algorithm's performance. The performance of the algorithms must be assessed using a set of typical issues or benchmark test functions with a range of complexity levels and various cohorts on a range of optimization tasks to solve the aforementioned issue. The initialization of the control parameters, the balance of the randomness in the algorithms, the computer performance used to implement the algorithms, the programming style of the algorithms, and the kinds of problems they are intended to solve are all factors that must be considered in an evaluation method to accurately compare WOA to its rival algorithms in the absence of experimental research on WOA in the literature. As a consequence, this part covers the experimental design, the list of benchmark problems utilized in the assessment, their control components, the statistical analysis of the evaluation technique, the pairwise statistical testing methods, and the evaluation outcomes.

12.3.1 Evaluation method

The evaluation method considers several factors, such as the initialization of parameters like problem dimensions, problem search space, and the number of iterations necessary to minimize the problem, the performance of the system used to implement the algorithms, the programming style of the algorithms, achieving a balance on the effect of randomization, and the use of various types of optimization problems to accurately compare WOA with BSA, PSO, FDO, and FF. This process compares WOA and its rival algorithms on 16 benchmark tasks with varying degrees of difficulty [8–10].

- A specific function with Nvar variables and the default search space for a population of 30 has to be optimized across several rounds. Nvars can have one of three values: 10, 30, or 60. Each algorithm is executed thirty times for each benchmark function, with 2000 iterations for Nvar values of 10, 30, and 60.

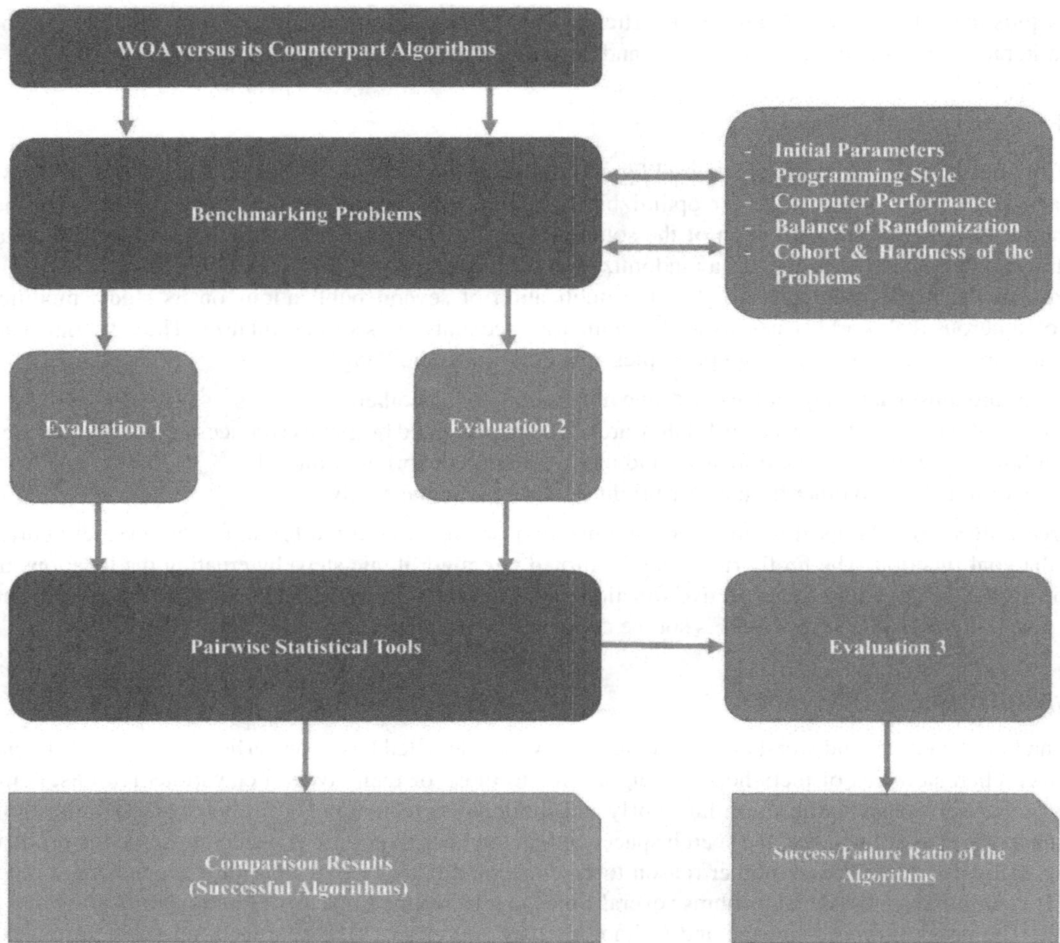

FIGURE 12.1 The performance evaluation framework of WOA.

- Many iterations are necessary to minimize functions with two variables in three different solution spaces with a population of thirty. Each algorithm is run 30 times with 2000 iterations for each of the three unique ranges (R1, R2, and R3): [-5, 5], [-250, 250], and [-500, 500].
- The proportion of successful function minimization for Evaluations 1 and 2 should be determined to compare the success rate of WOA to that of its competitors.

Fig. 12.1 displays the evaluation framework that comprises the procedures of Evaluation 1, 2, and 3.

12.3.2 Problems and initial parameters

In each of the three assessments, 16 benchmark issues were used to evaluate the associated performance of WOA against BSA, FDO, PSO, and FF in addressing a range of optimization problems based on their cohort and complexity level [8–10]. Regardless of how the test function problems were resolved, specific problems are more challenging. The chosen optimization strategies, their search space, global minimum, dimension, and percentage of difficulty, are shown in Table 12.1. Hardness percentages can range from 4.92 to 82.75 percent. The Table's most challenging problem is the Whitley function, while its most manageable problem is the sphere.

Also, for each benchmark problem, each algorithm is run thirty times with a population size of thirty for each of the three Evaluations, with a maximum of 2000 iterations. For the benchmark tasks, Evaluations 1, 2, and 3 each have different search regions and sizes. Each benchmark problem for Evaluation 1 receives the default dimension and has its search space separated into R1, R2, and R3. Nvar1, Nvar2, and Nvar3 are the search space dimensions for each benchmark function in evaluation 2. To evaluate the success rate of WOA to that of its rival techniques, it is necessary to determine the ratio

TABLE 12.1 Problems with a different success rate.

Problem ID, Name	Global min.	Space	Dimension	Success percentage
P1, Ackley	0	[-32, 32]	n	48.25
P2, Alpine01	0	[0, 10]	2	65.17
P3, Bird	-106.764	[4.701, 3.152], [-1.582, -3.13]	2	59.00
P4, Leon	0	[0, 10]	2	41.17
P5, CrossInTray	-2.062	[-10, 10]	2	74.08
P6, Easom	-1	[-100, 100]	2	26.08
P7, Whitley	0	[-10.24, 10.24]	2	4.92
P8, EggCrate	0	[-5, 5]	2	64.92
P9, Griewank	0	[-600, 600]	n	6.08
P10, HolderTable	-19.208	[-10, 10]	2	80.08
P11, Rastrigin	0	[-5.12, 5.12]	n	39.50
P12, Rosenbrock	0	[-5, 10]	n	44.17
P13, Salomon	0	[-100, 100]	2	10.33
P14, Sphere	0	[-1, 1]	2	82.75
P15, StyblinskiTang	-39.166	[-5, 5]	n	70.50
P16, Schwefel26	0	[-500, 500]	2	62.67

TABLE 12.2 Initial parameters of Evaluation 1.

Features	Dimension	Default
	Space	R1 [-5, 5]; R2 [-250, 250]; R3 [-500, 500]
	Hardness rate	4.92%–82.75%
Initial parameters	Number of executions	30
	iterations	2000
	Population size	30

TABLE 12.3 Initial parameters of Evaluation 2.

Features	Dimension	Nvar1: 10; Nvar2: 30 and Nvar3: 60
	Space	2
	Hardness rate	4.92%–82.75%
Initial parameters	Number of executions	30
	iterations	2000
	Population size	30

of successful function minimization for Evaluations 1 and 2. Tables 12.2 and 12.3 for Evaluations 1 and 2, respectively, illustrate the initializations of these parameters.

12.3.3 Statistical analysis and tool

Occasionally, meta-heuristic algorithms might present the worst and most relevant answers for a given problem; for instance, if an algorithm runs twice on an equivalent problem, it might receive the appropriate answer the first time and the worst solution the second time, and versa. In the studies [11] and [12], statistical methods were used to contrast WOA's success or failure of WOA, problem-solving algorithms, and others. The discussed experiment resolved numerical optimization problems using seven statistical measures: mean, standard deviation, best, worst, average computing time, number of effective minimizations, and the number of failed minimizations.

To evaluate two algorithms and determine whether one has a greater statistical success rate in solving a specific optimization issue, one can use the Wilcoxon signed-rank test and other Pairwise statistical testing tools [2], [13]. The Wilcoxon signed-rank test is used in the evaluation to compare WOA to other algorithms; the statistically significant value (H0) is set at 0.05, and Eq. (12.1) states the null hypothesis (H0) for a particular benchmark problem.

$$Median(Algorithm\ A) = Median(Algorithm\ B) \tag{12.1}$$

The Wilcoxon signed-rank test used R+, R-, and p-value to determine the approach that produced the statistically better response. Using GraphPad Prism, T+ and T- values ranging from 0 to 465 are determined for the same experiment. Comparable to the mathematical precision of today's application and software development tools, the P-value ranges between 4 and 6. In general, the precision used for Evaluations 1 and 2 was 6 since this level of precision may be necessary for real-world applications.

12.4 Result evaluation

This section consists of three sub-sections. The results of the three evaluations mentioned in the previous section are covered in the first sub-section. On the other hand, the computational cost of WOA compared to its comparative algorithms is described in sub-section two. Last but not least, the convergence curve of WOA is presented.

12.4.1 Results of three evaluations

Each technique decreased the four optimization issues, P3, P4, P5, and P8, in Evaluation 1. This evaluation solved optimization problems utilizing the algorithms' three dimensions and default search spaces: Nvars 1 through 3. However, since the algorithms could minimize four problems of varying dimensions, the dimensions of the numerical optimization problems did not affect the success rate of the algorithms used to minimize them. In addition, the shortened benchmark functions were of relatively low difficulty. The lowest success rate for these difficulties was 41.17 percent, while the best was 59 percent. As a result, none of the techniques effectively reduced difficulties with a high hardness score and changeable dimensions.

In contrast, problem P10, which had an overall success rate of 80.08 percent, could not be solved in any variable dimension by any approach. This indicates that a conclusive conclusion could not be made on the efficacy of WOA and other competing algorithms in reducing optimization issues based on their hardness of score. As a result, there is no intrinsic link between the difficulty of minimization optimization problems and the efficiency of these algorithms. Using problem-based statistical comparison techniques, it was established which of the experiment's algorithms could statistically resolve the benchmark functions. This method utilizes the computational time of the algorithms needed to complete 30 iterations and find the global minimum. Throughout the evaluation process, Using the Wilcoxon signed-rank test and an importance criterion of (0.05), WOA was compared to other algorithms. Also, Eq. (12.1) was investigated the null hypothesis (H0) for a specific benchmark problem. Tables 12.4, 12.5, and 12.6 show that the algorithms in Evaluation 1 statistically outperformed the other algorithms according to the Wilcoxon signed-rank test.

On the other hand, every method in Evaluation 2 resolved a subset of optimization problems in order of increasing difficulty. This evaluation resolved two-dimensional optimization issues in R1, R2, and R3—three distinct search spaces. In contrast to Evaluation 1, the number of search spaces impacted the success rates of the algorithms used to minimize

TABLE 12.4 Statistical evaluation utilizing the two-sided Wilcoxon Signed-Rank Test ($\alpha = 0.05$) to identify the best solution for the problems solved in Evaluation 1 (Nvar1).

Problems	WOA vs. BSA				WOA vs. FDO				WOA vs. PSO				WOA vs. FF			
	p-value	T+	T-	Winner	p-value	T+	T-	Winner	p-value	T+	T-	Winner	p-value	T+	T-	Winner
P3	<0.0001	0	465	+	0.0003	330	48	-	0.0003	269	196	+	0.0004	443	22	+
P4	<0.0001	465	0	+	0.0006	344	121	+	0.6023	157	308	-	0.0001	400	51	+
P5	0.0001	465	0	+	0.703	264	201	-	<0.0004	451	0	+	<0.0001	453	12	+
P8	0.6101	200	265	-	0.0001	465	0	+	0.0001	279	186	+	0.0006	307	158	+
+/=/-	**3/0/1**				**2/0/2**				**3/0/1**				**4/0/0**			

TABLE 12.5 Statistical evaluation utilizing the two-sided Wilcoxon Signed-Rank Test ($\alpha = 0.05$) to identify the best solution for the problems solved in Evaluation 1 (Nvar2).

Problems	WOA vs. BSA				WOA vs. FDO				WOA vs. PSO				WOA vs. FF			
	p-value	T+	T-	Winner	p-value	T+	T-	Winner	p-value	T+	T-	Winner	p-value	T+	T-	Winner
P3	0.0004	440	25	+	<0.0001	61	374	+	0.0001	465	0	+	0.0001	440	25	+
P4	0.0001	444	21	+	0.5001	214	0	=	<0.0001	460	5	+	0.0004	455	1	+
P5	0.5001	222	243	-	0.513	185	280	-	0.5263	210	255	-	<0.52001	315	148	-
P8	0.0001	433	32	+	0.0001	301	164	+	<0.0001	432	33	+	0.0001	355	110	+
+/=/1	**3/0/1**				**2/1/1**				**3/0/1**				**3/0/1**			

TABLE 12.6 Statistical evaluation utilizing the two-sided Wilcoxon Signed-Rank Test ($\alpha = 0.05$) to identify the best solution for the problems solved in Evaluation 1 (Nvar3).

Problems	WOA vs. BSA				WOA vs. FDO				WOA vs. PSO				WOA vs. FF			
	p-value	T+	T-	Winner	p-value	T+	T-	Winner	p-value	T+	T-	Winner	p-value	T+	T-	Winner
P3	0.0001	440	25	+	0.5101	249	216	-	0.6101	266	11	-	0.0001	425	40	+
P4	0.5816	250	210	=	0.602	211	254	-	0.786	273	192	+	0.0621	464	1	+
P5	<0.0001	315	0	+	0.0001	99	366	+	0.0001	465	0	+	0.0001	463	2	+
P8	0.0001	461	4	+	0.0631	462	3	+	0.0802	406	0	+	0.0761	453	12	+
+/=/-	**3/1/0**				**2/0/2**				**3/0/1**				**4/0/0**			

the problems since each approach may have removed a different number of problems in each search region. For instance, among 16 optimization tasks, the R1, R2, and R3 techniques removed 11, 9, and 8. The complexity of the solved benchmark functions ranged from low to high as well. For instance, the comparative success rates for identical problems in search space R3 were 6.08 percent and 82.75 percent. A significant difficulty score in search space R3 was also unable to decrease by any methods. In search area R3, P16 had an overall success rate of 62.67 percent and was unsolvable by any method.

The success rate of P7, which was 4.92 percent overall, could not be reduced by any of the algorithms operating in the same search space. This indicates that it is impossible to accurately determine how successful WOA and other algorithms are at reducing optimization problems based on their difficulty ratings. As a result, there is no inherent connection between the complexity of minimization optimization issues and the efficiency of these methods. Like Evaluation 1, Evaluation 2 used the Wilcoxon signed-rank test with a significance level (α) of 0.05 to compare WOA against other algorithms. The examined null hypothesis (H0) in this situation is given by Eq. (12.1) for a particular benchmark problem. Tables 12.7, 12.8, and 12.9 show the algorithms that outperformed the other algorithms in Evaluation 2 based on the results of the Wilcoxon signed-rank test.

In Tables 12.4 to 12.9, "-" denotes situations where the WOA was shown as statistically inferior performance, and the null hypothesis was rejected. '+' denotes situations in which the WOA was statistically superior to the null hypothesis, and the null hypothesis was rejected; '=' denotes situations in which there is no statistically significant difference between the two comparison algorithms when assessing the degree to which problems can be minimized. In pairwise problem-based statistical comparisons of the algorithms, the last rows of Tables 12.4–12.9 provide the (+/=/-) summation for statistically significant scenarios indicated by the symbols "+," "=," and "-." When the (+/=/-) data from Evaluations 1 and 2 are examined, it is possible to conclude that WOA outperformed the other comparative algorithms statistically regarding reducing numerical optimization problems.

For most evaluation parts, Evaluations 1 and 2 suggest that WOA is more efficient in resolving numerical optimization problems with varying complexity, variable sizes, and search areas. However, none of the approaches could minimize each of the sixteen benchmark functions successfully. The data presented in Tables 12.4, 12.5, and 12.6 are graphically represented in Fig. 12.2. Fig. 12.3 depicts the findings of Tables 12.7, 12.8, and 12.9.

Regarding the first and second evaluations, it was established that none of the algorithms could handle every benchmark issue. Concerning Evaluation 3, the percentage of successful minimization of each of the 16 varied benchmark functions for Nvar1, 2, and 3 with two variable dimensions as the default search space and three different search spaces (R1, R2, and

TABLE 12.7 Statistical evaluation utilizing the two-sided Wilcoxon Signed-Rank Test ($\alpha = 0.05$) to identify the best solution for the problems solved in Evaluation 2 (R1).

Problem	WOA vs. BSA				WOA vs. FDO				WOA vs. PSO				WOA vs. FF			
	p-value	T+	T-	Winner	p-value	T+	T-	Winner	p-value	T+	T-	Winner	p-value	T+	T-	Winner
P2	0.0001	461	4	+	0.5701	106	359	-	0.0001	459	6	+	0.0001	460	5	+
P4	0.5138	174	291	-	0.0001	445	20	+	0.541	167	298	-	0.5204	156	309	-
P5	0.0001	410	55	+	0.0001	423	42	+	0.0001	461	4	+	0.0003	451	14	+
P6	0.0001	444	21	+	0.5273	351	114	-	0.0001	464	1	+	0.6198	249	216	-
P8	0.5831	364	14	-	0.691	203	262	-	0.0711	324	141	+	0.0001	369	96	+
P9	0.0001	350	115	+	<0.5121	105	360	-	0.0001	145	320	+	0.0701	422	43	+
P11	<0.0001	461	4	+	0.0801	460	5	+	0.5004	369	96	-	0.0001	322	143	+
P12	0.7024	115	350	-	0.0001	465	0	+	<0.0001	154	311	+	<0.0001	231	234	+
P13	0.0001	425	40	+	0.551	166	299	-	0.0001	464	1	+	0.0001	371	94	+
P14	0.0593	264	210	+	0.0611	461	4	+	0.0801	16	284	+	0.0902	145	320	+
P15	0.0001	463	2	+	0.0001	261	204	+	0.0001	364	101	+	0.0321	458	7	+
+/=/-	**9/0/3**				**6/0/5**				**9/0/2**				**9/0/2**			

TABLE 12.8 Statistical evaluation utilizing the two-sided Wilcoxon Signed-Rank Test ($\alpha = 0.05$) to identify the best solution for the problems solved in Evaluation 2 (R2).

Problem	WOA vs. BSA				WOA vs. FDO				WOA vs. PSO				WOA vs. FF			
	p-value	T+	T-	Winner	p-value	T+	T-	Winner	p-value	T+	T-	Winner	p-value	T+	T-	Winner
P2	0.0301	401	34	+	0.672	220	245	=	<0.0001	407	11	+	0.0001	218	0	+
P4	0.0001	423	39	+	0.0001	201	133	+	0.5001	253	211	-	<0.0001	225	99	+
P8	0.5189	264	61	-	0.5501	465	0	-	0.0001	225	99	+	0.5211	265	143	-
P9	0.0201	465	0	+	<0.0001	147	61	+	0.5001	125	211	=	<0.0001	215	100	+
P11	0.5201	146	261	-	0.521	332	101	-	<0.0001	241	14	+	0.0003	178	217	+
P12	0.0001	378	0	+	0.2001	201	211	+	0.641	253	131	-	<0.0001	274	14	+
P13	0.6001	364	119	-	0.0001	228	29	+	<0.0001	208	211	+	0.0001	365	0	+
P14	0.0501	13	265	+	<0.0001	235	1	+	<0.0001	109	201	+	<0.0001	205	4	+
P15	0.3001	265	0	+	0.6201	165	249	-	0.0004	265	49	+	0.0001	465	0	+
+/=/-	**6/0/3**				**5/1/3**				**6/1/2**				**8/0/1**			

TABLE 12.9 Statistical evaluation utilizing the two-sided Wilcoxon Signed-Rank Test ($\alpha = 0.05$) to identify the best solution for the problems solved in Evaluation 2 (R3).

Problems	WOA vs. BSA				WOA vs. FDO				WOA vs. PSO				WOA vs. FF			
	p-value	T+	T-	Winner	p-value	T+	T-	Winner	p-value	T+	T-	Winner	p-value	T+	T-	Winner
P2	0.5001	233	220	=	0.6001	106	231	-	<0.0001	250	215	+	<0.0001	265	97	+
P4	0.0001	206	18	+	<0.0001	151	119	+	0.6511	178	177	-	0.0004	151	16	+
P8	0.6021	110	261	-	0.0003	460	5	+	0.0001	225	18	+	0.5619	176	201	-
P9	<0.0001	161	245	+	<0.0001	301	0	+	<0.0001	425	0	+	0.0001	376	11	+
P11	0.0001	232	143	+	0.2286	222	163	+	0.5130	204	219	-	0.0004	235	12	+
P12	<0.0001	250	99	+	0.5201	265	198	-	0.0198	28	68	+	<0.0001	205	200	+
P13	0.6001	177	218	-	0.6301	206	228	-	0.0004	151	128	+	0.0003	265	212	+
P14	0.0213	64	212	+	0.6098	165	218	-	<0.0001	231	200	+	0.0003	401	0	+
+/=/-	**5/1/2**				**4/0/4**				**6/0/2**				**7/0/1**			

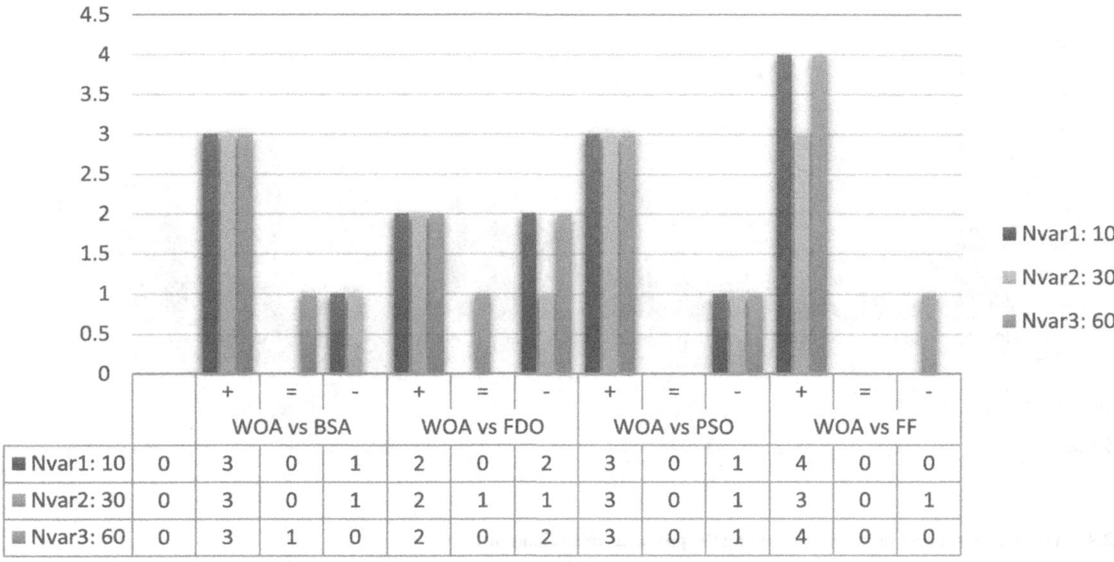

		WOA vs BSA			WOA vs FDO			WOA vs PSO			WOA vs FF		
		+	=	-	+	=	-	+	=	-	+	=	-
Nvar1: 10	0	3	0	1	2	0	2	3	0	1	4	0	0
Nvar2: 30	0	3	0	1	2	1	1	3	0	1	3	0	1
Nvar3: 60	0	3	1	0	2	0	2	3	0	1	4	0	0

FIGURE 12.2 A graphical form illustration of Evaluation 1.

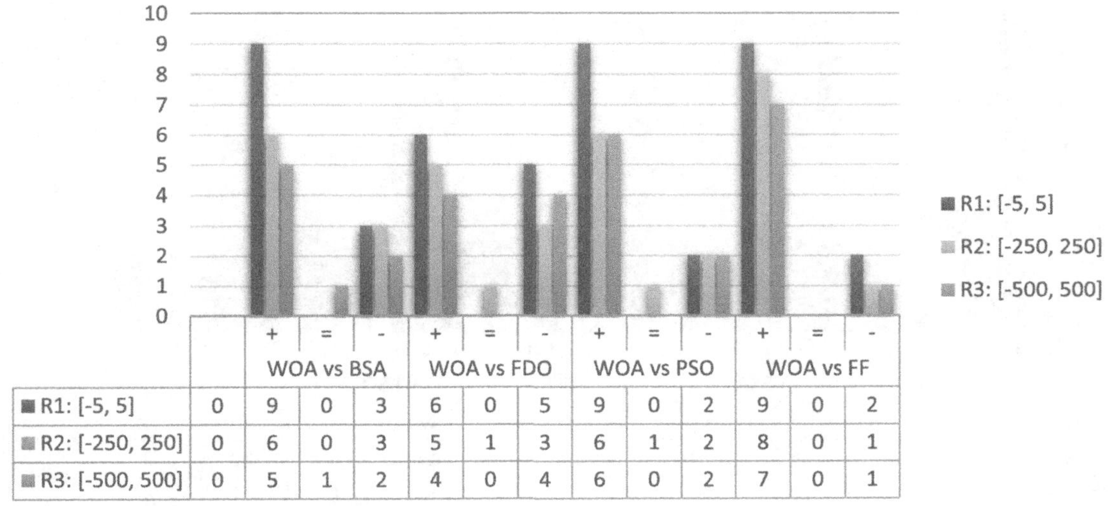

		WOA vs BSA			WOA vs FDO			WOA vs PSO			WOA vs FF		
		+	=	-	+	=	-	+	=	-	+	=	-
R1: [-5, 5]	0	9	0	3	6	0	5	9	0	2	9	0	2
R2: [-250, 250]	0	6	0	3	5	1	3	6	1	2	8	0	1
R3: [-500, 500]	0	5	1	2	4	0	4	6	0	2	7	0	1

FIGURE 12.3 A graphical form illustration of Evaluation 2.

R3). The success and failure rates for decreasing the 16 benchmark functions in Evaluations 1 and 2 are shown in Figs. 12.4 and 12.5, respectively.

Evaluation 3 determined that WOA was the most effective algorithm for decreasing the highest number of the 16 benchmark functions, while PSO was the least effective technique for attaining the same objective. It also showed that WOA could minimize the most optimization problems; nevertheless, the lowest ratio of successful P1-P16 functions minimization in Nvar1, 2, and 3 FDO uncovered variables that utilized the default search space after that PSO. In contrast, for the variable sizes of two and three different search spaces, WOA and DE had the highest ratio of successful minimization (R1, 2, and 3) of these optimization problems, respectively.

12.4.2 Computational cost

This section analyzes the total computational cost of the algorithms (running time/memory usage) based on the benchmarking problems. The running time and memory usage for the 30 solutions produced by WOA and other algorithms are shown in Table 12.10. We note that WOA has a faster rate of solving problems. The WOA used less memory than the

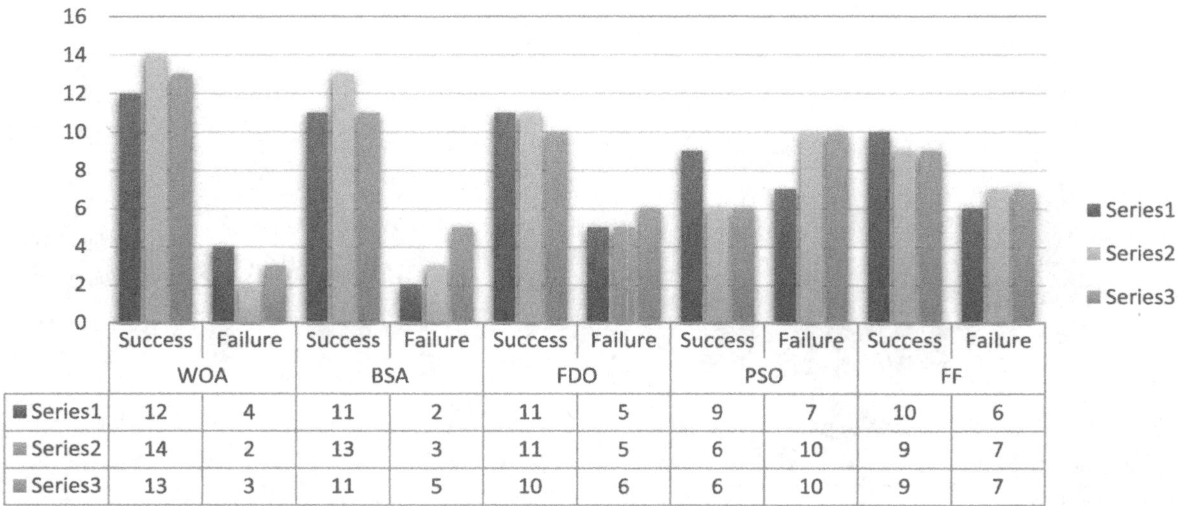

FIGURE 12.4 The failure-success ratio for minimizing the problems in Evaluation 1.

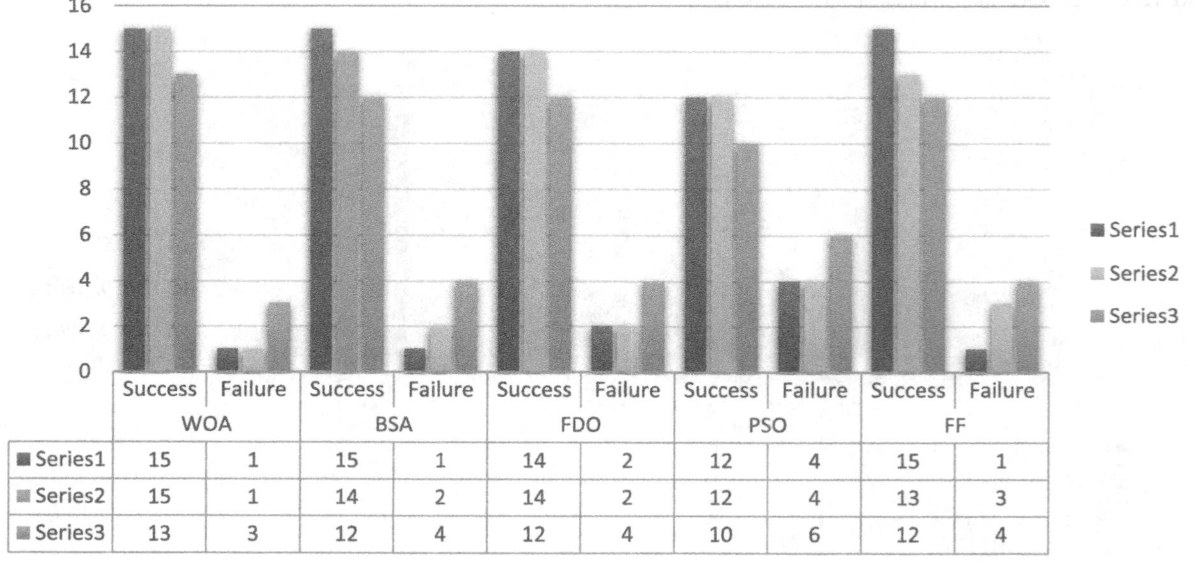

FIGURE 12.5 The failure-success ratio for minimizing the problems in Evaluation 2.

other methods, too. Surprisingly, FDO needs less running time than WOA for P6 and P12 and uses less memory than WOA for solving P16. At the same time, PSO needs less memory than WOA for P2 and less time for executing P4 than WOA. Referencing FF, it needs less running time than WOA for P11. Also, BSA uses less memory than WOA for solving P12. On average, the WOA executes more quickly and uses less memory than the other optimization techniques.

12.4.3 Convergence analysis

The convergence of WOA is assessed using sixteen benchmark issues (P1–P16). WOA has been independently performed using its comparing algorithms, and the best result has been shown after each repetition. Each algorithm has run 1000 times with a default search space of 30 and a population size of 30. As a result, Fig. 12.6 shows how quickly WOA and its rival algorithms converge. As observed, WOA improved the convergence compared to its comparison methods. Overall, based on the findings from the earlier sections, WOA performs better in exploitation and accelerating convergence. The advantage of WOA over other algorithms is its capacity to avoid local optima and get an optimum global result.

TABLE 12.10 Average running time and memory usage for the fifteen problems obtained by WOA and its counterpart algorithms.

Problems	Statistics	WOA	BSA	FDO	PSO	FF	Winner
P1	**Running time**	60.465	64.625	65.692	71.318	73.444	WOA
	Memory usage	107.610	123.882	117.958	122.546	142.770	WOA
P2	**Running time**	87.403	108.494	88.638	131.252	97.628	WOA
	Memory usage	168.778	190.248	187.935	150.853	187.574	PSO
P3	**Running time**	101.658	146.621	146.461	148.136	155.733	WOA
	Memory usage	102.490	110.601	107.337	154.050	137.452	WOA
P4	**Running time**	114.398	163.755	153.260	133.616	179.608	PSO
	Memory usage	129.673	142.252	355.075	301.916	197.261	WOA
P5	**Running time**	96.258	211.288	175.017	114.907	157.864	WOA
	Memory usage	139.250	168.400	161.505	196.943	155.040	WOA
P6	**Running time**	120.226	108.462	98.479	125.525	106.761	FDO
	Memory usage	115.785	142.418	134.172	131.066	163.013	WOA
P7	**Running time**	72.416	90.503	82.868	94.466	96.738	WOA
	Memory usage	107.937	148.801	119.474	146.541	161.103	WOA
P8	**Running time**	154.787	322.546	251.372	187.438	373.585	WOA
	Memory usage	218.852	265.149	304.041	308.048	329.304	WOA
P9	**Running time**	139.777	190.830	143.621	224.299	208.666	WOA
	Memory usage	150.697	424.252	447.155	432.043	287.968	WOA
P10	**Running time**	74.609	133.127	256.108	150.900	195.263	WOA
	Memory usage	117.771	126.280	207.945	122.908	219.695	WOA
P11	**Running time**	190.319	330.692	297.388	312.068	176.029	FF
	Memory usage	103.530	104.072	271.548	224.580	294.954	WOA
P12	**Running time**	197.392	205.910	132.140	137.415	337.746	FDO
	Memory usage	125.762	215.917	130.117	197.576	235.339	WOA
P13	**Running time**	125.196	145.005	170.095	133.859	173.290	WOA
	Memory usage	96.748	133.607	129.430	245.086	381.938	WOA
P14	**Running time**	206.919	222.166	295.787	232.808	230.426	WOA
	Memory usage	123.038	109.867	155.162	187.752	159.081	BSA
P15	**Running time**	213.554	246.275	223.216	267.662	258.846	WOA
	Memory usage	117.521	184.633	178.929	179.451	177.233	WOA
P16	**Running time**	116.488	181.630	173.427	137.377	151.673	WOA
	Memory usage	141.698	145.050	111.549	273.085	280.214	FDO

12.5 Summary

In this chapter, the performance of WOA was compared with its analogous algorithms for various real-world problems. The assessment results revealed that WOA is statistically better than competing approaches for reducing various numerical optimization problems without being unduly sensitive to issue dimensions. Although the statistically superior performance was shown between WOA and other algorithms in this study's evaluations, this does not suggest that WOA can be used for all optimization problems regarding hardness score, search space, and dimension problems. In this analysis, WOA could not minimize the sixteen benchmark functions utilized in this experiment. This failure is most likely the result of four factors: variable issue difficulty ratings, search spaces, and the number of dimensions with problem cohorts. The success and failure rates of WOA in resolving various problems with varying difficulty ratings, problem dimensions, and search areas cannot be precisely quantified. Consequently, this experiment illustrates that WOA is somewhat sensitive to the problem's difficulty score, problem dimensions and type, and search space. In addition, WOA was performing better on the sixteen problems in terms of convergence speed, running time, and memory utilization.

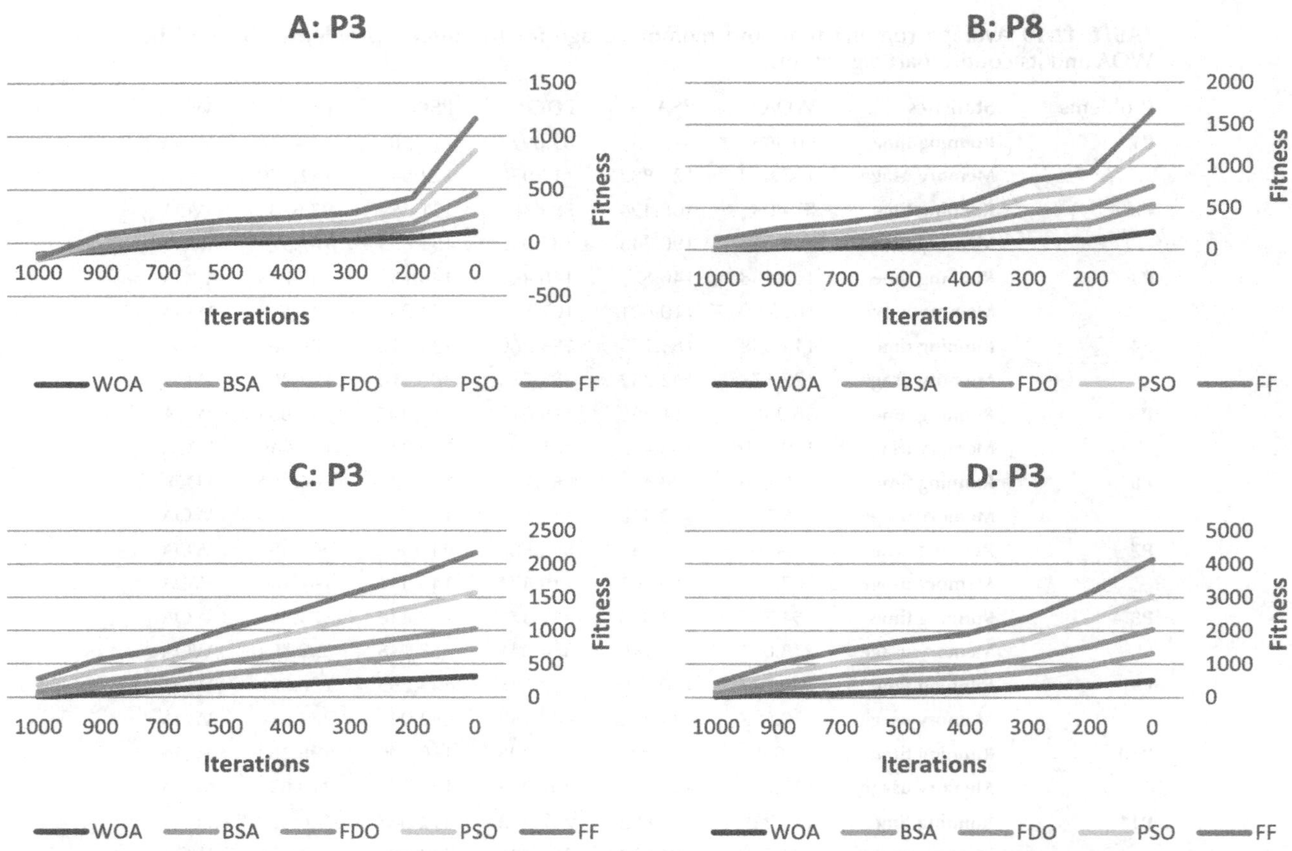

FIGURE 12.6 Convergence analysis curves of the WOA and its counterpart algorithms on four representative problems; (A) for P3, (B) for P8, (C) for P13 and (D) for P16.

References

[1] S. Mirjalili, A. Lewis, The whale optimization algorithm, Adv. Eng. Softw. 95 (2016) 51–67.

[2] P. Civicioglu, Backtracking search optimization algorithm for numerical optimization problems, Appl. Math. Comput. 219 (15) (2013) 8121–8144.

[3] J.M. Abdullah, T. Ahmed, Fitness dependent optimizer: inspired by the bee swarming reproductive process, IEEE Access 7 (2019) 43473–43486.

[4] G. Lindfield, J. Penny, Introduction to Nature-Inspired Optimization, Academic Press, 2017.

[5] X.-S. Yang, Firefly algorithms for multimodal optimization, in: International Symposium on Stochastic Algorithms, 2009, pp. 169–178.

[6] X.-S. Yang, Firefly algorithm source code, https://www.mathworks.com/matlabcentral/fileexchange/29693-firefly-algorithm, 2022. (Accessed 21 September 2021).

[7] B.A. Hassan, CSCF: a chaotic sine cosine firefly algorithm for practical application problems, Neural Comput. Appl. (2020) 1–20.

[8] M. Jamil, X.-S. Yang, A literature survey of benchmark functions for global optimization problems, arXiv:1308.4008, 2013.

[9] M.M. Ali, C. Khompatraporn, Z.B. Zabinsky, A numerical evaluation of several stochastic algorithms on selected continuous global optimization test problems, J. Glob. Optim. 31 (4) (2005) 635–672.

[10] Global Optimization Benchmarks and AMPGO, http://infinity77.net/global_optimization. (Accessed 24 November 2018).

[11] P. Civicioglu, Artificial cooperative search algorithm for numerical optimization problems, Inf. Sci. (NY) 229 (2013) 58–76.

[12] B.A. Hassan, T.A. Rashid, Operational framework for recent advances in backtracking search optimization algorithm: a systematic review and performance evaluation, Appl. Math. Comput. (2019) 124919.

[13] J. Derrac, S. García, D. Molina, F. Herrera, A practical tutorial on the use of nonparametric statistical tests as a methodology for comparing evolutionary and swarm intelligence algorithms, Swarm Evol. Comput. 1 (1) (2011) 3–18.

Chapter 13

Multi-objective archived-based whale optimization algorithm

Nima Khodadadi[a], Seyedeh Zahra Mirjalili[b], Seyed Mohammad Mirjalili[c], Mohammad H. Nadim-Shahraki[d], and Seyedali Mirjalili[b,e,f]

[a]*Department of Civil and Architectural Engineering, University of Miami, Coral Gables, FL, United States,* [b]*Centre for Artificial Intelligence Research and Optimisation, Torrens University Australia, Brisbane, QLD, Australia,* [c]*Department of Engineering Physics, Polytechnique Montréal, Montreal, QC, Canada,* [d]*Faculty of Computer Engineering, Najafabad Branch, Islamic Azad University, Najafabad, Iran,* [e]*Yonsei Frontier Lab, Yonsei University, Seoul, South Korea,* [f]*University Research and Innovation Center, Obuda University, Budapest, Hungary*

13.1 Introduction

Over the past decade, meta-heuristics have become a common technique for researchers and professionals seeking optimal solutions to various optimization problems. These algorithms complete the absence of the need for a derivative and have stochastic behaviors. The former term is a reference to random mechanisms and operators that are utilized in meta-heuristics. These components are responsible for generating random behaviors. The majority of these algorithms commence by using a set of arbitrary solutions. To achieve the desired level of precision, this set is then refined using a variety of mechanisms, many of which take cues from nature. Such compromises are necessary to find satisfactory solutions within a reasonable timeframe for problems with vast search spaces where exploring every potential solution is not feasible. The Chimp Optimization Algorithm (COA) [1], Whale Optimization Algorithm (WOA) [2], Stochastic Paint Optimizer (SPO) [3], Red Fox Optimization Algorithm (RFO) [4], Prairie Dog Optimization (PDO) [5], and Snake Optimizer (SO) [6] are just a few examples of researchers who have used these methods to resolve many meta-heuristic optimization problems. Problems such as truss structures [7], medical issues [8], and data mining [9] are the best instances of optimization problems.

Optimization problems can be categorized as either single-objective or multi-objective problems. In single-objective optimization (SOO), the primary objective is to identify the optimal solution that meets the problem requirements, such as minimizing or maximizing the objective function while adhering to predetermined constraints. In essence, the aim is to discover the most optimal solution. Real-world problems, on the other hand, are more complex and have more objectives. Multiple goals must be optimized at once in such situations. As a result, it is crucial to understand multi-objective optimization to attain the optimal solution (MOO). There is a fundamental difference between SOO and MOO. In most cases, SOO has an extremum point as the optimal solution, while obtaining such a point in MOO is complicated and often impossible [10].

The set, as mentioned earlier, is known as the Pareto optimal solutions or Pareto set [11]. Users must decide which option is best for them by considering their individual preferences and needs. In the absence of such data, all Pareto-optimal solutions are equivalent. Fig. 13.1 demonstrates the Pareto concept for a bi-objective optimization problem, with the feasible objective space represented in green (light gray in print version) and the Pareto set shown in black. As a result, all points marked from a to f with black circles are Pareto optimal, while points marked g to i with red circles (mid gray in print version) are not.

There are two techniques for solving multi-objective problems: Priori and Posteriori. In some cases, the designer's preference or constraints may determine the relative importance of objectives, and such problems can be converted to single-objective (SO) problems. Methods like lexicographic [12], goal programming [13], and weighted sum [14] are examples of a priori preference specification.

While a priori methods have their benefits, they also have some drawbacks. For example, not all multi-objective (MO) problems have preference data available. Additionally, because single-objective optimization (SOO) solutions are a single point, the designer may lack a complete understanding of the trade-offs between objectives. Consequently, posterior preference specification techniques have been proposed, with the primary goal of identifying the complete set of Pareto

Handbook of Whale Optimization Algorithm. https://doi.org/10.1016/B978-0-32-395365-8.00019-1

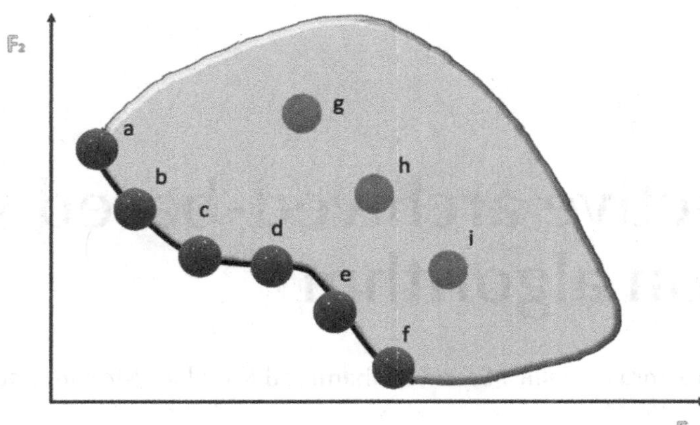

FIGURE 13.1 Pareto concept for multi-objective problems.

optimal solutions [15]. In the field of posterior methods, Evolutionary Algorithms (EAs) have produced some of the most well-known examples, including the Non-dominated Sorting Genetic Algorithm [16] and MO Particle Optimization Algorithm [17]. While the aforementioned algorithms have been around for over a decade, the more recent EAs inspired by nature are also being used in practice in MMOs. Researchers have found that by modeling natural behaviors, they can more effectively address practical issues. In recent years, metaheuristic algorithms that take inspiration from nature have proven effective in solving MOO problems such as the Multi-Objective Stochastic Paint Optimizer (MOSPO) [18], Multi-Objective Chimp Optimization Algorithm (MOChOA) [19], Multi-Objective Artificial Vultures Optimization Algorithm (MOAVOA) [20], Multi-Objective Search Group Algorithm (MOSGA) [21], Multi-Objective Non-dominated Advanced Butterfly Optimization Algorithm (MONSBOA) [22], Multi-Objective Material Generation Algorithm (MOMGA) [23], Multi-Objective Seagull Optimization Algorithm (MOSOA) [24]. In addition, various versions of multi-objective WOA with different methods were proposed in the last decade. Wang et al. proposed [25] the opposition-based multi-objective whale optimization algorithm with global grid ranking (MOWOA). One novel feature of the MOWOA algorithm is the use of a global grid ranking (GGR) mechanism, which is inspired by the grid mechanism. In addition, the algorithm incorporates opposition-based learning (OBL), which has been shown to improve the performance of optimization algorithms.

Non-Dominated Sorting Whale Optimization Algorithm (NSWOA) is another multi-objective version of WOA which was developed by Jangir et al. [26]. The proposed algorithm was designed for solving multi-objective optimization problems and works by collecting all non-dominated Pareto optimal solutions achieved during the course of the evolutionary process. Once the evolution process is complete, the best solutions are chosen from the collection of all Pareto optimal solutions using a crowding distance mechanism. The crowding distance mechanism is based on the coverage of solutions and allows for the selection of solutions that are not only non-dominated, but also well-distributed across the Pareto front. In addition, the proposed algorithm incorporates a bubble-net hunting strategy, which is used to guide humpback whales towards the dominated regions of multi-objective search spaces.

There are still challenges without effective solutions, although many metaheuristic algorithms have been proposed. According to the No Free Lunch theory [27], there is no universally applicable algorithm that can efficiently solve all optimization problems. As a result, nature-inspired algorithms help us find new ways to solve problems or enhance traditional methods.

Mirjalili and Lewis [2] developed an optimizer that goes by the name Whale Optimization Algorithm (WOA). Among its many advantageous properties are its low number of tuning parameters, simplicity of application, resistance to becoming stuck in local optima, and adequate convergence rate. Several engineering optimization projects used the WOA with great success. Therefore, this chapter's primary purpose is to use the WOA in MO problems to boost the understanding of the situation for decision-makers.

The following are some of the most important contributions to this study that were made:

- Applied archive to WOA to store the obtained Pareto optimal solutions.
- The algorithm can quickly navigate the multi-objective search space and reach Pareto optimality by incorporating a leader function.
- The distributed Pareto fronts Grid technique is added for spreading Pareto fronts.

- Eight well-known engineering problems are evaluated by MAWOA.
- Four performance metrics are used to find the efficiency of this method.

The rest of the chapter is organized as follows. Section 13.2 provides an overview of the original WOA algorithm. Section 13.3 elaborates on the multi-objective extension of the WOA, MOWOA. The experimental results are presented in Section 13.4. Finally, the chapter concludes with Section 13.5.

13.2 Whale optimization algorithm

The WOA is an intelligent algorithm based on swarm behavior that is utilized to solve continuous optimization problems. It has shown exceptional performance through the use of modern meta-heuristic techniques. In comparison to other swarm intelligence methods, it is simple to implement and robust, comparable to other nature-inspired algorithms. The algorithm utilizes only a reduced number of control parameters, with only a single parameter needing adjustment in practice. Humpback whales are used in WOA to explore a multi-dimensional search space for food, as illustrated in Fig. 13.2. The different decision variables represent the individual locations of the humpback whales, while the cost objective represents the distance between the whales and the food source. It is worth noting that the time-dependent location of each whale individual is determined by three operational processes: (1) shrinking encircling prey, (2) bubble-net attacking method (exploitation phase), and (3) search for prey (exploration phase) [2].

FIGURE 13.2 Whale Search space for food.

The proposed algorithm is based on the cooperative nature of humpback whales. The algorithm takes its cue from the bubble-net method of fishing.

These are the steps taken by the algorithm to arrive at the best possible answer.

- As part of the initialization process, we generate a random initial population of whales, determine each individual's fitness, and set the optimal whale location as the best position found so far.
- Bubble-net hunting strategy updates the location of the whale. The whale's position will be determined based on the position of the whale that has been determined to be the best so far. If the requirement isn't met, a whale at random will be selected.
- If more whales are discovered that are fitter, their fitness levels will be taken into account, and the best position will be updated.
- If the termination criteria have been met, the best position will be returned as the optimal solution for the problem that has been presented. If not, proceed to the steps in updating the whale position again.

It is worth mentioning that bubble-net feeding is a unique behavior only exhibited by humpback whales. This feeding technique involves the whales releasing bubbles in a spiral pattern to trap and concentrate their prey. The process has been mathematically modeled in reference [2] to enable optimization techniques to be applied to it.

13.3 Multi-objective whale optimization algorithm

We've added an extra dimension to the previously mentioned algorithm by adding three more parts to the WOA approach. These are very similar to the components used in MOPSO [17]. The No Free Lunch (NFL) [7] theorem proves that not all optimization problems can be solved using these techniques. To rephrase, an algorithm might fare well in one context but poorly in another. This highlights the importance of comparative studies across a spectrum of topics. The results of such a comparison study can also shed new light on the efficacy of meta-heuristics. In this chapter, we introduce the MAWOA algorithm and use it to solve eight different engineering problems. The followings are three mechanisms in the MAWOA algorithm:

i. **Archive Mechanism:** One of MAWOA's features is an archive that stores all of the non-dominated Pareto optimum solutions that have been discovered up until this point. If it is concave, convex, and disconnected, then it has the capability of spreading uniformly across the Pareto front. For example, this method only consists of a single component, which is the archive controller. This controller's primary responsibility is to decide if a given solution should be saved in the repository.
ii. **Grid Mechanism:** The grid mechanism is used to generate the distributed Pareto fronts. Several distinct regions make up the objective function space. If a new member is added to the population, the grid will require recalculation, and every member of the population will need to be repositioned. The grid provides an area in which solutions can be evenly dispersed via the use of hypercubes.
iii. **Leader Mechanism:** By using the leader function, the other search agents are directed towards new and promising regions of the search space where they have a greater chance of discovering a solution that is close to the global optimum. It can be challenging to compare outcomes in a multi-objective search space due to Pareto optimality, but it is possible in a single-objective search space. To address this, a mechanism for selecting a leader was introduced. A compilation of the most effective non-dominant solutions gathered so far is maintained. The leader selection module identifies the sparser regions of the search space from which non-dominated solutions are generated.

The probability for each hypercube is determined using the following hypercube [28]:

$$P_i = \frac{C}{N_i} \tag{13.1}$$

Following Eq. (13.1), hypercubes that contain a smaller total number of members are more likely to nominate new leaders. The probability of selecting a hypercube as a source from which to select leaders increases in direct correlation with the rate at which the variety of solutions can be obtained by solving a hypercube decrease.

The MAWOA algorithm is a modified version of the WOA procedure. A better consistency from the MAWOA algorithm is likely if we choose one of the solutions from the archive. When faced with a plethora of choices, it can be challenging to identify the Pareto-optimal responses. We applied the leader function collection and archive preservation to find a solution to this issue.

13.4 Simulation and results

In this section, a total of four multi-objective engineering problems [29] will be utilized to evaluate the effectiveness of MAWOA. The results will be analyzed and compared to the outcomes of other methodologies. MAWOA will be compared to four state-of-the-art multi-objective algorithms: Multi-Objective Ant Lion Optimizer (MOALO) [30], Multi-Objective Grey Wolf Optimizer (MOGWO) [28], Multi-Objective Harris Hawks Optimization (MOHHO) [31], Multi-Objective Moth Flame Optimization (MMFO) [32]. The MAWOA's efficacy is further demonstrated by a comparison to state-of-the-art algorithms. Experiments with the following designs have been selected for testing MAWOA:

- N_{pop} refers to the total number of particles in the population and is equivalent to 100.
- N_{grid} is the Number of Adaptive Grids, which equals 30.
- N_{rep} is the Archive Size which equals 100.
- γ is the probability of selecting parents from the neighborhood which equals 0.9.
- β is the leader selection pressure parameter which equals 4.
- The optimization problem has M objectives, where M here equals 2.
- In the MO problem, the total number of decision variables is D.
- 1000 is the highest possible number of times a function can be evaluated.
- Procedures are run 30 times.

Tables 13.1 and 13.2 display the IGD and GD performance metrics results for eight engineering design challenges. When compared to other algorithms, the MAWOA method performed better in six out of eight instances where IGD metrics were used. On the GD metric, the MAWOA performs better for all benchmarks except the WELDED BEAM and

TABLE 13.1 The IGD metric for different algorithms.

Functions		Algorithm				
		MOALO	*MOGWO*	*MOHHO*	*MMFO*	*MAWOA*
BNH	**Ave**	1.2198E-02	4.0494E-03	3.5815E-03	9.9928E-04	**8.4163E-04**
	SD	3.6455E-03	1.9951E-03	5.0573E-04	8.7345E-05	**8.6647E-05**
CONSTR	**Ave**	2.5041E-03	7.1841E-04	1.1642E-03	**5.8456E-04**	2.0968E-03
	SD	9.6528E-04	2.2443E-04	2.4569E-04	**7.8592E-05**	9.7900E-04
DISK BREAKE	**Ave**	2.2755E-03	6.7631E-04	1.3843E-03	5.9422E-04	**5.4859E-04**
	SD	1.2051E-03	1.2071E-04	1.9251E-04	8.1654E-05	**5.6420E-05**
4 BAR-TURSS	**Ave**	2.2004E-02	2.1258E-02	2.0264E-02	1.9977E-02	**1.9974E-02**
	SD	1.1024E-03	1.4265E-04	1.2763E-04	4.2378E-05	**2.6698E-05**
WELDED BEAM	**Ave**	6.1770E-03	1.9168E-03	1.6802E-03	6.1991E-04	**6.0255E-04**
	SD	2.8159E-03	1.8536E-03	6.9040E-04	**4.0206E-05**	5.6688E-05
OSY	**Ave**	8.1016E-03	9.1489E-03	1.0089E-02	7.0087E-03	**6.1173E-03**
	SD	2.2613E-04	3.3932E-03	4.4388E-03	1.7325E-03	**1.1274E-03**
SPEED REDUCER	**Ave**	1.7737E-02	1.4243E-02	2.9377E-02	1.2731E-02	**9.9519E-03**
	SD	3.3878E-03	3.2032E-03	1.4556E-02	1.0569E-02	**7.3369E-03**
SRN	**Ave**	6.2137E-03	2.4823E-03	1.5092E-03	**3.4377E-04**	5.4677E-04
	SD	1.8810E-03	1.0614E-03	4.0294E-04	**9.9166E-05**	1.4190E-04

TABLE 13.2 The GD metric for different algorithms.

Functions		Algorithm				
		MOALO	*MOGWO*	*MOHHO*	*MMFO*	*MAWOA*
BNH	**Ave**	5.3758E-02	5.8830E-02	3.1920E-02	3.3611E-02	**3.1514E-02**
	SD	1.7132E-02	1.1755E-02	2.7997E-03	1.9638E-03	**1.5608E-03**
CONSTR	**Ave**	1.5711E-03	8.2515E-04	8.8606E-04	8.0251E-04	**6.4151E-04**
	SD	7.7806E-04	8.5335E-05	9.3045E-05	1.0058E-04	**7.8527E-05**
DISK BREAKE	**Ave**	8.3694E-03	3.9801E-03	4.4411E-03	**1.6933E-03**	3.5732E-03
	SD	3.4518E-03	1.8173E-03	1.9495E-03	**1.6256E-04**	2.2308E-03
4 BAR-TURSS	**Ave**	2.5901E+00	1.2022E+01	1.3712E+01	1.4874E+01	**1.2053E+00**
	SD	2.2161E+00	2.8597E+00	1.0608E+00	1.0404E+00	**1.0016E+00**
WELDED BEAM	**Ave**	4.4399E-03	1.8035E-02	**1.2609E-02**	1.3382E-02	3.0169E-02
	SD	6.9956E-04	1.1288E-02	4.9246E-03	**3.2405E-03**	4.0992E-02
OSY	**Ave**	7.0760E-01	9.8497E-01	3.5636E+00	1.1503E+00	**6.7259E-01**
	SD	2.4932E-01	2.9382E-01	2.0479E+00	7.1836E-01	**1.7168E-01**
SPEED REDUCER	**Ave**	7.6817E+00	8.2889E+00	1.1628E+01	8.8962E+00	**6.8196E+00**
	SD	3.4697E+00	1.6129E+00	6.0902E+00	3.0599E+00	**2.7972E+00**
SRN	**Ave**	2.5798E-02	2.6743E-02	3.6360E-02	3.5736E-02	**2.1085E-02**
	SD	3.2892E-03	3.9924E-03	4.0201E-02	1.4179E-02	**2.1374E-03**

the DISK BRAKE. The SD values for both metrics showed that this method was able to consistently produce the same results throughout many separate experiments. When comparing MAWOA's results to the SD values, it got five out of eight and six out of eight for IGD and GD metrics, respectively. In comparison to the MOALO, MOGWO, HOHHO, and MMFO, it is evident that the proposed MAWOA achieves superior IGD and GD performances.

These results demonstrate the effectiveness of the proposed method for solving challenging multi-objective optimization problems in a mathematical setting. For the engineering design problems considered, Figs. 13.3 and 13.4 depict the algorithm's ability to generate better solutions that are closer to the Pareto front. This illustration shows that the proposed MAWOA algorithm converges excellently to all the actual Pareto-optimal fronts.

TABLE 13.3 The MS metric for different algorithms.

Functions		Algorithm				
		MOALO	*MOGWO*	*MOHHO*	*MMFO*	*MAWOA*
BNH	**Ave**	5.4090E-01	8.3839E-01	9.7375E-01	**1.0000E+00**	**1.0000E+00**
	SD	1.0692E-01	9.1795E-02	2.4069E-02	**0.0000E+00**	**0.0000E+00**
CONSTR	**Ave**	7.7822E-01	9.7506E-01	9.4665E-01	9.9271E-01	**9.9873E-01**
	SD	7.6343E-02	1.8575E-02	2.8600E-02	**7.8120E-03**	5.8106E-02
DISK BREAKE	**Ave**	8.1431E-01	9.9660E-01	9.9023E-01	**9.9950E-01**	9.9271E-01
	SD	1.1765E-01	1.6364E-02	9.1601E-03	**3.4466E-04**	1.4494E-02
4 BAR-TURSS	**Ave**	8.4793E-01	1.3470E+00	1.4598E+00	1.4873E+00	**1.4984E+00**
	SD	1.2454E-01	4.2641E-02	3.0207E-02	**4.9481E-04**	2.4608E-02
WELDED BEAM	**Ave**	6.2463E-01	1.0777E+00	1.0305E+00	1.0322E+00	**1.0868E+00**
	SD	6.9354E-02	1.1100E-01	1.1506E-01	7.8287E-02	**5.0601E-02**
OSY	**Ave**	5.7530E-01	5.8701E-01	7.9739E-01	6.7654E-01	**8.5664E-01**
	SD	2.6351E-02	1.4507E-01	2.0878E-01	5.9364E-02	**1.0308E-01**
SPEED REDUCER	**Ave**	6.6868E-01	7.6707E-01	4.0783E-01	7.8149E-01	**8.2273E-01**
	SD	6.0939E-02	2.4182E-02	2.1325E-01	8.6327E-02	**5.0578E-02**
SRN	**Ave**	3.9177E-01	7.0014E-01	8.3950E-01	**9.3385E-01**	8.6367E-01
	SD	8.2165E-02	8.0177E-02	5.3680E-02	3.3579E-02	**2.5807E-02**

TABLE 13.4 The S metric for different algorithms.

Functions		Algorithm				
		MOALO	*MOGWO*	*MOHHO*	*MMFO*	*MAWOA*
BNH	**Ave**	8.8402E-01	1.4365E+00	**7.9303E-01**	1.1807E+00	1.0305E+00
	SD	4.4648E-01	4.5168E-01	3.8938E-01	1.3487E-01	**1.3646E-01**
CONSTR	**Ave**	6.9071E-02	6.1162E-02	4.9323E-02	6.7608E-02	**3.2002E-02**
	SD	1.8623E-02	1.1995E-02	2.1079E-02	8.0015E-03	**1.1763E-03**
DISK BREAKE	**Ave**	2.5412E-01	1.4674E-01	9.8010E-02	1.0837E-01	**9.0736E-02**
	SD	4.3251E-02	3.6122E-02	1.9764E-02	2.4915E-02	**1.5571E-02**
4 BAR-TURSS	**Ave**	**4.6988E+00**	4.8379E+00	5.2590E+00	5.0172E+00	5.7093E+00
	SD	1.1721E+00	8.2413E-01	9.6304E-01	7.8121E-01	**4.3398E-01**
WELDED BEAM	**Ave**	2.2431E-01	3.2128E-01	**2.1478E-01**	2.4395E-01	3.9427E-01
	SD	1.3595E-01	1.2064E-01	8.1430E-02	**3.3963E-02**	3.8524E-01
OSY	**Ave**	1.4382E+00	1.2224E+00	1.3603E+00	1.3090E+00	**1.1617E+00**
	SD	4.8498E-01	4.7166E-01	7.7663E-01	2.1121E-01	**3.0824E-01**
SPEED REDUCER	**Ave**	3.6722E+01	2.0624E+01	9.3291E+00	2.0634E+01	**2.0021E+01**
	SD	9.3623E+00	3.1861E+00	5.5217E+00	5.5307E+00	**1.8301E+00**
SRN	**Ave**	**1.7386E+00**	2.6491E+00	1.8819E+00	2.1787E+00	2.0353E+00
	SD	8.6872E-01	1.2558E+00	9.9287E-01	2.9438E-01	**2.0348E-01**

Table 13.3 indicates that, on average, the MAWOA outperforms the MOALO, MOGWO, HOHHO, and MMFO algorithms for the MS metric in six out of the eight engineering design problems considered. According to Table 13.4, MAWOA's rank is second to none among other methods in terms of the S metric. MOALO and MOHHO outperform better just in two cases out of eight cases than other algorithms. As measured by standard deviation, the MAWOA performed better than any other method in both MS and S metrics. Through the use of the IGD, GD, S, and MS indices, this chapter shows that the proposed MAWOA algorithm is capable of achieving better results and outperforming existing approaches. The results obtained are highly competitive, and the proposed MAWOA method generates better solutions closer to the Pareto front for engineering problems when the actual and achieved Pareto fronts are considered.

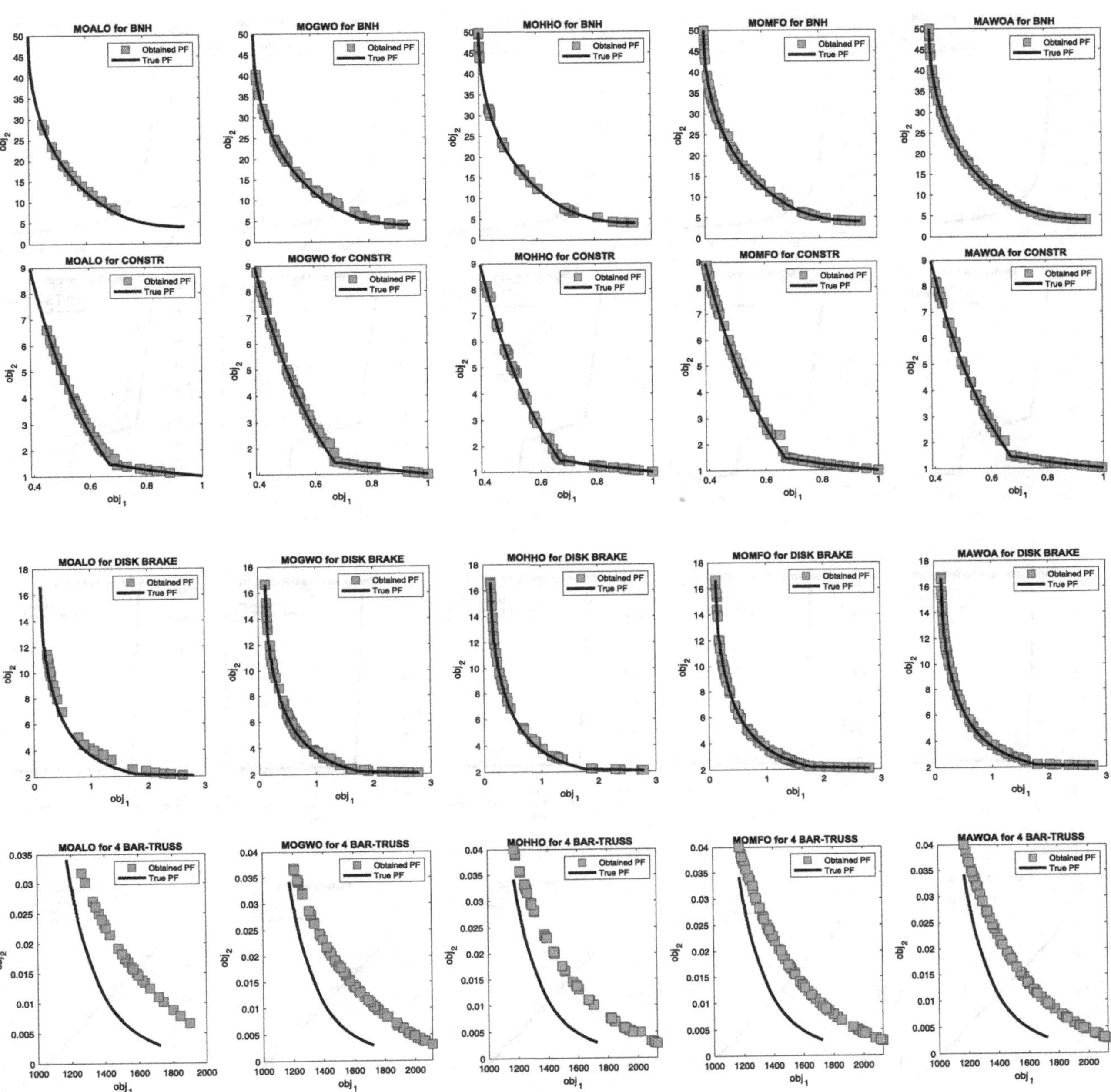

FIGURE 13.3 Results of different algorithms for BNH, CONSTR, DISK BRAKE, and 4 BAR-TRUSs.

13.5 Conclusion

In this chapter, the Multi-objective Whale Optimization Algorithm, also known as MAWOA, was presented as a means of optimizing for multiple objectives problems. Three different components, including leader selection, grid mechanism, and archive mechanism was added to WOA for solving the problem with multiple objectives. The MAWOA evaluation relied on the performance of benchmark tests on a total of eight well-known engineering problems. According to the findings, the MAWOA algorithm is superior to the MOALO, MOGWO, HOHHO, and MMFO. The approach that was suggested allows for the handling of a large number of restrictions, and the statistical findings indicate that it provides superior solutions to those that are currently offered by optimizers. From a computational standpoint, MAWOA is shown to be superior to the others. In the future, this strategy may be applied to other data clustering methods, and it was originally developed to deal with multi-objective problems in bioinformatics.

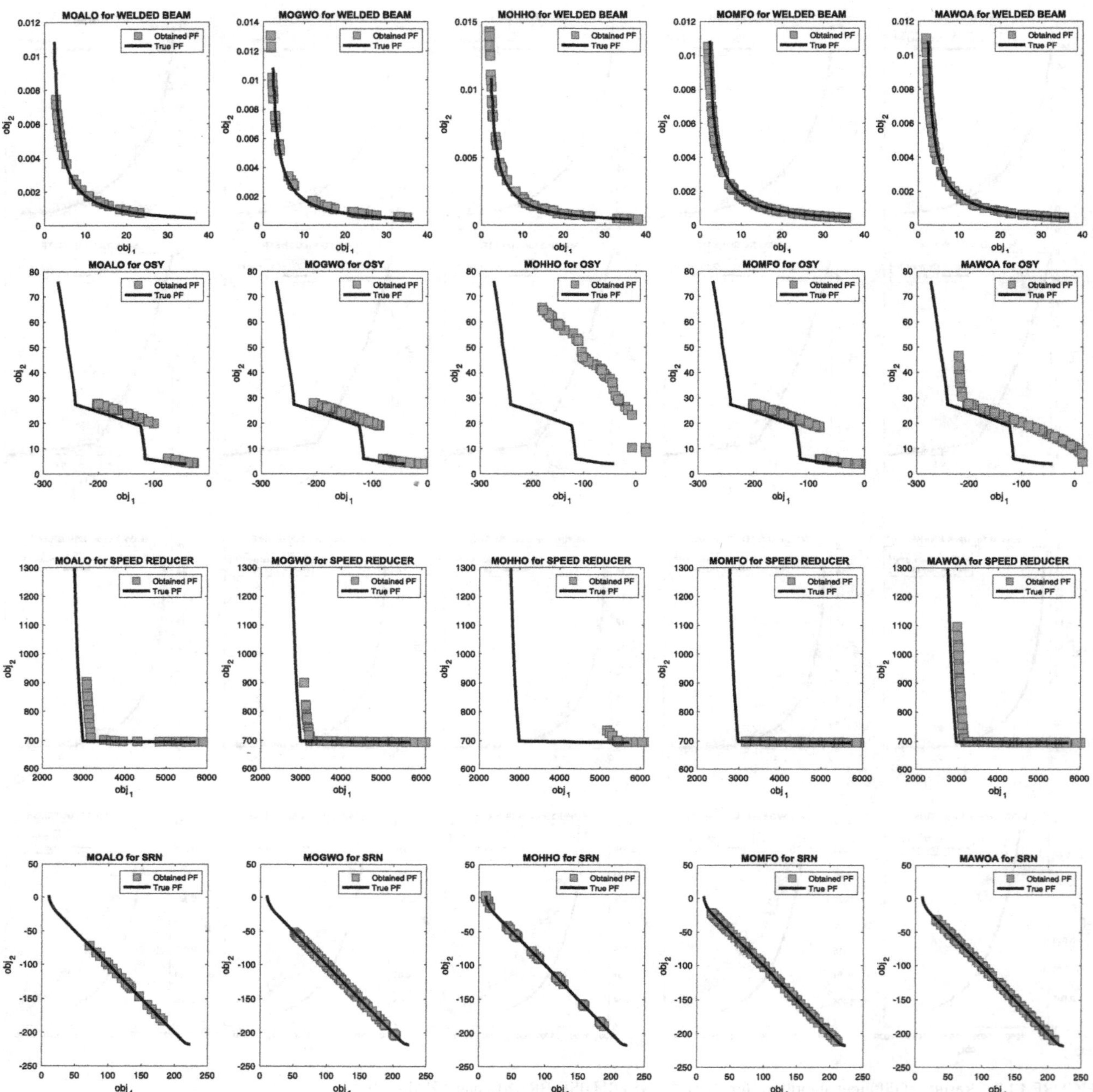

FIGURE 13.4 Results of different algorithms for WELDED BEAM, OSY, SPEED REDUCER, and SRN.

References

[1] M. Khishe, M.R. Mosavi, Chimp optimization algorithm, Expert Syst. Appl. 149 (2020) 113338.
[2] S. Mirjalili, A. Lewis, The whale optimization algorithm, Adv. Eng. Softw. 95 (2016) 51–67.
[3] A. Kaveh, S. Talatahari, N. Khodadadi, Stochastic paint optimizer: theory and application in civil engineering, Eng. Comput. (2020) 1–32.
[4] D. Połap, M. Woźniak, Red fox optimization algorithm, Expert Syst. Appl. 166 (2021) 114107.
[5] A.E. Ezugwu, J.O. Agushaka, L. Abualigah, S. Mirjalili, A.H. Gandomi, Prairie dog optimization algorithm, Neural Comput. Appl. (2022) 1–49.
[6] F.A. Hashim, A.G. Hussien, Snake optimizer: a novel meta-heuristic optimization algorithm, Knowl.-Based Syst. 242 (2022) 108320.
[7] N. Khodadadi, S. Talatahari, A.H. Gandomi, ANNA: advanced neural network algorithm for optimization of structures, Proc. Inst. Civ. Eng. Build. (2023) 1–59.

[8] A.A. Abdelhamid, et al., Classification of monkeypox images based on transfer learning and the Al-Biruni Earth radius optimization algorithm, Mathematics 10 (19) (2022) 3614.
[9] L. Abualigah, A.J. Dulaimi, A novel feature selection method for data mining tasks using hybrid sine cosine algorithm and genetic algorithm, Clust. Comput. (2021) 1–16.
[10] N. Khodadadi, S.M. Mirjalili, W. Zhao, Z. Zhang, L. Wang, S. Mirjalili, Multi-objective artificial hummingbird algorithm, in: Advances in Swarm Intelligence: Variations and Adaptations for Optimization Problems, Springer, 2022, pp. 407–419.
[11] P. Ngatchou, A. Zarei, A. El-Sharkawi, Pareto multi objective optimization, in: Proceedings of the 13th International Conference on, Intelligent Systems Application to Power Systems, 2005, pp. 84–91.
[12] L. Lai, L. Fiaschi, M. Cococcioni, Solving mixed Pareto-lexicographic multi-objective optimization problems: the case of priority chains, Swarm Evol. Comput. 55 (2020) 100687.
[13] R.W. Hanks, B.J. Lunday, J.D. Weir, Robust goal programming for multi-objective optimization of data-driven problems: a use case for the United States transportation command's liner rate setting problem, Omega 90 (2020) 101983.
[14] C. Lin, F. Gao, Y. Bai, An intelligent sampling approach for metamodel-based multi-objective optimization with guidance of the adaptive weighted-sum method, Struct. Multidiscip. Optim. 57 (3) (2018) 1047–1060.
[15] A. Messac, C.A. Mattson, Generating well-distributed sets of Pareto points for engineering design using physical programming, Optim. Eng. 3 (4) (2002) 431–450.
[16] N. Srinivas, K. Deb, Muiltiobjective optimization using nondominated sorting in genetic algorithms, Evol. Comput. 2 (3) (1994) 221–248.
[17] C.A.C. Coello, M.S. Lechuga, MOPSO: a proposal for multiple objective particle swarm optimization, in: Proceedings of the 2002 Congress on Evolutionary Computation. CEC'02 (Cat. No. 02TH8600), vol. 2, 2002, pp. 1051–1056.
[18] N. Khodadadi, L. Abualigah, S. Mirjalili, Multi-objective stochastic paint optimizer (MOSPO), Neural Comput. Appl. 34 (2022) 18035–18058, https://doi.org/10.1007/s00521-022-07405-z.
[19] M. Khishe, N. Orouji, M.R. Mosavi, Multi-objective chimp optimizer: an innovative algorithm for multi-objective problems, Expert Syst. Appl. 211 (2023) 118734.
[20] N. Khodadadi, F. Soleimanian Gharehchopogh, S. Mirjalili, MOAVOA: a new multi-objective artificial vultures optimization algorithm, Neural Comput. Appl. (2022) 1–39.
[21] T.H.B. Huy, P. Nallagownden, K.H. Truong, R. Kannan, D.N. Vo, N. Ho, Multi-objective search group algorithm for engineering design problems, Appl. Soft Comput. 126 (2022) 109287.
[22] S. Sharma, N. Khodadadi, A.K. Saha, F.S. Gharehchopogh, S. Mirjalili, Non-dominated sorting advanced butterfly optimization algorithm for multi-objective problems, J. Bionics Eng. (2022) 1–25.
[23] B. Nouhi, N. Khodadadi, M. Azizi, S. Talatahari, A.H. Gandomi, Multi-objective material generation algorithm (MOMGA) for optimization purposes, IEEE Access 10 (2022) 107095–107115.
[24] G. Dhiman, et al., MOSOA: a new multi-objective seagull optimization algorithm, Expert Syst. Appl. 167 (2021) 114150.
[25] W.L. Wang, W.K. Li, Z. Wang, L. Li, Opposition-based multi-objective whale optimization algorithm with global grid ranking, Neurocomputing 341 (2019) 41–59.
[26] P. Jangir, N. Jangir, Non-dominated sorting whale optimization algorithm (NSWOA): a multi-objective optimization algorithm for solving engineering design problems, Glob. J. Res. Eng. 17 (F4) (2017) 15–42.
[27] D.H. Wolpert, W.G. Macready, No free lunch theorems for optimization, IEEE Trans. Evol. Comput. 1 (1) (1997) 67–82.
[28] S. Mirjalili, S. Saremi, S.M. Mirjalili, L. dos, S. Coelho, Multi-objective grey wolf optimizer: a novel algorithm for multi-criterion optimization, Expert Syst. Appl. 47 (2016) 106–119.
[29] N. Khodadadi, L. Abualigah, E.-S.M. El-Kenawy, V. Snasel, S. Mirjalili, An archive-based multi-objective arithmetic optimization algorithm for solving industrial engineering problems, IEEE Access 10 (2022) 106673–106698, https://doi.org/10.1109/ACCESS.2022.3212081.
[30] S. Mirjalili, P. Jangir, S. Saremi, Multi-objective ant lion optimizer: a multi-objective optimization algorithm for solving engineering problems, Appl. Intell. 46 (1) (2017) 79–95.
[31] U. Yüzgeç, M. Kusoglu, Multi-objective Harris hawks optimizer for multiobjective optimization problems, BSEU J. Eng. Res. Technol. 1 (1) (2020) 31–41.
[32] N. Khodadadi, S.M. Mirjalili, S. Mirjalili, Multi-objective moth-flame optimization algorithm for engineering problems, in: Handbook of Moth-Flame Optimization Algorithm, CRC Press, 2022, pp. 79–96.

Chapter 14

U-WOA: an unsupervised whale optimization algorithm based deep feature selection method for cancer detection in breast ultrasound images

Payel Pramanik[a], Rishav Pramanik[a], Anurup Naskar[a], Seyedali Mirjalili[b,c], and Ram Sarkar[a]
[a]*Department of Computer Science and Engineering, Jadavpur University, Kolkata, India,* [b]*Centre for Artificial Intelligence Research and Optimisation, Torrens University Australia, Brisbane, QLD, Australia,* [c]*Yonsei Frontier Lab, Yonsei University, Seoul, South Korea*

14.1 Introduction

Breast Cancer is a significant health concern and a leading cause of premature death among women. According to statistics, a study released by the International Agency for Research on Cancer (IARC) in December 2020 shows that breast cancer has become the most commonly diagnosed form of cancer globally, surpassing lung disease [1]. Breast cancer is the most frequently diagnosed cancer among women globally, with a death rate of 12.7 per 100,000 women and a prevalence rate as high as 25.8 per 100,000 women. The estimated number of new cases of breast cancer is unarguably very high when compared to other cancerous diseases. A more favorable and optimistic clinical picture would result from increased health consciousness and the presence of screening programs and breast cancer treatment facilities [2]. Considering this is a life-threatening problem, it requires to be diagnosed and treated once the presence of a malignant tumor in the breast is manifested. Millions of women worldwide can potentially benefit greatly from early identification of breast cancer. One of the most effective methods for identifying cancerous tumors is breast image analysis. Radiologists employ a variety of imaging methods, such as diagnostic mammography, thermograms, ultrasound images, histopathological slides, and magnetic resonance imaging (MRI) scans, to determine the stage of breast cancer [3]. Radiologists are often tasked to grade tumors and evaluate morphological elements through the means of visual inspection. Considering clinicians with years of experience, the manual examination method is time-consuming, results in inter-observer variations, and more importantly, such examinations are highly prone to human-like errors. Considering the growth in population presently, the efforts by clinicians often amount to unnecessary biopsies that could be reduced by computer-aided diagnosis (CAD) frameworks. Notably, the advancement of machine learning (ML) and deep learning (DL) frameworks in particular has considerably boosted the reproducibility of the investigation process and reduced human-like errors, in addition to improvising the process by automation of breast cancer prediction [4–6]. To provide more suitable models for an incoming fresh batch of data, a variety of DL-based algorithms have recently been used to learn from previously obtained ultrasound images [7].

DL has significantly advanced the state-of-the-art in several medical imaging-based tasks for advancing the automated diagnosing process in several other tasks like cervical cancer screening [8], COVID-19 diagnosis [9], Parkinson's Disease [10], etc. One of the main contributing factors to this immense popularity of DL has to do with the good generalization traits DL has showcased in the past. DL leverages the use of derivatives or gradients as we commonly know it to optimize its internal weights. A typical DL method for image processing applications consists of two internal components: 1) a feature extractor and 2) a fully connected neural network-based classifier. Generally, both of these components work in synchronization for the model to work as an end-to-end model which is fed with an image and the desired class is obtained as the output. Majorly in the biomedical field, the process of knowledge distillation is used which is known as transfer learning (TL) to alleviate the need for datasets of larger size. In this process, the model is typically trained on a standard large-scale dataset and the internal weights are then transferred to a relatively smaller-sized dataset to be fine-tuned on [11]. Traditionally, researchers overlook one principal concept of eliminating redundant features which ultimately hamper the performance of the overall framework. For this very reason, recently several strategies like filter pruning [12], feature se-

Handbook of Whale Optimization Algorithm. https://doi.org/10.1016/B978-0-32-395365-8.00020-8

lection [13] are getting more popular these days. The objective of these methods is to reduce the presence of redundant and non-informative features. In this work, we use a deep feature selection strategy to perform this task, specifically, we perform feature selection using a popular meta-heuristic optimization algorithm. To sum up this chapter in a nutshell:

In the current study, we suggest a deep feature selection strategy based on the unsupervised whale optimization algorithm (U-WOA) for the identification of breast cancer from ultrasound images. We demonstrate a technique and achieved highest level of accuracy for the classification of breast tumors using a public dataset, called BUSI.

The remainder of the study is organized in the manner listed below. In Section 14.2, first, we go over some of the earlier techniques and their limitations with regard to ultrasound image-based breast cancer data and then the application of whale optimization algorithm (WOA) in neural network and feature selection. In Section 14.3, we first list the requirements for this study before going into further depth about our proposed approach. After that, in Section 14.4, we go over the metrics that we employ to assess the suggested model, together with the experimental data, analysis, and discussion. Finally, in Section 14.5, we draw a conclusion to our work and discuss some potential future directions.

14.2 Literature review

This section is divided into two subsections, the first of which discusses several recent approaches to breast cancer diagnosis using the BUSI database. In the following section, we discuss how WOA can be used in deep feature selection.

14.2.1 Methods for breast cancer detection using BUSI database

Numerous methods by a diverse group of researchers have been proposed which contribute significantly towards detecting breast cancer in ultrasound images. In this part, we go through a few of the most recent methods for identifying breast cancer in the BUSI dataset.

The work by Cai et al. [14] presented a classification method for breast ultrasound images where a feature extractor, namely phased congruence-based binary pattern (PCBP) was proposed. In this work, the authors devised a phase congruency (PC) approach to capture structural information through Fourier analysis. Additionally, the authors extracted textures using local binary patterns (LBP). Finally, the image classification was performed using a support vector machine (SVM). Even though this approach achieved an AUC index value of 0.894, it is observed that LBPs cannot capture translation invariant features at times thus making it a biased feature descriptor to work with [15]. In a work, Mishra et al. [16] presented a handcrafted feature extraction based methodology, where the authors extract handcrafted features using some popular techniques like histogram oriented gradients (HOG), gray-level co-occurrence matrix (GLCM) along with two other feature extractors. The authors then considered a feature selection method using the recursive feature elimination technique which was followed by an ML based classifier. The feature selection methodology consisted a hyper-parameter on how many features to be selected out of the entire feature set. It is to be noted that this approach may not be straightforward in all cases.

A CAD system for the detection of breast cancer was designed by Sadad et al. [17]. In this study, they extracted a lesion component from background tissues using a segmentation technique known as marker-controlled watershed transformation. Then, shape-based and texture features using GLCM were extracted to form a hybrid features set that were used for classification. This method achieved an accuracy of 96.6% using an ensemble classifier. Yang et al. [18] suggested another system based on texture analysis using multi-resolution gray-scale invariant features through the ranklet transformation.

Whereas, Moon et al. [19] proposed a methodology that combines an image fusion technique with various handcrafted feature representations and several convolutional neural network (CNN) architectures, the main of the study was aimed to ensemble various CNN architectures. In this, the authors selected the basis models from various CNN models that produce the best predicted accuracy for each type of image and then integrated the output from those base models using the weighted average ensemble method. Jabeen et al. [20] used deep learning techniques and then merged the best features utilizing optimization algorithms like the reformed Binary Grey Wolf Optimization (BGWO) and the reformed Differential Evaluation to classify breast cancer using ultrasound images. In this method, DarkNet-53 deep learning model was used for extraction of features.

Dhabyani et al. [21] looked at how the BUSI database for the diagnosis of breast cancer might be affected by the Generative adversarial network data augmentation approach. The CNN (AlexNet) and TL methodologies were two deep learning classification techniques that the authors employed in this work. In another work, Mishra et al. [22] proposed a feature fusion-based method to differentiate between benign and malignant tumors. The authors proposed a hybridized approach consisting of fusing handcrafted features with CNN-based features. In all evaluation measures taken into account, this method beat cutting-edge approaches on the classification test, with performance surpassing 95%. In order to address the problem of low specificity and inconsistent diagnosis, Li et al. [23] built a learning algorithm based on computational ultrasound image characteristics and selected a collection of clinically significant properties. In this, the radiomics features were extracted, which was then followed by mutual information based filtering and a recursive feature elimination-based feature selection procedure.

14.2.2 Applications of WOA

The objective of feature selection techniques is to minimize the amount of data and improve the effectiveness of learning algorithms by removing unrelated and redundant information. WOA is one of the efficient meta-heuristic algorithms for feature selection, and it has been used by various researchers in the medical domain.

For instance, Mafarja et al. [24] introduced a wrapper-based feature selection approach that utilized the WOA algorithm to identify the optimal set of minimum features. In this study, evolutionary operators such as selection, crossover, and mutation were combined with the binary form of WOA to improve exploration and exploitation. In another work, Nadimi-Shahraki et al. [25] proposed an enhanced version of WOA to address continuous optimization issues. The enhanced WOA incorporated a pooling mechanism and three search strategies, namely migrating, preferential selecting, and enriched encircling prey, to enhance both local and global search capabilities. Also, they introduced the binary version of enhanced WOA called BEWOA for the selection of optimal features from medical database. Kundu et al. [26] suggested a WOA-based feature selection method called AltWOA in DNA microarray gene expression datasets, where the WOA incorporates the idea of altruism into the population of candidate solutions. In this approach, altruism allows solutions with average fitness but the potential to converge to optimal solutions to survive through iterations, even at the cost of less promising alternatives. Another WOA-based feature selection method called quantum WOA (QWOA) was proposed by Agrawal et al. [27]. WOA and the quantum notion are combined to form QWOA. Each member of the population was shown in this as a quantum bit, and the quantum rotation gate operator served as a variation operator. For quantum-base represented exploring, shrinkage, and spiral movement of the whales in QWOA, the authors provided modified mutation and crossover operators. A chaotic based binary version of WOA called bWOA-S was introduced by Hussein et al. [28] for feature selection problem. In this, the S-shaped transfer function was used to convert the conventional WOA to the binary version. Whereas in [29], Hussien et al. presented a binary variation of the WOA, referred to as bWOA, that utilized a V-shaped design. The fitness function used in bWOA was a hyperbolic tangent function, which was employed to convert continuous values into binary values. In [30], an intelligent system for diagnosing lung tumors was introduced, in which the GLCM method was utilized to extract 19 texture and statistical features from lung computed tomography (CT) images. Here WOA was used to select the optimal set of prominent features.

The hybridization approaches for feature selection have also used WOA. For instance, In [31], the authors used the WOA and simulated annealing (SA) to create a hybrid WOA for feature selection. SA is employed as a component of the WOA algorithm to improve its exploitation capacity. Also, Zheng et al. [32] introduced a novel hybrid feature selection technique called the Maximum Pearson Maximum Distance Improved Whale Optimization Algorithm (MPMDIWOA). This method merges the filter approach of Maximum Pearson Maximum Distance (MPMD) with the wrapper approach of Improved Whale Optimization Algorithm (IWOA) to create a new algorithm.

Besides feature selection, WOA is also used in neural networks for optimizing the connection weights [33], hyperparameters [34] in neural networks.

14.3 Materials & methods

In this section, we discuss some prerequisites and the proposed method in details. First, we briefly describe the dataset used to evaluate the proposed method. Next, we describe the U-WOA method used for feature selection purpose and describe the entire pipeline in detail.

14.3.1 Dataset description

In the present study, we use the BUSI database which was generated in 2018 and made publicly accessible by Dhabyani et al. [35]. This database contains 780 breast ultrasound images (each 500 × 500 pixels) obtained from 600 people ranging in age from 25 to 75 years old utilizing two LOGIQ E9 ultrasound equipment. Along with the images, this database also includes the manually constructed ground truth mask generated by the Matlab® software. Fig. 14.1 displays instance images of three different categories and their corresponding masks.

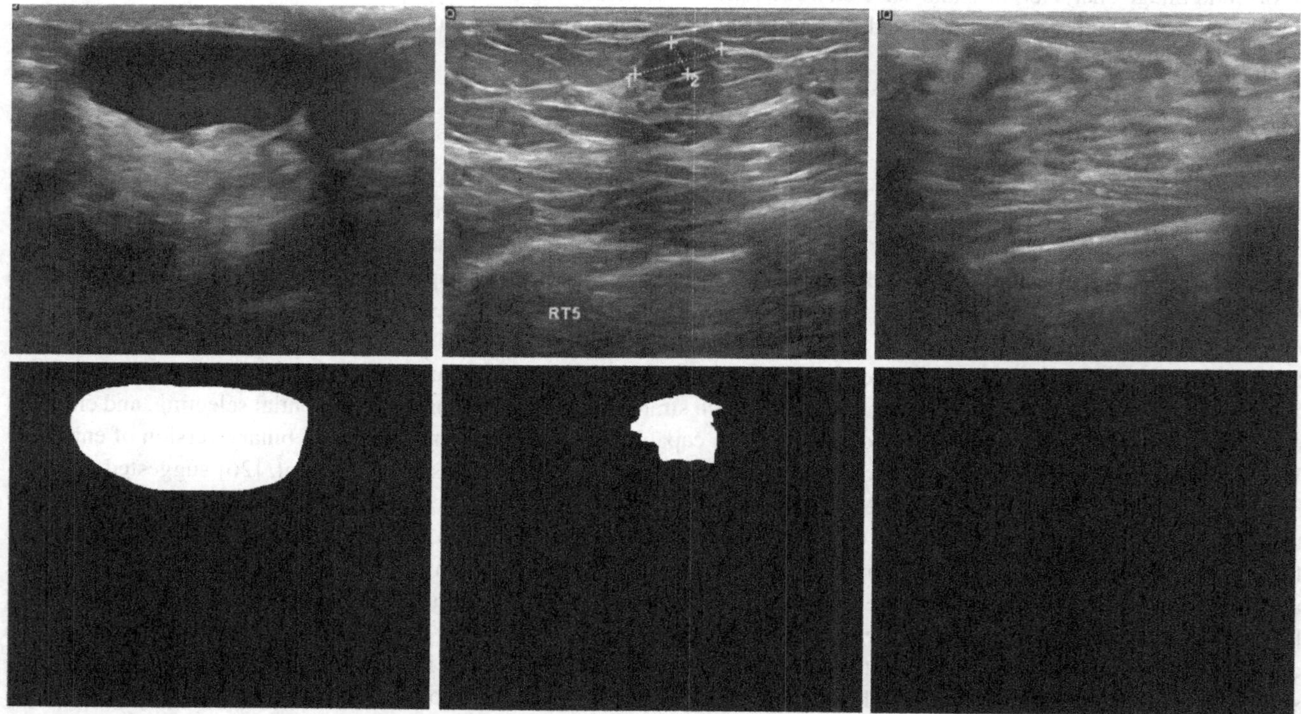

FIGURE 14.1 The top row from left to right depicts the Benign Ultrasound, Malignant Ultrasound, and Normal Ultrasound images, whereas the bottom row from left to right depicts the corresponding Benign, Malignant, and Normal masks.

14.3.2 Whale optimization algorithm

Whale optimization algorithm (WOA) by Mirjalili & Lewis [36] is a meta-heuristic optimization algorithm. The WOA meta-heuristic optimization algorithm imitates the humpback whales' bubble-net assault strategy, in which they plunge into the water and wrap a bubble-net around their victim. Three primary processes make up the WOA algorithm: encircling the prey, a bubble-net attacking phase (exploitation), and hunting for prey (exploration).

14.3.2.1 Encircling prey

In the initial stage, the most suitable search agent is selected with the help of a fitness function and then the distance between the top-performing (best) search agent and the remaining search agents is updated. The current prey is considered to be closer to the global optimum. The methods of spiral updating location and shrinking encircling have a 50% probability of success in the Bubble-net assaulting stage depending on the settings of specific restrictions. In the exploration phase known as "searching for prey," the search agent is free to conduct a haphazard search for prey without using the spiral updating location.

Eqs. (14.1) and (14.2) are used to update the position of the agent$\vec{P}$ where $\vec{P}'$ is the best agent. The coefficients E & F are evaluated as in Eqs. (14.3) and (14.4) respectively.

$$M = |F \cdot \vec{P}'(t) - \vec{P}(t)| \tag{14.1}$$

$$\vec{P}(t+1) = \vec{P}'(t) - \vec{E} * \vec{M} \tag{14.2}$$

$$E = 2 * m * x - m \tag{14.3}$$

$$F = 2 * x \tag{14.4}$$

m drops linearly from 2 to 0 over iterations, where x is an arbitrary variable lying between 0 to 1. We have updated the location of the best solution P' after each cycle if there is a better option or the search agent travels beyond the space of all possible physical solutions. Random variable x helps the search agent to choose the ideal search agent and carry out surrounding the prey.

14.3.2.2 Bubble-net attacking method (exploitation phase)

To model the spiral attacking mechanism, first, we calculate the distance as in Eq. (14.5) and further position is updated as in Eq. (14.6). Here, the form of the logarithmic spiral is determined by the constant b, and the value of $prob$ is chosen at random from (0, 1).

$$M' = |\vec{P}'(t) - \vec{P}(t)| \tag{14.5}$$

$$P(t+1) = \begin{cases} P'(t) - E \cdot M & \text{if } prob \leq 0.5 \\ \vec{M}' * e^{\text{bl}} * \cos(2\pi \text{l}) + \vec{P}' & \text{if } prob > 0.5 \end{cases} \tag{14.6}$$

14.3.2.3 Search for prey (exploration phase)

A randomized search is carried out by the agent in the search space during the algorithm's exploration phase. It serves to draw attention to the random search depending on the proximity of the agents. We have decided on the strategy for encircling based on the coefficient vector A's randomized values. To force the search agent to move away from the top ranked whale, we have used values of E in the range $[-1, 1]$. We have used values for which $|E| > 1$ for exploration. Here, the term "P" denotes a randomly selected agent from the population.

$$\vec{M} = |F \cdot \vec{P}''(t) - \vec{P}(t)| \tag{14.7}$$

$$\vec{P}(t+1) = \vec{P}''(t) - \vec{E} \cdot \vec{M} \tag{14.8}$$

14.3.3 Unsupervised WOA (U-WOA)

Being a population based algorithm, WOA heavily depends on the best evaluated position (P'). This position is updated with each iteration and acts as a guiding factor for the other agents, thereby making the entire process expensive to experiment with. Additionally, these approaches focus on a particular agent or a set of features. Notably, this approach also misses out on valuing each of the features. Taking this into account, we examine each of the features and assign importance scores to them using three popular filter-based feature ranking methods. To be specific, we consider the PCC [37], Spearman's Correlation Coefficient (SCC) [38] and ReliefF [39], which is discussed in the following subsection. In each iteration, we assign a pseudo-fitness value to each particle based on these importance scores, while the original fitness function used to rank the agents and evaluate the agents for the target solution remains unchanged. The objective of this pseudo fitness value is only to replace the leader guiding agent (P').

We construct the pseudo fitness function using Eq. (14.9).

$$Pseudo\_Fitness = \sum_{\forall i} S(\sigma(x_i)) \cdot (PCC_i + SCC_i + ReliefF_i) \tag{14.9}$$

In Eq. (14.9), i refers to the ith feature and S is the transfer function defined in Section 14.3.4.3. The feature importance scores for each feature are recorded. By doing this, we ensure that our solution is guided to select the best set of features obtained so far while at the same time exploring even better solutions through the steps of WOA.

14.3.3.1 Pearson's correlation coefficient (PCC)

PCC is a statistical measure used to evaluate the correlation or association between two continuous variables. Although it is based on the concept of covariance, it is recognized as one of the leading strategies for quantifying the relationship between variables of interest. It describes the amount of the association as well as the direction of the relationship. For a single feature or variable, the covariance is evaluated as per the following Eq. (14.10)

$$cov = \frac{\sum_{\forall i}(a_i - \bar{a})(b_i - \bar{b})}{\sqrt{\sum_{\forall i}(a_i - \bar{a})^2 \sum_{\forall i}(b_i - \bar{b})^2}} \tag{14.10}$$

Let us consider a feature set $\mathcal{X}$ for a particular sample, where the features can be represented as X_{ij} considering the ith sample and the jth feature. Following Eq. (14.10) we find the correlation coefficient as in Eq. (14.11) for a feature k.

$$cov = \sum_{\forall j} \frac{\sum_{\forall k}(a_k - \bar{a_k})(a_j - \bar{a_j})}{\sqrt{\sum_{\forall i}(a_k - \bar{a_k})^2 \sum_{\forall i}(a_j - \bar{a_j})^2}} \quad k \neq j \tag{14.11}$$

14.3.3.2 Spearman correlation coefficient (SCC)

SCC works in a similar fashion when compared to PCC. The only difference in SCC is that instead of using the features directly, in SCC the features are ranked based on their importance (in ascending order) before proceeding.

14.3.3.3 ReliefF

Kira and Rendell [40] developed the Relief algorithm, which was inspired by instance-based learning. Relief is a feature selection filtering technique that uses individual evaluations to calculate a score (or weight) for each feature, as a measure of its relevance to the prediction task. These scores, referred to as feature weights, range from −1 (worst) to +1 (best). However, the original Relief method had some limitations, such as being limited to binary classification problems and not having a mechanism for handling missing data. To overcome these limitations, the ReliefF algorithm was introduced in the literature. The detailed procedure of ReliefF is presented in Algorithm 14.1. ReliefF determines two nearest neighbor instances of the target: one of the same class, referred to as the nearhit, and one of the opposite class, referred to as the nearmiss.

Algorithm 14.1 ReliefF.

Input: The total number of samples n and number of features f. No. of neighbors for consideration N and features x.
Output: Feature importance scores I
$I \leftarrow$ Initialize all feature weights as 0
for $i \leftarrow$ to N **do**
    Select a random sample S
    Find a "nearhit" and "nearmiss" as instances
    for $j \leftarrow$ to f **do**
        $I[S] = I[S] - (x_i - nearhit_i)^2/N + (x_i - nearmiss_i)^2/N$
    end for
end for

14.3.4 Methodology

In the present study, we describe a technique for identifying breast tumors in ultrasound images that basically involves three steps. At first, the deep feature extraction step that involves extracting deep features from the ultrasound images used as input. Next, the feature selection step that involves choosing the most important and useful features to create the best feature subset. Lastly, the classification step that involves ML-based classifiers that have been trained on the test dataset to make predictions. Fig. 14.2 shows the entire pipeline of the proposed approach.

14.3.4.1 Deep feature extraction

We use a pre-trained transfer learning (TL) model, trained on the ImageNet database, for feature extraction and fine-tune it using the relevant dataset. To do this, we remove the last layer of the base TL model and replace it with a Global Average Pooling (GAP) layer followed by a dense layer with 256 units and a rectified linear activation unit (ReLU). Finally, we modify the output layer with a softmax activation function to match the categories of the target dataset, known as the BUSI database. The fine-tuned TL model structure for deep feature extraction is illustrated in Fig. 14.3.

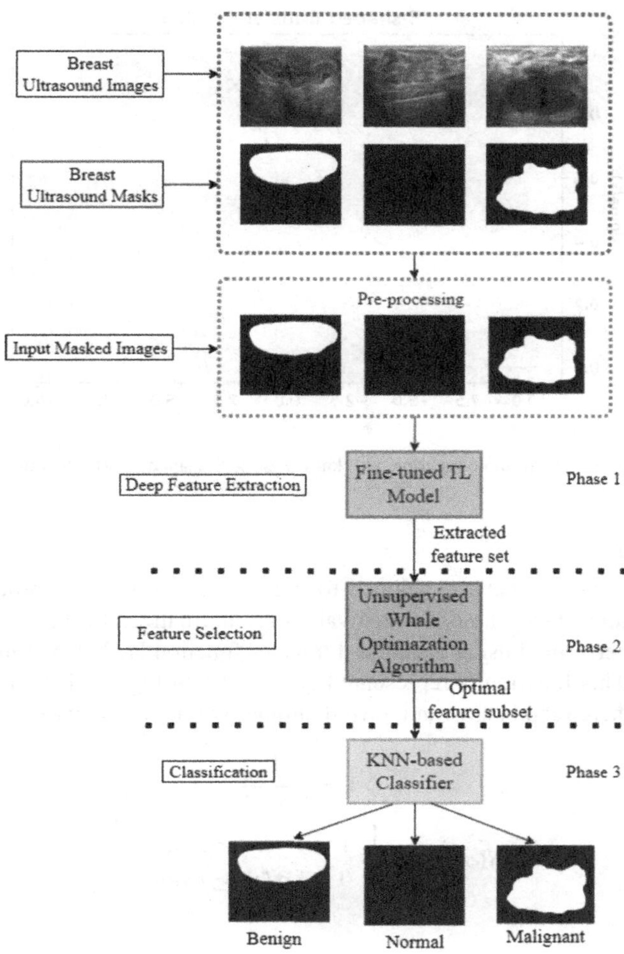

FIGURE 14.2 Pipeline of the proposed method used for predicting tumor in ultrasound images.

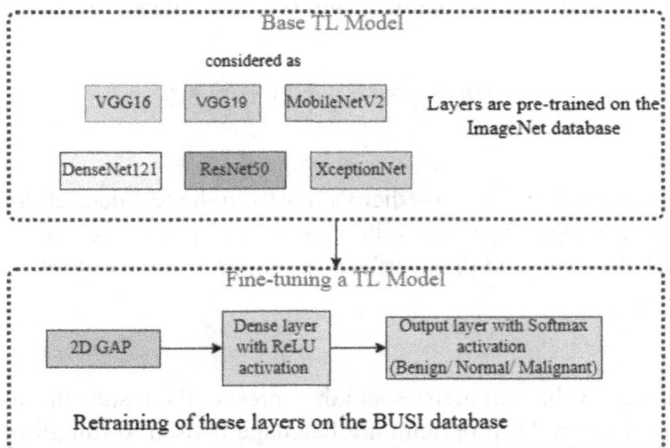

FIGURE 14.3 The TL framework used to extract deep features from the BUSI database consists of six different base TL models: VGG16, VGG19, ResNet50, MobileNetV2, DenseNet121, and XceptionNet. Each of these models is fine-tuned for the purpose of feature extraction in our experimentation.

14.3.4.2 Feature selection using U-WOA

We feed the extracted deep features into the U-WOA model.

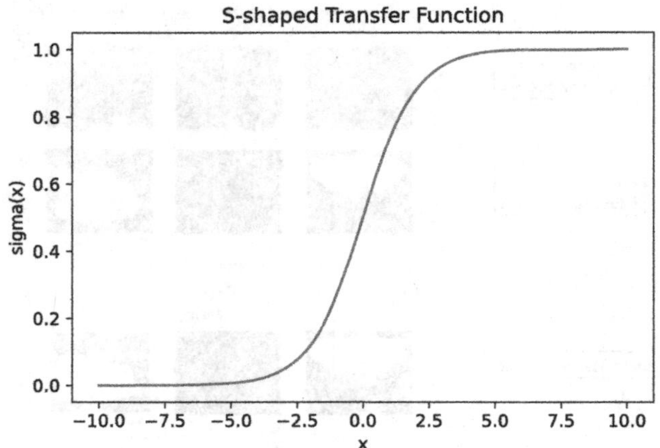

FIGURE 14.4 A graphical representation of the Sigmoid function, also known as the S-shaped transfer function.

14.3.4.3 Transfer function

As it is discussed above, WOA was originally developed for numerical optimization problems. WOA requires a step to discretize or binarize the continuous data. The optimized values are normalized to (0, 1) using a transfer function. In this work, this conversion has been performed using a standard transfer function [13]. This function is typically referred to as the S-shaped transfer function. This function is represented graphically in Fig. 14.4. In accordance with Eq. (14.13), we binarize continuous variables, where rand is any randomized number within the $sigma(x)$ range.

$$\sigma(x) = \frac{1}{1 + e^{-x}} \tag{14.12}$$

$$S(\sigma(x)) = \begin{cases} 1 & \text{if } \sigma(x) < rand \\ 0 & \text{if } \sigma(x) \geq rand \end{cases} \tag{14.13}$$

14.3.4.4 Fitness function

Using Eq. (14.14), we calculate the fitness value to evaluate candidate solutions. Here, α is a hyperparameter. The ratio between the total number of left-out features and the total number of features is represented by f and the classification accuracy is represented by a.

$$Fitness = \alpha * a + (1 - \alpha) * f \tag{14.14}$$

14.3.4.5 Classification

In this study, we employ a KNN-based classifier to predict values from the test dataset. We divide the optimal feature set, acquired in step 2, into 75 percent training, 10 percent validation, and 15 percent test sets. The training and validation sets are then used as input to the KNN classifier, and then performs prediction on the test set.

14.4 Results

At first, we describe the performance evaluation metrics and then present the results through figures and tables along with their corresponding analyses. The Python 3.6 programming language is used to run all of the experiments on a machine with a 12 GB NVIDIA Tesla T4 GPU. The Keras library is used to implement the deep learning models in the Tensorflow framework.

14.4.1 Performance metrics

This subsection covers the metrics used to evaluate our model. To understand these measures, we need to first define some crucial concepts.

The samples of the positive classes and negative classes that the classifier accurately identifies is indicated by True positives (T_{pos}) and True Negatives (T_{neg}) respectively. False Negatives (F_{neg}) and False Positives (F_{pos}) refer to the positive class samples and negative class samples respectively that are incorrectly labeled as the opposite by the classifier [41].

$$Accuracy = \frac{T_{pos} + T_{neg}}{T_{pos} + T_{neg} + F_{pos} + F_{neg}} \tag{14.15}$$

$$Precision = \frac{T_{pos}}{T_{pos} + F_{pos}} \tag{14.16}$$

$$Recall = \frac{T_{pos}}{T_{pos} + F_{neg}} \tag{14.17}$$

The weighted harmonic mean of the precision and recall values is known as the F1-score. Precision and recall are treated equally. It is determined as:

$$F1 - score = \frac{2 * Precision * Recall}{Precision + Recall} = \frac{2 * T_{pos}}{2 * T_{pos} + F_{pos} + F_{neg}} \tag{14.18}$$

14.4.2 Hyperparameters for TL models

We perform a few simulations to choose the ideal parameters for TL models. We initially use the grid search method with some common values in practice to find the best possible combination of initial learning rate and batch size for training the deep CNN models. In this current study, we consider the best possible combination of batch size (BS) and initial learning rate (L_0) as 64 and 0.001 respectively. We use the cross entropy loss and the commonly used Adam optimizer to optimize internal weights. The learning rate was lowered after the third epoch by a factor of two using a step learning rate scheduler to ensure smooth learning. In Table 14.1, we describe the values of different hyperparameters considered in this study.

TABLE 14.1 The hyperparameter specifications utilized in this work for the training of TL models.

Hyperparameter	*Value*
BS	64
Optimizer	Adam
L_0	0.001
Loss Function	Categorical Cross Entropy

14.4.3 Results and discussion

As previously stated, in our study, we use a standard DL model to extract features from ultrasound tumor images and after that, we choose an optimal set of features from this deep feature set using an unsupervised version of WOA. So, to extract deep features, we experiment with different DL based TL models and the obtained results are recorded in Fig. 14.5.

To train the TL model, we split the entire dataset into 75% training, 10% validation and 15% testing sets. Fig. 14.5 shows that the VGG16 model performs better than other TL models in accuracy over test datasets. So, we proceed with the deep features extracted by the VGG16 model. While training, the accuracy and loss values of the VGG16 model for BUSI dataset are noted and shown in Fig. 14.6 and Fig. 14.7 respectively.

Fig. 14.6 shows the training and validation accuracies for every epoch and it can be seen that the model does not overfit while Fig. 14.7 ensures smooth learning of the model during training. In the loss learning curve the popular categorical cross entropy loss function is used. The confusion matrix of the model is shown in Fig. 14.8. It is clear from this figure that the number of misclassified samples is very less and the model accurately predicts all three classes.

Lastly, we compare our method to previously established methods and present the comparison in Table 14.2. The comparison shows that our proposed method surpasses all the methods considered in this study.

One strong possible reason for our method outperforming the existing methods can be linked to the fact that our optimization algorithm selects the most optimal set of features on the basis of different types of statistical importance of features which was ignored in [43]. Specifically, the works reported in [16,17,42] consider several shape and texture based feature

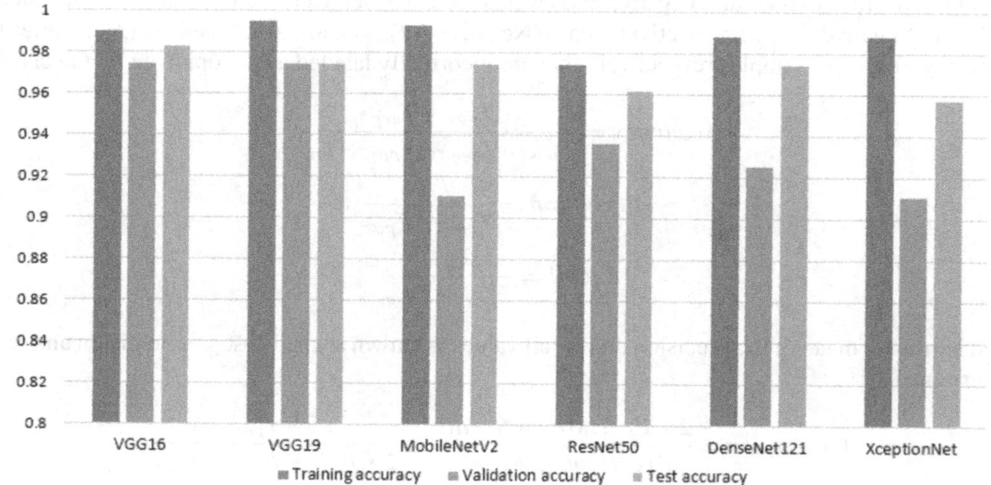

FIGURE 14.5 Performance of different TL models on the BUSI database. All the values are in %.

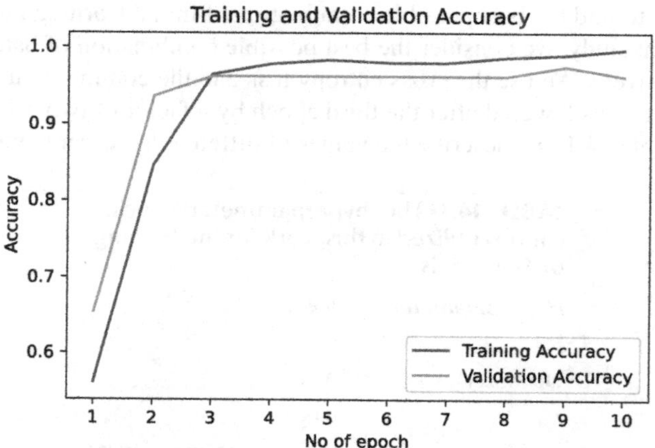

FIGURE 14.6 Learning curve (Accuracy) of the feature extraction model during training of the VGG16 model.

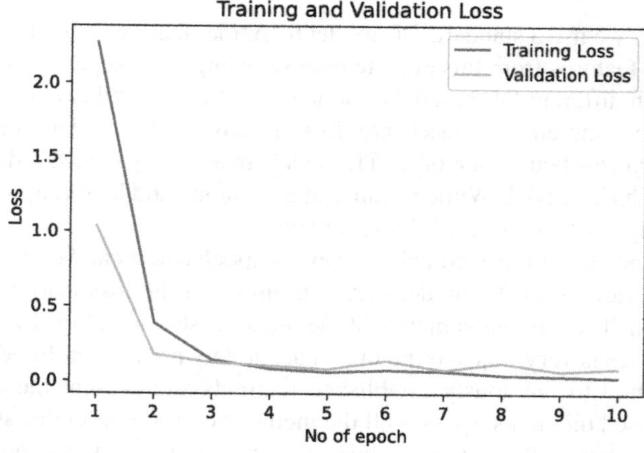

FIGURE 14.7 Learning curve (Loss) of the feature extraction model during training of the VGG16 model.

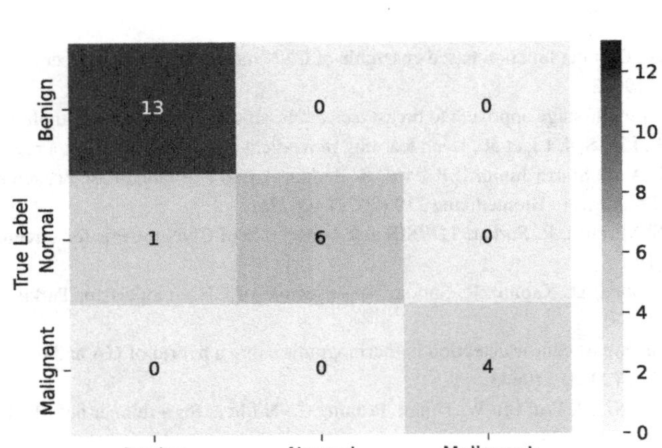

FIGURE 14.8 Confusion matrix of the proposed method.

TABLE 14.2 Performance of the proposed method is compared with previous methods on the BUSI database, with the highest values indicated in bold font.

Method	*Accuracy*	*Recall*	*Precision*	*F1-score*
Sadad et al. [17]	0.966	0.943	-	-
Moon et al. [42]	0.946	0.923	0.900	0.911
Mishra et al. [16]	0.974	0.96	0.958	0.959
Pramanik et al. [43]	0.987	0.987	0.987	0.987
Proposed method	**0.991**	**0.991**	**0.991**	**0.991**

extractors like GLCM to extract features and further use an ML based classification algorithm to classify the same. The fundamental disadvantage of utilizing such feature extractors is that they are not invariant to rotations and scaling-based transformations.

14.5 Conclusion

Breast cancer is a life-threatening disease that affects women worldwide. Early detection of malignancy in a woman's breast through ultrasound imaging can greatly aid in providing prompt care and treatment for breast cancer. In this work, we propose a novel deep feature selection method called U-WOA, for detecting cancer in breast ultrasound images. We assess our strategy using a variety of performance indicators on the openly available BUSI database. The results of the testing demonstrate that the proposed method generates promising outcomes on unseen test data and achieves state-of-the-art results, pointing to prospective applications of the suggested model in actual diagnostics for helping medical experts in order to provide better treatment. However, this approach also has certain limitations because it only focuses on the ground truth tumor masks, which is the RoI of the original ultrasound images. However, in real-world situations, such RoI masks are hard to come by. In the future, DL-based methods like U-net and its variants may be investigated for automatic segmentation from ultrasound images before applying any classification algorithms.

References

[1] H. Sung, J. Ferlay, R.L. Siegel, M. Laversanne, I. Soerjomataram, A. Jemal, et al., Global cancer statistics 2020: GLOBOCAN estimates of incidence and mortality worldwide for 36 cancers in 185 countries, CA: A Cancer Journal for Clinicians 71 (3) (2021) 209–249.

[2] S. Malvia, S.A. Bagadi, U.S. Dubey, S. Saxena, Epidemiology of breast cancer in Indian women, Asia-Pacific Journal of Clinical Oncology 13 (4) (2017) 289–295.

[3] S.V. Sree, E.Y.K. Ng, R.U. Acharya, O. Faust, Breast imaging: a survey, World Journal of Clinical Oncology 2 (4) (2011) 171.

[4] P. Pramanik, S. Mukhopadhyay, S. Mirjalili, R. Sarkar, Deep feature selection using local search embedded social ski-driver optimization algorithm for breast cancer detection in mammograms, Neural Computing & Applications (2022) 1–21.

[5] S. Majumdar, P. Pramanik, R. Sarkar, Gamma function based ensemble of CNN models for breast cancer detection in histopathology images, Expert Systems with Applications (2022) 119022.

[6] A. Bagchi, P. Pramanik, R. Sarkar, A multi-stage approach to breast cancer classification using histopathology images, Diagnostics 3 (1) (2022) 126.

[7] S. Liu, Y. Wang, X. Yang, B. Lei, L. Liu, S.X. Li, et al., Deep learning in medical ultrasound analysis: a review, Engineering 5 (2) (2019) 261–275.

[8] R. Pramanik, M. Biswas, S. Sen, L.A. de Souza Júnior, J.P. Papa, R. Sarkar, A fuzzy distance-based ensemble of deep models for cervical cancer detection, Computer Methods and Programs in Biomedicine 219 (2022) 106776.

[9] R. Pramanik, S. Dey, S. Malakar, S. Mirjalili, R. Sarkar, TOPSIS aided ensemble of CNN models for screening COVID-19 in chest X-ray images, Scientific Reports 12 (1) (2022) 1–19.

[10] A. Kurmi, S. Biswas, S. Sen, A. Sinitca, D. Kaplun, R. Sarkar, An ensemble of CNN models for Parkinson's disease detection using DaTscan images, Diagnostics 12 (5) (2022) 1173.

[11] R. Pramanik, P. Pramanik, R. Sarkar, Breast cancer detection in thermograms using a hybrid of GA and GWO based deep feature selection method, Expert Systems with Applications 219 (2023) 119643.

[12] J.H. Luo, H. Zhang, H.Y. Zhou, C.W. Xie, J. Wu, Lin W. Thinet, Pruning CNN filters for a thinner net, IEEE Transactions on Pattern Analysis and Machine Intelligence 41 (10) (2018) 2525–2538.

[13] R. Pramanik, S. Sarkar, R. Sarkar, An adaptive and altruistic PSO-based deep feature selection method for pneumonia detection from chest x-rays, Applied Soft Computing 128 (2022) 109464.

[14] L. Cai, X. Wang, Y. Wang, Y. Guo, J. Yu, Y. Wang, Robust phase-based texture descriptor for classification of breast ultrasound images, BioMedical Engineering Online 14 (1) (2015) 1–21.

[15] D. Huang, C. Shan, M. Ardabilian, Y. Wang, L. Chen, Local binary patterns and its application to facial image analysis: a survey, IEEE Transactions on Systems, Man and Cybernetics. Part C, Applications and Reviews 41 (6) (2011) 765–781.

[16] A.K. Mishra, P. Roy, S. Bandyopadhyay, S.K. Das, Breast ultrasound tumour classification: a machine learning—radiomics based approach, Expert Systems 38 (7) (2021) e12713.

[17] T. Sadad, A. Hussain, A. Munir, M. Habib, S. Ali Khan, S. Hussain, et al., Identification of breast malignancy by marker-controlled watershed transformation and hybrid feature set for healthcare, Applied Sciences 10 (6) (2020) 1900.

[18] M.C. Yang, W.K. Moon, Y.C.F. Wang, M.S. Bae, C.S. Huang, J.H. Chen, et al., Robust texture analysis using multi-resolution gray-scale invariant features for breast sonographic tumor diagnosis, IEEE Transactions on Medical Imaging 32 (12) (2013) 2262–2273.

[19] W.K. Moon, Y.W. Lee, H.H. Ke, S.H. Lee, C.S. Huang, R.F. Chang, Computer-aided diagnosis of breast ultrasound images using ensemble learning from convolutional neural networks, Computer Methods and Programs in Biomedicine 190 (2020) 105361.

[20] K. Jabeen, M.A. Khan, M. Alhaisoni, U. Tariq, Y.D. Zhang, A. Hamza, et al., Breast cancer classification from ultrasound images using probability-based optimal deep learning feature fusion, Sensors 22 (3) (2022) 807.

[21] W. Al-Dhabyani, M. Gomaa, H. Khaled, F. Aly, Deep learning approaches for data augmentation and classification of breast masses using ultrasound images, International Journal of Advanced Computer Science and Applications 10 (5) (2019) 1–11.

[22] A.K. Mishra, P. Roy, S. Bandyopadhyay, S.K. Das, Feature fusion based machine learning pipeline to improve breast cancer prediction, Multimedia Tools and Applications (2022) 1–29.

[23] Y. Li, W. Zhao, Accurate breast tumor identification using computational ultrasound image features, in: International Workshop on Computational Mathematics Modeling in Cancer Analysis, Springer, 2022, pp. 150–158.

[24] M. Mafarja, S. Mirjalili, Whale optimization approaches for wrapper feature selection, Applied Soft Computing 62 (2018) 441–453.

[25] M.H. Nadimi-Shahraki, H. Zamani, S. Mirjalili, Enhanced whale optimization algorithm for medical feature selection: a COVID-19 case study, Computers in Biology and Medicine 148 (2022) 105858.

[26] R. Kundu, S. Chattopadhyay, E. Cuevas, R. Sarkar, AltWOA: Altruistic Whale Optimization Algorithm for feature selection on microarray datasets, Computers in Biology and Medicine 144 (2022) 105349.

[27] R. Agrawal, B. Kaur, S. Sharma, Quantum based whale optimization algorithm for wrapper feature selection, Applied Soft Computing 89 (2020) 106092.

[28] A.G. Hussien, A.E. Hassanien, E.H. Houssein, S. Bhattacharyya, M. Amin, S-shaped binary whale optimization algorithm for feature selection, in: Recent Trends in Signal and Image Processing, Springer, 2019, pp. 79–87.

[29] A.G. Hussien, E.H. Houssein, A.E. Hassanien, A binary whale optimization algorithm with hyperbolic tangent fitness function for feature selection, in: 2017 Eighth International Conference on Intelligent Computing and Information Systems (ICICIS), IEEE, 2017, pp. 166–172.

[30] S. Vijh, D. Gaur, S. Kumar, An intelligent lung tumor diagnosis system using whale optimization algorithm and support vector machine, International Journal of System Assurance Engineering and Management 11 (2) (2020) 374–384.

[31] M.M. Mafarja, S. Mirjalili, Hybrid whale optimization algorithm with simulated annealing for feature selection, Neurocomputing 260 (2017) 302–312.

[32] Y. Zheng, Y. Li, G. Wang, Y. Chen, Q. Xu, J. Fan, et al., A novel hybrid algorithm for feature selection based on whale optimization algorithm, IEEE Access 7 (2018) 14908–14923.

[33] I. Aljarah, H. Faris, S. Mirjalili, Optimizing connection weights in neural networks using the whale optimization algorithm, Soft Computing 22 (1) (2018) 1–15.

[34] A. Brodzicki, M. Piekarski, J. Jaworek-Korjakowska, The whale optimization algorithm approach for deep neural networks, Sensors 21 (23) (2021) 8003.

[35] W. Al-Dhabyani, M. Gomaa, H. Khaled, A. Fahmy, Dataset of breast ultrasound images, Data in Brief 28 (2019) 104863.

[36] S. Mirjalili, A. Lewis, The whale optimization algorithm, Advances in Engineering Software 95 (2016) 51–67.

[37] K. Pearson, Notes on regression and inheritance in the case of two parents, Proceedings of the Royal Society of London 58 (1895) 240–242.
[38] C. Spearman, The proof and measurement of association between two things, The American Journal of Psychology 100 (3/4) (1987) 441–471.
[39] M. Robnik-Šikonja, I. Kononenko, Theoretical and empirical analysis of ReliefF and RReliefF, Machine Learning 53 (1) (2003) 23–69.
[40] K. Kira, L.A. Rendell, A practical approach to feature selection, in: Machine Learning Proceedings 1992, Elsevier, 1992, pp. 249–256.
[41] J. Han, J. Pei, M. Kamber, Data Mining: Concepts and Techniques, Elsevier, 2011.
[42] W.K. Moon, Y.W. Shen, C.S. Huang, L.R. Chiang, R.F. Chang, Computer-aided diagnosis for the classification of breast masses in automated whole breast ultrasound images, Ultrasound in Medicine & Biology 37 (4) (2011) 539–548.
[43] P. Pramanik, S. Mukhopadhyay, D. Kaplun, R. Sarkar, A deep feature selection method for tumor classification in breast ultrasound images, in: International Conference on Mathematics and Its Applications in New Computer Systems., Springer, 2022, pp. 241–252.

Chapter 15

Constraint optimization: solving engineering design problems using Whale Optimization Algorithm (WOA)

Syed Kumayl Raza Moosavi[a], Malik Naveed Akhter[b], Muhammad Hamza Zafar[c], and Majad Mansoor[d]
[a]School of Electrical Engineering and Computer Science, National University of Sciences and Technology, Islamabad, Pakistan, [b]School of Mechanical and Manufacturing Engineering, National University of Sciences and Technology, Islamabad, Pakistan, [c]Department of Engineering Sciences, University of Agder, Grimstad, Norway, [d]Dept. of Automation, University of Science and Technology of China, Hefei, China

15.1 Introduction

To maximize the life cycle and performance of an engineered product and minimize the cost has been a imperative subject for researchers in the last two decades in terms of technological and economic advancement [1]. The design optimization processes involve multiple interacting objects and constraints which contain many design variables to find the optimal objective solution. The constraints of the design can sometimes be non linear in nature such as the material behavior, properties, geometry, dynamics, and design cost [2]. Due to this non-linearity the optimum solution can not be accurately determined with any singular known programming solution. Global optima can be found only if the objective function and the associated constraints meet and satisfy certain properties.

Several mathematical models and approaches have been conceptualized using classical programming techniques to solve these design problems [3]. Traditionally, unconstrained problems are solved through direct method and gradient methods, whereas, constrained problem solving algorithms are classified under indirect and direct methods. The indirect technique tries to establish and solve more than one linear problem from the original one while the direct method aims to ascertain search points in succession.

Early Direct methods were simple and based on heuristics drawn in two dimensions. In 1991, research on this technique was revived in [4] in congruence with parallel computing accompanying the convergence analysis using multi directional search. Simulation based optimization methodology for direct methods produces results of computer simulations that design, analyze, control, and optimize complex physical systems [5]. Despite its ease of use, the issues with the direct search method are unavoidable. In 1960s, the Direct Search methods were applied widely but fell out of favor because it lacked coherent mathematical analysis. [6] summarized it as being developed heuristically with no proof of convergence and in most cases having slow rate of convergence.

On the other hand, indirect methods consist of two main challenges, difficulty in handling path inequality constraints and exponential increase in time and resources as the number of states and constraints of the design problem increases. Many applications have used techniques of indirect solutions to solve real world constraint applications. Indirect optimization was used to accurately estimate the Lagrange multipliers to compute minimum propellant spiral escapes and captures [7]. Fitting spline curves are a challenging issue in applications of CAD/CAM because of their dependence on nonlinear continuous variables. To solve this issue Galvez et al. [8] used an indirect approach to convert a non-linear problem to a convex optimization solution. [9] solved hybrid optimal control problems by proposing a min-H algorithm, which would be initialized intuitively and find a global optima solution.

These classical techniques have inherent shortcomings which trap the solution into a local minimum thereby causing the solution to be unable to obtain global optimum. To circumvent the issues arisen from the classical algorithms' weaknesses, researchers have proposed the use of meta-heuristic algorithms as an effective alternative to the deterministic techniques as they do not require the knowledge of the objective function gradients [10].

Handbook of Whale Optimization Algorithm. https://doi.org/10.1016/B978-0-32-395365-8.00021-X

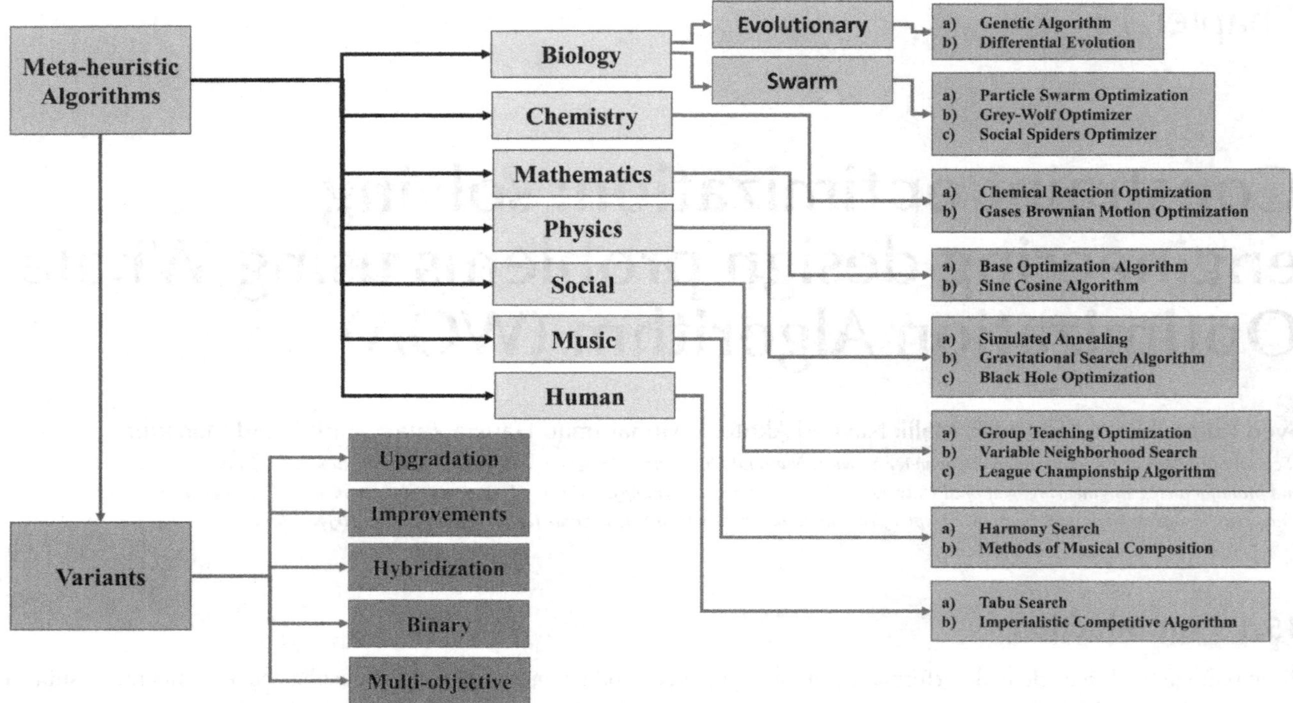

FIGURE 15.1 Types of Meta-heuristic Algorithms.

15.1.1 Meta heuristic techniques

Meta heuristic is a class of algorithms that is generally applied to a variety of optimization algorithms to provide a approximate global solution with few modifications. Fundamentally, meta-heuristics offers a domain-specific knowledge by employing upper level strategy of solution sharing between many nodes being implemented in parallel. Through incorporation of various mechanisms and simple mathematical functions, it extends from basic local search to advanced learning technique by preserving search experience. Typically, these algorithms functions are segmented into two phases, exploration and exploitation phases. While the exploration phase searches for a global solution around the search space using erratic movements, the exploitation phase converges towards the current iteration global solution. The trade off between the two phases is key to efficient search process, which is dictated by the hyper parameters set by each algorithm.

The optimization problems that attracted the attention of meta heuristic techniques have a large variance ranging from single to multi-objective, constrained to unconstrained, continuous to discontinuous. Solutions for NP-Hard problems are practically better executed with these techniques instead of exact polynomial solutions which although may provide best solutions, are unfeasible due to computational and time costs. Despite the achievements of the classical meta heuristic algorithms, developed before the year 2000, new, improved, and hybridized versions of these techniques are being introduced every day to solving benchmark problems sets. Over the years, researchers have divided the technique into many different types depending on the nature of the algorithm [11–13]. Fig. 15.1 shows the categorization of different types of meta-heuristic techniques found in literature.

The rest of the paper is organized as follows. First, the basic concepts of commonly used meta-heuristic algorithms with respect to solving engineering design problems are introduced. Then the details of the proposed Whale Optimization Algorithm (WOA) are elaborated. Then the usage of the WOA algorithm with constraint optimization problems is shown and compared with other similar meta-heuristic techniques. Moreover, a time complexity analysis of the WOA is evaluated. Finally, a discussion is outlined on the results generated and a conclusion is extracted.

15.2 Related work

Meta heuristics are probabilistic solvers belonging in the family of approximate optimization methods. Meta-heuristic algorithms are a desirable tool to employ in hard optimization problems due to its dual nature which include random-based and rule based techniques. In the real world, the design optimization problems are highly nonlinear with many design variables involving complicated constraints. Literature has shown that different optimization algorithms have been applied to structural design optimization tasks. For instance, a micro genetic algorithm was proposed for optimizing steel frames as an effective method with semi-redid joints [14]. Optimization of steel trusses was proposed in [15] by utilizing three techniques in combination namely; adaptive dimensional search, exponential big bang crunch optimization, and modified big bang crunch by exploring feasible solutions of the initial population. An effective numerical approach by integrating harmony search algorithm (HS) for reliability-based design optimization and failure probability analysis was proposed in [16] for nonlinear inelastic behaviors of frames. Considering the size, shape, and frame of truss structures, [17] proposed a variant meta heuristic technique, i.e., and improved simulated annealing algorithm to solving the dynamic constraint of natural frequency for truss sizing optimization.

Improved Differential evolutionary (DE) algorithm, belonging to a family of evolutionary algorithms, was proposed by [18] for optimizing non-linear inelastic steel frames with consideration on the constraint of panel zones. For the optimization of the objective function, i.e., the weight of the rigid-jointed steel frame structures [19] applied an array of seven population based meta-heuristic algorithms while satisfying the constraints put on displacement and stress limits. A novel meta-heuristic algorithm, i.e., Plasma Generation Optimization (PGO) was proposed in [20], the performance of which was assessed by finding optimization solutions of benchmark constrained design functions and truss problems. An efficient micro-organism inspired Barnacles Mating Optimizer (BMO) algorithm was proposed by [21] to solve the optimal reactive power dispatch (ORPD) problem equipped with the reactive compensator, equality and generator constraints.

The evidential reasoning approach to solving practical civil engineering design problems with constraints has been proposed in this work using the novel algorithm Whale Optimization Algorithm (WOA) [22]. The detailed work with practical examples is presented in Section 15.4.

15.3 Whale optimization algorithm

Swarm based techniques belonging to the family of nature-inspired algorithms are modeled by imitating the social mechanism of group of mammals and insects. A unique feature of such swarm algorithms is that the search space information is preserved over subsequent iterations, among multiple particles and usually include less operators and therefore are easier to implement. This section illustrates the inspiration and the mathematical model of the WOA algorithm.

15.3.1 Inspiration

Found in deep oceans and seas around the world, the humpback whale, a species of baleen whales, have a very distinctive body shape with long fins and a lumpy head. These whales live approximately 45-50 years and typically migrate up to 16,000 km every year. Whales are known to be the largest mammals in the world growing up to 30 m long and 180 tons in weight. Interestingly, whales are considered to be very intelligent mammals with emotion. According to [23], whale have spindle cells in the brain similar to that of humans which are responsible for emotions and analytical thinking. Whales have been proven to have judgment, learning abilities and communication where in some cases they learned to create their own dialect. The social behavior of these creatures is also unique which can be verified by their innovative hunting strategy towards smaller school fish close to the surface.

The foraging behavior of whales, the bubble net feeding mechanism, is done by creating distinctive bubbles in a circular fashion. The two main mechanisms observed by researchers in their hunting strategy was 'upward spiral' and 'double loops'. First the whales start from the bottom and start creating bubbles in a spiral pattern and move closer to the surface whilst doing so. The second stage involves lobtail and capture loop [24]. This approach, distinct to humpback whales, can be observed in Fig. 15.2. This paper aims to use the mathematical design of this very hunting method to solve real world problems.

15.3.2 Mathematical model

15.3.2.1 Multi-agent navigation

The central point of the encircling mechanism of humpback whale hunting technique is theoretically deduced as the optimal design function in the search space region. The WOA algorithm puts forth a supposition that the best solution from the multi-agents in the swarm is that same central point and hence is considered as the global optima solution. The rest of the agents follow a exploration and exploitation pattern to converge into that location whilst also traversing in the search space for other possible global optima solutions and then update their positions accordingly at each iteration. The updation is

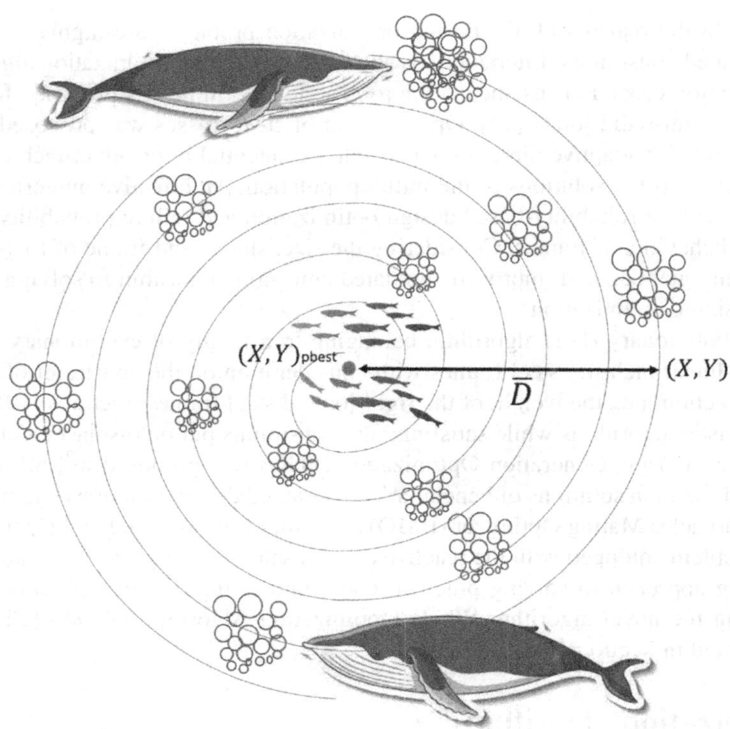

FIGURE 15.2 Spiral bubble-net hunting strategy of Whales.

represented with the following equations:

$$\vec{D} = |\vec{C}.\vec{X}_{pbest}(i) - \vec{X}(i)| \tag{15.1}$$

$$\vec{X}(i+1) = \vec{X}_{pbest}(i) - \vec{A}.\vec{D} \tag{15.2}$$

where i represents the current iteration, X is the current iteration's position vector, X_{pbest} is the position vector of best solution reach by the agent so far and $\vec{A}$ and $\vec{D}$ are the hyper parameters of the equation represented by the following:

$$\vec{A} = 2.a.r - a \tag{15.3}$$

$$\vec{X}(i+1) = \vec{X}_{pbest}(i) - \vec{A}.\vec{D} \tag{15.4}$$

where r represents a random value in the range [0, 1] and the parameter a linearly decreases from 2 to 0 over the course of the iterations. By introducing a random variable in the equation the convergence leads to searching solutions around and between keypoints of the global optima solution as well. This objectively simulates the encircling of the prey mechanism. This can be inferred the case for 2D, 3D, and n-dimensional search space. While the solution sharing mechanism is straightforward the attacking mathematical model of the humpback whales is divided into the exploration and the exploitation phases.

15.3.2.2 Exploitation phase

In order to mimic the bubble-net procedure, the model is mainly separated into two sections. Since the attack movement of the whale happens simultaneously between the two sections, the probabilistic likely-hood of the updation of the agent's position is 50% between Eqs. (15.4) and (15.5).

- Shrinking of the encircling mechanism: The effort of the parameter a that reduces from 2 to 0 in Eq. (15.3) ensures diminishing of search space area by which global optima is achieved.
- Spiral updation: As can be seen in Fig. 15.2, this approach focuses on the whale's movement towards the prey in a spiral equation depicting a helix-shape simplified in Eq. (15.5).

$$\vec{X}(i+1) = \overline{D}.e^{b.l}.cos(2.\pi.l) + \vec{X}_{pbest}(i) \tag{15.5}$$

where $\overline{D}$ indicates the distance of the agent to the global optima of the current iteration and is represented by the equation:

$$\overline{D} = \vec{X}_{pbest}(i) - \vec{X}(i) \tag{15.6}$$

l is a random variable in the range of $[-1, 1]$ and b is a constant in the equation for defining the spiral.

15.3.2.3 Exploration phase

In order to search for optimal solutions outside the convergence zone of the exploitation phase an approach similar to the exploitation phase is utilized except with minor changes. Instead of the value of A converging to zero a random value in the range above 1 is utilized to compel the agent to progress further away towards a possible bigger school of fish. This mechanism puts the emphasis of the algorithm towards a global search. Defined with the following equations, the mathematical model of the phase is defined:

$$\vec{D} = |\vec{C}.\vec{X}_{rand}(i) - \vec{X}(i)| \tag{15.7}$$

$$\vec{X}(i+1) = \vec{X}_{rand}(i) - \vec{A}.\vec{D} \tag{15.8}$$

where $\vec{X}_{rand}(i)$ is the location of a whale's position chosen randomly from the list of whales in the search space employed. The intuitive and detailed process of WOA is shown in Fig. 15.3. Due to its adaptive variation of exploitation and exploration phases the WOA algorithm can be considered a global optimizer useful to find solutions to nonlinear problems.

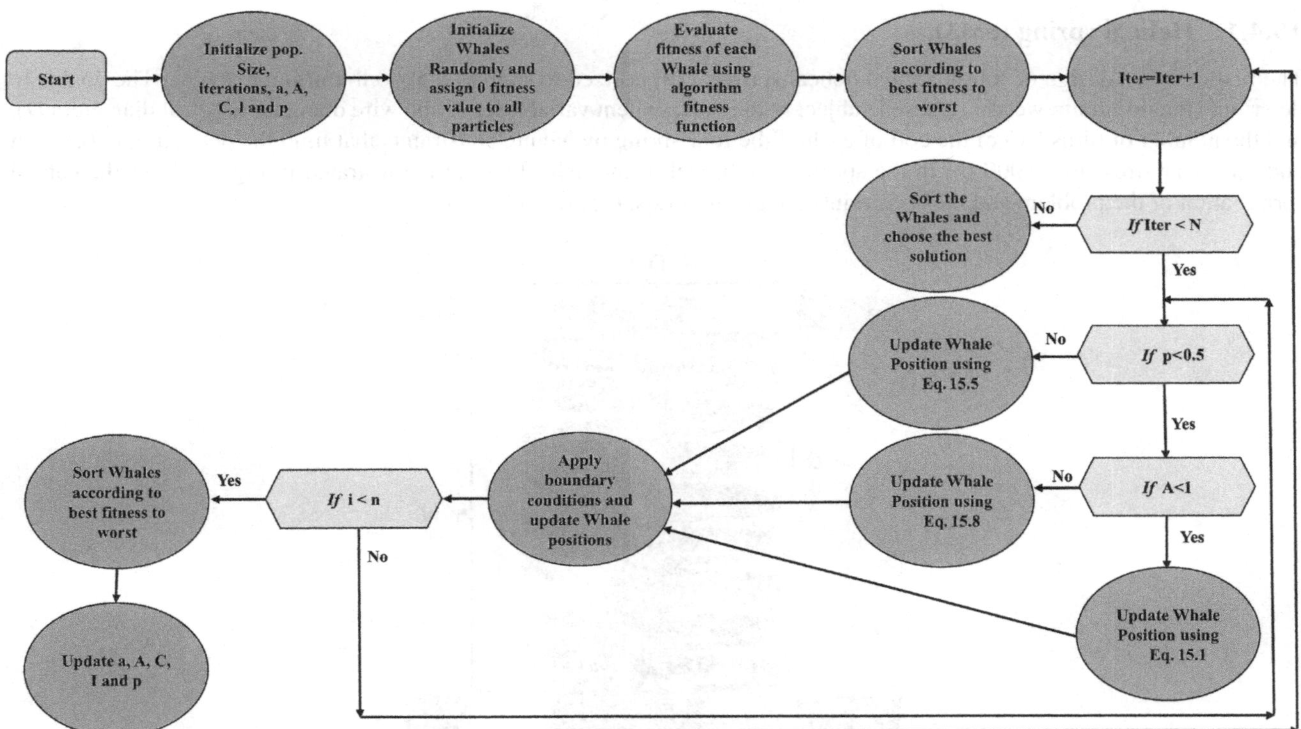

FIGURE 15.3 Process flow diagram.

15.4 Engineering design problems

Structural design problems are complex and sometimes when multiple independent variables with multiple design constraints are involved, the optimal solution of interest does not exist. In order to gauge how the WOA algorithm performs with such complicated problems, 12 standard structural engineering design test problems are formulated and run through the algorithm to find an optimal global solution. To show the efficacy of the present algorithm, the results are also compared with other meta-heuristic algorithms namely; Particle Swarm Optimization (PSO) algorithm [25], Grey Wolf Optimization (GWO) algorithm [26], Barnacles Mating Optimization (BMO) algorithm [21], and Arithmetic Optimization Algorithm (AOA) [27]. The manually defined parameters utilized for each meta-heuristic algorithm are tabulated in Table 15.1 and have been chosen according to the requisite papers in which they are introduced.

TABLE 15.1 Meta heuristic hyperparameters.

S.No.	Algorithm	Hyperparameter	value
1	WOA	$\vec{a}$	[2.0 - 0] over no. of iterations
2	GWO	a	[2.0 - 0] over no. of iterations
3	PSO	C1	1.5
		C2	2.0
4	AOA	a	5.0
		μ	0.6
5	BMO	q	2.0
		pl	0.7

15.4.1 Helical spring (FM1)

The first case is a design of four identical helical springs [28] utilized to support a 5000 lb milling machine. The goal is to determine the minimum weight required subject to the independent variables, i.e., the wire diameter (d), coil diameter (D), and the number of turns (N) of the coil of each of the four spring by having constraints that limit the deflection to 0.1 inch and the shear stress to 10,000 psi in the spring. The model of the helical spring is illustrated in Fig. 15.4. Mathematical formulation of the problem and the constraints are given in Eqs. (15.9) to (15.11).

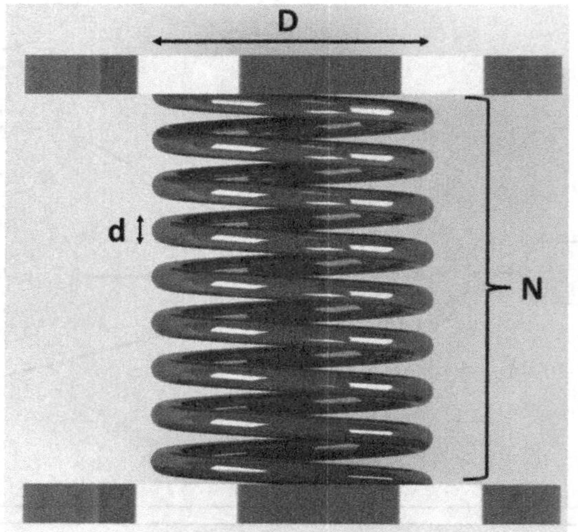

FIGURE 15.4 Helical Spring.

$$X = \begin{bmatrix} x_1 \\ x_2 \\ x_3 \end{bmatrix} = \begin{bmatrix} d \\ D \\ N \end{bmatrix} \quad (15.9)$$

$$d, x_1 > 0 \qquad D, x_2 > 0 \qquad N, x_3 > 0 \quad (15.10)$$

$$Minimize : f(X) = weight = \frac{\pi d^2}{4} \pi D N \rho \tag{15.11}$$

G is represented as the shear modulus ($12 \times 10^6 psi$), ρ is the density of spring ($0.3 lb/in^3$) and K_s is the shear stress correction factor (1.05). Table 15.2 shows the fitness values for the WOA algorithm for the Position value $X = [0.405841202 : 0.2 : 0.2]$.

TABLE 15.2 Fitness Values.

Best	Worst	S.D	Average
0.004876647	0.004876647	0	0.004876647

$$g_1(X) = 1 - \frac{d^4 G}{80 F D^3 N} = 120 x_1^4 x_2^{-3} x_3^{-1} \le 0, \quad g_2(X) = 1 - \frac{1250 \pi d^3}{K_s F D} = 2.992 x_1^3 x_2^{-1} \le 0$$

$$g_3(X) = 1 - \frac{\sqrt{G g d}}{200 \sqrt{2 \rho \pi} D^2 N} = 139.8388 x_1 x_2^{-2} x_3^{-1} \le 0$$

Table 15.3 shows the requisite constraints involved for the above mentioned value of X.

TABLE 15.3 Constraints.

G1	G2	G3
-2033.6317	-3E-12	-7093.0433

15.4.2 Tension/compression spring (FM2)

The goal in the problem introduced in [29] is to determine the weight of the compression spring by determining the optimal values of the independent variables, i.e., diameter of the wire (d), mean coil diameter (D), and the number of active coils (N) utilized. The model of the helical spring is illustrated in Fig. 15.5. Mathematical formulation of the problem and the constraints are given in Eqs. (15.12) to (15.14).

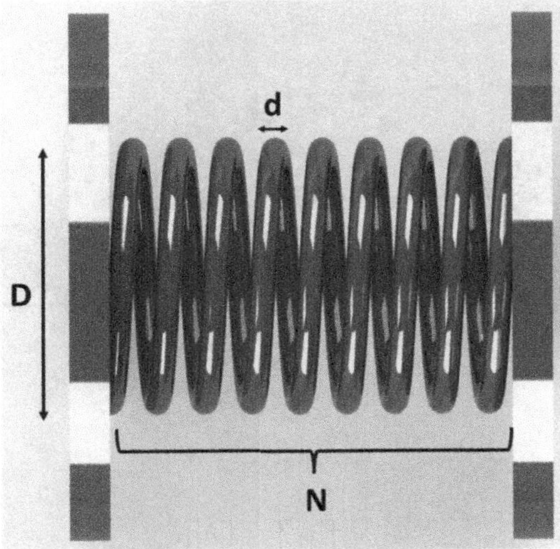

FIGURE 15.5 Tension/Compression Spring.

$$X = \begin{bmatrix} x_1 \\ x_2 \\ x_3 \end{bmatrix} = \begin{bmatrix} d \\ D \\ N \end{bmatrix} \tag{15.12}$$

$$0.5 < d < 2 \qquad 0.25 < D < 1.3 \qquad 2.0 < N < 15.0 \tag{15.13}$$

$$Minimize: f(X) = (N+2)Dd^2 \tag{15.14}$$

Table 15.4 shows the fitness values for the WOA algorithm for the Position value $X = [0.050000001 : 0.607614203 : 2]$.

TABLE 15.4 Fitness Values.

Best	Worst	S.D	Average
0.006076142	0.011544971	0.011544971	0.006642155

$$g_1(X) = 1 - \frac{D^3N}{71785d^4} \leq 0, \qquad g_2(X) = \frac{4D^2 - dD}{12566(Dd^3 - d^4)} + \frac{1}{5108d^2} - 1 \leq 0$$

$$g_3(X) = 1 - \frac{140.45d}{D^2N} \leq 0, \qquad g_4(X) = \frac{D+d}{1.5} - 1 \leq 0$$

Table 15.5 shows the requisite constraints involved for the above mentioned value of X.

TABLE 15.5 Constraints.

G1	G2	G3	G4
-3.0122E-09	-1.3589E+17	-8.5106	-0.5616

15.4.3 Welded beam design (FM3)

The goal of this structural design is to minimize the cost needed for fabrication of the welded beam [30] as illustrated in Fig. 15.6. The ensure the design satisfies the structural integrity, a few constraints are involved in the mechanism such as shear stress, buckling load of the bar, beam bending stress, deflection of the end beam, and side constraints. The decision variables for the design optimization include weld thickness (h), bar length (l), bar height (t), and the bar thickness (b). Mathematical formulation of the problem and the constraints are given in Eqs. (15.15) to (15.17).

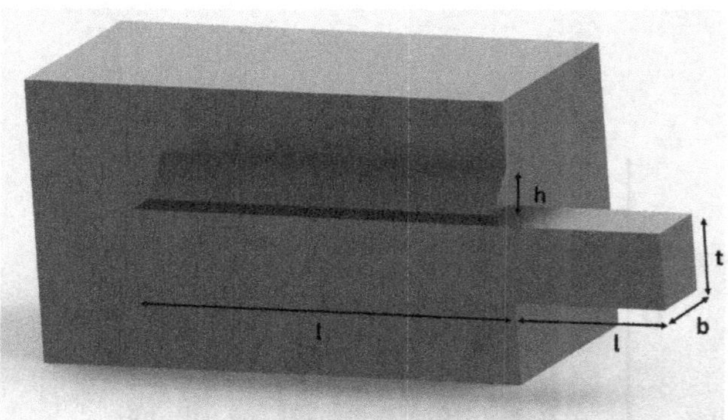

FIGURE 15.6 Welded Beam Problem.

$$X = \begin{bmatrix} x_1 \\ x_2 \\ x_3 \\ x_4 \end{bmatrix} = \begin{bmatrix} h \\ l \\ t \\ b \end{bmatrix} \tag{15.15}$$

$$0.1 < x_1, x_4 < 2.0 \qquad 0.1 < x_2, x_3 < 10.0 \tag{15.16}$$

$$Minimize: f(X) = 1.10471x_1^2x_2 + 0.04811x_3x_4(14.0 + x_2) \tag{15.17}$$

Table 15.6 shows the fitness values for the WOA algorithm for the Position value $X = [1.631522583 : 0.470963871 : 1.999941961 : 4.200243774]$.

TABLE 15.6 Fitness Values.

Best	Worst	S.D	Average
7.233132593	9.513480763	0.514501373	7.591820935

$$g_1(X) = \tau(X) - \tau_{max} \leq 0, \quad g_2(X) = \sigma(X) - \sigma_{max} \leq 0, \quad g_3(X) = x_1 - x_4 \leq 0$$
$$g_4(X) = 0.10471x_1^2 = 0.04811x_3x_4(14.0 + x_2) - 5.0 \leq 0, \quad g_5(X) = 0.125 - x_1 \leq 0$$
$$g_6(X) = \delta(X) - \delta_{MAX} \leq 0, \quad g_7(X) = P - P_C(X) \leq 0$$

where,

$$\tau(X) = \sqrt{(\tau')^2 + 2\tau'\tau''\frac{x_2}{2R} + (\tau'')^2}, \quad \tau' = \frac{P}{\sqrt{2}x_1x_2}, \quad \tau'' = \frac{MR}{J}, \quad M = P\left(L + \frac{x_2}{2}\right)$$

$$R = \sqrt{\frac{x_2^2}{4} + \left(\frac{x_1 + x_3}{2}\right)^2}, \quad J = 2\left\{\sqrt{2}x_1x_2\left[\frac{x_2^2}{12} + \left(\frac{x_1 + x_3}{2}\right)^2\right]\right\}, \quad \sigma(X) = \frac{6PL}{x_4x_3^2} \quad \delta(X) = \frac{4PL^3}{Ex_3^3x_4}$$

$$P_c = \frac{4.013E\sqrt{\frac{x_3^2x_4^6}{36}}}{L^2}\left(1 - \frac{x_3}{2L}\sqrt{\frac{E}{4G}}\right), \quad L = 14in, \quad E = 30 \times 10^6 psi, \quad G = 12 \times 10^6 psi$$

$$P = 6000lb, \quad \tau_{MAX} = 13600psi, \quad \sigma_{MAX} = 30000psi, \quad \delta_{MAX} - 0.25in$$

Table 15.7 shows the requisite constraints involved for the above mentioned value of X.

TABLE 15.7 Constraints.

G1	G2	G3	G4	G5	G6	G7
-1.3944E-07	-1.43E-13	-2.5687E-07	-3.8730E-07	-1.5065E-07	-1.8467E-08	-1.4309

15.4.4 Gear train design (FM4)

The problem of the gear train design [30] consists of calculating the minimum possible cost of the gear train for a gear ratio defined as:

$$gear\ ratio = \frac{\eta_B\eta_C}{\eta_D\eta_A} \tag{15.18}$$

where η_j is the number of teeth of the gear where $j = A, B, C, D$. Mathematical formulation of the problem and the constraints are given in Eqs. (15.19) to (15.21). (See Fig. 15.7.)

$$X = \begin{bmatrix} x_1 \\ x_2 \\ x_3 \\ x_4 \end{bmatrix} = \begin{bmatrix} \eta_A \\ \eta_B \\ \eta_C \\ \eta_D \end{bmatrix} \tag{15.19}$$

$$12 < x_i < 60 \qquad i = 1, ..., 4 \tag{15.20}$$

$$Minimize : f(X) = \left[\frac{1}{6.931} - \frac{x_3x_2}{x_1x_4}\right]^2 \tag{15.21}$$

Table 15.8 shows the fitness values for the WOA algorithm for the Position value $X = [60 : 24.73135053 : 12 : 44.54378108]$.

FIGURE 15.7 Gear Train.

TABLE 15.8 Fitness Values.

Best	Worst	S.D	Average
0.003104794	0.003104794	0	0.003104794

15.4.5 Pressure vessel design (FM5)

The problem aims to minimize the cost of a cylindrical pressure vessel [30] in terms of forming, material, and welding. The design variables involved include the shells thickness (T_s), thickness of the head (T_h), the inner radius of the shaft (R), and the length (L). Mathematical formulation of the problem and the constraints are given in Eqs. (15.22) to (15.24). (See Fig. 15.8.)

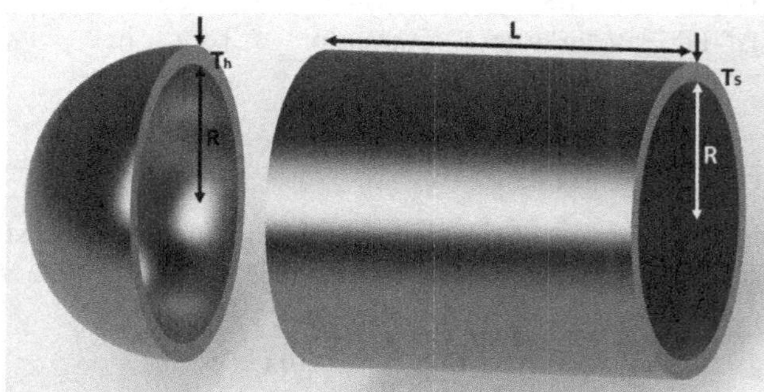

FIGURE 15.8 Pressure Vessel.

$$X = \begin{bmatrix} x_1 \\ x_2 \\ x_3 \\ x_4 \end{bmatrix} = \begin{bmatrix} T_s \\ T_h \\ R \\ L \end{bmatrix} \tag{15.22}$$

$$1 < x_1, x_2 < 99 \qquad 10.0 < x_3, x_4 < 200.0 \tag{15.23}$$

$$Minimize : f(X) = 0.6224x_1x_3x_4 + 1.778x_2x_3^2 + 3.166x_1^2x_4 + 19.84x_1^2x_3 \tag{15.24}$$

Table 15.9 shows the fitness values for the WOA algorithm for the Position value $X = [1 : 1 : 51.61687492 : 86.0134186]$.

TABLE 15.9 Fitness Values.

Best	Worst	S.D	Average
8797.097455	10336.21624	562.6438057	9130.421898

$$g_1(X) = -x_1 + 0.0193x_3 \leq 0, \qquad g_2(X) = -x_2 + 0.00954x_3 \leq 0$$

$$g_3(X) = -\pi x_3^2 x_4 - \frac{4}{3}\pi x_3^3 + 1296000 \leq 0, \qquad g_4(X) = x_4 - 240 \leq 0$$

Table 15.10 shows the requisite constraints involved for the above mentioned value of X.

TABLE 15.10 Constraints.

G1	G2	G3	G4
-0.0038	-0.5076	-0.0004	-153.9866

15.4.6 Three truss design (FM6)

The problem is a 3-bar planar truss design structure [31]. The independent variables involved in this design problem are A1 and A2. The objective is to minimize the volume of a 3 truss bar loaded statically. Mathematical formulation of the problem and the constraints are given in Eqs. (15.25) to (15.27). (See Fig. 15.9.)

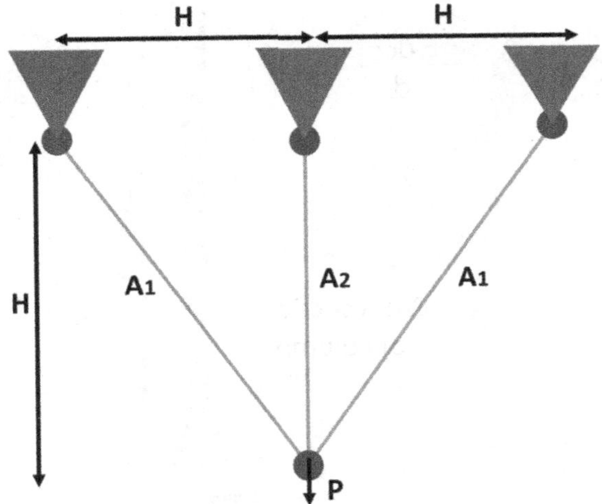

FIGURE 15.9 Three truss.

$$X = \begin{bmatrix} x_1 \\ x_2 \end{bmatrix} = \begin{bmatrix} A_1 \\ A_2 \end{bmatrix} \tag{15.25}$$

$$0 < A_1, A_2 < 1, \quad l = 100, \quad P = 2KN/cm^2, \quad \sigma = 2KN/cm^2 \tag{15.26}$$

$$Minimize : f(X) = (2\sqrt{2A_1} + A_2)xl \tag{15.27}$$

Table 15.11 shows the fitness values for the WOA algorithm for the Position value $X = [0.90477641 : 0.150982751]$.

TABLE 15.11 Fitness Values.

Best	Worst	S.D	Average
284.1375154	292.4250027	4.313382179	288.9727922

$$g_1(X) = Px\left(\frac{\sqrt{2}A_1 + A_2}{\sqrt{2}(A_1)^2 + 2A_1A_2}\right) - \sigma \le 0$$

$$g_2(X) = Px\left(\frac{A_2}{\sqrt{2}(A_1)^2 + 2A_1A_2}\right) - \sigma \le 0, \qquad g_3(X) = Px\left(\frac{1}{\sqrt{2}A_2 + A_1}\right) - \sigma \le 0$$

Table 15.12 shows the requisite constraints involved for the above mentioned value of X.

TABLE 15.12 Constraints.

G1	G2	G3
-0.0005	-1.7890	-0.2116

15.4.7 Tubular column design (FM7)

The target problem in this case is to design tubular cross-section based column with the capability for carrying a compressive load P with the objective function to minimize being the cost of fabrication. There are two independent variables involved in the design that is the mean diameter (d) and thickness (T) of the column. Mathematical formulation of the problem and the constraints are given in Eqs. (15.28) to (15.30). (See Fig. 15.10.)

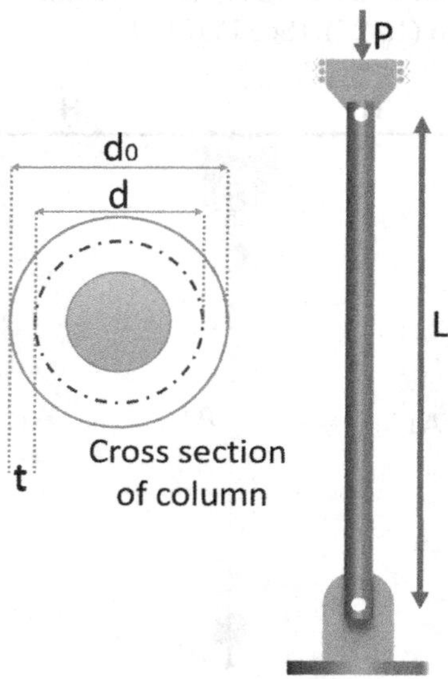

FIGURE 15.10 Tubular Column Design.

$$X = \begin{bmatrix} x_1 \\ x_2 \end{bmatrix} = \begin{bmatrix} d \\ T \end{bmatrix} \tag{15.28}$$

$$2 < d < 14, \quad 0.2 < T < 0.8 \tag{15.29}$$

$$Minimize : f(X) = 9.8dT + 2d \tag{15.30}$$

Table 15.13 shows the fitness values for the WOA algorithm for the Position value $X = [5.4511655 : 0.291964981]$.

TABLE 15.13 Fitness Values.

Best	Worst	S.D	Average
26.53134641	26.72766351	0.061980641	26.59822823

$$g_1(X) = \frac{P}{\pi dT} - \sigma_y \leq 0, \qquad g_2(X) = \frac{P}{\pi dT} - \frac{\pi^2 E(d^2 + T^2)}{8l^2} \leq 0, \qquad g_3(X) = -d + 2.0 \leq 0$$
$$g_4(X) = d - 14 \leq 0, \qquad g_5(X) = -T + 0.2 \leq 0, \qquad g_6(X) = T - 0.8 \leq 0$$

where,

$$P = 2500kgf, \quad \sigma_y = 500kgf/cm^2, \quad E = 8.5 \times 10^5 kgf/cm^2, \quad l = 250cm$$

Table 15.14 shows the requisite constraints involved for the above mentioned value of X.

TABLE 15.14 Constraints.

G1	G2	G3	G4	G5	G6
-1.2491E-07	-0.0017	-3.4512	-8.5488	-0.0920	-0.5080

15.4.8 Hydrodynamic thrust bearing design (FM8)

The problem defined in [32] is to design a hydrodynamic thrust such that the constraints involved be satisfied with minimal power loss. The four design variables included in this design's optimization are: bearing step radius (R), recess radius (R_o), oil viscosity (μ), and flow rate (Q). Mathematical formulation of the problem and the constraints are given in Eqs. (15.31) to (15.33). (See Fig. 15.11.)

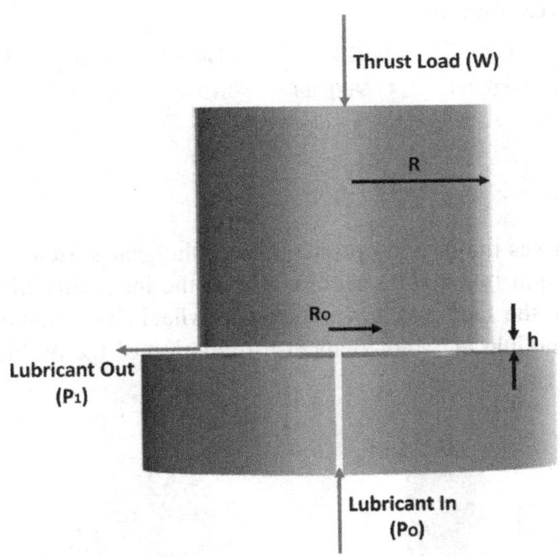

FIGURE 15.11 Hydrodynamic Thrust Bearing Design.

$$X = \begin{bmatrix} x_1 \\ x_2 \\ x_3 \\ x_4 \end{bmatrix} = \begin{bmatrix} R \\ R_o \\ \mu \\ Q \end{bmatrix} \tag{15.31}$$

$$1 < R, Q, R_o < 16, \quad 1 \times 10^{-6} < \mu < 16 \times 10^{-6} \tag{15.32}$$

$$Minimize: f(X) = \frac{QP_o}{0.7} + E_f \tag{15.33}$$

Table 15.15 shows the fitness values for the WOA algorithm for the Position value $X = [16 : 15.41386981 : 1 : 0.000016]$.

TABLE 15.15 Fitness Values.

Best	Worst	S.D	Average
7165.385742	7165.393759	0.001461699	7165.386024

$$g_1(X) = W_s - W \le 0, \qquad g_2(X) = P_o - P_{max} \le 0, \qquad g_3(X) = \Delta T - \Delta T_{max} \le 0$$
$$g_4(X) = h_{min} - h \le 0, \qquad g_5(X) = R_o - R \le 0$$
$$g_6(X) = \frac{\gamma}{g P_o}(\frac{Q}{2\pi Rh}) - 0.001 \le 0, \qquad g_7(X) = \frac{W}{\pi(R^2 - R_o^2)} - 5000 \le 0$$

where,

$$W = \frac{\pi P_o}{2}\frac{R^2 - R_o^2}{\ln \frac{R}{R_o}}, \qquad P_o = \frac{6\mu Q}{\pi h^3}\ln\frac{R}{R_o}, \qquad E_f = 9336Q\gamma C\Delta T, \qquad \Delta T = 2(10^P - 560)$$
$$P = \frac{\log_{10}\log_{10}(8.122 \times 10^6 \mu + 0.8) - C_1}{n}, \qquad h = \left(\frac{\pi N}{30}\right)^2 \frac{2\pi\mu}{E_f}\left(\frac{R^4}{4} - \frac{R_o^4}{4}\right)$$
$$\gamma = 0.0307, \quad C = 0.5, \quad n = -3.55, \quad C_1 = 10.04, \quad W_s = 101000, \quad P_{max} = 1000$$
$$\Delta T_{max} = 50, \quad h_{min} = 0.001, \quad g = 386.4, \quad N = 750$$

Table 15.16 shows the requisite constraints involved for the above mentioned value of X.

TABLE 15.16 Constraints.

G1	G2	G3	G4	G5	G6	G7
-101000	-999.9960	-1168.0588	-0.1956	-0.5861	0	-5000

15.4.9 Spur gear design (FM9)

The problem presented in [33] discusses the design optimization of the gear system containing many design variables in order to determine the minimum weight that can be acquired given the inequality of the constraints involved. The eight independent design variables include the face width b, pinion and wheel shaft diameter d_1 and d_2 respectively, number of pinion teeth Z_1, and the normal module m. Mathematical formulation of the problem and the constraints are given in Eqs. (15.34) to (15.36). (See Fig. 15.12.)

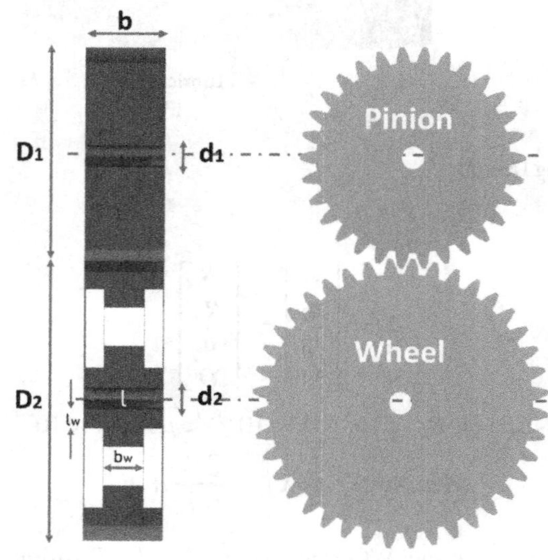

FIGURE 15.12 Spur Gear Design.

$$X = \begin{bmatrix} x_1 \\ x_2 \\ x_3 \\ x_4 \\ x_5 \\ x_6 \end{bmatrix} = \begin{bmatrix} b \\ Z_1 \\ d_1 \\ d_2 \\ m \\ H \end{bmatrix} \tag{15.34}$$

$$20 < b < 32, 10 < d_1 < 30, 30 < d_2 < 40, 18 < Z_1 < 25, 2.75 < m < 4, 200 < H < 400 \tag{15.35}$$

$$Minimize: f(X) = \left(\frac{\pi\rho}{4000}\right)\left[bm^2Z_1^2(1+\mu^2) - (D_i^2 - d_0^2)(l - b_w) - \eta d_p^2 b_w - b(d_1^2 + d_2^2)\right] \tag{15.36}$$

Table 15.17 shows the fitness values for the WOA algorithm for the Position value $X = [22 : 23.14932613 : 31.50882761 : 18 : 2.75 : 270.5384866]$.

TABLE 15.17 Fitness Values.

Best	Worst	S.D	Average
5918.386808	5918.386808	0	5918.386808

$$g_1(X) = b_1 - \frac{S_n C_x K_r K_{my} b - J - m}{K_v K_o K_m} \le 0, \quad g_2(X) = b_1 - \frac{D_1 I S_{fe}^2 C_l^2 C_r^2 b}{K_v K_m K_o C_p^2} \le 0$$

$$g_3(X) = (1 + D_2) - (1 + D_2) - \frac{D_1(2D_2 + D_1)sin^2\phi}{4m} \le 0, \quad g_4(X) = 8m - b \le 0$$

$$g_5(X) = b - 16m \le 0, \quad g_6(X) = b_3 - d_1^3 \le 0, \quad g_7(X) = b_4 - d_2^3 \le 0, \quad g_8(X) = \frac{Z_1(1+a)m}{2} - b_5 \le 0$$

where,

$$a = 4, \quad \rho = 8mg/m^3, \quad P = 7.5kW, \quad \eta = 6, \quad \tau = 19.62MPa, \quad N_1 = 1500rpm, \quad D_r = amZ_1 - 2.5m,$$

$$l_w = 2.5m, \quad b_w = 3.5m, \quad D_i = D_r - 2l_w, \quad D_1 = mZ_1, \quad D_2 = amZ_1, \quad d_o = d_2 + 25, \quad d_p = 0.25(D_i - d_o)$$

$$Z_2 = Z_1 D_2/D_1, N_2 = N_1/a, v = \pi D_1 N_1/60000, \quad b_5 = 151.5, \quad b_4 = 4868 \times 10^6 P/N_2\tau, \quad b_3 = 4868 \times 10^6 P/N_1\tau$$

$$b_1 = 1000P/v, \quad K_r = 0.814, \quad C_r = 1, \quad K_{ms} = 1.4, \quad K_m = 1.3, \quad K_o = 1, \quad C_l = 1, \quad C_p = 191$$

$$\phi = 25, \quad S_n = 1.7236H, \quad K_v = \frac{78 + \sqrt{196.85v}}{78}, \quad S_{fe} = 28H - 69, \quad I = \frac{asin\phi cos\phi}{2(a+1)}$$

Fig. 15.13 shows the standard values for AGMA information sheet 908-B89. This graph is used to obtain the Lewis geometry factor with respect to the number of teeth. The relation equation is supposed up to a 3rd polynomial order.

Table 15.18 shows the requisite constraints involved for the above mentioned value of X.

TABLE 15.18 Constraints.

G1	G2	G3	G4	G5	G6	G7	G8
-2334.6464	-144388.0357	-159.0608	-1.6E-09	-22	-0.1860	-18340.6739	-27.75

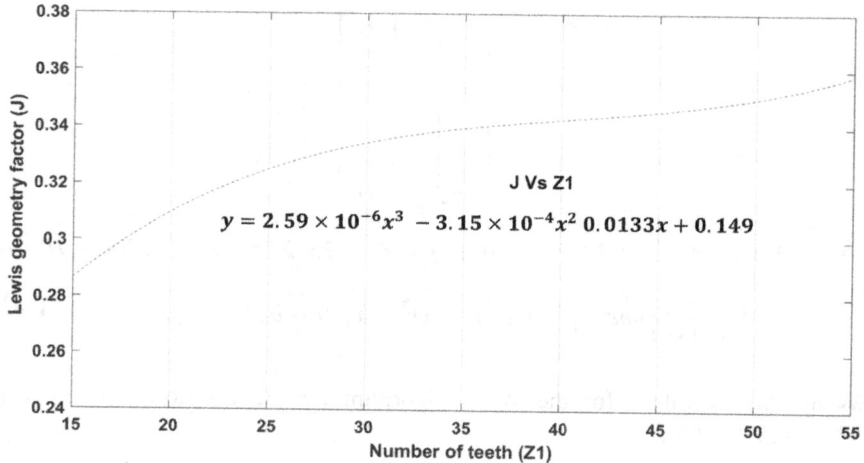

FIGURE 15.13 AGMA information sheet 908-B89.

15.4.10 Step cone pulley design (FM10)

The aim is to find the minimum weight of the four step cone pulley design [34]. Five decision variables determine the optimization problem, i.e., diameter of each step d_i, $i = 1, ..., 4$ and the pulley width ω. The step pulley transmits minimum of 0.75 hp given an input speed of 350 rpm. The output speeds for each step cone pulley are shown in Fig. 15.14. Mathematical formulation of the problem and the constraints are given in Eqs. (15.37) to (15.39).

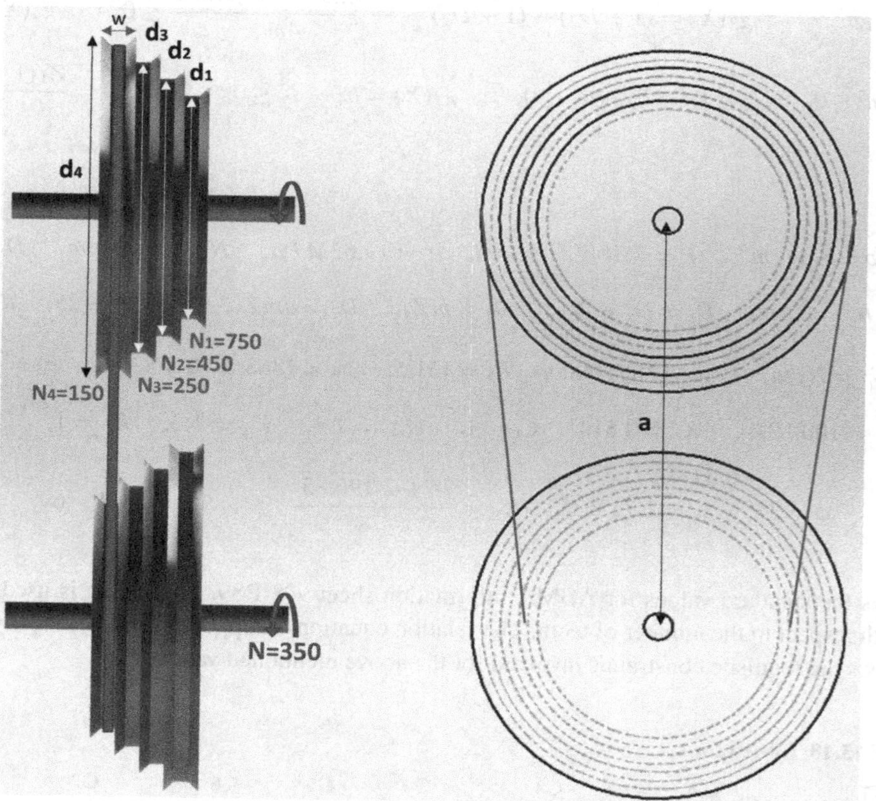

FIGURE 15.14 Step Cone pulley.

$$X = \begin{bmatrix} x_1 \\ x_2 \\ x_3 \\ x_4 \\ x_5 \end{bmatrix} = \begin{bmatrix} d_1 \\ d_2 \\ d_3 \\ d_4 \\ w \end{bmatrix} \tag{15.37}$$

$$16 \le w \le 100, 40 \le d_i \le 100m \text{ where } i = 1, 2, 3, 4 \tag{15.38}$$

$$Minimize: f(X) = \rho\omega\left[d_1^2\left(1+\left(\frac{N_1}{N}\right)^2\right)+d_2^2\left(1+\left(\frac{N_2}{N}\right)^2\right)+d_3^2\left(1+\left(\frac{N_3}{N}\right)^2\right)+d_4^2\left(1+\left(\frac{N_4}{N}\right)^2\right)\right] \tag{15.39}$$

Table 15.19 shows the fitness values for the WOA algorithm for the Position value $X = [40 : 40 : 85.82268898 : 44.60303996 : 16]$.

TABLE 15.19 Fitness Values.

Best	Worst	S.D	Average
3072399328	3272399328	17130894.3	3172399.328

$$g_i(X) = 2 - R_i \le 0, \qquad g_j(X) = (0.75x745.6998) - P_j \le 0$$

where,

$$C_i = \frac{\pi d_i}{2}\left(1+\frac{N_i}{N}\right)+\frac{\left(\frac{N_i}{N}-1\right)^2}{4a}+2a, \qquad R_i = exp\left[\mu\left(\pi - 2sin^{-1}\left(\frac{d_i}{2a}\left(\frac{N_i}{N}-1\right)\right)\right)\right]$$

$$P_i = stw\left[1 - exp\left[\pi - 2sin^{-1}\left[\left(\frac{N_i}{N}-1\right)\frac{d_i}{2a}\right]\right]\right]\frac{\pi d_i N_i}{60}$$

$$\rho = 7200lg/m^3, \quad a = 3m, \quad \mu = 0.35, \quad s = 1.75MPa, \quad t = 8mm, \quad i, j = 5, 6, 7, 8$$

Table 15.20 shows the requisite constraints involved for the above mentioned value of X.

TABLE 15.20 Constraints.

G1	G2	G3	G4	G5	G6	G7	G8
-1.6733	-2.6355	-2.9978	-2.7476	-466257.3811	-76385.0451	-6.715E-6	-19249.5984

15.4.11 Reinforced concrete beam design (FM11)

The goal of the problem defined in [35] contains a reinforced concrete beam that is supported with a 30 ft span. The beam is subject to a deal load of 1 klbf and live load of up to 2 klbf. The objective is to determine the fabrication cost's minimal value given the independent design variable: Cross sectional area of the reinforcement bar A_s, width of the beam w, and depth of the beam h. Mathematical formulation of the problem and the constraints are given in Eqs. (15.40) to (15.42). (See Fig. 15.15.)

$$X = \begin{bmatrix} x_1 \\ x_2 \\ x_3 \end{bmatrix} = \begin{bmatrix} A_s \\ b \\ h \end{bmatrix} \tag{15.40}$$

$$6 \le A_s \le 8.4, 28 \le b \le 40, 5 \le h \le 10 \tag{15.41}$$

$$Minimize: f(X) = 2.9A_s + 0.6bh \tag{15.42}$$

Table 15.21 shows the fitness values for the WOA algorithm for the Position value $X = [6.37916395 : 8.451865908 : 33.7832481]$.

$$g_1(X) = \frac{h}{b} - 4 \le 0, \qquad g_2(X) = 180 + 7.375\frac{A_s^2}{b} - A_s h \le 0$$

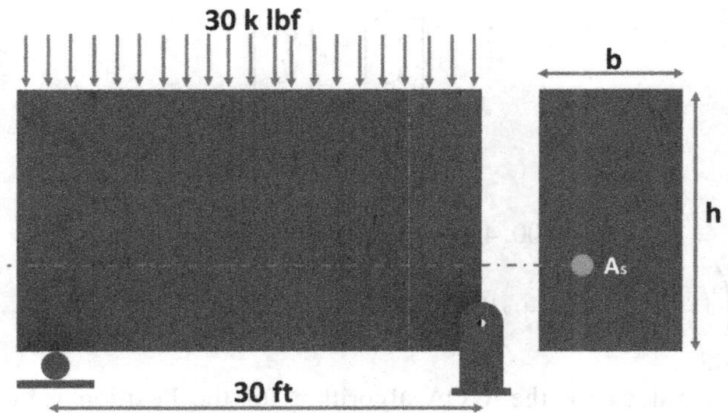

FIGURE 15.15 Reinforced Concrete beam.

TABLE 15.21 Fitness Values.

Best	Worst	S.D	Average
358.8663099	361.0237659	1.525551818	359.9450379

Table 15.22 shows the requisite constraints involved for the above mentioned value of X.

TABLE 15.22 Constraints.

G1	G2
-0.00287	-1.76619E-08

15.4.12 Piston lever design (FM12)

The objective of this design problem introduced in [36] is to determine the minimum value of the oil volume when the lever of the piston is lifted. The lift traverses from 0 deg to 45 deg. The design involves a total of four independent variables. Mathematical formulation of the problem and the constraints are given in Eqs. (15.43) to (15.45). (See Fig. 15.16.)

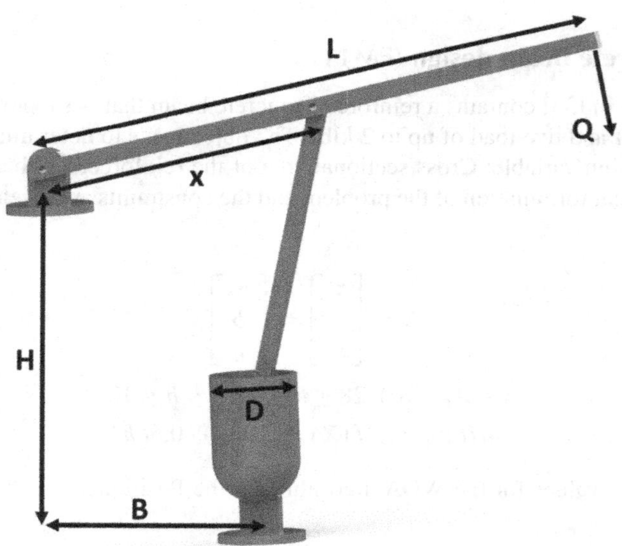

FIGURE 15.16 Piston Lever Design.

$$X = \begin{bmatrix} x_1 \\ x_2 \\ x_3 \\ x_4 \end{bmatrix} = \begin{bmatrix} H \\ B \\ D \\ X \end{bmatrix} \quad (15.43)$$

$$0.05 \leq B, D, H \leq 500, 0.05 \leq X \leq 120 \quad (15.44)$$

$$Minimize : f(X) = \frac{1}{4}\pi D^2(L_2 - L_1) \quad (15.45)$$

Table 15.23 shows the fitness values for the WOA algorithm for the Position value $X = [52.14320033 : 500 : 0.05 : 0.05]$.

TABLE 15.23 Fitness Values.

Best	Worst	S.D	Average
0.00003580	0.00003583	0.00000002	0.00003581

$$g_1(X) = QLcos\theta - RF \leq 0, \qquad g_2(X) = Q(L - X) - M_{max} \leq 0$$

$$g_3(X) = 1.2(L_2 - L_1) - L_1 \leq 0, \qquad g_4(X) = \frac{D}{2} - B \leq 0$$

where,

$$F = \pi PD^2/4, \quad L_1 = \sqrt{(X-B)^2 + H^2}, \quad L_2 = \sqrt{(Xsin\theta + H)^2 + (B - Xcos\theta)^2}, \quad \theta = 45\,\text{deg}$$

$$R = \frac{|-X(Xsin\theta + H) + H(B - Xcos\theta)|}{\sqrt{(X-B)^2 + H^2}}, \quad L = 240in, M_{max} = 1.8 \times 10^6, Q = 1500psi$$

Table 15.24 shows the requisite constraints involved for the above mentioned value of X.

TABLE 15.24 Constraints.

G1	G2	G3	G4
-7.1993E-09	-360.3	-502.6399	-499.975

15.4.13 Comparative analysis

The above mentioned structural design problems included in this study are useful for determining the efficacy of the meta-heuristic algorithm. This section shows the comparative study of the WOA with other similar techniques. For a fair comparison each algorithm uses 50 agents as the population size and a total of 30 runs are evaluated to determine the best, worst, standard deviation and average values for each design problem. The comparison of the results for other meta-heuristic approaches is shown in Table 15.25. The best, worst, and average values for the fitness solution are highlighted as bold for easy visualization. Evidently, the WOA algorithm performs better for most of the engineering design test problems compared to other meta-heuristic techniques. Particularly, in the case of FM4, each algorithm was able to find the single global optima solution despite having different combinations of input X in Eq. (15.19). The BMO algorithm, comparatively, performed worst for the optimization of structural design problems.

Table 15.26 illustrates the constraint values that are determined from the input X values from each meta-heuristic algorithm. Problem FM3 shows that the constraint condition $G1$, $G2$, and $G6$ have been violated by PSO algorithm. This means that during the iteration process the algorithm was unable to find a single local optima solution that would follow the constraint conditions. Similarly, GWO and AOA algorithms also violated constraints for problem FM5.

15.4.14 Computational complexity analysis

Computational complexity analysis provides a theoretical framework for understanding the behavior of metaheuristic algorithms. It can help to identify the key factors that affect the performance of the algorithms, such as the search space size,

TABLE 15.25 Statistical Results of fitness for each design by different methods.

	Fitness	WOA	PSO	AOA	GWO	BMO
FM1	Best	**0.004876647**	0.768023817	0.014377985	0.03096478	0.017078169
	Worst	**0.004876647**	0.768023817	0.014377985	0.03096478	0.017078169
	SD	0	0	2.93E-12	3.923E-12	1.256E-12
	Average	**0.004876647**	0.768023817	0.014377985	0.03096478	0.017078169
FM2	Best	**0.006076142**	0.006437839	0.05024129	0.00720571	0.05498987
	Worst	0.011544971	**0.01138261**	1.831659501	1.246502396	1.658080096
	SD	0.001102363	0.001394765	0.456877322	0.332539012	0.500024022
	Average	**0.006642155**	0.008572318	0.660690476	0.528933288	0.58758367
FM3	Best	**7.233132593**	9.33249971	7.261826712	10.83823042	8.50310902
	Worst	9.513480763	26.35602343	**9.030039521**	10.83823042	16.86309315
	SD	0.514501373	4.166377585	0.599931283	0	2.557469022
	Average	**7.591820935**	14.7666218	7.779617292	10.83823042	14.09403885
FM4	Best	0.003104794	0.003104794	0.003104794	0.003104794	0.003104794
	Worst	0.003104794	0.003104794	0.003104794	0.003104794	0.003104794
	SD	0	0	0	0	0
	Average	0.003104794	0.003104794	0.003104794	0.003104794	0.003104794
FM5	Best	**8797.097455**	8797.450208	8798.105689	8799.735961	9699.158396
	Worst	10336.21624	**10289.7857**	256039.5664	10566.38137	20684.48352
	SD	562.6438057	659.9734875	62557.2032	705.4554666	3296.384283
	Average	**9130.421898**	9273.557877	47353.69725	9399.369912	11511.67506
FM6	Best	**284.1375154**	286.2731515	289.6014254	286.9762808	285.6662847
	Worst	292.4250027	**286.2731515**	299.2342662	288.5139507	291.5231804
	SD	4.313382179	0	6.811701902	1.087296813	3.135358147
	Average	288.9727922	**286.2731515**	294.418026	287.7451158	287.9480322
FM7	Best	**26.53134641**	27.17977326	27.51312329	26.90432625	27.24043316
	Worst	**26.72766351**	34.92670765	30.84693854	34.08362105	33.90243684
	SD	0.061980641	2.456083972	1.314071664	2.065625162	2.111827297
	Average	**26.59822823**	29.89848276	29.22314003	29.02391402	30.37617983
FM8	Best	**7165.385742**	7246.165431	7165.385801	7165.396836	7165.393075
	Worst	7165.393759	12924.43842	7165.39677	**7165.366836**	7165.993075
	SD	0.001461699	0.001411506	0.004851816	4.18163E-05	0.004461699
	Average	**7165.386024**	9298.109901	7165.388098	7165.531836	7165.693075
FM9	Best	**5918.386808**	5930.872072	5970.100818	5926.513741	5983.261807
	Worst	**5918.386808**	6004.381687	6707.733187	6989.069233	5990.979746
	SD	0	29.84090906	228.5899323	232.6767978	2.403028635
	Average	**5918.386808**	5976.386754	6060.339185	6056.020656	5990.098094
FM10	Best	**3072399328**	3078113208	3075435904	3055092715	3094239847
	Worst	3272399328	3350210560	3564741235	3611527266	**3179427563**
	SD	17130894.3	110547546.2	199907517.3	172551102.5	60236812.06
	Average	3172399.328	3163632494	3232463487	3204110618	**3136833705**
FM11	Best	**358.8663099**	360.7092429	360.2494122	360.4985395	359.8035634
	Worst	361.0237659	366.0180618	**360.7092429**	360.8608767	361.7245789
	SD	1.525551818	0	0.21491107	0.25621107	1.358363097
	Average	**359.9450379**	360.7092429	360.433737	360.6797081	360.7640712
FM12	Best	**0.00003580**	0.00003583	**0.00003580**	**0.00003580**	0.000202858
	Worst	**0.00003583**	0.00003585	0.00003586	0.00003590	0.002244307
	SD	0.00000002	0.00000002	0.00000002	0.00000002	0.00060078
	Average	**0.00003581**	0.00003584	0.00003584	0.00003583	0.000825534

TABLE 15.26 Statistical Results of constraints for each design by different methods.

		WOA	PSO	GWO	AOA	BMO
FM1	G1	-2033.63169	-12918.10597	-689.0953714	-319.4344125	-579.9862465
	G2	-3E-12	-5E-11	-7E-12	-8E-12	-2E-12
	G3	-7093.04333	-45043.4674	-2405.119242	-1116.241816	-2024.694194
FM2	G1	-3.0122E-09	0	0	0	0
	G2	-1.3589E+17	**3.14353E+16**	-3.82276E+17	-2.33455E+16	-3.20148E+16
	G3	-8.51055638	-20	0	0	0
	G4	-0.56159053	0	0	0	0
FM3	G1	-1.3944E-07	**1611.287894**	-318.8623593	-9.6085957	-4136.825761
	G2	-1.43E-13	**115348.2442**	-7127.682111	-812.0100893	-1396.844946
	G3	-2.5687E-07	-2.389210025	-6.14584522	-5.36189897	-3.03429473
	G4	-3.873E-07	-7.259341814	-0.57158807	-2.11666808	-3.3825416
	G5	-1.5065E-07	-0.885	-0.82077162	-0.85663867	-1.36541047
	G6	-1.8466E-08	**0.376804309**	-0.19348468	-0.17294509	-0.18686876
	G7	-1.43086519	-3939242.318	-61149436.02	-41100250.85	-17665314.08
FM4	NA					
FM5	G1	-0.00379431	-0.006678206	-2.153506117	-0.016324322	-0.026479713
	G2	-0.50757501	-0.509000523	-4.356727667	**0.513768603**	-0.441541381
	G3	-0.0004388	-0.000810369	**1256853.571**	**0.000164059**	-21048.49489
	G4	-153.986581	-152.8869793	-190.4012644	-149.1512442	-195.7120343
FM6	G1	-0.00053874	-0.004648857	-0.054275539	-0.051058987	-0.021542078
	G2	-1.78897038	-1.676833943	-1.713694502	-1.890781153	-1.795565007
	G3	-0.21156836	-0.327814901	-0.340581037	-0.160277835	-0.225977071
FM7	G1	-1.2491E-07	-12.92878167	-20.71122032	-4.38087291	-6.893257648
	G2	-0.0016901	-34.53646859	-49.17372745	-26.05660491	-52.69234054
	G3	-3.4511655	-3.567951223	-3.604370277	-3.568581482	-3.696567993
	G4	-8.5488345	-3.432048777	-3.396297231	-8.431418518	-8.303532007
	G5	-0.09196498	-0.093428468	-0.096255248	-0.08833509	-0.083298016
	G6	-0.50803502	**0.506571532**	-0.503744752	-0.51166491	-0.516701984
FM8	G1	-101000	-101000	-101000	-101000	-101000
	G2	-999.995981	-610.3384216	-999.9959393	-999.9882148	-999.9908477
	G3	-1168.05877	-1168.060025	-1168.058778	-1168.059283	-1168.058768
	G4	-0.19563462	-0.001264491	-0.193647559	-0.101189419	-0.14846172
	G5	-0.58613019	-1.365282506	-0.592607227	-5.397513202	-4.735257627
	G6	0	-0.00099749	0	0	-7.40E-07
	G7	-5000	-5000	-5000	-5000	-5000
FM9	G1	-2334.64635	-4331.332124	-4410.706645	-4376.69497	-1223.106031
	G2	-144388.036	-313449.3216	-321618.3936	-319920.8227	-77522.08862
	G3	-159.06077	-159.0607697	-159.0607697	-159.0607697	-159.0607697
	G4	-1.6E-09	-0.131909681	-0.127155563	-0.005870559	-0.001221915
	G5	-22	-21.86809032	-21.87284444	-21.99412944	-21.99877808
	G6	-0.18598417	-3903.293852	-3218.009797	-9020.465527	-10448.14012
	G7	-18340.6739	-4283.773644	-20516.87734	-4626.7089	-22622.83384
	G8	-27.75	-27.25	-27.75	-27.75	-27.75

continued on next page

TABLE 15.26 *(continued)*

		WOA	PSO	GWO	AOA	BMO
FM10	G1	-1.67328253	-1.67328253	-1.67328253	-1.67328253	-1.67328253
	G2	-2.63553367	-2.635533665	-2.635533665	-2.635533665	-2.635533665
	G3	-2.99777753	-2.959193596	-2.993535167	-2.991792279	-2.909453426
	G4	-2.74756044	-2.864902074	-2.731437111	-2.99123949	-2.914305153
	G5	-466257.381	-466257.3811	-466257.3811	-466257.3811	-466257.3811
	G6	-76385.0451	-76385.04514	-76385.04514	-76385.04514	-76385.04514
	G7	-6.715E-06	-9770.883013	-1068.636805	-1508.07093	-22539.50483
	G8	-19249.5984	-9851.35085	-20567.35763	-102.7560765	-5994.135438
FM11	G1	-0.00287	-0.00100	-0.049477331	-0.028783578	-0.019268116
	G2	-1.7662E-08	-2.81858E-09	-6.88675E-08	-6.775E-09	-3.27501E-09
FM12	G1	-7.1993E-09	-3.783180612	-0.286687751	-0.091200912	-7.597014515
	G2	-360.3	-360.3	-360.3	-360.3	-360.3
	G3	-502.639949	-502.6602881	-502.6414878	-502.6404387	-37.44591648
	G4	-499.975	-499.975	-499.975	-499.975	-24.16223637

the problem structure, and the choice of parameters. This information can be used to design more efficient algorithms, and to guide the development of new metaheuristic methods. This section provides a time based computational cost investigation of the Whale Optimization algorithm using collection of 9 functions (C01 to C09) from CEC 2019 [37], as shown in Table 15.A.1. These functions are set as minimization problems with CEC01 to CEC03 having different dimensions while CEC04 to CEC09 functions, which are rotated and shifted, have a dimension value of 10. Four specific computational times, T_0, T_1, T_2, $\hat{T}_2$ are determined which establish the complexity of the algorithm, evaluation criteria for which is detailed in [38]. T_0 is the runtime of a specific mathematical algorithm, T_1 is the computational time of the CEC function which has been run 10,000 times, T_2 is the computational time of the technique used to solve the minimization problem of the CEC function in under 10,000 iterations and T_2 is the mean value of 5 repetitions of the $\hat{T}_2$ time analysis. It is executed 5 times to cater for the variations in the time of execution due to the probabilistic nature of the algorithms. The algorithms complexity is reflected by Eq. (15.46).

$$\vec{T} = \frac{\hat{T}_2 - T_1}{T_0} \tag{15.46}$$

The WOA is tested on these computational times and is compared with other similar metaheuristic algorithms on each CEC benchmark test function. The results of the experiment are shown in Table 15.27. The value of T_0 was determined to be 0.1162 seconds. Since the algorithms have multi-agent solution finding technique, a population size of 10 was used for each algorithm in comparison. Rest of the parameters are same as illustrated in Table 15.1. The machine used for the complete analysis is a Core i7 9750h with 16 GB RAM.

15.5 Conclusion

In the presented work, the WOA algorithm is utilized in the application of the optimization of structural design problems with multiple constraints. The validation of the algorithm has been conducted using several benchmark structural design problems found in literature ans is proven to be highly efficient. The extensive comparative analysis with other metaheuristic algorithms, traditional and new, reveals that WOA performs much superior. This is primarily due to the fact that the mathematical model of the spiral strategy and the decreasing value of the hyperparameter of the algorithm ensures equitable distribution of the exploitation and exploration phases in which the WOA algorithm traverses around the search space region as well as converge towards a global optima solution.

Taking in account the multi-agent technique, the dynamic strategies used for optimization can be further extended similar to the extension done with population based algorithms such as improvements, upgradation and hybridization to ascertain the application of multi-objective optimization burdened with various complicated constraints such as the NP hard problems. Most of the validation and proofs of meta-heuristic algorithms are highly dependent on empirical methods and thus the No Free Lunch (NFL) theorem posits an insightful conclusion that not any one single meta-heuristic algorithm

TABLE 15.27 Comparison of meta heuristic algorithms with WOA for $\vec{T}$ time complexity analysis.

Function	$\vec{T}_{WOA}$	$\vec{T}_{GWO}$	$\vec{T}_{PSO}$	$\vec{T}_{AOA}$	$\vec{T}_{BMO}$
CEC 01	**315.4728**	317.0034	317.4022	316.4658	329.9038
CEC 02	5.06911	5.268898	**4.918262**	5.81568	13.40771
CEC 03	**10.231**	10.58441	10.32667	10.60248	18.83066
CEC 04	**3.52374**	3.89401	3.723081	3.883752	12.11951
CEC 05	**4.309211**	4.970947	4.359501	4.397969	13.74307
CEC 06	111.012	111.3846	**110.3989**	110.049	115.1678
CEC 07	**4.1225602**	4.396979	4.235602	4.367849	11.9896
CEC 08	**4.189333**	4.388855	4.25469	4.507995	12.07964
CEC 09	4.11466	4.072453	4.292401	**3.904604**	12.05281

can be used for any defined problem, therefore, from an analytical standpoint, mathematical analysis and proves of why algorithms work is highly sought for. Progress in this area would surely help understand meta-heuristic algorithms to help design more efficient hybridization of such algorithms.

Appendix 15.A

Table 15.A.1 shows the CEC-2019 test functions in terms of description, dimension, range, and minimum value. These functions are used in the analysis for algorithm time complexity.

TABLE 15.A.1 CEC - 2019 Benchmark Test Functions.

Function	Description	Dimension	Range	f_min
CEC01	Storn's Chebyshev polynomial fitting problem	9	[-8192, 8192]	1
CEC02	Inverse Hilbert matrix problem	16	[-16384, 16384]	1
CEC03	Lennard–Jones minimum energy cluster	18	[-4, 4]	1
CEC04	Rastrigin's function	10	[-100, 100]	1
CEC05	Grienwank's function	10	[-100, 100]	1
CEC06	Weierstrass function	10	[-100, 100]	1
CEC07	Modified Schwefel's function	10	[-100, 100]	1
CEC08	Expanded Schaffer's F6 function	10	[-100, 100]	1
CEC09	Happy CAT function	10	[-100, 100]	1

References

[1] Z. Michalewicz, Genetic Algorithms + Data Structures = Evolution Programs, vol. 3, Springer, Berlin, 1995, https://doi.org/10.1007/978-3-662-03315-9.

[2] M. Abd Elaziz, A.H. Elsheikh, D. Oliva, L. Abualigah, S. Lu, A.A. Ewees, Advanced metaheuristic techniques for mechanical design problems, Archives of Computational Methods in Engineering 29 (1) (2022) 695–716.

[3] P. Drag, K. Styczeń, M. Kwiatkowska, A. Szczurek, A review on the direct and indirect methods for solving optimal control problems with differential-algebraic constraints, in: Recent Advances in Computational Optimization, 2016, pp. 91–105.

[4] V. Torczon, On the convergence of the multidirectional search algorithm, SIAM Journal on Optimization 1 (1) (1991) 123–145.

[5] T.G. Kolda, R.M. Lewis, V. Torczon, Optimization by direct search: new perspectives on some classical and modern methods, SIAM Review 45 (3) (2003) 385–482.

[6] R.M. Lewis, V. Torczon, M.W. Trosset, Direct search methods: then and now, Journal of Computational and Applied Mathematics 124 (1–2) (2000) 191–207.

[7] C.L. Ranieri, C.A. Ocampo, Indirect optimization of spiral trajectories, Journal of Guidance, Control, and Dynamics 29 (6) (2006) 1360–1366.

[8] A. Gálvez, A. Iglesias, From nonlinear optimization to convex optimization through firefly algorithm and indirect approach with applications to CAD/CAM, Scientific World Journal 2013 (2013) 283919, https://doi.org/10.1155/2013/283919.

[9] B. Passenberg, M. Leibold, O. Stursberg, M. Buss, A globally convergent, locally optimal min-h algorithm for hybrid optimal control, SIAM Journal on Control and Optimization 52 (1) (2014) 718–746.

[10] M.-Y. Cheng, D. Prayogo, Symbiotic organisms search: a new metaheuristic optimization algorithm, Computers & Structures 139 (2014) 98–112.

[11] M. Gendreau, J.-Y. Potvin, Metaheuristics in combinatorial optimization, Annals of Operations Research 140 (1) (2005) 189–213.

[12] I. Fister Jr., X.-S. Yang, I. Fister, J. Brest, D. Fister, A brief review of nature-inspired algorithms for optimization, arXiv preprint, arXiv:1307.4186.

[13] A.K. Sangaiah, Z. Zhiyong, M. Sheng, et al., Computational Intelligence for Multimedia Big Data on the Cloud with Engineering Applications, Academic Press, 2018.

[14] V.-H. Truong, P.-C. Nguyen, S.-E. Kim, An efficient method for optimizing space steel frames with semi-rigid joints using practical advanced analysis and the micro-genetic algorithm, Journal of Constructional Steel Research 128 (2017) 416–427.

[15] S. Kazemzadeh Azad, Seeding the initial population with feasible solutions in metaheuristic optimization of steel trusses, Engineering Optimization 50 (1) (2018) 89–105.

[16] V. Truong, S.-E. Kim, An efficient method for reliability-based design optimization of nonlinear inelastic steel space frames, Structural and Multidisciplinary Optimization 56 (2) (2017) 331–351.

[17] C. Millan-Paramo, et al., Exporting water wave optimization concepts to modified simulated annealing algorithm for size optimization of truss structures with natural frequency constraints, Engineering With Computers 37 (1) (2021) 763–777.

[18] M.-H. Ha, Q.-V. Vu, V.-H. Truong, Optimization of nonlinear inelastic steel frames considering panel zones, Advances in Engineering Software 142 (2020) 102771.

[19] A. Kaveh, K.B. Hamedani, S.M. Hosseini, T. Bakhshpoori, Optimal design of planar steel frame structures utilizing meta-heuristic optimization algorithms, Structures 25 (2020) 335–346, Elsevier.

[20] A. Kaveh, H. Akbari, S.M. Hosseini, Plasma generation optimization: a new physically-based metaheuristic algorithm for solving constrained optimization problems, Engineering Computations 38 (4) (2020) 1554–1606, https://doi.org/10.1108/EC-05-2020-0235.

[21] M.H. Sulaiman, Z. Mustaffa, M.M. Saari, H. Daniyal, Barnacles mating optimizer: a new bio-inspired algorithm for solving engineering optimization problems, Engineering Applications of Artificial Intelligence 87 (2020) 103330.

[22] S. Mirjalili, A. Lewis, The whale optimization algorithm, Advances in Engineering Software 95 (2016) 51–67.

[23] P.R. Hof, E. Van der Gucht, Structure of the cerebral cortex of the humpback whale, Megaptera novaeangliae (Cetacea, Mysticeti, Balaenopteridae), The Anatomical Record: Advances in Integrative Anatomy and Evolutionary Biology 290 (1) (2007) 1–31.

[24] A.S. Friedlaender, E. Hazen, J. Goldbogen, A. Stimpert, J. Calambokidis, B. Southall, Prey-mediated behavioral responses of feeding blue whales in controlled sound exposure experiments, Ecological Applications 26 (4) (2016) 1075–1085.

[25] Y. Zhang, S. Wang, G. Ji, et al., A comprehensive survey on particle swarm optimization algorithm and its applications, Mathematical Problems in Engineering 2015 (2015) 931256, https://doi.org/10.1155/2015/931256.

[26] S. Mirjalili, S.M. Mirjalili, A. Lewis, Grey wolf optimizer, Advances in Engineering Software 69 (2014) 46–61.

[27] L. Abualigah, A. Diabat, S. Mirjalili, M. Abd Elaziz, A.H. Gandomi, The arithmetic optimization algorithm, Computer Methods in Applied Mechanics and Engineering 376 (2021) 113609.

[28] S. Singiresu, et al., Engineering Optimization: Theory and Practice, John Wiley & Sons, 1996.

[29] J. Arora, Introduction to Optimum Design, McGraw-Hill, New York, 1989.

[30] E. Sandgren, Nonlinear integer and discrete programming in mechanical design optimization, Journal of Mechanical Design 112 (2) (1990) 223–229, https://doi.org/10.1115/1.2912596.

[31] T. Ray, P. Saini, Engineering design optimization using a swarm with an intelligent information sharing among individuals, Engineering Optimization 33 (6) (2001) 735–748.

[32] R.V. Rao, V.J. Savsani, D. Vakharia, Teaching–learning-based optimization: a novel method for constrained mechanical design optimization problems, Computer Aided Design 43 (3) (2011) 303–315.

[33] R.V. Rao, R.B. Pawar, Constrained design optimization of selected mechanical system components using Rao algorithms, Applied Soft Computing 89 (2020) 106141.

[34] S. Gupta, H. Abderazek, B.S. Yıldız, A.R. Yildiz, S. Mirjalili, S.M. Sait, Comparison of metaheuristic optimization algorithms for solving constrained mechanical design optimization problems, Expert Systems with Applications 183 (2021) 115351.

[35] H.M. Amir, T. Hasegawa, Nonlinear mixed-discrete structural optimization, Journal of Structural Engineering 115 (3) (1989) 626–646.

[36] P. Thanedar, G. Vanderplaats, Survey of discrete variable optimization for structural design, Journal of Structural Engineering 121 (2) (1995) 301–306.

[37] J.M. Abdullah, T. Ahmed, Fitness dependent optimizer: inspired by the bee swarming reproductive process, IEEE Access 7 (2019) 43473–43486.

[38] J.-J. Liang, B. Qu, D. Gong, C. Yue, Problem Definitions and Evaluation Criteria for the CEC 2019 Special Session on Multimodal Multiobjective Optimization, Computational Intelligence Laboratory, Zhengzhou University, 2019, https://doi.org/10.13140/RG.2.2.33423.64164.

Chapter 16

F-WOA: an improved whale optimization algorithm based on Fibonacci search principle for global optimization

Saroj Kumar Sahoo[a], Salpa Reang[a], Apu Kumar Saha[a], and Sanjoy Chakraborty[b,c]

[a]*Department of Mathematics, National Institute of Technology Agartala, Agartala, Tripura, India,* [b]*Department of Computer Science and Engineering, National Institute of Technology Agartala, Agartala, Tripura, India,* [c]*Department of Computer Science and Engineering, Iswar Chandra Vidyasagar College, Belonia, Tripura, India*

16.1 Introduction

In the process of optimization, an issue must be identified and resolved in a short amount of time. It is a procedure where we input something and attempt to acquire the greatest outcomes for the outputs, for instance, the number of inputs that we may alter for each issue, as well as the sorts of inputs that we can have, or the processing time required to obtain the results. In other words, cost-cutting and production maximization are the two main objectives of optimization. Numerous conventional techniques, including the random search method, the quasi-Newton method, the univariate approach, the steepest descent method, the pattern search method, the conjugate gradient method, etc. Traditional optimization has a few limitations, such as selecting a random solution to an optimization issue and the final answer being completely dependent on the original random solution. However, we cannot guarantee that the final solution will not become trapped in the local optimum, that it is not suited for discontinuous objective functions, that it is only suitable for one initial solution, and that it is not ideal for parallel computing [1]. To address these shortcomings, we have non-traditional optimization approaches known as metaheuristics in the current world.

To overcome these limitations, a bunch of algorithms have been developed by the different researchers called metaheuristic algorithms. We mainly have distinguished metaheuristic into two types: single solution based (SSB) in this search process the solution moves from a single require solution to another search space exploring the current solution of the neighborhood; population based (PB) this improves the algorithmic and heuristic. Broadly The population-based optimization methods are classified in four different groups: evolutionary algorithm, swarm intelligence algorithm, physical or chemical law-based algorithm and mathematical based algorithm few examples of these algorithms are differential evolution (DE) [2], particle swarm optimization (PSO) [3], symbiotic organisms search (SOS) [4], sine cosine algorithm (SCA) [5], JAYA algorithm [6], salp swarm algorithm (SSA) [7], butterfly optimization algorithm (BOA) [8], whale optimization algorithm (WOA) [9]. Except these novel algorithms, researchers worldwide are engaged in improving and modifying the existing algorithms for better performances. Also, these algorithms often found to be applied in several real-world problems. Apart from these modified and improved algorithms, some of the scientists have been trying to hybridize two or more algorithms to introduce balanced algorithms in respect of exploration and exploitation. Some of such improved, modified and hybrid algorithms can be found in [10–13].

In order for the population-based optimization strategy to work, two features, known as exploration and exploitation (or intensification and diversification, respectively), must be present. Generating a wide variety of solutions is what we mean when we talk about diversifying the search space. Intensification means focusing your efforts locally by sifting through the data uncovered by a currently viable solution. Higher exploration causes a disadvantage that the optimization speed is rather slow, and the convergence rate is very low for obtaining globally optimum solution. These two stages must be well-balanced for the final algorithm to be successful.

In this paper, WOA is considered to be studied and analyzed deeply. WOA algorithm was first discovered in 2016 [9]. Social behavior and bubble-net hunting method of whales (humpback whale) in the waters provided inspiration for WOA. The author of the WOA algorithm, showed that WOA acquired achievements that compare well to others in their

Handbook of Whale Optimization Algorithm. https://doi.org/10.1016/B978-0-32-395365-8.00022-1

field with other nature inspired meta-heuristic optimization algorithms. Further, WOA provides competitive results in the different sectors like medical, engineering, chemical science [14,15]. The demerits of WOA are (a) suffered as a result of entrapping at the local optima and (b) Slow convergence rate (c) not goodly maintained balanced between diversification and intensification. To this end, many strategies have been recommended to improve WOA's effectiveness; we'll examine some of these strategies in the next paragraph. By motivating from the literature work on WOA and according to 'no free lunch theorem' [16], Researchers from all over the globe are submitting different metaheuristic algorithms, as well as changes, upgrades, additions, hybrid variations, etc., since no one is suitable for use with an algorithm handling all kinds of optimization issues faced in the real world or in academia.

We have established an improved version algorithm of the WOA in this paper to address the aforementioned WOA shortcomings (F-WOA, in short) with the help of crossover weight and Fibonacci search method. Listed below are brief descriptions of the most significant contributions:

A newly crossover weight technique is embedded into the position update phase of the WOA algorithm, and it improves the capacity for intensification and diversification.

a) The Fibonacci search method is applied after the position update phase of the WOA algorithm which help the solutions of F-WOA converge to the optimal solution effectively.

b) The new F-WOA algorithm is very first evaluated on total set of 14 (fourteen) benchmark functions made up of unimodal and multimodal functions, then it is compared with six well-known metaheuristic algorithms to determine how effective it is. On the IEEE CEC'2019 test suite, the suggested F-WOA is also contrasted with six metaheuristic algorithms.

c) Convergence analysis and statistical experiments including Friedman rank test. Statistical analysis is then used to the results of the position update phase to assess how well the proposed method performed.

d) It has been put to the test in addressing two small-scale engineering design challenges in order to gauge its problem-solving capabilities.

The current article is organized in the given following manner: All the related works on the WOA algorithm are provided in Section 16.2. In Section 16.3, the WOA algorithm is summarized. In Section 16.4, the suggested F-WOA algorithm is displayed. Section 16.5 contains the simulation performance and findings. A brief explanation of benchmark functions, statistical tests, and convergence analysis is provided in Section 16.5 as well. In Section 16.6, real-world applications are demonstrated. Section 16.7 discusses conclusions.

16.2 Literature review

Improved WOA is a new innovative form of WOA developed by Xiong et al. [17] that fixes the problems of the original WOA (IWOA). The authors struck a decent balance between discovery and exploitation by employing a dual-prey method. The IWOA that was proposed has been used to solve the challenge of extracting parameters in a variety of PV models. Sahu et al. [18] was proposed the new novel variant of WOA called modified (MWOA). The author was added that this paper was mainly focused to design a new lead-lag structure for the SSSC-PSS coordinate structure. Tubishat et al. [19] developed a novel variant of WOA by using Elite Opposition-Based Learning (EOBL) and an evolutionary operator from DE. The author was added EOBL at the initial phase of the IWOA and added evolutionary operator of DE in last phase of IWOA. We have employed the heuristic of Information Gain (IG) with Support Vector Machines (SVM) in WOA's feature selection process in order to narrow the search space explored by WOA. Niu et al. [20] had improved upon an existing variant of WOA named SMWOA which is the combination of simplex method and WOA. The author added that the new proposed SMWOA boost the old WOA, make it more robust, and avoid premature convergence. Nonlinear adaptive weight and golden sine operator have been employed by Zhang et al. [21] to create a novel kind of WOA (NGS-WOA). The sine function and the unit circle have a unique relationship that allows for the author to add non-linear adaptive weight, which improves the proposed NGS-ability WOA's to perform global exploration by traversing the sine function in a way that carries out the same action as scanning the unit circle. Another improvement is the use of the golden sine operator integrated with WOA to update the best solution for global exploration and local exploitation. In addition, the author used the recommended NGS-WOA to resolve complex engineering design optimization issues. Du et al. [22] has proposed a new metaheuristic which was a novel variant of WOA called single-dimensional swimming WOA (SWWOA). The author has proposed this SWWOA to overcome the shortcoming. In the initial phase the tent is applied to increase a capability for searching. For the better results in the search space, the author has applied quasi-opposition after every iteration and logarithm function was used to trade a balance between the diversification and intensification and the last improvement the author added was that the new single dimensional was brought from the old WOA update position for further improvement. Heidari et al. [23] has combined the current WOA with the recent variant of hill climbing for better efficient in the exploitation phase called BMWOA and this new algorithm was proposed to conquer the deficit of the original WOA and to give better results in the tackling of the huge range of numerical optimization problems. A new hybrid unique WOA variant called ESSAWOA was developed by Fan et al. [24], which incorporated both SSA and LOBL. For the initial stage of the hybridization, the author used SSA on the fundamental WOA to refresh the population's location. Thirdly, LOBL was utilized to boost the population variety of the recommended optimizer, building on the second phase's introduction of SSA to WOA's surrounding prey and bubble-net attacking phases. It was suggested that this new hybrid strike a middle ground between exploration and exploitation in terms of trade. Liu et al. [25] was proposed modified Holt's exponential smoothing (MHES) algorithm by adding traditional exponential smoothing models with proposed MHES. MHES can predict better results compared to traditional method. Saafan et al. [26] have developed a new hybrid novel variant of WOA called hybrid improved whale optimization salp swarm algorithm (IWOSSA) is the combination of improved WOA (IWOA) and salp swarm algorithm (SSA). The author was used two modifications in the proposed IWOSSA. At the beginning, exponential relationships were used instead of linear relationships in WOA. Secondly, for a more equitable distribution of exploration and exploitation the author used a specific condition to choose one of the algorithms between IWOA and SSA. Fan et al. [27] has proposed a new metaheuristic called joint search mechanisms (JSWOA) by using tent chaotic map in order to maintain the diversity in the initial phase for global search of the improved algorithm and a new adaptive inertia weight is used to jump out of the local optimum and it was used to improve the speed and convergence accuracy. Lakshmi et al. [28] have developed a new hybrid metaheuristic called WOA-TLBO it's the combine of WOA and TLBO. TLBO was used by the author because it can easily jump out of the local optimum, and it has a speed convergence rate, without any efforts for sufficient preliminary constraints. Further, it is used to solve the Facial Emotion Recognition (FER) functional problem. Chakraborty et al. [29] was developed a new variant of WOA called enhanced WOA (WOAmM). The author integrates WOA with the modified SOS to overcome the demerits of WOA. And the modified SOS makes more efficient to search in the globally and avoiding the local trap in the exploitation phase. Chakraborty et al. [30] had developed a new hybrid novel variant of WOA called hunger search based WOA (HSWOA) by combining the Hunger Search Game (HGS) and WOA. The author proposed HSWOA to overcome the drawbacks of the WOA, and two weights are adapted in WOA by using the hunger level to strike a better balance between exploration and exploitation. Nadimi-Shahraki et al. [31] utilizing a pooling mechanism and three efficient search strategies presented a novel metaheuristic dubbed enhanced WOA (E-WOA). It was utilized in medicine to propose a binary E-WOA called BE-WOA to choose efficacy datasets from that field in order to validate the suggested E-WOA and it performed better when compared to the previous WOA. Additionally, the coronavirus disease 2019 (COVID-19) illness is recognized using the BE-WOA. Liu et al. [32] has created a new upgraded version of WOA (EGE-WOA) by the First and Lévy flights. The EGE-WOA incorporates Lévy flights for confined and unconstrained optimization problems to increase the effectiveness of its global exploration. By adding additional convergent dual adaptive weights, the EGE-WOA then enhances its convergence behavior. The EGE-WOA presents a novel technique for determining the condition of whale predation in accordance with the features of sperm whales that hunt by producing high-frequency ultrasound. Sun et al. [33] has developed a new type of metaheuristics called MWOA-CS which is the integration of modified WOA and crisscross optimization algorithm (CS). The author has added this new proposed MWOA-CS to overcome the drawbacks of the old WOA. To strike a better balance between diversification and exploitation, the new improved WOA has chosen the new non-linear convergence factor and nonlinear inertia weight. Liang et al. [34] was developed a new metaheuristic called OCDWOA to overcome the shortcomings of the WOA which easily fall in local optimum and cannot hit the global optimal solutions. The author has added four types of improvement in the WOA. Firstly, the opposition-based learning method was used to improve global search by avoiding the local trapping search, secondly, the nonlinear parameter has a good chance of achieving the procedure begins with widespread exploration and then progresses to exploitation later on. Thirdly, density peak clustering strategy is applied to make a group in small region to increase the variation of the population and differential evolution. Sun et al. [35] had developed a new novel variant of WOA called multi-strategy WOA (MSWOA), this algorithm was proposed to resolve the disadvantages of the WOA. The author was added some modified in the original WOA, Firstly, tent map function is embraced to minimize the optimization and to increase the search efficient capacity. Secondly, nonlinear strategy was used instead of linear coverage factor to trade a good balance between the diversification and intensification. Lastly, optimal feedback strategy is adopted to operate more on exploration phase and to improve the global search ability.

16.3 Whale optimization algorithm (WOA)

A new metaheuristic named WOA which is developed by Mirjalili in 2016 and it is based on the population based (PB) optimization. A global solution may be obtained by avoiding local optima using this swarm-based technique. WOA employs three random search techniques: looking for prey, surrounding prey, and assaulting prey. This can be described in the mathematically model for this WOA are described as follows:

Searching for the prey (exploration): In this strategy, the algorithm is applied to the whale's present location and the prey's known location to conduct a random search for food. This can be expressed mathematically with the following equation:

$$\vec{D} = |\vec{C} \cdot \vec{X}^*(t) - \vec{X}(t)| \tag{16.1}$$

$$\vec{X}(t+1) = \vec{X}^*(t) - \vec{A} \cdot \vec{D} \tag{16.2}$$

In the above equation, $\vec{D}$ be the distance between the present solution and the random one. $\vec{A}$ and $\vec{C}$ are the coefficient vectors, $\vec{X}^*$ vector is randomly chosen as the solution from the current solution and the best solution obtained, and t represent the present iteration number, $\vec{X}$ represent the position vector, here the dot $(\cdot)$ operator represents the multiplication process of the element, the above parameter $\vec{A}$ and $\vec{C}$ are the coefficient vectors and will be calculated as:

$$\vec{A} = 2\vec{a} \cdot \vec{r} - \vec{a} \tag{16.3}$$

$$\vec{C} = 2 \cdot \vec{r} \tag{16.4}$$

where the variable $\vec{a}$ decrease linearly from 2 to 0, and $\vec{r}$ is the random number whose interval is [0, 1].

Encircling the prey: It is similar to the searching for prey, but here for encircling the prey we replaced $\vec{X}^*$ by $\overrightarrow{X_{best}}$ as the best candidate solution and the formulated are as follows:

$$\vec{D} = |\vec{C} \cdot \overrightarrow{X_{best}}(t) - \vec{X}(t)| \tag{16.5}$$

$$\vec{X}(t+1) = \overrightarrow{X_{best}}(t) - \vec{A} \cdot \vec{D} \tag{16.6}$$

In the above equation $\overrightarrow{X_{best}}$ is the best fitness solution till the present iteration.

Attacking the prey (exploitation): Another name of attacking the prey we state as bubble-net attacking tactic, in this method humpback whale used two approaches viz.: 1) shrinking encircling technique (SET), and 2) spiral updating position (SUP).

a. SET: In this technique the variable $\vec{a}$ decrease linearly from 2 to 0 that we have used in Eq. (16.3) over the algorithm iteration, and here it means that the value of $\vec{A}$ vector is a random value over $[-a, a]$.

b. SUP: it is the method where humpback whale uses a spiral shaped path to find their prey, and they can determine the distance between the current whale position and the prey. This method can be mathematically expressed as:

$$\vec{X}(t+1) = \vec{D}' \cdot e^{bl} \cdot \cos(2\pi l) + \vec{X}^*(t) \tag{16.7}$$

In the above equation,
b represents the spiral path of the logarithmic and here b represent constant.
l represents the random number, and the equation that is used to arrive at this value is as follows:

$$l = (a_2 - 1)r + 1 \tag{16.8}$$

In the above equation a_2 decrease linearly from (-1) to (-2) with the iteration and $l \in [-1, 1]$, where $\vec{D}'$ represents the distance of the current whale between the position and the prey and it is formulated as: $\vec{D}' = |\vec{X}^*(t) - \vec{X}(t)|$. After the criss-cross between two phase diversification and intensification of the algorithm and again it is conducted by using of the coefficient parameter $\vec{A}$. Here we get two conditions: 1) the exploration process is selected when $\vec{A} \geq 1$, and the global search is proposed using Eqs. (16.1) and (16.2); 2) if $\vec{A} < 1$, depending on the probability value of a, the whale updates their position using Eqs. (16.6) and (16.7). Here the value of a is fixed as 0.5, and WOA can switch between either by surrounding the target or by using a net as an assault strategy. And their probability is 50%. The mathematically formulated can be represented as:

$$\vec{X}(t+1) = \begin{cases} \vec{X}^*(t) - \vec{A} \cdot \vec{D} & \text{if } \alpha < 0.5 \\ \vec{D}' \cdot e^{bl} \cdot \cos(2\pi l) + \vec{X}^*(t) & \text{if } \alpha \geq 0.5 \end{cases} \tag{16.9}$$

where α is the random number over the interval [0, 1].

16.4 Proposed F-WOA algorithm

In this section, the proposed F-WOA is introduced. The adjustment of crossover weight is discussed in Section 16.4.1 and adaption of Fibonacci search method is introduced in Section 16.4.2.

The two basic components of every generalized optimization technique are diversification and intensification. Diversification is the process of searching the whole search space, or doing a global search. Exploitation, on the other hand, is characterized as local search, which entails studying the limited portions of the large search space. The local minima stagnation problem (caused by exploitation) is avoided by the algorithm thanks to these events, which also improve convergence and variety of the solutions (from exploration). The harmony between these two occurrences is still another crucial component. Any algorithm that can accomplish these three goals can be categorized as a member of the family of cutting-edge algorithms. When compared to other cutting-edge algorithms, WOA is a very competitive optimization technique. According to the research, it has drawbacks including slow convergence, low solution accuracy, a lack of variation, as well as a tendency to stabilize at a local optimum, which is particularly problematic when dealing with complicated parameter optimization issues. Additionally, the algorithm may struggle to keep a balance between the two if it focuses on exploration rather than exploitation.

In this current study, a revised old version of the WOA algorithm is used called F-WOA is developed by combining crossover weight approach with Fibonacci search-based method in order to address the aforementioned problems. The details are discussed in the following subsections.

16.4.1 Adaption of crossover weight

In WOA, the current solution is updated using an alternative solution drawn at random from the whole population during the diversification phase (search prey phase). Due to the possibility of picking a solution that is close by, blind solution selection during exploration might diminish the variety of solutions. To an equilibrium of global and local search, we have added a nonlinear crossover weight in the position update phase of the WOA. The new updated equation is as follows:

$$\vec{X}(t+1) = \begin{cases} (\vec{X}^*(t) - \vec{A} \cdot \vec{D}).CR1 & \text{if } \alpha < 0.5 \\ \vec{D}' \cdot e^{bl} \cdot \cos(2\pi l).CR2 + \vec{X}^*(t).(1 - \text{CR2}) & \text{if } \alpha \geq 0.5 \end{cases} \tag{16.10}$$

where, $CR1 = exp(tan(rand(1, N)))$; and N is referred as size of population
$CR2 = 3.(0.5 - rand(1, N)).f$; and N is referred as size of population
$f = exp(-(Iteration/Max\_iteration))$

16.4.2 Improved solution technique

To give the better efficient quality of the solution as a whole and to avoid a stalemate in the pursuit of the local optimum solution in the proposed F-WOA, we have added an improve solution technique in WOA algorithm by using Fibonacci search method and average of three random solutions.

Fibonacci search method (FSM):

The FSM is a mathematical process that shifts and narrows down the search range by using Fibonacci numbers to obtain the extreme value of unimodal functions. The optimal point is always contained within the range being narrowed. Shifting can take place in both directions. The direction of the change is determined by the function values of that two different points in time different experiment points. The Fibonacci numbers, which can be understood by referring to the following definition, serve as the basis for the FSM.

$Fib = [F_1, F_2, F_3, \ldots, F_n]$, where, $F_i \; \forall \, i = 1, 2, \ldots n$ are Fibonacci numbers and generated by following equation.

$$F_0 = 1 = F_1, F_{no} = F_{no-1} + F_{no-2}, no = 2, 3, 4, \ldots .n \tag{16.11}$$

Let t_1 and t_2 be two experimental numbers of any finite length of interval with upper and lower limits as UL and LL, respectively. Calculate two initial points

$$t_1 = LL + \left(\frac{F_{no-2}}{F_{no}}\right)(UL - LL), t_2 = UL - \left(\frac{F_{no-2}}{F_{no}}\right)(UL - LL) \tag{16.12}$$

If the value of the function at t_2 is larger than that at t_1, then the range shifts to the right; however, if the given value of the function at t_1 is larger than that at t_2, then the range shifts to the left. The new values of t_3 and t_4 are computed by applying the Fibonacci search algorithm, which states that t_3 equals t_1 and t_4 equals t_2. If two functional values are not equal, then only one of them either t_3 or t_4 will be regarded as a totally a new experimental point. On the other hand, the other will be treated the same as either t_1 or t_2 depending on the direction in which the contracting is taking place. Both t_3 and t_4 are considered to be new experimental points whenever there are two function values that are identical to one another. A modified version of the Fibonacci search approach was utilized by the author of [36] for the partially shaded solar PV array. This was done to take advantage of the great computing efficiency of FSM. Recent research conducted by Yazici et al. [37] used an enhanced version of the Fibonacci search approach to investigate wind energy conversion systems. After the position update phase of WOA, the notion of the Fibonacci search technique was incorporated into the algorithm. In order to discover the new solution (X New) that will enhance the solution quality and prevent stagnation at the local optimal, the following method (Algorithm 16.1) is put into practice:

Algorithm 16.1 Process of obtaining new solution by Fibonacci search method.

If rand < 0.5
  $X_{New} = X_{avg}$
  Where, $X_{avg} = \frac{x_1+x_2+x_3}{3}$ and x_1, x_2 and x_3 are three random numbers from the population.
else
  Compute Fibonacci Search method using Eqs. (16.11)–(16.12)
End

The F-WOA algorithm's pseudocode is presented in Algorithm 16.2, and the specific stages are outlined below:

1st step: Put all of the parameters, including the maximum number of iterations, the number of populations, and the function evaluation, in a random order.
2nd step: Initializing population randomly.
3rd step: Update all required parameters.
4th step: Use Eq. (16.8) to update the number of flames after sorting the moth matrix and flame matrix according to fitness value.
5th step: Update whale position using Eq. (16.10).
6th step: Apply Fibonacci search process by the help of Algorithm 16.2 and discover the functional significance value of the new approaches, then update the solution.
7th step: If the stopping requirements are not met, move on to step two to obtain the greatest fitness value.

16.5 Simulation study and analysis

The benchmark functions are mathematical functions used to evaluate and compare the performance of optimization algorithms. Unimodal benchmark functions have a single optimal solution, while multimodal benchmark functions have multiple optimal solutions. And in this section it will be explained in a brief way. Interpretations of data obtained through these benchmark functions can provide valuable insights into the strengths and weaknesses of optimization algorithms, as well as the impact of different parameters on their performance. For example, an algorithm that performs well on unimodal functions but poorly on multimodal functions may indicate a limitation in its ability to handle complex, multi-peaked optimization problems. Section 16.5.1 delves deeply into the Benchmark features. The experimental setting for our suggested technique is explained in Section 16.5.2. In Section 16.5.3, a comparison of F-WOA with basic well-known optimization algorithms is presented. Statistical analysis and convergence performance functions are presented in Section 16.5.4. Experimental analysis of on IEEE CEC 2019 test problems is discussed in Section 16.5.5.

Algorithm 16.2 Pseudocode of F-WOA algorithm.

```
Input: Initialize the population Xi (i = 1, 2, ..., n), Objective function f(X), X = (X1X2...Xdim), Maximum iteration
(M_Itr), and other related parameters are determined;
Evaluate fitness value of each solution in the population;
Find the best solution BESTsol
while (Itr < M_Itr)
  Update a, A, C, l, and p for each individual;
        if1 (p<0.5)
            if2 (|A|< 1)
              update the current search agent's position using Eq. (16.1);
            else if2 (|A≥ 1|
              Apply the first part of Eq. (16.10) to update the position of the current search space;
            end if2
          elseif1 (p≥0.5)
        Apply the second part of the Eq. (16.10) to Update the position of the current search space;
    end if1
  end for
  Apply Algorithm 16.2;
  Verify and correct any search agents that stray outside the search area;
  Determine each search agent's fitness;
  Update BESTsol if there is a better solution;
  Current_iter = Current_iter + 1
end while
Output: return BESTsol
```

16.5.1 Benchmark functions

Benchmark functions are very important to verify the performance of the algorithms. The proposed Fibonacci WOA is tested to verify on different types of problems, 14 benchmark functions were taken and divided into the two categories, *viz.*, unimodal benchmark functions (F1 to F7), and multimodal benchmark functions (F8 to F14). And details are presented in Appendix 16.A in Table 16.A.1.

16.5.2 Experimental setup

The technique outlined in this paper is examined in this section by contrasting the evaluated outcomes with DE, PSO, JAYA, SSA, SCA, and WOA. To analyze the data, MATLAB® 2015a, Windows 10 (64-bit) is used, along with an Intel Core i5 CPU and 8 gigabytes of random-access memory. The efficiency of the novel technique is analyzed by using ten composite benchmark functions, fourteen unimodal benchmark functions, fourteen multimodal benchmark functions, and ten composite benchmark functions from the IEEE CEC 2019 test suites. A population size of fifty, with a maximum of four hundred and fifty iterations are used to test each algorithm. It is appropriate to compare the optimized outputs using the same benchmarks as before because the suggested technique is an improved version of WOA. To determine an average (A_{VRG}) and standard deviation, 30 runs were used (ST_D). Table 16.1 displays the parameter information for the algorithms under consideration.

16.5.3 Results and discussion on classical benchmark functions

In this part of section, we compare our proposed F-WOA's simulation results to those obtained using six (6) alternative meta-heuristics: WOA DE, PSO, SCA, JAYA, and SSA. 14 standard-setting operations, both unimodal and multimodal, have been analyzed.

The mean and standard deviation for optimized unimodal functions are shown in Table 16.2, F-WOA, and the other six techniques. The table makes it clear that, in comparison to other algorithms, F-WOA offered the fewest values. Table 16.2 shows the results of an investigation into the unimodal function optimization of the functions F1 to F7. Here the best quality

TABLE 16.1 The values of the parameters of the considered algorithms.

Algorithm	Parameters values
DE	Scaling Factor $(F) = 0.5 =$ Crossover probability
PSO	Inertia weight $(w) = 0.9$ to 0.4, Acceleration coefficient $c_1 = c_2 = 0.2$, Maximum velocity $(V_{max} = 4)$
SSA	Two random numbers (c_1 and c_2) where $0 \le c_1, c_2 \le 1$
SCA	Random numbers ($r_2 = rand * 2\pi$, $r_3 = rand * 2$, and $r_4 = rand$) and constant $a = 2$
JAYA	$r = rand\ (0, 1)$
WOA	Mix rate=1, two parameters 'a' and 'b' are uniformly random numbers between 0 and 1

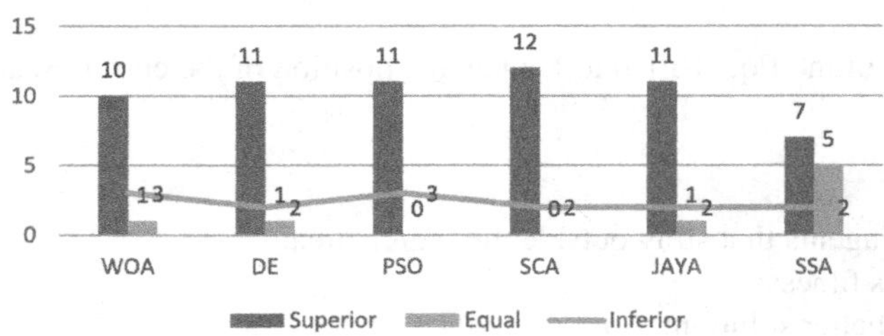

FIGURE 16.1 Performance comparison of F-WOA and other basic algorithms over basic benchmark functions.

outcomes for the functions F1 to F4, and F7 come from the suggested F-WOA algorithm. It provides results for functions F2 and F5 that are second-best. Table 16.2 also includes an investigation of functions F8 to F14 under multimodal function optimization. For the functions F8, F12, F13, and F14, it is obvious that F-WOA produces better results than alternative techniques. It offers the second- and third-best results possible for functions F9 and F10, respectively, and it performs worse than other methods for functions F14. As a result, it can be said that F-WOA is a better algorithm than the other six methods for optimizing multimodal functions.

Table 16.3 displays the frequency with which the performance of the F-mean WOA is better to, equivalent to, or worse than the comparison to the other six techniques. According to Table 16.3, it is abundantly obvious that F-WOA obtains greater performance levels than WOA, DE, PSO, SCA, JAYA, and BOA in the corresponding amounts of 10, 11, 12, 11, and 7 of the benchmark functions. The first, first, zero, and fifth benchmark functions, in that order, each include results that are comparable to one another. Fig. 16.1 illustrates the outcomes that are superior, comparable, and inferior respectively.

16.5.4 Discussion on statistical and convergence performance for basic benchmark problems

In recent years, computational approaches have increasingly used statistical tests as a standard tool to enhance performance evaluation [11,12]. Usually, they are employed in experimental studies to determine whether there is a certain algorithm that is better than others. This effort, which might be challenging, is required to ascertain whether a novel proposal for a solution significantly outperforms current approaches to a particular problem. The efficient functioning of the suggested F-WOA algorithm is evaluated in this study using Friedman rank tests and results were presented in this work using IBM-SPSS software.

Table 16.4 shows results from a Friedman-Rank test with 95% confidence that F-performance WOA's is substantial as well as being suitable with popular algorithms such as WOA, DE, PSO, SCA, JAYA, and SSA for a total of 14 benchmark functions. Table 16.4 shows this algorithm's mean level is lower than others. Fig. 16.2 shows Friedman rank test of F-WOA and other modern algorithms.

Some of the convergence graphs for benchmark functions like F1 (Sphere), F2 (Booth function), F6 (Lean function), F10 (Levy function), F12 (Alpine function), and F13 (Schaffer Function) were compared with DE, PSO, SCA, JAYA, SSA, and WOA in Fig. 16.3 to assess how quickly F-WOA converges in comparison to other algorithms. The vertical and horizontal

TABLE 16.2 Simulation of F-WOA with six traditional algorithms over basic benchmark problems.

Sl. No.	Performance measure	F-WOA	WOA	DE	PSO	SCA	JAYA	SSA
F1	A_{VRG}	1.32e-03	3.56E-01	7.49E-03	2.41E-03	7.24e-07	0	3.10E-02
	ST_D	1.34e-02	5.29E-01	1.32E-03	1.54E-02	1.22e-06	0	3.44E-02
F2	A_{VRG}	1.21e-04	1.88E-01	7.30E-05	2.20E-02	5.29e-02	0	2.05E-03
	ST_D	2.44e-04	2.08E-01	6.57E-05	1.47E-01	1.03e-01	0	2.12E-03
F3	A_{VRG}	0	2.10E-10	1.31E-04	2.87E-05	1.71e-187	8.45E-50	0
	ST_D	0	3.28E-11	1.21E-04	1.65E-03	0	6.01E-45	0
F4	A_{VRG}	0	4.72E-12	1.27E-08	4.40E-78	5.71e-196	2.55E-108	0
	ST_D	0	3.73E-10	4.74E-08	1.64E-74	0	0	0
F5	A_{VRG}	-3.79e-02	-3.79E-03	-3.79E-03	-3.79e-03	-3.79e-03	-3.79E-03	-3.30E-02
	ST_D	5.75e-01	4.11E-05	5.23E-06	4.10E-06	2.35e-08	3.37E-09	4.59E-06
F6	A_{VRG}	2.69e-07	1.73E-01	2.44E-02	6.62E-04	5.88e-03	3.64E-02	3.18E-04
	ST_D	4.85e-07	2.61E-01	2.29E-02	1.33E-03	1.42e-02	2.68E-02	2.84E-03
F7	A_{VRG}	0	7.47E-88	9.74E-15	5.50E-03	2.93e-03	1.14E-100	0
	ST_D	0	4.29E-77	3.78E-14	1.59E-03	1.35e-02	3.71E-112	0
F8	A_{VRG}	0	0	0	4.55E-10	8.16e-02	0	0
	ST_D	0	0	0	0	1.63e-01	0	0
F9	A_{VRG}	7.54e-01	6.26E-02	3.19E-15	2.39E-01	4.39e-03	3.60E-32	2.17E-02
	ST_D	6.65e-01	1.23E-01	5.47E-12	3.41E-03	1.40e-02	2.11E-47	3.24E-02
F10	A_{VRG}	2.55e-05	3.23E-50	3.22E-05	1.42E-10	1.55e-04	1.50E-09	0
	ST_D	2.22e-04	1.61E-50	2.10E-05	8.77E-12	7.49e-04	2.08E-07	0
F11	A_{VRG}	-4.50e-11	-1.17E-10	-1.12E-10	-7.67E-11	-1.14e-10	-1.10E-09	-8.40E-10
	ST_D	2.68e-11	1.23E-14	8.51E-14	4.41E-14	2.53e-13	1.56E-20	4.17E-10
F12	A_{VRG}	0	4.58E-29	2.45E-05	2.56E-10	1.00e-08	4.17E-25	0
	ST_D	0	2.69E-28	1.60E-05	3.13E-07	2.15e-08	9.74E-15	0
F13	A_{VRG}	7.12	1.61E+01	3.51	3.64	9.93	3.01	1.45E+02
	ST_D	1.44e+01	2.21E+01	2.62E-02	2.12E-01	2.86	1.03E-06	9.51
F14	A_{VRG}	4.21e-02	1.34E-01	2.93E-01	7.19E-01	1.59e-01	5.48E-02	7.31E-02
	ST_D	3.24e-01	1.24E-01	1.68E-01	5.43E-01	2.10e-01	3.08E-02	6.54E-02

TABLE 16.3 Performance valuation of F-WOA with six traditional methods.

	WOA	DE	PSO	SCA	JAYA	SSA
Superior to	10	11	11	12	11	7
Similar to	1	1	0	0	1	5
Inferior to	3	2	3	2	2	2

TABLE 16.4 Friedman Rank Test on F-WOA for basic benchmark function.

Algorithm	Mean rank	Rank
F-WOA	**2.98**	**1**
SSA	3.48	2
DE	4.51	3
SCA	5.17	4
PSO	5.44	5
JAYA	5.70	6

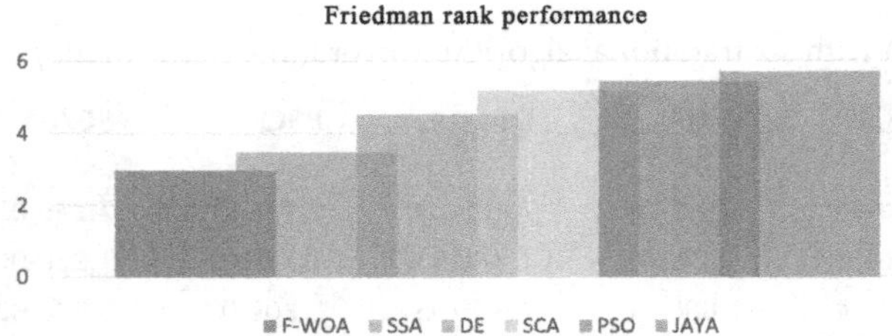

FIGURE 16.2 Friedman rank of F-WOA with considered algorithms over basic benchmark functions.

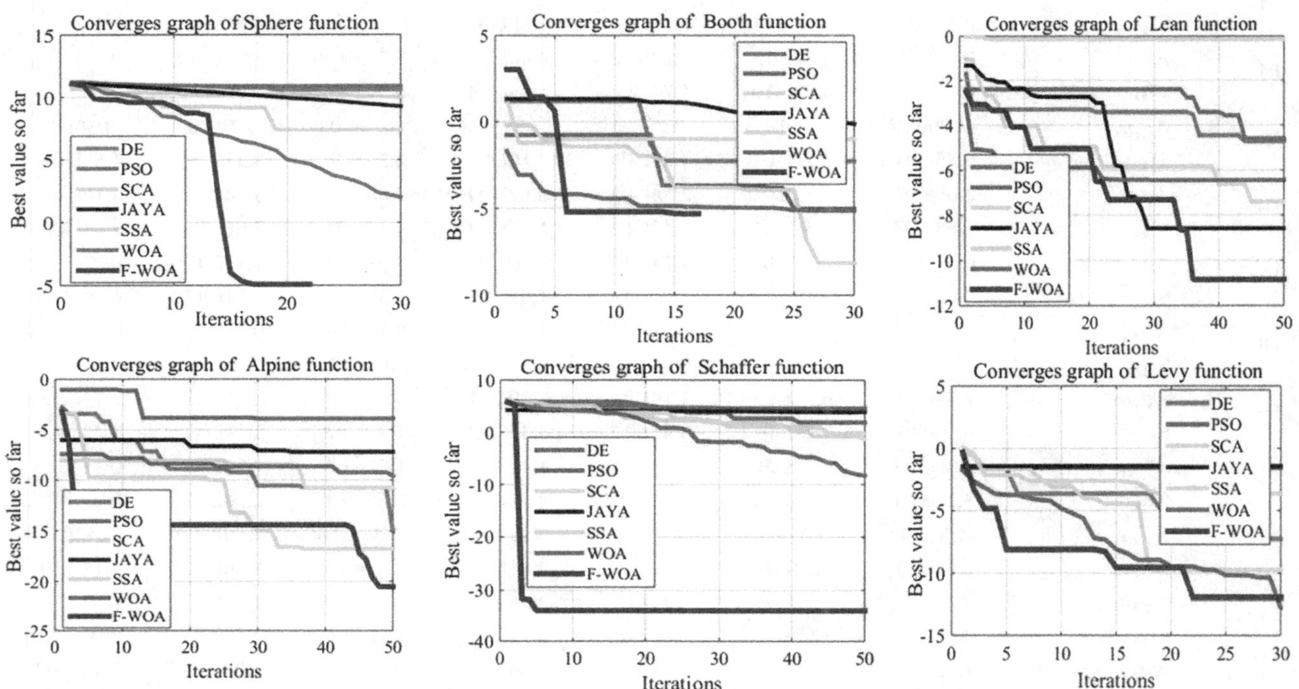

FIGURE 16.3 Convergence of F-WOA with other considered algorithms for different benchmark functions.

axes in these graphs display the function evaluation and the number of iterations, respectively. To make the curves easier to see, they were drawn at a scale of 100. In terms of the worldwide optimal, F-WOA converges more quickly than the other methods. On the other hand, other compared optimization algorithms have a moderate convergence rate because they get stuck in neighborhood optima.

16.5.5 Results and discussions on IEEE CEC 2019 benchmark function

In this study, we focus on the ten composite benchmark functions (SD1 - SD10) discussed in the IEEE CEC 2019 special session [38]. The complexity of those functions makes them better measures of performance than the simpler first 14 benchmark functions. The experiment compares F-WOA against the regular WOA as well as five additional Metaheuristics Algorithms which is in the terms of mean and standard deviation. Experiment findings are based on 30 separate trials, size of the population is 50, and the maximum iterations number is 30000.

16.5.5.1 Comparison of F-WOA over IEEE CEC 2019 benchmark problems

In this study, the functions SD1 through SD10 are solved using the suggested F-WOA method, and its performance is compared to that of DE, SOS, JAYA, BOA, WOA, and SCA. Table 16.5 displays a comparison of the mean (A_{VRG}) and standard deviation (STD) of ten benchmark functions across all of the algorithms.

From Table 16.5, we get some results are better than F-WOA, some are similar and very few Benchmark functions perform worse result. 8 Benchmark functions perform better results in F-WOA comparing with SCA, 2 are similar, where

TABLE 16.5 Comparison with other Swarm Intelligence Algorithms.

Algorithm	SD1		SD2		SD3		SD4		SD5	
	AVRG	STD	*AVRG*	STD	*AVRG*	STD	*AVRG*	STD	*AVRG*	STD
F-WOA	1	0	5	0	8.40	1.59	3.34e+01	3.39e+01	1.64	1.68
SCA	1	0	5	0	9.62	9.60e-01	1.05e+02	1.56e+01	1.00e+02	2.59e+01
DE	1.82e+06	8.25e+05	1.93e+03	3.40e+02	4.58	6.11e-01	8.54	1.99	1.77	2.64e-02
SOS	4.75e+03	6.88e+03	2.65e+02	1.96e+02	4.61	9.78e-01	2.08e+72	3.42e+72	1.19e+73	4.40e+73
JAYA	6.53e+06	2.83e+06	4.68e+03	7.78e+02	8.09	7.57e-01	4.10e+01	4.90	2.60	1.67e-01
BOA	1	0	5	0	6.79	1.15	1.26e+02	2.15e+01	1.09e+02	2.33e+01
WOA	1.45E+08	1.17E+08	1.10E+04	36E+03	8.16	1.60	6.62E+01	2.40E+01	3.83	1.43

Algorithm	SD6		SD7		SD8		SD9		SD10	
	AVRG	STD	*AVRG*	STD	*AVRG*	STD	*AVRG*	STD	*AVRG*	STD
F-WOA	3.07	1.21	8.22e+02	2.23e+02	4.25	2.41e-01	1.26	9.43e-02	2.15e+01	1.75e-01
SCA	1.21e+01	1.09	2.29e+03	2.51e+02	5.13	1.79e-01	3.86	6.23e-01	2.16e+01	1.43e-01
DE	1.11	1.48e-01	1.07e+03	1.24e+02	4.48	2.71e-01	1.29	4.20e-02	2.25e+01	2.52
SOS	2.15	2.22e-01	4.55e+71	4.09e+71	5.50	0	1.59e+71	2.34e+71	2.18e+01	8.74e04
JAYA	5.90	9.23e-01	1.51e+03	2.09e+02	4.28	2.08e-01	1.56	1.11e-01	2.16e+01	1.27e-01
BOA	1.36e+01	1.15	2.07e+03	2.89e+02	5.12	2.83e-01	4.14	5.51e-01	2.16e+01	1.85e-01
WOA	9.94	1.84	1.49E+03	2.85E+02	4.73	3.09E-01	1.48	1.69E-01	21.46	1.12E-01

TABLE 16.6 Performance assessment of F-WOA with compared algorithms on IEEE CEC-2019 functions.

	SCA	DE	SOS	JAYA	BOA	WOA
Superior to	8	7	8	9	7	8
Similar to	2	0	0	0	2	0
Inferior to	0	3	2	1	1	2

no benchmark functions show worse result.7 Benchmark functions perform better results in F-WOA comparing with DE, no one is similar, where 3 benchmark functions show worse result. 8 Benchmark functions perform better results in F-WOA comparing with SOS, no one is similar, where 2 benchmark functions show worse result. 9 Benchmark functions perform better results in F-WOA comparing with JAYA, no one shows similar, where 1 benchmark functions show worse result. 7 Benchmark functions perform better results in F-WOA comparing with BOA, 2 is similar, where 1 benchmark functions show worse result. 8 Benchmark functions perform better results in F-WOA comparing with WOA, 2 is similar, where no benchmark functions show worse result. Simulation Summary of results is shown in Table 16.6. The superior, similar and inferior results of the proposed F-WOA and other traditional optimization algorithms are illustrated in Fig. 16.4.

16.5.5.2 *Statistical performance of F-WOA over IEEE CEC 2019 test problems*

For statistical measurement, the Friedman rank test is conducted in this paper. The results are analyzed by Friedman test in terms of rank and average rank. Comparative study on rank and mean rank of F-WOA with other 6 metaheuristic algorithms which are tested on IBM-SPSS software. In addition to this, it can be observed from the average rank of the Friedman test in Table 16.7 that the F-WOA has achieved best rank as compared to DE, SOS, JAYA, BOA, WOA, and SCA.

16.5.6 Computational complexity of F-WOA algorithm

For a better evaluation of population-based algorithms, not only the solution quality but also the computational times should be investigated. Complexity of any algorithm is a function which provides the running time or space with respect to

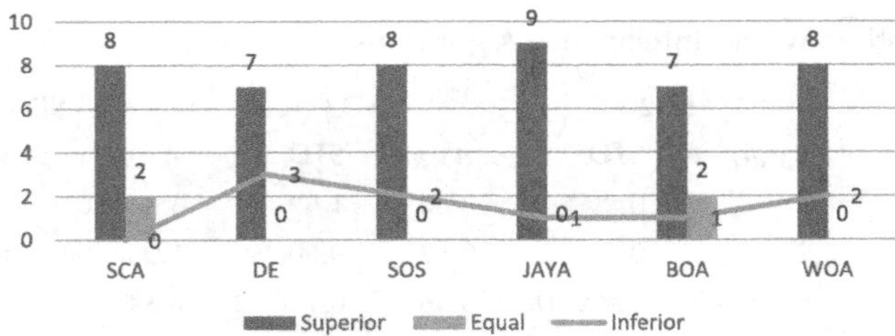

FIGURE 16.4 Performance comparison of F-WOA and compared algorithms over IEEE CEC 2019 functions.

TABLE 16.7 Friedman Rank Test of F-WOA and compared algorithms IEEE CEC 2019 benchmark functions.

Algorithm	Mean rank	Rank
F-WOA	**3.40**	**1**
BOA	4.08	3
SCA	4.31	4
DE	3.69	2
SOS	4.66	6
JAYA	5.01	7
WOA	4.51	5

input size. The process of finding a formula for total time required for successful execution of algorithm is known as time complexity. A big-O notation is used to analysis the computational complexity of the proposed F-WOA algorithm. The Complexity of F-WOA also depends on initialization of whale position (T_{IWP}), evaluation of whale position by exploration and exploitation operator (T_{EWP}), and update phase by improved solution technique (T_{UFIST}). Let maximum iterate number, variable number and moths' number are denoted by I, D, and N respectively. Here we will use time complexity for the comparison of both F-WOA and WOA algorithm. According to the quicksort algorithm, Computational complexity towards worst and best case is

$$\begin{aligned} T_{F-WOA} &= T_{IWP} + T_{EWP} + T_{UFIST} \\ &= O(2N) + O(N^2) + O(2N)I + O(N^2)I + O(N)I \end{aligned}$$

Further, from [32], the time complexity of MFO for the worst case is $O(2N) + O(N^2) + O(2N)I + O(N^2)I$. Therefore, the time complexity for F-WOA is greater than WOA. It is true that the complexity of computation has increased, but the increase is negligible in light of the enhancement.

16.6 Engineering design problems

In the part of this section, we apply the suggested F-WOA to the solution of two limited real-world engineering problems: the design considerations for both a tension-compression spring and a cantilever beam were considered. The appendix contains the mathematical formulation of all of the engineering design difficulties that have been encountered and each engineering problem features many restrictions of varying types. Constraints used to regulate these sorts of issues are typically dealt with using the death penalty functions approach, which is both straightforward and widely utilized. The maximum number of iterations for each, together with the population size of engineering problem is 500 and 30, respectively.

16.6.1 Tension-compression spring design (TSD) problem

Three restrictions in the widely are studied engineering design issue known as the Tension-Compression Spring Design (TSD): the wire diameter as (d), the number of active coils (N), and the mean coil diameter (D). Minimizing the cost to

FIGURE 16.5 Image of TSD problem.

TABLE 16.8 Comparison analysis of F-WOA with other algorithms for TSD problem.

Algorithm	Optimal variables			Optimal weight
	D	D	N	
F-WOA	**0.05100**	**0.54751**	**2.00000**	**0.0050716**
DE	0.051609	0.354714	11.410831	0.0126702
HS	0.051154	0.349871	12.076432	0.0126706
WOA	0.0518957	0.36410931	10.868421862	0.0126659
PSO	0.051728	0.357644	11.244543	0.0126747
ES	0.051989	0.363965	10.890522	0.0126810
GA	0.051480	0.351661	11.632201	0.0127048
Constraint correction	0.050000	0.315900	14.250000	0.0128334
Mathematical optimization	0.053396	0.399180	9.1854000	0.0128334

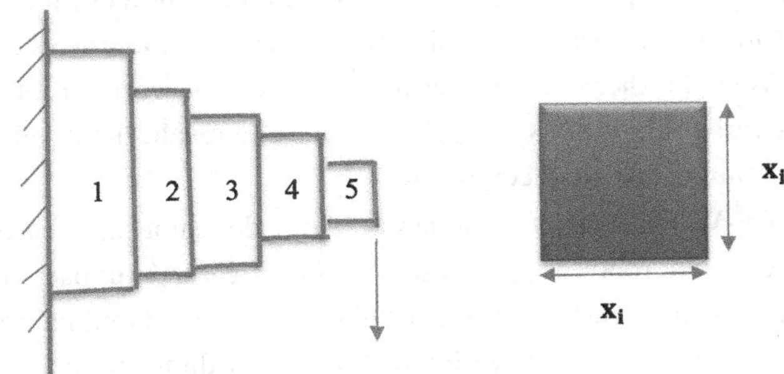

FIGURE 16.6 Cantilever beam design problem.

manufacture the spring while taking into account the aforementioned three structural factors is the target of this problem. Appendix 16.B and Fig. 16.5 present the mathematical expression and its diagram, respectively. Our proposed F-WOA algorithm solves this design problem, Table 16.8 presents the findings of the analysis together with those of the aforementioned techniques. It is clear from Table 16.8 that F-WOA is the most effective algorithm.

16.6.2 Cantilever beam design (CBD) problem

The F-WOA algorithm is applied so as to resolve the CBD problem and is then compared to a number of other metaheuristic algorithms that have been published in [10]. There are five square, hollow blocks in this problem, one constraint, and five parameters. To further understand the cantilever beam design problem, please refer to Appendix 16.C and Fig. 16.6. The top F-WOA values from the preceding methods are shown in Table 16.9, and from Table 16.9, F-WOA is the best of the bunch.

TABLE 16.9 Comparison analysis of F-WOA with other methods for CBD problem.

Algorithm	Optimal variables					Optimal weight
	x_1	x_2	x_3	x_4	x_5	
F-WOA	**5.97322**	**5.27141**	**4.46236**	**3.47657**	**2.13735**	**1.33159**
WOA	5.2982	5.6551	4.3305	3.6671	2.8271	1.3676
ALO	6.01812	5.31142	4.48836	3.49751	2.158329	1.33995
DE	5.5091	4.8861	4.2490	4.6868	3.0253	1.3761
SOS	6.01878	5.30344	4.49587	3.49896	2.15564	1.33996
CS	6.0089	5.3049	4.5023	3.5077	2.1504	1.33999
MMA	6.0100	5.3000	4.4900	3.4900	2.1500	1.3400
PSO	5.3791	6.9067	3.6337	4.9205	2.2560	1.37036
GCA-I	6.0100	5.30400	4.4900	3.4980	2.1500	1.3400
SSA	6.0783	4.0783	4.8385	5.7777	2.4201	1.5275
GCA-II	6.0100	5.3000	4.4900	3.4900	2.1500	1.3400

16.7 Conclusions and future extensions

In this study, we introduce F-WOA, a new and improved WOA algorithm. Here, we incorporate a novel crossover operator and the Fibonacci method's guiding principle into the WOA algorithm, helping the participants in F-WOA find the optimal solution and find a happy medium between broad exploration and targeted application. It has been evaluated using six cutting-edge metaheuristic algorithms on the IEEE CEC 2019 benchmark functions and fourteen traditional benchmark functions. In addition, the Friedman rank test was used to statistically evaluate F-WOA for significance. The recommended F-WOA has also been subjected to a convergence analysis. Additionally, two real-world optimization issues have been resolved using the suggested F-WOA. Furthermore, the results for these real-world problems show how much better the newly developed F-WOA is than earlier algorithms in the literature. As a result, the developed F-WOA is better than the state-of-the-art meta-heuristic algorithms that were compared.

Apart from the advantages of F-WOA, it has few demerits like it does not produce better optimal results for large number of constraints type problems like robot gripper design problem, car side impact design problems which leads to further enhancements of F-WOA algorithm. Moreover, it does not produce good optimal results as compared to the other new efficient optimization algorithms and improved version of MFO algorithms, these aids help researchers to developed new modification in F-WOA algorithm for more good solutions with extreme high coverage and convergence rate.

The F-WOA algorithm that was proposed, which builds on the present method and may be expanded to accommodate optimization of several or more objectives, could be used to investigate real-world optimization problems such as routing of vehicle, job shop planning, workflow planning, parameter estimation of the fuel cell problem, combined economic and emission dispatch problem, COVID-19 CT image segmentation problem, aircraft design problem, feature selection problems etc. Further, it can be used to solve higher dimensional optimization problems, used for theoretical investigation like convergence analysis with new statistical investigation.

Compliance with ethical standards

Conflict of interest: The authors of this study have stated unequivocally that they do not have any known conflicts of interest, either financial or interpersonal, that may be interpreted as having an effect on the research that was provided in this study.

Ethical approval: There is not a single study by any of the authors of this piece that used either people or animals as research subjects

Appendix 16.A Formulation of 14 benchmark functions

TABLE 16.A.1 Formulation of 14 benchmark functions.

S/N	Function	Formulation of objective functions	d	F_{min}	Search space
Unimodal Benchmark Functions					
F1	Sphere	$f(x) = \sum_{j=1}^{d} x_j^2$	30	0	[-100, 100]
F2	Booth	$f(x) = (2x_1 + x_2 - 5)^2 + (x_1 + 2x_2 - 7)^2$	2	0	[-10, 10]
F3	Matyas	$f(x) = 0.26\left(x_1^2 + x_2^2\right) - 0.48x_1x_2$	2	0	[-10, 10]
F4	Sumsquare	$f(x) = \sum_{i=1}^{D} x_i^2 \times i$	30	0	[-10, 10]
F5	Zettl	$f(x) = \left(x - 1^2 + x - 2^2 - 2x_1\right)^2 + 0.25x_1$	2	-0.00379	[-1, 5]
F6	Leon	$f(x) = 100\left(x_2 - x_1^3\right)^2 + (1 - x_1)^2$	2	0	[-1.2, 1.2]
F7	Zakhrov	$f(x) = \sum_{j=1}^{d} x_i^2 + \left(0.5\sum_{j=1}^{d} jx_j\right)^2 + \left(0.5\sum_{j=1}^{d} jx_j\right)^4$	2	0	[-5, 10]
Multimodal Benchmark Functions					
F8	Bohachevsky	$f(x) = x_1^2 + 2x_2^2 - 0.3\cos(3\pi x_1) - 0.3$	2	0	[-100, 100]
F9	Bohachevsky 3	$f(x) = x_1^2 + 2x_2^2 - 0.3\cos(3\pi x_1) - 0.3$	2	0	[-50, 50]
F10	Levy	$f(x) = sin^2(\pi x_1) + \sum_{i=1}^{D-1}(x_i - 1)^2\left[1 + 10sin^2(\pi x_i + 1)\right] + (x_D - 1)^2\left[1 + sin^2(2\pi x_D)\right]$ Where, $x_i = 1 + \frac{1}{4}(x_i - 1),\ i = 1, 2, \ldots\ldots\ldots D$	30	0	[-10, 10]
F11	Michalewicz	$f(x) = -\sum_{i=1}^{D} \sin(x_i) sin^{2m}(\frac{ix_i^2}{\pi})$, m=10	10	-9.66015	[0, π]
F12	Alpine	$f(x) = \sum_{i=1}^{D} \lvert x_i \sin(x_i) + 0.1x_i \rvert$	30	0	[-10, 10]
F13	Schaffers	$f(x) = 0.5 + \frac{sin^2\left(x_1^2 + x_2^2\right) - 0.5}{\left[1 + 0.001\left(x_1^2 + x_2^2\right)\right]^2}$	2	0	[-100, 100]
F14	Powersum	$f(x) = \sum_{i=1}^{D}\left[\left(\sum_{k=1}^{D}(x_k^i) - b_i\right)^2\right]$	30	0	[-10, 10]

Appendix 16.B Tension/compression spring design problem

Let us consider

$$\vec{x} = [x_1\ x_2\ x_3] = [d\ D\ N],$$

Minimize $f(\vec{x}) = x_1^2 x_2 (2 + x_3)$,
Subjected to

$$g_1(\vec{x}) = 71785x_1^4 - x_2^3 x_3 \le 0$$
$$g_2(\vec{x}) = \frac{4x_2^2 - x_1x_2}{12566\left(x_2x_1^3 - x_1^4\right)} - \frac{1}{5108x_1^2} \le 0$$
$$g_3(\vec{x}) = x_2^2 x_3 - 140.45x_1 \le 0$$
$$g_4(\vec{x}) = x_1 + x_2 - 1.5 \le 0$$

where, $0.05 \le x_1 \le 2.00$, $0.25 \le x_2 \le 1.30$, $2 \le x_3 \le 15.0$

Appendix 16.C Cantilever beam design problem

Let us consider

$$\vec{x} = [x_1\ x_2\ x_3\ x_4\ x_5] = [d\ D\ N],$$

Minimize $f(\vec{x}) = 0.6224\,(x_1 + x_2 + x_3 + x_4 + x_5)$,
Subjected to $g(\vec{x}) = \frac{61}{x_1^3} + \frac{27}{x_2^3} + \frac{19}{x_3^3} + \frac{7}{x_4^3} + \frac{1}{x_5^3}71785x_1^4 - x_2^3x_3 \leq 0$
where, $0.01 \leq x_1, x_2, x_3x_4, x_5 \leq 100$

References

[1] D.K. Pratihar, Soft Computing, Narosa Publishing House, New Delhi, 2008.

[2] K.V. Price, Differential evolution, in: Handbook of Optimization, Springer, Berlin, Heidelberg, 2013, pp. 187–214.

[3] J. Kennedy, R. Eberhart, Particle swarm optimization, in: Proceedings of ICNN'95. International Conference on Neural Networks, vol. 4, 1995, pp. 1942–1948.

[4] M.Y. Cheng, D. Prayogo, Symbiotic organisms search: a new metaheuristic optimization algorithm, Computers & Structures 139 (2014) 98–112.

[5] S. Mirjalili, SCA: a sine cosine algorithm for solving optimization problems, Knowledge-Based Systems 96 (2016) 120–133.

[6] W. Warid, H. Hizam, N. Mariun, N.I. Abdul-Wahab, Optimal power flow using the Jaya algorithm, Energies 9 (9) (2016) 678.

[7] S. Mirjalili, A.H. Gandomi, S.Z. Mirjalili, S. Saremi, H. Faris, S.M. Mirjalili, Salp Swarm Algorithm: a bio-inspired optimizer for engineering design problems, Advances in Engineering Software 114 (2017) 163–191.

[8] S. Arora, S. Singh, Butterfly optimization algorithm: a novel approach for global optimization, Soft Computing 23 (3) (2019) 715–734.

[9] S. Mirjalili, A. Lewis, The whale optimization algorithm, Advances in Engineering Software 95 (2016) 51–67.

[10] S.K. Sahoo, A.K. Saha, S. Nama, M. Masdari, An improved moth flame optimization algorithm based on modified dynamic opposite learning strategy, Artificial Intelligence Review (2022) 1–59.

[11] S.K. Sahoo, A.K. Saha, A hybrid moth flame optimization algorithm for global optimization, Journal of Bionics Engineering (2022) 1–22.

[12] S. Sharma, S. Chakraborty, A.K. Saha, S. Nama, S.K. Sahoo, mLBOA: a modified butterfly optimization algorithm with Lagrange interpolation for global optimization, Journal of Bionics Engineering (2022) 1–16.

[13] S.K. Sahoo, A.K. Saha, S. Sharma, S. Mirjalili, S. Chakraborty, An enhanced moth flame optimization with mutualism scheme for function optimization, Soft Computing 26 (6) (2022) 2855–2882.

[14] H. Chen, Y. Xu, M. Wang, X. Zhao, A balanced whale optimization algorithm for constrained engineering design problems, Applied Mathematical Modelling 71 (2019) 45–59.

[15] B. Alatas, ACROA: artificial chemical reaction optimization algorithm for global optimization, Expert Systems with Applications 38 (10) (2011) 13170–13180.

[16] D.H. Wolpert, W.G. Macready, No free lunch theorems for optimization, IEEE Transactions on Evolutionary Computation 1 (1) (1997) 67–82.

[17] G. Xiong, J. Zhang, D. Shi, Y. He, Parameter extraction of solar photovoltaic models using an improved whale optimization algorithm, Energy Conversion and Management 174 (2018) 388–405.

[18] P.R. Sahu, P.K. Hota, S. Panda, Modified whale optimization algorithm for coordinated design of fuzzy lead-lag structure-based SSSC controller and power system stabilizer, International Transactions on Electrical Energy Systems 29 (4) (2019) e2797.

[19] M. Tubishat, M.A. Abushariah, N. Idris, I. Aljarah, Improved whale optimization algorithm for feature selection in Arabic sentiment analysis, Applied Intelligence 49 (5) (2019) 1688–1707.

[20] Y. Niu, Z. Tang, Y. Zhou, Z. Wang, An enhanced whale optimization algorithm with simplex method, in: International Conference on Intelligent Computing, Springer, Cham, August 2019, pp. 729–738.

[21] J. Zhang, J.S. Wang, Improved whale optimization algorithm based on nonlinear adaptive weight and golden sine operator, IEEE Access 8 (2020) 77013–77048.

[22] P. Du, W. Cheng, N. Liu, H. Zhang, J. Lu, A modified whale optimization algorithm with single-dimensional swimming for global optimization problems, Symmetry 12 (11) (2020) 1892.

[23] A.A. Heidari, I. Aljarah, H. Faris, H. Chen, J. Luo, S. Mirjalili, An enhanced associative learning-based exploratory whale optimizer for global optimization, Neural Computing & Applications 32 (9) (2020) 5185–5211.

[24] Q. Fan, Z. Chen, W. Zhang, X. Fang, ESSAWOA: enhanced whale optimization algorithm integrated with salp swarm algorithm for global optimization, Engineering With Computers (2020) 1–18.

[25] L. Liu, L. Wu, Predicting housing prices in China based on modified Holt's exponential smoothing incorporating whale optimization algorithm, Socio-Economic Planning Sciences 72 (2020) 100916.

[26] M.M. Saafan, E.M. El-Gendy, IWOSSA: an improved whale optimization salp swarm algorithm for solving optimization problems, Expert Systems with Applications 176 (2021) 114901.

[27] Q. Fan, Z. Chen, Z. Li, Z. Xia, J. Yu, D. Wang, A new improved whale optimization algorithm with joint search mechanisms for high-dimensional global optimization problems, Engineering With Computers 37 (3) (2021) 1851–1878.

[28] A.V. Lakshmi, P. Mohanaiah, WOA-TLBO: whale optimization algorithm with teaching-learning-based optimization for global optimization and facial emotion recognition, Applied Soft Computing 110 (2021) 107623.

[29] S. Chakraborty, A.K. Saha, S. Sharma, S. Mirjalili, R. Chakraborty, A novel enhanced whale optimization algorithm for global optimization, Computers & Industrial Engineering 153 (2021) 107086.

[30] S. Chakraborty, A.K. Saha, R. Chakraborty, M. Saha, S. Nama, HSWOA: an ensemble of hunger games search and whale optimization algorithm for global optimization, International Journal of Intelligent Systems 37 (1) (2022) 52–104.

[31] M.H. Nadimi-Shahraki, H. Zamani, S. Mirjalili, Enhanced whale optimization algorithm for medical feature selection: a COVID-19 case study, Computers in Biology and Medicine 148 (2022) 105858.

[32] J. Liu, J. Shi, F. Hao, M. Dai, A novel enhanced global exploration whale optimization algorithm based on Lévy flights and judgment mechanism for global continuous optimization problems, Engineering With Computers (2022) 1–29.

[33] G. Sun, Y. Shang, R. Zhang, An efficient and robust improved whale optimization algorithm for large scale global optimization problems, Electronics 11 (9) (2022) 1475.

[34] X. Liang, S. Xu, Y. Liu, L. Sun, A modified whale optimization algorithm and its application in seismic inversion problem, Mobile Information Systems 2022 (2022) 1–18.

[35] G. Sun, Y. Shang, K. Yuan, H. Gao, An improved whale optimization algorithm based on nonlinear parameters and feedback mechanism, International Journal of Computational Intelligence Systems 15 (1) (2022) 1–17.

[36] R. Ramaprabha, M. Balaji, B.L. Mathur, Maximum power point tracking of partially shaded solar PV system using modified Fibonacci search method with fuzzy controller, International Journal of Electrical Power & Energy Systems 43 (1) (2012) 754–765, https://doi.org/10.1016/j.ijepes.2012.06.031.

[37] İ. Yazıcı, E.K. Yaylacı, B. Cevher, F. Yalçın, C. Yüzkollar, A new MPPT method based on a modified Fibonacci search algorithm for wind energy conversion systems, Journal of Renewable and Sustainable Energy 13 (1) (2021) 013304.

[38] P. Bujok, A. Zamuda, Cooperative model of evolutionary algorithms applied to CEC 2019 single objective numerical optimization, in: 2019 IEEE Congress on Evolutionary Computation (CEC), IEEE, June 2019, pp. 366–371.

Chapter 17

A random weight and random best solution based improved whale optimization algorithm for optimization issues

Sanjoy Chakraborty[a,b], Apu Kumar Saha[c], Saroj Kumar Sahoo[c], and Ashim Saha[a]
[a]*Department of Computer Science and Engineering, National Institute of Technology Agartala, Agartala, Tripura, India,* [b]*Department of Computer Science and Engineering, Iswar Chandra Vidyasagar College, Belonia, Tripura, India,* [c]*Department of Mathematics, National Institute of Technology Agartala, Agartala, Tripura, India*

17.1 Introduction

Despite its widespread use and significance, the optimization problem remains a formidable obstacle in the field of artificial calculation [1]. Engineering disciplines such as signal processing, machine learning, neural network optimization, and civil, mechanical, and electrical engineering, these fields all deal with optimization issues [2]. Due to the curse of dimensionality, conventional and gradient-based optimization approaches are frequently ineffective when confronted with such challenging problems. In addition, the gradient search is sensitive to the location of the starting point, making it particularly difficult to use gradient-based strategies in a problem involving local optimum solutions [1]. The optimization problems can be any one of the types, minimization or maximization.

Meta-heuristic algorithms, in contrast to conventional methods, often produce the best results on such problems. Advantages of these algorithms include their ease of use and execution, as well as their ability to avoid settling for suboptimal solutions. These methods were developed to quickly solve problems of large dimensionality and complexity [3]. They take into consideration concepts from the fields of biology, intelligence, mathematics, physics, neuroscience, and statistical mechanics. Metaheuristic algorithms can be divided into different categories by a number of common phenomena, including those connected to evolutionary, physics, swarm intelligence, human resistive frameworks, and human behavior. Different metaheuristics have various advantages, but the No Free Lunch (NFL) [4] hypothesis suggests that no single method is universally effective for solving all issues. Improved performance of an algorithm on one group of issues is no assurance of equivalent performance on another collection of functions. Furthermore, it is demonstrated that the same method yields varied results on the same problem depending on the parameters that are set. All of these factors motivate researchers all around the world to rethink existing methods and test out novel approaches.

Both global and local search are essential to the search process, yet they occur at different times. It is desirable to make as many random decisions as possible during the exploration phase of the algorithm. The evaluation of the promising regions uncovered during exploration marks the beginning of the exploitation phase. Therefore, the local search procedure is connected to the exploitation stage. An algorithm's performance can be attributed, in large part, to how well it strikes a balance between its first and second phases [5].

A meta-heuristic optimization method called the whale optimization algorithm (WOA) [6] leverages swarm intelligence to discover the best solution. Mirjalili and Lewis came up with the idea after observing the hunting techniques of humpback whales. As can be seen from the process, WOA employs a basic but effective mechanism with few regulating variables. Since its inception, WOA has been applied to tackle a wide range of optimization issues due to its effectiveness. Although being effective, it has many flaws, such as a lack of exploration capabilities, a slow rate at which solutions are produced, and a propensity to become stuck in the local solution [1].

Handbook of Whale Optimization Algorithm. https://doi.org/10.1016/B978-0-32-395365-8.00023-3

Over the years, a number of researchers have attempted to fix the problems with WOA. Below is a summary of some of WOA's most recent improvements and changes. Parameters were altered and the convergence rate of WOA was sped up with the help of chaotic maps in CWOA [5]. A nonlinear dynamic approach using the cosine function underpinned the introduction of MWOA [7]. The algorithm also made use of Levy flight and quadratic interpolation to avoid convergence to local optimum and boost the accuracy of the outcomes. The developers of WOAmM [8] By changing the mutualism phase of SOS and adding it to the solution space, WOAmM's creators expanded the original WOA. To speed up the algorithm's convergence during the modified mutualism phase, a greedy selection technique was used to pick the random solution. WAROA [9] has devised a military program that segregates the populace into cooperative crops and deep prey. The new strategy has also required significant adjustments to WOA's underlying characteristics. Tent map evaluation of the seed population and quasi-oppositional learning are two examples of how SWWOA [10] has been modified to increase variety. To strike a middle ground between discovery and exploitation, a novel method based on the logarithmic function for evaluating the co-efficient parameter (A) has been presented. Last but not least, the updated strategy, a one-dimensional swimming approach, replaces the prey behavior of basic WOA. ImWOA [11] was presented with more variety to fill the voids left by WOA. To encourage more mobility during the search for prey phase, designers changed the way random solutions were selected. The whale's cooperative hunting strategy was also incorporated into the algorithm's exploitation phase in order to achieve a healthy balance between the WOA exploration and exploitation phases. The total number of iterations was also split in two for the sake of investigation and exploitation. Through these tweaks, WOA was able to escape from local optima, yield better solutions, and converge more quickly.

Therefore, identifying the loopholes in the existing algorithms and their modifications to design an efficient variant is a common practice. In this work, RWbWOA, a novel version of WOA that has been developed. The modifications incorporated into WOA are as follows:

- A random weight is used throughout the exploration phase to add diversity in the search.
- In the exploitation phase of WOA another random weight is used to direct the search process nearby the present optimal result.
- To prevent the blind selection of the best solution utilized in WOA, a randomly chosen solution from a pool of best solutions is employed.
- Compared outcomes of twenty-five benchmark functions to examine efficacy of the proposed method.

The remainder of the study is structured in accordance with Section 17.2's description of the fundamental WOA, in Section 17.3, the modifications included in WOA are discussed. Section17.4 provides a discussion of the results of the evaluation. The study is ultimately concluded in Section 17.5.

17.2 Whale optimization algorithm

WOA is an algorithm inspired by swarm intelligence that accomplishes the search process in the same manner as humpback whale searches its food. The three stages of WOA are as follows.

17.2.1 Exploration phase

WOA provided accommodations for the hunt for prey technique to scout the search area. The whales explore the search space and update their position with respect to a random whale solution during this phase. This portion of the method can be expressed mathematically as:

$$D'' = |C.W_{rndm}^{(i)} - W^{(i)}| \tag{17.1}$$

$$W^{(i+1)} = W_{rndm}^{(i)} - \overline{A}.D'' \tag{17.2}$$

where W represents a population search agent, W_{rndm} represents a randomly selected solution within the present solutions, i represents the present value of iteration, and D'' shows the separation between the currently selected solution and the randomly chosen solution. Here, (.) operator represents the element wise reproduction operation, and the symbol | | signifies the absolute value.

To estimate the parameters $\overline{A}$ and C, utilize the ensuing equations:

$$\overline{A} = 2a_p \times rndm - a_p \tag{17.3}$$

$$C = 2 \times rndm \tag{17.4}$$

During the search process, $rndm$ was used to denote a random number in the range [0, 1], value of a_p moves from 2 to 0 during the search.

17.2.2 Exploitation phase

To conduct the local search phase, bubble-net attacking and encircling the target phases are employed. These phases are characterized as follows.

17.2.2.1 Encircling prey phase

The current global top solution is thought to be rather close to the most elegant one at this moment. Every other search agent in the population moves closer to the overall best result by changing where it is located in the search space. The scientific form of this step in the algorithm is as follows:

$$D' = \left| C.W_{gb}^{(i)} - W^{(i)} \right| \tag{17.5}$$

$$W^{(i+1)} = W_{gb}^{(i)} - \overline{A}.D' \tag{17.6}$$

where, P_{gb} is the global top solution obtained at that point of time.

17.2.2.2 Bubble-net attack

The whales' movement in this phase takes the shape of a spiral. To express this step of the method, we use the notation:

$$D^* = W_{gb}^{(i)} - W^{(i)} \tag{17.7}$$

$$W^{(i+1)} = D^* \cdot e^{bl} \cdot \cos(2\Pi l) + W_{gb}^{(i)} \tag{17.8}$$

While the constant variable b in Eq. (17.8) characterizes the logarithmic spiral shape, the arbitrary value l in this equation is defined by Eq. (17.9).

$$l = \left(a_q - 1\right) rndm + 1 \tag{17.9}$$

Throughout the search process, the value of a_q varies linearly between (−1) and (−2), and $rndm$ is a probability value between 0 and 1. Changing the value of $|A|$ allows you to move between the method's global and local search phases. When $|A|$ is lesser than 1, uses Eq. (17.2) is used to examine the region. If not, enables the exploiting procedure with Eq. (17.6) or (17.8). With a probability of 0.5, any one from Eq. (17.10) is chosen. Mathematical expressions for the process are as follows:

$$\begin{cases} W^{(i+1)} = W_{gb}^{i} - \overline{A}.D' & \text{when } P < 0.5 \\ W^{(i+1)} = D^* \cdot e^{bl} \cos(2\pi l) + W_{gb}^{i} & \text{when } P \geq 0.5 \end{cases} \tag{17.10}$$

where P is a fictional number that falls between 0 and 1.

17.3 Proposed RWbWOA

Traditional WOA relied on the co-efficient vector value $\overline{A}$ and a random value to decide between global and local search. As the process progressed into its second half, the arrangement restricted it to the exploitation phase, narrowing the possible solutions [1]. Less exploration of the search space leads to less diversity in the population. Local search and global search contribute to the accuracy of the solution, while the exploitation phase is responsible for the algorithm's speed of convergence. While in the exploration phase, WOA updates the current solution with another solution picked at random from the entire population. Due to the possibility of selecting a solution that is close to the current one, the diversity of the solutions can be reduced if they are selected blindly during the exploration phase. Another issue is that the same solution might be chosen multiple times to update solutions. Because of this, the search effort will be directed in the wrong way. In light of the foregoing, the RWbWOA uses a random weight during the search for prey phase, the whales are pushed far from the present and the selected random solution with a motive of searching a different location. The bar graph of the value evaluated in first 50 iterations of the search is given in Fig. 17.2(a). The modified solution can be represented as:

$$W^{(i+1)} = \omega_1 .* W^{(i)}_{rndm} - \overline{A}.D'' \tag{17.11}$$

where ω_1 is evaluated using the equation below:

$$\omega_1 = 0.5 + 0.5 * rand \tag{17.12}$$

While exploitation in classical WOA the best solution is blindly used to update other solutions. As WOA has low exploration ability it may stick into the chosen best solution or into any local solution nearby. Therefore, in RWbWOA instead of using the best solution a random best solution from a pool of first five best solutions is used with a random weight (ω_2). The blindness of WOA is rectified and ω_2 enables the search process to look for suitable solutions in neighboring areas. The bar graph of the value evaluated in last 50 iterations of the search is given in Fig. 17.2(b). The modified equations are as follows:

$$W^{(i+1)} = \omega_2 .* W^{(i)}_{rgb} - \overline{A}.D' \tag{17.13}$$

$$W^{(i+1)} = D^* \cdot e^{bl} \cdot \cos(2\Pi l) + \omega_2 .* W^{(i)}_{rgb} \tag{17.14}$$

$$\omega_2 = 0.1 + 0.1 * rand \tag{17.15}$$

In Fig. 17.1, the pseudo-code for RWbWOA is shown.

Initialize the Whale solution W^x $(x = 1,2,3, \ldots\ldots n)$

Evaluate fitness of every search agent

Find w_{rg_b} *and* w_{g_b}

While ($t <$ maximum iterations)

- For every whale solution

  Evaluate a, A, C, l and Ω

  - If (Ω<0.5)
    - When ($|A| < 1$)

      Evaluate new solution using Eqn. (17.13)
    - If ($|A| \geq 1$)

      Chose an arbitrary solution (WS_{rndm})

      Evaluate new solution using Eqn. (17.11)
    - End of If
  - Else If ($\Omega \geq$ 0.5)

    Evaluate new solution using Eqn. (17.14)
  - End of If
- End of For

Check boundary conditions

Evaluate fitness of each the present Whale solution

Update w_{rg_b} *and* w_{g_b}

Increment t by 1

- End while

Return w_{g_b}

FIGURE 17.1 Pseudo-code of RWbWOA.

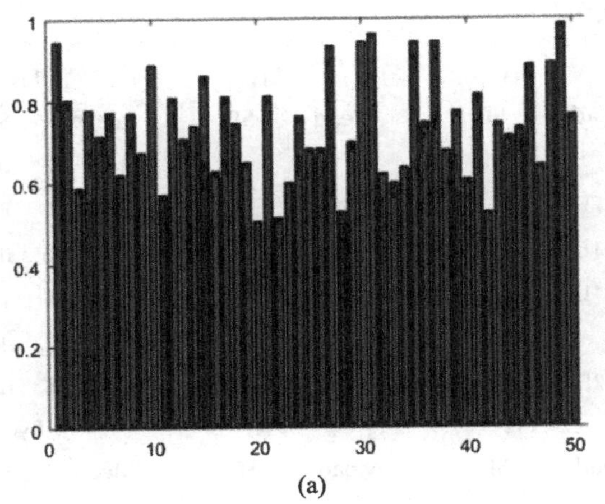

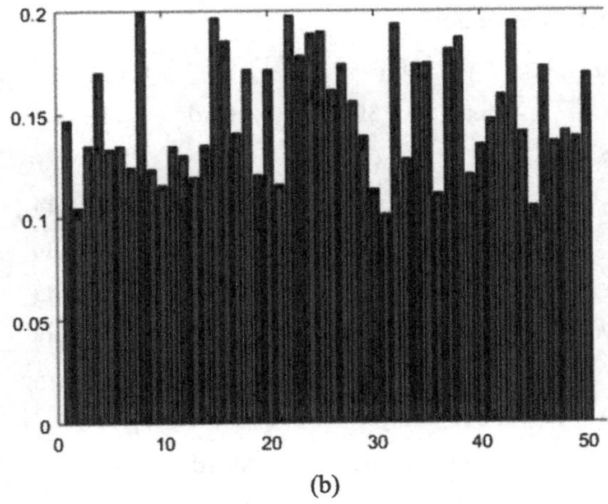

FIGURE 17.2 Evaluated values of ω_1 and ω_2.

17.4 Discussion of numerical results

Calculated outcomes from the selected functions have been compared to recent WOA variants, namely ESSAWOA [12], WOAmM [8], m-SDWOA [2], SHADE-WOA [13], HSWOA [14]. F1 through F13 are unimodal functions, while the remaining functions are multimodal. The details of the functions are given in Table 17.A.1 of Appendix 17.A. Whale population during evaluation is fixed to 30 and 15000 function evaluation is chosen as the termination criteria. Table 17.1 shows the average and standard deviation (SD) values assessed by the algorithms. Table 17.2 displays the pairwise comparison outcomes evaluated by RWbWOA. Similar values indicate the identical outcomes. RWbWOA found superior than the compared algorithms in both the unimodal and multimodal functions it confirms about the exploitation as well as exploration capacity of the algorithm. HSWOA is emerged as the 2nd best algorithm after RWbWOA.

TABLE 17.1 Comparison of RWbWOA results with the other WOAs.

Algorithm	F_1		F_2		F_3		F_4		F_5	
	Mead	SD	Mead	SD	Mead	SD	Mead	SD	Mead	SD
RWbWOA	0	0	3.31E-178	0	0	0	3.64E-165	0	0	0
ESAWOA	9.26E-20	1.56E-19	1.17E-10	1.29E-10	2.54E-17	5.77E-17	5.43E-11	5.52E-11	0	0
WOAmM	1.15E-55	4.13E-55	1.01E-28	2.22E-28	3.97E-41	1.60E-40	1.04E-27	2.50E-27	0	0
m-SDWOA	3.94E-46	1.10E-45	6.89E-24	1.88E-23	2.16E-25	2.38E-25	9.48E-23	1.58E-22	0	0
SHADE-WOA	1.90E-47	1.02E-46	1.24E-32	6.55E-32	1.34E+02	8.51E+01	2.76E+00	1.40E+00	3.67E-01	1.07E+00
HSWOA	2.56E-319	0	5.10E-216	0	3.21E-198	0	5.61E-226	0	0	0
Algorithm	F_6		F_7		F_8		F_9		F_{10}	
	Mead	SD	Mead	SD	Mead	SD	Mead	SD	Mead	SD
RWbWOA	4.31E-01	2.88E-01	0	0	0	0	0	0	0	0
ESAWOA	9.31E-05	7.46E-05	1.22E-10	2.41E-10	4.58E-12	1.04E-11	1.12E-13	1.51E-13	2.23E-19	4.24E-19
WOAmM	8.36E-04	7.85E-04	2.39E-12	7.46E-12	5.31E-48	1.86E-47	1.07E-12	2.98E-12	8.74E-95	4.79E-89
m-SDWOA	4.61E-04	3.57E-04	1.35E-22	2.31E-22	1.64E-39	2.79E-39	1.38E-46	3.47E-46	1.22E-97	6.69E-94
SHADE-WOA	9.92E-03	8.59E-03	5.73E-130	3.13E-129	7.79E-39	4.25E-38	3.28E-25	1.06E-24	1.13E-50	6.19E-50
HSWOA	7.06E-05	7.51E-05	6.11E-209	4.11E-128	4.29E-281	7.19E-103	8.94E-39	4.89E-38	0	0

continued on next page

TABLE 17.1 *(continued)*

Algorithm	F_{11}		F_{12}		F_{13}		F_{14}		F_{15}	
	Mead	SD	Mead	SD	Mead	SD	Mead	SD	Mead	SD
RWbWOA	0	0	0	0	0	0	0	0	8.88E-16	0
ESAWOA	6.59E-13	1.71E-12	1.09E-14	1.92E-14	1.17E-24	2.21E-24	0	0	6.31E-11	8.10E-11
WOAmM	1.43E-48	5.31E-48	2.14E-57	1.14E-56	9.41E-95	5.15E-94	0	0	8.88E-16	0
m-SDWOA	3.58E-40	7.68E-40	3.42E-49	9.55E-49	8.41E-102	2.93E-101	2.19E+00	6.77E+00	3.73E-15	1.45E-15
SHADE-WOA	2.56E-39	1.40E-38	2.01E-50	1.10E-49	0	0	2.98E+01	1.76E+01	5.15E-15	2.70E-15
HSWOA	6.58E-189	5.89E-107	3.25E-222	1.98E-107	7.12E-301	2.10E-117	0	0	8.88E-16	0
Algorithm	F_{16}		F_{17}		F_{18}		F_{19}		F_{20}	
	Mead	SD	Mead	SD	Mead	SD	Mead	SD	Mead	SD
RWbWOA	0	0	4.08E-182	0	0	0	0	0	0	0
ESAWOA	0	0	5.37E-13	5.16E-13	9.61E-69	3.59E-68	0	0	2.72E-11	2.51E-11
WOAmM	0	0	1.99E-09	4.85E-09	1.01E-160	5.52E-160	0	0	6.53E-14	3.44E-13
m-SDWOA	0	0	8.05E-54	4.41E-53	2.65E-140	1.42E-139	0	0	9.99E-02	1.10E-09
SHADE-WOA	1.15E-03	4.78E-03	2.02E-162	1.11E-161	5.56E-08	3.04E-07	0	0	4.07E-01	2.27E-01
HSWOA	0	0	0	0	1.09E-221	0	0	0	0	0
Algorithm	F_{21}		F_{22}		F_{23}		F_{24}		F_{25}	
	Mead	SD	Mead	SD	Mead	SD	Mead	SD	Mead	SD
RWbWOA	0	0	0	0	0	0	5.00E-01	0	1.00E+00	0
ESAWOA	0	0	0	0	0	0	5.00E-01	0	1.00E+00	2.44E-10
WOAmM	0	0	0	0	0	0	1.45E+01	0	1.00E+00	0
m-SDWOA	0	0	0	0	0	0	1.45E+01	0	1.00E+00	0
SHADE-WOA	0	0	0	0	0	0	1.45E+01	3.68E-03	1.00E+00	0
HSWOA	0	0	0	0	0	0	5.00E-01	0.00E+00	1.00E+00	0

TABLE 17.2 RWbWOA results pairwise comparison with the WOAs using Table 17.1 data.

EBWOA	ESSAWOA	WOAmM	m-SDWOA	SHADE-WOA	HSWOA
Superior to	16	16	17	16	11
Similar to	9	9	7	6	11
Inferior to	0	0	1	3	3

17.5 Conclusion

In order to increase the WOA's effectiveness, this research suggests a new variation. Due to the lack of weight given to the exploration phase during optimization, WOA mostly exploits the search space. The result is a propensity toward low solution accuracy and the pursuit of suboptimal solutions (local optima). In RWbWOA two random weights are used, one for exploration and another for exploitation. The weights guide the search process to move far while exploration and near by the optimal solution during the exploration. Moreover, during the exploration phase instead of the best solution, a random best solution is used to lessen the drawback of blind selection and to move slowly towards the optimal solution. The superiority of the suggested strategy is confirmed by a comparison of the evaluated results with other recently modified WOA variations.

Appendix 17.A

Here mathematical formulation of the benchmark functions is given in a table.

TABLE 17.A.1 Details of benchmark functions.

ID	Description	F^*
F_1	$f(y) = \sum_{k=1}^{D} y_k^2$	0
F_2	$f(y) = \sum_{k=1}^{D} \lvert y_k\rvert + \prod_{k=1}^{D} \lvert y_k\rvert$	0
F_3	$F(y) = \sum_{k=1}^{D} \left(\sum_{l=1}^{k} y_l\right)^2$	0
F_4	$f(y) = \max_k [\lvert y_k\rvert, 1 \le k \le D]$	0
F_5	$f(y) = \sum_{k=1}^{D} (\lvert y_k + 0.5\rvert)^2$	0
F_6	$f(y) = \sum_{k=1}^{D} y_k^4 + random(0, 1)$	0
F_7	$f(y) = \sum_{k=1}^{D} y_k^2 + \left(\sum_{k=1}^{D} 0.5ky_k\right)^2 + \left(\sum_{k=1}^{D} 0.5ky_k\right)^4$	0
F_8	$f(y) = y_1^2 + 10^6 \sum_{k-2}^{D} y_k^6$	0
F_9	$f(y) = \sum_{k=1}^{D/4}[(y_{4k-3} + 10y_{4k-2})^2 + 5(y_{4k-1} + y_{4k})^2 + (y_{4k-2} + 2y_{4k-1})^4 +$ $10(y_{4k-3} + 10y_{4k})^4]$	0
F_{10}	$f(y) = 10^6 y_1^2 + \sum_{k-2}^{D} y_k^6$	0
F_{11}	$f(y) = \sum_{k=2}^{D} \left(10^6\right)^{(k-1)(D-1)} \cdot y_k^2$	0
F_{12}	$f(y) = \sum_{k=1}^{n-1} \left(y_k^2\right)^{\left(y_{k+1}^2+1\right)} + \left(y_{k+1}^2\right)^{\left(y_k^2+1\right)}$	0
F_{13}	$f(y) = 0.26(\left(y_1^2 + y_2^2\right) - 0.48y_1y_2$	0
F_{14}	$f(y) = \sum_{k=1}^{D}[y_k^2 - 10\cos(2\Pi y) + 10]$	0
F_{15}	$f(y) = -20exp(-0.2\sqrt{\frac{1}{D}\sum_{k=1}^{D} y_k^2} - exp\left(\frac{1}{D}\sum_{k=1}^{D} \cos 2\Pi y_k\right)) + 20 + e$	0
F_{16}	$f(y) = \frac{1}{4000}\sum_{K=1}^{D} y_k^2 - \prod_{k=1}^{D} \cos(\frac{y_k}{\sqrt{k}}) + 1$	0
F_{17}	$f(y) = \sum_{k=1}^{D} \lvert y_k \sin(y_k) + 0.1y_k\rvert$	0
F_{18}	$f(y) = \sum_{k=1}^{D} y_k^6(2 + sin\frac{1}{y_k})$	0
F_{19}	$f(y) = 0.1D - (0.1\sum_{k=1}^{D} \cos(5\Pi y_k) - \sum_{k=1}^{D} y_k^2)$	0
F_{20}	$f(y) = 1 - \cos\left(2\Pi\sqrt{\sum_{k=1}^{D} y_k^2}\right) + 0.1\sqrt{\sum_{k=1}^{D} y_k^2}$	0
F_{21}	$f(y) = y_1^2 + 2y_2^2 - 0.3\cos(3\Pi y_1) - 0.4\cos(4\Pi y_2) + 0.7$	0
F_{22}	$f(y) = y_1^2 + 2y_2^2 - 0.3\cos(3\Pi y_1).0.4\cos(4\Pi y_2) + 0.3$	0
F_{23}	$f(y) = y_1^2 + 2y_2^2 - 0.3\cos(3\Pi y_1 + 4\Pi y_2) + 0.3$	0
F_{24}	$f(y) = 0.5 + \frac{\sin^2\left(y_1^2 - y_2^2\right) - 0.5}{\left[1 + 0.001\left(y_1^2 + y_2^2\right)\right]^2}$	0
F_{25}	$f(y) = \left\lvert y_1^2 + y_2^2 + y_1y_2\right\rvert + \lvert\sin(y_1)\rvert + \lvert cos(y_2)\rvert$	1

References

[1] S. Chakraborty, A.K. Saha, R. Chakraborty, M. Saha, An enhanced whale optimization algorithm for large scale optimization problems, Knowledge-Based Systems 233 (2021) 107543.

[2] S. Chakraborty, A.K. Saha, S. Sharma, R. Chakraborty, S. Debnath, A hybrid whale optimization algorithm for global optimization, Journal of Ambient Intelligence and Humanized Computing 14 (2023) 431–467.

[3] E.G. Talbi, Metaheuristics: From Design to Implementation, John Wiley & Sons, 2009, p. 74.

[4] D.H. Wolpert, W.G. Macready, No free lunch theorems for optimization, IEEE Transactions on Evolutionary Computation 1 (1) (1997) 67–82, https://doi.org/10.1109/4235.585893.

[5] G. Kaur, S. Arora, Chaotic whale optimization algorithm, Journal of Computational Design and Engineering 5 (3) (2018) 275–284.

[6] S. Mirjalili, A. Lewis, The whale optimization algorithm, Advances in Engineering Software 95 (2016) 51–67, https://doi.org/10.1016/j.advengsoft.2016.01.008.
[7] Y. Sun, X. Wang, Y. Chen, Z. Liu, A modified whale optimization algorithm for large-scale global optimization problems, Expert Systems with Applications 114 (2018) 563–577.
[8] S. Chakraborty, A.K. Saha, S. Sharma, S. Mirjalili, R. Chakraborty, A novel enhanced whale optimization algorithm for global optimization, Computers & Industrial Engineering 153 (2021) 107086.
[9] R. Jiang, M. Yang, S. Wang, T. Chao, An improved whale optimization algorithm with armed force program and strategic adjustment, Applied Mathematical Modelling 81 (2020) 603–623.
[10] P. Du, W. Cheng, N. Liu, H. Zhang, J. Lu, A modified whale optimization algorithm with single-dimensional swimming for global optimization problems, Symmetry 12 (11) (2020) 1892.
[11] S. Chakraborty, S. Sharma, A.K. Saha, A. Saha, A novel improved whale optimization algorithm to solve numerical optimization and real-world applications, Artificial Intelligence Review 55 (2022) 4605–4716.
[12] Q. Fan, Z. Chen, W. Zhang, X. Fang, ESSAWOA: enhanced whale optimization algorithm integrated with salp swarm algorithm for global optimization, Engineering With Computers 38 (2022) 797–814.
[13] S. Chakraborty, S. Sharma, A.K. Saha, S. Chakraborty, SHADE–WOA: a metaheuristic algorithm for global optimization, Applied Soft Computing 113 (2021) 107866.
[14] S. Chakraborty, A.K. Saha, R. Chakraborty, M. Saha, S. Nama, HSWOA: an ensemble of hunger games search and whale optimization algorithm for global optimization, International Journal of Intelligent Systems 37 (1) (2022) 52–104.

Chapter 18

Guided whale optimization algorithm (guided WOA) with its application

Abdelhameed Ibrahim[a], **El-Sayed M. El-kenawy**[b], **Nima Khodadadi**[c], **Marwa M. Eid**[d], and **Abdelaziz A. Abdelhamid**[e]

[a]*Computer Engineering and Control Systems Department, Faculty of Engineering, Mansoura University, Mansoura, Egypt,* [b]*Department of Communications and Electronics, Delta Higher Institute of Engineering and Technology, Mansoura, Egypt,* [c]*Department of Civil and Architectural Engineering, University of Miami, Coral Gables, FL, United States,* [d]*Faculty of Artificial Intelligence, Delta University for Science and Technology, Mansoura, Egypt,* [e]*Department of Computer Science, Faculty of Computer and Information Sciences, Ain Shams University, Cairo, Egypt*

18.1 Introduction

The act of determining which of several potential solutions offers the greatest benefit for resolving a certain issue is known as optimization. Taking into account the characteristics of each method, optimization algorithms may be loosely classified into two distinct groups: deterministic optimization algorithms and stochastic intelligent optimization algorithms. Because the initial starting values are always the same when utilizing deterministic algorithms, the solutions to the same problems are always the same as well [1,2]. In contrast to deterministic algorithms, probabilistic ones follow a series of random steps in order to achieve their objectives. The procedure of optimization can never again be carried out in such a manner.

In spite of this, in the vast majority of situations, the two of them are capable of arriving at the identical final optimal solutions. Stochastic algorithms can also be classified as either heuristic or metaheuristic, which provides a third classification option. A heuristic algorithm is a strategy for finding the best possible solutions through a process of iterative trial and error. On the other hand, metaheuristic algorithms handle optimization issues using a stochastic approach. These algorithms require some prior knowledge of the random search method. To put it another way, it is a technique for optimization that starts with a solution chosen at random. After that, during the subsequent iteration, it exploits and explores at random the accessible search space according to a predetermined probability [3,4].

The bio-inspired metaheuristic techniques are a novel methodology that takes their primary inspiration from the process of biological evolution in order to develop innovative and potent algorithmic competition. Over the course of the past two decades, there has been an increase in demand for these methods [3,5]. These methods are able to mine the crucial information that is hidden within the population in order to locate the approach that will prove to be the most successful. Numerous academics have published much research on these algorithms to this day, and several algorithms of nature-inspired metaheuristics, such as the Bat Algorithm (BA), the Grey Wolf Optimization (GWO) algorithm [6,7], the Dipper Throated Optimization (DTO) [8–10], and the Whale Optimization Algorithm (WOA) [11] have been proposed. According to some reports, there are approximately 150 different optimization algorithms that can be used to solve (get the best possible results in) optimization problems [12,13].

WOA is a more contemporary optimization algorithm that is based on swarm intelligence. It was inspired by the humpback whales behavior during the hunting phase. The WOA optimization technique incorporates three operators that simulate the behaviors of humpback whales: looking for a target, surrounding the prey, and attacking with a bubble net. Over the course of the past few years, the WOA algorithm has been successfully utilized in the resolution of a wide variety of problems involving optimization and feature selection [11].

The No Free Lunch Theorem is a concept in machine learning that essentially states that there is no one algorithm or approach that is universally better than others when it comes to solving all types of problems [14]. In other words, there is no "free lunch" when it comes to finding the most effective machine-learning solution. This theorem suggests that different algorithms and approaches may be more effective for different types of problems and that finding the optimal solution for a given problem requires careful consideration and experimentation. The No Free Lunch Theorem highlights the importance of understanding the problem at hand and the strengths and weaknesses of different algorithms in order to choose the best approach for a specific task. It also emphasizes the need for continued exploration and innovation in the field of machine

Handbook of Whale Optimization Algorithm. https://doi.org/10.1016/B978-0-32-395365-8.00024-5

learning, as there is always the potential for new approaches to emerging that may be better suited for certain types of problems.

The guided WOA is an optimization algorithm that combines the principles of swarm intelligence and evolutionary computing to find the optimal solution for a given problem [15]. This approach uses a guiding function to improve the performance of the standard WOA algorithm, which can sometimes struggle to converge on the optimal solution. The guiding function serves as an additional constraint that directs the search towards regions of the search space where the optimal solution is more likely to be found. By guiding the search process in this way, the guided WOA algorithm is able to improve its performance and find the optimal solution more efficiently than the standard WOA algorithm. The guided WOA algorithm has been applied to a range of optimization problems in various fields, including engineering and data science, and has demonstrated promising results in terms of improving the speed and accuracy of optimization processes.

To facilitate comprehension of the topic, this work is partitioned into the following sections. The first version of the WOA algorithm, as well as an examination of its level of complexity, can be found in Section 18.2. The guided WOA algorithm, as well as an examination of its complexity, may be found in the referenced Section 18.3. The binary-guided WOA for feature selection problems is introduced in Section 18.4. In Section 18.5, experimental findings, including more contemporary applications of guided WOA, are presented and discussed. Section 18.6 will contain a brief summary of the findings obtained from the study. Additionally, suggestions for future research will also be included in this section. The recommendations for further research will serve as a guide for scholars who wish to expand upon the research conducted in this study.

18.2 Whale optimization algorithm

The WOA algorithm was inspired by the hunting technique of whales, where they create a spiral-shaped pattern of bubbles to drive fish towards the surface and catch them. This foraging behavior of whales was used as a model to develop the WOA algorithm. The algorithm mimics the behavior of the whales in creating a spiral-shaped search pattern to find the best solutions for optimization problems. By utilizing this natural process, the WOA algorithm aims to efficiently find the optimal solution for a given problem, just as whales hunt for fish in an efficient manner [11,16]. In this behavior, whales use bubbles to trap their prey. The following equation serves as the mathematical foundation for the first optimization process implemented by this optimizer:

$$\mathbf{G}(t+1) = \mathbf{G}^*(t) - \mathbf{A}.\mathbf{D},\ \mathbf{D} = |\mathbf{C}.\mathbf{G}^*(t) - \mathbf{G}(t)| \tag{18.1}$$

where in each iteration of the process described in [17,18], vector $\mathbf{G}(t)$ represents a solution at that particular iteration, while $\mathbf{G}^*(t)$ indicates the prey position. The symbol "." in this context means that the vectors are multiplied pairwise. Vector $\mathbf{G}(t+1)$ represents the updated solution position. Additionally, $\mathbf{A}$ and $\mathbf{C}$ vectors are updated in each iteration. This is done by first changing the value of $\mathbf{a}$ linearly from 2 to 0, and then computing $\mathbf{A}$ and $\mathbf{C}$ as follows: $\mathbf{A} = 2\mathbf{a}.\mathbf{r}_1 - \mathbf{a}$, and $\mathbf{C} = 2.\mathbf{r}_2$, for $\mathbf{r}_1$ and $\mathbf{r}_2$ are updated randomly within [0, 1].

The second process involves two steps. The first step is a contracting circle that reduces the values of the $\mathbf{a}$ and $\mathbf{A}$ vectors. The second step involves a spiral process that updates the positions. More details will be provided in the following section.

$$\mathbf{G}(t+1) = \mathbf{D}'.e^{bl}.cos(2\pi l) + \mathbf{G}^*(t),\ \mathbf{D}' = |\mathbf{G}^*(t) - \mathbf{G}(t)| \tag{18.2}$$

where in the WOA mechanism, the distance between the ith whale, represented by $\mathbf{D}'$, and the best one is denoted by a variable called D'. The shape of the spiral is determined by a constant parameter b, while l is a random value that falls within the range of -1 to 1. It can be mathematically described using this equation.

$$\mathbf{G}(t+1) = \begin{cases} \mathbf{G}^*(t) - \mathbf{A}.\mathbf{D} & \text{if } \mathbf{r}_3 < 0.5 \\ \mathbf{D}'.e^{bl}.cos(2\pi l) + \mathbf{G}^*(t) & \text{otherwise} \end{cases} \tag{18.3}$$

where the symbol $\mathbf{r}_3$ represents a randomly generated value within [0, 1].

Final technique can be accomplished by utilizing the vector $\mathbf{A}$ as a foundation. To facilitate a search around the globe using the following equation, the search agent location is updated using the results of a random whale search $\mathbf{G}_{rand}$.

$$\mathbf{G}(t+1) = \mathbf{G}_{rand} - \mathbf{A}.\mathbf{D},\ \mathbf{D} = |\mathbf{C}.\mathbf{G}_{rand} - \mathbf{G}| \tag{18.4}$$

Therefore, the exploitation and exploration are under the control of the variable $\mathbf{A}$, while the spiraling or circular movement is under the control of the variable r_3.

The WOA algorithm is broken down into its component parts and presented in Algorithm 18.1. Firstly, the WOA algorithm initializes the population $\mathbf{G}_i$ $(i = 1, 2, ..., n)$ with a specified size and sets the maximum iteration number to be performed. Additionally, the algorithm sets the fitness function to be used and initializes the parameters of WOA, which include $\mathbf{a}$, $\mathbf{A}$, $\mathbf{C}$, l, $\mathbf{r}_1$, $\mathbf{r}_2$, $\mathbf{r}_3$, and t. The algorithm then calculates the fitness function for each search agent in the population and determines the best individual.

During the iterative search process, the algorithm uses a loop that executes while the current iteration t is less than or equal to the maximum iteration number specified. Within this loop, the algorithm updates the position of each search agent by iterating over the population, and checks if a specific condition is met using if-else statements. If $\mathbf{r}_3 < 0.5$ and $|\mathbf{A}| < 1$, the algorithm updates the current agent position using a specified equation. Otherwise, the algorithm selects a random agent and updates the current agent position by another equation. Additionally, if $\mathbf{r}_3 \geq 0.5$, the algorithm updates the current agent position using a third equation. After iterating over all search agents, the algorithm updates the WOA parameters and calculates the fitness function for each search agent again to find the best individual. Finally, the algorithm increments the iteration counter t and repeats the loop until the maximum iteration number is reached. The algorithm then returns the best search agent found during the search process.

Algorithm 18.1 Original WOA Pseudo-code.

```
Initialize population G_i (i = 1, 2, ..., n) with size n, fitness function F_n and iterations Max_iter.
Initialize parameters of a, A, C, l, r1, r2, r3, t = 1
Calculate fitness function F_n for each agent G_i
Find best individual G*
while t ≤ Max_iter do
    for (i = 1 : i < n + 1) do
        if (r3 < 0.5) then
            if (|A| < 1) then
                Update position of current agent from the search space by Eq. (18.1)
            else
                Select a random agent from the search space G_rand
                Update position of current agent from the search space by Eq. (18.4)
            end if
        else
            Update position of current agent from the search space by Eq. (18.2)
        end if
    end for
    Update parameters of a, A, C, l, r3
    Calculate fitness function F_n for each agent G_i
    Find best agent G*
    Increase counter t.
end while
Return best agent G*
```

This will discuss the computational complexity of the WOA algorithm, as presented in Algorithm 18.1. The number of agents in the populations will be denoted by n, while a total number of iterations will be represented by Max_{iter}. For each component of the algorithm, we will define its time complexity.

- Population initialization has a time complexity of $O(1)$.
- Initializing parameters: $O(1)$.
- Evaluating the fitness function: $O(n)$.
- Finding the best agent: $O(n)$.
- Updating the positions: $O(Max_{iter} \times n)$.
- Updating the parameters $\mathbf{a}$, $\mathbf{A}$, $\mathbf{C}$, l, $\mathbf{r}_3$: $O(Max_{iter})$.
- Evaluating the fitness function: $O(Max_{iter} \times n)$.
- Updating the best agent: $O(Max_{iter} \times n)$.
- Incrementing the iteration counter: $O(Max_{iter})$.

The WOA algorithm is a complex optimization algorithm that involves numerous computations. Taking into consideration the various intricacies involved in the algorithm, its overall computational complexity can be expressed as $O(Max_{iter} \times n)$. However, if we also factor in the number of variables (m) in the problem, then the computational complexity of the algorithm would be $O(Max_{iter} \times n \times m)$. This means that as the number of agents and variables increases, the computational complexity of the algorithm also increases, leading to longer computation times.

18.3 Guided WOA

The Guided WOA algorithm incorporates several modifications to the original WOA algorithm. One such modification involves replacing the search strategy for a randomly selected whale with a more sophisticated approach that can more efficiently move the whales towards the prey or best solution. This will allow you to get around the disadvantage of using this method. Eq. (18.4) from the original WOA has the whales wander around each other in a random pattern. In the improved WOA, also known as the Guided WOA, to improve its exploration performance, a whale is allowed to update based on three random whales. By substituting the following equation for the one in Eq. (18.4), this can motivate whales to engage in greater exploratory behavior without causing them to be influenced by the leader position.

$$\mathbf{G}(t+1) = \mathbf{w}_1 * \mathbf{G}_{rand1} + \mathbf{z} * \mathbf{w}_2 * (\mathbf{G}_{rand2} - \mathbf{G}_{rand3}) + (1 - \mathbf{z}) * \mathbf{w}_3 * (\mathbf{G} - \mathbf{G}_{rand1}) \tag{18.5}$$

where the equation involves several variables, including $\mathbf{G}_{rand1}$, $\mathbf{G}_{rand2}$, and $\mathbf{G}_{rand3}$, which represent three random solutions. Additionally, $\mathbf{w}_1$ is a randomly generated value between 0 and 0.5, while $\mathbf{w}_2$ and $\mathbf{w}_3$ are two other random values within [0, 1]. The variable $\mathbf{z}$ is used to gradually reduce the search radius, and its value decreases exponentially rather than linearly as follows:

$$\mathbf{z} = 1 - \left(\frac{t}{Max_{iter}}\right)^2 \tag{18.6}$$

where this equation describes the relationship between the iteration t and the most iterations possible Max_{iter}.

The Guided WOA algorithm, which can be found in (Algorithm 18.2), utilizes this equation. This is a pseudo-code for the Guided WOA algorithm. It begins by initializing the WOA population, parameters, and fitness function. Then, it calculates the fitness function for each individual and finds the best one. The algorithm enters a loop where it updates the position of each search agent based on certain conditions. In each iteration, it updates the WOA and Guided WOA parameters and calculates the fitness function for each individual. The algorithm terminates when it reaches Max_{iter} and returns the best individual.

We will discuss the computational complexity of the Guided WOA algorithm based on Algorithm 18.2. The time complexity for each section of the algorithm can be defined, where n is the number of populations.

- Population initialization has a time complexity of $O(1)$.
- Initializing parameters: $O(1)$.
- Evaluating the objective function: $O(n)$.
- Finding the best agent has a time complexity of $O(n)$.
- Updating the position of the whales: $O(Max_{iter} \times n)$.
- Updating **a**: $O(Max_{iter})$.
- Updating the parameters: $O(Max_{iter})$.
- Evaluating the objective function after each iteration: $O(Max_{iter} \times n)$.
- Updating the best individual after each iteration: $O(Max_{iter} \times n)$.
- Incrementing the iteration counter after each iteration: $O(Max_{iter})$.

Despite the additional steps involved in the Guided WOA algorithm, its computational complexity is still similar to that of the original WOA algorithm. Specifically, the overall computational complexity can be expressed as $O(Max_{iter} \times n)$, which is the same as that of the WOA algorithm. However, if we also take into account the number of variables involved in the problem, the final computational complexity can be expressed as $O(Max_{iter} \times n \times m)$. It is important to note that despite the modifications made to the WOA algorithm to create the Guided WOA algorithm, its overall computational complexity remains the same. This means that the Guided WOA algorithm can be used as a viable alternative to the original WOA algorithm without any significant increase in computational complexity.

Algorithm 18.2 Pseudo-code of the Guided WOA.

```
Initialize population G_i (i = 1, 2, ..., n) with size n, fitness function F_n, and iterations Max_iter.
Initialize parameters of a, A, C, l, r1, r2, r3, w1, w2, w3, t = 1
Calculate fitness function F_n for each agent G_i
Find best agent G*
while t ≤ Max_iter do
    for (i = 1 : i < n + 1) do
        if (r3 < 0.5) then
            if (|A| < 1) then
                Update position of current agent from the search space as
                G(t + 1) = G*(t) − A.D
            else
                Select three random agents from the search space G_rand1, G_rand2, and G_rand3
                Update (z) by the exponential form in Eq. (18.6)
                Update position of current agent from the search space as
                G(t + 1) = w1 * G_rand1 + z * w2 * (G_rand2 − G_rand3) + (1 − z) * w3 * (G − G_rand1)
            end if
        else
            Update position of current agent from the search space as
            G(t + 1) = D'.e^(bl).cos(2πl) + G*(t)
        end if
    end for
    Update parameters
    Calculate the fitness function F_n for each agent G_i
    Find best agent G*
    Increase counter t.
end while
Return best agent G*
```

18.4 Binary guided WOA algorithm

In recent years, the process of selecting features has evolved to become one of the most alluring steps in the whole procedure of data analysis. This is due to the fact that feature selection makes an effort to lower the high dimensionality of the data by removing features that are deemed to be unnecessary or redundant. Finding relevant characteristics that reduce classification mistakes is the major purpose of this feature selection optimization technique. As a consequence, they have been applied in a range of fields. The process of feature selection can be modeled mathematically as an optimization problem with the objective of achieving the minimum possible cost.

In the case that there are issues with the selection of features, the solutions provided by the Guided WOA algorithm will be binary, with values of either 0 or 1. The continuous values of the Guided WOA method that is being presented are going to be translated into binary values [0, 1] in order to make it simpler to select features from the dataset. This investigation makes use of the equation that is presented below, which is derived from the Sigmoid function [19].

$$x_d^{t+1} = \begin{cases} 1 & \text{if } Sigmoid(m) \geq 0.5 \\ 0 & \text{otherwise} \end{cases}, \qquad (18.7)$$

$$Sigmoid(m) = \frac{1}{1 + e^{-10(m-0.5)}}.$$

This equation represents the binary solution at t and dimension d: x_d^{t+1}. To convert the output solutions to binary, the $Sigmoid$ function can be used. The value changes to 1 if $Sigmoid(m) \geq 0.5$; otherwise, it remains at 0. The parameter m represents the selected features in the algorithm.

The binary version of the Guided WOA algorithm is shown in detail in Algorithm 18.3. The algorithm first initializes the population, objective function, and relevant parameters. It then converts the solution to a binary representation with values of either 0 or 1. After computing the fitness function for each agent, the algorithm identifies the best agent position.

The algorithm then enters a loop that continues until it reaches Max_{iter} specified by the user. Within the loop, the algorithm performs a series of operations on each search agent. If a specific condition is met, the algorithm updates the position of the current search agent based on various factors. Following each iteration, the updated solution is converted back to a binary representation, and the relevant parameters are updated. Finally, the algorithm returns the best agent position as the solution to the optimization problem. The computational complexity is determined by analyzing the Guided WOA algorithm, which is $O(Max_{iter} \times n)$, and for d dimensions, it becomes $O(Max_{iter} \times n \times d)$.

Algorithm 18.3 Binary Guided WOA Algorithm.

```
Initialize population, fitness function, and parameters
Convert solution to binary [0 or 1]
Calculate fitness function for each agents and get best agent position
while t ≤ Max_iter do
    for (i = 1 : i < n + 1) do
        if (r3 < 0.5) then
            if (|A| < 1) then
                Update current agent position
            else
                Select three random search agents
                Update (z) by the exponential form
                Update current agent position
            end if
        else
            Update current agent position
        end if
    end for
    Convert updated solution to binary
    Update parameters
    Calculate fitness function for agents and get the best agent position
end while
return best agent
```

The evaluation of a solution's quality in the binary Guided WOA algorithm is performed by applying the objective equation F_n. The error rate of the classifier, Err, selected features, s, and a total number of features, S, are involved in this formula.

$$F_n = \alpha Err + \beta \frac{|s|}{|S|} \tag{18.8}$$

The provided trait's population significance is denoted by α, while β is equal to $1-\alpha$. A popular classification technique that is simple to use and efficient is the k-nearest neighbor (k-NN) algorithm. This method's effectiveness mostly depends on its capacity to produce a feature subset that yields a low classification error rate. The k-NN classifier is used in this strategy to make sure the chosen features are accurate. The k-NN classifier is in charge of assessing the precision of the chosen feature subset and eliminating any characteristics that do not aid in precise categorization. The k-NN technique is a trustworthy and efficient way to choose features in classification situations.

18.5 Guided WOA applications

In this section, four different applications are presented that are based on the Guided WOA and binary Guided WOA algorithms. The first application focuses on the feature selection and classification of Computed Tomography (CT) images, where two algorithms based on Guided WOA, namely Stochastic Fractal Search (SFS)-Guided WOA, and Particle Swarm Optimization (PSO)-Guided-WOA, are proposed. The second application involves the transformer failure diagnosis feature selection and classification, where the AD-PRS-Guided WOA algorithm is recommended. For the third application, an additional method derived from Guided WOA is proposed for wind speed forecasting, and this method is called AD-PSO-Guided WOA. In the final application, the SFS-Guided WOA algorithm is used to optimize the deep neural network for speech emotion recognition.

Each of these applications demonstrates the versatility and effectiveness of the Guided WOA and binary Guided WOA algorithms in solving different problems across various domains. The use of these algorithms in feature selection and classification tasks shows their potential in improving the accuracy and efficiency of machine learning models. Overall, these applications provide insights into the diverse applications of Guided WOA and binary Guided WOA algorithms and their potential to enhance different fields of research.

18.5.1 First application: COVID-19

Research into the coronavirus, which has symptoms that are comparable to those of other types of pneumonia, requires a diagnosis as an essential first step in the prevention process. Imaging techniques such as CT scans and X-rays are quite helpful in this regard. However, the processing of chest CT images and utilizing processed images to effectively COVID-19 diagnoses is a task that requires a significant amount of computational resources. The application of approaches based on machine learning has the potential to solve this problem. The purpose of this study was to offer two optimization techniques for the selection of COVID-19 features and their subsequent categorization. The suggested structure was comprised of three sequential stages. In the first step of the process, a Convolutional Neural Network (CNN) by the name of AlexNet is used to extract characteristics from the CT images. The second step is to implement a suggested strategy for selecting features using the Guided WOA, which is based on the SFS algorithm. This step is then followed by balancing the characteristics that were chosen. Finally, a classifier using a voting technique called Guided WOA with PSO has been suggested [15].

This classifier combines the predictions of several distinct classifiers in order to select the class that has received the most votes. Because of this, the likelihood of individual classifiers, such as Neural Networks (NN), Decision Trees (DT), Support Vector Machine (SVM), and k-NN, showed large disparities increased. The effectiveness of the proposed SFS-Guided WOA algorithm as a feature selection method was tested on two datasets – one consisting of CT images with clinical findings positive for COVID-19 and the other consisting of CT images negative for COVID-19. The SFS-Guided WOA algorithm performance was compared against other optimization techniques that are commonly used in the recent research literature. Various statistical procedures such as the Wilcoxon rank-sum test, analysis of variance, and T-test were employed for evaluating the proposed algorithms' quality.

18.5.2 Second application: diagnostic accuracy of transformer faults

The identification of transformer defects allows for the prevention of the transformer's unwarranted removal from service and the maintenance of uninterrupted utility service. The study of dissolved gases is used to establish the diagnosis of defects in transformers (DGA). Conventional Dissolved Gas Analysis (DGA) techniques, such as Rogers' ratio method, Duval triangle method, and IEC code 60599 are often insufficient for detecting transformer defects. Even other DGA techniques, such as the Dornenburg method and the Key gas method, do not provide satisfactory results. Therefore, additional research is required to increase diagnostic accuracy by merging classic DGA approaches with techniques including artificial intelligence and optimization.

Researchers proposed the use of an algorithm known as Adaptive Dynamic (AD) Polar Rose (PRS)-Guided WOA (AD-PRS-Guided WOA) to optimize the parameters of classification approaches that are used to enhance the diagnostic accuracy of transformers [20]. The Guided WOA based AD-PRS algorithm optimizes the classification parameters to obtain the most accurate results. This approach has shown promising results in the diagnosis of transformer defects and could lead to significant improvements in the field. According to the findings, the AD-PRS-Guided WOA that was developed achieves a diagnostic accuracy of 97.1 percent for transformer faults, which was higher than the accuracy achieved by other DGA techniques described in the relevant research. The accuracy of the method had been verified through statistical analysis using a variety of tests, such as ANOVA and Wilcoxon's rank-sum.

18.5.3 Third application: wind speed forecasting

The creation and implementation of a reliable method for forecasting wind speed have the potential to enhance the security and stability of power networks that experience a substantial amount of wind penetration. It is extremely difficult to provide an accurate forecast of wind speed and power because of the unpredictability and instability exhibited by the wind. In order to achieve this goal and enhance the level of dependability of forecasting, a number of algorithms have been presented. The Long Short-Term Memory network, often known as LSTM, is a technology that is commonly utilized for the purpose of making accurate predictions of future occurrences based on time series data. Throughout the course of this investigation, a machine learning strategy was proposed for the purpose of wind speed ensemble forecasting. This strategy makes use of the Guided WOA-based AD-PSO algorithm in order to optimize the hyperparameters of an LSTM deep learning model [21]. The method was used for wind speed forecasting by integrating the aforementioned techniques.

To evaluate the effectiveness of the suggested approach, predictions were made regarding the hourly power generation up to 48 hours in advance using wind power forecasting data obtained from seven different wind farms. The wind forecasting section of the Kaggle Global Energy Forecasting Challenge in 2012 was where the dataset was collected. According to the findings, the accuracy attained by the suggested Guided WOA-based AD-PSO algorithm was superior to that achieved by existing optimization and deep learning strategies. In order to demonstrate that the suggested algorithm produces accurate results, a series of statistical tests, such as Wilcoxon's rank-sum and ANOVA, were carried out. In addition, the strategy that was suggested can be applied to other domains, such as the forecasting of load, stock price, and traffic, respectively. The Guided WOA-based AD-PSO method has the capability of optimizing the hyperparameters of a variety of deep learning models, such as CNN and Recurrent Neural Networks (RNN). The proposed method has tremendous potential to be implemented in a variety of contexts in order to enhance the accuracy of predictions, particularly in light of the growing availability of large amounts of data.

18.5.4 Fourth application: speech emotion recognition

One of the most significant challenges that the current techniques for voice emotion recognition need to surmount is the absence of a dataset that is sufficiently large to effectively train the deep learning models that are currently available. This is one of the most significant obstacles that must be overcome. The purpose of this research is to present a new data augmentation approach with the goal of enhancing the spoken emotions dataset by adding more samples via the precise incorporation of noise fractions. The deep learning models' hyperparameters are either constructed by hand or updated while the model is in the process of being trained. This is the case for all of the models. This can be accomplished in two ways: manually or automatically. Having said that, utilizing this strategy does not ensure that the optimal values will be determined for these parameters. As a result, we propose an improved version in which the hyperparameters of the deep learning model have been tweaked to obtain the optimal values for them, allowing for increased recognition accuracy as a result. This particular deep-learning model is made up of a CNN that has a layer of LSTM and comprises four local feature-learning blocks. While the CNN is intended to learn features from the log Mel-spectrogram of input speech samples, the LSTM layer is responsible for learning both short-term and long-term correlations. A recently developed stochastic fractal search (SFS)-guided whale optimization technique was applied [22]. This was done with the intention of improving the overall performance of this deep neural network. One of the SFS algorithm's strengths is that it is able to strike a balance between the exploration and exploitation of the positions held by the search agents. This ensures that the algorithm can locate the best potential global solution.

RAVDESS, Emo-DB, IEMOCAP, and SAVEE were the four speech emotion datasets that were utilized in the trials that were conducted for this research. Experiments like these were carried out so that the usefulness of the suggested strategy could be demonstrated. The SFS-guided WOA algorithm was utilized in order to optimize the deep-learning model parameters. The study highlights the possibility for employing advanced optimization approaches to increase the performance of deep learning models in voice emotion identification applications by utilizing advanced optimization techniques. The recognition accuracies that were achieved were, in order, 98.13 percent, 99.76 percent, 99.47 percent, and 99.50 percent, based on the four datasets. In addition, a statistical analysis of the accomplished goals is presented for the purpose of highlighting the dependability of the suggested methodology.

18.6 Conclusion

The most recent iteration of the Whale Optimization Algorithm (WOA) takes its cues from swarm intelligence, which is a form of artificial intelligence that mimics the behavior of animals that live in communities and cooperate with one another. This type of artificial intelligence was used to develop the WOA. This algorithm takes its cues from the hunting tactics of humpback whales, which use bubbles to trap their prey. The WOA optimization method replicates this technique by surrounding a target and then narrowing in on it until it can be captured. The advantage of this approach is that it can be used to solve complex optimization problems more effectively and efficiently than traditional optimization algorithms. Furthermore, because it is modeled after a biological process, it can be applied to a wide range of fields, including engineering, finance, and biology. In recent years, numerous problems relating to optimization and feature selection have been resolved thanks to the WOA algorithm. In comparison to the standard WOA format, the Guided WOA has undergone some changes in order to better suit its purpose. You will be able to use this to overcome the disadvantage of applying this strategy thanks to this. In the Guided WOA, also known as the Guided WOA, a whale is allowed to follow three random whales rather than just one other whale in order to improve its exploration performance. This is done so that the whale can find more food. This can inspire whales to engage in more adventurous action because it does not affect them in any way when they take on the role of leader. It has been demonstrated, through the use of a variety of applications and tests, that the Guided WOA-based approaches perform better than a wide variety of other comparative optimization algorithms and give higher levels of accuracy in comparison to other algorithms. This is the case in both terms of performance and levels of accuracy.

References

[1] X. Yang, Introduction to Mathematical Optimization: From Linear Programming to Metaheuristics, Cambridge International Science Publishing Ltd, 2008.

[2] J. Brownlee, Clever Algorithms: Nature-Inspired Programming Recipes, Jason Brownlee, 2011.

[3] X.-S. Yang, A.H. Gandomi, S. Talatahari, A.H. Alavi, Metaheuristics in Water, Geotechnical and Transport Engineering, Newnes, 2012.

[4] A.R. Yıldız, et al., An effective hybrid immune-hill climbing optimization approach for solving design and manufacturing optimization problems in industry, Journal of Materials Processing Technology 209 (6) (2009) 2773–2780.

[5] A.H. Gandomi, A.H. Alavi, Multi-stage genetic programming: a new strategy to nonlinear system modeling, Information Sciences 181 (23) (2011) 5227–5239.

[6] D.S. Khafaga, A.A. Alhussan, E.-S.M. El-Kenawy, A. Ibrahim, M.M. Eid, A.A. Abdelhamid, Solving optimization problems of metamaterial and double T-shape antennas using advanced meta-heuristics algorithms, IEEE Access 10 (2022) 74449–74471, https://doi.org/10.1109/access.2022.3190508.

[7] E.-S.M. El-kenawy, H.F. Abutarboush, A.W. Mohamed, A. Ibrahim, Advance artificial intelligence technique for designing double T-shaped monopole antenna, Computers, Materials & Continua 69 (3) (2021) 2983–2995, https://doi.org/10.32604/cmc.2021.019114.

[8] A.A. Alhussan, D.S. Khafaga, E.-S.M. El-Kenawy, A. Ibrahim, M.M. Eid, A.A. Abdelhamid, Pothole and plain road classification using adaptive mutation dipper throated optimization and transfer learning for self driving cars, IEEE Access 10 (2022) 84188–84211, https://doi.org/10.1109/access.2022.3196660.

[9] E.-S.M. El-Kenawy, S. Mirjalili, A.A. Abdelhamid, A. Ibrahim, N. Khodadadi, M.M. Eid, Meta-heuristic optimization and keystroke dynamics for authentication of smartphone users, Mathematics 10 (16) (2022) 2912, https://doi.org/10.3390/math10162912.

[10] E.-S.M. El-kenawy, F. Albalawi, S.A. Ward, S.S.M. Ghoneim, M.M. Eid, A.A. Abdelhamid, N. Bailek, A. Ibrahim, Feature selection and classification of transformer faults based on novel meta-heuristic algorithm, Mathematics 10 (17) (2022) 3144, https://doi.org/10.3390/math10173144.

[11] S. Mirjalili, A. Lewis, The whale optimization algorithm, Advances in Engineering Software 95 (2016) 51–67, https://doi.org/10.1016/j.advengsoft.2016.01.008, http://www.sciencedirect.com/science/article/pii/S0965997816300163.

[12] F. Valdez, O. Castillo, P. Melin, Bio-inspired algorithms and its applications for optimization in fuzzy clustering, Algorithms 14 (4) (2021) 122.

[13] E.-S.M. El-Kenawy, S. Mirjalili, F. Alassery, Y.-D. Zhang, M.M. Eid, S.Y. El-Mashad, B.A. Aloyaydi, A. Ibrahim, A.A. Abdelhamid, Novel meta-heuristic algorithm for feature selection, unconstrained functions and engineering problems, IEEE Access 10 (2022) 40536–40555, https://doi.org/10.1109/access.2022.3166901.

[14] D. Wolpert, W. Macready, No free lunch theorems for optimization, IEEE Transactions on Evolutionary Computation 1 (1) (1997) 67–82, https://doi.org/10.1109/4235.585893.

[15] E.-S.M. El-kenawy, A. Ibrahim, S. Mirjalili, M.M. Eid, S.E. Hussein, Novel feature selection and voting classifier algorithms for COVID-19 classification in CT images, IEEE Access 8 (2020) 179317–179335, https://doi.org/10.1109/access.2020.3028012.

[16] S. Mirjalili, S.M. Mirjalili, S. Saremi, S. Mirjalili, Whale Optimization Algorithm: Theory, Literature Review, and Application in Designing Photonic Crystal Filters, Springer International Publishing, Cham, 2020, pp. 219–238, https://doi.org/10.1007/978-3-030-12127-3_13.

[17] E. Cuevas, F. Fausto, A. González, Metaheuristics and Swarm Methods: A Discussion on Their Performance and Applications, Springer International Publishing, Cham, 2020, pp. 43–67, https://doi.org/10.1007/978-3-030-16339-6_2.

[18] F. Fausto, A. Reyna-Orta, E. Cuevas, Á.G. Andrade, M.A. Pérez-Cisneros, From ants to whales: metaheuristics for all tastes, Artificial Intelligence Review 53 (2019) 753–810, https://doi.org/10.1007/s10462-018-09676-2.

[19] E.-S.M. El-Kenawy, N. Khodadadi, S. Mirjalili, T. Makarovskikh, M. Abotaleb, F.K. Karim, H.K. Alkahtani, A.A. Abdelhamid, M.M. Eid, T. Horiuchi, et al., Metaheuristic optimization for improving weed detection in wheat images captured by drones, Mathematics 10 (23) (2022) 4421, https://doi.org/10.3390/math10234421.

[20] S.S.M. Ghoneim, T.A. Farrag, A.A. Rashed, E.-S.M. El-Kenawy, A. Ibrahim, Adaptive dynamic meta-heuristics for feature selection and classification in diagnostic accuracy of transformer faults, IEEE Access 9 (2021) 78324–78340, https://doi.org/10.1109/access.2021.3083593.

[21] A. Ibrahim, S. Mirjalili, M. El-Said, S.S.M. Ghoneim, M.M. Al-Harthi, T.F. Ibrahim, E.-S.M. El-Kenawy, Wind speed ensemble forecasting based on deep learning using adaptive dynamic optimization algorithm, IEEE Access 9 (2021) 125787–125804, https://doi.org/10.1109/access.2021.3111408.

[22] A.A. Abdelhamid, E.-S.M. El-Kenawy, B. Alotaibi, G.M. Amer, M.Y. Abdelkader, A. Ibrahim, M.M. Eid, Robust speech emotion recognition using CNN+LSTM based on stochastic fractal search optimization algorithm, IEEE Access 10 (2022) 49265–49284, https://doi.org/10.1109/access.2022.3172954.

Chapter 19

Optimal Power Flow with renewable power generations using hyper-heuristic technique

M.H. Sulaiman and Z. Mustaffa
Universiti Malaysia Pahang, Pekan, Pahang, Malaysia

19.1 Introduction

Optimal Power Flow (OPF) problem in power system operation is one of the well-known problems that have attracted numerous researchers and power engineers to propose better solutions since decades ago. The complexity of the Optimal Power Flow (OPF) solution makes it a challenging task. The solution of the OPF problem poses a challenge due to its complex nature, as it involves large-scale, non-linear, and non-convex optimization problems with various constraints, making it difficult to solve using traditional mathematical modeling techniques. To tackle this challenge, the paper proposes the use of metaheuristic algorithms as an alternative solution, which have been proven to be more effective compared to traditional methods [1,2].

Numerous metaheuristic algorithms, including Barnacles Mating Optimizer [3], Teaching Learning Based Optimization (TLBO) [4], Moth-Flame Optimizer (MFO) [5], Symbiotic Organisms Search Algorithm (SOSA) [6], Salp Swarm Algorithm (SSA) [7], and many others, have been proposed to date for solving OPF problems, in particularly for loss minimization. The modified, enhanced, and/or hybrid version of the metaheuristic algorithm, such as the hybrid modified Imperialist Competitive Algorithm and Sequential Quadratic Programming, have also been developed to improve the searching behavior of certain algorithms in solving OPF (ICA-SQP) [8], the Long-Term Memory Harris' Hawk Optimization (LTM-HHO) [9], the hybrid Artificial Bee Colony with Differential Evolution (ABC-DE) [10], the Improved Chaotic Electromagnetic Field Optimization algorithm (ICEFO) [1], the Grey Wolf Optimizer based on Crisscross Search (CS-GWO) [9], the Differential Evolution (DE) with variants [11–13], CS-GWO algorithm [14], Efficient Parallel GA (EPGA) [15], and Modified Grasshopper Optimization Algorithm (MGOA) [16].

This study proposes a new strategy to solve OPF problems together with the presence of wind-solar-small hydro power generations by implementing the hyper heuristic technique. Three recent metaheuristic algorithms are employed as low-level meta-heuristics (LLH) to discover the optimal solution of OPF problems namely Grey Wolf Optimizer (GWO) [17], Barnacles Mating Optimizer (BMO) [18–20], and Whale Optimization Algorithm (WOA) [21]. A high-level hyper heuristics (HHH) strategy called Exponential Monte Carlo with counter (EMCQ) is utilized as a selector when optimizing the control variables in order to get the best solutions. Related works on the HH approach to solve other optimization problems can be seen such as in t-way test suit generation [22,23], sorted-waste capacitated location routing problem [24], and image reconstruction [25].

The aim of this paper is to introduce a hyper-heuristic approach to address the optimal power flow (OPF) problem. Section 19.2 of the paper delves into the formulation of the OPF problem and provides a comprehensive explanation of the details required to comprehend the issue. In Section 19.3, the low-level heuristics (LLH) that have been selected for this study are briefly described. The hyper heuristic strategy is introduced and explained in detail in Section 19.4. In Section 19.5, the proposed hyper heuristic method for solving the OPF problem is put into action. The results and evaluation of the method are detailed in Section 19.6, and the conclusion, including key takeaways and future possibilities, is given in Section 19.7.

Handbook of Whale Optimization Algorithm. https://doi.org/10.1016/B978-0-32-395365-8.00025-7

19.2 Optimal Power Flow incorporating stochastic solar, wind, and small hydro power generation

The Optimal Power Flow (OPF) is a mathematical optimization problem that aims to determine the most efficient configuration of a power system, given certain constraints and objectives. The objective of OPF is to minimize a chosen cost or loss function, which is often a combination of real power losses, generation costs, and operational constraints. In a power system, the control variables can be the generator output levels, the tap positions on transformers, and the switch status of certain components. The OPF problem takes into account these control variables and the equality and inequality constraints, such as the power balance, voltage magnitude limits, and thermal limits of the components. This study specifically focuses on the application of OPF to a power system that includes both thermal generation sources and stochastic sources of wind, solar, and small hydro energy. The goal is to minimize losses in the system while taking into account the uncertainties and variability of the stochastic energy sources. By using OPF, the optimal control variables can be identified, leading to a more efficient and cost-effective power system operation.

In this work, the performance of the OPF solution will be analyzed on a revised IEEE 57-bus system, exhibited in Fig. 19.1. There are 57 buses in this system, 42 loads with 7 generators. There are eighty transmission lines for this system. From the seven generators, three generators are converted into solar power, solar with small hydro power as well as wind power generator located at bus 9, 6, and 12, respectively. In solving the OPF problem, the objective function, referred to as F_{Loss}, can be mathematically expressed as:

$$F_{Loss} = \sum_{i=j}^{nl} \sum_{j \neq i}^{nl} G_{ij} \left[V_i^2 + V_j^2 - 2V_i V_j \cos\left(\delta_i - \delta_j\right) \right] \tag{19.1}$$

where G_{ij} is the conductance of the *i*-*j* transmission line, V_i and V_j stand for the voltage at the sending and receiving ends of bus i and j, respectively, and nl stands for the total number of transmission lines in the power system. All workable solutions to the OPF problem must abide by the allowable constraints in order to be successful. Additionally, to account for the system's complexity, the power balance equations must also to be fulfilled, and can be mathematically expressed as follows:

$$P_{Gi} - P_{Di} - V_i \sum_{j=1}^{nB} V_j \left[G_{ij} \cos\left(\delta_{ij}\right) + B_{ij} \sin\left(\delta_{ij}\right) \right] = 0 \quad \forall i \in nB \tag{19.2}$$

$$Q_{Gi} - Q_{Di} - V_i \sum_{j=1}^{nB} V_j \left[G_{ij} \sin\left(\delta_{ij}\right) - B_{ij} \cos\left(\delta_{ij}\right) \right] = 0 \quad \forall i \in nB \tag{19.3}$$

The expression δ_{ij} stands for the variation in the angle of voltage between bus i and bus j. The real and reactive power generated at bus i, which encompasses energy from sources such as wind, solar and small hydro, are represented by P_{Gi} and Q_{Gi}. Meanwhile, the real and reactive power consumed at bus i is denoted by P_{Di} and Q_{Di}. Lastly, nB symbolizes the total number of bus systems in the system.

Moreover, to ensure the smooth operation of the power system components, these equations must also abide by the operating constraints, which are represented as:

$$P_{TGi}^{\min} \leq P_{TGi} \leq P_{TGi}^{\max} \quad i = 1, ..., N_{TG} \tag{19.4}$$

$$P_{WG,j}^{\min} \leq P_{WG,j} \leq P_{WG,j}^{\max} \quad j = 1, ..., N_{WG} \tag{19.5}$$

$$P_{SG,k}^{\min} \leq P_{SG,k} \leq P_{SG,k}^{\max} \quad k = 1, ..., N_{SG} \tag{19.6}$$

$$P_{SHG,k}^{\min} \leq P_{SHG,k} \leq P_{SHG,k}^{\max} \quad k = 1, ..., N_{SHG} \tag{19.7}$$

$$Q_{TGi}^{\min} \leq Q_{TGi} \leq Q_{TGi}^{\max} \quad i = 1, ..., N_{TG} \tag{19.8}$$

$$Q_{WG,j}^{\min} \leq Q_{WG,j} \leq Q_{WG,j}^{\max} \quad j = 1, ..., N_{WG} \tag{19.9}$$

$$Q_{SG,k}^{\min} \leq Q_{SG,k} \leq Q_{SG,k}^{\max} \quad k = 1, ..., N_{SG} \tag{19.10}$$

$$Q_{SHG,k}^{\min} \leq Q_{SHG,k} \leq Q_{SHG,k}^{\max} \quad k = 1, ..., N_{SHG} \tag{19.11}$$

$$V_{Gi}^{\min} \leq V_{Gi} \leq V_{Gi}^{\max} \quad i = 1, ..., N_G \tag{19.12}$$

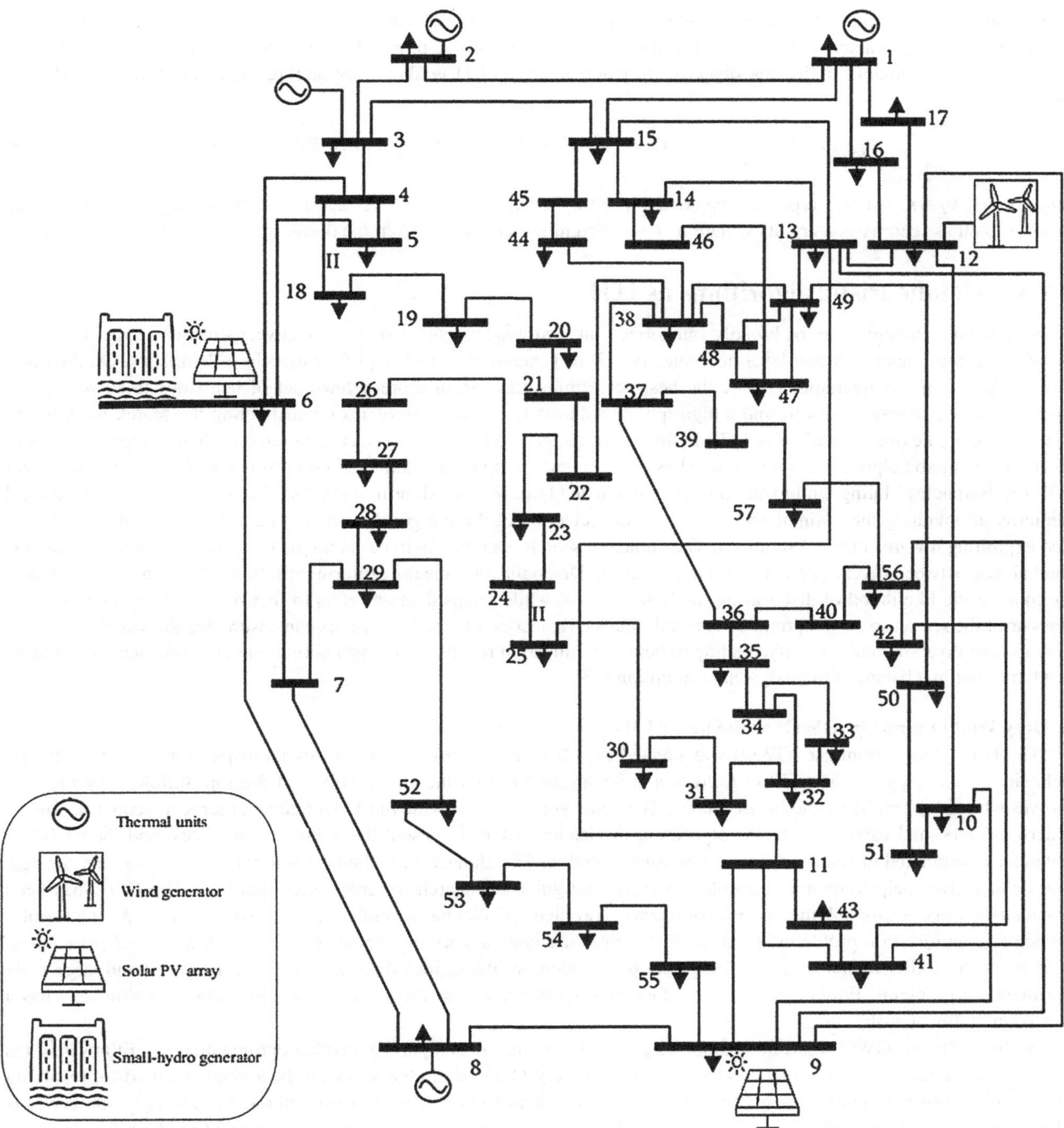

FIGURE 19.1 Modified IEEE 57-bus system [3].

It is noteworthy that all these restrictions can be met through the utilization of the MATPOWER [26] power flow program, guaranteeing the acquisition of precise results.

The OPF problem in the modified IEEE-57 bus system requires optimization of 13 variables for its solution. In this study, the following variables are optimized: the thermal power generation P_{TG2}, P_{TG3}, and P_{TG8}, the energy generated from renewable sources, P_{SHG} stand for combined solar-mini hydro generators, individual solar generator, P_{SG} and wind generators, P_{WG}, as well as the voltage magnitude at these power generators' buses including the swing bus. The permissible limits for these control and state variables are specified in [3]. To prevent any violations of the constraints previously described, and to guarantee that the state variables, including the swing power generator and reactive power generation, remain within acceptable limits, the objective function in Eq. (19.1) is altered by adding the penalty function P_{Fi}, as follows:

$$F_{Loss} = P_{Loss} + PF_1 * Qerr + PF_2 * V_L err + PF_3 * Serr + PF_4 * P_{G1} err \tag{19.13}$$

where *Qerr*, $V_L err$, and *Serr* represent the deviation from the target reactive power generation, the voltage violation at load buses as well as security constraints violation. $P_{G1} err$ signifies a deviation from the desired real power at the swing bus.

19.3 Metaheuristic algorithms as LLH

Hyper heuristic strategies can be broadly categorized into two types: generative and selective. Generative hyper heuristics involve the combination of low-level heuristics (LLH) to generate new and improved high-level heuristics. On the other hand, selective hyper heuristics choose the best performing LLH from a pre-defined set of heuristics to solve a given problem. These strategies aim to find a high-quality solution to complex problems by abstracting the search process and making it more flexible and adaptive [23]. In this study, the proposed solution is based on selective hyper-heuristics, where three metaheuristic algorithms are employed as LLH to find the optimal solution of OPF problems: Grey Wolf Optimizer (GWO), Barnacles Mating Optimizer (BMO), and Whale Optimization Algorithm (WOA). Each algorithm has merits and demerits in obtaining the optimal solution. Thus, the selection of these algorithms as LLH can be benefited in exploring and exploiting towards optimal solutions. The choice of search operators in hyper heuristic strategies is crucial as it affects the balance between diversification and intensification. Diversification means exploring different places in the search area to improve the likelihood of discovering the best solution, while intensification refers to focusing on the most promising regions of the search space to optimize it. A well-balanced selection of search operators can ensure that the search process is able to both diversify and intensify, leading to better optimization results. It is worth noting that choosing search operators randomly but in a balanced manner is also an option [23].

a. Grey Wolf Optimizer (GWO) as the first LLH

The Grey Wolf Optimizer (GWO) was created by Mirjalili in 2014 [17] and takes its inspiration from the hunting behavior of grey wolves. The GWO follows a strict hierarchy, with the alpha wolves at the top, followed by the beta wolves who aid the alpha in making decisions. The delta wolves make up the third hierarchy and serve as scouts, sentinels, elders, hunters, and caretakers. In the algorithm, the leader wolves represent the solution candidates, and their hunting behavior is simulated in search of the best possible solution. The alpha wolf is the best solution found so far and the beta and delta wolves help to explore the solution space and guide the search towards better solutions. The GWO has been shown to be effective for solving various optimization problems. GWO has several attractive features that make it a popular choice for optimization problems. It has a simple structure and easy to implement, with only a few control parameters. It also has the ability to handle a wide range of optimization problems, including non-linear, multi-modal and large-scale optimization problems. Furthermore, it has a fast convergence rate compared to other optimization algorithms and has a low computational cost.

Additionally, the GWO algorithm has shown good performance in comparison to other optimization algorithms in terms of solution accuracy and convergence speed. Overall, the GWO algorithm has shown to be a viable method for resolving optimization issues and has attracted the interest of researchers from a variety of disciplines. In this study, the GWO is utilized as a Local Search Heuristic (LLH) or search operator to produce new solutions, if selected by the hyper-heuristic algorithm. A diagram of the complete GWO algorithm as a search operator can be seen in Fig. 19.2.

b. Barnacles Mating Optimizer (BMO) as the second LLH

The barnacles' mating behavior was the foundation for the optimization technique known as the Barnacles Mating Optimizer (BMO) [18]. As a type of evolutionary-based algorithm, BMO uses the Hardy-Weinberg principle to ensure efficient exploitation of previously discovered solutions, while at the same time introducing the element of exploration to expand the search for even better solutions. The use of sperm cast, a key aspect of barnacle mating behavior, allows BMO to introduce a random component into the optimization process, making it a highly effective and versatile optimization method. With its unique blend of exploitation and exploration, the Barnacles Mating Optimizer is capable of finding solutions that would

Algorithm 1: GWO search operator

Input: The current population, pop_i {pop_1,...,pop_{dim}}, Current Best$_i$, T=max iteration,l=current iteration

Output: The updated current population, pop_i' {pop_1',...,pop_{dim}'}

```
Find Alpha, beta and delta positions (Alpha_pos, Beta_pos, Delta_pos)
a=2-l*(2/T)
    for j=1:dim
        r1=rand(),r2=rand()
        A1=2*a*r1-a; C1=2*r2;
        D_alpha=abs(C1*Alpha_pos(j)-popi(:,j));
        X1=Alpha_pos(j)-A1*D_alpha;
        r1=rand(),r2=rand()
        A2=2*a*r1-a; C2=2*r2;
        D_beta=abs(C2*Beta_pos(j)- popi(:,j));
        X2=Beta_pos(j)-A2*D_beta;
        r1=rand(), r2=rand()
        A3=2*a*r1-a; C3=2*r2;
        D_delta=abs(C3*Delta_pos(j)- popi(:,j));
        X3=Delta_pos(j)-A3*D_delta;
        popi(:,j)=(X1+X2+X3)/3;
    end for
Return popi'
```

FIGURE 19.2 GWO LLH.

Algorithm 2: BMO search operator

Input: The current population, pop_i {pop_1,...,pop_{dim}}, pop_{mum} {pop_1,...,pop_{dim}}, pop_{dad}{pop_1,...,pop_{dim}}

Output: The updated current population, pop_i' {pop_1',...,pop_{dim}'}

```
Set pl
        if selection of Dad and Mum ≤ pl
            for j=1:dim
                p=randn()
                popi'(:,j)= p*popmum(:,j)+(1-p)*popdad(:,j);
            end
        else
            for j=1:dim
                popi'(:,j)= rand()*popmum(:,j);
            end
        end if
Return popi'
```

FIGURE 19.3 BMO LLH.

be difficult to discover using traditional optimization algorithms. Fig. 19.3 in the referenced work shows how the BMO algorithm is used as a Local Search Heuristic (LLH) to find new solutions in an optimization problem.

c. Whale Optimization Algorithm (WOA) as the third LLH

WOA takes its inspiration from humpback whales' hunting habits and one of the swarm intelligence (SI) metaheuristic algorithms proposed by Mirjalili [21]. Humpback whales recognize the location of prey and encircle them. As a swarm intelligence (SI) algorithm, WOA leverages the remarkable hunting strategies of humpback whales to find optimal solutions to complex optimization problems. Humpback whales are known for their ability to recognize the location of prey and encircle it, effectively trapping it. In WOA, this concept is translated into an optimization algorithm through the use of a two-phase approach, with the exploitation phase being based on the humpback whale's bubble-net attacking method and the exploration phase relying on a random approach for diversified results. The exploitation phase focuses on refining previously discovered solutions, while the exploration phase broadens the search for new, potentially better solutions. By combining these two phases, WOA effectively balances exploitation and exploration in its search for optimal solutions. This makes it a highly effective optimization algorithm that is capable of delivering superior results, even in complex and challenging optimization problems. The LLH for WOA is depicted in Fig. 19.4.

Algorithm 3: WOA search operator

Input: The current population, pop_i {pop_1,...,pop_{dim}}, `Current Best_i, T=max iteration, l=current iteration, pop_no`

Output: The updated current population, pop_i' {pop_1',...,pop_{dim}'}

```
Leader_pos = Current Best_i
a=2-l*(2/T); a2=-1+l*((-1)/T);
r1=rand(),r2=rand()
A=2*a*r1-a; C1=2*r2; b=1;
t=(a2-1)*rand+1; p = rand();
      for j=1:dim
         if p<0.5
            if abs(A)>=1
               rand_leader_index = floor(pop_no*rand()+1);
               X_rand = pop_i(rand_leader_index, :);
               D_X_rand=abs(C*X_rand(j)-pop_i(i,j));
               pop_i(i,j)=X_rand(j)-A*D_X_rand;
            elseif abs(A)<1
               D_Leader=abs(C*Leader_pos(j)-pop_i(i,j));
               pop_i(i,j)=Leader_pos(j)-A*D_Leader;
            end
         elseif p>=0.5
            distance2Leader=abs(Leader_pos(j)-pop_i(i,j));
            pop_i(i,j)=distance2Leader*exp(b.*t).*cos(l.*2*pi)+Leader_pos(j);
         end
      end for
Return pop_i'
```

FIGURE 19.4 WOA LLH.

19.4 Hyper heuristic strategies for OPF solution

The idea of implementing hyper-heuristic strategies for solving OPF problem with the presence of stochastic wind-solar-small hydro power generations is illustrated in Fig. 19.5. Three LLH algorithms that have been presented in the previous section viz, GWO, BMO, and WOA are used to obtain the solution where the performance of the selected LLH algorithm will be evaluated at each population. In this study, the Exponential Monte Carlo with counter (EMCQ) method is used as a selector and acceptance mechanism, and it has been thoroughly discussed in [27]. The hyper heuristic approach is an approach that adapts and combines different low-level heuristics to solve optimization problems. It is frequently utilized in many different sectors, including power systems, due to its ability to handle complex and dynamic optimization problems. In the present study, the hyper heuristic approach is applied to solve the OPF problem with the presence of stochastic wind-solar-small hydro power generations. The proposed method demonstrates the benefits of the hyper heuristic approach by combining the strengths of different low-level heuristics and adapting to changing conditions in the optimization problem. This results in improved performance and solution quality compared to LLH optimization algorithms.

EMCQ is a parameter-free hyper-heuristic that uses probability density to accept lesser quality solutions for avoiding the local optima entrapment. The following expression represents the probability density, ψ:

$$\psi = e^{-\delta t/q} \tag{19.14}$$

where δ, t, and q are the difference value of objective evaluation function between current and the previous solutions, iteration number and counter for consecutive non-improving iterations, respectively. It is crucial to note that the value of q is raised after an unsuccessful move and reset to 1 following a successful move, in order to diversify the solution further [23,27]. The pseudo code for EMCQ HH strategy is exhibited in Fig. 19.6.

19.5 Implementation of HH into OPF solution

The usage of HH into OPF problems is started by providing the all the information regarding the problem dimension, test system used, boundaries limits, all the LLH to be employed as well as function evaluation. In solving the OPF with stochastic wind-solar-small hydro power generation, the set of control variables as well as state variables can be described

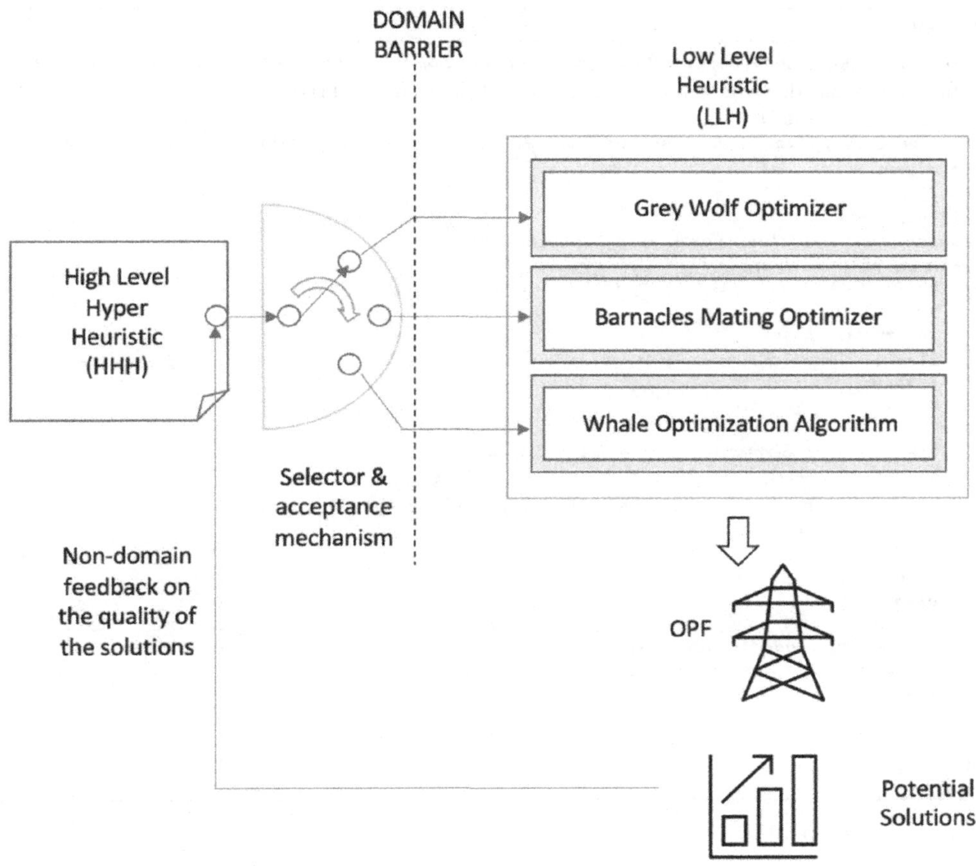

FIGURE 19.5 Hyper-heuristic strategies for selecting LLH and acceptance mechanism.

as follows:

$$x^T = \left[P_{G_2}, \ldots, P_{G_{NG}}, V_{G_1}, \ldots, V_{G_{NG}}\right] \tag{19.15}$$

$$u^T = \left[Q_{G_1}, \ldots, Q_{G_{NG}}, \ldots, V_{L_1}, \ldots, V_{L_p}, S_{L_1}, \ldots, S_{L_q}, P_{G_1}\right] \tag{19.16}$$

In order to optimize the modified IEEE-57 bus system, there is a total of 13 control variables that need to be optimized and 8 state variables of generations that must be operated within their specified limits. The optimization process begins with an initialization step, in which an initial solution is obtained with a fitness or objective function for each population. Then, a random LLH is selected for the first population and the Hyper-Heuristic (HH) determines whether to keep or switch to a different LLH for subsequent populations based on the performance of the selected LLH. The objective function is evaluated by converting the population into power flow data, running a power flow solution program, and then using Eq. (19.13) from the load flow solution to calculate the losses. This process is repeated for each population selected by the LLH and is iterated until the maximum number of iterations is reached. The implementation of the HH for solving the OPF problem is depicted in Fig. 19.7.

The computational complexity of a hyper heuristic approach is determined by several factors, including the dimension of the problem (d), the number of populations (n), the number of LLHs (L), as well as the cost of evaluating the objective function (Cof). Based on the equation $O(t(nd + L + Cof * n))$, we can determine the computational complexity of the hyper heuristic is considered low since t, n, d, and Cof are all constant. The termination criterion for the optimization process only involves a constant factor of $O(1)$.

Algorithm 5: EMCQ HH

Input: All population pop_{all} {$pop_{1\text{-}1},\ldots,pop_{1\text{-}dim}$; $popall_{\text{-}1}, \ldots, pop_{all\text{-}dim}$}, *N= no of population, Boundaries: ub, lb, Function evaluations: FUNC*

Output: Final solution: best population, pop_i' {$pop_1',\ldots,pop_{dim}'$}, best obj, percentage of LLH

```
Initialization, popall, Set q = 1; MAX_gen;
valall = FUNC (popall), Define H as a set of LLH, Hk= {MFO, BMO, TLBO, GBO}; Select randomly Hk
Select randomly Hk
for gen=1:MAX_gen
    for i=1: pop_size
        Generate new solution, popi' using selected Hk
        Evaluate vali' =FUNC(popi')
        if (vali' < vali) % better solution in terms of the objective value
            popi← popi' ;
            vali← vali';
            disp ("Improving Moves - Maintain current operator");
            Keep Hk
            q = 1;
        else
            disp ("Non-Improving Moves - Maintain current operator with probability");
            δ = vali' - vali;
            Generate a random number rand()[0,1], p_toss; ψ = e^(-δt/q); %probability from monte-carlo
            if p_toss < ψ
               Keep Hk
               q = 1;
            else
               disp ("Must change search operator");
                Select randomly new LLH, Hk
                q = q+1;
            end
        end
    end
    Record best solution so far
end
```

FIGURE 19.6 Pseudo code for EMCQ HH.

TABLE 19.1 Statistical results of loss minimization of OPF problem HH and LLH algorithms.

Algorithms	Min	Max	Avg.	Std.
GWO	19.8673	23.0340	20.7108	0.7831
BMO	19.7002	26.6108	21.0650	2.0995
WOA	20.2197	27.4457	23.4810	1.8649
HH	**19.6813**	**20.9314**	**19.9173**	**0.3583**

19.6 Results and discussion

The OPF problem solving simulations were carried out on a MacBook Pro with a 2.40 GHz Quad-Core Intel Core i5 processor and 8 GB RAM, using MATLAB® 2019b. The power flow solutions were obtained using the MATPOWER toolbox [26], a well-known and reliable tool for the assessment of function evaluations. A comparison of the performance of the algorithms GWO [17], BMO [18], WOA [21], and HH was done using 30 independent simulations with similar OPF problem settings. The results showed that the Hyper-Heuristic algorithm outperformed the other three algorithms (GWO, BMO, and WOA) in terms of performance. In Table 19.1, the algorithms' top outcomes are denoted in bold.

The results of optimizing 13 control variables and 8 state variables for all LLH algorithms and the HH are displayed in Table 19.2. From this table, it can be seen that the EMCQ HH algorithm resulted in the least total power loss of 19.6813 MW. The BMO followed with a power loss of 19.7002 MW, then the GWO with 19.8673 MW, and finally the WOA with 20.2197 MW. All of the optimized control and state variables were within the defined boundaries, as seen in columns 2 and 3 of the table. The results also indicate that all algorithms achieved maximum power generation from renewable energy sources, resulting in full utilization of renewable energy in comparison to thermal power generation. Incorporating

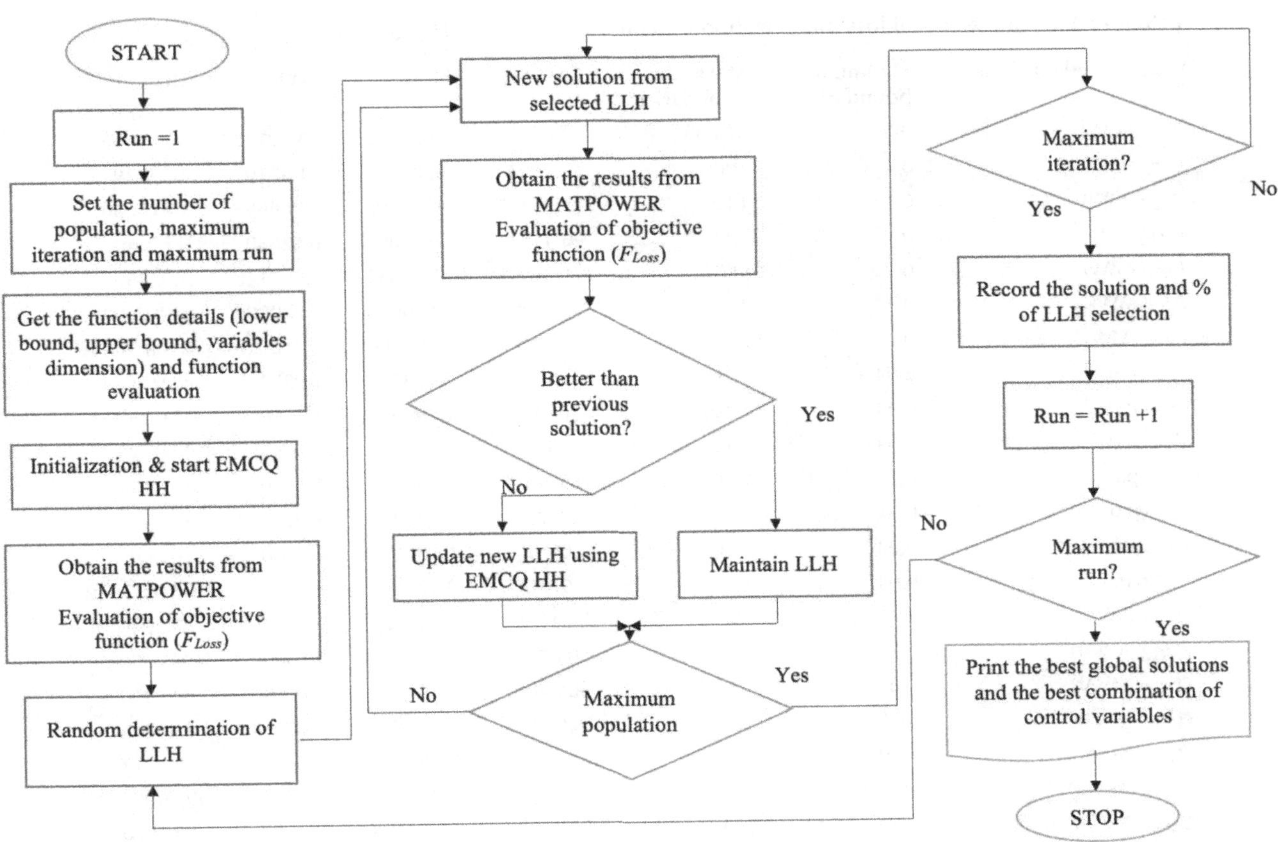

FIGURE 19.7 Flowchart of HH implementation into solving OPF problem.

renewable energy sources into existing power systems significantly reduces power loss and has a positive impact on the environment. From the convergence curve in Fig. 19.8, it is observed that all the algorithms converge within 200 iterations, with the exception of GWO which converged at 500 iterations. The best results from 30 simulations for all algorithms are shown in the figure, providing a visual representation of the optimization process.

Fig. 19.9 summarizes the percentage distribution of LLH by EMCQ HH. It can be noted that the selection of BMO is dominant as search operator in solving the stated problem, which is about 52% form total selection for generating the solution. From this figure also can be seen that WOA ranked as second LLH after BMO which is 27% from the total selection of HH to solve the loss minimization problem, even though as individual metaheuristic algorithm, WOA produced the worst results compared to BMO and GWO. This shows that the WOA is able to be produce much better results when incorporating with other algorithms, especially in this study.

Even though the results presented in this study show the effectiveness of the hyper heuristic technique, there are some drawbacks and limitation. In general, however, hyper-heuristic approaches can have limitations such as:

a. Sensitivity to parameters: the proposed hyper heuristic can be sensitive to the parameters used to control the low-level heuristics and finding the optimal combination of parameters can be a challenge.

b. Computational complexity: the proposed hyper heuristic can be computationally intensive, as they often involve multiple rounds of optimization and selection processes.

c. Difficulty in choosing low-level heuristics: choosing the right low-level heuristics for a problem can be challenging as their effectiveness may vary based on the specific attributes of the problem instance.

d. Performance variability: the proposed hyper heuristic can show variability in performance, as they rely on the performance of the low-level heuristics, which can be influenced by factors such as the size and structure of the problem instance, the quantity of solutions being assessed, and the starting solutions' quality.

TABLE 19.2 Detail results of loss minimization obtained by HH and LLH algorithms.

Variables\Algorithms	Minimum boundaries	Maximum boundaries	GWO	BMO	WOA	HH
P_{TG1} (MW)	0	575.88	299.7701	296.2552	287.9135	295.3856
P_{TG2} (MW)	0	100	15.7752	9.3801	31.4410	10.2729
P_{TG3} (MW)	0	140	135.5707	140.0000	129.6652	140.0000
P_{SHG} (MW)	0	100	96.4452	100.0000	89.2880	99.9658
P_{TG8} (MW)	0	550	313.1061	314.8649	322.7120	314.8570
P_{SG} (MW)	0	200	200.0000	200.0000	200.0000	200.0000
P_{WG} (MW)	0	210	210.0000	210.0000	210.0000	210.0000
V_1 in p.u.	0.94	1.10	1.0217	1.0232	1.0273	1.0234
V_2 in p.u.	0.94	1.10	1.0405	1.0986	1.1000	1.0525
V_3 in p.u.	0.94	1.10	1.0845	1.0158	1.1000	1.0152
V_6 in p.u.	0.94	1.10	0.9814	0.9501	1.1000	1.0201
V_8 in p.u.	0.94	1.10	1.0222	1.0282	1.0273	1.0257
V_9 in p.u.	0.94	1.10	1.0951	1.1000	1.1000	1.1000
V_{12} in p.u.	0.94	1.10	0.9770	0.9805	0.9665	0.9817
Q_{TG1} (MVAr)	-140	200	53.1499	62.9932	64.4605	63.5030
Q_{TG2} (MVAr)	-17	50	50.0000	50.0000	50.0000	50.0000
Q_{TG3} (MVAr)	-10	60	60.0000	36.0231	60.0000	29.9860
Q_{SHG} (MVAr)	-8	25	-8.0000	-8.0000	25.0000	4.4372
Q_{TG4} (MVAr)	-140	200	67.3071	76.6049	56.7254	65.4217
Q_{SG} (MVAr)	-3	9	9.0000	9.0000	9.0000	9.0000
Q_{WG} (MVAr)	-150	155	51.1770	55.0915	17.6582	59.1635
F_{Loss} (MW)			19.8673	19.7002	20.2197	**19.6813**

Hyper heuristics are especially useful for problems with high dimensionality and complex search spaces, such as the OPF problem, where it can be difficult to determine the best LLHs to use in advance. By using a flexible selection mechanism, hyper heuristics can dynamically adapt to the characteristics of the problem, which can result in faster convergence and better solutions. Additionally, hyper heuristic can handle multiple objectives and constraints, and can also handle uncertainty in the input data. These are important features for power system optimization problems, which often involve multiple objectives, constraints, and uncertainty in the input data. Overall, the hyper heuristic approach has shown promising results in power system optimization problems and has the potential to be a valuable tool for solving these types of problems. However, more research is needed to fully understand the potential of hyper heuristics in this field and to determine the best methods for applying them to different types of power system optimization problems.

19.7 Conclusion

A hyper-heuristic strategy to reducing losses in the OPF issue that uses renewable energy sources was proposed in this work. On a modified IEEE 57-bus system, the effectiveness of the suggested methodology and low-level metaheuristic algorithms was assessed. The outcomes demonstrated that the EMCQ hyper heuristic beat other low-level metaheuristics and reached exceptional performance. This shows that the suggested strategy could someday be a viable option for resolving OPF issues with various goals, such cost or emission reduction.

Acknowledgment

This study is supported by the Ministry of Higher Education Malaysia under Fundamental Research Grant Scheme Grant FRGS/1/2019/ICT02/UMP/02/4 (RDU1901130) & Universiti Malaysia Pahang under Distinguish Research Grant # RDU223003.

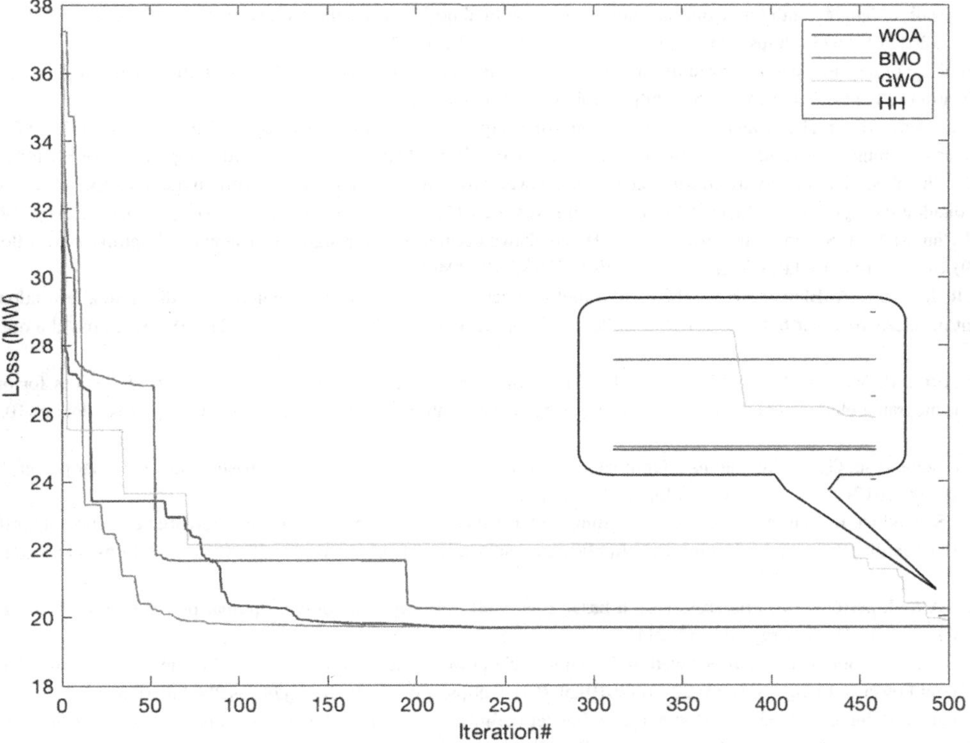

FIGURE 19.8 Convergence curve for all algorithms for loss minimization problem.

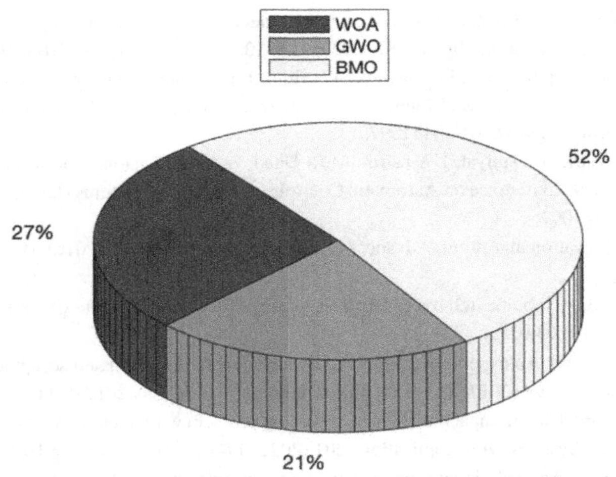

FIGURE 19.9 Search Operator Normalized Percentage Distribution for all LLH for all cases.

References

[1] H. Bouchekara, Solution of the optimal power flow problem considering security constraints using an improved chaotic electromagnetic field optimization algorithm, Neural Computing & Applications 32 (7) (2020) 2683–2703, https://doi.org/10.1007/s00521-019-04298-3.

[2] T.T. Nguyen, T.T. Nguyen, M.Q. Duong, A.T. Doan, Optimal operation of transmission power networks by using improved stochastic fractal search algorithm, Neural Computing & Applications 32 (13) (2020) 9129–9164, https://doi.org/10.1007/s00521-019-04425-0.

[3] M.H. Sulaiman, Z. Mustaffa, Solving optimal power flow problem with stochastic wind–solar–small hydro power using barnacles mating optimizer, Control Engineering Practice 106 (2021) 104672, https://doi.org/10.1016/j.conengprac.2020.104672.

[4] H.R.E.H. Bouchekara, M.A. Abido, M. Boucherma, Optimal power flow using Teaching-Learning-Based Optimization technique, Electric Power Systems Research 114 (2014) 49–59, https://doi.org/10.1016/j.epsr.2014.03.032.

[5] H. Buch, I.N. Trivedi, P. Jangir, Moth flame optimization to solve optimal power flow with non-parametric statistical evaluation validation, Cogent Engineering 4 (1) (2017) 1286731, https://doi.org/10.1080/23311916.2017.1286731.

[6] S. Duman, Symbiotic organisms search algorithm for optimal power flow problem based on valve-point effect and prohibited zones, Neural Computing & Applications 28 (11) (2017) 3571–3585, https://doi.org/10.1007/s00521-016-2265-0.

[7] M.Z. Islam, et al., Generation fuel cost and loss minimization using salp swarm algorithm based optimal power flow, in: 2020 International Conference on Computer Communication and Informatics (ICCCI), 22–24 Jan. 2020, 2020, pp. 1–6, https://doi.org/10.1109/ICCCI48352.2020.9104100.

[8] J. Ben Hmida, T. Chambers, J. Lee, Solving constrained optimal power flow with renewables using hybrid modified imperialist competitive algorithm and sequential quadratic programming, Electric Power Systems Research 177 (2019) 105989, https://doi.org/10.1016/j.epsr.2019.105989.

[9] K. Hussain, W. Zhu, M.N.M. Salleh, Long-term memory Harris' hawk optimization for high dimensional and optimal power flow problems, IEEE Access 7 (2019) 147596–147616, https://doi.org/10.1109/ACCESS.2019.2946664.

[10] J. Mahadevan, R. Rengaraj, A. Bhuvanesh, Application of multi-objective hybrid artificial bee colony with differential evolution algorithm for optimal placement of microprocessor based FACTS controllers, Microprocessors and Microsystems (2021) 104239, https://doi.org/10.1016/j.micpro.2021.104239.

[11] P.P. Biswas, P. Arora, R. Mallipeddi, P.N. Suganthan, B.K. Panigrahi, Optimal placement and sizing of FACTS devices for optimal power flow in a wind power integrated electrical network, Neural Computing & Applications 33 (12) (2021) 6753–6774, https://doi.org/10.1007/s00521-020-05453-x.

[12] P.P. Biswas, P.N. Suganthan, G.A.J. Amaratunga, Optimal power flow solutions incorporating stochastic wind and solar power, Energy Conversion and Management 148 (2017) 1194–1207, https://doi.org/10.1016/j.enconman.2017.06.071.

[13] P.P. Biswas, P.N. Suganthan, R. Mallipeddi, G.A.J. Amaratunga, Optimal power flow solutions using differential evolution algorithm integrated with effective constraint handling techniques, Engineering Applications of Artificial Intelligence 68 (2018) 81–100, https://doi.org/10.1016/j.engappai.2017.10.019.

[14] A. Meng, et al., A high-performance crisscross search based grey wolf optimizer for solving optimal power flow problem, Energy 225 (2021) 120211, https://doi.org/10.1016/j.energy.2021.120211.

[15] B. Mahdad, K. Srairi, T. Bouktir, Optimal power flow for large-scale power system with shunt FACTS using efficient parallel GA, International Journal of Electrical Power & Energy Systems 32 (5) (2010) 507–517, https://doi.org/10.1016/j.ijepes.2009.09.013.

[16] M.A. Taher, S. Kamel, F. Jurado, M. Ebeed, Modified grasshopper optimization framework for optimal power flow solution, Electrical Engineering 101 (1) (2019) 121–148, https://doi.org/10.1007/s00202-019-00762-4.

[17] S. Mirjalili, S.M. Mirjalili, A. Lewis, Grey wolf optimizer, Advances in Engineering Software 69 (2014) 46–61, https://doi.org/10.1016/j.advengsoft.2013.12.007.

[18] M.H. Sulaiman, Z. Mustaffa, M.M. Saari, H. Daniyal, Barnacles mating optimizer: a new bio-inspired algorithm for solving engineering optimization problems, Engineering Applications of Artificial Intelligence 87 (2020) 103330, https://doi.org/10.1016/j.engappai.2019.103330.

[19] M.H. Sulaiman, et al., Barnacles mating optimizer: a bio-inspired algorithm for solving optimization problems, in: 2018 19th IEEE/ACIS International Conference on Software Engineering, Artificial Intelligence, Networking and Parallel/Distributed Computing (SNPD), 27–29 June 2018, 2018, pp. 265–270, https://doi.org/10.1109/SNPD.2018.8441097.

[20] M.H. Sulaiman, Z. Mustaffa, M.M. Saari, H. Daniyal, I. Musirin, M.R. Daud, Barnacles mating optimizer: an evolutionary algorithm for solving optimization, in: 2018 IEEE International Conference on Automatic Control and Intelligent Systems (I2CACIS), 20-20 Oct. 2018, 2018, pp. 99–104, https://doi.org/10.1109/SNPD.2018.8441097.

[21] S. Mirjalili, A. Lewis, The whale optimization algorithm, Advances in Engineering Software 95 (2016) 51–67, https://doi.org/10.1016/j.advengsoft.2016.01.008.

[22] K.Z. Zamli, B.Y. Alkazemi, G. Kendall, A Tabu Search hyper-heuristic strategy for t-way test suite generation, Applied Soft Computing 44 (2016) 57–74, https://doi.org/10.1016/j.asoc.2016.03.021.

[23] K.Z. Zamli, F. Din, G. Kendall, B.S. Ahmed, An experimental study of hyper-heuristic selection and acceptance mechanism for combinatorial t-way test suite generation, Information Sciences 399 (2017) 121–153, https://doi.org/10.1016/j.ins.2017.03.007.

[24] C. Shang, L. Ma, Y. Liu, S. Sun, The sorted-waste capacitated location routing problem with queuing time: a cross-entropy and simulated-annealing-based hyper-heuristic algorithm, Expert Systems with Applications 201 (2022) 117077, https://doi.org/10.1016/j.eswa.2022.117077.

[25] N.R. Sabar, A. Turky, A. Song, A. Sattar, An evolutionary hyper-heuristic to optimise deep belief networks for image reconstruction, Applied Soft Computing 97 (2020) 105510, https://doi.org/10.1016/j.asoc.2019.105510.

[26] R.D. Zimmerman, C.E. Murillo-Sánchez, R.J. Thomas, MATPOWER: steady-state operations, planning, and analysis tools for power systems research and education, IEEE Transactions on Power Systems 26 (1) (2011) 12–19, https://doi.org/10.1109/TPWRS.2010.2051168.

[27] M. Ayob, G. Kendall, A Monte Carlo hyper-heuristic to optimise component placement sequencing for multi head placement machine, in: Proceedings of the International Conference on Intelligent Technologies, InTech, vol. 3, 2003, pp. 132–141.

Chapter 20

An efficient single image dehazing algorithm based on patch-wise transmission map estimation using Whale Optimization Algorithm

K. Ashwini[a], Hathiram Nenavath[b], and Ravi Kumar Jatoth[a]

[a]Department of Electronics and Communication Engineering, National Institute of Technology, Warangal, India, [b]Department of Electronics and Communication Engineering, National Institute of Technology, Jamshedpur, India

20.1 Introduction

Due to the presence of suspended particles like smoke, fog, dust particles, etc., images taken in a foggy environment have poor visibility and low contrast under present circumstances. However, small drops and other particles impede camera light the greatest on cloudy or foggy days. The lens of a camera acts as a filter, diminishing and scattering the light that enters it. As a result, the haze in the environment affects the images that were taken. As a result, the restoration of real haze-free images is a hard research topic with various applications, including surveillance and satellite photography. According to Koschmieder's law [1], the hazy scene is represented mathematically as:

$$I(z) = J(z)t(z) + A(1 - t(z)) \tag{20.1}$$

where $I(z)$ represents the foggy picture, $J(z)$ is an original scene, and z is location of the intensity value. A is the global atmospheric light and $t(z)$ is depth map at z, it can also be expressed in terms of the scene depth $d(z)$ as

$$t(z) = e^{-\psi d(z)} \tag{20.2}$$

where ψ is a scattering factor. By rearranging Eq. (20.1) to recover the true scene or haze-free image $J(z)$ as:

$$J(z) = \frac{I(z) - A}{t(z)} + A \tag{20.3}$$

From Eq. (20.3), Haze free image $J(z)$ can be restored by estimating the transmission map $t(z)$ and Atmospheric light A using haze input $I(z)$. Therefore, the estimation of accurate value to these parameters is a challenging task in image dehazing techniques.

Several methods are proposed to overcome the image dehazing problem. Classical prior based methods [2–5] are proposed to estimate global atmospheric light and depth map but When paired with a soft-mating, the dark channel prior suffers with halo effects, and it is also unable to recover color tones. Zhu et al. [6] implemented color attenuation prior using brightest and saturated pixels to estimate transmission map. These methods cannot estimate transmission map accurately in complex scenarios. To overcome that, physics based approaches [7], [8] are introduced based on certain assumptions and are not always applied to real-world haze models. To optimize the transmission appropriately and cut down on the halo effects, Y. Gao et al. [9] proposed an unique fast local Laplacian filtering with adaptive boundary restriction. Fusion-Based Methods [10], [11] are proposed based on patch-wise and pixel-wise transmission map estimation but require more computational complexity. According to Berman et al. [12], the intensities of a local patch of a haze-free image approximately lie in a line in the RGB color space, allowing for an estimate of the scene transmission. Salazar-Colores et al. [13] proposed image dehazing technique based on DCP and also uses morphological reconstruction for fast computing of transmission

Handbook of Whale Optimization Algorithm. https://doi.org/10.1016/B978-0-32-395365-8.00026-9

maps. Bai et al. [14] first generates the reference image using a deep pre-dehazer. It implements a progressive feature fusion module to fuse the foggy image and reference ground truth image to study guiding information. Finally, the picture restoration module uses the guiding information for clearer image restoration.

To solve these problems learning based models are proposed in image restoration applications. In learning based methods, deep learning models are trained to estimate the coefficients of Eq. (20.3) and restore the dehazed image from the haze input [15–18]. AOD-Net [19] generates dehazed images through a lightweight CNN architecture. But the dehazed images become darker in low-light regions. C. Wang et al. [20] introduced a unique weakly supervised method based on the multi-scale block that estimates the transmission map, atmospheric light, and intermediate dehazed image without any atmospheric light and actual transmission depth as supervision. RYF-Net [21] suggested YNet and RNet use the YCbCr and RGB color spaces, respectively, to accept the hazy scene as input and extract the haze-relevant features for dehazing. Kim and Kwon [22] developed a deep learning based dehazing approach, which enhance poor-light region needs to be applied to darkened images. Ren et al. [23] use temporal and semantic information to estimate the transmission map. Agrawal and Jalal [24] proposed dense haze removal using super pixel and nonlinear transform approach, which reduce halo artifacts in the dehazed images. Hence the learning-based approaches have made considerable progress, the restored images often have an overly saturated or overly smoothed appearance.

GAN-based methods [18], [25], [26] use a generator and a discriminator to implement image dehazing. Park et al. [27] have CycleGAN for natural color balancing and cGAN to recover local atmospheric light and transmission depth. Li et al. [28] has a dual-module structure (D-Net and T-Net). The T-Net would be connected to D-Net to transfer transmission depth. Two models can be coupled sequentially or in concurrently to communicate knowledge on different levels, but the generator adds artifacts that make the images look different from clear images. H. Zhu et al. [29] able to directly learn the physical features from inputs and remove the haze from hazy ones by utilizing a network consisting of an unique compositional generator and a novel deep-supervised discriminator. Y. Mo et al. [30] proposed dual GAN architecture based Dehazing model, which includes Cyclic Consistency Loss for training. LIGHT-Net [31] implemented two modules, color constancy module and haze reduction module. In the color constancy module, uses the GWA algorithm to assure colors in the obtained foggy image. The color constant foggy image is then processed by the suggested haze reducing module to restore the haze-free image. Chen et al. [32] proposed PDR-GAN. It is a GAN framework that incorporates both a physical-guided restoration stage and a depth-guided refinement stage.

To overcome the above problems, authors proposed a single image dehazing technique by using Whale Optimization Algorithm, called WOA-Dehaze. And estimated the Atmospheric light for given hazy image. We formulated fitness function in order to increase the image contrast and to minimize the Haze density by preserving edges data. Then, by minimizing the cost function, we are estimating optimal value of transmission map for every patch. Finally restored the dehazed image using estimated transmission depth map and local atmospheric light.

20.2 Whale optimization algorithm

WOA-Whale Optimization [33] Algorithm is an effective meta-heuristic optimization technique that has a high global search ability, fast convergence, and high efficiency. The whale optimization algorithm imitates the way humpback whales hunt. However, the most interesting facet of humpback whales is their unique hunting technique known as bubble-net feeding.

20.2.1 Encircle of the prey

The WOA algorithm considers that the target prey or a solution that is very close to best candidate now. After determining the best search agent, the other search agents will update their positions from the local best candidate. This process of updating search agents from local best is known as encircling prey.

$$\vec{Q}_i = \left|\vec{P}.\vec{X}_{lbest}(t) - \vec{X}_i(t)\right| \tag{20.4}$$

$$\vec{X}_i(t+1) = \vec{X}_{lbest}(t) - \vec{R}.\vec{Q}_i \tag{20.5}$$

Here $\vec{X}_{lbest}(t)$ is the best position vector of present iteration. $\vec{X}_i(t)$ is the position vector of ith candidate, $|.|$ is the absolute operator, and . is an element-by-element multiplication operation. $\vec{X}_{lbest}(t)$ should be calculated for each iteration if there exists a best solution. The coefficient vectors $\vec{R}$ and $\vec{P}$ are defined as follows:

$$\vec{R} = 2a_1.\vec{r}_v - a_1 \tag{20.6}$$

$$\vec{P} = 2.\vec{r}_v \tag{20.7}$$

where a_1 is a linearly decreasing variable from 2 to 0 for each iteration and $\vec{r}_v$ is a random vector whose values in the range of [0, 1].

20.2.2 Exploitation stage (bubble-net attacking method)

The humpback whales utilize the bubble-net technique to attack their prey. Shrinking encircling mechanism and Spiral are two strategies that are developed to mathematically simulate humpback whale activity in bubble nets with the probability of 50% each to updating position. Following was a mathematical model of these strategies:

$$\vec{X}_i(t+1) = \vec{X}_{lbest}(t) - \vec{R}.\vec{Q}_i \quad \text{if } \rho < 0.5 \tag{20.8}$$

$$\vec{X}_i(t+1) = \vec{Q}'_i.e^{bk}.\cos(2\pi k) + \vec{X}_{lbest}(t) \quad \text{if } \rho \geq 0.5 \tag{20.9}$$

where ρ is a random variable in the range of [0, 1], $\vec{Q}'_i = \left|\vec{X}_{lbest}(t) - \vec{X}_i(t)\right|$ specifies the difference between the ith whale to best whale. Express the logarithmic spiral shape with b being a constant, k is a random value between -1 to 1.

20.2.3 Exploration stage (search for prey)

During the exploitation phase, a candidate's location is updated not using the optimal position vector but rather a position vector picked at random. This method allows the WOA algorithm to execute a global search and $\left|\vec{A}\right| > 1$, which promotes exploration. The mathematical model is as follows:

$$\vec{Q}_i = \left|\vec{P}.\vec{X}_{rand} - \vec{X}_i(t)\right| \tag{20.10}$$

$$\vec{X}_i(t+1) = \vec{X}_{rand} - \vec{R}.\vec{Q}_i \tag{20.11}$$

where $\vec{X}_{rand}$ is a random agent selected randomly among the current population.

20.3 Proposed method

The proposed dehazing technique is implemented from the atmospheric scattering model. The proposed WOA-Dehazing technique predicts an accurate value of A, and we also estimated $t(x)$ for each patch by minimizing the fitness function. Finally reconstructed haze free image from the predicted data. Fig. 20.1 shows the block diagram of proposed technique.

Initially haze image is down sampled using single level Discrete Haar Wavelet Transform [34]. Discrete Haar Wavelet Transform converts the input haze image I^c, $c \in \{r, g, b\}$ into four sub modules, low-pass sub-module block $\hat{I}^c_A$ (i.e., approximation factors), numerous high-pass sub-band modules $\hat{I}^c_v$, $\hat{I}^c_h$, and $\hat{I}^c_d$ (i.e., vertical, horizontal and diagonal factors respectively). High-pass sub-module blocks are in gray color, which corresponds the edge information used to reconstruct a haze-free image.

Next the atmospheric light of haze image is estimated in whitish regions since most haze effected regions have white appearance. An atmospheric light [2] A^c, $c = 1, 2, 3$, to be obtained by choosing the minimal pixel intensity value among local patch μ of all pixels (m, n) within the two-dimensional $M \times N$ index η. At the end the highest pixel among the index set η is picked as atmospheric light A^c.

$$A^c = \max_{m,n \in \eta} \min_{k,l \in \mu(m,n)} I^c_A(k,l), \quad c = 1, 2, 3 \tag{20.12}$$

Finally, after approximating the atmospheric light A^c, the transmission depth $t(x)$ is predicted in order to reconstruct the dehazed image by minimizing the cost function using whale optimization algorithm (WOA). The cost function for image dehazing is formulated to maximize the contract of foggy image and to achieve the lowest possible haze density to predict best transmission depth $t(x)$ and recovers the $\hat{I}^c_A$ from Eq. (20.3) with estimated transmission map. Finally, single level Inverse Discrete Haar Wavelet Transform (IDHWT) is utilized to recover the dehazed scene.

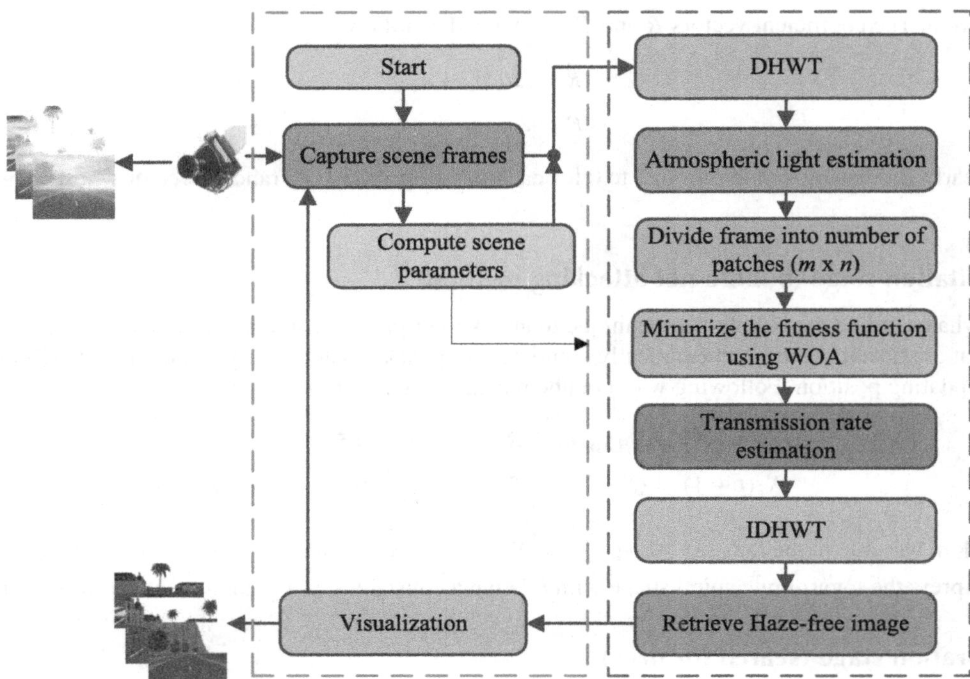

FIGURE 20.1 The proposed Image dehazing framework of WOA-Dehaze method.

20.3.1 Cost function

In general, densely hazy images have whitish or bright regions due to fog. It indicates that gradient component of the image is used to define quantitative value of haze density ($HD(x, y)$). The fog density ($HD(x, y)$) [35] of the hazy input image is inversely proportion to the gradient components, can be defined as follows:

$$HD(x, y) = \sum_{c\in\{r,g,b\}} \sum_{(x,y)\in\mu_A} \exp\left[-\Phi \times \left(\nabla \hat{J}_A^c(x, y)\right)^2\right] \tag{20.13}$$

Here μ_A is local patch of input image $\hat{J}_A^c$ and Φ is a non-negative density coefficient. The $HD(x, y)$ from Eq. (20.3):

$$HD(x, y) = \sum_{c\in\{r,g,b\}} \sum_{(x,y)\in\mu_A} \exp\left[-\Phi \times \left[\nabla\left(\frac{\hat{I}_A^c(x, y) - A^c}{\hat{t}(x, y)} + A^c\right)\right]^2\right] \tag{20.14}$$

In practice, the haze free image has lowest haze density. As a result $HD(x, y)$ is minimized by estimating the $\hat{t}(m, n)$ for each local patch $\hat{I}_A^c$. Therefore, fitness function for fog density is defined as

$$E^{HD} = \sum_{c\in\{r,g,b\}} \sum_{(x,y)\in\mu_A} \exp\left[-\Phi \times \left[\nabla\left(\frac{\hat{I}_A^{c,Q}(x, y) - A^c}{\hat{t}(x, y)} + A^c\right)\right]^2\right] \tag{20.15}$$

In general, a hazed block output typically has poor contrast, and as the estimated $\hat{t}(x, y)$ decreases, a de-hazed patch contrast increase. In order to ensure that the de-hazed block has the highest contrast, we optimize the ideal transmission map $\hat{t}(x, y)$ so that the de-hazed block has the maximum contrast. We used the contrast of Mean Squared Error to statistically assess the contrasts of dehazed blocks. The expression for the contrast of mean squared error (C_{MSE}) is given by

$$C_{MSE} = \sum_{c\in\{r,g,b\};(x,y)\in\mu_A} \frac{\left(\hat{J}_A^c(x,y) - \bar{J}_A^c\right)^2}{K_{\mu_A}} \tag{20.16}$$

$$= \sum_{c\in\{r,g,b\};(x,y)\in\mu_A} \frac{\left(\hat{I}_A^c(x,y) - \bar{I}_A^c\right)^2}{\left(\hat{t}(x,y)\right)^2 K_{\mu_A}} \tag{20.17}$$

where K_{μ_A} is the number of pixel intensities of each patch and $\bar{J}_A^c$, $\bar{I}_A^c$ are the mean pixel values of $\hat{J}_A^c$, $\hat{I}_A^c$ respectively. To remove haze, we increase the contrast of a dehazed image, hence higher mean squared error C_{MSE} [36] represents that high contrast image. From Eq. (20.17), the square of the transmission depth function $(\hat{t}(x,y))$ is reversely proportional to the mean squared error contrast (C_{MSE}). Therefore, the contrast of the hazy image can be changed by reducing the value of $(\hat{t}(x,y))^2$. Therefore, cost function for image contrast is defined as

$$E^{contrast} = -\sum_{c\in\{r,g,b\}}\sum_{(x,y)\in\mu_A} \frac{\left(\hat{J}_A^c(m,n) - \bar{J}_A^c\right)^2}{K_{\mu_A}} \tag{20.18}$$

$$= -\sum_{c\in\{r,g,b\}}\sum_{(x,y)\in\mu_A} \frac{\left(\hat{I}_A^c(x,y) - \bar{I}_A^c\right)^2}{\left(\hat{t}(x,y)\right)^2 K_{\mu_A}}. \tag{20.19}$$

The contrast cost, $E^{contrast}$ is described by taking the negative summation of the mean squared error contrast for each patch of $c \in \{R, G, B\}$ color channels, then we can maximize the mean squared error contrast, by minimizing C_{MSE}. But if the contrast is pushed too far, some pixel intensities are eliminated by underflow or overflow. In order to overcome this problem, we include one more parameter in the fitness function or cost function as the information loss. We describe the E^{Loss} (information loss fitness) [36] for the patch μ_A is the squared summation of truncated values,

$$E^{Loss} = \sum_{c\in\{r,g,b\}}\sum_{(x,y)\in\mu_A}\left\{\left(\min\left\{0, \hat{J}_A^c(x,y)\right\}\right)^2 + \left(\max\left\{0, \hat{J}_A^c(x,y) - 255\right\}\right)^2\right\} \tag{20.20}$$

$$= \sum_{c\in\{r,g,b\}}\left\{\sum_{i=0}^{\alpha}\left(\frac{i - A^c}{\hat{t}(x,y)} + A^c\right)h_c(i) + \sum_{i=\beta}^{255}\left(\frac{i - A^c}{\hat{t}(x,y)} + A^c - 255\right)h_c(i)\right\} \tag{20.21}$$

where $h_c(i)$ is the histogram of the local patch at intensity level i in the color channel c and (α, β) represent the intercepts at which the truncation occurs. The terms $\max\left\{0, \hat{J}_A^c(m,n) - 255\right\}$ and $\min\left\{0, \hat{J}_A^c(m,n)\right\}$ in Eq. (20.20) denote the truncated values due to overflow and underflow, respectively.

Finally, we formulated the overall cost function is a sum of contrast cost, haze density cost and information loss cost given by

$$E = E^{contrast} + \beta_1 E^{HD} + \beta_2 E^{Loss} \tag{20.22}$$

where β_1 and β_2 are weighting parameters that regulate the respective weights of cost of haze density and information loss. Last, we predict the optimum transmission $(\hat{t}(x,y))$ for each patch by reducing the total fitness function E using Whale Optimization algorithm (WOA).

20.4 Experimental results

In this section, we compared proposed technique with several state of art techniques both qualitatively and quantitatively. We adopted performance matrices like Structural Similarity Matrices (SSIM) [37], Weighted Peak Signal-To-Noise Ratio (WPSNR) [38], Feature Similarity Index (FSIM) [39], Mean Square Error (MSE), and Blur metrics (BM) [40] used to assess the efficiency of proposed technique. The comparison methods include He et al. [2], DehazeNet [41], Zhu et al. [6], AOD-Net [19], GridDehazeNet (CVPR 2019) [42], Zheng et al. (CVPR 2021) [43], and RefineDNet [17]. We also demonstrated the effect of our proposed method for different benchmark datasets which includes SOTS_Indoor and Outdoor [44], I-HAZE [45], REVIDE [46], D_HAZY [47], and Real life dataset. The dehazed images of various methods are shown

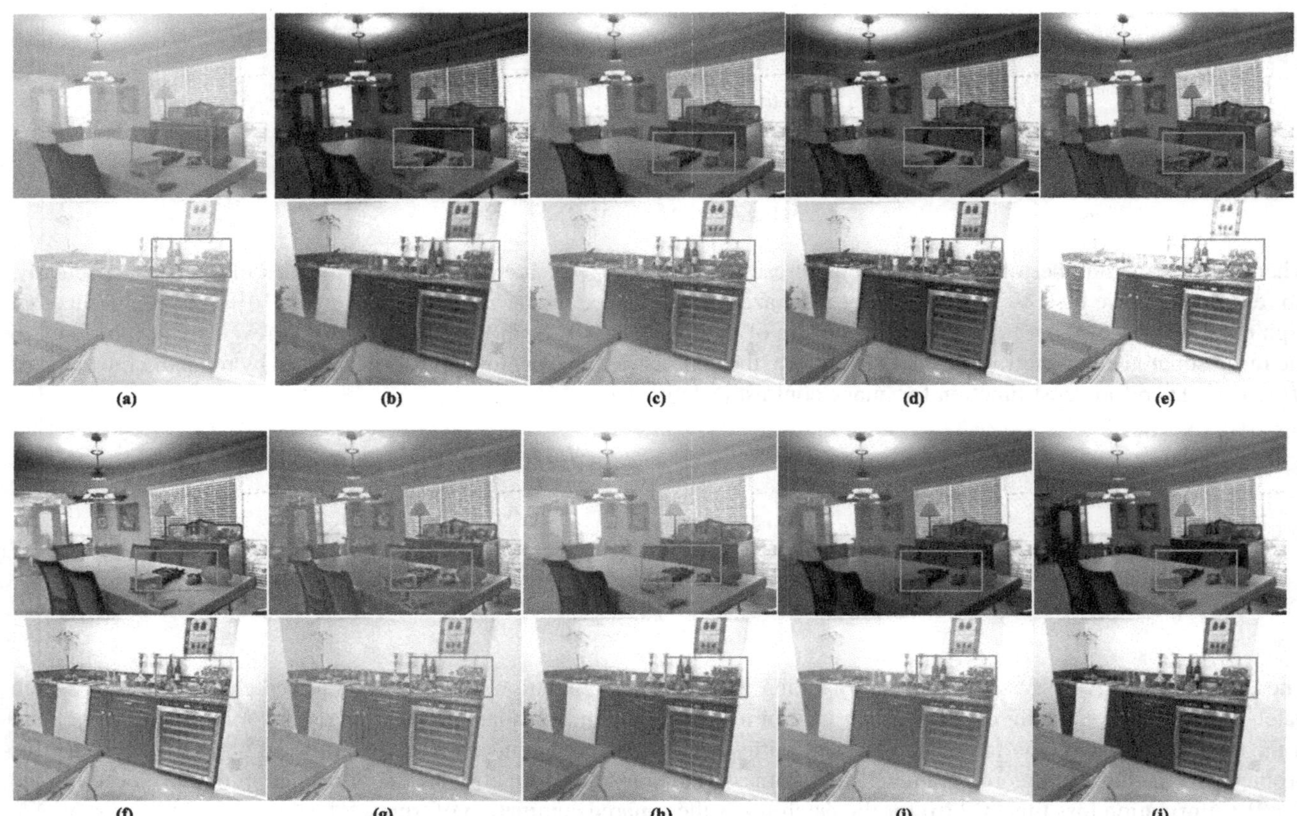

FIGURE 20.2 SOTS Indoor dataset image dehazing results: (a) is foggy input, (b)-(j) are the dehazed outputs of He et al. [2], DehazeNet [41], Zhu et al. [6], AOD-Net [19], GridDehazeNet (CVPR 2019) [42], Zheng et al. (CVPR 2021) [43], RefineDNet [17], WOA-Dehaze, and Ground truth.

in Fig. 20.2 to Fig. 20.6 for SOTS_Indoor, SOTS_Outdoor, REVIDE, I-HAZE, and D-HAZY respectively. Fig. 20.2 shows the dehazing results of two SOTS_Indoor dataset images for state of art methods. It can be observed that the result of He et al. [2] has color distortion and is over-saturated than the original truth image. And also we can observe that the dehazing images of DehazeNet [41] demonstrate haze. The dehazing results GridDehazeNet (CVPR 2019) [42] are brighter than the ground truth, and it indicates that it causes over-saturation.

Fig. 20.3 shows that dehazing results of two SOTS_Outdoor dataset images for state of art methods. From Fig. 20.3 (e), result of Zhu et al. [6]. We can clearly observe that the color alteration in sky region. And also in Fig. 20.3 (g), result of Zheng et al. (CVPR 2021) [43] the backside of car, tree areas, and building areas are become darker and not visible properly. Fig. 20.4 shows that dehazing results of REVIDE dataset image for various methods. From Figs. 20.4 (e) and 20.4 (f), we can observe that there is color distortion in wall regions. In Fig. 20.4, we can observe that the dehazing results of He et al. [2] and GridDehazeNet (CVPR 2019) [42] are oversaturated. Fig. 20.5 shows that dehazing result of D_HAZY dataset image for state of art methods. The Middlebury section of D_HAZY [47] holds 23 indoor paired images derived from the Middlebury dataset's images and superior depth maps and we can observe that the compared method failed to enhance the brightness of image without introducing the artifacts and our proposed method WOA-Dehaze can eliminate haze without introducing visible artifacts or distortions. Dehazed outputs of I-HAZE dataset are shown in Fig. 20.6. Although DehazeNet [41], Zhu et al. [6], and AOD-Net [19] gave accurate outputs in terms of fog removal, but some heavy haze zones, for example area around the toy in the first row and the red bounding box area in the second row, were not clearly cleaned.

The dehazing outcomes of real-world foggy images for various state of art approaches and WOA-Dehaze are shown in Fig. 20.7. Due to some realistic scenarios, original scene images of real-life foggy pictures are not possible to capture in nature. The dehaze results of He et al. [2] and Zhu et al. [6] are a little color distortion in the sky areas and the leaves area. And also, RefineDNet [17] does not affect the color in sky areas; mountain areas in the output image are slightly darker. Although these methods achieve remarkable dehazing performance, they usually fail in the above cases. However, from Fig. 20.2 to Fig. 20.7, it is proven that the proposed method, WOA-Dehaze, achieves satisfactory visual results.

FIGURE 20.3 SOTS Outdoor dataset image results: (a) is foggy input, (b)-(j) are the dehazed outputs of He et al. [2], DehazeNet [41], Zhu et al. [6], AOD-Net [19], GridDehazeNet (CVPR 2019) [42], Zheng et al. (CVPR 2021) [43], RefineDNet [17], WOA-Dehaze, and Ground truth.

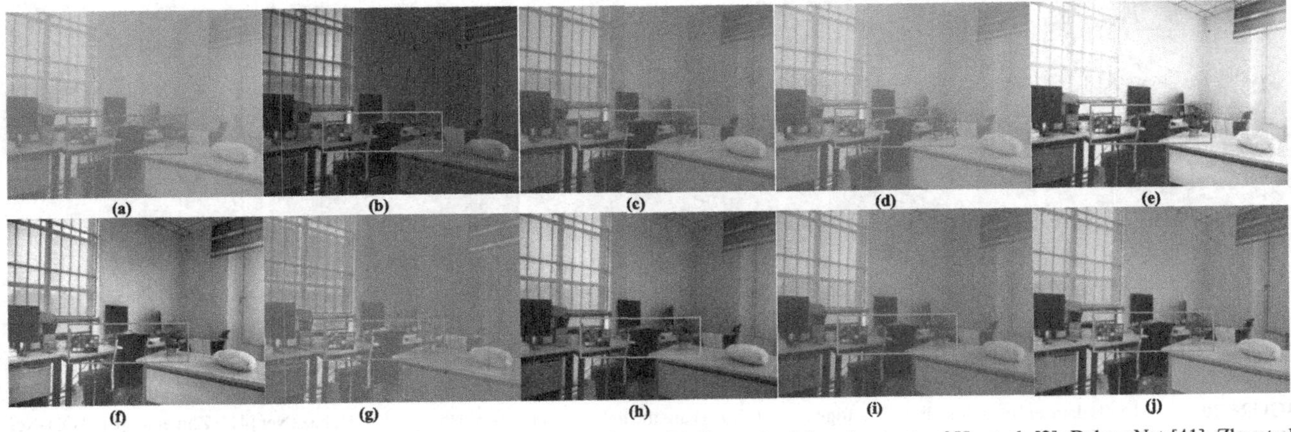

FIGURE 20.4 REVIDE dataset image dehazing results: (a) is foggy input, (b)-(j) are the dehazed outputs of He et al. [2], DehazeNet [41], Zhu et al. [6], AOD-Net [19], GridDehazeNet (CVPR 2019) [42], Zheng et al. (CVPR 2021) [43], RefineDNet [17], WOA-Dehaze, and Ground truth.

To prove the effectiveness of the proposed technique, we conducted quantitative assessments for proposed technique and state of art dehazing techniques using the assessment metrics. We adopted assessment metrics like Structural Similarity Matrices (SSIM) [37], Mean Square Error (MSE), Weighted Peak Signal-To-Noise Ratio (WPSNR) [38], Feature Similarity Index (FSIM) [39], and Blur metrics (BM) [40].

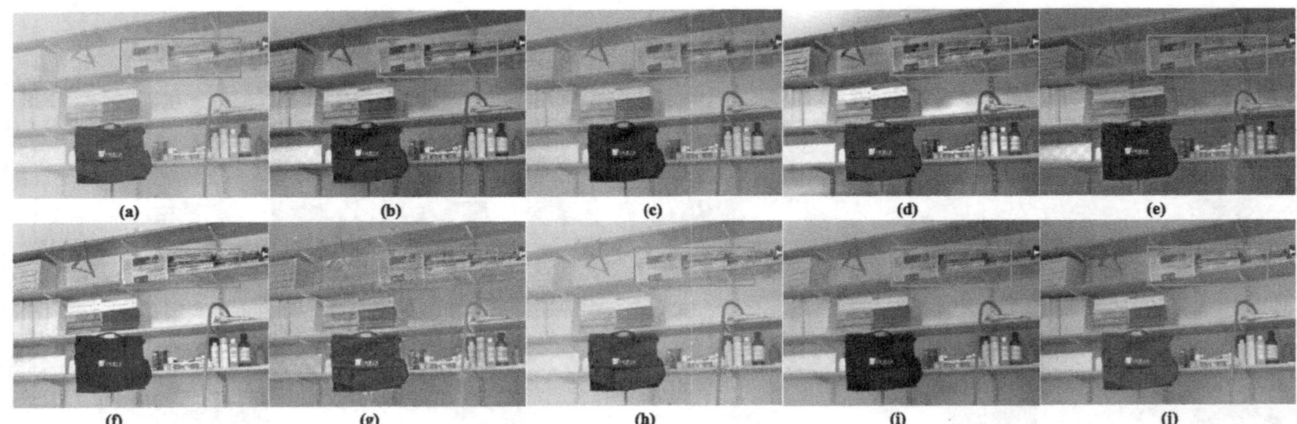

FIGURE 20.5 D-HAZY dataset image results: (a) is foggy input, (b)-(j) are the dehazed outputs He et al. [2], DehazeNet [41], Zhu et al. [6], AOD-Net [19], GridDehazeNet (CVPR 2019) [42], Zheng et al. (CVPR 2021) [43], RefineDNet [17], WOA-Dehaze, and Ground truth.

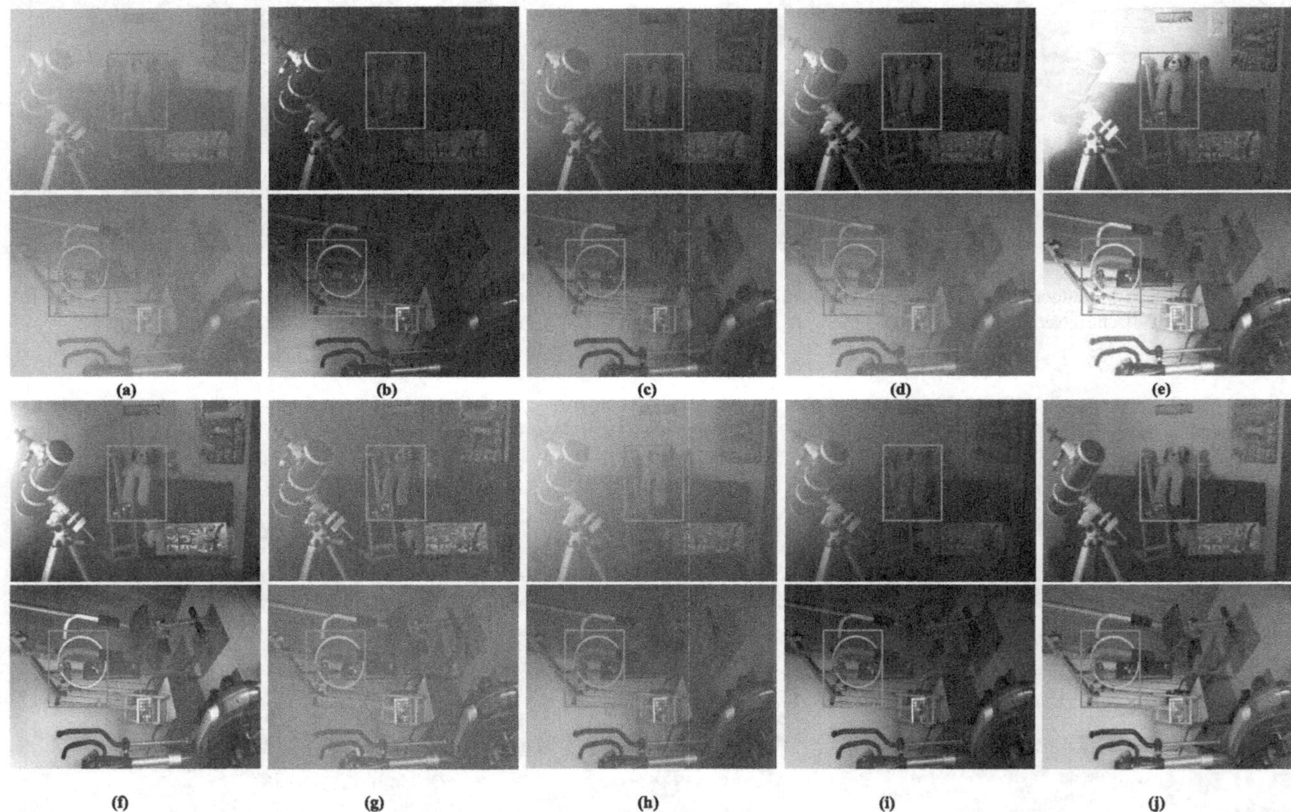

FIGURE 20.6 I-HAZE dataset image results: (a) is foggy input, (b)-(j) are the dehazed outputs of He et al. [2], DehazeNet [41], Zhu et al. [6], AOD-Net [19], GridDehazeNet (CVPR 2019) [42], Zheng et al. (CVPR 2021) [43], RefineDNet [17], WOA-Dehaze, and Ground truth.

20.4.1 Structural SIMilarity index

The Structural SIMilarity index [29] is a parameter used to assess the likeness among two photographs. The SSIM among two patches δ_1 and δ_2 of average range $N \times N$ is denoted by

$$SSIM(\delta_1, \delta_2) = \frac{(2\mu_{\delta 1}\mu_{\delta 2} + Z_1)(2\sigma_{\delta 12} + Z_2)}{\left(\mu_{\delta 1}^2 + \mu_{\delta 2}^2 + Z_1\right)\left(\sigma_{\delta 1}^2 + \sigma_{\delta 2}^2 + Z_2\right)} \tag{20.23}$$

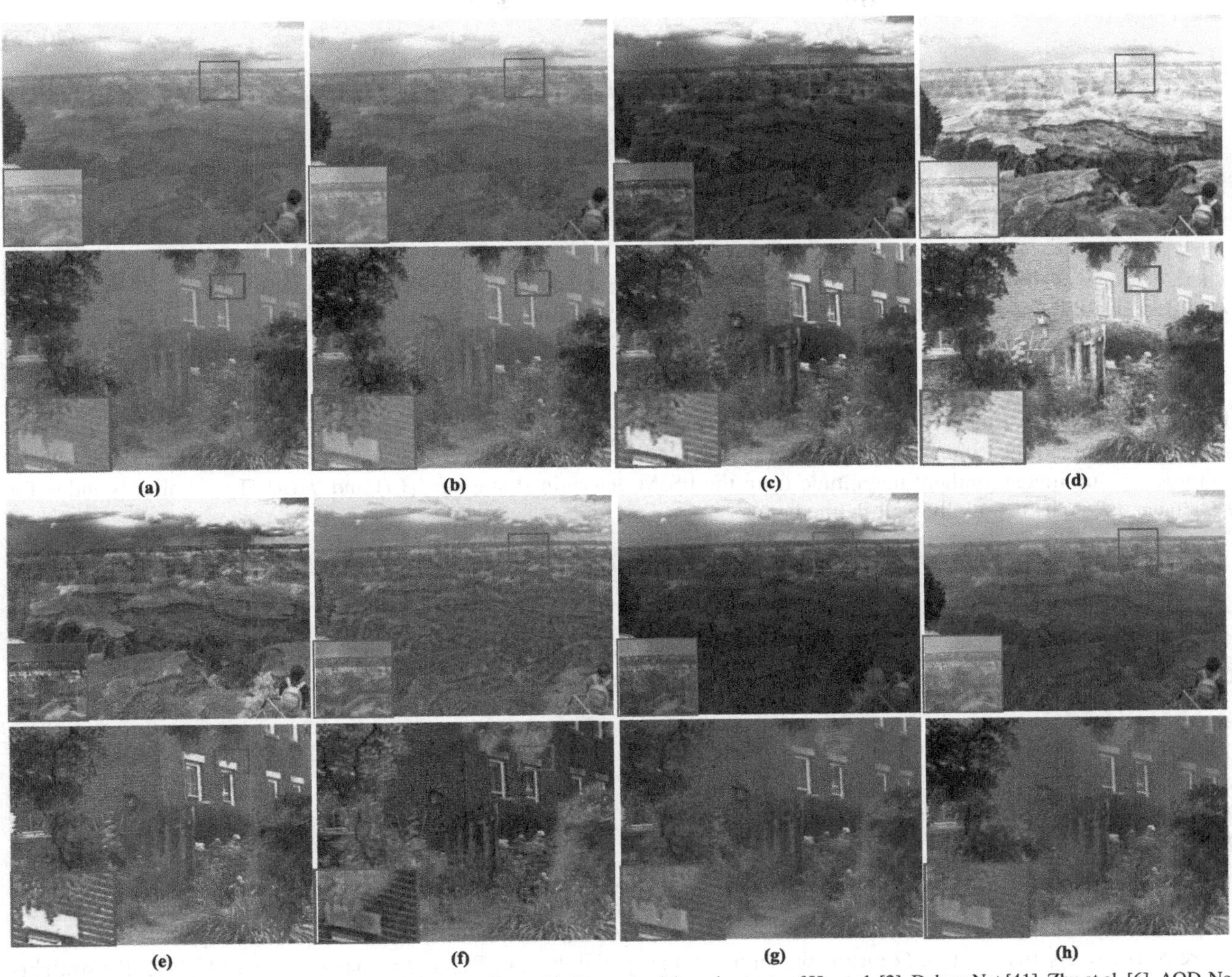

FIGURE 20.7 real-world hazy images results: (a) is foggy input, (b)-(h) are the dehazed outputs of He et al. [2], DehazeNet [41], Zhu et al. [6], AOD-Net [19], Zheng et al. (CVPR 2021) [43], RefineDNet [17], and WOA-Dehaze.

where $\mu_{\delta 1}$, $\mu_{\delta 2}$ are the averages of δ_1 and δ_2, $\sigma_{\delta 12}$ is the co-variance of δ_1 and δ_2, $\sigma_{\delta 1}^2$, $\sigma_{\delta 2}^2$ are the variances of δ_1 and δ_2. To restrict the division with a weak denominator, use two variables Z_1, Z_2.

20.4.2 WPSNR and MSE

The PSNR (peak signal-to-noise ratio) is easily distinguished over the MSE (mean squared error), however it performs poorly when representing vision from the human perspective. To overcome that Perceptually weighted PSNR (WPSNR) measures perceptual sensitivity of local image regions by calculating weighting factors. The MSE (mean squared error) of input image I and its noisy image J with the dimensions $p \times q$ is given by:

$$MSE = \frac{1}{pq} \sum_{a=0}^{p-1} \sum_{b=0}^{q-1} [I(a,b) - J(a,b)]^2 \tag{20.24}$$

The overall biased or weighted MSE distortion for an image MSE_k^w is given by

$$MSE_k^w = \sum_k w_k.MSE \tag{20.25}$$

$$MSE_k^w = \frac{1}{pq}\sum_k w_k.\sum_{a=0}^{p-1}\sum_{b=0}^{q-1}[I(a,b) - J(a,b)]^2 \tag{20.26}$$

$$WPSNR = 20\log_{10}\left(\frac{MAX_I}{\sqrt{MSE_k^w}}\right) \tag{20.27}$$

where MAX_I was the most probable pixel intensity of the scene and w_k was a weighted factor and if $w_k = 1$ for all then weighted mean squared error distortion MSE_k^w equals to the regular mean squared error MSE.

20.4.3 Feature similarity index

There are two steps to the Feature Similarity (FSIM) index calculation [39]. In the initial phase, we compute the local similarity map, and then integrate it into a single likeness score in the following phase. We isolate the two factors Phase congruency and image gradient magnitude from the FSIM determined among $f_1(u)$ and $f_2(u)$. The similarity index for $PC_1(u)$ and $PC_2(u)$ is initially well-defined as

$$S_{PC}(u) = \frac{2PC_1(u)\cdot PC_2(u) + \theta_1}{PC_1^2(u) + PC_2^2(u) + \theta_1} \tag{20.28}$$

where θ_1 is a constant having positive value that increases the stability of S_{PC}. Likewise, the gradient magnitude values $G_1(u)$ and $G_2(u)$ are linked, and the likeness measure is distinct as

$$S_G(u) = \frac{2G_1(u)\cdot G_2(u) + \theta_2}{G_1^2(u) + G_2^2(u) + \theta_2} \tag{20.29}$$

where θ_2 is a constant having positive value independent of the varying magnitudes of the gradients. The FSIM among $f_1(u)$ and $f_2(u)$ is as

$$FSIM = \frac{\sum_{u\in\Omega} S_L(u)\cdot PC_m(u)}{\sum_{u\in\Omega} PC_m(u)} \tag{20.30}$$

where $S_L(u) = S_{PC}(u)\cdot S_G(u)$, Ω means the entire image spatial domain $PC_m(u) = \max(PC_1(u), PC_2(u))$. By straightforwardly integrating the chromatic statistics, the FSIM index can be prolonged to $FSIM_C$ as:

$$FSIM_C = \frac{\sum_{u\in\Omega} S_{PC}(u)\cdot S_G(u)\cdot \left[S_I(u)\cdot S_Q(u)\right]^{\lambda}\cdot PC_m(u)}{\sum_{u\in\Omega} PC_m(u)} \tag{20.31}$$

where the reputation of the colors is limited by $\lambda > 0$ requirement (the Q and I are chrominance channels). Also, BM (Blur metrics) [40] is considered for performance comparison.

The quantitative analysis of proposed method and various dehazing techniques are shown in Table 20.1 to Table 20.4 for different datasets. The original scene image and the dehazed image have a higher WPSNR value, which suggests that there are less imperfections, and if the value of SSIM is high, then indicates that the method that was proposed was successful in maintaining the structure of the image. In Table 20.1 to Table 20.4 the first best three values are indicated in red (mid gray in print version), blue (dark gray in print version), and green (light gray in print version) colors respectively.

From Table 20.1 we can observe that our proposed technique WOA-Dehaze obtained best quantitative value than the other dehazing technique. The proposed technique obtained has second best value for Average Feature Similarity Index (FSIM) value of SOTS_Outdoor data set, which was marked blue in color. Table 20.2 demonstrates the quantitative analysis of REVIDE and D-HAZY dataset. The proposed method gained best values among all compared method in table. Table 20.3 the quantitative outputs of the different dehazing approaches for I-HAZE dataset. The proposed algorithm obtained best values than the other compared methods. Table 20.4 represents the superiority of the proposed technique for real world images. It is observed that from Table 20.1 to Table 20.4 the proposed method WOA_Dehaze obtained superior values than the state of art methods, which include He et al. [2], DehazeNet [41], Zhu et al. [6], AOD-Net [19], GridDehazeNet (CVPR 2019) [42], Zheng et al. (CVPR 2021) [43], and RefineDNet [17].

TABLE 20.1 Quantitative Comparison of various Dehazing methods using Average WPSNR, SSIM, MSE, FSIM, and BM on SOTS_Indoor and SOTS_Outdoor.

Data set		He et al.	DehazeNet	Zhu et al.	AOD-Net	GridDehazeNet (CVPR 2019)	Zheng et al. (CVPR 2021)	RefineDNet	WOA_Dehaze
SOTS_Indoor	WPSNR	16.2604	19.3721	16.4210	20.4357	21.3274	22.4226	22.7616	27.145
	SSIM	0.5004	0.7204	0.5050	0.4773	0.7734	0.8286	0.7668	0.9582
	MSE	0.1675	0.0733	0.1617	0.062	0.0585	0.0229	0.1004	0.0081
	FSIM	0.9857	0.9945	0.9879	0.9801	0.9977	0.9865	0.9799	1.0000
	BM	0.3710	0.3565	0.4418	0.3313	0.4252	0.3754	0.3951	0.435
SOTS_Outdoor	WPSNR	16.6881	16.4175	17.0411	24.1315	17.7954	25.8620	22.4412	26.8634
	SSIM	0.5155	0.5672	0.5833	0.5047	0.7559	0.8712	0.6471	0.9064
	MSE	0.0940	0.0944	0.0854	0.0259	0.0678	0.0106	0.0389	0.0052
	FSIM	0.9941	0.9965	0.9955	0.9852	0.9986	0.9789	0.9742	0.9981
	BM	0.2519	0.2281	0.2418	0.1846	0.2246	0.2173	0.1982	0.2872

TABLE 20.2 Quantitative Comparison of various Dehazing methods using WPSNR, SSIM, MSE, FSIM, and BM on REVIDE and D-HAZY.

Data set		He et al.	DehazeNet	Zhu et al.	AOD-Net	GridDehazeNet (CVPR 2019)	Zheng et al. (CVPR 2021)	RefineDNet	WOA_Dehaze
REVIDE	WPSNR	13.5397	20.5727	16.6836	19.3190	20.4939	24.4570	19.0703	38.0148
	SSIM	0.4186	0.8582	0.6193	0.4683	0.8412	0.8987	0.2580	0.9960
	MSE	0.1779	0.0351	0.0863	0.0536	0.0338	0.0144	0.0669	6.2800e-04
	FSIM	0.9757	0.9989	0.9964	0.9854	0.9995	0.9635	0.9809	1.0000
	BM	0.3257	0.2834	0.3117	0.2506	0.2844	0.2691	0.2724	0.2995
D-HAZY	WPSNR	14.9826	19.0980	16.9930	23.3899	21.1968	21.0683	17.7343	34.7065
	SSIM	0.5422	0.7563	0.6577	0.5969	0.8668	0.8449	0.6021	0.9901
	MSE	0.1427	0.0550	0.0909	0.0314	0.0326	0.0315	0.0742	0.0014
	FSIM	0.9677	0.9963	0.9899	0.9876	0.9989	0.9560	0.9709	1.0000
	BM	0.2888	0.3010	0.3143	0.2500	0.3050	0.2793	0.2845	0.3138

TABLE 20.3 Quantitative Comparison of various Dehazing methods using Average WPSNR, SSIM, MSE, FSIM, and BM on I-HAZE Dataset.

Data set		He et al.	DehazeNet	Zhu et al.	AOD-Net	GridDehazeNet (CVPR 2019)	Zheng et al. (CVPR 2021)	RefineDNet	WOA_Dehaze
I-HAZE	WPSNR	13.5188	16.5225	14.7926	20.4037	17.2227	21.6653	19.1127	32.2023
	SSIM	0.3734	0.6524	0.4631	0.4150	0.7044	0.8352	0.5918	0.9861
	MSE	0.1797	0.0894	0.1328	0.0436	0.0765	0.0273	0.0574	0.0024
	FSIM	0.9616	0.9933	0.9894	0.9864	0.9984	0.9620	0.9758	1.0000
	BM	0.2362	0.2846	0.3246	0.2284	0.2874	0.2422	0.2614	0.3121

TABLE 20.4 Quantitative Comparison of various Dehazing methods using Average WPSNR, SSIM, MSE, FSIM, and BM on Real life images.

Data set		He et al.	DehazeNet	Zhu et al.	AOD-Net	Zheng et al. (CVPR 2021)	RefineDNet	WOA_Dehaze
Real world images	WPSNR	16.0192	19.8661	19.9232	21.4476	23.1067	21.4986	33.2185
	SSIM	0.6283	0.6339	0.5680	0.6575	0.7911	0.8819	0.9871
	FSIM	0.9865	0.9873	0.9770	0.9643	0.9750	0.9979	0.9983
	MSE	0.0939	0.0964	0.0424	0.0404	0.0199	0.0269	0.0025
	BM	0.2491	0.2690	0.2040	0.2163	0.2256	0.2403	0.2414

20.5 Conclusion

We implemented a novel image dehazing technique, called WOA-Dehaze. Unlike traditional dehazing techniques, our technology may produce dehazed images with a wide range of illumination, color tones and atmosphere approaches. The proposed method performs dehazing operation by estimating transmission depth map and local atmospheric light. To address that, the fitness function is formulated using Haze density cost, Contrast cost and Information loss cost, and the proposed WOA-Dehaze method implemented using Whale Optimization Algorithm for optimizing the transmission map by minimizing the fitness function. Several qualitative and quantitative analyses show that the suggested WOA-Dehaze approach performs effectively on a variety of databases with respect to performance indices. And Experimental study demonstrates the superiority of our technology over the state-of-the-art techniques for dehazing.

References

[1] R.T. Tan, Visibility in bad weather from a single image, in: 2008 IEEE Conf. Comput. Vis. Pattern Recognit., 2008, pp. 1–8.

[2] K. He, J. Sun, X. Tang, Single image haze removal using dark channel prior, IEEE Trans. Pattern Anal. Mach. Intell. 33 (12) (2011) 2341–2353, https://doi.org/10.1109/TPAMI.2010.168.

[3] Y. Jiang, C. Sun, Y. Zhao, L. Yang, Image dehazing using adaptive bi-channel priors on superpixels, Comput. Vis. Image Underst. 165 (2017) 17–32, https://doi.org/10.1016/j.cviu.2017.10.014.

[4] M. Kaur, D. Singh, V. Kumar, K. Sun, Color image dehazing using gradient channel prior and guided L0 filter, Inf. Sci. (NY) 521 (2020) 326–342, https://doi.org/10.1016/j.ins.2020.02.048.

[5] J. Li, Q. Hu, M. Ai, Haze and thin cloud removal via sphere model improved dark channel prior, IEEE Geosci. Remote Sens. Lett. 16 (3) (2019) 472–476, https://doi.org/10.1109/LGRS.2018.2874084.

[6] Q. Zhu, J. Mai, L. Shao, A fast single image haze removal algorithm using color attenuation prior, IEEE Trans. Image Process. 24 (11) (2015) 3522–3533, https://doi.org/10.1109/TIP.2015.2446191.

[7] W. Liu, R. Yao, G. Qiu, A physics based generative adversarial network for single image defogging, Image Vis. Comput. 92 (2019) 103815, https://doi.org/10.1016/j.imavis.2019.10.001.

[8] J. Dong, J. Pan, Physics-based feature dehazing networks, in: ECCV 2020, in: LNCS, vol. 12375, 2020, pp. 188–204, https://doi.org/10.1007/978-3-030-58577-8_12.

[9] Y. Gao, Y. Su, Q. Li, J. Li, Single fog image restoration with multi-focus image fusion, J. Vis. Commun. Image Represent. 55 (2018) 586–595, https://doi.org/10.1016/j.jvcir.2018.07.004.

[10] D. Zhao, L. Xu, Y. Yan, J. Chen, L.Y. Duan, Multi-scale Optimal Fusion model for single image dehazing, Signal Process. Image Commun. 74 (2019) 253–265, https://doi.org/10.1016/j.image.2019.02.004.

[11] M. Zheng, G. Qi, Z. Zhu, Y. Li, H. Wei, Y. Liu, Image dehazing by an artificial image fusion method based on adaptive structure decomposition, IEEE Sens. J. 20 (14) (2020) 8062–8072, https://doi.org/10.1109/JSEN.2020.2981719.

[12] D. Berman, T. Treibitz, S. Avidan, Non-local image dehazing, in: Proc. IEEE Comput. Soc. Conf. Comput. Vis. Pattern Recognit., vol. 2016-Decem, 2016, pp. 1674–1682, https://doi.org/10.1109/CVPR.2016.185.

[13] S. Salazar-Colores, E. Cabal-Yepez, J.M. Ramos-Arreguin, G. Botella, L.M. Ledesma-Carrillo, S. Ledesma, A fast image dehazing algorithm using morphological reconstruction, IEEE Trans. Image Process. 28 (5) (2019) 2357–2366, https://doi.org/10.1109/TIP.2018.2885490.

[14] H. Bai, J. Pan, X. Xiang, J. Tang, Self-guided image dehazing using progressive feature fusion, IEEE Trans. Image Process. 31 (2022) 1217–1229.

[15] Z. Luan, Y. Shang, X. Zhou, Z. Shao, G. Guo, X. Liu, Fast single image dehazing based on a regression model, Neurocomputing 245 (2017) 10–22, https://doi.org/10.1016/j.neucom.2017.03.024.

[16] J.-L. Yin, Y.-C. Huang, B.-H. Chen, S.-Z. Ye, Color transferred convolutional neural networks for image dehazing, IEEE Trans. Circuits Syst. Video Technol. 30 (11) (2020) 3957–3967, https://doi.org/10.1109/tcsvt.2019.2917315.

[17] S. Zhao, L. Zhang, Y. Shen, Y. Zhou, RefineDNet: a weakly supervised refinement framework for single image dehazing, IEEE Trans. Image Process. 30 (2021) 3391–3404, https://doi.org/10.1109/TIP.2021.3060873.

[18] G. Kim, S.W. Park, J. Kwon, Pixel-wise Wasserstein autoencoder for highly generative dehazing, IEEE Trans. Image Process. 30 (2021) 5452–5462, https://doi.org/10.1109/TIP.2021.3084743.

[19] B. Li, X. Peng, Z. Wang, J. Xu, D. Feng, AOD-Net: all-in-one dehazing network, in: Proc. IEEE Int. Conf. Comput. Vis., vol. 2017-Octob, 2017, pp. 4780–4788, https://doi.org/10.1109/ICCV.2017.511.

[20] C. Wang, W. Fan, Y. Wu, Z. Su, Weakly supervised single image dehazing, J. Vis. Commun. Image Represent. 72 (2020) 102897, https://doi.org/10.1016/j.jvcir.2020.102897.

[21] A. Dudhane, S. Murala, RYF-Net: deep fusion network for single image haze removal, IEEE Trans. Image Process. 29 (2019) 628–640, https://doi.org/10.1109/tip.2019.2934360.

[22] G. Kim, J. Kwon, Deep illumination-aware dehazing with low-light and detail enhancement, IEEE Trans. Intell. Transp. Syst. 23 (3) (2022) 2494–2508, https://doi.org/10.1109/TITS.2021.3117868.

[23] W. Ren, et al., Deep video dehazing with semantic segmentation, IEEE Trans. Image Process. 28 (4) (2019) 1895–1908, https://doi.org/10.1109/TIP.2018.2876178.

[24] S.C. Agrawal, A.S. Jalal, Dense haze removal by nonlinear transformation, IEEE Trans. Circuits Syst. Video Technol. 32 (2) (2022) 593–607, https://doi.org/10.1109/TCSVT.2021.3068625.

[25] J. Zhu, L. Meng, W. Wu, D. Choi, J. Ni, Generative adversarial network-based atmospheric scattering model for image dehazing, Digit. Commun. Netw. 7 (2) (2021) 178–186, https://doi.org/10.1016/j.dcan.2020.08.003.

[26] K. Wang, S. Zhang, J. Chen, F. Ren, L. Xiao, A feature-supervised generative adversarial network for environmental monitoring during hazy days, Sci. Total Environ. 748 (2020) 141445, https://doi.org/10.1016/j.scitotenv.2020.141445.

[27] J. Park, D.K. Han, H. Ko, Fusion of heterogeneous adversarial networks for single image dehazing, IEEE Trans. Image Process. 29 (2020) 4721–4732, https://doi.org/10.1109/TIP.2020.2975986.

[28] Y. Li, Y. Liu, Q. Yan, K. Zhang, Deep dehazing network with latent ensembling architecture and adversarial learning, IEEE Trans. Image Process. 30 (2021) 1354–1368, https://doi.org/10.1109/TIP.2020.3044208.

[29] H. Zhu, et al., Single-image dehazing via compositional adversarial network, IEEE Trans. Cybern. 51 (2) (2021) 829–838, https://doi.org/10.1109/TCYB.2019.2955092.

[30] Y. Mo, C. Li, Y. Zheng, X. Wu, DCA-CycleGAN: unsupervised single image dehazing using dark channel attention optimized CycleGAN, J. Vis. Commun. Image Represent. 82 (2022) 103431, https://doi.org/10.1016/j.jvcir.2021.103431.

[31] A. Dudhane, P.W. Patil, S. Murala, An end-to-end network for image de-hazing and beyond, IEEE Trans. Emerg. Top. Comput. Intell. 6 (1) (2022) 159–170, https://doi.org/10.1109/TETCI.2020.3035407.

[32] X. Chen, Y. Li, C. Kong, L. Dai, Unpaired image dehazing with physical-guided restoration and depth-guided refinement, IEEE Signal Process. Lett. 29 (2022) 587–591, https://doi.org/10.1109/LSP.2022.3147434.

[33] S. Mirjalili, A. Lewis, The whale optimization algorithm, Adv. Eng. Softw. 95 (2016) 51–67, https://doi.org/10.1016/j.advengsoft.2016.01.008.

[34] M.J. Shensa, The discrete wavelet transform: wedding the a trous and Mallat algorithms, IEEE Trans. Signal Process. 40 (10) (1992) 2464–2482, https://doi.org/10.1109/78.157290.

[35] M. Ju, Z. Gu, D. Zhang, Single image haze removal based on the improved atmospheric scattering model, Neurocomputing 260 (2017) 180–191, https://doi.org/10.1016/j.neucom.2017.04.034.

[36] J.H. Kim, W.D. Jang, J.Y. Sim, C.S. Kim, Optimized contrast enhancement for real-time image and video dehazing, J. Vis. Commun. Image Represent. 24 (3) (2013) 410–425, https://doi.org/10.1016/j.jvcir.2013.02.004.

[37] E.P. Simoncelli, H.R. Sheikh, A.C. Bovik, Z. Wang, Image quality assessment: from error visibility to structural similarity, IEEE Trans. Image Process. 13 (4) (2004) 600–612.

[38] J. Erfurt, C.R. Helmrich, S. Bosse, H. Schwarz, D. Marpe, T. Wiegand, A study of the perceptually weighted peak signal-to-noise ratio (WPSNR) for image compression, in: 2019 IEEE Int. Conf. Image Process., 2019, pp. 2339–2343.

[39] L. Zhang, L. Zhang, X. Mou, D. Zhang, FSIM: a feature similarity index for image quality assessment, IEEE Trans. Image Process. 20 (8) (2011) 2378–2386, https://doi.org/10.1109/TIP.2011.2109730.

[40] F. Crete, T. Dolmiere, P. Ladret, M. Nicolas, The blur effect: perception and estimation with a new no-reference perceptual blur metric, in: Human Vision and Electronic Imaging XII, vol. 6492, 2007, p. 64920I, https://doi.org/10.1117/12.702790.

[41] B. Cai, X. Xu, K. Jia, C. Qing, D. Tao, DehazeNet: an end-to-end system for single image haze removal, IEEE Trans. Image Process. 25 (11) (2016) 5187–5198, https://doi.org/10.1109/TIP.2016.2598681.

[42] X. Liu, Y. Ma, Z. Shi, J. Chen, GridDehazeNet: attention-based multi-scale network for image dehazing, in: 2019 IEEE/CVF International Conference on Computer Vision (ICCV), 2019, pp. 7313–7322, https://doi.org/10.1109/ICCV.2019.00741.

[43] Z. Zheng, et al., Ultra-high-definition image dehazing via multi-guided bilateral learning, in: Proc. IEEE Comput. Soc. Conf. Comput. Vis. Pattern Recognit., 2021, pp. 16180–16189, https://doi.org/10.1109/CVPR46437.2021.01592.

[44] B. Li, et al., Benchmarking single-image dehazing and beyond, IEEE Trans. Image Process. 28 (1) (2019) 492–505, https://doi.org/10.1109/TIP.2018.2867951.

[45] C.O. Ancuti, C. Ancuti, R. Timofte, C. De Vleeschouwer, I-HAZE: a dehazing benchmark with real hazy and haze-free outdoor images, arXiv: 1804.05091v1, 2018, https://doi.org/10.1109/CVPRW.2018.00119.

[46] X. Zhang, et al., Learning to restore hazy video: a new real-world dataset and a new method, in: Proc. IEEE Comput. Soc. Conf. Comput. Vis. Pattern Recognit., 2021, pp. 9235–9244, https://doi.org/10.1109/CVPR46437.2021.00912.

[47] C. Ancuti, C.O. Ancuti, C. De Vleeschouwer, D-HAZY: a dataset to evaluate quantitatively dehazing algorithms, in: Proc. - Int. Conf. Image Process. ICIP, vol. 2016-Augus, 2016, pp. 2226–2230, https://doi.org/10.1109/ICIP.2016.7532754.

Chapter 21

An enhanced whale optimization algorithm with opposition-based learning for LEDs placement in indoor VLC systems

Abdelbaki Benayad[a], Amel Boustil[a], Yassine Meraihi[b], Seyedali Mirjalili[c,d], Selma Yahia[b], and Sylia Mekhmoukh Taleb[b]

[a]*LIMOSE Laboratory, University of M'Hamed Bougara Boumerdes, Boumerdes, Algeria,* [b]*LIST Laboratory, University of M'Hamed Bougara Boumerdes, Boumerdes, Algeria,* [c]*Centre for Artificial Intelligence Research and Optimisation, Torrens University Australia, Brisbane, QLD, Australia,* [d]*Yonsei Frontier Lab, Yonsei University, Seoul, South Korea*

21.1 Introduction

Currently, Light Communication (LC) technology is gaining more interest as a result of the massive bandwidth available in the uncontrolled optical spectrum [1]. Visible Light Communication (VLC) is among of such technology that can be considered as an alternative or complement to radio communications [2]. It can be used for many indoor applications such as localization [3,4], networking at home [5], and communication in locations like hospitals and airplane cabins where radio frequency (RF) radiation is restricted [6]. This technology is based on LEDs, which are regarded like a sustainable and energy-efficient source of light [7]. LEDs can both send data and provide illumination. Thus, achieving dual functionality. The placement of LEDs in indoor VLC systems is an important issue that affects the performance of the network. It is known to be an NP-hard problem [8–11]. Therefore, meta-heuristics are intelligent methods that can provide efficient solutions to solve it. Some studies based on meta-heuristics have been proposed for solving the issue of LEDs placement in indoor VLC communication systems. In this context, Rui et al. [10] utilized the PSO algorithm for solving the LED deployment problem in an indoor VLC communication system. The effect of changing the LEDs and user numbers were discussed. The results revealed that PSO gives good results compared to random models by obtaining the best-LEDs deployment and minimizing the average outage area rate.

In the work of Chaochuan et al. [11], an enhanced Cuckoo Search Algorithm, called VLP-IACS, was proposed for solving the 3D indoor LEDs positioning problem in an indoor VLC system. Simulations were carried out in a room size of $5 \text{ m} \times 5 \text{ m} \times 6 \text{ m}$ and the findings indicated that the VLP-IACS algorithm demonstrated strong performance.

In their study, Wen et al. [9] tackled the issue of LED placement in indoor VLC systems by proposing a Modified Artificial Fish Swarm Algorithm (MAFSA). Through experimentation, a positioning error of only 4.05 mm was achieved.

Kumawat et al. [8] used the Wale Optimization Algorithm for solving the LED panel placement Problem in the indoor VLC system. WOA was performed in a room with dimensions $5 \text{ m} \times 5 \text{ m} \times 3 \text{ m}$, having 4 to 6 LED panels considering the received power and SNR parameters. The findings that the WOA algorithm performed better than the PSO algorithm.

Wang et al. [12] investigated the layout of LEDs in indoor VLC communications based on evolutionary algorithm (EA) for optimizing the minimum SNR. Simulation results showed that the proposed EA approach obtained an optimal LEDs placement with maximization of the minimum SNR.

Whale Optimization Algorithm (WOA) is a novel swarm intelligence method introduced by Mirjalili and Lewis [13] in 2016. This meta-heuristic has some issues such as poor exploration and local optimum. To improve its solutions, WOA was mixed with some techniques like OBL mechanism and chaotic concept.

A. Ewees et al. [14], Abd Elaziz et al. [15], and Rohit Salgotra et al. [16], for example, combines the OBL strategy with WOA to improve its exploration. Yancang Li et al. [17], Hangqi Ding et al. [18], Gaganpreet Kaur et al. [19], Dharmbir Prasad et al. [20], Diego Oliva et al. [21], for example, combined some of chaotic maps with WOA to balance between exploitation and exploration, and take it away from local optima.

In this chapter, we proposed an enhanced version of WOA, called EWOA, for solving the LED deployment issue in an indoor VLC communication system. The proposed EWOA integrates two strategies including chaotic map concept and

Handbook of Whale Optimization Algorithm. https://doi.org/10.1016/B978-0-32-395365-8.00027-0

OBL method to improve the optimization performance of WOA. The performance of EWOA was assessed considering the user coverage and throughput parameters.

The rest of this chapter is arranged as follows. In Section 21.2, the formulation of the LEDs placement problem in indoor VLC communication systems is presented. Section 21.3 describes the structure of the WOA algorithm, chaotic map, OBL mechanism. Section 21.4 describes the EWOA model for solving the LEDs placement in indoor VLC systems. In Section 21.5, the simulation's findings are shown and discussed. Section 21.6 presents the final conclusion of the study.

21.2 LEDs placement problem formulation

21.2.1 System model

In this chapter, we consider a typical indoor VLC system (V) in a room of dimension, $X \times Y \times Z$ as illustrated in Fig. 21.1. This room is equipped with N Light Emitting Diodes (LEDs) installed on the ceiling to serve as optical access points (transmitters) and M receiving users inside this room. We further define d_{ij} the vertical distance between the i-th LED and the j-th receiver. Therefore, let:

- L is the set of N LEDs: $L = \{L_1, L_2, \ldots, L_N\}$ should be positioned for installation on the ceiling at locations (x_i, y_i, z_i), where $i \in (1, 2, 3, \ldots, N)$.
- U is the set of M Users $U = \{U_1, U_2, \ldots, U_M\}$, each user has a photo detector (PD) with him, used as wireless receiver. We assume that the users are randomly distributed in the $3D$ room. To say a LED covers the user U_j, if the value of the power received Pr from PD of the user U_j is positive, it should be noted that each user U_j can be associated with at most one LED (i.e., L_i). In other words, it is associated with the closest LED and the higher power (Pr).

21.2.2 Mathematical model

In our studies, we focus on the network's coverage and throughput as the two primary aspects of a VLC system that should be optimized.

- With the following Eqs. (21.1)–(21.2), the network's coverage is described:

$$Cov(V) = \sum_{j=1}^{M} max_{i\in\{1,\ldots,N\}}(Cov_{U_j}^{L_i}) \tag{21.1}$$

where $Cov_{U_j}^{L_i}$ demonstrates how each user U_j is covered by LED L_i. By (21.2), we can determine the value of coverage.

$$Cov_{U_j}^{L_i} = 1 \quad \text{if } Pr_j > 0 \tag{21.2}$$

- The throughput of the network is represented as follows:

$$Tr(V) = \sum_{j=1}^{M} max_{i\in\{1,\ldots,N\}}(Tr_{U_j}^{L_i}) \tag{21.3}$$

where with every user U_j covered by LED L_i has a throughput defined by $Tr_{U_j}^{L_i}$. It can be given as (21.4) [22]

$$Tr_{U_j}^{L_i} = B \times log_2(1 + SNR) \tag{21.4}$$

where B and SNR is the bandwidth, the signal-to-noise-ratio respectively. SNR can be expressed as (21.5)

$$SNR = \frac{(R.Pr)^2}{\sigma_t^2} \tag{21.5}$$

where R is the PD responsivity and σ_t^2 is the total noise variance. Pr is the power received by PD.

Therefore, here is how the objective function can be expressed:

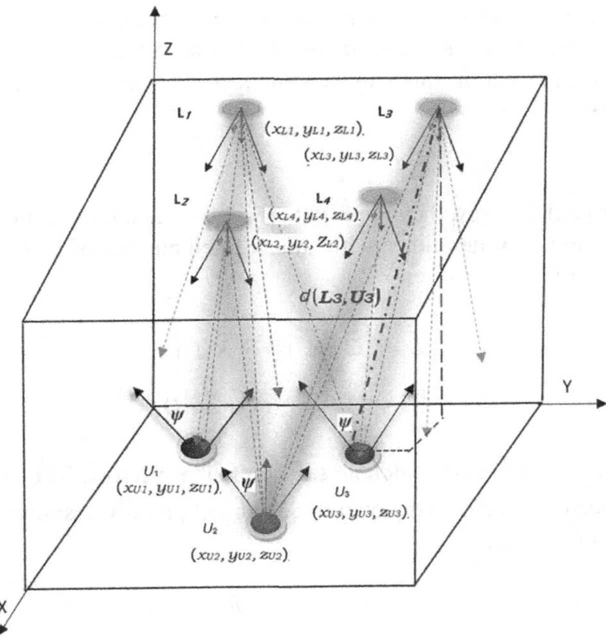

FIGURE 21.1 System model of our VLC system.

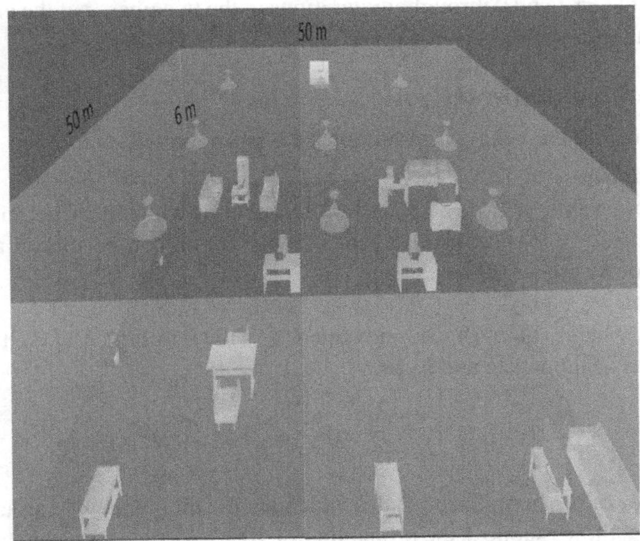

FIGURE 21.2 System model 3D.

$$f = (\lambda)(\frac{Cov(V)}{M}) + (1-\lambda)(\frac{Tr(V)}{M \times Tr_{max}}) \tag{21.6}$$

where $\lambda \in [0, 1]$.

21.3 Preliminaries

21.3.1 Whale optimizer algorithm (WOA)

Inspired from the hunting mechanism of humpback whales, Mirjalili et al. [13] proposed a special hunting mechanism and developed the meta-heuristic algorithm "Whale Optimizer Algorithm", in 2016.

Humpback whales use their natural behavior to surround prey by encircling them with flashing fins and creating a spiral-shaped bubble net, keep the prey contained, and finally block it from fleeing.

In the following section, we will present the mathematical model of [13] about search for prey, spiral bubble-net attacking technique, and encircling prey.

21.3.1.1 Encircling prey

When the humpback whales surround the prey, they modify their position and update it towards the best search agent from the first iteration to a maximum number of iterations, with an increased number of iterations. We describe mathematically this behavior in Eq. (21.7) and Eq. (21.8).

$$\vec{D} = \left|\vec{C} \cdot \overrightarrow{X^*}(t) - \vec{X}(t)\right| \tag{21.7}$$

$$\vec{X}(t+1) = \overrightarrow{X^*}(t) - \vec{A} \cdot \vec{D} \tag{21.8}$$

where t denotes the latest iteration and $\vec{C}$ and $\vec{A}$ denote the coefficient vectors. The best existing solution's placement vector is indicated by $\overrightarrow{X^*}$, the placement vector is denoted by $\vec{X}$, while $|\ |$ it only absolute value. We show in Eq. (21.9) and Eq. (21.10) how to get the vectors $\vec{C}$ and $\vec{A}$.

$$\vec{A} = 2.\vec{a}.\vec{r} - \vec{a} \tag{21.9}$$

$$\vec{C} = 2.\vec{r} \tag{21.10}$$

where $\vec{a}$ is decreased linearly between 2 and 0 throughout iterations in the two phases of the exploration and the exploitation. We mention that $\vec{r}$ could be a random vector between 0 and 1.

21.3.1.2 Bubble-net attacking method

There are two methods for simulating the behavior of humpback whales' bubble-net.

- **Shrinking encircling mechanism**
  for producing this result, the a value of Eq. (21.9) is lowered over an amount of iterations from 2 to 0. By choosing random numbers for $\vec{A}$ within $[-1, 1]$, the search agent's new position can be placed anywhere between the agent's initial location and the position of the actual best agent.
- **The spiral updating position**
  attempts to mimic humpback whales' helix shaped movement as shown in Eq. (21.11), it is determined by the equation of spiral between the positions of the whale and its prey.

$$\vec{X}(t+1) = \overrightarrow{D'} \cdot e^{bl} \cdot \cos(2\pi l) + \overrightarrow{X^*}(t) \tag{21.11}$$

In order to formalize the behavior of the humpback whales which swim simultaneously around the prey within a shrinking circle and along a spiral-shaped path, we suppose the probability of 50% to decide between the spiral model or the shrinking encircling strategy. To update whales' whereabouts, we use Eq. (21.12)

$$\vec{X}(t+1) = \begin{cases} \overrightarrow{X^*}(t) - \vec{A} \cdot \vec{D}\} & \text{if } p \leqslant 0.5 \\ \overrightarrow{D'} \cdot e^{bl} \cdot \cos(2\pi l) + \overrightarrow{X^*}(t)\} & \text{if } p \geqslant 0.5 \end{cases} \tag{21.12}$$

where $\overrightarrow{D'}$ represents how far away the i-th whale is from its prey (currently the most effective solution). For the purpose of describing the logarithmic spiral's form, b is a constant. The random values l and p are in the range of $[-1, 1]$ and $[0, 1]$ respectively.

Eq. (21.13) provides $\overrightarrow{D'}$.

$$\overrightarrow{D'} = \left|\overrightarrow{X^*}(t) - \vec{X}(t)\right| \tag{21.13}$$

21.3.1.3 Search for prey

In the exploration phase, in order to search for the prey, we exploit the variation of $\vec{A}$ vector, by using a random values larger than or less than 1 and -1 respectively to compel search agents to leave a reference whale. The exploration phase is formalized by Eq. (21.14).

$$\vec{D} = \left|\vec{C} \cdot \overrightarrow{X_{\text{rand}}} - \vec{X}\right| \tag{21.14}$$

$$\vec{X}(t+1) = \overrightarrow{X_{\text{rand}}} - \vec{A} \cdot \vec{D} \tag{21.15}$$

where $\overrightarrow{X_{\text{rand}}}$ is a random whale. This means a random location vector chosen from the present populace.

We present, the pseudo-code of WOA in Algorithm 21.1 [13].

Algorithm 21.1 The pseudo-code of WOA algorithm.

Initialize : the whale's population X_i $(i = 1, 2, \ldots, n)$.
Calculate : Calculate the fitness of each search agent.
X^* = the best search agent.
while ($t <$ maximum number of iterations) **do**
  for (each search agent) **do**
    Update a, A, C, l, and p.
    if ($p < 0.5$) **then**
      if ($|A| < 1$) **then**
        Update the position of the current search agent by Eq. (21.7)
      else if ($|A| >= 1$) **then**
        Select a random search agent (Xrand)
        Update the position of the current search agent by Eq. (21.15)
      end if
    else if ($|A| >= 0.5$) **then**
      Update the position of the current search by Eq. (21.11)
    end if
  end for
  Check if any search agent goes beyond the search space and amend it,
  Calculate the fitness of each search agent
  Update X^* if there is a better solution
  $t = t + 1$
end while
return X^*

21.3.2 Opposition-based learning (OBL)

The OBL approach was initially created in 2005 by Tizhoosh [23], to optimize the search space's exploring phase and to increase the convergence rates of many meta-heuristic algorithms Then, it was successfully combined with many meta-heuristic optimization algorithms. The OBL strategy focuses on calculating opposite numbers to approximate solutions.

In Eq. (21.16), we show how to identify the opposite of a real number $Pmo \in [lb, ub]$.

$$\bar{Pmo} = ub + lb - Pmo \tag{21.16}$$

A multidimensional search space can be added to the definition shown in Eq. (21.16) by Eq. (21.17).

$$\bar{Pmo}_i = ub_i + lb_i - Pmo_i, \quad i = 1, 2, \ldots, d \tag{21.17}$$

where the true vector Pmo is the one that $\bar{Pmo}$ is the opposite of.

21.3.3 Chaotic map concept

To improve also the searching capability and to increase the convergence rate, we used the Chaotic map concept, it's a deterministic approach to studying the behavior of dynamic and nonlinear systems. It has numerous crucial characteristics, including, non-converging, stochastic, ergodicity, bounded, regularity, non-repetitive, unpredictable, and non-periodic. These characteristics have been turned into a variety of chaotic maps, which are mathematical equations used to produce random parameters in meta-heuristics. Multiple chaotic maps have also been illustrated in the literature, including Sinusoidal map, Logistic map, Circle map, Gauss map, Piecewise map, Chebyshev map, Sine map, Piecewise map, Tent map, and Iterative map [24,25]. With straightforward operations and dynamic randomization, the sine map is among the most basic and often used chaotic maps Eq. (21.18) is the definition of the sine map equation.

$$SM_{k+1} = \frac{ac}{4}\sin(\pi SM_k), \quad 0 \le SM_O \le 1 \tag{21.18}$$

where SM_k is the value of Sine map at the k-th iteration. It is in [0, 1] and the control parameter ac between $0 < ac \le 4$.

21.4 The proposed EWOA for solving the LEDs placement problem

One of the main challenges with WOA is its slow convergence speed, as highlighted in [19]. In order to address this issue and improve overall performance, OBL strategy and chaos theory were incorporated into the WOA optimization phase. These modifications allowed for greater control over the exploitation and exploration phases, ultimately enhancing the algorithm's global convergence speed. Firstly, in order to increase WOA's convergence speed and explore effectively the search space, the OBL is implemented. Secondly, to speed up convergence rate and prevent being caught in local optima, chaotic maps are employed in the WOA algorithm.

The pseudo-code of EWOA is given in Algorithm 21.2 and presented in Fig. 21.3.

Algorithm 21.2 The pseudo-code of EWOA algorithm.

Initialize : the whales population randomly X_i $(i = 1, 2, \ldots, n)$.
Calculate : the Opposite $\bar{X}$ of the whale population.
Calculate : the fitness of each search agent X_i and Opposite $\bar{X}_i$ and select the n best from $X_i \bigcup \bar{X}_i$.
$X^* =$ the best current from the n best.
Initialize : the value of the sine chaotic map.
while ($t <$ maximum number of iterations) **do**
  Update : the chaotic value using Eq. (21.20).
  for (each search agent) **do**
    Update a, A, C, l.
    adjust the parameter 'p' using sine chaotic value Eq. (21.20)
    if ($p < 0.5$) **then**
      if ($|A| < 1$) **then**
        Update the position of the current search agent by Eq. (21.7)
      else if ($|A| >= 1$) **then**
        Select a random search agent (Xrand)
        Update the position of the current search agent by Eq. (21.15)
      end if
    else if ($|A| >= 0.5$) **then**
      Update the position of the current search by Eq. (21.11)
    end if
  end for
  Check if any search agent goes beyond the search space and amend it
  Calculate the fitness of each search agent
  Update X^* if there is a better solution
  $t = t + 1$
end while
return X^*

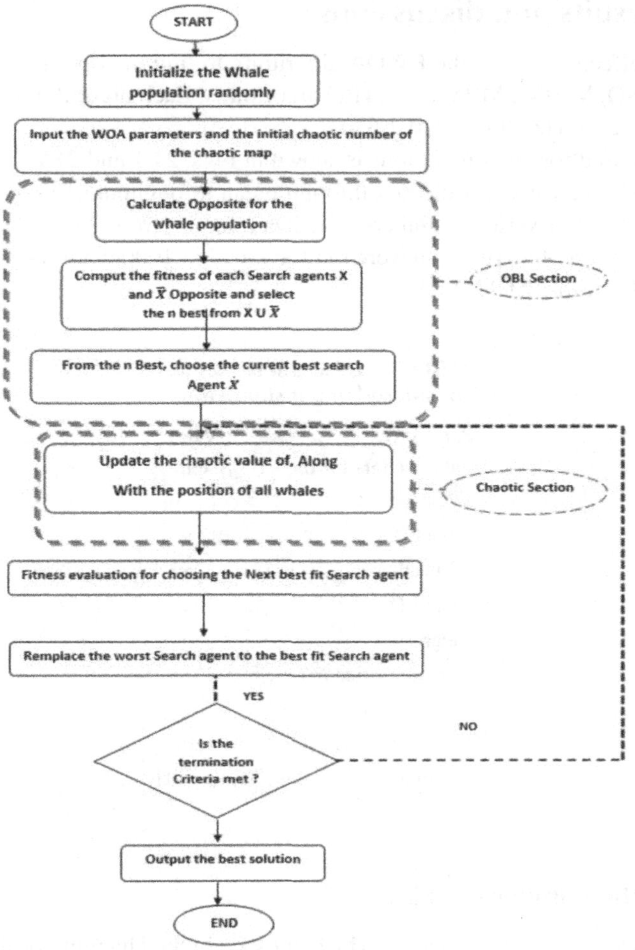

FIGURE 21.3 Flowchart of EWOA.

With a few adjustments, the EWOA follows the same processes as the WOA:

- In terms of the **OBL** technique, we determined the opposing $\overline{X}$ utilizing the OBL mechanism after the original solution X was generated randomly. The fitness value for both $\overline{X}$ and X is then calculated, and we keep just the best n from $\overline{X} \bigcup X$ to reuse it in residual iterations.
- Regarding the **Chaos** method, the parameter p of Eq. (21.12) is modified using Sine chaotic map during iterations, and the new frequency equation is written as:

$$\vec{X}(t+1) = \begin{cases} \overrightarrow{X^*}(t) - \vec{A} \cdot \vec{D}\} & \text{if } p_i \leqslant 0.5 \\ \vec{D'} \cdot e^{bl} \cdot \cos(2\pi l) + \overrightarrow{X^*}(t)\} & \text{if } p_i \geqslant 0.5 \end{cases} \tag{21.19}$$

$$p_{i+1} = \frac{ac}{4} \sin(\pi p_i), \tag{21.20}$$

where p_i is the chaotic map value of parameter p, in the i-th iteration, and it must fall among 0 and 1. $ac = 4$ and $p_0 = 0.7$.

21.5 Experimental results and discussions

This section will evaluate the effectiveness of the EWOA algorithm, to find the optimal LEDs placement and compare it with the original WOA, BA, PSO, MRFO, MPA, and CHIO algorithms. Each algorithm is coded using Matlab®. Runs all simulations in a Core i5-4310U 2.6 GHz-CPU RAM 12Go.

A typical scheme of a room used for our simulations is shown in Figs. 21.1 and 21.2.

The EWOA algorithm's performance was validated through mean throughput and coverage per user metrics using 16 scenarios. These scenarios consisted of various numbers of LEDs (ranging from 2 to 9) and users (ranging from 5 to 40). Each result in this chapter is the mean of 30 runs and were run for a total of 1000 iterations, while the simulation parameters used for the study can be found in Table 21.1.

TABLE 21.1 Parameter values considered in our simulations.

Parameter	Value
Parameters for the VLC system	
N° of users M	[5 40]
N° of LEDs N	[2 9]
Width W	50 m
Length D	50 m
Height H	6 m
Population size NS	30
N° of runs R	30
N° of iteration t_{max}	1000
l a random in	[−1, 1]

21.5.1 Effect of varying the number of LEDs

In this case, we'll discuss the results of eight scenarios which consists in the changing of the number of LEDs from 2 to 9 with a fixed number of users (30), on the coverage and throughput. The results are presented in Table 21.2 and Fig. 21.4. It is shown that as the number of LEDs in the room increases, the throughput increases and the coverage rises until covering approximately all users. When there are more than 7 LEDs, this is realized. In reality, as the number of LEDs rises, users have a greater probability of being covered, raising the metrics for throughput and coverage. In the same conditions, EWOA gives usually better performance in terms of throughput and coverage.

21.5.2 Effect of varying the number of users

In this case, we'll discuss the results of the eight remaining scenarios, which consists in the changing of the number of users from 5 to 40 with a fixed number of LEDs (30) on the coverage and throughput. The results are shown in Fig. 21.5 and Table 21.3. It is observed that as the number of users in the room increases, the coverage and the throughput decrease. We can easily see that in most cases, EWOA performs better than other algorithms in terms of coverage and throughput.

21.6 Conclusion

In this paper, we propose an efficient Whale Optimizer Algorithm, including chaotic map concept and Opposition-Based Learning (OBL) mechanism, for solving the LEDs placement problem in indoor visible light communication (VLC) system in an indoor room with dimensions 50 m × 50 m × 6 m. In our system, we have considered N LEDs and M users. The efficiency of the proposed EWOA is assessed considering the user coverage and throughput metrics in comparison with the original WOA, BA, PSO, MRFO, CHIO, and MPA approaches. The simulation results showed that EWOA obtained optimal LEDs layouts compared to other algorithms.

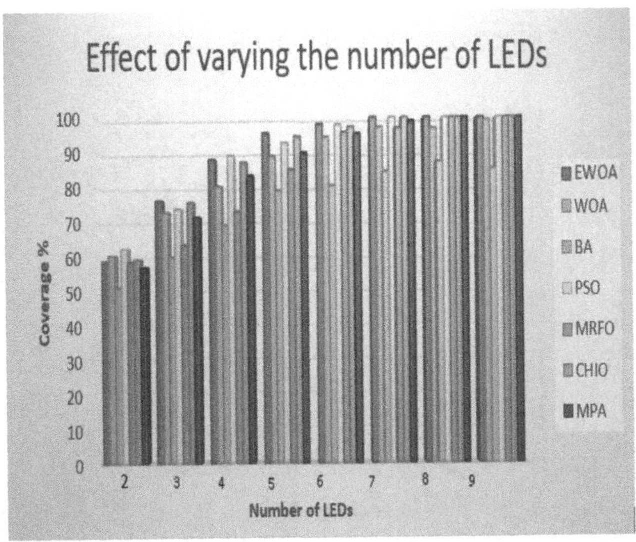

FIGURE 21.4 Effect of varying the number of LEDs.

TABLE 21.2 Coverage, mean throughput, and fitness under parameter various of number LED.

Number of LEDs	2	3	4	5	6	7	8	9
			Coverage (%)					
EWOA	58.8	**76.22**	88	**95.44**	**98.2**	**100**	**100**	**100**
WOA	60.44	72.88	80.22	88.88	94.44	97.22	96.88	99.22
BA	51.11	60.77	69.11	79	80.55	84.44	87.33	85.22
PSO	**62.4**	73.84	**89.1**	92.61	98	**100**	**100**	**100**
MRFO	58.7	63.45	73.1	85	95.8	97	100	100
CHIO	59.22	75.66	87.11	94.55	97.33	**100**	**100**	**100**
MPA	57.1	71.44	83.33	90	95.4	99	**100**	**100**
			Mean throughput per user (Mbps)					
EWOA	0.69	1.32	1.74	2.94	**4.23**	**6.1**	**7.14**	8.33
WOA	0.65	1.30	1.59	1.94	2.48	2.48	4.26	4.87
BA	0.60	1.27	1.57	1.63	2.20	2.28	2.30	2.35
PSO	**0.71**	**1.34**	1.59	2.15	2.84	3.23	4.87	5.1
MRFO	0.61	1.15	1.43	1.86	2.3	2.57	3.7	4.21
CHIO	0.52	1.28	1.68	**3.07**	4.03	5.89	7.05	**8.86**
MPA	0.62	1.21	**1.89**	2.63	3.8	4.2	4.33	4.83
			Fitness					
EWOA	0.36	**0.42**	0.45	**0.51**	**0.57**	**0.58**	**0.60**	**0.64**
WOA	0.30	0.30	0.41	0.45	0.48	0.50	0.51	0.52
BA	0.26	0.31	0.35	0.40	0.41	0.43	0.44	0.44
PSO	**0.39**	0.41	**0.47**	0.48	0.53	0.55	0.57	0.61
MRFO	0.27	0.33	0.37	0.44	0.46	0.47	0.49	0.51
CHIO	0.31	0.39	0.46	0.49	0.52	0.53	0.54	0.55
MPA	0.28	0.34	0.36	0.41	0.45	0.49	0.52	0.53

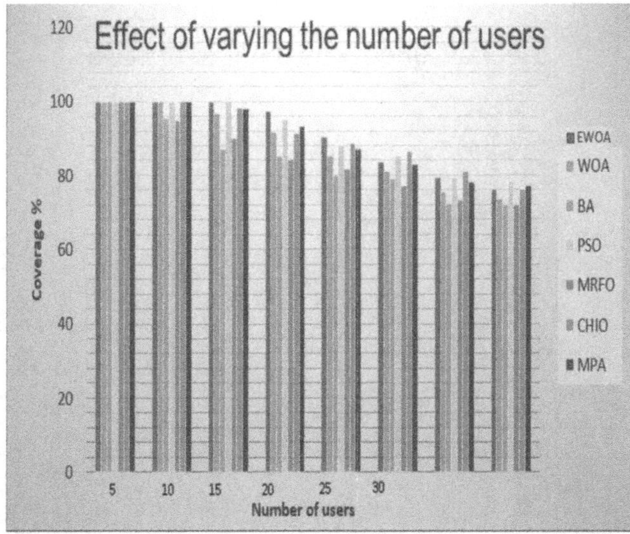

FIGURE 21.5 Effect of varying the number of Users.

TABLE 21.3 Coverage, mean throughput, and fitness under parameter various of number of users.

Number of clients	5	10	15	20	25	30	35	40
				Coverage (%)				
EWOA	**100**	**100**	**100**	**97.14**	**90.2**	83.45	79.33	76.11
WOA	**100**	**100**	96.74	91.64	85.14	80.85	75.54	73.46
BA	**100**	95.47	87.11	85.16	80	79.11	72.28	72
PSO	**100**	**100**	**100**	95	88.14	85.11	79.3	**78.44**
MRFO	**100**	94.8	90	84	81.5	77.14	73.11	72
CHIO	**100**	**100**	98.37	91.12	88.73	**86.44**	**81.33**	77.64
MPA	**100**	**100**	98.14	93.44	87.5	83.52	78.44	77.21
				Mean throughput per user (Mbps)				
EWOA	**18.3**	12.44	**10.84**	7.71	4.44	3.33	2.63	2.1
WOA	13.35	9.73	7.70	5.21	3.50	2.09	1.84	**2.62**
BA	6.50	5.26	4.27	3.31	2.64	1.68	1.48	1.41
PSO	17.4	**13.11**	10.2	7.31	4.14	3.11	2.87	2.24
MRFO	7.4	6.81	6.1	3. 62	2.91	2.14	1.81	1.44
CHIO	16.41	12.88	9.08	6.09	4.42	3.19	2.83	2.56
MPA	14.3	9.45	9.03	**8.32**	**5.74**	**3.51**	**2.92**	1.84
				Fitness				
EWOA	**0.69**	**0.65**	**0.59**	0.52	0.48	0.49	0.46	0.46
WOA	0.58	0.56	0.54	0.50	0.46	0.45	0.45	0.43
BA	0.54	0.53	0.46	0.44	0.41	0.40	0.40	0.38
PSO	0.67	0.61	0.53	**0.54**	0.51	**0.50**	0.48	0.46
MRFO	0.55	0.51	0.50	0.48	0.46	0.42	0.41	0.40
CHIO	0.60	0.58	0.56	**0.54**	**0.52**	0.49	**0.49**	**0.48**
MPA	0.63	0.59	0.53	0.50	0.49	0.47	0.47	0.45

References

[1] Svilen Dimitrov, Harald Haas, Principles of LED Light Communications: Towards Networked Li-Fi, Cambridge University Press, 2015.

[2] Ying-Dar Lin, Third quarter 2018 IEEE communications surveys and tutorials, IEEE Communications Surveys and Tutorials 20 (3) (2018) 1607–1615.

[3] Panagiotis Botsinis, Dimitrios Alanis, Simeng Feng, Zunaira Babar, Hung Viet Nguyen, Daryus Chandra, Soon Xin Ng, Rong Zhang, Lajos Hanzo, Quantum-assisted indoor localization for uplink mm-Wave and downlink visible light communication systems, IEEE Access 5 (2017) 23327–23351.

[4] Musa Furkan Keskin, Sinan Gezici, Orhan Arikan, Direct and two-step positioning in visible light systems, IEEE Transactions on Communications 66 (1) (2017) 239–254.

[5] Suseela Vappangi, Venkata Mani Vakamulla, Synchronization in visible light communication for smart cities, IEEE Sensors Journal 18 (5) (2017) 1877–1886.

[6] D. Tagliaferri, C. Capsoni, High-speed wireless infrared uplink scheme for airplane passengers' communications, Electronics Letters 53 (13) (2017) 887–888.

[7] Vaneet Aggarwal, Zhe Wang, Xiaodong Wang, Muhammad Ismail, Energy scheduling for optical channels with energy harvesting devices, IEEE Transactions on Green Communications and Networking 2 (1) (2017) 154–162.

[8] Ishwar Ram Kumawat, Satyasai Nanda, Ravi Maddila, Positioning LED panel for uniform illuminance in indoor VLC system using whale optimization, in: Optical and Wireless Technologies, 2018, pp. 131–139.

[9] Shangsheng Wen, Xiaoge Cai, W.P. Guan, Jiajia Jiang, Bangdong Chen, Mouxiao Huang, High-precision indoor three-dimensional positioning system based on visible light communication using modified artificial fish swarm algorithm, Optical Engineering 57 (2018) 106102.

[10] Rui Guan, Jin-Yuan Wang, Yun-Peng Wen, Jun-Bo Wang, Ming Chen, PSO-based LED deployment optimization for visible light communications, in: 2013 International Conference on Wireless Communications and Signal Processing, 2013, pp. 1–6.

[11] Jia Chaochuan, Yang Ting, Wang Chuanjiang, Sun Mengli, High-Accuracy 3D Indoor Visible Light Positioning Method Based on the Improved Adaptive Cuckoo Search Algorithm, Arabian Journal for Science and Engineering 47 (2022) 2479–2498.

[12] Lang Wang, Chunyue Wang, Xuefen Chi, Linlin Zhao, Xiaoli Dong, Optimizing SNR for indoor visible light communication via selecting communicating LEDs, Optics Communications 387 (2017) 174–181.

[13] Seyedali Mirjalili, Andrew Lewis, The whale optimization algorithm, Advances in Engineering Software 95 (2016) 51–67.

[14] Ahmed A. Ewees, Mohamed Abd Elaziz, Diego Oliva, A new multi-objective optimization algorithm combined with opposition-based learning, Expert Systems with Applications 165 (2021) 113844.

[15] Mohamed Abd Elaziz, Diego Oliva, Parameter estimation of solar cells diode models by an improved opposition-based whale optimization algorithm, Energy Conversion and Management 171 (2018) 1843–1859.

[16] Urvinder Salgotra, Rohit Singh, Sriparna Saha, On some improved versions of whale optimization algorithm, Arabian Journal for Science and Engineering 44 (2019) 9653–9691.

[17] Yancang Li, Muxuan Han, Qinglin Guo, Modified Whale Optimization Algorithm Based on Tent Chaotic Mapping and Its Application in Structural Optimization, KSCE Journal of Civil Engineering 24 (2020) 3703–3713.

[18] Hangqi Ding, Zhiyong Wu, Luchen Zhao, Whale optimization algorithm based on nonlinear convergence factor and chaotic inertial weight, Concurrency and Computation: Practice and Experience 32 (24) (2020) e5949.

[19] Gaganpreet Kaur, Sankalap Arora, Chaotic whale optimization algorithm, Journal of Computational Design and Engineering 5 (3) (2018) 275–284.

[20] Dharmbir Prasad, Aparajita Mukherjee, Gauri Shankar, Vivekananda Mukherjee, Application of chaotic whale optimisation algorithm for transient stability constrained optimal power flow, IET Science, Measurement & Technology 11 (8) (2017) 1002–1013.

[21] Diego Oliva, Mohamed Abd El Aziz, Aboul Ella Hassanien, Parameter estimation of photovoltaic cells using an improved chaotic whale optimization algorithm, Applied Energy 200 (2017) 141–154.

[22] Yusuf Said Eroglu, Alphan Sahin, Ismail Guvenc, Nezih Pala, Murat Yuksel, Multi-element transmitter design and performance evaluation for visible light communication, in: 2015 IEEE Globecom Workshops (GC Wkshps), IEEE, 2015, pp. 1–6.

[23] Hamid R. Tizhoosh, Opposition-based learning: a new scheme for machine intelligence, in: International Conference on Computational Intelligence for Modelling, Control and Automation and International Conference on Intelligent Agents, Web Technologies and Internet Commerce (CIMCA-IAWTIC'06), vol. 1, IEEE, 2005, pp. 695–701.

[24] Yassine Meraihi, Dalila Acheli, Amar Ramdane-Cherif, QoS multicast routing for wireless mesh network based on a modified binary bat algorithm, Neural Computing & Applications 31 (7) (2019) 3057–3073.

[25] Shahrzad Saremi, Seyedali Mirjalili, Andrew Lewis, Biogeography-based optimisation with chaos, Neural Computing & Applications 25 (5) (2014) 1077–1097.

Chapter 22

Adaptive bi-level whale optimization algorithm for maximizing the power output of hybrid wave-wind energy site

Mehdi Neshat[a], Nataliia Y. Sergiienko[b], Leandro S.P. da Silva[b], Erfan Amini[c], Mahdieh Nasiri[d], and Seyedali Mirjalili[a,e]

[a] *Centre for Artificial Intelligence Research and Optimisation, Torrens University Australia, Brisbane, QLD, Australia,* [b] *School of Mechanical Engineering, University of Adelaide, Adelaide, SA, Australia,* [c] *Department of Civil, Environmental and Ocean Engineering, Stevens Institute of Technology, Hoboken, NJ, United States,* [d] *Department of Mechanical Engineering, Stevens Institute of Technology, Hoboken, NJ, United States,* [e] *University Research and Innovation Center, Obuda University, Budapest, Hungary*

22.1 Introduction

The substantial increase in energy consumption over the past several decades has prompted communities, stakeholders, and authorities to address the scarcity of energy resources. Despite the substantial progress made in alternative energy development, the majority of consumable energy continues to come from oil, natural gas, and coal [1,2]. The generation of energy from these sources also poses a significant threat to the climate in the long term, leading to unavoidable and detrimental effects on both humans and the environment. Consequently, the outlook for energy consumption cannot be considered positive without the formulation of a clear plan and a robust framework for the energy transition. In accordance with the resolution agreed upon globally in Paris, decreasing global warming to less than 2°C and also in the pre-industrial levels at 1.5°C has been deemed a critical goal [3]. Meanwhile, the COVID-19 pandemic since 2019 and the Ukraine war since 2022 have only served to intensify the energy crisis.

The utilization of renewable energy sources, which are clean, sustainable, and limitless in nature, has gained significant prominence in recent years. The main sources of renewable energy, such as solar and wind, do not emit harmful greenhouse gases or pollutants, which are significant contributors to climate change and environmental degradation [4,5]. Moreover, the cost of renewable energy sources has been decreasing at a stable rate, while the cost of fossil fuels is escalating, and their prices are highly unstable [6,7]. Utilizing renewable energy converters offers numerous advantages in terms of climatic and economical advantages, including the reduction of greenhouse gas emissions and air pollution from fossil fuels, diversification of energy supply to reduce dependency on imported oil and gas, and promotion of economic growth through the creation of manufacturing, installation, and other job opportunities [6,7]. As depicted in Fig. 22.1, there are numerous types of renewable energy systems, with marine energy sources being one of the most promising globally [8,9]. The main sources of offshore energy, such as wave and wind, thermal, biomass, tidal stream, and tidal range, among others, offer immense potential [10,11].

The utilization of ocean wave energy and offshore wind energy as sources of marine renewable energy has emerged as two of the most prominent methods [12]. These renewable energy resources play a vital role in supplying power to coastal communities and offshore infrastructure. The concept of harnessing ocean waves as a source of energy dates back to 1799 in France [13], while commercial interest in wave energy didn't emerge until the 1970s. Similarly, the use of offshore wind turbines was not considered practical until the commissioning of the initial offshore wind farm globally, Vindeby, in Denmark in 1991 [14]. By 2021, the global generating capacity of all offshore wind energy projects had reached 368 GW and continues to grow at an exponential rate [15].

In the realm of renewable energy, offshore wind energy holds significant potential, particularly due to the abundance of wind in offshore locations, which surpasses that of onshore locations. Unlike onshore wind energy, which is a well-established technology, the recent related to offshore wind energy sites are in their nascent stage and continue to garner interest from researchers and practitioners alike [16,17].

Handbook of Whale Optimization Algorithm. https://doi.org/10.1016/B978-0-32-395365-8.00028-2

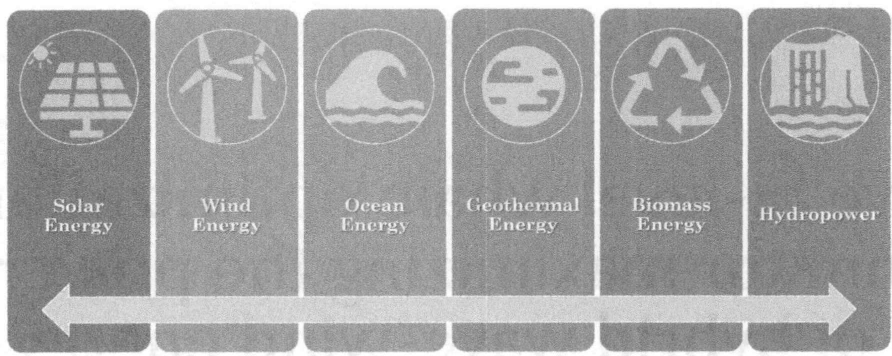

FIGURE 22.1 Schematic view of different types of sustainable energy systems.

The hybridization of two technologies, wave and offshore wind energy has emerged as a topic of research and has been explored to yield higher power output and a potentially lower levelized cost of energy (LCOE) through shared structures [18]. Several preliminary studies have investigated the possibility of combined wave energy harvesting methods, including point absorbers, oscillating physical structures, overtopping wave models, oscillating water column devices, etc. [19]. Moreover, the characteristics of combined wave-wind systems, such as technologies, water depths (deep or shallow), and location with regard to the shoreline (offshore, nearshore, onshore) differ, resulting in the categorization of hybrid, co-located and island systems based on the connectivity rate in the middle of wave energy converters (WECs) and offshore wind turbines [18].

The co-located systems, as a straightforward option, integrate the wind energy converters (WECs) with offshore wind turbines and share common features such as the operational area, electrical grid connection, and maintenance equipment [20,21]. Fig. 22.1 depicts two variants of co-located systems, namely, hybrid arrays and independent arrays. Co-located arrays with self-contained are located in close proximity and share a single electric grid connection [22]. On the other hand, co-located combined arrays incorporate a common area and relevant infrastructures, thereby constituting a single array. Within the realm of combined arrays, three variations exist: Peripherally Distributed Array (PDA) which arranges the WECs along the array's perimeter in accordance with the dominant wave direction, Uniformly Distributed Array (UDA) which uniformly distributes both WECs and seaward wind turbines all over the line-up, and Non-uniformly Distributed Array (NDA) where the WECs are arranged in a non-uniform manner within the offshore wind farm [18].

The hybrid system is another classification, in which offshore wind turbines and WECs are integrated towards a unit structure. There are two sub-classifications of hybrid systems, that revolve around the water depth: fixed-bottom and floating. Moreover, suitable devices based on shallow waters, and fixed-bottom foundations, such as monopile, are commonly used for offshore wind turbines. Recently, several fixed-bottom hybrid devices have been proposed, including the Wave Star by Danish Wave Energy Concepts [23], the Wave Treader by Scotland's Green Ocean Energy, and the WEGA, introduced by the company of Portugal's Sea For Life which is a WEC based on the gravity [24]. Schematics of these devices are depicted in Fig. 22.2.

The utilization of floating offshore wind turbines (FOWTs) has garnered growing attention in the offshore energy industry due to their economic viability in water depths that exceed 60 meters [26–28]. Furthermore, the abundance of wind energy sources in deep water regions has facilitated the growing popularity of FOWTs [29]. There have been numerous commercial hybrid projects utilizing FOWTs, including the Poseidon project by Floating Power Plant in the Netherlands, which is based on a triangular floating foundation [30]. The Pelagic Power AS introduced the W2Power power plant, which utilizes a triangular-shaped floating model with three wind turbines located at three corners, at the same time, collective oscillating body wind turbines are located on its edges, each generating 2–3 MW [31]. Another platform of note is the Kriso semi-submersible multi-unit floating offshore wind turbine, with a quadrangular shape-based and also embedded with four WECs and offshore wind turbines [32].

In the realm of wave-wind energy technologies, offshore multipurpose platforms, otherwise known as island systems, offer a unique and innovative solution. These systems are often larger in scale and afford the opportunity to harness multiple marine resources on a single platform. The two primary classifications of island systems are artificial and floating. The former is utilized for substantial energy storage, while the latter serves as smaller-scale floating multipurpose platforms that possess a higher capacity than most ships. Structurally, platforms in this category mainly consist of spars, semi-submersible, and tension leg platforms (TLPs), which are derived from offshore oil and gas industry structures. The versatility of these platforms enables the integration of wind turbines, water column overturning WECs, and point absorbers with heaving,

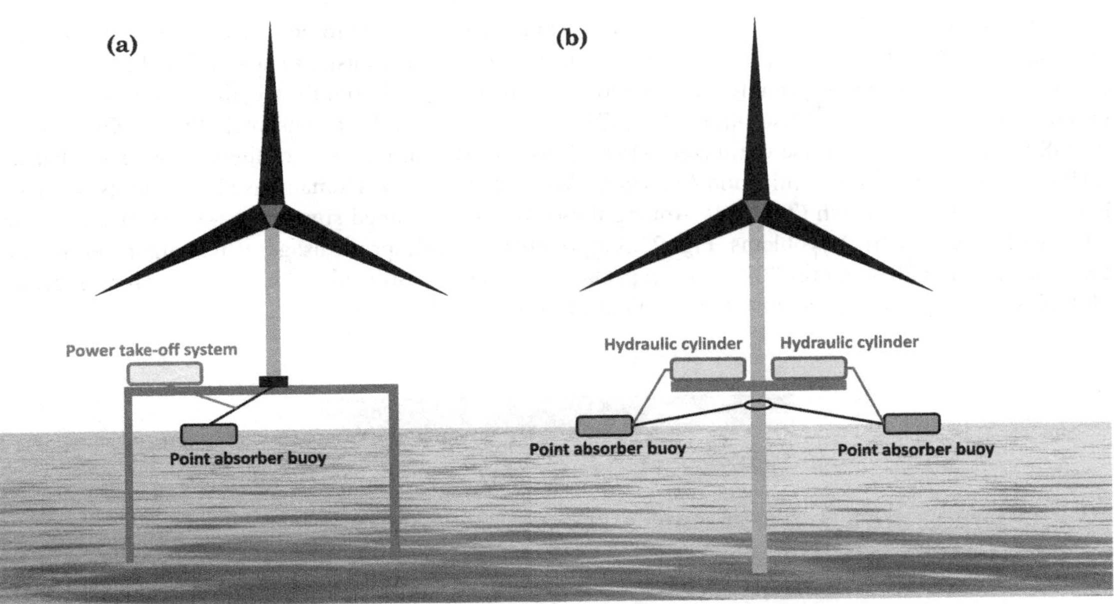

FIGURE 22.2 Schematics of hybrid devices, (a): WEGA, a gravitational wave energy absorber, (b): Wave Treader (Reproduced based on [25]).

pitching, or multiple degrees of freedom attachments. Moreover, there are numerous design variations that can be employed, as discussed in [33–36].

Recent research highlights the significance of optimizing the layout and size of co-located wave-wind systems, as demonstrated in [37]. Researchers have sought to maximize the structural and cost benefits of these systems by exploring the optimal layout for co-located wind and wave farms. For instance, an experimental model of an offshore wind farm consisting of 80 generators in the area of Wave Hub located in South West England assessed the impact of offshore wind farms on the environment [11,38]. Moreover, quantitative evaluations have been conducted to determine the most profitable farms by considering the correlation between wind and waves, as well as the most advantageous sizes for WECs as auxiliary hydrogen producers [39,40]. The results indicate that farms with a minimum relationship between waves and wind can recover a uttermost of 101.12 hours of equivalently rated manufacture [39]. The hybridization of offshore wind power and wave in co-located farms offers numerous advantages, including increased weather windows, popular electrical underpinning, shared operation, and maintenance equipment, and reduced carbon footprint, thus improving their economic viability [41].

The most recent development in wind-wave systems is the hybrid wind-wave farm, which integrates its capabilities more efficiently and effectively than conventional wind-wave co-located plants. However, the optimization of a hybrid wind-wave farm layout is subject to several constraints, such as the location of the farm, the number of extractors, the array layout, the separation distance between extractors, and the direction of incident winds and waves. To address these challenges, some optimization frameworks have been derived through simulations and validated in different case studies in the United States, Europe, and Africa [37,42]. Additionally, previous studies faced difficulties in applying wave wake analysis to hybrid wave-wind energy farms, as it requires excessive computational resources and a long iteration process, which makes the optimization process extremely slow [42]. As a result, innovative methods, such as the differential evolution (DE) optimization approach, have been proposed to optimize the device layout for actual environmental conditions using a multi-objective approach [43].

The optimization of various engineering fields, including the generation and conversion of energy, network grid design and supply, has been a topic of substantial attentiveness in recent times, particularly with the advent of information technology and the integration of AI-based approaches [44]. Metaheuristic optimization has gained popularity as a solution to address the hyper-parameter optimization problems faced in the design of renewable energy systems. This approach uses a combination of deterministic and stochastic components to find the best possible solution, which makes it a flexible and versatile method for solving a broad range of optimization problems. Unlike traditional heuristics, which were based on subjective experience, metaheuristics are derived from the objective function and can be applied to a much broader range of problems. Research has shown that metaheuristics can significantly improve the performance of renewable energy systems by reducing the number of iterations required to reach a solution, reducing the computational cost, and providing more accurate and reliable results [45]. In addition, the use of metaheuristics has led to new insights into the design of energy systems and has allowed engineers to explore new design space that was previously not accessible. [45].

Metaheuristic algorithms can be broadly categorized into four types: i) Evolutionary algorithms, such as the Genetic Algorithm (GA) [46] and Differential Evolution (DE) [47]; ii) Physics-based algorithms, including the Gravitational Search Algorithm (GSA) [48] and Multi-verse optimizer (MVO) [49]; iii) Swarm-based algorithms, such as the Particle Swarm Optimizer (PSO) [50] and Whale Optimization Algorithm (WOA) [51]; and iv) Human-based algorithms, such as Harmony Search (HS) [52] and Taboo Search (TS) [53]. Among these, WOA has gained significant popularity and is increasingly being used in various engineering problems. Fig. 22.3 highlights the widespread usage of this algorithm in research and academic documents from 2016 to mid-2022, with expectations of it being utilized and reviewed in an even greater number of research topics, leading to its appearance in more than 11,500 research outcomes.

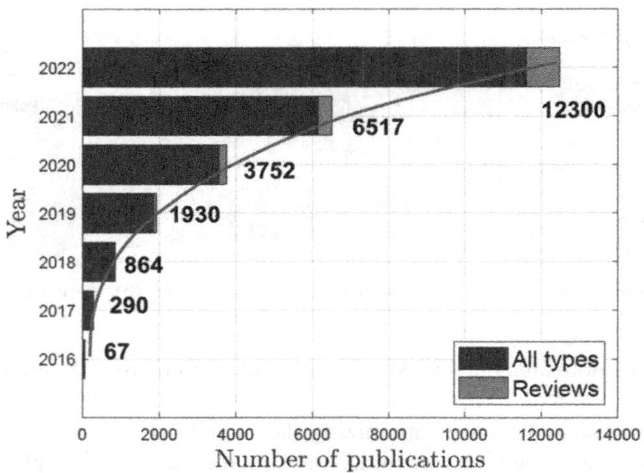

FIGURE 22.3 The number of research documents published each year included utilizing or reviewing the Whale Optimization Algorithm (WOA) and its applications, taking research articles, books, technical reports, case studies, and thesis into account (from 2016 to 2022, data credit: Google Scholar).

The WOA algorithm is one of the popular metaheuristics and optimization methods in the field of engineering, including the optimization of hybrid wind-wave systems. Its applications include site selection [54], installation, operation, and maintenance costs [55], dynamic analysis [56], and design and layout [57]. Despite its usefulness, research on the optimized power take-off (PTO) parameters of hybrid energy converters remains limited [58].

WOA has received increasing attention in the academic community due to its effectiveness and ease of adaptation. A significant body of literature has compared its performance to other metaheuristics and traditional optimization techniques [59]. To further enhance its applicability, various modifications have been proposed for the WOA algorithm, including its application in data mining and machine learning [60], hybridization in power systems [61] and electric photovoltaic (PV) systems [62,63], modification for IoT [64], applied mathematics [65], and aeronautics engineering [66]. The WOA algorithm has also been used to study the reliability of engineering systems [67].

In this book chapter, a novel Adaptive Bi-level Whale Optimization Algorithm (AWOA) is introduced to maximize the power output of a hybrid wave-wind energy site. This bi-level optimization consists of an outer level, where a geometry parameter tuning task is performed, and an inner level, where a PTO parameter optimization task is conducted. The proposed method is then compared to several state-of-the-art metaheuristic algorithms to develop a comprehensive evaluation of its effectiveness. The hybrid wind-wave system used in the study consists of a 5-MW OC4 semi-submersible platform coupled with three spherical WECs, which are designed to extract power solely from heaving motions. A spring-damper system is employed to implement the system of power take-off, with the constraint that the stiffness must be positive. The optimization problem on the design side pertains to the radius of the WECs and their distance from the main column.

This paper is organized as follows. Section 22.2 describes the system and modeling properties, including the hybrid wind-wave system, equations of motion, deployment site, and performance measures. Section 22.3 presents the optimization setup, the evaluation criteria, the algorithm procedure, and code development. In Section 22.4, we explain the properties of the whale optimization algorithm (WOA), the exploitation and exploration phase, and the development of the adaptive Bi-level Whale Optimization Algorithm (AWOA). In Section 22.5, we discuss the numerical results, visualize the research output, and assess the proposed methodology to achieve desired results. Finally, the findings and future scopes are listed in Section 22.6.

22.2 System description and modeling

22.2.1 Hybrid wind-wave system

The hybrid wind-wave system considered in this study is comprised of the 5-MW OC4 semi-submersible platform and three half-submerged spherical WECs mechanically coupled with the platform [68]. The WECs are designed to extract power from the heaving motion only. The power take-off system is modeled as a spring-damper system, where the stiffness can only take positive values. The radius of the WECs and their distance from the main column are a subject of optimization.

For the simplicity of this study, it has been assumed that the addition of WECs does not change the vertical location of the platform's center of gravity, and does not require a redesign of the mooring system. In addition, the platform, WECs, and tower are assumed to be rigid bodies. The physical dimensions, inertia properties and mooring configuration of the original 5-MW OC4 semi-submersible platform are specified in [69]. (See Fig. 22.4.)

FIGURE 22.4 The hybrid wind-wave system based on the OC4 semi-submersible platform and three half-submerged spherical WECs.

22.2.2 Equations of motion

The dynamics of the hybrid wind-wave platform is described in total by 8 degrees-of-freedom (DOFs): 3DOFs correspond to the platform's motion in surge (q_1^{FOWT}), heave (q_3^{FOWT}), and pitch (q_5^{FOWT}), 2DOFs correspond to the wind turbine dynamics including the blade pitch angle β, and the generator speed ϕ, and 3DOFs describe the heaving motion of three WECs ($q_3^{WEC,i}$, $i = 1\dots3$). The WECs extract power from the relative motion between the platform and each WEC as: $q_{rel}^i = q_3^{WEC,i} - q_3^{FOWT} + L_i q_5^{FOWT}$, where L_i is a vector that gives the horizontal distance from the vertical center line of the platform to the ith WEC's center of gravity. As a result, the vector of generalized coordinates of the hybrid wind-wave system is:

$$\mathbf{q}(t) = [q_1^{FOWT}, \quad q_3^{FOWT}, \quad q_5^{FOWT}, \quad \phi, \quad \beta, \quad q_3^{WEC,1}, \quad q_3^{WEC,2}, \quad q_3^{WEC,3}]^T. \tag{22.1}$$

Forces acting on the hybrid wind-wave system include:

- hydrodynamic forces, including radiation and excitation forces. The hydrodynamic coupling between the bodies is modeled using BEM solver NEMOH;
- aerodynamic forces (nonlinear) are modeled according to [70], which has been verified against OpenFAST output;
- mooring loads on the floating platform are represented as stiffness matrices derived using a quasi-static approach [71,72], where each stiffness term is a function of the platform's surge, heave and pitch displacements;

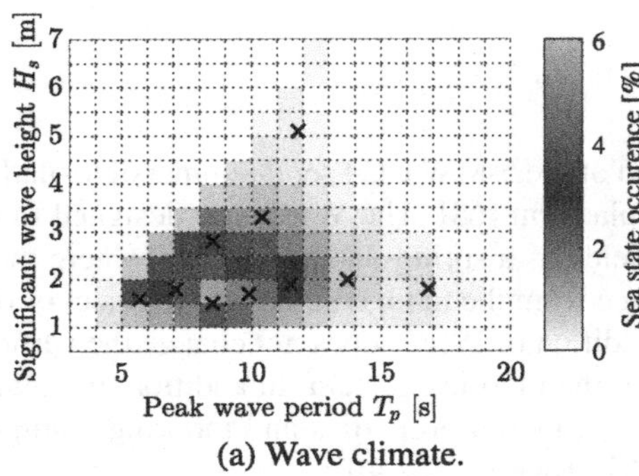

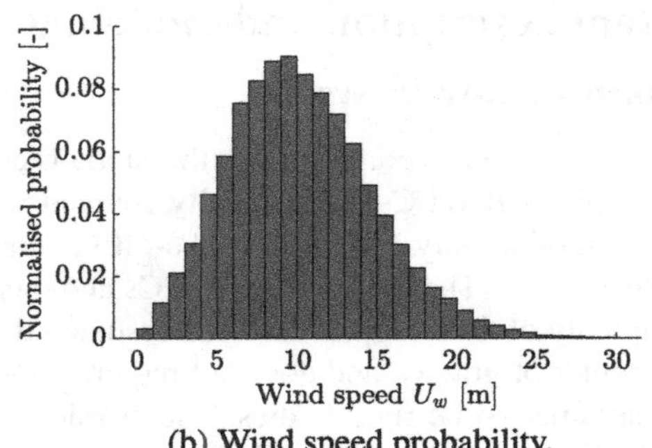

FIGURE 22.5 The wave and wind statistics (hindcast) for the Sydney site.

- hydrostatic forces on the FOWT in heave and pitch are modeled as a linearized stiffness matrix calculated using BEM solver NEMOH. For the WECs, the change in the cross-section of the half-submerged spherical WEC produces a non-linear hydrostatic restoring force which is modeled according to [73];
- viscous drag forces on the FOWT and WECs are represented according to the Morison equation [68,70];
- power take-off machinery is modeled as a linear spring-damper system, where power is extracted from the relative motion between each WEC and the platform. The mechanical coupling between the platform and the WECs is represented using stiffness and damping matrices.

The resultant equation of motion of the hybrid wind-wave system is written as:

$$\mathbf{M}\ddot{\mathbf{q}}(t) = \mathbf{F}_{exc}(t) + \mathbf{F}_{rad}(t) + \mathbf{F}_{aero}(t) + \mathbf{F}_{moor}(t) + \mathbf{F}_{hs}(t) + \mathbf{F}_{visc}(t) + \mathbf{F}_{pto}(t). \tag{22.2}$$

Eq. (22.2) contains non-linear forces as aerodynamic, hydrostatic, and viscous drag. Following the statistical linearization approach [74], Eq. (22.2) is linearized where all nonlinear terms are replaced by the equivalent linear matrices of damping and stiffness. The equivalent linear system of the hybrid wind-wave system is:

$$\left[-\omega^2(\mathbf{M} + \mathbf{A}_{rad}) + i\omega(\mathbf{B}_{rad} + \mathbf{B}_{aero}^{eq} + \mathbf{B}_{visc}^{eq} + \mathbf{B}_{pto}) + \mathbf{K}_{moor} + \mathbf{K}_{hs}^{eq} + \mathbf{K}_{pto}\right]\mathbf{q}(\omega) = \mathbf{Q}(\omega), \tag{22.3}$$

where $\mathbf{M}$ is a diagonal mass matrix (all equations are assembled around the center of gravity of each body), $\mathbf{A}_{rad}$ and $\mathbf{B}_{rad}$ are the radiation added mass and damping matrices, $\mathbf{B}_{pto}$ and $\mathbf{K}_{pto}$ are the PTO coupling matrices, $\mathbf{K}_{moor}$ is a mooring matrix, and $\mathbf{Q}$ is the generalized load vector that contains the wave excitation force and wind loads. All matrices with a superscript 'eq' correspond to the equivalent linear matrices that are calculated iteratively for each wind and wave condition following statistical linearization (for more details refer to [73]).

The input to Eq. (22.3) is the significant wave height, peak wave period and mean wind speed. It is assumed that the wave elevation is represented by the JONSWAP spectrum [63] with a peak enhancement factor of 3.3, and the wind speed fluctuations follow the Kaimal spectrum [75].

22.2.3 Deployment site

The potential for hybrid wind and wave exploration has been investigated in [76] for six different sites around Australia. Taking into account such factors as wind and wave energy potential, power availability and its seasonal variability, potential power smoothing effects and capacity factors, it has been found that Sydney area is one of the most desired locations for the installation of the hybrid wind and wave energy farm. In addition, the eastern coast of Australia is the most suitable for the use of floating offshore wind turbines due to its deep water. A wave-wind hindcast data for the Sydney site for the period from 2010 until 2014 is demonstrated in Fig. 22.5. This wave climate has a 19.7 kW/m wave power density with the dominant wave period of 8.5 s and a dominant wave height of 1.5 m. The wind potential is approximately 1.07 kW/m^2 with the most probable wind speed of 10 m/s. The water depth at this location is $>$100 m.

To evaluate the performance of the hybrid wind-wave system, the k-means clustering method has been used to extract 10 representative sea states. For each of these sea states, the most probable value of the mean wind speed at the hub height has been evaluated and shown in Table 22.1. The probability of occurrence of each environmental condition has been re-calculated to make sure that the total wave power flux at the deployment site remains unchanged.

TABLE 22.1 Environmental conditions.

T_p [sec]	H_s [m]	U_w [m/s]	Occurrence [%]
5.7	1.6	13	11.3
7.1	1.8	9	16.2
8.5	2.8	14	8.3
8.5	1.5	7	14.8
9.9	1.7	7	13.8
10.4	3.3	12	6.1
11.5	1.9	9	10.3
11.8	5.1	20	2.1
13.7	2.0	8	5.2
16.8	1.8	8	3

22.2.4 Performance measures

At this location, the 5-MW wind turbine will generate approximately 2 MW of power on average. Once the WECs are attached to the floating platform, there are two main questions that interest the hybrid system developers: (i) how much more power can WECs generate and (ii) how does the addition of WECs affect the wind turbine performance.

The average power generation of WECs in one sea state is evaluated as [73]:

$$P^{WEC,i} = \frac{1}{T}\int_0^T (\dot{q}^i_{rel})^2 B_{pto} dt = \sigma^2_{\dot{q}^i_{rel}} B_{pto}, \tag{22.4}$$

where B_{pto} is the PTO damping coefficient, T denotes the time interval, and $\sigma_{\dot{q}^i_{rel}}$ denotes the standard deviation of the relative velocity between the ith WEC and the platform.

The wind turbine performance is characterized by the horizontal component of the wind turbine nacelle acceleration [70,77]:

$$\ddot{q}^{nacelle} = \ddot{q}_1^{FOWT} + \ddot{q}_5^{FOWT}(z_{nacelle} - z_{cg}), \tag{22.5}$$

where $z_{nacelle}$ and z_{cg} are vertical coordinates of the nacelle and the platform's center of gravity. As nacelle acceleration fluctuates around zero, it is more representative to calculate its standard deviation for each environmental condition of interest. Once calculated for each environmental condition from Table 22.1, the annual average power production and average standard deviation of the nacelle acceleration are evaluated for the Sydney site using the probability of occurrence from Table 22.1.

22.3 Optimization setup

To set up an optimization function for the hybrid wind-wave system is not a trivial task. Ideally, it is desired to use WECs for the benefit of the floating platform by simultaneously increasing power generation and damping the motion of the platform where the PTO parameters K_{pto} and B_{pto} play a critical role in this task. However, as demonstrated in [68], these two objective functions (increase in power generation and reduction in platform motion) are conflicting by nature.

There are 22 parameters that require optimizing in this study including the radius of spherical WECs, the distance from the FOWT's main column to the WECs, and PTO stiffness and damping coefficients for each sea state. However, it is well known, that larger WECs generate more power (see Fig. 22.6), so the absolute value of the power production cannot be used as an objective function. Therefore, the q-factor can be introduced that demonstrates the gain in power generated by WECs attached to the floating platform as compared to WECs that operate in isolation (without the platform):

$$q = \frac{P^{hybrid}}{3P^{isolatedWEC}}. \tag{22.6}$$

The objective function should be formulated as a minimization problem, therefore, it is taken as $1/q$. In addition, the q-factor does not take into account the performance of the wind turbine, in particular, the nacelle acceleration. Therefore, the nacelle acceleration is introduced as a penalty factor into the objective function. So if the addition of WECs does not lead to an increase in the FOWT nacelle acceleration as compared to a stand-alone FOWT (without WECs), the penalty is zero. If the nacelle acceleration increases twice, the penalty takes a value of 1 (see Fig. 22.6).

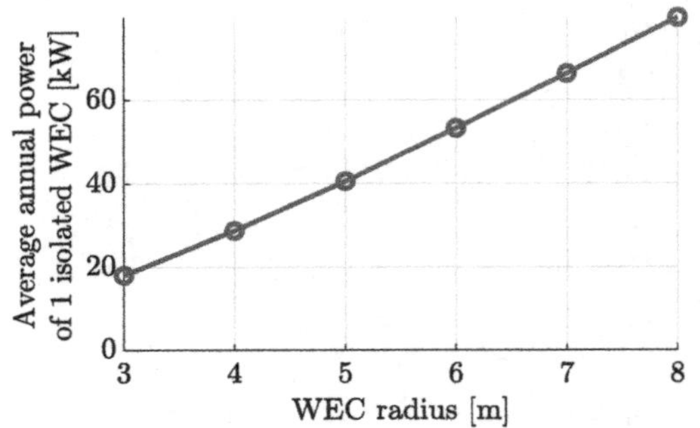

(a) Dependence of the WEC power production (Sydney site) on its radius.

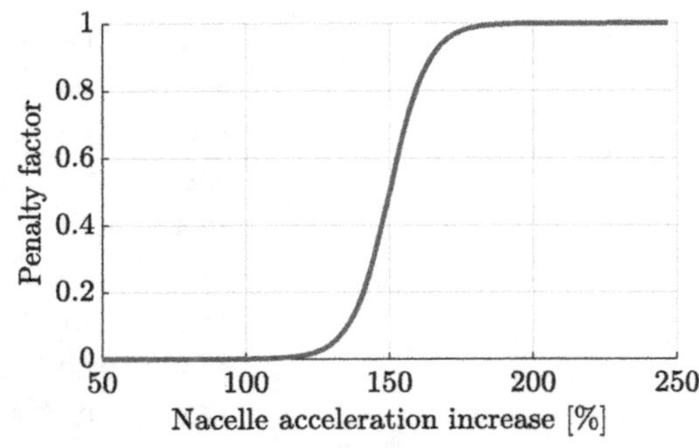

(b) Penalty factor.

FIGURE 22.6 Setting an objective function.

As a result, the objective function for this work is set as:

$$\text{Objective function} = 1/q + \text{penalty factor}. \tag{22.7}$$

So the optimization process in this work will attempt to find such WEC parameters (radius, distance to the platform and PTO coefficients) that the power gain is maximized while not significantly increasing the nacelle acceleration of the floating platform. The optimization variables can change within the limits specified in Table 22.2.

TABLE 22.2 Design variables.

Parameter	Unit	Values
WEC radius	m	[3, 4, 5, 6, 7, 8]
Distance from the main column to the WEC	m	[20, 25, 30, 35, 40, 45, 50]
PTO stiffness	N/m	$[0 \ldots 10^{10}]$
PTO damping	N/(m/s)	$[0 \ldots 10^{10}]$

22.4 Meta-heuristic optimization algorithms

22.4.1 Whale optimization algorithm (WOA)

The Whale Optimization Algorithm (WOA), a swarm-based approach, draws inspiration from the hunting behavior of humpback whales. These marine mammals are known for their intelligence and emotional intelligence, attributes that stem from their possession of spindle cells. Additionally, humpback whales have a distinct hunting technique known as bubble-net feeding, which has garnered significant attention in the scientific community [78].

This algorithm incorporates the principles of bubble-net feeding into its design [59]. In this hunting method, humpback whales produce bubbles in a circular or 9-shaped pattern to corral small fish near the surface of the water. There are two specific models of bubble-net hunting: upward spirals and double-loops [79]. The WOA optimization algorithm utilizes mathematical modeling of the spiral bubble-net feeding strategy, consisting of three key steps as described in subsequent sections.

22.4.1.1 Surrounding the prey

At the first step of their hunting, humpback whales encircle their prey and recognize where it is located. It is assumed by the WOA algorithm that the present best current particle plays the role of target prey or it is assumed that it can be near to the optimum since it is unknown in advance where the best configuration is in the search space. Following determining the best-found particle, other particles will try to adjust their coordination based on the best-found particle. This process is determined by following formulas [51]:

$$\vec{P}(iter+1) = \overrightarrow{P^*}(iter) - \vec{A}\cdot\vec{d}, \tag{22.8}$$

where

$$\vec{d} = \left|\vec{c}\cdot\overrightarrow{P^*}(iter) - \vec{P}(iter)\right|, \tag{22.9}$$

where the current iteration notifies by $iter$, $\vec{P}$ shows the position vector, $\overrightarrow{P^*}$ indicates the coordination vector of the best-found particle achieved so far, and $\vec{A}$ and $\vec{c}$ vectors are coefficient vectors as defined below:

$$\vec{A} = 2\vec{a}\cdot\vec{r} - \vec{a}, \tag{22.10}$$

$$\vec{c} = 2\vec{r}, \tag{22.11}$$

where $\vec{a}$ is reduced with a linear pattern between 2 and 0, and $\vec{r}$ is generated based on a uniform random number in [0, 1].

22.4.1.2 Bubble-net attacking method (exploitation phase)

As a next step of the hunting method, there are two different approaches for the bubble-net behavior model which are considered account simultaneously. It is worth mentioning that, the possibility of choosing between these two methods is 50%.

- Shrinking encircling mechanism:
  In this method the recent location of a particle can be determined everywhere between the initial location of the particle and the location of the current best-found particle. This behavior is obtained by decreasing the value of $\vec{a}$ in Eq. (22.10). In a 2D space, Fig. 22.7 depicts the feasible positions from (P, Q) towards (P^*, Q^*) that can be achieved by A in [0, 1].

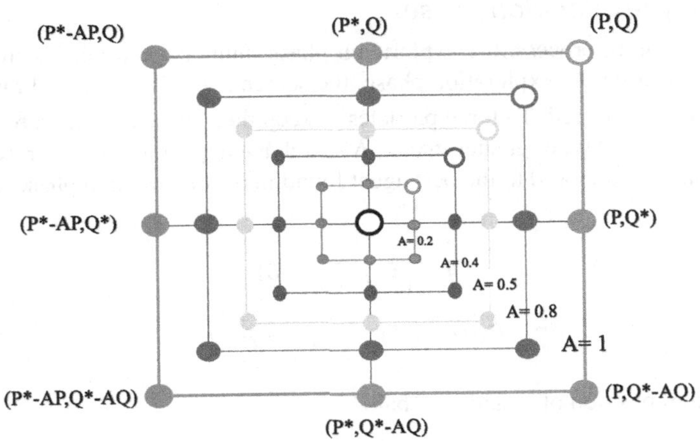

FIGURE 22.7 Shrinking encircling mechanism of bubble-net search mechanism, adapted from [51].

- Spiral updating position:
  which first calculates the distance between the whale located and the prey located, then a spiral equation is created between the position of the whale and prey to mimic the helix-shaped movement of humpback whales, as it is shown in Fig. 22.8. Here is the mathematical model of this process:

$$\vec{P}(iter+1) = \overrightarrow{d'}\cdot e^{bl}\cdot\cos(2\pi l) + \overrightarrow{P^*}(iter), \tag{22.12}$$

  where $\vec{d'}$ signifies the interspace of the ith particle to the prey (best-found particle achieved so far), b is constant, and l is a random number in [−1, 1].
  According to the fact that humpback whales surround the prey based on a circle model and across a spiral form path in parallel. The probability rate for switching between these two strategies is 50%. Thus, taking these two approaches into

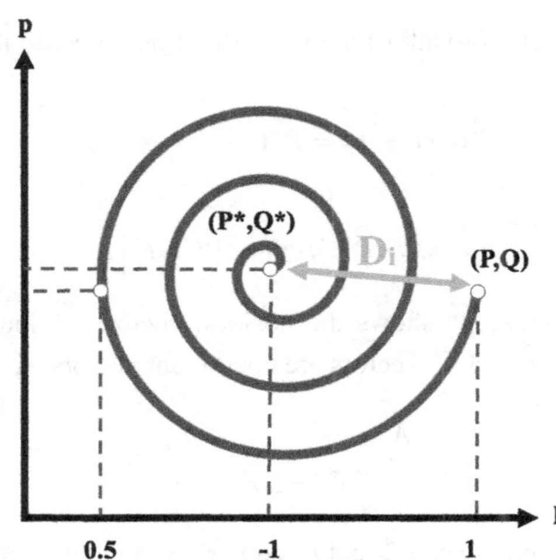

FIGURE 22.8 Spiral updating position of bubble-net search mechanism (P^* shows the best solution obtained), adapted from [51].

account, Eq. (22.13) describes the mathematical representation of the bubble-net search mechanism [51]:

$$\vec{P}(iter+1) = \begin{cases} \overrightarrow{P^*}(iter) - \vec{A} \cdot \vec{d} & \text{if } P < 0.5 \\ \overrightarrow{d'} \cdot e^{bl} \cdot \cos(2\pi l) + \overrightarrow{P^*}(iter) & \text{if } P \geq 0.5 \end{cases} \tag{22.13}$$

where P is a random number in [0, 1].

22.4.1.3 Explore the prey (exploration phase)

Finally, as a last part of the hunting process, the exploration phase, humpback whales search arbitrarily based on their locations in the search space. During the exploration phase, the search agent is positioned randomly instead of using the best search agent that has been discovered. To force particles to keep their distance from a reference whale, the $\vec{A}$ vector is generated based on random values less or greater than 1. A search agent position will be reset using a randomly selected particle in the exploration phase, as opposed to the best agent found in the exploitation phase. Following is a mathematical model [51]:

$$\vec{d} = \left| \vec{c} \cdot \overrightarrow{P_{\text{rand}}} - \vec{P} \right|, \tag{22.14}$$

$$\vec{P}(iter+1) = \overrightarrow{P_{\text{rand}}} - \vec{A} \cdot \vec{d}, \tag{22.15}$$

where $\overrightarrow{P_{\text{rand}}}$ represent a random position of a random whale.

22.4.2 Adaptive bi-level whale optimization algorithm (AWOA)

Although WOA is able to start the search process with a strong exploration ability and smoothly goes to the exploration phase, in some real-world optimization problems with heterogeneous decision variables encountered with premature convergence issues or the accuracy of best-found solutions is low [80]. In order to overwhelm this deficiency, various strategies have been proposed, such as applying a combination of self-adapting parameter adjustment and mixed mutation strategy [81], using the chaotic maps in order to adjust the parameter p [82], combining three sufficient search strategies with a pooling mechanism [83], developing WOA search mechanism by Levy and Brownian motion that cooperates with bubble-net attacking strategy [84].

This study proposes an enhanced bi-level whale optimization algorithm to solve a real-energy optimization method: a hybrid wave-wind energy system. The modified WOA is applied at the upper level to optimize both PTO and geometry parameters. At the lower level, we use a local search technique to speed up the convergence rate of the WOA by optimizing one of the heterogeneous setting parameters. The main contributions of the proposed adaptive WOA are including:

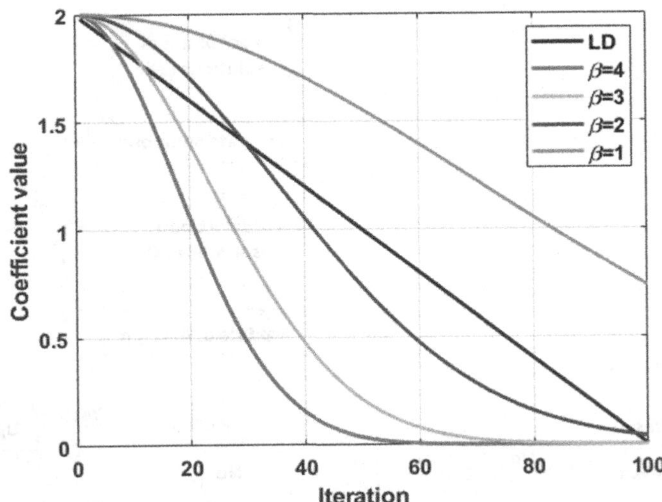

FIGURE 22.9 Various control parameter strategies to balance between the exploration and exploitation. The LD is related to a with a linearly decreased coefficient from 2 to 0 (standard a).

- Using a linearly decreased population size strategy rather than a fixed population size by Eq. (22.16)

$$NP_{iter+1} = \text{round}\left(\left[\left(\frac{NP_{min} - NP_{ini}}{Max_{iter}}\right) \times iter + NP_{ini}\right]\right). \tag{22.16}$$

- Proposing an effective control parameter coefficient to tune both exploration and exploitation phases properly (Eq. (22.17)). Fig. 22.9 indicates various exponential control parameters compared with the linear model applied in the standard WOA. As can be seen, increasing the coefficient of β results on concentrating more one the exploitation and search with small step sizes.

$$\vec{a} = 2 \times \exp\left(-\left(\beta \times \frac{iter}{Max_{iter}}\right)^2\right). \tag{22.17}$$

- Making a pool of initial random designs (three folds of the population size) and selecting the best ones to reinforce the exploration ability.
- Optimizing the PTO and geometry parameters of the hybrid model in two separate levels. Dividing the search space into homogeneous sections can improve the optimization algorithms performance [85].

The diagram of the technical details of the adaptive bi-level WOA can be seen in Fig. 22.10.

From Fig. 22.10, Im_{rate} computes the difference fitness value of the best-found solutions in two sequential generations. If this improvement rate is less than $\phi = 1\%$, the local search is applied to adjust PTO and geometry parameters in different levels. The $flag$ can be switched between 0 and 1, determining which setting parameters should be optimized. In order to select K_{pto} or B_{pto} parameters to optimize, a random variable (r) is used. It is noted that the local search is applied to improve the quality of the best-found solution in that population, so it is a single-solution based method.

Despite the considerable investigations to improve the WOA's performance [59], an adequate and efficient technique still needs to be developed to propose a fair balance between local and global search ability in the WOA. The primary reason may be related to enhancing the WOA's characteristics based on the case studies and is still a tricky research question that needs more investigation. To approach this goal, we focused on optimizing the performance of a hybrid wave-wind energy unit using WOA at two levels because the character of PTO parameters is intuitively different from geometry features. Furthermore, we tested five population sizes and compared their impact on the final optimal solutions and proposed adaptive population sizes, which performed better than other static sizes.

In the majority of real-engineering optimization problems, evaluating the new solutions using the fitness function is costly in terms of computational budget. Therefore, the computational complexity of the optimization methods can be estimated by the number of running fitness functions. As in each generation, WOA produced N new solutions and evaluated them, so with a fixed population size, the computational complexity is $O(N)$. However, in our proposed adaptive WOA, the population size linearly decreased, and the computational complexity is $O(\frac{N_{min}-N_{max}}{Max_i} \times i + N_{ini})$, where N_{min}, N_{max},

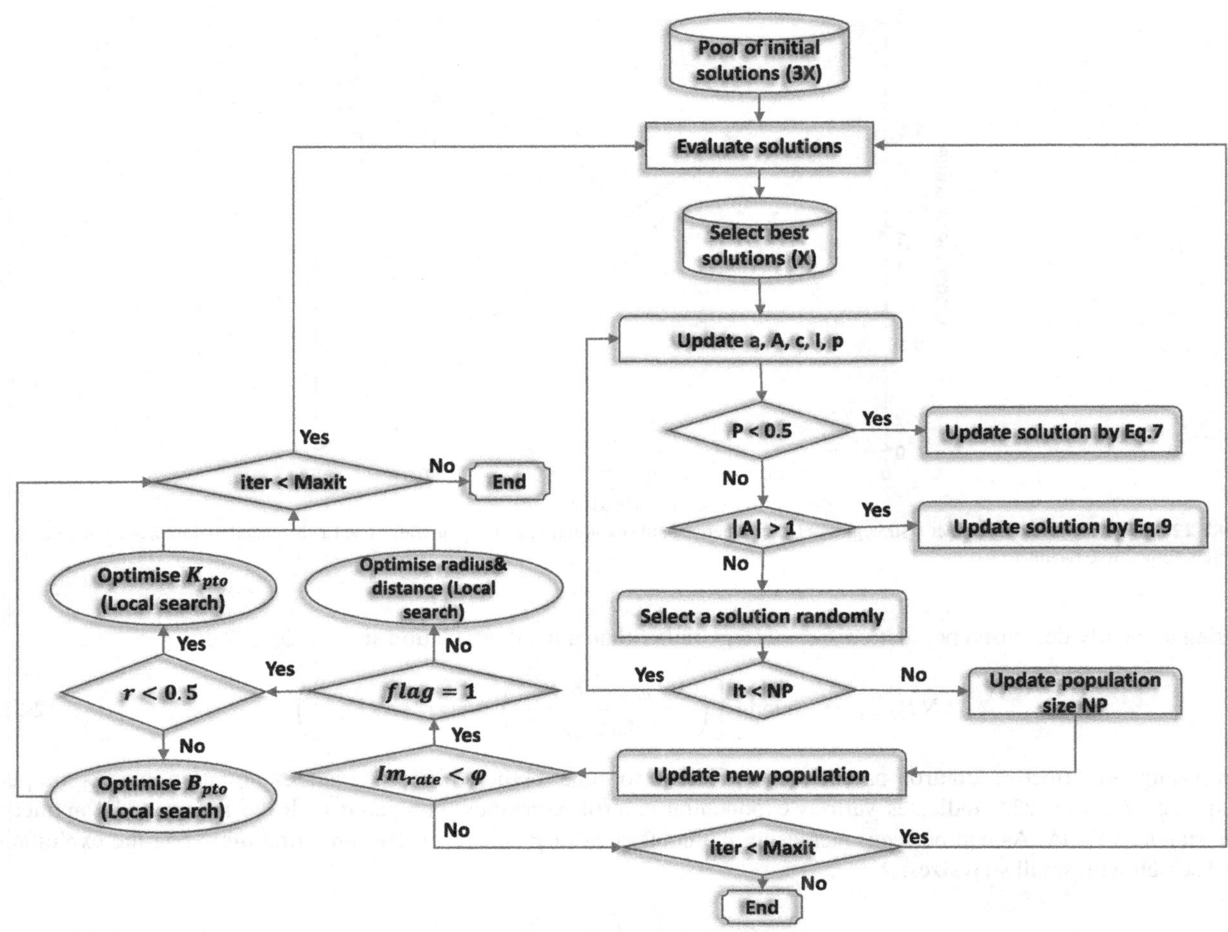

FIGURE 22.10 The flowchart of the Adaptive Bi-level Whale Optimization Algorithm (AWOA).

and N_{ini} are 30, 50, and 50. i and Max_i are iterations and the maximum number of generations. Therefore, the AWOA is 20% faster than the original WOA with a static population size with the same generation number.

22.5 Numerical results and discussions

In this section, to evaluate the efficiency of the proposed AWOA, a comparative framework is developed. Initially, using six popular swarm-based algorithms, including Grey Wolf Optimization (GWO), Grasshopper Optimization Algorithm (GOA), Moth Flame Optimizer (MFO), Sine Cosine Algorithm (SCA), Salp Swarm Algorithm (SSA), and Whale Optimization Algorithm (WOA). The control parameters in all optimization methods are based on the literature. The population size and the maximum number of evaluations are 50 and 1000 for all methods.

Fig. 22.11 shows the statistical optimization results of ten independent runs for six swarm-based algorithms, including the minimum, maximum, median, first quartile, third quartile, and outliers for the best-found solutions. In terms of median performance and the best-found solution, the WOA outperformed the other five meta-heuristics. Furthermore, it can be seen that the SCA's performance is also competitive and indicates a high level of robustness to converge appropriate solutions. The GWO, MFO, and SSA's efficiency approximately are the same; however, the median performance of the GWO is more promising than the other two methods.

In the second step in developing a comparative framework, we focused on one of the critical parameters of the WOA, which is population size (Np). This is mainly because small population sizes can achieve a faster convergence speed with the risk of struggling with local optimum. On the other hand, a large size of population leads to better exploration with a low convergence rate. Thus, finding a suitable number of agents can considerably improve the solutions' quality. Fig. 22.12 demonstrates how the WOA's performance shifts when Np varies (such as 12, 25, 50, and 100); in order to determine the

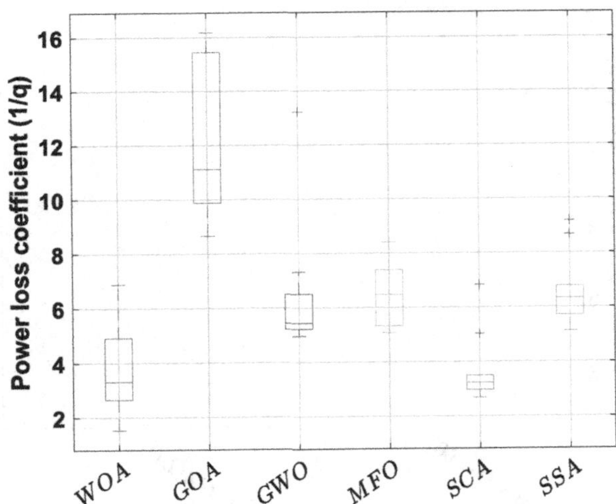

FIGURE 22.11 The statistical analysis of six optimization methods' performance to minimize the power loss coefficient. Each method runs 10 times independently and the best-found solution per each run is considered.

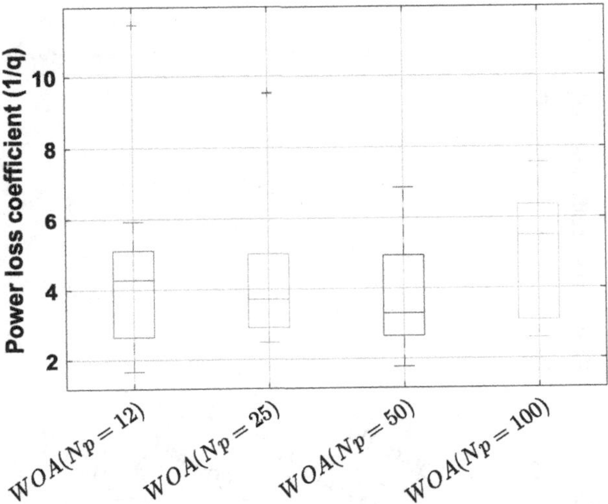

FIGURE 22.12 The impact of population size on the WOA's performance. Each population size runs 10 times independently and the best-found solution per each run is considered.

correct number of search agents for this problem. It is observed that the average quality of the best-found solutions with a population size of 50 is better than other sizes. However, both Np =12 and 50 could find solutions with power loss values less than 2.

To have a more comprehensive analysis of population size effect on the performance of WOA, we proposed a simple dynamic population size which is linearly reduced (LD) from 50 to 20 agents based on Eq. (22.16). We compared this adaptive population size scenario with a fixed size at 50 and ran each setting ten times. The best-found solutions for both fixed and dynamic measures can be seen in Fig. 22.13. As it can be seen that the WOA with a linearly decreased Np outweighs the WOA with $Np = 50$. This dynamic Np could provide more robust searchability than static population size.

In order to develop an effective balance between both the exploration and exploitation phases in the WOA, a control parameter (a) is introduced that adjusts the search step and is linearly reduced from 2 to 0 (can be seen in Fig. 22.9). To test the impact of other strategies for reducing the search coefficient, we proposed an exponentially decreased formula by Eq. (22.17). Fig. 22.14 shows a performance comparison between the WOA (linear population size (LD)) with linear control parameter, exponentially decreased a with various coefficients ($\beta = 2, 3,$ and 4) and the Adaptive Bi-level WOA (WOA-LD-NM).

At first glance of Fig. 22.14, we can see that WOA-LD with an exponentially control parameter ($\beta = 2$) performed best compared with other exponential coefficients and linear scenarios because as it can be seen in Fig. 22.9, the transforma-

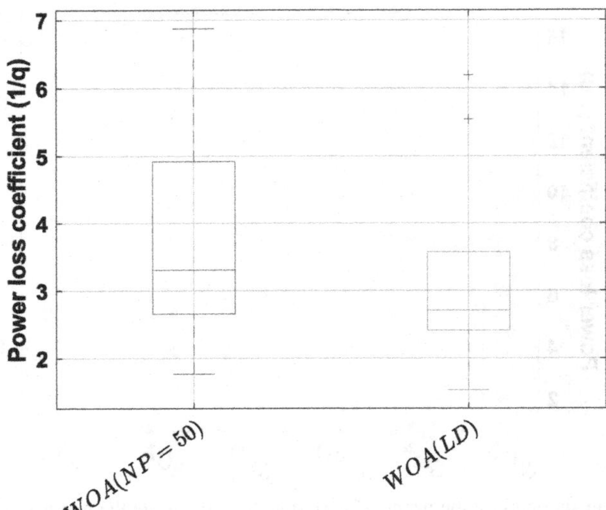

FIGURE 22.13 A comparison between a fixed population size with a linearly decreased from 50 to 20. Each configuration runs 10 times independently and the best-found solution per each run is considered.

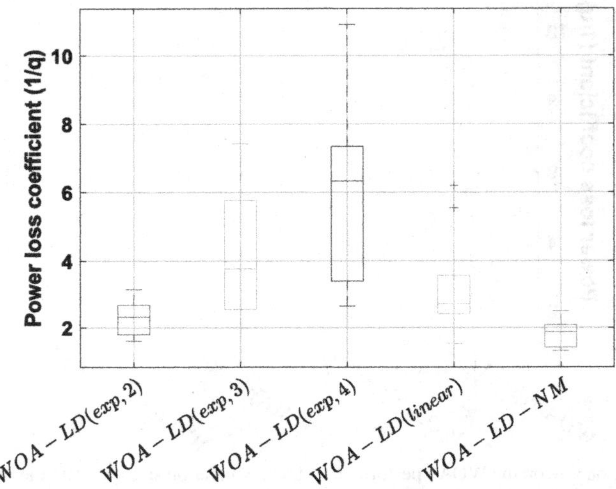

FIGURE 22.14 The performance statistical analysis of adaptive bi-level WOA (WOA-LD-NM) and four WOA with linearly decrease in population size and various exploration coefficient strategies. Each configuration runs 10 times independently and the best-found solution per each run is considered.

tion process from exploration ($a > 1$) to exploitation phase ($a < 1$) is smoother than the linear scenarios. Moreover, this exponentially reduced formula ($\beta = 2$) allocated 60% of the computation budget to the exploitation phase, and that leads to higher quality solutions. Furthermore, the main observation of box-plot above is that the AWOA could beat all other hybrid control parameters (population size plus search step coefficient) effectively.

To illustrate a convergence speed comparison between the proposed adaptive WOA with other meta-heuristics, the average best-so-far of the best-found solution found in each generation over the ten independent runs is plotted and can be seen in Fig. 22.15. As observed in this figure, the AWOA could converge the feasible solutions with higher quality faster than other popular meta-heuristics. In the initial 200 evaluation numbers, the AWOA's convergence is inclined to speed up due to the exponentially decreased control coefficient a strategy and removing poor solutions from the population. The convergence rate of WOA and SCA is the same and more considerable than SSA, GWO, MFO, and GOA.

Fig. 22.16 shows the best-found configuration of K_{pto} and B_{pto} variables achieved by the AWOA. From this figure, we can see the same fluctuation pattern for both PTO parameters with different scales. The B_{pto} parameters converged to high values more than 10^5; however, the K_{pto} settings tend to go down around 10^2, except for K_1, K_2, and K_4. The radius and distance variables of the best-performed configuration are 3 m and 20 m, respectively.

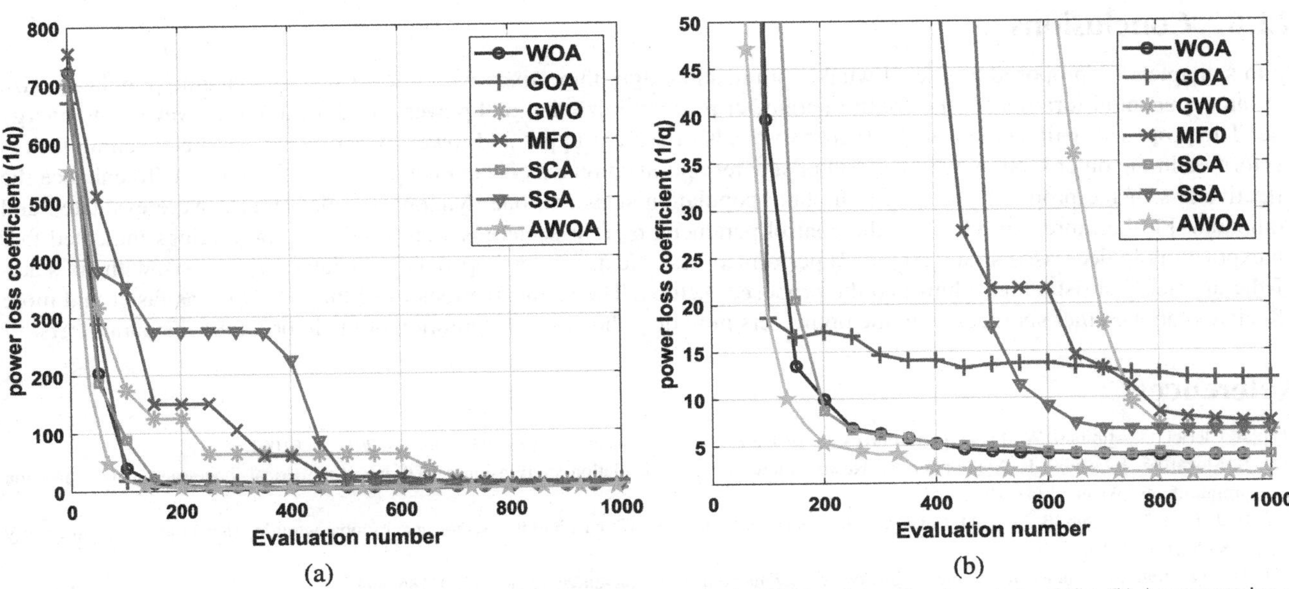

FIGURE 22.15 (a) The convergence speed comparison between the proposed adaptive WOA and the other six meta-heuristics. (b) A zoom version of the proposed method and others.

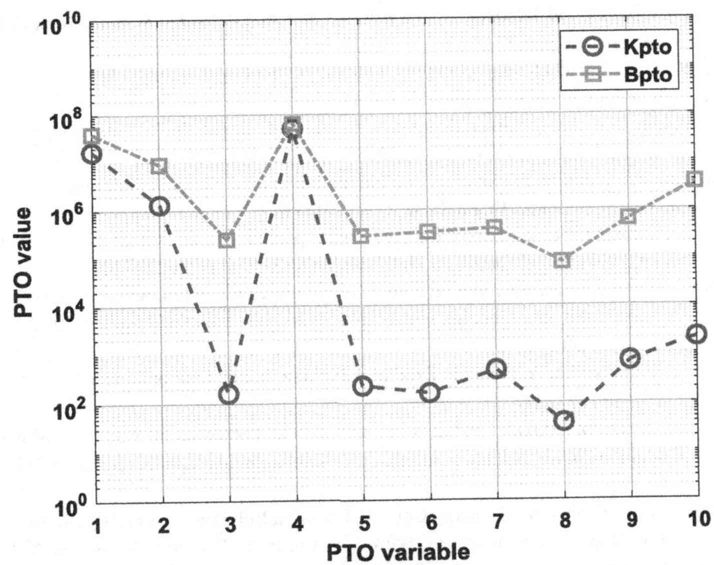

FIGURE 22.16 The best-configuration of K_{pto} and B_{pto} parameters proposed by AWOA (radius = 3 m, distance = 20 m, and power loss function = 1.32).

So the resultant hybrid wind-wave system includes three spherical WECs of a 3 m radius that are placed 20 m from the main column. A deeper analysis of the optimization results demonstrated that smaller WECs placed closer to the platform do not increase the nacelle acceleration, so the penalty factor for this case was approaching zero. Also, it seems that WECs attached to the floating platform absorb less power as compared to the WECs that operate in isolation, so WECs do not benefit from the hydrodynamic interaction with the FOWT. For this particular sea site, three WECs will produce approximately 40.5 kW, which comprises only 2% of the power production of the wind turbine.

According to the findings of this chapter, both original and adaptive WOA show a considerable performance in multi-modal with heterogeneous parameters. This is mainly because the WOA includes a strong exploration strategy compared with GOA, GWO, SSA, and MFO. Furthermore, a great balance mechanism was introduced in WOA to provide a fair chance for both exploration and exploitation during the optimization process.

22.6 Conclusions

In this chapter, we proposed a bi-level whale optimization algorithm mixed with a local search, dynamic population size and an exponential formula for the control parameter a to optimize the total power output of a hybrid wave-wind energy site. The proposed optimization method has two levels of search: upper and lower. A geometry parameter tuning task is performed in the outer section, while a PTO parameter optimization task is conducted in the inner section. To enhance the effectiveness of the proposed method, four static population sizes and one dynamic population size were evaluated and compared. Furthermore, for adjusting the control parameter α, four scenarios were used, and the findings indicated that an exponentially decreased strategy ($\beta = 2$) performed best. Next, we developed a comparative framework with six state-of-the-art meta-heuristic algorithms and the proposed method. The results demonstrated that AWOA was faster and more effective than the other six meta-heuristic optimizers in finding the best configuration of the hybrid model's parameters.

References

[1] A. Clerici, G. Alimonti, World energy resources, in: EPJ WEB of Conferences, vol. 98, EDP Sciences, 2015, p. 01001.

[2] S. Mozaffari, E. Amini, H. Mehdipour, M. Neshat, Flow discharge prediction study using a CDF-based numerical model and gene expression programming, Water 14 (4) (2022) 650.

[3] B.-J. Tang, Y.-Y. Guo, B. Yu, L.D. Harvey, Pathways for decarbonizing China's building sector under global warming thresholds, Applied Energy 298 (2021) 117213.

[4] The importance of renewable energies, https://www.acciona.com/renewable-energy/?_adin=02021864894.

[5] M. Neshat, S. Mirjalili, N.Y. Sergiienko, S. Esmaeilzadeh, E. Amini, A. Heydari, D.A. Garcia, Layout optimisation of offshore wave energy converters using a novel multi-swarm cooperative algorithm with backtracking strategy: a case study from coasts of Australia, Energy 239 (2022) 122463.

[6] United, Local Renewable Energy Benefits and Resources | US EPA — epa.gov, https://www.epa.gov/statelocalenergy/local-renewable-energy-benefits-and-resourcese, 2022. (Accessed 4 August 2022).

[7] K. Rahgooy, A. Bahmanpour, M. Derakhshandi, A. Bagherzadeh-Khalkhali, Distribution of elastoplastic modulus of subgrade reaction for analysis of raft foundations, Geomechanics and Engineering 28 (1) (2022) 89–105.

[8] J. Moccia, A. Arapogianni, J. Wilkes, C. Kjaer, R. Gruet, S. Azau, J. Scola, Pure power. Wind energy targets for 2020 and 2030, European Wind Energy Association EWEA, BRussels (Belgium) 43 (1) (2011) 1–77.

[9] E.O.E. Association, et al., European ocean energy roadmap 2010-2050, Ocean Energy Systems, Belgium, Brussels 1 (1) (2012) 1–36.

[10] A.G. Borthwick, Marine renewable energy seascape, Engineering 2 (1) (2016) 69–78.

[11] S. Filom, S. Radfar, R. Panahi, E. Amini, M. Neshat, Exploring wind energy potential as a driver of sustainable development in the southern coasts of Iran: the importance of wind speed statistical distribution model, Sustainability 13 (14) (2021) 7702.

[12] F. Taveira-Pinto, P. Rosa-Santos, T. Fazeres-Ferradosa, Marine renewable energy, Renewable Energy 150 (2020) 1160–1164.

[13] R. Mayon, D. Ning, B. Ding, N.Y. Sergiienko, Wave energy converter systems–status and perspectives, in: Modelling and Optimisation of Wave Energy Converters, CRC Press, 2022, pp. 3–58.

[14] C.L. Archer, A. Vasel-Be-Hagh, C. Yan, S. Wu, Y. Pan, J.F. Brodie, A.E. Maguire, Review and evaluation of wake loss models for wind energy applications, Applied Energy 226 (2018) 1187–1207.

[15] R. Zahedi, A. Ahmadi, R. Eskandarpanah, M. Akbari, Evaluation of resources and potential measurement of wind energy to determine the spatial priorities for the construction of wind-driven power plants in Damghan city, International Journal of Sustainable Energy and Environmental Research 11 (1) (2022) 1–22.

[16] J. Lee, F. Zhao, Global wind report 2019, Global Wind Energy Council, Brussels, Belgium 25 (1) (2019) 1–78.

[17] S. Watson, A. Moro, V. Reis, C. Baniotopoulos, S. Barth, G. Bartoli, F. Bauer, E. Boelman, D. Bosse, A. Cherubini, et al., Future emerging technologies in the wind power sector: a European perspective, Renewable & Sustainable Energy Reviews 113 (2019) 109270.

[18] C. Pérez-Collazo, D. Greaves, G. Iglesias, A review of combined wave and offshore wind energy, Renewable & Sustainable Energy Reviews 42 (2015) 141–153.

[19] E. Rusu, F. Onea, A review of the technologies for wave energy extraction, Clean Energy 2 (1) (2018) 10–19.

[20] E.D. Stoutenburg, N. Jenkins, M.Z. Jacobson, Power output variations of co-located offshore wind turbines and wave energy converters in California, Renewable Energy 35 (12) (2010) 2781–2791.

[21] E. Amini, D. Golbaz, R. Asadi, M. Nasiri, O. Ceylan, M. Majidi Nezhad, M. Neshat, A comparative study of metaheuristic algorithms for wave energy converter power take-off optimisation: a case study for eastern Australia, Journal of Marine Science and Engineering 9 (5) (2021) 490.

[22] C. Perez Collazo, S. Astariz, J. Abanades, D. Greaves, G. Iglesias, Co-located wave and offshore wind farms: a preliminary case study of an hybrid array, in: International Conference in Coastal Engineering (ICCE), 2014.

[23] H.R. Ghafari, H. Ghassemi, G. He, Numerical study of the Wavestar wave energy converter with multi-point-absorber around DeepCwind semisubmersible floating platform, Ocean Engineering 232 (2021) 109177.

[24] W. Jia, H. Yamamoto, K. Kuns, A. Effler, M. Evans, P. Fritschel, R. Abbott, C. Adams, R.X. Adhikari, A. Ananyeva, S. Appert, Point absorber limits to future gravitational-wave detectors, Physical Review Letters. 127 (24) (2021) 241102.

[25] Q. Gao, N. Ertugrul, B. Ding, M. Negnevitsky, Offshore wind, wave and integrated energy conversion systems: a review and future, in: 2020 Australasian Universities Power Engineering Conference (AUPEC), IEEE, 2020, pp. 1–6.

[26] J.M. Jonkman, Dynamics Modeling and Loads Analysis of an Offshore Floating Wind Turbine, University of Colorado at Boulder, 2007.

[27] C. Michailides, Z. Gao, T. Moan, Experimental study of the functionality of a semisubmersible wind turbine combined with flap-type wave energy converters, Renewable Energy 93 (2016) 675–690.

[28] Y. Ohya, T. Karasudani, T. Nagai, K. Watanabe, Wind lens technology and its application to wind and water turbine and beyond, Renewable Energy and Environmental Sustainability 2 (2017) 2.

[29] K. Thiagarajan, H. Dagher, A review of floating platform concepts for offshore wind energy generation, Journal of Offshore Mechanics and Arctic Engineering 136 (2) (2014) 020903.

[30] A. Yde, S.B. Bellew, R.S. Clausen, A.W. Nielsen, et al., Experimental and theoretical analysis of a combined floating wave and wind energy conversion platform, DTU Wind Energy, 2014.

[31] W2Power: Pelagic power, http://www.pelagicpower.no/.

[32] H.-C. Kim, K.-H. Kim, M.-H. Kim, K. Hong, Global performance of a KRISO semisubmersible multiunit floating offshore wind turbine: numerical simulation vs. model test, International Journal of Offshore and Polar Engineering 27 (01) (2017) 70–81.

[33] W. De Boer, F. Verheij, D. Zwemmer, R. Das, The energy island–an inverse pump accumulation station, in: EWEC 2007, 2007, pp. 7–10.

[34] W. Musial, P. Beiter, S. Tegen, A. Smith, Potential offshore wind energy areas in California: an assessment of locations, technology, and costs, Tech. Rep., National Renewable Energy Lab. (NREL), Golden, CO (United States), 2016.

[35] D. Zhang, Z. Chen, X. Liu, J. Sun, H. Yu, W. Zeng, Y. Ying, Y. Sun, L. Cui, S. Yang, et al., A coupled numerical framework for hybrid floating offshore wind turbine and oscillating water column wave energy converters, Energy Conversion and Management 267 (2022) 115933.

[36] D.M. Skene, N. Sergiienko, B. Ding, B. Cazzolato, The prospect of combining a point absorber wave energy converter with a floating offshore wind turbine, Energies 14 (21) (2021) 7385.

[37] A.M. Ragab, A.S. Shehata, A. Elbatran, M.A. Kotb, Numerical optimization of hybrid wind-wave farm layout located on Egyptian North Coasts, Ocean Engineering 234 (2021) 109260.

[38] S. Astariz, C. Perez-Collazo, J. Abanades, G. Iglesias, Towards the optimal design of a co-located wind-wave farm, Energy 84 (2015) 15–24.

[39] A. Saenz-Aguirre, J. Saenz, A. Ulazia, G. Ibarra-Berastegui, Optimal strategies of deployment of far offshore co-located wind-wave energy farms, Energy Conversion and Management 251 (2022) 114914.

[40] E. Amini, H. Mehdipour, E. Faraggiana, D. Golbaz, S. Mozaffari, G. Bracco, M. Neshat, Optimization study of hydraulic power take-off system for an ocean wave energy converter, arXiv preprint, arXiv:2112.09803.

[41] S. Astariz, C. Perez-Collazo, J. Abanades, G. Iglesias, Co-located wind-wave farm synergies (operation & maintenance): a case study, Energy Conversion and Management 91 (2015) 63–75.

[42] F. Haces-Fernandez, H. Li, D. Ramirez, A layout optimization method based on wave wake preprocessing concept for wave-wind hybrid energy farms, Energy Conversion and Management 244 (2021) 114469.

[43] Y. Wang, Z. Liu, H. Wang, Proposal and layout optimization of a wind-wave hybrid energy system using GPU-accelerated differential evolution algorithm, Energy 239 (2022) 121850.

[44] M. Abdel-Basset, L. Abdel-Fatah, A.K. Sangaiah, Metaheuristic algorithms: a comprehensive review, in: Computational Intelligence for Multimedia Big Data on the Cloud with Engineering Applications, 2018, pp. 185–231.

[45] X.-S. Yang, Metaheuristic optimization, Scholarpedia 6 (8) (2011) 11472.

[46] S. Katoch, S.S. Chauhan, V. Kumar, A review on genetic algorithm: past, present, and future, Multimedia Tools and Applications 80 (5) (2021) 8091–8126.

[47] S. Das, P.N. Suganthan, Differential evolution: a survey of the state-of-the-art, IEEE Transactions on Evolutionary Computation 15 (1) (2010) 4–31.

[48] N.M. Sabri, M. Puteh, M.R. Mahmood, A review of gravitational search algorithm, International Journal of Advances in Soft Computing and its Applications 5 (3) (2013) 1–39.

[49] S. Mirjalili, S.M. Mirjalili, A. Hatamlou, Multi-verse optimizer: a nature-inspired algorithm for global optimization, Neural Computing & Applications 27 (2) (2016) 495–513.

[50] J. Kennedy, R. Eberhart, Particle swarm optimization, in: Proceedings of ICNN'95-International Conference on Neural Networks, vol. 4, IEEE, 1995, pp. 1942–1948.

[51] S. Mirjalili, A. Lewis, The whale optimization algorithm, Advances in Engineering Software 95 (2016) 51–67.

[52] A. Ala'a, A.A. Alsewari, H.S. Alamri, K.Z. Zamli, Comprehensive review of the development of the harmony search algorithm and its applications, IEEE Access 7 (2019) 14233–14245.

[53] D. Cvijović, J. Klinowski, Taboo search: an approach to the multiple minima problem, Science 267 (5198) (1995) 664–666.

[54] Y.-H. Lin, M.-C. Fang, An integrated approach for site selection of offshore wind-wave power production, IEEE Journal of Oceanic Engineering 37 (4) (2012) 740–755.

[55] J. Izquierdo-Pérez, B.M. Brentan, J. Izquierdo, N.-E. Clausen, A. Pegalajar-Jurado, N. Ebsen, Layout optimization process to minimize the cost of energy of an offshore floating hybrid wind–wave farm, Processes 8 (2) (2020) 139.

[56] T.D. Pham, H. Shin, A new conceptual design and dynamic analysis of a spar-type offshore wind turbine combined with a moonpool, Energies 12 (19) (2019) 3737.

[57] J. Hu, B. Zhou, C. Vogel, P. Liu, R. Willden, K. Sun, J. Zang, J. Geng, P. Jin, L. Cui, et al., Optimal design and performance analysis of a hybrid system combing a floating wind platform and wave energy converters, Applied Energy 269 (2020) 114998.

[58] E. Amini, H. Mehdipour, E. Faraggiana, D. Golbaz, S. Mozaffari, G. Bracco, M. Neshat, Optimization of hydraulic power take-off system settings for point absorber wave energy converter, Renewable Energy 194 (2022) 938–954, https://doi.org/10.1016/j.renene.2022.05.164, https://www.sciencedirect.com/science/article/pii/S0960148122008242.

[59] N. Rana, M.S.A. Latiff, S.M. Abdulhamid, H. Chiroma, Whale optimization algorithm: a systematic review of contemporary applications, modifications and developments, Neural Computing & Applications 32 (20) (2020) 16245–16277.
[60] R.K. Saidala, N. Devarakonda, Improved whale optimization algorithm case study: clinical data of anaemic pregnant woman, in: Data Engineering and Intelligent Computing, Springer, 2018, pp. 271–281.
[61] B. Bentouati, L. Chaib, S. Chettih, A hybrid whale algorithm and pattern search technique for optimal power flow problem, in: 2016 8th International Conference on Modelling, Identification and Control (ICMIC), IEEE, 2016, pp. 1048–1053.
[62] C. Kumar, R.S. Rao, A novel global MPP tracking of photovoltaic system based on whale optimization algorithm, International Journal of Renewable Energy Development 5 (3) (2016) 225–232.
[63] K.F. Hasselmann, T.P. Barnett, E. Bouws, H. Carlson, D.E. Cartwright, K. Eake, J. Euring, A. Gicnapp, D. Hasselmann, P. Kruseman, A. Meerburg, P. Muller, D.J. Olbers, K. Richter, W. Sell, H. Walden, Measurements of wind-wave growth and swell decay during the Joint North Sea Wave Project (JONSWAP), Ergaenzungsheft zur Deutschen Hydrographischen Zeitschrift, Reihe A, 1973.
[64] T.A. Al-Janabi, H.S. Al-Raweshidy, Efficient whale optimisation algorithm-based SDN clustering for IoT focused on node density, in: 2017 16th Annual Mediterranean Ad Hoc Networking Workshop (Med-Hoc-Net), IEEE, 2017, pp. 1–6.
[65] H. Hu, Y. Bai, T. Xu, A whale optimization algorithm with inertia weight, WSEAS Transactions on Computers 15 (2016) 319–326.
[66] X. Huang, R. Wang, X. Zhao, K. Hu, Aero-engine performance optimization based on whale optimization algorithm, in: 2017 36th Chinese Control Conference (CCC), IEEE, 2017, pp. 11437–11441.
[67] K. Lu, Z. Ma, A modified whale optimization algorithm for parameter estimation of software reliability growth models, Journal of Algorithms & Computational Technology 15 (2021), https://doi.org/10.1177/17483026211034442.
[68] L.S.P. da Silva, N.Y. Sergiienko, B. Cazzolato, B. Ding, Dynamics of hybrid offshore renewable energy platforms: heaving point absorbers connected to a semi-submersible floating offshore wind turbine, Renewable Energy 199 (2022) 1424–1439.
[69] A. Robertson, J. Jonkman, M. Masciola, H. Song, A. Goupee, A. Coulling, C. Luan, Definition of the semisubmersible floating system for phase II of OC4, Tech. Rep., National Renewable Energy Lab. (NREL), Golden, CO (United States), 2014.
[70] L.S.P. da Silva, M. de Oliveira, B. Cazzolato, N. Sergiienko, G.A. Amaral, B. Ding, Statistical linearisation of a nonlinear floating offshore wind turbine under random waves and winds, Ocean Engineering 261 (2022) 112033.
[71] C.P. Pesce, G.A. Amaral, G.R. Franzini, Mooring system stiffness: a general analytical formulation with an application to floating offshore wind turbines, in: International Conference on Offshore Mechanics and Arctic Engineering, vol. 51975, American Society of Mechanical Engineers, 2018, p. V001T01A021.
[72] G.A. Amaral, P.C. Mello, L.H. do Carmo, I.F. Alberto, E.B. Malta, A.N. Simos, G.R. Franzini, H. Suzuki, R.T. Gonçalves, Seakeeping tests of a FOWT in wind and waves: an analysis of dynamic coupling effects and their impact on the predictions of pitch motion response, Journal of Marine Science and Engineering 9 (2) (2021) 179.
[73] L.S.P. da Silva, N.Y. Sergiienko, C.P. Pesce, B. Ding, B. Cazzolato, H.M. Morishita, Stochastic analysis of nonlinear wave energy converters via statistical linearization, Applied Ocean Research 95 (2020) 102023.
[74] J.B. Roberts, P.D. Spanos, Random Vibration and Statistical Linearization, Courier Corporation, 2003.
[75] J.C. Kaimal, J. Wyngaard, Y. Izumi, O. Coté, Spectral characteristics of surface-layer turbulence, Quarterly Journal of the Royal Meteorological Society 98 (417) (1972) 563–589.
[76] Q. Gao, S.S. Khan, N. Sergiienko, N. Ertugrul, M. Hemer, M. Negnevitsky, B. Ding, Assessment of wind and wave power characteristic and potential for hybrid exploration in Australia, Renewable & Sustainable Energy Reviews 168 (2022) 112747.
[77] P. Sclavounos, C. Tracy, S. Lee, Floating offshore wind turbines: responses in a seastate Pareto optimal designs and economic assessment, in: International Conference on Offshore Mechanics and Arctic Engineering, vol. 48234, 2008, pp. 31–41.
[78] W.A. Watkins, W.E. Schevill, Aerial observation of feeding behavior in four baleen whales: Eubalaena glacialis, Balaenoptera borealis, Megaptera novaeangliae, and Balaenoptera physalus, Journal of Mammalogy 60 (1) (1979) 155–163.
[79] J.A. Goldbogen, A.S. Friedlaender, J. Calambokidis, M.F. McKenna, M. Simon, D.P. Nowacek, Integrative approaches to the study of baleen whale diving behavior, feeding performance, and foraging ecology, BioScience 63 (2) (2013) 90–100.
[80] S. Mirjalili, S.M. Mirjalili, S. Saremi, S. Mirjalili, Whale optimization algorithm: theory, literature review, and application in designing photonic crystal filters, in: Nature-Inspired Optimizers, 2020, pp. 219–238.
[81] W. Tong, A new whale optimisation algorithm based on self-adapting parameter adjustment and mix mutation strategy, International Journal of Computer Integrated Manufacturing 33 (10–11) (2020) 949–961.
[82] G. Kaur, S. Arora, Chaotic whale optimization algorithm, Journal of Computational Design and Engineering 5 (3) (2018) 275–284.
[83] M.H. Nadimi-Shahraki, H. Zamani, S. Mirjalili, Enhanced whale optimization algorithm for medical feature selection: a COVID-19 case study, Computers in Biology and Medicine 148 (2022) 105858.
[84] M.H. Nadimi-Shahraki, S. Taghian, S. Mirjalili, L. Abualigah, M. Abd Elaziz, D. Oliva, EWOA-OPF: effective whale optimization algorithm to solve optimal power flow problem, Electronics 10 (23) (2021) 2975.
[85] M. Neshat, N.Y. Sergiienko, E. Amini, M. Majidi Nezhad, D. Astiaso Garcia, B. Alexander, M. Wagner, A new bi-level optimisation framework for optimising a multi-mode wave energy converter design: a case study for the Marettimo island, Mediterranean sea, Energies 13 (20) (2020) 5498.

Chapter 23

Sizing optimization of truss structures using hybrid whale optimization algorithm

Mohammed A. Awadallah[a,b], **Lamees Mohammad Dalbah**[c], **Malik Braik**[e], **Mohammed Azmi Al-Betar**[c,d], **Zaid Abdi Alkareem Alyasseri**[f], **and Seyedali Mirjalili**[g,h]

[a]*Department of Computer Science, Al-Aqsa University, Gaza, Palestine,* [b]*Artificial Intelligence Research Center (AIRC), Ajman University, Ajman, United Arab Emirates,* [c]*Artificial Intelligence Research Center (AIRC), College of Engineering and Information Technology, Ajman University, Ajman, United Arab Emirates,* [d]*Department of Information Technology, Al-Huson University College, Al-Balqa Applied University, Irbid, Jordan,* [e]*Department of Computer Science, Al-Balqa Applied University, As-Salt, Jordan,* [f]*Information Technology Research and Development Center (ITRDC), University of Kufa, Najaf, Iraq,* [g]*Centre for Artificial Intelligence Research and Optimisation, Torrens University Australia, Brisbane, QLD, Australia,* [h]*University Research and Innovation Center, Obuda University, Budapest, Hungary*

23.1 Introduction

Structural optimization is an important civil engineering problems [1]. Civil engineering is related to designing, constructing, operating, and maintaining buildings and infrastructures such as canals, dams, bridges, roads, etc. The main concern of structure optimization is the design of the most safety and maximum gains [2]. The safety of the design is measured by satisfying the problem constraints, while the maximum gains are reflected by the quality of the design defined by the objective function. In term of optimization context [1], structural optimization problem can be classified into four categories: sizing optimization, shape optimization, topology optimization, and multi-objective optimization. The sizing optimization class deals with the cross-sectional areas of structures, while the shape optimization class deals with the nodal coordinates of structures. The topology optimization finds the optimal design by deleting unnecessary structural members. This is achieved by focusing on how the joints are connected in the design. Finally, multi-objective optimization structure optimization class combines two or more classes of structural optimization.

Early days, the researchers in the domain of structure optimization introduced the classical mathematical theorems and programming techniques for small-sized problems like iterative 3D extended evolutionary structure optimization [3], cellular automata [4], and gradient-based algorithms [5]. Recently, metaheuristic algorithms are the most powerful techniques that are commonly used in the engineering optimization domain such as structure optimization [6]. This is due to the fact that these algorithms have the ability to intelligently explore and exploit the search space of the optimization problems and thus find the optimal solution that mostly satisfied the problem constraints.

Metaheuristic algorithms are divided into two main categories local-based and population-based algorithms [7,8]. Local-based algorithms begin with an initial solution and iteratively exploit this solution until the near-optimal solution is achieved. On the other hand, the population-based algorithms begin with a set of solutions and these solutions are enhanced using their operators (i.e., mutation and crossover) until the near-optimal solutions are achieved. The local-based algorithms more focus on exploiting the already visited regions with less concentration of exploring multi-regions, while the population-based algorithms more focus on exploring multi regions with less concentration on each region of search space. Therefore, the optimization community tends to hybridize the local search-based into population-based metaheuristic algorithms to complement their strengths in exploration and exploitation. In the literature, some of the local-based algorithms introduced for the sizing optimization of truss structures such as simulated annealing [9]. However, population-based algorithms for the sizing optimization of truss structures are the most commonly used such as flower pollination algorithm [2], gravitational search algorithm [10], harmony search [11], sine cosine algorithm [12], heat transfer search algorithm [13], grey wolf optimizer [14], political optimizer [15], mine blast algorithm [16], teaching-learning based optimization [17], artificial bee colony [18], and others reported in [1,6].

Due to the complexity of the sizing optimization of truss structures, some researchers have hybridized the local search algorithms as exploiter agents within the population-based algorithm to empower the exploitation ability of the algorithms and thus strike the right balance between exploration and exploitation abilities [19]. There are many hybrid algorithms

Handbook of Whale Optimization Algorithm. https://doi.org/10.1016/B978-0-32-395365-8.00029-4

introduced in the literature for the sizing optimization of truss structures [6]. Cheng et al. [20] introduced a new variant hybrid algorithm based on the global-best concept of PSO and the neighborhood search for the sizing optimization of truss structures. These modifications are suggested by the authors to omit randomness and empower the exploitation ability of the algorithm. In another study by Asl et al. [21], they combined the neighborhood search algorithm within the framework of the genetic algorithm in order to enhance its exploitation ability. The authors in [21] evaluated the effectiveness of their algorithm using seven cases of the sizing optimization of truss structures. Another hybrid algorithm based on harmony search and JAYA algorithm, called LSSO-HHSJA, is introduced by Degertekin et al. [22]. In LSSO-HHSJA, the JAYA algorithm is used to minimize the number of structural analyses required in the optimization process. The LSSO-HHSJA algorithm achieved satisfactory results when it was tested on large-scale structural optimization.

The Whale optimization algorithm (WOA) is one of the most successful population-based algorithms introduced by Mirjalili and Lewis [23]. It mimics the social behavior of humpback whales in finding food resources and hunting prey. Firstly, the whale is randomly searching for prey. Thereafter, the whale swims in a circle and creates bubbles along a circle if and only if finding prey to hunt. The main advantage of WOA is the lower opportunity to trap into local optima, it has the ability to navigate the problem search space without complicated operators, the output of the algorithm is not dependent on the initial solutions, and it has a few parameters that should be initialized before execution [24]. Therefore, WOA has successfully adopted for tackling several optimization problems such as resource allocation in wireless networks [25], parameter estimation of photovoltaic cells [26], feature selection [27], optimal power flow problem [28], dynamic economic dispatch [29], and others reported in [24,30].

However, there are some hybridized versions of WOA introduced in the literature, the authors in [31], integrated the tabu search algorithm as a local-search agent within the framework of WOA for green vehicle routing problems. Mafarja & Mirjalili [32] used the simulated annealing as a new local-search operator within the WOA to enhance its exploitation ability. Their hybrid algorithm was implemented for feature selection problems with superior results. In another study, the authors in [33] integrated the WOA with thermal exchange optimization, called GWOA-TEO, to enhance its exploitation capability. Furthermore, the crossover operator was utilized in GWOA-TEO to empower its exploration ability. Similarly, Tang et al. [34] hybridized the WOA with the artificial bee colony to enhance the WOA performance in terms of speeding up convergence and avoiding the problem of local optimum. However, there is still room for presenting a new hybrid variant of the WOA to enhance the WOA performance in terms of finding the right balance between the exploration and exploitation, convergence accuracy, and jumping out of local optimum.

The main aim of the research presented in this chapter is to tackle the sizing optimization of truss structures using a new variant of the hybrid whale optimization algorithm (HWOA). In HWOA, the adaptive β-hill climbing as a local search-based agent is integrated within the framework of the WOA to enhance its exploitation ability and thus make a right balance between exploitation and exploration abilities. The feasibility of the truss structure solution is maintained using a penalty assigned to each unsatisfied constraint in the objective function. The effectiveness of the proposed HWOA for solving the sizing optimization of truss structures is studied using four case studies (10-bar, 25-bar, 72-bar, and 200-bar) with ten variants according to the type of variables (continue or discrete). The performance of the proposed HWOA is compared with the classical WOA, as well as, other comparative methods published in the literature. In summary, the proposed HWOA is able to achieve viable results in comparison with those obtained by other state-of-the-art methods.

The remaining parts of this chapter are manipulated as follows: The mathematical formulation of the truss structure optimization problem is given in Section 23.2. The background of the procedural steps for the whale optimization algorithm is presented in Section 23.3, while the background of procedural steps for the adaptive β-hill climbing procedure is given in Section 23.4. The description of the proposed HWOA algorithm is modeled in Section 23.5. In Section 23.6, the test problems are discussed and the test the experimental results are comparatively analyzed. Finally, the conclusion and some future directions are illustrated in Section 23.7.

23.2 Truss structure problem

The sizing optimization problem of a truss structure aims to determine the optimum values of each member in the structure (i.e., cross-sectional areas), which leads to the minimum structural weight. However, the design with the minimum weight should satisfy the stress and deflection constraints. The objective function of this problem is formulated as follows:

$$\min W(X) = \sum_{i=1}^{m} X_i \sum_{k=1}^{gk} \rho_{k,i} \times L_{k,i} \tag{23.1}$$

Subject to

$$\sigma_i^{min} \leq \sigma_i \leq \sigma_i^{max} \quad i = 1, 2, \cdots, m \tag{23.2}$$

$$\gamma_j^{min} \leq \gamma_j \leq \gamma_j^{max} \quad j = 1, 2, \cdots, z \tag{23.3}$$

where $W(X)$ is the total weight of the structure. X is the solution, while X_i is the decision variable at the ith position. In the other words, X_i is the cross-sectional area of the ith group of bars. The value of X_i is in the range of $[X_i^{min}, X_i^{max}]$, where X_i^{min} and X_i^{max} are the lower and upper bounds of the ith decision variable, respectively. Furthermore, m is the total number of design variables of the solution, and the value of m is equal to the number of member groups included in the structure. $\rho_{k,i}$ is the mass density of the kth member in the ith group, while $L_{k,i}$ is the length of the kth member in the ith group. gk is the total number of members in the ith group.

The stress and deflection constraints are shown in Eq. (23.2) and Eq. (23.3), respectively. σ_i is the axial stress of the ith member, while its value within the range of the $[\sigma_i^{min}, \sigma_i^{max}]$. γ_j is the nodal displacement of the jth transnational degree of freedom, while its value within the range of $[\gamma_j^{min}, \gamma_j^{max}]$. z is the total number of transnational degrees of freedom.

Thereafter, the penalty function is used to handle the infeasible truss design caused by unsatisfying the stress and displacement constraints. Such that if the design is feasible then the penalty is zero, otherwise, the penalty is greater than zero. The nodal displacements and element stresses penalties are formulated as follows:

$$\varphi_\sigma(i) = \begin{cases} 0 & \sigma_i^{min} \leq \sigma_i \leq \sigma_i^{max}, \\ \left| \frac{\sigma_i - \sigma_i^{min/max}}{\sigma_i^{min/max}} \right| & \sigma_i < \sigma_i^{min} \text{ or } \sigma_i > \sigma_i^{max}, \end{cases} \tag{23.4}$$

$$\varphi_\gamma(j) = \begin{cases} 0 & \gamma_j^{min} \leq \gamma_j \leq \gamma_j^{max}, \\ \left| \frac{\gamma_j - \gamma_j^{min/max}}{\gamma_j^{min/max}} \right| & \gamma_j < \gamma_j^{min} \text{ or } \gamma_j > \gamma_j^{max}, \end{cases} \tag{23.5}$$

where $\varphi_\sigma(i)$ is the value of stress penalty of the ith member, and $\varphi_\gamma(j)$ is the value of the nodal deflection penalty of the jth node. Subsequently, the total penalty values of the stress and nodal deflection are shown in Eq. (23.6) and Eq. (23.7).

$$\varphi_\sigma = \sum_{i=1}^{m} \varphi_\sigma(i) \tag{23.6}$$

$$\varphi_\gamma = \sum_{j=1}^{z} \varphi_\gamma(j) \tag{23.7}$$

Finally, the objective function (see Eq. (23.1)) and optimization constraints handling are merged into a new objective function as follows:

$$\min F = \psi \times W \tag{23.8}$$

where F is the value of the objective function. ψ is the total penalty, and its value is equal to $(1 + \varphi_\gamma + \varphi_\sigma)$. W is the weight of the truss design calculated using Eq. (23.1).

23.3 Whale optimization algorithm (WOA)

The WOA is one of the most attractive swarm-based algorithm introduced in 2016 by Mirjalili and Lewis [23]. The feeding social behavior of the humpback whales is the main concept considered in this algorithm. The whales have two main strategies of foraging as follows: firstly, whales search randomly in the hope of finding prey for hunting. This represents the exploration ability in an optimization context. On the other hand, the whales hunt a group of fish or krill by swim within a circle and creating bubbles along a circle as shown in Fig. 23.1. This behavior reflects the exploiting capability in an optimization context. The mathematical formulation of the social behavior of humpback whales in optimization context is shown in the following procedural steps:

Step 1: Initialization of WOA's parameters. The two algorithmic parameters of the WOA should be initialized before execution. These two algorithmic parameters are the size of the population (N) and the maximum number of iterations (T_{max}). Fortunately, all control parameters of WOA are dynamically adapted during the search.

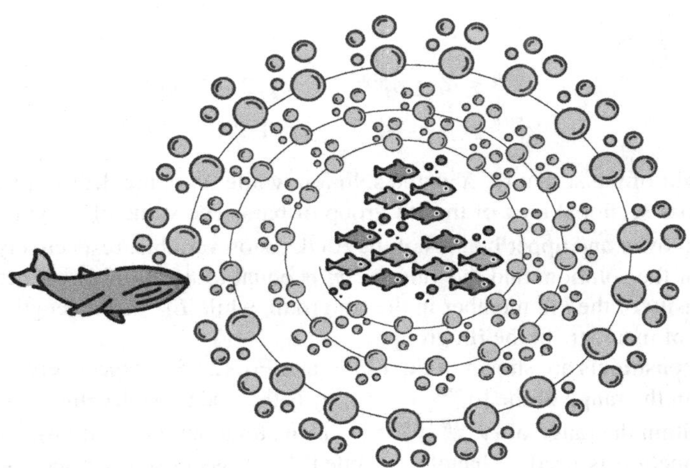

FIGURE 23.1 Bubble-net feeding social behavior of whales.

Step 2: Construct the initial of WOA's population. WOA is a population-based algorithm that begins with a set of solutions constructed and retained in the WOA Memory (**WOAM**). **WOAM** is represented as a 2D matrix with $N \times m$ dimensions as shown in Eq. (23.9). N is the number of solutions and m is the solution dimension. In **WOAM**, each row $\vec{X}_i = (X_{i,1}, X_{i,2}, \ldots, X_{i,m})$ represents the whale solution at the ith position. $X_{i,j}$ is the decision variable (i.e., element) of the ith whale at the j position. Each decision variable $X_{i,j}$ is assigned a random value in the range $[X_j^{min}, X_j^{max}]$. X_j^{min} and X_j^{max} are the lower and upper bounds of the jth decision variable.

$$\mathbf{WOAM} = \begin{bmatrix} X_{1,1} & X_{1,2} & \cdots & X_{1,m} \\ X_{2,1} & X_{2,2} & \cdots & X_{2,m} \\ \vdots & \vdots & \cdots & \vdots \\ X_{N,1} & X_{N,2} & \cdots & X_{N,m} \end{bmatrix} \tag{23.9}$$

Step 3: Fitness evaluation. In this step, the position of each whale (i.e., solution) in the population should be evaluated using the fitness function $f(\vec{X}_i)$, where $i = 1, 2, \ldots, N$. In addition, the whale with the best position (i.e., best solution) should be defined $\vec{X}^*$.

Step 4: Encircling the prey. Each whale in the population is updated its position using the position of the best whale. This behavior leads the whales in the population to update their position close to the best whale. In WOA, the target prey is the best whale. This behavior is formulated in Eq. (23.10).

$$\vec{X}_i(t+1) = \vec{X}^*(t) - \vec{D} \cdot \vec{A} \tag{23.10}$$

$$\vec{D} = |\vec{C} \cdot \vec{X}^*(t) - \vec{X}_i(t)| \tag{23.11}$$

$$\vec{A} = 2 \times a \times r - a \tag{23.12}$$

$$\vec{C} = 2 \times r \tag{23.13}$$

where $\vec{X}_i(t+1)$ is the position of the ith whale at the $(t+1)$ iteration. $\vec{X}^*(t)$ is the position of the best whale in the current population at the t iteration. D is the distance between the position of the current whale $\vec{X}_i(t)$ and the position of the best whale $\vec{X}^*(t)$. A and C are two coefficient vectors calculated in Eq. (23.12) and Eq. (23.13), respectively. r is random value in the range [0, 1]. a is a value linearly decreased from 2 to 0 at each iteration using Eq. (23.14). t is the current iteration and T is the maximum number of iterations. Finally, $(\cdot)$ reflects the multiplication process of two vectors and $|\ |$ is the absolute value.

$$a = 2 - t \times \frac{2}{T} \tag{23.14}$$

Step 5: Spiral updating position. The whales in the population swim follow a bubble-net where the whales move in a spiral-shaped path tracking prey. The formulation of this behavior is formulated in Eq. (23.15).

$$\vec{X}_i(t+1) = \vec{D}' \cdot e^{pl} \cdot \cos(2\pi l) + \vec{X}^*(t) \tag{23.15}$$

$$\vec{D}' = |\vec{X}^*(t) - \vec{X}_i(t)| \tag{23.16}$$

where $\vec{X}_i(t+1)$ is the position of the ith whale at the $(t+1)$ iteration. $\vec{D}'$ is the distance between the prey and the position of the current whale. b is a constant value used for defining the spiral shape. l is a random value within the range $[-1, 1]$. To model the previous two behavior of whales (i.e., encircling the prey and the spiral updating position), a probability of 50% is defined to choose between them during the search process as shown in Eq. (23.17). r is a random value between 0 and 1.

$$\vec{X}(t+1) = \begin{cases} \text{Encircling the prey (see Eq. (23.10))} & r < 0.5, \\ \text{Spiral updating position (see Eq. (23.15))} & r \geq 0.5. \end{cases} \tag{23.17}$$

Step 6: Search for prey. In this step, the position of the whale in the population is updated according to the positions of one of the whales in the population randomly chosen. This behavior is mathematically modeled using Eq. (23.18). It should be noted that this is the main source of exploration in the WOA algorithm.

$$\vec{X}_i(t+1) = \vec{X}_k(t) - \vec{D} \cdot \vec{A} \tag{23.18}$$

$$\vec{D} = |\vec{C} \cdot \vec{X}_k(t) - \vec{X}(t)| \tag{23.19}$$

where $\vec{X}_i(t+1)$ is the position of the ith whale at the $(t+1)$ iteration. $\vec{X}_k(t)$ is the position of the kth whale at the t iteration, which is randomly selected, where $k \in [1, N]$. D is the distance between the ith whale and the k whale. A and C are two coefficient vectors calculated in Eq. (23.12) and Eq. (23.13), respectively.

Step 7: Checking stop condition. Steps 3 to 6 are repeated until the maximum number of iterations is reached.

23.4 Adaptive β hill climbing (AβHC)

A β hill climbing (βHC) is one of the recent local search-based algorithms introduced by Al-Betar [35]. βHC has superior advantages over other similar algorithms such as it is simple in concepts, easy to implement, speedy in convergence, and powerfully avoids local optima [36]. Therefore it has been used for several optimization problems such as feature selection [37], multiple-reservoir scheduling [38], economic load dispatch [39], denoising ECG signals [40], text document clustering [41], etc.

A βHC is start with a single random solution, this solution is enhanced gradually using the three operators of βHC till the stop condition is met. $\mathcal{N}$-operator, β-operator, and $\mathcal{U}$-operator are the three operators of the βHC algorithm. The $\mathcal{N}$-operator is the main source of the exploitation by finding the neighbor solution, while the β-operator is the main source of the exploration by randomly modifying the current solution. $\mathcal{U}$-operator is the update operator used to replace the current solution with the new one if better. The first two operators (i.e., $\mathcal{N}$-operator, and β-operator) are controlled by N and β parameters, respectively. N and β parameters are assigned a value within the range [0, 1].

In order to simplify the original βHC, Adaptive β hill climbing (AβHC) is introduced as an extended version of βHC [42]. In AβHC, the values of N and β parameters are automatically updated during the search process to suitably strike the right balance between the exploration and exploitation operations. The procedural steps of AβHC are explained as follows:

Step 1: Initialization of AβHC's parameters. The parameters of the AβHC should be initialized before running. The settings of the AβHC's parameters are $\beta_{min} = 0.001$, $\beta_{max} = 0.6$, and $K = 20$. These settings are introduced in [42]. The β_{min} and β_{max} are used to control β parameter, while K used to control the N parameter during the search process. Finally, T_β is the maximum number of iterations.

Step 2: Construct the initial solution. In this step, the initial solution $\vec{X} = (X_1, X_2, \ldots, X_m)$ is randomly constructed. X_j is the decision variable (i.e., element) of the jth position. X_j is assigned a random value between the lower bound X_j^{min} and the upper bound X_j^{max}.

Step 3: The Nighbourhood operator (i.e., $\mathcal{N}$-operator). In this step, the current solution $\vec{X}$ is moved to it neighbor solution $\vec{X}'$ using Eq. (23.20).

$$X'_j = X_j \pm r \times \mathcal{N}(t) \tag{23.20}$$

where X'_j is the decision variable in the new solution $\vec{X}'$ at the jth position. X_j is the decision variable in the current solution $\vec{X}$ at the jth position. j is random number between 1 and m, where m is the solution dimension. r is a random value between 0 and 1. $\mathcal{N}(t)$ is the distance bandwidth between the current solution and the neighbor one. The value of $\mathcal{N}(t)$ is adopted using Eq. (23.21).

$$\mathcal{N}(t) = 1 - C(t) \tag{23.21}$$

$$C(t) = \frac{t^{\frac{1}{K}}}{T_\beta^{\frac{1}{K}}} \tag{23.22}$$

where t is the current iteration. T_β is the maximum number of iterations. K is a constant value equal to 20. The $C(t)$ is a value gradually decreased to 0 during the search process.

Step 4: The β-operator. One of the decision variables of the neighbor solution $\vec{X}'$ is assigned with a random value in the new solution $\vec{X}''$ as shown in Eq. (23.23).

$$X''_k = X_k^{min} + r \times (X_k^{max} - X_k^{min}) \tag{23.23}$$

where X''_k is the decision variable in the new solution $\vec{X}'$ at the kth position. It should be noted that the β-operator is controlled by the β parameter. The value of the β parameter is adopted using Eq. (23.24).

$$\beta(t) = \beta_{min} + t \times \frac{\beta_{max} - \beta_{min}}{T_\beta} \tag{23.24}$$

where $\beta(t)$ is the value of β parameter at the t iteration. T_β is the maximum number of iterations. β_{min} and β_{max} are the lower and upper value of β parameter. The value of β parameter gradually decreased from β_{max} to β_{min} during the search process.

Step 5: The update operator ($\mathcal{U}$-operator). In this step, the current solution $\vec{X}$ is replaced by the new one $\vec{X}''$ if it is better.

Step 6: Checking stop condition. Steps 3 to 5 are repeated until the maximum number of iterations (i.e., T_β) is meet.

23.5 Hybridizing the WOA with AβHC

Fig. 23.2 shows the procedural steps of the proposed hybrid WOA (HWOA) algorithm, while the pseudo-code of HWOA is given in Algorithm 23.1. In the proposed HWOA, the AβHC as a new local search operator is integrated within the framework of the WOA in order to enhance the exploitation of the proposed algorithm. In the proposed HWOA, each solution in the population is modified based on the three operators of the WOA (i.e., searching the prey, encircling the prey, and the Spiral updating position). Next, a solution is passed to AβHC with a probability of R. The solution passed to AβHC is considered the initial solution. This solution is enhanced locally using the three operators of AβHC (i.e., $\mathcal{N}$-operator, β-operator, and $\mathcal{U}$-operator). It should be noted that the higher rate of R leads to a higher rate of calling the AβHC operator and thus a higher rate of exploitation and higher computational time.

It should be noted that the AβHC is used as a local search-based agent within the structure of the population-based algorithms for solving several optimization problems. In [43], the AβHC is integrated with the harris hawks optimizer for economic load dispatch problems. Similarly, the sine cosine algorithm was combined with β-hill climbing optimizer for economic load dispatch problems [44]. In [36], the AβHC was integrated with six population-based algorithms (i.e., flower pollination algorithm, salp swarm algorithm, crow search algorithm, grey wolf optimization, particle swarm optimization, and JAYA algorithm) in six hybrid variants for the training of neural networks.

23.6 Experiments and results

For evaluation purposes, the performance of WOA and the Hybrid WOA (HWOA) was tested through four well-known truss optimization problems. The problems are (1) the 10-bar planar truss, (2) the 25-bar spatial truss, (3) the 72-bar spatial truss, and (4) the 200-bar planar truss. In which both the discrete and continuous search spaces were considered for the first three problems, and only the discrete for the 200-bar planar truss problem. The results obtained by the proposed algorithms (WOA & HWOA) were compared against well-established algorithms according to the implementation availability. Also, note that for easier illustration the best-achieved result for each case study is highlighted using bold fonts.

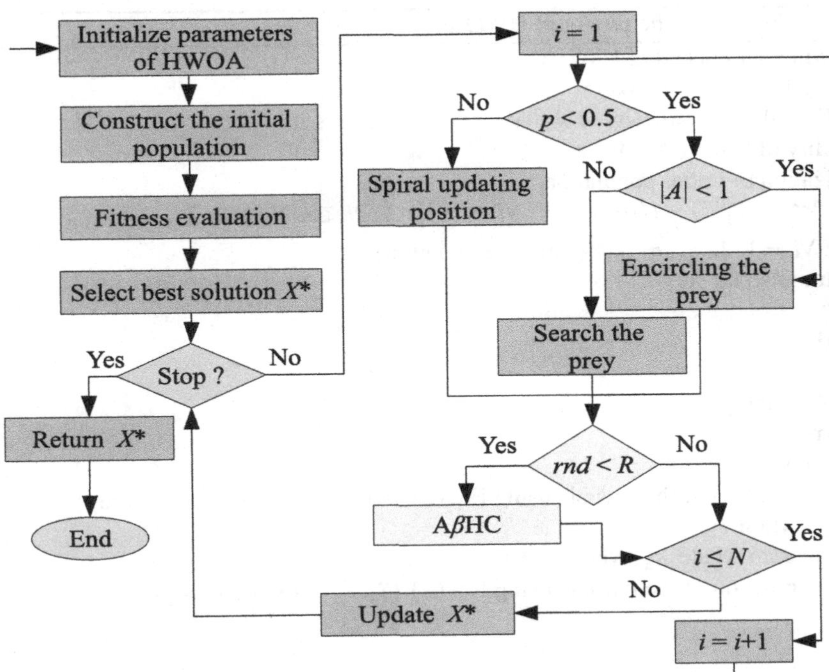

FIGURE 23.2 The flowchart of the proposed HWOA.

Regarding WOA and HWOA implementation environment, the algorithms were coded using Matlab® R2021a and run on a PC Core-i7 with 16 GB RAM. Furthermore, the parameters' specifications for WOA are as follows for all the study cases (1) the size of the population $(N) = 50$ and (2) the maximum number of iterations $(T_{max}) = 1000$. On the other hand, the parameters of HWOA are (1) the size of the population $(N) = 50$, (2) the maximum number of iterations $(T_{max}) = 200$, (3) the β_{min} and β_{max} are 0.001 and 0.6, respectively, (4) $T = 1000$, and (5) $K = 20$.

Lastly, the performance of the proposed HWOA is compared with the twenty-one other comparative algorithms published in the literature. These algorithms are ARCGA [45], MABC [46], WSA [47], TLBO [48], ALO [49], MBA [50], GA [51], CSP [52], FPA [2], HTS [13], FA [53], AFA [53], SOS [54], mSOS [54], PSO [55], HPSO [56], PSOPC [55], DHP-SACO [57], HS [58], HHS [20], and HACOHS-T [59]. The full names and the parameter settings of the other comparative algorithms can be found in the references mentioned besides the names of the algorithms.

23.6.1 Case study 1: 10-bar planar truss

Starting with the first case study, the planar truss shown in Fig. 23.3 represents the 10-bar truss structure. It consists of 6 nodes (joints) and 11 bars (members). The specifications of the structure are as follows; The Structure members' material has a Young modulus of $E = 10\,Msi$, and a density of $\rho = 0.1\,lb/in.^3$. Furthermore, a vertical load of $P = 100\,kip$ is acting on the 2nd and 4th joints. The displacement of joints in the structure is limited to $\gamma^{max} = 2\,in$ & $\gamma^{min} = -2\,in$ in both directions (x and y). In addition, the maximum allowable stress for all members is $\sigma^{max} = +25\,ksi$ and the compression stress is $\sigma^{min} = -25\,ksi$.

For this case study, three variants of cross-sectional areas were studied, where the first considers a continuous search space, such that the member's cross-sectional area can be a size between $0.1\,in^2$ and $35\,in^2$. On the other hand, the other two variants, define a set of available cross-sectional areas X^{Set}, and the member's cross-sectional area will be one of the available areas. Those variants are performed in discrete search spaces, where discrete case-1 has $X^{Set} = \{$ 1.62, 1.8, 1.99, 2.13, 2.38, 2.62, 2.88, 2.93, 3.09, 3.13, 3.38, 3.47, 3.55, 3.63, 3.84, 3.87, 3.88, 4.18, 4.22, 4.49, 4.59, 4.80, 4.97, 5.12, 5.94, 7.22, 7.97, 11.5, 13.50, 13.90, 14.2, 15.5, 16.0, 16.9, 18.8, 19.9, 22.0, 22.9, 28.5, 30.0, 33.5$\}$ $in.^2$ and discrete case-2 has $X^{Set} = \{$ 0.10, 0.50, 1, 1.50, 2, 2.50, 3, 3.50, 4, 4.50, 5, 5.50, 6, 6.50, 7, 7.50, 8, 8.50, 9, 9.50, 10, 10.5, 11, 11.5, 12, 12.5, 13, 13.5, 14, 14.5, 15, 15.5, 16, 16.5, 17, 17.5, 18, 18.5, 19, 19.5, 20, 20.5, 21, 21.5, 22, 22.5, 23, 23.5, 24, 24.5, 25, 25.5, 26, 26.5, 27, 27.5, 28, 28.5, 29, 29.5, 30, 30.5, 31, 31.5$\}$ $in.^2$.

Algorithm 23.1 The pseudo-code of the proposed HWOA.

```
Initialize the following parameters of HWOA
        $N$, $m$, and $T_{max}$ Parameters of WOA
        $\beta_{min}$, $\beta_{max}$, $K$, and $T_\beta$ parameters of A$\beta$HC
        $R$ the probability of calling A$\beta$HC
———— Initialization of initial population
$\vec{X}_{i,j} = X_j^{min} + (X_j^{max} - X_j^{min}) \times U(0,1)$    $\forall i = 1, 2, \ldots, N$, and $\forall j = 1, 2, \ldots, m$
Calculate $f(\vec{X}_i)$  $\forall i = 1, 2, \ldots, n$    {Fitness evaluation}
Select the best solution so far $\vec{X}^*$
$t$=1
while ($t \leq T_{max}$) do
  for i= 1 : N do
    Update $a$, $A$, $C$, $l$ and $p$.
    if $p < 0.5$ then
      if $|A| < 1$ then
        Update the position of the $i$ whale using Eq. (23.10).    {Encircling the prey}
      else if $|A| \geq 1$ then
        Select a random whale $X_k$, where $k \neq i$
        Update the position of the $i$ whale using Eq. (23.18).    {Search for prey}
      end if
    else
      Update the position of the $i$ whale using Eq. (23.15).    {Spiral updating position}
    end if
    —— Calling Adaptive $\beta$-hill climbing procedure.
    $rnd \in [0, 1]$
    if $rnd < R$ then
      $t2$=1
      while ($t2 \leq T_\beta$) do
        Update $\mathcal{N}(t2)$ using Eq. (23.21)
        Select $j$, where $j \in [1, m]$
        $X'_{i,j} = X_{i,j} \pm r \times \mathcal{N}(t2)$    {$\mathcal{N}$-operator}
        Select $k$, where $k \in [1, m]$
        $X''_{i,k} = X_k^{min} + r \times (X_k^{max} - X_k^{min})$    {$\beta$-operator}
        if $\vec{X}''_i$ is better than $\vec{X}_i$ then
          $\vec{X}_i = \vec{X}''_i$    {$\mathcal{U}$-operator}
        end if
        $t2 = t2 + 1$
      end while
    end if
    Calculate fitness of $\vec{X}_i$
    Update the $\vec{X}^*$
  end for
  $t = t + 1$
end while
Return the best solution $\vec{X}^*$
```

Tables 23.1–23.3 illustrate a comparison of the best-obtained cross-sectional areas by the WOA, the hybrid WOA (HWOA), and other literature algorithms for both continuous and discrete search spaces. It can be seen that the hybridization of WOA was fruitful, where it was able to achieve respectful results in comparison to the finding of the original WOA. Such that considering the continuous case, HWOA reported a total structure weight of 5060.873 which is very close to the best-achieved weight by HS, not as WOA which reported the worst W of 5220.440.

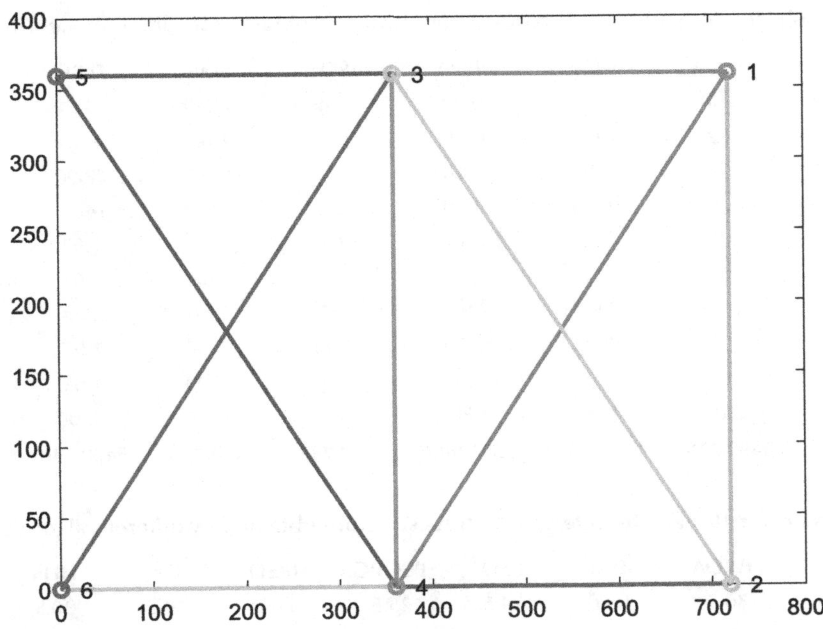

FIGURE 23.3 Planar 10-bar truss structure.

TABLE 23.1 Design results for 10-bar (Continuous case) truss structure obtained by different algorithms.

Algorithm	WOA	HWOA	HS	HPSO	WSA	TLBO	ALO
X_1	32.775	30.558	30.150	30.704	30.538	30.668	30.101
X_2	0.21	0.100	0.102	0.100	0.100	0.100	0.100
X_3	27.869	23.184	22.710	23.167	23.176	23.158	23.480
X_4	17.337	15.228	15.270	15.183	15.248	15.223	15.559
X_5	0.125	0.100	0.102	0.100	0.100	0.100	0.100
X_6	0.125	0.558	0.544	0.551	0.554	0.542	0.568
X_7	17.431	21.066	7.541	7.460	7.458	21.026	7.417
X_8	7.118	7.464	21.560	20.978	21.027	7.465	21.016
X_9	0.125	0.100	21.450	21.508	21.522	0.100	21.454
X_{10}	22.395	21.470	0.100	0.100	0.100	21.466	0.100
W(X)	5220.440	5060.873	**5057.880**	5060.920	5060.850	5060.973	5061.530

Moving to the discrete search space cases. In discrete case-1, HWOA was one of the leading algorithms that recorded a structured weight of 5490.738. In contrast, WOA was not able to optimize the structure members' cross-sectional areas and ended with a weight of 5696.942. Furthermore, in discrete case-2, HWOA achieved the second-best weight. Where the best was reported by GA. Once again, the WOA finding was not that good.

23.6.2 Case study 2: 25-bar spatial truss

The truss in Fig. 23.4 illustrates the 25-bar truss structure, it consists of 10 nodes (joints) and 25 bars (members). The structure specifications include a young modulus of $E = 10\,Msi$ for the members' material, and a density of $\rho = 0.1\,lb/in.^3$. Also, the displacement of joints in the structure is limited to $\gamma^{max} = 0.35\,in$ & $\gamma^{min} = -0.35\,in$ in all directions. In addition to a maximum allowable stress of $\sigma^{max} = +40\,ksi$ and a compression stress of $\sigma^{min} = -40\,ksi$ for all members. Moreover, the structure is subjected to two independent loading cases listed in Table 23.4. Here P_x, P_y, and P_z refer to the loads along the x, y, and z axes, respectively.

Furthermore, for this truss structure, the members are aggregated into 8 groups. Where the same cross-sectional area size is assigned to all the group's members. The structure groups are as follows: $GP_1 = \{X_1\}$; $GP_2 = \{X_2 - X_5\}$; $GP_3 =$

TABLE 23.2 Design results for 10-bar (Discrete case-1) truss structure obtained by different algorithms.

Algorithm	WOA	HWOA	MBA	HPSO	mSOS	SOS	TLBO	HHS	PSO
X_1	30.00	33.50	30.00	30.00	33.50	33.50	33.50	33.5	30.00
X_2	1.62	1.62	1.62	1.62	1.62	1.62	1.62	1.62	1.62
X_3	22.90	22.90	22.90	22.90	22.90	22.90	22.90	22.9	30.00
X_4	22.00	14.20	16.90	13.50	14.20	14.20	14.20	14.2	13.50
X_5	1.62	1.62	1.62	1.62	1.62	1.62	1.62	1.62	1.62
X_6	1.62	1.62	1.62	1.62	1.62	1.62	1.62	1.62	1.80
X_7	16.90	22.90	7.97	7.97	7.97	7.97	22.90	7.97	11.50
X_8	13.90	7.97	22.90	26.50	22.90	22.90	7.97	22.9	18.80
X_9	1.62	1.62	22.90	22.00	22.00	22.00	1.62	22	22.00
X_{10}	22.90	22.00	1.62	1.80	1.62	1.62	22.00	1.62	1.80
W(X)	5696.942	**5490.738**	5507.750	5531.980	**5490.738**	**5490.738**	5490.740	5490.74	5581.760

TABLE 23.3 Design results for 10-bar (Discrete case-2) truss structure obtained by different algorithms.

Algorithm	WOA	HWOA	MBA	PSO	PSOPC	HPSO	GA	HHS	FA	AFA
X_1	24.0	30.5	29.5	24.5	25.5	31.5	30.5	30.5	25.5	31
X_2	0.1	0.1	0.1	0.1	0.1	0.1	0.5	0.1	1.5	0.1
X_3	24.0	23.5	24.0	22.5	23.5	24.5	16.5	24	24.5	23
X_4	15.5	15.5	15.0	15.5	18.5	15.5	15	14	13.5	15
X_5	0.1	0.1	0.1	0.1	0.1	0.1	0.1	0.1	0.1	0.1
X_6	0.1	0.5	0.5	1.5	0.5	0.5	0.1	0.5	2	0.5
X_7	25.0	21.0	7.5	8.5	7.5	7.5	0.5	7.5	12	7.5
X_8	10.0	7.5	21.5	21.5	21.5	20.5	18	21.5	19	21
X_9	0.1	0.1	21.5	27.5	23.5	20.5	19.5	21.5	20	21.5
X_{10}	22.5	21.0	0.1	0.1	0.1	0.1	0.5	0.1	2.5	0.1
W(X)	5229.31	5052.42	5067.33	5243.71	5133.16	5073.51	**4217.30**	5067.33	5139.37	5059.87

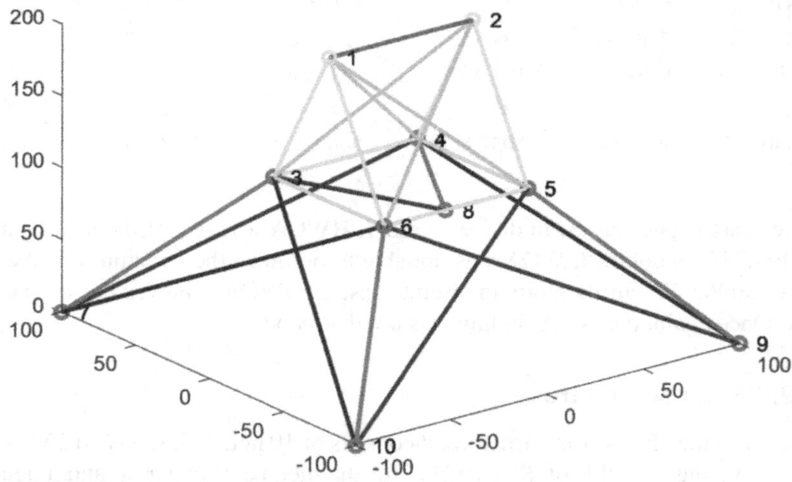

FIGURE 23.4 Planar 25-bar tower structure.

$\{X_6 - X_9\}$; $GP_4 = \{X_{10}, X_{11}\}$; $GP_5 = \{X_{12}, X_{13}\}$; $GP_6 = \{X_{14} - X_{17}\}$; $GP_7 = \{X_{18} - X_{21}\}$; $GP_8 = \{X_{22} - X_{25}\}$. Similar to the 10-bar truss structure, the 25-bar truss study case also includes variants. Where the lower and upper allowable cross-sectional areas for the continuous search space are $0.01\,in^2$ and $3.4\,in^2$, respectively. On the other hand, the discrete

TABLE 23.4 25-bar truss Loading cases.

Load Cases	Nodes	P_x (kips)	P_y (kips)	P_z (kips)
Case 1	1	1.0	10.0	-5.0
	2	0.0	10.0	-5.0
	3	0.5	0.0	0.0
	6	0.5	0.0	0.0
Case 2	1	0.0	20.0	-5.0
	2	0.0	-20.0	-5.0

TABLE 23.5 The cross-section areas following the AISC norm.

No.	in^2	No.	in^2	No.	in^2	No.	in^2
1	0.111	17	1.563	33	3.84	49	11.5
2	0.141	18	1.62	34	3.87	50	13.5
3	0.196	19	1.8	35	3.88	51	13.9
4	0.25	20	1.99	36	4.18	52	14.2
5	0.307	21	2.13	37	4.22	53	15.5
6	0.391	22	2.38	38	4.49	54	16
7	0.442	23	2.62	39	4.59	55	16.9
8	0.563	24	2.63	40	4.8	56	18.8
9	0.602	25	2.88	41	4.97	57	19.9
10	0.766	26	2.93	42	5.12	58	22
11	0.785	27	3.09	43	5.74	59	22.9
12	0.994	28	3.13	44	7.22	60	24.5
13	1	29	3.38	45	7.97	61	26.5
14	1.228	30	3.47	46	8.53	62	28
15	1.266	31	3.55	47	9.3	63	30
16	1.457	32	3.63	48	10.85	64	33.5

sets are $X^{Set} = \{0.1, 0.2, 0.3, 0.4, 0.5, 0.6, 0.7, 0.8, 0.9, 1.0, 1.1, 1.2, 1.3, 1.4, 1.5, 1.6, 1.7, 1.8, 1.9, 2.0, 2.1, 2.2, 2.3, 2.4, 2.6, 2.8, 3.0, 3.2, 3.4\}$ $(in._2)$ for discrete case-1, and $X^{Set} = \{0.01, 0.4, 0.8, 1.2, 1.6, 2.0, 2.4, 2.8, 3.2, 3.6, 4.0, 4.4, 4.8, 5.2, 5.6, 6.0\}$ $(in._2)$ for discrete case-2. While discrete case-3 considers the American Institute of Steel Construction (AISC) data listed in Table 23.5.

Consequently, the variants were implemented, and the algorithms' achievements were reported in Tables 23.6–23.9. In the case of continuous search space, the WOA suffers from optimizing the truss members' cross-sectional area sizes, whereas HWOA claimed a good enough total structure weight of 545.380 comparing it to the best weight of 544.380 reported by HS.

Next, for the discrete cases, the results of WOA and the HWOA were very close. Where in discrete case-1, proudly the HWOA was able to achieve the best weight of 482.214, and WOA reported the second best weight of 482.724. Overall, the performance of both the WOA and the HWOA was better than the other literature algorithms.

In the discrete case-2, the cross-sectional areas assigned by DHPSACO lead to the superiority of the algorithm. On the other hand, WOA and HWOA reported exactly the same truss weight and ranked as the second-best algorithms.

Lastly, in the third discrete case, the best claimed weight was 551.140 which was gained by HWOA, MBA, and HPSO. However, regarding WOA, its optimization process ended with a truss weight of 558.844. That is considered the third best weight comparing it with the literature results.

23.6.3 Case study 3: 72-bar spatial truss

The third case study considers the 72-bar truss structure, for illustration purposes refer to Fig. 23.5. This truss structure is built of 72 bars (members) connected through 20 nodes (joints). Where the Young modulus of members' material is $E = 10\,Msi$, and a density is $\rho = 0.1\,lb/in.^3$. In addition, the displacement of joints in the structure is limited to $\gamma^{max} =$

TABLE 23.6 Design results for 25-bar (Continuous case) truss structure obtained by different algorithms.

Algorithm	WOA	HWOA	HS	HTS	FPA	HPSO	WSA	CSP	ALO
GP_1	2.7275	0.0100	0.047	0.01	0.01	0.010	0.010	0.010	0.010
GP_2	1.9388	1.9867	2.022	2.0702	1.8308	1.970	1.983	1.910	1.994
GP_3	2.7837	2.9906	2.950	2.970031	3.1834	3.016	3.000	2.798	2.983
GP_4	0.0137	0.0100	0.010	0.01	0.01	0.010	0.010	0.010	0.010
GP_5	0.0368	0.0100	0.014	0.01	0.01	0.010	0.010	0.010	0.010
GP_6	0.7041	0.6902	0.688	0.67079	0.7017	0.694	0.683	0.708	0.684
GP_7	1.8935	1.6782	1.657	1.61712	1.7266	1.681	1.678	1.836	1.676
GP_8	2.6434	2.6548	2.663	2.6981	2.5713	2.643	2.661	2.645	2.665
W(X)	570.676	545.173	**544.380**	545.13	545.159	545.190	545.163	545.090	545.160

TABLE 23.7 Design results for 25-bar (Discrete case-1) truss structure obtained by different algorithms.

Algorithm	WOA	HWOA	MBA	GA	TLBO	HHS	PSO	PSOPC	HPSO
GP_1	0.1	0.1	0.1	0.1	0.1	0.1	0.4	0.1	0.1
GP_2	0.5	0.6	0.3	0.3	0.3	0.3	0.6	1.1	0.3
GP_3	3.4	3.4	3.4	3.4	3.4	3.4	3.5	3.1	3.4
GP_4	0.1	0.1	0.1	0.1	0.1	0.1	0.1	0.1	0.1
GP_5	1.7	2.3	2.1	2.0	2.1	2.1	1.7	2.1	2.1
GP_6	1	1	1	1	1	1	1	1	1
GP_7	0.4	0.2	0.5	0.5	0.5	0.5	0.3	0.1	0.5
GP_8	3.4	3.4	3.4	3.4	3.4	3.4	3.4	3.5	3.4
W(X)	482.724	**482.214**	484.850	483.354	484.850	484.85	486.540	490.160	484.850

TABLE 23.8 Design results for 25-bar (Discrete case-2) truss structure obtained by different algorithms.

Algorithm	WOA	HWOA	MBA	HHS	PSO	PSOPC	HPSO	DHPSACO
GP_1	0.01	0.01	0.01	0.01	0.01	0.01	0.01	0.01
GP_2	1.6	1.6	2	2	2	2	2	1.60
GP_3	3.2	3.2	3.60	3.6	3.60	3.60	3.60	3.20
GP_4	0.01	0.01	0.01	0.01	0.01	0.01	0.01	0.01
GP_5	0.01	0.01	0.01	0.01	0.40	0.01	0.01	0.01
GP_6	0.8	0.8	0.80	0.8	0.80	0.80	0.80	0.80
GP_7	2	2	1.60	1.6	1.60	1.60	1.60	2.00
GP_8	2.4	2.4	2.40	2.4	2.40	2.40	2.40	2.40
W(X)	553.813	553.813	560.590	560.59	566.440	560.590	560.590	**551.610**

TABLE 23.9 Design results for 25-bar (Discrete case-3) truss structure obtained by different algorithms.

Algorithm	WOA	HWOA	MBA	PSO	PSOPC	HPSO
GP_1	0.111	0.111	0.111	1	0.111	0.111
GP_2	1.563	2.13	2.130	2.620	1.563	2.130
GP_3	3.47	2.88	2.880	2.620	3.380	2.880
GP_4	0.111	0.111	0.111	0.250	0.111	0.111
GP_5	0.111	0.111	0.111	0.307	0.111	0.111
GP_6	0.563	0.766	0.766	0.602	0.766	0.766
GP_7	1.99	1.62	1.620	1.457	1.990	1.620
GP_8	2.62	2.62	2.620	2.880	2.380	2.620
W(X)	558.844	**551.140**	**551.140**	567.490	556.900	**551.140**

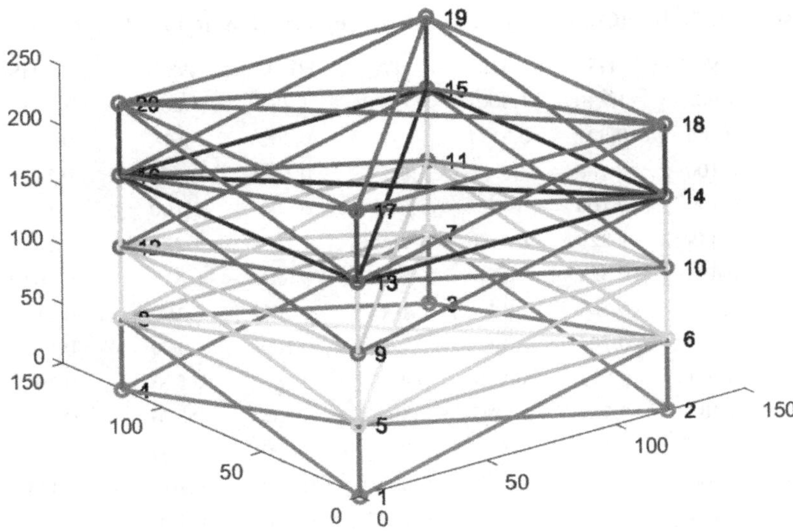

FIGURE 23.5 72-bar truss structure.

TABLE 23.10 72-bar truss Load cases.

Load Cases	Nodes	P_x (kips)	P_y (kips)	P_z (kips)
Case 1	17	0	0	-5
	18	0	0	-5
	19	0	0	-5
	20	0	0	-5
Case 2	17	5	5	-5

$0.25\,in$ & $\gamma^{min} = -0.25\,in$ in both x and y directions. Furthermore, the maximum allowable stress for the member is $\sigma^{max} = +25\,ksi$ and the compression stress is $\sigma^{min} = -25\,ksi$. Moreover, the structure is subjected to two loading cases listed in Table 23.10. Note that P_x, P_y, and P_z refer to the loads along the x, y, and z axes, respectively.

In terms of design variables, the truss members were assigned into 16 instead of 72 group. Subsequently, the members of one group have the same cross-sectional area size. The structure groups are as follows: $GP_1 = \{X_1 - X_4\}$; $GP_2 = \{X_5 - X_{12}\}$; $GP_3 = \{X_{13} - X_{16}\}$; $GP_4 = \{X_{17}, X_{18}\}$; $GP_5 = \{X_{19} - X_{22}\}$; $GP_6 = \{X_{23} - X_{30}\}$; $GP_7 = \{X_{31} - X_{34}\}$; $GP_8 = \{X_{35}, X_{36}\}$; $GP_9 = \{X_{37} - X_{40}\}$; $GP_{10} = \{X_{41} - X_{48}\}$; $GP_{11} = \{X_{49} - X_{52}\}$; $GP_{12} = \{X_{53}, X_{54}\}$; $GP_{13} = \{X_{55} - X_{58}\}$; $GP_{14} = \{X_{59} - X_{66}\}$; $GP_{15} = \{X_{67} - X_{70}\}$; $GP_{16} = \{X_{71}, X_{72}\}$; In addition, the cross-sectional area sizes considered in this study case are of two variants: (1) Continuous and (2) Discrete. Here in the continuous search space, the area sizes range between $0.1\,in^2$ and $3\,in^2$. On the other hand, the discrete sizes set includes X^{Set} = {0.10, 0.20, 0.30, 0.40, 0.50, 0.60, 0.70, 0.80, 0.90, 1, 1.1, 1.2, 1.3, 1.4, 1.5, 1.6, 1.7, 1.8, 1.9, 2, 2.1, 2.2, 2.3, 2.4, 2.5, 2.6, 2.7, 2.8, 2.9, 3, 3.1, 3.2}.

The optimized members' sizes for the 72-bar continuous and discrete case studies obtained by WOA, HWOA, and other literature algorithms are listed in Tables 23.11 & 23.12. Starting with the continuous case, WOA reported a truss weight of 383.807 which is acceptable but not that good. Given the same study scenario, HWOA was able to be ranked as the second-best algorithm comparing its findings to the literature. Where the best was achieved by HPSO, with a weight of 369.65.

Next is the discrete case. It can be seen that our proposed algorithm (HWOA) recorded the best truss weight of 385.540, which was also achieved by three other comparative methods (i.e., MBA, DHPSACO, and HHS). On the other hand, the optimized weight of WOA was 391.407. As a result, WOA was given the third-best rank.

23.6.4 Case study 4: 200-bar planar truss

The last studied case is the 200-bar planar truss, the truss structure of this problem consists of 200 bars connected utilizing 77 nodes as illustrated in Fig. 23.6. Where the structure specifications are as follows; the members' material has a Young

TABLE 23.11 Design results for 72-bar (Continuous case) truss structure obtained by different algorithms.

Algorithm	WOA	HWOA	HS	HTS	FPA	HPSO	WSA	HACOHS-T	TLBO	CSP
GP_1	1.957	1.902	1.790	1.9001	1.8758	1.857	1.885	1.563	1.881	1.945
GP_2	0.456	0.505	0.521	0.5131	0.516	0.505	0.514	0.563	0.514	0.503
GP_3	0.100	0.100	0.100	0.1	0.1	0.100	0.100	0.111	0.100	0.100
GP_4	0.105	0.100	0.100	0.1	0.1	0.100	0.100	0.111	0.100	0.100
GP_5	1.313	1.345	1.229	1.2456	1.2993	1.255	1.271	1.266	1.271	1.268
GP_6	0.549	0.496	0.522	0.508	0.5246	0.503	0.511	0.563	0.515	0.510
GP_7	0.106	0.100	0.100	0.1	0.1001	0.100	0.100	0.111	0.100	0.100
GP_8	0.100	0.100	0.100	0.1	0.1	0.100	0.100	0.111	0.100	0.100
GP_9	0.497	0.704	0.517	0.555	0.4971	0.496	0.526	0.391	0.532	0.507
GP_{10}	0.541	0.487	0.504	0.5227	0.5089	0.506	0.516	0.563	0.513	0.517
GP_{11}	0.142	0.100	0.100	0.1	0.1	0.100	0.100	0.111	0.100	0.108
GP_{12}	0.101	0.100	0.101	0.1	0.1	0.100	0.100	0.111	0.100	0.100
GP_{13}	0.157	0.133	0.156	0.1566	0.1575	0.100	0.156	0.196	0.157	0.156
GP_{14}	0.524	0.540	0.547	0.5407	0.5329	0.524	0.545	0.563	0.543	0.540
GP_{15}	0.360	0.404	0.442	0.4084	0.4089	0.400	0.412	0.391	0.408	0.422
GP_{16}	0.679	0.537	0.590	0.5669	0.5731	0.534	0.568	0.602	0.573	0.579
W(X)	383.807	377.962	379.270	379.73	379.095	**369.650**	379.618	390.180	379.632	379.970

TABLE 23.12 Design results for 72-bar (Discrete case) truss structure obtained by different algorithms.

Algorithm	WOA	HWOA	MBA	DHPSACO	HHS	PSO	PSOPC	HPSO
GP_1	1.6	1.9	2.0	1.9	1.9	2.6	3.0	2.1
GP_2	0.5	0.5	0.6	0.5	0.5	1.5	1.4	0.6
GP_3	0.1	0.1	0.4	0.1	0.1	0.3	0.2	0.1
GP_4	0.1	0.1	0.6	0.1	0.1	0.1	0.1	0.1
GP_5	1.2	1.3	0.5	1.3	1.4	2.1	2.7	1.4
GP_6	0.6	0.5	0.5	0.5	0.5	1.5	1.9	0.5
GP_7	0.1	0.1	0.1	0.1	0.1	0.6	0.7	0.1
GP_8	0.1	0.1	0.1	0.1	0.1	0.3	0.8	0.1
GP_9	0.6	0.5	1.4	0.6	0.5	2.2	1.4	0.5
GP_{10}	0.6	0.5	0.5	0.5	0.5	1.9	1.2	0.5
GP_{11}	0.1	0.1	0.1	0.1	0.1	0.2	0.8	0.1
GP_{12}	0.1	0.1	0.1	0.1	0.1	0.9	0.1	0.1
GP_{13}	0.2	0.2	1.9	0.2	0.2	0.4	0.4	0.2
GP_{14}	0.6	0.6	0.5	0.6	0.6	1.9	1.9	0.5
GP_{15}	0.4	0.4	0.1	0.4	0.4	0.7	0.9	0.3
GP_{16}	0.4	0.6	0.1	0.6	0.6	1.6	1.3	0.7
W(X)	391.407	**385.540**	**385.540**	**385.540**	**385.54**	1089.880	1069.790	388.940

modulus of $E = 30\,Msi$ and a density of $\rho = 0.268\,lb/in^3$. Also, the structure members must satisfy the stress constraints, such that the tensile and the compression stresses are $\sigma^{max} = +10\,ksi$ and $\sigma^{min} = -10\,ksi$, respectively. However, for this case study, the truss nodes are not subjected to any displacement constraint. Further, regarding the loading conditions, three independent loading conditions must be taken into consideration: i) 1.0 kip acting in the positive x-direction at nodes 1, 6, 15, 20, 29, 34, 43, 48, 57, 62, and 71; ii) 10 kips acting in the negative y-direction at nodes 1-6, 8, 10, 12, 14-20, 22, 24, 26, 28-34, 36, 38, 40, 42-48, 50, 52, 54, 56-62, 64, 66, 68, 70-75; and iii) the previous two loading conditions acting together.

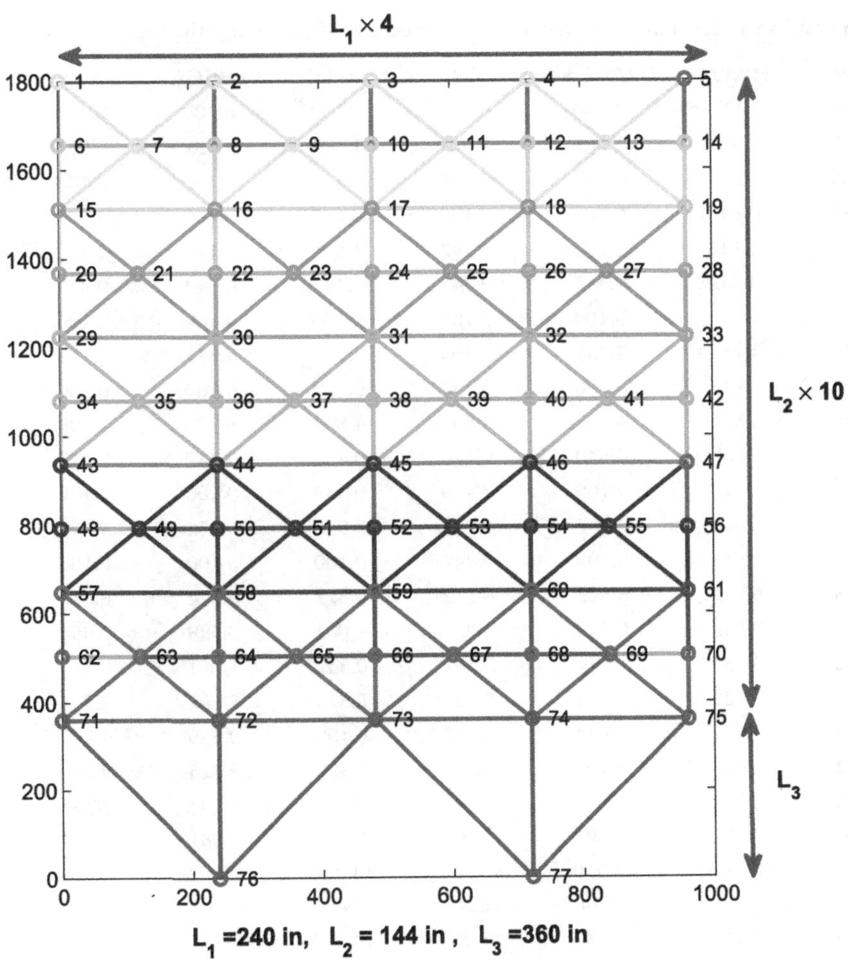

FIGURE 23.6 Planar 200-bar tower structure.

Furthermore, with respect to the design variable, the members are grouped into 29 groups as follows: $GP_1 = \{X_1 - X_4\}$; $GP_2 = \{X_5, X_8, X_{11}, X_{14}, X_{17}\}$; $GP_3 = \{X_{19} - X_{24}\}$; $GP_4 = \{X_{18}, X_{25}, X_{56}, X_{63}, X_{94}, X_{101}, X_{132}, X_{139}, X_{170}, X_{177}\}$; $GP_5 = \{X_{26}, X_{29}, X_{32}, X_{35}, X_{38}\}$; $GP_6 = \{X_6, X_7, X_9, X_{10}, X_{12}, X_{13}, X_{15}, X_{16}, X_{27}, X_{28}, X_{30}, X_{31}, X_{33}, X_{34}, X_{36}, X_{37}\}$; $GP_7 = \{X_{39} - X_{42}\}$; $GP_8 = \{X_{43}, X_{46}, X_{49}, X_{52}, X_{55}\}$; $GP_9 = \{X_{57} - X_{62}\}$; $GP_{10} = \{X_{64}, X_{67}, X_{70}, X_{73}, X_{76}\}$; $GP_{11} = \{X_{44}, X_{45}, X_{47}, X_{48}, X_{50}, X_{51}, X_{53}, X_{54}, X_{65}, X_{66}, X_{68}, X_{69}, X_{71}, X_{72}, X_{74}, X_{75}\}$; $GP_{12} = \{X_{77} - X_{80}\}$; $GP_{13} = \{X_{81}, X_{84}, X_{87}, X_{90}, X_{93}\}$; $GP_{14} = \{X_{95} - X_{100}\}$; $GP_{15} = \{X_{102}, X_{105}, X_{108}, X_{111}, X_{114}\}$; $GP_{16} = \{X_{82}, X_{83}, X_{85}, X_{86}, X_{88}, X_{89}, X_{91}, X_{92}, X_{103}, X_{104}, X_{106}, X_{107}, X_{109}, X_{110}, X_{112}, X_{113}\}$; $GP_{17} = \{X_{115} - X_{118}\}$; $GP_{18} = \{X_{119}, X_{122}, X_{125}, X_{128}, X_{131}\}$; $GP_{19} = \{X_{133} - X_{138}\}$; $GP_{20} = \{X_{140}, X_{143}, X_{146}, X_{149}, X_{152}\}$; $GP_{21} = \{X_{120}, X_{121}, X_{123}, X_{124}, X_{126}, X_{127}, X_{129}, X_{130}, X_{141}, X_{142}, X_{144}, X_{145}, X_{147}, X_{148}, X_{150}, X_{151}\}$; $GP_{22} = \{X_{153} - X_{156}\}$; $GP_{23} = \{X_{157}, X_{160}, X_{163}, X_{166}, X_{169}\}$; $GP_{24} = \{X_{171} - X_{176}\}$; $GP_{25} = \{X_{178}, X_{181}, X_{184}, X_{187}, X_{190}\}$; $GP_{26} = \{X_{158}, X_{159}, X_{161}, X_{162}, X_{164}, X_{165}, X_{167}, X_{168}, X_{179}, X_{180}, X_{182}, X_{183}, X_{185}, X_{186}, X_{188}, X_{189}\}$; $GP_{27} = \{X_{191} - X_{194}\}$; $GP_{28} = \{X_{195}, X_{197}, X_{198}, X_{200}\}$; $GP_{29} = \{X_{196}, X_{199}\}$. In this context, it must be noted that the members of the same group have the same cross-sectional area. Lastly, we studied this case from the discrete search space perspective, in which the members' size is limited to one of the following sizes $X^{set} = \{0.10, 0.347, 0.440, 0.539, 0.954, 1.081, 1.174, 1.333, 1.488, 1.764, 2.142, 2.697, 2.80, 3.13, 3.565, 3.813, 4.805, 5.952, 6.572, 7.192, 8.525, 9.30, 10.850, 13.330, 14.29, 17.17, 19.18, 23\ 28.08, 33.70\}$.

Consequently, WOA, HWOA, and the other seven literature algorithms results are recorded in Table 23.13. With the high dimensionality and the rugged search space of this truss structure, this case study is considered hard and requires a powerful algorithm that is able to cope with it. However, accordingly to the reported results, we can see that WOA was not able to escape from the local optima, subsequently it ranked as the worst algorithm. On the other hand, we can conclude

TABLE 23.13 Design results for 200-bar truss structure obtained by different algorithms.

Algorithm	WOA	HWOA	HACOHS-T	GA	MABC	ARCGA	SOS	mSOS	HHS
GP_1	3.131	0.1	0.100	0.347	0.100	0.100	0.100	0.100	0.1
GP_2	4.805	0.954	1.081	1.081	1.333	1.081	0.954	0.954	0.954
GP_3	0.539	0.1	0.347	0.100	0.100	0.100	0.100	0.440	0.1
GP_4	1.333	0.347	0.100	0.100	0.100	0.100	0.100	0.100	0.1
GP_5	19.18	2.142	2.142	2.142	2.697	2.142	2.142	2.142	2.142
GP_6	3.813	0.347	0.347	0.347	0.347	0.347	0.347	0.347	0.347
GP_7	10.85	0.1	0.100	0.100	0.100	0.100	0.100	0.100	0.1
GP_8	3.131	3.131	3.131	3.565	3.131	3.131	3.131	3.131	3.131
GP_9	3.813	0.1	0.100	0.347	0.100	0.100	0.100	0.100	0.1
GP_{10}	19.18	4.805	4.805	4.805	4.805	4.805	4.805	4.805	4.805
GP_{11}	6.572	0.44	0.440	0.440	0.440	0.347	0.440	0.440	0.44
GP_{12}	14.29	0.1	0.100	0.440	0.539	0.100	0.100	0.440	0.347
GP_{13}	3.131	5.952	5.952	5.952	5.952	5.952	5.952	5.952	5.952
GP_{14}	1.174	0.1	0.100	0.347	0.100	0.100	0.100	0.100	0.347
GP_{15}	3.565	6.572	6.572	6.572	6.572	6.572	6.572	6.572	6.572
GP_{16}	9.3	0.539	0.539	0.954	1.081	0.539	0.954	0.954	0.954
GP_{17}	7.192	0.539	1.174	0.347	0.347	1.081	0.347	0.347	0.347
GP_{18}	17.17	8.525	8.525	8.525	8.525	7.192	8.525	8.525	8.525
GP_{19}	17.17	0.1	0.100	0.100	0.100	0.539	0.100	0.100	0.1
GP_{20}	7.192	9.3	9.300	9.300	9.300	8.525	9.300	9.300	9.3
GP_{21}	5.952	1.174	1.333	0.954	0.954	1.333	0.954	0.954	1.081
GP_{22}	1.764	0.1	0.539	1.764	1.764	1.081	1.764	1.764	0.347
GP_{23}	14.29	13.33	13.330	13.330	13.330	10.850	13.330	13.330	13.33
GP_{24}	13.33	0.1	1.174	0.347	0.440	0.100	0.440	0.440	0.954
GP_{25}	8.525	13.33	13.330	13.330	13.330	13.330	13.330	13.330	13.33
GP_{26}	10.85	0.954	2.697	2.142	2.142	1.488	2.142	2.142	1.764
GP_{27}	9.3	5.952	3.565	4.805	3.813	5.952	3.813	3.813	3.813
GP_{28}	13.33	10.85	8.525	9.300	8.525	13.330	8.525	8.525	8.525
GP_{29}	17.17	14.29	17.170	17.170	19.180	14.290	17.170	17.170	17.17
W(X)	81840.08	**27113.326**	28030.200	28544.014	28366.365	28347.594	27544.191	27544.191	27163.59

that HWOA parameters were able to efficiently compromise between the exploration and exploitation phases, and report respective results, which outperform all the algorithms.

In summary, the performance of WOA and the proposed HWOA were evaluated utilizing four truss structure problems through discrete and continuous search space. In consequence, the results obtained were compared against a variety of literature algorithms. Generally speaking, the findings of the original WOA were the worst in all continuous cases, and somehow acceptable but not that good for the discrete cases (10-bar, 25-bar, and 72-bar problems). In addition, WOA performance in the 200-bar problem was unsatisfactory and its findings were not even in the range of other algorithms' results, which indicates the incapability of the algorithm in solving high-dimensional truss problems. On the other hand, the hybridization of adaptive β hill climbing (AβHC) operators with WOA (HWOA) empowers the algorithm, and eventually allows it to be ranked as either the best algorithm or the second-best algorithm (except in one case scenario 72-bar continuous). This performance is basically proofing the exploitation and exploration abilities gained from the adaptive β hill climbing (AβHC) to escape from the local optima, and as a result, achieve superior findings.

23.7 Conclusion and future work

This study presented a hybrid whale optimization algorithm (HWOA) for the sizing optimization of truss structures problems. In the proposed HWOA, the adaptive β-hill climbing optimizer as a local search-based procedure is injected into the main framework of the WOA as a new operator to enhance the convergence rate and the exploitation ability thus maintaining the balance between the exploration and exploitation abilities during all stages of the search process.

Four truss optimization problems with different study cases (10-bar, 25-bar, 72-bar, and 200-bar) were utilized to evaluate the performance of the proposed HWOA and compared to other comparative methods published in the literature. Two cases with discrete variables and one case with continues variables of 10-bar, three cases with discrete variables and one case with continues variables of 25-bar, one case with discrete variables and one case with continues variables of 71-bar, and one case with discrete variables of 200-bar were considered in this study.

The numerical results showed the effectiveness of the proposed HWOA by achieving the best results similar to some of the other competitors in one case of 10-bar problem with discrete variables. In addition, the proposed HWOA performs better than or is similar to the other comparative algorithms by getting the best results in two cases of 25-bar problem with discrete variables. The proposed HWOA achieved the best results similar to some of the other competitors in one case of 72-bar problem with discrete variables. Finally, the proposed WOA performs better than the classical WOA as well as the other comparative methods by achieving the best results for the 200-bar problem with discrete design variable. However, the proposed HWOA got very competitive results in the remaining cases of the different four truss optimization problems.

Clearly, the experimental results show the superiority of the proposed algorithm and this confirms the effectiveness of using the β-hill climbing optimizer to empower the local search ability of the proposed algorithm and thus making the right balance between the exploration and exploitation operators when navigating the search space of the truss structure problems.

Future research directions will focus on adapting the proposed hybrid algorithm for other variants of the truss optimization problem like shape optimization, topology optimization, and multi-objective optimization. In addition, the proposed algorithm can be applied to large-scale cases of the sizing optimization of truss structures.

References

[1] L. Mei, Q. Wang, Structural optimization in civil engineering: a literature review, Buildings 11 (2) (2021) 66.

[2] G. Bekdaş, S.M. Nigdeli, X.-S. Yang, Sizing optimization of truss structures using flower pollination algorithm, Applied Soft Computing 37 (2015) 322–331, https://doi.org/10.1016/j.asoc.2015.08.037.

[3] S. Białkowski, Structural optimisation methods as a new toolset for architects, in: Complexity & Simplicity - Proceedings of the 34th eCAADe Conference, CUMINCAD, 2016, pp. 255–264.

[4] M. Afshar, A. Faramarzi, Size optimization of truss structures by cellular automata, Journal of Computer Science and Engineering 3 (1) (2010) 1–9.

[5] U.T. Ringertz, A branch and bound algorithm for topology optimization of truss structures, Engineering Optimization 10 (2) (1986) 111–124, https://doi.org/10.1080/03052158608902532.

[6] A.R. Kashani, C.V. Camp, M. Rostamian, K. Azizi, A.H. Gandomi, Population-based optimization in structural engineering: a review, Artificial Intelligence Review 55 (1) (2022) 345–452.

[7] C. Blum, A. Roli, Metaheuristics in combinatorial optimization: overview and conceptual comparison, ACM Computing Surveys (CSUR) 35 (3) (2003) 268–308, https://doi.org/10.1145/937503.937505.

[8] M.A. Awadallah, M.A. Al-Betar, I.A. Doush, S.N. Makhadmeh, G. Al-Naymat, Recent versions and applications of sparrow search algorithm, Archives of Computational Methods in Engineering (2023) 1–28, https://doi.org/10.1007/s11831-023-09887-z.

[9] C. Millán-Páramo, Modified simulated annealing algorithm for discrete sizing optimization of truss structure, Jordan Journal of Civil Engineering 12 (4) (2018) 683–697.

[10] M. Khatibinia, H. Yazdani, Accelerated multi-gravitational search algorithm for size optimization of truss structures, Swarm and Evolutionary Computation 38 (2018) 109–119.

[11] S. Degertekin, Improved harmony search algorithms for sizing optimization of truss structures, Computers & Structures 92 (2012) 229–241.

[12] S. Gholizadeh, R. Sojoudizadeh, Modified sine-cosine algorithm for sizing optimization of truss structures with discrete design variables, Iran University of Science & Technology 9 (2) (2019) 195–212.

[13] S. Degertekin, L. Lamberti, M. Hayalioglu, Heat transfer search algorithm for sizing optimization of truss structures, Latin American Journal of Solids and Structures 14 (2017) 373–397.

[14] H. Alkhrisat, L.M. Dalbah, M.A. Al-Betar, M.A. Awadallah, K. Assaleh, M. Deriche, Size optimization of truss structures using improved grey wolf optimizer, IEEE Access 11 (2023) 13383–13397, https://doi.org/10.1109/ACCESS.2023.3243164.

[15] R. Awad, Sizing optimization of truss structures using the political optimizer (PO) algorithm, Structures 33 (2021) 4871–4894.

[16] A. Sadollah, H. Eskandar, A. Bahreininejad, J.H. Kim, Water cycle, mine blast and improved mine blast algorithms for discrete sizing optimization of truss structures, Computers & Structures 149 (2015) 1–16.

[17] A. Baghlani, M. Makiabadi, M. Maheri, Sizing optimization of truss structures by an efficient constraint-handling strategy in TLBO, Journal of Computing in Civil Engineering 31 (4) (2017) 04017004.

[18] F.K. Jawad, C. Ozturk, W. Dansheng, M. Mahmood, O. Al-Azzawi, A. Al-Jemely, Sizing and layout optimization of truss structures with artificial bee colony algorithm, Structures 30 (2021) 546–559.

[19] C. Blum, J. Puchinger, G.R. Raidl, A. Roli, Hybrid metaheuristics in combinatorial optimization: a survey, Applied Soft Computing 11 (6) (2011) 4135–4151.

[20] M.-Y. Cheng, D. Prayogo, Y.-W. Wu, M.M. Lukito, A hybrid harmony search algorithm for discrete sizing optimization of truss structure, Automation in Construction 69 (2016) 21–33.

[21] R. Najian Asl, M. Aslani, M. Shariat Panahi, Sizing optimization of truss structures using a hybridized genetic algorithm, arXiv e-prints, arXiv: 1306.1454, 2013, https://doi.org/10.48550/arXiv.1306.1454.

[22] S.O. Degertekin, M. Minooei, L. Santoro, B. Trentadue, L. Lamberti, Large-scale truss-sizing optimization with enhanced hybrid HS algorithm, Applied Sciences 11 (7) (2021) 3270.

[23] S. Mirjalili, A. Lewis, The whale optimization algorithm, Advances in Engineering Software 95 (2016) 51–67.

[24] F.S. Gharehchopogh, H. Gholizadeh, A comprehensive survey: whale optimization algorithm and its applications, Swarm and Evolutionary Computation 48 (2019) 1–24.

[25] Q.-V. Pham, S. Mirjalili, N. Kumar, M. Alazab, W.-J. Hwang, Whale optimization algorithm with applications to resource allocation in wireless networks, IEEE Transactions on Vehicular Technology 69 (4) (2020) 4285–4297.

[26] D. Oliva, M. Abd El Aziz, A.E. Hassanien, Parameter estimation of photovoltaic cells using an improved chaotic whale optimization algorithm, Applied Energy 200 (2017) 141–154.

[27] R. Agrawal, B. Kaur, S. Sharma, Quantum based whale optimization algorithm for wrapper feature selection, Applied Soft Computing 89 (2020) 106092.

[28] M.H. Nadimi-Shahraki, S. Taghian, S. Mirjalili, L. Abualigah, M. Abd Elaziz, D. Oliva, EWOA-OPF: effective whale optimization algorithm to solve optimal power flow problem, Electronics 10 (23) (2021) 2975.

[29] W. Yang, Z. Peng, Z. Yang, Y. Guo, X. Chen, An enhanced exploratory whale optimization algorithm for dynamic economic dispatch, Energy Reports 7 (2021) 7015–7029.

[30] H.M. Mohammed, S.U. Umar, T.A. Rashid, A systematic and meta-analysis survey of whale optimization algorithm, Computational Intelligence and Neuroscience (2019) 8718571.

[31] S.K. Dewi, D.M. Utama, A new hybrid whale optimization algorithm for green vehicle routing problem, Systems Science & Control Engineering 9 (1) (2021) 61–72.

[32] M.M. Mafarja, S. Mirjalili, Hybrid whale optimization algorithm with simulated annealing for feature selection, Neurocomputing 260 (2017) 302–312.

[33] C.-Y. Lee, G.-L. Zhuo, A hybrid whale optimization algorithm for global optimization, Mathematics 9 (13) (2021) 1477.

[34] C. Tang, W. Sun, M. Xue, X. Zhang, H. Tang, W. Wu, A hybrid whale optimization algorithm with artificial bee colony, Soft Computing 26 (5) (2022) 2075–2097.

[35] M.A. Al-Betar, β-hill climbing: an exploratory local search, Neural Computing & Applications 28 (1) (2017) 153–168, https://doi.org/10.1007/s00521-016-2328-2.

[36] M.A. Al-Betar, M.A. Awadallah, I.A. Doush, O.A. Alomari, A.K. Abasi, S.N. Makhadmeh, Z.A.A. Alyasseri, Boosting the training of neural networks through hybrid metaheuristics, Cluster Computing (2022) 1–23, https://doi.org/10.1007/s10586-022-03708-x.

[37] M.A. Al-Betar, A.I. Hammouri, M.A. Awadallah, I. Abu Doush, Binary β-hill climbing optimizer with S-shape transfer function for feature selection, Journal of Ambient Intelligence and Humanized Computing 12 (7) (2021) 7637–7665.

[38] E. Alsukni, O.S. Arabeyyat, M.A. Awadallah, L. Alsamarraie, I. Abu-Doush, M.A. Al-Betar, Multiple-reservoir scheduling using β-hill climbing algorithm, Journal of Intelligent Systems 28 (4) (2019) 559–570.

[39] M.A. Al-Betar, M.A. Awadallah, I. Abu Doush, E. Alsukhni, H. ALkhraisat, A non-convex economic dispatch problem with valve loading effect using a new modified β-hill climbing local search algorithm, Arabian Journal for Science and Engineering 43 (12) (2018) 7439–7456.

[40] Z.A.A. Alyasseri, A.T. Khader, M.A. Al-Betar, M.A. Awadallah, Hybridizing β-hill climbing with wavelet transform for denoising ECG signals, Information Sciences 429 (2018) 229–246.

[41] L.M. Abualigah, A.M. Sawaie, A.T. Khader, H. Rashaideh, M.A. Al-Betar, M. Shehab, β-hill climbing technique for the text document clustering, in: New Trends in Information Technology (NTIT), 2017, p. 60.

[42] M.A. Al-Betar, I. Aljarah, M.A. Awadallah, H. Faris, S. Mirjalili, Adaptive β-hill climbing for optimization, Soft Computing 23 (24) (2019) 13489–13512.

[43] M.A. Al-Betar, M.A. Awadallah, S.N. Makhadmeh, I.A. Doush, R.A. Zitar, S. Alshathri, M. Abd Elaziz, A hybrid Harris Hawks optimizer for economic load dispatch problems, Alexandria Engineering Journal 64 (2023) 365–389, https://doi.org/10.1016/j.aej.2022.09.010.

[44] M.A. Al-Betar, M.A. Awadallah, R.A. Zitar, K. Assaleh, Economic load dispatch using memetic sine cosine algorithm, Journal of Ambient Intelligence and Humanized Computing (2022) 1–29, https://doi.org/10.1007/s12652-022-03731-1.

[45] K. Koohestani, S. Kazemzadeh Azad, An adaptive real-coded genetic algorithm for size and shape optimization of truss structures, in: Proceedings of the First International Conference on Soft Computing Technology in Civil, Structural and Environmental Engineering, vol. 13, Civil-Comp Press, Stirlingshire, UK, 2009.

[46] A. Hadidi, S.K. Azad, S.K. Azad, Structural optimization using artificial bee colony algorithm, in: 2nd International Conference on Engineering Optimization, 2010, pp. 6–9.

[47] B. Adil, B. Cengiz, Optimal design of truss structures using weighted superposition attraction algorithm, Engineering With Computers 36 (3) (2020) 965–979.

[48] C.V. Camp, M. Farshchin, Design of space trusses using modified teaching–learning based optimization, Engineering Structures 62 (2014) 87–97.

[49] M. Salar, B. Dizangian, Sizing optimization of truss structures using ant lion optimizer, in: 2nd International Conference on Civil Engineering, Architecture and Urban Management in Iran, 2019.

[50] A. Sadollah, A. Bahreininejad, H. Eskandar, M. Hamdi, Mine blast algorithm for optimization of truss structures with discrete variables, Computers & Structures 102 (2012) 49–63.

[51] V. Toğan, A.T. Daloğlu, An improved genetic algorithm with initial population strategy and self-adaptive member grouping, Computers & Structures 86 (11–12) (2008) 1204–1218.

[52] A. Kaveh, R. Sheikholeslami, S. Talatahari, M. Keshvari-Ilkhichi, Chaotic swarming of particles: a new method for size optimization of truss structures, Advances in Engineering Software 67 (2014) 136–147.

[53] A. Baghlani, M. Makiabadi, M. Sarcheshmehpour, Discrete optimum design of truss structures by an improved firefly algorithm, Advances in Structural Engineering 17 (10) (2014) 1517–1530.

[54] D.T. Do, J. Lee, A modified symbiotic organisms search (mSOS) algorithm for optimization of pin-jointed structures, Applied Soft Computing 61 (2017) 683–699.

[55] L. Li, Z. Huang, F. Liu, A heuristic particle swarm optimization method for truss structures with discrete variables, Computers & Structures 87 (7–8) (2009) 435–443.

[56] L. Li, F. Liu, Harmony particle swarm algorithm for structural design optimization, in: Harmony Search Algorithms for Structural Design Optimization, Springer, 2009, pp. 121–157.

[57] A. Kaveh, S. Talatahari, A particle swarm ant colony optimization for truss structures with discrete variables, Journal of Constructional Steel Research 65 (8–9) (2009) 1558–1568.

[58] K.S. Lee, Standard harmony search algorithm for structural design optimization, in: Harmony Search Algorithms for Structural Design Optimization, Springer, 2009, pp. 1–49.

[59] M. Talebpour, A. Kaveh, V. Kalatjari, Optimization of skeletal structures using a hybridized ant colony-harmony search-genetic algorithm, Iranian Journal of Science and Technology. Transactions of Civil Engineering 38 (C1) (2014) 1.

Chapter 24

Whale Optimization Algorithm (WOA) for BIM-based resource trade-off in construction project scheduling

Milad Baghalzadeh Shishehgarkhaneh[a,b], Mahla Basiri[c], and Mahdi Azizi[c]

[a]*Department of Construction Management, Islamic Azad University of Tabriz, Tabriz, Iran,* [b]*Department of Civil Engineering, Faculty of Engineering, Monash University, Clayton, VIC, Australia,* [c]*Department of Civil Engineering, University of Tabriz, Tabriz, Iran*

24.1 Introduction

The planning phase of a construction project involves defining the activities that will be involved in the project, estimating the resources that will be needed, and determining the durations that will be necessary to carry out the activities that will be defined, followed by defining the interrelationships between the various tasks that will be involved in the project. With the growing complexity of construction projects, project managers need to find a balance between time, cost, quality, and risk at the outset of the project [1]. Research proposes that it may be possible to complete project activities ahead of schedule by allocating additional resources, but this would come at a higher cost. However, rushing the project timeline could compromise the quality and increase the risk associated with the project. Furthermore, shortening the operating time may reduce project quality and increase the project's risk. Since shorter project durations are often linked with greater construction costs, the time-cost trade-off (TCTP) is a problem defined to lower the cost of schedule compression [2]. The Time-Cost-Quality-Risk Trade-off Problems (TCQRTP) could be deemed as a major challenge in project management.

Furthermore, the construction industry is ultimately responsible for environmental issues brought on by building and using facilities. Building material manufacturing emits more carbon dioxide (CO_2) as compared to other industrial output, and construction activities play a substantial role in air pollution and greenhouse gas emissions [3]. To solve TCQRTP, over the last few years, researchers have most frequently used metaheuristic optimization algorithms, such as Atomic orbital search (AOS) [4,5], Grey Wolf Optimizer (GWO) [6], the Border Collie Optimization (BCO) [7], Material Generation Algorithm (MGA) [8,9], Artificial Bee Colony (ABC) [10,11], Stochastic Diffusion Search (SDS) [12], Black Widow Optimization Algorithm (BWO) [13,14], Charged System Search (CSS) [15,16], Fire Hawk Optimizer [17], Glowworm swarm optimization (GSO) [18], Energy Valley Optimizer (EVO) [19], Flower Pollination Algorithm (FPA) [20], and Crystal structure algorithm (CryStAl) [21].

Chassiakos, Samaras [22] represented a TCT model concerning the Critical Path Method (CPM) that can be employed in the discrete cost-time relationship of project activities. Feng, Liu [23] proposed a novel algorithm utilizing the Genetic Algorithm (GA) and Pareto method for TCT. In their other research work, the authors presented the TCT model under uncertainty using GA with a Simulation approach [24]. El-kholy [25] proposed a TCT model considering a linear programming model's budget variability and time uncertainty. Aziz, Hafez [26] presented a novel technique called the Smart Critical Path Method System (SCPMS), combining GA and CPM for resource optimization so as to diminish project time and cost and increase its quality. Considering three primary variables, including crashed costs, crashed durations, and resources number, Ballesteros-Pérez, Elamrousy [27] represented a non-linear model for TCT problems. Chen and Tsai [28] evaluated TCT problems with fuzzy parameters, a feasible approach for intricate networks in projects. Regarding the fact that the fuzzy environment in TCT problems contains just membership functions, with projects' uncertainty and execution time, Abdel-Basset, Ali [29] employed the neutrosophic method to puzzle out TCT problems. Albayrak [30] suggested a new hybrid version of particle swarm optimization (PSO) and a GA algorithm for solving TCT problems. Nevertheless, other different parameters like risk, safety, carbon dioxide emission (CO_2), etc., have been added to the projects' contracts in addition to time and cost. The new and evolving contracts in the construction industry are putting increased pressure on

Handbook of Whale Optimization Algorithm. https://doi.org/10.1016/B978-0-32-395365-8.00030-0

decision-makers to find the most optimal or near-optimal solutions while maximizing quality, reducing construction time, lowering costs, mitigating risk, and minimizing CO_2 emissions [31].

Concerning the time-cost-quality trade-off (TCQT), Babu and Suresh [32] proposed that quality ought to be deemed in the TCT problems, in which they created a linear programming approach for TCQT; Khang and Myint [37] applied it in a cement factory in Thailand, to endorse the proposed concept. El-Rayes and Kandil [31] proposed a 3D TCQT analysis instead of a conventional 2D analysis in order to minimize the cost and time of a highway construction project and maximize its quality. Tareghian and Taheri [33] resolved a TCQT problem by employing an electromagnetic scattering search that could be carried out on vast and huge projects. Furthermore, Kannimuthu, Raphael [34] developed a framework for Relaxed-Restricted Pareto filtering (RR-PARETO3)-based TCQT issues in a multi-state resource-constrained project planning context. Tran, Luong-Duc [35] proposed the Opposition multiple objective symbiotic organisms search (OMOSOS) strategy, a useful technique for dealing with challenges including trade-offs between time, cost, quality, and work continuity. Mohammadipour and Sadjadi [36] deemed risk in the CQTP. The authors properly supplied linear programming to reduce project risks, extra costs, and quality losses as much as possible. Safaei [37] creates a multi-objective mathematical programming model for the sake of TCQRT solved by the Multipurpose Genetic Algorithm (NSGAII). Furthermore, Baghalzadeh Shishehgarkhaneh, Moradinia [38] compared novel and classic metaheuristic algorithms' capabilities in dealing with solving the time–cost–quality–risk trade-off problems in dam construction projects.

When a computer eventually replaced the drawing board, the emergence of 2D CAD in the 1970s caused a revolution in the drawing process by making it possible to copy, electronically distribute, and, in some cases, automate the information [39]. Subsequently, object-oriented computer-aided design (CAD) systems replaced two-dimensional symbols with building elements (objects) capable of encapsulating the behavior of commonplace building components. These architectural features may be seen from numerous perspectives and given non-graphic properties. By incorporating parametric 3D geometry with changeable dimensions and assigned rules, these things gain "intelligence," allowing for the presentation of intricate geometric and functional interactions between architectural sections [40,41]. While van Nederveen and Tolman initially coined the term Building Information Modeling (BIM) in 1992, Eastman pioneered using virtual models in buildings in the 1970s [42]. The way architecture, engineering, and construction projects are typically designed has evolved in the last 20-30 years due to the use of Building Information Modeling (BIM). BIM allows for the effective management of resources throughout the entire process, from the initial planning phase to the final demolition stage. This has improved the efficiency of project planning, execution, and maintenance. Through the early design phase of any project, BIM could be an interesting opportunity for project management [43]. BuildingSMART Alliance developed the definition of BIM and its contribution in the following areas: BIM is a business procedure that enables all team members and stakeholders to produce and use asset data through the lifespan of a building for design, construction, and operation. Facilities' physical and functional aspects are shown digitally in BIM [44–46]. In order to make sure the model is as precise as possible before construction begins, team members are continually revising and updating their sections of the model in the BIM environment based on project requirements and design modifications [47,48].

BIM is a "richer repository" compared to a collection of CAD drawings; some stores multi-disciplinary can create and construct information and the attributes of buildings in a digital and graphical BIM model. BIM allows for the efficient use of information in the architectural model, eliminating the need to constantly recreate it and speeding up the design process. Additionally, it promotes collaboration and information sharing among the project team through the ability to export data. As a result, BIM use is supported by various government initiatives intended to increase the effectiveness of the construction sector [49]. In other words, BIM improves design and construction quality, speeds up project execution, and is more cost-effective from a management standpoint [50]. The Industry Foundation Classes (IFC) data model, which buildingSMART International created and maintained, is fundamental for cost-effectively supporting this interoperability without depending on any product or software-specific file formats [51]. According to the National Building Information Model Standard (NBIMS), BIM is defined as "*the process of creating an electronic representation of a facility for various purposes including visualization, engineering analysis, conflict resolution, code compliance analysis, cost estimation, documenting as-built conditions, budgeting, and more*" [52].

Furthermore, different definitions of BIM have been provided by prominent International Standards Organization (ISO), such as the ISO 29481-1:2016 standard defines BIM as "a method for mapping and describing the information processes involved in the life cycle of construction projects" [53] and ISO 19650-2:2018 as "the process of designing, constructing or operating a building or infrastructure asset using electronic object-oriented information" [54]. The use of BIM in construction offers several key benefits, including improved design quality and lifecycle management, more efficient maintenance, accurate cost estimation, streamlined information, integrated workflow, enhanced collaboration between stakeholders and project teams, and reduced energy consumption [1]. Additionally, for the entry-level user, the BIM-assisted estimate outperformed standard estimation approaches. Generally speaking, based on BS 1192-4:2014 (BSI, 2014a) and PAS 1192-2:2013 (BSI, 2013) standards, BIM maturity level could be divided into four distinct categories as follows [55–57]:

- **Level 0:** This is referred to as unmanaged CAD, which is likely to be two-dimensional, with the information provided through conventional paper drawings or, in several examples, digitally via PDF, which is an independent source of information covering fundamental asset information.
- **Level 1:** This is the level at which many businesses now operate, often accomplished via 3D CAD and 2D drawing for statutory approval documents and production information drafting.
- **Level 2:** This is characterized by collaborative modeling, in which all stakeholders use their 3D models but do not collaborate on a single, shared model. The design data is exchanged using a standard file format such as IFC or COBie.
- **Level 3:** This level embodies complete cooperation across all disciplines via a single, shared project model stored in a centralized repository; this level is referred to as "Open BIM."

In project planning problems, a variety of resources may be employed to execute project tasks. We are faced with a discrete decision-making situation where the resource allocations define the decisions. The number of resources devoted to the activity will determine when it is finished. We presume that each activity's and each project's resource allocation recognize the relationships between time, cost, quality, risk, and carbon dioxide (CO_2) emission. The Whale Optimization Algorithm (WOA) proposed by Mirjalili and Lewis [58] is used in this study. This chapter's major novelty is using a unique metaheuristic to address the time-cost-quality-risk-CO_2 trade-off (TCQRCT) problem in an actual construction project using the Building Information Modeling (BIM) process. For statistical analysis, Kruskal-Wallis (KW), Wilcoxon (W), Mann-Whitney (MW), and Kolmogorov-Smirnov (KS) were utilized. 30 independent optimization runs are used to calculate the mean, the worst, the standard deviation, and the needed number of objective function evaluations. A maximum of 5000 objective function evaluations are used to define a stopping condition. The advantages of the Whale Optimization Algorithm (WOA) include being parameter-free, demonstrating fast convergence behavior, and achieving the lowest possible objective function evaluation.

24.2 Problem statement

Three main components make up the framework as follows:

- The decision variables and initialization module;
- BIM module;
- The metaheuristic optimization algorithm (Whale Optimization Algorithm (WOA)) module.

24.2.1 Decision variables and initialization module

A building project's activity-on-node (AON) diagram consists of M nodes, and arrows represent the connections among the activities. Every task can be completed miscellaneously, and regarding resource number, technology, and equipment used, each has its own cost, time, risk, quality, and CO_2 emissions. By selecting the appropriate course of action for each activity, the TCRQC tradeoff problem optimization technique attempts to improve project quality while reducing project time, cost, risk, and carbon dioxide emissions. As a result, the first objective function in Eq. (24.1) is to shorten the project's duration, which is represented by T_p:

$$Minimum\ T_p = \min(\max(ST_i + D_i)) = \min(\max(FT_i)); \quad i = 1, \ldots, M \tag{24.1}$$

where D_i represents each activity's length; ST_i and FT_i show activities' start and finish times, respectively; M is project scheduling's overall number of nodes [59].

In addition, the overall cost of a project includes indirect costs (IC), direct costs (DC), and delay costs (DC). The cost of the materials, equipment, and resources needed to carry out tasks is included in the direct costs. The costs applied throughout the project, such as management, administration, and insurance, are referred to as indirect costs and are fixed [60]. Other methods exist for assessing the total project's cost differently; however, this research only considers direct, indirect, and delay costs. Eq. (24.2) illustrates the objective function that reduces the overall cost of the project as follows:

$$\text{Minimum}\ TC_p = D_{C_i}^j + I_{C_i}^j + DC \tag{24.2}$$

$$D_{C_i}^j = \sum_{i=1}^{n} C_i^j \tag{24.3}$$

$$I_{C_i}^j = C_{ic} \times T \tag{24.4}$$

$$DC = \begin{cases} C_1 (T_0 - T) & \text{if } T \leq T_0 \\ \left(e^{\frac{T-T_0}{T_0}} - 1\right)\left(D_{C_i}^j + I_{C_i}^j\right) & \text{if } T > T_0 \end{cases} \tag{24.5}$$

where TC_p shows the overall project's cost; $D_{C_i}^j$ and $I_{C_i}^j$ represent the direct and indirect costs in conjunction with the *j-th* mode of the *i-th* activity, respectively. C_i^j elucidates the cost in conjunction with *j-th* mode of *i-th* activity; DC shows the delay cost; C_{ic} is the indirect cost/time; T_0 is the contractual project's planned time; C_1 is the prize for early completion of the project, and T is the project's total duration [61,62].

The quality of the whole project is the total quality of all the individual activities since project resources may comprise various equipment, materials, and labor. The quality will increase as the activities are extended, but going beyond a certain point will result in a decline in quality. Consequently, the quality performance index (*QPIi*), which is determined as follows, serves to indicate the quality of each activity [62]:

$$QPI_i = a_i t_i^2 + b_i t_i + c_i \tag{24.6}$$

where t_i shows the length of *i-th* activity; a_i, b_i, and c_i represent coefficients determined using the quadratic function based on BD (Fig. 24.1). SD, BD, and LD elucidate the shortest, best, and longest duration. Nonetheless, BD is determined by Eq. (24.7). Hence, Eq. (24.8) represents the quality's objective function as follows:

$$BD = SD + 0.613\,(LD - SD) \tag{24.7}$$

$$\max Q = \sum_{i=1}^{M} \frac{QPI_i}{M} \tag{24.8}$$

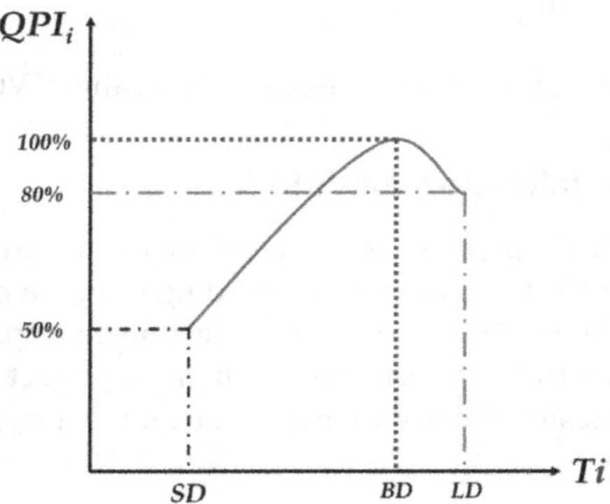

FIGURE 24.1 The quality performance index.

Nevertheless, using certain resources might devastate the environment by generating CO_2 during the project's phase of construction. There are two ways that CO_2 gas may be released during the on-site construction procedure: directly through energy and the burning of fuel and indirectly from the manufacture and transportation of building materials. Therefore, Eq. (24.9) can be used to identify the objective function of reducing the project's whole CO_2 emissions.

$$\min CE = \sum_{i=1}^{M} E_{dij} + \sum_{i=1}^{M} E_{inij} = (\sum_{i=1}^{M} Q_{ed} \times F_e + Q_{dd} \times F_d) + (\sum_{i=1}^{M} Q_k \times F_j + Q_{ek} \times F_e + Q_{dk} \times F_d) \tag{24.9}$$

where CE shows the total CO_2 emission in the project; E_{dij} and E_{inij} are project's direct and indirect CO_2 emissions, respectively; Q_{ed} is activity' electricity consumption; Q_{dd} elucidates activity's diesel consumption; Q_{ij} shows consumption of material k in activity; Q_{ek} is the amount of electricity used to transport the material k needed for the operation;

Q_{dk} displays the amount of diesel used to move the material k for the activity; the variables F_j, F_e, and F_d represent the per-unit production of material k, the carbon emission factor per unit of electricity, and the diesel consumption per unit, respectively.

The project's conditions, delivery methods, and contract terms all significantly determine the real project risk [63–65]. Consequently, Eq. (24.10) can be used to represent the objective function for risk:

$$\min R = w_1 \times \left(1 - \frac{TF_c + 1}{TF_{max} + 1}\right) + w_2 \times \left(\frac{\sum_{i=1}^{Pd}(R_t - \overline{R})^2}{P_d(\overline{R})^2}\right) + w_3 \times \left(1 - \frac{\overline{R}}{\max(R_t)}\right) \tag{24.10}$$

where the project's overall current float and flexible scheduling float are shown by TF_c and TF_{max}, respectively; $\overline{R}$ clarifies the level of uniform resources; R_t is the resource needed on day t; and w_i is the weights.

Eq. (24.11) is utilized to evaluate the capabilities of the WOA to concurrently optimize the time-cost-quality-risk-CO_2 (All) trade-off using the normalizing procedure:

$$F(x) = \frac{T - T_{min}}{T_{max} - T_{min}} + \frac{C - C_{min}}{C_{max} - C_{min}} + \frac{R - R_{min}}{R_{max} - R_{min}} + \frac{CO_2 - CO_{2(\min)}}{CO_{2(\max)} - CO_{2(\min)}} + \frac{Q_{min} - Q}{Q_{max} - Q_{min}} \tag{24.11}$$

24.2.2 BIM module

The case study used to test the model is a five-story residential dwelling with a basement which has a total floor area of 930 m^2. The case study is also used to verify the suggested algorithm concerning the following five factors: cost, time, risk, quality, and CO_2 emissions. Table 24.1 shows how the BIM procedure, project information, and expert judgments throughout the planning and designing stages elicit information about all 38 activities of the project. All actions follow a Finished to Start (FS) pattern, meaning they end before they begin. Architectural, Structural, and MEP (Mechanical, Electrical, and Piping) modeling were carried out using Autodesk Revit 2022; all components were modeled at LOD 350 according to BIMFourm 2019 standard. After that, a parametric model was made in Revit with the help of Dynamo visual programming. The next research step was using the Navisworks software for soft and hard clash detection. In the meanwhile, Lumion was used to create photorealistic images of the finished product. MATLAB® is then utilized for programming. Fig. 24.2 elucidates the modeling procedure based on BIM.

As seen in Table 24.1, the BIM process, the project data, and the experts' judgments are used in the designing and planning processes to extract all of the activity information. In other words, the information in this table was compiled using the experiences of a wide variety of highly accomplished individuals and specialists in the relevant sector. The amount of time and cost required for mode 1 is the first suggestions made by the contractor, and a majority of contractors proposed the lowest amount at the initial stage in order to win the auction. However, given the majority of contractors do not take into account rework, conflicts, hard or soft clashes, payment delays employers, and harsh weather conditions. Mode 3 is the results produced from the BIM procedure. Modes 2 and 4 were taken into consideration based on the recommendations made by specialists and experts working on this project. Mode 5 is the project's actual time and cost, which were derived from the final state of the construction. Then, a random risk percentage and carbon dioxide emissions for each activity are demonstrated regarding the opinions of top academicians and industry authorities.

24.2.3 Whale Optimization Algorithm (WOA)

The humpback is one of the most well-known and recognized huge whales (Fig. 24.3). Humpbacks belong to the family Balaenopteridae (rorquals) of baleen whales. As well as its occasional propensity to approach vessels, this species is well-known for its frequent acrobatic activity. Much has been learned about the biology and behavior of humpback whales in the past 40 years because of the individual identification of thousands of whales throughout the globe based on their natural markings [66,67].

Bubble-net feeding behavior is a term used to describe how they forage for food. As can be seen in Fig. 24.4, this foraging is carried out by releasing many bubbles in a circular or "9"-shaped path. For the purpose of optimization, WOA is based on a formulation based on the upward-spirals approach.

The behavior of whales in hunting, where they circle their target and adjust their position to determine the best strategy, can be modeled mathematically. The prey can be considered as the optimal solution, and the whale's behavior can be

TABLE 24.1 Project Data.

NO	Activity	Logical	Mode 1					Mode 2					Mode 3					Mode 4					Mode 5				
			Time	Cost $	Quality %	Risk	CO_2	Time	Cost $	Quality %	Risk	CO_2	Time	Cost $	Quality %	Risk	CO_2	Time	Cost $	Quality %	Risk	CO_2	Time	Cost $	Quality %	Risk	CO_2
1	Foundation	-	32	10125	93.3695	14.21833	281.6642	28	11140.1	83.09886	12.22777	222.4114	25	12352.5	95.23689	11.23248	253.8452	20	13162.5	73.76191	15.92453	194.3483	15	13930.89	89.63472	16.91982	266.5375
2	Retaining wall	1FS+1	19	2815	97.75215	12.15933	80.02299	15	3097.221	86.99941	10.45703	63.18881	13	3434.3	99.70719	9.605873	72.11941	11	3659.5	77.2242	13.61845	55.21586	10	3873.13	93.84206	14.46961	75.72538
3	Columns of ground	2FS	16	2518.75	94.8012	8.783333	133.1788	14	2771.271	84.37307	7.553667	105.1624	10	3072.875	96.69722	6.938833	120.0252	8	3274.375	74.89295	9.837333	91.89335	6	3465.523	91.00915	10.45217	126.0264
4	Beam and roof of ground	3FS+1	12	5406.25	96.579	14.93958	217.2037	10	5948.261	85.95531	12.84804	171.5112	8	6595.625	98.51058	11.80227	195.7512	7	7028.125	76.29741	16.73233	149.8705	5	7438.405	92.71584	17.7781	205.5388
5	Columns of 1st floor	4FS+2	16	1937.5	95.4771	4.40541	148.8838	12	2131.747	84.97462	3.788653	117.5636	10	2363.75	97.38664	3.480274	134.1791	8	2518.75	75.42691	4.934059	102.7298	6	2665.787	91.65802	5.242438	140.888
6	Beam and roof of 1st floor	5FS+1	12	4500	97.5375	8.333	294.1643	10	4951.153	86.80838	7.16638	232.2819	8	5490	99.48825	6.58307	265.1108	7	5850	77.05463	9.33296	202.9734	5	6191.505	93.636	9.91627	278.3663
7	Columns of 2nd floor	6FS+2	15	1937.5	95.7216	4.565584	175.9502	12	2131.747	85.19222	3.926402	138.9361	10	2363.75	97.63603	3.606811	158.5722	8	2518.75	75.62006	5.113454	121.4056	6	2665.787	91.89274	5.433045	166.5008
8	Beam and roof of 2nd floor	7FS+1	12	4500	99.5265	9.4605	204.4505	10	4951.153	88.57859	8.13603	161.4409	8	5490	93.55491	7.473795	184.2576	7	5850	78.62594	10.59576	141.0708	5	6191.505	95.54544	11.258	193.4705
9	Columns of 3rd floor	8FS+2	15	1937.5	95.8698	5.239164	198.253	12	2131.747	85.32412	4.505681	156.5472	10	2363.75	97.7872	4.13894	178.6723	8	2518.75	75.73714	5.867864	136.7946	6	2665.787	92.03501	6.234605	187.6059
10	Beam and roof of 3rd floor	9FS+1	12	4500	90.56025	4.427083	185.8913	10	4951.153	80.59862	3.807292	146.786	8	5490	92.37146	3.497396	167.5315	7	5850	71.5426	4.958333	128.265	5	6191.505	86.93784	5.268229	175.9081
11	Columns of 4th floor	10FS+2	15	1937.5	90.9685	8.366991	162.6227	12	2131.747	80.96197	7.195612	128.4123	10	2363.75	92.78787	6.609923	146.561	8	2518.75	71.86512	9.37103	112.2096	6	2665.787	87.32976	9.956719	153.8891
12	Beam and roof of 4th floor	11FS+1	12	4500	91.55625	10.08058	122.2504	10	4951.153	81.48506	8.669299	96.53301	8	5490	93.38738	7.963659	110.1762	7	5850	72.32944	11.29025	84.3528	5	6191.505	87.894	11.99589	115.685
13	Columns of 5th floor	12FS+2	15	1937.5	92.45555	13.32009	139.5912	12	2131.747	82.28544	11.45528	110.2258	10	2363.75	94.30466	10.52287	125.8043	8	2518.75	73.03988	14.9185	96.31791	6	2665.787	88.75733	15.85091	132.0945
14	Beam and roof of 5th floor	13FS+1	12	4500	94.6272	9.354555	95.98636	10	4951.153	84.21821	8.044917	75.79402	8	5490	96.51974	7.390098	86.50614	7	5850	74.75549	10.4771	66.23059	5	6191.505	90.84211	11.13192	90.83144
15	Columns of ridge roof	14FS+1	8	546	93.6513	4.99297	91.9894	6	600.7399	83.34966	4.293954	72.63789	5	666.12	95.52433	3.944446	82.90394	3	709.8	73.98453	5.592126	63.47269	2	751.2359	89.90525	5.941634	87.04914
16	Beam and roof of ridge floor	15FS+1	7	1387.5	96.6784	5.313047	222.5938	6	1526.606	86.04378	4.569221	175.7675	4	1692.75	98.61197	4.197307	200.609	3	1803.75	76.37594	5.950613	153.5897	2	1909.047	92.81126	6.322526	210.6395
17	Brickworks of ground	4FS+1	19	2025	91.21395	3.913667	207.7781	16	2228.019	81.18042	3.365753	164.0685	13	2470.5	93.03823	3.091797	187.2566	11	2632.5	72.05902	4.383307	143.3669	9	2786.177	87.56539	4.657263	196.6194
18	Mechanical installations of ground	17FS+2	13	1625	93.4479	3.797944	189.8038	11	1787.916	83.16863	3.266232	149.8754	9	1982.5	95.31686	3.000376	171.0576	7	2112.5	73.82384	4.253697	130.9646	5	2235.821	89.70998	4.519554	179.6105
19	Electrical installations of ground	17FS+2	18	1562.5	93.1515	11.248	157.7796	16	1719.15	82.90484	9.67328	124.5881	13	1906.25	95.01453	8.88592	142.1963	10	2031.25	73.58969	12.59776	108.868	6	2149.828	89.42544	13.38512	149.3061
20	Brickworks of 1st floor	6FS+1	17	2250	90.3658	7.512706	182.6314	15	2475.577	80.42556	6.460927	144.2118	13	2745	92.17312	5.935037	164.5935	11	2925	71.38898	8.41423	126.0156	9	3095.753	86.75117	8.94012	172.8232
21	Mechanical installations of 1st floor	20FS+2	13	2000	94.50173	9.886561	120.4436	11	2200.513	84.10654	8.502443	95.1063	9	2440	96.39176	7.810383	108.5479	7	2600	74.65636	11.07295	83.10611	5	2751.78	90.72166	11.76501	113.9753

continued on next page

TABLE 24.1 *(continued)*

NO	Activity	Logical	Mode 1					Mode 2					Mode 3					Mode 4					Mode 5				
			Time	Cost $	Quality %	Risk	CO_2	Time	Cost $	Quality %	Risk	CO_2	Time	Cost $	Quality %	Risk	CO_2	Time	Cost $	Quality %	Risk	CO_2	Time	Cost $	Quality %	Risk	CO_2
22	Electrical installations of 1st floor	20FS+2	11	1775	92.7675	11.34605	149.8917	10	1952.955	82.56308	9.757604	118.3595	8	2165.5	94.62285	8.96338	135.0875	6	2307.5	73.28633	12.70758	103.4253	4	2442.205	89.0568	13.5018	141.8418
23	Brickworks of 2nd floor	8FS+1	17	2250	90.69015	8.055757	216.5987	15	2475.577	80.71423	6.927951	171.0335	13	2745	92.50395	6.364048	195.206	11	2925	71.64522	9.022448	149.4531	10	3095.753	87.06254	9.586351	204.9663
24	Mechanical installations of 2nd floor	23FS+2	13	2100	96.8286	9.520074	108.9321	11	2310.538	86.17745	8.187264	86.01643	9	2562	98.76517	7.520858	98.1733	7	2730	76.49459	10.66248	75.16317	5	2889.369	92.95546	11.32889	103.082
25	Electrical installations of 2nd floor	23FS+2	11	1775	93.4755	9.204534	203.5518	10	1952.955	83.1932	7.915899	160.7313	8	2165.5	95.34501	7.271582	183.4477	6	2307.5	73.84565	10.30908	140.4507	4	2442.205	89.73648	10.9534	192.6201
26	Brickworks of 3rd floor	10FS+1	17	2250	93.2184	7.96893	157.2684	15	2475.577	82.96438	6.85328	124.1843	13	2745	95.08277	6.295455	141.7356	11	2925	73.64254	8.925202	108.5152	10	3095.753	89.48966	9.483027	148.8223
27	Mechanical installations of 3rd floor	26FS+2	13	2100	93.6564	2.93765	129.3102	11	2310.538	83.3542	2.526379	102.1077	9	2562	95.52953	2.320744	116.5387	7	2730	73.98856	3.290168	89.22406	5	2889.369	89.91014	3.495804	122.3657
28	Electrical installations of 3rd floor	26FS+2	11	1775	94.95675	11.1316	187.4805	10	1952.955	84.51151	9.573176	148.0408	8	2165.5	96.85589	8.793964	168.9637	6	2307.5	75.01583	12.46739	129.3615	4	2442.205	91.15848	13.2466	177.4119
29	Brickworks of 4th floor	12FS+1	17	2250	94.2319	11.01147	146.2233	15	2475.577	83.86639	9.469861	115.4628	13	2745	96.11654	8.699059	131.7814	11	2925	74.4432	12.33284	100.8941	10	3095.753	90.46262	13.10365	138.3704
30	Mechanical installations of 4th floor	29FS+2	13	2118.75	96.17625	11.2336	183.5407	11	2331.168	85.59686	9.660896	144.9299	9	2584.875	98.09978	8.874544	165.4131	7	2754.375	75.97924	12.58163	126.6431	5	2915.167	92.3292	13.36798	173.6838
31	Electrical installations of 4th floor	29FS+2	11	1775	91.7911	7.247047	127.1093	10	1952.955	81.69408	6.23246	100.3698	8	2165.5	93.62692	5.725167	114.5552	6	2307.5	72.51497	8.116692	87.70544	4	2442.205	88.11946	8.623986	120.283
32	Brickworks of 5th floor	14FS+1	17	2250	91.1596	8.528233	175.2401	15	2475.577	81.13204	7.33428	138.3754	13	2745	92.98279	6.737304	157.9323	11	2925	72.01608	9.551621	120.9157	10	3095.753	87.51322	10.1486	165.8289
33	Mechanical installations of 5th floor	32FS+2	13	2100	96.8406	6.975386	99.42853	11	2310.538	86.18813	5.998832	78.51208	9	2562	98.77741	5.510555	89.60834	7	2730	76.50407	7.812432	68.60569	5	2889.369	92.96698	8.30071	94.08876
34	Electrical installations of 5th floor	32FS+2	11	1775	99.81825	6.231033	202.6635	10	1952.955	88.83824	5.358689	160.0298	8	2165.5	94.82734	4.922516	182.6472	6	2307.5	78.85642	6.978757	139.8378	4	2442.205	95.82552	7.41493	191.7795
35	Rooftop	34FS	19	1168.75	96.4286	16.84735	211.2417	16	1285.925	85.82145	14.48872	166.8035	13	1425.875	98.35717	13.30941	190.3781	11	1519.375	76.17859	18.86903	145.7568	10	1608.071	92.57146	20.04835	199.897
36	Elevator	34FS+2	22	3000	96.2533	5.394279	128.5282	20	3300.769	85.66544	4.63908	101.4902	17	3660	98.17837	4.26148	115.834	15	3900	76.04011	6.041593	88.68447	10	4127.67	92.40317	6.419192	121.6257
37	Facade	34FS+5	73	7448	95.238	4.742953	221.6274	64	8194.709	84.76182	4.07894	175.0044	53	9086.56	97.14276	3.746933	199.7381	43	9682.4	75.23802	5.312108	152.9229	36	10247.63	91.42848	5.644114	209.725
38	Outdoors	35FS+1	54	3291.2	93.77538	11.02793	195.7652	48	3621.164	83.46009	9.484023	154.5827	39	4015.264	95.65089	8.712067	176.4302	27	4278.56	74.08255	12.35129	135.078	20	4528.329	90.02437	13.12324	185.2517

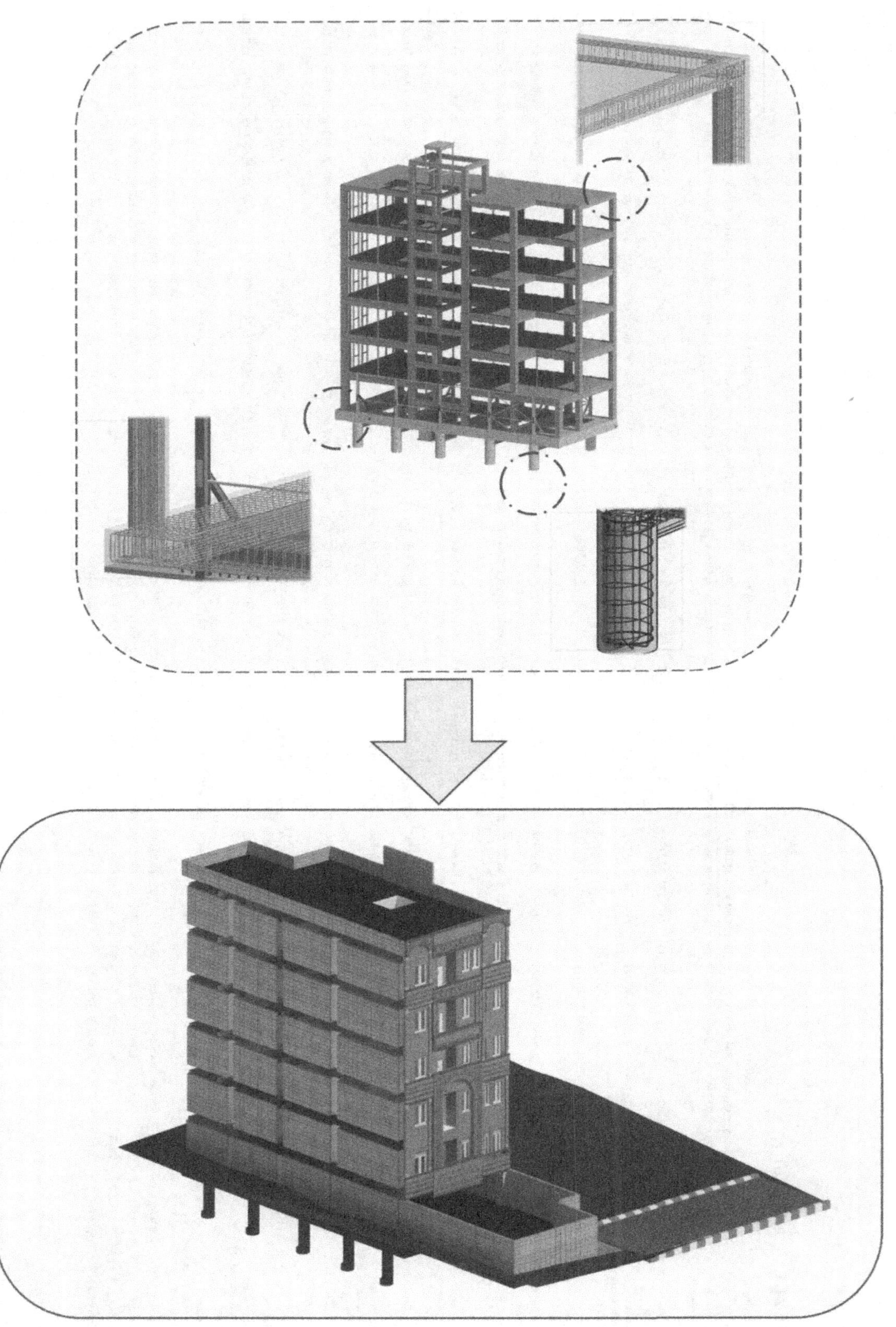

FIGURE 24.2 BIM-based modeling of the case study project.

FIGURE 24.3 Humpback whale.

FIGURE 24.4 Humpback whales are known to produce spiral-shaped bubbles.

represented by mathematical equations as follows:

$$\vec{D} = \left|\vec{C}.\vec{X}^*(t) - \vec{X}(t)\right| \tag{24.12}$$

$$\vec{X}(t+1) = \vec{X}^*(t) - \vec{A}.\vec{D} \tag{24.13}$$

$$\vec{A} = 2\vec{a}.\vec{r} - \vec{a} \tag{24.14}$$

$$\vec{C} = 2\vec{r} \tag{24.15}$$

where t is the current iteration, $\vec{X}^*$ shows the position vector of the best solution, $\vec{X}$ elucidates the position's vector, $\vec{D}$ shows the current iteration's position vector, | | is the absolute value, and . elucidates the element-by-element multiplication. For a better solution appears, $\vec{X}^*$ ought to be updated in each iteration. Here, vectors $\vec{A}$ and $\vec{C}$ show coefficient vectors, $\vec{a}$ elucidates a variable that decreases linearly from 2 to 0 as iterations progress, and $\vec{r}$ is a random vector with values in the range of [0, 1].

Humpback whales' bubble-net behavior is assumed to be controlled by spirals that update their positions and a mechanism for decreasing and circling. By lowering the value of a within the range of $\vec{A}$ in Eq. (24.14), the mechanism for contracting and encircling is provided. The reduction of $\vec{A}$ shows that the novel positions of search agents might be obtained anywhere between the best agent at the moment and the search agent's starting position. Fig. 24.5.a displays the different locations created by $0 \leq A \leq 1$ from (X, Y) to (X^*, Y^*). For the Spiral updating position, the distance between the whales at (X, Y) and the prey at (X^*, Y^*) is calculated (Fig. 24.5.b). An equation in spiral form is developed between

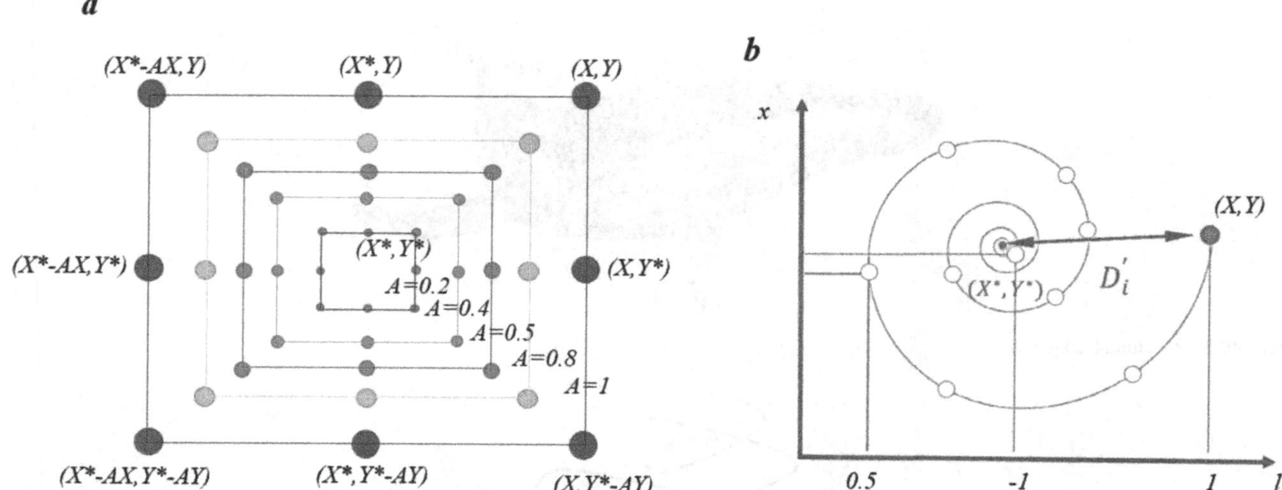

FIGURE 24.5 Bubble-net search mechanism in WOA: (a) shrinking encircling mechanism, and (b) spiral updating position.

the position of the prey and the whale in order to mimic the helix-like movement of humpback whales.

$$\vec{D}' = \left|\vec{X}^*(t) - \vec{X}(t)\right| \tag{24.16}$$

$$\vec{X}(t+1) = \vec{D}'.e^{bl}.(\cos 2\pi l) + \vec{X}^*(t) \tag{24.17}$$

where $\vec{D}'$ shows the current iteration's position vector, the constant value b is used to model the logarithmic spiral shape, and l is a random vector with uniform distribution in the range of $[-1, 1]$.

There is a 50% possibility that bubble-net searches will be employed to update the position of whales during optimization when humpback whales swim around their prey. The aforementioned mathematical model technique is shown in the following equations, where p stands for the probability for each looping route.

$$\vec{X}(t+) = \begin{cases} \vec{X}^*(t) - \vec{A}.\vec{D} & \text{if } p < 0.5 \\ \vec{D}'.e^{bl}.(\cos 2\pi l) + \vec{X}^*(t) & \text{if } p \geq 0.5 \end{cases} \tag{24.18}$$

To execute a global search, a position update for search agents is evaluated regarding randomly selected search agents, as opposed to the best search agent discovered to date. This method is used for considering random values greater than one for $\vec{A}$. The following equations characterize the mathematical model for the given procedure [58]:

$$\vec{D} = \left|\vec{C}.\vec{X}_{rand} - \vec{X}(t)\right| \tag{24.19}$$

$$\vec{X}(t+) = \vec{X}_{rand} - \vec{A}.\vec{D} \tag{24.20}$$

The variable $\vec{X}_{rand}$ represents a randomly selected whale or a random position vector from the current population. The pseudo-code for the Whale Optimization Algorithm (WOA) is depicted in Fig. 24.6.

24.3 Optimization results and discussion

To examine the effectiveness of the WOA algorithm in resolving resource trade-off problems in building projects, five alternative metaheuristic algorithms were used, including the Bees Algorithm (BA) [68], Biogeography-Based Optimization (BBO) [69], Cultural Algorithm (CA) [70], Charged System Search (CSS) [71], and Sine Cosine Algorithm (SCA) [72]. All optimization procedures were performed using MATLAB on a PC with 8 GB of RAM, a CORE i7 processor, and a 2.8 GHz frequency. Table 24.2 lists potential algorithms for each case and the best results of the WOA. However, 30 separate optimization runs are performed for the main statistical goals to calculate the mean, worst, standard deviation, and

```
Initialize the whales population Xi (i = 1, 2, ..., n)
Calculate the fitness of each search agent
X*=the best search agent
while (t < maximum number of iterations)
    for each search agent
    Update a, A, C, l, and p
        if1 (p<0.5)
            if2 (|A| < 1)
                Update the position of the current search agent by the Eq. (2.1)
            else if2 (|A| ≥ 1)
                Select a random search agent (Xrand)
                Update the position of the current search agent by the Eq. (2.8)
            end if2
        else if1 (p≥0.5)
                Update the position of the current search by the Eq. (2.5)
        end if1
    end for
    Check if any search agent goes beyond the search space and amend it
    Calculate the fitness of each search agent
    Update X* if there is a better solution
    t=t+1
end while
return X*
```

FIGURE 24.6 Pseudo-code of the WOA algorithm [58].

TABLE 24.2 The best findings of the WOA and other methods in the project.

	BA	BBO	CA	CSS	SCA	WOA
Time	423.00	362.00	391.00	351.00	424.00	342.00
Cost	213987.00	183185.51	192424.46	178197.08	209476.49	177082.12
Quality	93.23	95.53	95.43	94.06	94.65	94.59
Risk	7.49	6.77	6.84	6.76	7.60	6.79
CO_2	131.59	120.08	123.30	123.03	132.82	117.86
All	0.95	0.83	0.85	0.78	0.95	0.78

computation time. The maximum number of objective function evaluations and iterations determine the number of populations for each technique, with a stopping condition also set based on a maximum of 5000 evaluations. The convergence history of the WOA and other algorithms in addressing the trade-off problems is shown in Fig. 24.7.

Table 24.3 displays the statistical data obtained from the optimization performed on the case study. The WOA algorithm has the potential to outperform the vast majority of other metaheuristics in the first scenario of time optimization of the case study. This scenario estimates that 342 days is the best and most optimal time, so the WOA is likely to outperform other algorithms. The WOA algorithm produces the result with the smallest standard deviation (Std), followed by the BA method, which accounts for 8.01. The WOA algorithm is the one with the minimum outcome. In contrast, the SCA algorithm generates the highest value of Std, around 13.59. In addition, the WOA algorithm completed the time optimization process in the shortest possible time (0.15 seconds). On the other hand, the BA algorithm procured the longest computing time, necessitating a significantly longer amount of time to complete the optimization procedure in the mentioned scenario.

In the following scenario, which focuses on cost optimization, the WOA algorithm performs superior to other alternative metaheuristic algorithms. To put it another way, unlike the BA method, the WOA algorithm can identify the project's least cost, which determines the maximum optimal value of cost. Nevertheless, the BA method required the most time spent on calculation in this scenario, followed by the CSS. On the other hand, the WOA and SCA approach required the least amount of time spent on computation for the project's cost optimization described above. In addition, the WOA provided the minimum possible value for the standard deviation, which the CSS followed. In the meanwhile, the SCA came out on top with the highest standard deviation of all the algorithms that were investigated for this scenario. As a consequence of this, the WOA algorithm could be an appropriate metaheuristic for the optimization of costs associated with project and construction management.

The statistical results of the quality optimization performed on the case study demonstrate that the WOA approach can deliver acceptable quality. The CA came in second place behind the BBO algorithm regarding its remarkable quality value,

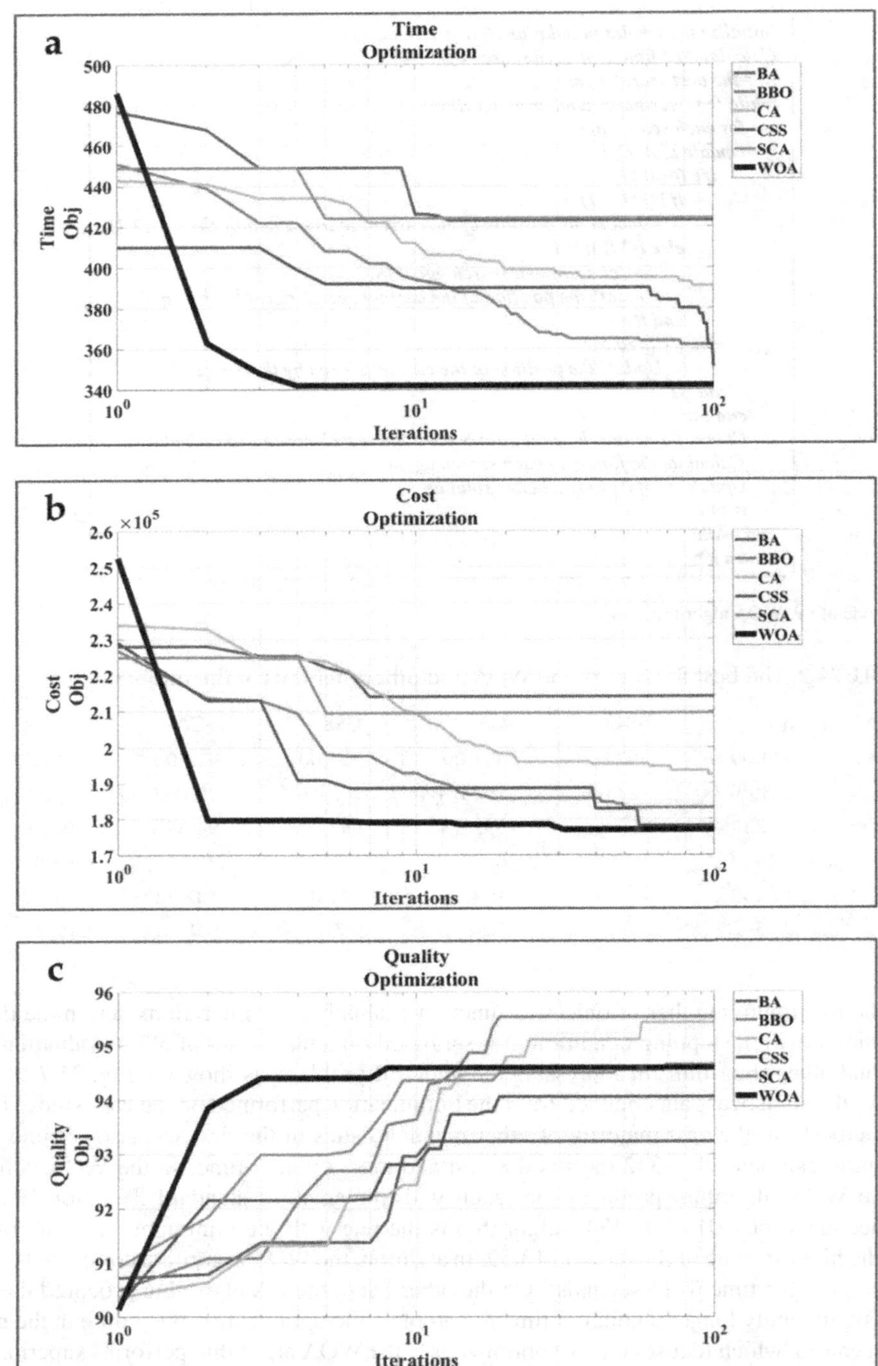

FIGURE 24.7 Convergence history of 30 independent optimization runs of WOA and other methods.

around 95.53. In addition, the WOA has the potential to deliver the least standard deviation, which in this instance is 0.04. In stark contrast, the CA has the highest Std. Compared to the BA, which needed around 1.00 seconds of computing time, the SCA method required much less time to achieve the same level of quality optimization as the latter. Consequently, even though the WOA algorithm can provide an acceptable quality, the BBO may be the more desirable option for project managers to go within this particular scenario. Despite this, CSS determined the lowest risk value within the case study scope, registered at almost 6.76. In addition, the SCA method demanded the least amount of computational time feasible

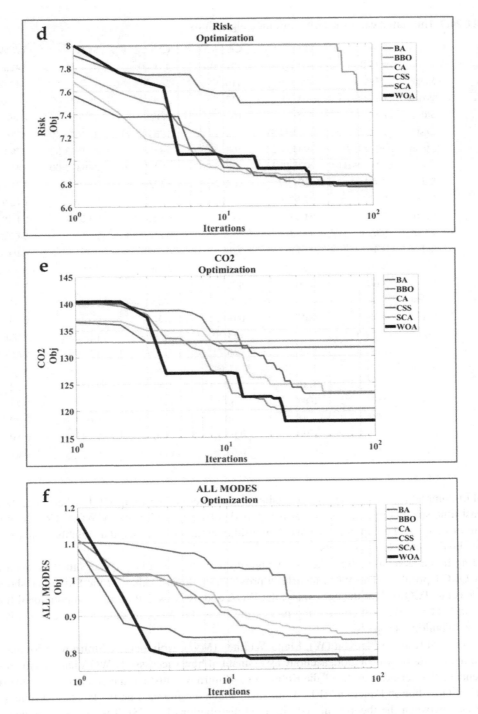

FIGURE 24.7 *(continued)*

in this case, followed by the WOA. As a result, the CSS and WOA can be excellent candidates for risk optimization in the context of project scheduling. During this time, the CSS and BBO are able to determine the least possible Std in the given scenario.

Regarding sustainable construction, the WOA could be feasible to reduce projects' carbon footprint. This is so that environmentally friendly construction could be done, as the WOA determined the least amount of CO_2 in the case study. On the other hand, the SCA gave the highest result for CO_2, which indicates that its performance was unfavorable in

TABLE 24.3 The statistical results of mentioned algorithms.

		BA	BBO	CA	CSS	SCA	WOA
Time	Best	423.00	362.00	391.00	351.00	424.00	**342.00**
	Mean	438.73	384.80	410.90	368.47	456.43	342.00
	Worst	451.00	402.00	437.00	389.00	484.00	342.00
	Std	8.01	9.70	10.47	9.32	13.59	**0.00**
Cost	Best	213987.00	183185.51	192424.46	178197.08	209476.49	**177082.12**
	Mean	222289.08	194121.23	208676.14	183183.79	232563.82	178351.29
	Worst	228275.31	207526.36	228854.67	192105.36	249372.00	179262.94
	Std	4073.16	5757.17	8224.95	3392.57	9355.17	**581.88**
Quality	Best	93.23	**95.53**	95.43	94.06	94.65	94.59
	Mean	92.21	94.52	94.18	93.25	94.58	94.48
	Worst	93.23	95.53	95.43	94.06	94.65	94.59
	Std	0.41	0.51	0.54	0.51	**0.04**	**0.04**
Risk	Best	7.49	6.77	6.84	**6.76**	7.60	6.79
	Mean	7.66	6.90	6.97	6.88	7.78	7.20
	Worst	7.82	7.05	7.13	7.02	8.02	7.58
	Std	0.08	**0.07**	0.08	**0.07**	0.13	0.23
CO_2	Best	131.59	120.08	123.30	123.03	132.82	**117.86**
	Mean	134.73	122.15	126.72	126.53	139.05	124.43
	Worst	137.07	125.19	131.64	129.60	141.92	132.87
	Std	1.57	**1.36**	2.03	1.45	1.92	4.72
All	Best	0.95	0.83	0.85	**0.78**	0.95	**0.78**
	Mean	0.99	0.86	0.90	0.82	1.06	0.79
	Worst	1.03	0.90	0.94	0.86	1.14	0.80
	Std	0.02	0.02	0.03	0.02	0.04	**0.01**

attaining the goal of completing the project with a smaller and smaller carbon footprint. Furthermore, the SCA technique resulted in the least time spent computing, which was reported at 0.15 (s), followed by WOA. Consequently, considering the typical amount of time spent computing, the WOA algorithm could be considered a suitable option for optimizing the quantity of carbon dioxide in building projects.

Using a dwelling home as a case study, the WOA algorithm can perform better than other metaheuristic algorithms in tackling the TCQRCT problem. The WOA algorithm provided the lowest value for the Std value, showing their better performance. To execute TCQRCT in the case study, the WOA and SCA used the least computational time, followed by the BBO with about 0.44 (s). Without considering processing time, the WOA method could be the only one that solves TCQRCT problems in building projects.

Four famous statistical tests—Wilcoxon (W), Mann-Whitney (MW), Kolmogorov-Smirnov (KS), and Kruskal-Wallis (KW)—were used in order to provide a more accurate assessment of how successfully WOA handled the resource trade-off based on BIM. Henceforth referred to as "KS," the Kolmogorov-Smirnov statistic is a member of the elite group of statistics known as "Empirical Distribution Function" (EDF) statistics, which are founded on the largest possible absolute value of the vertical difference between the theoretical and observed distributions [73–75]. The W test evaluates the comparison of the average rankings by examining pairs of rankings. The null hypothesis for this test is that there is no difference in the mean rankings of two variables selected randomly from two datasets. The dataset with the lower mean ranking is considered to have superior statistical behavior as per this metric [76]. The null hypothesis in the MW test indicates the difference between two randomly chosen variables from distinct datasets by considering the sum of the variables' rankings. This means that the statistical dataset with a lower rank summation has superior statistical behavior [77]. Furthermore, the KW test is a popular statistical test that examines the overall rankings of multiple variables across different datasets. It compares the mean of the rankings across multiple datasets, whereas the MW and W tests, which are based on the sum and mean of ranks, are performed between two datasets only [78,79].

TABLE 24.4 The p-values of different statistical tests.

Main Algorithm	*Statistical Test*	*Alternative Metaheuristic Algorithms*				
		BA	BBO	CA	CSS	SCA
WOA	*KS Test*	0.9996	0.9996	0.9996	0.9996	0.9996
	MW Test	0.69	1.00	0.84	0.84	0.84
	W Test	0.03	0.22	0.16	0.09	0.06

TABLE 24.5 The maximum difference of metaheuristics.

Main Algorithm	*Statistical Test*	*Alternative Metaheuristic Algorithms*				
		BA	BBO	CA	CSS	SCA
WOA	*MW Test*	30.00	28.00	29.00	29.00	29.00
		25.00	27.00	26.00	26.00	26.00
	W Test	21.00	17.00	18.00	19.00	20.00
		1.00	4.00	3.00	2.00	1.00

TABLE 24.6 The KW test results (mean of the ranks).

Rankings	*Algorithms*	*Mean of Ranks*
1	WOA	14.00
2	BBO	14.20
3	CSS	14.60
4	CA	15.60
5	SCA	17.20
6	BA	17.40
Chi-sq.	0.73	
Prob>Chi-sq.	0.98	

TABLE 24.7 The statistical results of mentioned algorithms.

	BA	BBO	CA	CSS	SCA	WOA
Time	1.37	0.41	0.77	0.98	0.16	**0.15**
Cost	1.45	0.38	0.76	0.97	**0.15**	**0.15**
Quality	1.36	0.44	0.80	1.00	**0.15**	0.16
Risk	1.20	0.46	0.76	0.97	**0.15**	0.18
CO_2	1.24	0.44	0.79	0.96	**0.15**	0.21
All	1.27	0.44	0.77	0.98	**0.15**	**0.15**

The p-values for the KS, MW, and W statistical tests are included in Table 24.4 for comparison. To put the capabilities of WOA into perspective, Table 24.5 shows the highest difference between several techniques, i.e., KS, MW, and W statistical tests. It is clear that the WOA can manage these problems since the mean (W test) and summation (MW test) of rankings in WOA are often lower than those of the other ways. Table 24.6 presents the overall rankings of the algorithms for dealing with BIM-based resource trade-offs, including the mean of ranks. It is also important to note that WOA now has the top rating, which is acceptable.

Regarding the complexity comparison of all comparative algorithms, the run-time analysis of different approaches is provided in Table 24.7 in which the mean values of run times for the conducted optimization runs for each algorithm in each case are provided. Regarding the fact that a total number of 100 iterations was utilized in all the optimization cases, the complexity of different algorithms may be different from each other; however, the initial number of population in all algorithms are determined in such a manner that all algorithms have equal number of objective function evaluations.

24.4 Conclusion

This chapter used a prominent metaheuristic algorithm and a unique framework, including building information modeling, to handle the resource trade-off problem in construction projects (BIM). As a famous and rigorous metaheuristic method, Whale Optimization Algorithm (WOA) is used in this situation. The case study's 3D BIM-based modeling was produced utilizing various programs, including Revit, Navisworks, Lumion, and Dynamo, for parametric modeling. The following is a summary of this research's major findings:

- Regarding the outcomes of the best optimization runs completed by various approaches for time optimization, the WOA algorithm was able to complete the case study in the least amount of time, 342 days.
- The WOA can give 177082.12 ($) for the case study cost, which is the greatest option among the others.
- The BBO algorithm produced the greatest results, while the WOA is also capable of offering better quality value.
- In comparison to alternative metaheuristics, the CSS can give the project the best outcomes for risk, and WOA gives the least CO_2.
- In comparison to other algorithms, the WOA method can deliver 0.78 concerning the best solutions to the TCQRCTP.

According to the findings and analyses, the WOA method has three key advantages over the other specified metaheuristics algorithms: quick convergence behavior, parameter-freeness, and the lowest possible objective function evaluation. Future research should assess the WOA algorithm's performance in more case studies, such as infrastructure construction projects like dams and tunnels, and compare with other metaheuristics; this research work's limitation is that only a residential building has been deemed as a case study. In addition, the WOA method has to be evaluated for future research involving complex optimization issues in various domains, such as large-scale engineering design problems like truss structures. Future research should aim to propose and develop a multi-objective version of the Whale Optimization Algorithm (MOWOA) to optimize time, cost, quality, risk, and CO_2 simultaneously, as these factors are often interrelated and in conflict with each other.

References

[1] A.H. Daqiqnia, S. Fard Moradinia, M. Baghalzadeh Shishehgarkhaneh, Toward nearly zero energy building designs: a comparative study of various techniques, AUT Journal of Civil Engineering 5 (2) (2021) 339–356, https://doi.org/10.22060/ajce.2021.20458.5771.

[2] M.B. Shishehgarkhaneh, et al., BIM-based resource tradeoff in project scheduling using Fire Hawk Optimizer (FHO), Buildings 12 (9) (2022) 1472.

[3] S. Abbasi, E. Noorzai, The BIM-based multi-optimization approach in order to determine the trade-off between embodied and operation energy focused on renewable energy use, Journal of Cleaner Production 281 (2021) 125359.

[4] M. Azizi, Atomic orbital search: a novel metaheuristic algorithm, Applied Mathematical Modelling 93 (2021) 657–683.

[5] M. Azizi, A.W. Mohamed, M.B. Shishehgarkhaneh, Optimum design of truss structures with atomic orbital search considering discrete design variables, in: Handbook of Nature-Inspired Optimization Algorithms: The State of the Art, Springer, 2022, pp. 189–214.

[6] S. Mirjalili, S.M. Mirjalili, A. Lewis, Grey wolf optimizer, Advances in Engineering Software 69 (2014) 46–61.

[7] T. Dutta, S. Bhattacharyya, S. Dey, J. Platos, Border Collie Optimization, IEEE Access 8 (2020) 109177–109197, https://doi.org/10.1109/ACCESS.2020.2999540.

[8] S. Talatahari, M. Azizi, A.H. Gandomi, Material generation algorithm: a novel metaheuristic algorithm for optimization of engineering problems, Processes 9 (5) (2021) 859.

[9] M. Azizi, M.B. Shishehgarkhaneh, M. Basiri, Optimum design of truss structures by Material Generation Algorithm with discrete variables, Decision Analytics Journal 3 (2022) 100043.

[10] D. Karaboga, An idea based on honey bee swarm for numerical optimization, Technical report-tr06, Erciyes University, Engineering Faculty, 2005, https://abc.erciyes.edu.tr/pub/tr06_2005.pdf.

[11] D. Karaboga, B. Basturk, Artificial Bee Colony (ABC) optimization algorithm for solving constrained optimization problems, in: Foundations of Fuzzy Logic and Soft Computing, Springer Berlin Heidelberg, Berlin, Heidelberg, 2007.

[12] S. Nasuto, J. Bishop, Stabilizing swarm intelligence search via positive feedback resource allocation, in: Nature Inspired Cooperative Strategies for Optimization (NICSO 2007), 2008, pp. 115–123.

[13] V. Hayyolalam, A.A. Pourhaji Kazem, Black Widow Optimization Algorithm: a novel meta-heuristic approach for solving engineering optimization problems, Engineering Applications of Artificial Intelligence 87 (2020) 103249.

[14] N. Jelodari, A. Asghar Pourhaji Kazem, Black widow optimization (BWO) algorithm in cloud brokering systems for connected internet of things, Journal of Computer & Robotics 15 (1) (2022) 33–45.

[15] S. Talatahari, et al., Optimization of large-scale frame structures using fuzzy adaptive quantum inspired charged system search, International Journal of Steel Structures 22 (3) (2022) 686–707.

[16] A. Kaveh, S. Talatahari, Optimal design of skeletal structures via the charged system search algorithm, Structural and Multidisciplinary Optimization 41 (2010) 893–911, https://doi.org/10.1007/s00158-009-0462-5.

[17] M. Azizi, S. Talatahari, A.H. Gandomi, Fire Hawk Optimizer: a novel metaheuristic algorithm, Artificial Intelligence Review 56 (2023) 287–363.

[18] K.N. Krishnanand, D. Ghose, Detection of multiple source locations using a glowworm metaphor with applications to collective robotics, in: Proceedings 2005 IEEE Swarm Intelligence Symposium, 2005. SIS 2005, IEEE, 2005.
[19] M. Azizi, et al., Energy valley optimizer: a novel metaheuristic algorithm for global and engineering optimization, Scientific Reports 13 (1) (2023) 226.
[20] X.-S. Yang, Flower Pollination Algorithm for Global Optimization, Springer Berlin Heidelberg, Berlin, Heidelberg, 2012.
[21] S. Talatahari, et al., Crystal Structure Algorithm (CryStAl): a metaheuristic optimization method, IEEE Access 9 (2021) 71244–71261.
[22] A.-P. Chassiakos, C.-I. Samaras, D.-D. Theodorakopoulos, An integer programming method for CPM time-cost analysis, Computer Modeling in Engineering & Sciences 1 (4) (2000) 9–18.
[23] C.-W. Feng, L. Liu, S. Burns, Using genetic algorithms to solve construction time-cost trade-off problems, Journal of Computing in Civil Engineering 11 (1997).
[24] C.-W. Feng, L. Liu, S.A. Burns, Stochastic construction time-cost trade-off analysis, Journal of Computing in Civil Engineering 14 (2) (2000) 117–126.
[25] A.M. El-kholy, Time–cost tradeoff analysis considering funding variability and time uncertainty, Alexandria Engineering Journal 52 (1) (2013) 113–121.
[26] R. Aziz, S. Hafez, Y. Abuol-Magd, Smart optimization for mega construction projects using artificial intelligence, Alexandria Engineering Journal 53 (2014).
[27] P. Ballesteros-Pérez, K.M. Elamrousy, M.C. González-Cruz, Non-linear time-cost trade-off models of activity crashing: application to construction scheduling and project compression with fast-tracking, Automation in Construction 97 (2019) 229–240.
[28] S.-P. Chen, M.-J. Tsai, Time–cost trade-off analysis of project networks in fuzzy environments, European Journal of Operational Research 212 (2) (2011) 386–397.
[29] M. Abdel-Basset, M. Ali, A. Atef, Uncertainty assessments of linear time-cost tradeoffs using neutrosophic set, Computers & Industrial Engineering 141 (2020) 106286.
[30] G. Albayrak, Novel hybrid method in time–cost trade-off for resource-constrained construction projects, Iranian Journal of Science and Technology, Transactions of Civil Engineering 44 (4) (2020) 1295–1307.
[31] K. El-Rayes, A. Kandil, Time-cost-quality trade-off analysis for highway construction, Journal of Construction Engineering and Management 131 (4) (2005) 477–486.
[32] A.J.G. Babu, N. Suresh, Project management with time, cost, and quality considerations, European Journal of Operational Research 88 (2) (1996) 320–327.
[33] H.R. Tareghian, S.H. Taheri, A solution procedure for the discrete time, cost and quality tradeoff problem using electromagnetic scatter search, Applied Mathematics and Computation 190 (2) (2007) 1136–1145.
[34] M. Kannimuthu, et al., Optimizing time, cost and quality in multi-mode resource-constrained project scheduling, Built Environment Project and Asset Management 9 (1) (2019) 44–63.
[35] D.-H. Tran, et al., Opposition multiple objective symbiotic organisms search (OMOSOS) for time, cost, quality and work continuity tradeoff in repetitive projects, Journal of Computational Design and Engineering 5 (2) (2018) 160–172.
[36] F. Mohammadipour, S.J. Sadjadi, Project cost–quality–risk tradeoff analysis in a time-constrained problem, Computers & Industrial Engineering 95 (2016) 111–121.
[37] M. Safaei, Sustainable survival pyramid model to balance four factors of cost, quality, risk and time limitation in project management under uncertainty, Pakistan Journal of Statistics and Operation Research (2020) 287–294.
[38] M. Baghalzadeh Shishehgarkhaneh, et al., Application of classic and novel metaheuristic algorithms in a BIM-based resource tradeoff in dam projects, Smart Cities 5 (4) (2022) 1441–1464.
[39] D. Cunz, D. Larson, Building information modeling, Under Construct (2006) 1–3, http://www.imageserve.com/naples2013/eunder_construction_12_06.pdf.
[40] I. Howell, B. Batcheler, Building information modeling two years later–huge potential, some success and several limitations, The Laiserin Letter 22 (4) (2005) 3521–3528.
[41] M. Baghalzadeh Shishehgarkhaneh, et al., Internet of Things (IoT), Building Information Modeling (BIM), and Digital Twin (DT) in construction industry: a review, bibliometric, and network analysis, Buildings 12 (10) (2022) 1503.
[42] A. Borrmann, et al., Building Information Modeling: Technologische Grundlagen und industrielle Praxis, Springer-Verlag, 2015.
[43] M.T.H. Khondoker, Automated reinforcement trim waste optimization in RC frame structures using building information modeling and mixed-integer linear programming, Automation in Construction 124 (2021) 103599.
[44] Y. Cao, S.N. Kamaruzzaman, N.M. Aziz, Building Information Modeling (BIM) capabilities in the operation and maintenance phase of green buildings: a systematic review, Buildings 12 (6) (2022).
[45] B. East, D. Smith, The United States national building information modeling standard: the first decade, in: 33rd CIB W78 Information Technology for Construction Conference (CIB W78 2016), 2016.
[46] J.C. Cheng, Q. Lu, A review of the efforts and roles of the public sector for BIM adoption worldwide, Journal of Information Technology in Construction (ITcon) 20 (27) (2015) 442–478.
[47] S. Azhar, Building Information Modeling (BIM): trends, benefits, risks, and challenges for the AEC industry, Leadership and Management in Engineering 11 (3) (2011) 241–252.
[48] M. Baghalzadeh Shishehgarkhaneh, S. Fard Moradinia, The role of Building Information Modeling (BIM) in reducing the number of project dispute resolution sessions, in: The 8th National Conference on Civil Engineering, Architecture and Sustainable Urban Development of Iran, 2020.

[49] R. Charef, H. Alaka, S. Emmitt, Beyond the third dimension of BIM: a systematic review of literature and assessment of professional views, Journal of Building Engineering 19 (2018) 242–257.
[50] R. Charef, et al., Building Information Modelling adoption in the European Union: an overview, Journal of Building Engineering 25 (2019) 100777.
[51] N.O. Nawari, Building Information Modeling: Automated Code Checking and Compliance Processes, CRC Press, 2018.
[52] S. Seyis, Mixed method review for integrating building information modeling and life-cycle assessments, Building and Environment 173 (2020) 106703.
[53] M. Poljanšek, Building information modelling (BIM) standardization, European Commission, 2017.
[54] M.Q. Huang, J. Ninić, Q.B. Zhang, BIM, machine learning and computer vision techniques in underground construction: current status and future perspectives, Tunnelling and Underground Space Technology 108 (2021) 103677.
[55] T. BSI, PAS 1192-2:2013: Specification for information management for the capital/delivery phase of construction projects using building information modelling, BSI, London, UK, 2013.
[56] G. Lea, et al., Identification and analysis of UK and US BIM standards to aid collaboration, https://doi.org/10.2495/BIM150411, 2015.
[57] R. Sacks, et al., BIM Handbook: A Guide to Building Information Modeling for Owners, Designers, Engineers, Contractors, and Facility Managers, John Wiley & Sons, 2018.
[58] S. Mirjalili, A. Lewis, The whale optimization algorithm, Advances in Engineering Software 95 (2016) 51–67.
[59] D.-T. Nguyen, J.-S. Chou, D.-H. Tran, Integrating a novel multiple-objective FBI with BIM to determine tradeoff among resources in project scheduling, Knowledge-Based Systems 235 (2022) 107640.
[60] L.M. Naeni, A. Salehipour, Optimization for project cost management, in: H. Golpîra (Ed.), Application of Mathematics and Optimization in Construction Project Management, Springer International Publishing, Cham, 2021, pp. 79–118.
[61] A. Panwar, K.N. Jha, Integrating quality and safety in construction scheduling time-cost trade-off model, Journal of Construction Engineering and Management 147 (2) (2021) 04020160.
[62] L. Zhang, J. Du, S. Zhang, Solution to the time-cost-quality trade-off problem in construction projects based on immune genetic particle swarm optimization, Journal of Management in Engineering 30 (2) (2014) 163–172.
[63] K.S. Al-Gahtani, Float allocation using the total risk approach, Journal of Construction Engineering and Management 135 (2) (2009) 88–95.
[64] J.M. de la Garza, A. Prateapusanond, N. Ambani, Preallocation of total float in the application of a critical path method based construction contract, Journal of Construction Engineering and Management 133 (11) (2007) 836–845.
[65] L.D. Long, D.-H. Tran, P.T. Nguyen, Hybrid multiple objective evolutionary algorithms for optimising multi-mode time, cost and risk trade-off problem, International Journal of Computer Applications in Technology 60 (3) (2019) 203–214.
[66] P.J. Clapham, Humpback whale: Megaptera novaeangliae, in: B. Würsig, J.G.M. Thewissen, K.M. Kovacs (Eds.), Encyclopedia of Marine Mammals, third edition, Academic Press, 2018, pp. 489–492.
[67] P.J. Palsbøll, et al., Genetic tagging of humpback whales, Nature 388 (6644) (1997) 767–769.
[68] D.T. Pham, et al., The bees algorithm—a novel tool for complex optimisation problems, in: Intelligent Production Machines and Systems, Elsevier, 2006, pp. 454–459.
[69] D. Simon, Biogeography-based optimization, IEEE Transactions on Evolutionary Computation 12 (6) (2008) 702–713.
[70] C.A. Coello Coello, R.L. Becerra, Efficient evolutionary optimization through the use of a cultural algorithm, Engineering Optimization 36 (2) (2004) 219–236.
[71] A. Kaveh, S. Talatahari, A novel heuristic optimization method: charged system search, Acta Mechanica 213 (3) (2010) 267–289.
[72] S. Mirjalili, SCA: a sine cosine algorithm for solving optimization problems, Knowledge-Based Systems 96 (2016) 120–133.
[73] N.M. Razali, Y.B. Wah, Power comparisons of Shapiro-Wilk, Kolmogorov-Smirnov, Lilliefors and Anderson-Darling tests, Journal of Statistical Modeling and Analytics 2 (1) (2011) 21–33.
[74] F.J. Massey Jr, The Kolmogorov-Smirnov test for goodness of fit, Journal of the American Statistical Association 46 (253) (1951) 68–78.
[75] V.W. Berger, Y. Zhou, Kolmogorov–Smirnov test: overview, in: Wiley StatsRef: Statistics Reference Online, 2014, https://doi.org/10.1002/9781118445112.stat06558.
[76] J. Cuzick, A Wilcoxon-type test for trend, Statistics in Medicine 4 (1) (1985) 87–90.
[77] S. Yue, C. Wang, The influence of serial correlation on the Mann–Whitney test for detecting a shift in median, Advances in Water Resources 25 (3) (2002) 325–333.
[78] A.C. Elliott, L.S. Hynan, A SAS macro implementation of a multiple comparison post hoc test for a Kruskal–Wallis analysis, Computer Methods and Programs in Biomedicine 102 (1) (2011) 75–80.
[79] E.F. Acar, L. Sun, A generalized Kruskal–Wallis test incorporating group uncertainty with application to genetic association studies, Biometrics 69 (2) (2013) 427–435.

Chapter 25

Applications of whale migration algorithm in optimal power flow problems of power systems

Mojtaba Ghasemi[a], Mohsen Zare[b], Soleiman Kadkhoda Mohammadi[c], and Seyedali Mirjalili[d,e]
[a]*Department of Electronics and Electrical Engineering, Shiraz University of Technology, Shiraz, Iran,* [b]*Department of Electrical Engineering, Faculty of Engineering, Jahrom University, Jahrom, Iran,* [c]*Department of Electrical and Electronic Engineering, Boukan Branch, Islamic Azad University, Boukan, Iran,* [d]*Centre for Artificial Intelligence Research and Optimisation, Torrens University Australia, Brisbane, QLD, Australia,* [e]*University Research and Innovation Center, Obuda University, Budapest, Hungary*

25.1 Introduction

One of the complex challenges in the world of engineering is OPF, which is of great importance in planning electrical networks, it has been attracting the attention of researchers in the field of electrical engineering for more than 60 years. During these years, the objectives considered in the problem have changed significantly. These changes can be seen as a result of the pliable structure of energy systems due to the presence of modern tools like storage elements, and renewable energy sources (RESs) [1] [2] [3] [4]. The first simplified OPF problem for the power systems was introduced in [3]. In the past, researchers used solving techniques based on mathematical methods like quadratic programming (QP) [5], and nonlinear programming (NLP) [6] to optimize these problems. In the following, heuristic solution methods were applied to optimize OPF. The complication of the practical OPF problem (because of its non-linear, non-derivative, and non-convex nature) is caused researchers to employ novel optimization methods to optimize OPF in recent years. Consequently, quite a few novel optimization methods have been proposed, which are mostly inspired via natural phenomena common in evolution and repetition in organisms (like birds, fish, ants). A brief description of several methods used in recent years includes the following:

Ant lion optimization [7–9], moth swarm algorithm (MSA) [10], Harris hawks optimization (HHO) [11], multi-objective dynamic OPF (MODOPF) [12], a grey wolf optimizer (DGWO) [13,14], salp swarm algorithm (SSA) [15], a particle swarm optimization (PSO) [17], an adaptive group search optimization (AGSO) [19], BAT algorithm [20], modified Jaya algorithm [21], a developing ABC algorithms [22–24], symbiotic organisms search (SOS) algorithm [25], an evolutionary algorithm (EA) [26], have been proposed for OPF problem including various objectives. A modified bacteria foraging algorithm (MBFA) [27], differential evolution (DE) [28–30], interior search algorithm (ISA) [31], chaotic invasive weed optimization algorithms (CIWO) [32], Jaya Algorithm [33], wind power to OPF [34], a new hybrid PSO [35], coronavirus herd immunity optimizer (CHIO) [36], multi-objective adaptive guided DE (MOAGDE) [37], a new parallel genetic algorithm (GA) (EPGA) [38], genetic teaching-learning-based optimization (TLBO) (G-TLBO) [39], a new improved adaptive DE [40–42], multi-objective mayfly algorithm (MOMA) [43], a new algorithm via Lévy mutation (LTLBO) [16], voltage stability constrained OPF (VSC-OPF) [44], manta ray foraging optimizer (MRFO) [45], glowworm swarm optimization (GSO) [46], social spider optimization (SSO) algorithms [47], multi-objective GSO (MOGSO) [48], a new Pareto evolutionary algorithm [49], sine-cosine algorithms (SCAs) [50,51], two-point estimate method (2PEM) [52], surrogate-assisted multi-objective probabilistic OPF [53], a combinatorial shuffle frog leaping algorithm (SFLA) with PSO [54], have been tested on the different IEEE systems. A novel hybrid firefly-bat algorithm (HFBA-COFS) [55], a hybrid PPSOGSA algorithm (hybrid of phasor PSO (PPSO) with gravitational search algorithm (GSA)) [56], PSO-GWO (the hybridization of PSO with GWO) [57], electromagnetism-like algorithm (ELA) [58], a turbulent flow of a water-based optimizer (TFWO) [59], tunicate swarm algorithm (TSA) [60], a cross entropy-cuckoo search algorithm (CE-CSA) [61], an effective CSA (ECSA) [62], chaotic Bonobo optimizer (CBO) [63], birds swarm algorithm (BSA) [64], slime mould-inspired algorithm (SMA) [65], a developed heap-based optimization (IHO) [66], CPSOGSA (a combinatorial PSOGSA with chaotic maps) [67],

Handbook of Whale Optimization Algorithm. https://doi.org/10.1016/B978-0-32-395365-8.00031-2

multi-objective OPF using GWO and DE algorithms [68], a developed honey bee mating algorithm (MHBMO) [69], for solving various OPF problems, etc.

25.2 Problem formulation

The purpose of solving the OPF problem is to determine and adjust a set of control variables in order to optimize the predetermined objectives in the operation of an electric network, which is the most important and basic goal of minimizing the cost of generation and transmission of electrical energy. The classic OPF objective function, $F(x,u)$, is formulated in the form of the following equations [10]:

$$Min\ F(x,u) \tag{25.1}$$

$$Subject\ to{:}\quad g(x,u)=0 \tag{25.2}$$

$$h(x,u)\leq 0 \tag{25.3}$$

In these equations, u and x are vectors of independent and decision parameters, respectively. Also, $h(x,u)$ and $g(x,u)$ are set of the inequality and equality constraints in OPF.

25.2.1 Decision parameters

The set of decision parameters in OPF equations include the following [10]:

1. $P_{G_2},\ldots,P_{G_{NG}}$: Real generation power on PV buses without the slack bus
2. $V_{G_1},\ldots,V_{G_{NG}}$: Voltage magnitudes of PV buses
3. $Q_{C_1},\ldots,Q_{C_{NC}}$: Compensation of shunt reactive VA
4. $T_1,\ldots,T_{NT}$: Setting transformer taps

According to decision parameters, u is stated as:

$$u^T=\left[P_{G_2},\ldots,P_{G_{NG}},T_1,\ldots,T_{NT},Q_{C_1},\ldots,Q_{C_{NC}},V_{G_1},\ldots,V_{G_{NG}}\right] \tag{25.4}$$

In Eq. (25.4), NC, NT, NG represent the number of reactive power compensators, transformers with taps, and units, respectively.

25.2.2 State variables

The set of state parameters in OPF equations include [10]:

1. P_{G_1}: Real generation power on the slack bus
2. $V_{L_1},\ldots,V_{L_{NPQ}}$: Voltage magnitudes of the load buses (QV)
3. $Q_{G_1},\ldots,Q_{G_{NG}}$: Reactive generation power of generators
4. $S_{l_1},\ldots,S_{l_{NTL}}$: Power loading on the lines

According to state parameters, x is stated as follows:

$$x^T=\left[P_{G_1},S_{l_1},\ldots,S_{l_{NTL}},Q_{G_1},\ldots,Q_{G_{NG}},V_{L_1},\ldots,V_{L_{NPQ}}\right] \tag{25.5}$$

In this equation, NPQ, NTL, and NG denote the number of QV, transmission lines, and generators, respectively.

25.2.3 Equality limits

Equality limits (or physical constraints) of OPF represent the technical status of the electrical network, which are stated using OPF formulations [10]:

$$Q_{Gi}=Q_{Di}+V_i\sum_{j=1}^{NB}V_j\left[-B_{ij}*\cos(\delta_{ij})+G_{ij}*\sin(\delta_{ij})\right] \tag{25.6}$$

$$P_{Gi}=P_{Di}+V_i\sum_{j=1}^{NB}V_j\left[B_{ij}*\sin(\delta_{ij})+G_{ij}*\cos(\delta_{ij})\right] \tag{25.7}$$

Here, V_i and V_j are voltage magnitudes on the i and j buses; Q_{Di} and P_{Di} show reactive and real demand; Q_{Gi} and P_{Gi} show reactive and real power generation; G_{ij}, B_{ij}, indicate conductance and susceptance of the line between buses j with i; δ_{ij} shows the difference of voltage phase angles between buses j with i; with NB buses.

25.2.4 Inequality limits

Technical limits of generators are stated as follows for $i = 1, 2, \ldots, NG$ [10]:

$$\begin{aligned} P_{Gi}^{\min} &\le P_{Gi} \le P_{Gi}^{\max} \\ V_{Gi}^{\min} &\le V_{Gi} \le V_{Gi}^{\max} \\ Q_{Gi}^{\min} &\le Q_{Gi} \le Q_{Gi}^{\max} \end{aligned} \tag{25.8}$$

In these equations, $V_{Gi}^{\max}$ and $V_{Gi}^{\min}$ are the allowed constraints of voltage magnitude of unit i; $P_{Gi}^{\max}$ and $P_{Gi}^{\min}$ are the allowed real power of generator i; and $Q_{Gi}^{\max}$ and $Q_{Gi}^{\min}$ express the allowed reactive power of generator i.

Technical constraints of transformers and VAR compensators have been stated here:

$$T_i^{\min} \le T_i \le T_i^{\max}, \quad i = 1, \ldots, NT \tag{25.9}$$

where $T_i^{\max}$ and $T_i^{\min}$ show the upper and lower limits of transformer taps.

$$Q_{Ci}^{\min} \le Q_{Ci} \le Q_{Ci}^{\max}, \quad i = 1, \ldots, NC \tag{25.10}$$

where $Q_{Ci}^{\max}$ and $Q_{Ci}^{\min}$ show the maximum and minimum amounts of VAR injection by compensator i.

Finally, network security limits applied to transmission line and bus voltages size is stated as follows:

i. Transmission line constraints
Power flow on the system lines for $i = 1, 2, \ldots, NTL$ is limited as:

$$S_{li} \le S_{li}^{\max} \tag{25.11}$$

where S_{li} with $S_{li}^{\max}$ show the apparent power flow with its maximum allowable boundary on line i.

ii. Bus constraints
The limit of voltage magnitude on the buses for $i = 1, 2, \ldots, NPQ$ is given as:

$$V_{Li}^{\min} \le V_{Li} \le V_{Li}^{\max} \tag{25.12}$$

25.2.5 Control constraints

A penalty function is considered to take into account the encroachment of the limits as follows [32]:

$$\begin{aligned} J = &\sum_{i=1}^{NG} F_i(P_{Gi}) + \lambda_P (P_{G1} - P_{G1}^{\lim})^2 + \lambda_S \sum_{i=1}^{NTL} (S_{li} - S_{li}^{\lim})^2 \\ &+ \lambda_Q \sum_{i=1}^{NG} (Q_{Gi} - Q_{Gi}^{\lim})^2 + \lambda_V \sum_{i=1}^{NPQ} (V_{Li} - V_{Li}^{\lim})^2 \end{aligned} \tag{25.13}$$

where λ_S, λ_Q, λ_V, and λ_P represent penalty factors, and $x^{\lim}$ denotes an auxiliary variable [32]:

$$x^{\lim} = \begin{cases} x^{\min}; & x < x^{\min} \\ x; & x^{\min} \le x \le x^{\max} \\ x^{\max}; & x > x^{\max} \end{cases} \tag{25.14}$$

25.3 Description of WMA

25.3.1 Initialization

In the first step, the WMA generates a population of immigrant whales, denoted by X_i, range from $X_{\min} = (X_{\min,1}, \ldots, X_{\min,D})$ to $X_{\max} = (X_{\max,1}, \ldots, X_{\max,D})$, where $X_i = (X_{i,1}, \ldots, X_{i,D})$ and $i = 1$ to N_{pop}. The dimension of the optimization function is also set D, where $d = 1 : D$:

$$X_i = X_{min} + rand\,(1, D) \times (X_{max} - X_{min})\,; \qquad i = 1 : N_{pop}. \tag{25.15}$$

25.3.2 Current local position

A group of migrating whales at any given moment is at one specific point in the ocean or along its journey. We consider this current position to be equivalent to the average current position of each member, i.e., X_{mean}. In each group of migratory whales, more experienced whales (here with a better position and fitness function) take over the task of leading the group and normal whales to the destination. Therefore, in the proposed WMA, there are N_L leaders, consisting of members with a better and superior position and fitness function. In this chapter, to speed up the convergence of WMA and also for WMA to escape from trapping in the local optima, through the experimental solutions obtained in the tests, we set X_{mean} equal to the average position of the current N_L leaders, that is:

$$X_{mean} = \frac{1}{N_L} \sum_{j=1}^{N_L} X_j \tag{25.16}$$

25.3.3 Movement of less-experienced whales

If the population have been set in terms of fitness function value or position (or, equivalently, whales' experience) from the best to the worst, the movement of a member is impacted by its nearest front member, i.e., X_k, which has a better fitness function. For instance, regarding the ith member, X_i, we have $X_1(Best)$, $X_2, \ldots, X_{i-1}$ (or X_k), X_i, X_{i+1}, ..., $X_{N_{pop}}(Worst)$; thus, an equation in the form of $rand * (X_k - X_i)$ is added to the motion equation of the ith member, X_i, where *rand* indicate a random value from 0 to 1 with a dimension equal to that of the test function. However, this equation is added only to the equation of less-experienced members (the number of such members is N_{pop}-N_L), not to that of leader whales.

25.3.4 Leader role in migrating the less experienced whales

As it was assumed before, the current position of migrating whales in the ocean is set to the average of the current positions of leader whales, X_{mean}. And, this current position of the group, X_{mean}, moves towards the position of the most experienced and powerful whale, as shown X_{best}, and this leads the less experienced whales toward the destination. Thus, an equation in the form of $rand * (X_{Best} - X_{mean})$ is added to the movement equation of less experienced whales.

Eventually, the movement and guidance or the novel location of the ith less experienced whale is written:

$$X_i^{new} = X_{mean} + rand * (X_k - X_i) + rand * (X_{Best} - X_{mean})\,; \qquad i = N_L + 1 : N_{pop} \tag{25.17}$$

Of course, it is noteworthy that in the case of substituting the current location X_i with the novel location X_i^{new}, the value of the fitness function $f\left(X_i^{new}\right)$ becomes less than $f(X_j)$.

25.3.5 Leaders discovering new space

In the group of migrating whales, the leaders are responsible for discovering and finding new places and more convenient routes toward the destination within the considered limits. The leaders perform this task according to Eq. (25.18):

$$X_j^{new} = X_j + rand * (X_{\min} + rand * (X_{\max} - X_{\min}))\,; \qquad j = 1 : N_L. \tag{25.18}$$

It is noted that in the case the new location X_j^{new} replaces the current location X_j, the value of $f(X_j^{new})$ becomes smaller than $f(X_j)$.

At the end of every generation of WMA, the population is sorted from the best to the worst and N_L best members are chosen as the leaders.

The optimization process of WMA has been shown in Fig. 25.1.

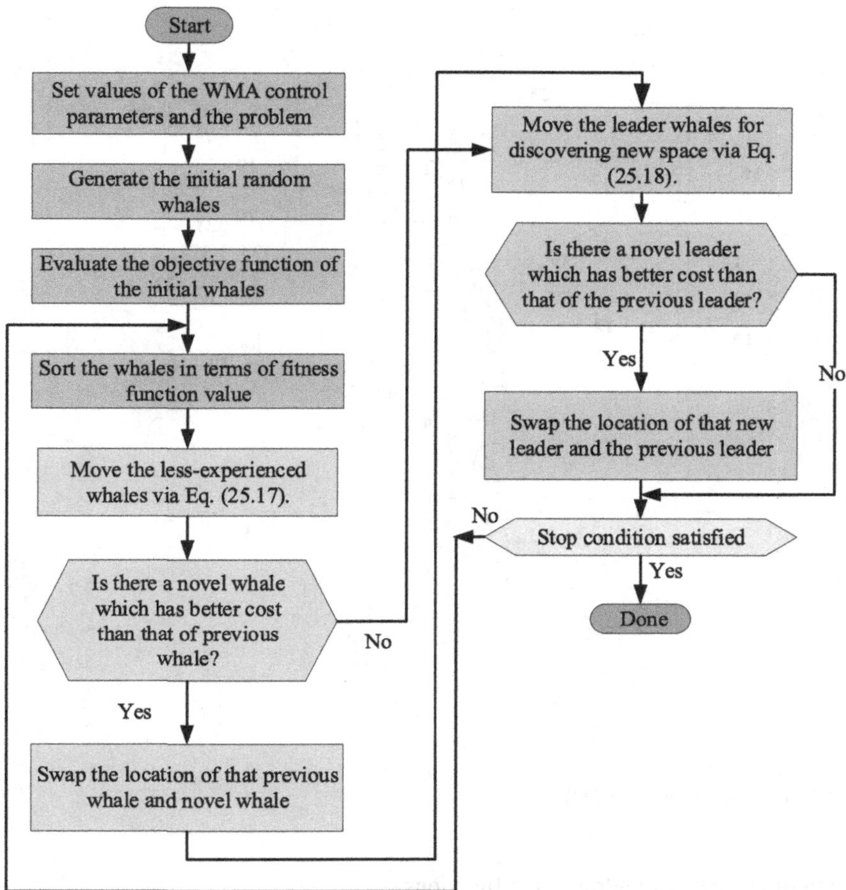

FIGURE 25.1 The optimization process of WMA.

25.3.6 Application of WMA in OPF problems

The proposed WMA was implemented in MATLAB® 2014a. Also, MATPOWER [70] was used for power flow analysis. All cases were executed on the IEEE 30-bus network [32]. All specifications of the system are based on the paper and are not repeated here to save space. (See Fig. 25.2.)

Six different OPF problems have been assumed and implemented on the studied system to indicate the successfulness of WMA. The problems are summarized as:

Problem I: Fuel cost minimization

In the present study, the objective function considered for the OPF consists of several sections. The main step is fuel cost minimization or the same traditional cost function of generators [32]:

$$J_1 = \sum_{i=1}^{NG} F_i(P_{Gi}) = \sum_{i=1}^{NG} (\alpha_i + b_i P_{Gi} + c_i P_{Gi}^2) \tag{25.19}$$

Here, α_i, b_i, and c_i indicate cost coefficients of generator i (refer to [32]).

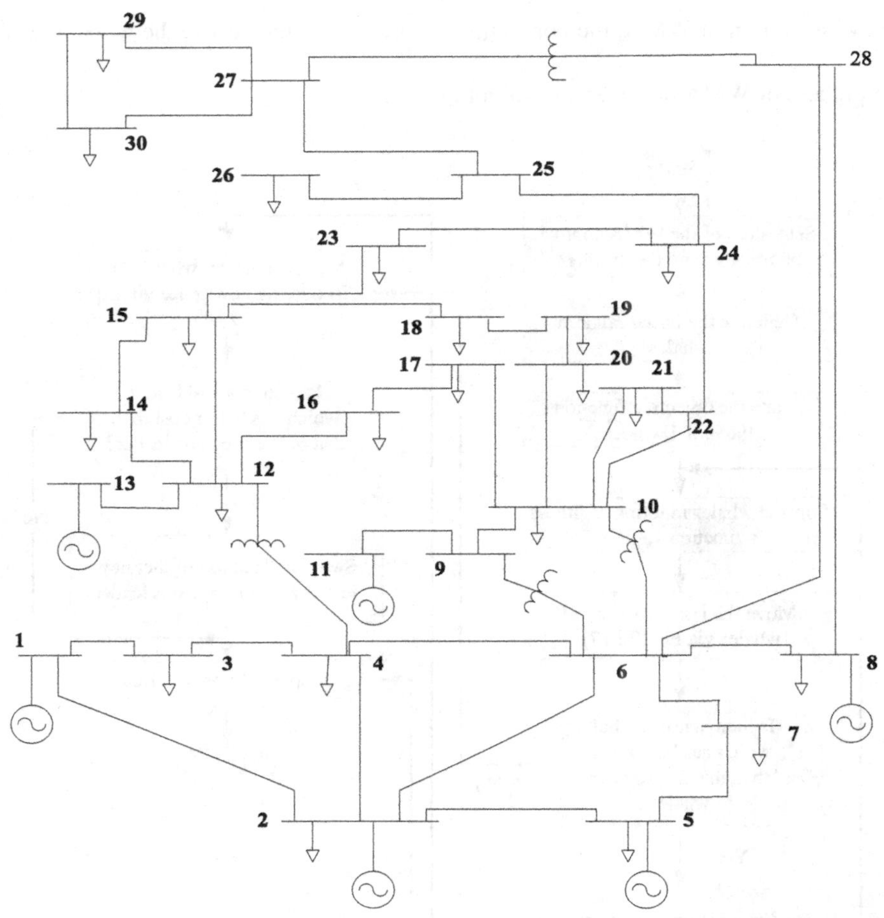

FIGURE 25.2 The studied IEEE 30-bus test network [32].

Problem II: Optimization of piecewise quadratic cost functions

Thermal generators can adopt oil, and coal depending on the performance of the network electrical. Consequently, theoretical analysis of the fuel cost curve of units 1 and 2 has been assumed as a sort of limits [32].

$$F(P_{Gi}) = \begin{cases} \alpha_{i1} + b_{i1}P_{Gi} + c_{i1}P_{Gi}^2 & P_{Gi}^{\min} \le P_{Gi} \le P_{Gi1} \\ \alpha_{i2} + b_{i2}P_{Gi} + c_{i2}P_{Gi}^2 & P_{Gi1} \le P_{Gi} \le P_{Gi2} \\ \dots & \\ \alpha_{ik} + b_{ik}P_{Gi} + c_{ik}P_{Gi}^2 & P_{Gik-1} \le P_{Gi} \le P_{Gi}^{\max} \end{cases} \tag{25.20}$$

Cost coefficients of unit i for the kth fuel type have been shown by α_{ik}, b_{ik}, and c_{ik}. It is evident that cost coefficients of other fuel source units are invariant and their values are similar to Problem I. So, the objective function for modeling the features of consumption fuel cost can be shown as [32]:

$$J_2 = \left(\sum_{i=3}^{NG} \alpha_i + b_i P_{Gi} + c_i P_{Gi}^2\right) + \left(\sum_{i=1}^{2} \alpha_{ik} + b_{ik} P_{Gi} + c_{ik} P_{Gi}^2\right) \tag{25.21}$$

Problem III: Minimization of fuel cost by taking into account valve point effects (VPEs)

Considering the efficacy of VPEs, the cost function becomes more accurate and real. VPEs occur as a result of the valves of thermal generators open in the case of accepting vapor, and this leads to an abrupt increase of losses and waves in the curve of the cost; it is described as [32]:

$$J_3 = \sum_{i=1}^{NG} \alpha_i + \left| d_i \sin\left(e_i \left(P_{Gi}^{\min} - P_{Gi}\right)\right)\right| + b_i P_{Gi} + c_i P_{Gi}^2 \tag{25.22}$$

Here, d_i with e_i indicate cost coefficients of generator i.

Problem IV: Considering the network losses

Transferring energy loss with minimized loss has always been one of the goals of engineers. Hence, the current scenario aims to reduce the costs associated with fuel costs and network loss simultaneously, this problem is described as [32]:

$$J_4 = \sum_{i=1}^{NG} \alpha_i + b_i P_{Gi} + c_i P_{Gi}^2 + \lambda p * P_{Loss} \tag{25.23}$$

Here, λp is set to 40, as was chosen in [10,71].

Network loss (P_{Loss}) is modeled as the following equation [10,71]:

$$P_{Loss} = \sum_{\substack{k=1 \\ k=(i,j)}}^{NTL} g_k (V_i^2 + V_j^2 - 2V_i V_j \cos\delta_{ij}) \tag{25.24}$$

In this equation, the conductance of the kth branch is given by g_k.

Problem V: Considering the voltage deviations (V.D.)

Voltage specification is the most essential criterion to determine the service manner of a power network. The voltage profile is modified via reducing the bus voltage deviation from the unity value. When only a cost-based OPF is considered, a suitable result with unfavorable V.D. is obtained. As a result, a multi-objective OPF problem has been presented below to simultaneously optimize V.D. and fuel cost, which is the objective function of scenario V of this paper's OPF [71].

$$J_5 = J_1 + \lambda v * \sum_{i=1}^{NPQ} |V_i - 1.0| \tag{25.25}$$

Here, λv has been set at 100 [71].

Problem VI: Minimization of fuel cost with considering the pollutant emission, V.D., and losses

The objective function of Problem VI aims to minimize the fuel cost with considering the pollutant emission, V.D., and losses across the system. The modeling of this function is given here [71]:

$$J_6 = J_1 + \lambda e * \sum_{i=1}^{NG} F_{Ei}(P_{Gi}) + \lambda p * P_{Loss} + \lambda v * \sum_{i=1}^{NPQ} |V_i - 1.0| \tag{25.26}$$

To provide a compromise among these different objective function terms, weight factors of $\lambda v = 21$, $\lambda p = 22$, and $\lambda e = 19$ are adopted [71]. The modeling of pollutant emission objective function is expressed as [71]:

$$F_E = \sum_{i=1}^{NG} \left(\alpha_i + \xi_i \exp(\lambda_i P_{Gi}) + \beta_i P_{Gi} + \gamma_i P_{Gi}^2\right) \tag{25.27}$$

Here, F_E is the amount of pollutant and γ_i, β_i, ξ_i, and λ_i indicate emission coefficients of unit i with β_i (ton/h MW), γ_i (ton/h MW^2), and α_i (ton/h) related to SO_X, and λ_i (1/MW) and ξ_i (ton/h) related to NO_X, respectively.

TABLE 25.1 Optimal values obtained by solving OPF problems using WMA.

Decision variables		Limits		Problems					
		Lower	Upper	I	II	III	IV	V	VI
P_{G1}	(MW)	50	250	177.0928	139.9995	198.8062	102.5005	176.1183	122.2923
P_{G2}		20	80	48.7218	54.9997	45.2316	55.6291	48.7819	52.3933
P_{G5}		15	50	21.3858	24.0363	18.0798	38.1146	21.6330	31.4753
P_{G8}		10	35	21.2801	34.9574	10.0000	35.0000	22.3681	35.0000
P_{G11}		10	30	11.9373	18.6487	10.0000	30.0000	12.3296	26.7609
P_{G13}		12	40	12.0000	17.5024	12.0000	26.6824	12.0000	21.0672
V_{G1}	(p.u.)	0.95	1.1	1.0832	1.0766	1.0815	1.0696	1.0422	1.0727
V_{G2}		0.95	1.1	1.0605	1.0589	1.0580	1.0575	1.0227	1.0572
V_{G5}		0.95	1.1	1.0345	1.0329	1.0301	1.0358	1.0144	1.0326
V_{G8}		0.95	1.1	1.0389	1.0424	1.0375	1.0438	1.0056	1.0409
V_{G11}		0.95	1.1	1.1000	1.0932	1.1000	1.0954	1.0733	1.0400
V_{G13}		0.95	1.1	1.0526	1.0457	1.0632	1.0567	0.9873	1.0227
T_{6-9}		0.9	1.1	1.0883	1.0951	1.0530	1.0816	1.0999	1.1000
T_{6-10}		0.9	1.1	0.9027	0.9029	0.9631	0.9004	0.9002	0.9598
T_{4-12}		0.9	1.1	0.9766	0.9784	0.9935	0.9903	0.9381	1.0317
T_{28-27}		0.9	1.1	0.9742	0.9706	0.9780	0.9750	0.9709	1.0042
Q_{C10}	(MVAR)	0.0	5.0	2.2594	4.9752	4.9870	0.8053	4.9638	4.9953
Q_{C12}		0.0	5.0	0.2553	4.8730	0.1084	0.6974	0.0186	0.7057
Q_{C15}		0.0	5.0	4.4550	5.0000	4.9235	4.4636	5.0000	3.9737
Q_{C17}		0.0	5.0	5.0000	5.0000	4.9992	5.0000	0	5.0000
Q_{C20}		0.0	5.0	4.2509	4.9961	5.0000	4.2527	5.0000	5.0000
Q_{C21}		0.0	5.0	5.0000	4.9741	5.0000	5.0000	4.9979	5.0000
Q_{C23}		0.0	5.0	3.2646	3.5526	3.0956	3.2610	5.0000	4.0727
Q_{C24}		0.0	5.0	5.0000	2.7829	4.9997	4.9995	5.0000	5.0000
Q_{C29}		0.0	5.0	2.6514	2.2717	2.6812	2.5555	2.6222	2.5319
Cost ($/h)		-	-	800.4789	646.5031	832.1707	859.1285	803.7059	830.2379
Emission (t/h)		-	-	0.3661	0.2835	0.4382	0.2288	0.3632	0.2530
Power losses (MW)		-	-	9.0178	6.7440	10.7176	4.5266	9.8309	5.5890
V.D. (p.u.)		-	-	0.9050	0.9123	0.8723	0.9323	0.0948	0.2981

Table 25.1 lists a summary of the optimal solutions obtained for WMA when used to the abovementioned six OPF problems in the studied electrical network.

25.3.7 Advantages of WMA

A comparison is made between the proposed WMA and its counterparts when used to the mentioned OPF of the previous section. Also, some OPF problems that are subject to real parameters are analyzed here. The following provides some novel optimization methods to perform a comparison study.

25.3.7.1 Aquila optimizer (AO)

This algorithm [72] is an imitation of the hunting behavior of Aquilas and it consists of four steps: 1) Choosing the desirable search environment, 2) Discovering the environment, 3) Attacking, and 4) Hunting the prey. Ref. [72] examines the capability of AO in finding the optimal solution by adopting several optimization problems and indicates that AO is superior to the compared technologies and it is shown that AO can be practically utilized for solving optimization problems.

25.3.7.2 Artificial rabbits optimization (ARO)

ARO [73] is relied on the attempt of rabbits to hide and survive and consists of two steps: random hiding and detour foraging. The former makes rabbits feed from near places and thus helping to stay hidden from predators. The latter assists rabbits to select a random place to hide. These are the bases of ARO. Ref. [73] examined the potential of this algorithm and compared it with some other methods, according to which it was demonstrated that ARO is successful in solving real-world complicated problems.

25.3.7.3 War strategy optimization (WSO)

The WSO [74] was developed by imitating the strategy of armies in war situations. This strategy was considered and modeled like an optimization problem, in which soldiers try to reach the optimal point. Two well-known war strategies were modeled: attack and defense. The strategy adopted for the war determines soldiers' positions. The algorithm incorporates a new weighting approach to update the positions [74]. The algorithm provides a suitable balance between the exploitation and exploration parts of the problem [74]. The authors in [74] examined the potential of WSO on fifty test functions and four practical problems. Also, a comparison was made between WSO and ten recent metaheuristic algorithms [74], demonstrating the acceptable performance of the WSO when applied to real-world problems.

25.3.7.4 Transit search (TS)

TS [75] was inspired by the exoplanet discovery approach. This approach helped to detect about 4000 planets using the provided data by telescopes. Astronomers believe that discovering planets is challenging due to their small size [75]. The TS has already proved its capability in astrophysics and cosmology [75] and has widely been adopted for these purposes. The basis of this algorithm is that it analyzes the light of stars when its luminosity changes and this helps to determine if a plant is in from of the star [75]. This algorithm was adopted and applied to around seventy problems and compared with its counterparts [75]. In [75] has been shown the TS obtained more accurate solutions than the rest of the algorithms.

25.3.7.5 Honey badger algorithm (HBA)

HBA [76] based on the foraging treatment of honey badgers and provides modeling of search strategy to solve optimization problems. Exploration and exploitation steps of HBA represent the digging and honey searching behavior of the honey badgers. One feature of HBA is its ability to preserve members diversity throughout the search [76]. HBA had been used to various functions and problems to demonstrate its capabilities [76]. The performed experiments and analyses revealed the superiority of HBA concerning convergence and the balance in the exploration and exploitation parts.

25.3.7.6 Remora optimization algorithm (ROA)

ROA [77] is based on bionics, nature, and metaheuristic approaches and imitates the parasitic behavior of remora. Two updating techniques are adopted in this algorithm depending on the size of the host. To be more specific, if the host is large, remora intends to feed on ectoparasites or the remaining food of the host, as is observed for large whales [77]. But, concerning small hosts, remora accompanies the host to reach a rich supply of prey, as is true for swordfish. Regardless of the updating technique, experience plays an essential duty in the action and decision of remora. In the ROA, the host can be substituted by another host, e.g., by ships, turtles, etc. Ref. [77] performed simulations on the behavior and treatment of ROA when applied to some engineering problem [77]. Also, a comparison was made between ROA and some heuristic algorithms [77]. According to the findings, ROA can surely be adopted as a reliable tool to solve problems.

25.3.7.7 Rat swarm optimizer (RSO)

RSO [78] imitates the chasing-attacking behavior of rats when trying to hunt prey. Ref. [78] provides a mathematical expression of chasing-attacking behavior and examines the algorithm on about forty problems. Also, a comparison was made between RSO and some popular optimizers [78]. As the results show in [78], RSO can be reliably incorporated into optimization problems in the real world.

25.3.7.8 Whale optimization algorithm (WOA)

WOA [79] based on the whales that use a bubble-net strategy for hunting prey. Ref. [79] adopted WOA to examine its potential when applied to optimization problems. WOA adopts prey searching, encircling, and bubble-net foraging treatment of humpback whales. According to the findings, this method performs far better than most metaheuristic algorithms [79].

25.4 Simulation

Six OPF problems are tested by adopting the suggested WMA and the algorithm was executed thirty times on each problem. The size of the swarm and generations for the suggested WMA was set at 90 and 400, respectively. As a result, the FES function was executed 36,000 times.

Simulation reports for thirty runs are displayed in Table 25.2, which shows the values of mean and standard deviation (Std.). As per the table, the results of the suggested WMA are preferable to those of the studied other metaheuristic algorithms. To compare these algorithms' convergence speeds, Fig. 25.3 illustrates the results. According to this, the superiority of the WMA is once again preferred. Moreover, it is evident that WMA is more robust and trustable than the analyzed modern algorithms and it should be selected as a reliable and new metaheuristic for future studies.

As shown in Table 25.2, the execution time of the WMA algorithm is only higher than WOA. Of course, it should be noted that all algorithms have almost the same computational complexity, so the simulation time of the algorithms is virtually in the same time frame.

TABLE 25.2 Statistical reports of WMA and new algorithms.

Algorithm	Min	Mean	Max	Std.	Time (s) (for the best solution)
			Problem I		
AO	801.3360	804.0073	806.0621	2.0317	539.25
ARO	801.2501	802.3285	803.5247	0.7489	548.73
WSO	801.8466	804.4639	806.6532	2.6532	586.92
TS	801.9005	804.1234	806.3890	1.9637	617.34
HBA	801.6710	803.6578	806.7329	1.5530	648.52
ROA	801.5436	802.9553	805.4536	1.6328	498.61
RSO	802.1682	804.7030	806.2563	2.1004	600.19
WOA	801.8374	803.1580	805.2798	1.1129	434.65
WMA	**800.4789**	**800.5208**	**801.0301**	**0.1314**	519.60
			Problem II		
AO	650.8516	706.4128	761.3912	63.1215	624.75
ARO	648.7201	697.3446	745.0086	57.6040	519.31
WSO	653.4912	722.6438	777.6820	81.6249	568.25
TS	652.7278	706.8126	780.6214	69.9004	712.88
HBA	653.5044	715.5734	788.9462	78.2341	684.64
ROA	653.7482	706.9516	769.2100	58.9784	574.52
RSO	651.5381	695.3478	780.5930	52.8237	523.76
WOA	650.7645	701.2135	755.7201	42.3585	455.40
WMA	**646.5031**	**679.9928**	**724.8688**	**41.6561**	450.38
			Problem III		
AO	834.5270	855.3274	889.6109	25.6438	562.49
ARO	833.8821	845.2003	850.4279	5.5929	574.00
WSO	837.8047	861.1238	897.9940	30.7128	515.74
TS	835.0087	858.4890	901.5281	42.1594	624.17
HBA	835.2469	849.2941	897.2226	39.8520	691.23
ROA	834.3256	847.0325	873.2615	23.9764	634.99
RSO	838.1795	863.7428	905.0087	41.4670	503.40
WOA	837.3284	852.4850	889.9094	13.2878	471.**01**
WMA	**832.1707**	**832.2148**	**833.5282**	**0.2841**	510.39

continued on next page

TABLE 25.2 *(continued)*

Algorithm	Min	Mean	Max	Std.	Time (s) (for the best solution)
			Problem IV		
AO	1052.6529	1061.7819	1074.9338	15.0940	608.24
ARO	1043.8314	1050.4161	1061.6846	9.8205	559.73
WSO	1049.2559	1058.8007	1069.5423	12.6237	646.17
TS	1050.5697	1060.2296	1071.8816	10.7462	612.48
HBA	1054.9885	1063.1136	1071.3346	11.8412	514.95
ROA	1049.6635	1052.8421	1069.7203	14.8637	493.06
RSO	1053.8534	1065.1948	1072.3895	10.6951	500.28
WOA	1052.2000	1060.5117	1070.2964	6.0733	485.31
WMA	**1040.1925**	**1040.3995**	**1041.5562**	**0.5102**	474.52
			Problem V		
AO	824.1608	836.4929	890.6201	52.6413	586.72
ARO	816.1733	826.9264	859.8250	14.3557	610.64
WSO	824.9046	830.7485	871.3209	36.6717	564.31
TS	823.7157	832.9081	875.1121	28.3446	672.15
HBA	825.1954	835.2717	892.9416	45.2517	623.46
ROA	820.3154	829.4836	864.2658	16.0058	584.37
RSO	822.3036	830.9326	888.7261	24.6301	465.92
WOA	821.2373	833.5007	886.9046	21.7344	421.86
WMA	**813.1859**	**813.9243**	**818.5932**	**1.6219**	470.65
			Problem VI		
AO	971.1222	977.6891	982.2607	5.4855	648.03
ARO	966.3601	972.5592	978.0045	3.1628	594.81
WSO	968.7921	975.0258	976.1479	1.9264	625.15
TS	970.6590	977.3619	981.5206	4.7125	643.75
HBA	968.8836	976.0745	979.3280	3.9846	516.86
ROA	967.0171	973.2516	978.4569	2.4921	523.93
RSO	967.2304	975.8635	978.1023	2.5830	468.10
WOA	969.0518	974.4014	977.2304	2.9229	495.55
WMA	**964.2630**	**964.2793**	**964.3003**	**0.0111**	549.81

25.5 Discussion

As previously shown, the suggested WMA metaheuristic is more suitable in terms of efficiency than the classic WOA, particularly in the case of testing them on OPF problems. A comparison between the optimal solutions of the suggested WMA metaheuristic and recent methods was made and the findings are reported in Tables 25.3 and 25.4. The exploitation and exploration features of the suggested new metaheuristic illustrate a more convincing treatment for both single- and multi-objective OPF problems.

25.6 Conclusion

The chapter presented a novel form of the WMA and applied it to six OPF problems implemented on a prototype system. Also, the chapter compared the performance of the designed WMA metaheuristic with its counterparts when problems are either single- or multi-objective. Global optimal solutions obtained by the suggested WMA are more acceptable than those of other methods. However, more future studies can still be conducted on the proposed method to improve its efficiency, especially by implementing it on some other practical optimization problems encountered in the engineering realm. Fur-

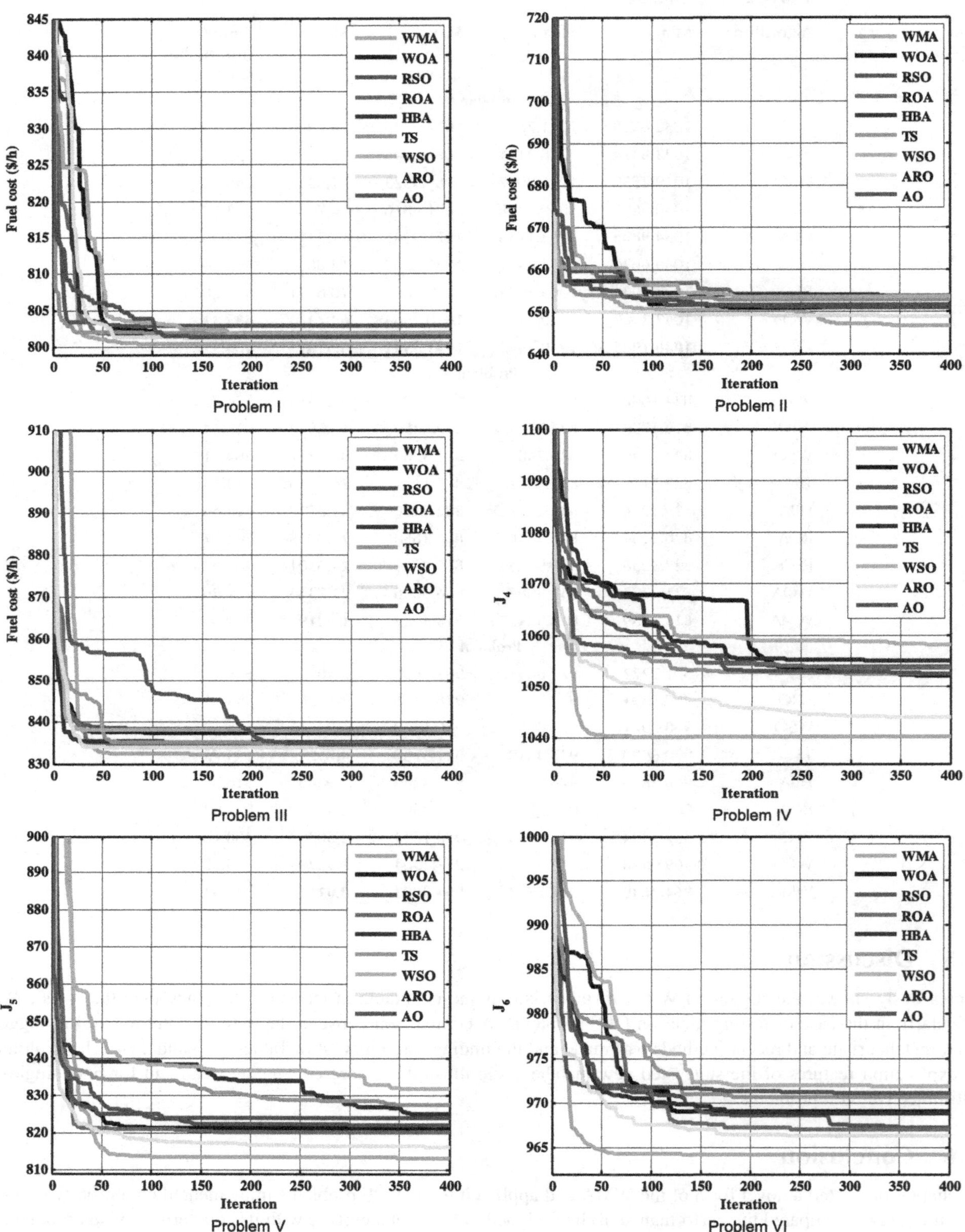

FIGURE 25.3 A comparison between convergence speeds of the algorithms.

TABLE 25.3 A comparison between the proposed WMA metaheuristic and some recent metaheuristics in case the problems are single-objective.

Algorithm	Fuel cost ($/h)	Emission (t/h)	Power losses (MW)	V.D. (p.u.)
Problem I				
MPSO-SFLA [54]	801.75	-	9.54	-
GWO [69]	801.41	-	9.30	-
FPA [80]	802.7983	0.35959	9.5406	0.36788
ARCBBO [85]	800.5159	0.3663	9.0255	0.8867
MGBICA [82]	801.1409	0.3296	-	-
IEP [83]	802.46	-	-	-
PSOGSA [84]	800.49859	-	9.0339	0.12674
MSA [80]	800.5099	0.36645	9.0345	0.90357
MICA-TLA [81]	801.0488	-	9.1895	-
EP [86]	803.57	-	-	-
MRFO [87]	800.7680	-	9.1150	-
AO [88]	801.83	-	-	-
SKH [89]	800.5141	0.3662	9.0282	-
MHBMO [68]	801.985	-	9.49	-
AGSO [17]	801.75	0.3703	-	-
TS [90]	802.29	-	-	-
JAYA [33]	800.4794	-	9.06481	0.1273
PPSOGSA [56]	800.528	-	9.02665	0.91136
MFO [80]	800.6863	0.36849	9.1492	0.75768
ABC [92]	800.660	0.365141	9.0328	0.9209
DE [93]	802.39	-	9.466	-
SFLA-SA [91]	801.79	-	-	-
WMA	**800.4789**	0.3661	9.0178	0.9050
Problem II				
LTLBO [16]	647.4315	0.2835	6.9347	0.8896
GABC [94]	647.03	-	6.8160	0.8010
MFO [80]	649.2727	0.28336	7.2293	0.47024
SSA [95]	646.7796	0.2836	6.5599	0.5320
FPA [80]	651.3768	0.28083	7.2355	0.31259
MDE [93]	647.846	-	7.095	-
SSO [47]	663.3518	-	-	-
IEP [83]	649.312	-	-	-
MICA-TLA [81]	647.1002	-	6.8945	-
MPSO-SFLA [54]	647.55	-	-	-
MSA [80]	646.8364	0.28352	6.8001	0.84479
WMA	**646.5031**	0.2835	6.7440	0.9123
Problem III				
HFAJAYA [1]	832.1798	0.4378	10.6897	0.8578
PSO [18]	832.6871	-	-	-
FA [1]	832.5596	0.4372	10.6823	0.8539
SP-DE [71]	832.4813	0.43651	10.6762	0.75042
WMA	**832.1707**	0.4382	10.7176	0.8723

TABLE 25.4 A comparison between the proposed WMA method and some recent metaheuristics in case the problems are multi-objective.

Algorithm	Fuel cost ($/h)	Emission (t/h)	Power losses (MW)	V.D. (p.u.)	*J*
			Problem IV		
MOALO [9]	826.4556	0.2642	5.7727	1.2560	1057.3636
MSA [80]	859.1915	0.2289	4.5404	0.92852	1040.8075
EMSA [96]	859.9514	0.2278	4.6071	0.7758	1044.2354
QOMJaya [97]	826.9651	-	5.7596	-	1402.9251
SpDEA [98]	837.8510	-	5.6093	0.8106	1062.223
MJaya [97]	827.9124	-	5.7960	-	1059.7524
WMA	859.1285	0.2288	4.5266	0.9323	**1040.1925**
			Problem V		
PSO [99]	804.477	0.368	10.129	0.126	817.0770
MPSO [80]	803.9787	0.3636	9.9242	0.1202	815.9987
MOMICA [100]	804.9611	0.3552	9.8212	0.0952	814.4811
BB-MOPSO [100]	804.9639	-	-	0.1021	815.1739
SSO [99]	803.73	0.365	9.841	0.1044	814.1700
EMSA [96]	803.4286	0.3643	9.7894	0.1073	814.1586
MFO [80]	803.7911	0.36355	9.8685	0.10563	814.3541
SpDEA [98]	803.0290	-	9.0949	0.2799	831.0190
TFWO [59]	803.416	0.365	9.795	0.101	813.5160
MNSGA-II [100]	805.0076	-	-	0.0989	814.8976
DA-APSO [101]	802.63	-	-	0.1164	814.2700
PSO-SSO [99]	803.9899	0.367	9.961	0.0940	813.3899
WMA	803.7059	0.3632	9.8309	0.0948	**813.1859**
			Problem VI		
I-NSGA-III [104]	881.9395	0.2209	4.7449	0.1754	994.2078
MODA [103]	828.49	0.265	5.912	0.585	975.8740
MSA [80]	830.639	0.25258	5.6219	0.29385	965.2907
J-PPS3 [102]	830.3088	0.2363	5.6377	0.2949	965.0228
J-PPS2 [102]	830.8672	0.2357	5.6175	0.2948	965.1201
PSO [99]	828.2904	0.261	5.644	0.55	968.9674
MOALO [9]	826.2676	0.2730	7.2073	0.7160	1005.0512
MFO [80]	830.9135	0.25231	5.5971	0.33164	965.8080
BB-MOPSO [100]	833.0345	0.2479	5.6504	0.3945	970.3379
J-PPS1 [102]	830.9938	0.2355	5.6120	0.2990	965.2159
MNSGA-II [100]	834.5616	0.2527	5.6606	0.4308	972.9429
SSO [99]	829.978	0.25	5.426	0.516	964.9360
WMA	830.2379	0.2530	5.5890	0.2981	**964.2630**

thermore, the proposed WMA can be modified and combined with other competitive algorithms to add even more features and achieve more satisfying results.

Declaration of competing interest

The authors declare that they have no conflict of interest.

References

[1] A.S. Alghamdi, A hybrid firefly–JAYA algorithm for the optimal power flow problem considering wind and solar power generations, Appl. Sci. 12 (2022) 7193.

[2] B. Mahdad, K. Srairi, Multi objective large power system planning under sever loading condition using learning DE-APSO-PS strategy, Energy Convers. Manag. 87 (2014) 338–350.

[3] H.W. Dommel, W.F. Tinney, Optimal power flow solutions, IEEE Trans. Power Appar. Syst. (1968) 1866–1876.

[4] R. Mota-Palomino, V.H. Quintana, Sparse reactive power scheduling by a penalty function-linear programming technique, IEEE Trans. Power Syst. 1 (1986) 31–39.

[5] R.C. Burchett, H.H. Happ, D.R. Vierath, Quadratically convergent optimal power flow, IEEE Trans. Power Appar. Syst. (1984) 3267–3275.

[6] O. Alsac, B. Stott, Optimal load flow with steady-state security, IEEE Trans. Power Appar. Syst. (1974) 745–751.

[7] A. Maheshwari, Y.R. Sood, S. Jaiswal, S. Sharma, J. Kaur, Ant lion optimization based OPF solution incorporating wind turbines and carbon emissions, in: 2021 Innovations in Power and Advanced Computing Technologies (i-PACT), 2021, pp. 1–6.

[8] A. Maheshwari, Y.R. Sood, Solution approach for optimal power flow considering wind turbine and environmental emissions, Wind Eng. 46 (2022) 480–502.

[9] O. Herbadji, L. Slimani, T. Bouktir, Optimal power flow with four conflicting objective functions using multiobjective ant lion algorithm: a case study of the Algerian electrical network, Iran. J. Electr. Electron. Eng. 15 (2019) 94–113, https://doi.org/10.22068/IJEEE.15.1.94.

[10] S. Duman, L. Wu, J. Li, Moth swarm algorithm based approach for the ACOPF considering wind and tidal energy, in: The International Conference on Artificial Intelligence and Applied Mathematics in Engineering, Springer, 2019, pp. 830–843.

[11] M.Z. Islam, N.I.A. Wahab, V. Veerasamy, H. Hizam, N.F. Mailah, J.M. Guerrero, et al., A Harris Hawks optimization based single- and multi-objective optimal power flow considering environmental emission, Sustainability 12 (2020) 5248.

[12] R. Ma, X. Li, Y. Luo, X. Wu, F. Jiang, Multi-objective dynamic optimal power flow of wind integrated power systems considering demand response, CSEE J. Power Energy Syst. 5 (2019) 466–473.

[13] M. Abdo, S. Kamel, M. Ebeed, J. Yu, F. Jurado, Solving non-smooth optimal power flow problems using a developed grey wolf optimizer, Energies 11 (2018) 1692.

[14] I.U. Khan, N. Javaid, K.A.A. Gamage, C.J. Taylor, S. Baig, X. Ma, Heuristic algorithm based optimal power flow model incorporating stochastic renewable energy sources, IEEE Access 8 (2020) 148622–148643.

[15] S. Kamel, M. Ebeed, F. Jurado, et al., An improved version of salp swarm algorithm for solving optimal power flow problem, Soft Comput. 25 (2021) 4027–4052.

[16] M. Ghasemi, S. Ghavidel, M. Gitizadeh, E. Akbari, An improved teaching–learning-based optimization algorithm using Lévy mutation strategy for non-smooth optimal power flow, Int. J. Electr. Power Energy Syst. 65 (2015) 375–384, https://doi.org/10.1016/j.ijepes.2014.10.027.

[17] J. Hazra, A.K. Sinha, A multi-objective optimal power flow using particle swarm optimization, Eur. Trans. Electr. Power 21 (2011) 1028–1045, https://doi.org/10.1002/etep.494.

[18] H.R.E.H. Bouchekara, A.E. Chaib, M.A. Abido, R.A. El-Sehiemy, Optimal power flow using an Improved Colliding Bodies Optimization algorithm, Appl. Soft Comput. 42 (2016) 119–131, https://doi.org/10.1016/j.asoc.2016.01.041.

[19] N. Daryani, M.T. Hagh, S. Teimourzadeh, Adaptive group search optimization algorithm for multi-objective optimal power flow problem, Appl. Soft Comput. 38 (2016) 1012–1024, https://doi.org/10.1016/j.asoc.2015.10.057.

[20] B. Venkateswara Rao, G.V. Nagesh Kumar, Optimal power flow by BAT search algorithm for generation reallocation with unified power flow controller, Int. J. Electr. Power Energy Syst. 68 (2015) 81–88, https://doi.org/10.1016/j.ijepes.2014.12.057.

[21] E.E. Elattar, S.K. ElSayed, Modified JAYA algorithm for optimal power flow incorporating renewable energy sources considering the cost, emission, power loss and voltage profile improvement, Energy 178 (2019) 598–609.

[22] K. Ayan, U. Kılıç, B. Baraklı, Chaotic artificial bee colony algorithm based solution of security and transient stability constrained optimal power flow, Int. J. Electr. Power Energy Syst. 64 (2015) 136–147, https://doi.org/10.1016/j.ijepes.2014.07.018.

[23] A. Khorsandi, S.H. Hosseinian, A. Ghazanfari, Modified artificial bee colony algorithm based on fuzzy multi-objective technique for optimal power flow problem, Electr. Power Syst. Res. 95 (2013) 206–213, https://doi.org/10.1016/j.epsr.2012.09.002.

[24] X. He, W. Wang, J. Jiang, L. Xu, An improved artificial bee colony algorithm and its application to multi-objective optimal power flow, Energies 8 (2015) 2412–2437, https://doi.org/10.3390/en8042412.

[25] S. Duman, J. Li, L. Wu, AC optimal power flow with thermal–wind–solar–tidal systems using the symbiotic organisms search algorithm, IET Renew. Power Gener. 15 (2021) 278–296.

[26] R.K. Avvari, V. Kumar D M, A novel hybrid multi-objective evolutionary algorithm for optimal power flow in wind, PV, and PEV systems, J. Oper. Autom. Power Eng. 11 (2) (2023) 130–143.

[27] A. Panda, M. Tripathy, A.K. Barisal, T. Prakash, A modified bacteria foraging based optimal power flow framework for Hydro-Thermal-Wind generation system in the presence of STATCOM, Energy 124 (2017) 720–740, https://doi.org/10.1016/j.energy.2017.02.090.

[28] M. Varadarajan, K.S. Swarup, Solving multi-objective optimal power flow using differential evolution, IET Gener. Transm. Distrib. 2 (2008) 720, https://doi.org/10.1049/iet-gtd:20070457.

[29] P.P. Biswas, P.N. Suganthan, G.A.J. Amaratunga, Optimal power flow solutions incorporating stochastic wind and solar power, Energy Convers. Manag. 148 (2017) 1194–1207.

[30] B.A. Kumari, K. Vaisakh, Integration of solar and flexible resources into expected security cost with dynamic optimal power flow problem using a Novel DE algorithm, Renew. Energy Focus 42 (2022) 48–69.

[31] S. Chandrasekaran, Multiobjective optimal power flow using interior search algorithm: a case study on a real-time electrical network, Comput. Intell. 36 (2020) 1078–1096, https://doi.org/10.1111/coin.12312.

[32] M. Ghasemi, S. Ghavidel, E. Akbari, A.A. Vahed, Solving non-linear, non-smooth and non-convex optimal power flow problems using chaotic invasive weed optimization algorithms based on chaos, Energy 73 (2014) 340–353, https://doi.org/10.1016/j.energy.2014.06.026.

[33] W. Warid, H. Hizam, N. Mariun, N.I. Abdul-Wahab, Optimal power flow using the Jaya algorithm, Energies 9 (2016) 678.

[34] L. Shi, C. Wang, L. Yao, Y. Ni, M. Bazargan, Optimal power flow solution incorporating wind power, IEEE Syst. J. 6 (2011) 233–241.

[35] K. Teeparthi, D.M.V. Kumar, Multi-objective hybrid PSO-APO algorithm based security constrained optimal power flow with wind and thermal generators, Eng. Sci. Technol. Int. J. 20 (2017) 411–426.

[36] Z.M. Ali, S.H.E.A. Aleem, A.I. Omar, B.S. Mahmoud, Economical-environmental-technical operation of power networks with high penetration of renewable energy systems using multi-objective coronavirus herd immunity algorithm, Mathematics 10 (2022) 1201.

[37] S. Duman, M. Akbel, H.T. Kahraman, Development of the multi-objective adaptive guided differential evolution and optimization of the MO-ACOPF for wind/PV/tidal energy sources, Appl. Soft Comput. 112 (2021) 107814.

[38] B. Mahdad, K. Srairi, T. Bouktir, Optimal power flow for large-scale power system with shunt FACTS using efficient parallel GA, Int. J. Electr. Power Energy Syst. 32 (2010) 507–517, https://doi.org/10.1016/j.ijepes.2009.09.013.

[39] M. Güçyetmez, E. Çam, A new hybrid algorithm with genetic-teaching learning optimization (G-TLBO) technique for optimizing of power flow in wind-thermal power systems, Electr. Eng. 98 (2016) 145–157.

[40] S. Li, W. Gong, L. Wang, X. Yan, C. Hu, Optimal power flow by means of improved adaptive differential evolution, Energy 198 (2020) 117314, https://doi.org/10.1016/j.energy.2020.117314.

[41] M. Basu, Multi-objective optimal power flow with FACTS devices, Energy Convers. Manag. 52 (2011) 903–910, https://doi.org/10.1016/j.enconman.2010.08.017.

[42] S. Duman, S. Rivera, J. Li, L. Wu, Optimal power flow of power systems with controllable wind-photovoltaic energy systems via differential evolutionary particle swarm optimization, Int. Trans. Electr. Energy Syst. 30 (2020) e12270.

[43] R. Kyomugisha, C.M. Muriithi, G.N. Nyakoe, Performance of various voltage stability indices in a stochastic multiobjective optimal power flow using mayfly algorithm, J. Electr. Comput. Eng. 2022 (2022), https://doi.org/10.1155/2022/7456333.

[44] R. Kyomugisha, C.M. Muriithi, M. Edimu, Multiobjective optimal power flow for static voltage stability margin improvement, Heliyon 7 (2021) e08631.

[45] H.T. Kahraman, M. Akbel, S. Duman, Optimization of optimal power flow problem using multi-objective manta ray foraging optimizer, Appl. Soft Comput. 116 (2022) 108334.

[46] S. Surender Reddy, C. Srinivasa Rathnam, Optimal power flow using glowworm swarm optimization, Int. J. Electr. Power Energy Syst. 80 (2016) 128–139, https://doi.org/10.1016/j.ijepes.2016.01.036.

[47] T.T. Nguyen, A high performance social spider optimization algorithm for optimal power flow solution with single objective optimization, Energy 171 (2019) 218–240, https://doi.org/10.1016/j.energy.2019.01.021.

[48] S.R. Salkuti, Optimal power flow using multi-objective glowworm swarm optimization algorithm in a wind energy integrated power system, Int. J. Green Energy 16 (2019) 1547–1561.

[49] X. Yuan, B. Zhang, P. Wang, J. Liang, Y. Yuan, Y. Huang, et al., Multi-objective optimal power flow based on improved strength Pareto evolutionary algorithm, Energy 122 (2017) 70–82, https://doi.org/10.1016/j.energy.2017.01.071.

[50] K. Dasgupta, P.K. Roy, V. Mukherjee, Power flow based hydro-thermal-wind scheduling of hybrid power system using sine cosine algorithm, Electr. Power Syst. Res. 178 (2020) 106018.

[51] A.-F. Attia, R.A. El Sehiemy, H.M. Hasanien, Optimal power flow solution in power systems using a novel Sine-Cosine algorithm, Int. J. Electr. Power Energy Syst. 99 (2018) 331–343, https://doi.org/10.1016/j.ijepes.2018.01.024.

[52] A. Saha, A. Bhattacharya, P. Das, A.K. Chakraborty, A novel approach towards uncertainty modeling in multiobjective optimal power flow with renewable integration, Int. Trans. Electr. Energy Syst. 29 (2019) e12136.

[53] C. Srithapon, P. Fuangfoo, P.K. Ghosh, A. Siritaratiwat, R. Chatthaworn, Surrogate-assisted multi-objective probabilistic optimal power flow for distribution network with photovoltaic generation and electric vehicles, IEEE Access 9 (2021) 34395–34414.

[54] M.R. Narimani, R. Azizipanah-Abarghooee, B. Zoghdar-Moghadam-Shahrekohne, K. Gholami, A novel approach to multi-objective optimal power flow by a new hybrid optimization algorithm considering generator constraints and multi-fuel type, Energy 49 (2013) 119–136, https://doi.org/10.1016/j.energy.2012.09.031.

[55] G. Chen, J. Qian, Z. Zhang, Z. Sun, Multi-objective optimal power flow based on hybrid firefly-bat algorithm and constraints-prior object-fuzzy sorting strategy, IEEE Access 7 (2019) 139726–139745.

[56] Z. Ullah, S. Wang, J. Radosavljević, J. Lai, A solution to the optimal power flow problem considering WT and PV generation, IEEE Access 7 (2019) 46763–46772.

[57] M. Riaz, A. Hanif, S.J. Hussain, M.I. Memon, M.U. Ali, A. Zafar, An optimization-based strategy for solving optimal power flow problems in a power system integrated with stochastic solar and wind power energy, Appl. Sci. 11 (2021) 6883.

[58] B. Jeddi, A.H. Einaddin, R. Kazemzadeh, Optimal power flow problem considering the cost, loss, and emission by multi-objective electromagnetism-like algorithm, in: 2016 6th Conference on Thermal Power Plants (CTPP), 2016.

[59] S. Sarhan, R. El-Sehiemy, A. Abaza, M. Gafar, Turbulent flow of water-based optimization for solving multi-objective technical and economic aspects of optimal power flow problems, Mathematics 10 (2022) 2106.

[60] R.A. El-Sehiemy, A novel single/multi-objective frameworks for techno-economic operation in power systems using tunicate swarm optimization technique, J. Ambient Intell. Humaniz. Comput. 13 (2022) 1073–1091.

[61] J. Sarda, K. Pandya, K.Y. Lee, Hybrid cross entropy—cuckoo search algorithm for solving optimal power flow with renewable generators and controllable loads, Optim. Control Appl. Methods 44 (2) (2023) 508–532, https://doi.org/10.1002/oca.2759.

[62] L.H. Pham, B.H. Dinh, T.T. Nguyen, Optimal power flow for an integrated wind-solar-hydro-thermal power system considering uncertainty of wind speed and solar radiation, Neural Comput. Appl. 34 (2022) 10655–10689, https://doi.org/10.1007/s00521-022-07000-2.

[63] M.H. Hassan, S.K. Elsayed, S. Kamel, C. Rahmann, I.B.M. Taha, Developing chaotic Bonobo optimizer for optimal power flow analysis considering stochastic renewable energy resources, Int. J. Energy Res. 46 (8) (2022) 11291–11325, https://doi.org/10.1002/er.7928.

[64] M. Ahmad, N. Javaid, I.A. Niaz, A. Almogren, A. Radwan, A bio-inspired heuristic algorithm for solving optimal power flow problem in hybrid power system, IEEE Access 9 (2021) 159809–159826.

[65] S. Mouassa, A. Althobaiti, F. Jurado, S.S.M. Ghoneim, Novel design of slim mould optimizer for the solution of optimal power flow problems incorporating intermittent sources: a case study of Algerian electricity grid, IEEE Access 10 (2022) 22646–22661.

[66] A.M. Shaheen, R.A. El-Sehiemy, H.M. Hasanien, A.R. Ginidi, An improved heap optimization algorithm for efficient energy management based optimal power flow model, Energy 250 (2022) 123795.

[67] S. Duman, J. Li, L. Wu, U. Guvenc, Optimal power flow with stochastic wind power and FACTS devices: a modified hybrid PSOGSA with chaotic maps approach, Neural Comput. Appl. 32 (2020) 8463–8492.

[68] A.A. El-Fergany, H.M. Hasanien, Single and multi-objective optimal power flow using grey wolf optimizer and differential evolution algorithms, Electr. Power Compon. Syst. 43 (2015) 1548–1559, https://doi.org/10.1080/15325008.2015.1041625.

[69] T. Niknam, M.R. Narimani, J. Aghaei, S. Tabatabaei, M. Nayeripour, Modified Honey Bee Mating Optimisation to solve dynamic optimal power flow considering generator constraints, IET Gener. Transm. Distrib. 5 (2011) 989, https://doi.org/10.1049/iet-gtd.2011.0055.

[70] R.D. Zimmerman, C.E. Murillo-Sanchez, R.J. Thomas, MATPOWER: steady-state operations, planning, and analysis tools for power systems research and education, IEEE Trans. Power Syst. 26 (1) (Feb. 2011) 12–19.

[71] P.P. Biswas, P.N. Suganthan, R. Mallipeddi, G.A.J. Amaratunga, Optimal power flow solutions using differential evolution algorithm integrated with effective constraint handling techniques, Eng. Appl. Artif. Intell. 68 (2018) 81–100, https://doi.org/10.1016/j.engappai.2017.10.019.

[72] L. Abualigah, D. Yousri, M. Abd Elaziz, A.A. Ewees, M.A.A. Al-Qaness, A.H. Gandomi, Aquila optimizer: a novel meta-heuristic optimization algorithm, Comput. Ind. Eng. 157 (2021) 107250.

[73] L. Wang, Q. Cao, Z. Zhang, S. Mirjalili, W. Zhao, Artificial rabbits optimization: a new bio-inspired meta-heuristic algorithm for solving engineering optimization problems, Eng. Appl. Artif. Intell. 114 (2022) 105082.

[74] T.S.L.V. Ayyarao, N.S.S. RamaKrishna, R.M. Elavarasan, N. Polumahanthi, M. Rambabu, G. Saini, et al., War strategy optimization algorithm: a new effective metaheuristic algorithm for global optimization, IEEE Access 10 (2022) 25073–25105.

[75] M. Mirrashid, H. Naderpour, Transit search: an optimization algorithm based on exoplanet exploration, Results Control Optim. 7 (2022) 100127.

[76] F.A. Hashim, E.H. Houssein, K. Hussain, M.S. Mabrouk, W. Al-Atabany, Honey Badger Algorithm: new metaheuristic algorithm for solving optimization problems, Math. Comput. Simul. 192 (2022) 84–110.

[77] H. Jia, X. Peng, C. Lang, Remora optimization algorithm, Expert Syst. Appl. 185 (2021) 115665.

[78] G. Dhiman, M. Garg, A. Nagar, V. Kumar, M. Dehghani, A novel algorithm for global optimization: rat swarm optimizer, J. Ambient Intell. Humaniz. Comput. 12 (2021) 8457–8482.

[79] S. Mirjalili, A. Lewis, The whale optimization algorithm, Adv. Eng. Softw. 95 (2016) 51–67.

[80] A.-A.A. Mohamed, Y.S. Mohamed, A.A.M. El-Gaafary, A.M. Hemeida, Optimal power flow using moth swarm algorithm, Electr. Power Syst. Res. 142 (2017) 190–206, https://doi.org/10.1016/j.epsr.2016.09.025.

[81] M. Ghasemi, S. Ghavidel, S. Rahmani, A. Roosta, H. Falah, A novel hybrid algorithm of imperialist competitive algorithm and teaching learning algorithm for optimal power flow problem with non-smooth cost functions, Eng. Appl. Artif. Intell. 29 (2014) 54–69, https://doi.org/10.1016/j.engappai.2013.11.003.

[82] M. Ghasemi, S. Ghavidel, M.M. Ghanbarian, M. Gitizadeh, Multi-objective optimal electric power planning in the power system using Gaussian bare-bones imperialist competitive algorithm, Inf. Sci. (NY) 294 (2015), https://doi.org/10.1016/j.ins.2014.09.051.

[83] W. Ongsakul, T. Tantimaporn, Optimal power flow by improved evolutionary programming, Electr. Power Compon. Syst. 34 (2006) 79–95, https://doi.org/10.1080/15325000691001458.

[84] J. Radosavljević, D. Klimenta, M. Jevtić, N. Arsić, Optimal power flow using a hybrid optimization algorithm of particle swarm optimization and gravitational search algorithm, Electr. Power Compon. Syst. 43 (2015) 1958–1970, https://doi.org/10.1080/15325008.2015.1061620.

[85] A. Ramesh Kumar, L. Premalatha, Optimal power flow for a deregulated power system using adaptive real coded biogeography-based optimization, Int. J. Electr. Power Energy Syst. 73 (2015) 393–399, https://doi.org/10.1016/j.ijepes.2015.05.011.

[86] Y. Sood, Evolutionary programming based optimal power flow and its validation for deregulated power system analysis, Int. J. Electr. Power Energy Syst. 29 (2007) 65–75, https://doi.org/10.1016/j.ijepes.2006.03.024.

[87] U. Guvenc, H. Bakir, S. Duman, B. Ozkaya, Optimal power flow using manta ray foraging optimization, in: Proceedings of the International Conference on Artificial Intelligence and Applied Mathematics in Engineering, 2020, pp. 136–149.

[88] A.K. Khamees, A.Y. Abdelaziz, M.R. Eskaros, A. El-Shahat, M.A. Attia, Optimal power flow solution of wind-integrated power system using novel metaheuristic method, Energies 14 (2021) 6117.

[89] H. Pulluri, R. Naresh, V. Sharma, A solution network based on stud krill herd algorithm for optimal power flow problems, Soft Comput. 22 (2018) 159–176, https://doi.org/10.1007/s00500-016-2319-3.

[90] M.A. Abido, Optimal power flow using tabu search algorithm, Electr. Power Compon. Syst. 30 (2002) 469–483, https://doi.org/10.1080/15325000252888425.

[91] T. Niknam, M. rasoul Narimani, M. Jabbari, A.R. Malekpour, A modified shuffle frog leaping algorithm for multi-objective optimal power flow, Energy 36 (2011) 6420–6432, https://doi.org/10.1016/j.energy.2011.09.027.
[92] K. Abaci, V. Yamacli, Differential search algorithm for solving multi-objective optimal power flow problem, Int. J. Electr. Power Energy Syst. 79 (2016) 1–10, https://doi.org/10.1016/j.ijepes.2015.12.021.
[93] S. Sayah, K. Zehar, Modified differential evolution algorithm for optimal power flow with non-smooth cost functions, Energy Convers. Manag. 49 (2008) 3036–3042, https://doi.org/10.1016/j.enconman.2008.06.014.
[94] R. Roy, H.T. Jadhav, Optimal power flow solution of power system incorporating stochastic wind power using Gbest guided artificial bee colony algorithm, Int. J. Electr. Power Energy Syst. 64 (2015) 562–578, https://doi.org/10.1016/j.ijepes.2014.07.010.
[95] L. Jebaraj, S. Sakthivel, A new swarm intelligence optimization approach to solve power flow optimization problem incorporating conflicting and fuel cost based objective functions, E-Prime - Adv. Electr. Eng. Electron. Energy 2 (2022) 100031.
[96] B. Bentouati, A. Khelifi, A.M. Shaheen, R.A. El-Sehiemy, An enhanced moth-swarm algorithm for efficient energy management based multi dimensions OPF problem, J. Ambient Intell. Humaniz. Comput. 12 (2021) 9499–9519, https://doi.org/10.1007/s12652-020-02692-7.
[97] W. Warid, H. Hizam, N. Mariun, N.I. Abdul Wahab, A novel quasi-oppositional modified Jaya algorithm for multi-objective optimal power flow solution, Appl. Soft Comput. 65 (2018) 360–373, https://doi.org/10.1016/j.asoc.2018.01.039.
[98] S.S.M. Ghoneim, M.F. Kotb, H.M. Hasanien, M.M. Alharthi, A.A. El-Fergany, Cost minimizations and performance enhancements of power systems using spherical prune differential evolution algorithm including modal analysis, Sustainability 13 (2021) 8113.
[99] R.A. El Sehiemy, F. Selim, B. Bentouati, M.A. Abido, A novel multi-objective hybrid particle swarm and salp optimization algorithm for technical-economical-environmental operation in power systems, Energy 193 (2020) 116817.
[100] M. Ghasemi, S. Ghavidel, M.M. Ghanbarian, M. Gharibzadeh, A. Azizi Vahed, Multi-objective optimal power flow considering the cost, emission, voltage deviation and power losses using multi-objective modified imperialist competitive algorithm, Energy 78 (2014) 276–289, https://doi.org/10.1016/j.energy.2014.10.007.
[101] C. Shilaja, K. Ravi, Optimal power flow using hybrid DA-APSO algorithm in renewable energy resources, Energy Proc. 117 (2017) 1085–1092.
[102] S. Gupta, N. Kumar, L. Srivastava, H. Malik, A. Pliego Marugán, F.P. García Márquez, A hybrid Jaya–Powell's pattern search algorithm for multi-objective optimal power flow incorporating distributed generation, Energies 14 (2021) 2831.
[103] H. Ouafa, S. Linda, B. Tarek, Multi-objective optimal power flow considering the fuel cost, emission, voltage deviation and power losses using Multi-Objective Dragonfly algorithm, in: International Conference on Recent Advances in Electrical Systems, Tunisia, 2017.
[104] J. Zhang, S. Wang, Q. Tang, Y. Zhou, T. Zeng, An improved NSGA-III integrating adaptive elimination strategy to solution of many-objective optimal power flow problems, Energy 172 (2019) 945–957.

Chapter 26

Optimizing CNN architecture using whale optimization algorithm for lung cancer detection

K. Sruthi[a], R.R. Rajalaxmi[b], R. Thangarajan[a], and C. Roopa[b]
[a]Department of IT, Kongu Engineering College, Perundurai, India, [b]Department of CSE, Kongu Engineering College, Perundurai, India

26.1 Introduction

Currently, despite huge advancements in the medical field worldwide, cancer remains one of the life-killer diseases to date. The fatality rate in the year 2020 is nearly 10 million (1). Breast cancer and lung cancer are the most prevalent cases in the year 2020, in terms of new cancer cases. Lung cancer is detected using a CT scan, which is one of the widely recognized tools for detecting cancer. The patients affected with lung cancer are detected in the middle or advanced stage of cancer. Detecting the lung nodule at an earlier stage and providing proper treatment is essential to reduce the mortality rate.

The development of malignant nodules in the lung's lobes is the primary cause of lung cancer. Lung nodules also known as pulmonary nodules, are small clumps of cells in the lungs. Detecting the nodule and differentiating it from non-nodules (non-cancerous cells) is a tedious task for the healthcare professional.

Deep learning tries to build artificial neural networks that can use algorithms to learn and intelligently make decisions. It uses neural networks with multiple layers of nodes between the input and output layers. The term "deep" in this case refers to the whole layers present in the network, i.e., the bigger number of layers, the deeper the network. A sequence of layers in the neural network present between input and output layer perform feature extraction. Deep learning plays a significant role in analyzing enormous amounts of data, carrying out complex algorithms, achieving the greatest performance with a huge amount of data, and performing efficient feature extraction. A larger range of issues is addressed by deep learning in the field of health care, from disease monitoring and cancer screening to offering individualized therapy recommendations.

The amount of data generated through radiological imaging methods such as MRI scans, X-rays, and CT are very huge. There aren't enough tools to transform all of this data into useful knowledge. In developed nations, one of the most terrible and deadly diseases is lung cancer. Medical imaging techniques like chest X-ray, CT scans, and MRI scans are employed in the early identification of lung cancer. In this case, identification entails categorizing the tumor into two classes: Malignant refers to cancerous tumors, while benign refers to non-cancerous tumors. When compared to a patient who is detected and treated for lung cancer at an early stage, a patient with advanced lung cancer has much lower survival odds.

Identification of lung cancer is a challenging and exceedingly complex problem. However, the patient has a good chance of survival if it is found in the early stages. According to diagnostic statistics, the majority of persons with this type of cancer who are diagnosed already have an advanced stage of the disease.

A crucial part of deep neural networks is their ability to distinguish between cancer cells and normal tissues, making them a useful tool for developing assistive AI-based cancer diagnosis. Only until the malignant cells are precisely distinguished from the healthy ones can a cancer treatment be effective. The classification of tumor cells is the cornerstone for deep learning-based cancer detection, followed by neural network training.

Convolutional Neural Networks (CNNs) is a subfield in deep learning and it is widely used for several applications, particularly in medical imaging. However, a wide range of factors influences system diagnosis precision. The use of computer-aided technology in this domain is becoming increasingly attractive to scientists in recent years. In the proposed methodology, network models are implemented using a novel method based on the CNN classifier. The parameters of the CNN model are also optimized using the whale optimization technique.

Handbook of Whale Optimization Algorithm. https://doi.org/10.1016/B978-0-32-395365-8.00032-4

26.2 Literature survey

Maier et al. [1] presented an introductory note about deep learning and its applications in medical image processing. This paper reviews the principles of perceptron and neural networks as well as the popularity of deep learning. Additionally, the work briefly illustrates medical image processing in detail, specifically image detection and recognition and segmentation computer-aided diagnostics and registration.

Nature provides numerous models for solving these challenges, hence the swarm intelligence optimization technique was developed. da Silva et al. [2] introduced a novel PSO Algorithm using Convolutional Neural Network to estimate the best hyper-parameters of CNN with decreased classification error. The experimental results exhibit 97.62% accuracy, 92.20% sensitivity, and 98.64% specificity. Due to the high potential ability of PSO optimal hyper-parameter values are obtained to show improved performance.

A deep learning-based technique for identifying lung nodules was introduced by Li et al. [3]. The characteristics are extracted using patch-based multi-resolution convolutional networks, which allowed four alternative combined models for classification. Comparing the suggested technique to other available traditional type technologies, the simulation test revealed improved robustness and effective performance.

Using various image-processing techniques, Vijh et al. [4] created an intelligent lung tumor diagnosis system. The steps include image enrichment, image segmentation, post-processing, feature abstraction, feature identification, and finally classification using a Support Vector Machine (SVM). WOA algorithm is used for the selection of the most predominant feature subset. WOA along with SVM is used in the automated assisted diagnosis scheme for concluding lung CT image as normal or abnormal.

In [5], different methods are developed to detect and classify pulmonary nodules with help of deep learning and swarm intelligence. The proposed approach utilizes swarm intelligence to optimize the weights in the CNN architecture. Seven different swarm algorithms are employed to test the performance of the model. Three of the algorithms exhibit superior performance with 25% faster learning than backpropagation-based learning. The experimental results show that the proposed models yield an accuracy of 93.71%. The sensitivity rate and specificity are 92.96% and 98.52%.

Rana et al. [6] introduced a novel approach to a deep learning assisted whale optimization algorithm for breast cancer prediction. CNN parameters are optimized with WOA to automatically adjust the CNN network structure with high detection accuracy. The main aim of this approach is to increase the accuracy of classifying breast cancer from histopathological images. The suggested technique outperforms other current algorithms when accuracy is compared to that of a standard CNN and other classifiers, according to the comparison.

A Multiscale CNN with Compound Fusions (MCNN-CF) architecture was suggested in this work [7] to lower the rate of false positive lung nodule detection. It performs feature fusion at two different levels. Class imbalance is handled with a novel training procedure. Experiments conducted on the LUNA16 dataset indicate the effectiveness of the method with a reduced false positive rate.

Murugan et al. [8] optimized Deep learning networks with the help of the Whale Optimization algorithm and backpropagation Algorithm for COVID-19 diagnosis. The weights and biases of ResNet-50 architecture are tuned to improve the performance of the network. The experiments conducted on the COVID-CT scan dataset reveal that the proposed method yields 98.78% accuracy, 98.37% sensitivity, 99.19% specificity, 99.18% precision, and 98.37% F1 score.

Ajai & Anitha [8] initially located the lung nodule regions using an entropy-based K-means clustering approach followed by segmentation. Next, an Improved Capuchin Search Algorithm (ICSA) is utilized to optimize the hybrid CNN-LSTM architecture. The experiments conducted on the LIDC–IDRI dataset produced an accuracy of 98% with the pulmonary nodule having a diameter greater than or equal to 3 mm.

In [9], an integrated Social Ski-Driver (SSD) algorithm and Shuffled Shepherd Optimization Algorithm (SSOA) is developed. Features are extracted from the CT images and a Multi-objective Rectification Network is used for classification. The proposed algorithm exhibits high performance with 89.6% accuracy, 10.4% Mean Absolute Error (MAE), 89.69% sensitivity, and 84.5% specificity.

A new method [10] is developed to identify non-small lung cancer for accurate diagnosis. The proposed method utilizes InceptionV3 focuses on pathological portions of the images by combining channel attention and spatial attention methods. The effectiveness of the technique is evaluated using a hospital-sourced, three-category dataset. Based on the tests, the accuracy on two open datasets is 95.24% and 98.14%.

The study in [11] proposed an artificial intelligence approach for lung cancer classification. The work uses a new modified Satin Bowerbird Optimization Algorithm to optimize the Alexnet architecture and identify relevant features. The proposed algorithm shows 95.96% accuracy when compared with other methods. Also, a higher F1-score and recall are obtained than other methods.

To locate the cancerous areas in the image, feature extraction is essential. A hybrid ensemble feature extraction model is created [12] for effectively identifying lung and colon cancer. The classification of the images is carried out using a combination of deep feature extraction and ensemble learning. Using histological lung and colon datasets, the model was assessed. Show that the suggested model has 99.05%, 100%, and 99.30% accuracy in detecting lung, colon, and (lung and colon) cancer, respectively.

Vijh et al. [13] presented a novel method to detect and classify lung cancer images using hybrid whale optimization and adaptive particle swarm optimization algorithm. Initially, pre-processing is applied over the images and segmented to separate the tumor regions. Different features like statistical, geometrical, texture, and structural features are extracted. The hybrid algorithm is applied to extract the relevant features and the CNN model is used for classification. The results indicate that the hybrid approach outperforms other methods with 97.18% accuracy, 97% sensitivity, and 98.66% specificity.

Automated lung nodule detection in CT images was suggested by Kumar et al. [14] using enhanced CNN and an improved whale optimization technique. The work is implemented in three stages pre-processing, segmentation, and feature extraction. Segmentation is carried out using Otsu thresholding and local binary pattern (LBP) features are extracted. An optimized convolutional neural network (CNN) model is used for classification and the experiments reveal that the proposed method outperforms other approaches.

To find the global best solution, Mirjalili & Lewis [15] suggested the Whale Optimization Algorithm (WOA), which is a kind of efficient method based on the swarm intelligence heuristic search algorithm. Due to its advantages of being straightforward, having fewer parameters, having good resilience, and being simple to apply, WOA has become increasingly popular. As a result, the WOA algorithm is helpful to tune the CNN model parameters, removing the need for a manual search to find the best hyper-parameters for classifying nodules and non-nodules.

26.3 Optimized convolutional neural network

The primary goal of the proposed work is to develop a customized CNN by tuning the hyper-parameters with a Whale optimization algorithm to categorize CT images into nodules and non-nodules.

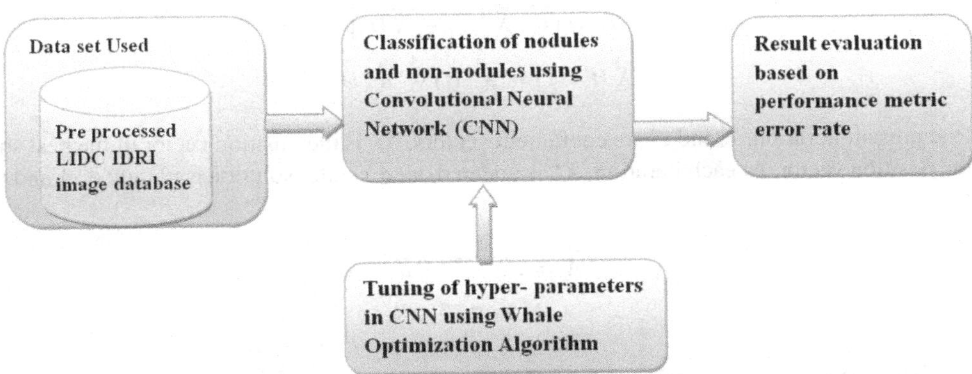

FIGURE 26.1 Proposed Methodology.

According to Fig. 26.1, the proposed methodology is broken down into three parts. The first stage involves preparing the images for identifying lung cancer. The Whale Optimization based CNN model is used in the second stage to classify data into nodules and non-nodules, followed by result evaluation in the final stage.

26.3.1 Convolutional neural network model

Fig. 26.2 depicts the CNN model used for the classification of lung cancer images. Four convolutional layers are used to extract the features from the images. Similarly, four max-pooling layers are used to identify the important features after each convolution operation. Finally, at the last layer flattening is performed followed by a classification task.

The main challenge faced in developing a CNN lies in selecting the suitable values for the hyper-parameters to yield accurate classification. The hyper-parameters associated with CNN include the number of filters, size of the filter, number of neurons, activation function, optimizers, etc. Metaheuristic techniques can be utilized to tune the hyper-parameters and improve the network performance thereby reducing the manual search of choosing optimal hyper-parameters. In the proposed work whale optimization algorithm is used to obtain optimal hyper-parameters of CNN.

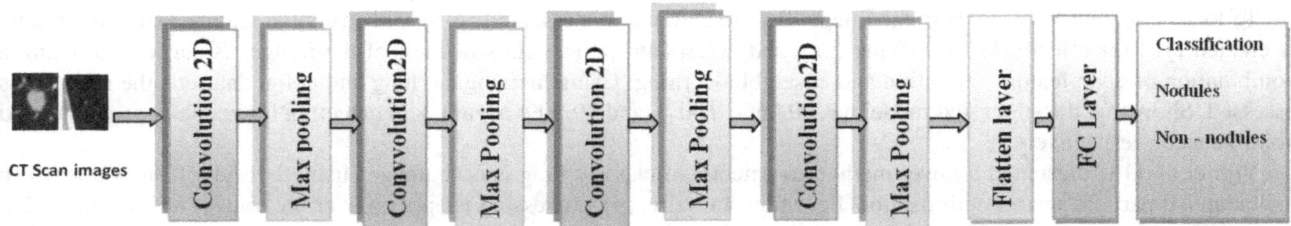

FIGURE 26.2 Convolutional Network Architecture.

26.3.2 Whale optimization algorithm

Whales are the largest intelligent mammal in the world. Whales come in a variety of species, including humpback, finback, and killer whales. The humpback whales hunting habits are the subject of this investigation. A new swarm-based optimization method namely Whale Optimization Algorithm (WOA) is developed by Mirjalili et al. [15]. It is developed based on the humpback whales hunting strategies. Three operators were used in this technique to mimic how humpback whales hunt by searching for prey, circling prey, and using bubble nets.

The prey for humpback whales is a group of small fishes or krill. This type of hunting approach is known as bubble net feeding. Bubble net feeding is the term for this kind of foraging behavior. To perform optimization this spiral bubble net feeding is mathematically modeled which includes encircling prey, spiral bubble net feeding, and search for prey.

Humpback whales need to locate the areas of prey and recognize and circle them. The WOA algorithm first chooses the current solution as the appropriate candidate solution for the target prey as the precise location of the prey is not known in advance. The best whale is identified that depicts the best solution. Based on the results, the other agents/whales update their positions to move closer to the best whale. The following equations depict such behavior:

$$\vec{D} = \left| \vec{C} . \vec{X}^*(t) - \vec{X}(t) \right| \tag{26.1}$$

$$\vec{X}(t+1) = \vec{X}^*(t) - \vec{A} . \vec{D} \tag{26.2}$$

where t denotes the present iteration, A and C are coefficient vectors, X^* is the position vector of the best solution obtained so far, and X is the position vector. At each iteration, X^* is updated once a better solution is obtained. A and C are computed as follows:

$$\vec{A} = 2\vec{a} . \vec{r} - \vec{a} \tag{26.3}$$

$$\vec{C} = 2 . \vec{r} \tag{26.4}$$

Here, for every iteration a is linearly reduced from 2 to 0 and a random vector r is in [0, 1]. The whales bubble-net behavior is performed using shrinking encircling and spiral updating approaches to locate the prey. Equation 26.3 models the shrinking encircling method. The humpback whales' movement in spiral space is represented as follows:

$$\vec{X}(t+1) = \vec{D'} . e^{bl} . \cos(2\pi l) + \vec{X}^*(t) \tag{26.5}$$

where $\vec{D'} = \vec{X}^*(t) - \vec{X}(t)$ denotes the distance between the whale and prey, b is used to define the logarithmic spiral shape and l depicts a random number in the range $[-1, 1]$. The whales swim within a shrinking circle and curved path simultaneously to locate the prey. This activity is modeled with a 50% probability to switch between shrinking encircling and spiral shapes. This is used to update the position of the whales as per the following equation:

$$\vec{X}(t+1) = \begin{cases} \vec{X}^*(t) - \vec{A} . \vec{D} & \text{if } p < 0.5 \\ \vec{D'} . e^{bl} . \cos(2\pi l) + \vec{X}^*(t) & \text{if } p \geq 0.5 \end{cases} \tag{26.6}$$

where p is a random number in [0, 1].

Also, the whales can search the prey randomly as per the equations:

$$\vec{D} = \left|\vec{C}.\vec{X}_{rand} - \vec{X}\right| \tag{26.7}$$

$$\vec{X}(t+1) = \vec{X}_{rand} - \vec{A}.\vec{D} \tag{26.8}$$

In this work, CNN hyper-parameters are tuned using WOA. Evaluating the fitness of whales is essential to identify the best hyper-parameters of CNN. The proposed work uses the error rate the model as a fitness function and is defined as follows:

$$Error\ Rate = \frac{\#misclassified\ images}{\#total\ images} \tag{26.9}$$

Fig. 26.3 depicts the flowchart of the proposed WOA method.

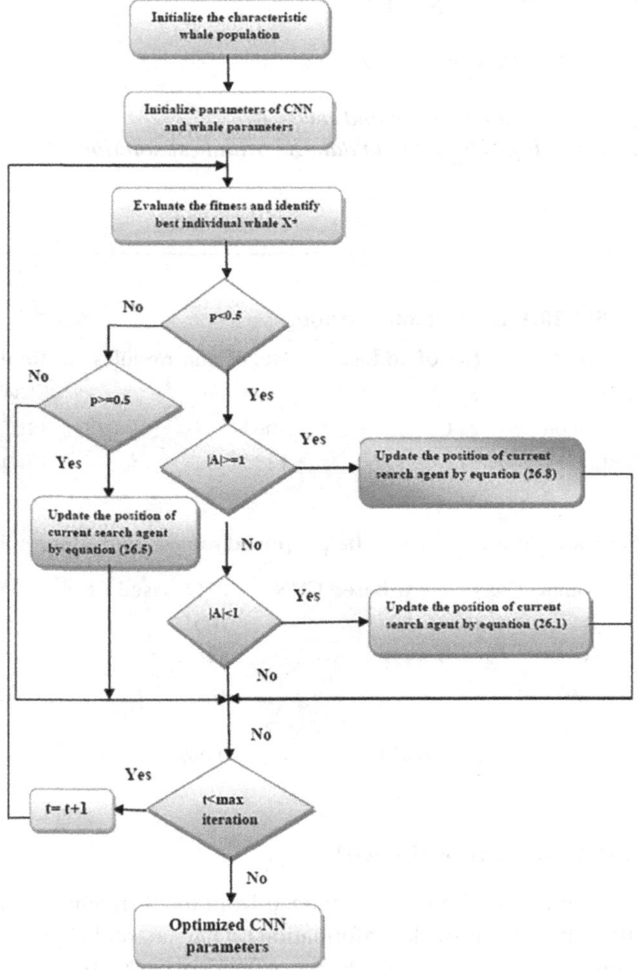

FIGURE 26.3 Proposed WOA.

The hyper-parameters of CNN optimized with WOA include, the number of filters in the first and second convolutional layer and also the number of neurons in the hidden layer.

Let #fc1 denote the number of filters in the first convolutional layer, #fc2 denote the number of filters in the second convolutional layer and #hn denote the number of neurons in hidden layers.

The pseudocode of the WOA based CNN algorithm is shown in Algorithm 26.1.

Algorithm 26.1 WOA algorithm.

Initialize CNN model parameters for lung cancer detection.
Initialize whale parameters and hyper-parameter values (#fc1, #fc2, #hn)
Initialize whale population X_i $(i = 1, 2 \ldots n)$
Evaluate fitness of whale using Eq. (26.9)
Identify the best individual whale X^*
while $(i <$ *number of iterations*)
for each whale
    update a, A, C, l, and p
    if $(p < 0.)$
        if $(|A| < 1)$
            Update current whale position using Eq. (26.1)
    else if $(|A| >= 1)$
            Choose a random whale (X_{rand})
            Update current whale position using Eq. (26.8)
    else
        Update current whale position using Eq. (26.5)
end
identify any whale that goes beyond the solution space and revise it
compute the fitness of each whale using Eq. (26.9) and update X^* *with best solution*
$i = i + 1$
end
return X^*

TABLE 26.1 Dataset information.

Dataset	No. of nodules	No. of non-nodules	Total
Train	845	4339	5184
Validation	224	1063	1287
Test	282	1298	1580

TABLE 26.2 Performance comparison of the proposed model with other approaches.

Parameters	Vanilla CNN	Bat-based CNN	PSO-based CNN	WOA-based CNN
# filters in C1	32	48	100	11
# filters in C2	32	90	44	43
# neurons in the hidden layer	128	78	100	53
Error Rate	0.0863	0.0518	0.0696	0.0500

26.4 Experimental results and discussion

This section presents a detailed discussion of the results obtained from the different methods. The entire methodology is implemented in Keras framework. Table 26.1 provides information on the dataset [16] utilized for the studies.

Table 26.2 shows the comparison of test accuracies between Vanilla CNN, Bat-based CNN, PSO-based CNN, and WOA-based CNN.

WOA-based CNN algorithm has achieved a lower error rate compared to Vanilla CNN [17], PSO-based CNN [2], and Bat-based CNN [18]. Although the proposed WOA method is implemented with smaller datasets and hyper-parameter optimization is limited to three parameters, the findings show that the CNN in conjunction with the WOA algorithm may produce good results for high-dimensional datasets. In some instances, the proposed model's features generated from the images are the same for nodule and non-nodule images. As a result, a few nodule images are classified as non-nodule images by the model.

TABLE 26.3 Statistical comparison of the proposed model versus other approaches.

Measure	Vanilla CNN	Bat-based CNN	PSO-based CNN	WOA-based CNN
Best	0.0863	0.0518	0.0696	0.0500
Worst	0.1157	0.0563	0.0810	0.0569
Mean	0.1012	0.0551	0.0731	0.0527
Median	0.1019	0.0557	0.0708	0.0512

TABLE 26.4 Run time comparison of proposed model with other approaches.

Runs	Run time comparison (in minutes)			
	Vanilla CNN	Bat-based CNN	PSO-based CNN	WOA-based CNN
1	2.33	77.56	124.76	76.19
2	2.28	77.41	123.02	76.30
3	2.36	76.95	123.28	76.27
4	2.33	77.63	123.55	76.45
5	2.34	77.54	123.56	76.61

Table 26.3 compares the algorithms' accuracy statistically. Since the meta-heuristic algorithms are stochastic in nature, the results are obtained from five different runs and compared. The proposed WOA-based CNN algorithm outperforms other algorithms in terms of best, worst, mean, and median values. When compared to Vanilla CNN, Bat based CNN, and PSO based CNN, the proposed WOA based CNN method outperforms them by 42.06%, 3.47%, and 28.16%, respectively. Furthermore, the results show that the mean error rate is 47.92%, 4.35%, and 27.90% lower than Vanilla CNN, Bat-based CNN, and PSO-based CNN. A comparison of median performance reveals that the proposed method outperforms Vanilla CNN, Bat-based CNN, and PSO-based CNN by 49.75%, 8.01%, and 27.68%, respectively. As a result, it is clear that the proposed model performs better comparatively to the other approaches.

The runtime analysis of all the algorithms is shown in Table 26.4. As Vanilla CNN model is trained with predefined parameters, it takes lesser time. However, meta-heuristic algorithms have to find the best solution by exploration and exploitation, they tend to take more time than standard CNN methods. From the results, it is noticed that WOA based CNN determines the best solution with minimum time than other meta-heuristic approaches.

26.5 Conclusion

The WOA algorithm is used to optimize the CNN parameters. WOA with CNN is employed in the proposed study to manage the hyper-parameters, resulting in superior solutions through exploration and exploitation. The result shows that the optimized CNN model achieved a lower error rate to classify nodules and non-nodules compared with other optimized CNN models. In the future, CNN architecture can be used with along with other nature inspiring methods to improve hyper-parameter optimization and reduce error rates. Also, in the proposed methodology only a few parameters in the CNN model have been optimized. In the future, more hyper-parameters can be optimized. The suggested method does not specify the stages of lung cancer; it only classifies CT images as nodules or non-nodules. Determining the stages, such as initial, moderate, and severe cases, can be the subject of future research.

References

[1] A. Maier, C. Syben, T. Lasser, C. Riess, A gentle introduction to deep learning in medical image processing, Z. Med. Phys. 29 (2) (May 2019) 86–101, https://doi.org/10.1016/J.ZEMEDI.2018.12.003.

[2] G.L.F. da Silva, T.L.A. Valente, A.C. Silva, A.C. de Paiva, M. Gattass, Convolutional neural network-based PSO for lung nodule false positive reduction on CT images, Comput. Methods Programs Biomed. 162 (Aug. 2018) 109–118, https://doi.org/10.1016/J.CMPB.2018.05.006.

[3] L. Li, Z. Liu, H. Huang, M. Lin, D. Luo, Evaluating the performance of a deep learning-based computer-aided diagnosis (DL-CAD) system for detecting and characterizing lung nodules: comparison with the performance of double reading by radiologists, Thorac. Cancer 10 (2) (Dec. 2018) 183–192, https://doi.org/10.1111/1759-7714.12931.

[4] S. Vijh, D. Gaur, S. Kumar, An intelligent lung tumor diagnosis system using whale optimization algorithm and support vector machine, Int. J. Syst. Assur. Eng. Manag. 11 (2) (Sep. 2019) 374–384, https://doi.org/10.1007/s13198-019-00866-x.

[5] C.A. de Pinho Pinheiro, N. Nedjah, L. de Macedo Mourelle, Detection and classification of pulmonary nodules using deep learning and swarm intelligence, Multimed. Tools Appl. 79 (21–22) (Jun. 2020) 15437–15465, https://doi.org/10.1007/s11042-019-7473-z.

[6] P. Rana, P. Kumar Gupta, V. Sharma, A novel deep learning-based whale optimization algorithm for prediction of breast cancer, Braz. Arch. Biol. Technol. 64 (2021) 21200221, https://doi.org/10.1590/1678-4324-2021200221.

[7] P.S. Mittapalli, T. V, Multiscale CNN with compound fusions for false positive reduction in lung nodule detection, Artif. Intell. Med. 113 (2021) 102017, https://doi.org/10.1016/j.artmed.2021.102017.

[8] R. Murugan, T. Goel, S. Mirjalili, D.K. Chakrabartty, WOANet: whale optimized deep neural network for the classification of COVID-19 from radiography images, Biocybern. Biomed. Eng. 41 (4) (Oct. 2021) 1702–1718, https://doi.org/10.1016/J.BBE.2021.10.004.

[9] A.K. Ajai, A. Anitha, Clustering based lung lobe segmentation and optimization based lung cancer classification using CT images, Biomed. Signal Process. Control 78 (2022) 103986, https://doi.org/10.1016/j.bspc.2022.103986.

[10] Z. Xu, H. Ren, W. Zhou, Z. Liu, ISANET: non-small cell lung cancer classification and detection based on CNN and attention mechanism, Biomed. Signal Process. Control 77 (2022) 103773, https://doi.org/10.1016/j.bspc.2022.103773.

[11] Y. Xu, Y. Wang, N. Razmjooy, Lung cancer diagnosis in CT images based on Alexnet optimized by modified Bowerbird optimization algorithm, Biomed. Signal Process. Control 77 (2022) 103791, https://doi.org/10.1016/j.bspc.2022.103791.

[12] M.A. Talukder, M.M. Islam, M.A. Uddin, A. Akhter, K.F. Hasan, M.A. Moni, Machine learning-based lung and colon cancer detection using deep feature extraction and ensemble learning, Expert Syst. Appl. 205 (2022) 117695, https://doi.org/10.1016/J.ESWA.2022.117695.

[13] S. Vijh, Prashant Gaurav, Hari M. Pandey, Hybrid bio-inspired algorithm and convolutional neural network for automatic lung tumor detection, in: Deep Neuro-Fuzzy Analytics in Smart Ecosystems, Neural Comput. Appl. (2020), https://doi.org/10.1007/s00521-020-05362-z.

[14] M.K. Kumar, A. Amalanathan, Automated lung nodule detection in CT images by optimized CNN: impact of improved whale optimization algorithm, Comput. Assist. Methods Eng. Sci. 29 (1–2) (2022) 7–31, https://doi.org/10.24423/cames.372.

[15] S. Mirjalili, A. Lewis, The whale optimization algorithm, Adv. Eng. Softw. 95 (May 2016) 51–67, https://doi.org/10.1016/J.ADVENGSOFT.2016.01.008.

[16] K.S. Mader, The Lung Image Database Consortium image collection (LIDC-IDRI), IEEE Dataport, https://doi.org/10.21227/zce3-jp96, 2021.

[17] H.F. Al-Yasriy, M.S. Al-Husieny, F.Y. Mohsen, E.A. Khalil, Z.S. Hassan, Diagnosis of lung cancer based on CT scans using CNN, IOP Conf. Ser., Mater. Sci. Eng. 928 (2) (Nov. 2020), https://doi.org/10.1088/1757-899X/928/2/022035.

[18] R.R. Rajalaxmi, K. Sruthi, S. Santhoshkumar, Bat algorithm with CNN parameter tuning for lung nodule false positive reduction, in: Third IFIP TC 12 International Conference on Computational Intelligence in Data Science, Chennai, India, Feb. 2020, pp. 131–142, https://doi.org/10.1007/978-3-030-63467-4_10.

Chapter 27

Multi-response optimization of plasma arc cutting on Monel 400 alloy through whale optimization algorithm

D. Rajamani, M. Siva Kumar, and E. Balasubramanian
Centre for Autonomous System Research, Department of Mechanical Engineering, Vel Tech Rangarajan Dr Sagunthala R&D Institute of Science and Technology, Chennai, India

27.1 Introduction

In today's cutthroat manufacturing sector, process optimization is crucial for lowering production costs, boosting product quality, increasing process performance while reducing human error, and fostering consistency. Furthermore, it helps the operator or machinist to determine the optimal set of process parameters with minimal effort and without having to consult any data handbooks. Machine malfunctions might occur when operators aren't aware of the best settings for the various controls they employ. If possible, it is best to apply multi-objective optimization methods, which can determine the optimal settings for the process parameters in order to optimize all of the relevant answers at once. This is because there are often competing goals during the machining process (like maximization of material removal rate and minimization of kerf characteristics, surface roughness, heat affected zone, and energy consumption, etc.) [1]. Researchers in the fields of additive manufacturing, machining, machine layout, selective assembly, etc., have recently adopted a number of multi-criteria decision-making techniques, such as desirability, gray relational analysis, TOPSIS, etc., and metaheuristic algorithms, such as NSGA, SA, PSO, ABC, TLBO, etc., to solve multi-objective problems.

Whilst metaheuristic algorithms are often successfully utilized for solving single-objective issues in a wide range of applications, there has been a lack of research into their usage for optimizing multi-response problems, especially in contemporary manufacturing processes. Recent studies suggest that the capabilities of metaheuristic algorithms for multi-performance optimization of atypical machining processes, in particular the PAC process [2], have not been fully established. Creating PAC components with desirable cutting qualities including enhanced surface quality, optimized substrate material removal, and refined kerf characteristics takes a long time because of the non-linear and complex cutting operations involved. This is due to the existence of a wide variety of process-related parameters, including but not limited to arc current, cutting speed, nozzle diameter, stand-off distance, etc. [3]. Thus, a powerful optimization method is necessary to attain optimum processing conditions for producing high-quality end-use components treated by the PAC method.

The humpback whale's bubble net hunting behavior served as inspiration for one of the first examples of nature-inspired metaheuristic algorithms, the whale optimization algorithm (WOA). It has seen extensive use in a number of fields for the purposes of streamlining multi-objective optimization problems and attaining optimum process conditions [4]. Because of its similarity to other nature-inspired algorithms and its simplicity and robustness in comparison to other swarm intelligence techniques, it is often used as an example of the latter. The method has fewer control parameters than most others, since just the time period needs to be changed. Optimizing process parameters with WOA has been employed by several researchers as of late to enhance the quality and performance aspects of traditional and non-traditional machining processes [5,6].

According to the extant literature, only a few researchers have used WOA to optimize machining parameters, and none of them have used WOA to optimize the PAC process. As a result, the current work focuses on the optimization of PAC parameters to improve the MRR, Kt, and HAZ of processed Monel 400 alloy sheets. The experiments were carried out

Handbook of Whale Optimization Algorithm. https://doi.org/10.1016/B978-0-32-395365-8.00033-6

in accordance with the BBD approach, and the response were measured for optimization. The statistical adequacy of the experiments was assessed using ANOVA, and the impact of PAC parameters on the specified response characteristics was investigated using response surface plots. WOA was used for the multi-response optimization, and its performance was compared to that of well-known metaheuristic approaches such as the particle swarm algorithm and the harmony search algorithm.

27.2 Methodologies

27.2.1 Response surface methodology

The optimization of complicated processes can be achieved by the application of RSM, an empirical statistical modeling tool [7]. It gives a simple and effective method for pinpointing the optimal design space, from the standpoint of performance. RSM, on the other hand, seeks to approximate it using a lower-order polynomial, namely a second-order polynomial equation. This article describes the quadratic response surface, which is used to establish a correlation between the independent variables (X_i) and the response (Y).

$$Y = b_0 + \sum_{i=1}^{n} b_i X_i + \sum_{i=1}^{n} b_{ii} X_i^2 + \sum_{i<j}^{n} b_{ij} X_i X_j \tag{27.1}$$

where n is the number of design variables, and b_0, b_i, b_{ii}, and b_{ij} represent the coefficient of constant, linear, quadratic, and cross product terms respectively.

In this investigation, the Box-behnken design (BBD) with four process variables at three levels was chosen as the appropriate model for optimizing the PAC process parameters. The BBD is commonly employed in non-sequential trials because it facilitates effective estimate of first and second order coefficients with fewer design points, hence reducing the number of experiments required. The current investigation followed the BBD experimental design to obtain a quadratic model of thirty experiments with six replicants in order to seek the optimum circumstances and investigate the impact of process factors on the PAC process. The model fitting method used to acquire the expected values is implemented within Design Expert software version 8.0.

27.2.2 Whale optimization algorithm

A swarm-based metaheuristic optimization algorithm called the Whale Optimization Algorithm (WOA) was developed in response to the distinctive bubble-net hunting strategy of humpback whales [8]. The three main stages of WOA include encircling the prey, using a spiral bubble-net strategy, and looking for prey. The following is a description of the WOA's later stages [9]:

Encircling: The humpback whale can locate the location of its prey, and it will surround that area since it is unsure of where to position itself inside the search area. The humpback will consequently presume that the prey's current location is closer to the ideal one. Encircling is a specific type of behavior and is defined as follows:

$$X(t+1) = X^*(t) - A.D \tag{27.2}$$

$$D = \left|C.X(t) - X^*(t)\right| \tag{27.3}$$

$$A = 2a.r - a \tag{27.4}$$

$$C = 2.r \tag{27.5}$$

where, X^* is denoted as the best solution of new position and it will update in each iteration if the new solution is better than existing result, A and C are the coefficient vectors, t indicates the current iteration number, r is random vector [0, 1], a is linearly decreases from 2 to 0 during the event of subsequent iterations. A hunt whale's new location can be anywhere between its previous location and the location of the best whale right now. In the WOA algorithm, this procedure is known as encircling.

Spiral bubble-net scheme: The humpback whale will swim in a spiral pattern around its prey as it is being hunted. The following is a description of the whale's route:

$$X(t+1) = D'.e^{bl}.\cos(2\pi l) + X^*(t) \tag{27.6}$$

$$D' = |X^*(t) - X(t)| \tag{27.7}$$

The shape of the spiral path is defined by b which is a constant, whereas the l is the random number between -1 to 1. Two behaviors, each of which has a 50% chance of occurring, are used in WOA and are described as follows:

$$X(t+1) = \left\{ \begin{array}{ll} X^*(t) - A.D & p < 0.5 \\ D'.e^{bl}.\cos(2\pi l) + X^*(t) & p \geq 0.5 \end{array} \right\} \tag{27.8}$$

where, p is randomly generated between 0 to 1. In addition to the spiral bubble-net strategy, humpback whales also haphazardly search for prey. In the section that follows, the method of random prey search is discussed.

Searching: The coefficient vector A must have values more than 1 or less than -1 to achieve the optimum global solutions. The humpback whales are randomly looking for new locations to be near one another. The mathematical representation of this mechanism, which is known as the exploration phase, is as follows:

$$X(t+1) = X_{rand}(t) - A.D \tag{27.9}$$

$$D = |C.X_{rand}(t) - X(t)| \tag{27.10}$$

where X_{rand} is the search whales' position vector, which was randomly chosen from the present population. Fig. 27.1 shows the process flow for executing the whale optimization algorithm for the proposed work.

27.3 Experimental details

The plasma arc cutting (PAC) experiments were performed on Monel 400 alloys sheet of thickness 3 mm and chemical composition of 63% Ni, 31.6% Cu, 2.5% Fe, 2% Mn, 0.5% Si, and 0.3% C based on the proposed BBD approach on a computerized controlled industrial use CNC plasma cutting machine. PlasmaCAM CNC software is included with the PAC system to ensure the precise motion of the plasma jet through the nozzle. A high-energy plasma is created using atmospheric air as a shield gas to thaw out and spew the melted metal over the workpiece surface. Through the use of a servo-operated torch with a copper nozzle designed with an air-cooled swirl, high precision cutting is achieved. The range of PAC parameters such as cutting speed (2200, 2400, and 2600 mm/min), gas pressure (3, 3.5, and 4 bar), arc current (45, 50, and 55 A) and stand-off distance (2, 2.5, and 3 mm) for performing the experiments and their corresponding response characteristics such as material removal rate, kerf taper, and heat affected zone is mentioned in Table 27.1.

27.4 Results and discussion

Statistical analysis of the conducted experiments and collected response characteristics is necessary to probe the efficiency of the experimental method. Multi parametric analysis of variance (ANOVA) was used to examine the correlation coefficient, sum of squares, goodness of fit, individual, interaction, and quadratic effects, and F statistics for the PAC treated alloy sheet at three different phases of manufacture. As a first step, analysis of variance (ANOVA) is performed to look at how a few different factors are affecting the reliability, validity, and precision of the response indices being employed. In the second step, we construct polynomial equations of the second order for each response, which we then use to analyze the impact of our selected independent variables on the final output. Lastly, the whale optimization process is used to find the optimal ranges for the parameters.

27.4.1 Statistical analysis of derived mathematical models

In order to further investigate the suggested RSM based experimental technique, the PAC experiments were conducted and response characteristics were assessed. Effectiveness of the established mathematical models was statistically examined

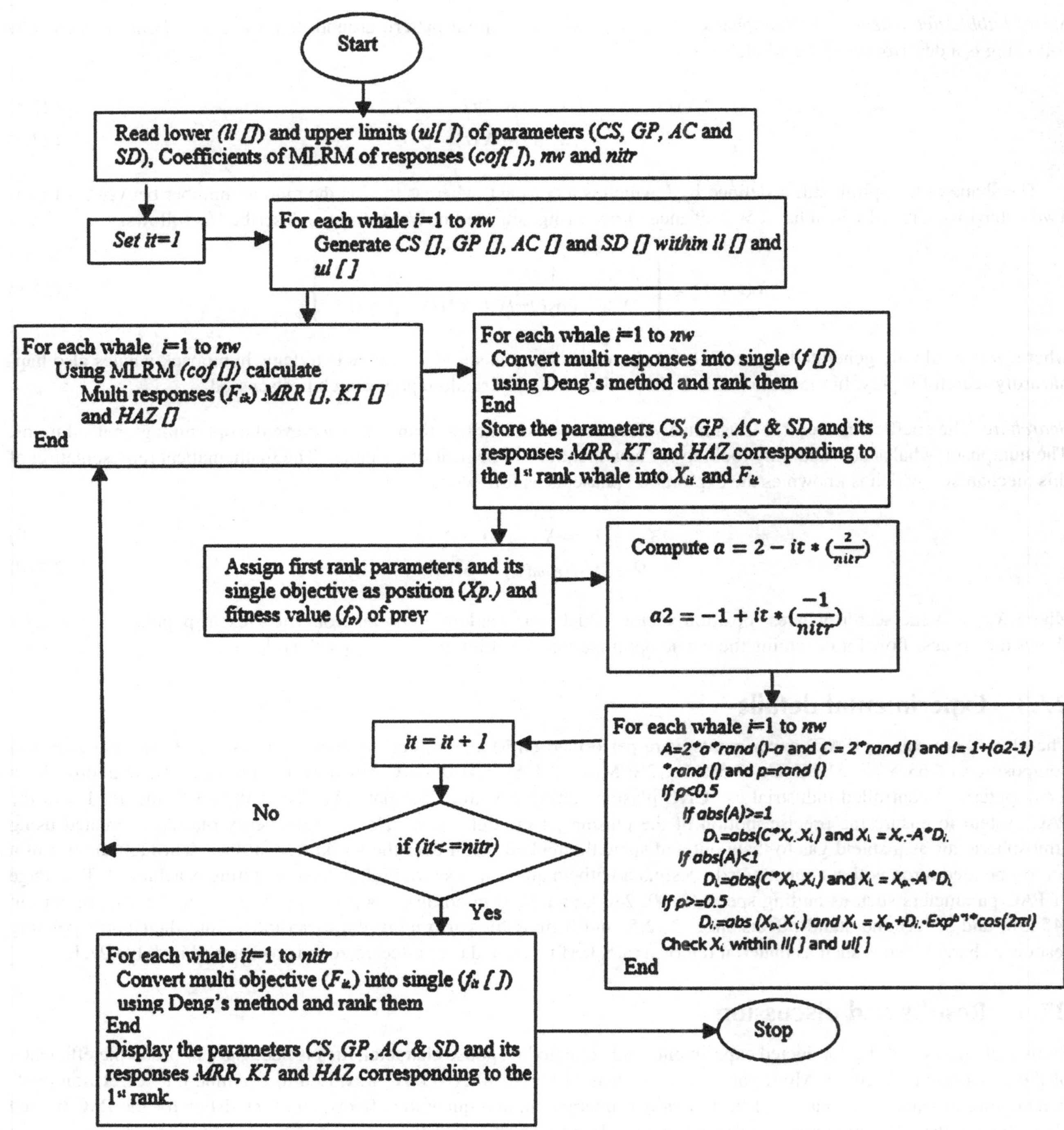

FIGURE 27.1 Flow chart for whale optimization algorithm to optimize PAC process.

with ANOVA. Response features including MRR, Kt, and HAZ were subjected to statistical analysis, and the outcomes were tabulated (Tables 27.2–27.4). Each response characteristic is subjected to an ANOVA with a 95% confidence interval, and the nonsignificant parameters (those with F-values greater than 0.05) are filtered out using a backward elimination method. According to the analysis of variance, after taking into account the effects of each subject, the chosen PAC parameters are significant. It was also determined that the gas pressure and stand-off distance had the greatest impact on MRR and HAZ, while cutting speed and stand-off distance had the greatest impact on Kt. The suitability of proposed regression

TABLE 27.1 Experimental design and measured response values.

Exp. No	Input parameters				Responses		
	Cutting speed (mm/min)	Gas pressure (Bar)	Arc current (A)	Stand-off distance (mm)	MRR (mm^3/min)	Kerf taper (Degree)	HAZ (mm)
1	2400	3.5	50	2.5	31.42	6.91	4.01
2	2400	3	50	3	33.84	5.79	4.74
3	2400	3.5	45	2	46.37	2.52	6.21
4	2400	3.5	50	2.5	32.55	7.59	4.26
5	2600	3.5	50	3	25.90	7.71	3.05
6	2200	3.5	50	3	28.74	7.07	3.84
7	2400	3	45	2.5	40.61	4.45	5.70
8	2400	4	55	2.5	29.47	4.66	3.81
9	2200	3	50	2.5	32.18	4.49	4.44
10	2400	3.5	55	2	34.11	6.63	4.56
11	2200	3.5	50	2	34.78	5.90	4.26
12	2600	3	50	2.5	33.57	8.56	4.50
13	2200	3.5	45	2.5	39.21	3.03	5.25
14	2200	3.5	55	2.5	27.48	5.62	3.68
15	2400	3.5	55	3	35.72	3.82	4.79
16	2200	4	50	2.5	27.75	6.91	3.21
17	2400	3.5	50	2.5	30.35	7.89	4.07
18	2400	3.5	50	2.5	29.36	7.44	4.23
19	2400	3	55	2.5	38.02	4.19	5.36
20	2400	3.5	50	2.5	30.26	7.40	4.05
21	2600	3.5	55	2.5	33.99	5.39	4.38
22	2400	3	50	2	40.28	7.08	5.40
23	2400	3.5	45	3	27.57	7.47	3.69
24	2400	4	50	2	36.35	4.55	4.34
25	2400	4	50	3	25.57	7.23	2.88
26	2600	3.5	50	2	32.76	7.46	4.38
27	2600	4	50	2.5	24.32	4.94	2.48
28	2600	3.5	45	2.5	25.60	6.87	3.42
29	2400	4	45	2.5	31.00	2.87	3.89
30	2400	3.5	50	2.5	30.63	7.15	3.95

models under experimental settings is demonstrated by the high values of the coefficient of determination (R2) at chosen design points (95.6% for MRR, 96.9% for Kt, and 98.1% for HAZ). Fig. 27.2(a-c) displays statistical residual plots, which, along with ANOVA, verify the statistical significance of developed models for further research by showing the ultimate distribution of data points and the vs plots at the center line.

27.4.2 Influence of PAC parameters

One of the key PAC process performance characteristics that significantly affects both production rate and cost is *MRR*. Higher *MRR* is typically desirable in machining operations to increase productivity while also lowering operating costs. Fig. 27.3(a-b) shows the impact of PAC variables on *MRR* and demonstrates the correlation that can be built between any two parameters while conserving mid-level values for the other two parameters. The effect of arc current and gas pressure on *MRR* is depicted in Fig. 27.3a, and it shows that when arc current and gas pressure increase from low to high levels, *MRR*

TABLE 27.2 Statistical analysis for *MRR*.

Source	DF	Adj. SS	Adj. MS	F-Value	P-Value
Model	14	742.063	53.005	45.96	0.000
Linear	4	375.689	93.922	81.44	0.000
CS	1	16.331	16.331	14.16	0.002
GP	1	161.634	161.634	140.15	0.000
AC	1	11.196	11.196	9.71	0.007
SD	1	186.527	186.527	161.73	0.000
Square	4	150.149	37.537	32.55	0.000
*CS*CS*	1	36.651	36.651	31.78	0.000
*GP*GP*	1	7.177	7.177	6.22	0.025
*AC*AC*	1	63.614	63.614	55.16	0.000
*SD*SD*	1	31.729	31.729	27.51	0.000
2-Way Interaction	6	216.226	36.038	31.25	0.000
*CS*GP*	1	5.811	5.811	5.04	0.040
*CS*AC*	1	101.143	101.143	87.70	0.000
*CS*SD*	1	0.169	0.169	0.15	0.707
*GP*AC*	1	0.280	0.280	0.24	0.629
*GP*SD*	1	4.681	4.681	4.06	0.062
*AC*SD*	1	104.142	104.142	90.30	0.000
Error	15	17.300	1.153		
Lack-of-Fit	10	11.252	1.125	0.93	0.571
Pure Error	5	6.048	1.210		
Total	29	759.363			
R^2	95.6%	Adj. R^2	90.32%		

TABLE 27.3 Statistical analysis for *Kt*.

Source	DF	Adj. SS	Adj. MS	F-Value	P-Value
Model	14	79.6412	5.6887	65.39	0.000
Linear	4	9.0056	2.2514	25.88	0.000
CS	1	5.2087	5.2087	59.87	0.000
GP	1	0.9667	0.9667	11.11	0.005
AC	1	0.7916	0.7916	9.10	0.009
SD	1	2.0386	2.0386	23.43	0.000
Square	4	37.0756	9.2689	106.55	0.000
CS*CS	1	0.0543	0.0543	0.62	0.442
GP*GP	1	8.5594	8.5594	98.39	0.000
AC*AC	1	31.4764	31.4764	361.82	0.000
SD*SD	1	0.2167	0.2167	2.49	0.135
2-Way Interaction	6	33.5600	5.5933	64.30	0.000
CS*GP	1	9.1295	9.1295	104.94	0.000
CS*AC	1	4.1453	4.1453	47.65	0.000
CS*SD	1	0.2121	0.2121	2.44	0.139
GP*AC	1	1.0476	1.0476	12.04	0.003
GP*SD	1	3.9363	3.9363	45.25	0.000
AC*SD	1	15.0893	15.0893	173.45	0.000
Error	15	1.3049	0.0870		
Lack-of-Fit	10	0.7170	0.0717	0.61	0.764
Pure Error	5	0.5879	0.1176		
Total	29	80.9461			
R^2	96.9%	Adj. R^2	93.8%		

TABLE 27.4 Statistical analysis for *HAZ*.

Source	DF	Adj. SS	Adj. MS	F-Value	P-Value
Model	14	19.6580	1.40414	109.99	0.000
Linear	4	11.4727	2.86818	224.67	0.000
CS	1	0.5105	0.51047	39.99	0.000
GP	1	7.5843	7.58430	594.09	0.000
AC	1	0.2107	0.21067	16.50	0.001
SD	1	3.1673	3.16727	248.10	0.000
Square	4	4.1514	1.03785	81.30	0.000
*CS*CS*	1	1.3655	1.36553	106.96	0.000
*GP*GP*	1	0.0087	0.00870	0.68	0.422
*AC*AC*	1	1.9581	1.95810	153.38	0.000
*SD*SD*	1	0.3024	0.30240	23.69	0.000
2-Way Interaction	6	4.0339	0.67231	52.66	0.000
*CS*GP*	1	0.1580	0.15801	12.38	0.003
*CS*AC*	1	1.6066	1.60656	125.84	0.000
*CS*SD*	1	0.2093	0.20931	16.40	0.001
*GP*AC*	1	0.0182	0.01823	1.43	0.251
*GP*SD*	1	0.1580	0.15801	12.38	0.003
*AC*SD*	1	1.8838	1.88376	147.56	0.000
Error	15	0.1915	0.01277		
Lack-of-Fit	10	0.1126	0.01126	0.71	0.697
Pure Error	5	0.0789	0.01579		
Total	29	19.8495			
R^2	98.1%	Adj. R^2	96.16%		

improves. A high concentration of plasma energy is transferred to the workpiece as arc current increases with gas pressure, causing the metal to melt quickly and vaporize, which in turn raises *MRR* [10]. The variation in *MRR* as a result of cutting speed and stand-off distance is shown in Fig. 27.3b. It has been noted that *MRR* declines with increasing cutting speed, whereas *MRR* declines with increasing stand-off distance. This is due to the fact that a higher stand-off distance attracts an absence of arc coherence, which causes the plasma arc to swerve, potentially increasing vulnerability to peripheral drag from the surrounding environment [11]. Therefore, when the torch's height increased, the plasma's kinetic energy at the obtrusion decreased, resulting in a decline in MRR.

Kerf taper is one of the most important quality requirements in metal cutting operations because it affects the dimensional accuracy of the final product (Kt). It is preferable to reduce the angle of Kt through proper process parameter selection in order to efficiently insert the components and omit post-processing. As seen in Fig. 27.3c, Kt decreases to a minimum as cutting speed and gas pressure are increased from low to high. Kt was increased by reducing gas pressure and increasing the cutting speed. The plasma stream from the nozzle induces intense melting and vaporisation, as well as an exothermic reaction, leading to uneven Kt at higher cutting speeds and low gas pressure. Reduced cutting speed and gas pressure produce a more uniform and precise cut surface [12]. Fig. 27.3d, which illustrates how gas pressure and arc current affect Kt, reveals that Kt shows a general upward trend at low to intermediate values before declining at high levels of both variables. At high cutting speeds, an unstable arc is created, and the plasma arc wanders off-axis from the torch in relation to the cutting edge, expanding Kt [13].

HAZ is an unfavorable geometrical deficit in the PAC process that can cause corrosion resistance and mechanical strength to decrease due to exposure to high operating temperatures and localized material melting. During the PAC process, the depth of HAZ is affected by the characteristics of the base material and the selection of appropriate cutting settings. While cutting metal, it's preferable to have a low HAZ so that the finished products last as long as possible. As can be seen in Fig. 27.3e, HAZ grows dramatically with increasing arc current and cutting speed, but decreases with decreasing arc current and cutting speed. The thickness of the HAZ is decreased at lower cutting speeds because to the limited heat transmission to the workpiece surface brought on by the inverse relationship between the cutting speed in

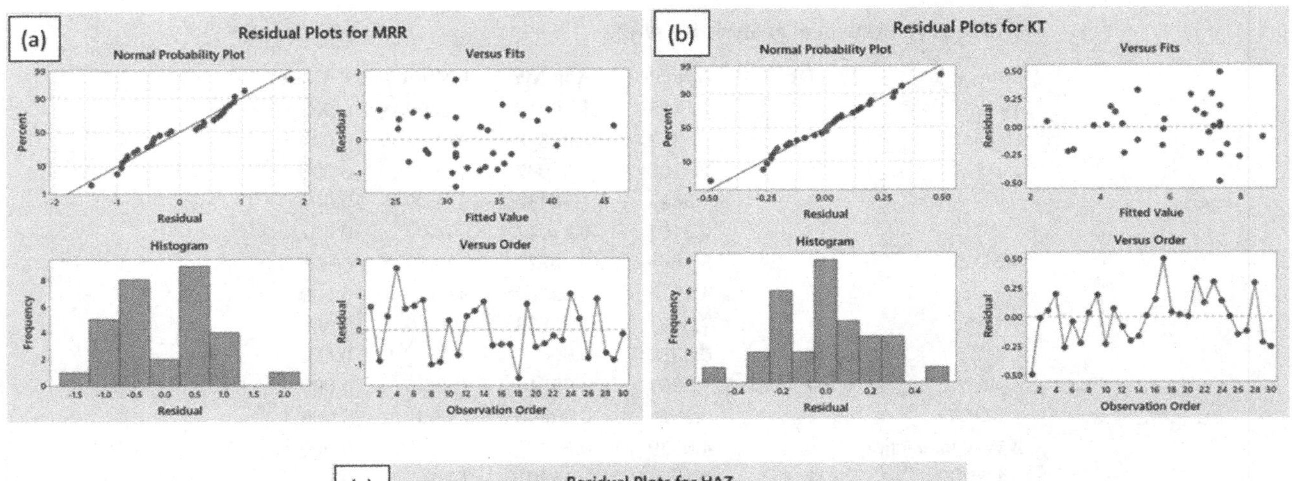

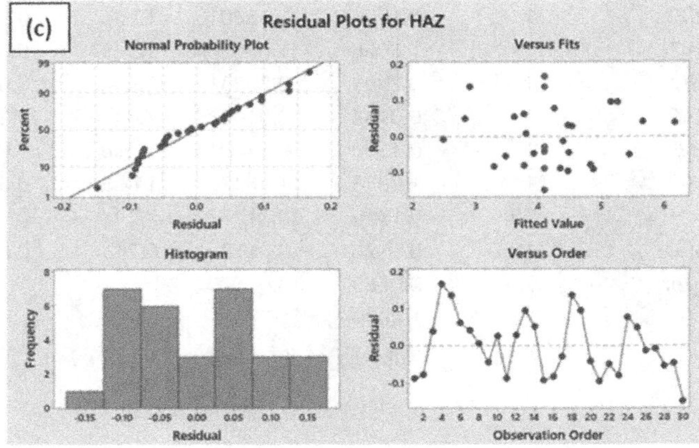

FIGURE 27.2 Statistical residual plots for: (a) *MRR*, (b) *Kt*, and (c) *HAZ*.

PAC and the energy density of the plasma jet [14]. Interaction of arc current and stand-off distance on HAZ, shown in Fig. 27.3f, reveals a dramatic increase in HAZ as arc current and stand-off distance are increased. More energy expenditure from the plasma jet leads to a higher heat affected zone (HAZ) [15] when a workpiece is impinged at a greater stand-off distance.

27.4.3 Multi-response optimization through WOA

The current work focuses on the multi-response optimization of PAC parameters to guarantee the enhanced cut quality features by optimizing for the maximum MRR and the minimum Kt and HAZ, respectively (WOA). Furthermore, the effectiveness of WOA was studied by contrasting the optimal solutions acquired via WOA with those produced via other, more conventional metaheuristic algorithms as the particle swarm optimization (PSO) algorithm and the harmony search algorithm (HSA) [16,17]. With its simple structure, ability to adapt to changing conditions, ability to solve low-dimensional and uni-model problems, ability to solve continuous and convex problems, ability to avoid local optima, and rapid convergence speed due to exploration and exploitation, the WOA has been widely adopted by researchers as a metaheuristic algorithm for solving complex engineering problems [18]. Mixing a process parameter set for one objective function with another objective function is generally frowned upon. As the goal functions employed in this investigation are incompatible, achieving a state of optimality is challenging. There are typically two methods employed when tackling such complex optimization issues. The first involves giving each goal some sort of weight or utility function in order to merge them into a single overarching goal. Secondly, we want to find the optimal values for all of the choice variables that aren't being dominated by any other variables.

Deng's method was used to determine the best solution among the acquired pareto optimum solutions, which yielded the optimal PAC parameters for individual answers. Tables 27.5–27.7 describe the various algorithms and the terms and factors involved in determining the best PAC settings. For the optimization to converge solutions effectively and to reach global

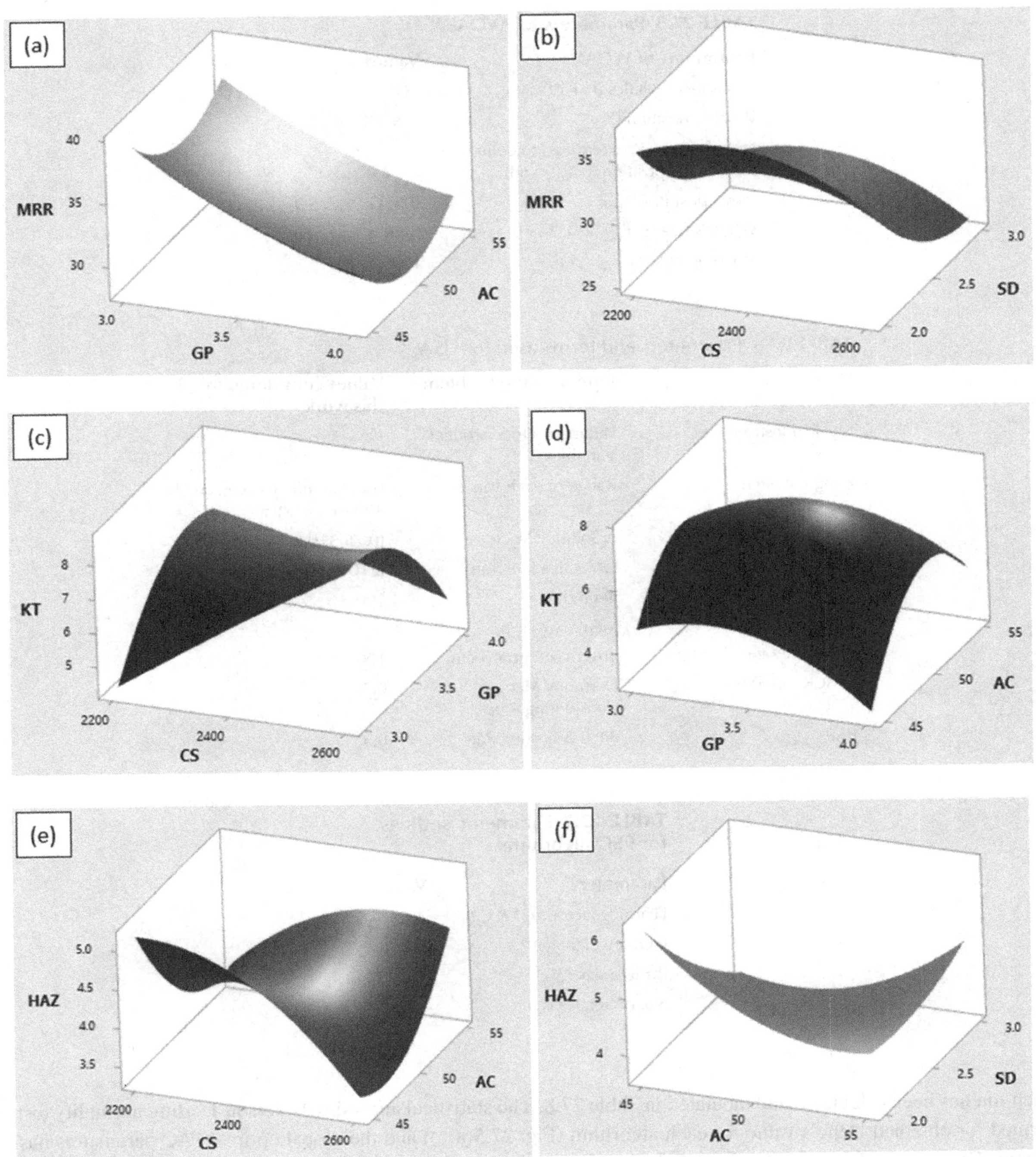

FIGURE 27.3 Influence of PAC process parameters on the response characteristics.

optimal solutions, the algorithm's parameters self-tuned at each iteration. Convergence graphs (Fig. 27.4(a)–(c)) illustrate the effectiveness of proposed optimization methods including WOS, PSO, and HSA for various response characteristics. The charts provide the impression that the WOA solutions converged efficiently compared to the other two algorithms for all of the chosen response features.

The algorithms were executed for twenty-seven times for obtaining pareto optimal PAC parameters. For each run, a set of optimal process parameters were attained for each optimization algorithms. Among the solutions, the best optimal processing variables for enhance response characteristics was obtained through Deng's statistical method. The Deng's value

TABLE 27.5 Parameters for WOA.

Parameters of WOA	Values
Quantity of whales (i=1,2,...nw)	100
Whale's position ($X_{i.}$)	[CS, GP, AC, SD]
No. of dimensions required to define the whale's position (j=1,2,...nd)	4
Prey's position ($X_{p.}$)	[CS, GP, AC, SD]
Whale's fitness (F_{ik}) (k=1,2...no)	MRR, KT, and HAZ
Stopping criteria	100

TABLE 27.6 Parameters and terms used for HSA.

HSA	Optimization Problem	Values considered in this work
Musical Instrument	Parameters/Independent Variables	-
Range of pitch	Range of each variable	For example, jet pressure 300 to 350 MPa
Harmony	A Solution Vector	[0.75 310 2.25 310]
Aesthetics	Objective function	[210 2.3 3.1]
Practice	Iteration	100
Experience	Memory Matrix	
Harmony Memory Size	Size of solution vector	100
HMCR (0.7-0.99)	Harmony Memory Considering Rate	0.7
PAR	Pitch Adjusting Rate	0.3

TABLE 27.7 Parameter settings for PSO algorithm.

Parameter	Value
Learning factors ($C1$ & $C2$)	2 & 2
Inertia weight (ω)	0.6
Particle size (N)	100
No. of iterations ($nitr$)	100

for each run has been calculated and tabulated in Table 27.8. The statistical analysis (Anderson-Darling normality test) was performed for obtained Deng's value for each algorithm (Fig. 27.5(a-c)) and the global optimal PAC parameters has been calculated. From the analysis, it indicates that the Deng's values were eventually distributed for all the algorithms. Whereas, the mean value at 95% confidence interval for WOA is 58.63% which is significantly higher than other two algorithms. Hence, it is proved that the WOA is outperformed the PSO and HSA algorithms in current study of the PAC process. Based on the Deng's values, the optimal PAC parameters and their corresponding response characteristics values were obtained and mentioned in Table 27.9. From the solutions, it is perceived that the improved MRR (46.69 mm^3/min) and Kt (0.503°) has been attained through WOA, whereas the improved HAZ (6.475 mm) was found at PSO algorithm. Moreover, the average computational time for obtaining optimized PAC parameters through different algorithms were calculated and mentioned in Table 27.10. It is perceived that the WOA is provided optimal PAC parameters at an average computational time of 3.287 s, which is 18.9% and 28.9% lesser than PSO and HSA, respectively. Therefore, it is determined that WOA is outperformed than PSO and HSA in terms of statistical analysis and computational time.

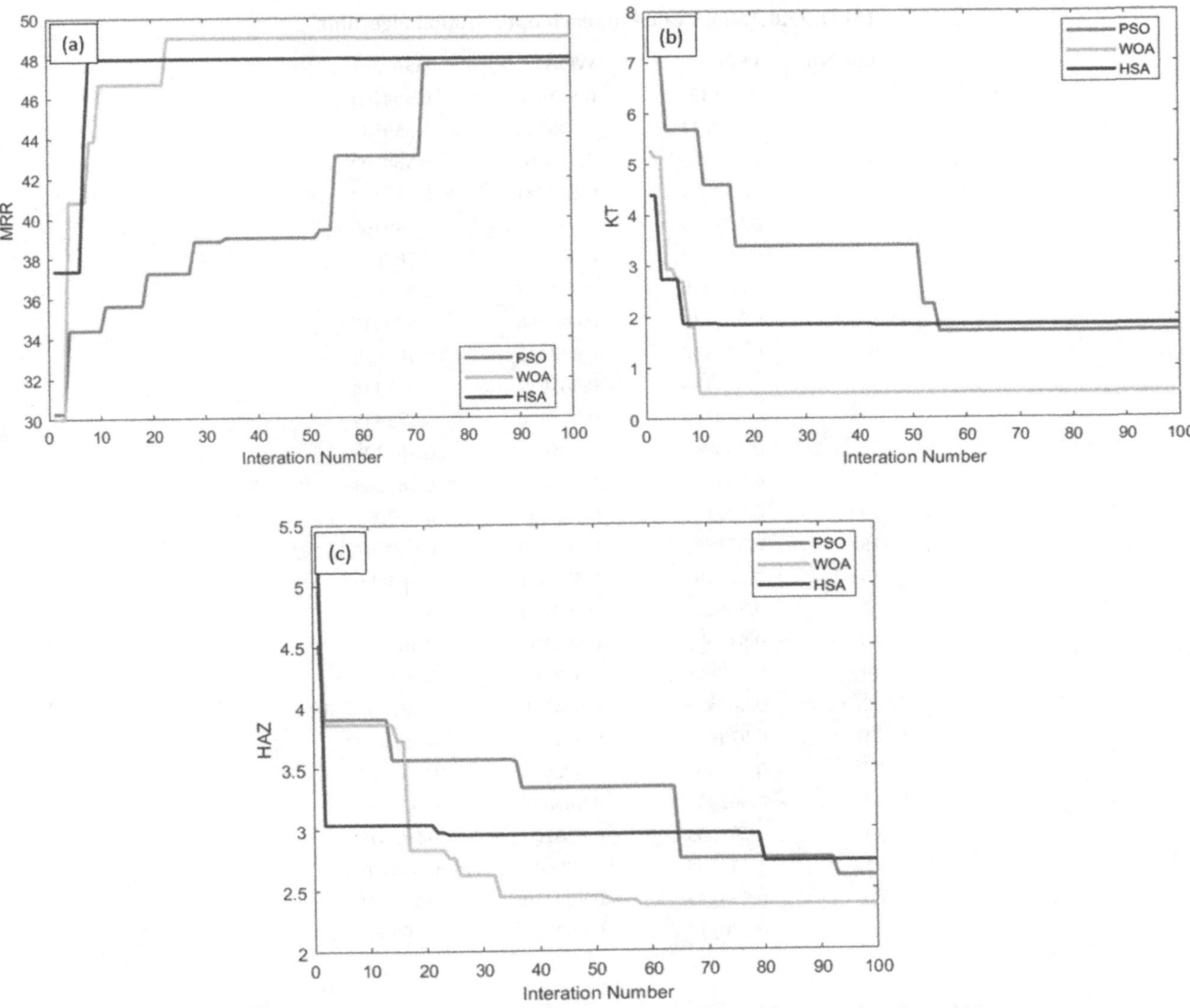

FIGURE 27.4 Convergence plots of different algorithms for: (a) *MRR*, (b) *Kt*, and (c) *HAZ*.

27.5 Conclusions

In this investigation, PAC cutting operations on Monel 400 were performed by modifying the cutting speed, gas pressure, arc current, and stand-off distance. The main effect of the PAC parameters on cut quality features was explored using RSM and WOA techniques. In addition, the performance of WOA in comparison to PSO and HSA was investigated. The significant findings of the work can be summarized as follows:

- The gas pressure and stand-off distance were shown to be the most significant parameters on MRR and HAZ, while the cutting speed and stand-off distance significantly influenced the Kt at a 95% confidence interval in the statistical ANOVA results.
- Cutting speed at 2206.87 mm/min, gas pressure at 3.99 Bar, arc current at 45.13 A, and stand-off distance at 2.02 mm are the ideal PAC settings for enhanced response characteristics for processed Monel 400 alloys, as determined by WOA.
- As compared to PSO and HSA, the results show that WOA is more efficient at optimizing PAC parameters, leading to more reliable outcomes with less time spent on computation. As a result, the WOA has been demonstrated to be an effective algorithm for resolving multi-objective manufacturing challenges due to its ease of use, adaptability, rapid convergence speed, and stochastic character.

TABLE 27.8 Deng's value for each optimization algorithm.

Run No.	PSO	WOA	HSA
1	0.571397	0.561098	0.564203
2	0.608211	0.535522	0.559731
3	0.53936	0.576162	0.568798
4	0.602011	0.573041	0.57529
5	0.597759	0.595796	0.549366
6	0.602854	0.645512	0.587605
7	0.548649	0.581107	0.567661
8	0.558233	0.646268	0.579647
9	0.541332	0.632443	0.584655
10	0.572196	0.563917	0.583338
11	0.569139	0.551658	0.573439
12	0.538043	0.604621	0.567395
13	0.564859	0.636789	0.581884
14	0.572159	0.597381	0.597091
15	0.57178	0.585059	0.568167
16	0.560691	0.572871	0.584496
17	0.56827	0.590754	0.55476
18	0.540673	0.59216	0.56238
19	0.557809	0.632167	0.564992
20	0.567458	0.604139	0.577795
21	0.573159	0.591444	0.560388
22	0.547563	0.6009	0.554252
23	0.562433	0.589636	0.576094
24	0.559083	0.628204	0.582819
25	0.555188	0.59964	0.558102
26	0.556125	0.609359	0.593101
27	0.570317	0.663128	0.566087

TABLE 27.9 Optimal PAC parameters obtained through PSO, WOA, and HSA.

Algorithms	*CS*	*GP*	*AC*	*SD*	*MRR*	*KT*	*HAZ*
PSO	2208.06	3.99	45.23	2.03	45.98	0.781	6.475
WOA	2206.87	3.99	45.13	2.02	46.69	0.503	6.559
HSA	2266.41	3.98	45.05	2.04	45.16	0.895	6.543

TABLE 27.10 Average computational time for proposed algorithms.

Algorithm	WOA	PSO	HSA
Execution time (s)	3.287	4.051	4.624

- This research will bring the optimal metaheuristic approach for the PAC procedure by determining the best parameter values. Moreover, comparable metaheuristic algorithms can be used to optimize complicated engineering problems during process design operations, hence reducing the number of experimental trials while maintaining a low computing complexity.

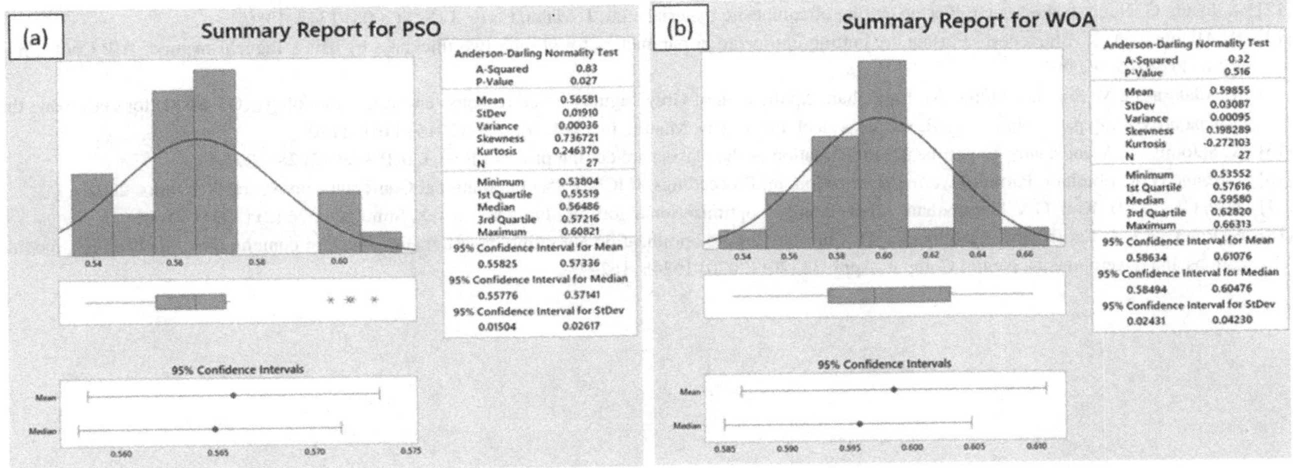

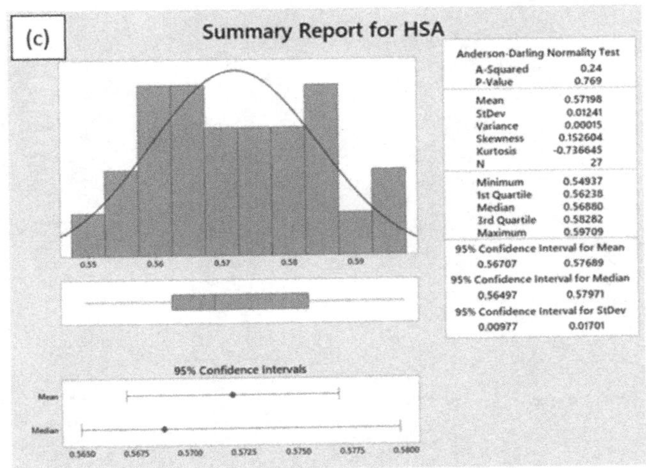

FIGURE 27.5 Statistical validation of proposed optimization algorithms: (a) PSO, (b) WOA, and (c) HSA.

References

[1] K. Kalita, S. Chakraborty, R.K. Ghadai, S. Chakraborty, Parametric optimization of non-traditional machining processes using multi-criteria decision-making techniques: literature review and future directions, Multiscale Multidiscip. Model. Exp. Des. 6 (2023) 1–40.

[2] M. Karthick, P. Anand, M. Siva Kumar, M. Meikandan, Exploration of MFOA in PAC parameters on machining Inconel 718, Mater. Manuf. Process. 37 (12) (2022) 1433–1445.

[3] K. Ananthakumar, D. Rajamani, E. Balasubramanian, J. Paulo Davim, Measurement and optimization of multi-response characteristics in plasma arc cutting of Monel 400 using RSM and TOPSIS, Measurement (Lond.) 135 (2019) 725–737.

[4] Y. Ling, Y. Zhou, Q. Luo, Lévy flight trajectory-based whale optimization algorithm for global optimization, IEEE Access 5 (2017) 6168–6186.

[5] M.H. Tanvir, et al., Multi-objective optimization of turning operation of stainless steel using a hybrid whale optimization algorithm, J. Manuf. Mater. Process. 4 (3) (2020) 64.

[6] M. Karthick, P. Anand, M. Meikandan, M. Siva Kumar, Machining performance of Inconel 718 using WOA in PAC, Mater. Manuf. Process. 36 (11) (2021) 1274–1284.

[7] M. Siva Kumar, et al., Intelligent modeling and multi-response optimization of AWJC on fiber intermetallic laminates through a hybrid ANFIS-salp swarm algorithm, Materials (Basel) 15 (20) (2022) 7216.

[8] S. Mirjalili, A. Lewis, The whale optimization algorithm, Adv. Eng. Softw. 95 (2016) 51–67.

[9] Y. Li, Y. He, X. Liu, X. Guo, Z. Li, A novel discrete whale optimization algorithm for solving knapsack problems, Appl. Intell. 50 (10) (2020) 3350–3366.

[10] R. Devaraj, E. Abouel Nasr, B. Esakki, A. Kasi, H. Mohamed, Prediction and analysis of multi-response characteristics on plasma arc cutting of Monel 400 alloy using Mamdani-fuzzy logic system and sensitivity analysis, Materials (Basel) 13 (16) (2020) 3558.

[11] D. Rajamani, K. Ananthakumar, E. Balasubramanian, J. Paulo Davim, Experimental investigation and optimization of PAC parameters on Monel 400 superalloy, Mater. Manuf. Process. 33 (16) (2018) 1864–1873.

[12] A. Iosub, G. Nagit, F. Negoescu, Plasma cutting of composite materials, Int. J. Mater. Form. 1 (S1) (2008) 1347–1350.
[13] B. Abdulnasser, R. Bhuvenesh, Plasma arc cutting optimization parameters for al with two thickness by using Taguchi method, AIP Conf. Proc. 1756 (1) (2016) 060002.
[14] R. Adalarasan, M. Santhanakumar, M. Rajmohan, Application of Grey Taguchi-based response surface methodology (GT-RSM) for optimizing the plasma arc cutting parameters of 304L stainless steel, Int. J. Adv. Manuf. Technol. 78 (5–8) (2015) 1161–1170.
[15] K. Salonitis, S. Vatousianos, Experimental investigation of the plasma arc cutting process, Proc. CIRP 3 (2012) 287–292.
[16] J. Kennedy, R. Eberhart, Particle swarm optimization, in: Proceedings of ICNN'95 - International Conference on Neural Networks, 2002.
[17] Z.W. Geem, J.H. Kim, G.V. Loganathan, A new heuristic optimization algorithm: harmony search, Simulation 76 (2) (2001) 60–68.
[18] N. Rana, M.S.A. Latiff, S.M. Abdulhamid, H. Chiroma, Whale optimization algorithm: a systematic review of contemporary applications, modifications and developments, Neural Comput. Appl. 32 (20) (2020) 16245–16277.

Chapter 28

Hybrid whale optimization algorithm for enhancing K-means clustering technique

Malik Braik[a], Mohammed A. Awadallah[b,c], Mohammed Azmi Al-Betar[d,e], Zaid Abdi Alkareem Alyasseri[f,g], Alaa Sheta[h], and Seyedali Mirjalili[i,j]

[a]*Department of Computer Science, Al-Balqa Applied University, As-Salt, Jordan,* [b]*Department of Computer Science, Al-Aqsa University, Gaza, Palestine,* [c]*Artificial Intelligence Research Center (AIRC), Ajman University, Ajman, United Arab Emirates,* [d]*Artificial Intelligence Research Center (AIRC), College of Engineering and Information Technology, Ajman University, Ajman, United Arab Emirates,* [e]*Department of Information Technology, Al-Huson University College, Al-Balqa Applied University, Irbid, Jordan,* [f]*Information Technology Research and Development Center (ITRDC), University of Kufa, Najaf, Iraq,* [g]*Department of Business Administration, College of Administrative and Financial Sciences, Imam Ja'afar Al-Sadiq University, Baghdad, Iraq,* [h]*Computer Science Department, Southern Connecticut State University, New Haven, CT, United States,* [i]*Centre for Artificial Intelligence Research and Optimisation, Torrens University Australia, Brisbane, QLD, Australia,* [j]*University Research and Innovation Center, Obuda University, Budapest, Hungary*

28.1 Introduction

Clustering analysis is one of the core methods for finding and comprehending underlying patterns embedded in data [1]. This is by dividing data objects into various clusters based on estimated or recognized internal qualities or similarities [1]. The clustering procedure groups data samples with a high degree of similarity into one cluster, whilst samples with differences are divided into various clusters. Numerous fields, including image segmentation [2], text mining [3], and bio-informatics [4], have widely utilized clustering analysis. Partitioning and hierarchical approaches are the two basic categories into which traditional clustering algorithms may be divided. The partitioning techniques separate the data samples into a number of clusters at once, each instance of which can only ever belong to a single cluster. The hierarchical approaches, on the other hand, create a hierarchy of clusters in either an agglomerated or an inharmonious manner. One of the most popular partitioning techniques is K-means (KM) clustering because it is simple, efficient, and easy to use [1].

Despite the advantages noted above, KM clustering method has certain drawbacks, including initialization sensitivity [1], noise sensitivity, and sensitivity to unfavorable sample distributions [5]. Real-world clustering activities, in particular, present KM clustering with a variety of challenges because of the complexity injected in datasets, such as disturbance from noise and outliers, enormous dimensionality, asymmetric, dispersed, clusters with overlap or cramped class margins and lopsided sample distributions [1]. Additionally, KM displays local optima traps and initiation sensitivity due to the fact that it's working mechanism comprises local search based on the arrangement of initial centroids [2].

Meta-heuristic algorithms have been extensively employed to help KM get out from local optima traps by scouting and finding more optimal cluster centroid configurations [6]. They are known for their mighty search aptitudes in respect of exploration and exploitation. As a result, the negative effects brought on by difficult real-life data can be lessened by hybridizing meta-heuristics with the KM technique due to better precise cluster determination produced from the optimized centroids of the KM technique. The efficacy of these hybrid clustering approaches has been thoroughly supported by experimental evidence, including Genetic Algorithms (GAs) [7], Fruit-fly Optimization Algorithm (FOA) [8], Crow Search (CS) algorithm [9], Harris Hawks Optimizer (HHO) [6], and many more algorithms [10,11].

In contrast to previous meta-heuristic algorithms, the Whale Optimization Algorithm (WOA) [12], one of the newest and promising developed meta-heuristic optimization methods, has a special capacity to conduct bubble-net feeding behavior. Due to its special nature, WOA is better able to handle multi-modal optimization problems like clustering analysis, which involve significant non-linearity and sub-optimal distraction [13,14]. The original WOA approach does, however, have limits regarding search variety and effectiveness. The diversity of the humpback whales of WOA would decline as they continue to forage, which implies that the localization of the area caused by the dispersal of solutions over the whole tractable area will restrict their ability to explore. As an illustration, when it comes to search diversification, the search conducts in WOA are almost constantly, in theory, constrained by a sluggish convergence rate, poor search capabilities and

Handbook of Whale Optimization Algorithm. https://doi.org/10.1016/B978-0-32-395365-8.00034-8

a propensity to settle into local optimum solutions. This is owing to its humble spiral and shrinkage encircling mechanisms to find prey in every possible position of the search space. This may result in stagnation because it makes it less likely for the humpback whales of WOA to locate more fruitful search directions. The existing search process, on the other hand, pushes humpback whales to navigate intensively in local search areas without taking into account the possibility that the global optimal solution may be far away, necessitating additional diversification of the search space. Since the exploitation is not completely effective in the local portions of the search guiding space, many navigation processes become pointless and ineffectual in directing the search to a more fruitful position. As a result, limited search variety compromises search efficiency. This necessitates more exploitation accompanied with further diversification of the search space. In this, expanding the variety of WOA's solutions can improve its capacity for exploration and shield it from falling victim to the local optimum's trap. In consequence, the main motivations for doing this work are the shortcomings of WOA, the limitations of KM clustering technique, and the many difficulties encountered in practical clustering tasks.

These defects with the native WOA model are addressed in this chapter, together with the local optima traps and initialization sensitivity of the traditional KM clustering method. Here, Hybrid WOA with Chameleon Swarm Algorithm (CSA), referred to as HWOA, was evolved in this chapter as an optimization approach for the KM clustering technique to act as a clustering approach for clustering tasks (i.e., various datasets) based on determining the centroids of a preset number of clusters, in which each cluster comprises with comparable features, where each feature was handled as a member of the dataset under study.

The proposed HWOA is collaborated with the KM clustering method to improve the efficiency of the clustering process. The proposed clustering method based on HWOA and KM seeks to optimize the partitioning decisions based on a user-specified initial set of clusters as input that is updated after each iteration. To demonstrate the efficacy and robustness of the hybridization of HWOA and KM clustering method, it was tested using a total of 10 datasets from UCI, each with a different amount of features and samples. The effectiveness of the proposed clustering model was evaluated using intra-cluster distances measure. The potentiality of the clustering computational results of the proposed HWOA was then contrasted with that of the well-known KM clustering method. Finally, this work makes the following contributions to the literature:

1. The first contribution consists of proposing of new enhancements to the current WOA through the adoption of a few modifications to its mathematical model by fusing it with the mathematical model of CSA.
2. The rotation mechanism of the CSA's search agents is used by the WOA's search agents at every level of iteration, which is the second contribution.
3. The third contribution is the investigation of the efficiency of HWOA in optimizing the KM clustering technique in performing clustering tasks on different UCI datasets of varying levels of complexity.
4. The fourth contribution is a comprehensive assessment of the proposed clustering-based HWOA compared to popular as well as other meta-heuristics-based KM-based clustering technique.

The remainder of this work is organized as follows: in Section 28.2, the k-means clustering technique, the incorporation of meta-heuristics with clustering methods, and the WOA and CSA algorithms are all presented. The proposed HWOA is thoroughly presented in Section 28.3. The experimental results and assessment tests of HWOA and a comparison with alternative algorithms are given in Section 28.4. Finally, terminations and ideas for more research are presented in Section 28.5.

28.2 Related works

In this section, we first provide an overview of the traditional KM clustering, traditional WOA, and traditional CSA algorithms. The literature on clustering models that used meta-heuristic methods is then reviewed.

28.2.1 K-means clustering

This clustering method splits the data samples into several clusters on the basis of distance measurements, and locates a division where the squared error between a cluster's points and empirical mean is kept to a minimum [1]. Let $s = s_1, s_2, \ldots, s_n$ be a collection of n data samples that will be bundled into a group of M clusters, $c = c_i, i = 1, 2, 3, \ldots, M$. Minimizing the sum of the squared error over all M clusters is the aim of KM clustering, which is presented as shown below:

$$J(c) = \sum_{i=1}^{M} \sum_{O_l \in c_i} (O_l - z_i)^2 \tag{28.1}$$

where c_i, o_l, z_i, and M stand for the ith cluster, its associated data samples, its centroid, and the overall number of clusters, respectively.

Cluster centroids are randomly initialized in the KM clustering technique. The distance lengths between the data samples and the associated centroid are used to identify which cluster has the closest set of data samples. By determining the mean value of each cluster's data samples, it is simple to update the centroid of each cluster. The procedure of partitioning data samples into the appropriate clusters is then repeated in accordance with the updated cluster centroids until the preset termination conditions are met.

The KM clustering technique exhibits outstanding capabilities in many applications such as pattern recognition [15], computer vision [16], and information retrieval [17]. It frequently provides a basic configuration for other sophisticated models as a fundamental pre-step technique. Despite its benefits and acceptance by researchers in image processing computer vision fields, KM clustering technique has certain drawbacks because of its constrained underlying assumptions and working principles. Initialization sensitivity is one of the main issues of the KM clustering method [1,5]. The total of intra-cluster distances in the KM clustering technique is especially minimized via local searches around the initial centroids. As a consequence, the performance of the KM clustering technique relies strongly on the initial design of the cluster centroids. In addition, because to its operational principles and the randomization during centroid initialization, this clustering technique is more likely to have local optima traps. One of the primary motives behind this study is this flaw of the KM clustering algorithm.

28.2.2 Integration of meta-heuristics with clustering

The local optimum traps and initialization sensitivity of traditional clustering methods have been addressed by a variety of meta-heuristic optimization techniques. Karaboga and Ozturk [18] devised a clustering strategy based on Artificial Bee Colony (ABC) algorithm by combining it with the KM clustering technique. The ABC-based clustering method was evaluated on Thirteen UCI datasets. The results demonstrated how competitive the ABC with KM clustering technique is at handling clustering tasks when compared to different classification methods. By fusing Gravitational Search Algorithm (GSA) with KM clustering algorithm, Hatamlou et al. [19] created a hybrid clustering approach known as GSA-KM. The performance of the GSA-KM clustering algorithm was assessed on 5 UCI datasets. This hybrid clustering method revealed promising benefits in respect of convergence speed and solution quality over the KM clustering method and a set of 6 mainstream meta-heuristic algorithms. The Black Hole (BH) algorithm was then evolved by Hatamlou [20] to boost the performance level of the KM clustering method. Six datasets from UCI were used to assess the BH-based clustering technique. By including both the cloning and fairness ideas, Das et al. [21] developed a Modified Bee Colony Optimization (MBCO) algorithm. Fairness was included, giving bees with soft probability an opportunity to be chosen to increase search diversification. The used cloning idea made it possible to keep the global best solution in the following iteration for the purpose of speeding up the convergence process. A promising hybrid clustering model, referred to as FAPSO-ACO-K, was introduced by Niknam and Amiri [22] by fusing three conventional algorithms, namely fuzzy adaptive PSO algorithm (FAPSO), KM clustering, and ACO algorithm. Six UCI datasets and four artificial datasets were used to evaluate the proposed approach. The issue of initialization sensitivity in KM clustering was solved by FAPSO-ACO-KM. The HWOA with the chameleon swarm algorithm was presented in this study to improve the KM clustering technique with its application on clustering tasks. The next two subsections present the WOA and CSA algorithms.

28.2.3 Whale optimization algorithm

The WOA's mathematical model is introduced and explained in this section. One of the largest whales in the oceans is the humpback whale that WOA relied on. The most amazing aspect of humpback whales is their characteristic bubble-net feeding method, which is how they hunt. Fig. 28.1 [12] illustrates how this foraging is carried out via the creation of unique bubbles along a circular or '9'-shaped course.

It is crucial to understand that humpback whales are the only species known to feed using bubble-nets. The mathematical model of prey encirclement, the spiral bubble-net feeding technique, and the hunt for prey are all described here, together with the underlying WOA processes.

28.2.4 Encircling prey

Humpback whales may realize their prey while hunting and circle around them. The WOA approach makes the assumption that the best candidate solution at any given time is either the target prey or substantially near to it since the position of the

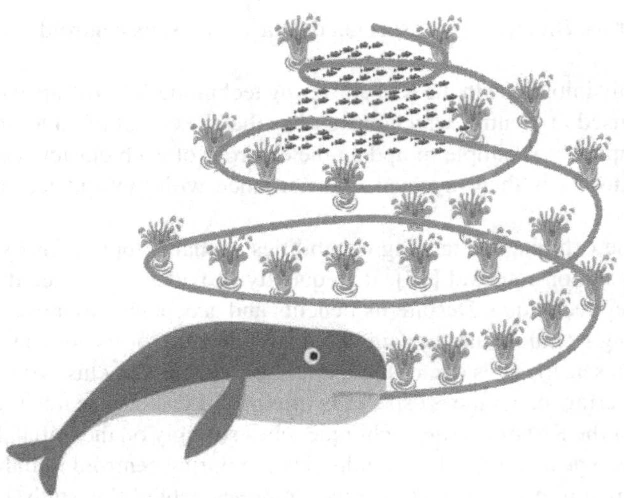

FIGURE 28.1 Bubble-net feeding techniques for humpback whales [12].

optimum design in the search space is unknown a priori. When the best humpback whale has been determined, the other humpback whales will make an effort to move closer to that humpback whale. The following formula demonstrates this conduct:

$$\vec{W}(t+1) = \vec{W}^*(t) - \vec{A} \cdot \vec{D} \tag{28.2}$$

where $\vec{D}$ can be defined as shown below:

$$\vec{D} = \left|\vec{C}\vec{W}^*(t) - \vec{W}(t)\right| \tag{28.3}$$

where t denotes the current iteration count, $\vec{C}$ and $\vec{A}$ are two coefficient vectors, $\vec{W}$ and $\vec{W}^*$ respectively represent the vectors of the current position and best position found thus far, | | denotes the absolute value, and · indicates a component-by-component multiplication.

It is crucial to note that each cycle should change $\vec{W}^*$ if a better solution materializes. At each iteration cycle of the WOA algorithm, the vectors $\vec{A}$ and $\vec{C}$ are updated with the following mathematical formulas:

$$\vec{A} = 2\vec{a}\vec{r} - \vec{a} \tag{28.4}$$

$$\vec{C} = 2 \cdot \vec{r} \tag{28.5}$$

where $\vec{r}$ is a random vector of values each with a range of [0, 1] and $\vec{a}$ is decreased linearly from a value of 2 to a value of 0 throughout the course of iterations.

Fig. 28.2 demonstrates the goal underlying Eq. (28.2) for a 2D situation. The position of a humpback whale (W, V) may be amended to reflect the location of the most recent best record (W^*, V^*). It is feasible to reach alternative places surrounding the optimum humpback whale relative to its current location by altering the values of the vectors $\vec{C}$ and $\vec{A}$.

It should be highlighted that any place inside the search area between the key points in Fig. 28.2 may be reached by constructing the random vector $\vec{r}$. As a result, Eq. (28.2) simulates the environment around the prey and enables any humpback whale to update its location close to the best current solution. The same conception may be used to a search space with a dimension of n, in which case the humpback whales will circle the optimum solution so got in hyper-cubes. The same bubble-net method described in the section above is used by humpback whales to attack their prey. This approach is represented mathematically by the following section.

28.2.4.1 Bubble-net attack technique

Two strategies are developed, as presented below, to mathematically model humpback whale bubble-net behavior:

1. Shrinkage encircling strategy: This behavior is achieved by lowering the value of $\vec{a}$ in Eq. (28.4). Notably, $\vec{a}$ also contributes to a narrowing of the variation range of $\vec{A}$. Alternatively, $\vec{A}$ is a random number selected from the range

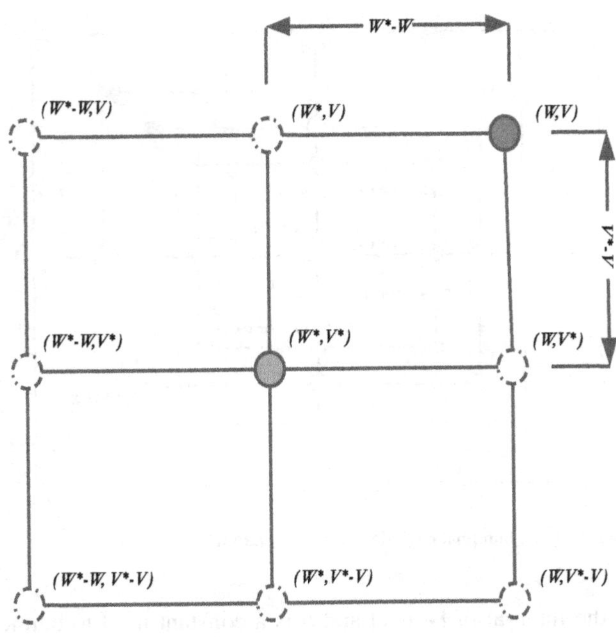

FIGURE 28.2 2D position vectors and their possible next places (W^* is the best solution thus far).

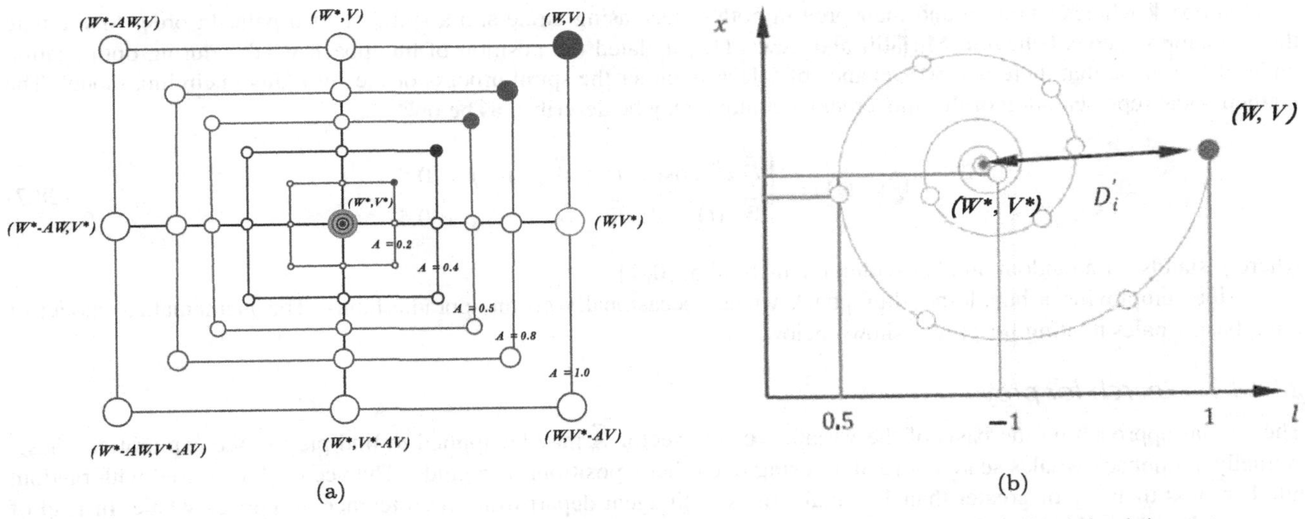

FIGURE 28.3 Bubble-net hunting technique (W^* is the best outcome thus far): (a) shrinking encircling strategy and (b) spiral updating strategy.

$[-a, a]$, where a is lessened from 2 to 0 throughout the iteration path. Using random numbers for $\vec{A}$ in the range of $[-1, 1]$, the new location of a humpback whale can be determined in any place between the humpback whale's initial position and the position of the best humpback whale at the moment. Fig. 28.3 (a) shows the several points in a 2D space that $0 \leq A \leq 1$ may reach from (W, V) in the direction of (W^*, V^*).

2. Spiral updating position strategy: As illustrated in Fig. 28.3 (b), this strategy first calculates the distance between the whale, which is positioned at (W, V), and the prey, which is positioned at (W^*, V^*).
A spiral model connecting the location of the whale and the prey is generated as defined in (28.6) in order to imitate the helix-shaped motion of humpback whales.

$$\vec{W}(t+1) = \vec{D}' \cdot e^{bl} \cdot \cos(2\pi l) + \vec{W}^*(t) \tag{28.6}$$

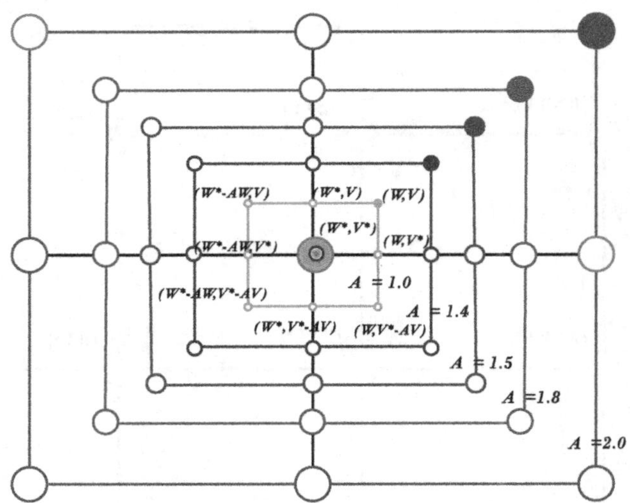

FIGURE 28.4 WOA's exploration model (W^* is a humpback whale chosen at random).

where l is a random value in the interval of $[-1, 1]$ and b is a constant used to define the logarithmic spiral's shape. The ith humpback whale's distance from its prey is shown by $\vec{D}' = \left|\vec{W}^*(t) - \vec{W}(t)\right|$, which represents the best solution obtained so far.

Humpback whales swim around their prey in both a decreasing circle and a spiral-shaped path. In order to simulate this contemporaneous behavior, Mirjalili and Lewis [12] updated the position of humpback whales during optimization under the premise that there is a 50% chance of selecting either the spiral process or the shrinking encircling model. The mathematical representation of this movement technique may be described as below:

$$\vec{W}(t+1) = \begin{cases} \vec{D}' e^{bl} \cos(2\pi l) + \vec{W}^*(t) & p \geq 0.5 \\ \vec{W}^*(t) - \vec{A} \cdot \vec{D} & p < 0.5 \end{cases} \tag{28.7}$$

where p stands for a random number within the interval of $[0, 1]$.

Besides employing a bubble-net, humpback whales occasionally go on sporadic hunts. The mathematical model of humpback whales hunting for prey is shown below.

28.2.4.2 Search for prey

The similar approach on the basis of the variation of the vector $\vec{A}$ may be applied to find prey in the exploration phase. Actually, humpback whales search at random using one other's positions as a guide. The vector $\vec{A}$ was used with random numbers less than -1 or greater than 1 to make the search agent depart from the reference humpback whale. Instead of using the best humpback whale so far, the position of a humpback whale might be changed throughout the exploration phase as opposed to the exploitation phase using a randomly chosen humpback whale. This method and $\left|\vec{A}\right| \geq 1$ enable WOA to do a global search process. The mathematical model of this mechanism can be modeled as shown below:

$$\vec{D} = \left|\vec{C}\vec{W}_r - \vec{W}\right| \tag{28.8}$$

$$\vec{W}(t+1) = \left|\vec{W}_r - \vec{A} \cdot \vec{D}\right| \tag{28.9}$$

where $\vec{W}_r$ stands for s position vector of a randomly chosen humpback whale from the current population of whales.

Fig. 28.4 shows a few of the several places close to a specific solution with $\vec{A} > 1$.

The WOA method starts with a pool of solutions generated at random. The humpback whales update their locations with reference to either the current best solution or a randomly chosen humpback whale each time WOA iterates. The a parameter is decreased from a value of 2 to a value of 0 to aid in diversification and intensification conducts. When $\left|\vec{A}\right| < 1$, the best solution is selected to update the position of the whales, and when $\left|\vec{A}\right| \geq 1$, a random whale is picked. On the basis of the value of the parameter p, WOA can alternate between a spiral and a circular movement. The WOA is eventually terminated when a termination condition is satisfied. The WOA pseudo-code can be found in Algorithm 28.1.

Algorithm 28.1 A pseudo-code outlining the WOA algorithm's key steps.

Initialize population of humpback whales, $W_i (i = 1, 2, \ldots, n)$
Evaluate the fitness of all humpback whales
T ← Largest number of iterations
$\vec{W}^*$ = the best humpback whale
while $(t < T)$ **do**
    for *each humpback whale* **do**
        Update l, p, $\vec{A}$, $\vec{a}$, and $\vec{C}$ using the correct formulas
        if $p < 0.5$ **then**
            if $|\vec{A}| < 1$ **then**
                Update the humpback whales' positions using Eq. (28.3)
            else if $|\vec{A}| \geq 1$ **then**
                Select a humpback whale at random ($\vec{W}_{rand}$)
                Update the humpback whales' positions using Eq. (28.9)
            end if
        else if $p \geq 0.5$ **then**
            Update the humpback whales' positions using Eq. (28.6)
        end if
    end for
    Amend if any humpback whale is outside the proper space area
    Evaluate the fitness of all humpback whales
    Update the vector $\vec{W}^*$ if a better solution exists.
    $t = t + 1$
end while
return the best solution $\vec{W}^*$

The adaptive variation of $\vec{A}$ enables WOA to easily switch between diversification and intensification actions. By lowering $\vec{A}$, some iterations are dedicated to intensification action ($|A| \geq 1$) while the remaining iterations are devoted to diversification action ($|A| < 1$).

28.2.5 Chameleon swarm algorithm

The most captivating information about chameleons' social behavior in foraging activity, according to Braik [23], is that they use three different methods of searching while looking for prey: they constantly keep an eye on and search for prey to stay alive, searching for prey with the help of the ability to roll their eyes, and shooting the prey with their tongues. These three basic concepts of CSA are actively used by chameleons in their hunt for and capture of prey as described below.

28.2.5.1 Hunt for prey

As chameleons come closer to the prey, the chameleons' position in the search area changes due to the prey's continually fluctuating location. How a chameleon could update its location in CSA while foraging is shown by the following mathematical formula:

$$y_{i,j}^{t+1} = \begin{cases} y_{i,j}^t + c_1 \dot{y}_{i,j}^t r_1 + c_2 \ddot{y}_{i,j}^t r_2 & rand \geq q \\ y_{i,j}^t + \lambda r_j^* \cdot \mathrm{sgn}(r - 0.5) & \text{otherwise} \end{cases} \tag{28.10}$$

where $y_{i,j}^{t+1}$ and $y_{i,j}^t$ are the ith chameleon's current and subsequent locations, respectively, $\ddot{y}_{i,j}^t = G_j^t - y_{i,j}^t$ represents the distance between the global best position reached thus far by any chameleon and the present location of the ith chameleon, $\dot{y}_{i,j}^t = P_{i,j}^t - y_{i,j}^t$ denotes the distance between the best position got by chameleon i and the best solution globally found so far, $r_j^* = \left[(u_j - l_j) r_3 + l_j\right]$ stands for the generation of random solutions in the search domain, u_j and l_j stand for the upper and lower boundaries of the search domain in the jth dimension, the parameters c_1 and c_2 have values of 2.0 and

2.0, respectively, $rand$, r_1, r_2, r_3, and r refer to random values generated in the interval [0, 1], q is the probability that a chameleon will capture prey that has a value of 0.1, sgn$(r - 0.5)$ produces values that can be either -1 or 1, and λ is defined as shown in Eq. (28.11).

$$\lambda = k_1 e^{(-k_2 t/T)^{k_3}} \tag{28.11}$$

where T and t are the maximum and current number of iterations, respectively, k_1, k_2, and k_3 are three positive parameters with values of 1.0, 3.5, and 3.5, respectively, where these values are advised in [23].

28.2.5.2 Chameleon eyes' rotation

According to Braik [23], chameleons can locate prey by rotating their gaze to the place where it is most likely to be in a 360° area. In this case, utilizing the ocular rotation attribute, it is feasible to imitate the location update of chameleons during hunting behavior as follows:

$$\vec{y}^{t+1} = \vec{y_r}^t + \vec{\bar{y}}^t \tag{28.12}$$

where $\vec{y}^{t+1}$ denotes the chameleon's new location. $\vec{\bar{y}}^t$ denotes its current position's midpoint, and $\vec{y_r}^t$ denotes the chameleon's rotating focused coordinates, which may be found using Eq. (28.13).

$$\vec{y_r}^t = m \times \vec{y_c}^t \tag{28.13}$$

where m identifies a rotation matrix that considers the rotation of the chameleon's eyes as illustrated in Eq. (28.15), and $\vec{y_c}^t$ stands in the centering coordinates as specified in Eq. (28.14).

$$\vec{y_c}^t = \vec{y}^t - \vec{\bar{y}}^t \tag{28.14}$$

$$m = R(\theta, \vec{z}_1, \vec{z}_2) \tag{28.15}$$

where $\vec{z}_2$ and $\vec{z}_1$ are orthogonal vectors in a search area with n dimension, θ is the rotational angle of the chameleon eyes, which is given by Eq. (28.16), and R identifies the rotation matrices in the respective axes as presented below:

$$R_\phi = \begin{bmatrix} 1 & 0 & 0 \\ 0 & \cos\phi & -\sin\phi \\ 0 & \sin\phi & \cos\phi \end{bmatrix}$$

where ϕ refers to the rotational angle in regard to the x-axis.

$$R_\theta = \begin{bmatrix} \cos\theta & 0 & \sin\theta \\ 0 & 1 & 0 \\ -\sin\theta & 0 & \cos\theta \end{bmatrix}$$

where θ refers to the rotation's angle about the y-axis.

$$\theta = r \cdot \text{sgn}(rand - 0.5) \cdot 180° \tag{28.16}$$

where r signifies a random value in the range of [0, 1], signifying a rotational angle between 0° and 180°, $rand$ identifies a random number created between 0 and 1, and sgn$(rand - 0.5)$ identifies the rotation's orientation, which can be -1 or 1.

28.2.5.3 Hunting prey

When a chameleon uses its tongue to attack a prey, its optimum position is somewhat altered since it may drop twice as far. The following formula describes how quickly a chameleon's tongue moves as it approaches its prey:

$$v_{i,j}^{t+1} = \omega v_{i,j}^t + a_1 \dot{y}_{i,j}^t r_1 + a_2 \ddot{y}_{i,j}^t r_2 \tag{28.17}$$

where $v_{i,j}^t$ and $v_{i,j}^{t+1}$ are the ith chameleon's current and next speeds, respectively, r_1 and r_2 are two random numbers generated in the interval of [0, 1], a_1 and a_2 are two positive numbers, which are equal to 2.0 and 1.8, respectively, and ω stands for the velocity's inertia weight which is identified in Eq. (28.18).

$$\omega = (1 - t/T)^{(\rho\sqrt{(t/T)})} \tag{28.18}$$

where ρ is a positive parameter with a value of 1.0.

Eq. (28.19) can be employed to determine the chameleon's position as it lowers its tongue near its food.

$$y_{i,j}^{t+1} = y_{i,j}^{t} + \left(\left(v_{i,j}^{t}\right)^2 - \left(v_{i,j}^{t-1}\right)^2\right)/(2a) \tag{28.19}$$

where $v_{i,j}^{t-1}$ denotes the prior velocity of the ith chameleon and a signifies the rate of acceleration of the projection of chameleon's tongue, which grows progressively until it arrives at a maximum value of 2590 meters per second squared, which can be as identified as presented in Eq. (28.20).

$$a = 2590 \times \left(1 - e^{-log(t)}\right) \tag{28.20}$$

The pseudo code for the basic CSA may be summed up as in Algorithm 28.2.

Algorithm 28.2 A pseudo-code characterizing the main steps of the standard CSA.

Initialize the CSA's main parameters
Randomly initialize and evaluate each chameleon's position.
Initialize the velocity of the chameleons' dropping tongues
while ($t \leq T$) **do**
  Update ω, λ, and a using Eqs. (28.18), (28.11), and (28.20), respectively
  Update the chameleons' position using Eq. (28.10)
  Update the chameleons' position using the property of eye rotation using Eq. (28.12)
  Calculate the speed of the chameleon's tongue using Eq. (28.17)
  Calculate the chameleon's position as it lowers its tongue toward its prey using Eq. (28.19)
  Calculate the chameleon's position as its tongue falls toward its target using Eq. (28.19)
  Amend the chameleons' position to match the lower and top bounds of the search space
  t = t+1
end while

28.3 Hybrid whale optimization algorithm

A hybrid clustering model is built on the foundation of WOA due to its distinct characteristics of bubble-net attack, shrinkage encircling and spiral updating position strategies. These strategies demonstrated how well they performed in WOA while dealing with multi-modal optimization problems [14]. The original WOA model's established search diversity and search efficiency limits, however, may place certain restrictions on the ability of WOA to identify the best centroids using clustering analysis. In order to solve the shortcomings of the original WOA algorithm and reduce the problems of initialization sensitivity and local optima traps of the KM clustering approach, Hybrid WOA with CSA algorithm, or HWOA, is proposed in this work involving a number of enhancements in the mathematical model of WOA. The proposed hybrid model loosens the restrictions of the mathematical models in the original WOA algorithm and strengthens exploratory diversity both locally and globally, and also evolves a balance between the exploration and exploitation aspects. The following subsections provide a detailed description of the proposed hybrid WOA algorithm.

28.3.1 Encirclement of prey

As indicated before, during hunting, humpback whales may become aware of their target and circle around them in all directions. Therefore, the target prey or a nearby prey is presupposed to be the best potential solution by the WOA algorithm at any given time. In HWOA, once a humpback whale realizes a prey and surround it, all other humpback whales will move closer to that humpback whale, which is the best one. In accordance with what was just described, the WOA algorithm alternates between spiral and circular movements depending on the values of the random parameter p and whether or not $|\vec{A}|$ is less than one. This behavior is updated and amended in HWOA to enable humpback whales to search for prey in all directions with the assistance of a new flag parameter. This behavior was defined as presented in Eqs. (28.21) and (28.24), respectively.

$$\vec{W}(t+1) = \vec{W}^*(t) + f \cdot \vec{A} \cdot \vec{D} \tag{28.21}$$

where the parameter f is defined as presented in Eq. (28.22).

$$f = \vec{F}(f_{in}) \tag{28.22}$$

where $\vec{F} = [-1, 1]$, which means that it can be either -1 or 1 and the parameter f_{in} is defined in Eq. (28.23).

$$f_{in} = \lfloor 2 \times rand + 1 \rfloor \tag{28.23}$$

where $rand$ denotes a value chosen at random and falling between [0, 1].

$$\vec{W}(t+1) = \vec{W}^*(t) + f \cdot \vec{A} \cdot \vec{W}^*(t) \tag{28.24}$$

28.3.2 Spiral and shrinkage encircling mechanisms

Hybridization of two meta-heuristic algorithms at the iteration loops of the dominant algorithm is a possible way to enhance the performance degree of the hybridized algorithms. This boost in the performance of the dominant algorithm is typically attributed to the reinforcement of the diversification and intensification features of the embrace algorithm (i.e., WOA) with those corresponding features of the corroborative algorithm (i.e., CSA). In this, the spiral and shrinkage encircling movement models of the humpback whales in WOA are hybridized with the movement updating models of chameleons in CSA. These movement models are swapped out for one another based on the values of p and $|\vec{A}|$. In this regard, the mathematical model of the spiral and shrinkage encircling strategies of the humpback whales during surrounding, moving around and close to prey is defined in the proposed HWOA algorithm as shown below:

$$\vec{W}(t+1) = \begin{cases} \vec{D}' e^{bl} \cos(2\pi l) + \vec{W}^*(t) & p \geq 0.5,\ r \geq 0.1 \\ \vec{W}^*(t) + \lambda \times f \times r_j^* & p \geq 0.5,\ r < 0.1 \\ \vec{W}^*(t) + f \cdot \vec{A} \cdot \vec{D} & p < 0.5,\ |\vec{A}| \geq 1 \\ \vec{W}^*(t) + f \cdot \vec{A} \cdot \vec{W}^*(t) & p < 0.5,\ |\vec{A}| < 1 \end{cases} \tag{28.25}$$

where r denotes a uniformly generated random number with the range [0, 1].

Eq. (28.25) was drafted to mathematically represent the movement strategies of humpback whales in HWOA using helical and shrinkage encircling mechanisms that benefits from the locomotion strategies of chameleons in CSA. To put it another way, humpback whales continually alter their directions of search in pursuit of possible prey, particularly when they hear, smell or see prey. These movement mechanisms are updated iteratively during the execution of HWOA with the help of the parameter f. Accordingly, as the exploration and exploitation facets of HWOA are fostered over those of the parent algorithms (i.e., WOA and CSA), there is a significant chance that HWOA will locate the global optimal solutions during optimization. These improvements have a significant practical influence on the native WOA's performance and may aid in its ability to efficiently and globally look for prey.

28.3.3 Humpback whales' rotation

The humpback whales in the HWOA algorithm employ a rotational mechanism approved in CSA [23] to revolve in a 360° and change their orientations in order to concurrently scan the entirety of the search area for prey. In this situation, by making use of the rotation feature used by humpback whales in HWOA, it is possible to mimic the location update of humpback whales when they are hunting for prey by doing the following:

$$\vec{W}(t+1) = \begin{cases} \vec{W}_r(t) + \vec{\bar{X}}(t) & rand \geq 0.1 \\ \vec{W}^*(t) + \lambda \times f \times r_j^* & rand < 0.1 \end{cases} \tag{28.26}$$

where $\vec{\bar{X}}(t)$ indicates the midpoint position of the humpback whales and $\vec{W}_r(t)$ stands for the humpback whale's rotating focused coordinates, which is defined as follows:

$$\vec{W}_r(t) = m \times \vec{W}_c(t) \tag{28.27}$$

where m denotes a rotation matrix that takes the rotation of humpback whales into account as presented in Eq. (28.15), and $\vec{W}_c$ represents the centering coordinates as indicated in Eq. (28.28).

$$\vec{W}_c(t) = \vec{W}(t) - \bar{\vec{X}}(t) \tag{28.28}$$

While the aforementioned mathematical models of HWOA definitely improve WOA, its performance might yet be enhanced. This can be accomplished if some auxiliary motions are carried out adaptively with systematic processing in accordance with the demands made by exploration and exploitation operations. Here, humpback whales' rotational motion in the search area may increase its bubble-net behavior toward additional global exploration and local exploitation aspects, which may provide extra benefits in dealing with optimization problems. The humpback whales in HWOA now have a promising feature that might aid in locating new candidate possible solutions that could be close to or distant from the current best ones.

28.3.4 Search for prey

As a matter of fact, humpback whales explore the search space most of their time while randomly searching for prey. Accordingly, the humpback whales' positions constantly vary during the exploration aspect as opposed to the exploitation aspect using a randomly search technique. To locate prey during the exploration phase, a similar strategy based on the variations of the vectors $\vec{A}$ and $\vec{D}$, and the parameter f may be applied. This method enables the proposed HWOA algorithm to do a global search process. The mathematical formula of this mechanism can be identified as shown in Eq. (28.29).

$$\vec{W}(t+1) = \left|\vec{X}_r + f \cdot \vec{A} \cdot \vec{D}\right| \tag{28.29}$$

To put it briefly, it is expected that changing the direction of humpback whale searches over the path of iterations of HWOA using Eq. (28.29) would have a promising convergence path towards the global solution in a timely way. They could also provide HWOA with a strong ability to avoid stagnation in local optima regions and find the overall optimum solution.

28.3.5 Implementation of the proposed HWOA

The exploration and exploitation features of the dominant WOA algorithm were integrated with those of the proponent CSA in the initial and final stages of HWOA to increase its potentiality to find the global optimal solutions. The pseudo-code of HWOA is presented in Algorithm 28.3.

As illustrated in Algorithm 28.3, the exploration and exploitation features acquired by the dominant WOA from CSA as well as the additional targeted improvements made to it heighten the balance between exploration and exploitation in HWOA. As a result, it is expected that HWOA would perform better than the parent algorithms (i.e., WOA and CSA).

28.3.6 The proposed clustering approach based-HWOA

The original KM clustering method's initialization sensitivity and local optima traps are then addressed using the proposed HWOA algorithm to build a promising clustering model. The proposed clustering method's pseudo-code is presented in Algorithm 28.4. The first humpback whale in the population is replaced with a seed solution for cluster centroids created using the original KM clustering approach in order to improve convergence and search effectiveness. During the partitioning process, the Euclidean distance is used to judge how similar the data samples are to one another. Depending on the sum of intra-cluster distance values, the quality of each humpback whale's representation of the centroid solution is assessed. The bubble-net behavior, spiral, shrinkage encircling and search for prey mechanisms of the population are governed and controlled by the proposed HWOA model. Through the intensified neighboring and global search processes, which are made possible by the increased variety of the search scopes and directions in HWOA, a cluster centroid solution of higher quality is found, and the likelihood of becoming stuck in local optima is greatly decreased. In the proposed HWOA, a cluster centroid solution with a superior quality is identified through the augmented neighboring and global search processes, and the chance of being caught in local optima is greatly decreased as a result of the strengthened diversification of the search directions and scopes.

The next section provides a thorough analysis and discussion of the results of the proposed clustering approach.

Algorithm 28.3 A pseudo-code summarizing the key steps of the proposed HWOA.

```
Initialize the parameters of HWOA
Initialize the population of HWOA at random
Assess the fitness of each humpback whale
while (k < K) do
    for each humpback whale do
        Update the parameters of HWOA using the correct formulas
        if p < 0.5 then
            if |A| ≥ 1 then
                Update the humpback whale's position using Eq. (28.21)
            else if |A| < 1 then
                Update the humpback whale's position using Eq. (28.24)
            end if
        else if p ≥ 0.5 then
            if rand ≥ 0.1 then
                Update the humpback whale's position using the first part of Eq. (28.25)
            else
                Update the humpback whale's position using the second part of Eq. (28.25)
            end if
        end if
    end for
    for each humpback whale do
        if rand ≥ 0.1 then
            Update the humpback whale's position using the first part of Eq. (28.26)
        else
            Update the humpback whale's position using the second part of Eq. (28.26)
        end if
    end for
    Evaluate and update the current, best and global solutions
    k = k + 1
end while
Return the global best solution
```

Algorithm 28.4 The main procedural steps of the proposed HWOA-based clustering method.

```
Set up the initial settings and import datasets
Create a swarm of humpback whales called S and initialize it as potential cluster centroids
Create the initial cluster centroid C0 as a seed solution by running KM on the dataset
Substitute the first humpback whale in the swarm S with C0
while (t < T) or other termination requirement not being satisfied do
    As the centroids, use each humpback whale to cluster the data samples using Euclidean distance
    Assess the fitness value of each humpback whale utilizing the sum of intra-cluster distance values as presented in
    Eq. (28.30)
    Update humpback whale positions using the proposed HWOA model
    t = t + 1
end while
Return the optimal fitness value and optimal global position in S
```

28.4 Evaluation and discussion of the results

The proposed HWOA is assessed and contrasted with other eight meta-heuristic algorithms in addition to one traditional clustering method in order to conduct a thorough and objective investigation of the clustering performance. GA and PSO are two of the most well-known meta-heuristic techniques due to their novelty and contributions to the creation of several meta-

TABLE 28.1 Parameter settings of the competitive meta-heuristic algorithms.

Algorithm	Parameter settings
Common Settings	Number of iterations: 200
	Population size: 50
	Number of runs: 30
DE	Crossover: 0.9
	Scale factor (*F*): 0.5
GA	Selection: Roulette wheel
	Crossover: 0.9
	Mutation: 0.05
WSO	$f_{min} = 0.01$, $f_{max} = 0.98$
	$a_0 = 10$, $a_1 = 100$, $a_2 = 0.001$
CS	Flight length (fl): 2
	Awareness probability (AP): 0.1
CSA	$c_1 = 2.0$, $c_2 = 2.0$, $\rho = 1.0$
	$a_1 = 2.0$, $a_2 = 1.80$, $q = 0.1$
	$k_1 = 1.0$, $k_2 = 3.5$, $k_3 = 3.5$
PSO	Inertia weight (ω): [0.9, 0.4]
	Cognitive factor (c_1): 1.8
	Social factor (c_2): 2.0
FPA	Switch Probability (p): 0.8
WOA	l: [−1, 1]
	$\vec{r}$: [0, 1]
	$\vec{a}$: Linear decrease from 2.0 to 0.0

heuristic algorithms. As such, the proposed HWOA was assessed and contrasted with the parent related meta-heuristics (i.e., WOA [12] and CSA [23]) as well as one popular conventional clustering method, the KM clustering method. The clustering strategy utilized by the proposed HWOA was adopted by other meta-heuristic algorithms, including DE [24] algorithm, GA [25], White Shark Optimizer (WSO) [26], Crow Search (CS) algorithm [27,28], PSO [29], and Flower Pollination Algorithm (FPA) [30].

To compare performance with fairness between the optimization algorithms, KM clustering technique is amalgamated into each optimization model. The sum of intra-cluster distances (also known as fitness scores) was used to assess ten datasets with a variety of dimensions. Further, the same number of function evaluations, as a stopping condition, was used for all meta-heuristics in order to secure a fair comparison between them. In the carried out experiments, the maximum number of iterations and the population size were set to 200 and 50, respectively. To lessen the impact of results' fluctuation and to ensure the stability of the results, we additionally conduct 30 independent runs in each experiment.

28.4.1 Parameter settings

The meta-heuristic methods used in this chapter have the same parameter settings as those described in the corresponding original works. As a result, an experimental study is used to pick the initial parameters that are applied to HWOA [12,23]. Table 28.1 contains a list of the specific parameter settings for each meta-heuristic method.

28.4.2 UCI benchmark datasets

The features of data samples, such as data distribution, disturbance, and dimensionality, have a substantial effect on the performance degree of clustering methods. Therefore, in order to judge the potency of the proposed clustering method, the following datasets from diverse areas with various attributes are employed. For assessment, a set of 10 datasets from the UCI machine learning repository was employed. Table 28.2 provides an illustration of the key traits of the used datasets.

TABLE 28.2 Ten selected datasets for evaluation.

Dataset	Number of features	No. of classes	Missing values	No. of instances
Hill-Valley	101	2	No	606
Dermatology	34	6	No	366
Iris	4	3	No	150
Wine	13	3	No	178
Balance	4	2	No	625
E. coli	7	3	No	336
TAE	5	3	No	151
Seeds	7	3	No	210
CMC	10	3	No	1473
Hungarian	14	2	No	294

In Table 28.2, Hill-Valley and Dermatology have relatively high feature dimensionality among the chosen datasets, with values of 101 and 34, respectively. Comparatively lower feature dimensions may be found in the other datasets (that is, 4 for Iris, 5 for TAE, and 10 for CMC). Other datasets have medium feature dimensions such as Wine and Hungarian with values of 13 and 14, respectively. Furthermore, due to significant sample imbalance between classes in some datasets, such as the E. coli dataset, only those classes with an adequate amount of samples were chosen for comparison of the performance of clustering methods.

28.4.3 Performance comparison metrics

The proposed clustering method uses the sum of intra-cluster distances as the fitness function, which is lessened during the iterative process, much like the KM clustering technique. Eq. (28.30) defines the sum of the distances evaluation method between the data samples and their respective centroids.

$$f(O,C) = \sum_{i=1}^{k} \sum_{O_l \in C_i} (O_l - Z_i)^2 \tag{28.30}$$

where C_i and Z_i stand for the ith cluster and its centroid, respectively, while O_l and k stand for the ith cluster's data and the total number of clusters.

The partitioned clusters are more compact, the lower the sum of intra-cluster distances. Each meta-heuristic algorithm is combined with the KM clustering algorithm for a total of 30 independent runs for each dataset. The average performance for each performance measure is computed and applied as the key evaluation criterion for comparison.

28.4.4 Performance evaluation of clustering

The results of clustering before feature selection for each dataset displayed in Table 28.3.

HWOA obtained the minimal distance measurements for the intra-cluster distance measure for many datasets, as shown in Table 28.3. According to the average performance over 30 runs, HWOA produced the lowest intra-cluster measures in seven datasets, including, Hill-Valley, Dermatology, Wine, CMC, Balance, and Seeds, while GA produced the lowest fitness scores with two datasets, including Iris and E. coli, and DE produced the lowest fitness score in the TAE and Hungarian datasets. In general, the HWOA-based clustering model showed significantly greater dominance over other algorithms-based clustering in locating enhanced centroids that result in more compact clusters when compared to the KM classical clustering method and other meta-heuristic-based clustering methods, namely DE, GA, WSO, CS, CSA, PSO, and WOA. The very small values of STD of the HWOA algorithm reveal the high level of stability that this algorithm has.

As a matter of fact, how long it takes an algorithm to cluster a dataset is directly related to the complexity of that algorithm. For the goal of thoroughly evaluating the performance of the proposed HWOA, the processing time must be measured, where this time must be within the acceptable range. The PC that served as the test platform for the clustering methods described in this chapter was equipped with an Intel Core i7-5200U CPU running at 2.3 GHz and 8.0 GB of RAM,

TABLE 28.3 The average clustering outcomes of HWOA and other clustering techniques before feature selection over 30 independent runs.

Dataset	Criterion	Features	KM	DE	GA	WSO	CS	CSA	PSO	FPA	WOA	HWOA
Iris	AVE	4	131.37	107.29	**99.78**	112.69	142.32	106.7	99.94	167.05	108.53	101.17
	STD		14.597	2.5905	**0.011234**	14.97	13.286	13.709	0.26154	10.031	8.5651	7.6842
Wine	AVE	13	16404	16312	16296	16371	16537	16353	16327	16833	16411	**16293**
	STD		52.533	4.1746	1.449	37.693	163.62	31.189	9.2502	116.79	23.2343	**0.39976**
Balance	AVE	4	1480.9	1467.1	1459.3	1461.6	1477	1459.1	1459.7	1504.3	1460.4	**1458.2**
	STD		9.0299	2.805	3.6199	3.7445	8.0414	3.0262	2.6032	6.9345	9.9182.4	**1.9207**
E. coli	AVE	7	185.72	158.43	**152.38**	185.77	230.26	160.14	152.87	246.32	167.83	152.73
	STD		14.799	1.5349	**0.17754**	9.8235	12.57	3.5084	0.63725	9.935	0.9845	0.58893
TAE	AVE	5	1538.6	**1506.3**	1522.8	1524.6	1547.8	1523.1	1511.1	1600.6	1522.7	1513.0
	STD		18.053	**2.942**	17.69	18.876	17.912	18.73	18.103	18.401	24.54	19.025
Seeds	AVE	7	417.99	345.04	320.73	339.4	397.85	330.17	322.64	461.6	343.6	**320.56**
	STD		33.685	5.403	0.090008	8.3746	18.621	4.9318	1.6654	16.365	1.7686	**0.000266**
CMC	AVE	9	6243	5751.7	5696.9	5945.3	6283.9	5885.6	5748	6702.8	5981.3	**5693.7**
	STD		132.47	13.031	1.5476	66.056	163.71	66.98	23.972	126.97	67.544	**0.006019**
Dermatology	AVE	34	3040.9	2775.2	2718.2	3130.9	3275.1	2948.3	2757.1	3545.7	3091	**2634.6**
	STD		55.841	14.091	14.504	43.948	35.978	49.348	19.562	44.589	13.541	**1.7755**
Hungarian	AVE	13	16069	**13001**	13088	14522	16150	13806	13367	17527	14132	13323
	STD		1095.5	**130.095**	491.77	1033.9	845.73	163.39	169.11	498.71	143.5453	133.95
Hill-Valley	AVE	101	4.92E+08	8.94E+07	1.37E+08	2.29E+08	2.74E+08	9.54E+07	7.47E+07	3.03E+08	1.06E+08	**5.52E+07**
	STD		4.66E+07	2.48E+07	6.91E+06	1.06E+07	8.11E+06	3.07E+07	9.79E+06	3.79E+06	3.12E+07	**1.63E+06**

TABLE 28.4 The average computation times of HWOA and other clustering techniques before feature selection over 30 independent runs.

Dataset	Criterion	KM	DE	GA	WSO	CS	CSA	PSO	FPA	WOA	HWOA
Iris	AVE	3.6504	2.8678	3.7877	**2.0778**	2.6326	3.1485	3.3904	3.731	3.484	4.6307
	STD	0.92193	0.3067	0.73962	0.28553	0.81267	0.63895	0.65437	0.75638	**0.2669**	5.5742
Wine	AVE	5.1157	5.0132	5.2473	**2.8720**	3.5883	3.9608	4.0766	5.4657	6.9542	5.3839
	STD	0.83716	1.6492	0.53748	0.4939	0.26574	0.2992	0.38288	1.7829	1.1761	**0.2598**
Balance	AVE	6.1596	4.009	4.7599	**3.1572**	4.0196	4.796	4.2831	4.357	4.5664	4.5038
	STD	1.5088	**0.3528**	0.57748	0.5411	0.67531	0.78425	0.45911	0.43665	0.49006	1.0365
E. coli	AVE	4.2747	3.3109	3.9481	**2.5224**	2.8707	3.4461	3.431	3.6469	4.297	4.7341
	STD	0.32922	0.2127	0.15043	0.51948	0.19936	0.16681	0.19181	0.20719	**0.1475**	1.1159
TAE	AVE	3.3666	3.7035	5.7281	3.6596	**2.8268**	3.1314	5.035	5.815	5.7625	4.7901
	STD	0.19248	0.88934	0.24729	0.9881	0.2859	**0.13494**	0.59488	0.89327	0.30068	0.48021
Seeds	AVE	4.7957	3.7400	4.7262	**2.483**	3.6168	3.7065	4.1256	4.0852	4.6794	4.8125
	STD	0.59539	0.33034	0.56532	**0.29561**	0.50038	0.52182	0.5484	0.43111	0.52314	1.1012
CMC	AVE	7.5378	5.4395	7.1205	**3.2306**	5.9738	5.5308	5.8572	5.8832	5.3631	5.48
	STD	1.0054	0.22175	1.2841	**0.1684**	1.2886	0.29262	0.96654	0.27172	0.47476	0.9556
Dermatology	AVE	**3.753**	4.9404	5.7423	3.8745	4.1915	5.117	5.1928	5.8423	7.942	7.993
	STD	**0.2629**	0.45567	0.39872	0.62312	0.29482	0.50065	0.49728	0.7362	0.81022	4.2425
Hungarian	AVE	**1.5854**	2.1499	2.8933	1.7463	2.7136	2.2422	2.76	2.3883	4.135	4.1429
	STD	0.1113	0.14252	0.32924	**0.089066**	0.52788	0.22329	0.11806	0.18962	0.32495	0.8747
Hill-Valley	AVE	3.8285	7.8682	8.5025	**3.7368**	6.7404	7.2216	7.607	8.5812	16.337	16.86
	STD	0.77866	0.54452	0.60833	0.26587	**0.4969**	0.52065	0.57134	0.61584	1.2812	0.8422

where MATLAB® 2020 A environment was used to run each of the evaluation tests. The average run times of HWOA and other rival clustering algorithms before feature selection, over 30 separate runs, are shown in Table 28.4.

Even though the proposed HWOA algorithm has a higher level of complexity than WOA, CSA, and some other competing techniques, its execution time is only slightly increased, which is acceptable because it is within the range of other competing methods. In all tests conducted, the maximum number of iterations of the meta-heuristic algorithms used for clustering was fixed at 200 iterations. The convergence curves in Fig. 28.5 show that each clustered dataset converges before the 100th iteration. This denotes that the number of iteration loops might be lowered even more, cutting the processing time in half. These convergence curves are presented over 30 independent runs as per the fitness results.

The best algorithm in Fig. 28.5 is the one that achieves the lowest fitness outcomes in the fewest iterations while also avoiding local optima. Examination of the plots in Fig. 28.5 reveals that the proposed HWOA method has superior convergence conducts compared to all other competitors in several datasets, including Hill-Valley, Dermatology, Wine, CMC, Balancing, and Seeds. Moreover, this figure shows that the nine compared methods behaved slightly differently when exploring the search space of each dataset as can be observed from the convergence curves of E. coli and TAE datasets.

28.4.5 Performance evaluation of feature selection and clustering

In this chapter, Minimum Redundancy Maximum Relevance (mRMR) [31] as a Feature Selection (FS) is utilized to reduce the dimensionality of the feature space on the considered datasets and to look into its fundamental effects on the performance of the clustering process due to the high dimensionality of Hill-Valley and Dermatology datasets and the potential inclusion of redundant features. The number of selected features from the two high-dimensional datasets, namely Hill-Valley and Dermatology datasets, are 52 and 11 from the initial 101 and 34 features, respectively. The sizes of these features were discovered through trial and error, and they produced the best results for almost all assessed clustering methods. The results on feature selection are also in line with previous works [32,33], where the ranges of chosen feature numbers are 50-54 [33] and 10-15 [32] for Hill-Valley and Dermatology datasets, respectively. This confirms the effectiveness of the mRMR-based feature selection algorithm used in this chapter. According to the empirical findings, feature selection improved the performance of the clustering methods for the majority of test instances. As an example, for the Hill-Valley

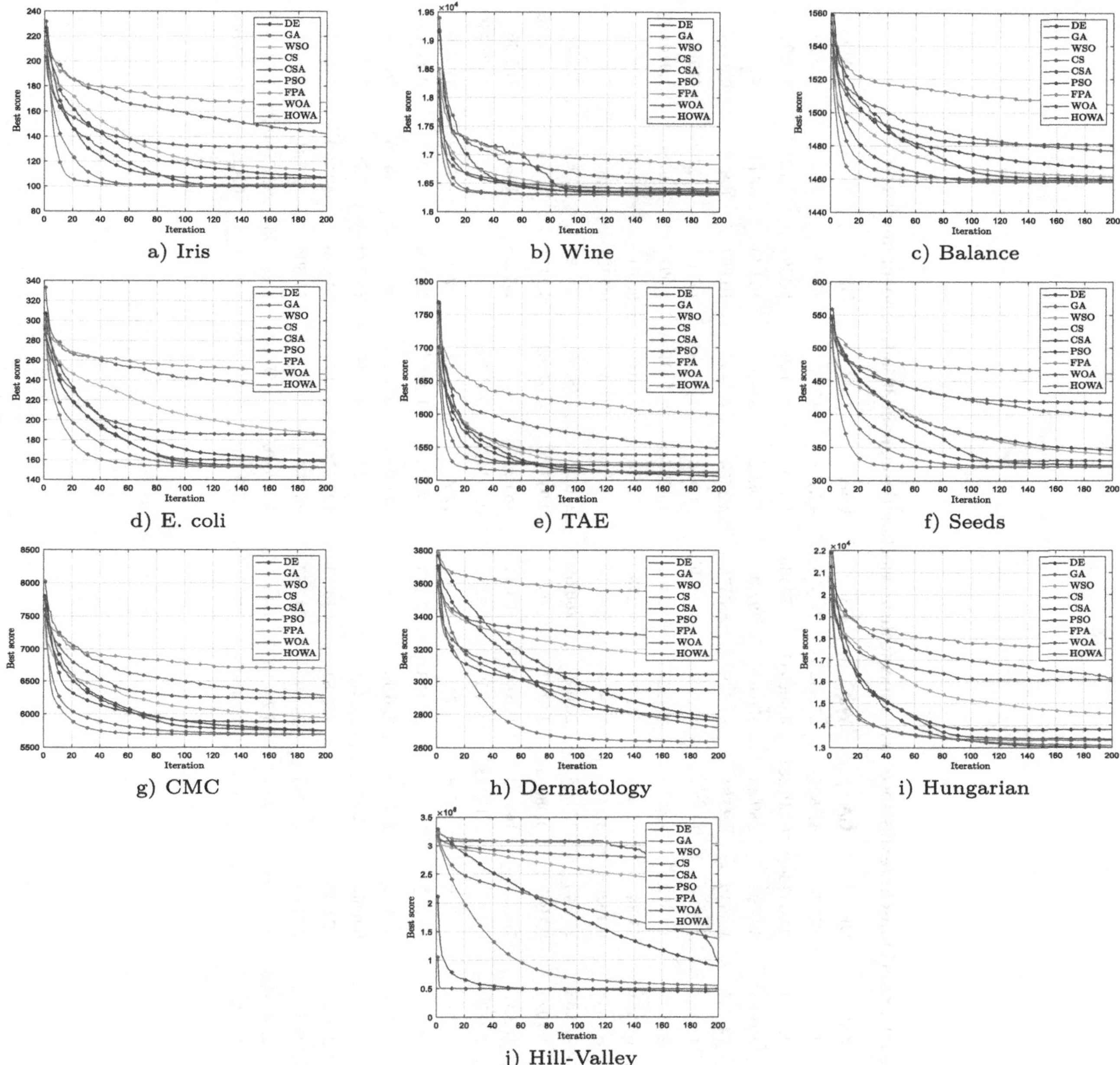

FIGURE 28.5 Convergence curves of the HWOA-based clustering before feature selection and all other competing methods for all studied datasets.

dataset demonstrated in Table 28.3, the number of features is lessened from the original 101 to 52. The comprehensive evaluation findings obtained over 30 runs for each evaluation metric measure after mRMR-based feature selection are displayed in Table 28.5.

According to the average performance over 30 runs as shown in Table 28.5, the proposed HWOA obtained the minimal distance measurements for the intra-cluster distance measure, or fitness scores, in 6 out of 10 datasets, including Hill-Valley, CMC, Seeds, Wine, Balance, and TAE, while DE produced the lowest fitness scores in three datasets, namely E. coli, Dermatology and Hungarian, and GA produced the lowest fitness score in the Iris dataset. In general, the HWOA-based clustering model showed significantly greater dominance over other algorithms-based clustering in locating enhanced centroids that result in more compact clusters when compared to the KM classical clustering method and other meta-heuristic-based clustering methods, namely DE, GA, WSO, CS, CSA, PSO, and WOA. The expanded global exploration capabilities of HWOA in contrast to those of other meta-heuristic optimization algorithms in obtaining the global best

TABLE 28.5 The average outcomes of the minimum intra-cluster distance metric after feature selection throughout 30 independent runs.

Dataset	Criterion	Features	KM	DE	GA	WSO	CS	CSA	PSO	FPA	WOA	HWOA
Iris	AVE	2	48.587	47.738	**47.629**	47.636	48.456	47.979	47.979	52.607	47.642	47.979
	STD		2.8064	0.089334	0.001261	**0.006555**	0.74588	1.9199	1.9189	1.339	2.109	1.919
Wine	AVE	9	408.84	377.92	369.88	376.83	397.04	374.92	377.64	462.43	384.6	**368.52**
	STD		14.761	1.5535	1.1781	3.0549	6.8375	3.3764	6.5495	10.893	0.879	**0.05989**
Balance	AVE	3	1126	1123.4	1121.1	1122.1	1125.9	1120.8	1121.8	1131.8	1121.1	**1120**
	STD		2.5527	1.1898	1.5076	1.9807	2.2461	1.8257	2.1001	3.091	2.875	**1.0141**
E. coli	AVE	5	65.551	**55.34**	57.432	71.064	82.38	61.998	57.033	88.619	63.051	60.732
	STD		8.4638	**0.10268**	3.3427	11.438	7.2224	6.8141	3.4477	3.1321	7.894	7.764
TAE	AVE	3	277.87	275.48	275.25	281.48	288.65	276.27	275.26	288.94	277.63	**275.21**
	STD		1.1979	0.19171	0.068633	2.6127	3.0334	0.91311	0.07567	3.2171	1.383	**0.011866**
Seeds	AVE	4	168.17	166.3	165.98	166.45	169.34	165.94	166	178.36	166.62	**165.86**
	STD		1.257	0.15	0.18343	0.38301	1.8539	0.13443	0.15365	3.161	1.325	**0.078456**
CMC	AVE	5	1379.5	1257.9	1255.8	1274.4	1461.2	1259.9	1255.2	1494.8	1261.2	**1255.6**
	STD		76.592	1.9861	0.5247	23.178	47.378	8.4293	1.2361	42.046	34.232	**0.05388**
Dermatology	AVE	11	643.58	**574.75**	707.69	826.36	966.05	611.05	611.24	953.05	738.36	602.72
	STD		31.029	**6.6002**	52.196	42.321	37.886	30.721	26.314	26.799	34.764	51.212
Hungarian	AVE	6	376.28	**272.38**	356	438.83	628.33	392.32	320.77	548.96	404.19	378.75
	STD		46.109	**2.5786**	28.088	24.399	94.752	24.403	30.229	41.804	32.531	29.574
Hill-Valley	AVE	52	2.64E+08	2.42E+07	2.88E+07	3.24E+07	1.40E+08	3.90E+07	1.68E+07	1.48E+08	3.37E+07	**1.54E+07**
	STD		1.06E+07	1.91E+06	2.12E+06	1.18E+07	7.03E+06	1.11E+07	1.17E+06	5.43E+06	6.42E+06	**1.11E+06**

TABLE 28.6 The average computation times of HWOA and other clustering techniques after feature selection throughout 30 independent runs.

Dataset	Criterion	KM	DE	GA	WSO	CS	CSA	PSO	FPA	WOA	HWOA
Iris	AVE	2.3788	2.9104	3.8606	**2.3013**	2.361	2.7926	3.1538	3.1446	2.8997	3.9209
	STD	0.33466	0.52066	0.51635	0.41117	**0.23773**	0.46983	0.45143	0.45225	0.33138	1.8601
Wine	AVE	2.7166	2.9831	3.553	**2.3835**	3.0673	3.3385	2.7607	3.5276	6.6081	3.925
	STD	**0.14758**	0.36383	0.40954	0.25732	1.0504	0.49322	0.3333	0.35135	1.946	0.72151
Balance	AVE	2.3911	4.06	4.9136	**2.3844**	3.6187	3.3516	4.2743	3.4466	3.8839	3.383
	STD	**0.22812**	0.76681	1.6711	0.30073	0.80781	0.37582	1.7461	0.41238	0.59776	0.59219
E. coli	AVE	3.0715	2.6822	3.1452	2.4466	**2.3791**	2.7144	2.8488	2.7912	2.8969	3.2997
	STD	0.56265	0.23019	0.39208	0.29666	0.21831	0.22387	**0.20013**	0.28836	0.27089	0.27087
TAE	AVE	2.2993	2.8295	3.5285	**2.1127**	2.5044	2.9329	2.9399	3.1494	2.9212	3.6989
	STD	0.20179	0.18464	0.22778	**0.14703**	0.12365	0.22301	0.18354	0.17216	0.18148	0.6252
Seeds	AVE	2.5428	2.9772	3.7033	3.7233	**2.5199**	2.9667	5.2929	3.6198	5.5809	3.691
	STD	0.51493	0.27991	**0.08285**	0.24301	0.12237	0.14829	1.261	0.28238	0.89748	0.27115
CMC	AVE	3.3999	5.7849	7.467	**3.3978**	5.8692	6.3728	5.7582	6.6053	6.7898	6.4267
	STD	1.401	0.53907	0.84699	0.62856	0.90275	0.82402	0.53563	0.76464	1.1386	**0.48057**
Dermatology	AVE	2.4723	3.4844	4.0767	**2.3669**	3.9122	3.5472	3.5557	3.8478	5.5216	4.8678
	STD	**0.03665**	0.2438	0.15049	0.26443	0.78331	0.13549	0.18205	0.11238	1.0132	0.31981
Hungarian	AVE	**1.9713**	4.6336	4.4977	2.1113	2.9979	3.8882	5.3594	3.8423	4.2358	5.0987
	STD	**0.28276**	0.76501	0.92848	0.30288	0.32364	0.70277	1.4929	0.54305	0.59935	0.49337
Hill-Valley	AVE	**3.4116**	4.2671	5.2323	1.8098	3.9979	4.1822	4.3157	4.9728	5.5036	5.521
	STD	**0.12119**	0.34675	0.3994	0.18036	0.28776	0.34431	0.3203	0.16222	0.71123	1.8214

solutions are explained by the developed search mechanisms of both WOA and CSA. The standard deviations of HWOA in clustering the aforementioned datasets are very small, which assert that the performance of HWOA in the target study is settled.

The average run times of HWOA and all other competitors before feature selection over 30 independent runs are shown in Table 28.6.

It is clearly seen from Table 28.6 that the computational time of HWOA is not large and is within the range of other competing algorithms.

Fig. 28.6 shows the convergence curves of the proposed HWOA and all other competing clustering methods as per the fitness results.

The convergence conducts of the proposed HWOA method are almost superior than all other comparison algorithms in various datasets, as evidenced by the convergence curves presented in Fig. 28.6. In the Iris, Wine, and Dermatology datasets, the convergence behavior of the nine competing meta-heuristic algorithms becomes equal after half of the iterations. As seen in the curves of the CMC and Seeds datasets, the convergence behavior of HWOA, GA, and PSO is not similar in most iterations but becomes indiscernible in the last few iterations. Moreover, in the complex Hill-Valley dataset, the convergence behavior of HWOA outperforms the others.

28.4.6 Statistical tests

This subsection used Friedman's test [34], which necessitates the computation of the average ranking value, to determine if the differences in performance of all meta-heuristic-based clustering algorithms are statistically significant. To establish if the null hypothesis is rejected using Friedman's test, it is necessary to compare the critical values obtained for the studied significance level ($\alpha = 0.05$). A post-hoc statistical analysis of Holm's test technique [35] was also taken into account in this chapter in order to determine which clustering algorithms are superior to or inferior to the proposed HWOA algorithm and at what level of significance.

The poorest behaving algorithm receives the greatest rank when Friedman's test is applied, while the best performing algorithm receives the lowest rank and is utilized as a control method for post-hoc analysis. The intra-cluster distance metric measure for the proposed HWOA and other comparative methods is a one key comprehensive performance indicator that is

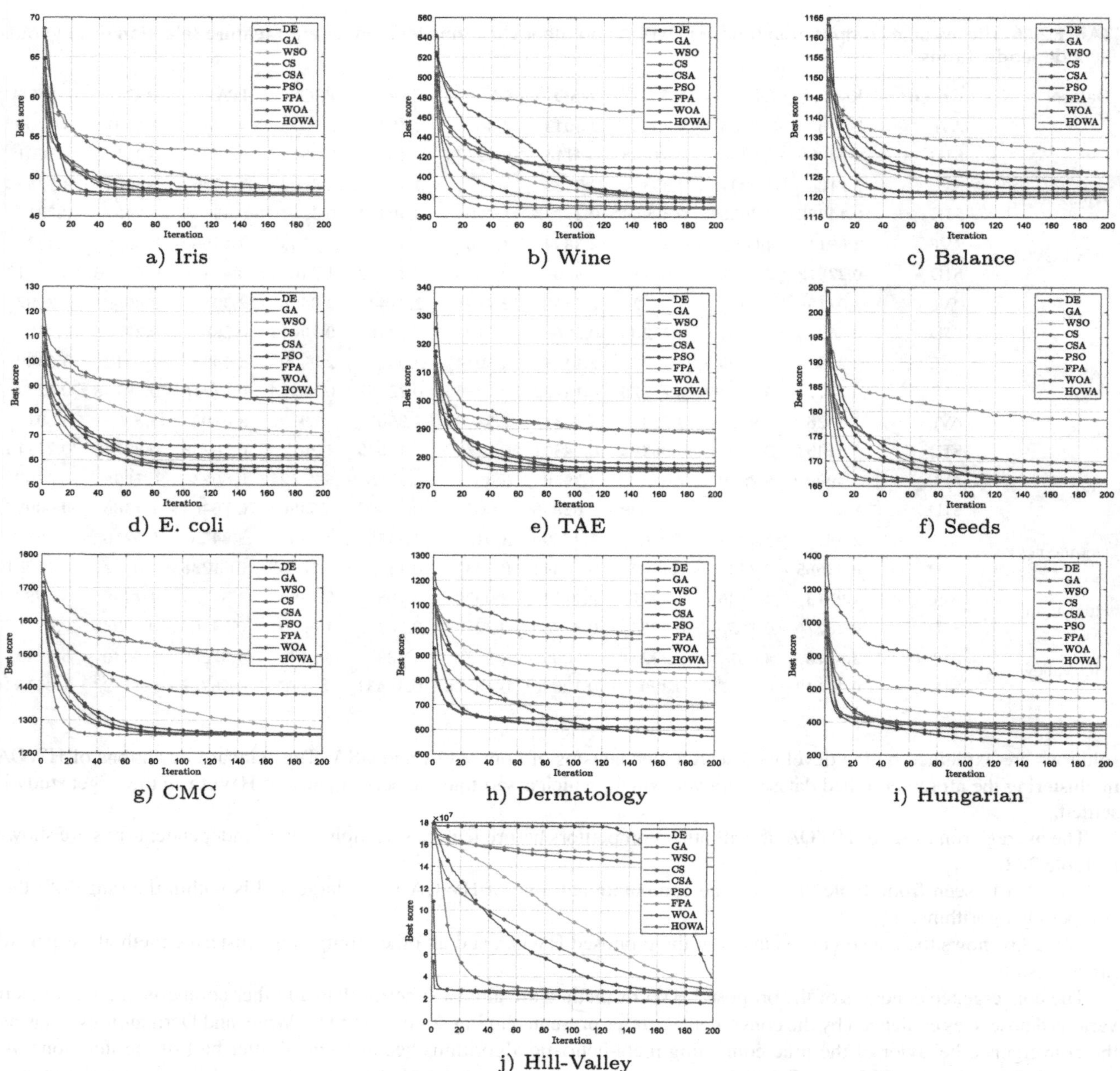

FIGURE 28.6 Convergence curves of the HWOA-based clustering after feature selection and all other competing methods for all studied datasets.

used while performing Friedman's and Holm's tests. The mean ranking results for this performance indicator used in this test for the proposed HWOA-based clustering method before feature selection are presented in Table 28.7.

Based on the findings of the intra-cluster distance test provided in Table 28.7, Friedman's test produced a p-value of 4.2574E-11. As a consequence, HWOA had the highest rank of 1.90, which was the highest. It is specifically followed in this order by GA, PSO, DE, CSA, WOA, WSO, KM, CS, and FPA came last.

Friedman's test was followed by Holm's test to see whether there were any statistically significant differences between HWOA and the competing techniques shown in Table 28.7. Table 28.8 provides the specifics of the results of the application of Holm's test.

Table 28.8 removes hypotheses with a p-value ≤ 0.0125 based on Holm's test. According to the Friedman's and Holm's tests, Tables 28.7 and 28.8 demonstrate that the proposed HWOA significantly outperforms all rival clustering techniques in terms of intra-cluster distance measure.

TABLE 28.7 A statistical comparison of the proposed HWOA and other competing techniques using the intra-cluster distance metric before feature selection.

Method	Rank	Method	Rank
KM	7.20	CSA	4.60
DE	4.00	PSO	2.90
GA	2.70	FPA	10.0
WSO	6.80	WOA	6.10
CS	8.80	HWOA	1.90

TABLE 28.8 Results of Holm's method between HWOA and every other competing technique before feature selection.

i	Method	$z = (R_0 - R^i)/SE$	p-value	$\alpha \div i$	Hypothesis
9	FPA	5.9822464	2.200809E-09	0.00555555	Rejected
8	CS	5.09598772	3.469276E-07	0.00625	Rejected
7	KM	3.91430941	9.066331E-05	0.00714285	Rejected
6	WSO	3.61888983	2.958695E-04	0.00833333	Rejected
5	WOA	3.10190557	0.00192279	0.01	Rejected
4	CSA	1.99408215	0.04614307	0.0125	Rejected
3	DE	1.55095278	0.12091299	0.01666666	Not rejected
2	PSO	0.73854894	0.46018093	0.025	Not rejected
1	GA	0.59083915	0.55462819	0.05	Not rejected

TABLE 28.9 A statistical comparison of the proposed HWOA and other competing techniques using the intra-cluster distance metric after feature selection.

Method	Rank	Method	Rank
KM	6.90	CSA	4.40
DE	3.50	PSO	3.20
GA	3.15	FPA	9.80
WSO	6.29	WOA	5.95
CS	8.90	HWOA	2.90

Table 28.9 displays the mean ranking results based on Friedman's test for the proposed HWOA-based clustering approach and other competing methods.

In Table 28.7, Friedman's test produced a p-value of 1.0465E-09 on the basis of the results of intra-cluster distance measure. With a significance level of 5%, HWOA received the best rank with a value of 2.90, which is followed in order by GA, PSO, DE, CSA, WOA, WSO, KM, CS, and FPA.

The results of applying Holm's test on the ranking results generated by Friedman's test in Table 28.9 are shown in Table 28.10.

A comparison of HWOA with the other rival clustering methods was performed on the basis of Holm's test as presented in Table 28.10, where this test suppresses hypotheses with p-value $\leq 0.0.008333$. Tables 28.7 and 28.8 affirm that HWOA is the best clustering method among all other algorithms, which is far superior to other competing clustering methods, which demonstrate the advantages of design search strategies for HWOA.

28.5 Conclusion and future work

This chapter has proposed Hybrid Whale Optimization Algorithm (WOA) with Chameleon Swarm Algorithm (CSA), referred to as HWOA, to solve the issues with initialization sensitivity and local optima traps of the traditional K-Means (KM) clustering method. The rotation mechanism in CSA was utilized by the humpback whales in the HWOA to broaden search

TABLE 28.10 Results of Holm's method between HWOA and every other competing technique after feature selection.

i	Method	$z = (R_0 - R^i)/SE$	p-value	$\alpha \div i$	Hypothesis
9	FPA	5.09598772	3.469276E-07	0.00555555	Rejected
8	CS	4.43129367	9.366941E-06	0.00625	Rejected
7	KM	2.95419578	0.00313485	0.00714285	Rejected
6	WSO	2.51106641	0.01203670	0.00833333	Rejected
5	WOA	2.25257428	0.024286004	0.01	Rejected
4	CSA	1.10782341	0.26793808	0.0125	Not rejected
3	DE	0.44312936	0.65767216	0.01666666	Not rejected
2	PSO	0.22156468	0.82465277	0.025	Not rejected
1	GA	0.18463723	0.85351357	0.05	Not rejected

diversification and efficiency. The mathematical model of HWOA contributes to enhancing exploitation and exploration features in the local and global areas of the search space. A set of 10 UCI datasets with varying complexity, number of features and instances, and problems in both low and high dimensions are used to investigate the efficacy of the proposed HWOA in enhancing the KM clustering algorithm. In terms of clustering performance with respect to the intra-cluster distance measure over 30 independent runs, the proposed HWOA-based KM clustering method outperforms the conventional KM clustering algorithm, and eight meta-heuristic algorithms, including the parent algorithms (i.e., WOA and CSA). The enhanced cluster centroids of the KM clustering technique, which in turn avoided the problems of KM, are attributed to benefits of the proposed approach. The obtained results confirm the usefulness and benefits of the proposed technique over the baseline methods in handling low and high-dimensional clustering tasks, by virtue of its improved exploration capabilities and wider search space. Future study will use additional objective functions that take into account both intra- and inter-cluster measurements to improve the proposed method for addressing complicated and atypical data distribution problems. It might be possible to hybridize the basic WOA with other meta-heuristics than CSA. This demands additional study. The extension of the HWOA algorithm to address other complicated datasets would be valuable. Various domains of optimization problems, including discriminatory feature selection, image enhancement, and image segmentation might potentially be assessed using the proposed HWOA algorithm.

References

[1] A.K. Jain, Data clustering: 50 years beyond k-means, in: Joint European Conference on Machine Learning and Knowledge Discovery in Databases, Springer, 2008, pp. 3–4.

[2] R. GeethaRamani, L. Balasubramanian, Macula segmentation and fovea localization employing image processing and heuristic based clustering for automated retinal screening, Computer Methods and Programs in Biomedicine 160 (2018) 153–163.

[3] L.M. Abualigah, A.T. Khader, M.A. Al-Betar, O.A. Alomari, Text feature selection with a robust weight scheme and dynamic dimension reduction to text document clustering, Expert Systems with Applications 84 (2017) 24–36.

[4] I. Triguero, S. Del Río, V. López, J. Bacardit, J.M. Benítez, F. Herrera, ROSEFW-RF: the winner algorithm for the ECBDL'14 big data competition: an extremely imbalanced big data bioinformatics problem, Knowledge-Based Systems 87 (2015) 69–79.

[5] A. Likas, N. Vlassis, J.J. Verbeek, The global k-means clustering algorithm, Pattern Recognition 36 (2) (2003) 451–461.

[6] L.-G. Zhang, X. Xue, S.-C. Chu, Improving K-means with Harris hawks optimization algorithm, in: Advances in Intelligent Systems and Computing, Springer, 2022, pp. 95–104.

[7] F.B. Ashraf, A. Matin, M.S.R. Shafi, M.U. Islam, An improved K-means clustering algorithm for multi-dimensional multi-cluster data using meta-heuristics, in: 2021 24th International Conference on Computer and Information Technology (ICCIT), IEEE, 2021, pp. 1–6.

[8] T. Bezdan, C. Stoean, A.A. Naamany, N. Bacanin, T.A. Rashid, M. Zivkovic, K. Venkatachalam, Hybrid fruit-fly optimization algorithm with k-means for text document clustering, Mathematics 9 (16) (2021) 1929.

[9] M. Braik, A. Sheta, S. Aljahdali, Diagnosis of brain tumors in MR images using metaheuristic optimization algorithms, in: International Conference Europe Middle East & North Africa Information Systems and Technologies to Support Learning, Springer, 2019, pp. 603–614.

[10] S. Harifi, M. Khalilian, J. Mohammadzadeh, S. Ebrahimnejad, Using metaheuristic algorithms to improve k-means clustering: a comparative study, Revue d'Intelligence Artificielle 34 (3) (2020) 297–305.

[11] A.F. Jahwar, A.M. Abdulazeez, Meta-heuristic algorithms for k-means clustering: a review, PalArch's Journal of Archaeology of Egypt/Egyptology 17 (7) (2020) 12002–12020.

[12] S. Mirjalili, A. Lewis, The whale optimization algorithm, Advances in Engineering Software 95 (2016) 51–67.

[13] L. Zhang, W. Srisukkham, S.C. Neoh, C.P. Lim, D. Pandit, Classifier ensemble reduction using a modified firefly algorithm: an empirical evaluation, Expert Systems with Applications 93 (2018) 395–422.
[14] M. Braik, Hybrid enhanced whale optimization algorithm for contrast and detail enhancement of color images, Cluster Computing (2022) 1–37, https://doi.org/10.1007/s10586-022-03920-9.
[15] H.H. Ali, L.E. Kadhum, K-means clustering algorithm applications in data mining and pattern recognition, International Journal of Science and Research (IJSR) 6 (8) (2017) 1577–1584.
[16] A. Risheh, P. Tavakolian, A. Melinkov, A. Mandelis, Infrared computer vision in non-destructive imaging: sharp delineation of subsurface defect boundaries in enhanced truncated correlation photothermal coherence tomography images using K-means clustering, NDT & E International 125 (2022) 102568.
[17] S.H. Toman, M.H. Abed, Z.H. Toman, Cluster-based information retrieval by using (K-means)-hierarchical parallel genetic algorithms approach, arXiv preprint, arXiv:2008.00150.
[18] D. Karaboga, C. Ozturk, A novel clustering approach: Artificial Bee Colony (ABC) algorithm, Applied Soft Computing 11 (1) (2011) 652–657.
[19] A. Hatamlou, S. Abdullah, H. Nezamabadi-Pour, A combined approach for clustering based on K-means and gravitational search algorithms, Swarm and Evolutionary Computation 6 (2012) 47–52.
[20] A. Hatamlou, Black hole: a new heuristic optimization approach for data clustering, Information Sciences 222 (2013) 175–184.
[21] P. Das, D.K. Das, S. Dey, A modified Bee Colony Optimization (MBCO) and its hybridization with k-means for an application to data clustering, Applied Soft Computing 70 (2018) 590–603.
[22] T. Niknam, B. Amiri, An efficient hybrid approach based on PSO, ACO and K-means for cluster analysis, Applied Soft Computing 10 (1) (2010) 183–197.
[23] M.S. Braik, Chameleon swarm algorithm: a bio-inspired optimizer for solving engineering design problems, Expert Systems with Applications 174 (2021) 114685.
[24] R. Storn, K. Price, Differential evolution–a simple and efficient heuristic for global optimization over continuous spaces, Journal of Global Optimization 11 (4) (1997) 341–359.
[25] X.-S. Yang, Nature-Inspired Metaheuristic Algorithms, Luniver Press, 2010.
[26] M. Braik, A. Hammouri, J. Atwan, M.A. Al-Betar, M.A. Awadallah, White shark optimizer: a novel bio-inspired meta-heuristic algorithm for global optimization problems, Knowledge-Based Systems 243 (2022) 108457.
[27] A. Askarzadeh, A novel metaheuristic method for solving constrained engineering optimization problems: crow search algorithm, Computers & Structures 169 (2016) 1–12.
[28] M. Braik, H. Al-Zoubi, M. Ryalat, A. Sheta, O. Alzubi, Memory based hybrid crow search algorithm for solving numerical and constrained global optimization problems, Artificial Intelligence Review 56 (2023) 27–99, https://doi.org/10.1007/s10462-022-10164-x.
[29] J. Kennedy, R. Eberhart, Particle swarm optimization (PSO), in: Proc. IEEE International Conference on Neural Networks, Perth, Australia, 1995, pp. 1942–1948.
[30] X.-S. Yang, Flower pollination algorithm for global optimization, in: International Conference on Unconventional Computing and Natural Computation, Springer, 2012, pp. 240–249.
[31] H. Peng, F. Long, C. Ding, Feature selection based on mutual information criteria of max-dependency, max-relevance, and min-redundancy, IEEE Transactions on Pattern Analysis and Machine Intelligence 27 (8) (2005) 1226–1238.
[32] Y. Wan, M. Wang, Z. Ye, X. Lai, A feature selection method based on modified binary coded ant colony optimization algorithm, Applied Soft Computing 49 (2016) 248–258.
[33] T. Bhadra, S. Bandyopadhyay, Supervised feature selection using integration of densest subgraph finding with floating forward–backward search, Information Sciences 566 (2021) 1–18.
[34] D.G. Pereira, A. Afonso, F.M. Medeiros, Overview of Friedman's test and post-hoc analysis, Communications in Statistics. Simulation and Computation 44 (10) (2015) 2636–2653.
[35] S. Holm, A simple sequentially rejective multiple test procedure, Scandinavian Journal of Statistics 6 (1979) 65–70.

Chapter 29

Whale optimization algorithm based controller design for air-fuel ratio system

Serdar Ekinci and Davut Izci
Department of Computer Engineering, Batman University, Batman, Turkey

29.1 Introduction

Since the usage of the fossil fuels, the vehicles have played a considerable role in climate change which has brutally affected the planet earth. On the hand, the engine power has a vital importance for today's modern society, therefore, sustainable solutions must be investigated such that the adverse effects can effectively be dealt with [1]. Due to its significant role of increasing the efficiency of spark-ignition engines and reducing the harmful emissions, it is worth to investigate the adjustment of the air-fuel mixture ratio [2]. In these engines, three-way catalytic converter is used for converting the products of the combustion into less harmful pollutants by oxidizing hydrocarbons to carbon dioxide and water, oxidizing carbon monoxide to carbon dioxide, and reducing nitrogen oxide to nitrogen and oxygen [3].

The air-fuel ratio dictates the efficiency of the above-mentioned conversion. To explain this from a more meaningful perspective, the following example can be used; a one percent higher deviation from the stoichiometric level would almost cause fifty percent higher nitrogen oxide discharge, on the other hand, a one percent less deviation would lead to higher emissions of carbon monoxide, hydrocarbons, and carbon oxide [4]. The air-fuel ratio, therefore, must be adjusted around the stoichiometric level in order to achieve a maximum conversion efficiency.

The performance of a spark-ignited engine can be improved with the utilization of appropriate control techniques in an air-fuel ratio system. The control of this system, however, is relatively complicated due to its time-delayed and nonlinear nature [5–7]. In this context, different control schemes such as neural network control [8], fault-tolerant control [9], model predictive control [4], observer model control [10,11], adaptive control [12], sliding mode control [13], and fuzzy control [14] have so far been used for transient performance improvement of an air-fuel ratio system. Those studies have not sufficiently addressed the control efficiency although they showed promising capabilities.

In this chapter, a proportional-integral controller with a feedforward mechanism is proposed for controlling the air-fuel ratio system as an efficient technique for a lean-burn spark-ignition engine by considering the previously adopted techniques. Due to its flexibility, gradient-free mechanism, and high local optima avoidance ability, a stochastic optimization technique named whale optimization algorithm [15] is also used in this chapter to tune the parameters of the proportional-integral controller with the feedforward mechanism. Besides, an appropriate cost function is also adopted for the purpose of obtaining optimal parameters.

To demonstrate the good promise of the whale optimization algorithm tuned proportional-integral controller with the feedforward compensated mechanism, the ant lion optimization algorithm [16], cuckoo search algorithm [17], and particle swarm optimization algorithm [18] are used in this chapter for comparisons. For the latter purpose, statistical, computational time, converging ability, transient response, input signal tracking ability, disturbance rejection, and robustness analyses are presented. Moreover, a performance evaluation based on several different cost functions is also provided. All those analyses confirm the more promising capability of the proposed whale optimization algorithm based and feedforward mechanism compensated proportional-integral controller for the air-fuel ratio system.

29.2 Whale optimization algorithm

The whale optimization algorithm (WOA) is a widely used stochastic optimization algorithm proposed by Mirjalili and Lewis [15] and mimics the bubble-net feeding behavior of humpback whales. The WOA algorithm is basically a population-based metaheuristic approach, thus, uses search agents for the determination of global optimum. The following equation is

Handbook of Whale Optimization Algorithm. https://doi.org/10.1016/B978-0-32-395365-8.00035-X

the main mathematical model of the WOA algorithm:

$$X(t+1)=\begin{cases} X^*(t)-AD & \text{if } p<0.5 \\ D'e^{bl}\cos(2\pi t)+X^*(t) & \text{if } p\geq 0.5 \end{cases} \tag{29.1}$$

In here, $D'=|X^*(t)-X(t)|$ indicating the distance of the ith agent and the current best solution, $D=|CX^*(t)-X(t)|$ where $C=2r$ and $A=2ar-a$. p, r, and l represent a random number within [0, 1], [0, 1], and [−1, 1], respectively, b stands for a constant that defines the shape of the logarithmic spiral, t is the current iteration. a is a linearly decreasing (from 2 to 0) parameter throughout the iterations. The encircling mechanism of the humpback whales is represented by the first component of Eq. (29.1) whereas the second component of this equation mimics the bubble-net feeding behavior. In WOA algorithm an equal probability is specified between the two components by switching the variable p. The exploration and exploitation phases are guaranteed in through adaptive adjustment of a and C parameters. The best solution in WOA algorithm plays a significant role to update the position of search agents, thus, guarantee the exploration and convergence whereas the currently obtained best solution has the importance for exploitation. The flexibility, gradient-free mechanism, and high local optima avoidance ability of the WOA algorithm motivates the work presented in this chapter for efficient operation of an air-fuel ratio system.

29.3 Problem definition and proposed design methodology

An air-fuel ratio (AFR) system, illustrated in Fig. 29.1, is constructed from the components named heat exhaust gas oxygen (HEGO) sensor, air-fuel path, universal exhaust gas oxygen (UEGO) sensor, throttle, lean nitrogen oxide trap (LNT), and three-way catalytic (TWC) [19]. Large time-varying delays, occurring due to gas transportation ($\tau_g=\alpha/\dot{m}_a$) and cycle ($\tau_c=120/N$), are the main difficulties of an AFR system control as they lead to instability. In here, $\dot{m}_a$ denotes the airflow rate and N stands for the engine speed whereas α is a constant. The overall time delay can be computed as $\tau=\tau_c+\tau_g$ [10] considering the sources of the delay. $k\dot{y}(t)+y(t)=u(t-\tau)$ can be used to model the UEGO sensor (first-order lag system), where $u(t)$ is the control input, k is the time constant and $y(t)$ is the output [20]. The dynamic model of the AFR system can be expressed as $P(s)=Y(s)/U(s)=e^{-\tau s}/(ks+1)$ where $\tau=1.5$ s, $k=0.2$ s [10]. The configuration of an AFR system is provided in Fig. 29.1.

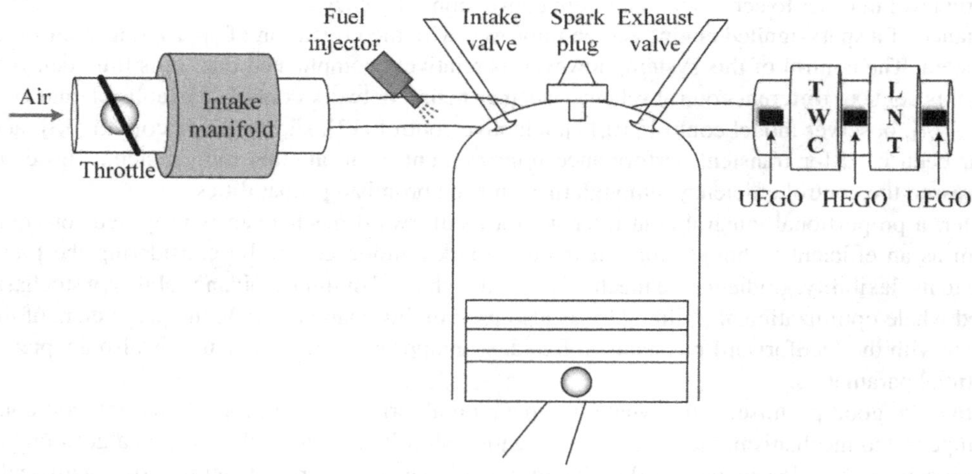

FIGURE 29.1 System configuration of an air-fuel ratio.

An air-fuel ratio (AFR) system has a time-delayed structure which needs to be tackled appropriately for efficient control [21]. To make the system capable of reacting changes efficiently, a feedforward (FF) control mechanism with a proportional-integral (PI) controller is used for this study [22]. The following equations respectively provide the transfer functions used for the PI controller, $C(s)$, and FF mechanism, $F(s)$.

$$C(s)=K_P+\frac{K_I}{s} \tag{29.2}$$

$$F(s)=\frac{K_F}{sT_F+1} \tag{29.3}$$

In here, K_P and K_I are PI related parameters whereas K_F and T_F are the FF related parameters. Appropriate adjustment of those parameters can significantly improve the performance of the system. Fig. 29.2 illustrates the block diagram of the PI and FF controllers used in this chapter for the control the AFR system.

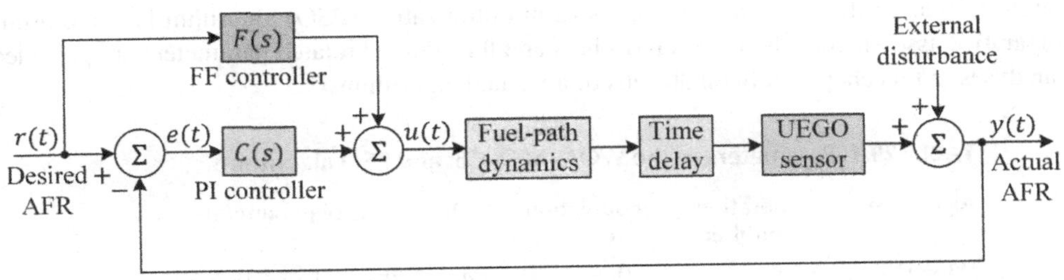

FIGURE 29.2 Closed loop AFR system with PI controller and FF mechanism.

For the AFR system, low or no overshoot with fast settling time is desirable. In this chapter the following cost function, J, is proposed to meet the aforementioned criteria:

$$J = \rho \times \int_0^{\infty} |e(t)|\, dt + (1 - \rho) \times T_s \tag{29.4}$$

where $e(t)$ is the error between the input and output signals, $e(t) = r(t) - y(t)$, and ρ is a balancing coefficient which is set as $\rho = 0.80$ in this chapter and T_s is the settling time. The parameter limitation is considered as $0.01 \leq K_P, K_I, K_F, T_F \leq 0.5$ in this chapter.

The WOA algorithm-based implementation procedure to adjust the PI controller with FF mechanism is illustrated in Fig. 29.3. As demonstrated, the parameter initialization of the WOA algorithm starts the optimization procedure and the parameters of K_P, K_I, K_F, and T_F are being updated using the proposed cost function. The optimization procedure ends after total number of iterations (t_{max}) are completed. Then the best obtained parameters are used.

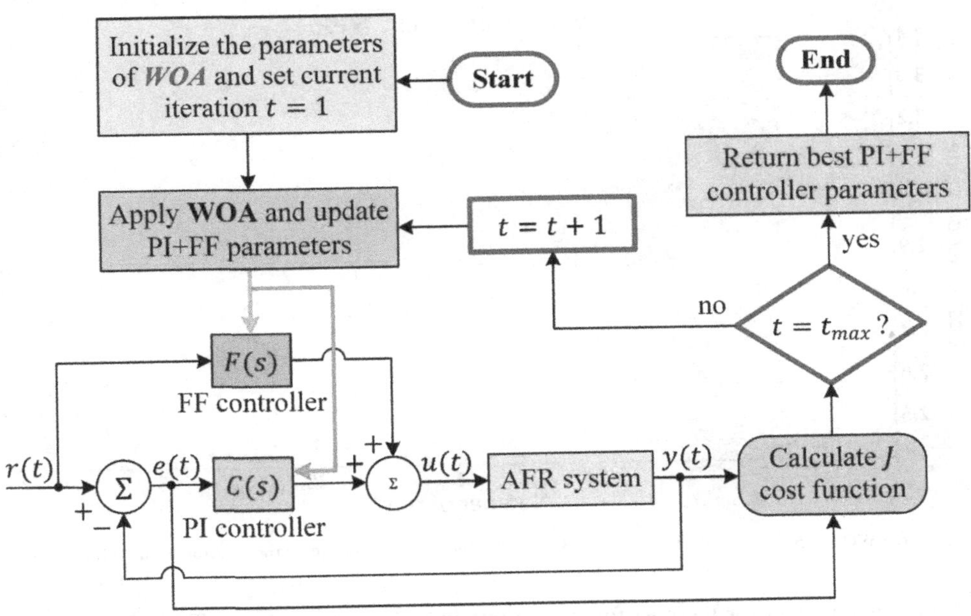

FIGURE 29.3 Application procedure for the WOA algorithm-based PI controller with an FF mechanism.

29.4 Simulation results

This section of the chapter provides the simulation results from a comparative perspective. In this regard, the statistical, computational time, converging ability, transient response, input signal tracking ability, disturbance rejection and robustness analyses are demonstrated. Besides, a performance evaluation based on several different cost functions is also provided.

29.4.1 Compared algorithms

To demonstrate not only the promise of the whale optimization algorithm (WOA) [15] but also the common efficiency of the metaheuristic optimization techniques for this specific engineering problem, the ant lion optimization (ALO) algorithm [16], cuckoo search (CS) algorithm [17], and particle swarm optimization (PSO) algorithm [18] are used in this chapter to provide a comparative assessment. The listed algorithms and the adopted related parameters are provided in Table 29.1. The performed analyses in this chapter rely on 30 runs of all stated algorithms.

TABLE 29.1 Parameters of the WOA, ALO, CS, and PSO algorithms.

Algorithm	Iteration number	Population size	Other control parameters
WOA [15]	50	30	$a=[0,2]$; $a_2=[-2,-1]$; $b=1$
ALO [16]	50	30	$I=10^w$; $w=[2,6]$
CS [17]	50	30	$p_a=0.25$
PSO [18]	50	30	$c_1=2$; $c_2=2$; inertia weight decreases linearly from 0.9 to 0.4

29.4.2 Statistical analysis

The initial assessment of the WOA [15] is comparatively performed from a statistical point of view by analyzing its ability of minimizing the cost function. Fig. 29.4 provides a comparative illustration of the ability of WOA [15], ALO [16], CS [17], and PSO [18] algorithms in terms of minimizing the cost function values with respect to number of runs. As shown in this figure, the WOA [15] algorithm is quite capable to better minimize the cost function value since it reaches the lowest value for all runs.

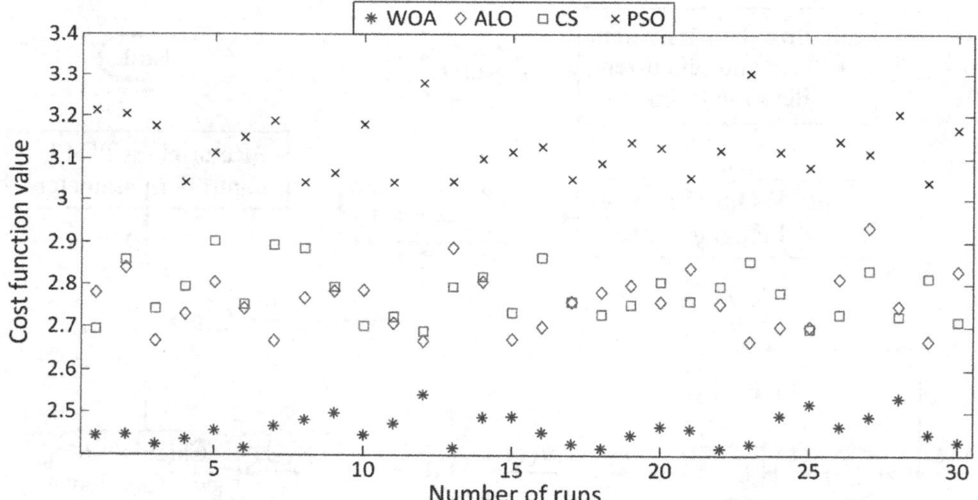

FIGURE 29.4 Ability of the WOA [15], ALO [16], CS [17], and PSO [18] algorithms to minimize the cost function with respect to number of runs.

The statistical metrics obtained for the cost function minimization are provided in Table 29.2. The numerical results in this table show that the WOA [15] algorithm is also capable to reach better statistical values of minimum, maximum, average, median, standard deviation and variance.

To provide the statistical performance from an illustrative perspective, a boxplot analysis is also demonstrated as can be seen in Fig. 29.5. The respective analysis shows that the WOA [15] algorithm reaches the lowest value for the cost function compared to ALO [16], CS [17], and PSO [18] algorithms. The performance of the WOA [15] algorithm is quite significant such that even the worst value achieved by the WOA algorithm is still lower than that of the lowest values achieved by ALO [16], CS [17], and PSO [18] algorithms.

TABLE 29.2 Statistical metrics for the cost function values obtained using WOA [15], ALO [16], CS [17], and PSO [18] algorithms.

Metric	WOA [15]	ALO [16]	CS [17]	PSO [18]
Best (minimum)	2.4150	2.6676	2.6912	3.0420
Worst (maximum)	2.5425	2.9347	2.9011	3.3048
Average	2.4606	2.7584	2.7795	3.1274
Median	2.4543	2.7589	2.7693	3.1186
Standard deviation	0.0351	0.0692	0.0627	0.0700
Variance	0.0012	0.0048	0.0039	0.0049

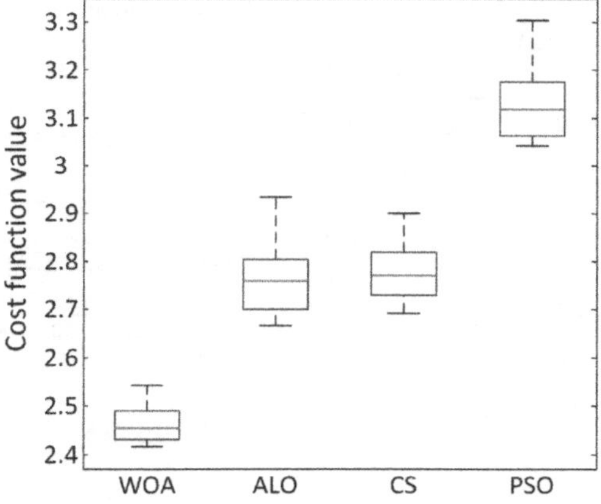

FIGURE 29.5 Boxplot analysis.

The last effort in terms of the statistical analysis is to perform a non-parametric statistical test such that the more excellent promise of the WOA [15] algorithm can be demonstrated from a more convincing perspective. In this regard, the Wilcoxon's test [23] is performed. As can be observed from the comparative results provided in Table 29.3, the WOA [15] algorithm becomes the winner with respect to ALO [16], CS [17], and PSO [18] algorithms as the p values smaller than 0.05 are achieved. That means the better effectiveness of the WOA algorithm is not by chance.

TABLE 29.3 Non-parametric statistical analysis.

Cost function	WOA [15] versus ALO [16]		WOA [15] versus CS [17]		WOA [15] versus PSO [18]	
	p-value	Winner	p-value	Winner	p-value	Winner
J	1.7344E−06	WOA	1.7344E−06	WOA	1.7344E−06	WOA

29.4.3 Computation time

The average computational time for each run of the WOA [15], ALO [16], CS [17], and PSO [18] algorithms is provided in Table 29.4 in order to demonstrate the efficiency in terms of the time taken to perform the optimization task. As shown in the presented numerical data, the ALO [16] algorithm reaches the best time efficiency. The WOA [15] algorithm follows the ALO [16] as the second-best algorithm for computational time efficiency. Meanwhile, there is not a significant difference between the ALO [16] and WOA [15] algorithms.

TABLE 29.4 Average computation times for per run of WOA, ALO, CS, and PSO algorithms.

Algorithm	WOA [15]	ALO [16]	CS [17]	PSO [18]
Time (s)	47.6801	44.2752	54.1855	57.0653

29.4.4 Convergence curve profile and obtained controller parameters

The convergence profiles of the WOA [15], ALO [16], CS [17], and PSO [18] algorithms for the minimization of the cost function are illustrated in Fig. 29.6. As shown in the related comparative convergence curves, the WOA [15] algorithm presents good capability of reaching the lowest cost function value with respect to ALO [16], CS [17], and PSO [18] algorithms. The best obtained controller parameters obtained via WOA [15], ALO [16], CS [17], and PSO [18] algorithms are provided in Table 29.5. Those parameters are used to perform the analysis of the system provided in Fig. 29.2.

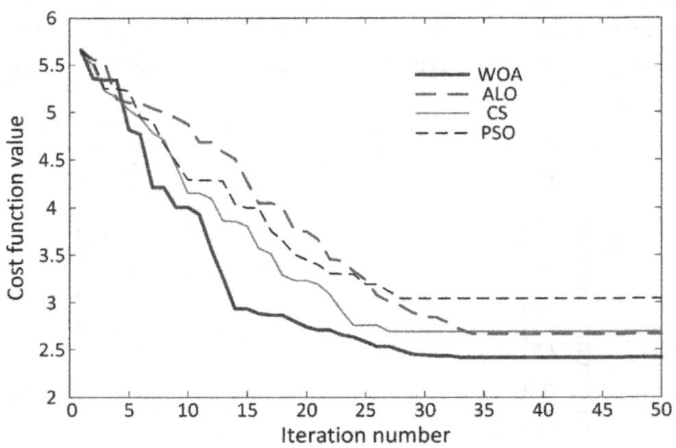

FIGURE 29.6 Convergence curves for best runs.

TABLE 29.5 The best obtained controller parameters.

Controller parameter	WOA [15]	ALO [16]	CS [17]	PSO [18]
K_P	0.20091	0.18463	0.23665	0.19655
K_I	0.32647	0.34908	0.31545	0.37916
K_F	0.29322	0.19521	0.33289	0.24987
T_F	0.21364	0.36290	0.46024	0.11796

29.4.5 Transient response

A comparative step response of the AFR system designed with the WOA [15], ALO [16], CS [17], and PSO [18] algorithms is provided in Fig. 29.7 in order to observe the transient response capability. The related figure demonstrates more desirable ability of the WOA [15] algorithm for AFR system control. To further support the demonstrated ability, the numerical results for the transient response performance metrics are also provided in Table 29.6. The related table numerically confirms the more excellent promise of the WOA [15] algorithm for the AFR system.

TABLE 29.6 Transient response performance metrics.

Performance indicator	WOA [15]	ALO [16]	CS [17]	PSO [18]
Rise time (s)	1.3572	1.6519	1.2597	1.3114
Settling time (s)	3.3188	3.8294	4.5898	6.0923
Overshoot (%)	1.4168	1.9182	4.2040	7.3100
Peak time (s)	4.2700	4.7280	3.8680	4.3230
Steady-state error	0	0	0	0

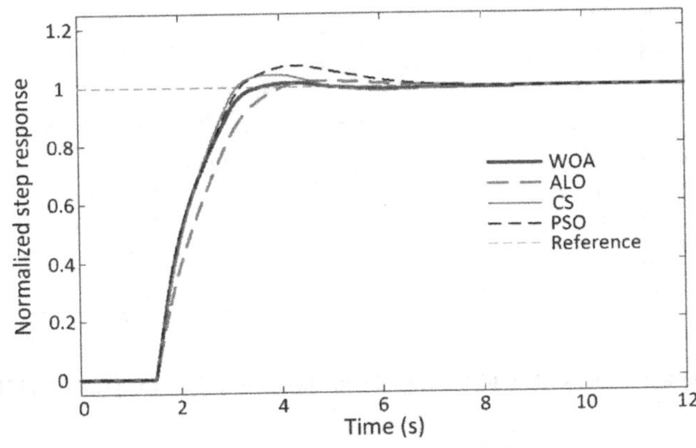

FIGURE 29.7 Step responses of the system designed with WOA [15], ALO [16], CS [17], and PSO [18] algorithms.

29.4.6 Input signal tracking ability

The input signal tracking performance of the system designed with the WOA [15], ALO [16], CS [17], and PSO [18] algorithms is provided in Fig. 29.8. At first glance, all the algorithms can be seen to track the input signal with a time delay. However, the tracking behavior of the WOA [15] algorithm based AFR system for the input signal is more efficient as can be observed from the fall or rise of the input signal.

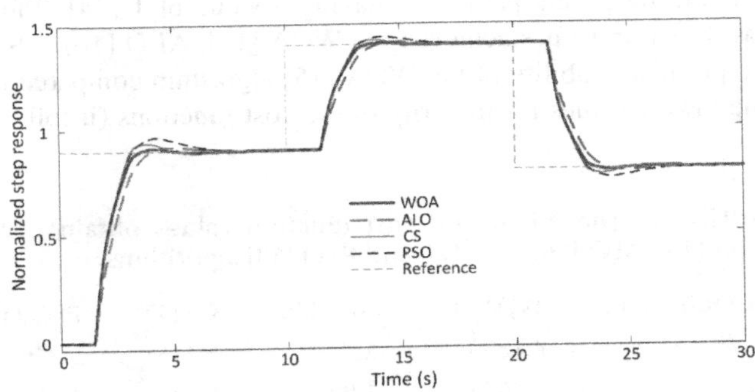

FIGURE 29.8 Input signal tracking behavior of the WOA [15], ALO [16], CS [17], and PSO [18] algorithms.

29.4.7 Disturbance rejection

A disturbance having 50% value of the reference signal is applied at 15 seconds in order to observe the disturbance rejection capability of the system designed with the WOA [15], ALO [16], CS [17], and PSO [18] algorithms. Fig. 29.9 demonstrates the disturbance rejection of the AFR system designed with WOA [15], ALO [16], CS [17], and PSO [18] algorithms. From this figure, the WOA [15] based system design can be seen to have the best disturbance rejection profile.

29.4.8 Performance evaluation on various cost functions

The performance of the WOA [15] based design is also investigated by using other widely utilized cost functions. In this context, integral of squared error, J_{ISE}, the integral of absolute error, J_{IAE}, integral of time-weighted squared error, J_{ITSE}, integral of time-weighted absolute error, J_{ITAE}, and ZLG, J_{ZLG}, are used to further provide the comparative assessment from a wider point of view. The mathematical definitions of J_{IAE}, J_{ISE}, J_{ITAE}, J_{ITSE} and J_{ZLG} are provided as follows [24–26].

$$J_{IAE} = \int_0^{\infty} |e(t)|\,dt \tag{29.5}$$

$$J_{ISE} = \int_0^{\infty} e^2(t)dt \tag{29.6}$$

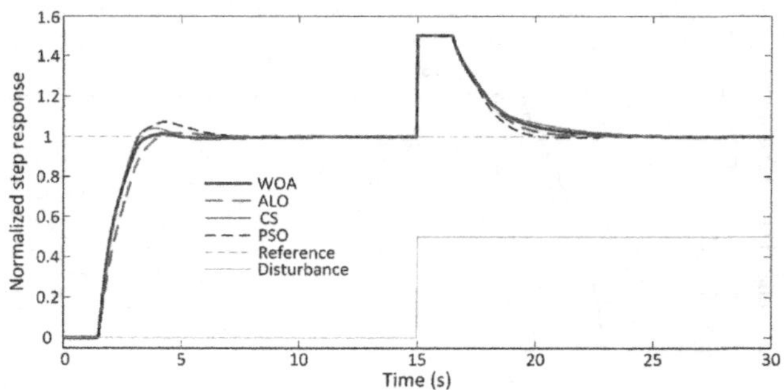

FIGURE 29.9 Disturbance rejection profile of the AFR system designed with the WOA [15], ALO [16], CS [17], and PSO [18] algorithms.

$$J_{ITAE} = \int_0^{\infty} t\,|e(t)|\,dt \tag{29.7}$$

$$J_{ITSE} = \int_0^{\infty} te^2(t)dt \tag{29.8}$$

$$J_{ZLG} = \left(1 - e^{-\sigma}\right)\left(\frac{\%OS}{100} + E_{ss}\right) + e^{-\sigma}(T_s - T_r) \tag{29.9}$$

In here, e represents the error, E_{ss} is the steady-state error, $\%OS$ is the maximum percent overshoot, T_s is the settling time, T_r is the rise time and σ is the weighting parameter having a value of 1 [24]. Table 29.7 presents the minimized numerical values of different cost function values achieved via WOA [15], ALO [16], CS [17], and PSO [18] algorithms. One can easily observe the more promising ability of the WOA [15] algorithm compared to ALO [16], CS [17], and PSO [18] algorithms as it achieves the lowest values for majority of the cost functions (it follows the PSO [18] algorithm only for J_{ISE}).

TABLE 29.7 The minimized cost function values obtained via WOA [15], ALO [16], CS [17], and PSO [18] algorithms.

Cost function	WOA [15]	ALO [16]	CS [17]	PSO [18]
J_{IAE}	2.1890	2.3772	2.2165	2.2794
J_{ISE}	1.8683	1.9851	1.8731	1.8611
J_{ITAE}	2.6678	3.1357	2.8200	3.1124
J_{ITSE}	1.8022	2.0629	1.8112	1.8067
J_{ZLG}	0.7306	0.8132	1.2517	1.8050

29.4.9 Robustness analysis under parameter changes

To assess the robustness comparatively, two different parameter uncertainty cases are considered. The respective cases (I and II) are provided in Table 29.8. The numerical values in Table 29.8 show the competitive performance of the WOA [15] based design for the AFR system in terms of overshoot, rise time, settling time, peak time, and steady state error. The two cases are further demonstrated via normalized step responses presented in Figs. 29.10 and 29.11.

29.5 Conclusion

This chapter discusses the more excellent promise of whale optimization algorithm as an efficient metaheuristic approach to deal with efficient operation of a feedforward compensated mechanism based proportional-integral controller-based air-fuel ratio system. An efficient cost function is also used to better optimize the parameters of the feedforward compensated proportional-integral controller adopted in an air-fuel ratio system. The more excellent capability of the whale optimization algorithm-based technique is demonstrated by using the ant lion optimization algorithm, cuckoo search algorithm and particle swarm optimization algorithm as competitive optimization techniques. The comparisons in terms of statistical, computational time, converging ability, transient response, input signal tracking ability, disturbance rejection and robustness

TABLE 29.8 Comparative robustness assessment against parameter uncertainty.

Parameter changes	Performance indicator	WOA [15]	ALO [16]	CS [17]	PSO [18]
Case (I) $\tau = 2.7$ s $k = 0.15$ s	Rise time (s)	1.3245	1.5620	1.2235	1.2784
	Settling time (s)	19.5386	20.8250	14.7282	25.2025
	Overshoot (%)	39.0890	42.1749	41.1194	51.1155
	Peak time (s)	6.5820	6.9370	5.9200	6.6310
	Steady-state error	0	0	0	0
Case (II) $\tau = 1.35$ s $k = 0.40$ s	Rise time (s)	1.5905	1.8942	1.4587	1.5095
	Settling time (s)	3.6153	6.5200	5.0943	6.8654
	Overshoot (%)	1.9897	3.1776	3.8740	7.7070
	Peak time (s)	4.6530	5.2690	4.1880	4.5950
	Steady-state error	0	0	0	0

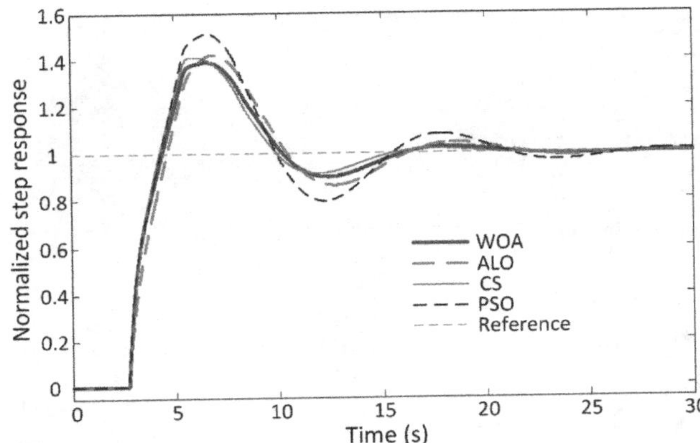

FIGURE 29.10 Robustness profile for Case I.

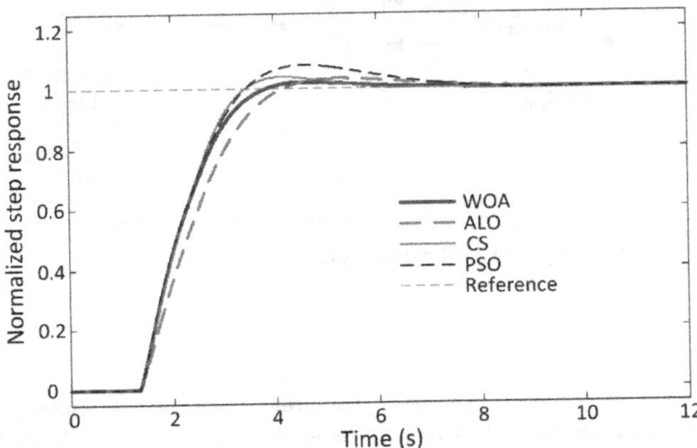

FIGURE 29.11 Robustness profile for Case (II).

analyses are presented. Furthermore, the widely used cost functions-based evaluation also demonstrates the more promising capacity of the whale optimization algorithm for the air-fuel ratio system control.

Appendix 29.A

The MATLAB® codes and the developed Simulink model for analysis of the air-fuel ratio system are provided in Table 29.A.1 and Fig. 29.A.1, respectively.

TABLE 29.A.1 MATLAB codes for WOA algorithm based AFR system.

```
%% AFR system
T=1.5; % 0.3<T<2.7
k=0.2;
% k=0.15; T=2.7; % Case (I)
% k=0.40; T=1.35; % Case (II)
s=tf('s');
P=exp(-T*s)/(1+k*s);
%% WOA-based PI+FF controller
Kp=0.20091; Ki=0.32647; Kf=0.29322; Tf=0.21364; % WOA
C = Kp+Ki/s; F = Kf/(Tf*s+1);
P.InputName = 'u';
P.OutputName = 'y';
C.InputName = 'e';
C.OutputName = 'uc';
F.InputName = 'r';
F.OutputName = 'uf';
Sum1 = sumblk('e','r','y','+-');    % e = r-y
Sum2 = sumblk('u','uf','uc','++');  % u = uf+uc
Twoa = connect(P,C,F,Sum1,Sum2,'r','y'); % stepinfo(Twoa)
t=0:0.001:12; y=step(Twoa, t); stepinfo(y, t, 1);
stepplot(Twoa)
legend('WOA')
xlabel('Time (s)'), ylabel('Normalized step response')
```

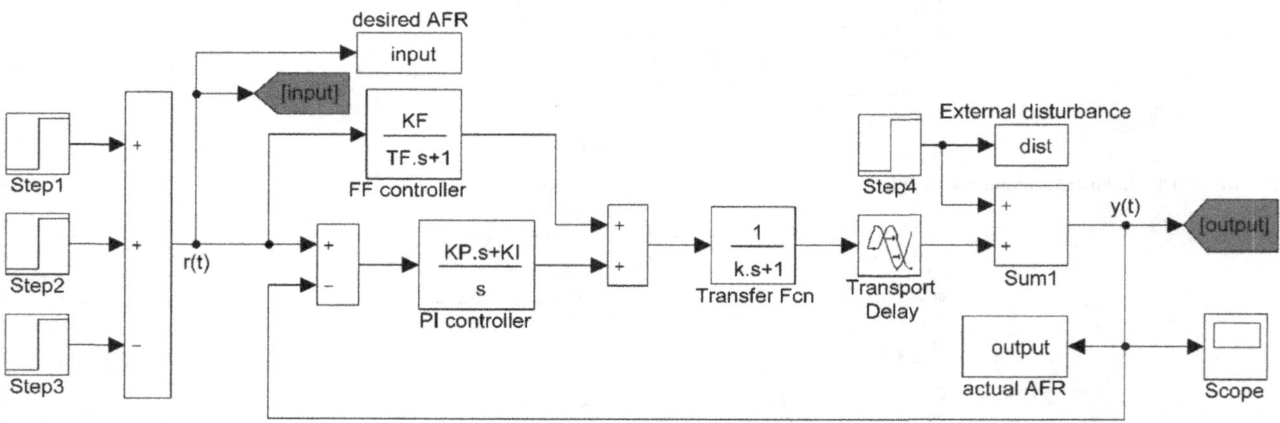

FIGURE 29.A.1 Developed Simulink model for analysis of AFR control system.

References

[1] A. Buonomano, G. Barone, C. Forzano, Advanced energy technologies, methods, and policies to support the sustainable development of energy, water and environment systems, Energy Reports 8 (2022) 4844–4853, https://doi.org/10.1016/j.egyr.2022.03.171.

[2] Z. Li, Q. Zhou, Y. Zhang, J. Li, H. Xu, Enhanced intelligent proportional-integral-like fuzzy knowledge–based controller using chaos-enhanced accelerated particle swarm optimization algorithm for transient calibration of air–fuel ratio control system, Proceedings of the Institution of Mechanical Engineers. Part D, Journal of Automobile Engineering 234 (2020) 39–55, https://doi.org/10.1177/0954407019862079.

[3] M. Postma, R. Nagamune, Air-fuel ratio control of spark ignition engines using a switching LPV controller, IEEE Transactions on Control Systems Technology 20 (2012) 1175–1187, https://doi.org/10.1109/TCST.2011.2163937.

[4] C. Manzie, M. Palaniswami, D. Ralph, H. Watson, X. Yi, Model predictive control of a fuel injection system with a radial basis function network observer, Journal of Dynamic Systems, Measurement, and Control 124 (2002) 648–658, https://doi.org/10.1115/1.1515328.

[5] K. Song, T. Hao, H. Xie, Disturbance rejection control of air–fuel ratio with transport-delay in engines, Control Engineering Practice 79 (2018) 36–49, https://doi.org/10.1016/j.conengprac.2018.06.009.

[6] B. Ebrahimi, R. Tafreshi, M. Franchek, K. Grigoriadis, J. Mohammadpour, A dynamic feedback control strategy for control loops with time-varying delay, International Journal of Control 87 (2014) 887–897, https://doi.org/10.1080/00207179.2013.861612.

[7] R. Tafreshi, B. Ebrahimi, J. Mohammadpour, M.A. Franchek, K. Grigoriadis, H. Masudi, Linear dynamic parameter-varying sliding manifold for air–fuel ratio control in lean-burn engines, IET Control Theory & Applications 7 (2013) 1319–1329, https://doi.org/10.1049/iet-cta.2012.0823.

[8] Y.-J. Zhai, D.-L. Yu, Neural network model-based automotive engine air/fuel ratio control and robustness evaluation, Engineering Applications of Artificial Intelligence 22 (2009) 171–180, https://doi.org/10.1016/j.engappai.2008.08.001.

[9] M.S. Iqbal, A.A. Amin, Genetic algorithm based active fault-tolerant control system for air fuel ratio control of internal combustion engines, Measurements & Control 55 (2022) 703–716, https://doi.org/10.1177/00202940221115233.

[10] H.-M. Wu, R. Tafreshi, Observer-based internal model air–fuel ratio control of lean-burn SI engines, IFAC Journal of Systems and Control 9 (2019) 100065, https://doi.org/10.1016/j.ifacsc.2019.100065.

[11] T. Alsuwian, M.S. Iqbal, A.A. Amin, M.B. Qadir, S. Almasabi, M. Jalalah, A comparative study of design of active fault-tolerant control system for air–fuel ratio control of internal combustion engine using particle swarm optimization, genetic algorithm, and nonlinear regression-based observer model, Applied Sciences 12 (2022) 7841, https://doi.org/10.3390/app12157841.

[12] X. Jiao, J. Zhang, T. Shen, J. Kako, Adaptive air-fuel ratio control scheme and its experimental validations for port-injected spark ignition engines, International Journal of Adaptive Control and Signal Processing 29 (2015) 41–63, https://doi.org/10.1002/acs.2456.

[13] Z. Salehi, S. Azadi, A. Mousavinia, Sliding mode air-to-fuel ratio control of spark ignition engines in comprehensive powertrain system, in: 2021 7th International Conference on Control, Instrumentation and Automation (ICCIA), IEEE, 2021, pp. 1–5, https://doi.org/10.1109/ICCIA52082.2021.9403601.

[14] Z. Li, J. Li, Q. Zhou, Y. Zhang, H. Xu, Intelligent air/fuel ratio control strategy with a PI-like fuzzy knowledge–based controller for gasoline direct injection engines, Proceedings of the Institution of Mechanical Engineers. Part D, Journal of Automobile Engineering 233 (2019) 2161–2173, https://doi.org/10.1177/0954407018779180.

[15] S. Mirjalili, A. Lewis, The whale optimization algorithm, Advances in Engineering Software 95 (2016) 51–67, https://doi.org/10.1016/j.advengsoft.2016.01.008.

[16] S. Mirjalili, The ant lion optimizer, Advances in Engineering Software 83 (2015) 80–98, https://doi.org/10.1016/j.advengsoft.2015.01.010.

[17] A.H. Gandomi, X.S. Yang, A.H. Alavi, Cuckoo search algorithm: a metaheuristic approach to solve structural optimization problems, Engineering Computations 29 (2013) 17–35, https://doi.org/10.1007/s00366-011-0241-y.

[18] Y. Shi, R.C. Eberhart, Empirical study of particle swarm optimization, in: Proceedings of the 1999 Congress on Evolutionary Computation-CEC99 (Cat. No. 99TH8406), IEEE, 1999, pp. 1945–1950, https://doi.org/10.1109/CEC.1999.785511.

[19] H.-M. Wu, R. Tafreshi, Air–fuel ratio control of lean-burn SI engines using the LPV-based fuzzy technique, IET Control Theory & Applications 12 (2018) 1414–1420, https://doi.org/10.1049/iet-cta.2017.0063.

[20] H.-M. Wu, R. Tafreshi, Fuzzy sliding-mode strategy for air-fuel ratio control of lean-burn spark ignition engines, Asian Journal of Control 20 (2018) 149–158, https://doi.org/10.1002/asjc.1544.

[21] M. Mohammadzaheri, R. Tafreshi, An enhanced Smith predictor based control system using feedback-feedforward structure for time-delay processes, The Journal of Engineering Research [TJER] 14 (2017) 156, https://doi.org/10.24200/tjer.vol14iss2pp156-165.

[22] A.P. Montoya-Ríos, F. García-Mañas, J.L. Guzmán, F. Rodríguez, Simple tuning rules for feedforward compensators applied to greenhouse daytime temperature control using natural ventilation, Agronomy 10 (2020) 1327, https://doi.org/10.3390/agronomy10091327.

[23] S. Ekinci, D. Izci, R. Abu Zitar, A.R. Alsoud, L. Abualigah, Development of Lévy flight-based reptile search algorithm with local search ability for power systems engineering design problems, Neural Computing & Applications 34 (2022) 20263–20283, https://doi.org/10.1007/s00521-022-07575-w.

[24] S. Ekinci, D. Izci, M. Kayri, An effective controller design approach for magnetic levitation system using novel improved manta ray foraging optimization, Arabian Journal for Science and Engineering 47 (2022) 9673–9694, https://doi.org/10.1007/s13369-021-06321-z.

[25] D. Izci, S. Ekinci, S. Mirjalili, Optimal PID plus second-order derivative controller design for AVR system using a modified Runge Kutta optimizer and Bode's ideal reference model, International Journal of Dynamics and Control 11 (2023) 1247–1264, https://doi.org/10.1007/s40435-022-01046-9.

[26] D. Izci, B. Hekimoğlu, S. Ekinci, A new artificial ecosystem-based optimization integrated with Nelder-Mead method for PID controller design of buck converter, Alexandria Engineering Journal 61 (2022) 2030–2044, https://doi.org/10.1016/j.aej.2021.07.037.

Chapter 30

Application of whale optimization algorithm to infinite impulse response system identification

Davut Izci and Serdar Ekinci
Department of Computer Engineering, Batman University, Batman, Turkey

30.1 Introduction

It is feasible to classify the digital systems as finite impulse response (FIR) and infinite impulse response (IIR) systems. The main difference of those systems is that the FIR systems rely on the past and present input signals whilst the IIR systems considers the past outputs, as well, in addition to present and past input signals [1]. Over the past few decades, there has been an increasing trend towards investigation of adaptive IIR filters. The rising trend can mainly be attributed to their applicability in system identification and modeling problems in the signal processing field [2]. Meanwhile, the IIR systems can be used to describe several other phenomena such as parameter estimation [3], robotics [4], and radar processing [5] along with communication and control systems [6]. Due to such a wider applicability scope, the rising trend of the IIR filtering keeps its continuation. Another fact that makes the IIR filter models attractive is that they can effectively model an unknown system with fewer number of coefficients compared to adaptive FIR systems [7]. Therefore, the adaptive IIR systems are also computationally efficient models.

The IIR system identification can be implemented in two stages. In the first stage, an appropriate plant is chosen for the identification. Then, the computation of the optimal filter coefficients is carried out using a convenient optimization technique. The main objective of such an approach is to determine an optimal set of coefficients which would allow the output of the adaptive IIR system to track the output of the unknown plant if they are subjected to the same input signal [8]. In another word, the IIR system identification problem relies on minimization of the error between the outputs of the adaptive filter and the unknown plant. Nevertheless, there are certain issues that must be tackled despite the advantages of the IIR systems. The structures of IIR models tend to generate errors with multimodal surfaces that leads to difficulties in terms of minimization of cost functions [9].

The minimization process is traditionally performed via gradient-based search algorithms, e.g., least mean square and Quasi-Newton approaches [10]. However, the minimization of the stated error related fitness functions using gradient-based search algorithms is difficult since such algorithms get stuck in local minima and cannot converge to the global minima [11]. To deal with such difficulties, metaheuristic approaches have recently been used as promising candidates as they can converge to a global optimal solution faster [12]. Because they have random search and selection capabilities. Therefore, it is feasible to reach more accurate and robust solutions [7]. In this regard, several metaheuristic techniques such as average differential evolution algorithm with local search [11], adaptive simulated annealing algorithm [13], particle swarm optimization algorithm [14], ant colony optimization algorithm [15], slime mold algorithm [16], and its enhanced version [17], teacher learner based optimization algorithm [18], firefly algorithm [19], grey wolf optimization algorithm [20], harmony search algorithm [21], modified imperialist competitive algorithm [22], tabu search algorithm [23], artificial bee colony algorithm [24], cat swarm optimization algorithm [25], and seeker optimization algorithm [26] can be encountered as the reported works regarding the IIR system identification problem.

In this chapter, efficiency of whale optimization algorithm is investigated in terms of adaptive IIR system identification problem. The whale optimization algorithm is a widely used metaheuristic algorithm mimicking the bubble-net feeding behavior of humpback whales [27]. It is basically a population-based metaheuristic approach that uses search agents to determine global optimum. The ability of the whale optimization algorithm is evaluated for the adaptive IIR model identification. In this context, four different examples with both reduced order and same order cases are considered for the IIR

Handbook of Whale Optimization Algorithm. https://doi.org/10.1016/B978-0-32-395365-8.00036-1

system identification problem. In terms of the comparisons, sine-cosine algorithm [28], gravitational search algorithm [29], and artificial bee colony optimization algorithm [30] are used for the comparative evaluation. The proposed whale optimization algorithm based IIR filtering for system identification is comparatively analyzed by computing the mean squared error as the cost function. The obtained statistical and convergence profile analyses indicate highly competitive ability of the whale optimization algorithm for the adaptive IIR system identification problem.

30.2 Whale optimization algorithm

This algorithm is a widely used stochastic optimization technique proposed by Mirjalili and Lewis [27] and mimics the humpback whales' bubble-net feeding behavior. The whale optimization algorithm (WOA) is basically a population-based metaheuristic approach using search agents to determine global optimum. Eq. (30.1) can be used as the main mathematical model of the WOA algorithm.

$$X(t+1) = \begin{cases} X^*(t) - AD & \text{if } p < 0.5 \\ D'e^{bl}\cos(2\pi t) + X^*(t) & \text{if } p \geq 0.5 \end{cases} \tag{30.1}$$

In Eq. (30.1), $D' = |X^*(t) - X(t)|$ indicating the distance of the ith agent and the current best solution, $D = |CX^*(t) - X(t)|$ where $C = 2r$ and $A = 2ar - a$. p, r, and l represent a random number within [0, 1], [0, 1], and [−1, 1], respectively, b stands for a constant that defines the shape of the logarithmic spiral, t is the current iteration. a is a linearly decreasing (from 2 to 0) parameter throughout the iterations. The encircling mechanism of the humpback whales is represented by the first component of Eq. (30.1) whereas the second component of this equation mimics the bubble-net feeding behavior. In WOA algorithm an equal probability is specified between the two components by switching the variable p. The exploration and exploitation phases are guaranteed in through adaptive adjustment of a and C parameters. The best solution in WOA algorithm plays a significant role to update the position of search agents, thus, guarantee the exploration and convergence whereas the currently obtained best solution has the importance for exploitation. High local optima avoidance ability, gradient-free mechanism, and flexibility of the WOA algorithm motivates the work presented in this chapter for efficient IIR system identification.

30.3 Problem formulation for IIR system identification

It is feasible achieve more accurate representation of physical plants for real world applications and to meet the performance specifications by using infinite impulse response models with fewer number of parameters [31]. Fig. 30.1 is used to provide an illustrative intuition via an arbitrary plant's infinite impulse response identification model.

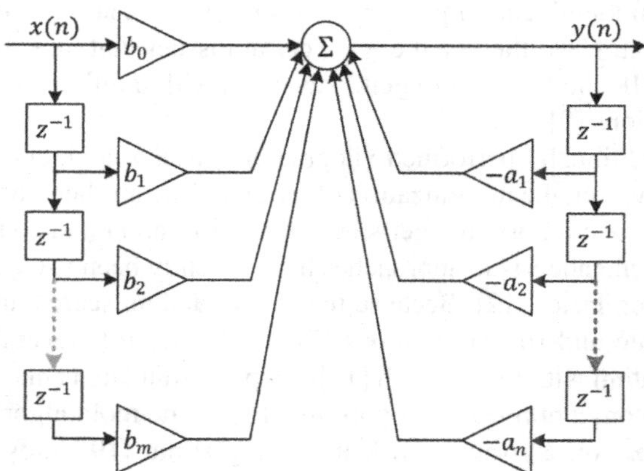

FIGURE 30.1 Detailed structure of IIR filter.

In the respective illustration, $y(n)$ represent the output of the IIR filter and $x(n)$ stands for the applied input signal. b_j and a_i are respectively the coefficients of the numerator and denominator where $i = 1, 2, \ldots, n$ and $j = 0, 1, \ldots, m$. The following transfer function can be obtained for an IIR filter system considering the above-mentioned information.

$$H(z) = \frac{b_0 + b_1 z^{-1} + b_2 z^{-2} + \cdots + b_m z^{-m}}{1 + a_1 z^{-1} + a_2 z^{-2} + \cdots + a_n z^{-n}} \tag{30.2}$$

The aim of an IIR system identification is to determine an unknown system using the IIR filter. Since the coefficients of the IIR filter attempt to approach to that of the unknown system, this can be considered as an optimization problem. Therefore, it is feasible to use an optimization technique to minimize the error between the outputs of the IIR filter model and the unknown plant. This case can illustratively be represented as shown in Fig. 30.2.

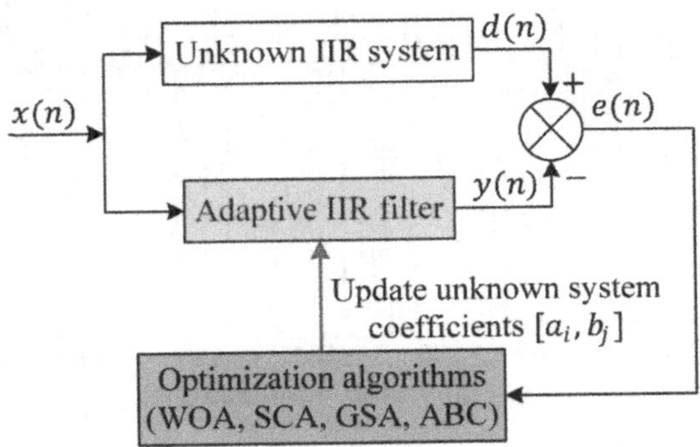

FIGURE 30.2 Block diagram of WOA, SCA, GSA, and ABC algorithms based IIR system identification problem.

In this figure, $e(n)$ represents the error between $d(n)$, unknown system, and $y(t)$ adaptive IIR filter. In this way, a minimization problem is obtained which can be described with the following mean squared error (MSE) function:

$$MSE = \frac{1}{W}\sum_{n=1}^{W}(d(n) - y(n))^2 \tag{30.3}$$

where W represents the number of input samples. To represent the MSE in dB form, Eq. (30.3) can be rewritten as $10\log_{10}(MSE)$.

30.4 Simulation results

In this section of the chapter, the comparative performance assessment of the WOA algorithm, for the IIR system identification problem, is discussed. A uniformly distributed white-noise sequence, taking values from $(-1, 1)$ and $W = 200$ is used as an input signal for the simulations. The related input signal is presented in Fig. 30.3 which is applied to the transfer functions of the adaptive IIR filter and the unknown IIR system. In this work, the difference of those models is obtained, and the adaptive filter model is minimized using the employed metaheuristic optimization algorithms. Table 30.1 presents the respective parameters of the WOA algorithm [27] along with the sine-cosine algorithm (SCA) [28], gravitational search algorithm (GSA) [29], and artificial bee colony (ABC) optimization algorithm [30] that are used for the comparative evaluation.

TABLE 30.1 The parameter settings of different algorithms for performance evaluation.

Optimization algorithm	Maximum iterations	Population size	Other control parameters
WOA [27]	500	30	$a = [0, 2]$; $a_2 = [-2, -1]$; $b = 1$
SCA [28]	500	30	$A = 2$
GSA [29]	500	30	$G_0 = 100$, $\alpha = 20$
ABC [30]	500	30	$Limit = 3$

In this chapter, four different examples with both reduced order and same order cases are considered for the IIR system identification problem. These examples and the respective cases are provided in the following subsections.

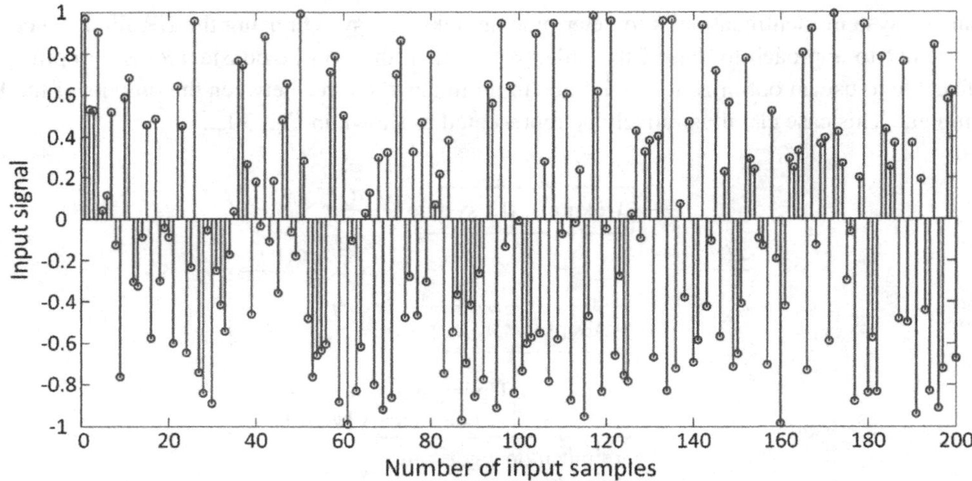

FIGURE 30.3 The input signal used for simulations.

30.4.1 Example I and related results

The initial assessment of the WOA algorithm for the IIR system identification problem is demonstrated through modeling a second order IIR system, given in Eq. (30.4), using second-order, Case (a), and first-order, Case (b), IIR filters.

$$H_P(z) = \frac{0.05 - 0.4z^{-1}}{1 - 1.1314z^{-1} + 0.25z^{-2}} \tag{30.4}$$

30.4.1.1 Case (a)

The same-order IIR model of the unknown IIR system in Example I is presented in Eq. (30.5). The comparative statistical results for this case are presented in numerical form as shown in Table 30.2 whereas the related obtained optimal coefficients are listed in Table 30.3. As can be seen from Table 30.2, the best MSE value, shown in bold, (−38.0416 in dB form) is obtained with the WOA algorithm indicating its better capacity for this type of IIR system identification problem.

$$H_S = \frac{b_0 + b_1 z^{-1}}{1 - a_1 z^{-1} - a_2 z^{-2}} \tag{30.5}$$

TABLE 30.2 MSE based statistical results for Example I (same order system).

Optimization algorithm	MSE				MSE (in dB)		
	Best	Worst	Mean	SD	Best	Worst	Mean
WOA [27]	**1.5698E−04**	4.7674E−02	1.8792E−02	1.4030E−02	−38.0416	−13.2172	−17.2603
SCA [28]	5.6896E−04	2.4368E−02	3.2316E−03	4.2461E−03	−32.4492	−16.1318	−24.9058
GSA [29]	2.0453E−03	2.6152E−02	8.4352E−03	5.1973E−03	−26.8924	−15.8250	−20.7390
ABC [30]	3.3885E−04	5.1514E−03	1.7066E−03	1.0655E−03	−34.6999	−22.8807	−27.6787

TABLE 30.3 Optimal coefficients of Example I in case of same order system.

Coefficients	Actual values	WOA [27]	SCA [28]	GSA [29]	ABC [30]
a_1	1.1314	1.0834	1.1352	0.9302	1.1460
a_2	−0.2500	−0.2055	−0.2447	−0.0644	−0.2608
b_0	0.0500	0.0564	0.0420	0.0554	0.0733
b_1	−0.4000	−0.4234	−0.4057	−0.4610	−0.4140

To provide an illustrative intuition, the convergence curves of the compared algorithms for Example I and Case (a) are provided in Fig. 30.4 in terms of dB form of MSE values. As demonstrated, this figure further indicates the good capability of the WOA algorithm since it converges to the lowest value with respect to other algorithms.

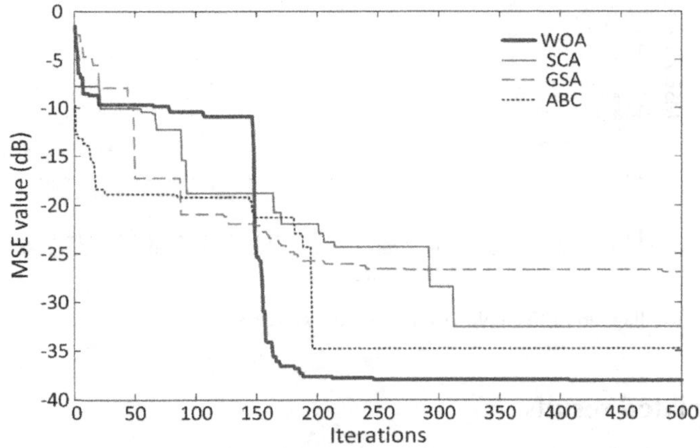

FIGURE 30.4 Comparative convergence diagram of Example I (same order system).

30.4.1.2 Case (b)

The reduced-order IIR model of the unknown IIR system in Example I is presented in Eq. (30.6). The comparative statistical results for this case are presented in numerical form as shown in Table 30.4 whereas the related obtained optimal coefficients are listed in Table 30.5. As can be seen from Table 30.4, the best MSE value, shown in bold, (−13.3721 in dB form) is obtained with the WOA algorithm indicating its better capacity for this type of IIR system identification problem.

$$H_R = \frac{b_0}{1 - a_1 z^{-1}} \tag{30.6}$$

TABLE 30.4 MSE based statistical results for Example I (reduced order system).

Optimization algorithm	MSE				MSE (in dB)		
	Best	Worst	Mean	SD	Best	Worst	Mean
WOA [27]	**4.6003E−02**	5.4014E−02	5.1819E−02	1.6569E−03	−13.3721	−12.6749	−12.8551
SCA [28]	5.4464E−02	6.2863E−02	5.9121E−02	1.9648E−03	−12.6389	−12.0160	−12.2826
GSA [29]	4.7545E−02	8.2473E−02	6.0596E−02	1.0738E−02	−13.2290	−10.8369	−12.1756
ABC [30]	4.9212E−02	5.5430E−02	5.3206E−02	1.6589E−03	−13.0793	−12.5626	−12.7404

TABLE 30.5 Optimal coefficients of Example I in case of reduced order system.

Coefficients	WOA [27]	SCA [28]	GSA [29]	ABC [30]
a_1	0.9102	0.9019	0.9129	0.9127
b_0	−0.2862	−0.3652	−0.3081	−0.3048

To provide an illustrative intuition, the convergence curves of the compared algorithms for Example I and Case (b) are provided in Fig. 30.5 in terms of dB form of MSE values. As demonstrated, this figure further indicates the more promising ability of the WOA algorithm since it converges to the lowest value with respect to other algorithms.

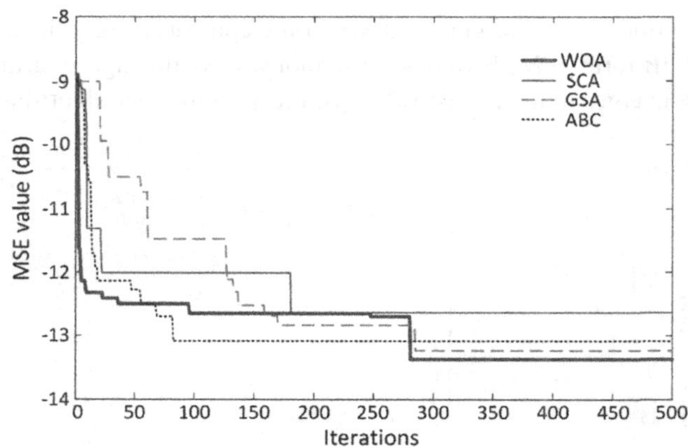

FIGURE 30.5 Comparative convergence diagram of Example I (reduced order system).

30.4.2 Example II and related results

The second assessment of the WOA algorithm for the IIR system identification problem is demonstrated through modeling a third order IIR system, given in Eq. (30.7), using third-order, Case (a), and second-order, Case (b), IIR filters.

$$H_P(z) = \frac{-0.2 - 0.4z^{-1} + 0.5z^{-2}}{1 - 0.6z^{-1} + 0.25z^{-2} - 0.2z^{-3}} \tag{30.7}$$

30.4.2.1 Case (a)

The same-order IIR model of the unknown IIR system in Example II is presented in Eq. (30.8). The comparative statistical results for this case are presented in numerical form as shown in Table 30.6 whereas the related obtained optimal coefficients are listed in Table 30.7. Although the best MSE value is obtained by the GSA algorithm for this type of IIR system identification problem, shown in bold (−184.1515 in dB form), the WOA algorithm still indicates the competitive performance with respect to the rest of the compared algorithms.

$$H_S = \frac{b_0 + b_1 z^{-1} + b_2 z^{-2}}{1 - a_1 z^{-1} - a_2 z^{-2} - a_3 z^{-3}} \tag{30.8}$$

TABLE 30.6 MSE based statistical results for Example II (same order system).

Optimization algorithm	MSE				MSE (in dB)		
	Best	Worst	Mean	SD	Best	Worst	Mean
WOA [27]	8.6994E−05	3.4255E−03	1.7333E−03	8.7323E−04	−40.6051	−24.6528	−27.6113
SCA [28]	1.1101E−03	5.5231E−03	2.8491E−03	1.0537E−03	−29.5464	−22.5782	−25.4529
GSA [29]	**3.8446E−19**	1.1734E−03	2.8440E−04	3.4927E−04	−184.1515	−29.3055	−35.4607
ABC [30]	1.9267E−05	1.3060E−03	5.0564E−04	3.3089E−04	−47.1519	−28.8406	−32.9616

To provide an illustrative intuition, the convergence curves of the compared algorithms for Example II and Case (a) are provided in Fig. 30.6 in terms of dB form of MSE values. This figure complies with the presented numerical values as GSA algorithm indicates better performance and the WOA algorithm has a competitive capability compared to the rest of the algorithms.

30.4.2.2 Case (b)

The reduced-order IIR model of the unknown IIR system in Example II is presented in Eq. (30.9). The comparative statistical results for this case are presented in numerical form as shown in Table 30.8 whereas the related obtained optimal coefficients are listed in Table 30.9. As can be seen from Table 30.8, the best MSE value, shown in bold, (−26.2195 in dB

TABLE 30.7 Optimal coefficients of Example II in case of same order system.

Coefficients	Actual values	WOA [27]	SCA [28]	GSA [29]	ABC [30]
a_1	0.6000	0.6046	0.4758	0.6000	0.5713
a_2	−0.2500	−0.2770	−0.3078	−0.2500	−0.2522
a_3	0.2000	0.2055	0.1174	0.2000	0.1872
b_0	−0.2000	−0.1886	−0.1685	−0.2000	−0.1970
b_1	−0.4000	−0.4021	−0.4183	−0.4000	−0.4058
b_2	0.5000	0.4913	0.4083	0.5000	0.4856

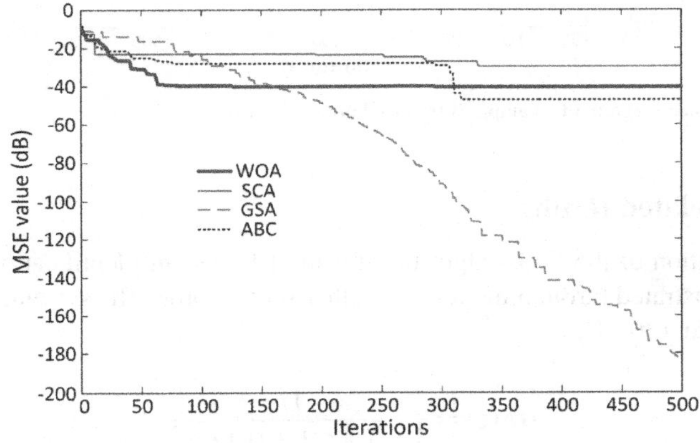

FIGURE 30.6 Comparative convergence diagram of Example II (same order system).

TABLE 30.8 MSE based statistical results for Example II (reduced order system).

Optimization algorithm	MSE				MSE (in dB)		
	Best	Worst	Mean	SD	Best	Worst	Mean
WOA [27]	**2.3881E−03**	4.5106E−03	3.1445E−03	4.4902E−04	−26.2195	−23.4577	−25.0245
SCA [28]	3.0711E−03	6.8610E−03	4.6661E−03	6.7150E−04	−25.1271	−21.6361	−23.3105
GSA [29]	2.4481E−03	3.0344E−03	2.6585E−03	1.2396E−04	−26.1117	−25.1793	−25.7536
ABC [30]	2.6008E−03	3.5803E−03	3.2281E−03	2.4197E−04	−25.8489	−24.4608	−24.9105

TABLE 30.9 Optimal coefficients of Example II in case of reduced order system.

Coefficients	WOA [27]	SCA [28]	GSA [29]	ABC [30]
a_1	−0.1574	−0.1816	−0.1622	−0.1416
a_2	−0.3416	−0.3532	−0.3893	−0.3499
b_0	−0.2256	−0.2128	−0.2200	−0.2192
b_1	−0.6029	−0.5738	−0.5755	−0.5860

form) is obtained with the WOA algorithm indicating its better capacity for this type of IIR system identification problem.

$$H_R = \frac{b_0 + b_1 z^{-1}}{1 - a_1 z^{-1} - a_2 z^{-2}} \tag{30.9}$$

To provide an illustrative intuition, the convergence curves of the compared algorithms for Example II and Case (b) are provided in Fig. 30.7 in terms of dB form of MSE values. As demonstrated, this figure further indicates the more promising ability of the WOA algorithm since it converges to the lowest value with respect to other algorithms.

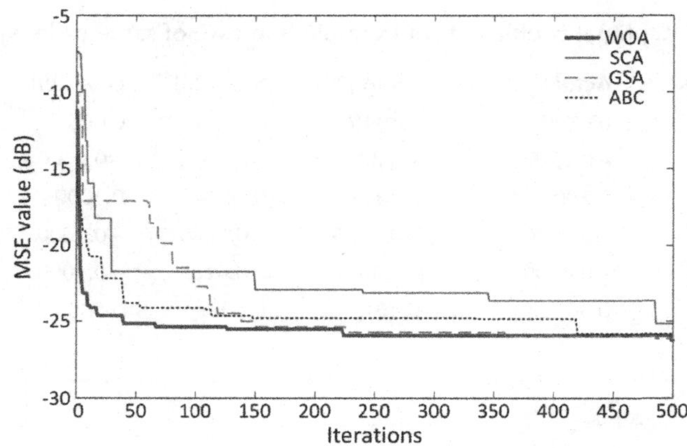

FIGURE 30.7 Comparative convergence diagram of Example II (reduced order system).

30.4.3 Example III and related results

In the third phase of the evaluation of the WOA algorithm for the IIR system identification problem, another IIR system identification problem is demonstrated through modeling another second order IIR system, given in Eq. (30.10), using the same order IIR filter, given in Eq. (30.11).

$$H_P(z) = \frac{1}{1 - 1.4z^{-1} + 0.49z^{-2}} \tag{30.10}$$

$$H_S = \frac{b_0}{1 - a_1 z^{-1} - a_2 z^{-2}} \tag{30.11}$$

The comparative statistical results for this case are presented in numerical form as shown in Table 30.10 whereas the related obtained optimal coefficients are listed in Table 30.11. Although the best MSE value is obtained by the ABC algorithm for this type of IIR system identification problem, shown in bold (−53.8731 in dB form), the WOA algorithm still indicates the competitive performance with respect to the rest of the compared algorithms.

TABLE 30.10 MSE based statistical results for Example III.

Optimization algorithm	MSE				MSE (in dB)		
	Best	Worst	Mean	SD	Best	Worst	Mean
WOA [27]	1.1444E−03	3.7848E−01	1.2083E−01	1.0894E−01	−29.4142	−4.2196	−9.1783
SCA [28]	1.1846E−03	3.3270E−02	1.0198E−02	7.8893E−03	−29.2643	−14.7795	−19.9148
GSA [29]	2.4243E−02	5.3904E−01	2.5161E−01	1.3789E−01	−16.1541	−2.6838	−5.9927
ABC [30]	**4.0991E−06**	8.9782E−03	1.4746E−03	1.9120E−03	−53.8731	−20.4681	−28.3133

TABLE 30.11 Optimal coefficients of Example III.

Coefficients	Actual values	WOA [27]	SCA [28]	GSA [29]	ABC [30]
a_1	1.4000	1.3833	1.4163	1.3021	1.3992
a_2	−0.4900	−0.4732	−0.5063	−0.3915	−0.4893
b_0	1.0000	0.9957	0.9950	0.9819	1.0000

To provide an illustrative intuition, the convergence curves of the compared algorithms for Example III are provided in Fig. 30.8 in terms of dB form of MSE values. This figure complies with the presented numerical values as ABC algorithm indicates better performance and the WOA algorithm has a competitive capability compared to the rest of the algorithms.

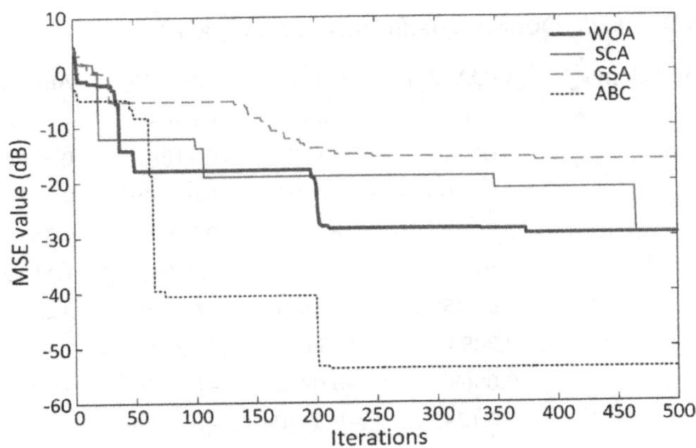

FIGURE 30.8 Comparative convergence diagram of Example III.

30.4.4 Example IV and related results

In the last phase of the evaluation of the WOA algorithm for the IIR system identification problem, a high order IIR system identification problem is demonstrated through modeling a sixth order IIR system, given in Eq. (30.12), using a fourth order IIR filter, given in Eq. (30.13).

$$H_P(z) = \frac{1 - 0.4z^{-2} - 0.65z^{-4} + 0.26z^{-6}}{1 - 0.77z^{-2} - 0.8498z^{-4} + 0.6486z^{-6}} \tag{30.12}$$

$$H_R = \frac{b_0 + b_1 z^{-1} + b_2 z^{-2} + b_3 z^{-3} + b_4 z^{-4}}{1 - a_1 z^{-1} - a_2 z^{-2} - a_3 z^{-3} - a_4 z^{-4}} \tag{30.13}$$

The comparative statistical results for this case are presented in numerical form as shown in Table 30.12 whereas the related obtained optimal coefficients are listed in Table 30.13. Although the best MSE value is obtained by the GSA algorithm for this type of IIR system identification problem, shown in bold (−24.2288 in dB form), the WOA algorithm still indicates the competitive performance with respect to the rest of the compared algorithms.

TABLE 30.12 MSE based statistical results for Example IV.

Optimization algorithm	MSE				MSE (in dB)		
	Best	Worst	Mean	SD	Best	Worst	Mean
WOA [27]	1.4071E−02	2.7312E−01	1.3882E−01	6.2572E−02	−18.5168	−5.6365	−8.5755
SCA [28]	4.2265E−03	1.2939E−01	4.8350E−02	3.5835E−02	−23.7402	−8.8810	−13.1560
GSA [29]	**3.7768E−03**	1.2815E−01	3.1727E−02	3.2812E−02	−24.2288	−8.9228	−14.9857
ABC [30]	5.5327E−02	1.8341E−01	1.0881E−01	3.8047E−02	−12.5706	−7.3658	−9.6333

To provide an illustrative intuition, the convergence curves of the compared algorithms for Example IV are provided in Fig. 30.9 in terms of dB form of MSE values. This figure complies with the presented numerical values as GSA algorithm indicates better performance and the WOA algorithm has a competitive capability compared to the rest of the algorithms.

30.4.5 Comparison of elapsed times

In this section of the chapter, a comparative assessment regarding the computational times of the algorithms for all provided examples and cases are presented. Considering the presented elapsed times in Table 30.14 and the overall performance of the WOA algorithm discussed for each example and case, one can conclude the better efficiency of the WOA algorithm both for computational times and efficiency of reaching promising results for IIR system identification problem.

TABLE 30.13 Optimal coefficients of Example IV.

Coefficients	WOA [27]	SCA [28]	GSA [29]	ABC [30]
a_1	0.0165	−0.0004	0.0464	−0.2671
a_2	0.0600	0.0137	0.1188	0.2248
a_3	−0.0706	−0.0019	−0.0239	0.3317
a_4	0.7686	0.8229	0.7300	0.5519
b_0	0.8150	0.9422	1.0494	1.1162
b_1	−0.0350	−0.0041	−0.0850	0.2207
b_2	0.2893	0.3532	0.2239	0.1308
b_3	0.0963	−0.0222	−0.0076	−0.1711
b_4	−0.1242	−0.3420	−0.2626	−0.2473

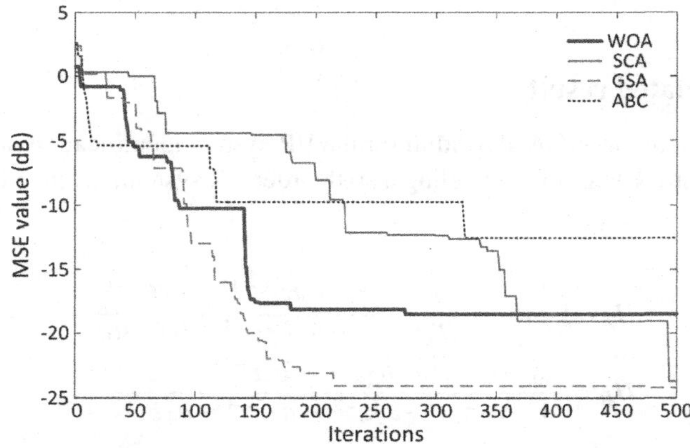

FIGURE 30.9 Comparative convergence diagram of Example IV.

TABLE 30.14 Comparison of average elapsed times (in seconds) per run for WOA, SCA, GSA, and ABC algorithms.

System	WOA [27]	SCA [28]	GSA [29]	ABC [30]
Example I, Case (a)	0.5473	0.4809	1.2744	1.3255
Example I, Case (b)	0.3953	0.3964	1.1069	1.2368
Example II, Case (a)	0.7766	0.6929	1.5160	1.8059
Example II, Case (b)	0.6890	0.6492	1.4507	1.9136
Example III	0.5119	0.4711	1.2631	1.3682
Example IV	0.9395	0.8467	1.7513	1.9746

30.5 Conclusion

In this chapter, the promise of the whale optimization algorithm has been investigated for the adaptive infinite impulse response system identification problem. The whale optimization algorithm has been used to model an unknown plant and its performance was evaluated on four different infinite impulse response filtering benchmarks. The promise of the whale optimization algorithm has been demonstrated through modeling the plant using the same and reduced order models. For the comparative purposes, the sine-cosine, gravitational search and artificial bee colony optimization algorithms have been used as other competitive metaheuristic algorithms. The comparisons have been carried out using mean squared error as the cost function. The whale optimization algorithm based infinite impulse response system identification has been demonstrated to be highly competitive compared to other good performing metaheuristic structures based identification problems.

Declaration of competing interests

All authors declare that they have no conflict of interest.

References

[1] M. Kumar, A. Aggarwal, T.K. Rawat, Bat algorithm: application to adaptive infinite impulse response system identification, Arab. J. Sci. Eng. 41 (2016) 3587–3604, https://doi.org/10.1007/s13369-016-2222-3.

[2] A. Mohammadi, S.H. Zahiri, S.M. Razavi, Infinite impulse response systems modeling by artificial intelligent optimization methods, Evolving Syst. 10 (2019) 221–237, https://doi.org/10.1007/s12530-018-9218-z.

[3] J. Lin, C. Chen, Parameter estimation of chaotic systems by an oppositional seeker optimization algorithm, Nonlinear Dyn. 76 (2014) 509–517, https://doi.org/10.1007/s11071-013-1144-9.

[4] M.R. Soltanpour, M.H. Khooban, A particle swarm optimization approach for fuzzy sliding mode control for tracking the robot manipulator, Nonlinear Dyn. 74 (2013) 467–478, https://doi.org/10.1007/s11071-013-0983-8.

[5] H.P. Phyu, K.S. Lwin, T.T. Oo, P.T. Aung, Analysis on ground clutter mitigation using IIR filter and frequency domain filters for C-band phased array weather radar, in: 2020 International Conference on Advanced Information Technologies (ICAIT), IEEE, 2020, pp. 48–52.

[6] S.H. Pauline, S. Dhanalakshmi, A robust low-cost adaptive filtering technique for phonocardiogram signal denoising, Signal Process. 201 (2022) 108688, https://doi.org/10.1016/j.sigpro.2022.108688.

[7] R. Zhao, Y. Wang, C. Liu, P. Hu, H. Jelodar, C. Yuan, Y. Li, I. Masood, M. Rabbani, H. Li, B. Li, Selfish herd optimization algorithm based on chaotic strategy for adaptive IIR system identification problem, Soft Comput. 24 (2020) 7637–7684, https://doi.org/10.1007/s00500-019-04390-9.

[8] P. Upadhyay, R. Kar, D. Mandal, S.P. Ghoshal, IIR system identification using differential evolution with wavelet mutation, Int. J. Eng. Sci. Technol. 17 (2014) 8–24, https://doi.org/10.1016/j.jestch.2014.02.002.

[9] Q. Luo, Y. Ling, Y. Zhou, Modified whale optimization algorithm for infinitive impulse response system identification, Arab. J. Sci. Eng. 45 (2020) 2163–2176, https://doi.org/10.1007/s13369-019-04093-1.

[10] Y. Niu, X. Yan, Y. Wang, Y. Niu, Dynamic opposite learning enhanced artificial ecosystem optimizer for IIR system identification, J. Supercomput. 78 (2022) 13040–13085, https://doi.org/10.1007/s11227-022-04367-w.

[11] B. Durmuş, Infinite impulse response system identification using average differential evolution algorithm with local search, Neural Comput. Appl. 34 (2022) 375–390, https://doi.org/10.1007/s00521-021-06399-4.

[12] M. Kumar, T.K. Rawat, A. Aggarwal, Adaptive infinite impulse response system identification using modified-interior search algorithm with Lèvy flight, ISA Trans. 67 (2017) 266–279, https://doi.org/10.1016/j.isatra.2016.10.018.

[13] S. Chen, R. Istepanian, B.L. Luk, Digital IIR filter design using adaptive simulated annealing, Digit. Signal Process. 11 (2001) 241–251, https://doi.org/10.1006/dspr.2000.0384.

[14] D.J. Krusienski, W.K. Jenkins, Particle swarm optimization for adaptive IIR filter structures, in: Proceedings of the 2004 Congress on Evolutionary Computation (IEEE Cat. No. 04TH8753), IEEE, 2004, pp. 965–970.

[15] N. Karaboga, A. Kalinli, D. Karaboga, Designing digital IIR filters using ant colony optimisation algorithm, Eng. Appl. Artif. Intell. 17 (2004) 301–309, https://doi.org/10.1016/j.engappai.2004.02.009.

[16] D. İzci, S. Ekinci, M. Güleydin, Application of slime mould algorithm to infinite impulse response system identification problem, Comput. Sci. IDAP-2022 (2022) 45–51, https://doi.org/10.53070/bbd.1172833.

[17] X. Liang, D. Wu, Y. Liu, M. He, L. Sun, An enhanced slime mould algorithm and its application for digital IIR filter design, Discrete Dyn. Nat. Soc. 2021 (2021) 1–23, https://doi.org/10.1155/2021/5333278.

[18] S. Singh, A. Ashok, M. Kumar, T.K. Rawat, Adaptive infinite impulse response system identification using teacher learner based optimization algorithm, Appl. Intell. 49 (2019) 1785–1802, https://doi.org/10.1007/s10489-018-1354-4.

[19] P. Upadhyay, R. Kar, D. Mandal, S.P. Ghoshal, A new design method based on firefly algorithm for IIR system identification problem, J. King Saud Univ., Eng. Sci. 28 (2016) 174–198, https://doi.org/10.1016/j.jksues.2014.03.001.

[20] S. Zhang, Y. Zhou, Grey wolf optimizer with ranking-based mutation operator for IIR model identification, Chin. J. Electron. 27 (2018) 1071–1079, https://doi.org/10.1049/cje.2018.06.008.

[21] S.K. Saha, R. Kar, D. Mandal, S.P. Ghoshal, Harmony search algorithm for infinite impulse response system identification, Comput. Electr. Eng. 40 (2014) 1265–1285, https://doi.org/10.1016/j.compeleceng.2013.12.016.

[22] M.A. Sharifi, H. Mojallali, A modified imperialist competitive algorithm for digital IIR filter design, Optik (Stuttg) 126 (2015) 2979–2984, https://doi.org/10.1016/j.ijleo.2015.07.022.

[23] A. Kalinli, N. Karaboga, A new method for adaptive IIR filter design based on tabu search algorithm, AEÜ, Int. J. Electron. Commun. 59 (2005) 111–117, https://doi.org/10.1016/j.aeue.2004.11.003.

[24] N. Karaboga, A new design method based on artificial bee colony algorithm for digital IIR filters, J. Franklin Inst. 346 (2009) 328–348, https://doi.org/10.1016/j.jfranklin.2008.11.003.

[25] G. Panda, P.M. Pradhan, B. Majhi, IIR system identification using cat swarm optimization, Expert Syst. Appl. 38 (2011) 12671–12683, https://doi.org/10.1016/j.eswa.2011.04.054.

[26] Chaohua Dai, Weirong Chen, Yunfang Zhu, Seeker optimization algorithm for digital IIR filter design, IEEE Trans. Ind. Electron. 57 (2010) 1710–1718, https://doi.org/10.1109/TIE.2009.2031194.

[27] S. Mirjalili, A. Lewis, The whale optimization algorithm, Adv. Eng. Softw. 95 (2016) 51–67, https://doi.org/10.1016/j.advengsoft.2016.01.008.

[28] S. Mirjalili, SCA: a sine cosine algorithm for solving optimization problems, Knowl.-Based Syst. 96 (2016) 120–133, https://doi.org/10.1016/j.knosys.2015.12.022.

[29] E. Rashedi, H. Nezamabadi-pour, S. Saryazdi, GSA: a gravitational search algorithm, Inf. Sci. (NY) 179 (2009) 2232–2248, https://doi.org/10.1016/j.ins.2009.03.004.

[30] D. Karaboga, B. Akay, A comparative study of Artificial Bee Colony algorithm, Appl. Math. Comput. 214 (2009) 108–132, https://doi.org/10.1016/j.amc.2009.03.090.

[31] E. Cuevas, J. Gálvez, S. Hinojosa, O. Avalos, D. Zaldívar, M. Pérez-Cisneros, A comparison of evolutionary computation techniques for IIR model identification, J. Appl. Math. 2014 (2014) 1–9, https://doi.org/10.1155/2014/827206.

Chapter 31

Optimization of SHE problem with WOA in AC-AC choppers

Satılmış Ürgün[a] and Halil Yiğit[b]

[a]Faculty of Aeronautics and Astronautics, Kocaeli University, Kocaeli, Türkiye, [b]Department of Information Systems Engineering, Kocaeli University, Kocaeli, Türkiye

31.1 Introduction

Power converters are devices that produce AC or DC power at different levels or frequencies using a regulated AC or DC source through semiconductor switches. They are widely exploited in many applications such as consumer electronics, industrial electronics, photovoltaic systems, electric vehicles, energy storage systems and energy distribution systems. AC-AC choppers in these converters convert constant amplitude AC voltage or current into AC voltage or current of different amplitude [1]. While AC-AC choppers are mainly used in industrial heating, lighting control and variable frequency motor drives, preventing AC voltage spikes and the correction of power factor are other application areas [2–5]. The phase control principle was first employed to control the output voltage. However, this method has disadvantages such as low-quality power factor and undesirable low frequency harmonics in the output voltage.

PWM method is deployed in AC-AC converters to eliminate these drawbacks. Classical PWM, Sinusoidal PWM, and Space-Vector PWM are the prominent control techniques [6,7]. When classical PWM techniques are utilized for harmonic elimination, it is necessary to increase the operating frequency. On the other hand, increasing frequency causes switching losses, electromagnetic interference (EMI), and switching stresses [8]. In order to overcome these problems, selective harmonic elimination (SHE) PWM method has been applied in recent years.

The SHE technique provides a significant advantage by eliminating the desired harmonic components at low switching frequency in applications where switching losses are disadvantageous. Obtaining analytical solutions of nonlinear transcendental equations in the SHE problem is quite troublesome. The difficulty increases even more, especially if a high number of harmonics are tried to be eliminated [9]. Iterative techniques such as Newton Raphson have been used to solve these equations, but the challenge of estimating the initial values and convergence to more than one solution is one of the most important limitations [10].

Metaheuristic algorithms are nature-based search procedures designed to solve problems that are difficult to optimize. Most important advantages of metaheuristic algorithms compared to classical search methods are simplicity, the availability of strategies to avoid getting trapped at the local optima, and being easily applied to different problems [11,12]. Therefore, metaheuristic optimization algorithms are widely used to search for an optimal solution in engineering problems [13–16].

Metaheuristic algorithms such as Genetic Algorithm (GA), Bee Colony Optimization (BCO), and Particle Swarm Optimization (PSO) have been employed to solve the SHE problem in AC-AC choppers [17–21]. Whale Optimization Algorithm (WOA) is a metaheuristic optimization algorithm based on the social behavior of humpback whales [22]. In the literature, there are studies in which the WOA method is applied to solve the SHE problem in multi-level inverters (MLI) [23–25], but there is no research on AC-AC choppers.

In this chapter, the optimization of the SHE problem in AC-AC choppers, which are widely used in power electronics, is presented with WOA, one of the current metaheuristic algorithms. The optimum switching moments required by the AC-AC chopper were analyzed using WOA. In this context, firstly, the fitness function is proposed and the switching moments are optimized. The output current harmonic elimination in AC-AC choppers was investigated by performing FFT analysis.

31.2 PWM AC-AC chopper

AC-AC choppers are power electronic converters with many topologies and the amplitude of the load voltage can be changed with frequency. In this chapter, a one-phase topology controlled by PWM will be considered.

Handbook of Whale Optimization Algorithm. https://doi.org/10.1016/B978-0-32-395365-8.00037-3

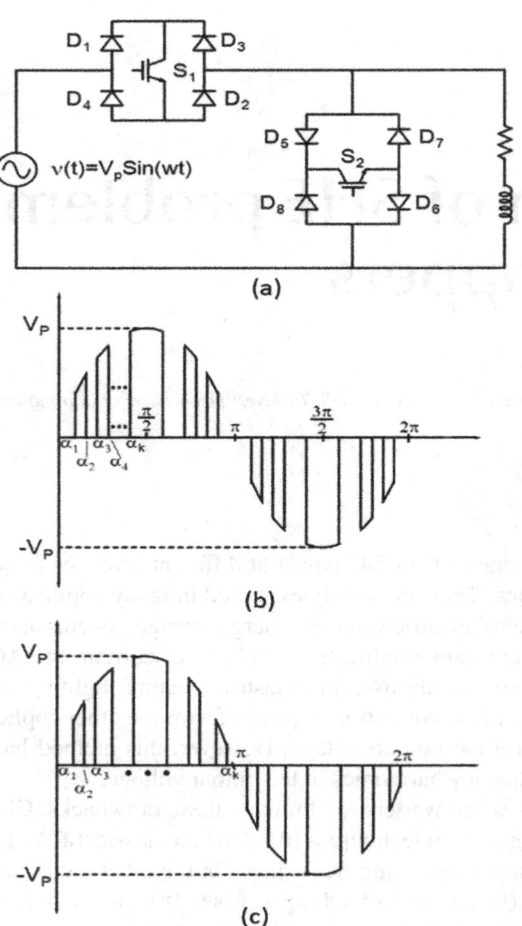

FIGURE 31.1 AC-AC chopper topology and chopped output voltage.

The main task of AC-AC choppers is to adjust the average power at the output by chopping the constant amplitude input voltage. However, when adjusting the output voltage, it should be considered especially in the elimination of low frequency harmonics. Therefore, it is important to determine the switching moments during chopping. SHE-PWM technique is used to achieve both purposes at the same time.

The structure of a SHE-PWM controlled one-phase AC-AC chopper is shown in Fig. 31.1. In this circuit, switches S_1 and S_2 operate at opposite moments in the same period. Diodes are used to direct the current. The voltage on the load is controlled by the conduction and cutting of the switches. The switching angles are indicated by the index α and switch S_1 turns on at odd values and turns off at even values. These switching moments are produced using Fourier series.

The chopper output voltage is written using the Fourier series as given by (31.1).

$$V_0(\omega t) = \sqrt{2}V_{in}\left(a_0 + \sum_{1}^{n}[a_n\cos(n\omega t) + b_n\sin(n\omega t)]\right) \tag{31.1}$$

where V_{in} is the input voltage of the chopper and $n = 1, 2, 3, \ldots$.

As seen in Fig. 31.1.b, there is symmetry in the output voltage and therefore even order harmonics will not be present in the output voltage. As a result, the coefficients a_0 and a_n become zero and (31.1) is simplified as follows:

$$V_{out}(\omega t) = \sqrt{2}V_{in}\sum_{1}^{n}[b_n\sin(n\omega t)] \tag{31.2}$$

where $n = 1, 3, 5, \ldots$.

The value of b_n is computed by (31.3),

$$b_n = \frac{2V_p}{\pi}\left[\frac{\sin(n-1)\omega t}{(n-1)} - \frac{\sin(n+1)\omega t}{(n+1)}\right]_{\alpha_1,\alpha_3,\ldots,\alpha_k}^{\alpha_2,\alpha_4,\ldots,\frac{\pi}{2}} \tag{31.3}$$

where $n \neq 1$ and V_p is the maximum value of the output sine wave. The fundamental component, b_1, shown in (31.4) is derived from (31.3).

$$b_1 = \frac{2V_p}{\pi}\left[\omega t - \frac{\sin(2\omega t)}{2}\right]_{\alpha_1,\alpha_3,\ldots,\alpha_k}^{\alpha_2,\alpha_4,\ldots,\frac{\pi}{2}} \tag{31.4}$$

31.3 Whale optimization algorithm (WOA)

WOA is a recently developed, nature-inspired, meta-heuristic optimization algorithm that simulates the bubble hunting behavior of humpback whales [26]. The bubble-net search principle of humpback whale is as follows: after the whale finds its prey, it creates a bubble-net along the spiral path and moves upward towards its prey. This predatory behavior includes three phases to simulate the search for prey, encircling prey, and bubble-net foraging behavior of humpback whales.

31.3.1 Surrounding prey stage

In the WOA algorithm, the humpback whale first identifies the location of the prey and then surrounds it. However, the whale has no prior knowledge of the location of the hunt. Therefore, assuming the current optimum position is the target prey, all other individuals in the group move to the optimum position. The surrounding stage is expressed by the following mathematical model:

$$X(t+1) = X^*(t) - A \cdot D, \tag{31.5}$$

$$D = \left|C \cdot X^*(t) - X(t)\right|, \tag{31.6}$$

where t is the current iterations, $X^*(t)$ is the prey position vector (the position vector of the best solution obtained), $X(t)$ is the position vector at iteration t. A and D is the surrounding step size, and

$$A = 2a \cdot rand - a, \tag{31.7}$$

$$C = 2 \cdot rand, \tag{31.8}$$

where $rand$ is a random number between [0, 1], a is a control parameter and decreases linearly from 2 to 0 with increasing iterations. The expression is as follows:

$$a = 2 - \frac{2t}{T_{max}}, \tag{31.9}$$

where T_{max} is the maximum number of iterations.

31.3.2 Bubble-net attack stage

After the prey is located, the humpback whale moves along the spiral path toward the prey in the shrinking encirclement. This behavior of whales is described with two methods which are shrinking and surrounding mechanism and spiral update position, respectively.

31.3.2.1 Shrinking and surrounding mechanism

Eqs. (31.7), (31.8), and (31.9) indicate the shrinking and surrounding mechanism. This mechanism is achieved by reducing the convergence factor a.

When the control parameter a is linearly reduced from 2 to 0, variable A takes values in the range $[-a, a]$. Based on the value taken by A, the whale individual's new position can be defined anywhere between the original position and the best available position, as represented in (31.5) and (31.6).

31.3.2.2 Spiral update position mechanism

In the spiral update position mechanism, first the distance between each whale and the current optimal position (prey) is calculated. The whales are then simulated to catch prey inside the spiral. (31.10) and (31.11) depict the logarithmic spiral equations between the position of the whales and the current optimum solution.

$$X(t+1) = D' \cdot e^{bl} \cdot \cos(2\pi l) + X^*(t), \tag{31.10}$$

$$D' = \left| X^*(t) - X(t) \right| \tag{31.11}$$

where D is the distance between the ith whale and the current optimal position, b is a constant coefficient utilized to control the logarithmic spiral shape, and l takes the random number in the range $[-1, 1]$. In the process of hunting, as the humpback whales attacks its prey, they move along a spiral path while they are encircling. Therefore, the shrinking encircling mechanism and the spiral updating position mechanism are performed with the same probability of 0.5 to obtain this synchronous model. (See Fig. 31.2.)

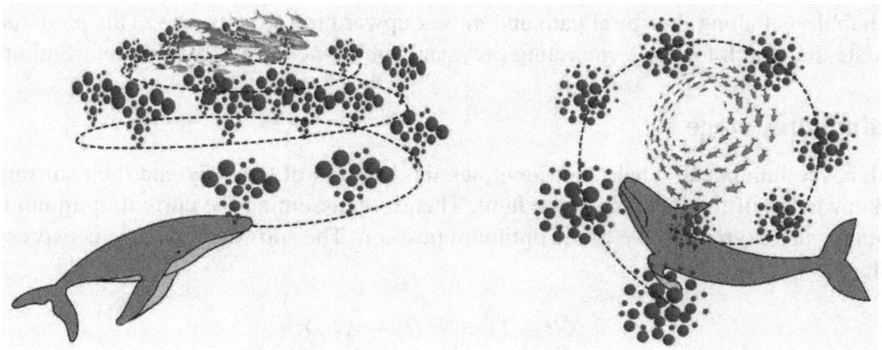

FIGURE 31.2 Schematic diagram of the whale bubble-net attack.

31.3.3 Hunting prey stage

During hunting of humpback whales, each whale randomly searches for prey based on each other's location. In case of $|A| \geq 1$, individual whales randomly selected to replace the current optimal solution can be kept away from reference whales to find another more suitable target. This action allows the WOA algorithm to search in a global scope. The mathematical model is presented in the following equations.

$$X(t+1) = X_{rand} - A \cdot D, \tag{31.12}$$

$$D = |C \cdot X_{rand} - X(t)| \tag{31.13}$$

where X_{rand} is the position vector of a randomly selected whale individual from the current whale population. Fig. 31.3 illustrates WOA flowchart.

31.4 Problem formulation and simulation result

Using Eq. (31.3), the 3rd, 5th, 7th, and 9th harmonic components can be computed as given by (31.14)–(31.18).

$$b_1 = \frac{2V_p}{\pi}\left[-\alpha_1 - \alpha_3 + \alpha_2 - \frac{-\sin(2\alpha_1) + \sin(2\alpha_2) - \sin(2\alpha_3)}{2}\right] \tag{31.14}$$

$$b_3 = \frac{2V_p}{\pi}\left[\frac{-\sin 2\alpha_1 + \sin 2\alpha_2 - \sin 2\alpha_3}{2} - \frac{-\sin 4\alpha_1 + \sin 4\alpha_2 - \sin 4\alpha_3}{4}\right] \tag{31.15}$$

$$b_5 = \frac{2V_p}{\pi}\left[\frac{-\sin 4\alpha_1 + \sin 4\alpha_2 - \sin 4\alpha_3}{4} - \frac{-\sin 6\alpha_1 + \sin 6\alpha_2 - \sin 6\alpha_3}{6}\right] \tag{31.16}$$

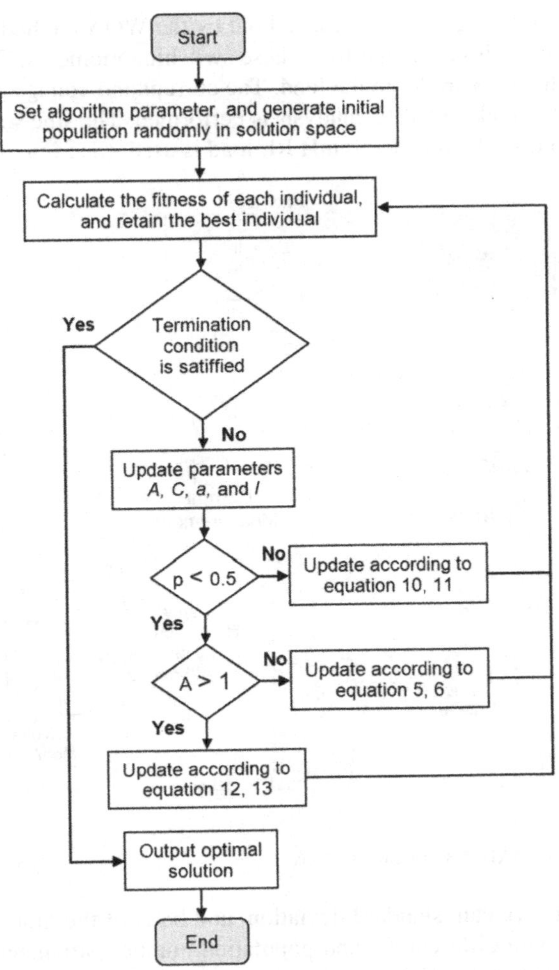

FIGURE 31.3 WOA flowchart.

$$b_7 = \frac{2V_p}{\pi}\left[\frac{-\sin 6\alpha_1 + \sin 6\alpha_2 - \sin 6\alpha_3}{6} - \frac{-\sin 8\alpha_1 + \sin 8\alpha_2 - \sin 8\alpha_3}{8}\right] \tag{31.17}$$

$$b_9 = \frac{2V_p}{\pi}\left[\frac{-\sin 8\alpha_1 + \sin 8\alpha_2 - \sin 8\alpha_3}{8} - \frac{-\sin 10\alpha_1 + \sin 10\alpha_2 - \sin 10\alpha_3}{10}\right] \tag{31.18}$$

Here, the aim of the optimization process is to eliminate the selected harmonic components and to produce the output voltage of the desired amplitude. In this case, the fitness function includes two components, output voltage and harmonics.

$$hc = |b_3| + |b_5| + |b_7| + \cdots \tag{31.19}$$

$$err = V_{out} - b_1 \tag{31.20}$$

$$Fa = |err| + hc \tag{31.21}$$

$$Mi = \frac{V_{des}}{V_{in}} \tag{31.22}$$

where Fa is the fitness function, hc is the sum of harmonic components to be eliminated, V_{des} is the peak value of output voltage, and Mi is the modulation index.

This study consists of two stages. The first stage is to calculate the optimum switching moments using the WOA algorithm, and the other is to run the circuit simulation utilizing these switching moments in MATLAB® Simulink Toolbox. The bridge consisting of insulated-gate bipolar transistor (IGBT) and four diodes depicted in Fig. 31.1 is given as a block. The

PWM pulse generator applies the switching moments carried out by the WOA method to the IGBT block in milliseconds. In addition, the duty cycle of the pulses is calculated from these switching moments. The inversion of the PWM signals is also applied to the IGBT block, which is parallel to the load. The current and voltage data captured by the scopes are sent to the Fast Fourier Transform (FFT) block and THD analysis is performed. The sine source supplied the circuit is at 300 V peak value and 50 Hz frequency, and a 20 ohm and 30 mH RL load is used. (See Fig. 31.4.)

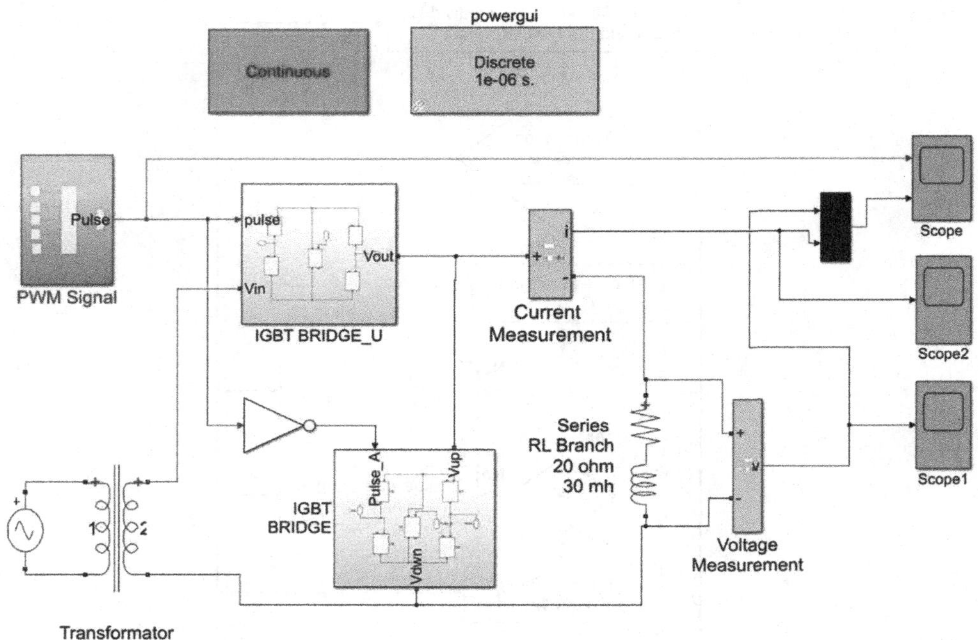

FIGURE 31.4 AC chopper circuit created in MATLAB Simulink toolbox.

Table 31.1 lists statistical features (mean, standard deviation, and best) of the fitness function obtained by WOA. The performance of WOA is evaluated with different Mi and population number parameters. These performance values have been obtained by taking the mean, standard deviation and optimum result of the fitness function values acquired as a result of 1000 runs. When Table 31.1 is analyzed, it is revealed that the increase in the number of population does not have a significant effect on the fitness value. As the number of population increases, the search space also increases, as can be seen from Table 31.2, the time to perform a single iteration of the WOA algorithm increases. A single iteration time is the time, in seconds, for the WOA algorithm to generate a fitness value in the search space. While this time depends on the computational complexity of the method, it also depends on the hardware and software features of the computer used in the simulation. When Table 31.2 is investigated in terms of Mi, it is concluded that the modulation index has no effect on the fitness value. Mi value adjusts the desired output voltage value of the AC-AC chopper in the fitness function.

In Fig. 31.5, the variation of switching moments throughout the modulation range, Mi, is given. While the switching moments change according to the fitness function demand throughout the modulation range, the differences between the angles at the end of the range (Mi=1) decrease to zero, resulting in the formation of a complete sine sign. In the following sections, different fitness functions will be examined. The switching signals for each case give visually similar results.

In Figs. 31.6, 31.7, and 31.8, current and voltage waveforms are illustrated for Mi=0.3, 0.6, and 0.9, respectively. The current value is multiplied by 10 so that the relationship between current and voltage can be seen easily. In cases where the modulation index approaches 1, the output voltage and current are completely similar to the sinus sign.

Fig. 31.9 plots the FFT analysis screen, where data on harmonic components and output voltage are achieved. A single screen is provided showing the working process performed for a single modulation index (Mi=0.8). FFT analysis is executed for each modulation index. Afterwards, the data obtained in the FFT analysis are visualized using the MATLAB Curve Fitting Tool.

Determining the fitness function is as important as optimizing the WOA parameters. Therefore, different fitness functions are considered. In this study, five switching moments are used in a quarter period. This offers five degrees of freedom. In other words, the desired four harmonic components can be eliminated while keeping the output voltage constant. However, as can be seen from Eqs. (31.9)–(31.11), increasing the number of harmonic components can affect the performance.

TABLE 31.1 Statistical features (a) mean, (b) standard deviation, and (c) best of the fitness function obtained by WOA.

(a)

	Mean	Population number		
		20	50	100
Modulation index, Mi	0.1	8,81487495E-03	2,48375055E-02	4,77687086E-02
	0.2	3,29254044E-01	1,44277103E-01	9,74035434E-02
	0.3	1,58822909E-01	6,95013323E-02	2,28543822E-01
	0.4	6,98103519E-02	1,41658849E-01	3,75771403E-01
	0.5	1,02241728E-01	5,33268913E-02	1,66765170E-01
	0.6	6,14940202E-02	1,18996192E-01	2,42480949E-01
	0.7	2,05567069E-01	1,42727337E-01	1,75375715E-01
	0.8	6,38995220E-02	6,39177362E-02	1,63709003E-01
	0.9	2,00094096E-01	2,65632645E-01	3,36717206E-02
	1.0	2,46378768E-03	2,00019439E-03	2,00001663E-03

(b)

	Std	Population number		
		20	50	100
Modulation index, Mi	0.1	4,48006353E-02	4,42805643E-02	4,32469587E-02
	0.2	4,91064754E-02	5,63185961E-02	5,54008521E-02
	0.3	4,52667555E-02	6,31963370E-02	4,56715867E-02
	0.4	4,33207241E-02	3,87825252E-02	2,80536950E-01
	0.5	6,36887915E-02	5,44422144E-02	1,20322743E-01
	0.6	5,14686051E-02	4,26427462E-02	1,27692965E-01
	0.7	1,29378877E-01	8,56596701E-02	1,16732040E-01
	0.8	4,34109998E-02	5,05757443E-02	4,27413158E-02
	0.9	7,90643142E-02	9,36746471E-02	4,76013801E-02
	1.0	4,58877816E-02	4,47213510E-02	4,47213588E-02

(c)

	Best	Population number		
		20	50	100
Modulation index, Mi	0.1	3,18014302E-03	1,86331742E-02	4,47766532E-02
	0.2	3,13611530E-01	1,31672044E-01	7,27797568E-02
	0.3	1,50050448E-01	5,23183255E-02	2,21701272E-01
	0.4	6,32670912E-02	1,39475841E-01	9,41494415E-02
	0.5	7,38605391E-02	3,16994163E-02	1,04153530E-01
	0.6	4,42459005E-02	1,13444871E-01	1,36415358E-01
	0.7	1,28721170E-01	5,15645307E-02	1,39409986E-01
	0.8	6,02536582E-02	5,28328569E-02	1,57314395E-01
	0.9	1,40434380E-01	1,87900432E-01	1,77067034E-02
	1.0	4,62488805E-09	2,88892680E-08	1,66683089E-08

TABLE 31.2 Comparison of single iteration time depending on population number.

Population number	20	50	100
Single iteration time (s)	0.5066626073	1.1157343596	2.3321748066

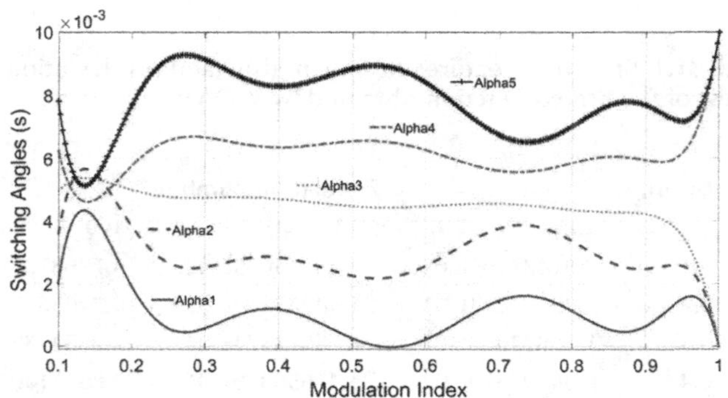

FIGURE 31.5 Switching moments throughout the modulation range.

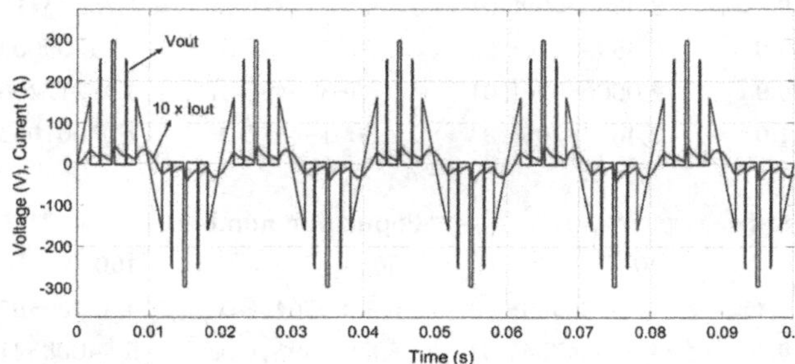

FIGURE 31.6 The current and voltage signals for the 0.3 value of the modulation index.

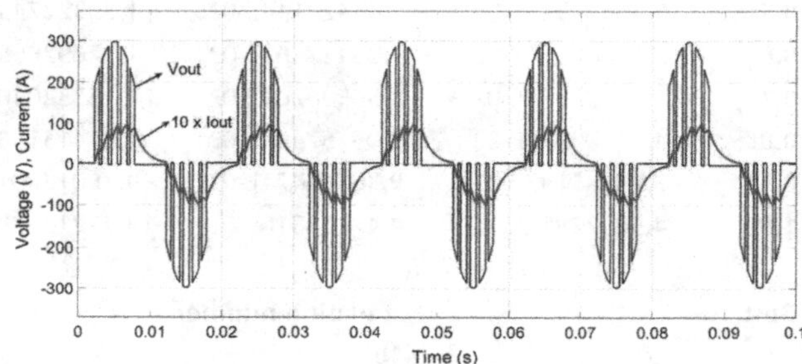

FIGURE 31.7 The current and voltage signals for the 0.6 value of the modulation index.

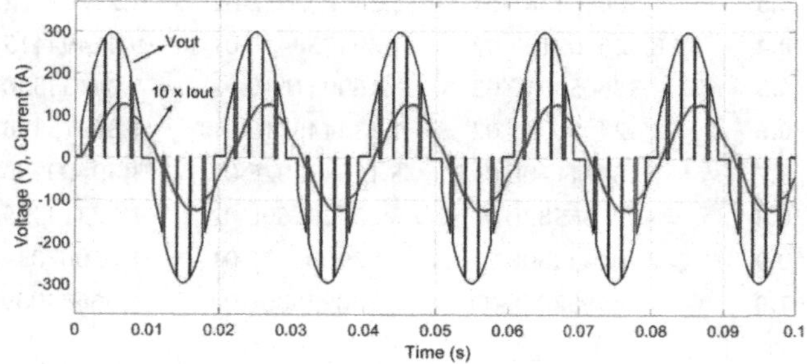

FIGURE 31.8 The current and voltage signals for the 0.9 value of the modulation index.

31.4.1 Case I

In the first scenario, which refers Eq. (31.23), the amplitude of the fundamental output voltage is tried to be kept constant while the 3rd and 5th harmonic components are tried to be eliminated. Therefore, only 3 of the five degrees of freedom

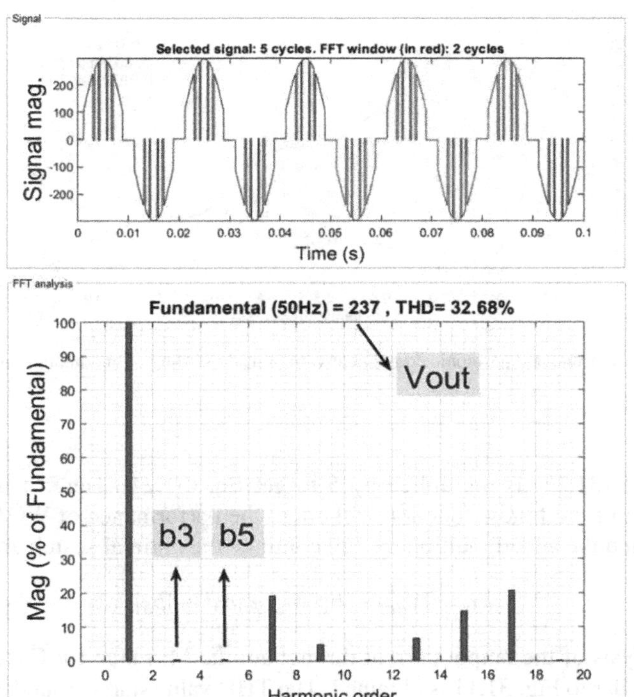

FIGURE 31.9 FFT analysis screen for modulation index 0.8.

are used.

$$Fa = |V_{des} - b_1| + |b_3| + |b_5| \tag{31.23}$$

Fig. 31.10 shows the output voltage relationships obtained throughout the modulation index (Mi) range. V_{des} is the desired output voltage depending on Mi and V_{out} is the voltage value measured on the load in the chopper circuit. From the data obtained for each case, a difference of 3 V was observed between these voltages. The reason for this is the voltage drop in the IGBT block connected in series between the source and the load.

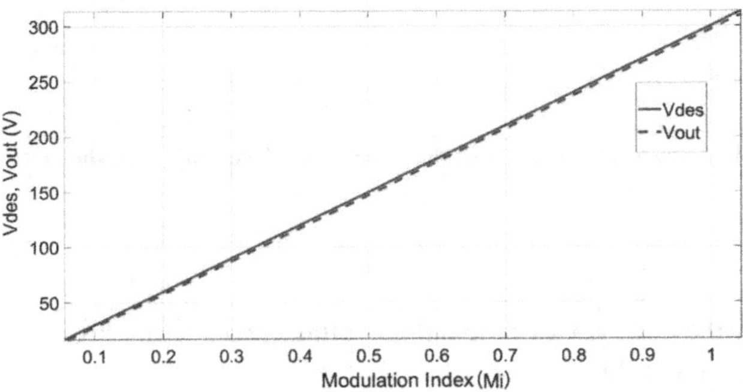

FIGURE 31.10 The output voltages obtained throughout the Mi for Case 1.

The THD analysis of the output current is given in Fig. 31.11. The THD-I plot gives the THD variation of the output current. It is clear seen from the figure that it started from 50% at low modulation indices and approached zero at the end of the modulation index. It is an expected result. Because when the modulation index is 1, the output current approaches the ideal sine sign. In Fig. 31.11, graphs of the variations obtained from the division of the 3rd and 5th components of the output current harmonics into the fundamental component are given. As can be seen from the results, the 3rd component value is slightly higher at small Mi values. Although the fitness function generally reduced the THD value, it is not able to eliminate the low components as desired.

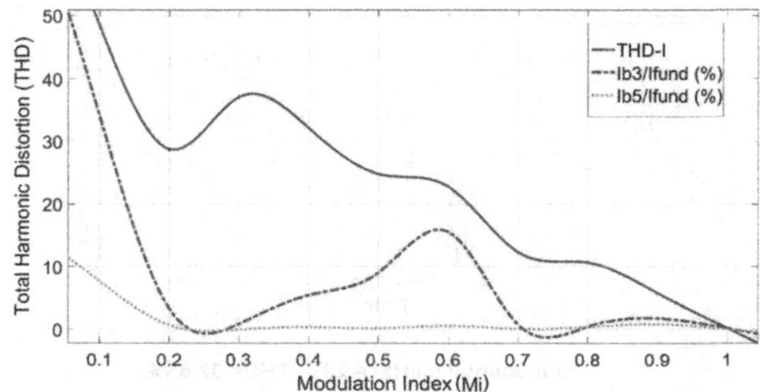

FIGURE 31.11 The output current THD value and harmonic component variations depending on modulation index for Case 1.

31.4.2 Case II

In Case 2, the b_3 coefficient in (31.23) is multiplied by 5 to get Eq. (31.24). Here, it is aimed to investigate how the coefficients to be added in front of the fitness function will affect the performance of WOA. After all, the fitness function consists of discrete functions, and the added multipliers will continue the optimal search around this function.

$$Fa = |V_{des} - b_1| + 5 * |b_3| + |b_5| \tag{31.24}$$

In Fig. 31.12, the FFT analysis of the output current throughout the Mi range for Case 2 is shown. From the point of view of THD, a THD graph similar to Fig. 31.11 is obtained. The THD value starts around 50% and drops to zero as the Mi value approaches 1. However, the remarkable result is that the ratio of the b_3 coefficient to the amplitude of fundamental component has decreased to a great extent. The multiplier added in front of b_3 in Eq. (31.24) depicts that WOA concentrates more on the minimization of the b_3 coefficient.

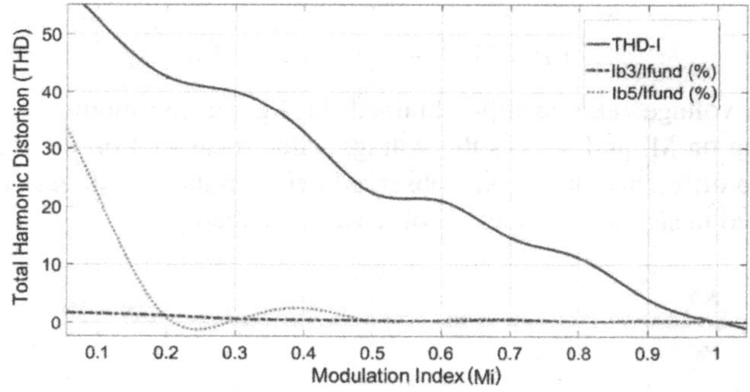

FIGURE 31.12 The output current THD value and harmonic component variations depending on modulation index for Case 2.

31.4.3 Case III

In Case 3, four functions, namely the peak value of the output voltage and 3 harmonic components, are tried to be optimized in the fitness function as seen in Eq. (31.25).

$$Fa = |V_{des} - b_1| + |b_3| + |b_5| + |b_7| \tag{31.25}$$

Fig. 31.13 illustrates the output voltage relationships obtained throughout the modulation index (Mi) range. V_{des} is the desired output voltage depending on Mi and V_{out} is the voltage value measured on the load in the chopper circuit. Considering the voltage drops on the circuit elements, the WOA method has perfectly optimized the peak value of the output voltage.

In Fig. 31.14, the FFT analysis of the output current throughout the Mi range for Case 3 is given. It is clearly seen that The THD value decreases even more than in Case 1 and Case 2. The THD value starts around 40% and drops to zero as the Mi value approaches 1. The amplitude of the 3rd component of the output current is also significantly reduced. However,

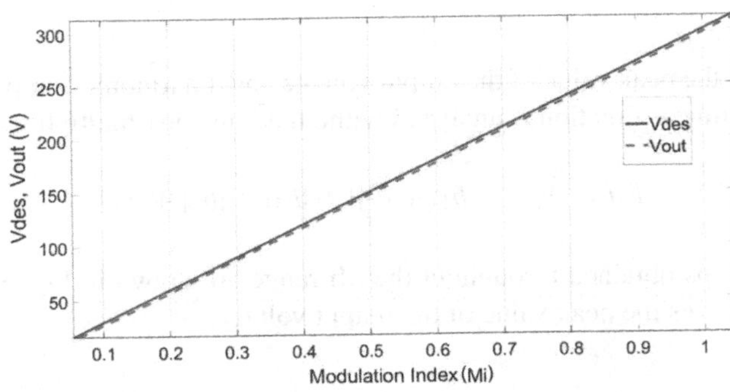

FIGURE 31.13 The output voltages obtained throughout the Mi for Case 3.

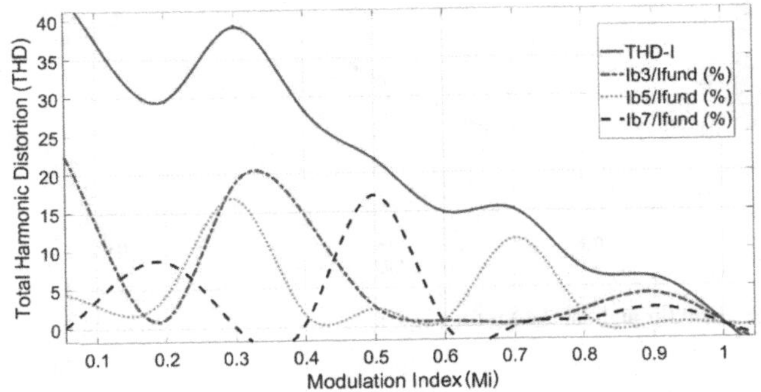

FIGURE 31.14 The output current THD value and harmonic component variations depending on modulation index for Case 3.

the 3rd and 5th components could not be completely eliminated. Therefore, there is a need to add a coefficient in front of the harmonic components, as in the case of Case 2.

31.4.4 Case IV

In the case of Case 4, Eq. (31.26) is obtained by adding the factor of 5 in front of the coefficient b_3 in (31.25).

$$Fa = |V_{des} - b_1| + 5 * |b_3| + |b_5| + |b_7| \tag{31.26}$$

As can be seen from the FFT analysis given in Fig. 31.15, the coefficient added in front of the b_3 component in the fitness function ensures that the third component is largely eliminated. It even takes the value of zero after Mi=0.5.

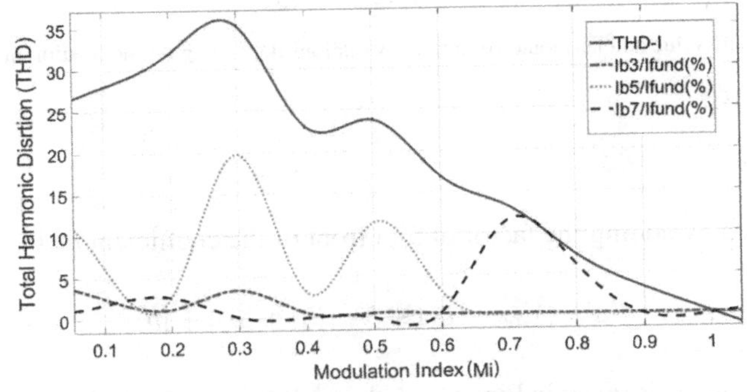

FIGURE 31.15 The output current THD value and harmonic component variations depending on modulation index for Case 4.

31.4.5 Case V

In Case 5, five functions, namely the peak value of the output voltage and 4 harmonic components, are tried to be optimized in the fitness function. First, the fitness function is analyzed without adding any multiplier as seen in Eq. (31.27).

$$Fa = |V_{des} - b_1| + |b_3| + |b_5| + |b_7| + |b_9| \tag{31.27}$$

The output voltage relationships obtained throughout the Mi range are drawn in Fig. 31.16. As in the previous cases, the WOA method perfectly optimizes the peak value of the output voltage.

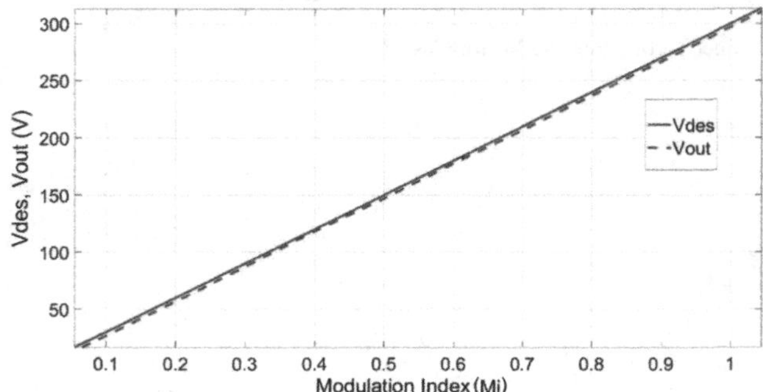

FIGURE 31.16 The output voltages obtained throughout the Mi for Case 5.

In Fig. 31.17, the FFT analysis of the output current throughout the Mi range for Case 5 is given. THD value exhibited similar performance to Case 1 and Case 2. It is clearly seen that The THD value decreases even more than in Case 1 and Case 2. The THD value starts around 50% and drops to zero as the Mi value approaches 1. Although the fitness function contains more harmonic component equations, the THD value has not decreased, and the 3rd component is still high. For this reason, it is necessary to add a multiplier in front of the harmonic components, as in Case 2 and Case 4.

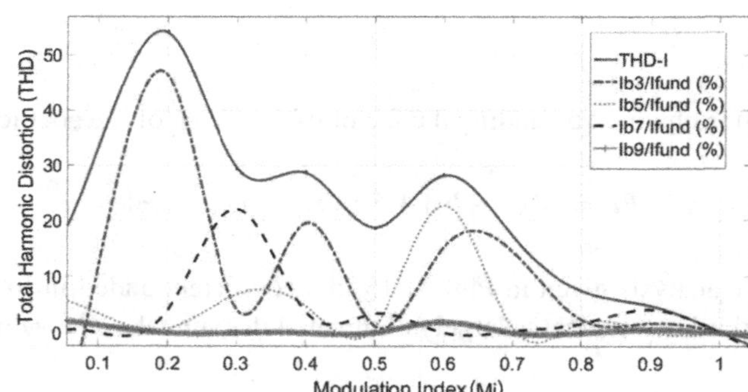

FIGURE 31.17 The output current THD value and harmonic component variations depending on modulation index for Case 5.

31.4.6 Case VI

In Case 6, Eq. (31.28) is obtained by adding the factor of 5 in front of the coefficient b_3 in (31.27).

$$Fa = |V_{des} - b_1| + 5 * |b_3| + |b_5| + |b_7| \tag{31.28}$$

As can be seen from the FFT analysis given in Fig. 31.18, the multiplier added in front of the b_3 component in the fitness function provides that the third component is largely eliminated. However, the amplitude of the other harmonic components increases as the focus is on eliminating the 3rd harmonic component.

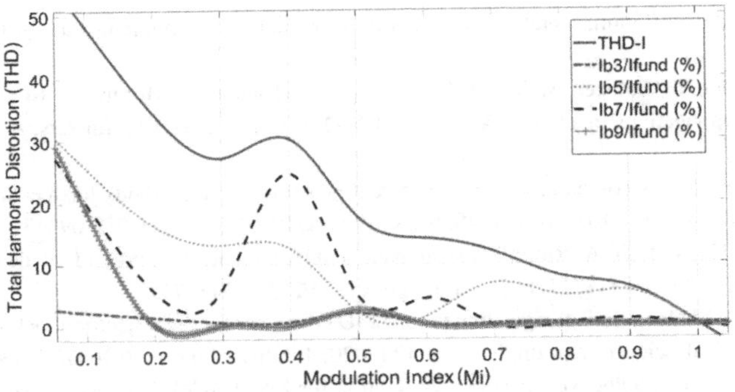

FIGURE 31.18 The output current THD value and harmonic component variations depending on modulation index for Case 5.

31.5 Conclusion

In this chapter, the harmonic elimination of the output voltage of AC-AC choppers is carried out using the WOA method. The selection of the fitness function, which is as important as the optimization method itself, has been made. The THDs of the output voltage have been examined with the FFT analysis. The results of this study can be classified as follows:

- Determining the fitness function is very important when metaheuristic algorithms are exploited for optimization in a system.
- In the fitness function consisting of discrete equations, a multiplier can be added in front of the discrete equations so that the WOA algorithm concentrates the search on this region.
- Using all five degrees of freedom that occurs in AC choppers using five switching moments may not eliminate all low-component harmonics.
- The most optimal THD minimization is to use the fitness function consisting of four parameters in the circuit using five switching angles. This corresponds to the situation where an output voltage and 3 harmonic components are eliminated.

As a result, harmonic elimination in AC-AC choppers can be performed successfully with WOA method.

References

[1] J.C. Rosas-Caro, F. Mancilla-David, J.M. Ramirez, A. Gonzalez-Rodriguez, E.N. Salas-Cabrera, P.A. Rojas-Molina, AC chopper topology with multiple steps switching capability, in: 2010 IEEE Energy Conversion Congress and Exposition, 2010, pp. 1808–1815, https://doi.org/10.1109/ECCE.2010.5618138.

[2] S. Jothibasu, M.K. Mishra, An improved direct AC–AC converter for voltage sag mitigation, IEEE Transactions on Industrial Electronics 62 (1) (Jan. 2015) 21–29, https://doi.org/10.1109/TIE.2014.2334668.

[3] T.B. Soeiro, C.A. Petry, J.C. Fagundes, I. Barbi, Direct AC–AC converters using commercial power modules applied to voltage restorers, IEEE Transactions on Industrial Electronics 58 (1) (Jan. 2011) 278–288, https://doi.org/10.1109/TIE.2010.2045320.

[4] M.K. Metwaly, H.Z. Azazi, S.A. Deraz, M.E. Dessouki, M.S. Zaky, Power factor correction of three-phase PWM AC chopper fed induction motor drive system using HBCC technique, IEEE Access 7 (2019) 43438–43452, https://doi.org/10.1109/ACCESS.2019.2907791.

[5] S. Subramanian, M.K. Mishra, Interphase AC–AC topology for voltage sag supporter, IEEE Transactions on Power Electronics 25 (2) (Feb. 2010) 514–518, https://doi.org/10.1109/TPEL.2009.2027601.

[6] Jong-Hyun Kim, Byung-Duk Min, Bong-Hwan Kwon, Sang-Chul Won, A PWM buck-boost AC chopper solving the commutation problem, IEEE Transactions on Industrial Electronics 45 (5) (Oct. 1998) 832–835, https://doi.org/10.1109/41.720341.

[7] Jang-Hyoun Youm, Bong-Hwan Kwon, Switching technique for current-controlled AC-to-AC converters, IEEE Transactions on Industrial Electronics 46 (2) (April 1999) 309–318, https://doi.org/10.1109/41.753769.

[8] S. Natarajan, T. Sudhakar Babu, K. Balasubramanian, U. Subramaniam, D.J. Almakhles, A state-of-the-art review on conducted electromagnetic interference in non-isolated DC to DC converters, IEEE Access 8 (2020) 2564–2577, https://doi.org/10.1109/ACCESS.2019.2961954.

[9] M.G. Cimoroni, M. Tinari, C. Buccella, C. Cecati, A high efficiency Selective Harmonic Elimination technique for multilevel converters, in: 2018 International Symposium on Power Electronics, Electrical Drives, Automation and Motion (SPEEDAM), 2018, pp. 673–677, https://doi.org/10.1109/SPEEDAM.2018.8445285.

[10] T.R. Sumithira, A. Nirmal Kumar, Elimination of harmonics in multilevel inverters connected to solar photovoltaic systems using ANFIS: an experimental case study, Journal of Applied Research and Technology (ISSN 1665-6423) 11 (1) (2013) 124–132, https://doi.org/10.1016/S1665-6423(13)71521-9.

[11] X.S. Yang, G. Bekdaş, S.M. Nigdeli, Review and applications of metaheuristic algorithms in civil engineering, in: X.S. Yang, G. Bekdaş, S. Nigdeli (Eds.), Metaheuristics and Optimization in Civil Engineering, in: Modeling and Optimization in Science and Technologies, vol. 7, Springer, Cham, 2016, https://doi.org/10.1007/978-3-319-26245-1_1.

[12] Ashkan Memari, Robiah Ahmad, Abdul Rahim, Metaheuristic algorithms: guidelines for implementation, Journal of Soft Computing and Decision Support Systems 4 (2017) 1–6.

[13] A.D. Boursianis, M.S. Papadopoulou, M. Salucci, A. Polo, P. Sarigiannidis, K. Psannis, S. Mirjalili, S. Koulouridis, S.K. Goudos, Emerging swarm intelligence algorithms and their applications in antenna design: the GWO, WOA, and SSA optimizers, Applied Sciences 11 (2021) 8330, https://doi.org/10.3390/app11188330.

[14] A.R. Yildiz, H. Abderazek, S. Mirjalili, A comparative study of recent non-traditional methods for mechanical design optimization, Archives of Computational Methods in Engineering 27 (2020) 1031–1048, https://doi.org/10.1007/s11831-019-09343-x.

[15] Hardi M. Mohammed, Shahla U. Umar, Tarik A. Rashid, A systematic and meta-analysis survey of whale optimization algorithm, Computational Intelligence and Neuroscience 2019 (2019) 8718571, https://doi.org/10.1155/2019/8718571.

[16] F. Loucif, S. Kechida, A. Sebbagh, Whale optimizer algorithm to tune PID controller for the trajectory tracking control of robot manipulator, Journal of the Brazilian Society of Mechanical Sciences and Engineering 42 (2020) 1, https://doi.org/10.1007/s40430-019-2074-3.

[17] N. Kumar, S. Chhawchharia, S.K. Sahoo, Selective harmonic elimination for PWM AC/AC voltage converter using real coded genetic algorithm, in: 2017 Innovations in Power and Advanced Computing Technologies (i-PACT), 2017, pp. 1–5, https://doi.org/10.1109/IPACT.2017.8245026.

[18] S. Mahendran, I. Gnanambal, A. Maheswari, FPGA-based genetic algorithm implementation for AC chopper fed induction motor, International Journal of Electronics 103 (12) (2016) 2029–2041, https://doi.org/10.1080/00207217.2016.1175034.

[19] K. Sundareswaran, A.P. Kumar, Voltage harmonic elimination in PWM AC chopper using genetic algorithm, IEE Proceedings. Electric Power Applications 151 (2004) 26–31, https://doi.org/10.1049/ip-epa:20040061.

[20] Wanchai Khamsen, Apinan Aurasopon, Chanwit Boonchuay, Optimal switching pattern for PWM AC-AC converters using bee colony optimization, Journal of Power Electronics 14 (2014), https://doi.org/10.6113/JPE.2014.14.2.362.

[21] A. Kouzou, M.O. Mahmoudi, M.S. Boucherit, Application of SHE-PWM for seven-level inverter output voltage enhancement based on Particle Swarm Optimization, in: 2010 7th International Multi-Conference on Systems, Signals and Devices, 2010, pp. 1–6, https://doi.org/10.1109/SSD.2010.5585588.

[22] S. Mirjalili, A. Lewis, The whale optimization algorithm, Advances in Engineering Software (ISSN 0965-9978) 95 (2016) 51–67, https://doi.org/10.1016/j.advengsoft.2016.01.008.

[23] Aala Kalananda Vamsi Krishna Reddy, Komanapalli Venkata Lakshmi Narayana, Optimal total harmonic distortion minimization in multilevel inverter using improved whale optimization algorithm, International Journal of Emerging Electric Power Systems 21 (3) (2020) 20200008, https://doi.org/10.1515/ijeeps-2020-0008.

[24] G. Nalcaci, M. Ermis, Selective harmonic elimination for three-phase voltage source inverters using whale optimizer algorithm, in: 2018 5th International Conference on Electrical and Electronic Engineering (ICEEE), 2018, pp. 1–6, https://doi.org/10.1109/ICEEE2.2018.8391290.

[25] M.A. Shahmi Bin Bimazlim, B. Ismail, M.Z. Aihsan, S. Khodijah Mazalan, M.S. Muhammad Azhar Walter, M.N. Khairul Hafizi Rohani, Comparative study of optimization algorithms for SHEPWM five-phase multilevel inverter, in: 2020 IEEE International Conference on Power and Energy (PECon), 2020, pp. 95–100, https://doi.org/10.1109/PECon48942.2020.9314491.

[26] W.A. Watkins, W.E. Schevill, Aerial observation of feeding behavior in four baleen whales: Eubalaena glacialis, Balaenoptera borealis, Megaptera novaeangliae, and Balaenoptera physalus, Journal of Mammalogy 60 (1) (1979) 155–163.

Chapter 32

A WOA-based path planning approach for UAVs to avoid collisions in cluttered areas

Mehmet Enes Avcu[a], Harun Gökçe[b], and İsmail Şahin[b]

[a] *Unmanned Systems Engineering Department, Titra Technology, Ankara, Turkey,* [b] *Faculty of Technology, Department of Industrial Design Engineering, Gazi University, Ankara, Turkey*

32.1 Introduction

This chapter has focused on the path planning algorithm for UAVs and solving with the whale optimization algorithm. Path planning of UAV systems has been an excellent area of research over the past decades. Path planning consists operations to find the route that passes through all of the points of interest in a given area. And also, there are many interesting areas about UAV systems. Examples include a collision avoidance [1], decision making [2], formation control [3], and control [4]. On the other side, optimization algorithms are extremely useful in the robotics area. There is a many optimization techniques that meta-heuristic and deterministic algorithms such as ant colony optimization [5], mixed integer programming [6], sequential convex optimization [7], and grey wolf optimization [8]. These algorithms are not guaranteed to give full performance in each path planning cases but each of them has theirs own specification which makes them suitable in sophisticated situation

In the case of path planning, that is the cognitive level of UAVs and also is used for different purposes such as task assignment, collision avoidance, etc. In the literature, the optimization algorithms mentioned above are widely used as decision makers in solving similar problems. The initial stage of cognitive autonomous vehicles is path planning. Typically, the law of motion, which is produced by a specific module for motion planning, serves as the input for the robot's control system [9]. To control spellbindingly, the limits of UAV flying must be considered to generate flyable pathways that connect specified sites [10]. The fixed-wing UAVs can be described as nonholonomic [11]. Therefore, we consider the dynamic limit of fixed-wing UAVs to design paths. Bank angle is one of the most important and the dynamic of fixed-wing UAV is mentioned in Section 32.3.1. And also, the dynamic of the rotary wing is mentioned in Section 32.3.2.

In [12], UAV path planning approaches are presented and categorized into five classes including sampling-based, node-based, mathematical-based, nature-based, and multi-fusion-based approaches. On the other hand, UAV path planning objectives and presented UAV path planning approaches classified into classical and heuristic techniques [13]. And also, UAV path planning techniques classified into three categories including representative techniques, coordinate techniques, and non-coordinate techniques in [14]. In the case of whale optimization, this algorithm is a nature-inspired algorithm [15]. However, Nature-inspired algorithms are extremely useful for robotics problems such as path planing, and job scheduling [16]. High-level techniques are known as meta-heuristics direct the exploration agents to gradually enhance the overall answer [16]. Therefore, these algorithms can bypass local optima solutions and whale optimization can find global optima [15]. And also, in comparison to other meta-heuristic optimization approaches, WOA is simpler in terms of idea and code implementation [17] and implemented at [18].

32.2 The whale optimization algorithm

In this section, the principles of the whale optimization algorithm (WOA) are explained and mathematical equations are given. The whale optimization algorithm (WOA) is one of the swarm-based algorithms that subtopic of meta-heuristic algorithms [15]. The hunting strategy of humpback whales is an inspiration for the whale optimization algorithm. The bubble-net hunting strategy is identical to that method only used by humpback whales [15], [17]. The WOA starts with initializing random solution like other swarm-based algorithms [19]. And also, the law of WOA is the key feature that develops candidate solutions in every step and that features distinguish the WOA from other swarm-based algorithms [15].

Handbook of Whale Optimization Algorithm. https://doi.org/10.1016/B978-0-32-395365-8.00038-5

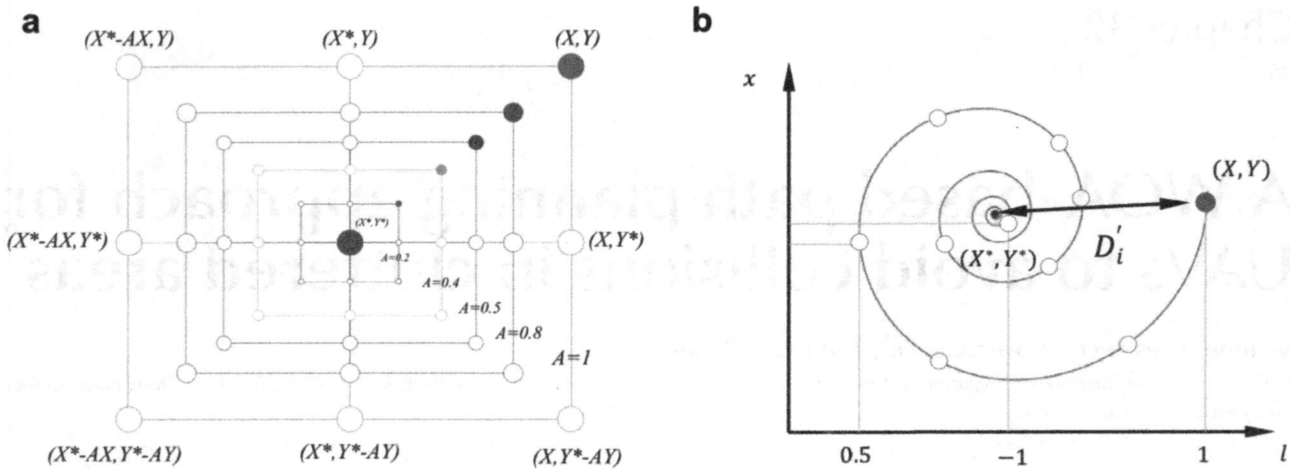

FIGURE 32.1 Bubble-net search mechanisms that use in WOA a-) The WOA Shrinking Encircling Mechanism b-) The WOA Spiral Updating Position.

32.2.1 Prey encircling

Humpback whales may first locate their prey and then encircle them. To begin, the WOA assumes that the current best optimal solution is the best and that it can be located near or on the prey. Furthermore, other search agents vary their location concerning the top search agent. The following equations are used to model the behavior of algorithms.

$$\vec{D} = |\vec{C}.\vec{X}^*(t) - \vec{X}(t)| \tag{32.1}$$

$$\vec{X}(t+1) = \vec{X}^*(t) - \vec{A}.\vec{D} \tag{32.2}$$

where t denotes the current step. $\vec{A}$ and $\vec{C}$ are coefficient vectors. $\vec{X}^*$, $\vec{X}$ is the best solution's position vector and the best solution's position vector, respectively. | | and . are the symbols denoting absolute value and component by component multiplication, respectively. If a better solution is discovered for each phase, $\vec{X}^*$ is updated. The vectors $\vec{A}$ and $\vec{C}$, on the other hand, are computed using the following formulas;

$$\vec{A} = 2\vec{a}.\vec{r} - \vec{a} \tag{32.3}$$

$$\vec{C} = 2.\vec{r} \tag{32.4}$$

where $\vec{a}$ is an augmented value ranging from 0 to 2 and $\vec{r}$ is a random vector in the range [0, 1].

32.2.2 Bubble-net attacking method

Two methodologies are used to represent the humpback whale's hunting style.

In this strategy, the a value in Eq. (32.3) is reduced, and therefore the value of A is reduced. As can be seen from this phrase, $\vec{a}$ is assigned a randomized value within the range $[-a, a]$ and $\vec{a}$ is reduced from 2 to 0.

32.2.2.1 Spiral updating position

In this strategy, the gap between both whale and prey is calculated, and a spiral formula is produced across whale and prey to replicate humpback whale helix-shaped behavior using the following equation;

$$\vec{X}(t+1) = \vec{D}'.e^{bl}.\cos(2\pi) + \vec{X}^*(t) \tag{32.5}$$

Humpback whales draw spiral-shape around their prey. This is simultaneously movement. To mathematically model this movement we assume % 50 probability to choose between two approaches and shown in Fig. 32.1. This behavior of algorithms can be expressed by the following equation and where p is a random value in the interval [0, 1];

$$\vec{X}(t+1) = \begin{cases} \vec{X}^*(t) - \vec{A}.\vec{D} & \text{if } p < 0.5 \\ \vec{D}'.e^{bl}.\cos(2\pi l) + \vec{X}*(t) & \text{if } p \geq 0.5 \end{cases} \tag{32.6}$$

32.2.3 Search for prey (exploration phase)

Humpback whales explore indiscriminately in nature, regardless of their position [19], [20]. WOA has achieved global optimization at this point, [20]. A vector in the interval $[-1, 1]$ is produced to drive the explore agent to travel away from the reference whale, as seen in the figure. As a result, the mathematical modeling of this step is expressed by the following formula and shown in Fig. 32.2;

$$\vec{D} = |\vec{C}.\vec{X}_{rand} - \vec{X}| \tag{32.7}$$

$$\vec{X}(t+1) = \vec{X}_{rand} - \vec{A}.\vec{D} \tag{32.8}$$

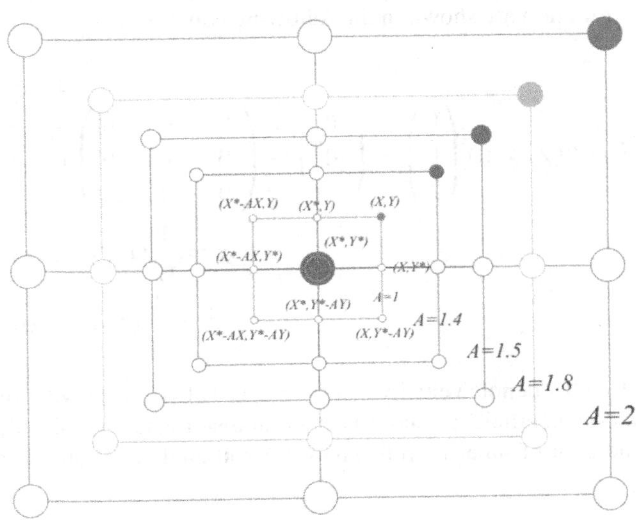

FIGURE 32.2 The WOA Exploration Mechanism.

32.3 Dynamics of agents and constraints

32.3.1 Dynamic of fixed-wing UAV

In the following, we accept dynamic of fixed wing agent as point-mass model [21]. Point mass model widely used in aircraft re-planning algorithms.

$$\dot{x} = V \cos\gamma \cos\chi \tag{32.9}$$

$$\dot{y} = V \cos\gamma \sin\chi \tag{32.10}$$

$$\dot{z} = V \sin\gamma \tag{32.11}$$

$$\dot{\gamma} = (g/V)(n \cos\phi - \cos\gamma) \tag{32.12}$$

$$\dot{\chi} = (g/V)((n \sin\phi)/(\cos\gamma)) \tag{32.13}$$

$$\dot{V} = (T - D)/m - g \sin\gamma \tag{32.14}$$

where x, y, and z denote the position of aircraft in the Cartesian system. And also, γ, χ, and ϕ denote respectively flight path angle, yaw angle, and bank angle of the vehicle. This angle can use for motion planning. T represents the aircraft trust vector, D is the aerodynamic drag force, and m and g denote respectively mass of the aircraft and gravitational acceleration. For safe motion planning, these values are extremely important. The velocity of nonholonomic aircraft systems has to be greater than the stall velocity. Because of that, velocity calculation is important for re-planning algorithms. And, n represent load factor (n = L/gm) and L denotes aerodynamic lift force. The other important constraint of nonholonomic aircraft systems is bank angle. The bank angle depends on the airspeed and turn radius of the vehicle and formulated (32.15).

$$\phi = \arctan(V^2/\rho g) \tag{32.15}$$

$$\rho_{min} = V^2/(g \tan(\phi_{max})) \tag{32.16}$$

where ρ denotes the turn radius of the aircraft. The minimum turn radius comes from the kinematic equation of vehicle [22] and depends on airspeed and bank angle.

32.3.2 Dynamic of rotary-wing UAV

Rotary-wing UAV is 6 degrees of freedom vehicle [23]. Therefore, this vehicle is defined by a fixed and body frame coordinate system. W and B denote fixed and body frames respectively as shown in Fig. 32.3. p denotes the position of the UAV in the B frame. Furthermore, we designate the vehicle's roll, pitch, and yaw angles by rotation axis. In this design, we consider a low-level attitude controller with first-order behavior that really can control desired roll and pitch angles [23], [24]. Classical system identification approaches may be used to identify the inner-loop first order parameters [23]. The airframe's rigid body formulas for motion are shown in the following equations;

$$\dot{\xi} = \nu \tag{32.17}$$

$$\dot{\nu} = R(\chi, \phi, \gamma)\begin{pmatrix} 0 \\ 0 \\ T \end{pmatrix} + \begin{pmatrix} 0 \\ 0 \\ -g \end{pmatrix} - \begin{pmatrix} A_x & 0 & 0 \\ 0 & A_y & 0 \\ 0 & 0 & A_z \end{pmatrix} \nu + d \tag{32.18}$$

$$\dot{\chi} = \frac{1}{\tau_\chi}(K_\chi \chi d - \chi) \tag{32.19}$$

$$\dot{\phi} = \frac{1}{\tau_\phi}(K_\phi \phi_d - \phi) \tag{32.20}$$

where ξ denote position of vehicle and ν denote velocity of vehicle. And also, g is gravitational acceleration, T is the thrust, A_x, A_y, A_z are mass normalized drag coefficients, and d is external disturbance. γ, ϕ, χ denote respective pitch, roll, and yaw angle. The time constant and gain of inner-loop behavior for roll angle and pitch angle are denoted by K_ϕ, τ_ϕ and K_γ, τ_γ, respectively.

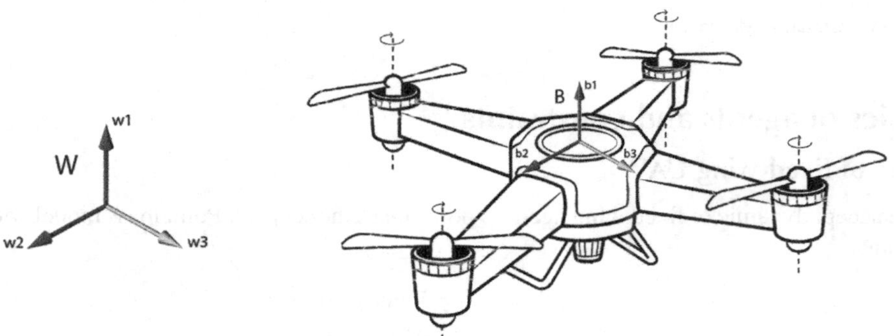

FIGURE 32.3 The UAVs coordinate systems.

32.4 Path planning

This section explains the principles of path planning in cluttered areas, and the cost function of path planning is given. The cluttered area is channeling and hard. Path planning of a UAV in cluttered areas has a few cost functions, for example, the distance between the start and end point, smoothness of path and time, etc. [25]. The distance between the start and end points will be calculated as an Euler or Hamilton distance. If cluttered areas are modeled as a city, distance will be Hamilton distance, or if cluttered areas are modeled as a forest or mountain range, distance will be Euler distance. The smoothness of the path can be provided by giving aircraft modeling as a constraint.

32.4.1 Path planning preliminaries

The position of the way-point is represented by P. The way-point matrix is $P = [w_{ij}] \in R^{n \times m}$. The number of way-points represented by n and the dimension of the path represented by m. Here i means the x-axis position of the path and

j means the y-axis position of the path. These representations can be seen in the figure. However, w_{ij} represent way-point at path. The purposes of path planning are the shortest and collision-free path. The one of the input of algorithm start point that represented by $S = [w_{11}, w_{21}]$ and also endpoint that represented by $F = [w_{1j}, w_{2j}]$. All of the position of UAVs is represented by;

$$P = \begin{pmatrix} w_{1,1} & w_{1,2} & \cdots & w_{1,m} \\ w_{2,1} & w_{2,2} & \cdots & w_{2,m} \\ \vdots & \vdots & \ddots & \vdots \\ w_{n,1} & w_{n,2} & \cdots & w_{n,m} \end{pmatrix}$$

We design a path planning algorithm that can use for 2D and 3D environments. To define the parameters and constraints of the path planning problem, we have to define two coordinate systems. First, the Cartesian coordinate system must be used to reduce algorithm solving time. This coordinate system represents the location of a point, for example X, Y, Z, its distance to the point that is Origin. The second coordinate system is the northeast-down (NED) coordinate system. And, origin of this coordinate system is located at the UAV's center of gravity. However, this is not flying at a constant altitude means flying at a constant z above the Earth's surface. The local coordinate system is represented by latitude, longitude, and altitude. Path planning of UAV is written in the Cartesian coordinate system, but to use this algorithm in commercial autopilot systems, such as PX4 [26], we convert the Cartesian system to the NED system. To convert the Cartesian coordinate system to NED coordinate system, we use the haversine formula. The haversine formula gives the shortest distance between two points over the Earth's surface. Distance between two points is represented by the following formula [27];

$$d_{NED} = 2r \arcsin(\sqrt{\sin^2(\frac{\alpha_n - \alpha_{n+1}}{2}) + \cos(\alpha_n).\cos(\alpha_{n+1}).\sin^2(\frac{\lambda_n - \lambda_{n+1}}{2})}) \tag{32.21}$$

where d_{NED} denotes the distance between two points at NED coordinate system. And, α_n and α_{n+1} denote respectively latitude position of waypoint n and latitude position of waypoint $n+1$. λ_n and λ_{n+1} denote respectively longitude position of waypoint n and longitude position of waypoint $n+1$. Furthermore, r represents the radius of the Earth which is approximately 6371 km. The bearing angle between two points is calculated by the following formula;

$$\theta_{NED} = \arctan(\sin(\Delta\lambda)\cos(\alpha_{n+1})\,,\ \cos(\alpha_1)\sin(\alpha_2) - \sin(\alpha_1)\cos(\alpha_2)\cos(\Delta\lambda)) \tag{32.22}$$

where θ denotes the bearing angle between waypoint n and waypoint $n+1$. $\Delta\lambda$ is equal to $\lambda_1 - \lambda_2$ that means difference in longitude positions. In this algorithm, we use the start position as the home position of the UAV. The other formula is to calculate the next waypoint location. Given the initial location of the waypoint, distance to the next waypoint, and bearing to calculate the next waypoint location by the following formula;

$$\alpha_{n+1} = \arcsin(\sin(\alpha_n)\sin(\Gamma) + \cos(\alpha_n)\sin(\Gamma)\cos(\theta) \tag{32.23}$$

$$\lambda_{n+1} = \lambda_n + \arctan(\sin(\theta)\sin(\Gamma)\cos(\alpha_1), \cos(\Gamma) - \sin(\alpha_n)\sin(\alpha_{n+1})) \tag{32.24}$$

where Γ denote angular distance between n'th waypoint and $n+1$'th waypoint and equal to d/r. Eqs. (32.21), (32.22), (32.23), and (32.24) are using to calculate path at NED coordinate system. However, we use Cartesian coordinate system for local path planning. Therefore, the distance between two waypoints at the Cartesian coordinate system is calculated by following formula [8];

$$d_{CAR} = \sqrt{(P_{x_n} - P_{x_{n+1}})^2 + (P_{y_n} - P_{y_{n+1}})^2} \tag{32.25}$$

where d_{CAR} is distance between n'th waypoint and $n+1$'th waypoint at the Cartesian coordinate system. The angle between two points can be calculated by using vector calculation. The angle between two vector vectors is calculated by the following equation [28];

$$\theta_{CAR} = \arccos(\frac{\vec{v_n}.\vec{v_{n+1}}}{|\vec{v_n}|.|\vec{v_{n+1}}|}) \tag{32.26}$$

where θ_{CAR} is angle between two vectors at Cartesian coordinate system. $\vec{v}_n$ represent $(P_{\vec{x}_n} - P_{\vec{x}_{n-1}}, P_{\vec{y}_n} - P_{\vec{y}_{n-1}})$, $\vec{v}_{n+1}$ represent $(P_{\vec{x}_{n+1}} - P_{\vec{x}_n}, P_{\vec{y}_{n+1}} - P_{\vec{y}_n})$ [28]. $P_{\vec{x}}$ and $P_{\vec{y}}$ represent respectively x-axis coordinate of position and y-axis coordinate of position.

Algorithm 32.1 Local position to Global Position.

Data: $P_l \leftarrow Local\ path\ list$
$n \leftarrow Number\ of\ way-point$
$[I_{lat}, I_{lon}] \leftarrow Initial\ way-point\ global\ position$
$i = 0$
$P_g = [\ ] \leftarrow Global\ path\ list$
$P_o \leftarrow previous\ way-point$
$P_o \leftarrow take\ initial\ way-point\ as\ a\ previous\ way-point$
while $i \leq n$ **do**
$\theta_i \leftarrow calculate\ angle\ between\ (i)'th\ way-point\ and\ (i+1)'th\ way-point\ by\ using$ (32.26)
$d_i \leftarrow calculate\ distance\ between\ (i)'th\ way-point\ and\ (i+1)'th\ way-point\ by\ using$ (32.25)
$P_i \leftarrow calculate\ new\ position\ by\ using\ \theta_i\ and\ d_i$
$P_g \leftarrow append\ P_i\ position\ to\ P_g\ list$
i = i + 1
end

32.4.2 Interpolation

To sample the generated pathways in a way that allows modern guidance controllers to track such a planned path, a sampling technique is required [29]. The path of UAVs should be smooth for tracking [30]. In order to smooth path we use interpolation algorithms.

32.4.2.1 Quadratic interpolation

In order to solve quadratic interpolation of function that denote $f(x)$ we define following formula;

$$z(y) = ay^2 + by + c \tag{32.27}$$

where a, b, and $c \in R$. The function $z(y)$ has three unknown parameters because of that we define three function with given points y_1 and y_2 [31]. We can find three parameters by solving following formula;

$$f(y_1) = ay_1^2 + by_1 + c = z(y_1) \tag{32.28}$$

$$f(y_2) = ay_2^2 + by_2 + c = z(y_2) \tag{32.29}$$

$$f'(y_1) = 2ay_1 + b = z'(y_1) \tag{32.30}$$

For solving these equations we can find a and b parameters. We use open source library to take interpolation which is SciPy [32].

32.4.2.2 Linear interpolation

Linear interpolation is easy to implement. Thus, we can reduce calculation time. A technique for roughly calculating the value of a function $f(y)$ and $z(y)$ can be represented by following formula [33];

$$z(y) = a(y - y_1) + b \tag{32.31}$$

where a and $b \in R$. To choose a and b value we use given two point y_1 and y_2 following formula;

$$z(y_1) = f(y_1) \tag{32.32}$$

$$z(y_2) = f(y_2) \tag{32.33}$$

With using these conditions, the unique function is evaluated;

$$z(y) = \frac{f(y_2) - f(y_1)}{y_2 - y_1}(y - y_1) + f(y_1) \tag{32.34}$$

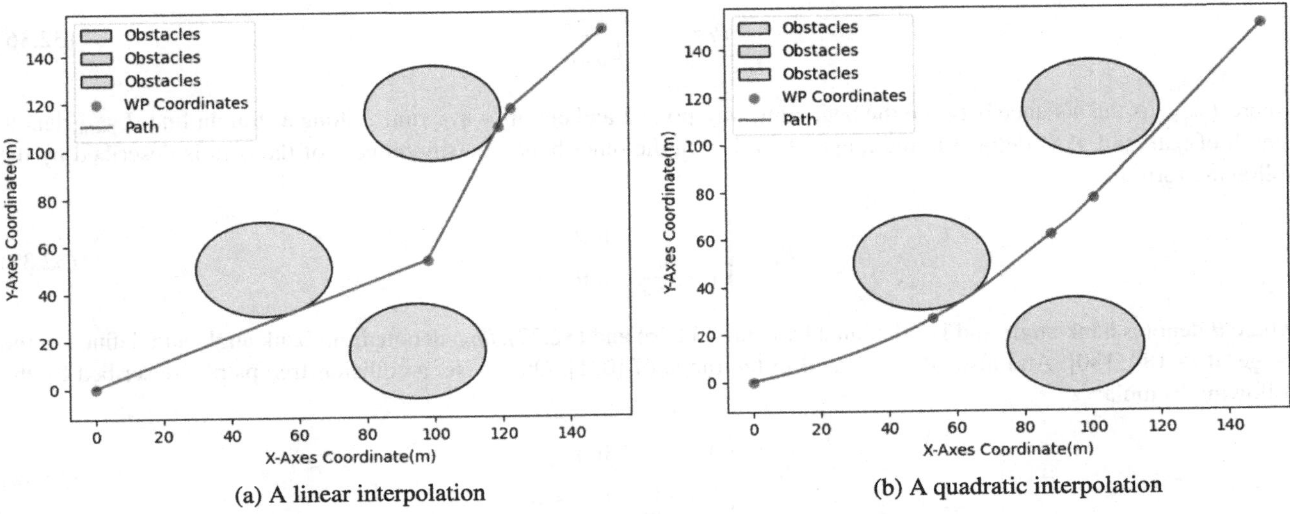

(a) A linear interpolation (b) A quadratic interpolation

FIGURE 32.4 The example map 1 of path planning via obstacle by using linear interpolation 32.4a and quadratic interpolation 32.4b.

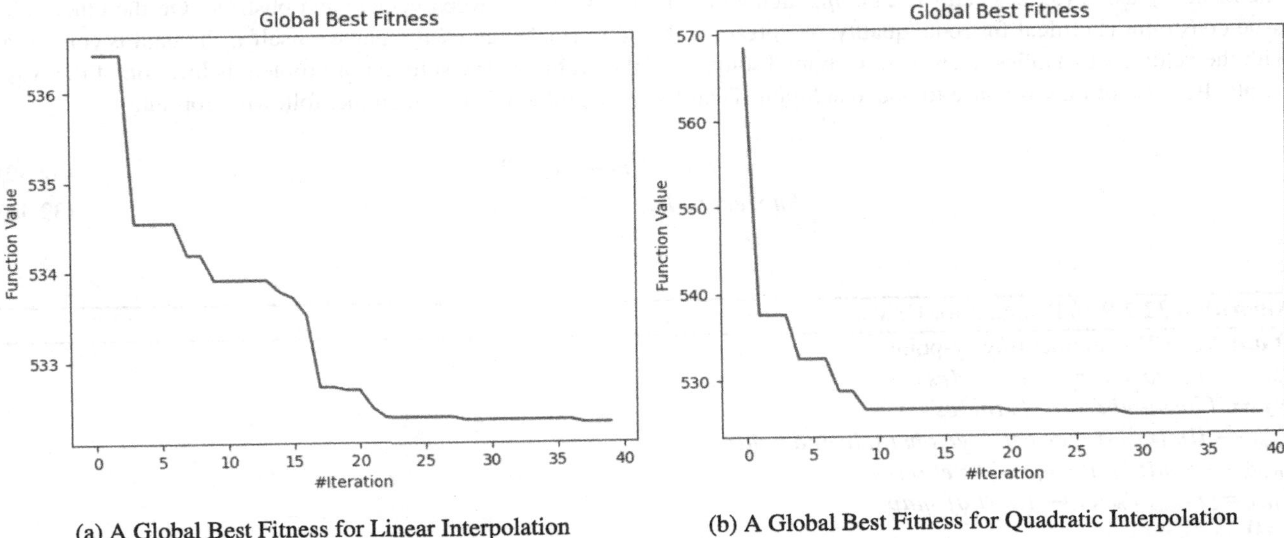

(a) A Global Best Fitness for Linear Interpolation (b) A Global Best Fitness for Quadratic Interpolation

FIGURE 32.5 The Result of Local and Global Best Fitness Function with 3 obstacles like Fig. 32.4.

32.4.3 Cost functions and boundary

The cost function is the length of the path between the start and final points. The best path in the scenario of UAV motion planning is far more complicated and comprises several semantic categories. The cost function is the same for both fixed and rotary wing UAVs. However, the constraints of the problem will change for both of them. The best path in the scenario of UAV motion planning is far more complicated and comprises several semantic categories. To satisfy the desired necessity, the cost function is used and developed. These necessities can be defined as the length of the path, the smoothness of the path, and the forbidden zone. However, the cost function of the problem can be represented by the following formula;

$$F_{cost} = P_l + P_s + P_f \tag{32.35}$$

where F_{cost} denote cost function of problem. And also, P_l, P_s, P_f respectively penalize the length of the path, smoothness of path, and paths going through collision-free path. These criteria improve the quality of trajectory. In the cost function, the length of the path is described by following formula [34];

$$P_l = 1 - \frac{L_{w_1 w_n}}{L_{path}} \tag{32.36}$$

where $L_{w_1 w_n}$ is the distance between the beginning way-point 1 and ending way-point 2 along a straight line. L_{path} denote length of path and P_l is defined in the range of [0, 1]. On the other hand, the smoothness of the path is described by the following formula;

$$P_s = \begin{cases} \infty & \text{if } \theta \geq \theta_{lim} \\ 1 - \frac{\theta}{\theta_{lim}} & \text{if } \theta < \theta_{lim} \end{cases} \tag{32.37}$$

where θ denotes bank angle and is calculated by using (32.26) and (32.22). θ_{lim} denote limit bank angle and defined in the range of [−180, 180]. And also, P_s is defined in the range of [0, 1]. Otherwise, a collision-free path is described by the following formula;

$$P_f = \begin{cases} 0 & \text{if } d \geq d_{lim} \\ 1 - \frac{d}{d_{lim}} & \text{if } d < d_{lim} \end{cases} \tag{32.38}$$

where d denotes the distance between the position of the obstacle and the position of the vehicle. However, d can be calculated by using (32.25) and (32.21). d_{lim} denote the limit distance between vehicle and obstacle. On the other side, issue constraint is critical for route quality. We presume that the vehicle can easily manage itself if the path is consistent with the vehicle's dynamics. However, we bound solution of the problem. The solution of problem is location of the way-points. Because of that, we have to bound solution. Therefore, the problem is rewritten like following formula;

$$F_{cost} = P_l + P_s + P_f \tag{32.39}$$

$$Subject\ to\ P_{min} \leq P \leq P_{max} \tag{32.40}$$

Algorithm 32.2 Path Planning for UAVs.

Data: N_w = The number of way-point
$P_o = The\ position\ of\ obstacles$
$N_o = The\ number\ of\ obstacles$
$d_{lim} \leftarrow limit\ distance\ between\ obstacle\ and\ vehicle$
$\theta_{bank} \leftarrow limit\ bank\ angle\ of\ vehicles$
$P_{lim} = [P_{min}, P_{max}] \leftarrow limit\ of\ map$
i = 0
$F_{cost} = 0$
$C_{constraint} = [\,]$
while $i \leq n$ **do**
  In this pseudocode, all of the calculation made in Cartesian coordinate system
  if *i = 0* **then**
    $L_{path} \leftarrow calculate\ lenght\ of\ path\ by\ using$ 32.4.2.1 or 32.4.2.2
    $\rho_{min} \leftarrow calculate\ limit\ turn\ radius\ of\ vehicles$ (32.16)
  end
  $P_l \leftarrow calculate\ length\ of\ path\ cost\ function\ by\ using$ (32.36)
  $\phi_{limit} \leftarrow calculate\ limit\ bank\ angle\ of\ vehicle\ by\ using$ (32.15)
  $P_s \leftarrow calculate\ smoothness\ of\ path\ cost\ function\ byusing$ (32.37)
  $P_f \leftarrow calculate\ free\ path\ cost\ function\ by\ using$ (32.38)
  $F_{cost} = F_{cost} + P_l + P_s + P_f$
  Solve Eqs. (32.40)
  i = i + 1
end

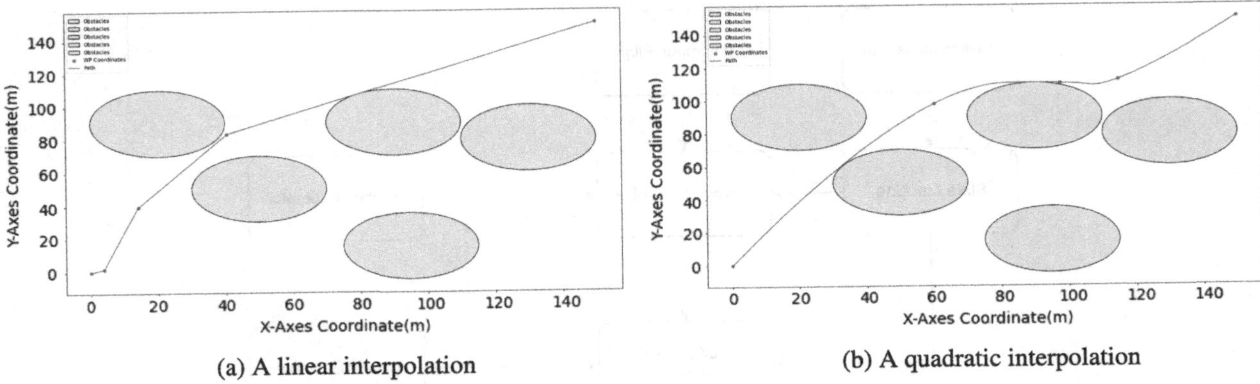

(a) A linear interpolation (b) A quadratic interpolation

FIGURE 32.6 The example map 2 of path planning via obstacle by using linear interpolation 32.6a and quadratic interpolation 32.6b.

32.5 Simulation environment

32.5.1 PX4 autopilot system

PX4 is commercial and open source autopilot system [26]. PX4 have been used to control rotary and fixed-wing. And also, we can use PX4 for rovers and unmanned underwater vehicles. PX4 has a modular software and node-based architecture [26]. PX4 software architecture can be shown in Fig. 32.8.

FIGURE 32.7 The Overall control scheme of vehicle.

32.5.2 QGroundControl and MAVLink messages

To see and upload mission QGroundControl is used. QGroundControl is ground control station for commercial drones. We use MAVLink messages to upload mission to PX4 autopilot system. In order to control vehicles, there is a lat of MAVLink messages which you can reach from this website https://mavlink.io/en/messages/common.html. In this project, we use mission item MAVLink messages. And also, we can show relationship between QGroundControl and PX4 in Fig. 32.7.

32.5.3 Simulation results

To test the performance of the algorithm Software In the loop is used with QGroundControl and PX4. And also, we use Matplotlib to visualize algorithm output [35]. For interpolation, we use the SciPy python library [32]. And also, a whale optimization algorithm is implemented successfully at the mealy python library [18]. With this environment, we test our algorithm. To test the performance of our algorithm we use Python programming language and a PC with an Intel i7 and 2.9 GHz CPU.

We use whale optimization as a solver that solves (32.35). Time is an important parameter for path planning algorithms. We show time that elapsed time at Fig. 32.9. In this cart, we show solving time at every iteration. The average elapsed time is about 0.147 and this result is admissible for global path planning. To solve the problem, the procedure is used in pseudo-code that shows at 32.2 This chart shows that solving time of Fig. 32.4. As you can see in Fig. 32.9, linear interpolation has a slower solving time than quadratic interpolation.

The parameters of a whale optimization algorithm are shown in Table 32.1. These parameters affect solving time of the problem. And also, an iteration value of the problem affects total solving time and finding a global optimum. In Fig. 32.5, the line graph shows the global best fitness value of quadratic and linear interpolation at every iteration. As you can see in Fig. 32.5, a WOA algorithm solves an easier quadratic interpolation method than linear interpolation.

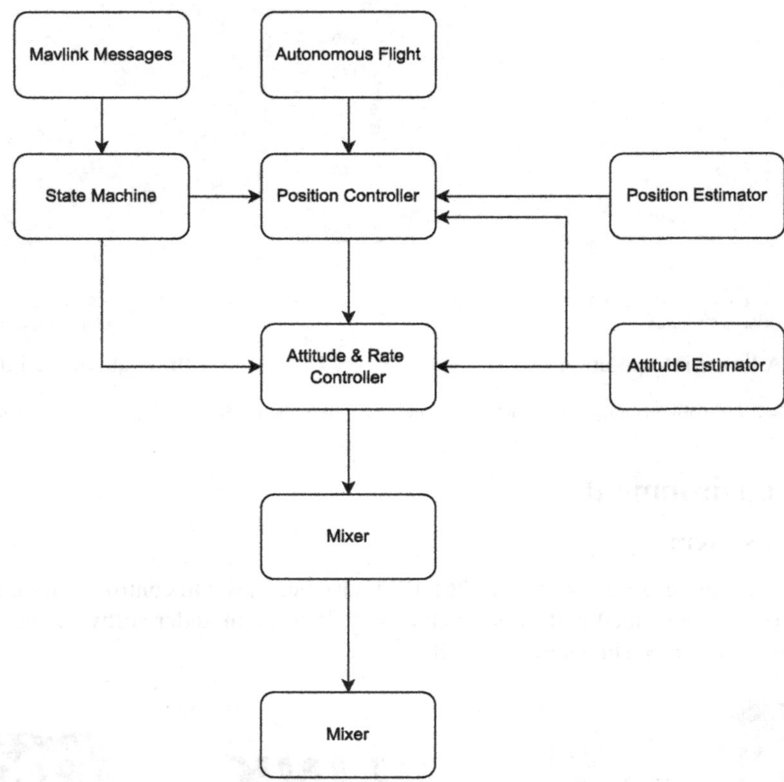

FIGURE 32.8 The Controller Scheme of PX4.

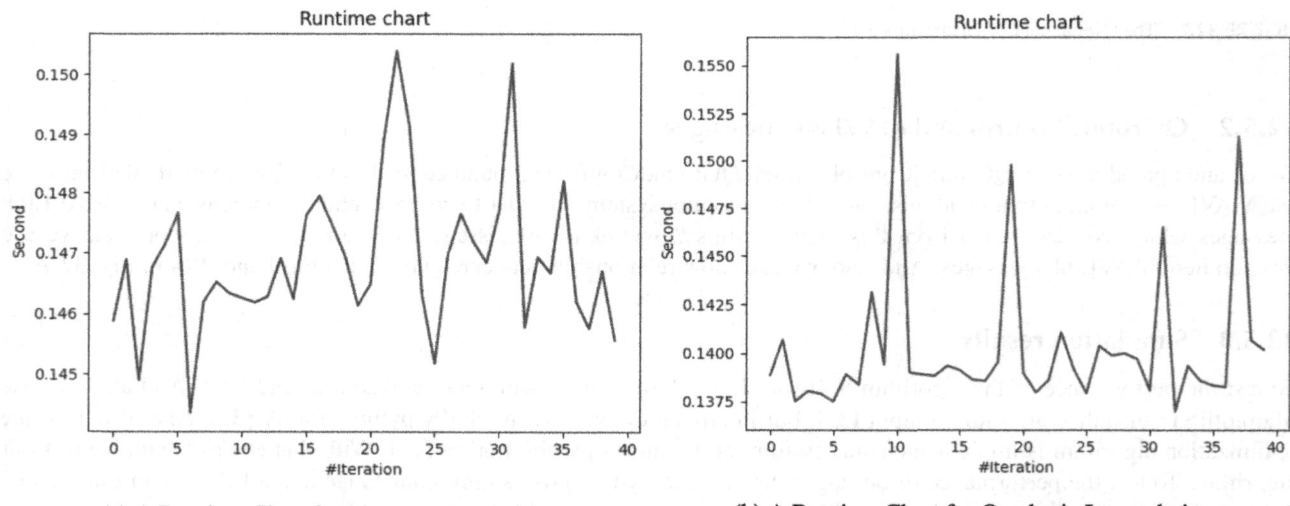

(a) A Runtime Chart for Linear Interpolation

(b) A Runtime Chart for Quadratic Interpolation

FIGURE 32.9 The Solving Time of the Problem at Fig. 32.4.

As shown in Figs. 32.9, 32.5, and 32.10, a whale optimization algorithm is appropriate for global path planning. As shown in Fig. 32.5, a whale optimization reach global optimum after 20-25 iteration. If there are more obstacles in the flight path a whale optimization algorithm require more iteration to reach global optimum solution. And also, a whale optimization algorithm can also solve the more complex area like map-2 at Fig. 32.6. Hence, a whale optimization algorithm is useful real-world global optimization.

TABLE 32.1 Parameters of the Whale Optimization Algorithm.

Name of Algorithm	Name of Parameters	Value
A Whale Optimization Algorithm	Parameter (a)	[2, 0]
	Population Size	10-50
	Iteration Range	10-100

In order to use a WOA-based path planning algorithm with PX4 autopilot system, we use the Algorithm 32.1. This algorithm converts local coordinates to global coordinates. PX4 autopilot system uses MAVLink messages to get and handle the mission. As shown in the examples Fig. 32.4 and 32.6, all of the coordinate is calculated in unit meter. However, MAVLink mission messages only accept coordinate at latitude and longitude coordination. After calculation way point by using WOA-based path planning algorithm, we use algorithm that given pseudo-code 32.1.

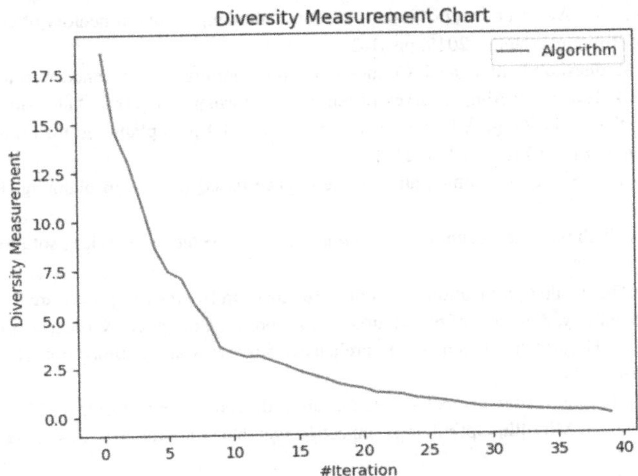

FIGURE 32.10 The Diversity of Problem at Fig. 32.4.

32.6 Conclusion

This section represents a path planning algorithm by using a whale optimization algorithm for UAVs in 2D environments. This algorithm can be used in 3D environments as well as 2D. This algorithm uses the 3 main functions to use in the cost function. One of them is the length of the path which is important for reaching the target as soon as possible. The other function is about a collision with obstacles, to avoid a collision, we set a limited distance between an obstacle and the vehicle. The last function is the smoothing function. When the algorithm avoids a collision, the angle between the way-point is not dynamically suitable for UAVs. Therefore, we use the smoothing function to handle this situation. In the end, we can easily say a whale optimization algorithm is useful for UAV's path planning algorithms.

32.7 Future works

The path planning is a challenging area in the real world. This chapter presented a whale optimization algorithm that is a good fit for challenging robotics path planning problems. In future work, we plan to use whale optimization for more complex path planning problems such as terrain flights, forest flights, etc with UAV. In addition to flights path planning problems, the whale optimization algorithm plan to use for other challenging robotics problems like controller design, decision-making algorithms, formation control of UAVs, etc.

References

[1] Yucong Lin, Srikanth Saripalli, Sampling-based path planning for UAV collision avoidance, IEEE Transactions on Intelligent Transportation Systems 18 (11) (2017) 3179–3192.

[2] Jesús Capitan, Luis Merino, Aníbal Ollero, Cooperative decision-making under uncertainties for multi-target surveillance with multiples UAVs, Journal of Intelligent & Robotic Systems 84 (1) (2016) 371–386.

[3] Javier Alonso-Mora, Stuart Baker, Daniela Rus, Multi-robot formation control and object transport in dynamic environments via constrained optimization, The International Journal of Robotics Research 36 (9) (2017) 1000–1021.

[4] Björn Lindqvist, Sina Sharif Mansouri, Ali-akbar Agha-mohammadi, George Nikolakopoulos, Nonlinear MPC for collision avoidance and control of UAVs with dynamic obstacles, IEEE Robotics and Automation Letters 5 (4) (2020) 6001–6008.

[5] Michael Brand, Michael Masuda, Nicole Wehner, Xiao-Hua Yu, Ant Colony Optimization Algorithm for Robot Path Planning, in: 2010 International Conference on Computer Design and Applications, vol. 3, IEEE, 2010, V3–436.

[6] Tom Schouwenaars, Bart De Moor, Eric Feron, Jonathan How, Mixed integer programming for multi-vehicle path planning, in: 2001 European Control Conference (ECC), IEEE, 2001, pp. 2603–2608.

[7] John Schulman, Yan Duan, Jonathan Ho, Alex Lee, Ibrahim Awwal, Henry Bradlow, Jia Pan, Sachin Patil, Ken Goldberg, Pieter Abbeel, Motion planning with sequential convex optimization and convex collision checking, The International Journal of Robotics Research 33 (9) (2014) 1251–1270.

[8] Ram Kishan Dewangan, Anupam Shukla, W. Wilfred Godfrey, Three dimensional path planning using grey wolf optimizer for UAVs, Applied Intelligence 49 (6) (2019) 2201–2217.

[9] Alessandro Gasparetto, Paolo Boscariol, Albano Lanzutti, Renato Vidoni, Path planning and trajectory planning algorithms: a general overview, in: Motion and Operation Planning of Robotic Systems, 2015, pp. 3–27.

[10] Antonios Tsourdos, Brian White, Madhavan Shanmugavel, Cooperative Path Planning of Unmanned Aerial Vehicles, John Wiley & Sons, 2010.

[11] Kwang-Kyo Oh, Myoung-Chul Park, Hyo-Sung Ahn, A survey of multi-agent formation control, Automatica 53 (2015) 424–440.

[12] Liang Yang, Juntong Qi, Jizhong Xiao, Xia Yong, A literature review of UAV 3D path planning, in: Proceeding of the 11th World Congress on Intelligent Control and Automation, IEEE, 2014, pp. 2376–2381.

[13] Yijing Zhao, Zheng Zheng, Yang Liu, Survey on computational-intelligence-based UAV path planning, Knowledge-Based Systems 158 (2018) 54–64.

[14] Shubhani Aggarwal, Neeraj Kumar, Path planning techniques for unmanned aerial vehicles: a review, solutions, and challenges, Computer Communications 149 (2020) 270–299.

[15] Seyedali Mirjalili, Andrew Lewis, The whale optimization algorithm, Advances in Engineering Software 95 (2016) 51–67.

[16] Simon Fong, Suash Deb, Ankit Chaudhary, A review of metaheuristics in robotics, Computers & Electrical Engineering 43 (2015) 278–291.

[17] Farhad Soleimanian Gharehchopogh, Hojjat Gholizadeh, A comprehensive survey: whale optimization algorithm and its applications, Swarm and Evolutionary Computation 48 (2019) 1–24.

[18] Nguyen Van Thieu, A collection of the state-of-the-art meta-heuristics algorithms in Python, Mealpy, 2020.

[19] Ibrahim Aljarah, Hossam Faris, Seyedali Mirjalili, Optimizing connection weights in neural networks using the whale optimization algorithm, Soft Computing 22 (1) (2018) 1–15.

[20] Marwa Sharawi, Hossam M. Zawbaa, Eid Emary, Feature selection approach based on whale optimization algorithm, in: 2017 Ninth International Conference on Advanced Computational Intelligence (ICACI), IEEE, 2017, pp. 163–168.

[21] Antonio Bicchi, Lucia Pallottino, On optimal cooperative conflict resolution for air traffic management systems, IEEE Transactions on Intelligent Transportation Systems 1 (4) (2000) 221–231.

[22] Zijie Lin, Lina Castano, Edward Mortimer, Huan Xu, Fast 3D collision avoidance algorithm for fixed wing UAS, Journal of Intelligent & Robotic Systems 97 (3) (2020) 577–604.

[23] Mina Kamel, Thomas Stastny, Kostas Alexis, Roland Siegwart, Model predictive control for trajectory tracking of unmanned aerial vehicles using robot operating system, in: Robot Operating System (ROS), Springer, 2017, pp. 3–39.

[24] Robert Mahony, Vijay Kumar, Peter Corke, Multirotor aerial vehicles: modeling, estimation, and control of quadrotor, IEEE Robotics & Automation Magazine 19 (3) (2012) 20–32.

[25] Purushothaman Raja, Sivagurunathan Pugazhenthi, Optimal path planning of mobile robots: a review, International Journal of Physical Sciences 7 (9) (2012) 1314–1320.

[26] Lorenz Meier, Dominik Honegger, Marc Pollefeys, PX4: a node-based multithreaded open source robotics framework for deeply embedded platforms, in: 2015 IEEE International Conference on Robotics and Automation (ICRA), IEEE, 2015, pp. 6235–6240.

[27] Nitin R. Chopde, Mangesh Nichat, Landmark based shortest path detection by using A* and haversine formula, International Journal of Innovative Research in Computer and Communication Engineering 1 (2) (2013) 298–302.

[28] Wei Jiang, Yongxi Lyu, Yongfeng Li, Yicong Guo, Weiguo Zhang, UAV path planning and collision avoidance in 3D environments based on POMPD and improved grey wolf optimizer, Aerospace Science and Technology 121 (2022) 107314.

[29] Thomas Bucher, Robust wind-aware path optimization on board small fixed-wing UAVs: Towards a more robust path planning in wind using trochoids and clothoids, Master's thesis, ETH Zurich, 2021.

[30] Dmitri Dolgov, Sebastian Thrun, Michael Montemerlo, James Diebel, Practical search techniques in path planning for autonomous driving, Ann Arbor 1001 (48105) (2008) 18–80.

[31] Wenyu Sun, Ya-Xiang Yuan, Optimization Theory and Methods: Nonlinear Programming, vol. 1, Springer Science & Business Media, 2006.

[32] Pauli Virtanen, Ralf Gommers, Travis E. Oliphant, Matt Haberland, Tyler Reddy, David Cournapeau, Evgeni Burovski, Pearu Peterson, Warren Weckesser, Jonathan Bright, Stéfan J. van der Walt, Matthew Brett, Joshua Wilson, K. Jarrod Millman, Nikolay Mayorov, Andrew R.J. Nelson, Eric Jones, Robert Kern, Eric Larson, C.J. Carey, İlhan Polat, Yu Feng, Eric W. Moore, Jake VanderPlas, Denis Laxalde, Josef Perktold, Robert Cimrman, Ian Henriksen, E.A. Quintero, Charles R. Harris, Anne M. Archibald, Antônio H. Ribeiro, Fabian Pedregosa, Paul van Mulbregt, SciPy 1.0 Contributors, SciPy 1.0: fundamental algorithms for scientific computing in Python, Nature Methods 17 (2020) 261–272.

[33] Philip J. Davis, Interpolation and Approximation, Courier Corporation, 1975.
[34] Vincent Roberge, Mohammed Tarbouchi, Gilles Labonté, Comparison of parallel genetic algorithm and particle swarm optimization for real-time UAV path planning, IEEE Transactions on Industrial Informatics 9 (1) (2012) 132–141.
[35] John D. Hunter, Matplotlib: a 2D graphics environment, Computing in Science & Engineering 9 (03) (2007) 90–95.

Chapter 33

Application of an Improved Whale Optimization Algorithm for optimal design of shell and tube heat exchanger☆

Diab Mokeddem[a], Seyedali Mirjalili[b,c], and Dallel Nasri[a]

[a]*Department of Electrical Engineering, Faculty of Technology, University of Ferhat Abbas Setif-1, Setif, Algeria,* [b]*Centre for Artificial Intelligence Research and Optimisation, Torrens University Australia, Brisbane, QLD, Australia,* [c]*University Research and Innovation Center, Obuda University, Budapest, Hungary*

33.1 Introduction

Heat exchangers are elements used to transfer heat between two fluids or between a solid and a fluid. In industrial societies, the heat exchanger is an essential element of any energy management policy. They are mainly used in the industrial sector (chemicals, petrochemicals, iron and steel, food processing, energy production, etc.), in the transport sector (automotive, aeronautics), but also in the residential sectors (heating, air conditioning, etc.) [1,2].

Shell and tube heat exchangers (STHE), in its different architectural variations, are the most likely the most prevalent and widely utilized fundamental heat exchanger structure in the process industries. A good selection of the appropriate heat exchanger design permits the process to gain energy and lower costs. For that, many techniques were used for optimization design of shell and tube heat exchangers [3].

A number of strategies for solving the design optimization problem have been presented, including numerical resolution of the stationary point equations of a nonlinear objective function [4], and schematic illustration of the search process [5]. Gaddis [6] provided a novel approach for determining shell-side pressure loss that is primarily based on the Delaware technique. Saunders [7] suggested a very practical way in which basic design elements are supplied, allowing the Bell method to be employed quickly for a predefined set of geometrical parameters and more in [8]. Traditional design methods based on mathematical programming techniques have several drawbacks. Therefore, some research on the optimization of shell and tube heat exchangers using artificial intelligence approaches has been used. Mariani et al. [9] used a combination of quantum particle swarm optimization (QPSO) and Zaslavskii chaotic pattern sequencing (QPSOZ) to develop an ideal shell and tube heat exchanger. Caputo et al. [10] used GA to build a heat exchanger focused on economical optimization. Patel and Rao [11] used Particle swarm optimization (PSO) to reduce the overall yearly cost of STHEs. Selbas et al. [12] employed a genetic algorithm (GA), using pressure drop as a constraint for reaching optimum parameters.

In this study, the Improved Whale Optimization Algorithm (IWOA) developed by Mokeddem and Mirjalili [13] is employed to solve the design problem of STHE. The performance of IWOA is compared with some well-known meta-heuristics, including genetic algorithm (GA) [14,15], Particle swarm optimizer (PSO) [11], Artificial bee colony (ABC) algorithm [16,17], Biogeography-based optimization (BBO) [18], Intelligent tuned harmony search algorithm (ITHS) [19], and improved intelligent tuned harmony search algorithm (I-ITHS) [20].

The rest of the chapter is organized as follows. Section 33.2 discusses briefly the Mechanism of Whale Optimization Algorithm (WOA). In Section 33.3 we describe the improved whale optimization algorithm. The mathematical model of SHTE is described in Section 33.4. The results discussion are presented in Section 33.5. Finally, Section 33.6 concludes the chapter.

☆. The contents of Section 33.2 and Section 33.3 have been taken from Mokeddem, D., & Mirjalili, S. (2020). Improved whale optimization algorithm applied to design PID plus second-order derivative controller for automatic voltage regulator system. Journal of the Chinese Institute of Engineers, 43(6), 541-552 by permission of the publisher (Taylor & Francis Ltd, http://www.tandfonline.com).

Handbook of Whale Optimization Algorithm. https://doi.org/10.1016/B978-0-32-395365-8.00039-7

33.2 Mechanism of Whale Optimization Algorithm (WOA)

The WOA algorithm is a new nature-inspired algorithm developed by Mirjalili and Lewis in 2016 [21]. It mimics the hunting behavior of humpback whales; their favorite food is small fish and krill herds close to the surface. The humpback whales foraging behavior is called the bubble-net feeding method, in which they create distinctive bubbles from about twelve meters down in spiral shape around the prey, encircling prey and then swim up to the surface to start hunting them. The hunting behavior of the whale optimizer is described by three main steps: encircling prey, bubble-net attacking and search for prey.

33.2.1 Phase of encircling

Once the prey location is defined, WOA starts encircling them. The prey here is the best optimal solution found so far [21]. After finding the best candidate solution in each iteration, the other agents update their positions in the neighborhood of the best search agent following the equations:

$$\vec{X}(t+1) = \vec{X}^*(t) - A \cdot \vec{D}, \tag{33.1}$$

$$\vec{D} = \left| C \cdot \vec{X}^*(t) - \vec{X}(t) \right|, \tag{33.2}$$

$$A = 2 \cdot a \cdot r - a, \tag{33.3}$$

$$C = 2 \cdot r, \tag{33.4}$$

where $\vec{X}^*$, $\vec{X}$ indicates the position vector of the optimal solution and the position vector of a solution, respectively, t represents the current step. The values of a are dropped from 2 to 0 throughout the duration of iterations, while r is a random number between 0 and 1.

33.2.2 Phase of bubble-net attacking (exploitation)

Mainly obtained by shrinking encircling mechanism, mathematically defined by Eq. (33.3) where the value of A is decreasing. As such, the value of A is set to be a random value in the interval $[-a, a]$ through iterations. Between original position and position of the current best agent we find the new position of A.

To mimic the movement of the humpback whales towards the prey, a spiral updating position equation is given [21]:

$$\vec{X}(t+1) = \vec{X}^*(t) + e^{bl} \cdot \cos(2\pi l) \cdot \vec{D}', \tag{33.5}$$

where $\vec{D}' = \left| \vec{X}^*(t) - \vec{X}(t) \right|$ is the distance between the prey (best solution) and the ith whale, l is a random value uniformly distributed in the range of $[-1, 1]$, and b is a constant for defining the shape of the logarithmic spiral.

The two foraging mechanisms, shrinking encircling and spiral updating position discussed above are mathematically modeled as follows:

$$\vec{X}(t+1) = \begin{cases} \vec{X}^*(t) - A \cdot \vec{D} & \text{if } p < 0.5 \\ \vec{X}^*(t) + e^{bl} \cdot \cos(2\pi l) \cdot \vec{D}' & \text{if } p > 0.5 \end{cases}, \tag{33.6}$$

p is a random number in [0, 1] switching between these two mechanisms Eq. (33.6) with an equal probability (50%) to update whale position.

33.2.3 Phase of searching (exploration)

The search agents (humpback whales) look for the best solution (prey) randomly and change their positions according to other agent positions. A take the values > 1 or < -1 in order to force the search agent to move far away from target whale. It is used to make the transition between exploration and exploitation phase easier by decreasing A.

When the value of $|A| \geq 1$, the WOA algorithm performs a global search and emphasizes exploration. For $|A| < 1$ the algorithm performs a local search emphasizing exploitation.

The mathematical model of the exploration phase can be given as:

$$\vec{X}(t+1) = \vec{X}_{rand} - A \cdot \vec{D''}, \tag{33.7}$$

$$\vec{D''} = \left| C \cdot \vec{X}_{rand} - \vec{X}(t) \right|, \tag{33.8}$$

where $\vec{X}_{rand}$ is randomly selected from whales (agents) in the current iteration.

33.3 Improved Whale Optimization Algorithm (IWOA)

This section proposes an improved version of WOA to improve the accuracy of global solution and avoid near global solution stagnation using crossover operator.

33.3.1 Arithmetic crossover

Genetic algorithm is a powerful evolutionary stochastic search technique, which is inspired from biological processes in nature. It relies on selection and the operators of crossover and mutation [22].

In this work, the standard WOA algorithm is integrated with crossover operator from genetic algorithm. The advantage of incorporating the crossover operator is to promote the exploratory behavior of search agents (whales). The crossover operator is involved in the production of new children (offsprings) by combining the genes of two parents (agents); to improve reproduction, producing better offspring.

In this study, we used an arithmetic crossover operator [13,23] that creates new offspring from two parent chromosomes using a linear combination and following the equations:

$$\begin{cases} Offspring1 = a * Parent1 + (1-a) * Parent2 \\ Offspring2 = (1-a) * Parent1 + a * Parent2 \end{cases} \tag{33.9}$$

where a is the weight factor with a random value between 0 and 1 which control individual dominance in reproduction. For each gene this operation is executed separately.

Consider two parents selected for crossover consisting of four floats genes for each:

$$\begin{array}{ll} Parent\ 1: & |\,0.2\,|\,1.4\,|\,0.2\,|\,7.5\,| \\ Parent\ 2: & |\,0.5\,|\,3.5\,|\,0.1\,|\,6.6\,| \end{array}.$$

If $a = 0.6$, two offspring are produced as follows:

$$\begin{array}{ll} Offspring\ 1: & |\,0.32\,|\,2.24\,|\,0.16\,|\,7.14\,| \\ Offspring\ 2: & |\,0.38\,|\,2.66\,|\,0.14\,|\,6.96\,| \end{array}.$$

By integrating arithmetic crossover operator into the WOA, the improved whale optimization algorithm (IWOA) is summarized in Fig. 33.1. The pseudo code of the IWOA algorithm begins with a random initialization of population. Then, it evaluates the fitness of each search agent to assign the best one and update other agent positions in the neighborhood of this best search agent, in the phase of encircling. This phase is followed by exploration and exploitation stages. Crossover is applied to emphasize global search by producing new offspring in the search space and re-calculate the fitness to update the current best solution. These procedures are repeated until the end of iterations.

33.4 Mathematical models of SHTE

The objective is to minimize the total cost by optimizing the design parameters of STHE from an economic standpoint. The next subsections present the mathematical models for heat transfer, pressure drop in a heat exchanger and the formulation of objective function.

```
Create an initial population of whales
Initialize the main controlling parameters

while the end condition is not satisfied
      Evaluate each whale using the objective function
      Update X*
      Update a, A, and C

      for all whales in the population
            if p < 0.5
                  if A < -1 or A > 1
                        Update the whale using Eq. (33.1)
                  else
                        Update Xrand
                        Update the whale using Eq. (33.7)
                  end if
            else
                  Update the whale using Eq. (33.5)
            end if
      end for

      Perform crossover using Eq. (33.9)
end while

Return X*
```

FIGURE 33.1 Pseudo code of the IWOA algorithm.

33.4.1 Heat exchanger design formulation

The surface area of the heat exchanger can be calculated according to the following formula [24,25]:

$$S = \frac{Q}{U \cdot \Delta T_{LM} \cdot F} \tag{33.10}$$

where Q indicates the rate of heat transmission, ΔT_{LM} is the logarithmic average temperature difference for the counter current arrangement, U is the heat transfer coefficient, and F is the adjustment factor. The heat transfer rate Q is computed as follows:

$$Q = m_s C_{ps} (T_{is} - T_{os}) = m_t C_{pt} (T_{ot} - T_{it}) \tag{33.11}$$

where

$$d_i = 0.8 d_o \tag{33.12}$$

The heat transfer coefficient for the shell side h_s is calculated based on the segmented baffle tube heat exchanger formulation proposed by Kern [24]:

$$h_s = 0.36 \frac{K_s}{D_e} Re_s^{0.55} Pr_s^{\frac{1}{3}} \left(\frac{\mu_s}{\mu_w}\right)^{0.14} \tag{33.13}$$

where D_e represents the hydraulic diameter of the shell [24,25]:

- In case of a square tube:

$$D_e = \frac{4\left(P_t^2 - \left(\pi d_o^2/4\right)\right)}{\pi d_o} \tag{33.14}$$

- In case of a triangular tube:

$$D_e = \frac{4\left(0.43 P_t^2 - \left(0.5\pi d_o^2/4\right)\right)}{0.5\pi d_o} \tag{33.15}$$

The Prandtl number (Pr_s) and Reynolds number (Re_s) for the shell side are given by:

$$Pr_s = \frac{\mu_s \cdot C_{ps}}{K_s} \tag{33.16}$$

$$Re_s = \frac{\rho_s . \vartheta_s \cdot D_e}{\mu_s} \tag{33.17}$$

where ϑ_s is the velocity of the fluid for the shell side, expressed by [24,25]:

$$\vartheta_s = \frac{m_s}{a_s.\rho_s} \tag{33.18}$$

and a_s shows the cross sectional area normal to the direction of flow [24,25]:

$$a_s = \frac{D_s \cdot b \cdot C_1}{\rho_s} \tag{33.19}$$

with C_1 is the clearance of the shell given by:

$$C_1 = P_t - d_o \tag{33.20}$$

The heat transfer coefficient on the tube side h_t is measured using the following equations [10]:

$$h_t = \frac{K_t}{d_i}\left[3.657 + \frac{0.0677\left(Re_t Pr_t \frac{d_i}{L}\right)^{1.33}}{1+0.1P_r\left(Re_t\frac{d_i}{L}\right)^{0.3}}\right] \quad (Re_t < 2300) \tag{33.21}$$

$$h_t = \frac{K_t}{d_i}\left\{\frac{\frac{f_t}{8}(Re_t - 1000)\,Pr_t}{1+12.7\sqrt{\frac{f_t}{8}}\left(Pr_t^{\frac{2}{3}} - 1\right)}\left[1+\left(\frac{d_i}{L}\right)^{0.67}\right]\right\} \quad (2300 < Re_t < 10000) \tag{33.22}$$

$$h_t = 0.027\frac{K_t}{D_t}Re_t^{0.8}\,Pr_t^{\frac{1}{3}}\left(\frac{\mu_t}{\mu_w}\right)^{0.14} \quad (Re_t > 10000) \tag{33.23}$$

where f_t is the factor of Darcy friction given by [1]:

$$f_t = 1/(1.82\log_{10} Re_t - 1.64)^2 \tag{33.24}$$

The numbers Pr_t and Re_t on the tube side given by Prandtl and Reynolds respectively are calculated by:

$$Pr_t = \frac{\mu_t.C_{p_t}}{k_t} \tag{33.25}$$

$$Re_t = \frac{\rho_t \cdot \vartheta_t \cdot D_t}{\mu_t} \tag{33.26}$$

and the flow velocity for tube side ϑ_t is found by:

$$\vartheta_t = \frac{m_t}{\frac{\pi d_t^2}{4}\rho_s} \cdot \frac{n_t}{N_t} \tag{33.27}$$

where n_t represents the number of tube passes and N_t denotes the number of tubes [24,25]:

$$N_t = K_1\left(\frac{D_s}{d_o}\right)^{n1} \tag{33.28}$$

where K_1 and n_1 are coefficients with values dependent on the number of passes and the flow arrangement. Based on total heat exchanger surface area S, the necessary tube length L is determined by the following equation:

$$L = \frac{S}{\pi \cdot d_o.N_t} \tag{33.29}$$

Considering the cross-flow between adjacent baffles, the logarithmic mean temperature difference ΔT_{LM} in Eq. (33.10) is determined according to the equation:

$$\Delta T_{LM} = \frac{(T_{is} - T_{ot}) - (T_{os} - T_{it})}{\ln\left[\frac{(T_{is}-T_{ot})}{(T_{os}-T_{it})}\right]} \tag{33.30}$$

In Eq. (33.10) the correction factor F for the flow configuration is a function of dimensionless temperature ratio, for most flow configuration of interest, and can be formulated as follows:

$$F = \frac{\sqrt{R^2-1}}{R-1} \cdot \frac{\ln\left(\frac{1-G}{1-GR}\right)}{\ln\left[\frac{2-G\left(R+1-\sqrt{R^2-1}\right)}{2-G\left(R+1+\sqrt{R^2-1}\right)}\right]} \tag{33.31}$$

where R represents the correction coefficient given by:

$$R = \frac{T_{is} - T_{os}}{T_{ot} - T_{it}} \tag{33.32}$$

and η is the pumping power calculated using the formula:

$$\eta = \frac{T_{ot} - T_{it}}{T_{is} - T_{it}} \tag{33.33}$$

33.4.2 Pressure drop

The tube side pressure drop in heat exchanger is the static fluid pressure, which may be expended to drive the fluid through the exchanger. To calculate the pressure drop, we consider the sum of distributed pressure drop along the tubes length and concentrated pressure losses in elbows and in the inlet and outlet nozzles from Kern [24]:

$$\Delta P_t = \Delta P_{tubelength} + \Delta P_{tubeelbow} = \frac{\rho_t \vartheta_t^2}{2}\left(\frac{L}{d_t} f_t + p\right) \cdot n \tag{33.34}$$

Kern [24] considered the value of the constant $p = 4$ meanwhile, Sinnot et al. [25] assumed $p = 2.5$.

The shell side pressure drop can be expressed by:

$$\Delta P_s = f_s \frac{\rho_s \vartheta_s^2}{2}\left(\frac{L}{B}\right) \cdot \left(\frac{D_s}{D_e}\right) \tag{33.35}$$

where the friction factor (f_s) is given by:

$$f_s = 2b_o Re_s^{-0.15} \tag{33.36}$$

and $b_o = 0.72$ [26] valid for $Re_s < 40000$.

33.4.3 Objective function

In this study the objective function considered is the total cost C_{total}, which includes capital investment (C_{inv}), annual operating cost (C_{annual}), energy cost (C_E), and total discounted operating cost (C_{total_disc}) [10]. This optimization function is subjected to design variables as the tubes outside diameter, shell inside diameters, baffles spacing, and number of tubes.

$$Minimize\ C_{total} = C_{inv} + C_{total\_disc} \tag{33.37}$$

As suggested by Taal et al. [27], the Hall's correlation is used adopted and the capital investment (C_{inv}) is calculated as a function of the heat exchanger surface according to:

$$C_{inv} = a_1 + a_2 S^{a_3} \tag{33.38}$$

where $a_1 = 8000$, $a_2 = 259.2$, and $a_3 = 0.91$ are numerical constants for shells and tubes made of stainless steel [27].

The net discounted operating cost (C_{total_disc}) is dependent on the pumping power to overcome frictional losses and is computed as follows:

$$C_{total\_disc} = \sum_{k=1}^{n_y} \frac{C_{annual}}{(I+1)^k} \tag{33.39}$$

TABLE 33.1 The process input and physical properties for different case studies [10].

	Mass flow (kg/s)	T_{input} (°C)	T_{output} (°C)	ρ (kg/m³)	C_p (kJ/kg K)	μ (Pas)	k (W/m K)	Rf (m² K/W)
Case 1:								
Shell side: methanol	27.80	95.00	40.00	750.00	2.84	0.00034	0.19	0.00033
Tube side: sea water	68.90	25.00	40.00	995.00	4.20	0.0008	0.59	0.0002
Case 2:								
Shell side: kerosene	5.52	199.00	93.30	850.00	2.47	0.0004	0.13	0.00061
Tube side: crude oil	18.80	37.80	76.70	995.00	2.05	0.00358	0.13	0.00061
Case 3:								
Shell side: distilled water	22.07	33.90	29.40	995.00	4.18	0.0008	0.62	0.00017
Tube side: raw water	35.31	23.90	26.70	999.00	4.18	0.00092	0.62	0.00017

and

$$C_{annual} = G \cdot C_E \cdot A \tag{33.40}$$

where $C_E = 0,12\ e/kWh$, $A = 7000\ h/an$, $G = \frac{1}{\eta}(\frac{m_t}{\rho_t}\Delta P_t + \frac{m_s}{\rho_s}\Delta P_s)$, $\eta = 80\%$, $n_y = 10$ years, and annual discounted rate $I = 10\%$ [10,24,28].

33.5 Results and discussion

In this section, the performance of the proposed IWOA is evaluated by solving three case studies considered for the Economic Design and Optimization of STHE. In the first case [24] the heat exchanger works between the operating fluids methanol and brackish water, with an operating heat load of 4.34 MW. For the second case [24], the heat exchanger operates between kerosene and crude oil with an operating heat load of 1.44 MW. In the last case [10], the heat exchanger operates between distilled water and raw water with an operating heat load of 0.46 MW. Table 33.1 shows the original design specifications [10] taken as an input for the optimization algorithm IWOA. In each case, there are four optimization variables: tube outside diameter d_0, shell inside diameter D_s, baffle spacing b, and number of tube passes n. The values of the variables are bounded as follows:

$$\begin{aligned} 0.01 \le d_o \le 0.032 \\ 0.15 \le D_s \le 1.2 \\ 0.2 \le b \le 0.45 \\ 1 \le n \le 8 \end{aligned} \tag{33.41}$$

When realizing the experiment, the simulation parameters of the IWOA algorithm are 30 agents and 100 iterations. In order to check the efficiency of the IWOA algorithm its results are compared with some of those of best algorithms in the literature: GA [10], PSO [11], ABC [17], BBO [18], ITHS [19], and I-ITHS [20] as well as the design of the original study [24]. The following subsections report comparison results for each case study.

33.5.1 Case 1

In this case, the heat exchanger operates between two fluids, methanol and sea water, with a charge heat equal to 4.34 MW. The results obtained by IWOA algorithm and a comparison with the solutions available in the literature from the aforementioned algorithms for capital investment (C_{inv}), annual operating cost (C_{annual}), operated discounted cost (C_{total_disc}), the total cost (C_{total}), and some other design variables are depicted in Table 33.2.

From the results of Table 33.2, it is noticed that the IWOA algorithm obtained the best solutions compared to the other algorithms. IWOA algorithm was able to achieve an optimum design with the lowest total cost (C_{total}) than other existing algorithms. In addition, the heat exchanger area (S) was significantly reduced and so is the length L of exchanger, which results in a decrease in the tube diameter (d_o). When compared to the original design [24], GA [10], PSO [11], ABC [17], BBO [18], ITHS [19], and I-ITHS [20], the capital investment (C_{inv}) reduced by 21.41%, 17.82%, 12.85%, 9.15%, 9.10%, 8.62%, 8.53%, and 2.91% respectively. The annual operating cost (C_{annual}) was also reduced by 97.1% compared to the

TABLE 33.2 Comparison of IWOA with other existing algorithms solving Case 1 for design and economic optimization of STHE.

Parameters	Original study	GA	PSO	ABC	BBO	ITHS	I-ITHS	IWOA
D_s (m)	0.894	0.83	0.81	1.3905	0.801	0.762	0.7635	0.92036
L (m)	4.83	3.379	3.115	3.963	2.04	2.0791	2.0391	1.1959
b (m)	0.356	0.5	0.424	0.4669	0.5	0.4988	0.4955	0.4333
d_o (m)	0.02	0.016	0.015	0.0104	0.01	0.0101	0.01	0.010
P_t (m)	0.025	0.02	0.0187	–	0.0125	0.1264	0.0125	0.0125
C_1 (m)	0.005	0.004	0.0037	–	0.0025	0.0253	0.0025	0.0025
n	2	2	2	2	2	2	2	2
N_t	918.0	1567.0	1658.0	1528.0	3587	3454	3558	5378.4401
v_t (m/s)	0.75	0.69	0.67	0.36	0.77	0.782	0.7744	0.51227
Re_t	14925.0	10936.0	10503.0	–	7642.49	7842.52	7701.29	5097.0985
Pr_t	5.7	5.7	5.7	–	5.7	5.7	5.7	5.6949
h_t (W/m^2 K)	3812.0	3762.0	3721.0	3818.0	4314	4415.918	4388.79	12282.8419
f_t	0.028	0.031	0.0311	–	0.034	0.0354	0.03555	0.038337
ΔP_t (Pa)	6251.0	4298.0	4171.0	3043.0	6156	6998.7	6887.63	2540.7597
a_s (m^2)	0.032	0.083	0.0687	–	0.0801	0.07602	0.07567	0.079759
D_e (m)	0.014	0.011	0.0107	–	0.007	0.00719	0.00711	0.0071092
v_s (m/s)	0.58	0.44	0.53	0.118	0.46	0.48755	0.48979	0.46474
Re_s	18381.0	11075.0	12678.0	–	7254 12	7736.89	7684.054	7287.9633
Pr_s	5.1	5.1	5.1	–	5.1	5.08215	5.08215	5.0821
h_s (W/m^2 K)	1573.0	1740.0	1950.8	3396.0	2197	2213.89	2230.913	2075.8019
f_s	0.33	0.357	0.349	–	0.379	0.3759	0.37621	0.37929
ΔP_s (Pa)	35789.0	13267.0	20551.0	8390.0	13799	14794.94	14953.91	10976.0297
U (W/m^2 K)	615.0	660.0	713.9	832.0	755	760.594	761.578	859.4687
S (m^2)	278.6	262.8	243.2	–	229.95	228.32	228.03	202.0624
C_{inv} (€)	51507.0	49259.0	46453.0	44559.0	44536	44301.66	44259.01	40480.8034
C_{annual} (€/year)	21111.0	947.0	1038.7	1014.5	984	964.164	962.4858	611.922
C_{total_disc} (€)	12973.0	5818.0	6778.2	6233.8	6046	5924.343	5914.058	3759.9959
C_{total} (€)	64480.0	55077.0	53231.0	50793.0	50582	50226	50173	44240.7992

original study and decreased by 35.38%, 41.09%, 39.68%, 37.81%, 36.53%, and 36.42%, compared to GA, PSO, ABC, BBO, ITHS, and I-ITHS, respectively.

Due to the reduction in capital investment (C_{inv}) and discounted operation cost (C_{total_disc}), the total cost (C_{total}), improved when compared to the original design [24], GA [10], PSO [11], ABC [17], BBO [18], ITHS [19], and I-ITHS [20] by 31.39%, 19.67%, 16.90%, 12.90%, 12.54%, 11.92%, and 11.82%, respectively.

For Case 1, Fig. 33.2 illustrates a comparison of the different costs calculated by the proposed IWOA algorithm and the other competitive algorithms. Comparing IWOA solutions to those of the original study [24], a considerable reduction of 21.41% in capital investment (C_{inv}), 97.1% in annual operating cost (C_{annual}), 71.02% in total discounted operating cost (C_{total_disc}), and 31.39% in total cost (C_{total}) is observed. The evolution of the total cost objective function (C_{total}) is shown in Fig. 33.3.

33.5.2 Case 2

The second case considers a heat exchanger operating between two fluids, methanol and sea water, with a charge heat equal to 1.44 MW. Table 33.3 depicted the results obtained by IWOA algorithm and other competitive algorithms for capital investment (C_{inv}), annual operating cost (C_{annual}), operated discounted cost (C_{total_disc}), the total cost (C_{total}), and some other design variables. As demonstrated in Table 33.3, the IWOA algorithm was able to develop a design with a significantly

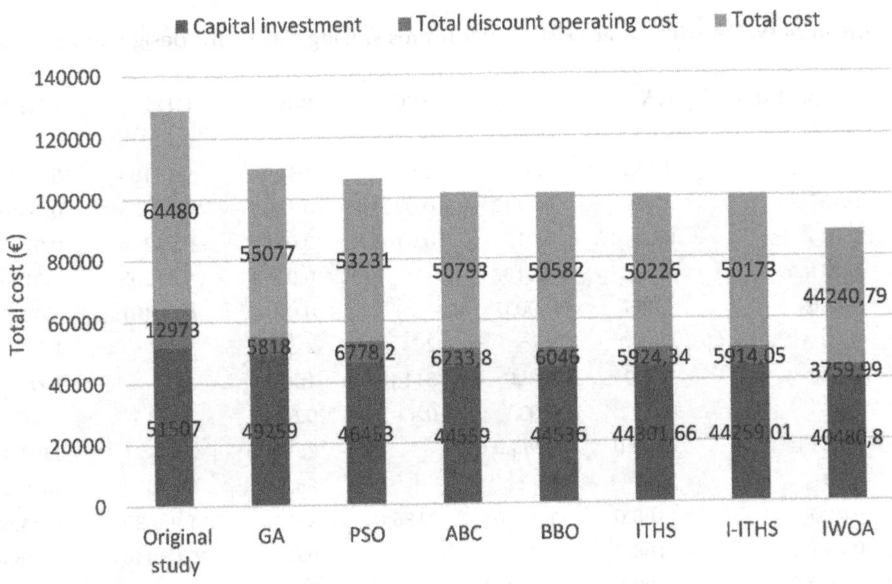

FIGURE 33.2 Different costs comparison for Case 1.

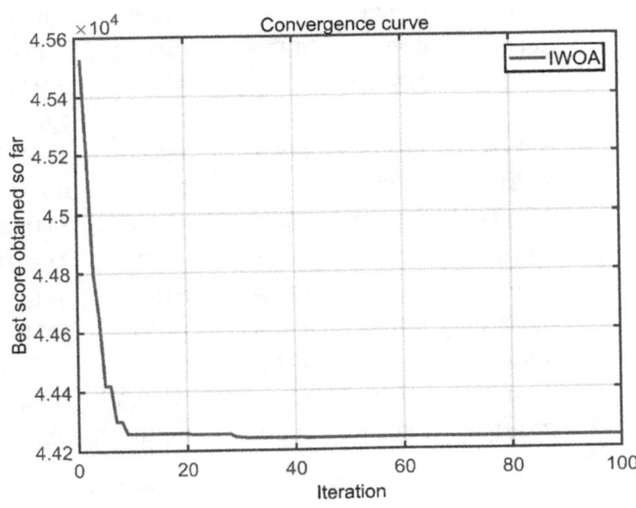

FIGURE 33.3 Convergence curve of IWOA for Case 1.

lower cost (C_{total}) than other existing techniques. The total cost (C_{total}) obtained using IWOA algorithm decreased from the original design [24], GA [10], PSO [11], ABC [17], BBO [18], ITHS [19], and I-ITHS [20] by 31.97%, 9.46%, 7.73%, 9.11%, 7.20%, 9.05%, and 8.92%, respectively. The IWOA method produced superior results than the other algorithms tested.

Fig. 33.4 represents a comparison of the different costs calculated by the proposed IWOA algorithm and the other competitive algorithms for Case 2. Comparing IWOA solutions to those of the original study [24], a considerable reduction of 10.53% in capital investment (C_{inv}), 83.25% in annual operating cost (C_{annual}), 83.25% in total discounted operating cost (C_{total_disc}), and 31.97% in total cost (C_{total}) is observed. The evolution of the total cost objective function (C_{total}) is shown in Fig. 33.5.

33.5.3 Case 3

In this last case, the heat exchanger operates between distilled water and raw water, with a charge heat equal to 0.46 MW. Table 33.4 resumes the results obtained by IWOA algorithm and other algorithms for some design variables, capital in-

TABLE 33.3 Comparison of IWOA with other existing algorithms solving Case 2 for design and economic optimization of STHE.

Parameters	Original study	GA	PSO	ABC	BBO	ITHS	I-ITHS	IWOA
D_s (m)	0.539	0.63	0.59	0.3293	0.74	0.32079	0.31619	0.47562
L (m)	4.88	2.153	1.56	3.6468	1.199	5.15184	5.06235	3.2526
b (m)	0.127	0.12	0.1112	0.0924	0.1066	0.24725	0.24147	0.021952
d_o (m)	0.025	0.02	0.015	0.0105	0.015	0.01204	0.01171	0.15703
P_t (m)	0.031	0.025	0.0187	–	0.0188	0.01505	0.01464	0.027441
C_1 (m)	0.006	0.005	0.0037	–	0.0038	0.00301	0.00293	0.0054788
n	4	4	2	2	2	1	1	2
N_t	158.0	391.0	646.0	511.0	1061	301	309	220.9339
v_t (m/s)	1.44	0.87	0.93	0.43	0.69	0.8615	0.8871	0.7061
Re_t	8227.0	4068.0	3283.0	–	2298	2306.77	2303.46	3446.5049
Pr_t	55.2	55.2	55.2	–	55.2	56.4538	56.4538	56.4538
h_t (W/m^2 K)	619.0	1168.0	1205.0	2186.0	1251	1398.85	1435.68	2836.4832
f_t	0.033	1168.0	0.044	–	0.05	0.04848	0.04854	0.043438
ΔP_t (Pa)	49245.0	14009.0	16926.0	1696.0	5109	10502.45	11165.45	5975.3039
a_s (m^2)	0.0137	0.0148	0.0131	–	0.0158	0.01585	0.01527	0.015003
D_e (m)	0.025	0.019	0.0149	–	0.0149	0.01188	0.01157	0.015606
v_s (m/s)	0.47	0.43	0.495	0.37	0.432	0.40948	0.42526	0.40956
Re_s	25281.0	18327.0	15844.0	–	13689	10345.29	10456.39	13582.2982
Pr_s	7.5	7.5	7.5	–	7.5	7.6	7.6	7.6
h_s (W/m^2 K)	920.0	1034.0	1288.0	868.0	1278	1248.86	1290.789	1065.9379
f_s	0.315	0.331	0.337	–	0.345	0.35987	0.35929	0.34547
ΔP_s (Pa)	24909.0	15717.0	21745.0	10667.0	15275	14414.26	15820.74	15546.6353
U (W/m^2 K)	317.0	376.0	409.3	323.0	317.75	326.071	331.358	363.4609
S (m^2)	61.5	52.9	47.5	61.566	60.35	58.641	57.705	49.5588
C_{inv} (€)	19007.0	17599.0	16707.0	19014.0	18799	18536.55	18383.46	17040.5554
C_{annual} (€/year)	1304.0	440.0	523.3	197.139	164.414	272.576	292.7937	218.4095
C_{total_disc} (€)	8012.0	2704.0	3215.6	1211.3	1010.25	1674.86	1799.09	1342.0318
C_{total} (€)	27020.0	20303.0	19922.6	20225.0	19810	20211	20182	18382.5872

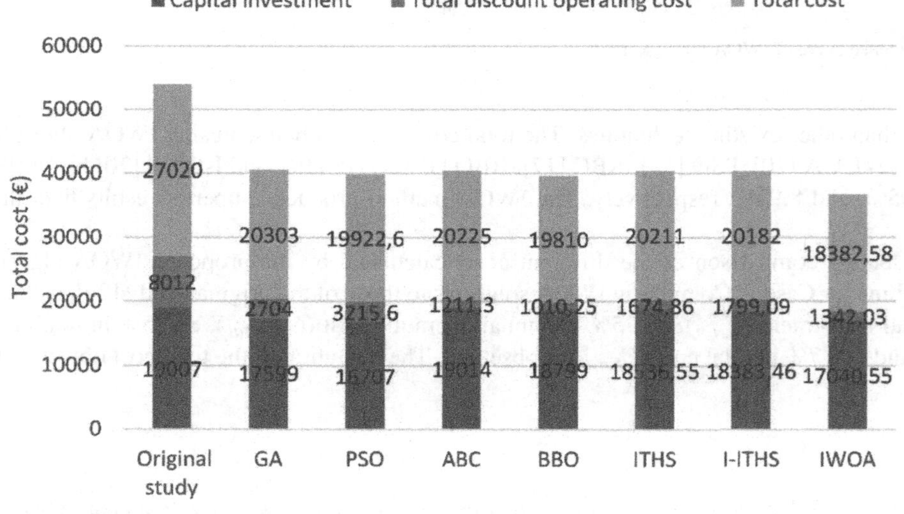

FIGURE 33.4 Different costs comparison for Case 2.

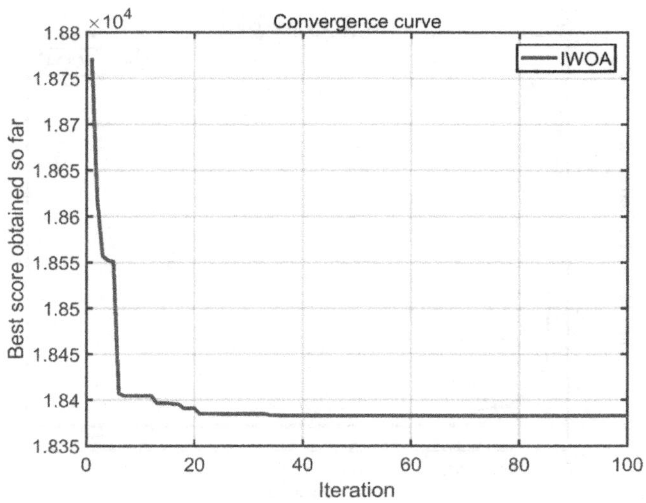

FIGURE 33.5 Convergence curve of IWOA for Case 2.

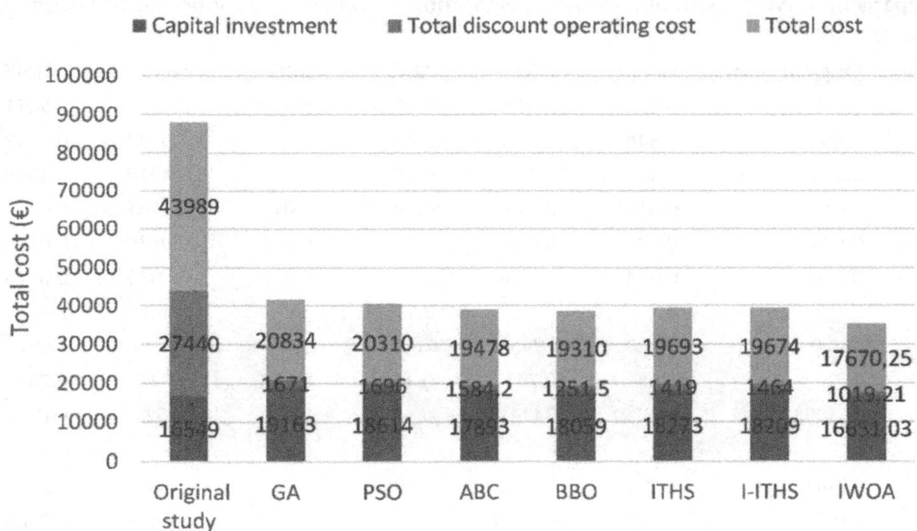

FIGURE 33.6 Different costs comparison for Case 3.

vestment (C_{inv}), annual operating cost (C_{annual}), operated discounted cost (C_{total_disc}), and the total cost (C_{total}). From Table 33.4 it is clear that IWOA algorithm was able to find the best design with the lowest cost (C_{total}) than other existing techniques. It is observed that the shell diameter (Ds) increased by 66.06% which causes a considerable reduction in the tube's length (L) by 87.35% when compared to original study [24]. The total cost (C_{total}) obtained using IWOA algorithm decreased from the original design [24], GA [10], PSO [11], ABC [17], BBO [18], ITHS [19], and I-ITHS [20] by 59.83%, 15.18%, 13.0%, 9.28%, 8.49%, 10.27%, and 10.18% respectively. The IWOA method produced superior results than the other algorithms tested.

Fig. 33.6 represents a comparison of the different costs calculated by the proposed IWOA algorithm and the other competitive algorithms for Case 3. Analyzing Fig. 33.6 it is found that the capital investment (C_{inv}) increased by 0.62%, the operational cost (C_{annual}) reduced by 96.29%, the total discounted operating cost (C_{total_disc}) decreased by 96.28%, and the total cost (C_{total}) improved by 59.83%. The convergence plot achieved in the simulation for minimizing the total cost (C_{total}) is shown in Fig. 33.7.

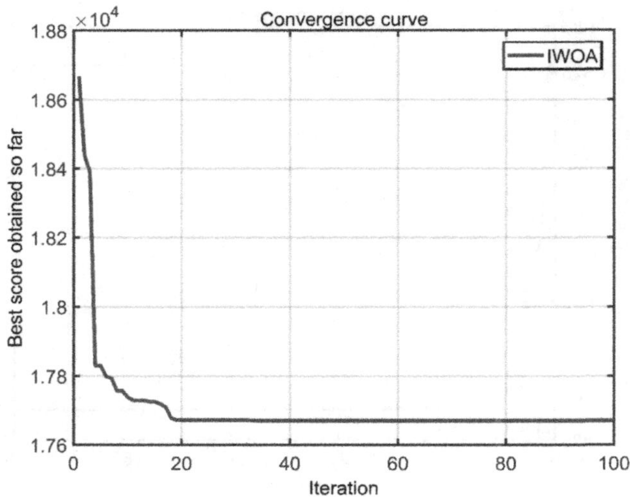

FIGURE 33.7 Convergence curve of IWOA for Case 3.

TABLE 33.4 Comparison of IWOA with other existing algorithms solving Case 3 for design and economic optimization of STHE.

Parameters	Original study	GA	PSO	ABC	BBO	ITHS	I-ITHS	IWOA
D_s (m)	0.387	0.62	0.0181	1.0024	0.55798	0.5726	0.5671	0.64267
L (m)	4.88	1.548	1.45	2.4	1.133	0.9737	0.9761	0.61728
b (m)	0.305	0.44	0.423	0.354	0.5	0.4974	0.4989	0.01
d_o (m)	0.019	0.016	0.0145	0.103	0.01	0.0101	0.01	0.46953
P_t (m)	0.023	0.02	0.0187	–	0.0125	0.0126	0.0125	0.0125
C_1 (m)	0.004	0.004	0.0036	–	0.0025	0.0025	0.0025	0.0025
n	2	2	2	2	2	2	2	2
N_t	160.0	803.0	894.0	704.0	1565	11845	1846	2434.8781
v_t (m/s)	1.76	0.68	0.74	0.36	0.898	0.747	0.761	0.57758
Re_t	36409.0	9487.0	9424.0	–	7804	6552	6614	5017.4471
Pr_t	6.2	6.2	6.2	–	6.2	6.2	6.2	6.2026
h_t (W/m^2 K)	6558.0	6043.0	5618.0	4438.0	9180	5441	5536	13093.579
f_t	0.023	0.031	0.0314	–	0.0337	0.0369	0.0368	0.038524
ΔP_t (Pa)	62812.0	3673.0	4474.0	2046.0	4176	3869	4049	2323.7134
a_s (m^2)	0.0236	0.0541	0.059	–	0.0558	0.0569	0.0565	0.060354
D_e (m)	0.013	0.015	0.01	–	0.0071	0.0071	0.0071	0.0071
v_s (m/s)	0.94	0.41	0.375	0.12	0.398	0.3893	0.3919	0.36751
Re_s	16200.0	8039.0	4814.0	–	3515	3473	3461	3249.5513
Pr_s	5.4	5.4	5.4	–	5.4	5.4	5.4	5.3935
h_s (W/m^2 K)	5735.0	3476.0	4088.3	5608.0	4911	4832	4871	4994.9289
f_s	0.337	0.374	0.403	–	0.423	0.4238	0.4241	0.42814
ΔP_s (Pa)	67684.0	4365.0	4271.0	27166.0	5917	4995	5062	3419.2171
U (W/m^2 K)	1471.0	1121.0	1177.0	1187.0	1384	1220	1229	1474.5572
S (m^2)	46.6	62.5	59.2	54.72	55.73	57.3	56.64	47.2174
C_{inv} (€)	16549.0	19163.0	18614.0	17893.0	18059	18273	18209	16651.0393
C_{annual} (€/year)	4466.0	272.0	276.0	257.82	203.68	231	238	165.8725
C_{total_disc} (€)	27440.0	1671.0	1696.0	1584.2	1251.5	1419	1464	1019.2145
C_{total} (€)	43989.0	20834.0	20310.0	19478.0	19310	19693	19674	17670.2538

33.6 Conclusion

In this study, an improved whale optimization algorithm (IWOA) is employed to design shell and tube heat exchanger (STHE) from an economic viewpoint. IWOA benefits from good convergence speed to reach the global optima due to the crossover operator. For shell and tube heat exchanger design, the total cost is taken as the objective function to minimize. Three case studies are solved using the proposed IWOA and compared to current approaches from literature. Comparison of the results revealed that IWOA achieved the lowest costs for capital investment, annual operating cost and total discounted operating cost. When comparing to the original study, the total cost reduces significantly for Case 1, Case 2, and Case 3 by: 31.39%, 31.97%, and 59.83% respectively. The proposed algorithm IWOA has proved good performance to find the optimal design for shell and tube heat exchanger with low cost and high efficacy.

References

[1] G.F. Hewitt, Heat Exchanger Design Handbook, Begell House, 1998.

[2] K. Thulukkanam, Heat Exchanger Design Handbook, CRC Press, 2000.

[3] M. Serna, A. Jimenez, An efficient method for the design of shell and tube heat exchangers, Heat Transfer Engineering 25 (2) (2004) 5–16.

[4] M. Reppich, S. Zagermann, A new design method for segmentally baffled heat exchangers, Computers & Chemical Engineering 19 (1995) 137–142.

[5] T. Poddar, G. Polley, Heat exchanger design through parameter plotting, Chemical Engineering Research and Design 74 (8) (1996) 849–852.

[6] E.S. Gaddis, V. Gnielinski, Pressure drop on the shell side of shell-and-tube heat exchangers with segmental baffles, Chemical Engineering and Processing: Process Intensification 36 (2) (1997) 149–159.

[7] E.A.D. Saunders, Heat Exchangers: Selection, Design and Construction, Longman Scientific & Technical, John Wiley & Sons, Harlow, New York, 1988.

[8] Y.A. Kara, O. Güraras, A computer program for designing of shell-and-tube heat exchangers, Applied Thermal Engineering 24 (13) (2004) 1797–1805.

[9] V.C. Mariani, A.R.K. Duck, F.A. Guerra, L. dos Santos Coelho, R.V. Rao, A chaotic quantum-behaved particle swarm approach applied to optimization of heat exchangers, Applied Thermal Engineering 42 (2012) 119–128.

[10] A.C. Caputo, P.M. Pelagagge, P. Salini, Heat exchanger design based on economic optimisation, Applied Thermal Engineering 28 (10) (2008) 1151–1159.

[11] V. Patel, R. Rao, Design optimization of shell-and-tube heat exchanger using particle swarm optimization technique, Applied Thermal Engineering 30 (11–12) (2010) 1417–1425.

[12] R. Selbaş, O. Kızılkan, M. Reppich, A new design approach for shell-and-tube heat exchangers using genetic algorithms from economic point of view, Chemical Engineering and Processing: Process Intensification 45 (4) (2006) 268–275.

[13] D. Mokeddem, S. Mirjalili, Improved whale optimization algorithm applied to design PID plus second-order derivative controller for automatic voltage regulator system, Journal of the Chinese Institute of Engineers 43 (6) (2020) 541–552.

[14] S. Sivanandam, S. Deepa, Genetic algorithms, in: Introduction to Genetic Algorithms, Springer, 2008, pp. 15–37.

[15] H. Najafi, B. Najafi, P. Hoseinpoori, Energy and cost optimization of a plate and fin heat exchanger using genetic algorithm, Applied Thermal Engineering 31 (10) (2011) 1839–1847.

[16] D. Karaboga, B. Gorkemli, C. Ozturk, N. Karaboga, A comprehensive survey: Artificial Bee Colony (ABC) algorithm and applications, Artificial Intelligence Review 42 (1) (2014) 21–57.

[17] A.Ş. Sahin, B. Kılıc, U. Kılıc, Design and economic optimization of shell and tube heat exchangers using Artificial Bee Colony (ABC) algorithm, Energy Conversion and Management 52 (11) (2011) 3356–3362.

[18] A. Hadidi, A. Nazari, Design and economic optimization of shell-and-tube heat exchangers using biogeography-based (BBO) algorithm, Applied Thermal Engineering 51 (1–2) (2013) 1263–1272.

[19] P. Yadav, R. Kumar, S.K. Panda, C. Chang, An intelligent tuned harmony search algorithm for optimisation, Information Sciences 196 (2012) 47–72.

[20] O.E. Turgut, M.S. Turgut, M.T. Coban, Design and economic investigation of shell and tube heat exchangers using improved intelligent tuned harmony search algorithm, Ain Shams Engineering Journal 5 (4) (2014) 1215–1231.

[21] S. Mirjalili, A. Lewis, The whale optimization algorithm, Advances in Engineering Software 95 (2016) 51–67.

[22] M. Furqan, H. Hartono, E. Ongko, M. Ikhsan, Performance of arithmetic crossover and heuristic crossover in genetic algorithm based on alpha parameter, IOSR Journal of Computer Engineering (IOSR-JCE) 19 (1) (2017) 31–36.

[23] O. Köksoy, T. Yalcinoz, Robust Design using Pareto type optimization: a genetic algorithm with arithmetic crossover, Computers & Industrial Engineering 55 (1) (2008) 208–218.

[24] D.Q. Kern, Process heat transfer, Tech. Rep., 1950.

[25] R.K. Sinnott, J.M. Coulson, J.F. Richardson, Chemical Engineering Design, vol. 6, Elsevier Butterworth-Heinemann, Oxford, 2005.

[26] M.S. Peters, K.D. Timmerhaus, Plant Design and Economics for Chemical Engineers, McGraw-Hill, New York, 1991.

[27] M. Taal, I. Bulatov, J. Klemes, P. Stehlik, Cost estimation and energy price forecast for economic evaluation of retrofit projects, Applied Thermal Engineering 23 (14) (2003) 1819–1835.

[28] V.H. Iyer, S. Mahesh, R. Malpani, M. Sapre, A.J. Kulkarni, Adaptive Range Genetic Algorithm: a hybrid optimization approach and its application in the design and economic optimization of Shell-and-Tube Heat Exchanger, Engineering Applications of Artificial Intelligence 85 (2019) 444–461.

33.6 Conclusion

References

Chapter 34

Whale-optimized convolutional neural network for potato fungal pathogens disease classification

D.N. Kiran Pandiri, R. Murugan, and Tripti Goel
Department of Electronics and Communication Engineering, National Institute of Technology Silchar, Assam, India

34.1 Introduction

Agriculture is the practice of cultivating plants and livestock. For many years, agriculture has been practiced by our ancestors and us until now. Agriculture, with its allied sectors, is the largest source of livelihood in many countries across the globe. Across the globe, food consumption is increasing with the increase in population around the world and food production through crops and plantations is reducing due to the diseases caused by insecticides, pesticides, and fungal pathogens. Every year 20–40 per cent of crop loss occurs due to the damage wrought by pests and diseases. To meet the demand and supply of the food chain across the globe, crop yield losses due to plant leaf diseases must be reduced [1]. This chapter concentrated on potato plant leaves diseases among various crops.

The potato has become a staple food in backwards nations and is considered the most important vegetable crop globally. India is the second largest producer nation of potatoes, followed by China. The potato crop is affected by bacterial and fungal pathogens at different stages of crop growth [2]. Considering the importance of potato crops, it is essential to study the fungal pathogens that cause EB [3] and LB [4] leaf diseases. Continuous monitoring of the crop to identify diseases using traditional approaches or by agricultural professionals has become a financial burden to the farmers and consumes a lot of time and resources. To aid the farmers in monitoring the leaf diseases, computer vision techniques [5] and DL algorithms can help detect the plant leaf diseases at a low cost and with fewer resources.

The book chapter discusses the EB, LB, and healthy leaves images of the potato plant that are publicly available. This chapter explains the convolutional neural network (CNN) used in classifying potato plant leaf diseases caused by fungal pathogens. The chapter discusses the application of the Whale optimization algorithm (WOA) for optimizing the performance of the CNN model in the training process. To achieve this, the hyperparameter of the CNN is optimized using WOA.

34.2 Fungal pathogens

Approximately 160 diseases influence the potato crop yield during different crop growth stages. They may harm tubers, foliage, or both. Pathogens that thrive in certain environments might destroy a crop. Among these diseases, fungal pathogens like EB and LB are major fungal diseases that affect potato crops [6].

Globally, LB cuts potato crop production by 15%, and Phytophthora infestans cause it. Initially, LB appears pale green and yellow at the margins and tips of the potato leaves and spreads to the stems of the plant. The tubers in the soil are affected by LB diseases. EB appears earlier than the LB, affects the crop at the earlier stages, and often affects the completely grown plants. EB is severe in environmental conditions where the climate changes from dry to wet more often. The loss caused due to EB may exceed the loss due to LB in these climatic conditions. These fungal pathogens can be controlled by using fungicides, but more often, usage of the fungicides leads to environmental pollution and threat to the human population. Fig. 34.1 shows the sample images of EB and LB of potato leaves.

Handbook of Whale Optimization Algorithm. https://doi.org/10.1016/B978-0-32-395365-8.00040-3

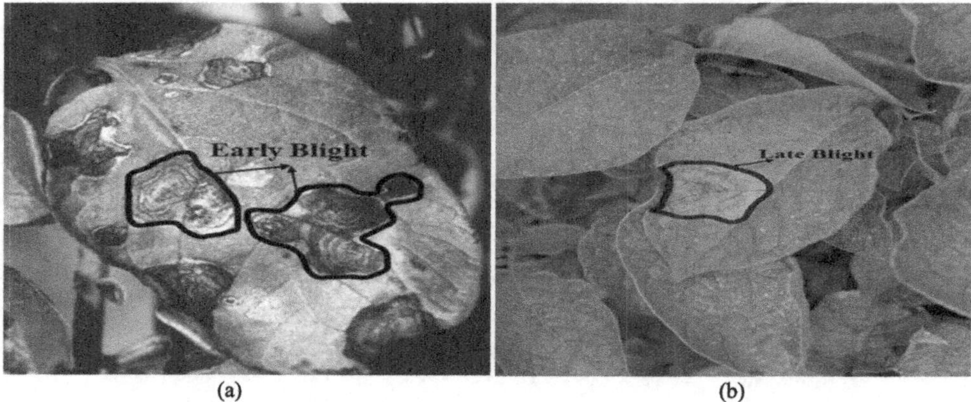

FIGURE 34.1 Fungal pathogen diseases of potato leaves (a) Early Blight. (b) Late Blight. (Source: [7].)

34.3 Database

To detect the EB and LB diseases of potato leaves in the early stages using images of the diseased leaves, publicly available potato leaves images database where the images were captured in the laboratory in certain lighting conditions [8]. Fig. 34.2 shows the sample images of the potato disease leaves database. The potato leaves dataset consists of 1500 images with three classes: EB, LB, and Healthy. The dataset is partitioned into 70:30 for training and testing the CNN model. The total number of images used is 1500, of which the number of images for EB, LB, and healthy is 500 for each label.

FIGURE 34.2 Database images of potato diseased leaves. (a) Early blight images at different stages. (b) Late blight images at different stages.

34.4 Artificial intelligence (AI)

AI is the ability of machines to replicate or enhance human intellect, such as reasoning and learning from experience. With the help of AI, researchers extend and expand human intelligence by developing technologies, methods, and applications for systems [9]. For the past two decades, AI has been introduced in various fields like automobiles, industries, medical, scientific research, and many others for better performance in detecting, identifying, and classifying objects with low cost and high precision and accuracy.

The sub-field of AI is deep learning (DL), where the approaches are based on artificial neural networks (ANN). ANNs [10] are inspired the by the neuron structure of the human brain. ANNs are comprised of three layers: input, hidden, and output layers, whereas DL models consist of multiple layers like convolutional layers, activation layers, pooling layers, softmax, and fully connected layers [11] [12]. DL models automatically learn the mathematical relations between the data by using hidden layers. Conventional and machine learning techniques depend on hand-crafted features, whereas DL models extract useful features from the input data. This reduces the time and costs of developing the model. The DL models perform better than the ML algorithms when the dataset of the model is increased. In this chapter, the general model in classifying the fungal pathogen diseases of the potato leaves by optimizing the CNN hyperparameters using the Whale optimization algorithm is shown in Fig. 34.3. The model to classify the potato fungal diseased leaves comprises the database, deep CNN architecture, a Whale optimizer to optimize the performance of the CNN architecture and the classification.

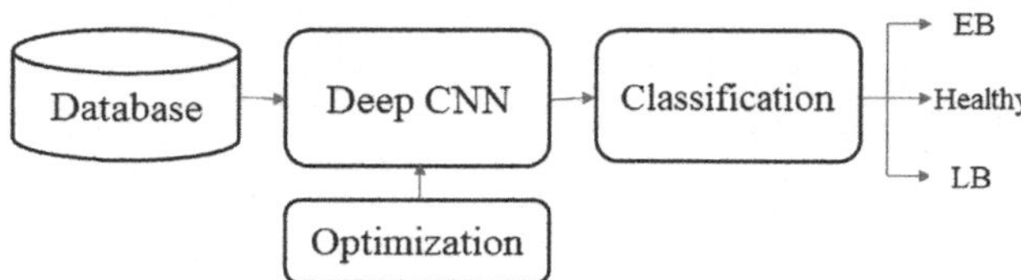

FIGURE 34.3 The general model for potato disease classification using optimized deep CNN.

34.5 Convolutional neural network

CNN's are widely utilized in object detection, classification, segmentation, image enhancement, image generation, and many other applications. The number of layers and filters of the CNN is decided by the researchers based on the application required to perform effectively. The architecture of CNN has layers of distinct functions. The convolutional layer (CL) converts the input data into deep feature maps. The CL operation is performed by sliding the filter over the input data [13]. To avoid over-fitting and to normalize the data, which accelerates the training process done by using the Batch Normalization layer [14]. The pooling layer is used to reduce the feature map dimensions and the number of parameters [15]. The fully connected layer flattens the features that represent the entire input data. CNN architecture is ended with the softmax layer in classification operation. Each class is identified based on the probability of the classes by the softmax activation layer. Several deep pre-trained CNN architectures were developed and trained on the ImageNet database. Some pre-trained networks are ResNet [16], VGG19 [17], AlexNet [18], EfficientNet [19], Inception [20], MobileNet [21], etc. Researchers used transfer learning [22], extreme learning [23], and ensemble learning techniques using pre-trained models to speed up the training process. At the same time, researchers designed their own CNN network to classify the objects based on the application requirement.

To improve the performance of the CNNs and pre-trained DL networks, the hyperparameters of the model have to be optimized. Optimizing the hyperparameters of the DL model by manual trial and error method is a challenging task and computationally cost-effective for researchers.

In this chapter, a deep CNN architecture has been designed and trained with the potato leaves disease images database. The deep CNN architecture consists of 22 layers with various filter sizes to create the feature map of the input data for classifying the images. The hyperparameters of the CNN architecture are optimized by using the WOA. The Whale optimized CNN architecture used in the classification of potato leaves is shown in Fig. 34.4.

34.6 Whale optimization algorithm

Real-world optimization problems involve varying complicated factors, including non-linearity, discontinuity, and other mixed variable types. Solving them using a classical algorithm approach is impractical and inefficient. To achieve optimal solutions by optimizing the parameters, alternate methods are to implement the stochastic methods. An approach which belongs to stochastic algorithms and uses a meta-heuristic algorithm is the Whale Optimization Algorithm (WOA) presented by Mirjalili and Lewis [24]. The WOA mimics nature by using the principles of biological or physical. Because of ease of implantation, high-speed computation, improved system performance, and effective utilization of resources, meta-heuristic

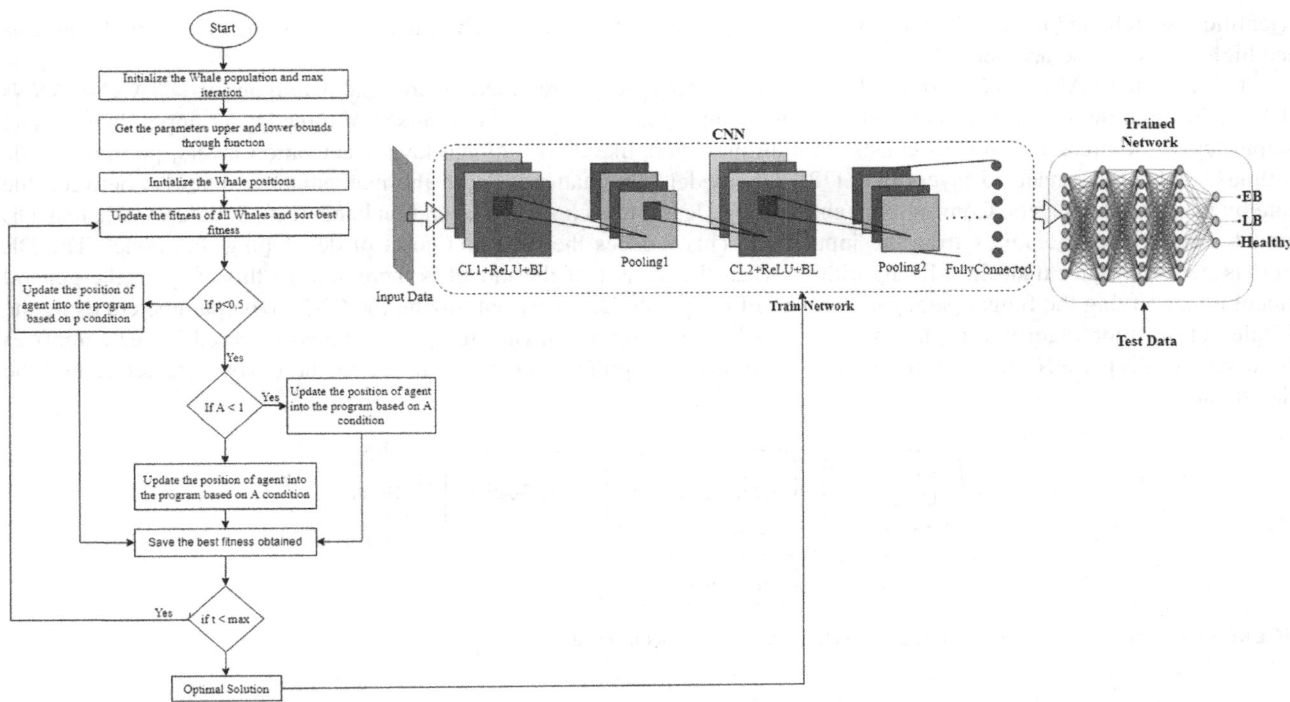

FIGURE 34.4 The schematic diagram of CNN architecture optimized the hyperparameters using WOA to classify the fungal pathogens of potato leaves.

algorithms are significant methods in industry applications, image processing, research science, and many other disciplines [25]. WOA is one of the meta-heuristic algorithms that is growing in multiple disciplines. During the process of exploration and exploitation, the WOA does not require additional tuning parameters, but it depends on a linear vector. At each iteration, the linear vector decreases from value 2 to 0.

WOA working: The WOA is inspired by the hunting behavior of humpback whales, where it creates a bubble net around the targeted prey called a bubble net prey feeding method. The bubble net attacking method is mathematically represented and implemented using a spiral equation to simulate the model. The whale population acts as search agents and searches for prey randomly. The steps involved in following the bubble net feeding method in obtaining the optimal solution for the CNN architecture are searching for the prey, encircling, and attacking the prey.

Encircling the prey: Once the humpback whale reaches the prey location, it encircles the prey. Initially, the whale agents assume that their current location is the best optimal position, and the other search agents update their position towards the best search agent to the optimal solution. Once the optimal solution is found, the search agents move towards the prey and explore the area around the prey for an optimal position than the previous position. The mathematical representation of this behavior is given by

$$P_{WA} = \left| A \cdot Z^*(t) - Z(t) \right| \tag{34.1}$$

$$Z(t+1) = Z^*(t) - B \cdot P_{WA} \tag{34.2}$$

where, P_{WA} is the whale search agent's current position, A and B are coefficients for the present iteration, Z is the position vector, and Z^* is the local optimal solution. The coefficients A and B are obtained from Eq. (34.3) and Eq. (34.4), respectively.

$$A = 2 \cdot r \tag{34.3}$$

$$B = (2 \cdot d \cdot r) - d \tag{34.4}$$

$$d = 2 - \frac{2 * t}{Max_{it}} \tag{34.5}$$

where, t is the current iteration and Max_{it} is the maximum number of iterations, r is the random number ranging from 0 to 1, and d is the parameter decremented linearly from 2 to 0. If $|B| \geq 1$, the whale search agent is far from the optimal solution and search agents explore the new position. If $|B| < 1$, then the optimal solution is achieved by the search agent.

Bubble net attacking: During bubble net attacking, the search agents choose either the exploitation or exploration phase.

34.6.1 Exploitation phase

In the exploitation phase, the search agents choose either to shrink the encircle or update the spiral position in each iteration to reach the prey or optimal solution. Based on the probability values, chooses any one of the approaches. In the shirking strategy, the encircling spiral shape is reduced as the search agent progress towards the optimal solution in each iteration. The value of d decreases to zero using Eq. (34.5). The positions of the search agents are updated based on $|B|$ value. This is done such that the search agent position is present between the random position and optimal position and is given by Eq. (34.6)

$$Z(t+1) = Z^*(t) - B \cdot P_{WA} \quad \text{if } prob < 0.5 \ (Shrinking\ encircle) \tag{34.6}$$

The search agents move in a helix shape in the spiral position updating approach. The mathematical form of the spiral equation is represented in Eq. (34.7).

$$Z(t+1) = P'_{WA} \cdot e^{s\alpha} \cdot \cos(2\pi s) + Z^*(t); \quad \text{if } prob \geq 0.5 \ (spiral\ update\ position) \tag{34.7}$$

where, $P'_{WA} = |Z^*(t) - Z(t)|$ represents the distance between ith iteration position to the best position or prey of the search agent, s indicates the shape of the spiral, which is constant, α denotes the random value in $[-1, 1]$. (See Figs. 34.5 and 34.6.)

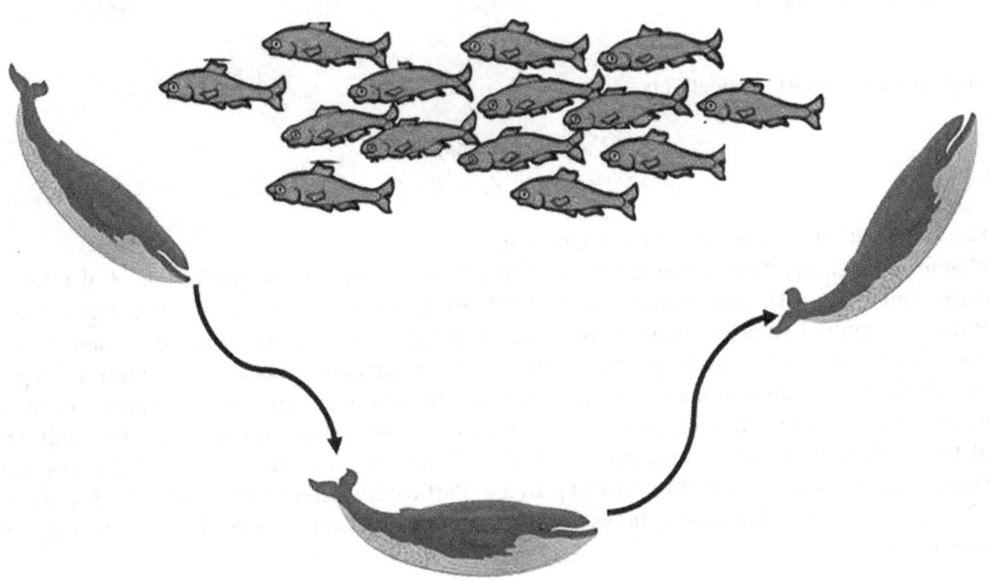

FIGURE 34.5 Whale search agents randomly searching for prey.

34.6.2 Exploration phase

To improve the search of prey (exploration) towards the refinement of the optimal solution, randomization of $|B|$ with values less than 1 or greater than or equal to 1. In the exploration phase, the whale search agents move towards the solution or away from the previous whale position, opposite to shrinking encircles (exploitation). This ensures diversification among the solutions obtained and reaches the global optimum solution. The exploration process is represented in equation form

FIGURE 34.6 Creating spiral shape bubbles by shrinking the circles to catch the prey.

TABLE 34.1 CNN architecture hyperparameter with non-optimized and optimized values.

Hyperparameters	Epochs	Learning rate	L2-regularization	Momentum
CNN	8	0.001	0.0001	0.9
CNN optimized using WOA	7	0.0045	0.0003	0.952

given by Eq. (34.8) and Eq. (34.9), respectively.

$$P_{WA} = |(A \cdot Z_r) - Z| \tag{34.8}$$

$$Z(t+1) = Z_r - B \cdot P_{WA} \tag{34.9}$$

where, Z_r is a random position chosen among the population.

This book chapter concentrated on the application of WOA in optimizing the performance of CNN architecture to classify the potato plants' EB, LB, and healthy leaves. CNN hyperparameters are optimized during the training phase with WOA to obtain the optimal values to increase the overall performance in terms of accuracy and precision. The CNN architecture is trained with 30 whale search agents to optimize the parameters: epoch, momentum, L2 regularization, and learning rate. The CNN architecture is trained 120 times to obtain the best fit values. For each epoch, the maximum number of iterations considered is 32. The performance of the CNN architecture is improved by using the optimized parameters. The accuracy of the CNN architecture is enhanced to 98.66%. The performance of the model is compared with other optimized algorithms: Slime Mould Algorithm (SMA) [26] and Particle Swarm Optimizer (PSO) [27], where the accuracy is 96.67% and 97.3%, respectively. Table 34.1 shows the hyperparameters that train the CNN architecture with and without optimization using WOA.

34.7 Performance analysis

The Whale optimized CNN architecture in classifying the fungal pathogens of potato leaves from healthy leaves is evaluated by using the performance metrics precision (P), recall (R), F1-score (F), and accuracy [28]. The performance metrics are calculated using the confusion matrices generated while testing the CNN architecture. Fig. 34.7 shows the confusion matrix of CNN without and with optimization using WOA and other optimized models. The performance metrics of CNN models are shown in Table 34.2.

By observing the precision, recall, and F1-score values in Table II, we can see that the performance of the CNN after optimizing the hyperparameters using WOA is increased in classifying the EB, LB, and healthy leaves. The overall perfor-

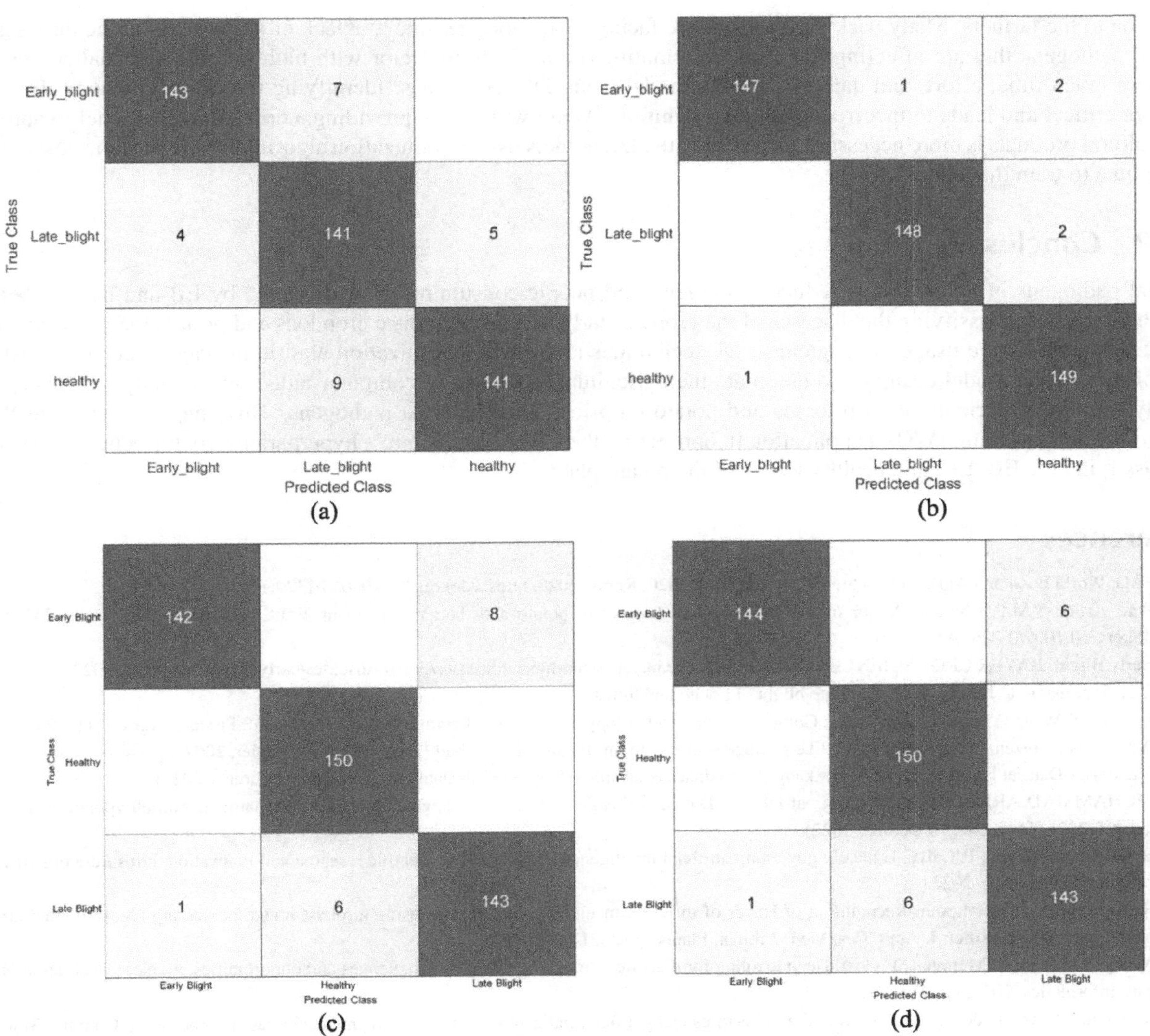

FIGURE 34.7 Confusion Matrix. (a) CNN. (b) CNN optimized with WOA. (c) CNN optimized using SMA. (d) CNN optimized using PSO.

TABLE 34.2 Performance metrics.

Method	EB			LB			Healthy		
	P	R	F	P	R	F	P	R	F
CNN	0.953	0.972	0.962	0.94	0.898	0.918	0.94	0.965	0.952
Optimized CNN with WOA	0.98	0.993	0.986	0.986	0.993	0.989	0.965	0.973	0.982
CNN optimized using SMA	0.946	0.993	0.969	0.953	0.947	0.95	1	0.961	0.98
CNN optimized using PSO	0.96	0.993	0.976	0.953	0.959	0.956	1	0.961	0.98

mance in classifying the fungal pathogens is increased by 4.22% after optimizing the hyperparameters of CNN using WOA and more efficient than SMA and PSO optimized algorithms.

34.8 Challenges

The world population is increasing, and at the same time, the agricultural crop is being affected by pests, insects, bacteria, and fungal pathogens. Monitoring and identifying diseases of the crop at every stage with the help of expertise have become a burden to the farmers. Many backward nations are facing food shortages due to a lack of knowledge on the diseases and fungal pathogens that are affecting the crop. Automating the agriculture sector with high precision is challenging as it requires much time, effort, and data to train the model using DL techniques. Identifying the infected among the leaves is more critical and leads to incorrect pattern recognition. Along with these, providing a cost-effective model to automate agricultural products is more necessary. Optimizing the DL models using optimization algorithms to reduce the cost requires more time to train the model.

34.9 Conclusion

Fungal pathogens in potato plants reduce crop yield, and people consuming potato affected by EB and LB has become health hazardous. Classifying the diseases of the crops at early stages can reduce crop loss and protect the environment by reducing the excessive usage of fungicides. DL techniques along with optimization algorithms can reduce the depth and complexity of the model designed to automate the agriculture sector using computer-aided image analysis. This chapter briefly introduces agriculture crop losses and potato crop loss due to fungal pathogens. This chapter covers the Whale Optimization Algorithm (WOA) application in optimizing the CNN architecture's hyperparameters for better performance in classifying the EB, LB, and healthy leaves of the potato plant.

References

[1] FAO, World Food and Agriculture - Statistical Yearbook 2020, Rome, 2020, https://doi.org/10.4060/cb1329en.

[2] R.K. Arora, S.M.P. Khurana, Major fungal and bacterial diseases of potato and their management, Fruit Veg. Dis. 1 (2006) 189–231, https://doi.org/10.1007/0-306-48575-3_6.

[3] Early Blight, BAYER CROP SCIENCE, https://cropscience.bayer.co.uk/threats/diseases/potato-diseases/early-blight-potatoes/, 2022.

[4] G.L. Schumann, C.J. Arcy, G.H. Cai, Late blight of potato and tomato, 2005.

[5] H. Tian, T. Wang, Y. Liu, X. Qiao, Y. Li, Computer vision technology in agricultural automation—a review, Inf. Process. Agric. 7 (1) (2020) 1–19.

[6] A.J. Termorshuizen, Fungal and fungus-like pathogens of potato, in: Potato Biology and Biotechnology, Elsevier, 2007, pp. 643–665.

[7] PlantifyDr Dataset | Kaggle, https://www.kaggle.com/datasets/lavaman151/plantifydr-dataset. (Accessed 9 February 2023).

[8] MUHAMMAD ARDI PUTRA, Potato Leaf Disease Dataset | Kaggle, https://www.kaggle.com/datasets/muhammadardiputra/potato-leaf-disease-dataset, 2021. (Accessed 10 October 2022).

[9] E. González Esteban, P. Calvo, Ethically governing artificial intelligence in the field of scientific research and innovation, https://doi.org/10.1016/j.heliyon.2022.e08946, 2022.

[10] R. Azadnia, K. Kheiralipour, Recognition of leaves of different medicinal plant species using a robust image processing algorithm and artificial neural networks classifier, J. Appl. Res. Med. Aromat. Plants 25 (2021) 100327.

[11] N. Jagan Mohan, R. Murugan, T. Goel, Deep learning for diabetic retinopathy detection: challenges and opportunities, in: Next Generation Healthcare Informatics, 2022, pp. 213–232.

[12] S. Belciug, Learning deep neural networks' architectures using differential evolution. Case study: medical imaging processing, Comput. Biol. Med. 146 (2022) 105623.

[13] S. Mostafa, F.-X. Wu, Diagnosis of autism spectrum disorder with convolutional autoencoder and structural MRI images, in: Neural Engineering Techniques for Autism Spectrum Disorder, Elsevier, 2021, pp. 23–38.

[14] H. Shenai, J. Gala, K. Kekre, P. Chitale, R. Karani, Combating COVID-19 using object detection techniques for next-generation autonomous systems, in: Cyber-Physical Systems, Elsevier, 2022, pp. 55–73.

[15] Q. Sellat, S.K. Bisoy, R. Priyadarshini, Semantic segmentation for self-driving cars using deep learning: a survey, in: Cognitive Big Data Intelligence with a Metaheuristic Approach, Elsevier, 2022, pp. 211–238.

[16] B.K. Durga, V. Rajesh, A ResNet deep learning based facial recognition design for future multimedia applications, Comput. Electr. Eng. 104 (2022) 108384, https://doi.org/10.1016/j.compeleceng.2022.108384.

[17] N. Dey, Y.-D. Zhang, V. Rajinikanth, R. Pugalenthi, N.S.M. Raja, Customized VGG19 architecture for pneumonia detection in chest X-rays, Pattern Recognit. Lett. 143 (2021) 67–74.

[18] P. Sabitha, G. Meeragandhi, A dual stage AlexNet-HHO-DrpXLM archetype for an effective feature extraction, classification and prediction of liver cancer based on histopathology images, Biomed. Signal Process. Control 77 (2022) 103833.

[19] M. Tan, Q. Le, EfficientNet: rethinking model scaling for convolutional neural networks, in: International Conference on Machine Learning, 2019, pp. 6105–6114.

[20] C. Szegedy, V. Vanhoucke, S. Ioffe, J. Shlens, Z. Wojna, Rethinking the inception architecture for computer vision, in: Proceedings of the IEEE Conference on Computer Vision and Pattern Recognition, 2016, pp. 2818–2826.

[21] N. Jagan Mohan, R. Murugan, T. Goel, S. Mirjalili, P. Roy, A novel four-step feature selection technique for diabetic retinopathy grading, Phys. Eng. Sci. Med. 44 (4) (2021) 1351–1366, https://doi.org/10.1007/s13246-021-01073-4.

[22] L.T. Duong, N.H. Le, T.B. Tran, V.M. Ngo, P.T. Nguyen, Detection of tuberculosis from chest X-ray images: boosting the performance with vision transformer and transfer learning, Expert Syst. Appl. 184 (2021) 115519.

[23] N.J. Mohan, R. Murugan, T. Goel, P. Roy, Fast and robust exudate detection in retinal fundus images using extreme learning machine autoencoders and modified KAZE features, J. Digit. Imag. 35 (3) (2022) 496–513.

[24] S. Mirjalili, A. Lewis, The whale optimization algorithm, Adv. Eng. Softw. 95 (2016) 51–67.

[25] N. Rana, M.S.A. Latiff, S.M. Abdulhamid, H. Chiroma, Whale optimization algorithm: a systematic review of contemporary applications, modifications and developments, Neural Comput. Appl. 32 (20) (2020) 16245–16277.

[26] D.N.K. Pandiri, R. Murugan, T. Goel, ODNet: optimized deep convolutional neural network for classification of Solanum tuberosum leaves diseases, in: 2022 IEEE Region 10 Symposium (TENSYMP), 2022, pp. 1–6.

[27] T.M. Shami, A.A. El-Saleh, M. Alswaitti, Q. Al-Tashi, M.A. Summakieh, S. Mirjalili, Particle swarm optimization: a comprehensive survey, IEEE Access 10 (2022) 10031–10061.

[28] M. Grandini, E. Bagli, G. Visani, Metrics for multi-class classification: an overview, arXiv preprint, arXiv:2008.05756, 2020.

Chapter 35

Whale optimization algorithm for scheduling and sequencing

Muhammad Najeeb Khan and Amit Kumar Sinha
School of Mechanical Engineering, Shri Mata Vaishno Devi University, Katra, Jammu & Kashmir, India

35.1 Introduction

An optimization algorithm is a process that is carried out repeatedly while evaluating numerous solutions in search of the best or most optimum one. In optimization problems, it is essential to quickly and efficiently identify the best solution to a particular issue while balancing a number of challenging constraints. Modern intelligent methods are typically used to address such optimization challenges [20]. Various approaches are recommended to address these problems, but they are insufficient to produce better results [19].

Meta-heuristic optimization algorithms received a lot of attention from researchers in recent years in scientific communities with major advancements, mostly for handling numerous complex optimization issues [12]. Prior to meta-heuristic algorithms, Hill-Climbing, Random Search, and Simulated Annealing were considered the standard techniques to handle optimization problems. Because they rely on relatively simple concepts and are simple to implement, do not require gradient information, can bypass local optima, and can be applied to a wide range of problems spanning various disciplines; meta-heuristic optimization algorithms are becoming increasingly popular in engineering applications [6]. Additionally, there are three categories of algorithms based on meta-heuristic multiple solutions: swarm-based, physics-based, and evolutionary-based techniques. Recombination, mutation, and eventually selection are general population-based meta-heuristics that are anchored in biological evolution. Regarding the fundamental fitness landscape, there are no presumptions for these approaches. The principle of Darwinian evolution is used by the class of well-known optimization techniques known as Genetic Algorithms (GA) [5,10].

The biogeography-based optimization (BBO) algorithm is a novel sort of optimization strategy based on the biogeography idea. Differential evolution (DE), which has achieved excellent results in both benchmark functions and real applications, is one of the most popular and successful evolutionary techniques for numerical optimization. Those algorithms that are based on physical principles belong to the second class of algorithms [18]. Each search agent of this sort can communicate with others and move around the search area in accordance with established physics. The laws of gravity, inertia, electromagnetic force, and other forces can be mentioned. For instance, Snell's law of light refraction served as the inspiration for the algorithms Ray Optimization (RO), Electromagnetic Field Optimization (EFO), and Gravity Search Algorithm (GSA) [21]. The metallurgical annealing procedure served as another inspiration for Simulated Annealing (SA). The third section is made up of swarm-based algorithms, which are based on the social behavior of creatures [4]. Its collective intelligence is built on the interactions between swarm members. Swarm-based algorithms are simpler to implement than evolutionary-based algorithms since they have fewer operators (i.e., selection, crossover, mutation).

These methods attempt to increase the quality of the group by starting with a set of potential solutions to the optimization problem that have been randomly chosen. A set of simple, mostly natural criteria are repeatedly adjusted to the current solutions to increase the values of their cost functions. The methods indicated above efficiently search the search space and minimize its size to discover the solution. According to Hammouche et al. [9], Katebi et al. [11], and Meng et al. [14], algorithms differ from one another principally in how they balance exploitation and exploration (global search capability) and local search capability around the nearly optimal answer.

Whale optimization based robot scheduling mathematical model has been proposed by Petrović et al. [17] while considering altogether seven fitness functions. Petrović et al. [17] single mobile robot scheduling problem has been successfully optimized using whale optimization algorithm. NP hard nature of the scheduling problem is optimized using whale optimization algorithm [17]. Al-qaness et al. [3] developed solution of parallel machining scheduling problem based on the modified whale optimization algorithm. In this modification they used beauty of two ombined algorithms namely whale

Handbook of Whale Optimization Algorithm. https://doi.org/10.1016/B978-0-32-395365-8.00041-5

optimization and firefly algorithm (FA). They also concluded they the modified whale optimization algorithm gives better results than simple whale optimization algorithm of any parallel machine scheduling problem. Ala et al. [2] used the whale optimization for solving health care scheduling problem. They also conclude that they got better scheduling solution as compared with NSGA-II bases optimization of health scheduling model.

Recently, Mirjalili and Lewis [15] created a brand-new meta-heuristic optimization technique named WOA. Whales are thought to be the most intellectual animals capable of motion. The WOA algorithm was driven by the distinctive humpback whale hunting habits. The humpback whales prefer to hunt fish that are swimming close to the water's surface, like krill or small fish. The bubble net feeding method is a distinctive hunting strategy used by humpback whales. In figure, you can see how the whales use this technique to swim around their prey while producing distinctive bubbles along a circle- or "9" shaped course.

35.2 Whale optimization algorithm (WOA)

A swarm intelligence technique called WOA is recommended for problems demanding continuous optimization. It has been shown that this algorithm outperforms some of the currently used algorithmic techniques, if not better. The hunting strategies used by humpback whales served as inspiration for WOA. In WOA, every solution is viewed as a whale. As the best member of the group, the whale in this response tries to fill in a new spot in the search area [16]. The whales use two distinct strategies to hunt and attack their prey [15]. In the first, the prey are contained, while bubble nets are created in the second. Whales explore their surroundings in pursuit of prey, and they take advantage of it when they are attacking. In the former method, humpback whales descend to a depth of around 12 meters, start to spiral-shaped bubble around their food, and then swim up towards the surface. The coral loop, lob-tail, and capture loop stages make up the later [8]. A mathematical model is first created using actions like circling prey, searching for prey, and spiral bubble-net feeding [15]. The flow chart of the WOA is illustrated in Fig. 35.1.

Encircling prey

Humpback whales can find and circle their prey when they are engaged in hunting. Because the position of the optimal design in the search space is unknown a priori, the WOA assumes that the best candidate solution at this time is either the target prey or very close to the optimum. The best search agent will be sought out, and other search agents will update their positions in close proximity to the greatest search agent. The behavior is expressed by the following equations (see Eqs. (35.1) and (35.2)) as stated by Mirjalili and Lewis [15].

$$\vec{B} = \left|\vec{T}\vec{X}^*(t) - \vec{X}(t)\right| \tag{35.1}$$

$$\vec{X}(t+1) = \vec{X}^* - \vec{E} \bullet \vec{B} \tag{35.2}$$

where,

$\vec{X} \rightarrow$ Position Vector
$X^* \rightarrow$ Position Vector of the best solution
$t \rightarrow$ Current Iteration
$\vec{E} \rightarrow$ Coefficient Vector $= 2\vec{e} \bullet \vec{s} - \vec{e}$
$\vec{T} \rightarrow$ Coefficient Vector $= 2 \bullet \vec{s}$
$\vec{e} \rightarrow$ Linear decreases from 2 to 0 over the course of iteration
$\vec{r} \rightarrow$ Random Vector $\in [0, 1]$

Bubble-net attacking method

A spiral, which mimics the helix-shaped movement of humpback whales, is used to numerically describe the position of the whale in respect to the prey. According to Mirjalili and Lewis [15], the behavior is represented by the following equations (see Eqs. (35.3), (35.4), and (35.5)).

$$\vec{X}(t+1) = \vec{B}' \bullet e^{gf} \bullet \mathrm{Cos}\,(2\pi f) + \vec{X}^*(t) \tag{35.3}$$

where,

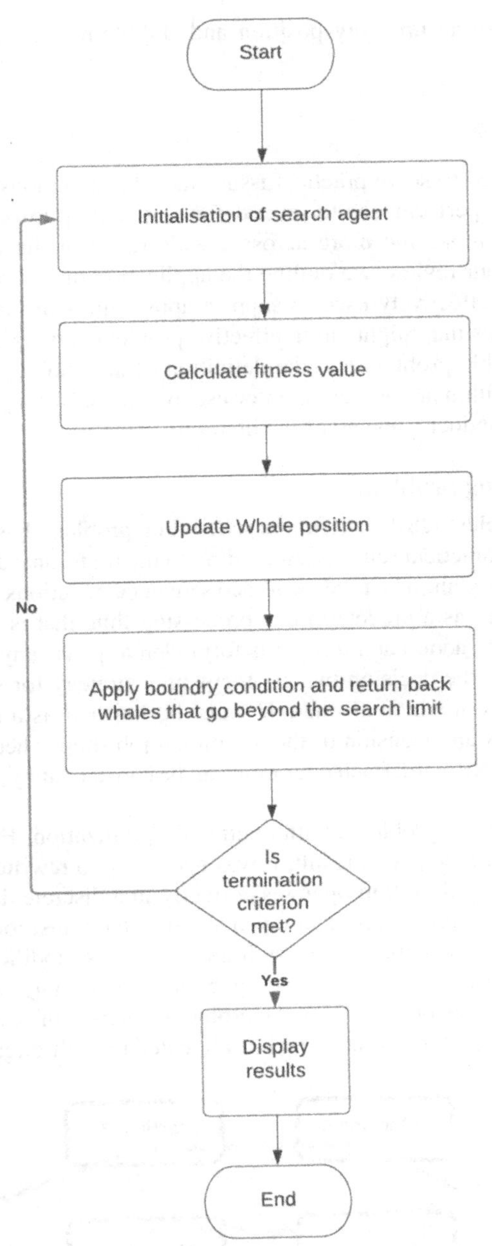

FIGURE 35.1 Flow chart of WOA.

$$\vec{B}' = \left|\vec{X}^*(t) - \vec{X}(t)\right| \tag{35.4}$$

$\vec{B}' \rightarrow$ Distance of the ith whale to the prey

$g \rightarrow$ Constant for defining the shape of the logarithmic spiral

f $\rightarrow$ Random No $[-1, 1]$

$\bullet \rightarrow$ Element Matrix Multiplication

$$\vec{X}(t+1) = \left\{ \begin{array}{ll} \vec{X}^*(t) - \vec{E} \bullet B & \text{if } p < 0.5 \\ \vec{B}' \bullet e^{gf} \bullet \text{Cos}\,(2\pi f) + \vec{X}^*(t) & \text{if } p \geq 0.5 \end{array} \right\} \tag{35.5}$$

$p \rightarrow$ Random No $\in [0, 1]$

The detailed information about Spiral updating position and Bubble-net search mechanism of whale can be found in [15].

35.3 Applications of WOA

Numerous ways that WOA can be used to solve practical issues have been demonstrated in literary works [7,13,22]. The majority of the research compared its performance to several conventional optimization techniques and other benchmark meta-heuristics. WOA is being used more and more across a wide range of disciplines thanks to its effectiveness and adaptability. As the primary goal of our review, we outline the application of the WOA in several technical domains and subdomains in this section. WOA is effectively used to support applications in many disciplines, including engineering optimization issues. The main concerns that might aid in effective problem solving when working with engineering applications are the formulation of a suitable problem and the definition of acceptable variables and objective functions. The key issues with the optimization algorithm are proper operator use, parameter setting, and population illustration. Here, the application areas are categorized to conduct a review of the literature.

WOA for scheduling and sequencing problem

The classical job shop scheduling problem (JSP), a common scheduling problem, has received a lot of attention in a variety of domains due to its extensive use in practical settings. According to the traditional JSP, a set of jobs must be processed by a number of machines, with each job containing a predetermined sequence of actions that must be completed on a particular machine. Additionally, each operation has a predetermined processing time that is fixed. At time zero, every machine is available and only capable of doing one action at a time. It is forbidden to pause any machine action that is in progress. In order to achieve a predetermined goal, the decision-makers focus on a strategy for sequence permutation for all machine activities. Makespan, or the time point at which all the jobs must be finished, is a typical JSP requirement. The flexible job shop scheduling problem (FJSP) is an extension of the traditional job shop scheduling problem (JSP), and it has been demonstrated that the FJSP is strongly NP-hard. Each operation can be carried out by one machine in an alternative machine set.

The first WOA was created to address problems with continual optimization. FJSP, on the other hand, is a common discrete combinatorial optimization problem. As a result, it was necessary to rewrite the WOA using a discrete encoding strategy for the issues. To make sure the algorithm operates correctly in a discrete domain, a discrete individual updating approach was created. Six adjustment curves of were employed to adjust the transition between the algorithm's exploration and exploitation while a hybrid variable neighborhood search approach was modified to create the population with good quality. An enhanced variable neighborhood search was incorporated into the suggested technique and used to the current optimal individual to further enhance its performance. For the problem structure of scheduling and sequencing, each colored line representing a job that has to go to each machine, which is allocated to it, through which its line goes. (See Fig. 35.2.)

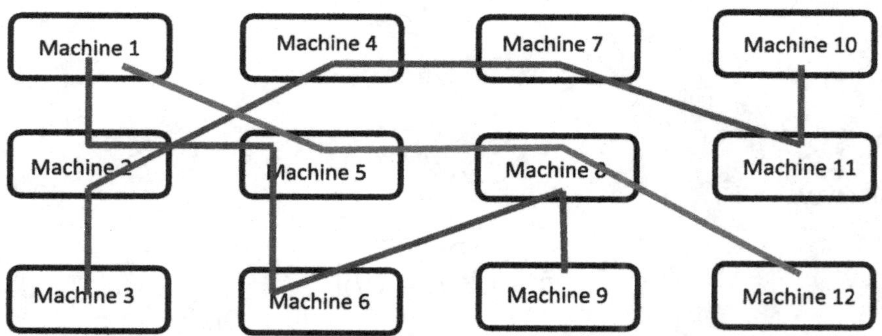

FIGURE 35.2 Problem structure of S&S.

Methodology of WOA for scheduling and sequencing problem

The Whale Optimization Algorithm (WOA), a distinctive intelligence algorithm frequently used to address optimization problems, has been shown to be effective by the majority of academics. WOA has undergone a number of beneficial improvements that will improve its ability to handle the job shop issue (JSP) [1,23,24]. Whereas WOA is an optimization strategy created to address the continuous optimization problem, JSP is a discrete combinatorial optimization challenge.

TABLE 35.1 Generated random number.

rand	0.62	0.31	0.37	0.92	0.27	0.15
rand0	6	5	2	3	1	4
OS	3	3	1	2	1	2
MS	2	3	4	2	1	3

Thus, WOA needs to be changed via a conversion technique in order to address the FJSP issue. The operation sequencing problem and the machine assignment problem are organized using a two-layer coding technique that combines operation sequence (OS) and machine sequence (MS) in accordance with the characteristics of the FJSP problem. The following method is used to explain the principle: The task number's location index is then determined by items in the mapped rand'. A set of random integers with a size equal to O operations, for instance, could be created and mapped from small to large. Rand, for instance, is equal to [0.62, 0.31, 0.37, 0.92, 0.27, 0.15] and [6, 5, 2, 3, 1] respectively. Specific information on random numbers is shown in Table 35.1. The encoding schematic's pseudo code is shown in Fig. 35.3.

$G \rightarrow$ Population Size
A $\rightarrow$ n×m Matrix
p $\rightarrow$ Smallest Prime Number
$p \geq 2m+3$

for $n = 1:G$
  for $k = 1:m$

$$r_k = \left\{2Cos\left(\frac{2\pi k}{p}\right)\right\}; \quad Where, \quad 1 \leq k \leq m$$

$$p_n(k) = \left\{\left\{r_1^{(m)} * k\right\}\left\{r_2^{(m)} * k\right\} \quad \ldots\left\{r_{1n}^{(m)} * k\right\}, \quad Where, k = 1,2, \ \ldots..n\right\}$$

  end
end

FIGURE 35.3 Pseudo code for the schematic of encoding.

In order to distribute the starting population more evenly and provide greater coverage in the first stage, the theory of good point set (GPS) was first created. A nonlinear convergence factor (NCF) is proposed to coordinate the global search and local search in WOA in order to improve the coordination role that the convergence factor in WOA plays. Ultimately, a completely new multi-neighborhood structure (MNS) with a total of three new neighborhoods is suggested. WOA's capacity to solve the job shop issue has been increased thanks to a number of beneficial developments (JSP). In order to distribute the starting population more evenly and provide greater coverage in the first stage, the theory of good point set (GPS) was first created. A nonlinear convergence factor (NCF) is proposed to coordinate the global search and local search in WOA in order to improve the coordination role that the convergence factor in WOA plays. For the visual representation of the time span taken by the collection of jobs performed on specified machines through a specified path see Fig. 35.4.

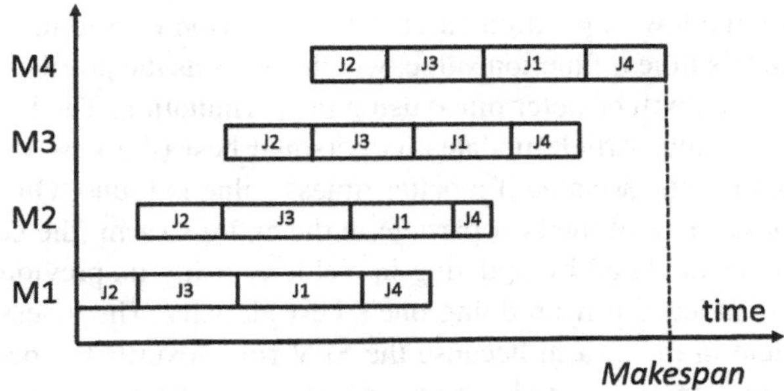

FIGURE 35.4 Visual representation of Makespan.

Ultimately, a completely new multi-neighborhood structure (MNS) with a total of three new neighborhoods is suggested. Three more neighborhoods, N1, N2, and N3, as well as a new multi-neighborhood structure that employs a multi-neighborhood approach to FJSP optimization are described. In terms of various machine pairing options, the N1 neighborhood structure optimizes the scheduling solution. The scheduling solution is optimized for various machine selection pairings while the machine selection is updated using a gene mutation approach. The neighborhood structure for the N2 operation update technique, the scheduling solution is kept from settling into a local optimum by utilizing a perturbation method to partially shift the locations of whale individuals. The scheduling solution's operation update method has been improved. Furthermore, the N3 neighborhood structure is used to improve the local search capabilities of WOA. By increasing the capacity of local search, WOA is better able to surpass the local optimal limit and lock the core function for updating as the iteration goes on. The population diversity reception mechanism (DRM), which aids populations in maintaining some of their diversity across time, is the following stage.

Comparative algorithms used for optimization of scheduling and sequencing problems

While we've seen the whale optimization algorithm for optimizing scheduling and sequencing problems, there are other metaheuristic algorithms that are also employed for addressing these NP-hard problems and produce efficient solutions. Genetic algorithms are one illustration of this kind of algorithm. Although GA is a nature-inspired algorithm like WOA, it has also been demonstrated through history to be extremely effective for optimization. The scheduling and sequencing problems can be resolved using a genetic algorithm with binary values. Priority criteria can be used to define random paths for our problem, and each path can be represented as a binary string. By applying GA to these binary strings, we can obtain the best solution for the given fitness function. Model parameters representing a solution to the optimization problem in a binary-coded GA are encoded on a chromosome of binary strings of 0s and 1s. A population of chromosomes drawn at random from the search space serves as the method's initial population, which represents a path jobs' should follow to complete. A new population that is superior to the previous one is created after the population is subjected to genetic processes including selection, cross-over, and mutation. Instead of focusing on a single spot, GA searches for data from many different locations. GA uses payoff (objective function) data rather than derivatives. GA is particularly robust in terms of local maxima and minima. For the representation of genes and chromosomes in the genetic algorithm see Fig. 35.5.

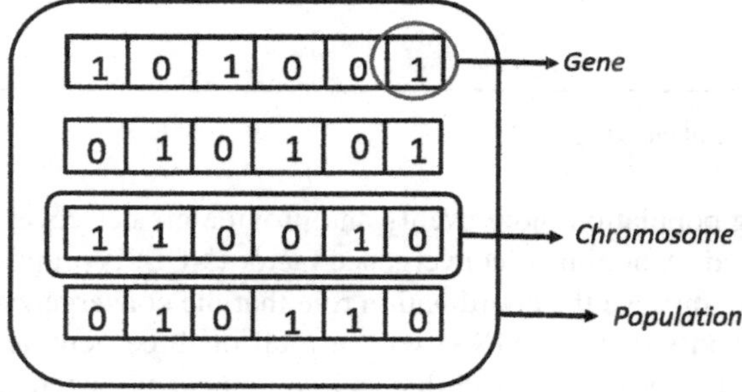

FIGURE 35.5 Genetic representation.

Another illustration of such an algorithm is particle swarm optimization (PSO). When PSO was initially created, it was capable of handling objective functions in continuous search spaces. After initializing the parameters, the initial population is generated at random. Since the evaluation of each particle in the swarm requires the identification of the permutation of jobs for scheduling and sequencing, the lowest position value (SPV) criterion is used to each particle to find its associated permutation. As a result, each particle's fitness function value, which serves as the problem's objective function with regard to the makespan/total flowtime criterion, will be determined using the permutation. The PSO algorithm iteratively performs the following steps after evaluation: Each particle updates its personal best (the best value achieved by each individual thus far) with its position, velocity, and fitness value if a better fitness value is found. On the other hand, the global best is updated using the position and fitness value of the best particle in the entire swarm (the best particle in the whole swarm). After that, each particle's position is calculated by updating its velocity using its previous velocity, lessons learned from doing one's personal best, and lessons learned from doing one's best globally. The assessment is performed to determine the fitness of each individual particle in the swarm because the SPV rule governs the permutation. To expand the search space investigation, a local search or mutation can also be conducted to a particular subset of the swarm's particles. This procedure is terminated by a specified halting circumstance. (See Fig. 35.6.)

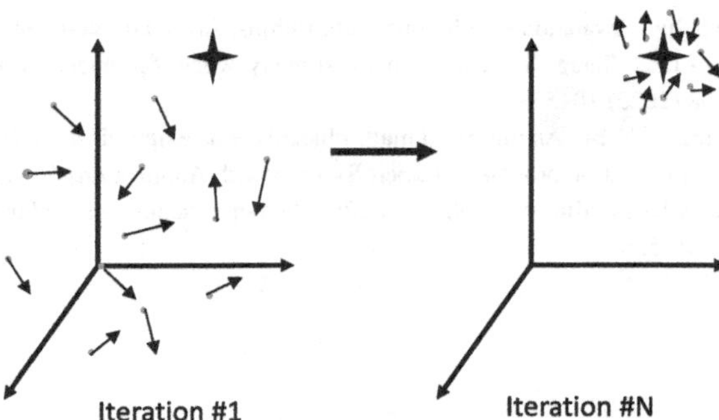

FIGURE 35.6 PSO visualization.

35.4 Conclusion

The basic concept of WOA is illustrated in this chapter. The basic engineering application of WOA has been explored. Scheduling and sequencing problem has been given more attention. Step by step procedures for implementing WOA in flexible job shop scheduling problem (FJSP) has been demonstrated. This chapter just enlightens the use of WOA for researchers and practitioners. The detailed information about WOA can be seen from the basic research paper of Mirjalili & Lewis [15]. In different real life problem like feature selection, fault diagnosis, breast cancer diagnosis, training of neural network, image segmentation problem, vehicle fuel consumption problem, etc., can be solved using WOA.

References

[1] M. Abdel-Basset, G. Manogaran, D. El-Shahat, S. Mirjalili, A hybrid whale optimization algorithm based on local search strategy for the permutation flow shop scheduling problem, Future Generation Computer Systems 85 (03) (2018).

[2] A. Ala, F.E. Alsaadi, M. Ahmadi, S. Mirjalili, Optimization of an appointment scheduling problem for healthcare systems based on the quality of fairness service using whale optimization algorithm and NSGA-II, Scientific Reports 11 (1) (2021) 19816.

[3] M.A. Al-qaness, A.A. Ewees, M. Abd Elaziz, Modified whale optimization algorithm for solving unrelated parallel machine scheduling problems, Soft Computing 25 (14) (2021) 9545–9557.

[4] D. Bertsimas, J. Tsitsiklis, Simulated annealing, Statistical Science 8 (1) (1993) 10–15.

[5] L.B. Booker, D.E. Goldberg, J.H. Holland, Classifier systems and genetic algorithms, Artificial Intelligence 40 (1–3) (1989) 235–282.

[6] O. Bozorg-Haddad, M. Solgi, H.A. Loáiciga, Meta-Heuristic and Evolutionary Algorithms for Engineering Optimization, John Wiley & Sons, 2017.

[7] S. Chakraborty, S. Sharma, A.K. Saha, A. Saha, A novel improved whale optimization algorithm to solve numerical optimization and real-world applications, Artificial Intelligence Review 55 (2022) 4605–4716.

[8] F.S. Gharehchopogh, H. Gholizadeh, A comprehensive survey: Whale Optimization Algorithm and its applications, Swarm and Evolutionary Computation 48 (2019) 1–24.

[9] K. Hammouche, M. Diaf, P. Siarry, A comparative study of various meta-heuristic techniques applied to the multilevel thresholding problem, Engineering Applications of Artificial Intelligence 23 (5) (2010) 676–688.

[10] J.H. Holland, Genetic algorithms, Scientific American 267 (1) (1992) 66–73.

[11] J. Katebi, M. Shoaei-parchin, M. Shariati, N.T. Trung, M. Khorami, Developed comparative analysis of metaheuristic optimization algorithms for optimal active control of structures, Engineering With Computers 36 (4) (2020) 1539–1558.

[12] A. Kaveh, A. Dadras, A novel meta-heuristic optimization algorithm: thermal exchange optimization, Advances in Engineering Software 110 (2017) 69–84.

[13] X. Lin, X. Yu, W. Li, A heuristic whale optimization algorithm with niching strategy for global multi-dimensional engineering optimization, Computers & Industrial Engineering 171 (2022) 108361.

[14] Z. Meng, G. Li, X. Wang, S.M. Sait, A.R. Yıldız, A comparative study of metaheuristic algorithms for reliability-based design optimization problems, Archives of Computational Methods in Engineering 28 (3) (2021) 1853–1869.

[15] S. Mirjalili, A. Lewis, The whale optimization algorithm, Advances in Engineering Software 95 (2016) 51–67.

[16] S. Mostafa Bozorgi, S. Yazdani, IWOA: an improved whale optimization algorithm for optimization problems, Journal of Computational Design and Engineering 6 (3) (2019) 243–259.

[17] M. Petrović, Z. Miljković, A. Jokić, A novel methodology for optimal single mobile robot scheduling using whale optimization algorithm, Applied Soft Computing 81 (2019) 105520.

[18] A.K. Qin, V.L. Huang, P.N. Suganthan, Differential evolution algorithm with strategy adaptation for global numerical optimization, IEEE Transactions on Evolutionary Computation 13 (2) (2008) 398–417.

[19] R.V. Rao, V.J. Savsani, Mechanical Design Optimization Using Advanced Optimization Techniques, Springer London, 2012.

[20] G. Venter, Review of optimization techniques, https://doi.org/10.1002/9780470686652.eae495, 2010.

[21] D. Whitley, S. Rana, J. Dzubera, K.E. Mathias, Evaluating evolutionary algorithms, Artificial Intelligence 85 (1–2) (1996) 245–276.

[22] W. Yang, K. Xia, S. Fan, L. Wang, T. Li, J. Zhang, Y. Feng, A multi-strategy whale optimization algorithm and its application, Engineering Applications of Artificial Intelligence 108 (2022) 104558.

[23] W. Yankai, W. Shilong, L. Dong, S. Chunfeng, Y. Bo, An improved multi-objective whale optimization algorithm for the hybrid flow shop scheduling problem considering device dynamic reconfiguration processes, Expert Systems with Applications 174 (2021) 114793.

[24] J. Zhu, Z.H. Shao, C. Chen, An improved whale optimization algorithm for job-shop scheduling based on quantum computing, International Journal of Simulation Modelling 18 (3) (2019) 521–530.

Chapter 36

Tuning SVMs' hyperparameters using the whale optimization algorithm

Sunday O. Oladejo[a], Stephen O. Ekwe[b], Adedotun T. Ajibare[c], Lateef A. Akinyemi[d], and Seyedali Mirjalili[e,f]

[a]School for Data Science and Computational Thinking, Stellenbosch University, Stellenbosch, South Africa, [b]Department of Electrical, Electronic and Computer Engineering, Cape Peninsula University of Technology, Cape Town, South Africa, [c]Faculty of Information and Communications Technology, Rosebank College, Cape Town, South Africa, [d]Department of Electronic and Computer Engineering, Lagos State University, Epe, Nigeria, [e]Centre for Artificial Intelligence Research and Optimisation, Torrens University Australia, Brisbane, QLD, Australia, [f]University Research and Innovation Center, Obuda University, Budapest, Hungary

36.1 Introduction

The Support Vector Machine (SVM) is a kernel-based supervised Machine Learning (ML) algorithm [1–3], widely used for classification and sometimes for regression problems. It possesses high generalization and discriminative capabilities, and strong statistical and theoretical foundations [2,4,5]. Its popularity also stems from its ease and simplicity of implementation and deployment, coupled with its high scalability [6]. When compared to other classification algorithms, it has a high level of performance accuracy and can compete with all current alternatives.

The SVM's mode of operation stems from the use of kernels (such as linear, nonlinear, polynomial, and radial basis function (RBF)) to transform the training datasets into a higher dimensionality in order to create an optimal separation boundary among the respective classes embedded in the considered dataset. Data points closest to the hyperplanes are called 'Support Vectors' (SV), and a 'margin' exists between the support vectors and the hyperplane. The primary goal of the SVM is to ensure that the margin between the SV and the hyperplane is as large as possible. The positioning of the hyperplane determines the flexibility, sensitivity, and accuracy of SVMs in a classification task or activity.

The kernel-based dimensionality transformation plays a critical role in the positioning of the hyperplane, the hyperparameters of the kernel, and by extension the SVM. Hyperparameters are the parameters of an algorithm or model whose values are not directly learnt during the learning process of ML algorithms. In other words, the way that the values of the hyperparameters are determined is totally separate from the learning process of the ML. One of the major challenges of the SVM is the choice of optimal values for the respective hyperparameters [3,7].

A widely agreed-upon solution to the SVM's hyperparameter conundrum is the use of metaheuristics, owing to their simplicity and ease of implementation, local optima avoidance, no need for gradient-based information, and flexibility [8]. Thanks to these advantages, metaheuristics have found applications in several fields of endeavors such as engineering design, production, resource allocation in wireless networks, and scheduling [9–15]. Fig. 36.1 illustrates some of these applications.

Recently, there has been increased interest in the development of new metaheuristics due to the No Free Lunch (NFL) theorem [16]. Popular metaheuristics include the Genetic Algorithm (GA) [17,18], Particle Swarm Optimization (PSO) [19], Grey Wolf Optimizer (GWO) [20], and Whale Optimizer Algorithm (WOA) [8]. This chapter's primary goal is to address the SVM's hyperparameter tuning challenges by the adoption of different variants of the WOA. First, we take a look at the history and a general overview of SVMs, before we tune their hyperparameters with the WOA.

36.2 Whale optimization algorithm and improved versions

In tuning the hyperparameters of the SVM, we deploy a common metaheuristic, the WOA [8], and some its improved versions [21] from the literature. These include the multi-strategy ensemble WOA (MSWOA) [22], Levy flight trajectory-based WOA (LWOA) [23], Elite-opposition-based-Golden-sine WOA (EGolden-SWOA) [24], improved WOA based on nonlinear adaptive weight and golden sine operator (NGS-WOA) [25], and Whale optimization algorithm with a modified mutualism phase (WOAmM) [26].

Handbook of Whale Optimization Algorithm. https://doi.org/10.1016/B978-0-32-395365-8.00042-7

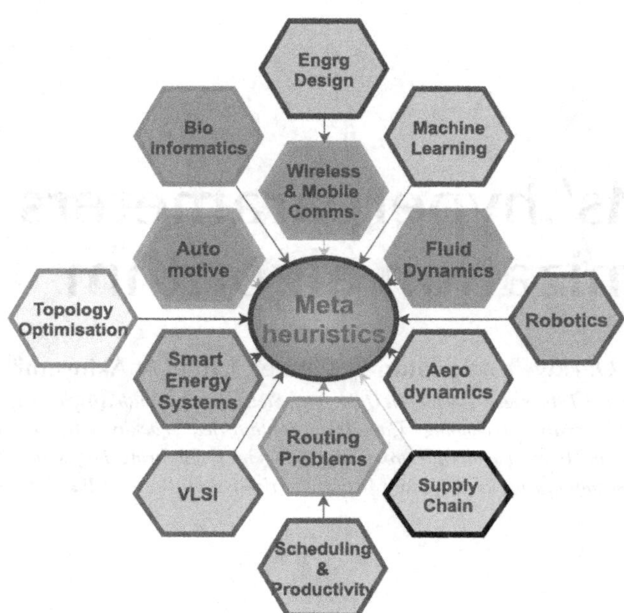

FIGURE 36.1 Some areas of application of metaheuristic algorithms.

36.2.1 Whale optimization algorithm

WOA draws its inspiration from the hunting behaviors of humpback whales. This behavioral pattern can be categorized into three stages [8,21]: (i) searching, (ii) encircling, and (iii) attacking the prey. The whales search for their prey in the ocean, which can be likened to the land search space in optimization and the hyperparameter space in hyperparameter optimization. Upon locating prey, the whales encircle it via multiple successive helix-shaped movements, and thereafter attack it by the creation of bubble nets.

The WOA process of finding an optimal or near optimal solution to an optimization problem mimics a whale's hunting processes of encircling its prey (i.e. for exploration) and bubble net creation (i.e. for exploitation). The WOA is premised on swarm intelligence and is suitable for continuous optimization problems [8]. Other than their large physical size, whales exhibit social behavior by living and hunting in groups and have been observed as being emotionally complex [27]. Owing to the social behavior of whales, the WOA is also categorized as a population-based metaheuristic (otherwise referred to as 'multiple-solution based metaheuristics') [21]. To this end, the WOA is well suited for global optimization. Additionally, the WOA's framework is designed such that the mathematical modeling for the behavioral patterns of the humpback whale is the basis of the WOA.

36.2.1.1 Mathematical modeling of the WOA

In this subsection, the WOA's mathematical background is described. On location of a prey by a pod (i.e. a group of whales), the whales encircle their prey. This is achieved by members of the pod updating their position based on the position or location of the whale that sighted the prey first. The encircling process is given by [8]:

$$\vec{\rho} = \left|\vec{\varphi} \cdot \vec{X}^*(t) - \vec{X}(t)\right|, \tag{36.1}$$

$$\vec{X}(t+1) = \vec{X}^*(t) - \vec{\alpha} \cdot \vec{\rho}, \tag{36.2}$$

where $\vec{X}(t)$ is the current position vector of a whale at iteration or time t. $\vec{X}^*(t)$ is the position of the best placed whale. Since the best placed or situated whale is stochastic, $\vec{X}^*(t)$ is checked and updated every iteration. The whales update their position using expressions (36.1) and (36.2). Likewise, $\vec{\varphi}$ and $\vec{\alpha}$ are randomly generated coefficients vectors given by [8]:

$$\vec{\alpha} = 2\vec{\epsilon} \cdot \vec{\kappa} - \vec{\epsilon}, \tag{36.3}$$

$$\vec{\varphi} = 2 \cdot \vec{\kappa}, \tag{36.4}$$

where $\vec{\kappa}$ is a uniformly distributed random vector between 0 and 1, whereas $\vec{\epsilon}$ linearly decreases from 2 to 0 over the course of the iteration. $\vec{\epsilon}$ enables the whales to shrink the encircling process in (36.1) and (36.2). To this end, in (36.4) $\vec{\varphi}$ lies in the range $-\vec{\epsilon} \leq \vec{\varphi} \leq \vec{\epsilon}$.

As a whale spirally encircles the prey in a helix-shaped movement, the spiral position is updated by [8]:

$$\vec{X}(t+1) = \vec{\rho}' \cdot e^{bl} \cdot \cos(2\pi l) + \vec{X}^*(t), \tag{36.5}$$

where b defines the logarithmic spiral's shape of the encircling process and l represents the stochastic characteristic of the logarithmic spiral's shape and is given as a random number in $[-1, 1]$. The '$\cdot$' in (36.5) indicates the element-wise multiplication of the first term in the right-hand-side of the expression. Additionally, $\vec{\rho}'$ denotes the distance of a whale to the best position of the prey found and it is simply expressed as $|\vec{X}^*(t) - \vec{X}(t)|$.

As stated earlier, for a whale to attack its prey, she encircles the prey in a helix-shaped movement and also moves closer to the prey in the shrinking process. Hence, a whale decides whether to update it helix-shaped movement (i.e. spirally) or draw closer to the prey. This decision process is premised on a probability given by [8]

$$\vec{X}(t+1) = \begin{cases} \vec{X}^*(t) - \vec{\alpha} \cdot \vec{\rho}, & \text{if } p < 0.5, \\ \vec{\rho}' \cdot e^{bl} \cdot \cos(2\pi l) + \vec{X}^*(t), & \text{if } p \geq 0.5, \end{cases} \tag{36.6}$$

where p is the movement decision variable, and being a probability value, it is a random number in $[0, 1]$.

Similar to the exploitation process (i.e. the encircling and shrinking mechanisms modeled in expression (36.1)–(36.6)), the exploration process of the WOA is given as follows:

$$\vec{\rho} = \left|\vec{\varphi} \cdot \vec{X}_r(t) - \vec{X}(t)\right|, \tag{36.7}$$

$$\vec{X}(t+1) = \vec{X}_r(t) - \vec{\alpha} \cdot \vec{\rho}, \tag{36.8}$$

where variable definitions in (36.7) and (36.8) are similar to expressions in (36.1) and (36.2), however, with slight changes in the limits. $\vec{X}_r(t)$ is a random position vector of a whale chosen from the pod (i.e. population). To ensure global search in the search process in the WOA, whales move far away from a reference whale by ensuring $\vec{\alpha}$ having values such that $\vec{\alpha} > 1$ or $\vec{\alpha} < -1$. (See Fig. 36.2.)

36.2.2 Multi-strategy ensemble whale optimization algorithm

The MSWOA improves the search deficiencies of the WOA by employing chaotic initialization to enhance the quality of the initialization population. Chaotic initialization is premised on the ergodicity, randomness, and regularity embedded in chaos theory [22,28,29]. Furthermore, the chaos initialization ensures the whales, or agents in general, are evenly distributed in the search space or hyperspace in our case. Chaos is a dynamic behavior though unstable, it is bounded and its dependent on the initial conditions with the features of infinite unstable periodic motions. The chaotic initialization is based on the mathematical expression given by [22]:

$$z(t+1) = 1 - m(\cos(n \cos^{-1} z(t)))^2, \tag{36.9}$$

where $z^i \in [-1, 1]$ is random generated. Additionally, m and n are the control variables, such that the entire system becomes chaotic with m and n having values of 2 and 4.

In addition to the improvement of the population generation using chaos, the MSWOA also improves the search strategy of the WOA by incorporating the 'best individual' concept as a reference for whale position updating, which can be expressed by [22]:

$$\vec{X}(t+1) = \vec{X}(t) \cdot rand_3 + K \cdot rand_4 \cdot \left|\vec{X}^*(t) - \vec{X}(t)\right|, \tag{36.10}$$

where K denotes the correction factor with a value of z while $rand_3$ and $rand_4$ are random numbers in $[0, 1]$.

To mitigate the loss of diversity inherent in the spiral updating process of the WOA [22], the MSWOA adopts the Levy flight [30,31] strategy, which is a random walk in which the step-lengths are characterized by a heavy-tailed probability distributed widely referred as Levy distribution. The Levy flight strategy balances the drive between local search and global search capabilities of the WOA. This enhances the exploration process of the MSWOA. The Levy distribution is given by:

$$L(s) \approx s^{-\theta}, \quad 1 \leq \theta \leq 3, \tag{36.11}$$

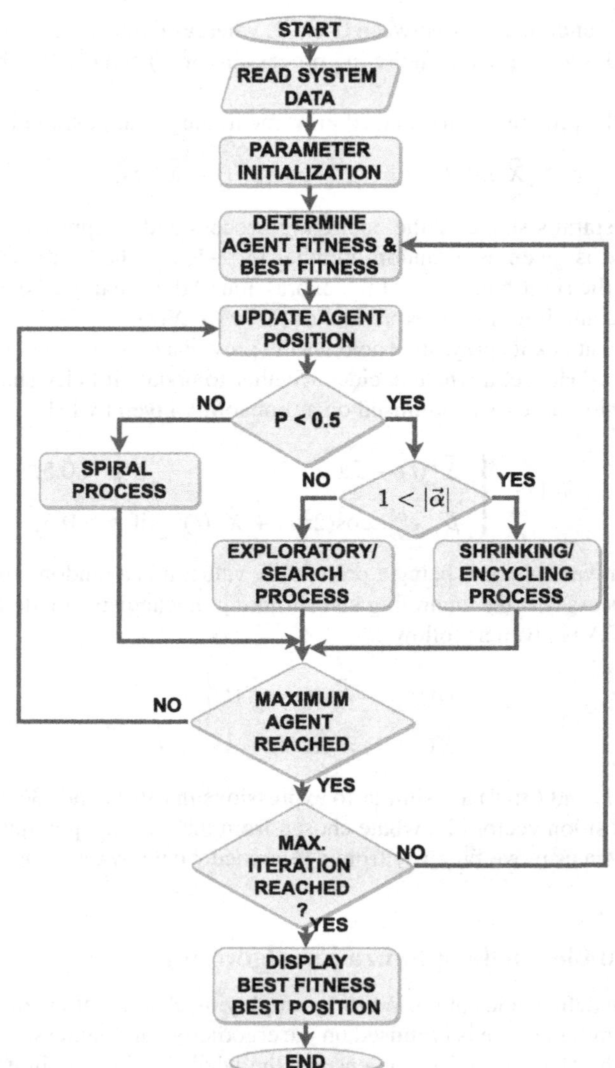

FIGURE 36.2 Flowchart for the WOA.

where s denotes the Levy flight's step size. Therefore, from [22], s is given by:

$$s = \frac{\mu}{|\nu|^{\frac{1}{\theta}}}, \quad \mu \sim \mathcal{N}(0, \sigma_\mu^2), \quad \nu \sim \mathcal{N}(0, \sigma_\nu^2) \tag{36.12}$$

$$\sigma_\mu^2 = \left\{ \frac{\Gamma(1+\theta)}{\theta \cdot \Gamma(1+\theta)/2} \cdot \frac{\sin(\pi \cdot \theta/2)}{2^{(\theta-1)/2}} \right\}, \ \sigma_\nu^2 = 1, \tag{36.13}$$

where $\mathcal{N}$ represents the Gaussian distribution. To this end, the spiral updating process of the WOA is improved by the MSWOA by including the Levy distribution as follows:

$$\vec{X}(t+1) = \begin{cases} \vec{X}(t) + D \cdot e^{bl} \cdot \cos(2\pi l) \cdot L(s), & \text{if } |\vec{\alpha}| \geq 1, \\ \vec{X}^*(t) + D \cdot e^{bl} \cdot \cos(2\pi l) \cdot L(s), & \text{if } |\vec{\alpha}| < 1. \end{cases} \tag{36.14}$$

Unlike the WOA where the solutions are constrained in the hyperparameter space by the upper and lower limits of the hyperparameter, in the MSWOA, the solutions are bounded by near boundary walks of the whales. This contributes to the exploratory capabilities of the MSWOA. The new positions of the whales if they exceed the boundaries of the hyperparameter space are given by:

$$\vec{X}(t+1) = \begin{cases} UB + rand_5 \cdot \frac{UB - \vec{X}(t)}{\vec{X}(t)} \cdot UB, & \text{if } \vec{X}(t) > UB, \\ LB + rand_6 \cdot \left| \frac{LB - \vec{X}(t)}{\vec{X}(t)} \cdot LB \right|, & \text{if } \vec{X}(t) < LB, \end{cases} \tag{36.15}$$

where $rand_5$ and $rand_6$ denote randomly generated numbers [0, 1]. Besides LB and UB represent the upper and lower limits of the hyperparameters. The MSWOA is described in the flowchart presented in Fig. 36.3.

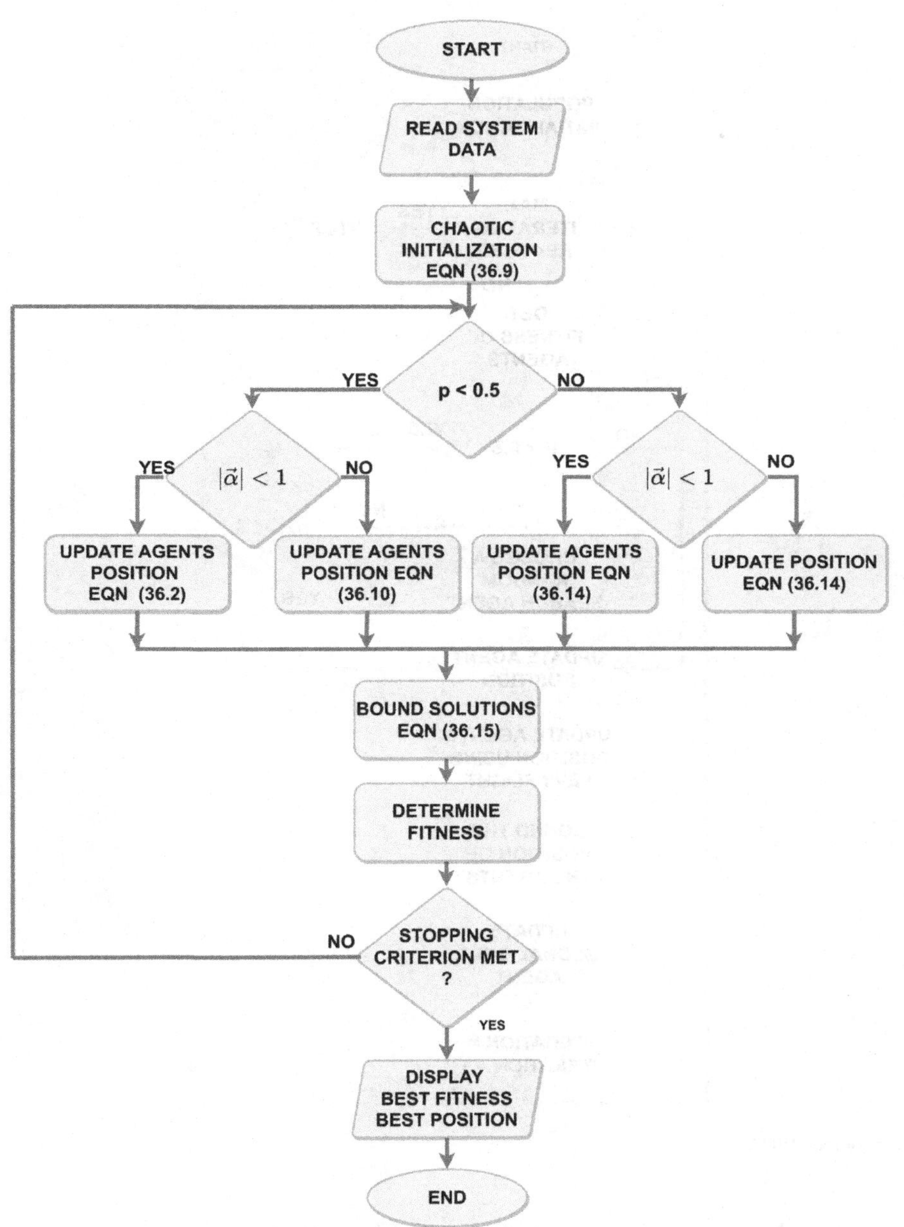

FIGURE 36.3 Flowchart for the MSWOA.

36.2.3 Levy flight trajectory WOA

The LWOA was developed to address the inefficiencies of the WOA in solving high-dimensional and multi-modal problems [23]. The LWOA is based on the Levy distribution given in expressions (36.11)–(36.13). As stated earlier in Section 36.2.2, the Levy flight enhances the search capabilities of the WOA, and therefore avoids local optima entrapment. The whale updating expression for LWOA is given by [23]:

$$\vec{X}(t+1) = \vec{X}(t) + \mu\, sign[rand_7 - 1/2] \cdot L(s) \tag{36.16}$$

where $rand_7$ is a random number in [0, 1] and the evaluation of $sign[rand_7 - 1/2]$ has only three values (i.e. 0, 1, and −1). Moreover, the μ and $L(s)$ were described in expressions (36.11)–(36.13). The flowchart of the LWOA is illustrated in Fig. 36.4.

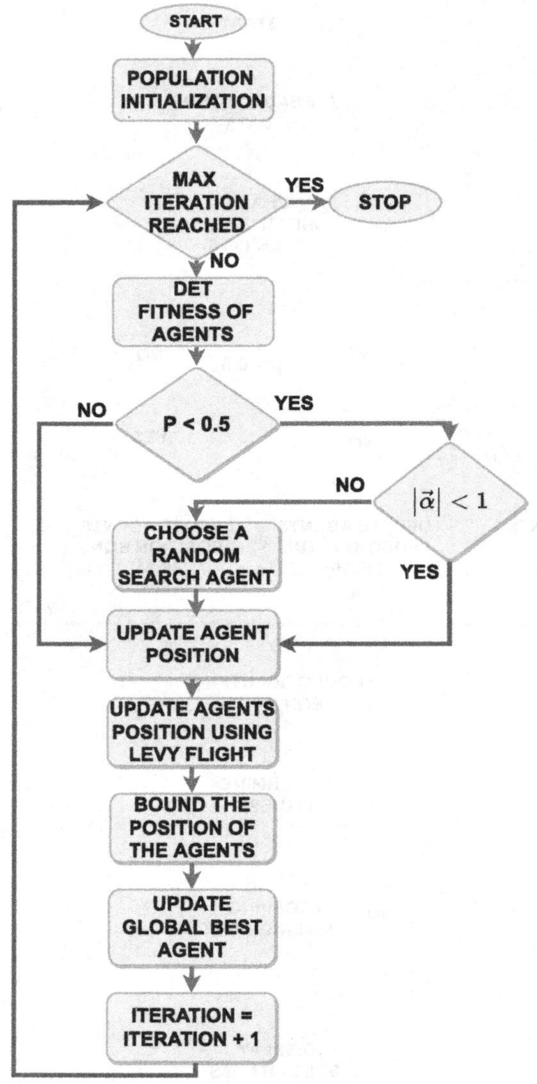

FIGURE 36.4 Flowchart for the LWOA.

36.2.4 Elite opposition-based Golden-sine WOA (EGolden-SWOA)

To improve the convergence rate and global search capabilities of the WOA, the authors in [24] proposed the EGolden-SWOA. This modification of the WOA is a combination of two widely used techniques in improving metaheuristics: (i) the

Opposition-Based Learning (OBL) mechanism [32], and (ii) the Golden Sine Algorithm [33]. The elite reverse learning strategy improves the diversity of the whale population by employing a group selection mechanism. The OBL mechanism is premised on calculating both a feasible solution and the opposite solution, and choosing the better of the two to be part of the next generation agent. For instance, if the opposite solution or position is given as $\vec{X}'(t) \in [m, n]$ be a real number defined in the interval, then the opposite solution is given [32,34] by:

$$\vec{X}'(t) = rand_8 * \Big(m(t) + n(t)\Big) - \vec{X}(t), \tag{36.17}$$

where $rand_8$ is a random number in [0, 1]. The variables $m(t)$ and $n(t)$ are defined by:

$$m(t) = min(\vec{X}(t)), \tag{36.18}$$

$$n(t) = max(\vec{X}(t)). \tag{36.19}$$

In the EGolden-SWOA, the GSA improves the global search of the whales. The GSA is a mathematical heuristic which is based on the periodic variation of the sine function in trigonometry. In other words, the GSA incorporates the movement along (i.e. points) on the unit circle [35] i.e. $[0, 2\pi]$. In the EGolden-SWOA, the position of the whales is updated by [33,35]:

$$\vec{X}(t+1) = \vec{X}(t) + \Big|\sin(r_1)\Big| - r_2 * \sin(r_1) * \Big|m_1\vec{X}^*(t) - m_2 * \vec{X}(t)\Big|, \tag{36.20}$$

where r_1 and r_2 are random numbers between $[0, 2\pi]$ and $[0, \pi]$, respectively. Additionally, m_1, and m_2 are coefficients defined by [33,35]:

$$m_1 = -\pi + (1 - \tau) * 2\pi, \tag{36.21}$$

$$m_2 = -\pi + \tau * 2\pi, \tag{36.22}$$

where τ is the golden ratio defined by $\tau = (\sqrt{5} - 1)/2$. m_1 and m_2 help to reduce the hyperparameter space and consequently improve the search efficiencies of the agents (i.e. whales, in this case). The flowchart in Fig. 36.5 describes the operational process of the EGolden-SWOA.

36.2.5 Improved WOA-based on non-linear adaptive weight and Golden sine operator (NGS-WOA)

The NGS-WOA [25] introduced a non-linear adaptive weight variable into the WOA search mathematical expressions to improve the WOA's search capabilities. The non-linear adaptive weight variable iteratively guides the search agents in the exploratory process. Moreover, the NGS-WOA also incorporates the GSA [33] described in 36.2.4. In designing the NGS-WOA, modifications were made to expressions (36.2), (36.8), and (36.5), respectively, by the inclusion of a non-linear variable, C_1, and rewritten as:

$$\vec{X}(t+1) = \vec{X}^*(t) - C_1 \cdot \vec{\alpha} \cdot \vec{\rho}, \tag{36.23}$$

$$\vec{X}(t+1) = \vec{X}_r(t) - C_1 \cdot \vec{\alpha} \cdot \vec{\rho}, \tag{36.24}$$

$$\vec{X}(t+1) = \vec{\rho}' \cdot e^{bl} \cdot C_1 \cos(2\pi l) + \vec{X}^*(t) \tag{36.25}$$

The introduction of C_1 into (36.23)–(36.25) enables $\vec{\alpha}$ and l to be adaptively controlled by the number of iterations. This reduces the influences of randomness on the performance of the NGS-WOA. The value of C_1 is determined by:

$$C_1 = \begin{cases} 0.5\left[1 + \cos(\frac{\pi t}{T})\right]^k, & \text{if } t < 0.5T \\ 0.5\left[1 - \cos(\pi + \frac{\pi t}{T})\right]^k, & \text{if } t \geq 0.5T, \end{cases} \tag{36.26}$$

where k, T, and t denote the convergence adjustment factor (i.e. $k = 0.5$ for optimum performance), the total number of iterations, and the current iteration number [25], respectively. The flowchart in Fig. 36.6 describes the operational process of the NGS-WOA.

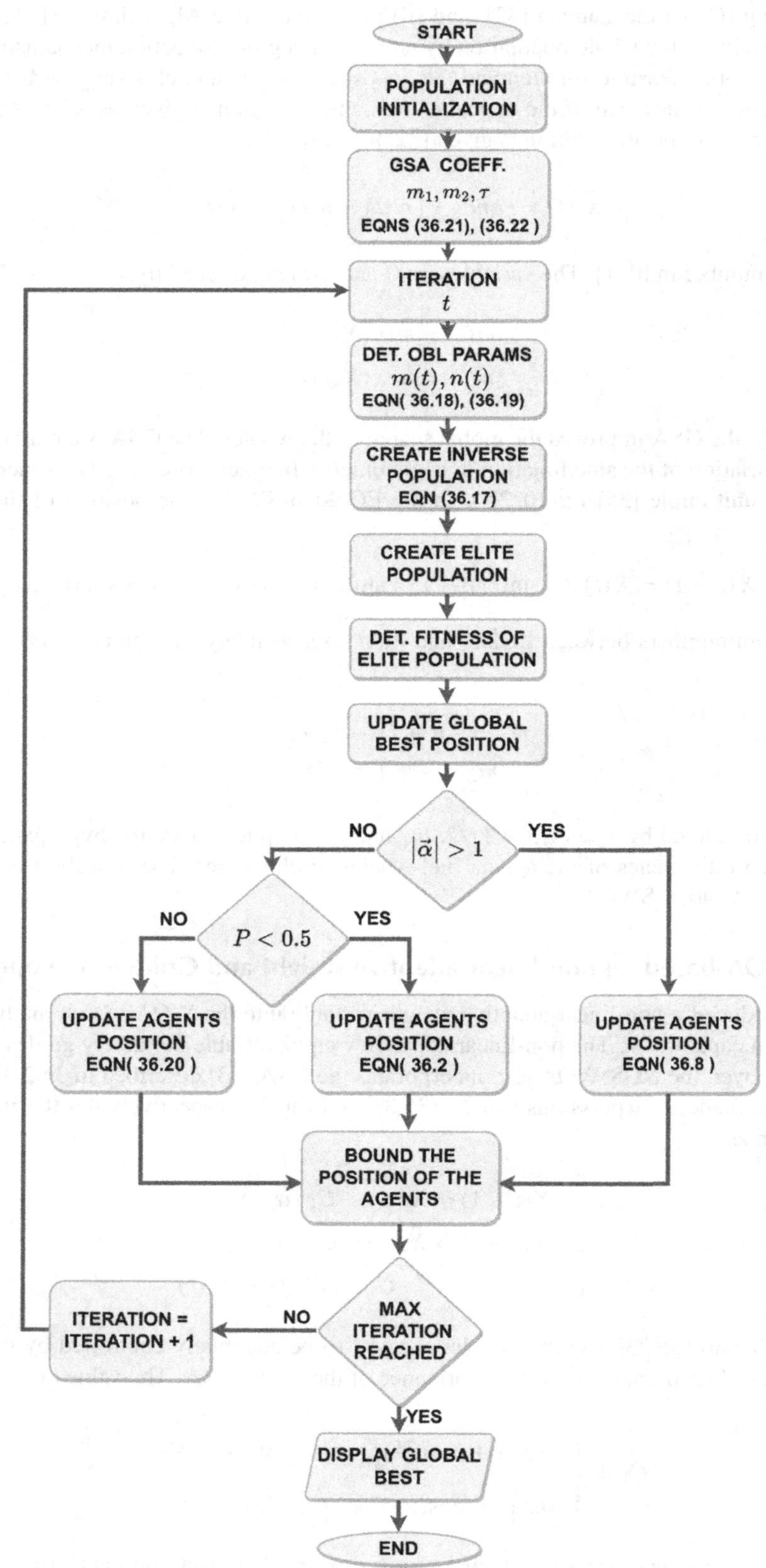

FIGURE 36.5 Flowchart for the EGOLDEN-SWOA.

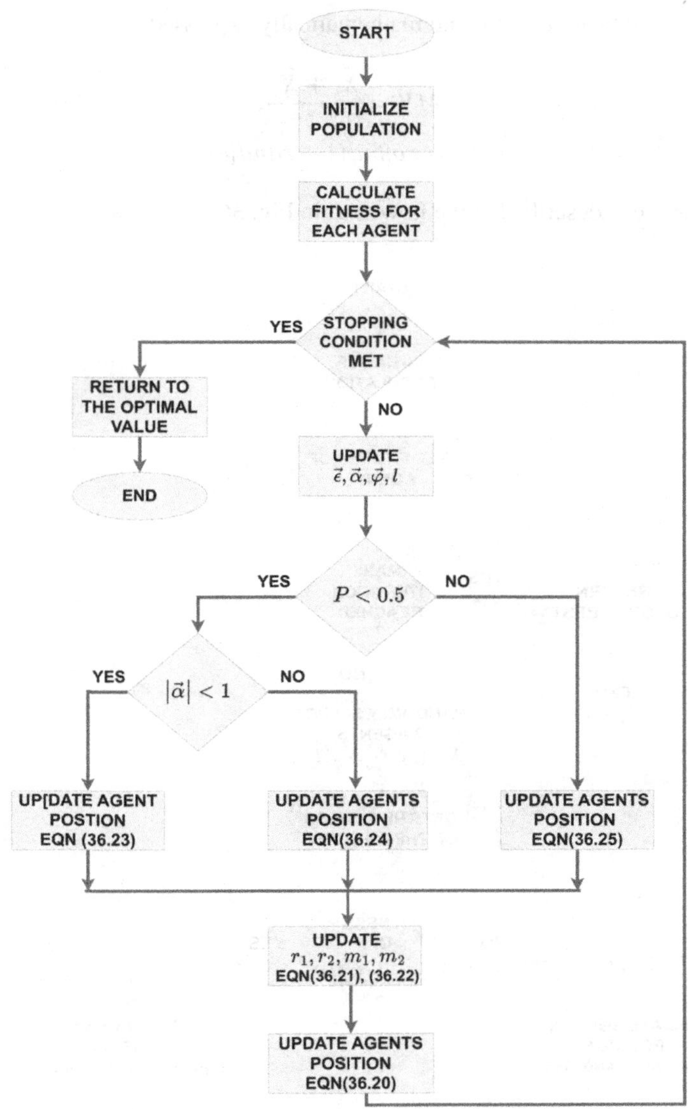

FIGURE 36.6 Flowchart for the NGS-WOA.

36.2.6 Whale optimization algorithm with a modified mutualism phase

The Whale Optimization Algorithm with a Modified Mutualism Phase (WOAmM) [26] was proposed to address the premature convergence of the WOA by the inclusion of a modified mutualism stage of the Symbiotic Organisms Search (SOS) [36]. The SOS uses the symbiotic relationship between the two distinct species. Common symbiotic relationships are mutualism, commensalism, and parasitism. However, in [36], the emphasis is on the modified mutualism phase of the SOS.

Mutualism, simply put, is a beneficial, inter-dependable relationship between two organisms. A common example of mutualism is the relationship between bees and flowers in the pollination process. The process is expressed mathematically as:

$$\vec{X}_i(t+1) = \vec{X}_i(t) + rand_9 * \left(\vec{X}^*(t) - MV * BF1\right), \tag{36.27}$$

$$\vec{X}_j(t+1) = \vec{X}_j(t) + rand_{10} * \left(\vec{X}^*(t) - MV * BF2\right), \tag{36.28}$$

where $\vec{X}_i$ is the ith element or population in the population vector, whereas $\vec{X}_j$ is selected randomly from the population vector and it is otherwise referred to as the organism. Moreover, $\vec{X}_i$ and $\vec{X}_j$ interact in a mutual relationship. MV and the

BF's denote the mutual vector and benefit vector and mathematically expressed as

$$MV = \frac{\vec{X}_i + \vec{X}_j}{2}, \tag{36.29}$$

$$BF = round(1 + rand_{10}) \tag{36.30}$$

The modified mutualism phase is described in the flowchart in Fig. 36.7.

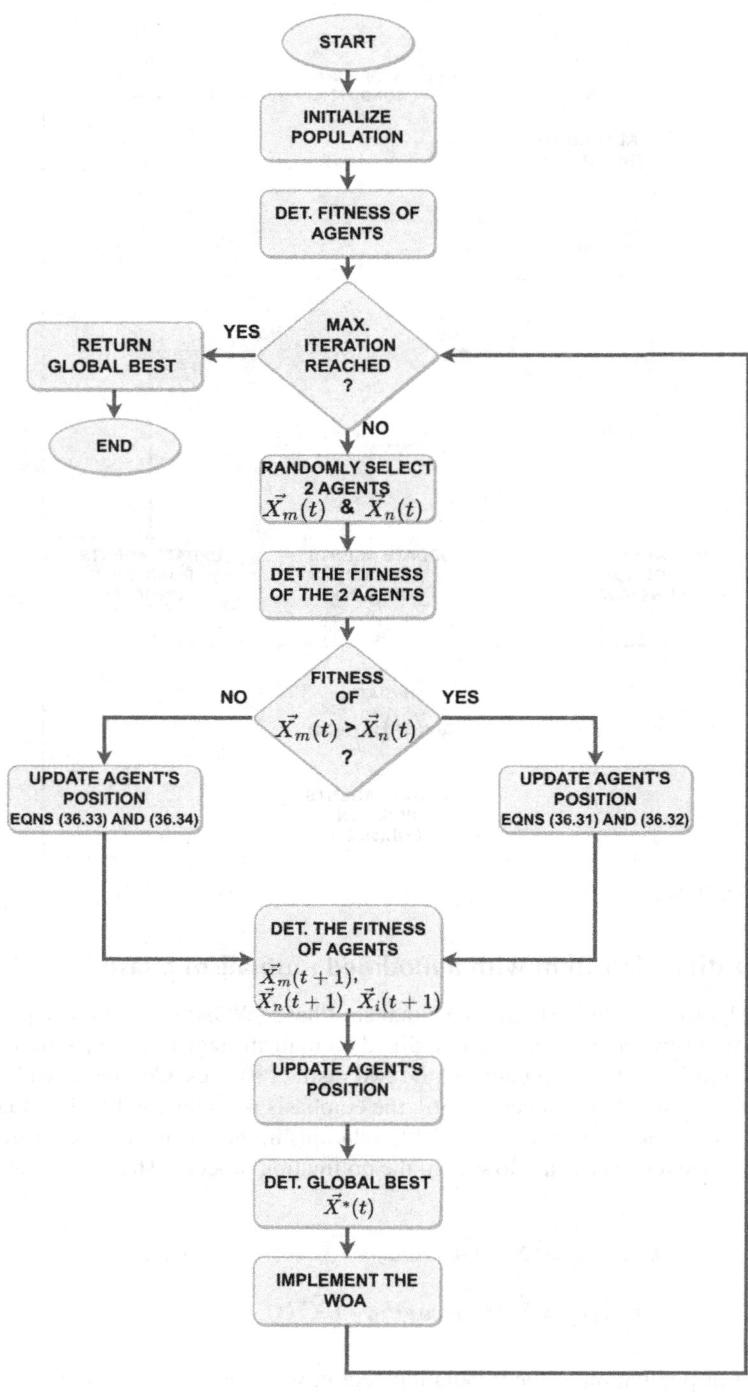

FIGURE 36.7 Flowchart for the WOAmM.

In the modified mutualism phase of an agent $\vec{X}_i$, two search agents are randomly selected from the population for each iteration. The fitness of randomly selected agents $\vec{X}_m$ and $\vec{X}_n$, determine the updating process of $\vec{X}_i$. If the fitness value of $\vec{X}_m$ is less than that of $\vec{X}_n$, hence, the mathematical expressions of the updating process is given by (36.31) and (36.32). However, if the fitness value of $\vec{X}_m$ is greater than that of $\vec{X}_n$, then the position updating process will follow (36.33) and (36.34).

$$\vec{X}_i(t+1) = \vec{X}_i(t) + rand_{11} * \left(\vec{X}_m(t) - MV * BF1\right), \tag{36.31}$$

$$\vec{X}_n(t+1) = \vec{X}_n(t) + rand_{12} * \left(\vec{X}_m(t) - MV * BF2\right), \tag{36.32}$$

$$\vec{X}_i(t+1) = \vec{X}_i(t) + rand_{13} * \left(\vec{X}_n(t) - MV * BF1\right), \tag{36.33}$$

$$\vec{X}_m(t+1) = \vec{X}_m(t) + rand_{14} * \left(\vec{X}_n(t) - MV * BF2\right). \tag{36.34}$$

36.3 SVM: a brief history and recent developments

In 1907, Stanislaw Zaremba, then a mathematics professor at Jagiellonian University in Krakow, Poland, introduced the idea of reproducing kernels with to respect boundary value problems for harmonic functions, a central concept in statistical learning [37,38] (and the bedrock of SVMs several years later). Ronald A. Fisher, in 1936, in his seminal work on pattern recognition [39], using statistical methods, classified three plants species; namely, Iris setosa, Iris virginica, and Iris veriscolor [40]. The idea of reproducing kernels was not used again for several years, until Eliakim H. Moore, Nachman Aronszajn, and Stefan Bergman revisited it. In 1950, Nachman Aronszajn systematically developed the reproducing kernels Hilbert space [37,41].

Vladimir Vapnik and Alexander Larner, in 1963, proposed a novel algorithm named the 'Generalized Portrait Algorithm' (GPA) [42]. The GPA was a pattern recognition algorithm. The SVM's framework relies on the non-linear implementation of the GPA. Alexey Chervonenkis and Vladimir Vapnik improved the GPA in 1963 based on Frank Rosenblatt's perceptron [43] and went on to develop the Vapnik-Chervonenkis (VC) theory [44], which is also referred to as 'the fundamental theory of learning'. In the following years, i.e. 1973 and 1974, Richard Duda and Peter Hart both worked on improving the margins of the hyperplanes [45]. SVMs gained prominence between 1974 and 1982, coinciding with the translations of Vladimir Vapnik's work [46–48] in Russian to English [49] and German [48].

In the 1990s, Vapnik collaborated with his colleagues at AT&T. The SVM's margin maximization was shown to minimize the maximum loss by Bernard Boser, Isabelle Guyon, and Vladimir Vapnik [50]. Moreover, 1995, Corinna Cortes and Vapnik developed the idea of the soft margin classifier [51]. In 1998, Vapnik improved the soft margin classifier to address regression problems [1]. In that same year, Peter Bartlett, John Shawe-Taylor, Robert Williamson, and Martin Anthony examined the SVM's hard margin statistical bound [52,53]. John Shawe-Taylor and Nello Christianini went on to examine the SVM's soft-margin's statistical bound [54].

36.4 SVMs: a general overview

SVMs have found uses in fields such as disease diagnosis and prognosis; bioinformatics; financial prediction; handwritten character recognition; text and hypertext categorization; encryption and decryption; geoinformatics; image segmentation; image recognition; and speech and emotion recognition. Fig. 36.8 illustrates domains in which SVMs have been applied.

36.4.1 The hard-margin SVM optimization problem

SVMs are supervised ML models capable of traditionally solving two-class classification problems. The SVM completes the classification by creating an optimal decision line or boundary that divides m-dimensional space into classes. Consequently, a data point can be correctly categorized using the divided m-dimensional space. The optimal decision line or boundary is called the 'hyperplane'. In the m-dimensional space, the data points closest to the hyperplane are referred to as 'support vectors' (SV). The SVM's goal is to ensure that the SVs are as far away from the hyperplane as possible. The perpendicular distance from the SVs to the hyperplane is called the 'margin'. SVs affect and determine the position of the hyperplane and, resultantly, the accuracy of the SVM. Fig. 36.9 illustrates these critical components of the SVM classifier.

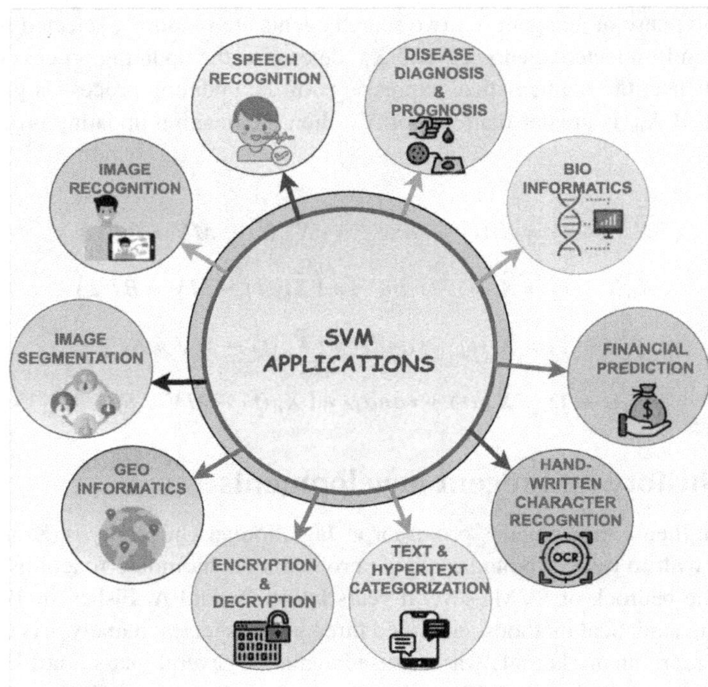

FIGURE 36.8 Some areas of application of the SVM.

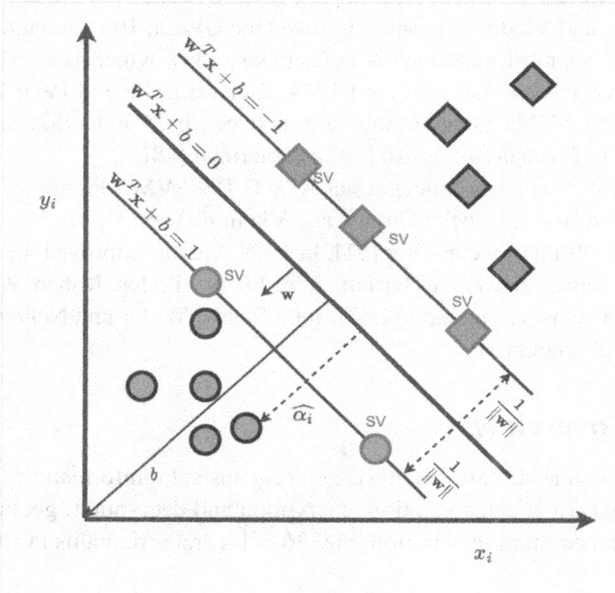

FIGURE 36.9 Illustration of the SVM classifier.

Assuming we have a dataset $\mathcal{G}$ of n m-dimension with input vector $\mathbf{x}_i$ associated with classes y_i, the dataset $\mathcal{G}$ is defined by:

$$\mathcal{G} = \left\{ (\mathbf{x}_i, y_i) \mid \mathbf{x}_i \in \mathbb{R}^m,\ y_i \in \{-1, 1\} \right\}_{i=1}^{n} \tag{36.35}$$

From geometry of the equation of a line defined by:

$$y = a\mathbf{x} + b \tag{36.36}$$

Renaming the two variables y and x as $\mathbf{x}_1$ and $\mathbf{x}_2$. Therefore, (36.36) is rewritten as:

$$a\mathbf{x}_1 - \mathbf{x}_2 + b = 0 \tag{36.37}$$

Assuming we combine $\mathbf{x}_1$ and $\mathbf{x}_2$ into a 2-dimensional vector, then $\mathbf{x} = (\mathbf{x}_1, \mathbf{x}_2)$, $\mathbf{w} = (a, -1)$, and b is the bias. Hence, by rewriting (36.37), the hyperplane is defined by:

$$\mathbf{w}^T\mathbf{x} + b = 0 \tag{36.38}$$

A comparison of the SVM classifier prediction given by $\mathbf{w}^T\mathbf{x} + b$ and the actual class y_i yields the functional margin defined by:

$$\alpha_i = y_i(\mathbf{w}^T\mathbf{x} + b) \tag{36.39}$$

Owing to the scalability weakness of the functional margin, (36.39) is normalized using the norm of $\mathbf{w}$ and giving rise to the geometric margin $\widehat{\alpha_i}$ given by:

$$\widehat{\alpha_i} = \frac{\alpha_i}{\|\mathbf{w}\|} \tag{36.40}$$

With the functional margin α_i set to 1, therefore, $1/\|\mathbf{w}\|$. It is important to note that the geometric margin is scaling-invariant. The goal of the SVM is to maximize the geometric margin.

$$\max_{b,w} \quad \widehat{\alpha_i} \tag{36.41}$$

s.t.

$$y_i(\mathbf{w}^T\mathbf{x} + b) \geq 1, \quad 1, 2, 3, \ldots, n$$

A careful inspection of the objective function shows that the minimization of the denominator of $\widehat{\alpha_i}$ equals the maximization of $\widehat{\alpha_i}$. Hence, we rewrite (36.41):

$$\min_{b,w} \quad \|\mathbf{w}\| \tag{36.42}$$

s.t.

$$y_i(\mathbf{w}^T\mathbf{x} + b) \geq 1, \quad 1, 2, 3, \ldots, n$$

However, (36.42) is non-convex and may be difficult to solve. Owing to the invariant characteristics of optimizing under a square function transformation and since $\mathbf{x}^2$ is monotonic for values of $\mathbf{x} \geq 0$, optimizing over x is the same as optimizing over $\mathbf{x}^2$. We can rewrite (36.42) as:

$$\min_{b,w} \quad \frac{1}{2}\|\mathbf{w}\|^2 \tag{36.43}$$

s.t.

$$y_i(\mathbf{w}^T\mathbf{x} + b) \geq 1, \quad 1, 2, 3, \ldots, n$$

The addition of the multiplier factor (i.e. 1/2) to the objective function in (36.43) is for mathematical convenience and manipulation in later steps. The SVM optimization in (36.43) can be solved using the Lagrangian multipliers [55]. The Lagrangian of (36.43), which is a dual problem, is defined as:

$$\mathcal{L}(\mathbf{w}, b, \gamma) = \frac{1}{2}\|\mathbf{w}\|^2 - \sum_{i=1}^{n} \gamma_i[y_i(\mathbf{w}^T\mathbf{x} + b) - 1] \tag{36.44}$$

where γ_i is the Lagrangian multiplier. In this case, the Lagrangian multiplier is the SV. From the duality standpoint, an optimization problem can be viewed from two perspectives: (i) as a primal problem, which entails the minimization problem; and (ii) as a dual problem, which focuses on the maximization problem [56,57]. The two perspectives lead to the same solution [3,58]. The Lagrangian primal problem of (36.44) is given as:

$$\min_{b,w}\max_{\gamma} \quad \mathcal{L}(\mathbf{w}, b, \gamma) \tag{36.45}$$

s.t.

$$\gamma_i \geq 0, \quad 1, \ldots, n$$

In solving, the derivative of Lagrangian function is taken wrt $\mathbf{w}$ and b [59]:

$$\frac{\partial \mathcal{L}(\mathbf{w}, b, \gamma)}{\partial \mathbf{w}} = \mathbf{w} - \sum_{i=1}^{n} \gamma_i y_i \mathbf{x}_i = 0 \tag{36.46}$$

$$\frac{\partial \mathcal{L}(\mathbf{w}, b, \gamma)}{\partial b} = -\sum_{i=1}^{n} \gamma_i y_i = 0 \tag{36.47}$$

From (36.46), $\mathbf{w}$ is defined as:

$$\mathbf{w} = \sum_{i=1}^{n} \gamma_i y_i \mathbf{x}_i \tag{36.48}$$

Since $\|\mathbf{w}\|^2$ in (36.44) is the same as $\mathbf{w}.\mathbf{w}$, and with (36.48), we can rewrite (36.44) as:

$$\mathcal{L}(\mathbf{w}, b, \gamma) = \frac{1}{2}\left(\sum_{i=1}^{n} \gamma_i y_i \mathbf{x}_i\right) \cdot \left(\sum_{j=1}^{n} \gamma_j y_j \mathbf{x}_j\right) - \sum_{i=1}^{n} \gamma_i \left[y_i \left(\left(\sum_{j=1}^{n} \gamma_j y_j \mathbf{x}_j \right) .\mathbf{x}_i + b \right) - 1 \right] \tag{36.49}$$

$$= \frac{1}{2}\sum_{i=1}^{n}\sum_{j=1}^{n} \gamma_i \gamma_j y_i y_j \mathbf{x}_i.\mathbf{x}_j - \sum_{i=1}^{n} \gamma_i y_i \left(\left(\sum_{j=1}^{n} \gamma_j y_j \mathbf{x}_j \right) .\mathbf{x}_i + b \right) + \sum_{i=1}^{n} \gamma_i \tag{36.50}$$

$$= \sum_{i=1}^{n} \gamma_i - \frac{1}{2}\sum_{i=1}^{n}\sum_{j=1}^{n} \gamma_i \gamma_j y_i y_j \mathbf{x}_i.\mathbf{x}_j - b\sum_{i=1}^{n} \gamma_i y_i \tag{36.51}$$

However, from (36.47) $\sum_{i=1}^{n} \gamma_i y_i = 0$, therefore (36.51) yields:

$$= \sum_{i=1}^{n} \gamma_i - \frac{1}{2}\sum_{i=1}^{n}\sum_{j=1}^{n} \gamma_i \gamma_j y_i y_j \mathbf{x}_i.\mathbf{x}_j \tag{36.52}$$

Hence, with (36.52) we can rewrite (36.44) as a Wolfe-Lagrangian dual problem, which is given as:

$$\max_{\gamma} \sum_{i=1}^{n} \gamma_i - \frac{1}{2}\sum_{i=1}^{n}\sum_{j=1}^{n} \gamma_i \gamma_j y_i y_j \mathbf{x}_i.\mathbf{x}_j \tag{36.53}$$

s.t.

$$\gamma_i \geq 0, \quad 1, \ldots, n$$

$$\sum_{i=1}^{n} \gamma_i y_i = 0$$

This is the hard margin SVM optimization problem which is suitable for linearly separable data [60,61] and would not work with non-linearly separable data. 'Linearly separable data' refers to simple datasets that can easily be classified using simple linear classifiers. In other words, there exists a hyperplane that can classify the dataset. A detailed mathematical explanation on linear separability of datasets is given in [62,63].

36.4.2 The soft margin SVM optimization problem

Real-world datasets are not only linearly separable but also non-linearly separable [64]. Unlike hard margin SVMs, which do not make mistakes, soft margin SVMs are designed to make a few mistakes, as illustrated in Fig. 36.10. This is achieved by the addition of a slack variable (or a vector) β_i to the objective function in (36.43). With a large value of β_i, the constraint will be met at the expense of an optimal hyperplane. To avoid this situation, the objective function of (36.43) is regularized by the addition of the sum of the individual β_i. Moreover, to control the soft margin classifier, a new variable, C, is introduced. When $C = 0$, the soft margin SVM becomes a hard margin SVM.

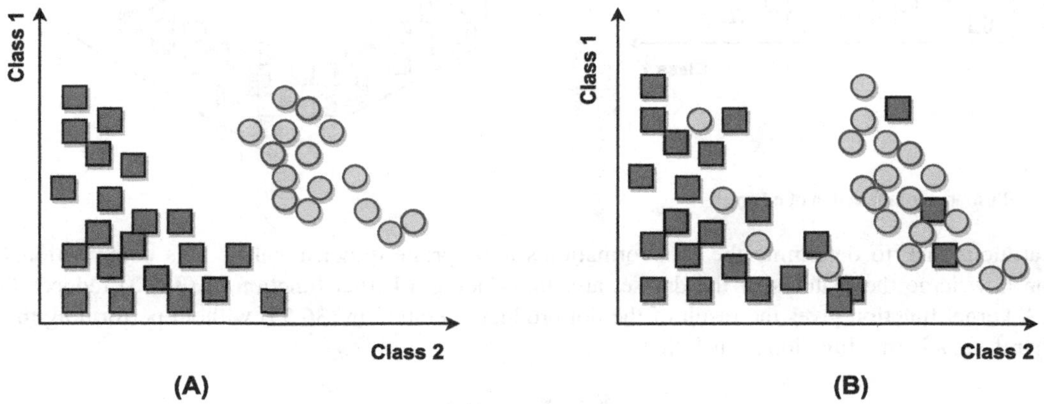

FIGURE 36.10 Illustration of the hard and soft margin classifiers in (A) and (B), respectively.

To this end, the soft margin SVM optimization is formulated as:

$$\min_{b,w,\beta} \frac{1}{2}\|\mathbf{w}\|^2 + C\sum_{i=1}^{n} \beta_i \tag{36.54}$$

s.t.

$$y_i(\mathbf{w}^T\mathbf{x} + b) \geq 1 - \beta_i, \quad i = 1, \ldots, n$$
$$\beta_i \geq 0, \quad i = 1, \ldots, n$$

The soft margin SVM optimization in (36.54) is the 1-norm soft margin [1]. The objective function of the 2-norm soft margin [54] optimization is given as:

$$\min_{b,w,\beta} \frac{1}{2}\|\mathbf{w}\|^2 + C\sum_{i=1}^{n} \beta_i^2 \tag{36.55}$$

A detailed Lagrangian solution of the 1-norm, 2-norm soft margin SVM optimization problem is given in [51] as:

$$= \sum_{i=1}^{n} \gamma_i - \frac{1}{2}\sum_{i=1}^{n}\sum_{j=1}^{n} \gamma_i\gamma_j y_i y_j \mathbf{x}_i.\mathbf{x}_j - \frac{1}{4C}\sum_{i=1}^{n} \gamma_i^2 \tag{36.56}$$

36.4.3 The kernel trick

When SVMs are used to classify non-linearly separable data, the data is first transformed into a higher dimensionality and then fed into the SVM as inputs such that [59] $\mathbf{x} \in \mathbb{R}^m \rightarrow \Phi(\mathbf{x})$, where $\mathbb{R}^m \rightarrow \mathbb{R}^k$; that is, $\mathbb{R}^m$ is transformed into $\mathbb{R}^k$ on the condition that $k \gg m$. We illustrate what a kernel does in Fig. 36.11.

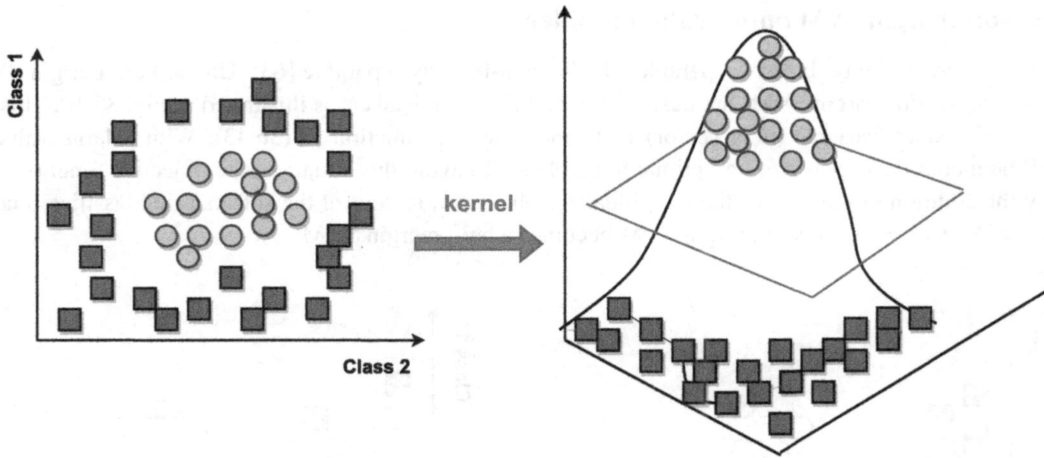

FIGURE 36.11 Illustration of the action of a kernel.

It is an arduous task to determine the transformation's appropriate dimensionality. It is computationally complex, depending on how large the features of the dataset are. Introducing a kernel function in (36.53) reduces the solution's complexity. A kernel function gives the result of the dot product operation in (36.53) without performing the transformation [65]. Therefore, a kernel function is such that:

$$\mathbf{F}(\mathbf{x}_i, \mathbf{x}_j) = \mathbf{x}_i.\mathbf{x}_j \tag{36.57}$$

To this end, the dot product in the above equations is replaced by a kernel function.

There are four widely used kernels in the literature: (i) the linear kernel, (ii) the polynomial kernel, (iii) the Gaussian Radial Basis Function (RBF), and (iv) the sigmoid kernel. Linear kernels are applied when the dataset is linearly separable. With linear kernels, the SVM classifier applies a straight line to differentiate the classes. It is a simple kernel and defined by:

$$\mathbf{F}(\mathbf{x}_i, \mathbf{x}_j) = \mathbf{x}_i.\mathbf{x}_j \tag{36.58}$$

Unlike linear kernels, polynomial kernels are applied when the data is non-linearly separable and defined by:

$$\mathbf{F}(\mathbf{x}_i, \mathbf{x}_j) = (\mathbf{x}_i.\mathbf{x}_j + e)^d \tag{36.59}$$

where d denotes the degree of the polynomial kernel and e represents a constant factor. Upon careful inspection of (36.59), we see that (36.59) is the same as (36.58) if $e = 0$ and $d = 1$. Simply put, with a polynomial degree of 1 and a constant factor of zero, the polynomial kernel becomes a linear kernel. As the value of d increases, the performance of the polynomial kernel improves; however, this may lead to overfitting.

In scenarios where the polynomial kernels' performance is poor, the RBF kernel is employed. The RBF kernel is suitable for multivariate functions and is defined by:

$$\mathbf{F}(\mathbf{x}_i, \mathbf{x}_j) = e^{-\alpha(\mathbf{x}_i - \mathbf{x}_j)^2} \tag{36.60}$$

where α is the tuning variable. As α becomes very small, the RBF kernel operates in a similar way with a linear kernel. As α increases, the RBF's performance greatly depends on the SVs. The Gaussian Kernel is a regularized RBF and defined by:

$$\mathbf{F}(\mathbf{x}_i, \mathbf{x}_j) = e^{-\frac{\|\mathbf{x}_i - \mathbf{x}_j\|}{2\sigma^2}} \tag{36.61}$$

The sigmoid kernel, otherwise known as the 'hyperbolic tangent kernel', is similar to a two-layer perceptron model in a neural network. It is defined by:

$$\mathbf{F}(\mathbf{x}_i, \mathbf{x}_j) = \tanh(\alpha.\mathbf{x}_i + \mathbf{x}_j + e) \tag{36.62}$$

where α denotes the slope, and it is generally given as the inverse of the data dimension. Additionally, e denotes the intercept constant [66]. A list of kernels is given in [65]. Table 36.1 summarizes the kernel types and their demerits.

TABLE 36.1 Merits of some widely employed kernels.

Kernel	Formula	Merits
Linear	$\mathbf{F}(\mathbf{x}_i, \mathbf{x}_j) = \mathbf{x}_i.\mathbf{x}_j$	• Memory Efficient. • Faster than other kernels. • Effective with higher dimensional datasets.
Polynomial	$\mathbf{F}(\mathbf{x}_i, \mathbf{x}_j) = (\mathbf{x}_i.\mathbf{x}_j + e)^d$	• Effective with normalized training data. • Great global performance.
RBF	$\mathbf{F}(\mathbf{x}_i, \mathbf{x}_j) = e^{-\alpha(\mathbf{x}_i - \mathbf{x}_j)^2}$	• Effective in no prior knowledge of data. • Less tuning parameter. • Great local performance.

36.5 Hyperparameter tuning

Like other ML algorithms, the SVM's parameters and hyperparameters greatly affect its performance. Hyperparameters are parameters of ML algorithms that are not directly learnt in the model's learning process. Resultantly, hyperparameter values and thresholds are specified outside the learning or training process, as illustrated in Fig. 36.12.

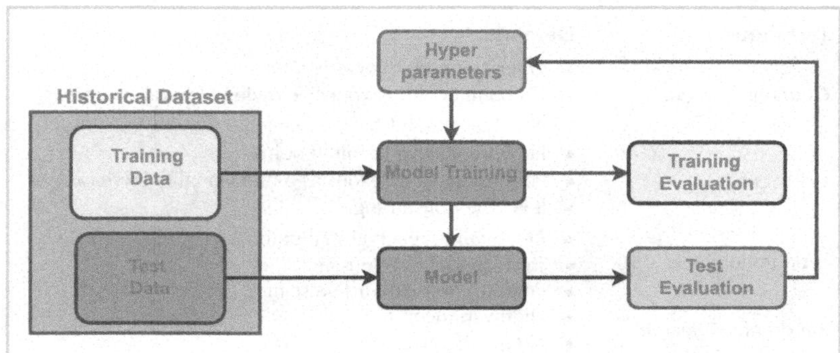

FIGURE 36.12 Illustration of an ML model creation process.

Hyperparameters control the model's flexibility and capacity to fit the training data. The SVM's hyperparameters are: (i) the kernel type, (ii) the regularization strength, C, (iii) the width of the RBF, α, and (iv) the degree of the polynomial kernel, d.

Though hyperparameters have a substantial impact on the training algorithm's performance, the SVM's hyperparameter settings differ for different datasets. It is, therefore, critical to determine the hyperparameter combinations to maximize the SVM's performance while reducing computational cost.

The hyperparameter tuning process follows the determination of the hyperparameters' space, which entails the hyperparameters' threshold. Additionally, a method of sampling the hyperparameter candidates is chosen before the training data is cross-validated. 'Cross-validation' is the resampling of the data in small folds to train the model over several iterations to establish a more accurate generalized fit. The hyperparameters' performance metric is established to determine the SVM classifier's best hyperparameter settings. Fig. 36.13 describes the stages in the hyperparameter tuning process.

Traditional optimization techniques such as gradient descent, grid search, randomized search, and experimental methods have been employed in the literature to obtain the SVM's parameters. However, these traditional techniques frequently fail in their search for the global optima. Table 36.2 lists some of these techniques' demerits. In [67], several strategies applied to the SVM optimization problem are discussed. Among them are evolutionary algorithms, i.e. metaheuristics [68]. In this work, we focus extensively on the application of the WOA, a widely used metaheuristic, in improving the performance of SVMs.

36.6 Empirical analysis of metaheuristic-based SVM training

In this section, we present and discuss the empirical analysis of the performance of SVMs when tuned by the WOA and improved versions. The simulation and empirical analysis were carried out in a Python Jupyter Notebook [69] on a computer with a configuration of Intel(R) Core (TM) i7-7600U CPU at 2.80 GHz 2.90 GHz (2 processors), installed

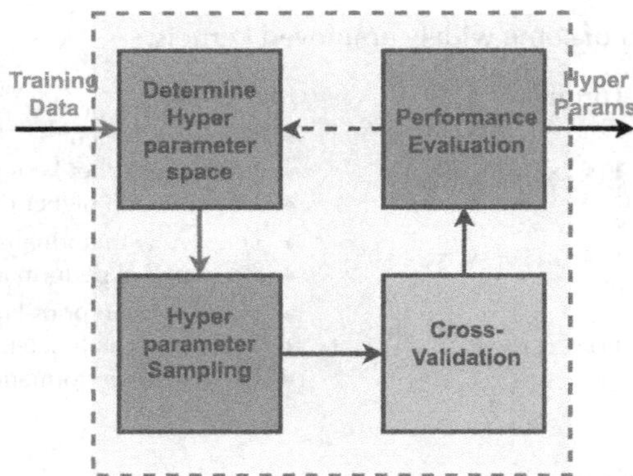

FIGURE 36.13 Hyperparameter tuning process.

TABLE 36.2 Demerits of some traditional techniques employed in solving optimization problems.

Technique	Demerits
Gradient Descent	• Often trapped in local optima. • Computationally expensive owing to frequent updates. • Slow convergence.
Grid Search	• Highly sensitive to initial point. • Performances degrades as dimensionality increases. • It is time-consuming.
Experimental	• Often fail to reach global optima. • Time and labor intensive. • Prone to non-exhaustive search.
Randomized Search	• High variance. • No intelligence.

memory of 16.0 GB, and an NVIDIA Graphics Processor. The SVM's performance was assessed on some widely employed benchmark datasets in the literature, such as: (i) the Breast Cancer Wisconsin (Diagnostic) dataset [70], (ii) the Iris dataset [71], and (iii) the Modified National Institute of Standards and Technology (MNIST) dataset [72], (iv) the Wine Recognition dataset [73], (v) the Glass Identification dataset [74], (vi) the Car Evaluation dataset [75], and (vii) the Indian Liver Patient dataset [71].

36.6.1 Dataset employed for empirical analysis

The Breast Cancer Wisconsin dataset is intended to classify whether a patient's diagnosis is benign or malignant. The dataset has 30 attributes and 569 data points or instances. The Iris dataset contains three classes of Iris plants: Setosa, Veriscolor, and Virginica. The dataset has 150 data points and four attributes, namely: Sepal Length, Sepal Width, Petal Length, and Petal Width, with all four attributes measured in centimeters. The MNIST dataset contains 70000 instances of 28×28 pixel grayscale images. These images are handwritten single digits from 0 to 9. The training dataset comprises 60000 instances/data points and the remaining 10000 instances are the test set. The Wine Recognition dataset comprises 178 data instances with 13 unique numeric features (such as flavonoids, color intensity, and alkalinity of ash) and three classes. The dataset is as a result of chemical analysis of three types of wine grown in Italy. The Glass Identification data set is an imbalanced data set based on a comparison test a rule-based system, called BEAGLE, the nearest-neighbor algorithm, and discriminant analysis. It is premised on the need to identify glass types in a crime scene for evidence purpose in criminology. The data set has nine unique features (such as refractive index, silicon, and magnesium) indicating the oxide content of the glass material, six classes, and 214 data points or instances. The car data sets comprises six numeric and categorical features, with 1728 instances and four classes of cars. A brief description of the seven datasets is depicted in Table 36.3.

TABLE 36.3 Description of the datasets.

No.	Dataset	Instances	Attributes	Classes	Sources
1.	Breast Cancer Wisconsin	569	30	2	[70]
2.	Iris	150	4	3	[71]
3.	MNIST	70000	784	10	[72]
4.	Wine Recognition	178	13	3	[73]
5.	Glass Identification	214	9	6	[74]
6.	Cars Evaluation	1728	6	4	[75]
7.	Indian Liver Patients	583	10	2	[71]

TABLE 36.4 SVM's test accuracy performance evaluation with hyperparameters tuned by WOA and its improved versions for seven datasets.

Data set	Metric	WOA	MSWOA	LWOA	EGOLDEN	NGS-WOA	WOAmM
Breast Cancer	AVG.	0.9220	0.8622	0.9220	0.6257	0.8035	0.9220
	STD.	0.0000	0.1183	0.0000	0.0000	0.1452	0.0000
	WORST	NA	0.6257	NA	NA	0.6257	NA
	BEST	0.9220	0.9220	0.9220	0.6257	0.9220	0.9220
	MODE	0.9220	0.9220	0.9220	0.6257	0.9220	0.9220
Iris	AVG.	0.9810	0.9829	0.9848	0.9629	0.9829	0.9838
	STD.	0.0000	0.0071	0.0047	0.0124	0.0071	0.0044
	WORST	NA	0.9714	0.9810	0.9524	0.9714	0.9810
	BEST	0.9810	0.9905	0.9905	0.9810	0.9905	0.9905
	MODE	0.9810	0.9810	0.9810	0.9524	0.9810	0.9810
Wine Recognition	AVG.	0.8073	0.8042	0.8057	0.4316	0.6803	0.8074
	STD.	0.0000	0.0042	0.0034	0.1118	0.1872	0.0001
	WORST	NA	0.7990	0.7990	0.3943	0.3943	0.8073
	BEST	0.8073	0.8077	0.8077	0.7670	0.8077	0.8077
	MODE	0.8073	0.8077	0.8073	0.3943	0.7993	0.8073
Glass Identification	AVG.	0.7106	0.7119	0.7120	0.6635	0.7106	0.7139
	STD.	0.0020	0.0061	0.0052	0.0310	0.0057	0.0033
	WORST	0.7099	0.6961	0.7032	0.6294	0.6961	0.7099
	BEST	0.7166	0.7168	0.7168	0.7032	0.7168	0.7168
	MODE	0.7099	0.7099	0.7099	0.6294	0.7099	0.7166
Car Evaluation	AVG.	0.9866	0.9864	0.9866	0.9471	0.9224	0.9868
	STD.	0.0005	0.0007	0.0003	0.0860	0.1124	0.0000
	WORST	0.9851	0.9851	0.9859	0.7006	0.7006	NA
	BEST	0.9868	0.9868	0.9868	0.9859	0.9868	0.9868
	MODE	0.9868	0.9868	0.9868	0.9851	0.9868	0.9868
Indian Liver Patient	AVG.	0.7257	0.7257	0.7257	0.7257	0.7257	0.7257
	STD.	0.0000	0.0000	0.0000	0.0000	0.0000	0.0000
	WORST	NA	NA	NA	NA	NA	NA
	BEST	0.7257	0.7257	0.7257	0.7257	0.7257	0.7257
	MODE	0.7257	0.7257	0.7257	0.7257	0.7257	0.7257
MNIST	AVG.	0.8621	0.8613	0.8592	0.0838	0.5502	
	STD.	0.0012	0.0000	0.0016	0.0000	0.3809	
	WORST	0.8613	NA	0.8575	NA	0.0838	
	BEST	0.8637	0.8613	0.8613	0.0838	0.8613	
	MODE	0.8613	0.8613	0.8575	0.0838	0.8613	

TABLE 36.5 Performance evaluation of the computational time taken to tune the SVM's hyperparameters in seconds.

Data set	Metric	WOA	MSWOA	LWOA	EGOLDEN	NGS-WOA	WOAmM
Breast Cancer	AVG.	688.89	755.82	776.98	1762.21	781.55	3565.62
	STD.	3.64	81.73	3.49	15.03	109.72	5.07
	WORST	695.13	918.77	781.19	1784.21	915.84	3572.34
	BEST	684.56	710.06	773.10	1744.98	677.98	3558.97
	MODE	684.56	710.06	773.10	1744.98	677.98	3558.97
Iris	AVG.	39.97	35.39	36.02	105.99	35.82	184.14
	STD.	8.23	2.44	2.29	7.63	2.73	5.66
	WORST	59.15	40.07	39.59	118.01	41.00	195.74
	BEST	32.76	31.66	32.28	92.80	31.57	176.83
	MODE	32.76	31.66	32.28	92.80	31.57	176.83
Wine Recognition	AVG.	92.40	95.77	107.13	229.27	101.99	451.38
	STD.	4.84	3.85	4.86	5.84	19.26	7.45
	WORST	99.54	99.97	115.97	239.99	143.94	472.05
	BEST	86.27	88.56	100.06	221.52	84.16	445.37
	MODE	86.27	88.56	100.06	221.52	84.16	445.37
Glass Identification	AVG.	180.17	188.66	196.20	425.84	172.70	924.84
	STD.	9.66	7.08	6.15	24.01	6.70	23.94
	WORST	194.04	199.51	210.29	479.70	188.97	973.81
	BEST	168.62	174.82	187.15	396.18	166.76	880.05
	MODE	168.62	174.82	193.00	396.18	166.76	880.05
Car Evaluation	AVG.	1649.39	1769.82	2536.61	9804.55	3337.51	7264.25
	STD.	97.79	548.87	129.38	1270.78	2064.32	188.99
	WORST	1777.96	3410.68	2843.97	11869.78	7102.17	7548.50
	BEST	1499.70	1514.08	2404.04	7884.70	1643.42	7024.02
	MODE	1499.70	1514.08	2404.04	7884.70	1643.42	7024.02
Indian Liver Patient	AVG.	779.66	778.20	776.73	4056.56	804.24	4382.42
	STD.	1.40	0.80	2.63	98.92	2.67	18.14
	WORST	782.48	779.86	781.97	4122.98	808.95	4411.09
	BEST	777.02	777.35	772.71	3779.64	799.68	4341.80
	MODE	777.02	777.38	772.71	3779.64	799.68	4341.80
MNIST	AVG.	5650.70	5493.76	5925.95	13768.20	5955.66	
	STD.	180.93	60.74	18.65	63.47	79.86	
	WORST	5904.60	5574.71	5943.28	13819.67	7135.23	
	BEST	5496.28	5428.41	5900.06	13644.12	5107.25	
	MODE	5496.28	5428.41	5900.06	13644.12	5107.25	

36.6.2 Performance evaluation

In Tables 36.4–36.6, the SVM's hyperparameter tuning process performance evaluation is examined in terms of accuracy, number of support vectors, and computational time for the tuning process of the WOA and its improved versions. For the purposes of fairness in the benchmarking and performance evaluation process, all the metaheuristics had 20 search agents and a maximum iteration of 200. To ensure the accuracy of the results, we adopt the Monte Carlo simulation technique, which relies on the law of large numbers.

In Table 36.4, the EGolden-SWOA tuned SVM had the lowest average classification accuracy rate for most of the examined data sets. In Table 36.4, the LWOA and WOAmM tuned SVM performed better than the other versions for almost

TABLE 36.6 SVM's number of support vectors performance evaluation with hyperparameters tuned by WOA and its improved versions for seven datasets.

Data set	Metric	WOA	MSWOA	LWOA	EGOLDEN	NGS-WOA	WOAmM
	AVG.	317	333	317	398	349	317
	STD.	0	33	0	0	40	0
Breast Cancer	MIN	317	317	317	398	317	317
	MAX	317	398	317	398	398	317
	MODE	317	317	317	398	317	317
	AVG.	17	22	18	41	22	17
	STD.	1	11	1	15	11	1
Iris	MIN	15	15	16	16	14	16
	MAX	18	46	19	57	50	18
	MODE	15	15	18	57	17	16
	AVG.	94	98	94	122	104	94
	STD.	1	3	2	7	14	2
Wine Recognition	MIN	93	92	92	102	93	93
	MAX	97	101	99	124	124	100
	MODE	94	100	93	124	94	94
	AVG.	100	106	101	120	108	100
	STD.	1	8	4	6	6	4
Glass Identification	MIN	99	98	98	112	98	98
	MAX	100	116	109	129	116	112
	MODE	100	99	99	129	98	99
	AVG.	304	305	299	502	564	299
	STD.	15	13	0	355	372	0
Car Evaluation	MIN	299	299	299	267	299	299
	MAX	347	334	299	1209	1209	299
	MODE	299	299	299	267	299	299
	AVG.	404	404	404	404	404	404
	STD.	0	0	0	0	0	0
Indian Liver Patient	MIN	404	404	404	404	404	404
	MAX	404	404	404	404	404	404
	MODE	404	404	404	404	404	404
	AVG.	638	636	632	800	702	
	STD.	3	0	14	0	80	
MNIST	MIN	636	636	612	800	636	
	MAX	642	636	642	800	800	
	MODE	636	636	612	800	800	

all the examined data sets. The LWOA and WOAmM tuned SVM the had highest average classification accuracy rate. The WOA, MSWOA, and NGS-WOA can also tune the hyperparameters to achieve near optimal or optimal classification accuracy for the SVM. Moreover, it is pertinent to examine the classification accuracy of the tuned SVM against the backdrop of the computational time of the tuning process.

Table 36.5 shows the time taken (in seconds) to tune the SVM's hyperparameters by the respective metaheuristics for each of the examined datasets. It is observed that though the SVM tuned by the improved versions of the WOA may achieve a higher classification accuracy than the WOA tuned SVM, this comes at the expense of longer computational time

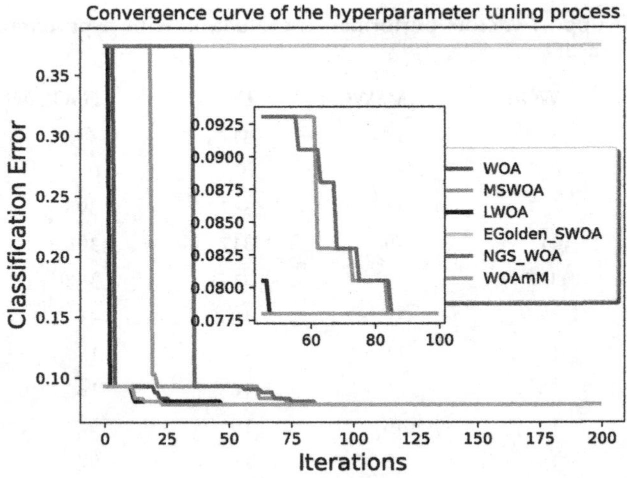

(a) Breast Cancer Wisconsin data set.

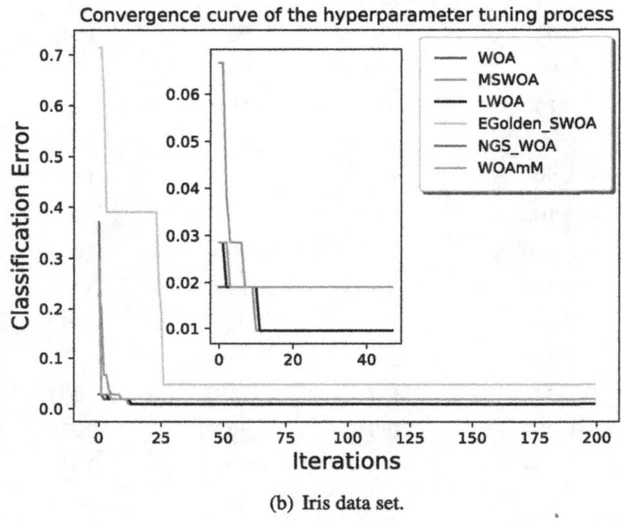

(b) Iris data set.

FIGURE 36.14 Instances of the convergence curves of WOA and its improved versions for Breast Cancer Wisconsin and Iris data sets.

in tuning the hyperparameters. We see that the WOAmM and EGolden-SWOA took the longest time in tuning the SVM hyperparameters, which may be attributed to the complexity order of the two algorithms. Moreover, it is observed that the WOA and MSWOA had the shortest possible average time taken to tune the hyperparameters for all the data sets. In general, the number of attributes, classes, and instances greatly affect the hyperparameter tuning time; this is observed in the time taken to tune the SVM for data sets such as the MNIST, Car Evaluation, Indian Liver Patient, and Breast Cancer Wisconsin. A careful comparison of Tables 36.3 and 36.5 gives more insight into this. Table 36.6 highlights the performance of the tuning process in identifying the support vectors.

Figs. 36.14–36.16, show the convergence curves of the WOA and its improved version in the finding optimal hyperparameters of the SVM. The inset plots in Figs. 36.14 and 36.15 give greater detail of the main chart where some or all of the WOA variants have the same or very close classification error values. In Fig. 36.14(a), the inset plot is to capture the main chart within the classification error values from 0 to 0.1, for iterations number 50 to 100. Similarly, in Fig. 36.14(b), the inset chart captures the main chart for iterations 0 to 50. The inset charts in Fig. 36.15 capture the main chart for iterations 0 to 100, and 0 to 6, respectively. In Figs. 36.14–36.16, the convergence curves of six of the examined data sets are depicted. It is observed that WOA, LWOA, and WOAmM, respectively converge faster than the other versions for most of the data sets. Additionally, they converge at about the 25th iterations. The MSWOA, NGS-WOA, and EGolden-SWOA exhibit

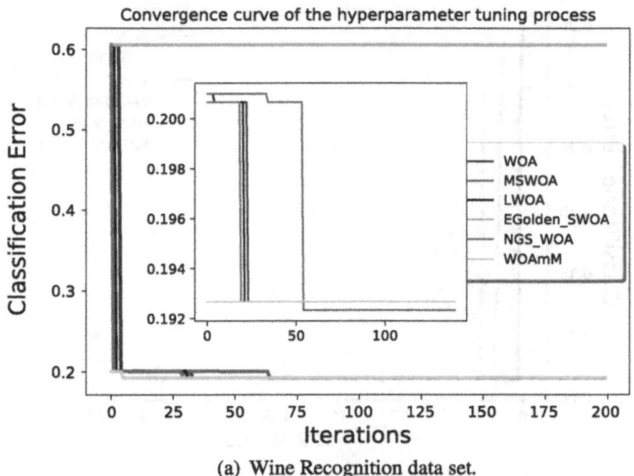

(a) Wine Recognition data set.

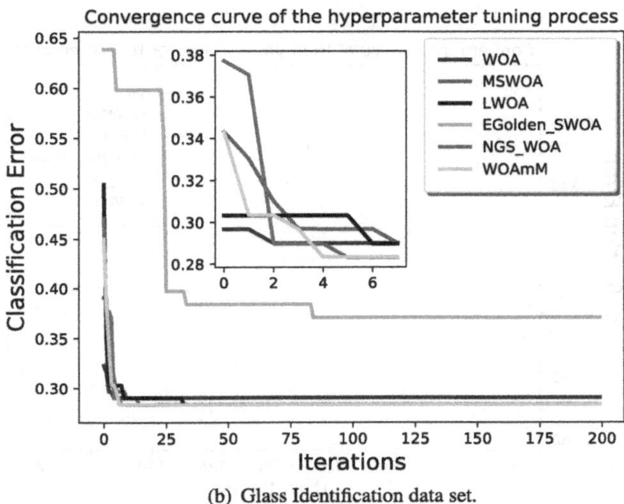

(b) Glass Identification data set.

FIGURE 36.15 Instances of the convergence curves of WOA and its improved versions for Wine Recognition and Glass Identification data sets.

great exploratory capabilities and therefore tend to converge later than the others for most of the data sets involved. It is observed that for the Indian Liver Patient, all the metaheuristics converge at the first iteration and this may be attributed to the characteristics of the dataset.

In Table 36.7, the impact of the number of search agents of the WOA and its improved version on the classification accuracy of the SVM is presented. The SVM's hyperparameters are tuned by a varying number of agents; 5-25 for the classification of the Iris data set. Generally, we observe that the SVM's classification accuracy increases as the number of agents increases for the WOA and its improved version. This may be attributed to the improved exploratory capabilities of the search agents in finding the SVM's optimal hyperparameters as its number increases.

Similar to the results in Table 36.7, in Table 36.8 the impact of increasing the number of agents on the computational time taken to tune the SVM's hyperparameters is examined. Specifically, in Table 36.8, we employ the Iris data set for the sake of spacial limitations. We note that as the number of agents increases, the computational time also increases for all six metaheuristics. Additionally, similar to Table 36.5, the computational times of the six metaheuristics differ in Table 36.8.

36.7 Conclusion

In this chapter, the capabilities of the WOA and five of its improved versions, namely the MSWOA, LWOA, EGolden-SWOA, NGS-WOA, and WOAmM, were deployed for hyperparameter tuning in SVMs. The hyperparameter tuning problem in SVMs was formulated as an optimization problem to enable the use of the WOA and its improved versions. The

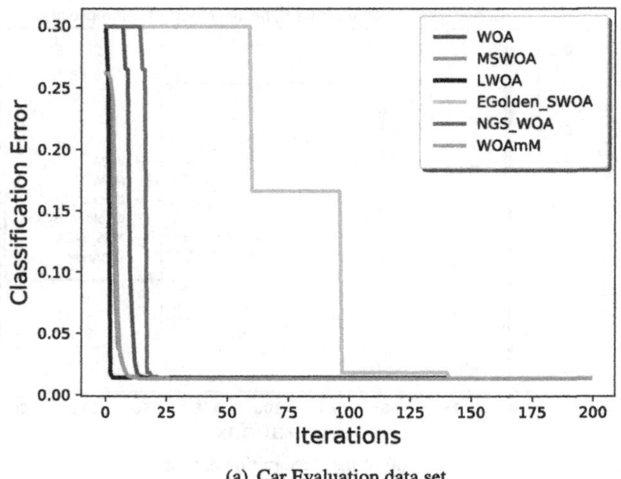

(a) Car Evaluation data set.

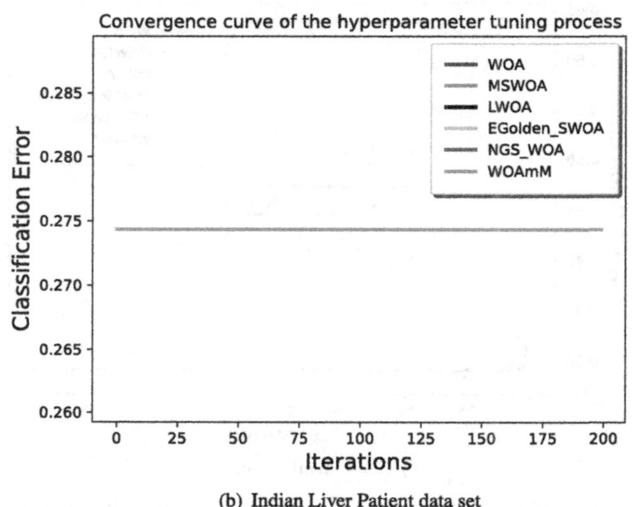

(b) Indian Liver Patient data set

FIGURE 36.16 Instances of the convergence curves of WOA and its improved versions for the Car Evaluation and Indian Liver Patient data sets.

hyperparameter tuning was carried out in an intensive Monte Carlo simulation environment for different scenarios involving seven widely used classification data sets; namely, Breast Cancer Wisconsin, Iris, Wine Recognition, Glass Identification, Car Evaluation, Indian Liver Patient, and MNIST, respectively. These data sets all have unique characteristics which make them suitable for hyperparameter tuning.

To evaluate the performance of the hyperparameter tuning process, we consider the classification accuracy, computational time, number of support vectors, and convergence rate metrics. We observed that the SVM had higher classification accuracy when its hyperparameters were tuned by the LWOA and WOAmM, and the lowest accuracy when tuned by the EGolden-SWOA. Additionally, in terms of computational time taken to complete the tuning process, the WOA and MSWOA achieved the fastest time of the six metaheuristics, whereas the WOAmM and EGolden-SWOA had the longest time to completion. The convergence rates of the WOA and others were also examined as they carried out the hyperparameter optimization. We see that WOA, LWOA, and WOAmM had fast convergence rates, while MSWOA, NGS-WOA, and EGolden-SWOA had better exploratory capacities, which ultimately delays their convergence rates. The impact of the number of search agents on the SVM classification accuracy was also considered and for the most part, the classification accuracy increases with the number of agents. Lastly, the WOA and its other versions can be employed for hyperparameter tuning of SVMs and by extension other machine learning algorithms.

TABLE 36.7 Impact of increasing the number of agents of the metaheuristics on the classification accuracy of the SVM on the Iris data set.

Number of Agents	Metric	WOA	MSWOA	LWOA	EGOLDEN	NGS-WOA	WOAmM
5	AVG.	0.9819	0.9743	0.9857	0.9829	0.9762	0.9800
	STD.	0.0029	0.0086	0.0048	0.0103	0.0077	0.0067
	WORST	0.9810	0.9619	0.9810	0.9619	0.9714	0.9714
	BEST	0.9905	0.9905	0.9905	0.9905	0.9905	0.9905
	MODE	0.9810	0.9714	0.9810	0.9905	0.9714	0.9810
10	AVG.	0.9829	0.9829	0.9848	0.9838	0.9781	0.9857
	STD.	0.0038	0.0057	0.0047	0.0074	0.0105	0.0048
	WORST	0.9810	0.9714	0.9810	0.9714	0.9619	0.9810
	BEST	0.9905	0.9905	0.9905	0.9905	0.9905	0.9905
	MODE	0.9810	0.9810	0.9810	0.9905	0.9810	0.9810
15	AVG.	0.9810	0.9867	0.9838	0.9657	0.9876	0.9829
	STD.	0.0000	0.0047	0.0044	0.0122	0.0061	0.0038
	WORST	0.9810	0.9810	0.9810	0.9524	0.9714	0.9810
	BEST	0.9810	0.9905	0.9905	0.9905	0.9905	0.9905
	MODE	0.9810	0.9905	0.9810	0.9524	0.9905	0.9810
25	AVG.	0.9810	0.9838	0.9848	0.9571	0.9857	0.9810
	STD.	0.0000	0.0061	0.0047	0.0064	0.0077	0.0000
	WORST	0.9810	0.9714	0.9810	0.9524	0.9714	0.9810
	BEST	0.9810	0.9905	0.9905	0.9714	0.9905	0.9810
	MODE	0.9810	0.9810	0.9810	0.9524	0.9905	0.9810

TABLE 36.8 Impact of increasing the number of agents of the metaheuristics on the computational time of the tuning the SVM's hyperparameter for the Iris data set.

Number of Agents	Metric	WOA	MSWOA	LWOA	EGOLDEN	NGS-WOA	WOAmM
5	AVG.	12.67	14.76	13.83	31.64	14.85	53.06
	STD.	0.26	1.78	0.81	2.92	1.40	5.19
	WORST	13.41	16.97	15.19	37.55	16.36	64.69
	BEST	12.48	12.46	12.70	28.65	12.65	48.43
	MODE	12.48	12.46	12.70	28.65	15.73	48.43
10	AVG.	26.03	26.00	26.72	70.19	29.31	125.00
	STD.	0.47	1.34	0.91	5.75	3.01	1.51
	WORST	26.92	29.53	28.90	76.73	33.69	128.44
	BEST	25.49	24.86	25.81	57.89	25.98	123.40
	MODE	25.49	25.00	25.81	57.89	25.98	123.40
15	AVG.	39.41	38.70	40.95	114.92	40.37	196.72
	STD.	0.72	0.79	2.01	5.04	3.09	2.69
	WORST	40.26	40.72	45.49	122.94	49.14	200.00
	BEST	37.92	37.76	38.88	108.15	37.96	190.88
	MODE	37.92	37.76	38.88	108.15	37.96	190.88
25	AVG.	60.71	63.42	66.47	199.90	67.56	337.66
	STD.	0.20	3.99	3.49	5.73	6.23	1.51
	WORST	61.04	75.25	73.08	208.45	79.93	340.26
	BEST	60.36	61.27	63.22	189.51	61.18	333.85
	MODE	60.66	61.27	63.22	189.51	61.18	333.85

References

[1] V.N. Vapnik, The Nature of Statistical Learning Theory, Springer Science & Business Media, New York, USA, 1999.

[2] N. Christianini, J. Shawe-Taylor, An Introduction to Support Vector Machines and Other Kernel-Based Learning Methods, Cambridge University Press, Cambridge, ISBN 0-521-78019-5, 2000.

[3] J. Cervantes, F. Garcia-Lamont, L. Rodríguez-Mazahua, A. Lopez, A comprehensive survey on support vector machine classification: applications, challenges and trends, Neurocomputing 408 (2020) 189–215.

[4] S. Maldonado, J. Merigó, J. Miranda, IOWA-SVM: a density-based weighting strategy for SVM classification via OWA operators, IEEE Transactions on Fuzzy Systems 28 (9) (2019) 2143–2150.

[5] L. Diosan, A. Rogozan, J.-P. Pecuchet, Improving classification performance of support vector machine by genetically optimising kernel shape and hyper-parameters, Applied Intelligence 36 (2) (2012) 280–294.

[6] J. Liang, Z. Qin, J. Ni, X. Lin, X.S. Shen, Practical and secure SVM classification for cloud-based remote clinical decision services, IEEE Transactions on Computers 70 (10) (2021) 1612–1625.

[7] A. Kirchner, C.S. Signorino, Using support vector machines for survey research, Survey Practice 11 (1) (2018) 1–14.

[8] S. Mirjalili, A. Lewis, The whale optimisation algorithm, Advances in Engineering Software 95 (2016) 51–67.

[9] S.O. Ekwe, S.O. Oladejo, L.A. Akinyemi, N. Ventura, A socially-inspired energy-efficient resource allocation algorithm for future wireless network, in: 2020 16th International Computer Engineering Conference (ICENCO), IEEE, 2020, pp. 168–173.

[10] S.O. Ekwe, L.A. Akinyemi, S.O. Oladejo, N. Ventura, Social-aware joint uplink and downlink resource allocation scheme using genetic algorithm, in: 2021 IEEE AFRICON, IEEE, 2021, pp. 1–6.

[11] S.O. Oladejo, S.O. Ekwe, L.A. Akinyemi, Multi-tier multi-domain network slicing: a resource allocation perspective, in: 2021 IEEE AFRICON, IEEE, 2021, pp. 1–6.

[12] S.O. Oladejo, S.O. Ekwe, L.A. Akinyemi, Multi-tier multi-tenant network slicing: a multi-domain games approach, ITU-Journal of Future and Evolving Technologies (ITU-JFET) 2 (6) (2021) 57–82.

[13] S.O. Oladejo, O.E. Falowo, Latency-aware dynamic resource allocation scheme for 5G heterogeneous network: a network slicing-multitenancy scenario, in: 2019 International Conference on Wireless and Mobile Computing, Networking and Communications (WiMob), IEEE, 2019, pp. 1–7.

[14] S.O. Oladejo, O.E. Falowo, Latency-aware dynamic resource allocation scheme for multi-tier 5G network: a network slicing-multitenancy scenario, IEEE Access 8 (2020) 74834–74852.

[15] S.O. Oladejo, Efficient radio resource management for the fifth generation slice networks, Ph.D. thesis, University of Cape Town, Cape Town, South Africa, 2021, http://hdl.handle.net/11427/35992.

[16] D.H. Wolpert, W.G. Macready, No free lunch theorems for optimization, IEEE Transactions on Evolutionary Computation 1 (1) (1997) 67–82.

[17] J.H. Holland, et al., Adaptation in Natural and Artificial Systems: An Introductory Analysis with Applications to Biology, Control, and Artificial Intelligence, MIT Press, 1992.

[18] J.H. Holland, Genetic algorithms, Scientific American 267 (1) (1992) 66–73.

[19] J. Kennedy, R. Eberhart, Particle swarm optimization, in: Proceedings of ICNN'95-International Conference on Neural Networks, vol. 4, IEEE, 1995, pp. 1942–1948.

[20] S. Mirjalili, S.M. Mirjalili, A. Lewis, Grey wolf optimizer, Advances in Engineering Software 69 (2014) 46–61.

[21] F.S. Gharehchopogh, H. Gholizadeh, A comprehensive survey: whale optimization algorithm and its applications, Swarm and Evolutionary Computation 48 (2019) 1–24.

[22] X. Yuan, Z. Miao, Z. Liu, Z. Yan, F. Zhou, Multi-strategy ensemble whale optimization algorithm and its application to analog circuits intelligent fault diagnosis, Applied Sciences 10 (11) (2020), https://www.mdpi.com/2076-3417/10/11/3667.

[23] Y. Ling, Y. Zhou, Q. Luo, Lévy flight trajectory-based whale optimization algorithm for global optimization, IEEE Access 5 (2017) 6168–6186.

[24] X. Ziya, L. Sheng, Elite reverse golden sine whale algorithm and its engineering optimization research, Electronic Journal 47 (10) (2019) 2177–2186.

[25] J. Zhang, J.S. Wang, Improved whale optimization algorithm based on nonlinear adaptive weight and golden sine operator, IEEE Access 8 (2020) 77013–77048.

[26] S. Chakraborty, A.K. Saha, S. Sharma, S. Mirjalili, R. Chakraborty, A novel enhanced whale optimization algorithm for global optimization, Computers & Industrial Engineering 153 (2021) 107086.

[27] P.R. Hof, E. Van der Gucht, Structure of the cerebral cortex of the humpback whale, Megaptera novaeangliae (Cetacea, Mysticeti, Balaenopteridae), The Anatomical Record: Advances in Integrative Anatomy and Evolutionary Biology 290 (1) (2007) 1–31.

[28] R.M. Rizk-Allah, A.E. Hassanien, S. Bhattacharyya, Chaotic crow search algorithm for fractional optimization problems, Applied Soft Computing 71 (2018) 1161–1175.

[29] C. Li, X. An, R. Li, A chaos embedded GSA-SVM hybrid system for classification, Neural Computing & Applications 26 (3) (2015) 713–721.

[30] B.B. Mandelbrot, B.B. Mandelbrot, The Fractal Geometry of Nature, vol. 1, WH Freeman, New York, 1982.

[31] A.F. Kamaruzaman, A.M. Zain, S.M. Yusuf, A. Udin, Levy flight algorithm for optimization problems- a literature review, Applied Mechanics and Materials 421 (2013) 496–501.

[32] H.R. Tizhoosh, Opposition-based learning: a new scheme for machine intelligence, in: International Conference on Computational Intelligence for Modelling, Control and Automation and International Conference on Intelligent Agents, Web Technologies and Internet Commerce (CIMCA-IAWTIC'06), vol. 1, IEEE, 2005, pp. 695–701.

[33] E. Tanyildizi, G. Demir, Golden sine algorithm: a novel math-inspired algorithm, Advances in Electrical and Computer Engineering 17 (2) (2017) 71–78.

[34] H.-H. Xu, R.-L. Tang, Particle swarm optimization with adaptive elite opposition-based learning for large-scale problems, in: 2020 5th International Conference on Computational Intelligence and Applications (ICCIA), IEEE, 2020, pp. 44–49.
[35] W. Xie, J.-S. Wang, Y. Tao, Improved black hole algorithm based on golden sine operator and Levy flight operator, IEEE Access 7 (2019) 161459–161486.
[36] M.-Y. Cheng, D. Prayogo, Symbiotic organisms search: a new metaheuristic optimization algorithm, Computers & Structures 139 (2014) 98–112.
[37] E. Barletta, S. Dragomir, F. Esposito, Weighted Bergman kernels and mathematical physics, Axioms 9 (2) (2020) 48.
[38] S. Zaremba, Léquation biharmonique et une classe remarquable de fonctions fondamentales harmoniques, Imprimerie de l'Universite, 1907.
[39] E. Hannan, Ronald Aylmer Fisher: 1890–1962, Economic Record 38 (84) (1962) 507.
[40] R.A. Fisher, The use of multiple measurements in taxonomic problems, Annals of Eugenics 7 (2) (1936) 179–188.
[41] N. Aronszajn, Theory of reproducing kernels, Transactions of the American Mathematical Society 68 (3) (1950) 337–404.
[42] V. Vapnik, Pattern recognition using generalized portrait method, Automation and Remote Control 24 (1963) 774–780.
[43] F. Rosenblatt, Principles of neurodynamics. Perceptrons and the theory of brain mechanisms, Tech. Rep., Cornell Aeronautical Lab Inc, Buffalo, NY, 1961.
[44] V. Vapnik, A note one class of perceptrons, Automation and Remote Control 25 (1964) 103–109.
[45] R.O. Duda, P.E. Hart, et al., Pattern Classification and Scene Analysis, vol. 3, John Wiley, New York, 1973.
[46] V.N. Vapnik, A.Â. Červonenkis, Teoriâ raspoznavaniâ obrazov: Statističeskie problemy obučeniâ, Izdatel'stvo "Nauka" Glavnaâ Redakciâ Fiziko-Matematičeskoj Literatury, Moscow, Russia, 1974.
[47] V. Vapnik, Estimation of Dependences Based on Empirical Data, Nauka, 1979.
[48] W.N. Wapnik, A. Tscherwonenkis, Theorie der Zeichenerkennung, De Gruyter, Berlin, 1979.
[49] V. Vapnik, Estimation of Dependencies Based on Empirical Data, Springer, Information and Control, 1982.
[50] B.E. Boser, I.M. Guyon, V.N. Vapnik, A training algorithm for optimal margin classifiers, in: Proceedings of the 5th Annual ACM Workshop on Computational Learning Theory, 1992, pp. 144–152.
[51] C. Cortes, V. Vapnik, Support-vector networks, Machine Learning 20 (3) (1995) 273–297.
[52] P.L. Bartlett, The sample complexity of pattern classification with neural networks: the size of the weights is more important than the size of the network, IEEE Transactions on Information Theory 44 (2) (1998) 525–536.
[53] J. Shawe-Taylor, P.L. Bartlett, R.C. Williamson, M. Anthony, Structural risk minimization over data-dependent hierarchies, IEEE Transactions on Information Theory 44 (5) (1998) 1926–1940.
[54] J. Shawe-Taylor, N. Cristianini, Margin distribution and soft margin, in: Advances in Large Margin Classifiers, 2000, p. 349.
[55] B.T. Smith, Lagrange Multipliers Tutorial in the Context of Support Vector Machines, Memorial University of Newfoundland St. John's, Newfoundland, Canada, 2004, p. 17.
[56] A. Gretton, Support vector machines, http://www.gatsby.ucl.ac.uk/~gretton/coursefiles/Slides5A.pdf, 2021. (Accessed 9 February 2022).
[57] K.P. Bennett, E.J. Bredensteiner, Duality and geometry in SVM classifiers, in: International Conf. on Machine Learning (ICML), vol. 2000, Morgan Kaufman, San Francisco, 2000, pp. 57–64.
[58] A. Beck, A. Ben-Tal, Duality in robust optimisation: primal worst equals dual best, Operations Research Letters 37 (1) (2009) 1–6.
[59] C.J. Burges, A tutorial on support vector machines for pattern recognition, Data Mining and Knowledge Discovery 2 (2) (1998) 121–167.
[60] A. Statnikov, A Gentle Introduction to Support Vector Machines in Biomedicine: Theory and Methods, vol. 1, World Scientific Publishing Co. Inc., River Edge, NJ, USA, 2011.
[61] D. Hardin, I. Tsamardinos, C.F. Aliferis, A theoretical characterization of linear SVM-based feature selection, in: Proceedings of the Twenty-First International Conference on Machine Learning, 2004, p. 48.
[62] D. Chen, Q. He, X. Wang, On linear separability of data sets in feature space, Neurocomputing 70 (13–15) (2007) 2441–2448.
[63] D. Elizondo, The linear separability problem: some testing methods, IEEE Transactions on Neural Networks 17 (2) (2006) 330–344.
[64] S. Tasoulis, N.G. Pavlidis, T. Roos, Nonlinear dimensionality reduction for clustering, Pattern Recognition 107 (2020) 107508.
[65] C.R. Souza, Kernel functions for machine learning applications, http://crsouza.com/2010/03/17/kernel-functions-for-machine-learning-applications/, 2010.
[66] H.-T. Lin, C.-J. Lin, A study on sigmoid kernels for SVM and the training of non-PSD kernels by SMO-type methods, Neural Computation 3 (1–32) (2003) 16.
[67] J. Nalepa, M. Kawulok, Selecting training sets for support vector machines: a review, Artificial Intelligence Review 52 (2) (2019) 857–900.
[68] L. Yang, A. Shami, On hyperparameter optimisation of machine learning algorithms: theory and practice, Neurocomputing 415 (2020) 295–316.
[69] T. Kluyver, B. Ragan-Kelley, F. Pérez, B. Granger, M. Bussonnier, J. Frederic, K. Kelley, J. Hamrick, J. Grout, S. Corlay, P. Ivanov, D. Avila, S. Abdalla, C. Willing, Jupyter notebooks – a publishing format for reproducible computational workflows, in: F. Loizides, B. Schmidt (Eds.), Positioning and Power in Academic Publishing: Players, Agents and Agendas, IOS Press, 2016, pp. 87–90.
[70] K.P. Bennett, O.L. Mangasarian, Robust linear programming discrimination of two linearly inseparable sets, Optimisation Methods and Software 1 (1) (1992) 23–34.
[71] D. Dua, C. Graff, UCI machine learning repository, http://archive.ics.uci.edu/ml, 2017.
[72] Y. Lecun, L. Bottou, Y. Bengio, P. Haffner, Gradient-based learning applied to document recognition, Proceedings of the IEEE 86 (11) (1998) 2278–2324.
[73] S. Aeberhard, D. Coomans, O. de Vel, Comparative analysis of statistical pattern recognition methods in high dimensional settings, Pattern Recognition 27 (8) (1994) 1065–1077, https://www.sciencedirect.com/science/article/pii/0031320394901457.
[74] I.W. Evett, E.J. Spiehler, Rule induction in forensic science, in: Knowledge Based Systems, Halsted Press, Div. of John Wiley & Sons, New York, United States, 1989, pp. 152–160.
[75] M. Bohanec, V. Rajkovic, Knowledge acquisition and explanation for multi-attribute decision making, in: 8th Intl Workshop on Expert Systems and Their Applications, Citeseer, 1988, pp. 59–78.

Chapter 37

Gene selection for microarray data classification based on mutual information and binary whale optimization algorithm

Maha Nssibi[a,b], **Ghaith Manita**[a,c], **and Ouajdi Korbaa**[a,d]

[a]*Laboratory MARS, LR17ES05, ISITCom, University of Sousse, Sousse, Tunisia,* [b]*ENSI, University of Manouba, Manouba, Tunisia,* [c]*ESEN, University of Manouba, Manouba, Tunisia,* [d]*ISITCom, University of Sousse, Sousse, Tunisia*

37.1 Introduction

DNA microarray technology is considered an essential asset in bioinformatics and biotechnology. Breakthroughs with this technology, along with the gene expression technique, uncovered several enigmas of the biological aspects of all living cells and have allowed the analysis and the treatment of such endogenous expression of genes. Thus, dealing with thousands of genes that can be present in a single experiment [1], remains a challenging task, not only due to the immense number of genes but also the redundancy and the irrelevance of a part of them, which decreases and impacts the performance of the microarray technology. Therefore, the gene selection problem is tackled by the Feature Selection (FS) [2] approach. The purpose of feature selection is to reduce the number of genes to reduce the time and the computational requirements, increase accuracy, and gain data interpretability efficiency. Feature selection methods are assembled into three sets: filter, wrapper, and embedded [3]. Wrapper and embedded methods [4,5] are based on classification techniques, and the selection of the genes is a part of the training phase of the learning algorithms. In contrast, filter methods are based on ranking techniques to select genes that are ranked above the fixed threshold [6]. Considering the large space of solutions and the incremental number of genes, selecting the most relevant subset of genes is regarded as an NP-complete combinatorial optimization problem [7]. Wherefore, metaheuristics are applied to solve the feature selection problem in a wrapper-based approach, using the classification algorithms as a fitness evaluation. Wrapper approaches for feature selection based on gene expression data are modeled as a combinatorial optimization, where the combinatorial variables are the genes, and the objective function is whether to minimize or maximize the fitness function based on the classification accuracy of a given classifier. Since the objective is whether to select a feature, the feature selection problem is modeled as a binary optimization problem. Many studies have proposed a binary version of different metaheuristics to tackle this problem [8–10].

One of the recent metaheuristic optimization algorithms, the Whale Optimization Algorithm (WOA) [11], is based on how humpback whales hunt. WOA uses artificial whales as search agents in the search process. The humpback whales use a distinctive method of hunting known as bubble-net feeding. Compared to other metaheuristic algorithms, it has been demonstrated that this metaheuristic produces competitive results. Indeed, WOA was introduced to address the problems associated with feature selection as well as continual real search space optimization challenges [12,13].

WOA provides the benefits of a balanced application of the global and local search methods and a successful execution with fewer parameters. However, it also has certain disadvantages, such as local optima stagnation and a failure to apply the global search thoroughly. These limitations are the motivation of this study, where in this chapter, we propose an Improved Binary Whale Optimization Algorithm based on Mutual Information (MI). The MI-based feature selection approach is transformed into global optimization. The binary WOA is used to select features effectively, avoid being trapped in local optima, and maximize the MI between features and class labels. This chapter proposes a new feature selection algorithm for microarray data based on a binary whale optimization algorithm and mutual information (M-BWOA).

The rest of the chapter is organized as follows. Section 37.2 introduces the WOA algorithm. The proposed M-BWOA algorithm is proposed in Section 37.3. Section 37.4 depicts the experimental setting and the analysis of the results. Finally, Section 37.5 presents the summary of this chapter.

Handbook of Whale Optimization Algorithm. https://doi.org/10.1016/B978-0-32-395365-8.00043-9

37.2 Whale Optimization Algorithm (WOA)

The authors of [11] presented the Whale Optimization Algorithm (WOA), which was modeled after the bubble-net feeding technique used by whales. In WOA, the target prey or a location close to it is the current best candidate solution. Then, the other whales will make an effort to improve their positions. The WOA models the swarming behavior mathematically as follows:

$$D = |CX^*(t) - X(t)| \tag{37.1}$$

$$X(t+1) = X^*(t) - A \cdot D, \tag{37.2}$$

where t represents the current iteration, C and A are coefficient vectors, X^* represents the position vector of the best solution, and X is the position vector. A and C values are calculated by the following equations:

$$A = 2 \cdot a \cdot r - a \tag{37.3}$$

$$C = 2 \cdot r, \tag{37.4}$$

where a is linearly decreased from two to zero over iterations and $r \in [0, 1]$. The exploitation phase is simulated mathematically as follows. (1) Shrinking encircling: obtained by decreasing the values according to Eq. (37.4). Remark that a is a random value in $[-a, a]$. (2) Spiral updating: calculates the distance between the prey and the whale. Eq. (37.5), calculates the spiral that mimics the spiral movement as follows:

$$X(t+1) = D^l e^{bl} \cdot \cos(2\pi l) + X^*(t) \tag{37.5}$$

where b is fixed and l is a random number in $[-1, 1]$. To choose either the spiral model or the shrinking encircling mechanism model, a probability of 50% is assumed as follows:

$$X(t+1) = \begin{cases} X^*(t) - A \cdot D & \text{if } p < 0.5 \\ D^l \cdot e^{bl} \cdot \cos(2\pi l) + X^*(t) & \text{otherwise} \end{cases} \tag{37.6}$$

where p is a random number in a uniform distribution. On the other hand, in the exploration phase, $1 < A < -1$ is used to force the agent to move away from this location. Eqs. (37.7) and (37.8) represent the exploration phase mathematically as follows:

$$D = |C \cdot X_{\text{rand}} - X| \tag{37.7}$$

$$X(t+1) = X_{\text{rand}} - A \cdot D. \tag{37.8}$$

37.3 Binary Whale Optimization Algorithm (BWOA)

The Gene selection problem tackled by the feature selection approach is considered a binary optimization problem since it is formulated as to whether to select a gene or not. Consequently, the WOA metaheuristic is transformed into a binary version to adapt the feature selection approach. In the continuous version of the WOA algorithm, the whales change their position in a continuous search space. Thus, in binary optimization, the search agents (the whales) are restricted to {0, 1} values, and a feature subset is represented as a single-dimensional vector, where the length of the vector is equal to the number of features. Each feature is set to 1 for the selected feature or 0 for the non-selected feature. To convert the continuous optimization algorithm to the binary version, transfer functions (TF) are utilized, which has been demonstrated as a successful tool for this conversion [14]. In this work, a recent class of new transfer functions, Taper-shaped TF [15], are applied. Taper-shaped TF function has an attractive property, unlike the existing transfer functions, Taper-shaped TF has a unified calculation formula of $T(x) = (|\frac{x}{A}|)^{\frac{1}{n}}$, $x \in [-A, A]$, $n \geq 1$. In the interest of this unified formula and the calculation simplicity, a high gain in the computational requirements and the algorithm robustness is achieved. The calculation formulas of the four Taper-shaped TF are given in Table 37.1, and their curves on $[-6, 6]$ are shown in Fig. 37.1. According to authors in [15], $T3$ and $T4$ exhibited the best performance. Therefore in this study, we adopt the third and fourth taper shape functions.

37.3.1 Proposed fitness function

Filter and wrapper techniques are used to select the most representative features generally. Based on the qualities of the data itself, filters reduce attributes. Filters are applied outside of the learning procedure by measuring the utility of rates.

TABLE 37.1 Mathematical formula of Taper-shape transfer functions.

Taper-shaped TF
T1: $TF(x) = \frac{\sqrt{\lvert x \rvert}}{\sqrt{\lvert A \rvert}}$
T2: $TF(x) = \frac{\lvert x \rvert}{\lvert A \rvert}$
T3: $TF(x) = \frac{\sqrt[3]{\lvert x \rvert}}{\sqrt[3]{\lvert A \rvert}}$
T4: $TF(x) = \frac{\sqrt[4]{\lvert x \rvert}}{\sqrt[4]{\lvert A \rvert}}$

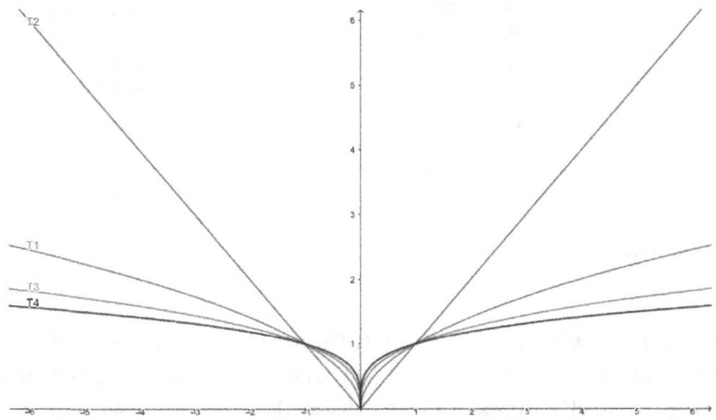

FIGURE 37.1 Taper shape transfer function.

Before training begins, attributes not anticipated to offer helpful information for classification are filtered out of the dataset. When using the wrapper strategy, the attribute space is combed to identify a feature subset motivated by the classification success of individual attribute subsets. Therefore, intelligent search space exploration is never easy because only one fitness function assessment requires a lot of time. This method is more time-consuming because the classifier has to be retrained on all potential attribute sets, and its efficiency has to be evaluated.

Regardless of the classifier employed, filter approaches always perform poorly in attribute reduction since they only consider gauging attributes' relevance based on the data's qualities. In contrast, the wrapper technique examines a significantly more extensive set of attribute combinations. This fact may be wasteful, but it can perform better with careful application because it is heavily classifier-guided.

The suggested approach is a wrapper technique that is guided by filter-based principles to make use of both the effectiveness of the filter-based techniques and the classification performance of the wrapper-based ones. Finding attribute combinations that use both filter-methods principles and wrapper-based methods concepts is done using a two-stage Whale Optimization Algorithm; see Fig. 37.2.

In the first stage, WOA is used to search the most appropriate set of features that maximizes the following fitness function based on mutual information index (MI).

Formally, the main goal of MI-based feature section methods is to identify the subset of features $S = \{f_1, \ldots, f_k\}$ from a given set F of n features that have the greatest degree of dependency with the target class C, or to identify the features S that maximize the value of $I(S; C)$ as formulated in (37.9).

$$\begin{aligned} I(S; C) &= I(f_1, \ldots, f_k; C) \\ &= \sum_{f_1, \ldots, f_k} \sum_{C} P(f_1, \ldots, f_k; C) \log \frac{P(f_1, \ldots, f_k; C)}{P(f_1, \ldots, f_k) P(C)} \end{aligned} \tag{37.9}$$

The majority of approaches recommend adding features to S sequentially since it is computationally costly to compute the joint MI, I(S; C) in (37.9). As mentioned before, we decide to use WOA to deal with the problem.

By the end of the first step, a population is obtained that optimizes the mutual information equation in (37.9). The following are the first solutions for the second level optimization that utilized BWOA to enhance classification performance using the obtained population.

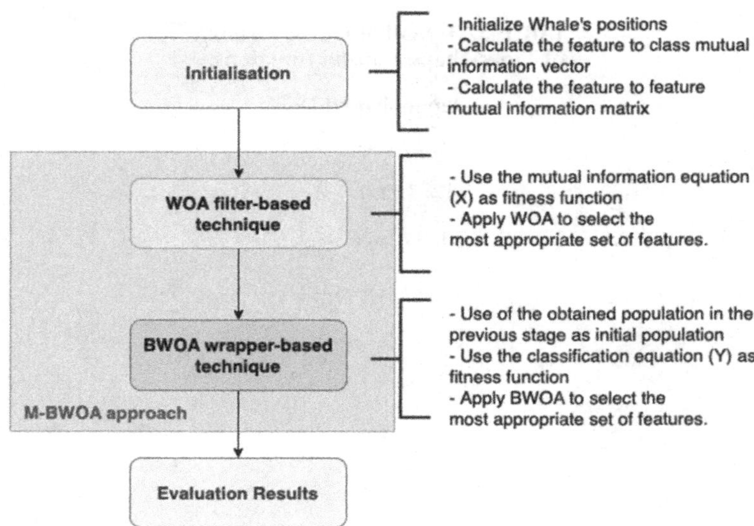

FIGURE 37.2 The proposed M-BWOA flowchart.

The fitness function, which assesses whether a subset fulfills the objectives, is the core component of BWOA. Researchers have employed accuracy and feature number as key criteria in their evaluations of each feature subset and the fitness function. The accuracy of a method's illness classification determines how well it performs overall. The detection rate or classification accuracy (P) and the number of attributes substantially impact the gene selection problem. The fitness function is used in the model in the manner described below:

$$Fitness = P \cdot a + \left(\frac{1}{\mathrm{NF}}\right) \cdot b \tag{37.10}$$

where NF is the number of subset features for classification, while P, a, and b depend on the empirical scope.

37.4 Experimental results

The proposed approaches were implemented using MATLAB® R2020a and run on an Intel Core i7 machine, 2.6 GHz CPU and 16 GB of RAM. Experiments were repeated 30 independent times to obtain statistically meaningful results.

37.4.1 Datasets description and parameter settings

For the evaluation of the proposed M-BWOA, nine biological gene expression datasets are used for experiments that are detailed in Table 37.2. For performance comparison, the M-BWOA will be compared against the basic version of WOA. State-of-the-art metaheuristics that tackled the gene selection problem are chosen namely: Particle Swarm Optimization (PSO) [16], Genetic Algorithm (GA) [17], Grey Wolf Optimizer (GWO) [18], Harris Hawks Algorithm (HHO) [19], and Atom Search Optimization (ASO) [20]. The parameters of these algorithms are selected based on their reference paper, where they studied the gene selection problem. These parameters are detailed in Table 37.3. During the experiments, the k-fold cross-validation [21] is used to assess the robustness of the generated results. The data is divided into K equal folds, the $K - 1$ folds are used for the classifier's training during the optimization process, and the remaining fold is used for testing and validating the classifier's performance. The procedure is repeated K times by using different folds as the testing set. The standard form of this technique in the wrapper feature selection algorithms is based on the k-nearest neighbor (KNN) classifier [22], with $K = 5$, which is applied during this study.

37.4.2 Results and discussion

The performance of the proposed algorithm variants, including M-BWOA-1 and M-BWOA-2, referring to T3 and T4 transfer functions, respectively, will be evaluated compared to the other algorithms. At first, the evaluation is conducted on statistical measures, including the best fitness value (best), the mean fitness value (avg), and the standard deviation (std).

TABLE 37.2 Details of datasets.

Dataset	No. of instances	No. of features
CLL_SUB_111 [23]	111	11340
colon [24]	62	2000
leukemia [25]	72	7070
lung [26]	203	3312
lung_discrete [27]	73	325
lymphoma [28]	96	4026
nci9 [29]	60	9712
Prostate_GE [30]	102	5966
SMK_CAN_187 [31]	187	19993

TABLE 37.3 Algorithms parameter settings.

Algorithm	Parameter	Value
PSO	c1, c2	2
	ω	0.1
GA	crossover rate	0.9
	mutation rate	0.1
GWO	α	[2,0]
HHO	α	[0,1]
	μ	
ASO	α	50
	β	0.2
	Vmax	6
AVOA	L_1	0.8
	L_2	0.2
	k	2.5
	P_1	0.6
	P_2	0.4
	P_3	0.6
WOA	a	[2,0]
All of them	search agents (particles, atoms, vultures, whales, ...)	30
	maximum iterations	1000

After that, the evaluation is based on the average accuracy and the average number of the selected features. Furthermore, the evaluation is based on the nonparametric Wilcoxon rank sum test [32], to check whether the obtained results of the algorithms are significantly different or not.

The listed results of M-BWOA variants in Table 37.4, are significantly better than those achieved by AVOA, PSO, GA, GWO, HHA, and ASO algorithms. Whereas M-BWOA is ranked the best algorithm, GWO outperforms the whole set only on the Lung_discrete dataset. Therefore, the fitness optimization results reported in Table 37.4, confirm the efficiency and robustness of the proposed algorithm. For further evaluation, the performance comparison is validated in terms of classification accuracy in Table 37.5 and the average number of selected features in Table 37.6. As reported in Table 37.5, where the best results are highlighted in **bold**, the proposed M-BWOA-2 corresponding to the T4 TF variant can be ranked the best, considering it achieved the highest accuracy value at four datasets, including colon, lymphoma, Prostate_GE, and SMK_CAN_187. AVOA scored the best result on one of nine datasets, namely lung_discrete. Likewise, WOA exhibits the same response as AVOA in the case of nci9, along with ASO for only one dataset. Meanwhile, PSO shows to be the best for the two datasets. The reported results confirm that the proposed approach is capable of maintaining the right balance between exploration and exploitation during the optimization of the FS problem. To sustain the obtained results, the results of the average number of selected features/genes are presented in Table 37.6. Where it can be observed that M-BWOA-2

TABLE 37.4 Comparison of fitness optimization results obtained by the algorithm variants, standard WOA and other state of the art metaheuristics.

Dataset		M-BWOA-1	M-BWOA-2	WOA	AVOA	PSO	GA	GWO	HHA	ASO
CLL_SUB_111	best	**1.3743248**	1.27877296	0.96436496	1.08025332	1.07306067	1.15662849	1.2008585	1.04237217	1.19574759
	avg	1.00971912	**1.18505224**	0.94100357	1.06296935	1.00436891	0.98926097	1.09576796	0.89106439	1.0970509
	std	0.06782328	0.02369878	**0.01269984**	0.0201787	0.02571917	0.02162362	0.02266634	0.02123002	0.02283814
colon	best	**1.39817119**	1.48865281	1.23370549	1.17504308	1.14920773	1.30626218	1.47801015	1.18665296	1.25919157
	avg	1.23036316	**1.48777521**	1.17220016	1.16517437	1.03160424	1.29074666	1.4774784	1.12683063	1.23429508
	std	0.05057689	0.00892654	**0.00209217**	0.00340003	0.00292122	0.01085551	0.00251832	0.00679585	0.00421995
leukemia	best	1.41611125	**1.43829054**	1.26136404	1.21371755	1.28452995	1.27448296	1.40478397	1.15910125	1.21131753
	avg	1.40450239	**1.40960115**	1.17612744	1.16283023	1.24653574	1.14506653	1.40381336	1.02625109	1.19416237
	std	0.02906808	0.00396534	0.00722952	0.00561986	0.00997707	0.00289441	**0.00064836**	0.00900891	0.00144809
lung	best	1.36425884	**1.37264377**	1.25503703	1.10851065	1.2506919	1.31910366	1.38813643	1.19366887	1.05570218
	avg	**1.39648822**	1.27189625	1.19424208	1.02009503	1.18405058	1.22877465	1.38342265	1.18199936	1.0979579
	std	0.00809461	0.00482431	0.00449598	0.00203174	0.00575898	0.00541519	0.0081018	0.00948634	**0.00138208**
lung_discrete	best	1.29011938	1.37198797	1.18183717	1.26161551	1.27519302	1.32150821	**1.46978642**	1.24778867	1.13987029
	avg	1.28711733	1.35418009	1.15975381	1.25365462	1.22207717	1.21178219	**1.42041832**	1.15195527	1.09618814
	std	0.04852272	0.02061378	0.00501576	0.00467198	0.00283071	0.00372168	**0.00111037**	0.01334876	0.00601429
lymphoma	best	1.28094412	**1.39679683**	1.15446792	1.15949125	1.02162029	1.10568557	1.29567435	1.00152708	1.02468364
	avg	1.22402695	**1.38608924**	1.09389518	0.94006145	0.99013828	1.03115639	1.28181744	0.99577954	0.93036123
	std	0.03240048	0.01036192	0.00754532	0.01125517	**4.9531E-06**	0.00592058	0.00429726	0.00286504	0.00572038
nci9	best	**1.31793174**	1.16942407	1.15742923	0.91378101	1.10083742	1.03177329	1.22963438	0.99710549	0.94954832
	avg	0.88603706	0.99990226	1.0606673	0.8869253	0.87245826	0.85226529	**1.17956533**	0.93546506	0.92810748
	std	0.06134346	0.04969804	0.03086962	0.03771434	0.03069184	0.02976959	0.03414012	**0.02431834**	0.03260979
Prostate_GE	best	1.3101027	**1.40548247**	1.18610861	1.07061578	1.14445445	1.27513978	1.32900073	1.23448423	1.17171262
	avg	1.29486464	**1.40028197**	1.1668322	1.031855	1.1070187	1.26738616	1.31745972	1.14311545	1.10390338
	std	0.03639641	0.00775107	0.00238939	**0.00067249**	0.00498682	0.00953512	0.00177775	0.00539038	0.00660185
SMK_CAN_187	best	**1.37406662**	1.3623046	1.05724616	1.05267296	1.08474464	1.24549307	1.31372838	0.96018679	1.12852369
	avg	1.20523488	**1.32235292**	0.99183224	0.90712398	0.96278468	1.05315301	1.17817344	0.82643456	1.01104592
	std	0.02873458	0.01317575	0.01114278	0.02139391	0.01017348	0.01906291	0.01611149	0.0140111	**0.00715141**

TABLE 37.5 The average accuracy of the proposed algorithm variants compared to other metaheuristics.

Dataset	M-BWOA-1	M-BWOA-2	WOA	AVOA	PSO	GA	GWO	HHA	ASO
CLL_SUB_111	0.5279	0.6994	0.6766	0.8233	0.7396	0.6710	0.6340	0.6940	**0.8349**
colon	0.7758	**0.9990**	0.8553	0.8704	0.7677	0.9084	0.9864	0.8199	0.8922
leukemia	0.9273	0.9175	0.8911	0.8877	**0.9722**	0.7612	0.9110	0.7646	0.9034
lung	0.9127	0.8150	0.9110	0.6959	**0.9206**	0.8147	0.8991	0.8901	0.7721
lung_discrete	0.8752	0.9167	0.8601	**0.9950**	0.9619	0.8606	0.9607	0.8980	0.8169
lymphoma	0.7530	**0.90124**	0.7394	0.6400	0.6886	0.6300	0.7816	0.6991	0.6489
nci9	0.4789	0.5486	**0.7959**	0.6418	0.6002	0.5374	0.7330	0.7382	0.6621
Prostate_GE	0.8216	**0.9186**	0.8927	0.7840	0.8145	0.9125	0.8402	0.9014	0.8090
SMK_CAN_187	0.7109	**0.8363**	0.7345	0.6829	0.7105	0.7264	0.7233	0.6075	0.7561

TABLE 37.6 The average number of selected features of the proposed algorithm variants compared to other metaheuristics.

Dataset	M-BWOA-1	M-BWOA-2	WOA	AVOA	PSO	GA	GWO	HHA	ASO
CLL_SUB_111	257.3	**187.62**	5585.41	6384.14	5623.11	4399.46	1103.75	6901.91	5630.84
colon	42.34	**31.03**	907.2	979.4	944.51	626.94	125.54	1032.07	924.18
leukemia	209.45	**42.16**	3240.73	3470.59	3404.03	1961.36	339.51	3498.41	3356.92
lung	**83.61**	109.25	1463.4	1615.38	1569.27	915.67	192.23	1629.41	1533.21
lung_discrete	45.69	31.56	121.74	149.94	138.7	94.9	**25.31**	158.4	129.2
lymphoma	152.41	**71.52**	1574.2	1931.39	1903.45	1022.39	155.81	1959.52	1844.89
nci9	1627.87	**785.56**	4720.81	5232.5	4811.22	3932.79	1062.24	6128.74	4799.88
Prostate_GE	**170.27**	171.27	2833.86	3032.71	2883.41	1841.66	424.93	3086.76	2853.46
SMK_CAN_187	**222.56**	371.68	9922.05	11069.31	9945.36	7199.64	1945.67	11519.65	9945.69

outperforms the other algorithms and selects the smallest subset of genes for five datasets, including CLL_SUB_11, colon, leukemia, lymphoma, nci9. At the same time, the second best approach seems to be the M-BWOA-1, where the number of genes is efficiently selected for three datasets. As for the lung_discrete dataset, the GWO algorithm achieves the best small number, followed by M-BWOA-2 and M-BWOA-1.

Considering the fitness optimization results, the obtained accuracy and the average selected number of genes, it can be stated that the proposed approach, M-BWOA performed well compared to the state of art optimization approaches. More specifically, the second variant, M-BWOA-2, that is based on the T4 taper shape transfer function. In addition and to provide a significant comparison, the nonparametric Wilcoxon rank sum test is utilized to verify if the obtained results of the proposed approach are significantly different from the other approaches with a p-value set to 5% of the significance level. A significant difference is recognized when the results of the other versions are equal or higher than the p-value. To compare the effectiveness of M-BWOA-2 as it showed superior results among the different algorithms and to detect significant improvements, Wilcoxon rank sum test results are reported in Table 37.7. The results indicate that M-BWOA-2 performance is significantly different from the other algorithms on most datasets. Overall, the obtained results validate that the proposed approach can provide promising and superior results to the existing wrapper-based optimization approaches in handling the gene selection problem.

37.5 Conclusion

This chapter has applied a two-stage approach named M-BWOA to select the most appropriate features to classify gene diseases. The proposed M-BWOA uses mutual information (MI) as a fitness function to reduce the number of attributes in the first step. In addition, it aims to filter out features that do not have any helpful information for the classification. We use the conventional Whale Optimization Algorithm (WOA) to do it. In the second step, we implement a wrapper-based technique to select the most relevant features from those chosen in the first step. Hence, a binary version of WOA (BWOA) is used. In addition, a recent transfer function, Taper-shaped TF, was applied to transform the continuous search space into a binary space to tackle the gene selection problem better. Experiments on nine biological datasets demonstrate the

TABLE 37.7 p-values of the Wilcoxon test of M-BWOA-2 versus other metaheuristics ($p \geq 0.05$ are underlined).

Dataset	M-BWOA-1	WOA	AVOA	PSO	GA	GWO	HHA	ASO
CLL_SUB_111	1.83E-06	0.0365	6.15E-10	2.80E-09	1.77E-09	8.27E-10	1.05E-09	5.14E-10
colon	5.14E-10	5.14E-10	5.13E-10	5.14E-10	5.14E-10	5.13E-10	5.46E-10	5.14E-10
leukemia	5.14E-10	5.14E-10	5.13E-10	5.14E-10	5.14E-10	5.14E-10	6.53E-10	5.14E-10
lung	5.14E-10	5.13E-10	5.14E-10	5.14E-10	5.14E-10	5.13E-10	5.14E-10	5.13E-10
lung_discrete	5.04E-10	5.04E-10	5.00E-10	5.07E-10	5.09E-10	4.98E-10	5.09E-10	5.08E-10
lymphoma	<u>0.0608</u>	5.14E-10	5.14E-10	5.14E-10	5.13E-10	5.13E-10	4.17E-08	5.14E-10
nci9	<u>0.0608</u>	0.0001	8.77E-10	6.92E-09	5.15E-10	5.15E-10	2.44E-08	5.15E-10
Prostate_GE	5.14E-10	5.14E-10	5.14E-10	5.14E-10	5.14E-10	5.14E-10	5.79E-10	5.14E-10
SMK_CAN_187	<u>6.08E-02</u>	<u>0.7785</u>	5.15E-10	2.23E-09	5.15E-10	1.05E-09	4.42E-09	6.93E-10

proposed approach's excellent classification accuracy and the selection of the most relevant and minimum subset of genes. For further studies, the proposed method can be investigated for multi-objective optimization problems and combined with feature attribution methods. In addition, it will allow an accurate explication for selecting relevant genes that can be applied to a real-world problem, such as performing cancer classification based on gene selection.

References

[1] K. Yang, Z. Cai, J. Li, G. Lin, A stable gene selection in microarray data analysis, BMC Bioinformatics 7 (1) (2006) 228.

[2] M. Dash, H. Liu, Feature selection for classification, Intelligent Data Analysis 1 (3) (1997) 131–156.

[3] G. Chandrashekar, F. Sahin, A survey on feature selection methods, Computers & Electrical Engineering 40 (2014) 16–28.

[4] G. Chen, J. Chen, A novel wrapper method for feature selection and its applications, Neurocomputing 159 (2015) 219–226.

[5] T.N. Lal, O. Chapelle, J. Weston, A. Elisseeff, Embedded methods, in: I. Guyon, M. Nikravesh, S. Gunn, L.A. Zadeh (Eds.), Feature Extraction: Foundations and Applications, in: Studies in Fuzziness and Soft Computing, Springer, Berlin, Heidelberg, 2006, pp. 137–165.

[6] N. Sánchez-Maroño, A. Alonso-Betanzos, M. Tombilla-Sanromán, Filter methods for feature selection – a comparative study, in: H. Yin, P. Tino, E. Corchado, W. Byrne, X. Yao (Eds.), Intelligent Data Engineering and Automated Learning - IDEAL 2007, in: Lecture Notes in Computer Science, Springer, Berlin, Heidelberg, 2007, pp. 178–187.

[7] X. Wang, J. Yang, X. Teng, W. Xia, R. Jensen, Feature selection based on rough sets and particle swarm optimization, Pattern Recognition Letters 28 (4) (2007) 459–471.

[8] M. Nssibi, G. Manita, O. Korbaa, Binary Giza pyramids construction for feature selection, Procedia Computer Science 192 (2021) 676–687.

[9] G. Manita, O. Korbaa, Binary political optimizer for feature selection using gene expression data, Computational Intelligence and Neuroscience 2020 (2020) 8896570.

[10] V. Kumar, A. Kaur, Binary spotted hyena optimizer and its application to feature selection, Journal of Ambient Intelligence and Humanized Computing 11 (2020) 2625–2645, Springer, Berlin, Heidelberg.

[11] S. Mirjalili, A. Lewis, The whale optimization algorithm, Advances in Engineering Software 95 (2016) 51–67.

[12] S. Mirjalili, S.M. Mirjalili, S. Saremi, S. Mirjalili, Whale optimization algorithm: theory, literature review, and application in designing photonic crystal filters, in: Nature-Inspired Optimizers, 2020, pp. 219–238.

[13] N. Rana, M.S.A. Latiff, S.M. Abdulhamid, H. Chiroma, Whale optimization algorithm: a systematic review of contemporary applications, modifications and developments, Neural Computing & Applications 32 (20) (2020) 16245–16277.

[14] S. Mirjalili, A. Lewis, S-shaped versus V-shaped transfer functions for binary particle swarm optimization, Swarm and Evolutionary Computation 9 (2013) 1–14.

[15] Y. He, F. Zhang, S. Mirjalili, T. Zhang, Novel binary differential evolution algorithm based on taper-shaped transfer functions for binary optimization problems, Swarm and Evolutionary Computation 69 (2022) 101022.

[16] F. Han, C. Yang, Y.-Q. Wu, J.-S. Zhu, Q.-H. Ling, Y.-Q. Song, D.-S. Huang, A gene selection method for microarray data based on binary PSO encoding gene-to-class sensitivity information, IEEE/ACM Transactions on Computational Biology and Bioinformatics 14 (1) (2015) 85–96.

[17] S. Li, X. Wu, X. Hu, Gene selection using genetic algorithm and support vectors machines, Soft Computing 12 (7) (2008) 693–698.

[18] S. Manikandan, R. Manimegalai, M. Hariharan, Gene selection from microarray data using binary grey wolf algorithm for classifying acute leukemia, Current Signal Transduction Therapy 11 (2) (2016) 76–83.

[19] A. Dabba, A. Tari, S. Meftali, A new multi-objective binary Harris hawks optimization for gene selection in microarray data, Journal of Ambient Intelligence and Humanized Computing (2021) 1–20.

[20] J. Too, A. Rahim Abdullah, Binary atom search optimisation approaches for feature selection, Connection Science 32 (4) (2020) 406–430.

[21] T. Fushiki, Estimation of prediction error by using K-fold cross-validation, Statistics and Computing 21 (2011) 137–146.

[22] J. Friedman, T. Hastie, R. Tibshirani, et al., The Elements of Statistical Learning, vol. 1, Springer Series in Statistics, Springer, New York, 2001.

[23] T.R. Golub, D.K. Slonim, P. Tamayo, C. Huard, M. Gaasenbeek, J.P. Mesirov, H. Coller, M.L. Loh, J.R. Downing, M.A. Caligiuri, C.D. Bloomfield, E.S. Lander, Molecular classification of cancer: class discovery and class prediction by gene expression monitoring, Science (New York, N.Y.) 286 (5439) (1999) 531–537.

[24] A. Bhattacharjee, W.G. Richards, J. Staunton, C. Li, S. Monti, P. Vasa, C. Ladd, J. Beheshti, R. Bueno, M. Gillette, M. Loda, G. Weber, E.J. Mark, E.S. Lander, W. Wong, B.E. Johnson, T.R. Golub, D.J. Sugarbaker, M. Meyerson, Classification of human lung carcinomas by mRNA expression profiling reveals distinct adenocarcinoma subclasses, Proceedings of the National Academy of Sciences of the United States of America 98 (24) (2001) 13790–13795.

[25] Hanchuan Peng, Fuhui Long, C. Ding, Feature selection based on mutual information criteria of max-dependency, max-relevance, and min-redundancy, IEEE Transactions on Pattern Analysis and Machine Intelligence 27 (8) (2005) 1226–1238.

[26] A.A. Alizadeh, M.B. Eisen, R.E. Davis, C. Ma, I.S. Lossos, A. Rosenwald, J.C. Boldrick, H. Sabet, T. Tran, X. Yu, J.I. Powell, L. Yang, G.E. Marti, T. Moore, J. Hudson, L. Lu, D.B. Lewis, R. Tibshirani, G. Sherlock, W.C. Chan, T.C. Greiner, D.D. Weisenburger, J.O. Armitage, R. Warnke, R. Levy, W. Wilson, M.R. Grever, J.C. Byrd, D. Botstein, P.O. Brown, L.M. Staudt, Distinct types of diffuse large B-cell lymphoma identified by gene expression profiling, Nature 403 (6769) (2000) 503–511.

[27] D.T. Ross, U. Scherf, M.B. Eisen, C.M. Perou, C. Rees, P. Spellman, V. Iyer, S.S. Jeffrey, M. Van de Rijn, M. Waltham, A. Pergamenschikov, J.C.F. Lee, D. Lashkari, D. Shalon, T.G. Myers, J.N. Weinstein, D. Botstein, P.O. Brown, Systematic variation in gene expression patterns in human cancer cell lines, Nature Genetics 24 (3) (2000) 227–235.

[28] U. Scherf, D.T. Ross, M. Waltham, L.H. Smith, J.K. Lee, L. Tanabe, K.W. Kohn, W.C. Reinhold, T.G. Myers, D.T. Andrews, D.A. Scudiero, M.B. Eisen, E.A. Sausville, Y. Pommier, D. Botstein, P.O. Brown, J.N. Weinstein, A gene expression database for the molecular pharmacology of cancer, Nature Genetics 24 (3) (2000) 236–244.

[29] D. Singh, P.G. Febbo, K. Ross, D.G. Jackson, J. Manola, C. Ladd, P. Tamayo, A.A. Renshaw, A.V. D'Amico, J.P. Richie, E.S. Lander, M. Loda, P.W. Kantoff, T.R. Golub, W.R. Sellers, Gene expression correlates of clinical prostate cancer behavior, Cancer Cell 1 (2) (2002) 203–209.

[30] J.B. Welsh, L.M. Sapinoso, A.I. Su, S.G. Kern, J. Wang-Rodriguez, C.A. Moskaluk, H.F. Frierson, G.M. Hampton, Analysis of gene expression identifies candidate markers and pharmacological targets in prostate cancer, Cancer Research 61 (16) (2001) 5974–5978.

[31] A. Spira, J.E. Beane, V. Shah, K. Steiling, G. Liu, F. Schembri, S. Gilman, Y.-M. Dumas, P. Calner, P. Sebastiani, S. Sridhar, J. Beamis, C. Lamb, T. Anderson, N. Gerry, J. Keane, M.E. Lenburg, J.S. Brody, Airway epithelial gene expression in the diagnostic evaluation of smokers with suspect lung cancer, Nature Medicine 13 (3) (2007) 361–366.

[32] M. Neuhäuser, Wilcoxon–Mann–Whitney Test, Springer Berlin Heidelberg, Berlin, Heidelberg, 2011, pp. 1656–1658.

Chapter 38

A new hybrid whale optimization algorithm and golden jackal optimization for data clustering

Farhad Soleimanian Gharehchopogh[a], Seyedali Mirjalili[b,c], Gültekin Işık[d], and Bahman Arasteh[e]
[a]*Department of Computer Engineering, Urmia Branch, Islamic Azad University, Urmia, Iran,* [b]*Centre for Artificial Intelligence Research and Optimisation, Torrens University Australia, Brisbane, QLD, Australia,* [c]*University Research and Innovation Center, Obuda University, Budapest, Hungary,* [d]*Department of Computer Engineering, Igdir University, Igdir, Turkey,* [e]*Department of Computer Engineering, Faculty of Engineering and Natural Science, Istinye University, Istanbul, Turkey*

38.1 Introduction

Optimization is an important aspect of engineering because it helps find the most optimal or preferred solution. To address optimization problems, engineers will utilize both accurate and approximate approaches. However, deterministic methods, which require complex calculations, often have a high risk of failure [1]. On the other hand, approximate algorithms provide a more efficient way to find the optimal or appropriate solution to optimization problems. These algorithms can produce good results in less time, and meta-heuristic algorithms are a popular example of such approximate methods. One area where these algorithms can be applied is clustering, which can also be viewed as an optimization problem that can benefit from meta-heuristic algorithms. Clustering is a type of unsupervised learning that automatically groups samples into categories based on their similarities. These categories are referred to as clusters, each of which is a collection of data samples that share common characteristics not found in other clusters. The K-means algorithm is a widely used clustering technique that helps identify the clusters and group similar data points together. The K-means algorithm first selects k random centers and then assigns each data sample to the nearest center based on a defined similarity measure. Centers in each cluster are recomputed in several iterations. Despite being a simple and fast algorithm, the K-means algorithm has several limitations in achieving optimal clustering results [2]. For example, it relies heavily on the initial values, making it susceptible to noise and unbalanced clustering. In addition, the algorithm may converge to local minima in the process of minimizing the objective function [3].

In the past two decades, the practice of employing meta-heuristic and evolutionary algorithms to solve clustering problems has become increasingly popular as a means of meeting the challenges indicated in the previous paragraph. Solving clustering problems with optimization algorithms is a challenging issue [4]. No algorithm can completely solve all clustering problems. It means that a method may work well on a set of problems and may not be able to achieve optimal solutions for some different problems [5]. Meta-heuristic algorithms and population-based optimization methods have been widely used to address such issues.

Meta-heuristic algorithms reach the optimal solution due to their simplicity, low number of parameters, lack of dependence on the type of problems, and being away from the usual complications [6]. Meta-heuristic algorithms are intelligent techniques inspired by nature and the evolutionary behaviors of animals and plants. Over the past few decades, these algorithms have made significant advances in solving complex and optimization problems with remarkable success. Although these methods may not guarantee the optimal solution, their mechanisms, and operators increase the probability of reaching the optimal global solution or a solution close to it by guiding the search process toward creating desirability [7]. The most crucial factor that controls the efficiency and accuracy of a meta-heuristic algorithm is the balance between exploration (freely searching the entire space regardless of its achievements during the search) and exploitation (paying attention to the accomplishments of the algorithm during the search). Exploration refers to the ability of a meta-heuristic algorithm to search different regions of the search space to discover the appropriate optimal solution. On the other hand, exploitation refers to the ability of a search algorithm to focus on the optimal range of values to obtain the desired solution. A good meta-heuristic algorithm balances these two conflicting goals. Every meta-heuristic algorithm, or the improvement version,

Handbook of Whale Optimization Algorithm. https://doi.org/10.1016/B978-0-32-395365-8.00044-0

tries to improve efficiency by controlling these two parameters. Usually, in the initial iterations, the power of exploration should be increased, and over time, the power of exploitation should become more prominent.

To find the best solutions, meta-heuristic algorithms need to find a happy medium between exploring new territories and exploiting existing ones. This involves conducting a broad and diverse search of the problem space during the initial iterations and then focusing on the identified areas more precisely in the later iterations. By finding the best balance between exploration and exploitation, the algorithm can increase its chances of discovering the most suitable solution to the optimization problem. To avoid getting trapped in local convergence and discover new solutions, meta-heuristic algorithms employ exploration, while exploitation is used to find the best solutions. The best solutions are identified during the exploitation phase by utilizing the solutions discovered during the exploration phase. However, most meta-heuristic algorithms are typically designed to perform well in either the exploration or exploitation phase.

This book chapter uses the hybridization of the WOA [8] and the GJO algorithm [9] for data clustering. The WOA is a nature-inspired meta-heuristic algorithm that mimics the collective behavior of humpback whales. The bubble network strategy inspires this algorithm. The GJO algorithm is a novel optimization method inspired by the hunting behavior of golden jackals in nature. The algorithm consists of three basic stages – searching for prey, surrounding it, and then pouncing on it – which are mathematically modeled and implemented accordingly. The purpose of the proposed model is to improve the effectiveness of the WOA by incorporating the techniques of the GJO.

WOA has serious flaws that, to some extent, prevent the application of optimization: (1) Unsatisfactory performance as a result of stochastic operating equation selection. (2) Longer execution times brought on by arbitrary condition selection techniques. A prospective hybrid method to increasing WOA's capacity to trade-off between global and local searches is inspired by the simplicity and logic of the GJO architecture. This hybrid approach will increase processing effectiveness and accelerate convergence while improving global best results. In general, the proposed model possesses the following strengths: a reasonable trade-off between global and local searches, which contributes to higher global best results and faster convergence; high simplicity, which shortens execution time; low computational complexity, which offers excellent processing efficiency; excellent clustering features and a powerful search capability, both of which increase the algorithm's dependability and favor the correctness of search operations; and excellent processing efficiency.

The rest of this book chapter is structured as follows. In Section 38.2, some research in terms of data clustering is taken into account. In Section 38.3, research on fundamental topics such as data clustering and the GJO algorithm was discussed. In Section 38.4, the proposed model is introduced. In Section 38.5, the performance and operation of the proposed model are tested, and the results are illustrated in the table and graphs. In the final section of the book chapter, the conclusion and future works are discussed.

38.2 Related works

Data clustering is a widely used statistical analysis technique in various domains, including image analysis, bioinformatics, data mining, pattern recognition, and machine learning. The primary goal of data clustering is to represent large amounts of data as a smaller set of primitives or primitive clusters, which simplifies the data modeling process. Therefore, it plays an essential role in knowledge discovery and data mining. Meta-heuristic algorithms and different versions of these algorithms have been used to use data clustering. Some of these are mentioned below.

Habib et al. presented a hybridization Artificial Bee Colony (ABC) with Differential Evolution (DE) [10]. It can serve as a suitable platform for the development of meta-heuristic algorithms that have a higher convergence rate and a more optimal balance between exploration and exploitation. An automatic fuzzy clustering method based on the new version of Variable string length ABC (VABC) is presented for data clustering [11]. By using VABC, the number of variable clusters is possible. This method makes the C-Means fuzzy clustering technique based on VABC-FCM not need the predetermined number of clusters. In addition, VABC-FCM has a powerful global search ability under reasonable parameter settings.

In [12], the improved Krill herd algorithm creates a global search capability for data clustering. The improvement is based on adding a global search operator to explore the optimal search region; thus, Krill particles move toward the best global solution. The elitism strategy is also applied to keep the best krill in the krill update stages.

Mao et al. have proposed a modified version of the ABC algorithm that incorporates equation adaptive search and extended memory to achieve global optimization [13]. The proposed algorithm has been evaluated through experiments on benchmark numerical functions to determine its effectiveness. The result indicates that the algorithm can achieve a suitable trade-off between exploration and exploitation, resulting in improved accuracy of the optimal solution and better convergence speed compared to other meta-heuristic algorithms.

In [14], a multi-objective artificial immune algorithm (AIA) is used for fuzzy clustering based on multiple kernels. The multi-objective AIA is used to improve the classic fuzzy clustering algorithm and solve some crucial limitations, such as vulnerability in local convergence. By using the multi-objective AIA and achieving an optimal solution set, the Pareto method is used. At first, through the process of primary antibody population, clone propagation, non-uniform mutation, degradation problems, and premature convergence are avoided. Ultimately, the best solution is selected from the Pareto optimal solutions. The evaluation of the UCI dataset has shown a reduction in error and optimal clustering. Xiang et al. introduced a new algorithm called the grey ABC algorithm in their research, which utilizes a unique neighbor selection method based on the degree of gray connection [15]. The results of relevant tests show the ABC algorithm's efficiency and superiority.

A Hybrid ABC (HABC) algorithm is proposed for data clustering [16]. The main approach of the hybrid algorithm is to improve the information exchange (social learning) between the bees using the mutation operator of the genetic algorithm. And then, the HABC algorithm was used for data clustering. The experimental results indicate that the HABC algorithm outperforms other methods and represents a viable alternative for clustering data. Another study introduced a new global search and ABC algorithm to improve the performance of feature selection and data clustering. The goal was to explore the relationship between the rapid growth of ABC and the speed of convergence using elite solution-guided search equations [17]. An enhanced version named EABC_elite is proposed, confirming the proposed algorithm's superior results. Also, in [18], an ABC based on MapReduce for clustering large amounts of data is presented by Banharnsakun. The experimental results indicate that the proposed algorithm is suitable for clustering large data while maintaining a good quality level of the clustering results.

Ashish et al. [19] presented a fast and efficient parallel bat algorithm for data clustering using map-reduce architecture. The parallel bat algorithm is very efficient and valuable due to the use of an evolutionary approach for clustering instead of other algorithms, such as k-means. It also has high speed due to the use of the Hadoop architecture. Various experiments have shown that the parallel bat algorithm outperforms particle swarm optimization and exhibits faster performance than other similar algorithms as the number of nodes increases.

Saida et al. [20] have presented a quantum chaotic cuckoo search algorithm. Various experiments have shown that the quantum chaotic cuckoo search algorithm outperforms other comparative algorithms in terms of both internal and external clustering quality across all datasets. Zhou et al. [21] have also introduced an enhanced version of the algorithm by incorporating the symbiotic organism search (SSO) algorithm to solve clustering problems. The SSO algorithm models the symbiotic interactions between organisms in an ecosystem to optimize their survival and reproduction. Experiments have shown that the SSO algorithm outperforms other algorithms in terms of accuracy, precision, and overall performance.

Qaddoura et al. showed a modified evolutionary behavior genetic algorithm and an advanced version of the nearest neighbor search technique based on allocation and selection mechanisms for the clustering problem [22]. The ability of evolutionary algorithms to solve various machine learning problems, including clustering, has been demonstrated. The purpose of the algorithm that has been proposed is to improve the overall quality of the clustering results by locating a solution that optimizes the differentiation between the various clusters as well as the coherence between the data points that are contained within the same cluster. The suggested algorithm outperforms other algorithms on the offered data sets, according to the results of several different tests, and it has good performance when utilizing the silhouette coefficient objective function.

Rahnema and Gharehchopogh presented an improved ABC based on a WOA for data clustering [5]. The ABC algorithm has been modified in this research to address the issues of late convergence and discovery. Two types of memory, namely random memory and the memory of elites, are used in the proposed algorithm. During the phase in which it searches for prey, the WOA makes use of its random memory. Convergence will also be improved because of the utilization of exceptional memory. Moreover, there is dynamic control over the amount of RAM that elites use. The proposed algorithm was evaluated on ten standard datasets from the UCI machine learning repository and compared with other algorithms using statistical criteria and the ANOVA test. The experiment results indicate that the proposed algorithm outperforms several meta-heuristic algorithms.

In this section, several research papers on the use of meta-heuristic algorithms for clustering purposes were reviewed, and in most of these studies, a meta-heuristic algorithm was used to solve the clustering problem as an optimization problem, and also the objective function of the intra-cluster set as they use the objective function.

38.3 Fundamental research

38.3.1 Data clustering

Unsupervised learning involves data clustering, which is an important problem in which data is divided into clusters or groups to maximize the similarity between data within each cluster and minimize the similarity between data from different clusters. The clustering process can be categorized into supervised and unsupervised learning. In supervised learning, the categories are clear from the beginning, and each training data is assigned to specific data called terms. An observer provides the learner with information (such as the number of clusters, etc.) in addition to the educational data during training. In unsupervised learning, no information is available to the learner without educational data, and the learner must look for a specific structure in the data. Data and the patterns that can be extracted from them are among the most critical indicators of the information world, and data clustering is one of the best methods that have been presented to work with data and these patterns. Clustering is a powerful technique that can effectively analyze large amounts of data by identifying patterns in the data space. It works by dividing a set of data into clusters, which are groups of similar data points. These clusters are unique and different from each other, and the data within each cluster share similar characteristics. Overall, clustering is an automated process that can efficiently categorize data into groups based on their similarities. Different similarity criteria can be considered for the similarity of data samples in a cluster, the most important of which is the distance criterion. This criterion finds objects closer to each other as a cluster, called distance-based clustering. Calculating the distance between two data is very important in clustering. Distance is a measure of dissimilarity that allows us to navigate through data space and create clusters. When we calculate the distance between two data points, it provides us with information about their similarity or dissimilarity, which helps us to group similar points to form clusters. Several mathematical functions are available for measuring the distance between data objects. The most common distance measures used to determine the similarity between N data objects are the Euclidean distance and the cosine distance.

38.3.2 Golden jackal optimization

The GJO algorithm is a novel optimization technique based on the cooperative hunting behavior of golden jackals. It aims to provide a new method for solving practical engineering problems. The algorithm consists of three basic steps, namely prey search, prey enclosure, and prey pounce, which are mathematically formulated and used in the algorithm.

This section will guide you through the processes involved in developing the GJO algorithm, which is a method of optimization that is both simple and effective.

38.3.2.1 Formulating the search space

Like other meta-heuristic optimization algorithms, GJO is based on a population of solutions. The first attempt to find a solution involves a uniform distribution of the initial solution across the search space (Eq. (38.1)).

$$Y_0 = Y_{min} + rand \times (Y_{max} - Y_{min}) \tag{38.1}$$

where Y_{max} and Y_{min} represent the upper and lower limits of the range for the variables, while $rand$ refers to a random vector that is uniformly distributed between 0 and 1.

The process of setting up the initial state of the algorithm involves creating a matrix called Prey, from which the two best solutions are selected to form a pair of jackals. Eqs. (38.2) and (38.3) represent the form of the prey matrix.

$$Prey = \begin{bmatrix} Y_{1.1} & Y_{1.2} & \cdots & Y_{1.d} \\ Y_{2.1} & Y_{2.2} & \cdots & Y_{2.d} \\ \vdots & \vdots & \vdots & \vdots \\ Y_{n.1} & Y_{n.2} & \cdots & Y_{n.d} \end{bmatrix} \tag{38.2}$$

$$F_{OA} = \begin{bmatrix} f(Y_{1.1}.Y_{1.2}\ldots Y_{1.d}) \\ f(Y_{2.1}.Y_{2.2}\ldots Y_{2.d}) \\ \vdots \\ f(Y_{n.1}.Y_{n.2}\ldots Y_{n.d}) \end{bmatrix} \tag{38.3}$$

The algorithm initializes the Prey matrix, which represents the population of prey, with the first and second fittest individuals as male and female jackals. The fitness of each prey is stored in the F_{OA} matrix, and Y_{ij} represents the jth dimension of the ith prey. The objective function f is applied to each prey, and the fittest prey is called the male jackal, while the second fittest is called the female jackal. The positions of the male and female jackals correspond to their respective prey positions.

38.3.2.2 Phase of exploration or search for the prey

This section discusses the exploration strategy of the GJO algorithm, which is inspired by the hunting behavior of jackals. In the wild, jackals are adept at tracking and pursuing prey, but sometimes the prey is difficult to catch and escape. In such cases, the jackals will temporarily stop pursuing and look for other prey. The male jackal leads the hunt while the female jackal follows the male.

$$Y_1(t) = Y_M(t) - E.|Y_M(t) - rl.Prey(t)| \tag{38.4}$$

$$Y_2(t) = Y_{FM}(t) - E.|Y_{FM}(t) - rl.Prey(t)| \tag{38.5}$$

The variables used in these equations include t, which represents the current iteration, $Prey(t)$ is the position of the prey, $Y_M(t)$ and $Y_{FM}(t)$ are the current positions of the male and female jackals. The new positions of the male and female jackals are represented by $Y_1(t)$ and $Y_2(t)$. Additionally, the Evading energy of the prey, denoted by E, is calculated using Eq. (38.6).

$$E = E_1 - E_0 \tag{38.6}$$

The value E_1 represents the energy reduction of the prey, while E_0 denotes the initial energy state of the prey, as shown in Eqs. (38.7) and (38.8).

$$E_0 = 2 \times r - 1 \tag{38.7}$$

$$E_1 = c_1 \times \left(1 - \frac{t}{T}\right) \tag{38.8}$$

The value of r is a random number between 0 and 1, T is the maximum number of iterations, c_1 is a constant value of 1.5, and t represents the current iteration. The energy value E_1 decreases linearly from 1.5 to 0 as the number of iterations increases. $|Y_M(t) - rl.Prey(t)|$ in Eqs. (38.4), (38.5) compute the distance between jackal and prey. Depending upon the prey's evading energy, this distance gets subtracted or added to the jackal's current position. In Eqs. (38.4) and (38.5), rl is a vector consisting of random numbers following the Lévy distribution, which is used to represent the Lévy motion. By multiplying rl by the prey, the prey motion is simulated in a Lévy fashion, and this calculation is shown in Eq. (38.9).

$$rl = 0.05 \times LF(y) \tag{38.9}$$

LF is the levy flight function, which is calculated using Eq. (38.10).

$$L_F = 0.01 \times \frac{u \times \sigma}{|v|^{1/\beta}}; \quad \sigma = \left(\frac{\Gamma(1+\beta) \times \sin(\pi\beta/2)}{\Gamma((1+\beta)/2) \times \beta \times 2^{(\beta-1)/2}}\right)^{1/\beta} \tag{38.10}$$

The variables u and v represent random values between 0 and 1, and β is a constant with a default value of 1.5. The jackal positions are updated by computing the means in Eq. (38.11).

$$Y(t+1) = \frac{Y_1(t) + Y_2(t)}{2} \tag{38.11}$$

38.3.2.3 Exploitation stage, or the enclosing and pouncing on prey

The reduction in the prey's evading energy means that the jackals are closing in on the prey, and once the prey is located, the jackal pair surrounds it, and then they attack and consume it. This collective hunting behavior of the male and female jackals is expressed through mathematical Eqs. (38.12) and (38.13).

$$Y_1(t) = Y_M(t) - E.|rl.Y_M(t) - Prey(t)| \tag{38.12}$$

$$Y_2(t) = Y_{FM}(t) - E.|rl.Y_{FM}(t) - Prey(t)| \tag{38.13}$$

The variables used in these equations include t, which represents the current iteration, $Prey(t)$ is the position of the prey, $Y_M(t)$ and $Y_{FM}(t)$ are the current positions of the male and female jackals. The new positions of the male and female jackals are represented by $Y_1(t)$ and $Y_2(t)$. The evading energy of the prey (E) is calculated using Eq. (38.6), and the position of the male and female jackals is updated using Eq. (38.11).

The purpose of rl in Eqs. (38.12) and (38.13) is to introduce a random behavior in the exploitation phase, which encourages exploration and helps avoid getting stuck in local optima. rl is computed based on Eq. (38.9) and serves to avoid getting stuck in local optima, especially in the last iterations. The role of the rl element in the exploitation phase can be understood as a way to mimic obstacles that could hinder the jackals' path to the prey in natural environments. These obstacles can cause difficulties and slow down the pursuit, and rl helps to incorporate this aspect into the algorithm to avoid local optima.

38.3.2.4 Transition from exploration to exploitation

The GJO algorithm uses the energy of the prey to switch from the exploration phase to the exploitation phase. During the process of evading the predator, the prey gradually loses its energy. Therefore, the evading energy of the prey is modeled using Eq. (38.6). The initial energy value denoted as E_0, is assigned an arbitrary value between -1 and 1 for each iteration. In the GJO algorithm, the energy of the prey decreases as it escapes from the jackal pair. The initial energy E_0 is randomly chosen from -1 to 1 in each iteration. If E_0 decreases from 0 to -1, it means that the prey's energy is weakening, but if it increases from 0 to 1, it means that the prey's strength is improving. When the absolute value of E is greater than 1, the jackal pair searches for prey in different areas. However, if the absolute value of E is less than 1, the GJO attacks the prey to exploit it.

Finally, the GJO algorithm starts by generating a random set of prey (possible solutions). In each iteration, a male and female jackal pair tries to predict the position of the prey. Then, each candidate solution in the population updates its distance from the jackal pair based on their predicted prey position. In GJO, the E_1 parameter decreases from 1.5 to 0 to prioritize exploration and exploitation. The hunting pair of golden jackals moves away from the prey when E is greater than 1 and moves towards it when E is less than 1. The algorithm terminates when the end criterion is met.

38.3.2.5 Execution time complexity

The execution time complexity of the GJO algorithm is based on three steps: initialization, finesse evaluation, and golden jackal updating. Note that if there are n jackals, the computational complexity of the initialization process is $O(n)$. The updating mechanism has a computational complexity of $O(T \times N) + O(T \times N \times D)$, which includes searching for the best location and updating the location vector of all jackals. T is the maximum number of iterations, and D is the dimension of specific problems. Therefore, the runtime complexity of the GJO algorithm is $O(N \times (T + TD + 1))$.

38.4 Proposed model

This section explains the steps of the proposed model to solve the clustering problem. In the clustering problem, a set with N samples $C = \{c_1.c_2.c_3 \ldots c_n\}$ and d dimensions are divided into clusters (K) based on standard features. WOA and GJO algorithms have several operators to move the search agents toward the optimal solution in the search space. The WOA algorithm faces problems such as local search entrapment, premature convergence, and balance between the search process, and the proposed model aims to deal with these weaknesses. The combination of different operators leads to the power of exploration and exploitation in the proposed model, resulting in better results. GJO operators are used in the exploration phase to enhance WOA. After the update process is completed, random operations will be applied to improve the diversity of solutions. An array of variables is formulated in the clustering problem to find the best cluster centers. The dimensions of each solution depending on the number of features considered in the clustering process. The flowchart of the proposed model is shown in Fig. 38.1.

38.4.1 Initial population

The search process in the proposed model starts by determining the initial value for N factors (which represent the solutions) according to Eq. (38.14). Each agent has a set of D variables representing the data set's dimensions.

$$X = \begin{bmatrix} x_1^1 & x_2^1 & \cdots & x_d^1 \\ x_1^2 & x_2^2 & \cdots & x_d^2 \\ \vdots & \vdots & \vdots & \vdots \\ x_1^n & x_2^n & \cdots & x_d^n \end{bmatrix} \tag{38.14}$$

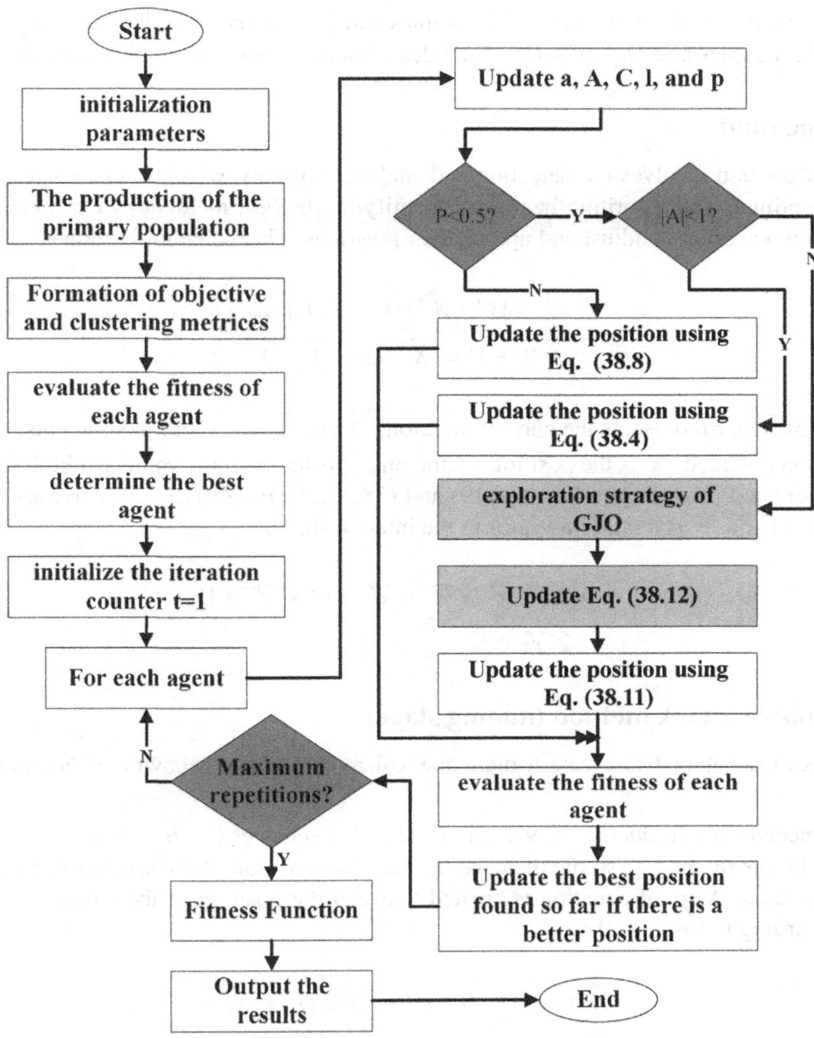

FIGURE 38.1 Flow chart of the proposed model for solving the clustering problem.

The next step is to determine the quality of each factor using the fitting function according to Eq. (38.15) and discover the best factor. The d-dimensional matrix of the variables used in the clustering optimization problem is shown in Eq. (38.14).

$$F = \begin{bmatrix} f(x_1^1.x_2^1 \ldots x_d^1) \\ f(x_1^2.x_2^2 \ldots x_d^2) \\ \vdots \\ f(x_1^n.x_2^n \ldots x_d^n) \end{bmatrix} \tag{38.15}$$

Based on the fitness of the vectors, the best vectors are considered to increase the better position and achieve a fast convergence rate. The Euclidean distance criterion is utilized to match the vectors, and it is a measure of similarity between samples. Eq. (38.16) calculates the distance between the samples and the cluster heads.

$$D_{ij} = \sqrt{\sum_{d=1}^{D} (X_{jd} - X_{id})^2} \tag{38.16}$$

In Eq. (38.16), i is the index of the members of the samples and j is the index of the cluster center of the samples, and D is the dimension of the samples and D_{ij} shows the Euclidean distance between samples and cluster centers.

38.4.2 Surround the hunt

The Humpback whale algorithm involves the detection and enclosure of prey, which in this context represents the goal or a solution close to the optimal. The algorithm begins by identifying the best member of the population, which the other whales then use as a reference point to adjust and update their positions. This behavior is shown according to Eq. (38.17).

$$\vec{D} = |\vec{C} \cdot \vec{X}^*(t) - \vec{X}(t)| \tag{38.17}$$

$$\vec{X}(t+1) = \vec{X}^*(t) - \vec{A} \cdot \vec{D} \tag{38.18}$$

In Eqs. (38.17) and (38.18), t represents the current iteration, $\vec{A}$ and $\vec{C}$ are vector coefficients, X is the position vector of the current best solution obtained, $\vec{X}$ is the position vector, and $|\,|$ is the absolute value. Multiplying element by element. A and C vectors are calculated according to Eqs. (38.19) and (38.20). In Eq. (38.6), $\vec{a}$ decreases linearly from 2 to zero during the repetition period and $\vec{r}$ is a random vector in the interval [0, 1].

$$\vec{A} = 2.\vec{a}.\vec{r} - \vec{a} = \vec{a} \times (2 \times \vec{r} - 1) \tag{38.19}$$

$$\vec{C} = 2.\vec{r} \tag{38.20}$$

38.4.3 Network-bubble attack method (mining stage)

Two techniques have been developed to create a mathematical model of the behavior of humpback whales forming a network bubble:

- Blockage reduction mechanism: Reducing the value in Eq. (38.19) achieves this behavior.
- Spiral update status: In this method, first, the distance between the position of the whale (X, Y) and the position of the prey (X^*, Y^*) is calculated. A spiral equation is created between the position of the whale and the prey to mimic the spiral movement, according to Eq. (38.21).

$$\vec{X}(t+1) = \vec{D'} \cdot e^{bl} \cdot \cos(2\pi l) + \vec{X}^*(t) \tag{38.21}$$

In Eq. (38.21), $\vec{D'} = |\vec{X^*}(t) - \vec{X}(t)|$ represents the distance of the ith whale to the prey, b is a constant to define the shape of the logarithmic spiral, l is a random number in the interval [0, 1], and element-by-element multiplication.

To simulate the circling behavior of humpback whales around the prey in a spiral path, the optimization algorithm assumes a 50% chance of using the spiral model to update the positions of the whales. The mathematical model is according to Eq. (38.22).

$$\vec{X}(t+1) = \begin{cases} \vec{X}^*(t) - A.D & \text{if } p < 0.5 \\ \vec{D'} \cdot e^{bl} \cdot \cos(2\pi l) + \vec{X}^*(t) & \text{if } p \geq 0.5 \end{cases} \tag{38.22}$$

In Eq. (38.22), p is a random number in the interval [0, 1]. In addition to the net-bubble method, humpback whales search for prey randomly.

38.4.4 Hunting (exploration phase)

This method is based on the changes in vector $\vec{A}$ can be used to search for prey (random search). $\vec{A}$ is used with random values greater than 1 or less than -1 to force the search agent to move away from a reference whale. In the random search stage, the position of a search agent is obtained by random selection instead of choosing the best search agent obtained during the non-random search stage. This emphasizes random search and allows the Whale Optimization Algorithm to conduct a global search. The condition $|\vec{A}| > 1$ is used to support this mechanism. The mathematical model is defined according to Eq. (38.24).

$$\overrightarrow{D} = |\overrightarrow{C}.\overrightarrow{X}_{rand} - \overrightarrow{X}| \tag{38.23}$$

$$\overrightarrow{X}(t+1) = |\overrightarrow{X}_{rand} - \overrightarrow{a} \times (2 \times \overrightarrow{r} - 1).\overrightarrow{D}| \tag{38.24}$$

In Eq. (38.23), the vector $\overrightarrow{X}_{rand}$ corresponds to the position of a randomly chosen whale from the current group of whales. A is the most important parameter in the WOA algorithm, which has a direct relationship with the speed of convergence and accuracy of the algorithm. From Eq. (38.19), we can conclude that A is mainly determined by the value of a. As the number of iterations increases, the value of a gradually decreases, which leads to an increase in the ability of local search and an acceleration of the convergence rate, but the probability of getting stuck in the local optimum increases. However, although the global detection ability of WOA is strong in early iterations, the convergence speed is slow. To improve the convergence rate in the early stages, the speed of reducing a should be increased. Therefore, in the proposed model, a new type of convergence coefficient is defined linearly by the GJO algorithm according to Eq. (38.25).

$$\overrightarrow{a} = a_1 - a_0 \tag{38.25}$$

a_1 represents the balance between exploration and exploitation. Therefore, a_1 is reduced from 1.5 to 0 to emphasize exploration and exploitation. a_0 represents the random position of whales in the search space, where "r" is an arbitrary number between 0 and 1.

$$a_0 = 2 \times r - 1 \tag{38.26}$$

$$a_1 = c_1 \times \left(1 - \frac{t}{T}\right) \tag{38.27}$$

The variable T stands for the maximum number of iterations, and c_1 is a fixed value of 1.5. The variable t represents the current iteration. As the iterations progress, the value of E_1 decreases linearly from 1.5 to 0.

38.4.5 Objective function

The Mean squared error is the most common objective function used to evaluate clustering. The Mean squared error is the square of the sum of the total distances of all samples of the data set with the center of the cluster in which they are located. The Mean squared error evaluation function is defined according to Eq. (38.28), where k is equal to the number of clusters and z_j is the center of cluster j [23].

$$SSE = \sum_{i=1}^{k} \sum_{x_i \in z_j} \|x_i - z_j\|^2 \tag{38.28}$$

38.4.6 Computational complexity

The main elements that contribute to the computational complexity of the proposed model are the initialization of the population and the update of positions. To simplify this, the population size is represented by N, the maximum number of iterations is T, and the dimension is D. Hence, the critical steps and their corresponding computational complexity can be expressed as follows:

Population initialization: To create a population of N individuals in a D-dimensional space, the computational complexity is represented by $O(N \times D)$. The time complexity of sorting the population and finding the best whale in the worst-case scenario can be expressed as $O(N^2 \times T)$. The updating mechanism has a complexity of $O(T \times N) + O(T \times N \times D)$ for finding the best location and updating the location vector of all jackals. The termination condition must be determined, and that is the last step. This part's computational complexity is $O(1)$, making OGSWOA's overall computational complexity $O(N \times D) + O(N^2 \times T) + O(T \times N) + O(T \times N \times D) + O(1)$.

38.5 Result and discussion

Blockage reduction mechanism: This behavior is achieved by reducing the value in Eq. (38.19). The number **of** iterations and the initial population in all algorithms are equal to 200 and 50, respectively. All algorithms have been evaluated in

TABLE 38.1 Characteristics of the datasets.

Datasets	No. of classes	Categorical attributes	Size
Glass	6	10	214
Vowel	6	3	871
CMC	3	10	1473
Iris	3	4	150
Wine	3	13	178
Cancer	2	9	699
Seeds	3	7	210
Heart	2	13	270

the MATLAB® 2019 environment on a system with 8 GB of memory, an Intel Core i5 processor, and Windows 10 with a 64-bit operating system. Table 38.1 shows the specifications of different datasets taken from the UCI website (https://archive.ics.uci.edu/ml/datasets.php).

In Table 38.2, the results of the models are shown based on the Best, Mean, and Worst criteria. The findings show that the proposed model incorporates a greater number of optimal values when compared to WOA and GJO. The performance of the proposed model on five datasets is better than GJO. The optimal value for Glass by the proposed model and GJO is 201.56 and 203.17, respectively. The optimal value for CMC by the proposed model and GJO is equal to 5645.33 and 5633.51, respectively. The optimal value for Iris by the proposed model and GJO is equal to 95.21 and 94.75, respectively. The proposed model can achieve better results in terms of the worst, average, and best criteria, which are ranked first. WOA and GJO algorithms provide better results than PSO and k-means algorithms. The proposed model ranked first in the Vowel dataset in the best benchmark. The results of the proposed model indicate optimal solutions, acceptable trade-offs between exploration and exploitation, and improved convergence rates.

38.5.1 Convergence rate

In this section, the convergence rate of the models in reaching the optimal solution is examined. Figures 38.2 to 38.5 show the convergence diagram of the models on 8 data sets. In Fig. 38.2, the Glass diagram shows that the proposed model has achieved the optimal solution in the initial iterations. GJO, WOA, and PSO models are in the first to third categories. The Vowel diagram shows that there is strong competition between the WOA and GJO models to find the best solution, which has reached the optimal value in the final iterations of the GJO model. In Fig. 38.3, the CMC diagram shows that there is competition between the proposed model and GJO to find the best solution, which has reached the optimal value in the final iterations of the GJO model. The Iris diagram shows that there is strong competition between the proposed model and WOA to find the best solution, which has reached the optimal value in the final iterations of the proposed model. In Fig. 38.4, the Wine and Cancer diagrams show that the proposed model has converged to the optimal points in the early stages of implementation, and the distance between the solutions discovered by the proposed model is large compared to GJO and WOA. Compared with WOA and GJO, the PSO model is unable to find optimal solutions and is stuck in the local optimum. In Fig. 38.5, the Seeds diagram shows that there is competition between the proposed model and GJO to find the best solution, which has reached the optimal value in the final iterations of the GJO model. Also, the PSO model has achieved optimal solutions close to WOA in the final iterations. The heart diagram shows that the proposed model has a better performance compared to other models.

The results prove that the combination of the WOA algorithm with GJO has led to finding optimal solutions. In most cases, the proposed model has obtained better and more promising results compared to other algorithms. The accurate results of the proposed model are due to the best exploitation of WOA and GJO. The reliability and robustness of the proposed model are due to the high discovery and escape from the local optimum using the GJO search method. Also, the results showed that the worst performances were recorded by PSO and k-means.

38.6 Conclusion and future works

This book chapter proposes a new model for data clustering using the hybridization of GJO and WOA algorithms. In the proposed model, the centers of the clusters were determined using an eagle search among the data points. Evaluation of eight different datasets showed that the proposed model had a lower error rate than the K-means, PSO, WOA, and GJO.

TABLE 38.2 Comparison of the proposed model with other algorithms on different datasets.

Datasets	Criterion	k-means	PSO	WOA	GJO	Proposed model
Glass	Best	215.72	216.26	211.68	203.17	**201.56**
	Mean	235.46	216.88	211.59	203.77	201.83
	worst	253.17	216.94	212.49	203.92	202.46
Vowel	Best	149415.31	149011.55	148489.23	148321.15	**148308.15**
	Mean	159256.94	156229.16	148826.57	148586.62	148359.43
	worst	160046.76	159389.74	148849.81	148861.55	148649.68
CMC	Best	5840.62	5745.29	5738.29	**5633.51**	5645.33
	Mean	5892.49	5722.85	5741.64	5635.47	5646.47
	worst	5911.23	5766.32	5743.08	5691.56	5695.36
Iris	Best	97.65	96.27	96.23	**94.75**	95.21
	Mean	107.19	96.75	96.59	95.39	95.65
	worst	121.24	102.35	96.72	95.56	95.85
Wine	Best	16553.64	16365.91	16304.24	16311.36	**16298.25**
	Mean	18062.12	16386.68	16384.95	16310.49	16295.66
	worst	18561.38	16552.49	16391.38	16315.75	16302.23
Cancer	Best	2998.48	2945.54	2922.55	2913.52	**2874.26**
	Mean	3253.62	3071.26	2948.42	2926.15	2882.65
	worst	3522.67	3214.79	3038.76	2935.78	2932.42
Seeds	Best	485.16	312.38	310.43	**309.26**	310.14
	Mean	487.96	312.79	311.56	310.21	310.23
	worst	560.25	313.64	312.91	311.05	311.65
Heart	Best	10681.44	10626.44	10622.25	10576.19	**10565.42**
	Mean	10687.68	10623.46	10639.36	10568.42	10531.58
	worst	10701.75	10657.75	10640.92	10592.35	10587.35

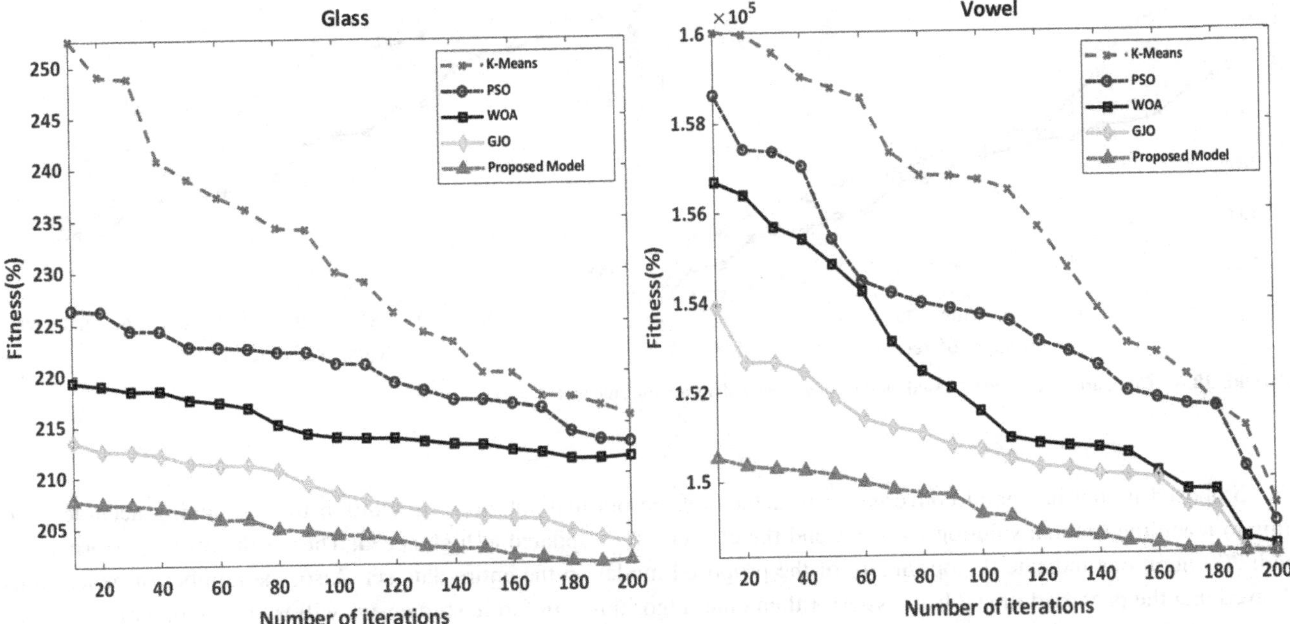

FIGURE 38.2 Comparison of models based on the convergence diagram on Glass and Vowel.

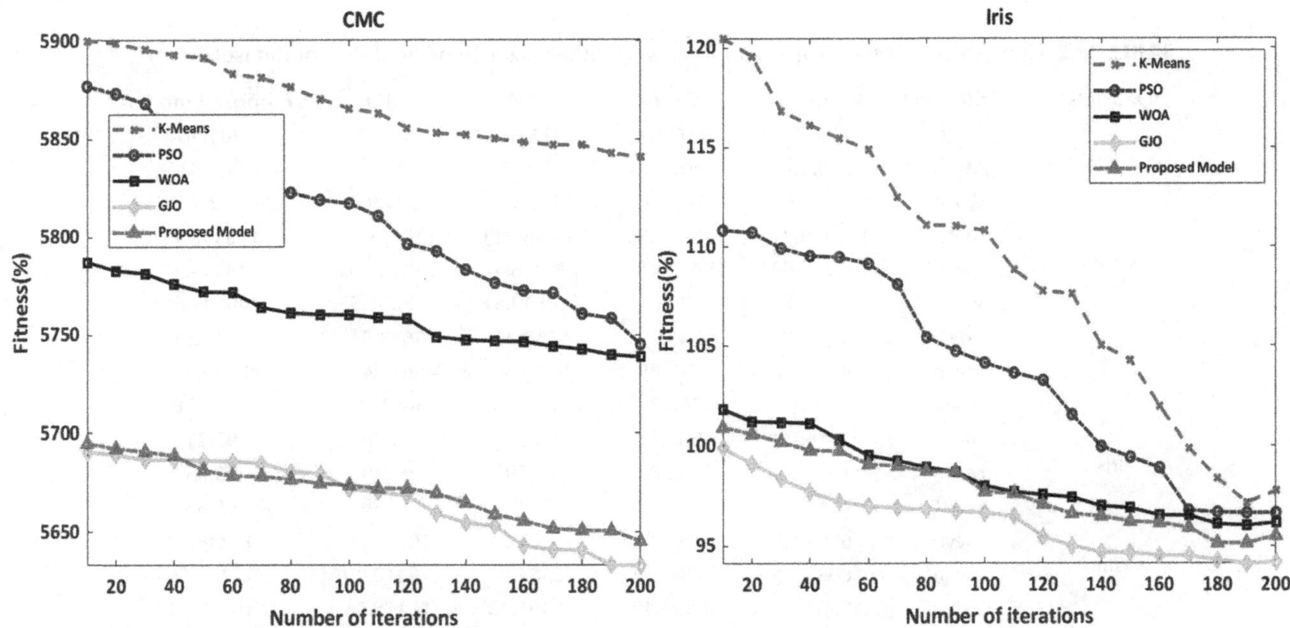

FIGURE 38.3 Comparison of models based on the convergence diagram on CMC and Iris.

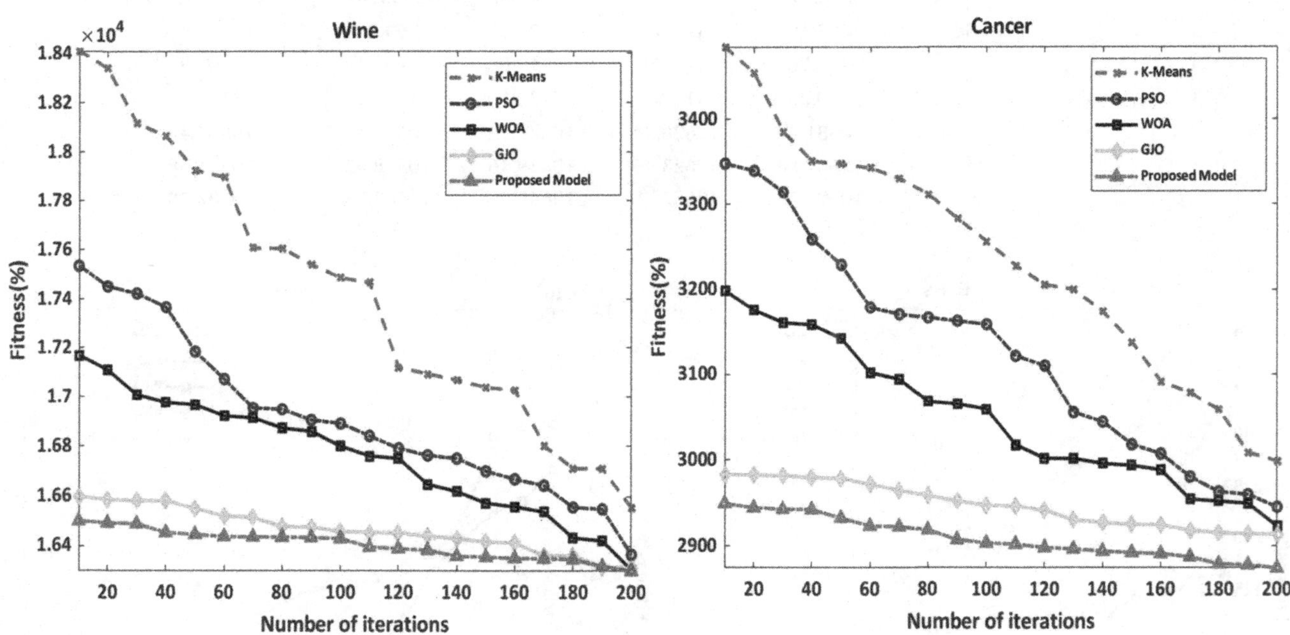

FIGURE 38.4 Comparison of models based on the convergence diagram on Line and Cancer.

The proposed model has good convergence in achieving the optimal solution. In addition to maintaining accuracy, the time to reach the optimal solution was low, and the clusters were updated at high speed. The results of the performances with 200 iterations indicate the optimality of the proposed model on the entire data set. Also, the number of generations showed that the proposed model had less error than other algorithms. In future studies, we will try to use other population-based meta-heuristic algorithms, and inspiring clustering methods in different types of samples in high volume will be clustered.

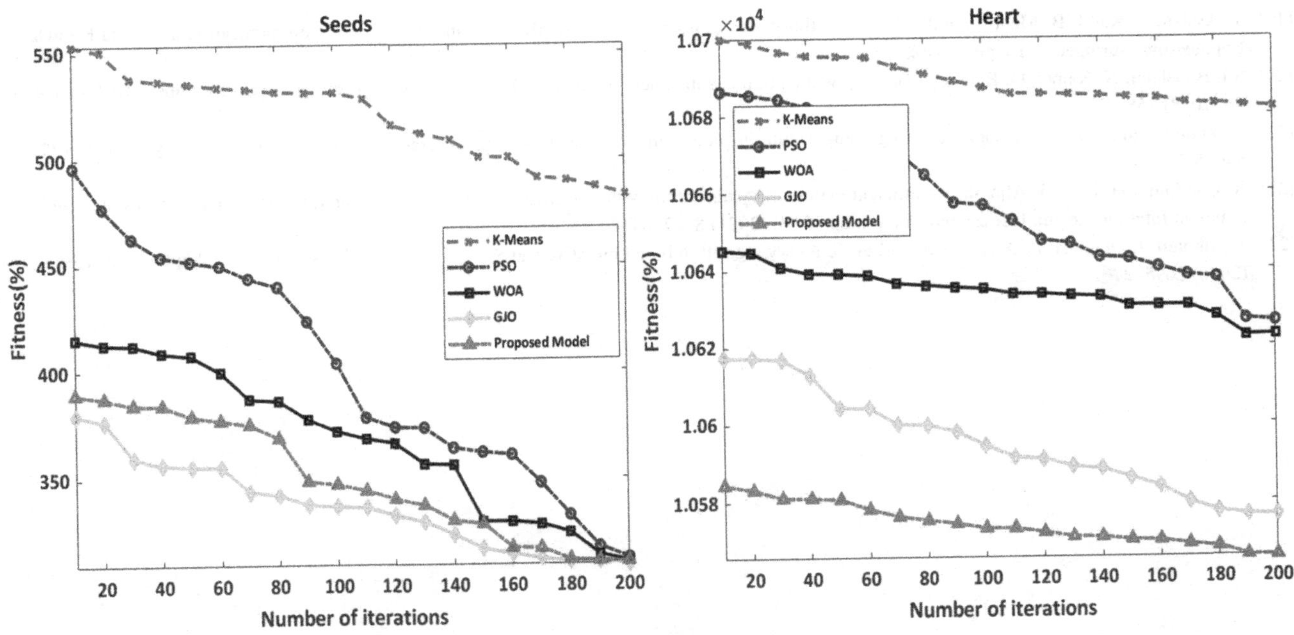

FIGURE 38.5 Comparison of models based on the convergence diagram on Seeds of Heart.

References

[1] Y. Shen, et al., An improved whale optimization algorithm based on multi-population evolution for global optimization and engineering design problems, Expert Systems with Applications 215 (2023) 119269.

[2] M.H. Nadimi-Shahraki, et al., Binary approaches of quantum-based avian navigation optimizer to select effective features from high-dimensional medical data, Mathematics 10 (15) (2022) 2770.

[3] F.S. Gharehchopogh, et al., CQFFA: a chaotic quasi-oppositional farmland fertility algorithm for solving engineering optimization problems, Journal of Bionic Engineering 20 (2023) 158–183.

[4] M.H. Nadimi-Shahraki, H. Zamani, DMDE: diversity-maintained multi-trial vector differential evolution algorithm for non-decomposition large-scale global optimization, Expert Systems with Applications 198 (2022) 116895.

[5] N. Rahnema, F.S. Gharehchopogh, An improved artificial bee colony algorithm based on whale optimization algorithm for data clustering, Multimedia Tools and Applications 79 (43) (2020) 32169–32194.

[6] M.H. Nadimi-Shahraki, et al., Hybridizing of whale and moth-flame optimization algorithms to solve diverse scales of optimal power flow problem, Electronics 11 (5) (2022) 831.

[7] M.H. Nadimi-Shahraki, et al., An improved moth-flame optimization algorithm with adaptation mechanism to solve numerical and mechanical engineering problems, Entropy 23 (12) (2021) 1637.

[8] S. Mirjalili, A. Lewis, The whale optimization algorithm, Advances in Engineering Software 95 (2016) 51–67.

[9] N. Chopra, M.M. Ansari, Golden jackal optimization: a novel nature-inspired optimizer for engineering applications, Expert Systems with Applications 198 (2022) 116924.

[10] S. Farshidpour, F. Keynia, Using artificial bee colony algorithm for MLP training on software defect prediction, Oriental Journal of Computer Science & Technology 5 (2) (2012) 231–239.

[11] Z.-g. Su, et al., Automatic fuzzy partitioning approach using Variable string length Artificial Bee Colony (VABC) algorithm, Applied Soft Computing 12 (11) (2012) 3421–3441.

[12] R. Jensi, G.W. Jiji, An improved krill herd algorithm with global exploration capability for solving numerical function optimization problems and its application to data clustering, Applied Soft Computing 46 (2016) 230–245.

[13] M. Mao, Q. Duan, L. Zhang, Artificial bee colony algorithm based on adaptive search equation and extended memory, Cybernetics and Systems 48 (5) (2017) 459–482.

[14] R. Shang, et al., Multi-objective artificial immune algorithm for fuzzy clustering based on multiple kernels, Swarm and Evolutionary Computation 50 (2019) 100485.

[15] W.-l. Xiang, et al., A grey artificial bee colony algorithm, Applied Soft Computing 60 (2017) 1–17.

[16] X. Yan, et al., A new approach for data clustering using hybrid artificial bee colony algorithm, Neurocomputing 97 (2012) 241–250.

[17] Z. Du, D. Han, K.-C. Li, Improving the performance of feature selection and data clustering with novel global search and elite-guided artificial bee colony algorithm, Journal of Supercomputing 75 (8) (2019) 5189–5226.

[18] A. Banharnsakun, A MapReduce-based artificial bee colony for large-scale data clustering, Pattern Recognition Letters 93 (2017) 78–84.

[19] T. Ashish, S. Kapil, B. Manju, Parallel bat algorithm-based clustering using MapReduce, in: Networking Communication and Data Knowledge Engineering, Springer, 2018, pp. 73–82.
[20] S.I. Boushaki, N. Kamel, O. Bendjeghaba, A new quantum chaotic cuckoo search algorithm for data clustering, Expert Systems with Applications 96 (2018) 358–372.
[21] Y. Zhou, et al., Automatic data clustering using nature-inspired symbiotic organism search algorithm, Knowledge-Based Systems 163 (2019) 546–557.
[22] R. Qaddoura, H. Faris, I. Aljarah, An efficient evolutionary algorithm with a nearest neighbor search technique for clustering analysis, Journal of Ambient Intelligence and Humanized Computing 12 (8) (2021) 8387–8412.
[23] T. Niknam, J. Olamaei, B. Amiri, A hybrid evolutionary algorithm based on ACO and SA for cluster analysis, Journal of Applied Sciences 8 (15) (2008) 2695–2702.

Chapter 39

Feature selection based on dataset variance optimization using Whale Optimization Algorithm (WOA)

Hassaan Bin Younis[a], Syed Kumayl Raza Moosavi[a], Muhammad Hamza Zafar[b], Shahzaib Farooq Hadi[a], and Majad Mansoor[c]

[a]*National University of Sciences and Technology, Islamabad, Pakistan,* [b]*Department of Engineering Sciences, University of Agder, Grimstad, Norway,* [c]*Dept. of Automation, University of Science and Technology of China, Hefei, China*

39.1 Introduction

Technologists are widely exploring subject of data mining in IT sector as this field has got a huge amount of data from which it is easier to fetch useful information [1]. Data mining, pattern classification, and information processing is essentially needed knowledge discovery procedure. Data preprocessing contributes largely towards the performance of data mining [2]. During preprocessing, feature selection is one of the preliminary tasks as it eradicates laid off, extraneous variables within a dataset. Feature categorization methods are segregated into wrappers and filters [3]. Filter prototype analyze the article subset based on the data with the help of nominated techniques while wrappers analyze selected feature subsets during a search process through a learning algorithm (e.g. Classification or regression models) [4]. Filters are faster as compared to wrappers in performance as they give results using information gain, feature distance, and their dependence which is cheaper method [3] while wrappers give better performance during classification in terms of accuracy [5]. In a wrapper feature selection method, three factors must be considered, i.e., classifier, subset evaluation criteria, and searching mechanism for best combination of feature [6].

Many researchers have proposed feature selection techniques based on matrix computation. Among them, recent methods [7–9] utilize theoretical bounds for such feature selection. Previously, these techniques have been applied on K-means [10], Support Vector Machine (SVM) [11,12], and ridge regression [13] which gave reasonable results which packed down existing models.

The challenging task in this problem is exploring an optimal subset. Meta-heuristic methods have handled this issue quite satisfactorily for the past twenty years [14]. For example, for N feature, 2^N solutions must be evaluated which requires high computational cost [15]. In this situation, meta-heuristics have been widely applied for feature selection. Tabu Search (TS) [16], Simulated Annealing (SA) [17], Record-to-Record Travel Algorithm (RRT) [18,19], Genetic Algorithm (GA) [20], Particle Swarm Optimization (PSO) [21], Ant Colony Optimization (ACO) [22], Differential Evolution (DE) [23], and Artificial Bee Colony (ABC) [24] are some of the prominent algorithms applied in such applications.

Meta-heuristic involves two stages, i.e., exploration and exploitation. Keeping this in view, meta-heuristics methods can be classified into population based and one solution based algorithms. Population based algorithms are exploration focused while single solution based algorithms are exploitation focused. Best strategy in designing an algorithm is maintaining a balance between the two searching techniques, i.e., exploration and exploitation. So, a hybrid model is designed for such classification which is called *memetic algorithm*.

Talbi et al. [14] presented a hybrid model combining conventional meta-heuristic with complementary meta-heuristic. In this way, a low level and a high level hybridization models are proposed. In a low level model, local search replaces a particular function. This is usually done by local optimization while global optimization is undertaken by population based algorithms [25]. In contrast to it, independent meta-heuristics are calculated in an order. At each stage, two levels of hybridization techniques are executed, termed as relay based and teamwork hybridization. In a relay based hybridization model, meta-heuristics subsets are executed in a serial fashion where every succeeding algorithm is fed with output of preceding algorithm while in teamwork hybridization model, meta-heuristic agents work in parallel. Each agent works

Handbook of Whale Optimization Algorithm. https://doi.org/10.1016/B978-0-32-395365-8.00045-2

in its workspace [25]. In order to choose the best feature subset for diagnosing coronary artery disease, a novel wrapper feature selection technique is suggested by Tashi et al. using a combination of GWO and SVM [26]. Xie et al. [27] proposed enhanced Particle Swarm Optimization for feature selection of classification model.

It is the feature selection approach that is proposed in this paper, a hybrid WOA algorithm is used. The proposed method intends to improve the WOA algorithm's use though variance based optimization. Tournament selection is used to choose search agents from the public because it provides every person a chance to be chosen, protecting the algorithm's variety.

The remainder of this article is formatted as follows: The related works are presented in Section 39.2. In Section 39.3, background theory is presented. The specifics of the suggested strategy are presented in Section 39.4. The experimental findings are presented and discussed in Section 39.5. Section 39.6 offers findings and suggestions for future work.

39.2 Related work

Many authors applied meta-heuristic hybridization models in their research work [14]. In comparison to conventional models, hybrid models have shown better performance in the recent past [25]. Oh et al. [28] used hybrid model for feature selection for the first time back in 2004 by combining local searching models with GA.

Martin et al. [29] combined Markov Chain model with SA to design a new hybrid algorithm and validated by implementing it on a travel salesman problem. Lenin et al. [30] solved reactive power problem with the combination of Tabu Search and SA. The algorithm was also applied on a symmetrical travel salesman problem. SA has been meshed with GA in different ways to design hybrid models for specific problems [31–34]. In these studies, hybrid models have outpaced conventional models for feature selection applications.

In the field of feature assortment, many researchers proposed their models successfully. Mafarja et al. [35] proposed a hybrid feature selection technique by combining SA and GA. The algorithm showed good performance upon testing it on UCI dataset in terms of number of selected attributes in comparison with state of the art approaches. Azmi et al. [36] proposed their hybrid model using GA and SA and tested it on hand written Farsi character. Another wrapper feature selection hybrid algorithm was also proposed by Wu [37]. He merged megalopolis approval benchmark of SA with crossover operative of GA. Afterwards, a hybrid model of SA and GA to produce a hybrid wrapper feature selection classification algorithm to predict power disruption in power quality issue and for optimization of SVM parameters for the same issue. Olabiyisi et al. [38] solved the local optima problem in timetabling feature selection using a hybrid GA-SA model in 2012. Tang et al. [39] used Fuzzy Artmap Neural Network for classification in wrapper feature selection manner in hybridization of GA and TS. Mafarja et al. [40] proposed two feature selection algorithms. In these models, major parameters were controlled by employing fuzzy logic which was merged with GA later on.

Moradi et al. [41] presented a local exploration model to guide PSO in selecting marginal deducts with respect to their correlation data. Moreover, in [42] SVM classifier was hybridized with combined model of GA and PSO and applied on micro-array dataset for verification. Yong et al. [43] presented a method to handle feature selection of unpredictable data using a multi-objective bare-bones PSO algorithm by incorporating dual operator technique. One is reinforced memory strategy and other is hybrid mutation. In 2012, Jona et al. [44] proposed a hybrid algorithm called Genetical Swarm Optimization (GSO) with the combination of PSO and Genetic Algorithm (GA) for feature optimization of CAD based digital mammogram system. The algorithm showed better results as compared to GA and PSO. Two other techniques were proposed based on wrapper feature selection method with a hybrid combination of ACO and GA [45,46]. Following the same pattern, Jona et al. [47] combined ACO with Cuckoo Search (CS) algorithm for feature selection in digital mammograms. Harmony Search (HS) and Stochastic Local Search (SLS) algorithms were combined to form a new technique for wrapper based feature exploitation by Nekkaa [48]. Readers are encouraged to study review papers on meta-heuristics and feature selection to have a better insight on these topics [49,50].

In presence of so much literature in meta-heuristics as mentioned above, a question might be raised if there is a space for new memetic algorithm. The answer to this question is a famous theorem in optimization called No Free Lunch (NFL) according to which no algorithm can solve all optimization problems with greater accuracy. In our scope, it can be rephrased as none of the above proposed wrapper feature selection algorithms are able to solve all of the feature selection problems. Since, there is always a room for better algorithms to solve optimization problems with greater accuracy, there is a big motivation to contribute in this field. The following section explains our algorithm for feature selection.

39.3 Method

39.3.1 Whale optimization algorithm

Mirjalili et al. [51] presented novel meta-heuristic algorithm, called Whale Optimization Algorithm (WOA), imitates the foraging behavior of humpback whales. The humpback whales swim in a constricted circle around the school of krill or small fishes as they pursue them near the surface while emitting unique bubbles, either in the form of a circle or a "9" (see Fig. 39.1). Attacking using a spiral bubble-net and encircling the prey. The first part of the algorithm represented the method which is called exploitation phase and the second phase where you look for prey randomly (exploration phase). The subsections that follow discuss the detailed mathematical representation of each stage. Observe that a uniform distribution is used in the equations is going to be used to produce random numbers.

Humpback whales first circle their target before beginning to hunt. This behavior can be mathematically modeled using Eqs. (39.1) and (39.2).

$$D = \left| C.\overrightarrow{X^*}(t) - \overrightarrow{X}(t) \right| \tag{39.1}$$

$$\overrightarrow{X}(t+1) = \overrightarrow{X^*}(t) - \overrightarrow{A}.D \tag{39.2}$$

where X^* denotes the best solution thus far, t denotes the current iteration, X represents the position vector, $|\,|$ represents mod and is a multiplication done element by element. Additionally, Eqs. (39.3) and (39.4) are used to calculate the coefficient vectors A and C, respectively:

$$\overrightarrow{A} = 2\overrightarrow{a}.\overrightarrow{r} - \overrightarrow{a} \tag{39.3}$$

$$\overrightarrow{C} = 2.\overrightarrow{r} \tag{39.4}$$

where, during the course of iterations, a drops linearly from 2 to 0 (in exploration and exploitation phases) and r is evenly segregated random vector created in the range [0, 1]. The search agents (whales), in accordance with Eq. (39.2), modify their positions in accordance with the location of the best-known answer (prey). The locations in which a whale can really be found near its prey are controlled by changing the values of the A and C vectors. The value of a in Eq. (39.3) must be decreased in accordance with Eq. (39.5) to produce the shrinking encircling behavior.

$$a = 2 - t\frac{2}{MaxIter} \tag{39.5}$$

where t is the increment and $MaxIter$ is the maximum number of iterations that can be performed. The distance between a search agent (X) and the best known search agent up to this point (X^*) is calculated to simulate the spiral-shaped path (see Fig. 39.1), after which a spiral equation is employed to generate the position of the neighbor search agent as in Eq. (39.6).

$$\overrightarrow{X}(t+1) = D'.e^{bl}.\cos(2\pi l) + \overrightarrow{X^*}(t) \tag{39.6}$$

The distance between the i-th whale and the prey is indicated by $D' = \left|\overrightarrow{X^*}(t) - \overrightarrow{X}(t)\right|$ and (best solution found thus far), b is a constant for determining the geometry of the logarithmic loop, and l is a random value in the range $[-1, 1]$.

A probability of 50% is considered to select between the two methods, shrinking surrounding and the spiral path, throughout the optimization procedure as it is in Eq. (39.7).

$$\overrightarrow{X}(t+1) = \begin{cases} \text{Shrinking Encircling (Eq. (39.2))} & \text{if } (p < 0.5) \\ \text{Spiral Shaped Path (Eq. (39.6))} & \text{if } (p \geq 0.5) \end{cases} \tag{39.7}$$

where p falls between [0, 1] randomly.

39.3.2 Exploration phase (search for prey)

In order to improve the exploration in WOA, a random search agent is chosen to lead the search rather than modifying the locations of the search agents in accordance with the position of the fastest one thus far. In order to have the search agent

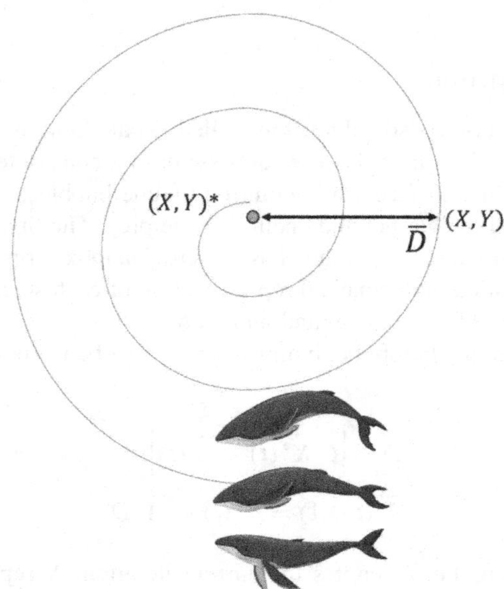

FIGURE 39.1 Humpback whales' distinctive bubble-net feeding techniques.

wander far away from the most well-known search agent, a vector A with random values higher than 1 or less than 1 is utilized. Eqs. (39.8) and (39.9) provide mathematical models for this system.

$$\vec{D} = \left| \vec{C}.\overrightarrow{X_{rand}} - \vec{X} \right| \tag{39.8}$$

$$\vec{X}(t+1) = \overrightarrow{X_{rand}} - \vec{A}.\vec{D} \tag{39.9}$$

where X_{rand} is a whale selected randomly out from the present population.

The WOA algorithm's pseudo code is displayed in Fig. 39.2. As can be seen, when the optimization process begins, WOA generates a random initial population and assesses it through a fitness function. The algorithm then repeatedly performs the subsequent steps until the end criterion is satisfied after determining the optimal solution. The primary coefficients are updated first. Next, a random value is produced. The algorithm alters the location of an answer using either Eqs. (39.2), (39.9), or Eq. (39.1) depending on this random vector (39.6). Furthermore, the solutions are restrained from straying too far from the search space. The method then gives the best result discovered as a close approximation to the global optimum.

As previously indicated, WOA is a population-based stochastic algorithm. The use of the best solution so far acquired to update the position of the other solutions is what ensures the convergence of this algorithm. This method, however, might direct solutions to global optimum. To update the positions of the solutions, three equations are switched between utilizing random variables. The adaptive parameter a is used to strike a compromise between convergence and avoiding local optima (exploration and exploitation). This parameter enhances convergence and exploitation proportionally to the number of iterations while smoothly reducing the size of alterations in the solutions.

The time complexity of Whale Optimization Algorithm is summarized as follows:

- WOA needs $O(N \times m)$ time in initialization phase that is surrounding the prey, where N is the population size and m represents dimension of the problem.
- Calculation of control parameters over the entire population in second phase of algorithm that is bubble net attack stage requires $O(N \times m)$ time.
- The next stage of algorithm is hunting the prey in which positions of agents are updated. This phase also requires $O(N \times m)$ time.

Provided the analysis, the total time complexity of WOA is $O(N \times m) + O(2 \times N \times m \times MaxIter)$, where $MaxIter$ indicates maximum number of iterations.

```
Generate Initial Population X_i = (i = 1,2, ..., n)
Calculate fitness of each solution
X* = the best search agent
While (t < Max Iteration)
        for each solution
                update a, A, C, l and p
                if (p < 0.5)
                        if (|A| < +1)
                        upgrade the position of current solution through Eq. (39.2)
                else if (|A| > +1)
                        select a random search agent (X_rand)
                        update the position of current search agent using Eq. (39.9)
                end
                else if (p ≥ 0.5)
                        update the position of current search using Eq. (39.6)
                end
        end for
        check if any solution goes beyond search space and repair it
        Calculate the fitness of each solution
        Update X* if there is a better solution t=t+1
end while
return X*
```

FIGURE 39.2 Pseudo-code of the Native WOA algorithm.

39.3.3 Limitations of standard whale optimization algorithm

Since its introduction, WOA has been applauded by many technologists but similar to other swarm-based optimization algorithms, WOA is also prone to some limitations that are encountered in various datasets and complex problems. Following are some of the limitations as per recent studies found in literature:

- WOA can be slow in converging to the global optimum solution, especially in complex and high-dimensional optimization problems.
- The WOA algorithm is prone to getting stuck in local optimum solutions, due to which its ability to find the global optimum solution is limited. This is quite crucial especially in complex optimization problems with many local optima. In a recent study, researchers presented an improved version of WOA to avoid local optima [26].
- In addition, WOA can have difficulty in handling multimodal optimization problems. Sometimes, it can become computationally intensive for large-scale optimization problems, which may limit its applicability in real-world scenarios.
- Each meta-heuristic optimization algorithm is highly dependent on the control parameter selection for a specific problem. Improper selection of which can result in slow convergence or premature convergence to a suboptimal solution.

39.4 Proposed approach

The binary [1] values are the only ones that can be used as solutions to the binary optimization issue of feature selection. When applied to the feature selection problem, a binary version must be devised for it. This work uses a one-dimensional vector to represent a solution, with the length of the vector depending on the number of attributes in the original dataset. A "1" or a "0" is used to indicate each value in a vector (cell). If the value is "1," then the associated attribute is picked; if otherwise, it is set to "0."

It is possible to think about feature selection as a multi-objective optimization issue where the two opposing goals of lesser picked features and better classification accuracy both need to be met. The better the solution, the fewer features there will be in it and the higher the classification accuracy. Each solution is weighed in accordance with the fitness function that has been proposed, which depends on the variance based classifier, in order to determine the solution's classification accuracy, quantity of chosen features it contains, time taken to find optimal solution and precision. The fitness function in Eq. (39.10) is utilized WOA algorithm to evaluate search agents in order to strike a compromise between the amount of selected features in each solution (minimum) and the classification accuracy (maximum).

$$Fitness = \alpha\gamma_R(D) + \beta\frac{|R|}{|N|} \tag{39.10}$$

where $\gamma_R(D)$ denotes the classification failure rate of a particular classifier (in this case, variance based classifier). Additionally, $|R|$ is the adjacency matrix of the chosen subset and $|N|$ is the total amount of features in the dataset. The parameters α and β, $\alpha \in [0, 1]$ chosen randomly and $\beta = (1 - \alpha)$, represent the significance of classification quality and subset length, respectively.

As seen in the WOA algorithm, exploitation (like in Eqs. (39.2) and (39.6)) hinges on figuring out how far away the search agent is from the whale that is now known to exist. We predict that the outcomes will be enhanced by using an effective local search algorithm to look for nearby solutions. Additionally, since the native WOA algorithm's exploration (as shown in Eq. (39.9)) depends on shifting each search agent's position in accordance with a randomly chosen solution, we think that employing a different selection mechanism, such as tournament selection, may enhance the algorithm's ability to explore.

The predicted fitness of a design is a significant quantity because it tells us whether a WOA would be able to locate optimal or nearly optimal solutions by recombining different building blocks in a given situation. To determine how much of sampling necessary to accept or reject a building block relative to one of its competitors, we must take into account both the statistical variance of fitness and the schema average fitness because most WOAs rely on statistical sampling. To do this, we must compute the variance of model fitness, often known as collateral noise.

The following formula determines how to calculate the default fitness of feature selection:

$$Fitness = \alpha \times accuracy(agent) + \beta \times \frac{tot\_feat - selected\_feat}{tot\_feat} \tag{39.11}$$

where α and β values are movable. These values are chosen to be 0.5 for this study. In this conventional method the objective is to maximize the fitness thereby increasing weighted accuracy and number of unselected features. The agent's accuracy is calculated using the dataset's accuracy and the features that the agent chose for that iteration. The accuracy is attained by applying the KNN method to the new, smaller dataset for each iteration.

For this round of Dataset Variance Optimization, the accuracy is swapped out for the variance of the new, smaller dataset and the fitness function becomes a minimization problem. Minimum variance and minimum number of selected features as per Eq. (39.12).

$$Fitness = \alpha \times Variance\,(Agent_{Dataset}) + \beta \times \frac{tot\_feat - unselected\_feat}{tot\_feat} \tag{39.12}$$

By doing this, the feature selection process is freed from the requirement for a classification algorithm. This drastically cuts down on the amount of time the algorithm needs to run.

39.5 Experimentation

Utilizing MATLAB®, the suggested algorithm is implemented. The tests are run on 18 FS benchmark datasets from the UCI data repository in order to evaluate the performance of the suggested techniques [52]. The dataset is split into two sections, training and testing dataset. Table 39.1 provides information about the datasets that were used, including the number of attributes and instances in each dataset.

39.6 Results and comparative analysis

This section reports all of the findings from the suggested methods. The two feature selection approaches for WOA algorithm namely KNN based and variance based are compared for each algorithm. All of the recommended options are evaluated to see which the best approach is. The proposed methods are contrasted with up-to-date feature selection techniques, including Grey Wolf Optimizer (GWO) [53], Particle Swarm Optimizer (PSO) [27], Binary Bat Algorithm (BBA) [54], Genetic Algorithm (GA) [55], and Equilibrium Optimizer (EO) [56] according to the following measures:

- Percentage accuracy, precision and F1-Score of classification using the test dataset with the chosen features. The average scores improved over 30 runs are tabulated, as seen in Table 39.2, Table 39.5 and Table 39.6 respectively.
- Table 39.3 presents the comparison on the basis of number of features selected from the training dataset.
- Time taken to optimize features in the training dataset is compared for all state-of-the-art approaches in seconds, as seen in Table 39.4.
- Finally, confusion matrices for WOA based on two optimization methods under study are presented for the entire dataset collection.

TABLE 39.1 Dataset List for the Experiments.

S. No.	Dataset	No. of Attributes	No. of Instances
1.	Ad	1558	3279
2.	Adult	14	48842
3.	Car	6	1728
3.	CTG	23	2126
4.	Voting	16	435
5.	Tic Tac Toe	9	958
6.	Gestures	50	9900
7.	HCV	14	615
8.	King-Rook Vs King-Pawn	36	3196
9.	Nursery	8	12960
10.	Obesity	17	2111
11.	Promoters	58	106
13.	Biodeg	41	1055
14.	Room	16	10129
15.	Seismic	19	2584
16.	Splice	61	3190

TABLE 39.2 Comparison of Percentage Accuracy of WOA for KNN and Variance based Classification.

Dataset	Method	%age Accuracy					
		WOA	GWO	PSO	BBA	GA	EO
Ad	KNN	94.49	94.70	92.37	94.92	94.70	95.13
	Variance	95.13	94.07	94.28	95.13	94.70	94.49
Adult	KNN	32.52	81.07	73.81	76.42	71.81	71.92
	Variance	72.13	71.67	72.19	71.92	71.83	71.49
Car	KNN	56.07	56.07	71.68	71.68	71.68	71.68
	Variance	71.68	58.96	58.96	58.96	71.68	71.68
CTG	KNN	89.67	80.75	82.16	85.92	83.57	85.92
	Variance	84.51	82.86	84.98	80.75	83.10	86.15
Voting	KNN	89.36	97.87	97.87	97.87	97.87	93.62
	Variance	97.87	97.87	91.49	82.98	53.19	89.36
Tic Tac Toe	KNN	60.94	60.94	69.27	60.94	50.00	69.27
	Variance	60.94	60.94	58.33	60.42	60.94	60.94
Gestures	KNN	85.25	95.90	95.90	95.90	95.90	95.90
	Variance	95.90	95.90	95.90	95.90	95.90	95.90
HCV	KNN	24.55	26.35	20.22	22.74	25.63	24.91
	Variance	24.55	22.02	26.35	22.74	22.02	22.74
King-Rook Vs King-Pawn	KNN	91.88	71.25	95.16	90.16	84.22	93.44
	Variance	54.69	66.41	58.91	71.88	68.75	65.78
Nursery	KNN	82.25	82.25	66.86	82.25	82.25	87.27
	Variance	40.82	40.82	40.82	40.82	40.82	40.82
Obesity	KNN	77.78	82.51	78.72	81.56	74.47	81.56
	Variance	73.52	80.85	76.12	67.38	73.76	78.25
Promoters	KNN	77.27	86.36	86.36	81.82	77.27	81.82
	Variance	81.82	54.55	77.27	63.64	77.27	68.18
Biodeg	KNN	78.67	78.20	76.30	81.99	80.09	81.99
	Variance	79.15	76.30	80.09	72.99	77.25	79.15

continued on next page

TABLE 39.2 *(continued)*

Dataset	Method	%age Accuracy					
		WOA	GWO	PSO	BBA	GA	EO
Room	KNN	92.94	99.31	96.64	99.51	98.86	98.52
	Variance	97.09	98.47	87.36	92.94	98.72	90.28
Seismic	KNN	94.78	94.39	94.97	94.39	94.97	94.39
	Variance	94.20	94.78	94.20	94.78	94.20	94.78
Splice	KNN	65.99	59.87	72.10	71.94	67.40	60.03
	Variance	57.99	53.92	56.74	64.73	65.52	53.13

TABLE 39.3 Comparison of Number of Features of WOA for KNN and Variance based Classification.

Dataset	Method	No. of Features					
		WOA	GWO	PSO	BBA	GA	EO
Ad	KNN	404	473	543	489	475	727
	Variance	407	479	563	495	475	718
Adult	KNN	1	4	1	4	1	2
	Variance	2	2	2	2	3	2
Car	KNN	1	1	1	1	1	1
	Variance	1	1	1	1	1	1
CTG	KNN	3	5	3	6	4	2
	Variance	8	8	9	3	10	9
Voting	KNN	2	4	2	4	3	3
	Variance	4	2	1	4	2	2
Drug	KNN	2	3	2	3	2	2
	Variance	1	2	1	3	2	2
Tic Tac Toe	KNN	1	1	1	2	1	1
	Variance	1	1	1	2	1	1
Gestures	KNN	10	15	15	15	15	17
	Variance	15	15	15	8	15	17
HCV	KNN	4	8	5	8	7	7
	Variance	8	8	5	2	8	9
King-Rook Vs King-Pawn	KNN	15	11	12	14	12	14
	Variance	8	10	6	10	10	9
Nursery	KNN	2	2	1	2	2	3
	Variance	1	1	1	1	1	1
Obesity	KNN	3	5	4	4	3	4
	Variance	2	5	3	2	3	4
Promoters	KNN	20	17	19	18	17	20
	Variance	9	19	12	19	17	17
Biodeg	KNN	9	12	9	12	10	10
	Variance	12	12	12	5	13	13
Room	KNN	2	4	2	4	3	3
	Variance	3	4	3	2	3	3
Seismic	KNN	3	4	1	5	2	3
	Variance	4	5	4	1	4	4
Splice	KNN	12	18	16	19	16	17
	Variance	19	20	22	21	20	15

TABLE 39.4 Comparison of Time(s) of WOA for KNN and Variance based Classification.

Dataset	Method	Time (s)					
		WOA	GWO	PSO	BBA	GA	EO
Ad	KNN	167.535	195.740	126.730	208.844	82.002	251.201
	Variance	38.098	72.009	40.778	48.016	22.677	67.585
Adult	KNN	2958.402	2859.682	966.541	2090.048	1270.714	2996.454
	Variance	103.481	135.085	182.614	46.749	169.468	265.392
Car	KNN	76.139	77.339	11.870	49.598	35.883	74.483
	Variance	8.248	8.536	1.262	3.312	8.357	8.107
CTG	KNN	107.728	111.714	35.954	67.023	49.084	103.320
	Variance	10.764	11.135	1.816	3.749	10.185	10.732
Voting	KNN	13.716	13.519	5.277	9.264	7.076	13.075
	Variance	3.140	2.667	1.541	1.437	2.515	3.077
Drug	KNN	100.305	101.688	34.552	59.644	42.251	96.212
	Variance	9.795	10.708	3.156	3.671	9.029	10.091
Tic Tac Toe	KNN	46.224	47.561	16.938	29.181	21.542	45.911
	Variance	5.499	5.187	2.078	2.234	5.171	5.764
Gestures	KNN	123.489	123.519	56.876	90.684	57.207	121.566
	Variance	12.216	12.934	5.875	3.640	12.247	13.294
HCV	KNN	75.342	77.527	27.500	43.241	31.759	73.249
	Variance	8.357	8.639	3.109	2.781	7.483	8.029
King-Rook Vs King-Pawn	KNN	284.669	290.940	126.349	225.873	131.725	283.794
	Variance	27.025	27.166	12.219	11.482	25.510	27.197
Nursery	KNN	805.836	786.390	281.817	383.867	331.324	754.779
	Variance	66.204	72.174	23.006	22.608	68.454	68.641
Obesity	KNN	100.698	76.187	35.667	65.486	46.709	100.899
	Variance	9.295	10.360	4.255	3.515	9.873	10.498
Promoters	KNN	7.826	5.514	3.507	5.233	4.187	8.248
	Variance	2.249	1.750	1.193	1.140	2.078	2.874
Biodeg	KNN	63.564	45.443	26.880	42.413	27.682	62.314
	Variance	7.061	4.952	3.302	2.406	6.467	7.467
Room	KNN	526.632	324.361	218.797	373.049	252.529	598.581
	Variance	52.129	28.634	19.480	18.043	48.927	51.942
Seismic	KNN	82.340	66.078	42.584	77.374	55.675	117.036
	Variance	7.030	6.686	4.530	4.187	11.716	11.747
Splice	KNN	295.808	209.311	141.657	244.776	128.942	283.591
	Variance	27.540	19.683	13.560	14.512	26.135	28.103

TABLE 39.5 Comparison of Precision of WOA for KNN and Variance based Classification.

Dataset	Method	Class	Precision					
			WOA	GWO	PSO	BBA	GA	EO
Ad	KNN	0	96.77	95.38	98.00	92.86	91.55	92.96
		1	94.15	94.59	91.71	95.27	95.26	95.51
	Variance	0	95.52	96.67	89.04	94.20	91.55	90.28
		1	95.06	93.69	95.24	95.29	95.26	95.25

continued on next page

TABLE 39.5 (*continued*)

Dataset	Method	Class	Precision					
			WOA	GWO	PSO	BBA	GA	EO
Adult	KNN	0	70.48	84.06	77.09	81.63	93.50	77.23
		1	23.04	66.18	39.33	52.13	45.76	35.16
	Variance	0	77.51	77.07	77.39	77.23	77.27	77.25
		1	36.57	34.11	36.29	35.16	35.15	34.51
Car	KNN		78.36	78.36	51.37	51.37	51.37	51.37
	Variance		51.37	54.95	54.95	54.95	51.37	51.37
CTG	KNN	0	89.40	78.71	80.00	84.18	82.78	84.04
		1						
	Variance	0	83.83	81.76	83.98	75.45	80.71	85.11
		1						
Voting	KNN	0	90.48	100.00	100.00	100.00	100.00	95.24
		1	88.46	96.15	96.15	96.15	96.15	92.31
	Variance	0	100.00	100.00	87.50	75.00	50.00	90.48
		1	96.15	96.15	95.65	94.74	63.64	88.46
Tic Tac Toe	KNN	0	0.00	0.00	61.11	0.00	40.19	61.11
		1	60.94	60.94	74.17	60.94	62.35	74.17
	Variance	0	0.00	0.00	46.38	49.06	0.00	0.00
		1	60.94	60.94	65.04	64.75	60.94	60.94
Gestures	KNN		85.03	95.93	95.93	95.93	95.93	95.93
	Variance		95.93	95.93	95.93	95.93	95.93	95.93
HCV	KNN		24.98	27.08	21.22	22.82	27.52	25.81
	Variance		25.32	22.69	27.69	22.82	22.69	23.91
King-Rook Vs King-Pawn	KNN	0	97.08	69.44	96.97	95.24	87.73	93.79
		1	87.98	73.10	93.59	86.38	81.54	93.11
	Variance	0	53.94	66.21	63.91	75.19	62.17	62.89
		1	55.18	66.57	57.11	69.63	84.57	69.34
Nursery	KNN		80.95	80.95	50.10	80.95	80.95	85.63
	Variance		26.99	26.99	26.99	26.99	26.99	26.99
Obesity	KNN		78.14	83.21	79.28	82.21	74.57	81.81
	Variance		73.73	81.51	76.34	67.22	73.93	78.73
Promoters	KNN	0	80.00	92.31	86.67	81.25	90.91	85.71
		1	71.43	77.78	85.71	83.33	63.64	75.00
	Variance	0	100.00	75.00	76.47	75.00	90.91	70.59
		1	66.67	42.86	80.00	50.00	63.64	60.00
Biodeg	KNN	0	85.42	87.41	88.19	88.65	86.71	88.65
		1	64.18	61.84	58.33	68.57	66.18	68.57
	Variance	0	85.52	84.89	88.32	79.22	84.62	88.15
		1	65.15	59.72	64.86	56.14	61.76	63.16
Room	KNN		92.85	99.31	96.56	99.51	98.86	98.54
	Variance		97.12	98.48	86.24	92.85	98.72	89.41
Seismic	KNN	0	94.96	95.29	94.97	94.94	94.97	95.12
		1	0.00	28.57	0.00	0.00	0.00	20.00
	Variance	0	94.93	94.96	94.93	94.96	94.93	94.96
		1	0.00	0.00	0.00	0.00	0.00	0.00
Splice	KNN		68.51	67.01	74.21	76.11	70.80	65.55
	Variance		61.79	58.51	60.57	67.50	66.70	55.19

TABLE 39.6 Comparison of F-1 Score of WOA for KNN and Variance based Classification.

Dataset	Method	Class	F1-Score					
			WOA	GWO	PSO	BBA	GA	EO
Ad	KNN	0	82.19	83.22	73.13	84.42	83.87	85.16
		1	96.74	96.86	95.56	96.96	96.83	97.08
	Variance	0	84.77	80.56	82.80	84.97	83.87	83.33
		1	97.10	96.50	96.57	97.09	96.83	96.70
Adult	KNN	0	29.44	88.09	84.31	85.07	78.35	82.78
		1	35.34	53.87	20.72	43.86	59.60	24.01
	Variance	0	82.85	82.63	82.94	82.78	82.69	82.42
		1	25.65	23.22	24.82	24.01	24.39	24.60
Car	KNN		57.39	57.39	59.85	59.85	59.85	59.85
	Variance		59.85	56.88	56.88	56.88	59.85	59.85
CTG	KNN		89.01	79.42	80.74	84.30	82.91	84.49
	Variance		83.71	81.75	83.97	76.82	81.61	85.19
Voting	KNN	0	88.37	97.67	97.67	97.67	97.67	93.02
		1	90.20	98.04	98.04	98.04	98.04	94.12
	Variance	0	97.67	97.67	91.30	84.00	62.07	88.37
		1	98.04	98.04	91.67	81.82	38.89	90.20
Tic Tac Toe	KNN	0	0.00	0.00	59.86	0.00	47.25	59.86
		1	75.73	75.73	75.11	75.73	52.48	75.11
	Variance	0	0.00	0.00	44.44	40.62	0.00	0.00
		1	75.73	75.73	66.67	70.31	75.73	75.73
Gestures	KNN		84.81	95.89	95.89	95.89	95.89	95.89
	Variance		95.89	95.89	95.89	95.89	95.89	95.89
HCV	KNN		24.20	25.71	20.08	22.10	25.38	24.72
	Variance		24.22	21.70	26.29	22.10	21.70	22.54
King-Rook Vs King-Pawn	KNN	0	91.10	70.98	94.89	89.19	82.79	93.18
		1	92.53	71.52	95.39	90.96	85.43	93.67
	Variance	0	48.58	64.34	45.09	68.31	73.75	66.97
		1	59.50	68.24	67.17	74.72	61.39	64.51
Nursery	KNN		81.30	81.30	55.73	81.30	81.30	86.30
	Variance		32.31	32.31	32.31	32.31	32.31	32.31
Obesity	KNN	0	77.78	82.24	78.47	81.43	74.16	80.66
		1						
	Variance	0	73.11	80.82	75.98	67.02	73.39	78.12
		1						
Promoters	KNN	0	82.76	88.89	89.66	86.67	80.00	85.71
		1	66.67	82.35	80.00	71.43	73.68	75.00
	Variance	0	83.33	54.55	83.87	69.23	80.00	77.42
		1	83.33	54.55	83.87	69.23	80.00	77.42
Biodeg	KNN	0	84.54	83.69	81.75	86.81	85.52	86.81
		1	65.65	67.14	66.22	71.64	68.18	71.64
	Variance	0	84.93	82.52	85.21	81.06	83.45	84.40
		1	66.15	63.24	69.57	52.89	63.64	68.57
Room	KNN	0	92.88	99.30	96.59	99.51	98.86	98.53
		1						
	Variance	0	97.10	98.47	86.58	92.88	98.71	89.66
		1						

continued on next page

TABLE 39.6 *(continued)*

Dataset	Method	Class	F1-Score					
			WOA	GWO	PSO	BBA	GA	EO
Seismic	KNN	0	97.32	97.10	97.42	97.11	97.42	97.11
		1	0.00	12.12	0.00	0.00	0.00	6.45
	Variance	0	97.01	97.32	97.01	97.32	97.01	97.32
		1	0.00	0.00	0.00	0.00	0.00	0.00
Splice	KNN	0 1	66.49	61.12	72.18	72.77	67.92	61.34
	Variance	0 1	58.89	54.58	57.21	64.85	65.34	53.67
Drug	KNN		25.71	28.81	27.83	27.45	27.84	29.30
	Variance		27.53	30.17	29.57	25.41	25.71	27.47

In the case of precision and F1-score the class is also specified in cases where the dataset has binary classes.

Following is a comparison of confusion matrices of WOA for KNN and variance based classification for the entire dataset collection. (See Figs. 39.3–39.19.)

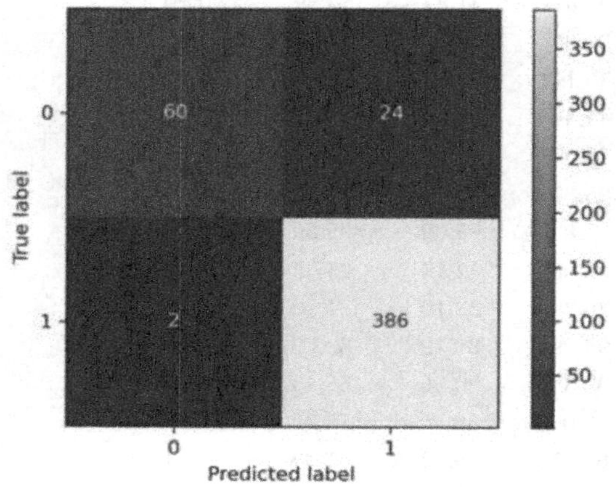

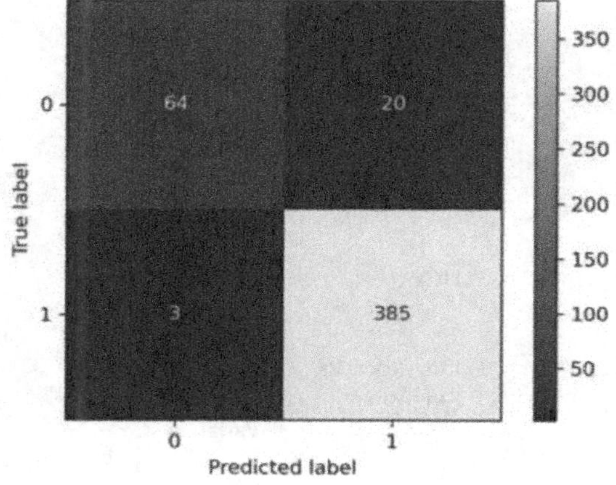

FIGURE 39.3 Confusion Matrix – Ad – KNN (left) Variance (right).

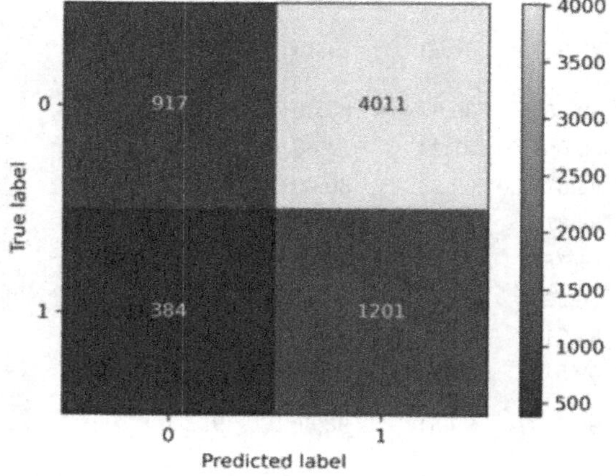

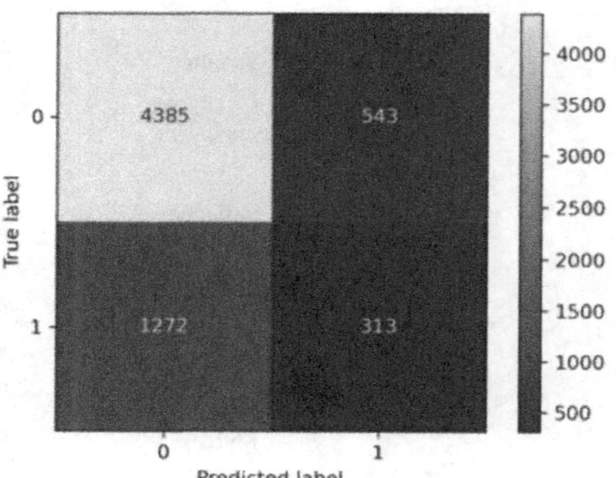

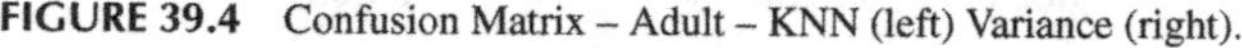

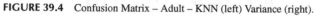

FIGURE 39.4 Confusion Matrix – Adult – KNN (left) Variance (right).

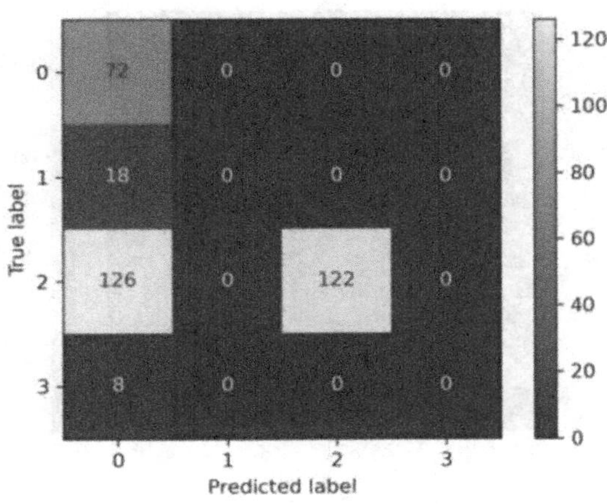

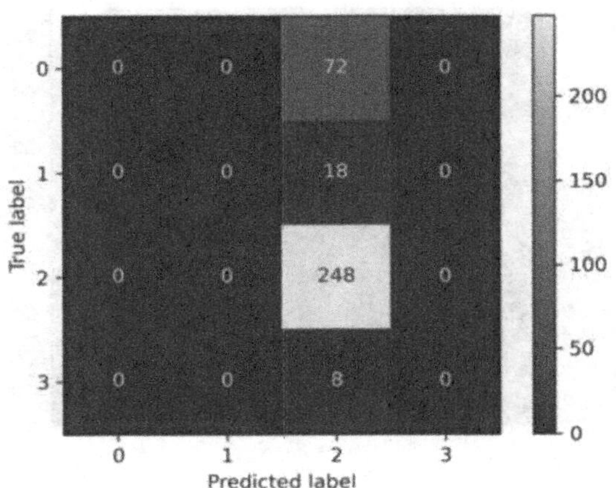

FIGURE 39.5 Confusion Matrix – Car – KNN (left) Variance (right).

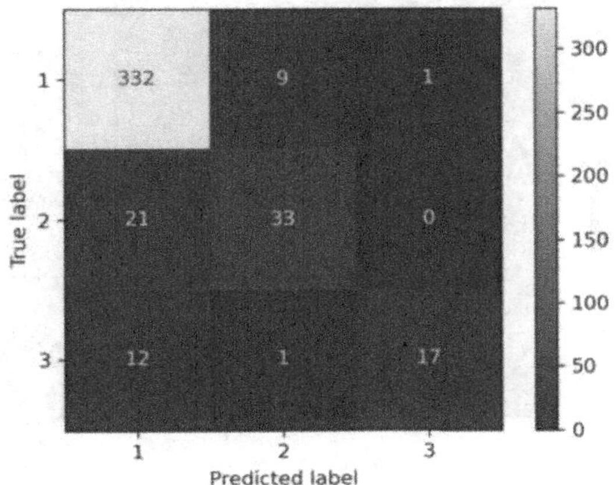

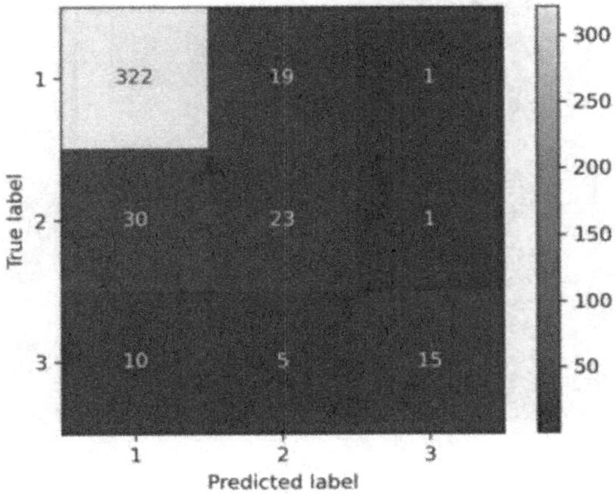

FIGURE 39.6 Confusion Matrix – CTG – KNN (left) Variance (right).

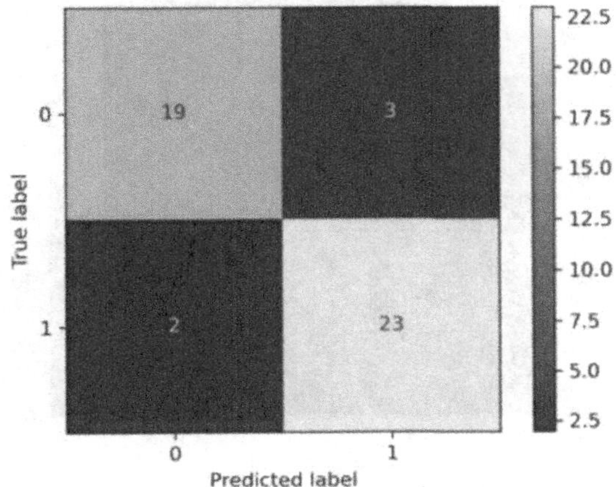

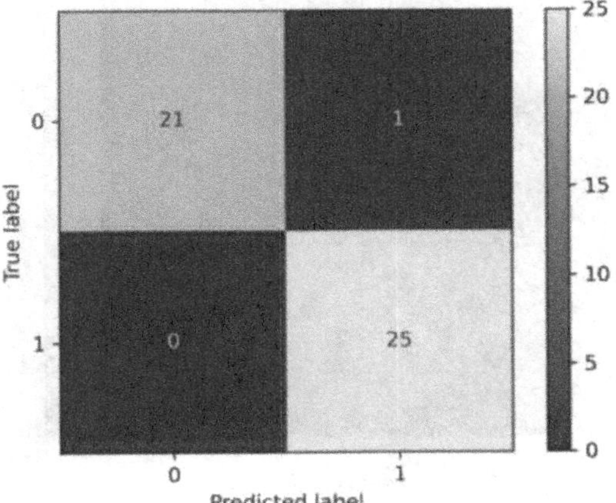

FIGURE 39.7 Confusion Matrix – Voting – KNN (left) Variance (right).

39.6.1 Discussion

In order to evaluate the impact of enhancing exploration, exploitation, and both of them taken together, this subsection analyzes the six suggested options. In the WOA algorithm, variance based optimization complements WOA, which improves

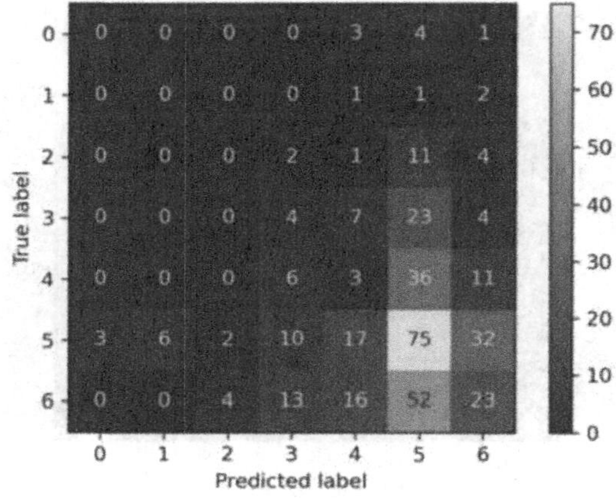
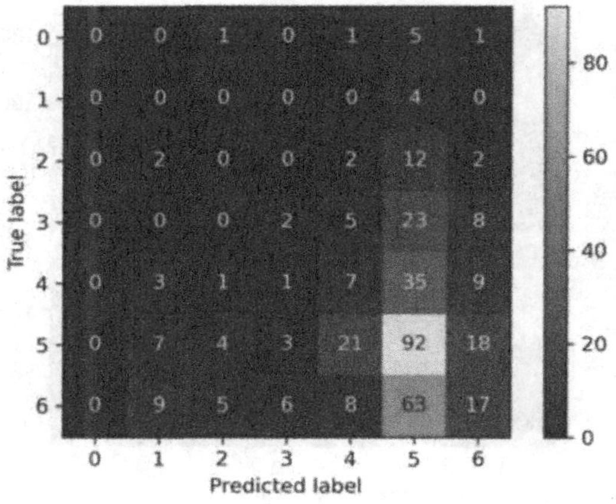

FIGURE 39.8 Confusion Matrix – Drug – KNN (left) Variance (right).

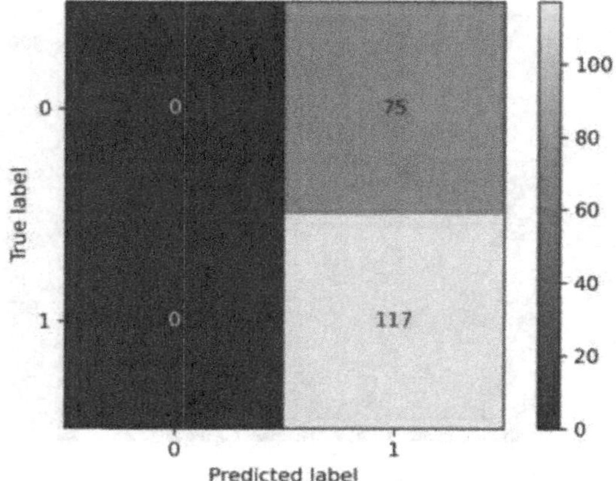
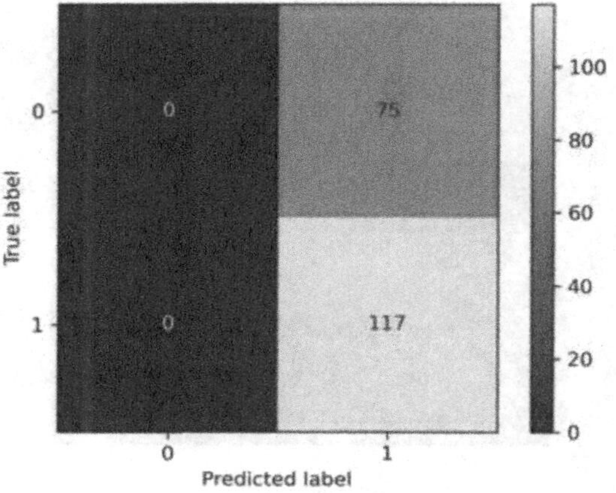

FIGURE 39.9 Confusion Matrix – Tic Tac Toe – KNN (left) Variance (right).

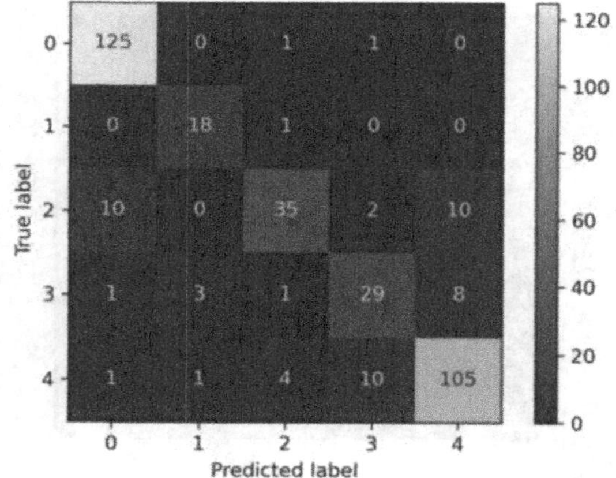
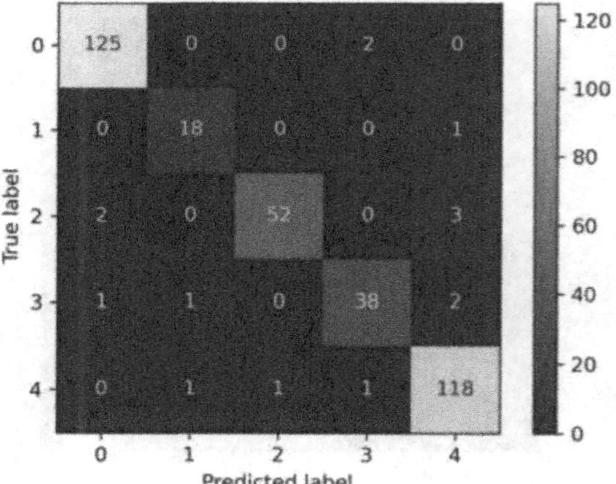

FIGURE 39.10 Confusion Matrix – Gestures – KNN (left) Variance (right).

exploitation, by improving exploration. Analysis of Table 39.2 and Table 39.3 results reveals that WOA with variance based optimization out passes other algorithms in terms of percentage accuracy and number of selected features. Time is the most critical factor in any classification problem. It can be seen from numbers in Table 39.4 that time taken to reach global target is least for variance based approach in almost all datasets. KNN based approach takes time for the process of the classifi-

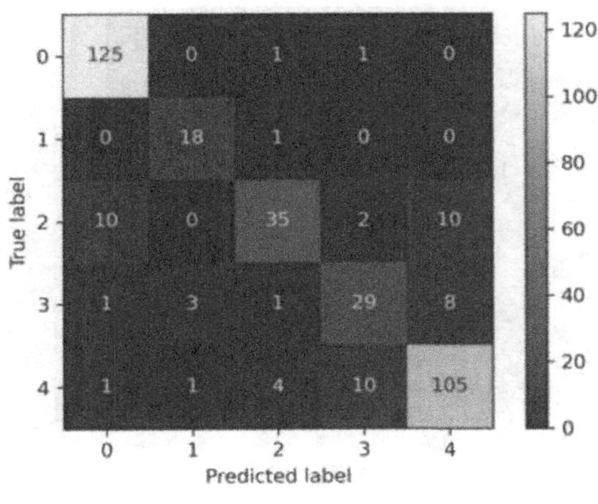

FIGURE 39.11 Confusion Matrix – HCV – KNN (left) Variance (right).

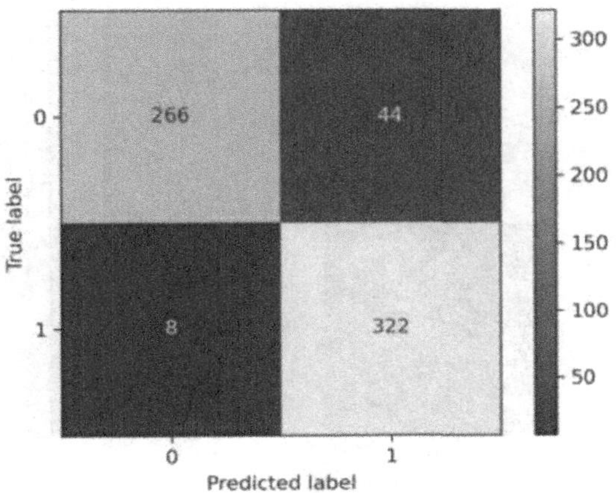

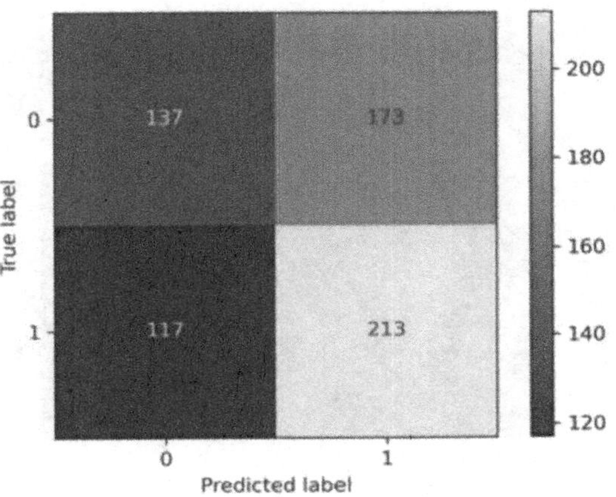

FIGURE 39.12 Confusion Matrix – King Rook vs King Pawn – KNN (left) Variance (right).

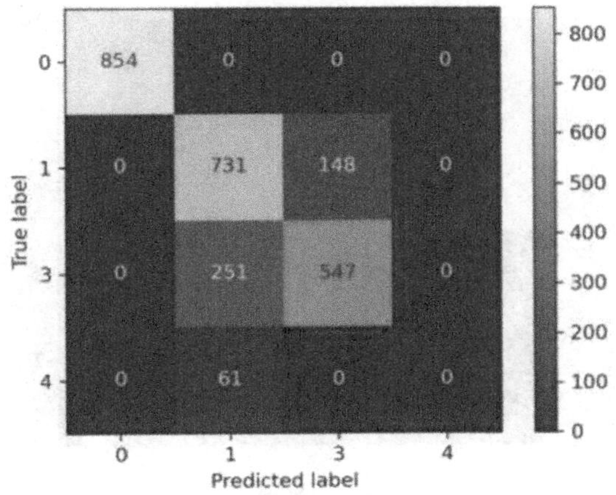

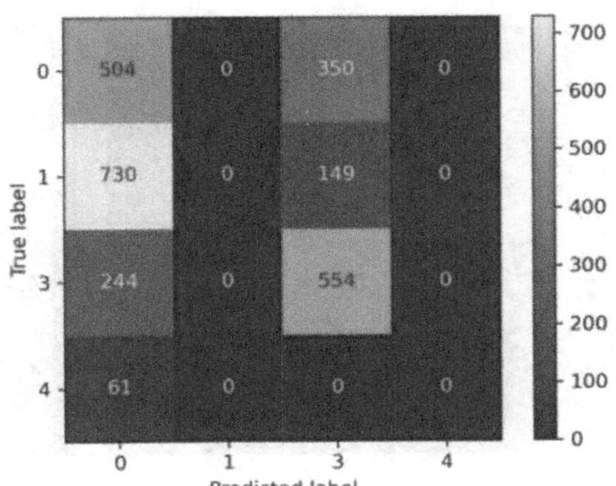

FIGURE 39.13 Confusion Matrix – Nursery – KNN (left) Variance (right).

cation task while the variance based approach uses the information in the dataset to determine dataset variance. It may be said that WOA is a reliable algorithm that strikes a balance between exploitation and exploration when looking for global best solution while variance based optimization enhances exploitation and decreases time taken to reach optimal solution. Selecting fewer features decreases the search space by removing duplicated or irrelevant attributes from a dataset.

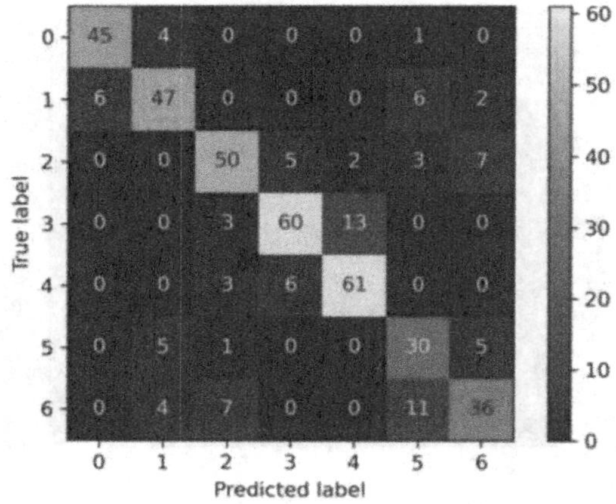

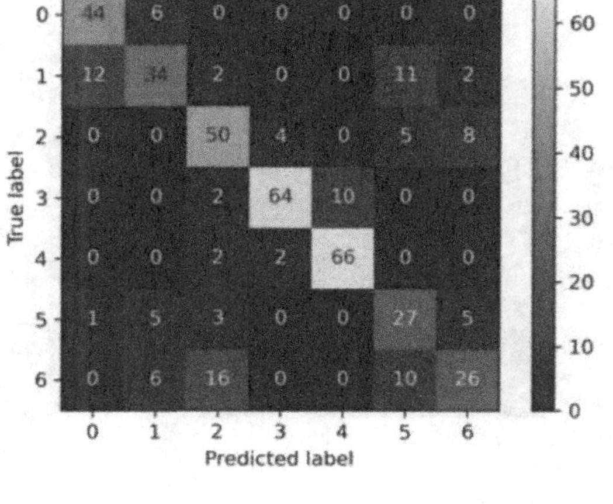

FIGURE 39.14 Confusion Matrix – Obesity – KNN (left) Variance (right).

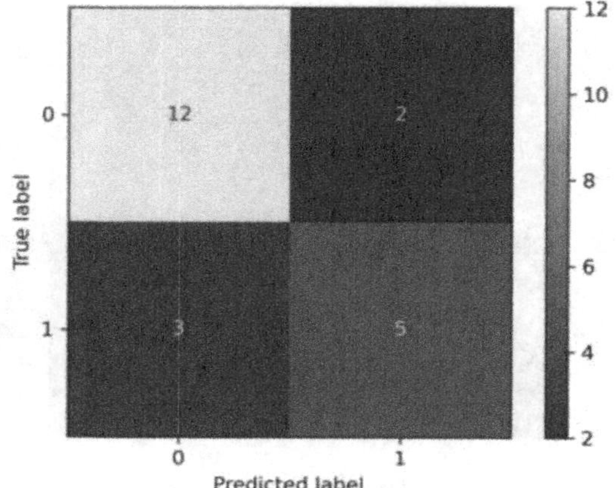

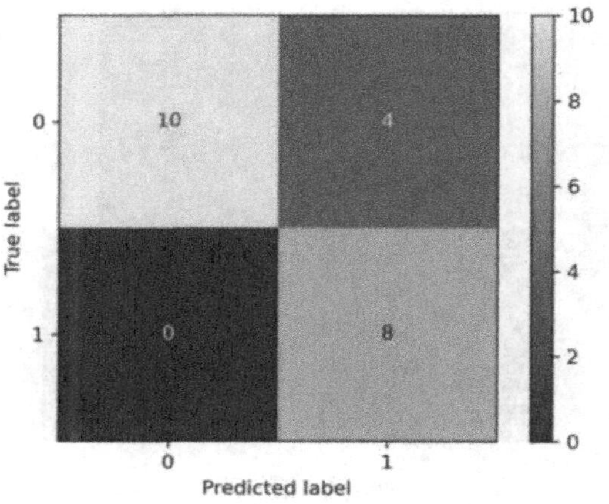

FIGURE 39.15 Confusion Matrix – Promoters – KNN (left) Variance (right).

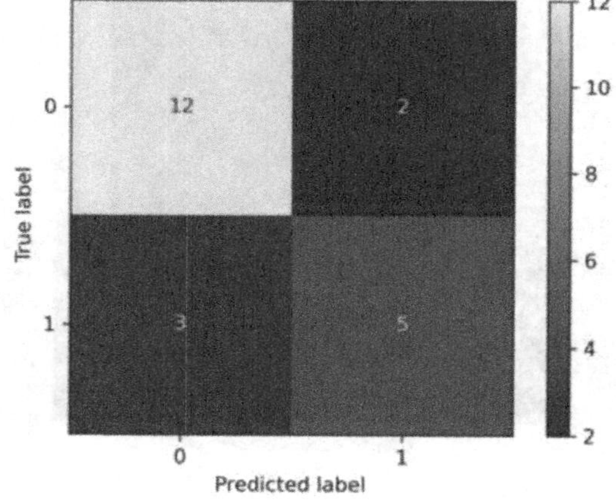

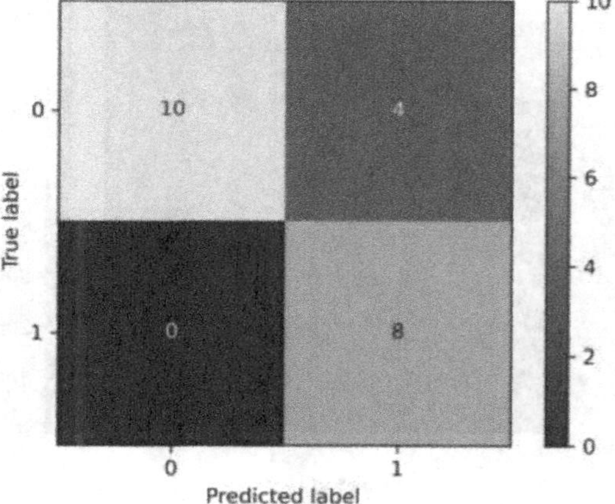

FIGURE 39.16 Confusion Matrix – Biodeg – KNN (left) Variance (right).

39.7 Conclusion and future work

One of the most important elements in improving the classifier's performance in the classification problem is feature selection. This research offered optimization of WOA algorithm based on the variance based optimization method. The method

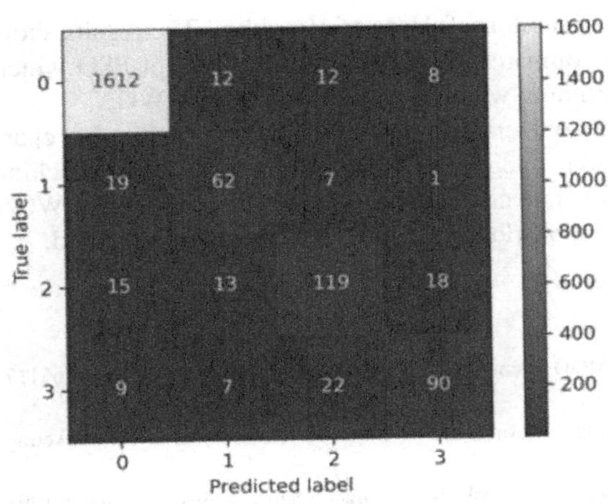

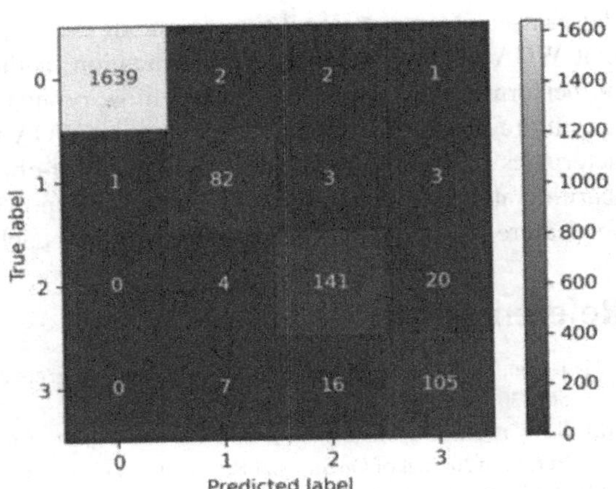

FIGURE 39.17 Confusion Matrix – Room – KNN (left) Variance (right).

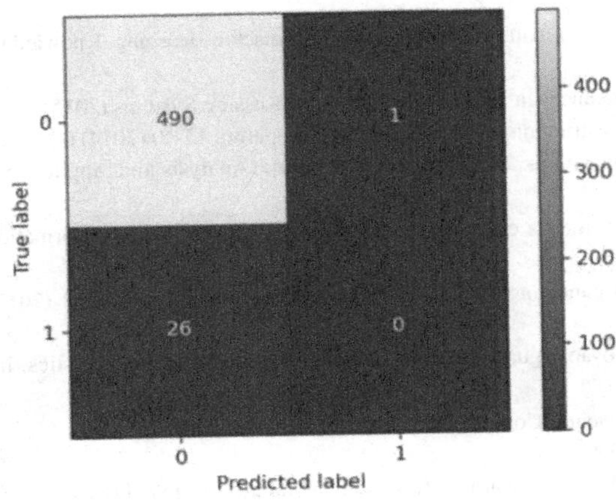

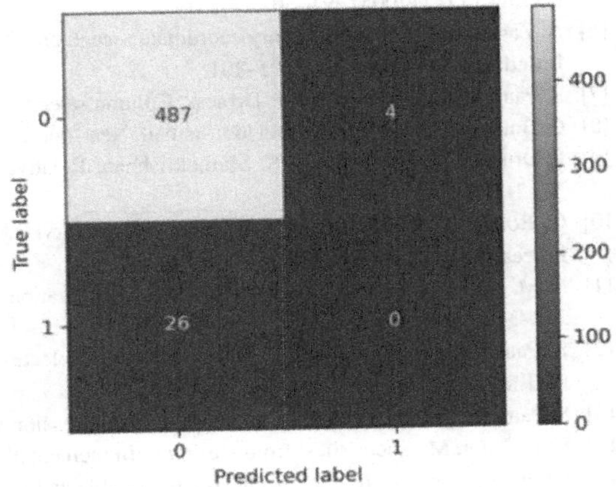

FIGURE 39.18 Confusion Matrix – Seismic – KNN (left) Variance (right).

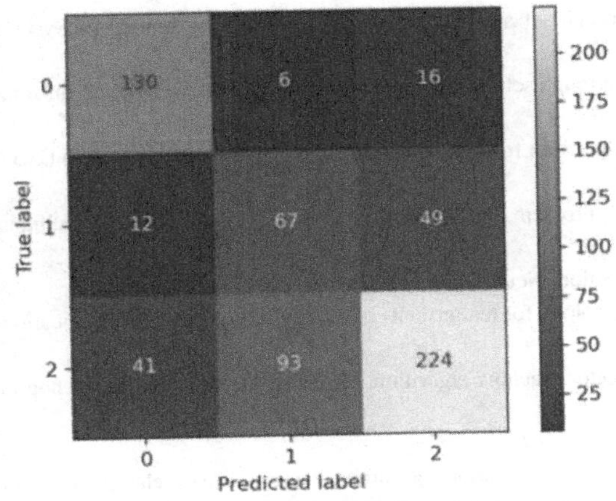

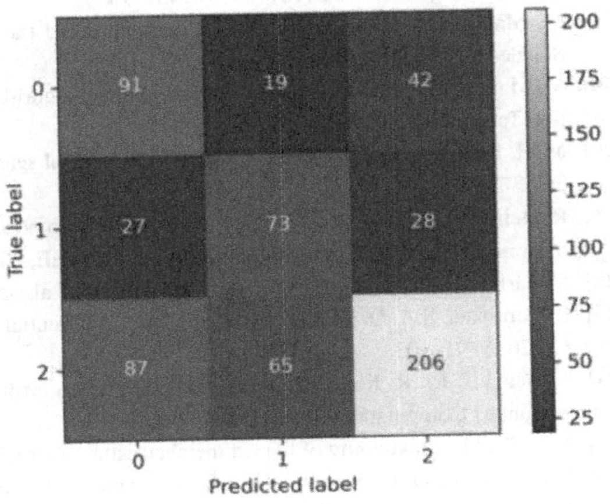

FIGURE 39.19 Confusion Matrix – Splice – KNN (left) Variance (right).

was tested on different datasets of UCI repository. The results are compared with benchmark KNN based optimization approach in order to test the validity of algorithm. The results are also compared with other wrapper based feature selection methods like GWO, PSO, BBA, GO, and EO in terms of percentage accuracy, precision, number of selected features and computational time. It was found that WOA along with our proposed approach outperforms other techniques in 90% of the datasets. Computational time is one of the most critical factor to assess the usefulness of algorithm. The results show that WOA with variance based optimization method is quite faster as compared to other algorithms except PSO which outperforms other algorithms but it is more prone to be stuck in local optimum which is suitably handled in WOA.

Future work directions for applying the WOA include exploring new objective functions that better capture the characteristics of the problem, investigating multi-objective feature selection, integrating the WOA with different machine learning algorithms such as deep learning, support vector machines (SVM), decision trees. The performance of the WOA for feature selection has been evaluated using benchmark datasets, but real-world applications have yet to be explored.

References

[1] J. Han, M. Kamber, F. Berzal, N. Marín, Data mining: concepts and techniques, SIGMOD Record 31 (2002) 66–68, https://doi.org/10.1145/565117.565130.

[2] S.F. Crone, S. Lessmann, R. Stahlbock, The impact of preprocessing on data mining: an evaluation of classifier sensitivity in direct marketing, European Journal of Operational Research 173 (3) (2006) 781–800.

[3] H. Liu, H. Motoda, Feature Extraction, Construction and Selection: A Data Mining Perspective, vol. 453, Springer Science & Business Media, 1998.

[4] R. Kohavi, G.H. John, Wrappers for feature subset selection, Artificial Intelligence 97 (1–2) (1997) 273–324.

[5] H. Liu, L. Yu, Toward integrating feature selection algorithms for classification and clustering, IEEE Transactions on Knowledge and Data Engineering 17 (4) (2005) 491–502.

[6] A. Zarshenas, K. Suzuki, Binary coordinate ascent: an efficient optimization technique for feature subset selection for machine learning, Knowledge-Based Systems 110 (2016) 191–201.

[7] S. Paul, M. Magdon-Ismail, P. Drineas, Column selection via adaptive sampling, Advances in Neural Information Processing Systems (2015) 28.

[8] C. Boutsidis, P. Drineas, M. Magdon-Ismail, Near-optimal column-based matrix reconstruction, SIAM Journal on Computing 43 (2) (2014) 687–717.

[9] P. Drineas, M.W. Mahoney, S. Muthukrishnan, Relative-error CUR matrix decompositions, SIAM Journal on Matrix Analysis and Applications 30 (2) (2008) 844–881.

[10] C. Boutsidis, P. Drineas, M.W. Mahoney, Unsupervised feature selection for the k-means clustering problem, Advances in Neural Information Processing Systems (2009) 22.

[11] M.M. Mafarja, S. Mirjalili, Hybrid whale optimization algorithm with simulated annealing for feature selection, Neurocomputing 260 (2017) 302–312.

[12] S. Paul, M. Magdon-Ismail, P. Drineas, Feature selection for linear SVM with provable guarantees, in: Artificial Intelligence and Statistics, in: PMLR, 2015.

[13] S. Paul, P. Drineas, Feature selection for ridge regression with provable guarantees, Neural Computation 28 (4) (2016) 716–742.

[14] E.-G. Talbi, Metaheuristics: from Design to Implementation, John Wiley & Sons, 2009.

[15] I. Guyon, A. Elisseeff, An introduction to variable and feature selection, Journal of Machine Learning Research 3 (Mar 2003) 1157–1182.

[16] A.-R. Hedar, J. Wang, M. Fukushima, Tabu search for attribute reduction in rough set theory, Soft Computing 12 (9) (2008) 909–918.

[17] R. Jensen, Q. Shen, Semantics-preserving dimensionality reduction: rough and fuzzy-rough-based approaches, IEEE Transactions on Knowledge and Data Engineering 16 (12) (2004) 1457–1471.

[18] M. Mafarja, S. Abdullah, A fuzzy record-to-record travel algorithm for solving rough set attribute reduction, International Journal of Systems Science 46 (3) (2015) 503–512.

[19] M. Mafarja, S. Abdullah, Record-to-record travel algorithm for attribute reduction in rough set theory, Journal of Theoretical and Applied Information Technology 49 (2) (2013) 507–513.

[20] M.M. Kabir, M. Shahjahan, K. Murase, A new local search based hybrid genetic algorithm for feature selection, Neurocomputing 74 (17) (2011) 2914–2928.

[21] R. Bello, et al., Two-step particle swarm optimization to solve the feature selection problem, in: Seventh International Conference on Intelligent Systems Design and Applications (ISDA 2007), IEEE, 2007.

[22] S. Kashef, H. Nezamabadi-pour, An advanced ACO algorithm for feature subset selection, Neurocomputing 147 (2015) 271–279.

[23] E. Zorarpacı, S.A. Özel, A hybrid approach of differential evolution and artificial bee colony for feature selection, Expert Systems with Applications 62 (2016) 91–103.

[24] J. Wang, T. Li, R. Ren, A real time IDSs based on artificial bee colony-support vector machine algorithm, in: Third International Workshop on Advanced Computational Intelligence, IEEE, 2010.

[25] E.-G. Talbi, A taxonomy of hybrid metaheuristics, Journal of Heuristics 8 (5) (2002) 541–564.

[26] Qasem Al-Tashi, Helmi Rais, Said Jadid, Feature selection method based on grey wolf optimization for coronary artery disease classification, in: Recent Trends in Data Science and Soft Computing: Proceedings of the 3rd International Conference of Reliable Information and Communication Technology (IRICT 2018), Springer International Publishing, 2019.

[27] Hailun Xie, et al., Feature selection using enhanced particle swarm optimisation for classification models, Sensors 21 (5) (2021) 1816.

[28] I.-S. Oh, J.-S. Lee, B.-R. Moon, Hybrid genetic algorithms for feature selection, IEEE Transactions on Pattern Analysis and Machine Intelligence 26 (11) (2004) 1424–1437.

[29] O.C. Martin, S.W. Otto, Combining simulated annealing with local search heuristics, Annals of Operations Research 63 (1) (1996) 57–75.

[30] K. Lenin, B.R. Reddy, M. Suryakalavathi, Hybrid Tabu search-simulated annealing method to solve optimal reactive power problem, International Journal of Electrical Power & Energy Systems 82 (2016) 87–91.

[31] P. Vasant, Hybrid simulated annealing and genetic algorithms for industrial production management problems, International Journal of Computational Methods 7 (02) (2010) 279–297.
[32] Z. Li, P. Schonfeld, Hybrid simulated annealing and genetic algorithm for optimizing arterial signal timings under oversaturated traffic conditions, Journal of Advanced Transportation 49 (1) (2015) 153–170.
[33] Y. Li, et al., A hybrid genetic-simulated annealing algorithm for the location-inventory-routing problem considering returns under E-supply chain environment, The Scientific World Journal 2013 (2013) 125893.
[34] L. Junghans, N. Darde, Hybrid single objective genetic algorithm coupled with the simulated annealing optimization method for building optimization, Energy and Buildings 86 (2015) 651–662.
[35] M. Mafarja, S. Abdullah, Investigating memetic algorithm in solving rough set attribute reduction, International Journal of Computer Applications in Technology 48 (3) (2013) 195–202, https://doi.org/10.1504/IJCAT.2013.056915.
[36] R. Azmi, et al., A hybrid GA and SA algorithms for feature selection in recognition of hand-printed Farsi characters, in: 2010 IEEE International Conference on Intelligent Computing and Intelligent Systems, IEEE, 2010.
[37] J. Wu, Z. Lu, L. Jin, A novel hybrid genetic algorithm and simulated annealing for feature selection and kernel optimization in support vector regression, in: 2012 IEEE 13th International Conference on Information Reuse & Integration (IRI), IEEE, 2012.
[38] Stephen O. Olabiyisi, et al., Hybrid metaheuristic feature extraction technique for solving timetabling problem, International Journal of Scientific and Engineering Research 3 (8) (2012) 1–6.
[39] S. Ferchichi, K. Laabidi, S. Zidi, Genetic algorithm and tabu search for feature selection, Studies in Informatics and Control 18 (2) (2009) 181–187.
[40] M. Majdi, S. Abdullah, N.S. Jaddi, Fuzzy population-based meta-heuristic approaches for attribute reduction in rough set theory, International Journal of Computer, Electrical, Automation, Control and Information Engineering 9 (12) (2015) 2462–2470, http://hdl.handle.net/20.500.11889/4244.
[41] P. Moradi, M. Gholampour, A hybrid particle swarm optimization for feature subset selection by integrating a novel local search strategy, Applied Soft Computing 43 (2016) 117–130.
[42] E.-G. Talbi, et al., Comparison of population based metaheuristics for feature selection: application to microarray data classification, in: 2008 IEEE/ACS International Conference on Computer Systems and Applications, IEEE, 2008.
[43] Z. Yong, G. Dun-wei, Z. Wan-qiu, Feature selection of unreliable data using an improved multi-objective PSO algorithm, Neurocomputing 171 (2016) 1281–1290.
[44] J. Jona, N. Nagaveni, A hybrid swarm optimization approach for feature set reduction in digital mammograms, WSEAS Transactions on Information Science and Applications 9 (11) (2012) 340–349.
[45] M.E. Basiri, S. Nemati, A novel hybrid ACO-GA algorithm for text feature selection, in: 2009 IEEE Congress on Evolutionary Computation, IEEE, 2009.
[46] R.S. Babatunde, S.O. Olabiyisi, E.O. Omidiora, R.A. Ganiyu, Feature dimensionality reduction using a dual level metaheuristic algorithm, International Journal of Applied Information Systems 7 (1) (April 2014) 49–52, https://doi.org/10.5120/ijais14-451134.
[47] J. Jona, N. Nagaveni, Ant-cuckoo colony optimization for feature selection in digital mammogram, Pakistan Journal of Biological Sciences: PJBS 17 (2) (2014) 266–271.
[48] M. Nekkaa, D. Boughaci, Hybrid harmony search combined with stochastic local search for feature selection, Neural Processing Letters 44 (1) (2016) 199–220.
[49] I. Boussaïd, J. Lepagnot, P. Siarry, A survey on optimization metaheuristics, Information Sciences 237 (2013) 82–117.
[50] G. Chandrashekar, F. Sahin, A survey on feature selection methods, Computers & Electrical Engineering 40 (1) (2014) 16–28.
[51] S. Mirjalili, A. Lewis, The whale optimization algorithm, Advances in Engineering Software 95 (2016) 51–67.
[52] C. Blake, CJ Merz UCI Repository of Machine Learning Databases, University of California at Irvine, 1998.
[53] E. Emary, H.M. Zawbaa, A.E. Hassanien, Binary grey wolf optimization approaches for feature selection, Neurocomputing 172 (2016) 371–381.
[54] R.Y. Nakamura, L.A. Pereira, K.A. Costa, D. Rodrigues, J.P. Papa, X.S. Yang, BBA: a binary bat algorithm for feature selection, in: 2012 25th SIBGRAPI Conference on Graphics, Patterns and Images, IEEE, August 2012, pp. 291–297.
[55] O.H. Babatunde, L. Armstrong, J. Leng, D. Diepeveen, A genetic algorithm-based feature selection, International Journal of Electronics Communication and Computer Engineering 5 (4) (2014) 899–905.
[56] Y. Gao, Y. Zhou, Q. Luo, An efficient binary equilibrium optimizer algorithm for feature selection, IEEE Access 8 (2020) 140936–140963.

Chapter 40

Whale optimization algorithm for Covid-19 detection based on ECG

Imene Latreche[a], Mohamed Akram Khelili[a,b], Sihem Slatnia[a], Okba Kazar[c], and Saad Harous[d]

[a]*Department of Computer Science, University of Biskra, Biskra, Algeria,* [b]*Numidia Institute of Technology, Algies, Algeria,* [c]*University of Kalba, Sharjah, United Arab Emirates,* [d]*Department of Computer Science, College of Computing and Informatics, University of Sharjah, Sharjah, United Arab Emirates*

40.1 Introduction

COVID-19 is the name of a global pandemic that developed in China in 2019. Covid-19 is a disease that affects the respiratory system. By 2022, this pandemic had afflicted around 635 million individuals, causing in 6 million fatalities [1]. The World Health Organization (WHO) recognized the virus's danger and quick spread, necessitating the development of a multitude of remedies to battle and limit the pandemic.

Among these options, the World Health Organization (WHO) suggested reverse-transcription polymerase chain reaction (RT-PCR) as the standard test for detecting COVID-19. Despite its status as the gold standard, this test has several drawbacks, including the time necessary to produce findings, the quality of the laboratory employees, the necessity for specialists to check the results, and the risk of providing false results. Scientists, on the other hand, developed alternate ways for diagnosing viral infection, such as chest X-rays and computed tomography (CT) scans. Many cutting-edge works, such as those done by Khelili et al. [2] these procedures could properly detect the virus in a patient's lungs within a short amount of time after the virus's infection was discovered. However, in the presence of other lung infections, such as pneumonia, the involvement of radiologists as specialists is essential to confirm the source of the infection in the lungs.

With the introduction of artificial intelligence (AI), several sectors, including medicine, have embraced it. Researchers take the advantages of these techniques and apply it on detecting the infections using image sources of the lungs. However, the application these techniques require specialized and qualified scientist.

As a result, fresh ways are needed to aid COVID-19 detection as the pandemic progresses [3]. The novel coronavirus predominantly affects the heart and other essential human organs [4]. COVID-19 individuals have been found to have a variety of cardiovascular abnormalities, including QRST anomalies, arrhythmias [5] [6], ST-segment modifications [7] [8], and conduction issues. These changes were seen in the electrocardiograms (ECGs) of COVID-19 individuals. These cardiac changes [9] [10] have suggested the use of ECG in the diagnosis of COVID-19. ST alterations on the ECG can be used to detect the virus and may suggest preclinical or symptomatic cardiovascular damage. Aside from PCR testing and chest X-rays or CT images, because of its mobility, availability, ease of use, cheap cost, safety, and real-time inspection, COVID-19 automated diagnosis from ECG may be of significant help.

Numerous studies have used a 2D representation of 1D ECG signals [11] [12] to train CNN architecture, such as ECG time-amplitude images [13] [14], time-frequency representations using the Short-Time Fourier Transform (STFT) [15] [16] and Continuous Wavelet Transform (CWT) [17], higher-order spectral representations [18], and dual beat coupling matrices [19]. Because paper-based ECG readings are so common [20], special care is needed to fill the gap of automated identification for cardiac problems requiring special attention.

The rest of the paper's structure is: Section 40.2 showed the most recent advances in COVID-19 diagnosis using ECG pictures. Section 40.3 discusses the dataset, CNN model, and WOA utilized for hyperparameter auto-selection. Section 40.4 for showing the results and comments. Section 40.5 of our report finishes with a review of the system's important results, limitations, and future research.

40.2 Related work

Some researchers have used artificial intelligence approaches-based ECG data to diagnose the heart diseases. The works done in [11] [12] used ECG data to treat problems. Other investigations [17] used wavelet transformations to convert ECG

Handbook of Whale Optimization Algorithm. https://doi.org/10.1016/B978-0-32-395365-8.00046-4

data to pictures in order to detect irregularities. Previous strategies that used publicly accessible ECG signal datasets had substantial success, but their real time application still challenging topic. This is due to the nature of ECG measurements that accumulated over time. This is not often the case in clinical settings, when ECG data is collected and stored as images [21].

Image of ECG data comprises of multiple distinct and carefully separated lead signals. Furthermore, there is some overlap in an ECG picture between the waveforms of distinct leads and the surrounding solid secondary axis, making it difficult to properly pinpoint useful properties. Furthermore, reducing the sample rate of ECG digital signals to ECG pictures leads in considerable information loss, affecting the effectiveness of AI systems [22].

This process, however, is computationally consuming, and the quality of data is restricted [23]. This is due to the fact that the noise created by this conversion has a significant influence on the performance of the used DL techniques. To detect a heart issue, the differences between virtually all sorts of heart abnormalities are often tiny, and these little variances are critical components for abnormality diagnosis. Despite its great learning capacity, deep learning algorithms cannot consistently identify the distinguishing features.

All of the restrictions outlined above prevent the application of ECG data as pictures. There have been several studies that have used ECG trace pictures to identify a range of cardiac diseases; however, the number of these research is restricted. As a result, a number of research papers used AI algorithms to detect a variety of cardiac abnormalities, such as normal and irregular heartbeats, myocardial infarction (MI), and Positive ECG (COVID-19).

The authors of [3] described the ECG-BiCoNet pipeline, which diagnoses COVID-19 using a 2D ECG trace picture and five deep learning techniques. DWT, symmetrical uncertainty (SU) technique, and three ML classifiers, are the suggested techniques in the system. The classification results of the five evaluated models show a promising outcome of accuracy in a binary and a multiclass classification 98.8% and 91.73% respectively.

Another study [24] employed an ECG-based diagnostic tool with eleven DL models to diagnose COVID-19, including ten models. According to the data obtained, the accuracy for multiclass levels is 98.2% and 91.6% for binary levels.

Ref. [25] introduced an innovative and successful approach to transform an ECG signal to images in order to detect COVID-19. The scientists extracted features and constructed hexaxial mapping pictures, which they then fed into the CNN model. The findings show promising outcomes, such as 96.20% of accuracy and 96.30% of F1-Score. To assess COVID-19's diagnostic skills, ECG data were categorized using the multiclass technique. The achieved values are 93.00% accuracy and 93.20% F1-score.

Rahman et al. [26] presented a new transfer learning-based system, which incorporates six deep learning models. The authors used multiple preprocessing techniques, such as gamma correction and standardization, to convert the dataset to their proposed models. According to the classification findings, both binary and multiclass classification worked well, with 99.1% and 97.83% accuracy, respectively.

Another research, [27], employed 12-lead ECG pictures for detecting COVID-19. The authors used a 3D CNN model with residual connections, and the average accuracy were 99.0% and 92.0% for binary and multiclass classifications respectively.

The authors in [28] applied various approaches on the EfficientNet deep learning technique's COVID-19 classification performance. The authors came to the conclusion that augmentation tactics can improve performance to some extent. However, overuse of improvement methods may reduce performance. The maximum degree of precision obtained was 81.8%.

These works, however, are subject to various restrictions. The first restriction is the inadequate training/validation database for deep learning with hundreds of hyperparameters. Furthermore, this research did not address the issue of class inequality.

Furthermore, neither the baseline rhythm nor the impact of the baseline rhythm is reported for each patient. Furthermore, optimization methodologies for selecting deep learning hyperparameters were lacking in these investigations. Furthermore, the dataset employed in the study is made up of verified COVID-19 instances. Asymptomatic infections may not be detectable with the same sensitivity.

As a result, the breadth of the results is relatively limited. As a result, we provide the Whale Optimization Algorithm for choosing the model's hyperparameters to diagnose COVID-19 using the traced ECG picture.

40.3 Material and methods

We will discuss in detail each component of our proposed approach, beginning with a description of the dataset before and after the preprocessing and augmentation steps, followed by a detailed presentation of the implemented model and an explanation of the WOA's process optimization algorithm, which will be fused with the used model.

40.3.1 Dataset

The suggested diagnostic technique makes use of a new public dataset [29] that incorporates pictures of COVID-19 and other heart diseases' ECG recordings. This dataset is one of the big and publicly used dataset for COVID-19 ECG recordings. This collection contains 1937 ECG pictures on a range of subjects. As indicated in Table 40.1, the collection includes 250 scans of patients infected with the new virus, 280 recordings with a current or previous myocardial infarction (MI), 548 ECG records of irregular heartbeats, and 859 normal pictures with no cardiac abnormalities.

Three extra electrodes were implanted on the two arms and left leg to simulate the following six limb leads: AVR, AVL, AVF, Lead I, II, and III. Using a telemedicine ECG diagnostic method, medical specialists assessed the images in the collection. A team of professional cardiologists with extensive experience in ECG annotation and research oversaw this study. (See Fig. 40.1.)

TABLE 40.1 Data detail.

Case	ECG images
COVID-19	250
Normal	859
Myocardial Infarction	77
Patients have past of Myocardial Infraction	203
Abnormal Heartbeat	548

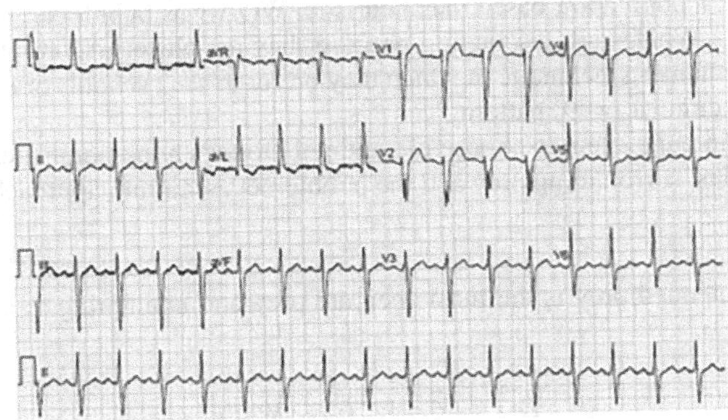

FIGURE 40.1 ECG image of Covid-19 patient.

40.3.2 Preprocessing

Gamma correction enhancement algorithm [30] were used to improve the quality of the ECG pictures. Individual pixels are typically exposed with mathematical operations during picture normalization. Gamma correction is a nonlinear approach that uses the internal map to project and connect the pixel value and the gamma value. If A displays a pixel value between 0 and 255, it signifies an angle value. Eqs. (40.1)–(40.5) are accurate if X reflects the pixel's gray-scale value (A). Suppose X_m is the middle of the range [0, 255].

P is the group's linear map, which comprises the following elements:

$$\varphi : A \rightarrow \Omega,\ \Omega = \{\omega \mid \omega = \varphi(x)\},\ \varphi(x) = \frac{\pi x}{2X_m} \tag{40.1}$$

The equation of the translation from Ω to Γ is:

$$j : \Omega \rightarrow \Gamma, \Gamma = \{\gamma \mid \gamma = j(X)\} \tag{40.2}$$

$$j(X) = 1 + f_1(X) \tag{40.3}$$

$$f_1(X) = a\cos(\varphi(x)) \tag{40.4}$$

Group A can be correlated with group pixel values based on this map. A predefined gamma value is used to determine the arbitrary pixel value. Suppose that (X)=j(X) and the Gamma correction function is:

$$s(K) = 255(\frac{k}{255})^{1/\gamma(k)} \tag{40.5}$$

The grayscale output pixel correction value is represented by the s(K) value. After gamma correction, the dataset is processed to resize the ECG images to meet CNN network input image-size criterion (e.g., 224 by 224 for residual and dense networks, and 299 by 299 for the inception network). The image's Z-score was normalized using the average and standard deviation. [26].

40.3.3 Augmentation

Training using an uneven dataset may result in a biased model since the dataset is unbalanced and does not contain a proportional number of photos for each category. Data augmentation for the training sample can assist guarantee that each class has an equal number of photographs, resulting in more accurate results. To balance the training pictures in this work, three technics were used: rotation, scaling, and translation. The picture enhancement procedure needed 7 to 12 degrees clockwise and counterclockwise rotations of the photos.

Scaling is the increase or reduction of an image's frame size; picture magnifications varied from 2.5% to 12% in this investigation. For image translation, images were horizontally and vertically translated by 3% to 22%.

40.3.4 Whale optimization algorithm (WOA)

Authors in [31] proposed WOA, a population-based metaheuristic. WOA was bubble-net hunting-based strategy, that emulate humpback whales' actions. Whales are fascinating because they are clever and emotionally complex. Furthermore, humpback whales have a unique hunting technique in which they prefer to seek schools of krill or microscopic fish near the surface by blowing bubbles in a circular or '9' pattern.

We employed the algorithm in our system to choose the optimal model's hyperparameters based on the accuracy of the produced models using two tactics, surrounding prey and the bubble-net assaulting approach.

40.3.4.1 Encircling prey

Consists of assuming that the intended prey is the most accurate prey and attempting to change the whale's place using Eqs. (40.6) and (40.7):

$$\overrightarrow{Dis} = \left|\overrightarrow{Co}.\overrightarrow{W^*}(i) - \overrightarrow{W}(i)\right| \tag{40.6}$$

$$\overrightarrow{W}(i+1) = \overrightarrow{W^*}(i) - \overrightarrow{A}c.\overrightarrow{Dis} \tag{40.7}$$

where $\overrightarrow{Ac}$ and $\overrightarrow{Co}$ are coefficient vectors, $\overrightarrow{W}$ is the vector's actual location, and W^* is the best solution's prior position vector. If a better solution exists, the location of W^* will be adjusted in each iteration (i). The vectors $\overrightarrow{Ac}$ and $\overrightarrow{Co}$ are determined as follows:

$$\overrightarrow{Ac} = 2\overrightarrow{p}.\overrightarrow{ra} - \overrightarrow{p} \tag{40.8}$$

$$\overrightarrow{Co} = 2.\overrightarrow{ra} \tag{40.9}$$

The $\overrightarrow{p}$ metric is linearly reduced in [2-0] over the course of iterations and $\overrightarrow{ra}$ is a random vector [0, 1].

40.3.4.2 Bubble-net attacking method

Two tactics are necessary at this period.

40.3.4.2.1 Shrinking encircling mechanism

The first strategy is based on reducing the value of $\overrightarrow{p}$ according to Eq. (40.8), which also decreases $\overrightarrow{Ac}$ because they are generally between $-p$ and p. In our system, we generated $\overrightarrow{Ac}$ between -1 and 1.

40.3.4.2.2 Spiral updating position

At this phase, the distance between the whale's (W) and prey's (Y) positions done with Eq. (40.10):

$$\vec{W}(t+1) = \overrightarrow{Dis'}.e^{bl}.\cos(2\pi l) + \overrightarrow{W^*}(i) \tag{40.10}$$

where $\overrightarrow{Dis'} = \left|\overrightarrow{W^*}(i) - \vec{W}(i)\right|$, b is a logarithmic spiral, and $l \in [-1, 1]$.

The whale has the same possibility (50%) to use either shrinking circle or spiral shaped based on v, where v is random from [0, 1] as it presented in Eq. (40.11).

$$\vec{W}(t+1) = \begin{Bmatrix} \overrightarrow{W^*}(i) - \overrightarrow{Ac}.\overrightarrow{Dis} & \text{if } v < 0.5 \\ \overrightarrow{Dis'}.e^{bl}.\cos(2\pi l) + \overrightarrow{W^*}(t) & \text{if } v \geq 0.5 \end{Bmatrix} \tag{40.11}$$

Furthermore, the whale undertakes another random search for prey using the following equations to assure exploration and avoid the potential of filling up the local optimum:

$$\overrightarrow{Dis} = \left|\overrightarrow{Co}.\overrightarrow{W_{rand}} - \vec{W}\right| \tag{40.12}$$

$$\vec{W}(t+1) = \overrightarrow{W_{rand}} - \vec{A}.\overrightarrow{Dis} \tag{40.13}$$

The $\overrightarrow{W_{rand}}$ is whale's position that chosen randomly.

40.3.4.3 *The pseudo algorithm of the proposed WOA-CNN*

```
Initialization of W_i (i = 1, 2, ..., n)
Compute the fitness
W*=the best search agent
While (k < max)
    for each search agent
    Re-compute a, A, Co, l, and p
        if (p<0.5)
            if (|Ac|< 1)
                Apply Eq. (40.6)
            else if (|Ac|>= 1)
                Select a random search agent ()
                Apply Eq. (40.13)
            end if
        else if (v >= 0.5)
            Apply Eq. (40.10)
        end if
    end for
Re-compute W*
k=k+1
end While
Get W*
```

40.3.5 Model description

We utilized CNN for training in our system because of its capability in automated feature extraction from pictures and deep analyses. We choose the CNN model, which comprises convolution and pooling layers, since it is beneficial in image processing. In addition, deep neurons with narrow kernel windows are used to increase feature learning and minimize model complexity. COVID-19 diagnostic systems use ECG images to create a deep model that extracts the most information from the image.

In our model, there are variety of layers (Figs. 40.2 and 40.3). The categorization results are stored in the final fully linked layer.

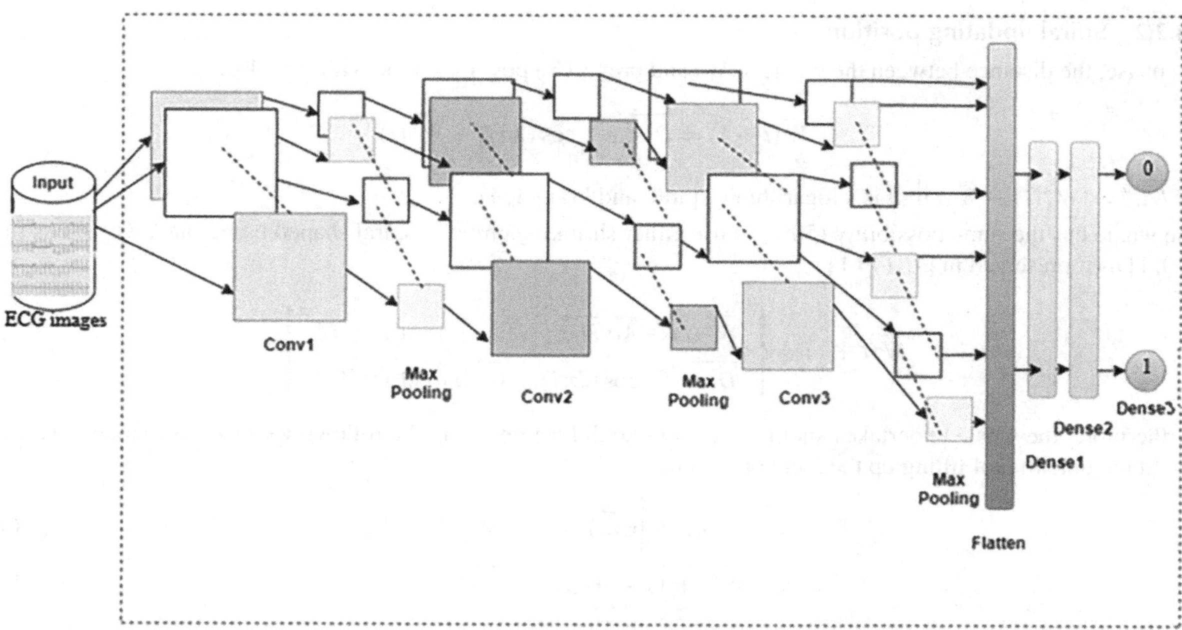

FIGURE 40.2 CNN model for binary classification.

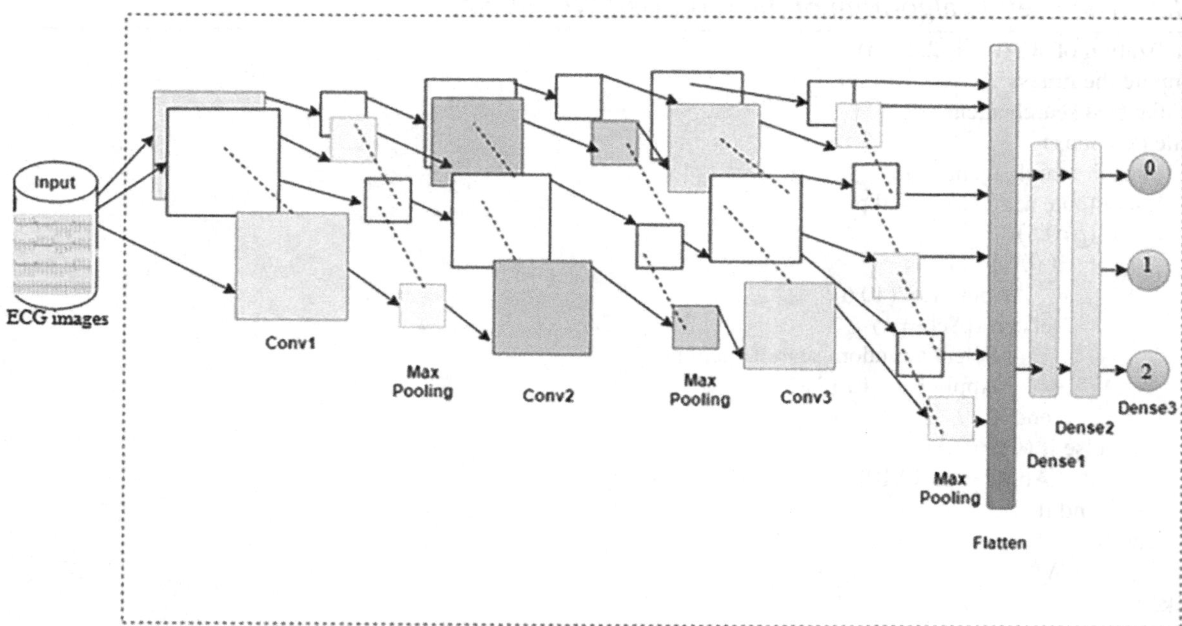

FIGURE 40.3 CNN model for 3class classification.

40.3.6 Model preparation

In the first step, we must establish our model's hyperparameters and build the starting population for our WOA. In our scenario, we choose 10 hyperparameters: optimizers, learning rate, number of filters, activation function, maxpooling layer, dropout rate, kernel size, number of epochs, random state, and loss function. The initial step in this phase is to produce n vectors, each of which contains random Hyperparameter values required to build n models. Fig. 40.4 shows that these models must be trained using a preprocessed dataset of ECG pictures in order to acquire the accuracy of each model, which will be regarded its fitness and will constitute the starting population in our system.

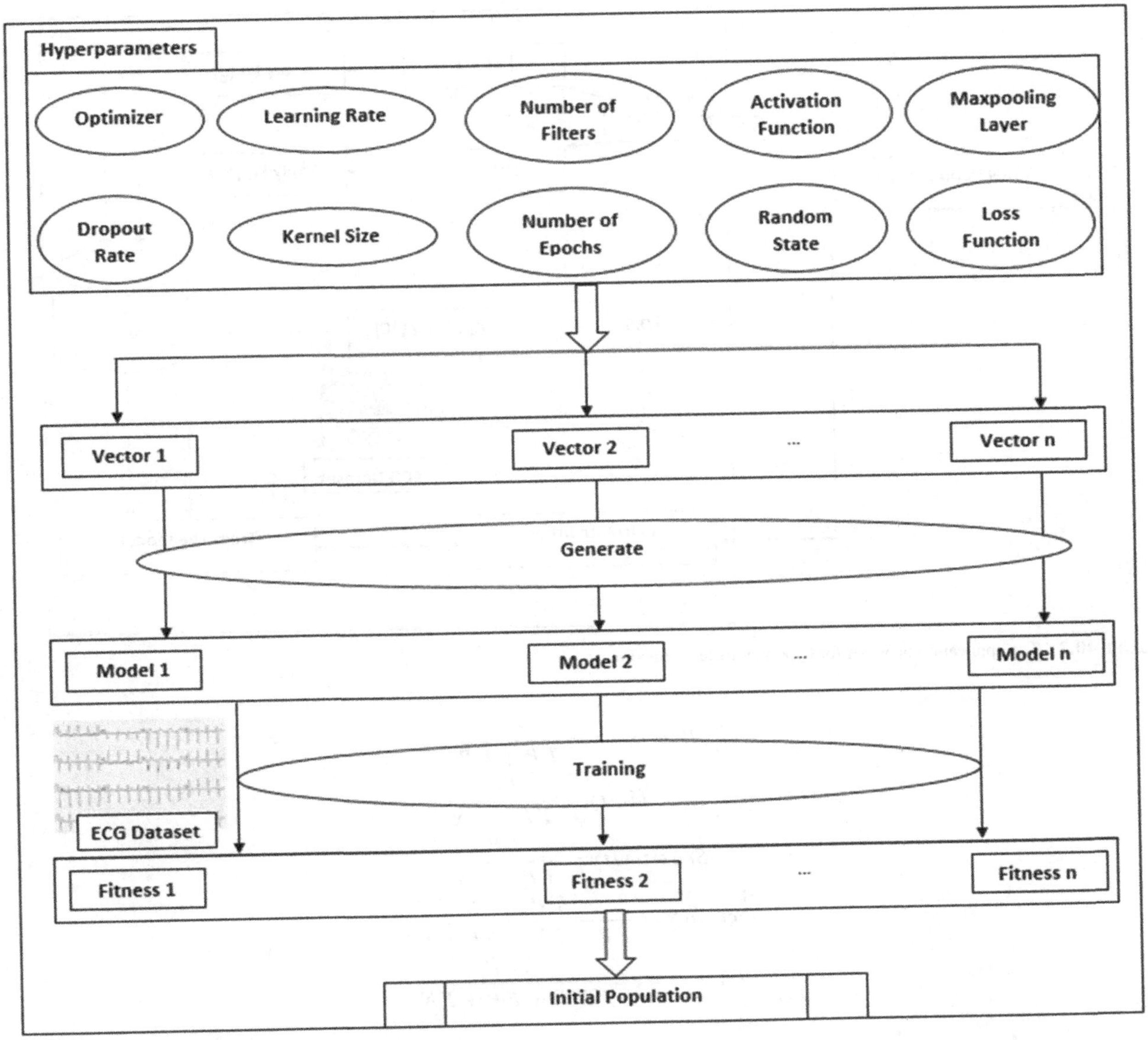

FIGURE 40.4 Generation of the first population.

In the second step, we must apply the WOA to our system using the original population created, and we must choose the best candidate from the population to apply for the WOA. As a result, we must first compute the probability (P) that each strategy will be adopted. If P = 0.5, we apply Eqs. (40.6) and (40.13) to determine if the vector A >= 1 or A 1 to decide the technique that will be used to update the location of the real solution, and then store the new position. In the alternative scenario, when P >= 0.5, we must use Eq. (40.13) and store the new location. Then, as illustrated in Fig. 40.5, we must retrain the model using the new hyperparameters' vector and continue the procedure until we reach the target or satisfy the stop criterion.

40.3.7 Classification performance evaluation

Various classification metrics were used such as:

$$Recall = \frac{TP}{TP + FN}$$

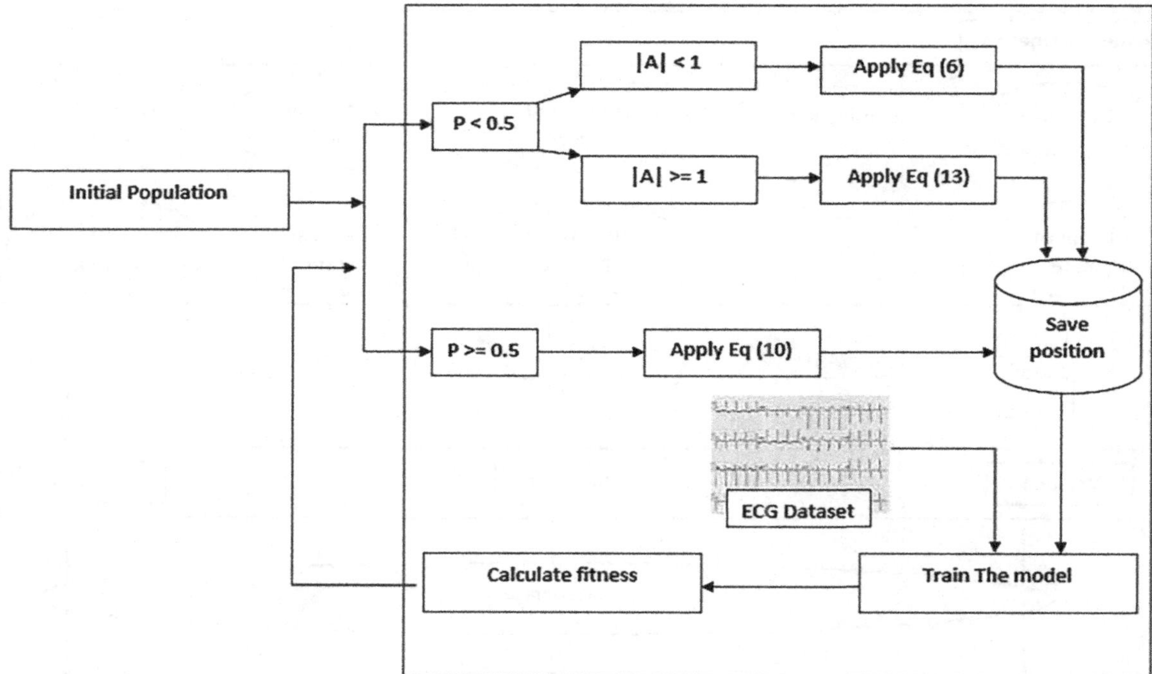

FIGURE 40.5 The application of WOA for hyperparameters selection.

$$Precision = \frac{TP}{TP + FP}$$

$$Specificity = \frac{TN}{FP + TN}$$

$$Sensitivity = \frac{TP}{TP + FN}$$

$$Accuracy = \frac{TP + TN}{TP + TN + FP + FN}$$

$$F1 - score = \frac{2TP}{2TP + FP + FN}$$

40.4 Results and description

In our system, we classified things in two ways.

40.4.1 Binary classification

We used a binary categorization that included both normal and COVID-19 patients. In this situation, we employed data augmentation to equalize the sample counts of the two classes. We then feed these into our training model. According to Table 40.2, the classification result reflects the performance of the proposed method, where class 0 represents COVID-19 samples and class 1 represents normal samples. In addition, as shown in Fig. 40.6, we display the training and validation accuracy to better describe and assure the results of our system in binary classification.

TABLE 40.2 Classification results.

Binary classification	Precision	Recall	F1-score	Sensitivity	Specificity	Accuracy	Loss
Class 0	1	0.96	0.98	0.96	1	0.99	0.056
Class 1	0.99	1	0.99	1	0.96		

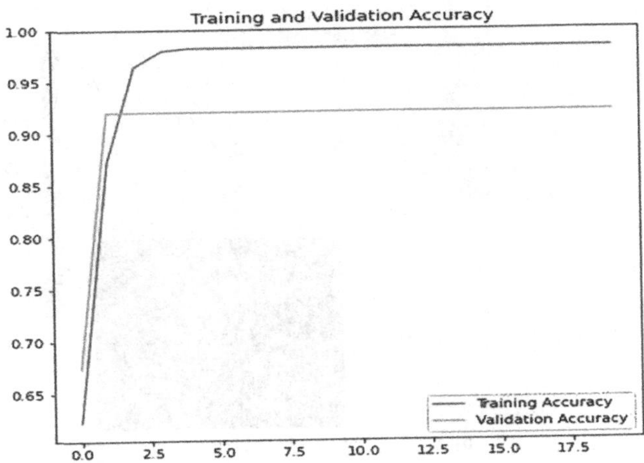

FIGURE 40.6 Training and validation accuracy of the system.

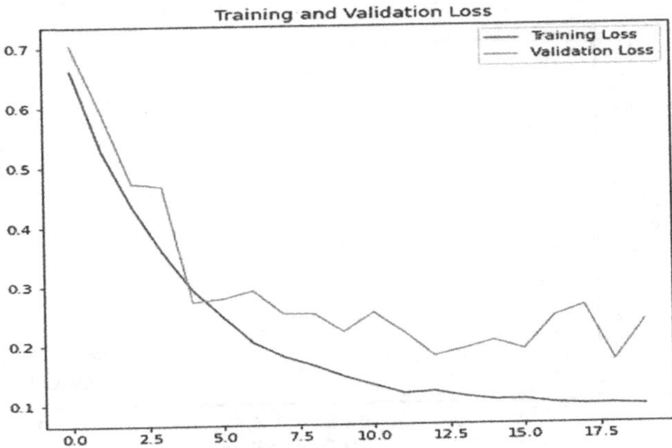

FIGURE 40.7 Training and validation loss of the system.

Furthermore, we included loss measurements during the training and validation phases to confirm the model's overfitting (see Fig. 40.7). All of these amazing results were acquired after only 20 training epochs, demonstrating WOA's efficiency in hyperparameter selection.

In addition, we generate the confusion matrix displayed in Fig. 40.8. The system's accuracy is proved by the fact that it correctly predicted all COVID-19 classes with only two exceptions.

Furthermore, the ROC curves of the model, as shown in Fig. 40.9, help to show the effectiveness of the suggested model.

40.4.2 Three-class classification

An extra class, abnormal heart beats, was added to the binary classification procedure in this form of classification to boost the system's accuracy and specificity in discriminating between COVID-19 and abnormal heart beats data. The classification findings show that precision, recall, F1-score, accuracy, and loss are all good measures (see Table 40.3). In addition, for a better understanding of the process, we provide the training and validation accuracy and loss using plots, as illustrated in Figs. 40.10 and 40.11.

TABLE 40.3 Classification results.

3-class classification	Precision	Recall	Sensitivity	Specificity	F1-score	Accuracy	Loss
Class 0	0.91	0.96	0.96	0.98	0.93	0.83	0.3
Class 1	0.86	0.66	0.66	0.95	0.74		
Class 2	0.81	0.91	0.91	0.76	0.86		

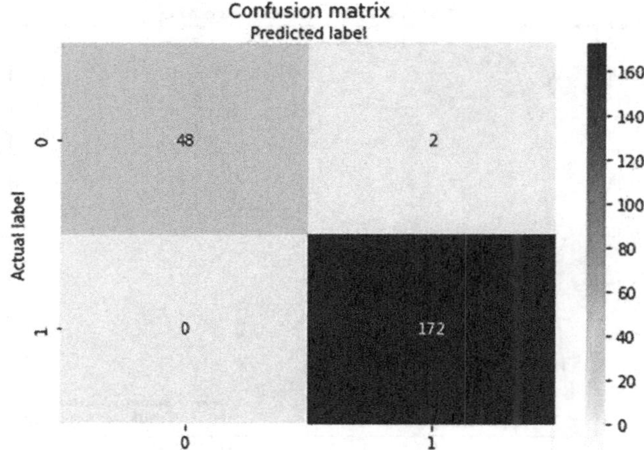

FIGURE 40.8 Confusion matrix of the model.

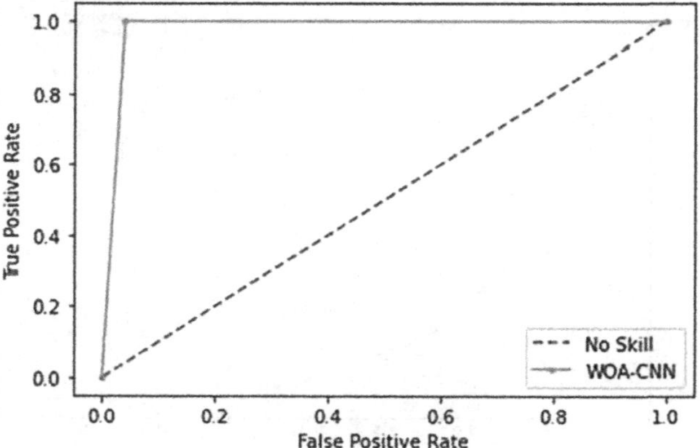

FIGURE 40.9 The ROC curve of the model.

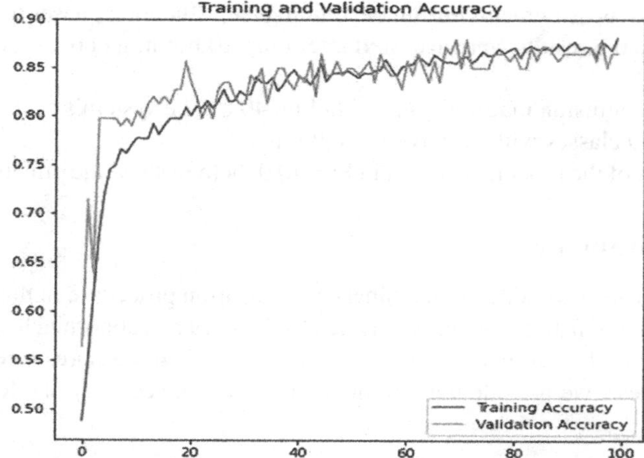

FIGURE 40.10 The accuracy of the model.

Furthermore, we validate the system's confusion using the confusion matrix, which shows that the model correctly predicts 83% of the samples and fails to predict 16% of the samples, as seen in Fig. 40.12.

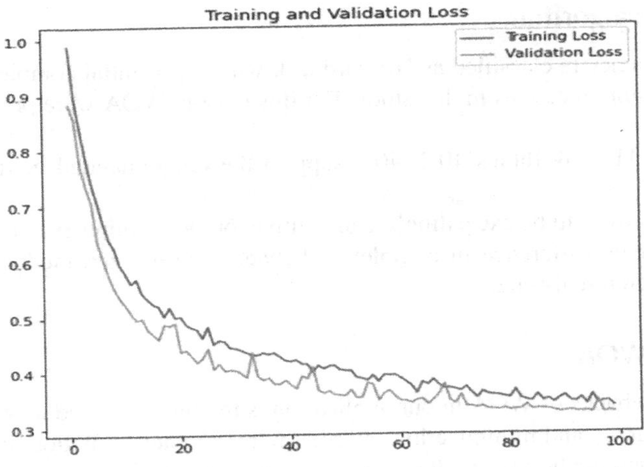

FIGURE 40.11 The loss of the model.

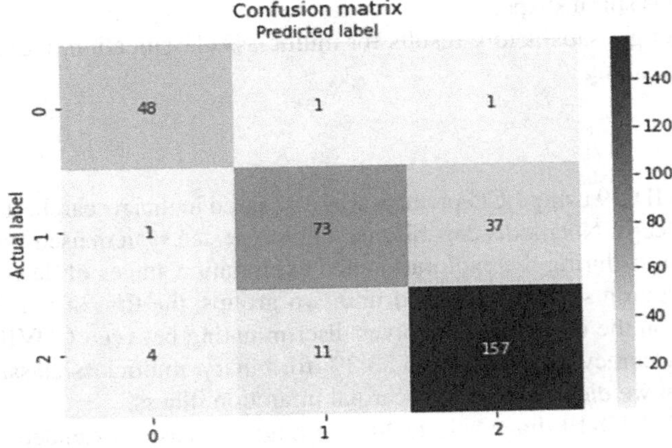

FIGURE 40.12 Confusion matrix of the model.

40.4.3 Comparative study

Our suggested system attained the greatest accuracy in the binary class among the state-of-the-art given studies, as indicated in Table 40.4. Unfortunately, in the multiclass setting, our suggested approach could not attain the desired outcomes and continues to be limited.

TABLE 40.4 Comparative study.

The Work	Technique	Accuracy (Binary/Multiclass)
	DWT + SU	98.8% / 91.73%
[25]	GLCM	96.20% / 93.00%
[27]	ResNet18, ResNet50, ResNet101, InceptionV3, DenseNet201, and MobileNetv2	99.1% / 97.83%
[28]	EfficientNet + Adam + AdamW	81.8% / 81.8%
Proposed system	WOA + CNN	99.99% / 83.3%

40.4.4 Complexity of the algorithm

Since the selection of hyperparameters classified as NP-hard task with exponential complexity $O(n^p)$ [32] [33], reducing this complexity is one of the main objectives in this study. For this reason, WOA were proposed to reduce the complexity to $O(n)$.

Figs. 40.7, 40.9, 40.10, 40.11, and Tables 40.2, 40.3 support the recommended model's strong performance in the majority of the criteria assessed.

The proposed technique is proven to be exceedingly competitive, outperforming even unlabeled or missing data. Based on the findings and the significant difference in complexity between the utilized models and those without WOA, the proposed model may be considered real-time.

40.4.5 Limitation of the WOA

One of the most significant drawbacks of WOA in our application is the time required to identify precise hyperparameters that would increase system accuracy and minimize loss. Furthermore, another limitation that may challenge our system is the randomness of selecting which method to employ between decreasing circle or spiral shaped that may alter a lot in our system's situation. The system could not function as planned because to the deferential specificities of each strategy, since when it was supposed to update the coefficient vector $\overrightarrow{A}$, the produced random p indicated that the strategy that will be used to update the prey position is spiral shaped.

Furthermore, WOA could not get satisfactory results for multiclass classification, indicating that their implementation in our system requires further tweaking.

40.5 Conclusion

The prospect of identifying COVID-19 using ECG pictures was discussed in this research. It developed a unique ECG-based automated diagnostic tool with deep CNN model capabilities. The suggested system used WOA to choose hyperparameters, which produced outstanding results during the exploration and exploitation stages of determining the ideal model configuration. The suggested categorization system is divided into two groups, the first of which differs between infected and not-infected cases. The second, on the other hand, involves discriminating between COVID-19, normal, and pathological heartbeats. The system had an accuracy of 99.99% and 83.3% for binary, multiclass classification respectively. The most serious flaw in our method is that we did not study myocardial infarction illness.

The disclosed ECG-based COVID-19 diagnostic technique might be easily extended to real-time cloud systems and done on mobile device-based decision-making apps in future research. As a result, it may help healthcare practitioners by providing a rapid and accurate method for diagnosing COVID-19, as well as reduce contamination and hospital load by minimizing unnecessary hospital visits. We planned to apply the WOA to more multiclassification situations in the future to identify and improve the WOA's shortcomings. We also want to compare WOA to another metaheuristic technique, such as the Marine Predator Algorithm (MPA), utilizing the same dataset. Furthermore, we are interested in reviewing the ECG records of the patient who previously received Covid-19 to see whether there is a chronic influence on the heart beat or not.

References

[1] COVID-19 CORONAVIRUS PANDEMIC, Worldmeter, https://www.worldometers.info/coronavirus/. (Accessed 1 November 2022).

[2] Khelili Mohamed Akram, Slatnia Sihem, Kazar Okba, Saad Harous, IoMT-fog-cloud based architecture for Covid-19 detection, Biomed. Signal Process. Control 76 (2022) 13.

[3] O. Attallah, ECG-BiCoNet: An ECG-based pipeline for COVID-19 diagnosis using Bi-Layers of deep features integration, Comput. Biol. Med. 142 (2022) 105210.

[4] R.S. Soumya, T. Govindan Unni, K.G. Raghu, Impact of COVID-19 on the cardiovascular system: a review of available reports, Cardiovasc. Drugs Ther. 35 (2021) 411–425.

[5] Andrea Denegri, Giuseppe Pezzuto, Matteo D'arienzo, et al., Clinical and electrocardiographic characteristics at admission of COVID-19/SARS-CoV2 pneumonia infection, Int. Emerg. Med. 16 (6) (2021) 1451–1456.

[6] Sripal Bangalore, Atul Sharma, Alexander Slotwiner, et al., ST-segment elevation in patients with Covid-19 - a case series, N. Engl. J. Med. 382 (25) (2020) 2478–2480.

[7] Behzad B. Pavri, Juergen Kloo, Darius Farzad, et al., Behavior of the PR interval with increasing heart rate in patients with COVID-19, Heart Rhythm 17 (9) (2020) 1434–1438.

[8] Saud Ahmed Khawaja, Poornima Mohan, Richard Jabbour, et al., COVID-19 and its impact on the cardiovascular system, Open Heart 8 (1) (2021) e001472.

[9] Rafael Bellotti Azevedo, Bruna Gopp Botelho, João Victor Gonçalves de Hollanda, et al., Covid-19 and the cardiovascular system: a comprehensive review, J. Hum. Hypertens. 35 (2) (2021) 4–11.

[10] Bruna Predabon, Arthur Zanfrilli Marques Souza, Gustavo Henrique Sumnienski Bertoldi, et al., The electrocardiogram in the differential diagnosis of cardiologic conditions related to the Covid-19 pandemic, J. Cardiac Arrhythmias 33 (3) (2020) 133–141.

[11] S. Kiranyaz, T. Ince, M. Gabbouj, Real-time patient-specific ECG classification by 1-D convolutional neural networks, IEEE Trans. Biomed. Eng. 63 (3) (2015) 664–675.

[12] O. Attallah, M. Sharkas, H. Gadelkarim, Deep learning techniques for automatic detection of embryonic neurodevelopmental disorders, Diagnostics 10 (1) (2020) 27.

[13] E. Izci, M. Ozdemir, M. Degirmenci, A. Akan, Cardiac arrhythmia detection from 2D ECG images by using deep learning technique, in: Proceedings of the 2019 Medical Technologies Congress (TIPTEKNO), Izmir, Turkey, 2019, pp. 1–4.

[14] G. Bortolan, I. Christov, I. Simova, Potential of rule-based methods and deep learning architectures for ECG diagnostics, Diagnostics 11 (9) (2021) 1678.

[15] Z. Zhu, X. Lan, T. Zhao, Y. Guo, P. Kojodjojo, Z. Xu, Z. Liu, S. Liu, H. Wang, X. Sun, Identification of 27 abnormalities from multi-lead ECG signals: an ensembled SE_ResNet framework with sign loss function, Physiol. Meas. 42 (6) (2021) 065008.

[16] Y.-Y. Jo, J. Kwon, K.-H. Jeon, Y.-H. Cho, J.-H. Shin, Y.-J. Lee, M.-S. Jung, J.-H. Ban, K.-H. Kim, S. Lee, Detection and classification of arrhythmia using an explainable deep learning model, J. Electrocardiol. 67 (2021) 124–132.

[17] X. Yang, X. Zhang, M. Yang, L. Zhang, 12-lead ECG arrhythmia classification using cascaded convolutional neural network and expert feature, J. Electrocardiol. 67 (2021) 56–62.

[18] H. Zhang, C. Liu, Z. Zhang, Y. Xing, X. Liu, R. Dong, Y. He, L. Xia, F. Liu, Recurrence plot-based approach for cardiac arrhythmia classification using inception-ResNet-V2, Front. Physiol. 12 (2021) 648950.

[19] V. Krasteva, I. Christov, S. Naydenov, T. Stoyanov, I. Jekova, Application of dense neural networks for detection of atrial fibrillation and ranking of augmented ECG feature set, Sensors 21 (2021) 6848.

[20] H. Dai, H.-G. Hwang, V. Tseng, Convolutional neural network based automatic screening tool for cardiovascular diseases using different intervals of ECG signals, Comput. Methods Programs Biomed. 203 (2021) 106035.

[21] Fabio Badilini, Tanju Erdem, Wojtek Zareba, et al., ECGscan: a method for digitizing paper ECG printouts, J. Electrocardiol. 36 (2003) 39.

[22] Siddharth Mishra, Gaurav Khatwani, Rupali Patil, et al., ECG paper record digitization and diagnosis using deep learning, J. Med. Biol. Eng. 41 (4) (2021) 422–432.

[23] Lawrence E. Widman, Gregory L. Freeman, A-to-D conversion from paper records with a desktop scanner and a microcomputer, Comput. Biomed. Res. 22 (4) (1989) 393–404.

[24] O. Attallah, An intelligent ECG-based tool for diagnosing COVID-19 via ensemble deep learning techniques, Biosensors 12 (5) (2022) 299.

[25] A.O. Mehmet, D.O. Gizem, G. Onan, Classification of COVID-19 electrocardiograms by using hexaxial feature mapping and deep learning, BMC Med. Inform. Decis. Mak. 21 (2021) 170.

[26] R. Tawsifur, A. Alex, E.H.C. Muhammad, et al., COV-ECGNET: COVID-19 detection using ECG trace images with deep convolutional neural network, Health Inf. Sci. Syst. 10 (1) (2022).

[27] S. Nebras, S. Abdulkadir, T. Ru-San, A.U. Rajendra, Attention-based 3D CNN with residual connections for efficient ECG-based COVID-19 detection, Comput. Biol. Med. 143 (2022) 105335.

[28] S.Z.T. Anwar, Effect of image augmentation on ECG image classification using deep learning, in: 2021 International Conference on Artificial Intelligence (ICAI), IEEE, 2021, pp. 182–186.

[29] H.K. Ali, H. Muzammil, K.M. Muhammad, ECG Images dataset of Cardiac and COVID-19 Patients, Data in Brief 34 (2021) 106762.

[30] Tawsifur Rahman, Amith Khandakar, Yazan Qiblawey, et al., Exploring the effect of image enhancement techniques on COVID-19 detection using chest X-ray images, Comput. Biol. Med. 132 (2021) 104319.

[31] S. Mirjalili, A. Lewis, The whale optimization algorithm, Adv. Eng. Softw. 95 (2016) 51–67.

[32] N. Immerman, Descriptive Complexity, Springer Science & Business Media, 2012, https://doi.org/10.1007/978-1-4612-0539-5.

[33] W. Salet, L. Bertolini, M. Giezen, Complexity and uncertainty: problem or asset in decision making of mega infrastructure projects?, Int. J. Urban Reg. Res. 37 (6) (2013) 1984–2000.

Chapter 41

Whale optimization algorithm for optimization of truss structures with multiple frequency constraints

Nima Khodadadi[a], El-Sayed M. El-kenawy[b], Marwa M. Eid[c], Ziad Azzi[d], Abdelaziz A. Abdelhamid[e], and Seyedali Mirjalili[f,g]

[a]*Department of Civil and Architectural Engineering, University of Miami, Coral Gables, FL, United States,* [b]*Department of Communications and Electronics, Delta Higher Institute of Engineering and Technology, Mansoura, Egypt,* [c]*Faculty of Artificial Intelligence, Delta University for Science and Technology, Mansoura, Egypt,* [d]*Department of Civil and Environmental Engineering, Florida International University, Miami, FL, United States,* [e]*Department of Computer Science, Faculty of Computer and Information Sciences, Ain Shams University, Cairo, Egypt,* [f]*Centre for Artificial Intelligence Research and Optimisation, Torrens University Australia, Brisbane, QLD, Australia,* [g]*University Research and Innovation Center, Obuda University, Budapest, Hungary*

41.1 Introduction

Analytical methods are founded on mathematics. Compared to only approximative methods, these techniques can locate the correct answer to the problem while requiring significantly less effort to compute. However, the quality of the solution achieved by using such methods depends on many factors, the most important of which is choosing the right starting point. On top of that, when applied to complex optimization problems, such as the optimal design of truss structures under frequency constraints, these techniques only manage to reach local minima rather than the global minimum. The majority of approximate techniques are global search techniques known as metaheuristics. The purpose of these techniques is to make up for the shortcomings of traditional methods. Using the wisdom of natural phenomena, metaheuristic algorithms can locate global or nearly global solutions. The metaheuristic algorithms employ the intelligence of natural phenomena such as evolution and social behaviors to solve optimization problems. For instance, the Stochastic Paint Optimizer (SPO) [1], Lion optimization algorithm (LOA) [2], Farmland Fertility Algorithm (FFA) [3], Moth Search Algorithm (MSA) [4], Mountain Gazelle Optimizer (MGO) [5], Grey Wolf Optimizer (GWO) [6], and Special Relativity Search (SRS) [7] are some well-known optimization algorithms.

Population-based metaheuristic algorithms consist of two main stages: initialization and searching. A population of agents is generated randomly in the initialization phase, and then those agents are used in the problem-solving phase following the procedure of imitated phenomena. Metaheuristic algorithms also use this process. Furthermore, algorithms' capabilities in responding to problems vary according to their unique histories. So, some algorithms may work better than others. It's an intriguing concept to combine the fundamental principles of algorithms. The similarities between algorithms have led researchers to explore the potential benefits of hybridizing multiple algorithms into a single one, such as the hybrid algorithm of African Vulture Optimization Algorithm and Harmony Search [8], the hybrid algorithm of Aquila Optimizer and Particle Swarm Optimizer [9], and the hybrid algorithm of Genetic Algorithm and ant Colony Optimization [10].

When it comes to building structures like bridges, arches, domes, etc., trusses are the structural system of choice for engineers [11]. An important topic in structural vibration analysis is the design of truss structures with frequency limitations. The vibrational response of the structures is significantly affected by the fundamental frequency and the corresponding mode shape. It is possible to improve a structure's responsiveness to dynamic excitation by manipulating the values of its natural frequency field [12].

Such engineering structures require optimal designs due to limited resources and high construction costs. According to the published research [13], the earliest attempts to optimize truss structures within frequency constraints primarily considered sizing variables. Sergeyev and Mroz [14], however, claimed that more efficient designs could be created by considering the structural shape and the size variables. This effort considerably complicates the original optimization problem. Finding

Handbook of Whale Optimization Algorithm. https://doi.org/10.1016/B978-0-32-395365-8.00047-6

the best possible solution for designing truss structures that meet multiple frequency constraints is challenging due to the problem being non-convex and highly non-linear.

In the process of solving complex structural optimization problems, metaheuristic algorithms demonstrated performance levels that were considered satisfactory. Because of this, these methods are used to deal with the issue of optimizing the size and shape of truss structures within the constraints of the natural frequency [15].

Mirjalili and Lews [16] introduced the Whale Optimization Algorithm (WOA) in 2016, which is a novel population-based algorithm. This algorithm imitates the way that humpback whales interact with one another in social environments. WOA is a population-based algorithm, meaning it starts with a random pool of candidates (the population) and then uses three rules—"encircle the prey," "spiral update the position," and "search for prey"—to improve the position of the population's best solutions iteratively. This paper presents WOA to be employed for designing the truss structures under frequency constraints.

The structure of the remaining parts of this chapter is as follows: Section 41.2 defines the problem of optimizing structural design. The outcomes of the study are presented in Section 41.3, while Section 41.4 serves as the concluding section.

41.2 Problems definition

In this subsection, the numerical equations for optimizing the size of truss structures are provided, along with several constraints on natural frequencies. The optimization process ensures that the truss structures have optimal cross-sectional values (A_i), leading to a decrease in the overall weight of the structure (W). Moreover, the resulting design must also fulfill the following set of minimum requirements:

$$Find \qquad [x_1, x_2, \ldots, x_{ng}] \tag{41.1}$$

$$Minimize \qquad W(\{x\}) = \sum_{i=1}^{nm} \gamma_i . A_i . L_i(x) \tag{41.2}$$

$$Subjected\ to \qquad \begin{cases} x_{\min} \le x_i \le x_{\max} \\ \omega_j \le \omega_j^* \\ \omega_k \ge \omega_k^* \end{cases} \tag{41.3}$$

where $W(\{x\})$ and ng represent the design variable vector and number, respectively. In this context, nm represents the quantity of structural components, while $W(\{x\})$ represents the weight of the structure. The material density, length, and cross-sectional area of members are denoted by γ_i, L_i, and x_i, respectively [17]. Fig. 41.1 presents a flowchart of the definition of truss problems with frequency constraints.

The jth natural frequency of the truss is represented as ω_j, and its upper bound is denoted as ω_j^*. Similarly, the kth natural frequency of the truss is denoted by ω_k, and its lower bound is represented as ω_k^*.

$$f_{penalty}(X) = (1 + \varepsilon_1 . \nu)^{\varepsilon_2}, \quad \nu = \sum_{i=1}^{n} \max[0, \nu_i] \tag{41.4}$$

where ν is the total number of design constraints violated. The value of the constant ε_1 is set to 1, and the value of ε_2 begins at 1.5 and increases linearly until it reaches 3.

41.3 Optimization benchmark with results

This section offers explanations for the examples that are presented. In order to demonstrate that the WOA is more effective than for structures with frequency constraints, the following well-known benchmark design cases are solved utilizing the mentioned algorithm compared with Arithmetic Optimization Algorithm (AOA) [18] and Material Generation Algorithm (MGA) [19]. The truss problems were run on a MacBook Pro with a 2.3 GHz CPU (8-Core an Intel Core i9 computer platform), 16 GB RAM, and MATLAB® code (macOS Monterey). Some examples will demonstrate the WOA algorithm's efficacy. Table 41.1 displays the material properties, cross-sectional area, and natural frequency constraints used for three

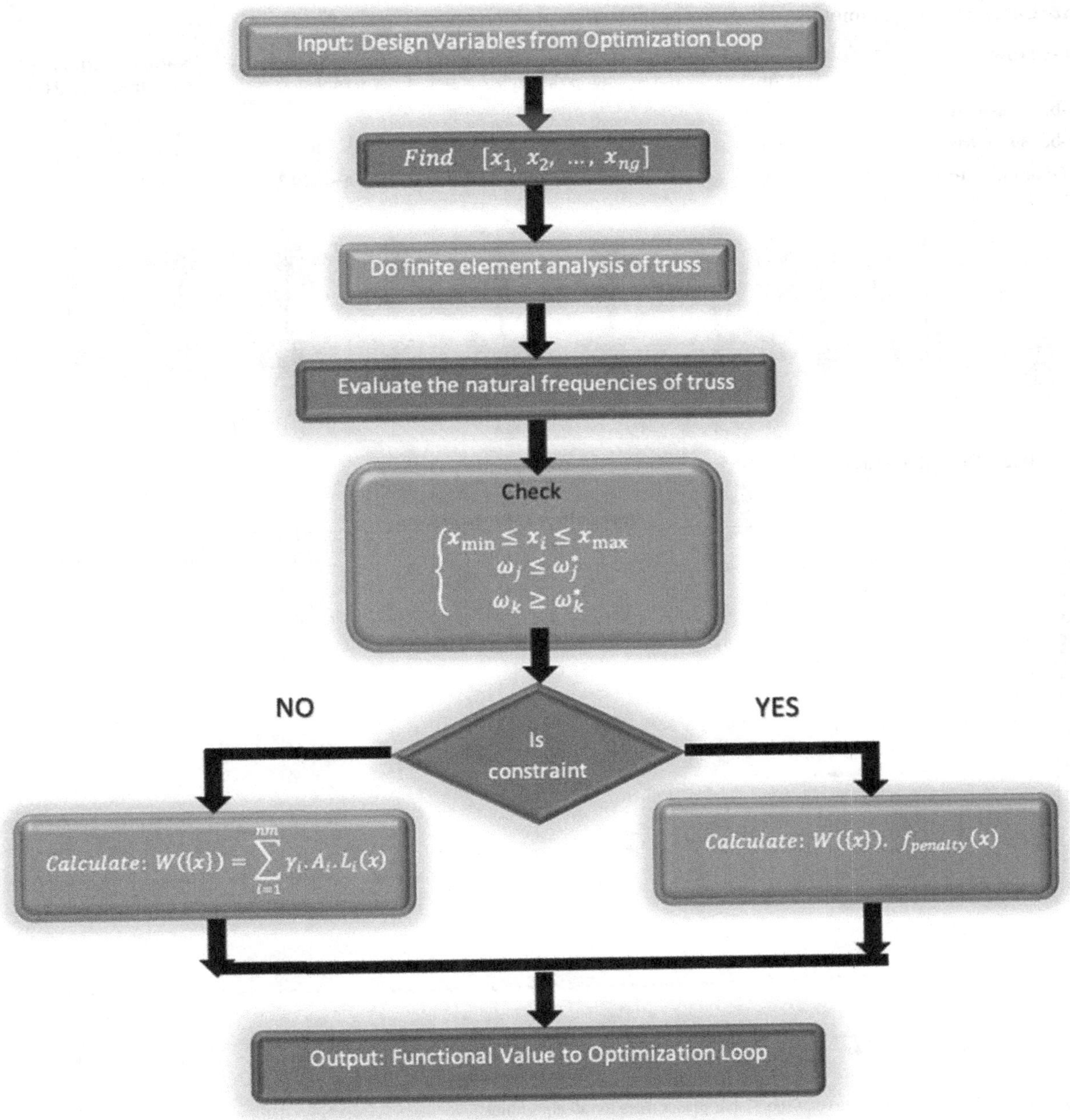

FIGURE 41.1 The flowchart of truss problems definition with frequency constraints.

examples: a 37-bar planar truss, a 72-bar space truss, and a 120-bar dome truss. The results of each example are compared to the outcomes achieved by several other optimization methods that were also used to solve each problem. The proposed process involves a total of 10000 evaluations for all examples. In addition, each example uses a population of 50 particles for ten runs.

41.3.1 The 37-bar planar truss

The first example presented in this chapter involves optimizing the 37-bar planar truss, which is depicted in Fig. 41.2. Design Parameters for this design problem are considered in Table 41.1. The design parameters for this example include the nodal coordinates of the upper chord and the members. The symmetry of the structure is also preserved through the

TABLE 41.1 Design Parameters for various optimization problems.

Truss Problems	Modulus of Elasticity E (N/m²)	Material Density ρ (kg/m³)	Cross-Sectional A (cm²)	Natural Frequency Constraints ω (Hz)
37-bar planer truss	2.1×10^{11}	7800	$1 \le A$	$\omega_1 \ge 20,\ \omega_2 \ge 40,\ \omega_3 \ge 60$
72-bar space truss	6.98×10^{11}	2770	$0.645 \le A$	$\omega_1 = 4,\ \omega_3 \ge 6$
120-bar dome truss	2.1×10^{11}	7971.810	$1 \le A \le 129.3$	$\omega_1 \ge 9,\ \omega_2 \ge 11$

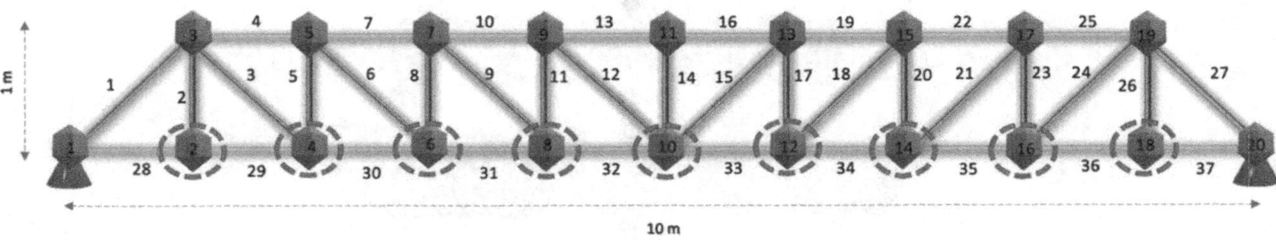

FIGURE 41.2 The 37-Bar Planar Truss.

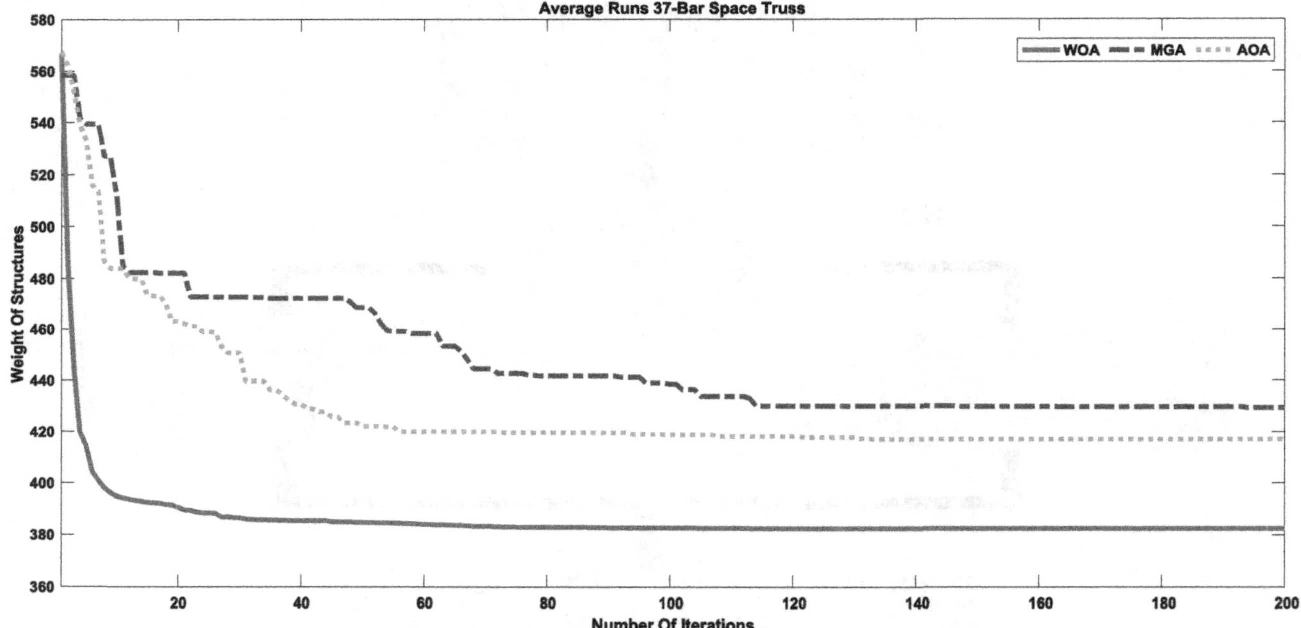

FIGURE 41.3 The convergence curve of the 37-bar planar truss for all algorithms.

connection between nodal coordinates and member areas. Every free node on the lower chord has a nonstructural mass of 10 kg attached to it. In this case, we take into account as limitations the first three of the structure's natural frequencies ($f_1 \ge 20$ Hz, $f_2 \ge 40$ Hz, and $f_3 \ge 60$ Hz). Consequently, this optimization problem involves a total of 19 design variables, including 14 sizing variables and 5 layout variables, as well as 3 frequency constraints.

Table 41.2 presents a comparison between the optimal designs reported in the literature and the results obtained using AOA and MGA in this chapter. The results demonstrate that WOA outperforms other metaheuristics in terms of achieving the best weight, as well as generating an average optimized weight and standard deviation. The optimized structural frequencies (Hz) for the different approaches are displayed in Table 41.3. All terms and conditions regarding allowable frequencies have been met, while the frequency results for WOA are near to minimum permissible frequency. Regarding Fig. 41.3, other methods, such as AOA and MGA, require more structural analyses than WOA to find the best solution.

TABLE 41.2 The design results of different methods for 37-bar planar truss.

Member Group	MGA	AOA	WOA
(Y_3-Y_{19}) m	1.2351	1.0000	1.3653
(Y_5-Y_{17}) m	1.4950	1.0000	1.7004
(Y_7-Y_{15}) m	1.6484	1.3583	1.8952
(Y_9-Y_{13}) m	2.0909	1.5923	2.0418
(Y_{11}) m	2.5067	1.5879	2.2837
(A_1-A_{27}) cm^2	2.9766	5.0000	1.6253
(A_2-A_{26}) cm^2	3.4396	4.2071	1.2070
(A_3-A_{24}) cm^2	3.9871	5.0000	2.0926
(A_4-A_{25}) cm^2	2.9688	5.0000	2.4774
(A_5-A_{23}) cm^2	1.3464	1.7929	1.3552
(A_6-A_{21}) cm^2	3.7166	1.7929	1.8578
(A_7-A_{22}) cm^2	2.6472	5.0000	2.1626
(A_8-A_{20}) cm^2	3.5585	3.0160	1.2318
(A_9-A_{18}) cm^2	2.1988	1.4052	2.3467
(A_{10}-A_{19}) cm^2	2.7868	3.3193	2.3843
(A_{11}-A_{17}) cm^2	4.3535	5.0000	1.4854
(A_{12}-A_{15}) cm^2	1.6328	1.0000	1.5520
(A_{13}-A_{16}) cm^2	5.0000	5.0000	1.5979
(A_{14}-A_{16}) cm^2	3.6339	2.2253	1.4949
Best weight	414.5971	404.7790	372.1079
Average weight	429.4048	417.0927	382.4203
Standard deviation	11.9675	9.1162	6.5135
No. analyses	5100	6650	4250

TABLE 41.3 The frequency of different methods for 37-bar planar truss.

No. Frequency	MGA	AOA	WOA
f_1	21.4571	22.069	20.000
f_2	44.9642	41.036	40.000
f_3	77.9348	69.505	60.047

41.3.2 The 72-bar space truss

The second example in this chapter involves optimizing the 72-bar space truss, and the specific details of this example are provided in Table 41.1. The truss is divided into sixteen groups, and as shown in Fig. 41.4, four nonstructural masses, weighing 2268 kg, are attached to the upper nodes. The optimization results achieved using WOA for this structure are compared with the outcomes generated by other methods, with a population size of 50 and 10000 function number evaluations.

AOA and MGA are 345.9465 kg, 461.8752 kg, and 413.8412 kg. According to Table 41.4, WOA's rank is second to none. WOA needs 5200 analyses to get the best results which are low compared with AOA and MGA. Statistical results for WOA, such as standardbred deviation and mean of weight, are the lowest among all other methods. The findings indicate that compared to other reported results, WOA is a more dependable and better option. Fig. 41.5 demonstrates the convergence patterns of the average weight for a 72-bar space truss, while Table 41.5 displays the occurrence rates of various methods used for the same truss. All terms and conditions regarding allowable frequencies have not been met.

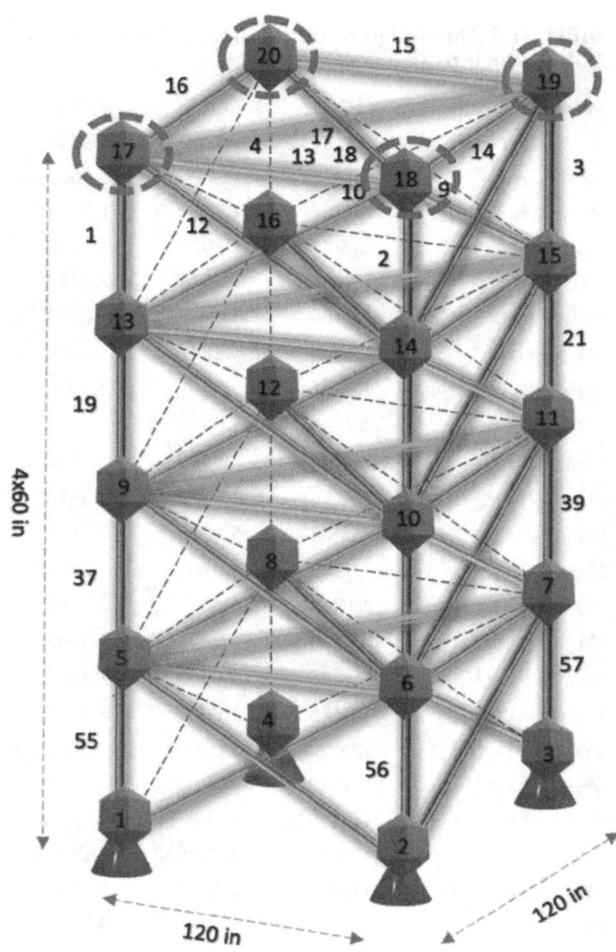

FIGURE 41.4 The 37-Bar Space Truss.

41.3.3 The 120-bar dome truss

The 120-bar space truss is the 3rd example presented in Fig. 41.6. Zolghadr and Kaveh [20] optimized this example initially. Considerations of this example are reported in Table 41.1. According to Fig. 41.6, nonstructural masses are added to this example. The masses can be described in terms of their respective weights in the following manner:

- 3000 kg at node 1 (red dash node; dark gray in print version)
- 1500 kg at nodes 2:13 (yellow dash circle; light gray in print version)
- 100 kg at nodes 14 to 37 (green dash circle; mid gray in print version)

These elements are divided into seven groups. 1 cm^2 and 129.3 cm^2 are the minimum and maximum cross-sectional areas for this example. Table 41.6 shows each group's best weight, average weight, standard deviation, number of analyses, and optimum cross-sectional area. As depicted in this table, WOA ranked first among other methods with a weight of 9463.8663 kg. According to Table 41.6, WOA gets the best results within 4800 function evaluation which is the lightest (see Fig. 41.7). These results show the WOA is more reliable and effective. Table 41.7 shows that the presented method fully met all the imposed limitations on natural frequency constraints.

The results of this example demonstrate that WOA is superior to the compared algorithms concerning both performance and accuracy. A statistical analysis of ten separate runs shows WOA's superiority over other metaheuristic algorithms. The algorithm escapes from local optima and performs exceptionally well because of its features. The study's findings are significant, as they suggest that the WOA algorithm is a viable and efficient solution to structural optimization problems. This may have important implications for a range of industries and fields that rely on structural optimization, including civil engineering, mechanical engineering, and architecture.

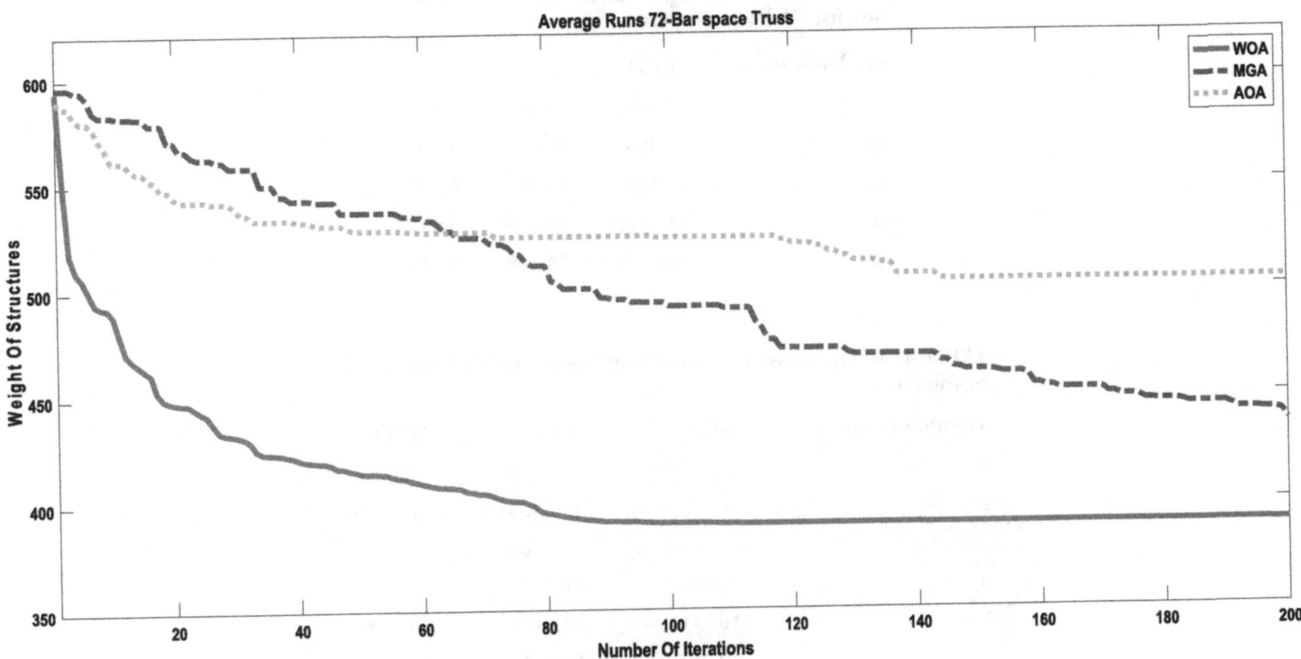

FIGURE 41.5 The convergence curve of the 72-bar space truss for all algorithms.

TABLE 41.4 The design results of different methods for 72-bar space truss.

Member Group	MGA	AOA	WOA
1 (A_1-A_4)	5.9530	10.6507	7.1617
2 (A_5-A_{12})	6.8826	9.1490	7.2909
3 (A_{13}-A_{16})	1.5543	2.2506	0.6450
4 (A_{17}-A_{18})	2.4483	12.6340	1.9046
5 (A_{19}-A_{22})	11.4453	20.0000	9.8425
6 (A_{23}-A_{30})	9.4307	8.3545	10.8259
7 (A_{31}-A_{34})	6.0880	2.7050	0.6450
8 (A_{35}-A_{36})	3.2847	1.9563	0.6857
9 (A_{37}-A_{40})	11.2135	6.7091	14.6872
10 (A_{41}-A_{48})	10.5151	7.7288	6.5532
11 (A_{49}-A_{52})	0.8822	2.3157	0.6450
12 (A_{53}-A_{54})	2.5903	18.6301	0.6450
13 (A_{55}-A_{58})	20.0000	18.6301	13.2706
14 (A_{59}-A_{66})	7.5589	7.2839	8.2663
15 (A_{67}-A_{70})	6.4759	0.6450	1.1531
16 (A_{71}-A_{72})	0.6450	2.5980	1.3171
Best weight	413.8412	461.8752	345.9465
Average weight	436.0913	503.2495	389.6596
Standard deviation	43.3850	37.2501	30.8039
No. analyses	10000	71150	5200

TABLE 41.5 The frequency of different methods for 72-bar space truss.

No. Frequency	MGA	AOA	WOA
f_1	4.194	3.932	4.003
f_2	4.194	3.932	4.004
f_3	6.149	6.030	6.000
f_4	10.426	12.356	9.095
f_5	11.533	14.169	9.324

TABLE 41.6 The design results of different methods for 120-bar dome truss.

Member Group	MGA	AOA	WOA
1	20.4173	49.1694	20.4054
2	56.6231	43.0303	36.4325
3	20.6543	11.7906	11.5168
4	23.8124	21.5781	20.7986
5	16.6761	12.1151	14.3718
6	19.7874	17.2590	11.4430
7	41.4071	26.1526	19.6669
Best weight	14085.7871	12624.1241	9463.8663
Average weight	15901.1846	16288.4596	11785.3724
Standard deviation	1494.0559	2276.6673	1319.7432
No. analyses			

TABLE 41.7 The frequency of different methods for 120-bar dome truss.

No. Frequency	MGA	AOA	WOA
f_1	9.028	10.242	9.000
f_2	12.452	11.047	11.097
f_3	12.458	11.060	11.141
f_4	12.469	11.815	11.153
f_5	12.504	11.873	11.167

41.4 Conclusion and future work

In this chapter, the whale optimization algorithm (WOA) method was utilized to determine the optimal weight design for truss structures while adhering to frequency constraints. The effectiveness of this method was compared to other commonly used meta-heuristic algorithms across three structural benchmarks of increasing size. Analysis of the results from the design examples showed that the WOA outperformed the AOA and MGA in terms of efficiency, accuracy, and overall performance. However, the WOA's fast convergence to an optimal solution could pose difficulties for larger-scale problems, as demonstrated by the comparison of convergence speeds. Therefore, it is suggested that future work should incorporate appropriate mechanisms to address these challenges. The WOA algorithm is recommended as a research direction, as it can be further optimized for use in optimization problems through the implementation of fitness-driven techniques.

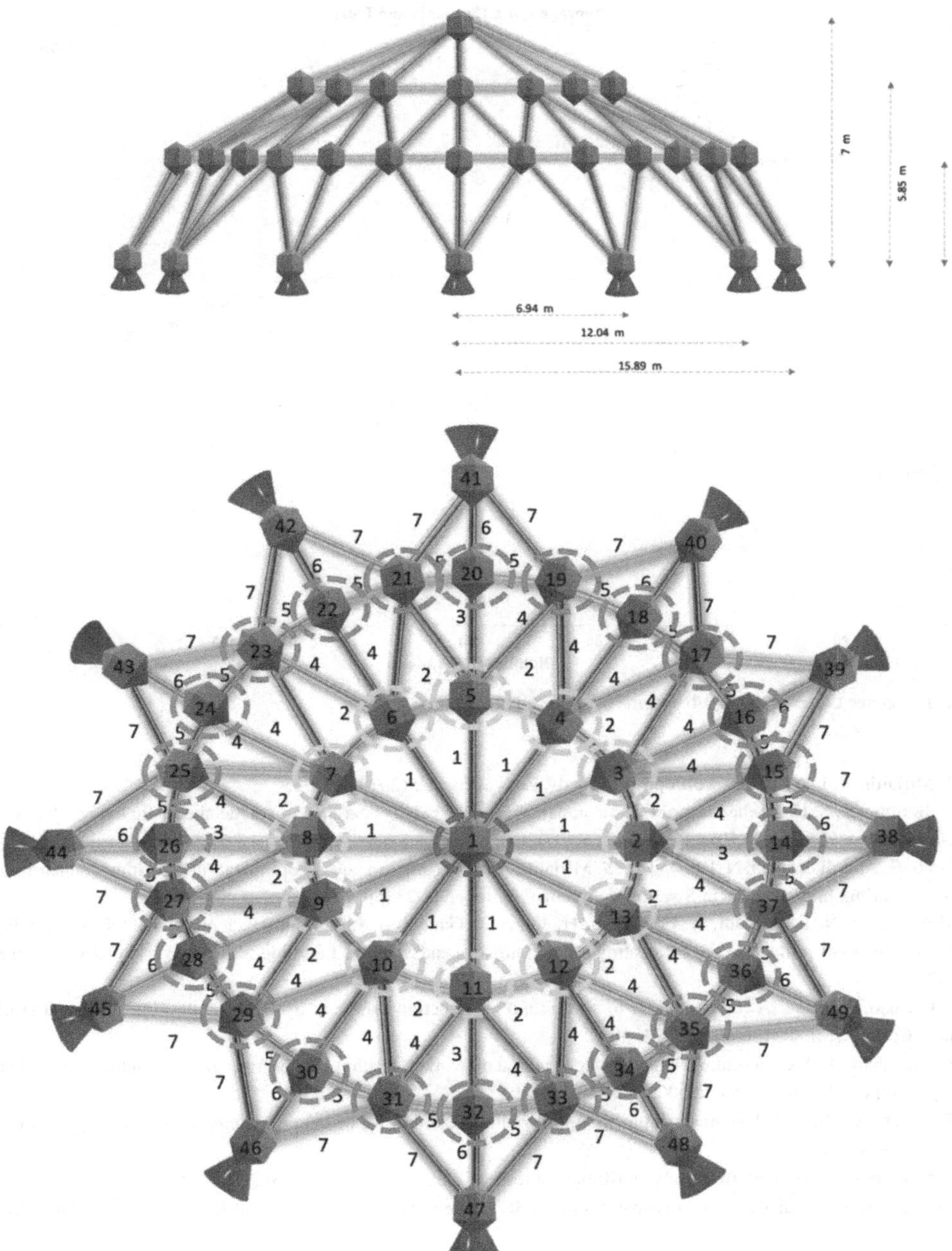

FIGURE 41.6 The 120-Bar dome Truss.

References

[1] A. Kaveh, S. Talatahari, N. Khodadadi, Stochastic paint optimizer: theory and application in civil engineering, Eng. Comput. 38 (2022) 1921–1952, https://doi.org/10.1007/s00366-020-01179-5.

[2] M. Yazdani, F. Jolai, Lion optimization algorithm (LOA): a nature-inspired metaheuristic algorithm, J. Comput. Des. Eng. 3 (1) (2016) 24–36.

[3] H. Shayanfar, F.S. Gharehchopogh, Farmland fertility: a new metaheuristic algorithm for solving continuous optimization problems, Appl. Soft Comput. 71 (2018) 728–746.

[4] G.-G. Wang, Moth search algorithm: a bio-inspired metaheuristic algorithm for global optimization problems, Memetic Comput. 10 (2) (2018) 151–164.

[5] B. Abdollahzadeh, F.S. Gharehchopogh, N. Khodadadi, S. Mirjalili, Mountain gazelle optimizer: a new nature-inspired metaheuristic algorithm for global optimization problems, Adv. Eng. Softw. 174 (2022) 103282.

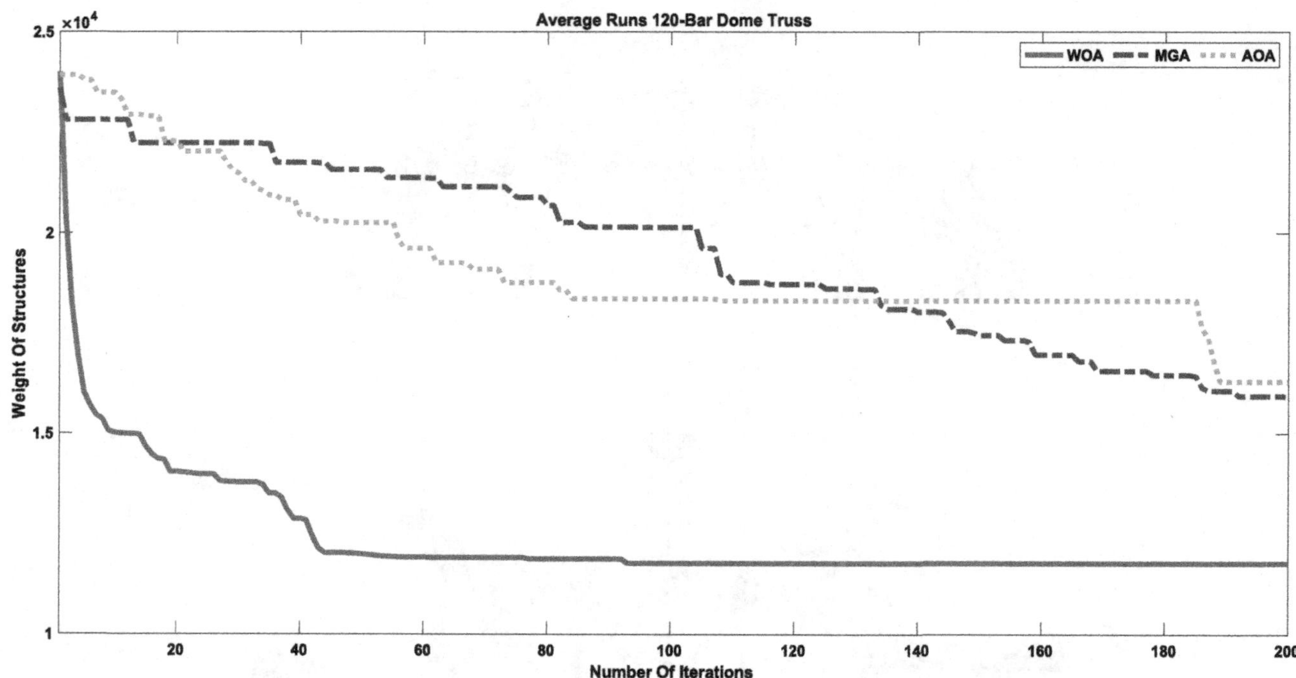

FIGURE 41.7 The convergence curve of the 120-bar dome truss for all algorithms.

[6] S. Mirjalili, S.M. Mirjalili, A. Lewis, Grey wolf optimizer, Adv. Eng. Softw. 69 (2014) 46–61.

[7] V. Goodarzimehr, S. Shojaee, S. Hamzehei-Javaran, S. Talatahari, Special relativity search: a novel metaheuristic method based on special relativity physics, Knowl.-Based Syst. 257 (2022) 109484.

[8] F.S. Gharehchopogh, B. Abdollahzadeh, N. Khodadadi, S. Mirjalili, A hybrid African vulture optimization algorithm and harmony search: algorithm and application in clustering, in: Advances in Swarm Intelligence, Springer, 2023, pp. 241–254.

[9] L. Abualigah, M.A. Elaziz, N. Khodadadi, A. Forestiero, H. Jia, A.H. Gandomi, Aquila optimizer based PSO swarm intelligence for IoT task scheduling application in cloud computing, in: Integrating Meta-Heuristics and Machine Learning for Real-World Optimization Problems, Springer, 2022, pp. 481–497.

[10] J. Luan, Z. Yao, F. Zhao, X. Song, A novel method to solve supplier selection problem: hybrid algorithm of genetic algorithm and ant colony optimization, Math. Comput. Simul. 156 (2019) 294–309.

[11] N. Khodadadi, S. Talatahari, A.H. Gandomi, ANNA: advanced neural network algorithm for optimisation of structures, Proc. Inst. Civ. Eng., Struct. Build. (2023), https://doi.org/10.1680/jstbu.22.00083.

[12] N. Khodadadi, S. Vaclav, S. Mirjalili, Dynamic arithmetic optimization algorithm for truss optimization under natural frequency constraints, IEEE Access 10 (2022) 16188–16208.

[13] R. Grandhi, Structural optimization with frequency constraints–a review, AIAA J. 31 (12) (1993) 2296–2303.

[14] O. Sergeyev, Z. Mroz, Sensitivity analysis and optimal design of 3D frame structures for stress and frequency constraints, Comput. Struct. 75 (2) (2000) 167–185.

[15] N. Khodadadi, S.M. Mirjalili, S. Mirjalili, Optimal design of truss structures with continuous variable using moth-flame optimization, in: Handbook of Moth-Flame Optimization Algorithm, CRC Press, 2022, pp. 265–280.

[16] S. Mirjalili, A. Lewis, The whale optimization algorithm, Adv. Eng. Softw. 95 (2016) 51–67.

[17] N. Khodadadi, S. Mirjalili, Truss optimization with natural frequency constraints using generalized normal distribution optimization, Appl. Intell. 52 (2022) 10384–10397, https://doi.org/10.1007/s10489-021-03051-5.

[18] L. Abualigah, A. Diabat, S. Mirjalili, M. Abd Elaziz, A.H. Gandomi, The arithmetic optimization algorithm, Comput. Methods Appl. Mech. Eng. 376 (2021) 113609.

[19] S. Talatahari, M. Azizi, A.H. Gandomi, Material generation algorithm: a novel metaheuristic algorithm for optimization of engineering problems, Processes 9 (5) (2021) 859.

[20] A. Kaveh, A. Zolghadr, Shape and size optimization of truss structures with frequency constraints using enhanced charged system search algorithm, Asian J. Civ. Eng. 12 (4) (2011) 487–509.

Chapter 42

A novel version of whale optimization algorithm for solving optimization problems

Nima Khodadadi[a], El-Sayed M. El-kenawy[b], Sepehr Faridmarandi[c], Mansoureh Shahabi Ghahfarokhi[c], Abdelhameed Ibrahim[d], and Seyedali Mirjalili[e,f]

[a]*Department of Civil and Architectural Engineering, University of Miami, Coral Gables, FL, United States,* [b]*Department of Communications and Electronics, Delta Higher Institute of Engineering and Technology, Mansoura, Egypt,* [c]*Department of Civil and Environmental Engineering, Florida International University, Miami, FL, United States,* [d]*Computer Engineering and Control Systems Department, Faculty of Engineering, Mansoura University, Mansoura, Egypt,* [e]*Centre for Artificial Intelligence Research and Optimisation, Torrens University Australia, Brisbane, QLD, Australia,* [f]*University Research and Innovation Center, Obuda University, Budapest, Hungary*

42.1 Introduction

Since the world is rapidly becoming more technological, problems in engineering and scientific research are of paramount importance. High-dimensional, complex global optimization problems constitute a significant portion of these issues. Finding a solution is difficult, as the search space grows exponentially with each additional dimension added to the problem. In addition to the fact that they cannot escape the "curse of dimensionality," traditional optimization techniques and those based on gradients are ineffective in resolving the complex issues that plague modern society for various reasons. In addition, the gradient search is highly sensitive to the location of the starting point, making it challenging to use in a problem with a local solution. As a result of these considerations, researchers were able to create optimization methods that effectively address practical optimization issues. One of the given algorithms relies on natural law and utilizes a heuristic search to produce its results. Heuristic methods aim to solve problems by understanding the problem first and then utilizing that understanding intelligently while minimizing the computation required to find a reasonable solution. In contrast to the more precise methods of classical heuristics, meta-heuristics are approximations based on a wide variety of theories and principles, including those of evolution, AI, natural phenomena, neural systems, and statistical mechanics. The optimization problems that nature-inspired meta-heuristics are meant to solve stem from the natural world. In order to function, these algorithms require values to be generated at random. The top-performing individuals of the present generation are utilized to produce the subsequent generation, which helps in enhancing the population's optimization over time. They can efficiently handle multiple local optimums at once and effectively solve high-dimensional global optimization problems. Such features have contributed to the widespread adoption of these algorithms for global optimization problems. There are many different types of nature-inspired optimization algorithms, but they can be roughly categorized into the following categories:

I. Evolutionary, which includes techniques like Genetic Programming (GP) [1] and Differential Evolution (DE) [2] that are inspired by the natural laws of evolution [3].
II. Laws of physics, chemistry, and art, replicate the rules of these fields in the cosmos; for example, the Water Evaporation Optimization (WEO) [4], Atom Search Optimization (ASO) [5], and the Stochastic Paint Optimizer (SPO) [6].
III. Human and animal behavior investigates facets of the human and animal condition, such as the Neural Network Algorithm (NNA) [7] and the Mountain Gazelle Optimizer (MGO) [8].

Mirjalili and Lewis [9] provide a more logical framework for categorizing meta-heuristic search algorithms.

Many improved versions of meta-heuristics have been proposed lately in the literature at an unprecedented acceleration pace. Some of the more popular and recent algorithms are the Improved Butterfly Optimization Algorithm (IBOA) [10], Chaotic Stochastic Paint Optimizer (CSPO) [11], Improved Whale Algorithm (IWA) [12], Evolutionary Population Dynamics Harmony Search (EPDHS) [13], and Advanced neural network algorithm (ANNA) [14].

Handbook of Whale Optimization Algorithm. https://doi.org/10.1016/B978-0-32-395365-8.00048-8

The recent introduction of the No Free Lunch (NFL) [15] theorem is considered an exciting development in numerical optimization. The theorem justifies the claim that no single optimization algorithm can effectively address all possible optimization scenarios. As a result, researchers all over the world are looking into various algorithms in various forms, including original, modified, improved, and hybrid versions. In addition, it is seen that applying the same algorithm to a problem with different sets of parameters generates different outcomes.

WOA was used for different applications, and it is well-known among other methods [16]. This chapter proposes an advanced version of the Whale Optimization Algorithm (AWOA). The benefits of this algorithm and the modification implemented here are shown in a comparison study with other meta-heuristics. The remaining chapter is structured in the following manner: Section 42.2 presents an overview of the WOA algorithm, while Section 42.3 delves into the specifics of the advanced version of WOA. In Section 42.4, problems are defined and accompanied by experimental results. Finally, Section 42.5 serves as a conclusion to the chapter.

42.2 Whale optimization algorithm (WOA)

Mirjalili and Lewis [9] introduced the whale optimization algorithm (WOA) as a more recent addition to the meta-heuristic algorithms. The bubble-net hunting strategy, utilized by humpback whales when targeting schools of krill or small fish near the surface, served as the inspiration for WOA. To accomplish this hunting strategy, humpback whales encircle their prey by swimming around it in a shrinking circle and simultaneously follow a spiral-shaped path to create distinguishable bubbles in a circular or "9"-shaped pattern. WOA simulates this behavior by allowing a 50% probability for choosing between the shrinking encircling mechanism and the spiral model for updating the whale's position during optimization. The formulas for these mechanisms are presented in the following section [17].

Shrinking encircling prey: The WOA algorithm treats the most promising solution candidate as the ultimate goal and then has the other search agents adjust their positions to correspond with that one. The following formula describes this pattern of behavior:

$$\vec{X}(t+1) = \vec{X}^*(t) - A.\vec{D} \tag{42.1}$$

$$\vec{D} = |C.\vec{X}^*(t) - \vec{X}(t)| \tag{42.2}$$

$$A = 2.a.r - a \tag{42.3}$$

$$C = 2.r \tag{42.4}$$

In order to accomplish this, the current iteration is represented by t, $\vec{X}^*$ denotes the best position recorded, and $\vec{X}$ represents the position of a whale. To gradually reduce the value of a from 2 to 0 in a linear manner, a uniformly distributed random number between 0 and 1 is assigned to r in each iteration.

Spiral bubble-net feeding maneuver: In order to imitate the helix-shaped path followed by humpback whales, a spiral equation is employed to calculate the positions of the whale and its prey at each moment.

$$\vec{X}(t+1) = e^{bk}.\cos(2\pi k).\vec{D}' + \vec{X}^*(t) \tag{42.5}$$

$$\vec{D}' = |\vec{X}^*(t) - \vec{X}(t)| \tag{42.6}$$

Here, b is responsible for defining the shape of the logarithmic spiral, while k is a random number uniformly distributed between the values of -1 and 1.

When the value of A exceeds 1 or is less than -1, the search agent is updated utilizing a randomly selected search agent rather than the optimal search agent. This facilitates the implementation of a global optimizer.

$$\vec{X}(t+1) = \vec{X}_{rand} - A.\vec{D}'' \tag{42.7}$$

$$\vec{D}'' = |C.\vec{X}_{rand} - \vec{X}(t)| \tag{42.8}$$

where in the current iteration $\vec{X}_{rand}$ is a random whale.

42.3 Advanced whale optimization algorithm (AWOA)

The WOA is a straightforward approach to the problem of finding solutions on a global scale. This research introduces the advanced whale optimization algorithm (AWOA) to increase its solution accuracy, search reliability, and convergence speed. An important aspect of algorithm improvement is retaining the approach's inherent simplicity.

The search direction of the population is exclusively reliant on the current and optimal position as determined by WOA. When the best member of a population gets stuck in a local optimum, the rest of the members of that population have a better chance of arriving at a global optimum.

The absolute value symbol is crucial when it comes to maintaining WOA's ethnically and racially diverse population. It is crucial to keep in mind that, in engineering problems, the upper and lower limits of the design variables are not zero. This means that the absolute symbol cannot be utilized when applying WOA to address such optimization problems. To eliminate the potential risks associated with the absolute value symbol and the local exploitation strategy in AWOA, Eqs. (42.5) and (42.6) are redefined in the following manner:

$$N = rand \times \left(r \times \vec{X}^*(t)\right) + (1 - r) \times M \tag{42.9}$$

$$\vec{X}(t+1) = N - rand \times (A \times \vec{X}^w(t) + (1 - rand) \times M - \vec{X}(t)) \tag{42.10}$$

where $\vec{X}^w$ is the worst position ever recorded, and M is the current average solution. Fig. 42.1 displays the pseudo-code for the AWOA algorithm.

AWOA has a computational complexity of $O(T \times P \times D)$, where T is the maximum number of iterations, P is the number of pop, D is the dimension of the problem being optimized.

It is important to consider the potential weaknesses of the Adaptive Whale Optimization Algorithm (AWOA) as well as its strengths when assessing its performance and effectiveness. The first potential flaw mentioned is that the AWOA involves a large number of random parameters, which can lead to increased uncertainty in the algorithm. This means that the performance of the algorithm may be difficult to predict or control, as it depends on the values of many random parameters. The second potential flaw mentioned is that the parameters used in the AWOA are designed to balance the exploitation and exploration phases of the algorithm, but are random and do not adaptively change with each iteration. The random parameters may not be well-suited to the specific problem being optimized, and may not adjust appropriately over time to improve the search process.

```
Initialize the whales' population Xi (i = 1,2,...,n)
Calculate the fitness of each search agent
X* is the best search agent
Xw is the worst search agent
  while t < Maxit
     for each search agent
      Update a, A, C, l, M and p
       if p <0.5
          if |A| < 1
          Update the position of the current search agent by Eq. (42.2)
          else if |A| <1
          Select a random search agent (Xrand )
          Update the position of the current search agent by Eq. (42.7)
          end if
       else if p ≥0.5
       Update the position of the current search by Eq. (42.10)
       end if
     end for
     Check if any search agent goes beyond the search space and amend it
     Calculate the fitness of each search agent
     Update X* if there is a better solution
     t = t + 1
  end while
return X*
```

FIGURE 42.1 Pseudo code of the AWOA algorithm.

42.4 Engineering problems

To demonstrate the efficacy of AWOA, this section analyzes several common engineering design problems. Results from the BBO [18], GWO [19], SCA [20], CS [21], and WOA [9] algorithms are reported so that comparisons can be made. The parameters of the algorithm have been adjusted such that their default values are used. Arcuri and Fraser [22] found that leaving all algorithms at their default values lowered the possibility of better parametrization bias. The referenced papers were also used to collect the control parameters of the comparing algorithms. Here, we employ the optimal parameter for each method based on their respective literature. Take note that for each and every one of the studies, a maximum of one 1000 iterations and one 100 populations were utilized.

42.4.1 Tension/compression spring

The problem of minimizing the weight of the tension/compression spring is a widely recognized optimization problem introduced by Belegundu and Arora [23] which is illustrated in Fig. 42.2.

FIGURE 42.2 The tension/compression spring.

Additional limitations, such as shear stress, frequency, and minimum deflection, are necessary to accomplish this reduction. The cost function can be formulated by taking into account three decision variables:

$$
\begin{aligned}
&\textit{Consider} && \vec{X} = [x_1, x_2, x_3] = [d, D, N] \\
&\textit{Minimize} && f_{cost}(\vec{X}) = (2 + x_3) \times x_2 x_1^2 \\
&\textit{Subject to} && g_1(\vec{X}) = 1 - \frac{x_3 x_2^3}{71785 x_1^4} \leq 0 \\
& && g_2(\vec{X}) = \frac{4x_2^2 + x_2 x_1}{12566(x_2 x_1^3 - x_1^4)} + \frac{1}{5108 x_1^2} - 1 \leq 0 \\
& && g_3(\vec{X}) = 1 - \frac{140.45 x_1}{x_2^2 x_3} \leq 0 \\
& && g_4(\vec{X}) = \frac{x_1 + x_2}{1.5} - 1 \leq 0 \\
& && g_6(\vec{X}) = \delta(\vec{X}) - \delta_{max} \\
& && g_7(\vec{X}) = P - P_c(\vec{X}) \leq 0 \\
&\textit{Variable Range} && 0.05 \leq x_1 \leq 2, \quad 0.25 \leq x_1 \leq 1.3, \quad 2 \leq x_1 \leq 15
\end{aligned}
\tag{42.11}
$$

Table 42.1 displays the outcomes of the AWOA algorithm along with other algorithms. The average of 30 independent results by AWOA is the best among other methods. Table 42.2 shows that the minimum value for both the worst result and the standard deviation is achieved by the AWOA method compared to all other methods. Table 42.1 displays the optimal variable values for various algorithms. As evident from the tables, the AWOA algorithm has provided a solution that surpasses any other approach.

42.4.2 Compound gear

The given problem illustrated in Fig. 42.3 is an example of a discrete design problem in mechanical engineering. The gear ratio, which is the proportion of the output shaft's angular speed to the input shaft's angular speed, needs to be reduced. The specifics of this problem can be found in reference [24], where the number of teeth in the gears is determined by four variables. The aim is to minimize the gear ratio.

TABLE 42.1 The results of a tension/compression spring.

Algorithm	$x_1 = d$	$x_2 = D$	$x_3 = N$	Optimal cost
AWOA	**0.0517562**	**0.3583359**	**11.1947206**	**0.012665**
WOA	0.051207	0.345215	12.004032	0.012676
BBO	0.051989	0.363965	10.890522	0.012681
SCA	0.051209	0.34522	12.00399	0.01267
GWO	0.05169	0.356737	11.28885	0.012666
CS	0.050279	0.32359	13.52539	0.012699

TABLE 42.2 The statistical results of a tension/compression spring.

Algorithm	Best	Mean	Worst	STD
AWOA	**0.012665**	**0.012683**	**0.012712**	**4.23E-05**
WOA	0.012676	0.0127	0.012832	3.00E-03
BBO	0.012681	0.012745	0.012953	8.06E-05
SCA	0.01267	0.013342	0.016293	1.32E-03
GWO	0.012666	0.012742	0.012842	5.32E-05
CS	0.012699	0.012711	0.012863	4.53E-05

TABLE 42.3 The results of compound gear.

Algorithm	$x_1 = T_a$	$x_2 = T_b$	$x_3 = T_d$	$x_3 = T_f$	Optimal cost
AWOA	**43**	**16**	**19**	**49**	**2.70E-12**
WOA	51	30	13	53	2.31E-11
BBO	53	26	15	51	2.31E-11
SCA	**43**	**16**	**19**	**49**	**2.70E-12**
GWO	**49**	**19**	**16**	**43**	**2.70E-12**
CS	**43**	**16**	**19**	**49**	**2.70E-12**

FIGURE 42.3 The compound gear.

The results of the BBO [18], GWO [19], SCA [20], CS [21], and WOA [9] are presented in Table 42.3. According to Table 42.3, displays the optimum configurations obtained by different permutations on four variables.

The following is the mathematical formula for compound gear:

$$\begin{aligned} &\textbf{\textit{Consider}} && \vec{X} = [x_1, x_2, x_3, x_4] = [T_a, T_b, T_d, T_f] \\ &\textbf{\textit{Minimize}} && f_{cost}(\vec{X}) = \left(\frac{1}{6.931} - \frac{x_3 x_2}{x_1 x_4}\right)^2 \\ &\textbf{\textit{Discerte Variable Range}} && 12 \le x_1, x_2, x_3, x_4 \le 60 \end{aligned} \tag{42.12}$$

TABLE 42.4 The statistical results of a compound gear.

Algorithm	Best	Mean	Worst	STD
AWOA	**2.70E-12**	**1.58E-10**	**1.24E-09**	**3.09E-10**
WOA	2.31E-11	2.30E-09	1.22E-08	2.40E-06
BBO	2.31E-11	4.54E-08	4.20E-07	7.30E-08
SCA	**2.70E-12**	8.81E-10	2.36E-09	6.45E-10
GWO	**2.70E-12**	3.38E-10	9.92E-10	4.10E-10
CS	**2.70E-12**	1.98E-09	2.35E-09	3.55E-09

In Table 42.4, it is evident that AWOA, SCA, GWO, and CS techniques have produced substantially improved outcomes than the BBO and WOA algorithms regarding optimal cost. The AWOA method, in particular, has outperformed other algorithms with respect to the mean, standard deviation, and worst-case scenarios for the final results.

42.4.3 Welded beam

The benchmark design problem was originally introduced by Coello [25] and has been explored by numerous researchers. Fig. 42.4 displays a diagram of the vertical force acting on a beam. The objective is to obtain a design that results in the lowest possible objective function, while also satisfying constraints related to stress, deflection, welding, and geometry.

$$
\begin{aligned}
&\textit{Consider} && \vec{X} = [x_1, x_2, x_3, x_4] = [h, l, t, b] \\
&\textit{Minimize} && f_{\text{cost}}(\vec{X}) = 1.10471 x_1^2 x_2 + 0.04811 x_3 x_4 (14.0 + x_2) \\
&\textit{Subject to} && g_1(\vec{X}) = \tau(\vec{X}) - \tau_{max} \\
& && g_2(\vec{X}) = \sigma(\vec{X}) - \sigma_{max} \\
& && g_3(\vec{X}) = x_1 - x_4 \leq 0 \\
& && g_4(\vec{X}) = 0.10471 x_1^2 + 0.04811 x_3 x_4 (14.0 + x_2) - 5.0 \leq 0 \\
& && g_6(\vec{X}) = \delta(\vec{X}) - \delta_{max} \\
& && g_7(\vec{X}) = P - P_c(\vec{X}) \leq 0 \\
&\textit{Where} && \tau(\vec{X}) = \sqrt{(\tau')^2 + 2\tau'\tau''\frac{x_2}{2R} + (\tau'')^2} \\
& && \tau' = \frac{P}{\sqrt{2} x_1 x_2}, \quad \tau'' = \frac{MR}{J} \\
& && M = P\left(L + \frac{x_2}{2}\right), \quad R = \sqrt{\frac{x_2^2}{4} + \left(\frac{x_1 + x_3}{2}\right)^2} \\
& && J = 2\left\{\sqrt{2} x_1 x_2 \left[\frac{x_2^2}{12} + \left(\frac{x_1 + x_3}{2}\right)^2\right]\right\} \\
& && P_c(\vec{X}) = \frac{4.013E\sqrt{\frac{x_3^2 x_4^6}{36}}}{L^2}\left(1 - \frac{x_3}{2L}\sqrt{\frac{E}{4G}}\right) \\
& && \sigma(\vec{X}) = \frac{6PL}{x_4 x_3^2}, \quad \delta(\vec{X}) = \frac{4PL^3}{E x_3^3 x_4} \\
& && P = 6000lb, \quad L = 14in, \quad E = 30 \times 10^6 psi \\
& && G = 12 \times 10^6 psi \\
&\textit{Variable Range} && 0.1 \leq x_1, x_4 \leq 2, \quad 0.1 \leq x_2, x_3 \leq 10
\end{aligned}
\tag{42.13}
$$

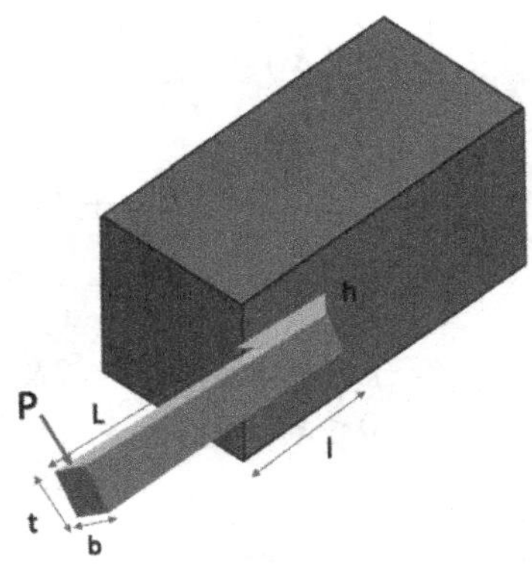

FIGURE 42.4 The welded beam.

TABLE 42.5 The results of the welded beam.

Algorithm	$x_1 = h$	$x_2 = l$	$x_3 = t$	$x_4 = b$	Optimal cost
AWOA	**0.205732**	**3.4704891**	**9.0366244**	**0.205733**	**1.724852**
WOA	0.205396	3.856979	10.00000	0.202376	1.730499
BBO	0.185486	4.3129	8.439903	0.235902	1.918055
SCA	0.205977	3.805191	8.99305	0.208803	1.786863
GWO	0.205676	3.478377	9.03681	0.205778	1.72624
CS	0.205799	3.469634	9.03495	0.205806	1.725135

TABLE 42.6 The statistical results of a welded beam.

Algorithm	Best	Mean	Worst	STD
AWOA	**1.724852**	**1.726179**	**1.765670**	**0.00068**
WOA	1.730499	1.7320	1.78965	0.0226
BBO	1.918055	2.630412	3.606933	0.410914
SCA	1.786863	1.849364	1.925162	0.034689
GWO	1.72624	1.72631	1.728487	0.000771
CS	1.725135	1.79433	2.237755	0.109935

The best possible result of AWOA and other processes is listed in Table 42.5. In addition, AWOA discovers the best variables for solving this problem, as is demonstrated in Table 42.5. The statistical evaluation of AWOA is shown in Table 42.6 in regard to other algorithms. Best, worst, and average results with zero standard deviation were all obtained by the present algorithm, as seen in Table 42.6.

42.4.4 Three-bar truss

This particular issue is another frequently cited topic in the literature. [26]. Fig. 42.5 depicts the symmetric configuration of the truss's three bar components. The problem is extremely constrained, and the objective function is extremely simple. Structural design issues face a wide variety of limitations, including stress, deflection, and buckling constraints. Mathematically, this issue is stated as:

$$\textit{Consider} \quad \vec{X} = [x_1, x_2] = [A_1 = A_3, A_2]$$

$$\textit{Minimize} \quad f_{cost}(\vec{X}) = (2\sqrt{2}X_1 + X_2) \times l$$

$$\textit{Subject to} \quad g_1(\vec{X}) = \frac{\sqrt{2}x_1 + x_2}{\sqrt{2}x_1^2 + 2x_1x_2} P - \sigma \leq 0$$

$$g_2(\vec{X}) = \frac{\sqrt{2}x_1 + x_2}{\sqrt{2}x_1^2 + 2x_1x_2}P - \sigma \leq 0$$

$$g_3(\vec{X}) = \frac{x_2}{\sqrt{2}x_1^2 + 2x_1x_2}P - \sigma \leq 0$$

$$g_4(\vec{X}) = \frac{1}{\sqrt{2}x_2 + x_1}P - \sigma \leq 0$$

$$\textit{Where} \quad l = 100cm, \quad P = 2KN/cm^3, \quad \sigma = 2KN/cm^3$$

$$\textit{Variable Range} \quad 0 \leq x_1, x_2 \leq 1 \tag{42.14}$$

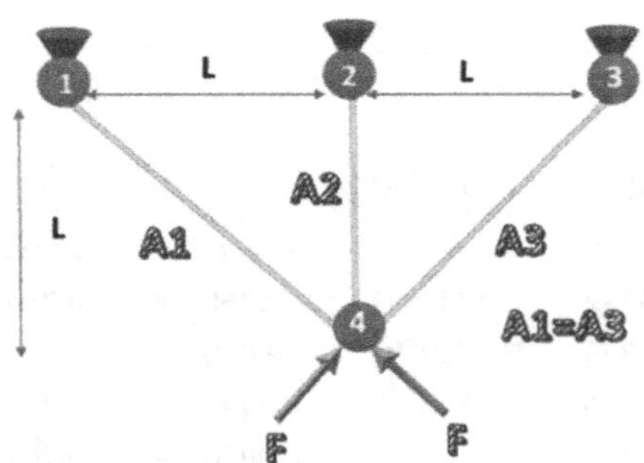

FIGURE 42.5 The three-bar truss.

Even though there are only a few variables, the number of constraints makes solving optimization problems difficult. Table 42.7 displays the outcomes of the six comparative algorithms used in this study. This table demonstrates that AWOA obtains optimal design with a cost of 263.895891.

TABLE 42.7 The results of the three-bar truss.

Algorithm	$x_1 = A_1 = A_3$	$x_2 = A_2$	Optimal cost
AWOA	**0.7889036**	**0.40760249**	**263.895891**
WOA	0.788091	0.409903	263.896097
BBO	0.789066	0.408227	263.896653
SCA	0.787854	0.41059	263.897786
GWO	0.788648	0.408325	263.896006
CS	0.78867	0.40902	263.97156

TABLE 42.8 The statistical results of the three-bar truss.

Algorithm	Best	Mean	Worst	STD
AWOA	**263.895891**	**263.89680**	**263.90381**	**0.0000843**
WOA	263.896097	264.186596	265.082636	0.35863425
BBO	263.896653	264.502146	267.748846	0.88058464
SCA	263.897786	264.752041	282.842712	3.73097657
GWO	263.896006	263.897955	263.904218	0.00161422
CS	263.97156	264.0669	266.4352	0.00009

Table 42.8 displays the results of a comparison between the best, average, Standard Deviation (SD), and worst values. AWOA finds the most reliable design compared to other approaches, with an SD of 8.43E-05.

The simulation experiments conducted in the study involved the use of engineering problems to assess the performance of the advanced version of the Whale Optimization Algorithm (AWOA). The results of the experiments demonstrate that the proposed optimization strategy effectively improves the performance of AWOA, leading to better optimization results.

Furthermore, the study found that AWOA outperforms other algorithms, making it a more competitive option for solving complex optimization problems. In summary, the results showed that AWOA's performance was enhanced by the addition of new formula and that it was able to outperform its competitors in the vast majority of the case studies in this chapter. The experimental findings indicate that NAWOA is capable of efficiently addressing complex, high-dimensional optimization and engineering problems. These results establish a solid theoretical basis for utilizing AWOA in practical applications.

42.5 Conclusion and future work

This research introduced an enhanced version of the Whale Optimization Algorithm. Since AWOA does not involve an additional internal parameter, it retains the same ease of use as WOA. Four classical optimization problems in mechanical engineering are used to demonstrate the effectiveness and suitability of AWOA. This algorithm is compared with other algorithms such as BBO, GWO, SCA, CA, and WOA. In all of the test problems, AWOA eventually converged on better designs. The statistical findings from the independent optimization trials in all benchmark instances demonstrated that AWOA generated lower results, indicating that the proposed method has superior search reliability. The AWOA algorithm was found to outperform other algorithms in the majority of case studies. The further evaluation studies are needed to determine the ability of the proposed optimization algorithm to solve specific problems, such as steel frame structures and other structural engineering problems. Further research and experimentation are needed to determine its effectiveness in solving specific problems, and to improve its performance in multi-objective optimization applications.

References

[1] P.J. Angeline, Genetic programming and emergent intelligence, Adv. Genet. Program. 1 (1994) 75–98.

[2] K.V. Price, Differential evolution, in: Handbook of Optimization, Springer, 2013, pp. 187–214.

[3] R. Storn, K. Price, Differential evolution–a simple and efficient heuristic for global optimization over continuous spaces, J. Glob. Optim. 11 (4) (1997) 341–359.

[4] A. Kaveh, T. Bakhshpoori, Water evaporation optimization: a novel physically inspired optimization algorithm, Comput. Struct. 167 (2016) 69–85.

[5] W. Zhao, L. Wang, Z. Zhang, Atom search optimization and its application to solve a hydrogeologic parameter estimation problem, Knowl.-Based Syst. 163 (2019) 283–304.

[6] A. Kaveh, S. Talatahari, N. Khodadadi, Stochastic paint optimizer: theory and application in civil engineering, Eng. Comput. 38 (2022) 1921–1952, https://doi.org/10.1007/s00366-020-01179-5.

[7] A. Sadollah, H. Sayyaadi, A. Yadav, A dynamic metaheuristic optimization model inspired by biological nervous systems: neural network algorithm, Appl. Soft Comput. 71 (2018) 747–782.

[8] B. Abdollahzadeh, F.S. Gharehchopogh, N. Khodadadi, S. Mirjalili, Mountain gazelle optimizer: a new nature-inspired metaheuristic algorithm for global optimization problems, Adv. Eng. Softw. 174 (2022) 103282.

[9] S. Mirjalili, A. Lewis, The whale optimization algorithm, Adv. Eng. Softw. 95 (2016) 51–67.

[10] G. Li, F. Shuang, P. Zhao, C. Le, An improved butterfly optimization algorithm for engineering design problems using the cross-entropy method, Symmetry (Basel) 11 (8) (2019) 1049.

[11] N. Khodadadi, S.M. Mirjalili, S.Z. Mirjalili, S. Mirjalili, Chaotic stochastic paint optimizer (CSPO), in: Proceedings of 7th International Conference on Harmony Search, Soft Computing and Applications, 2022, pp. 195–205.

[12] F. Jiang, L. Wang, L. Bai, An improved whale algorithm and its application in truss optimization, J. Bionic Eng. 18 (3) (2021) 721–732.

[13] S.Z. Mirjalili, S. Sajeev, R. Saha, N. Khodadadi, S.M. Mirjalili, S. Mirjalili, Evolutionary population dynamic mechanisms for the harmony search algorithm, in: Proceedings of 7th International Conference on Harmony Search, Soft Computing and Applications, 2022, pp. 185–194.

[14] N. Khodadadi, S. Talatahari, A.H. Gandomi, ANNA: advanced neural network algorithm for optimisation of structures, Proc. Inst. Civ. Eng., Struct. Build. (2023), https://doi.org/10.1680/jstbu.22.00083.

[15] D.H. Wolpert, W.G. Macready, No free lunch theorems for optimization, IEEE Trans. Evol. Comput. 1 (1) (1997) 67–82.

[16] S.M. Mirjalili, S.Z. Mirjalili, N. Khodadadi, V. Snasel, S. Mirjalili, Grey wolf optimizer, whale optimization algorithm, and moth flame optimization for optimizing photonics crystals, in: Advances in Swarm Intelligence, Springer, 2023, pp. 169–179.

[17] A. Kaveh, M.I. Ghazaan, Enhanced whale optimization algorithm for sizing optimization of skeletal structures, Mech. Based Des. Struct. Mach. 45 (3) (2017) 345–362.

[18] D. Simon, Biogeography-based optimization, IEEE Trans. Evol. Comput. 12 (6) (2008) 702–713.

[19] S. Mirjalili, S.M. Mirjalili, A. Lewis, Grey wolf optimizer, Adv. Eng. Softw. 69 (2014) 46–61.

[20] S. Mirjalili, SCA: a sine cosine algorithm for solving optimization problems, Knowl.-Based Syst. 96 (2016) 120–133.

[21] A.H. Gandomi, X.-S. Yang, A.H. Alavi, Cuckoo search algorithm: a metaheuristic approach to solve structural optimization problems, Eng. Comput. 29 (1) (2013) 17–35.

[22] A. Arcuri, G. Fraser, Parameter tuning or default values? An empirical investigation in search-based software engineering, Empir. Softw. Eng. 18 (3) (2013) 594–623.

[23] A.D. Belegundu, J.S. Arora, A study of mathematical programming methods for structural optimization. Part I: theory, Int. J. Numer. Methods Eng. 21 (9) (1985) 1583–1599.

[24] B.K. Kannan, S.N. Kramer, An augmented Lagrange multiplier based method for mixed integer discrete continuous optimization and its applications to mechanical design, J. Mech. Des. 116 (2) (1994) 405–411, https://doi.org/10.1115/1.2919393.
[25] C.A.C. Coello, Use of a self-adaptive penalty approach for engineering optimization problems, Comput. Ind. 41 (2) (2000) 113–127.
[26] M.-Y. Cheng, D. Prayogo, Symbiotic organisms search: a new metaheuristic optimization algorithm, Comput. Struct. 139 (2014) 98–112.

Chapter 43

Binary whale optimization algorithm for topology planning in wireless mesh networks

Sylia Mekhmoukh Taleb[a], Yassine Meraihi[a], Seyedali Mirjalili[b,c], Selma Yahia[a], and Amar Ramdane-Cherif[d]

[a]*Systems Engineering and Telecommunications Laboratory, University of Boumerdes, Boumerdes, Algeria,* [b]*Centre for Artificial Intelligence Research and Optimisation, Torrens University Australia, Brisbane, QLD, Australia,* [c]*University Research and Innovation Center, Obuda University, Budapest, Hungary,* [d]*Systems Engineering Laboratory of Versailles (LISV), University of Paris-Saclay, Velizy, France*

43.1 Introduction

Wireless Mesh Networks (WMNs) represent a new generation of multi-hop wireless networks, used to connect various wireless devices by building an unwired mesh [1,2]. They seek to offer mobile and fixed consumers high-speed internet access everywhere, anytime, and are gaining more and more real interest from the R&D community, network operators, and service providers [2,3]. Self-configuration, self-organization, and the capacity to autonomously create and sustain connectivity between nodes in locations and surroundings devoid of the internet are all traits that WMNs display [2]. These traits offer a number of benefits, including low implementation costs, simple network management, robustness, etc.

The WMN is composed of three different types of nodes, including: Mesh Gateway (MG), Mesh Router (MR), and Mesh Client (MC). MCs can be mobile phones, desktop computers, laptops, Pocket PCs, Personal Digital Assistants (PDAs), and many other devices. They connect to the internet through MRs, which relay traffic to and from MGs, which in turn are linked to the internet infrastructure. The typical architecture of such networks is depicted in Fig. 43.1.

Despite the benefits of Wireless Mesh Networks (WMNs), these networks present many challenges for network operators, such as the problem of optimizing the location of mesh nodes in the deployment region. Poor placement of mesh nodes (MRs/MGs) can result in interference and congestion, leading to high levels of packet loss, delay, and low throughput. The topology planning problem has been shown to be NP-hard [4] and cannot be solved using traditional exact methods. As a result, meta-heuristics have been shown to be the most effective approach for solving the problem. Several meta-heuristics have been applied, such as Tabu Search (TS) [5], simulated annealing (SA) [6], Hill Climbing (HC) [7], Genetic Algorithm (GA) [8], Particle Swarm Optimization (PSO) [9], Firefly Algorithm (FA) [10], Honey Badger Algorithm (HBA) [11], Coyote Optimization Algorithm (COA) [12], and Moth Flame Optimization (MFO) [13].

Based on the literature review presented in [14], when solving the mesh nodes placement problem, the metrics taken into account are namely coverage and connectivity. On the other hand, there are few studies that have taken into account the cost metric [4,15,16]. This lack motivates our attempt to address the mesh router placement problem with cost minimization, while satisfying full coverage and full connectivity requirements.

This chapter proposes a BWOA for tackling the topology planning problem in WMNs. BWOA is obtained using S-shaped and V-shaped families of transfer functions. BWOA was validated using five generated in under different numbers of mesh clients.

This essay has the following organizational structure. We give a general overview of the system model and problem formulation in Section 43.2. The Whale Optimization Algorithm (WOA) is introduced in Section 43.3. The Binary Whale Optimization Algorithm is described in Section 43.4. The effectiveness of BWOA is assessed in Section 43.5 along with the findings. Section 43.6 wraps up the essay.

43.2 Problem formulation

In this study, the Wireless Mesh Network (WMN) is modeled as a fully connected graph denoted by $G = (V, E)$, where V represents the set of MRs and E represents the set of links between them. Each MR is equipped with a radio interface

Handbook of Whale Optimization Algorithm. https://doi.org/10.1016/B978-0-32-395365-8.00049-X

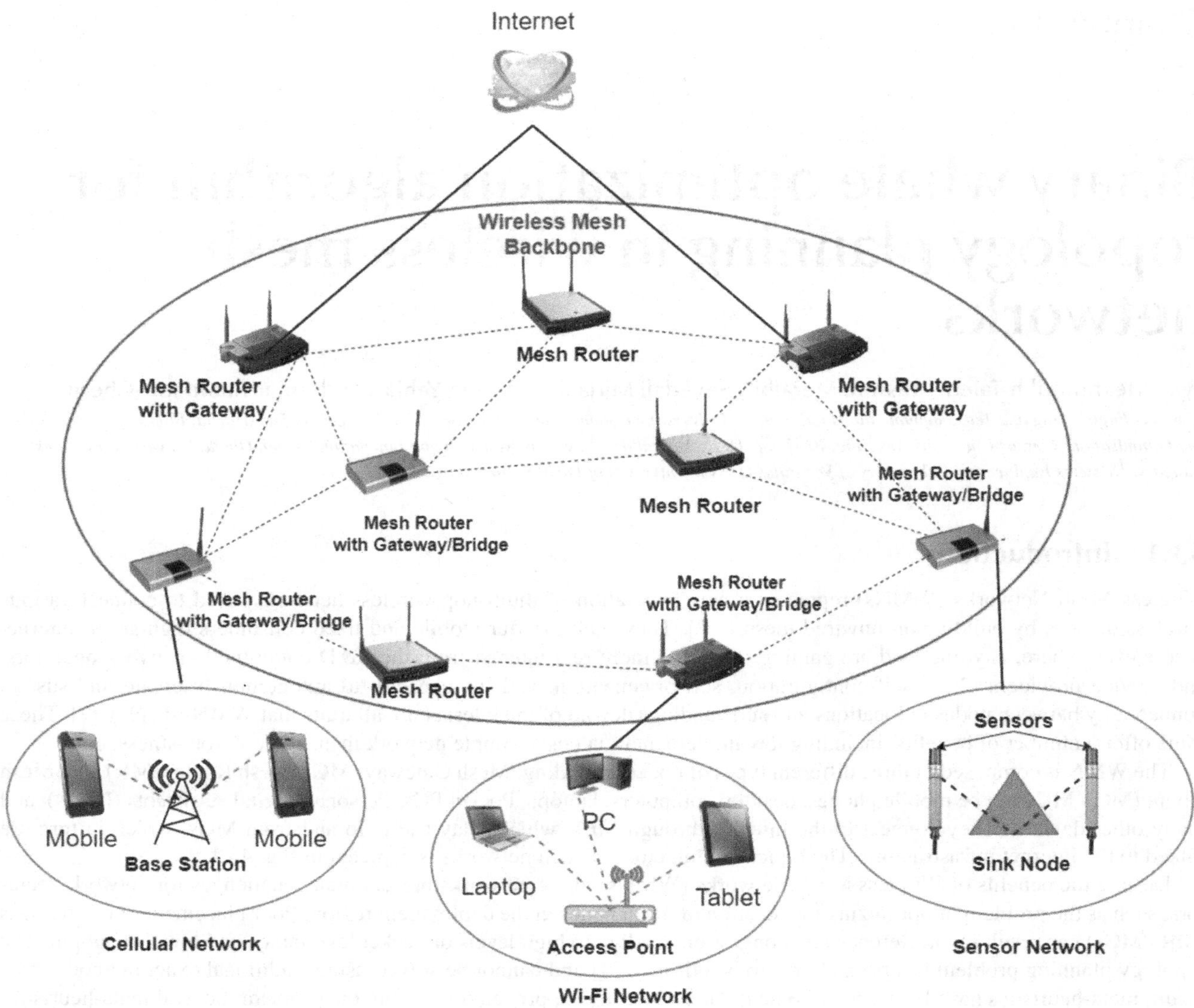

FIGURE 43.1 WMNs architecture.

that has an identical coverage radius R. The set of MCs is denoted as $MC = MC_1, ..., MC_n$, and the set of Candidate Sites (CS) to host an MR is denoted as $CS = CS_1, ..., CS_s$. An MC is considered to be covered by an MR if it is within the MR's transmission range and is assigned to the closest one. Two installed MRs can establish a link between them if their transmission range is greater than the Euclidean distance between the two MRs.

The coverage variable a_{ij} is given as follows:

$$a_{ij} = \begin{cases} 1 & \text{if MC}_\text{i} \text{ is covered by CS}_\text{j} \\ 0 & \text{otherwise} \end{cases} \tag{43.1}$$

Let $i \in MC$, $j \in CS$, the assignment variable x_{ij} is described as follows:

$$x_{ij} = \begin{cases} 1 & \text{if MC}_\text{i} \text{ is assigned to CS}_\text{j} \\ 0 & \text{otherwise} \end{cases} \tag{43.2}$$

The installation variable r_j is given in Eq. (43.3):

$$r_j = \begin{cases} 1 & \text{if MR is installed in } \mathrm{CS_j} \\ 0 & \text{otherwise} \end{cases} \quad (43.3)$$

Let j and $l \in CS$, $j \neq l$, the connectivity variable A_{jl} (adjacency matrix) is specified as defined in Eq. (43.4).

$$A_{jl} = \begin{cases} 1 & \text{if } \mathrm{CS_j} \text{ and } \mathrm{CS_l} \text{ can be connected} \\ 0 & \text{otherwise} \end{cases} \quad (43.4)$$

Our planning problem aims to identify a group of potential locations for installing MRs that can fulfill the requirements of full coverage and full connectivity. To frame the problem, we can express it in the following manner:

$$m = \mathrm{Min} \sum_{j \in CS} r_j \quad (43.5)$$

Subject to:

$$\sum_{j \in CS} x_{ij} = 1 \quad \forall i \in MC \quad (43.6)$$

$$x_{ij} \leqslant r_j a_{ij} \quad \forall i \in MC \quad \forall j \in CS, \quad (43.7)$$

$$\sum_{k=1}^{m-1} \mathrm{A}^k \neq 0 \quad (43.8)$$

The objective function in Eq. (43.5) reduces the overall number of installed MRs m in the network. Eq. (43.6) ensures the full coverage of all MCs. Inequality (43.7) implies that a MC_i is assigned and covered by a deployed MR in CS_j. Eq. (43.8) imposes to have at least one path between each pair of MRs. Consequently, the graph G is fully connected.

43.3 Whale optimization algorithm (WOA)

Based on the hunting strategies of humpback whale, Mirjalili and Lewis introduced a novel swarm intelligence method called Whale Optimization Algorithm (WOA) in 2016 [17]. WOA's exploitation phase is modeled based encircling prey mechanism and spiral updating position approach. The description is given as follows:

- **Encircling prey: if ($P < 0.5$ and $|A| < 1$)**
  Eqs. (43.9) and (43.10) are utilized to modify the position of the $X(t+1)$ solution:

$$\vec{D'} = \left| C \vec{X}_{best}(t) - \vec{X}(t) \right| \quad (43.9)$$

$$\vec{X}(t+1) = \vec{X}_{best}(t) - \vec{A} \cdot \vec{D'} \quad (43.10)$$

where t denotes the current iteration, $\vec{X}$ and $\overrightarrow{X_{best}}$ represent the current and the best solutions, respectively. The vectors of coefficients $\vec{A}$ and $\vec{C}$ are calculated as in Eqs. (43.11) and (43.12).

$$\vec{A} = 2\vec{a} \cdot \vec{rl} - \vec{a} \quad (43.11)$$

$$\vec{A} = 2\vec{r} \quad (43.12)$$

where $\vec{a}$ drops linearly from 2 to 0 over iterations (simulating the shrinking encircling behavior as in Eq. (43.11)) and $\vec{rl}$ is a random vector in the range [0, 1].
Let $t_{\max}$ be the total number of iterations. a is determined using the formula given in Eq. (43.13).

$$a = 2\left(1 - t/t_{\max}\right) \quad (43.13)$$

- **Spiral updating position: if $P \geq 0.5$**
  Let l be a random number in the interval $[-1, 1]$, b is a constant, and $\vec{D}$ represents the distance between the current solution and $\vec{X}_{best}$ at iteration t. The spiral-shaped path followed by the whales is modeled using the spiral rule in Eq. (43.14).

$$\vec{X}(t+1) = \vec{D} \cdot e^{bl} \cdot \cos(2\pi l) + \vec{X}_{best}(t) \tag{43.14}$$

$$\vec{D} = \left| \vec{A} \cdot \vec{X}_{best}(t) - \vec{X}(t) \right| \tag{43.15}$$

To simulate the exploration phase in WOA (if $Pr < 0.5$ and $|A| \geq 1$), the current whale position is updated by a search agent chosen at random from the population, as shown in Eq. (43.17):

$$\vec{D} = \left| \vec{C} \vec{X}_{rand}(t) - \vec{X}(t) \right| \tag{43.16}$$

$$\vec{X}(t+1) = \vec{X}_{rand}(t) - \vec{A} \cdot \vec{D} \tag{43.17}$$

where the vector $\vec{X}_{rand}$ represents a search agent randomly selected from the current population, while the vector $\vec{A}$ is defined as a vector containing random values that fall within the range of $[-1, 1]$.

The primary stages of the WOA method are illustrated in Algorithm 43.1.

Algorithm 43.1 The WOA Algorithm.

```
Initialize WOA parameters
Randomly initialize the population of solutions
Evaluate the population and determine X_best
for t = 1 to t_max do
    for i = 1 to N do
        Update a, A, C, l, and P
        if P < 0.5 then
            if |A| < 1 then
                Calculate X_i(t + 1) using Eq. (43.10)
            else
                Calculate X_i(t + 1) using Eq. (43.17)
            end if
        else
            Calculate X_i(t + 1) using Eq. (43.14)
        end if
    end for
    Evaluate the population and determine X_best
    t = t + 1
end for
Return the best solution
```

43.4 Binary whale optimization algorithm (BWOA)

WOA technique has been utilized for solving various optimization problems [18–26] and Machine Learning (ML) problems [27–32]. However, like any other optimization algorithm, WOA has its limitations when applied to ML methods. Some of these limitations are:

- Convergence speed: Although WOA has been shown to converge quickly on some optimization problems, it may struggle to find the global optimum in high-dimensional ML problems. This is due to the fact that the search space in ML problems is often complex and multi-modal, making it difficult for the algorithm to converge efficiently.
- Sensitivity to parameters: The performance of WOA is dependent on the choice of parameters, such as the step size and population size. The performance of the algorithm can be significantly influenced by these parameters, which might require careful tuning for each machine learning problem to achieve optimal results.

TABLE 43.1 S-shaped, V-shaped, and U-shaped, and Traper-shaped families of transfer functions.

Name	Transfer function	Name	Transfer function
	S-shaped		**V-shaped**
S1 [34]	$S1(x) = \frac{1}{1+e^{-2x}}$	V1 [34]	$V1(x) = \left\|\text{erf}\left(\frac{\sqrt{\pi}}{2}x\right)\right\|$
S2 [35]	$S2(x) = \frac{1}{1+e^{-x}}$	V2 [36]	$V2(x) = \|\tanh(x)\|$
S3 [34]	$S3(x) = \frac{1}{1+e^{(-x/2)}}$	V3 [34]	$V3(x) = \left\|(x)/\sqrt{1+x^2}\right\|$
S4 [34]	$S4(x) = \frac{1}{1+e^{(-x/3)}}$	V4 [34]	$V4(x) = \left\|\frac{2}{\pi}\arctan\left(\frac{\pi}{2}x\right)\right\|$
	U-shaped		**Taper-shaped**
U1 [37]	$U1(x) = \|x\|^{1.5}$	T1 [38]	$T1(x) = \frac{\sqrt{\|x\|}}{\sqrt{\|A\|}}$
U2 [37]	$U2(x) = \|x\|^2$	T2 [38]	$T2(x) = \frac{\|x\|}{\|A\|}$
U3 [37]	$U3(x) = \|x\|^3$	T3 [38]	$T3(x) = \frac{\sqrt[3]{\|x\|}}{\sqrt[3]{\|A\|}}$
U4 [37]	$U4(x) = \|x\|^4$	T4 [38]	$T4(x) = \frac{\sqrt[4]{\|x\|}}{\sqrt[4]{\|A\|}}$

- Local minima: Like many other optimization algorithms, WOA may get stuck in local minima, preventing it from locating the global optimum solution. This can be a particular challenge in ML problems, where the objective function is often complex and non-linear.
- Computational complexity: WOA is a computationally intensive algorithm and may require a large amount of computational resources, especially for high-dimensional ML problems. This can be a limiting factor for real-time or online applications where the processing time is critical.

Despite these limitations, WOA has shown promising results on a variety of ML problems and has the potential to be a valuable tool for optimizing complex ML models. However, it is important to carefully consider the limitations of WOA when applying it to ML problems and to choose the appropriate optimization algorithm based on the specific requirements of each problem.

The topology planning problem in WMNs is not a machine learning (ML) problem in the traditional sense, as it does not involve training a model to make predictions based on input data. Rather, it is an optimization problem that can be solved successfully using meta-heuristic approaches such as WOA. This chapter addresses a design problem that must be solved in a discrete space, therefore, a binary version of WOA is required to tackle this issue.

The binarization of WOA requires the use of transfer functions in order to obtain '0' or '1' values. The V-shaped, S-shaped, and U-shaped families of transfer functions are the three forms that are most frequently utilized.

A novel X-shaped transfer function was put out in [33] in order to enhance the binary social mimic optimization's (BSMO) capabilities for exploration and exploitation. By merging two elements ($S2$ and $S2'$) and applying a crossover operation, the new solution is created. A further novel category of transfer functions known as Taper-shaped transfer functions was introduced in [33]. Although they were influenced by the U-shaped family of transfer functions, these functions were constructed employing the power function over a symmetric interval of positive real numbers. Table 43.1 lists the calculation formulas for the S-shaped, V-shaped, U-shaped, and Taper-shaped families of transfer functions, and Fig. 43.2 shows their graphical representations.

In this chapter, we use V-shaped and S-shaped families of transfer functions for discretizing WOA. The binary versions of WOA based on V1-V4 functions are denoted by BWOAV1, BWOAV2, BWOAV3, and BWOAV4, respectively. BWOAS1, BWOAS2, BWOAS3, and BWOAS4 are the binary versions of WOA based on S1, S2, S3, and S4, respectively.

Based on the S-shaped functions, the binarization method is specified in Eq. (43.18).

$$X_i^j(t+1) = \begin{cases} 1 & \text{if rand()} < S\left(X_i^j(t+1)\right) \\ 0 & \text{if rand()} \geq S\left(X_i^j(t+1)\right) \end{cases} \tag{43.18}$$

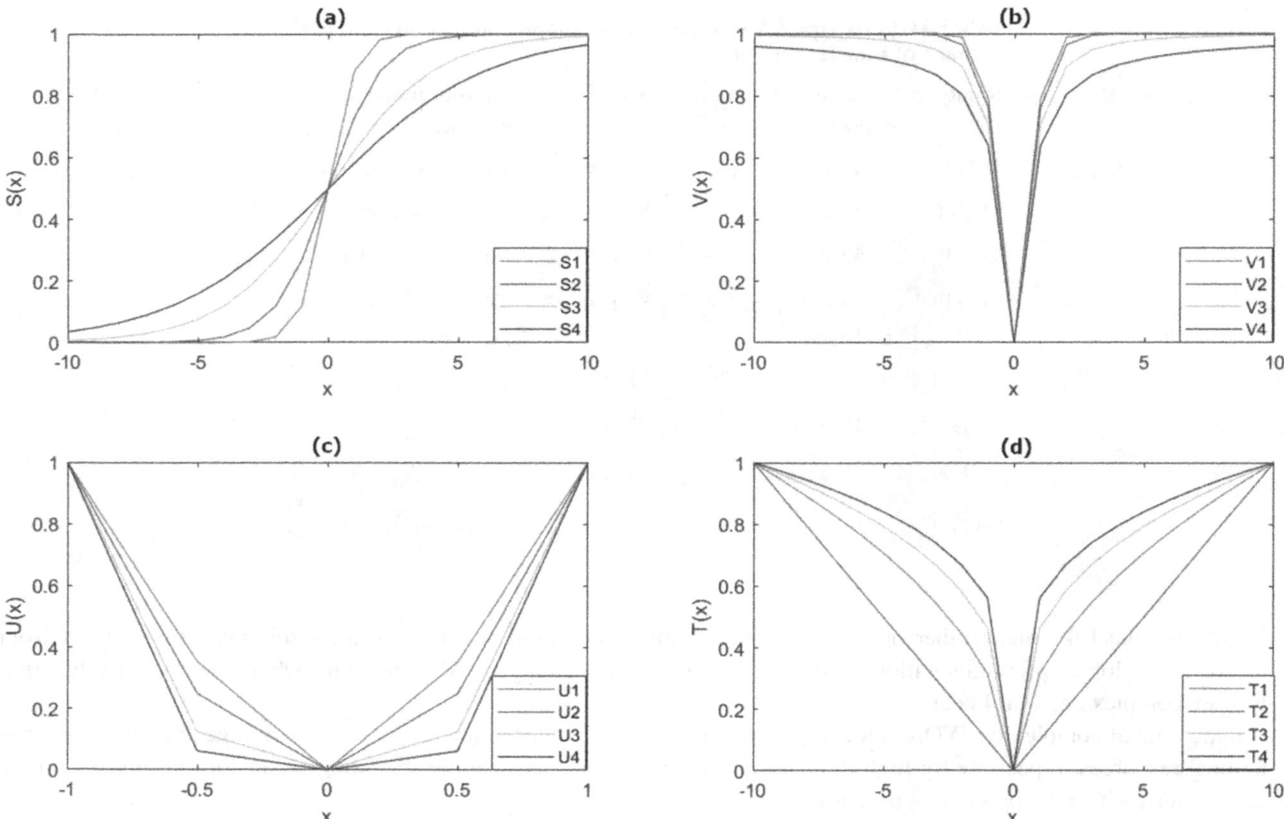

FIGURE 43.2 (a) S-shaped (b) V-shaped (c) U-shaped (d) Traper-shaped families of transfer functions.

On the other hand, based on V-shaped functions, the binarization method is specified in Eq. (43.19).

$$X_i^j(t+1) = \begin{cases} \sim X_i^j(t) & \text{if rand} < V\left(X_i^j(t+1)\right) \\ X_i^d(t) & \text{else} \end{cases} \tag{43.19}$$

43.5 Simulation results

This section assesses the performance of the eight binary versions of Whale Optimization Algorithm (WOA). All binary versions of WOA were implemented using Matlab®. The simulations were conducted on a Core i7 machine, with a population size of 20 solutions. 1000 iterations were used as the total number of iterations. Each result provided in this section is the mean value obtained from 10 simulations.

The evaluation process is done in terms of three metrics including minimum, maximum, and average number of MRs, taking into account various numbers of MCs (from 20 to 100). A transmission range of $250m$ is assigned for each installed MR. A grid topology of 10×10 was considered for the candidate sites.

Figs. 43.3.a, 43.3.b, 43.3.c, 43.3.d, 43.3.e, 43.3.f, 43.3.g, and 43.3.h report an example of planned network using BWOAS1, BWOAS2, BWOAS3, BWOAS4, BWOAV1, BWOAV2, BWOAV3, and BWOAV4, respectively. The planned network is a resolution for a network scenario comprising of 100 randomly dispersed MCs within a deployment region of $1000m \times 1000m$, represented by blue points. Each installed MR is represented by the Bold black point. It is clearly seen that binary versions of WOA based V-shaped functions are more appropriate for addressing the challenge of topology planning in WMNs.

The simulation outcomes, which are displayed in Table 43.2 and Fig. 43.4, demonstrated that an increase in the number of MCs requires an increase in the number of MRs to provide the full coverage. In fact, more routers are required to cover the additional Mcs. Furthermore, BWOA algorithms utilizing V-shaped transfer functions outperform S-shaped transfer

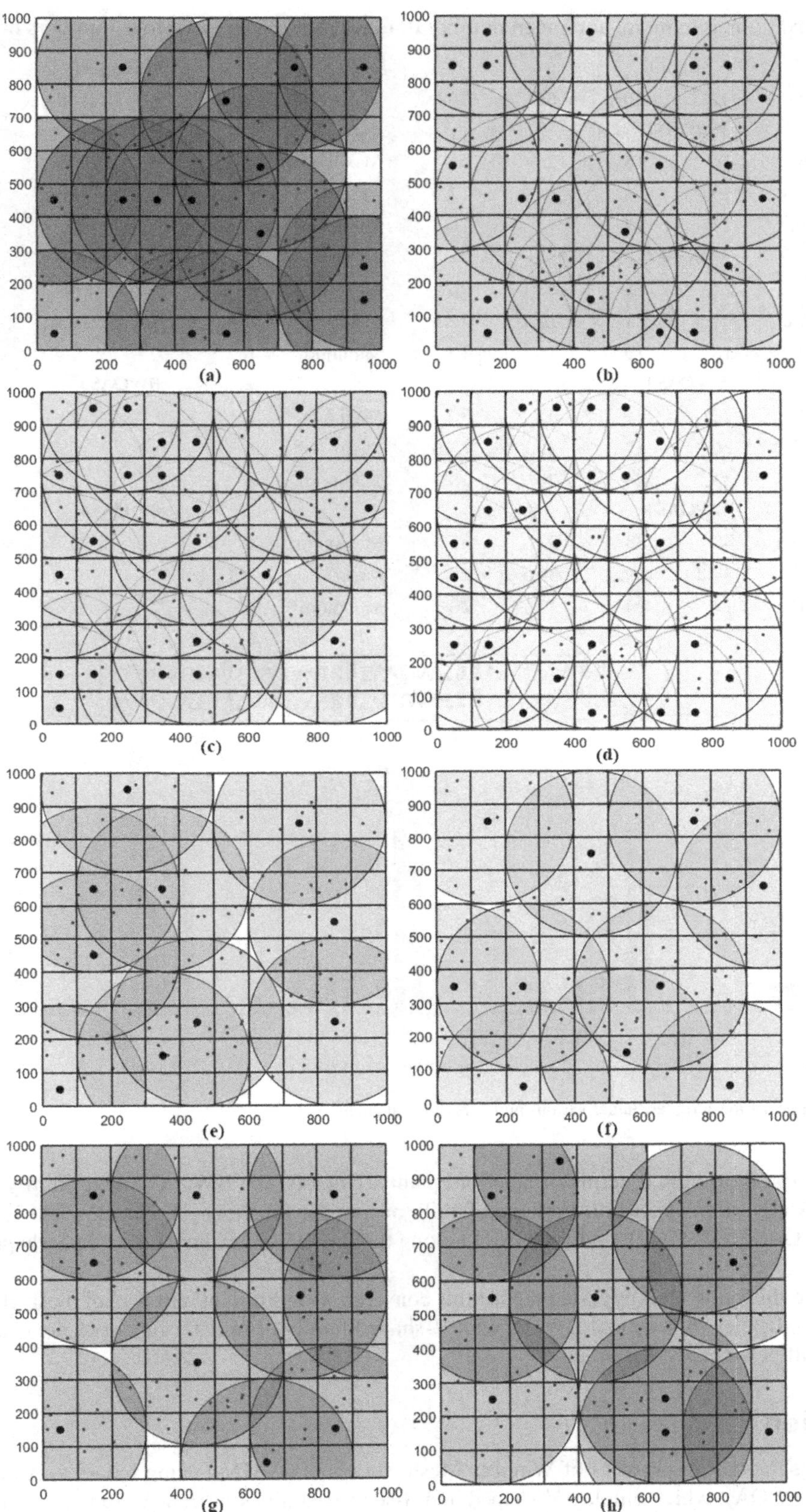

FIGURE 43.3 Obtained placements using: (a) BWOAS1 (b) BWOAS2 (c) BWOAS3 (d) BWOAS4 (e) BWOAV1 (f) BWOAV2 (g) BWOAV3 (h) BWOAV4.

TABLE 43.2 Maximum, minimum, and mean number of mesh routers under various numbers of mesh clients.

S-shaped						V-shaped					
n	20	40	60	80	100	*n*	20	40	60	80	100
		BWOAS1						BWOAS1			
Maximum	16	18	19	20	21	Maximum	8	9	10	10	10
Mean	14.8	15.7	18	17.3	17.9	Mean	6.8	7.6	8.6	9.3	9.2
Minimum	12	13	17	14	16	Minimum	6	7	8	8	7
		BWOAS2						BWOAS2			
Maximum	24	24	25	25	26	Maximum	8	9	10	10	10
Mean	20.8	21.5	21.8	22.1	22.5	Mean	6.8	7.9	8.4	9	9.1
Minimum	17	16	19	18	19	Minimum	6	7	7	8	8
		BWOAS3						BWOAS3			
Maximum	30	28	31	31	30	Maximum	8	9	9	11	10
Mean	27.5	26.7	27.6	28.2	27.6	Mean	6.7	8.1	8.6	9.7	9
Minimum	24	23	24	24	25	Minimum	6	6	8	8	8
		BWOAS4						BWOAS4			
Maximum	32	32	33	33	31	Maximum	8	8	9	11	11
Mean	29	28.2	29.9	30.2	29	Mean	**6.5**	**7.5**	**7.9**	**8.7**	9.7
Minimum	24	24	24	27	26	Minimum	6	7	7	8	8

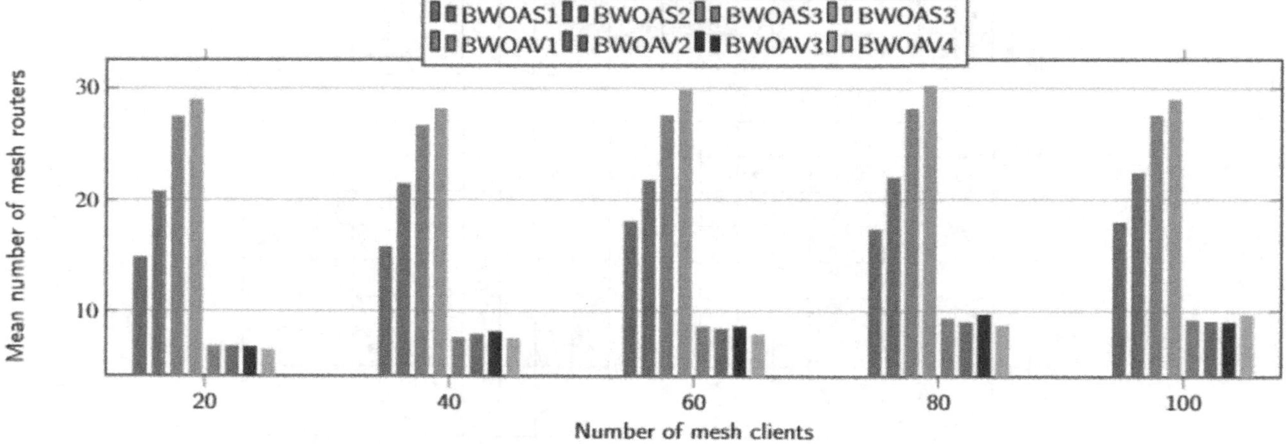

FIGURE 43.4 Mean number of mesh routers under various numbers of mesh clients.

functions-based BWOA algorithms in terms of solution quality. Notably, the BWOAV4 algorithm yields the best results among all the BWOA algorithms based on V-shaped family of transfer functions. Again, BWOAS1, which uses S1, performs better than BWOAS2, BWOAS3, and BWOAS4 among the BWOA algorithms based on S-shaped family of transfer functions.

Figs. 43.5.a–43.5.e show how the suggested algorithms converge with different numbers of mesh clients (20, 40, 60, 80, and 100, respectively). Its clearly seen that BWOA with V-shaped family of transfer functions are capable of determining the global solution with few needed iterations

43.6 Conclusion

This chapter addresses the design problem of Wireless Mesh Networks (WMNs) using binary versions of the Whale Optimization Algorithm (WOA), which employ V-shaped and S-shaped families of transfer functions to convert the original WOA into a binary form. The primary objective of the study is to reduce the number of MRs required to achieve full coverage and full connectivity. The study evaluates the algorithm's performance using three metrics such minimum, maximum,

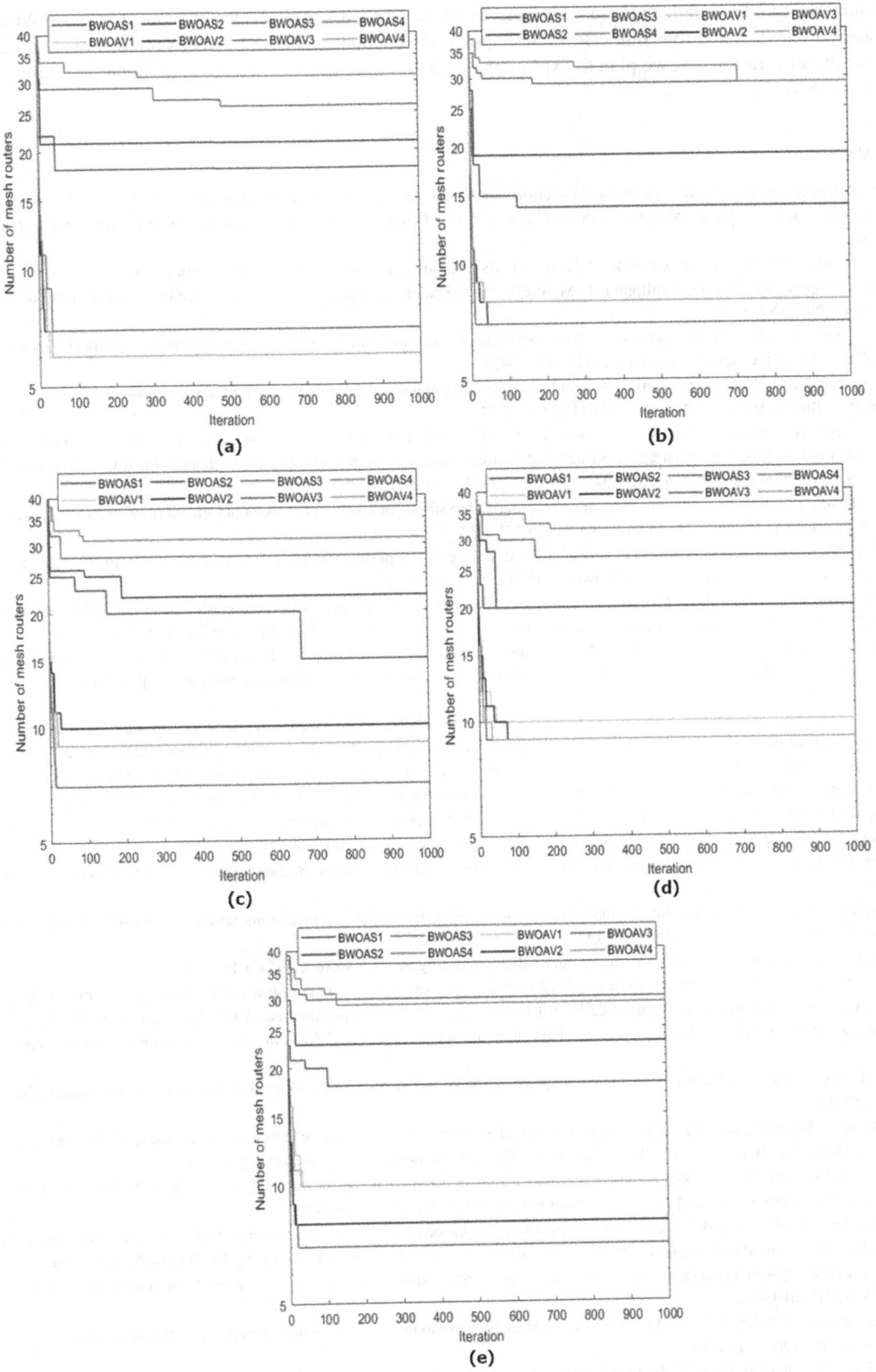

FIGURE 43.5 Example of algorithms convergence using (a) 20 MCs (b) 40 MCs (c) 60 MCs (d) 80 MCs (e) 100 MCs.

and average number of MRs, under various numbers of MCs. Experimental results indicate that the BWOA algorithms utilizing V-shaped transfer functions outperform S-shaped transfer functions-based BWOA algorithms in terms of solution quality. As part of our future work, we plan to explore the application of multi-objective approaches to address the topology planning issue in WMNs.

References

[1] A. Ouni, Optimisation de la capacité et de la consommation énergétique dans les réseaux maillés sans fil, Ph.D. thesis, INSA de Lyon, 2013.

[2] I.F. Akyildiz, X. Wang, W. Wang, Wireless mesh networks: a survey, Computer Networks 47 (2005) 445–487, https://doi.org/10.1016/j.comnet.2004.12.001.

[3] I. Akyildiz, X. Wang, Wireless Mesh Networks, Advanced Texts in Communications and Networking, Wiley, 2009.

[4] E. Amaldi, A. Capone, M. Cesana, I. Filippini, F. Malucelli, Optimization models and methods for planning wireless mesh networks, Computer Networks 52 (2008) 2159–2171.

[5] F. Xhafa, C. Sánchez, A. Barolli, M. Takizawa, Solving mesh router nodes placement problem in wireless mesh networks by tabu search algorithm, Journal of Computer and System Sciences 81 (2015) 1417–1428.

[6] F. Xhafa, A. Barolli, C. Sánchez, L. Barolli, A simulated annealing algorithm for router nodes placement problem in wireless mesh networks, Simulation Modelling Practice and Theory 19 (2011) 2276–2284.

[7] A. Hirata, T. Oda, N. Saito, Y. Nagai, K. Toyoshima, L. Barolli, A CCM-based HC system for mesh router placement optimization: a comparison study for different instances considering normal and uniform distributions of mesh clients, in: International Conference on Network-Based Information Systems, Springer, 2021, pp. 329–340.

[8] F. Xhafa, C. Sanchez, L. Barolli, E. Spaho, Evaluation of genetic algorithms for mesh router nodes placement in wireless mesh networks, Journal of Ambient Intelligence and Humanized Computing 1 (2010) 271–282.

[9] V.T. Le, N.H. Dinh, N.G. Nguyen, A novel PSO-based algorithm for gateway placement in wireless mesh networks, in: 2011 IEEE 3rd International Conference on Communication Software and Networks, IEEE, 2011, pp. 41–45.

[10] L. Sayad, D. Aissani, L. Bouallouche-Medjkoune, Placement optimization of wireless mesh routers using firefly optimization algorithm, in: 2018 International Conference on Smart Communications in Network Technologies (SaCoNeT), IEEE, 2018, pp. 144–148.

[11] S.M. Taleb, Y. Meraihi, S. Mirjalili, D. Acheli, A. Ramdane-Cherif, A.B. Gabis, Enhanced honey badger algorithm for mesh routers placement problem in wireless mesh networks, in: 2022 International Conference on Advanced Aspects of Software Engineering (ICAASE), IEEE, 2022, pp. 1–6.

[12] S.M. Taleb, Y. Meraihi, A.B. Gabis, S. Mirjalili, A. Zaguia, A. Ramdane-Cherif, Solving the mesh router nodes placement in wireless mesh networks using coyote optimization algorithm, IEEE Access 10 (2022) 52744–52759, https://doi.org/10.1109/ACCESS.2022.3166866.

[13] S.M. Taleb, Y. Meraihi, S. Mirjalili, D. Acheli, A. Ramdane-Cherif, A.B. Gabis, Mesh router nodes placement for wireless mesh networks based on an enhanced moth–flame optimization algorithm, Mobile Networks and Applications (2023) 1–24, https://doi.org/10.1007/s11036-022-02059-6.

[14] S.M. Taleb, Y. Meraihi, A.B. Gabis, S. Mirjalili, A. Ramdane-Cherif, Nodes placement in wireless mesh networks using optimization approaches: a survey, Neural Computing & Applications 34 (2022) 5283–5319, https://doi.org/10.1007/s00521-022-06941-y.

[15] M. da Silva, E.L.F. Senne, N.L. Vijaykumar, Wireless mesh networks planning based on parameters of quality of service, in: ICORES, 2012, pp. 441–446.

[16] A. So, B. Liang, Optimal placement and channel assignment of relay stations in heterogeneous wireless mesh networks by modified bender's decomposition, Ad Hoc Networks 7 (2009) 118–135.

[17] S. Mirjalili, A. Lewis, The whale optimization algorithm, Advances in Engineering Software 95 (2016) 51–67.

[18] T.-K. Dao, T.-S. Pan, J.-S. Pan, A multi-objective optimal mobile robot path planning based on whale optimization algorithm, in: 2016 IEEE 13th International Conference on Signal Processing (ICSP), IEEE, 2016, pp. 337–342, https://doi.org/10.1109/ICSP.2016.7877851.

[19] J. Wu, H. Wang, N. Li, P. Yao, Y. Huang, H. Yang, Path planning for solar-powered UAV in urban environment, Neurocomputing 275 (2018) 2055–2065.

[20] S.V. Kumar, R. Jayaparvathy, B. Priyanka, Efficient path planning of AUVs for container ship oil spill detection in coastal areas, Ocean Engineering 217 (2020) 107932.

[21] Z. Yan, J. Zhang, Z. Yang, J. Tang, Two-dimensional optimal path planning for autonomous underwater vehicle using a whale optimization algorithm, Concurrency and Computation: Practice and Experience 33 (2021) e6140, https://doi.org/10.1002/cpe.6140.

[22] A. Chhillar, A. Choudhary, Mobile robot path planning based upon updated whale optimization algorithm, in: 2020 10th International Conference on Cloud Computing, Data Science & Engineering (Confluence), IEEE, 2020, pp. 684–691.

[23] S. Nasrollahzadeh, M. Maadani, M.A. Pourmina, Optimal motion sensor placement in smart homes and intelligent environments using a hybrid WOA-PSO algorithm, Journal of Reliable Intelligent Environments 8 (2022) 345–357, https://doi.org/10.1007/s40860-021-00157-y.

[24] P. Singh, S. Prakash, Optical network unit placement in Fiber-Wireless (FiWi) access network by whale optimization algorithm, Optical Fiber Technology 52 (2019) 101965.

[25] A. Al-Moalmi, J. Luo, A. Salah, K. Li, L. Yin, A whale optimization system for energy-efficient container placement in data centers, Expert Systems with Applications 164 (2021) 113719.

[26] K.M.S. Alzaidi, O. Bayat, O.N. Uçan, Multiple DGs for reducing total power losses in radial distribution systems using hybrid WOA-SSA algorithm, International Journal of Photoenergy 2019 (2019), https://doi.org/10.1155/2019/2426538.

[27] M. Sharawi, H.M. Zawbaa, E. Emary, H.M. Zawbaa, E. Emary, Feature selection approach based on whale optimization algorithm, in: 2017 Ninth International Conference on Advanced Computational Intelligence (ICACI), IEEE, 2017, pp. 163–168, https://doi.org/10.1109/ICACI.2017.7974502.
[28] M.M. Mafarja, S. Mirjalili, Hybrid whale optimization algorithm with simulated annealing for feature selection, Neurocomputing 260 (2017) 302–312.
[29] M. Tubishat, M.A. Abushariah, N. Idris, I. Aljarah, Improved whale optimization algorithm for feature selection in Arabic sentiment analysis, Applied Intelligence 49 (2019) 1688–1707.
[30] J. Nasiri, F.M. Khiyabani, A whale optimization algorithm (WOA) approach for clustering, Cogent Mathematics & Statistics 5 (2018) 1483565.
[31] B.M. Sahoo, H.M. Pandey, T. Amgoth, A whale optimization (WOA): meta-heuristic based energy improvement clustering in wireless sensor networks, in: 2021 11th International Conference on Cloud Computing, Data Science & Engineering (Confluence), IEEE, 2021, pp. 649–654.
[32] M. Kumar, A. Chaparala, OBC-WOA: opposition-based chaotic whale optimization algorithm for energy efficient clustering in wireless sensor network, Intelligence 250 (2019).
[33] K.K. Ghosh, P.K. Singh, J. Hong, Z.W. Geem, R. Sarkar, Binary social mimic optimization algorithm with X-shaped transfer function for feature selection, IEEE Access 8 (2020) 97890–97906.
[34] S. Mirjalili, A. Lewis, S-shaped versus V-shaped transfer functions for binary Particle Swarm Optimization, Swarm and Evolutionary Computation 9 (2013) 1–14.
[35] S. Mirjalili, H. Zhang, S. Mirjalili, S. Chalup, N. Noman, A novel u-shaped transfer function for binary particle swarm optimisation, in: Soft Computing for Problem Solving 2019: Proceedings of SocProS 2019, vol. 1, Springer, 2020, pp. 241–259.
[36] Z.W. Geem, J.H. Kim, G.V. Loganathan, A new heuristic optimization algorithm: harmony search, Simulation 76 (2001) 60–68.
[37] L. Wang, X. Fu, M.I. Menhas, M. Fei, A modified binary differential evolution algorithm, in: Life System Modeling and Intelligent Computing, Springer, 2010, pp. 49–57.
[38] Y. He, F. Zhang, S. Mirjalili, T. Zhang, Novel binary differential evolution algorithm based on taper-shaped transfer functions for binary optimization problems, Swarm and Evolutionary Computation 69 (2022) 10102.

Chapter 44

A survey of different Whale Optimization Algorithm applications in water engineering and management

Yashar Dadrasajirlou and Hojat Karami
Civil Engineering, Semnan University, Semnan, Iran

44.1 Application of WOA in lake water level (LWL) modeling

A lake refers to a space that contains a large but limited volume of water. Another feature of the lake is that it is surrounded by land [1]. However, ponds and very small spaces in terms of volume cannot be included in this classification. On the other hand, it is possible to build a lake, in other words, in order to provide water, electricity, agricultural purposes, and recreational purposes, humans collect water in a place that is called an artificial lake [2]. Knowing the level of the lake is essential for water supply and planning for water distribution. This information (Fig. 44.1) is generally closely related to time series related to hydrological parameters such as rainfall, volume of runoff entering the lake over time, evaporation rate, temperature changes [3]. So far, the use of intelligent data-related methods has shown acceptable results in the estimation of LWL fluctuations.

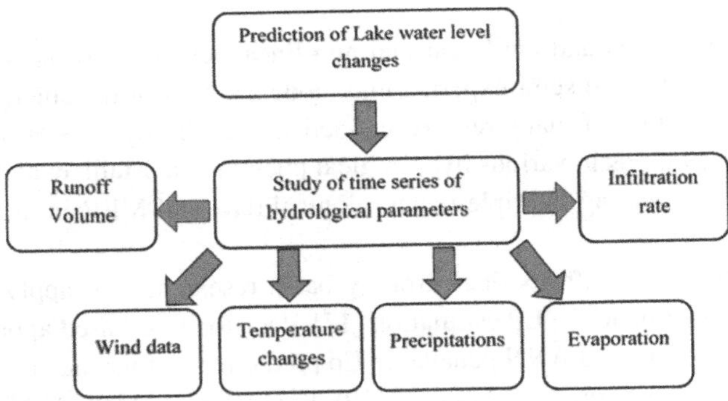

FIGURE 44.1 Parameters required for lake water prediction level.

Some studies have been conducted during last 20 years to integrate the intelligent methods for better understanding of LWL fluctuations. The investigation of water level of the Caspian Lake in a monthly period by Vaziri [4] using ANN approach is among them. Altunkaynak [5] has also used ANN in the simulation of LWL in Van, Turkey. In [6], Talebizadeh and Meridenjad studied the LWL of Urmia Lake in Iran using ANN and ANFIS. Aytek et al. [7] used the GEP to model the changes in the LWL of Van Lake in Turkey. Many other smart models were considered in LWL studies. For example, using chaos models, SARIMA, and MLR for LWL modeling [9,11], RF model, SVM [8], ANLR, DT, FFNN, GRNN, and RBFNN [10].

Nevertheless, water science researchers are always trying to find newer, stronger and more reliable methods of intelligent methods so that they can more accurately simulate the fluctuations of the water level of lakes. These efforts are related to the comprehensiveness of the variety of influential data in this modeling. Because the lack of stability in the water level of the lake is closely related to various hydrological (water discharge in the lake), climatic (evaporation, rainfall) and morphological (infiltration and bed changes) processes. Therefore, it is important to improve the development of more efficient intelligent methods in the studies of the above cases. Despite the proper efficiency of intelligent models in

Handbook of Whale Optimization Algorithm. https://doi.org/10.1016/B978-0-32-395365-8.00050-6

simulations related to LWL changes, their inability to optimize their calculation parameters [12] has caused researchers to solve this problem by using evolutionary optimization algorithms [13]. One of these evolutionary optimization algorithms is the WOA [14]. By creating a powerful training optimization space in the MLP model, the WOA has been able to increase its effectiveness in predicting the LWL. WOA has done this by speeding up convergence and breaking out of local optimal stagnation. Also, in the mentioned research, the performance of MLP-WOA is compared with other smart methods, such as CCNN model, SOMNN, DTR, RFR which showed the positive effect of using this optimization algorithm.

44.2 Application of WOA in pan evaporation estimation

Due to the high importance of evaporation estimation studies, researchers have invented several methods, which can be mentioned as water balance method, energy balance method, mass transfer method, Penman method and evaporation pan method (Fig. 44.2).

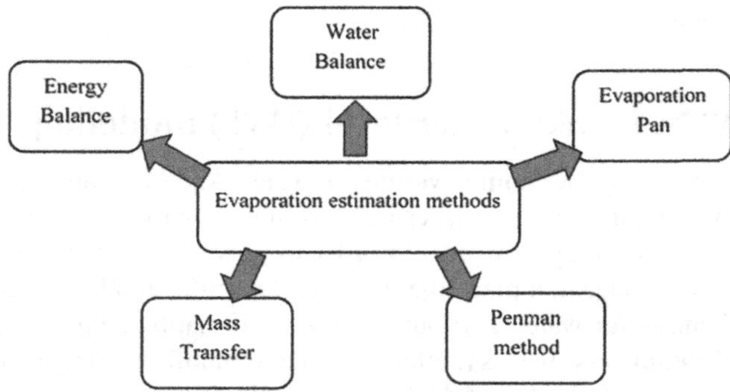

FIGURE 44.2 Evaporation estimation methods.

But the multiplicity of these variables and the linear and non-linear relationships between them have not stopped researchers from producing experimental and semi-experimental models [15]. But the emergence of intelligent data mining methods was able to attract the attention of many researchers active in hydrology, because several studies have been observed in the application of these methods in various hydrological phenomena. Malik et al. [16] evaluated the performance of RBFNN and SOMNN models by having multiple meteorological data and MLR in daily Ep estimation of Pantnagar, India.

Several examples can be given in the efforts of hydrology basin researchers to apply smart methods in evaporation estimation. For example, ANN performance in EP estimation [17], decision tree-based approaches to calculate daily Ep in Poyang Lake [18], Evaluation of ANFIS and ANN penatsil in Ep prediction in four meteorological stations in two cases of presence and absence of sufficient data to predict Milky Ep [19], Evaluation of M5T, MARS and MLP models in Ep estimation [20], DNN performance in Ep prediction [21], mentioned. Furthermore, to increase the ability of these algorithms in estimating the amount of Ep, researchers took the approach of using intelligent models with data pre-processing procedure [22,23].

However, most data mining models can hardly adjust the model parameters by themselves. But this method could not help in the inefficiency of data-mining methods in optimizing modeling parameters. Therefore, to solve this problem, the use of meta-heuristic algorithms became common, because these algorithms showed the ability to optimize and solve the parameters in data mining methods in Ep prediction. One of the newest meta-heuristic algorithms used to increase the speed of solving data mining methods has been introduced, the WOA, which was introduced by Mirjalili and Lewis in [14]. Various studies have been conducted to study the performance of this optimization algorithm in the presence of data mining methods. Wu et al. [24] combined WOA with ELM in order to predict the monthly amount of Ep and compared the obtained results with DE-ELM, M5T and ANN. Azar et al. [25] also estimated the daily evaporation from the reservoir of Qaleh Chay Ajab Shir dam by combining the WOA with ANN. His study showed that the WOA improves ANN estimation capability of the daily evaporation rate. Based on above investigation, Table 44.1 illustrates some of the positive and negative point in both experimental and intelligent methods for evaporation estimation.

TABLE 44.1 Pros & Cons for Experimental and Intelligent evaporation estimation methods.

	Experimental Method	Intelligent Method
Positive	• Having well examined in different climate conditions	• Fast & Cheap • Capable of establishing complicated relations between evaporation parameters
Negative	• Time & Money Consuming • using more parameters with complicated relationships	• Need for optimizing the algorithm parameters using optimization methods such as WOA

44.3 Application of WOA in modeling reference evapotranspiration

The contribution of evapotranspiration (ET) to the hydrological cycle is significant, so that more than 60% of the world's precipitation returns to the atmosphere in this way [26] and this shows the importance of evaporation and transpiration [27]. Accurate calculation of evaporation and transpiration rate of individual plants is challenging in terms of complexity and cost. For this reason, reference evaporation and transpiration (ET0) is usually calculated and attributed as plant evaporation and transpiration by applying coefficients. Researchers have done many researches in order to invent low-cost and practical methods for accurate estimation of ET0. These attempts to introduce experimental methods have faced problems such as the lack of universality of the proposed method due to the high variability of ET0 and the lack of sufficient data. For this reason, the acceptance of AI methods was considered due to its non-linear nature and ability to simulate the complex behavior of ET0. For example, using GEP in six regions of Burkina Faso [28], using ANFIS and ANN to model monthly mean ET0 at 275 stations in Turkey [29], using ANFIS-GP, ANFIS-SC, ANN and GEP in modeling monthly mean TE0 in 50 stations in Iran [30], using ANN and GEP to model daily ET0 in 19 stations in Saudi Arabia [31], using ANN, LSSVM, and ELM to calculate weekly ET0 in Jodhpur and Pali, India [32], using RF and GRNN models in daily ET0 estimation for 2 meteorological stations in China [33] using ANN to model daily ET0 at Aminheto meteorological station in Greece [15], using MARS, SVM and GEP to estimate monthly ET0 in 44 station in Iran [34], using MARS and GEP to estimate daily ET0 in six stations with different climates in Iran [34], using M5P, SVR and RF to calculate ET0 in Florida, America [35], using FNNN, ANFIS and SVR to model ET0 in 14 stations in Iran, Iraq, Turkey, Libya and Cyprus [36].

These numerous studies showed the existence of problems in the computational procedure of AI approaches in ET0 modeling, so attention was paid to the use of optimization algorithms in developing intelligent mode methods. Based on this, several studies have been conducted to improve and improve the performance of common AI methods in estimating ET0, by combining them with optimization algorithms. One of the most successful of these algorithms is the WOA [14]. For example, Mohammadi and Mehdizadeh [37] were able to improve the simulation capacity of ET0 in the SVR algorithm by combining the whale algorithm in SVR training. In this study, 3 stations in arid (Isfahan), semi-arid (Urmia) and very arid (Yazd) climates were investigated. Tikhamarine et al. [38] combined WOA with ANN to calculate the monthly ET0 in two stations RANICHAURI (India) and El Beida station (Algeria). Wu et al. [39] calculated daily ET0 in different regions of China with diverse climates by optimizing experimental models using WOA. Gao et al. [40] combined WOA and ANN to calculate the parameters of ANN in order to estimate ET0 in 6 stations in Loess Plateau of north China. Zhao et al. [41] by optimizing the parameters of the ELMAN prediction model, using WOA, calculated ET0 in different regions of China with different climates. Ye et al. [42] used a combination of WOA in DENFIS and MARS to calculate daily ET. Fig. 44.3 briefly explains this section.

44.4 Application of WOA in rainfall & runoff modeling estimation

Runoff estimation is another important part oh hydrological studies. In addition, time series analysis and rainfall modeling also play an important role in scientific studies. This importance is clearly evident in the wide application of runoff in the urban drinking water supply and agricultural water supply sectors [43]. Having sufficient knowledge and information for accurate modeling of runoff and precipitation is very essential [44]. Like other hydrological phenomena, predicting the amount of precipitation presents many challenges [45]. There are a number of conducted studies about runoff prediction. For example, the use of ANN and LSTM [46], the comparison of ENN and ANN in runoff modeling [47] Feasibility of using ANN to simulate runoff in 11 basins [48]. Also, common intelligent methods are used to predict rainfall. Such as ANN, FL, GP, SVR [12]. However, their results show that ML models are not successful in predicting rainfall over long periods of time

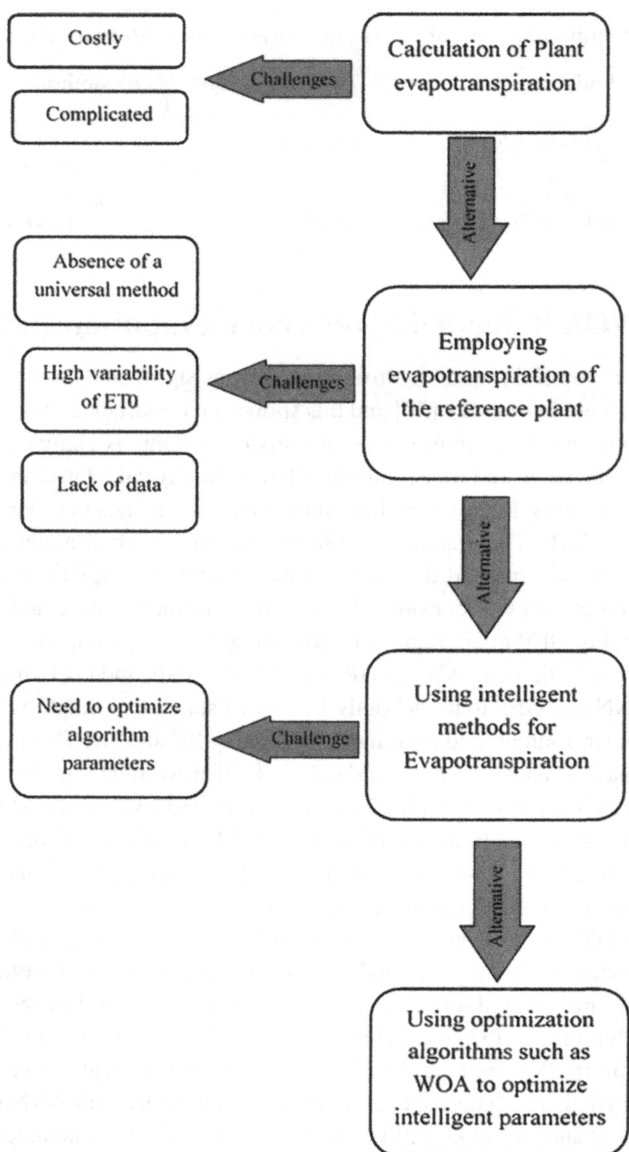

FIGURE 44.3 Application of WOA in evapotranspiration estimation.

[49]. Therefore, the combination of intelligent methods was used in rainfall forecasting. For example, using WAVELET-SVR and WAVELET-ANN [50], combining GA and ANN (Solgi et al., 2014), using wavelet-ANN [51]. However, the researchers paid attention to the approach of using optimization algorithms to increase the accuracy of long-term rainfall forecast. Diop et al. [49] by combining MLP and WOA has made an annual forecast of rainfall. In another study (0028) by combining smart methods (SVM-WOA MLP-WOA) with conceptual methods (IHACRES, GR4J, and MISD) they tried to introduce an efficient model in predicting rainfall-runoff. The presence of the WOA in the MLP structure showed that the combination of this smart calculator with the IHACRES tool shows a good ability in rainfall-runoff modeling. Valikhan Anaraki [52] used LSSVM-WOA to estimate the amount of precipitation in six stations of Karun3 basin under climate change conditions. Farpour and KhozeymeNezhad [53] used the WOA to calibrate the parameters of Hymod's rainfall-runoff model and concluded that the WOA is highly efficient [54]. Lv et al. [55] improved the accuracy of runoff calculated using SWAT and ANN methods using WOA. Fig. 44.4 explains the process of survey in using application of WOA in rainfall-runoff estimation.

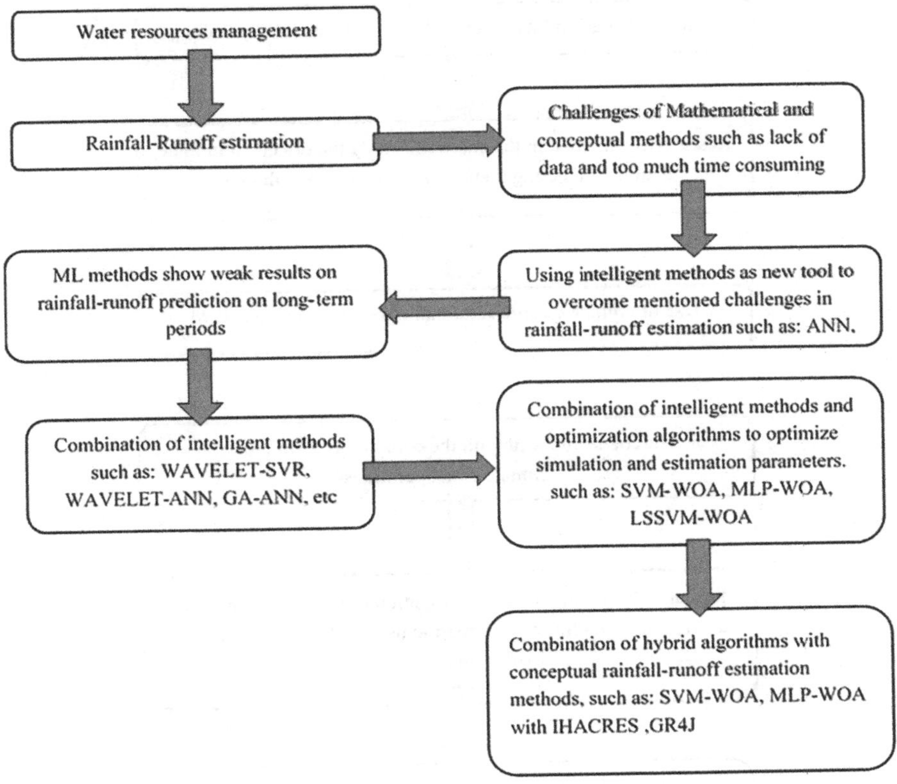

FIGURE 44.4 The flowchart of rainfall-runoff estimation.

44.5 Application of WOA in flood frequency analysis and daily water level

Undoubtedly, global warming and its severe effects on the hydrology of the planet are not hidden from anyone. These tremendous effects have led to the approach of hydrology basin researchers to adapt in the future and deal with extreme events [56]. Global climate models (GCM) are among the most well-known methods of predicting climate change. However, due to the universality of this model's performance, it is necessary to use the spatial and temporal microscales of GCM information in regional studies [57]. Both dynamic and statistical approaches are the most common downscaling measures. Dynamic downscaling methods use high-resolution regional climate models (RCMs). The basis of scale reduction in the statistical method is based on the relationship between large-scale climatic features and small-scale hydrological features [58]. In recent years, researchers have investigated the performance of smart methods in reducing the scale of meteorological variables. For example, using SSVM and ANN [59] concluded that SSVM has a good performance in downscaling precipitation. Using ANN, KNN, and SVM in order to determine the scale of rainfall in Zayandeh Rood Dam Basin [60], using LSSVM for downscaling variations in precipitation, temperature, and reference evapotranspiration [61], comparing the performance of SSVM and the performance of the statistical downscaling method (SDSM) in reducing the scale of temperature and daily precipitation [56] are such researches.

Also, several types of research have been conducted on the application of intelligent algorithms in flow simulation. For example, predicting river flow based on data extracted from the HadCM3 model and using SVM, GEP, and ANN [62]. Prediction of the daily and monthly flow of four rivers using ANN, SVR, and ELM are part of [63] research. The mentioned researches show the proper performance of smart methods in reducing the scale of hydrological phenomena, but adjusting the parameters of smart algorithms is highly effective in their performance. For this purpose, the use of meta-engineering algorithms with the approach of optimizing the parameters of smart algorithms has been suggested [64]. One of the algorithms used to adjust the factors of smart methods is the WOA. Du et al. [65] have used WOA in setting LSSVM parameters and Bui et al. [66] have used WOA in setting ANFIS parameters. Valikhan Anaraki et al. [67] analyzed flood frequency considering climate change impact, by using data pre-processing, ML, and WOA. The purpose of using WOA is to adjust the parameters of LSVM for more accurate reducing the scale of precipitation and temperature in the flood simulation process. The sequence of what has been written above is briefly indicated in Fig. 44.5.

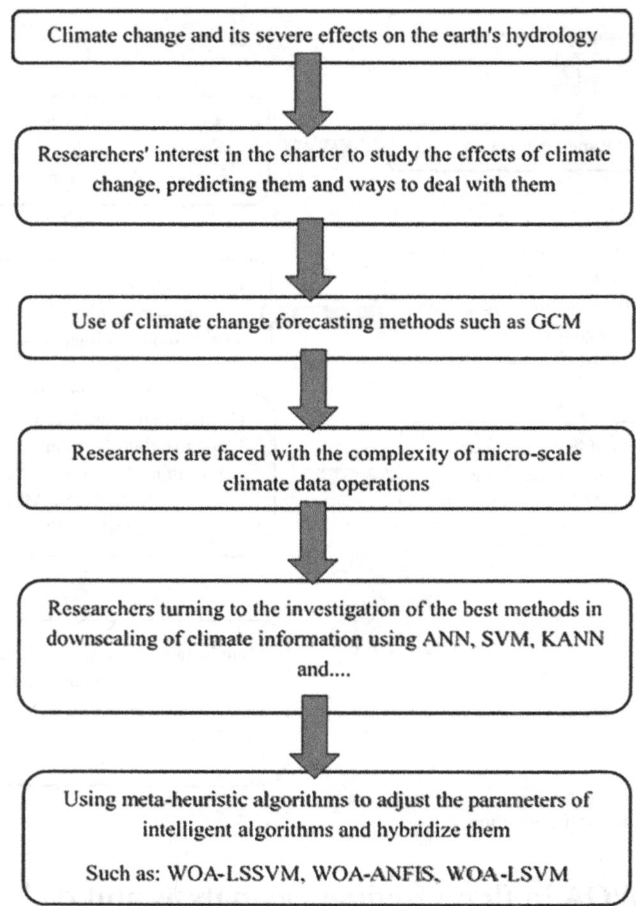

FIGURE 44.5 Using WOA in regulating the Intelligent methods' parameters for flood frequency estimation.

44.6 Application of WOA in groundwater level modeling

Modeling underground water levels and water balance situation is critical in hydrological studies of catchment. Modeling using common methods has simulation errors and uncertainty in validation, so it is necessary to use new methodologies to forecast the underground water level [68]. While conceptual methods are considered powerful measures for simulating hydrological phenomena, to increase their accuracy, there is a need for very complete information about the watershed. This leads to the ineffectiveness of these tools in all regions of the world. Therefore, researchers have proposed intelligent models in order to solve the mentioned obstacles [69]. While ANNs use limited and little information for simulation, the powerful performance of ANN in simulation has been confirmed [70]. The high ability to simulate the groundwater level (GWL) has been proven in several studies [71,72]. The network structure, algorithm parameters, and training method are determining factors of the quality of simulation. Therefore, it is recommended to used optimization to increase the simulation accuracy [73–75]. One of the main problems of ANN is the low rate of convergence and falling into local minima, and to solve this problem, the use of meta-heuristic algorithms is recommended [76,77]. WOA is one of the meta-heuristic algorithms that has shown good results in solving optimization problems. [14,78,79]. The proper performance of WOA in optimizing ANN, ANFIS, and SVR parameters has been proven [80,81]. The combination of WOA and ANN has shown good performance in simulating the phenomenon of evaporation [82]. In [83], the combination of WOA and ANN optimization algorithm was used to study the underground water level of the aquifer. The obtained results showed that the application of the WOA algorithm was able to simulate the ability of ANN in simulating the underground water level of the aquifer by considering temperature, evaporation, precipitation, aquifer viscosity, and aquifer input. Fig. 44.6 briefly indicates the above information.

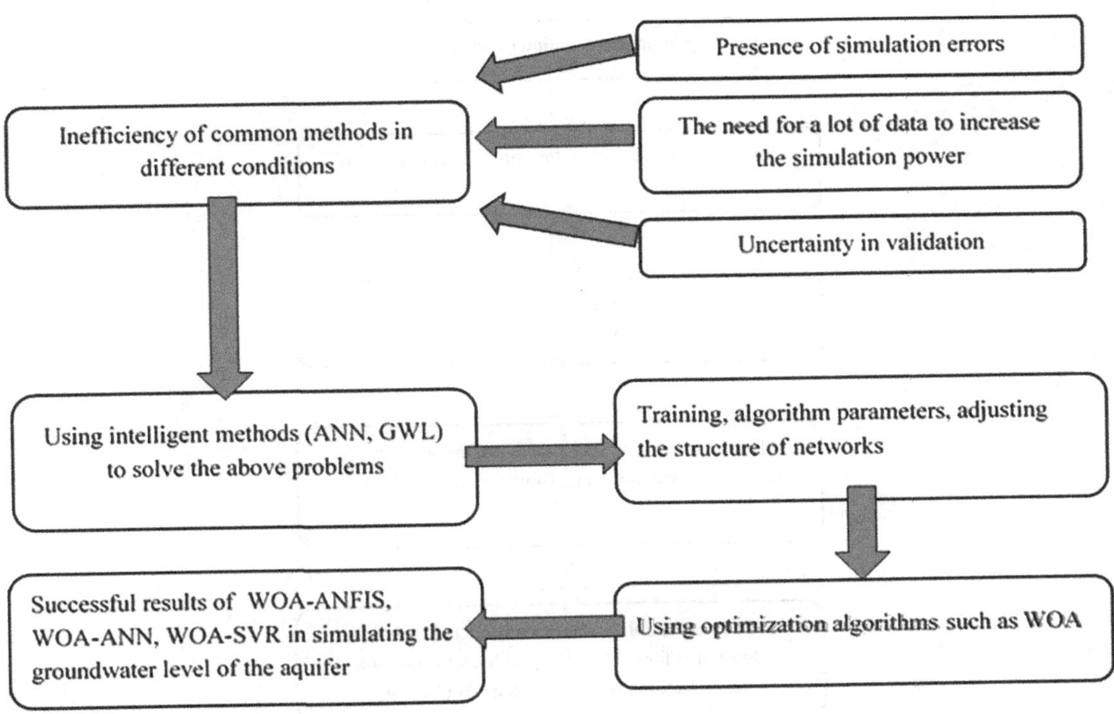

FIGURE 44.6 Using WOA in regulating the Intelligent methods' parameters for groundwater level modeling.

44.7 Application of WOA in reservoirs operation

Undoubtedly, water resource management is one of the sciences in which optimization plays an important role and this approach is used to plan and control water resources systems. But due to the existence of complex non-linear behavior in water resources, they need to apply optimal methods to achieve the work solution. One of the parts related to water resources that are usually optimized with mathematical models are reservoirs. This optimization is usually proposed as a multi-objective optimization because the optimization of the reservoir has several requirements. One of the reasons that challenge the comprehensiveness of conventional optimization methods is that the approaches are limited in their ability to solve optimization problems, they cannot have sufficient strength in getting stuck in local minima, the lack of unique objective functions, and excessive sensitivity to the starting point of the solution region and suffer from low accuracy [84]. One of the methods of reservoir optimization is the use of dynamic programming, which was used in the study of Ben Alaya et al. [85] to optimize the reservoir. However, using this method in multiple reservoir optimization problems is considered a fundamental challenge. Therefore, creating alternative and efficient methods in response to multi-dimensional problems in the reservoir is completely felt. For this reason, meta-heuristic algorithms have been introduced to optimize a wide range of problems. In recent decades, meta-heuristic algorithms have been modeled on natural systems. These algorithms have been used in extensive studies in the optimization of male reservoirs [86,87]. Ashofteh et al. [88] used genetic programming to optimize reservoir activity and showed that it outperforms GAs in meeting downstream demand. Amirkhani et al. [89] presented a method to optimize the activity of Karaj reservoir in Iran using NSGA-II. Lai et al. [90] With the combination of LEVY and WOA, they were able to optimize the operation system of the KLANG Gate dam reservoir. Mohammadi et al. [91] optimized the operation problems of Salman Farsi dam reservoir by using the combination of WOA and genetic algorithm and considering the multi-objective approach. Donyaii et al. [92] based on the non-linear programming model solution, they used the whale algorithm to optimize the operation of the reservoir dam of Golestan Dam. Fig. 44.7 briefly illustrates the process mentioned above.

44.8 List of abbreviations

WOA Whale Optimization Algorithm
LWL Lake Water Level

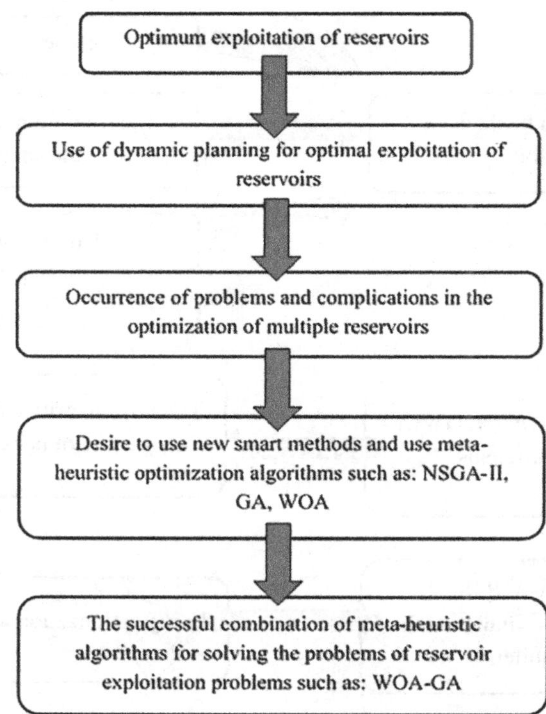

FIGURE 44.7 Using WOA in regulating the Intelligent methods' parameters for reservoir operation.

ANN Artificial Neural Network
ANFIS Adaptive Neuro Fuzzy Inference System
GEP Gene Expression Programming
SARIMA Seasonal Autoregressive Integrated Moving Average
MLR Multiple Linear Regression
RF Random forest
SVM Support Vector Machine
ANLR Additive Non-Linear Regression
DT Decision Tree
FFNN Feed-Forward Neural Networks
GRNN Generalized Regression Neural Networks
RBFNN Radial Basis Function Neural Networks
CCNN Cascade Correlation Neural Network
SOMNN Self-Organizing Map Neural Network
DTR Decision Tree Regression
RFR Random Forest Regression
MARS Multivariate Adaptive Regression Spline
DNN Deep Neural Network
DENFIS Dynamic Evolving Neural-Fuzzy Inference System
ENN Emotional Neural Network
FL Fuzzy Logic
GP Genetic Programming
SSVM Smooth Support Vector Machine
ML Machine Learning

References

[1] G.J. Van Geest, H. Wolters, F.C.J.M. Roozen, et al., Water-level fluctuations affect macrophyte richness in floodplain lakes, in: Hydrobiologia, 2005.

[2] P.A. Gantzer, L.D. Bryant, J.C. Little, Lake and reservoir management, Water Environ. Res. 81 (2009) 1854–1956, https://doi.org/10.2175/106143009X12445568400494.

[3] B. Yadav, K. Eliza, A hybrid wavelet-support vector machine model for prediction of lake water level fluctuations using hydrometeorological data, Measurement 103 (2017) 294–301, https://doi.org/10.1016/j.measurement.2017.03.003.

[4] M. Vaziri, Predicting Caspian Sea surface water level by ANN and ARIMA models, J. Waterw. Port Coast. Ocean Eng. 123 (1997) 158–162.

[5] A. Altunkaynak, Forecasting surface water level fluctuations of lake van by artificial neural networks, Water Resour. Manag. 21 (2007) 399–408, https://doi.org/10.1007/s11269-006-9022-6.

[6] M. Talebizadeh, A. Moridnejad, Uncertainty analysis for the forecast of lake level fluctuations using ensembles of ANN and ANFIS models, Expert Syst. Appl. 38 (2011) 4126–4135, https://doi.org/10.1016/j.eswa.2010.09.075.

[7] A. Aytek, O. Kisi, A. Guven, A genetic programming technique for lake level modeling, Hydrol. Res. 45 (2014) 529–539, https://doi.org/10.2166/nh.2013.069.

[8] B. Li, G. Yang, R. Wan, X. Dai, Y. Zhang, Comparision of random forests and other statistical methods for the prediction of lake water level: a case study of the Poyang Lake in China, Hydrol. Res. 47 (S1) (2016) 69–83, https://doi.org/10.2166/nh.2016.264.

[9] C-C. Young, W-C. Liu, W-L. Hsieh, Predicting the water level fluctuation in an Alpine lake using physically based, artificial neural network, and time series forecasting models, Math. Probl. Eng. 2015 (2015) 1–11, https://doi.org/10.1155/2015/708204.

[10] B. Vaheddoost, H. Aksoy, H. Abghari, Prediction of water level using monthly lagged data in Lake Urmia, Iran, Water Resour. Manag. 30 (2016) 4951–4967, https://doi.org/10.1007/s11269-016-1463-y.

[11] R. Khatibi, M.A. Ghorbani, L. Naghipour, V. Jothiprakash, T.A. Fathima, M.H. Fazelifard, Inter-comparison of time series models of lake levels predicted by several modeling strategies, J. Hydrol. 511 (2014) 530–545, https://doi.org/10.1016/j.jhydrol.2014.01.009.

[12] Z.M. Yaseen, I. Ebtehaj, H. Bonakdari, et al., Novel approach for streamflow forecasting using a hybrid ANFIS-FFA model, J. Hydrol. 554 (2017) 263–276, https://doi.org/10.1016/j.jhydrol.2017.09.007.

[13] Z.M. Yaseen, R.C. Deo, I. Ebtehaj, H. Bonakdari, Hybrid data intelligent models and applications for water level prediction, in: Handbook of Research on Predictive Modeling and Optimization Methods in Science and Engineering, 2018.

[14] S. Mirjalili, A. Lewis, The whale optimization algorithm, Adv. Eng. Softw. 95 (2016) 51–67, https://doi.org/10.1016/j.advengsoft.2016.01.008.

[15] V.Z. Antonopoulos, A.V. Antonopoulos, Daily reference evapotranspiration estimates by artificial neural networks technique and empirical equations using limited input climate variables, Comput. Electron. Agric. 132 (2017) 86–96, https://doi.org/10.1016/j.compag.2016.11.011.

[16] A. Malik, A. Kumar, O. Kisi, Daily pan evaporation estimation using heuristic methods with gamma test, J. Irrig. Drain. Eng. 144 (9) (2018) 04018023.

[17] A. Jain, T. Roy, Evaporation modelling using neural networks for assessing the self-sustainability of a water body, Lakes Reserv. Res. Manag. 22 (2) (2017) 123–133.

[18] X. Lu, Y. Ju, L. Wu, J. Fan, F. Zhang, Z. Li, Daily pan evaporation modeling from local and cross-station data using three tree-based machine learning models, J. Hydrol. 566 (2018) 668–684.

[19] J. Shiri, P. Marti, S. Karimi, G. Landeras, Data splitting strategies for improving data driven models for reference evapotranspiration estimation among similar stations, Comput. Electron. Agric. 162 (2019) 70–81.

[20] O. Kisi, S. Heddam, Evaporation modelling by heuristic regression approaches using only temperature data, Hydrol. Sci. J. 64 (6) (2019) 653–672.

[21] B. Majhi, D. Naidu, A.P. Mishra, S.C. Satapathy, Improved prediction of daily pan evaporation using Deep-LSTM model, Neural Comput. Appl. 32 (2020) 7823–7838.

[22] S.N. Qasem, S. Samadianfard, S. Kheshtgar, S. Jarhan, O. Kisi, S. Shamshirband, K.W. Chau, Modeling monthly pan evaporation using wavelet support vector regression and wavelet artificial neural networks in arid and humid climates, Eng. Appl. Comput. Fluid Mech. 13 (1) (2019) 177–187.

[23] L. Pammar, P.C. Deka, Daily pan evaporation modeling in climatically contrasting zones with hybridization of wavelet transform and support vector machines, Paddy Water Environ. 15 (4) (2017) 711–722.

[24] L. Wu, G. Huang, J. Fan, X. Ma, H. Zhou, W. Zeng, Hybrid extreme learning machine with meta-heuristic algorithms for monthly pan evaporation prediction, Comput. Electron. Agric. 168 (2020) 105115, https://doi.org/10.1016/j.compag.2019.105115.

[25] N.A. Azar, N. Kardan, S.G. Milan, Developing the artificial neural network–evolutionary algorithms hybrid models (ANN–EA) to predict the daily evaporation from dam reservoirs, Eng. Comput. 39 (2023) 1375–1395, https://doi.org/10.1007/s00366-021-01523-3.

[26] S. Wang, J. Lian, Y. Peng, B. Hu, H. Chen, Generalized reference evapotranspiration models with limited climatic data based on random forest and gene expression programming in Guangxi, China, J. Hydrol. 221 (2019) 220–230, https://doi.org/10.1016/j.agwat.2019.03.027.

[27] S. Mehdizadeh, H. Saadatnejadgharahassanlou, J. Behmanesh, Calibration of Hargreaves–Samani and Priestley–Taylor equations in estimating reference evapotranspiration in the Northwest of Iran, Arch. Agron. Soil Sci. 63 (7) (2017) 942–955, https://doi.org/10.1080/03650340.2016.1249474.

[28] S. Traore, A. Guven, Regional-specific numerical models of evapotranspiration using gene-expression programming interface in Sahel, Water Resour. Manag. 26 (15) (2012) 4367–4380, https://doi.org/10.1007/s11269-012-0149-3.

[29] H. Citakoglu, M. Cobaner, T. Haktanir, O. Kisi, Estimation of monthly mean reference evapotranspiration in Turkey, Water Resour. Manag. 28 (1) (2014) 99–113, https://doi.org/10.1007/s11269-013-0474-1.

[30] O. Kisi, H. Sanikhani, M. Zounemat-Kermani, F. Niazi, Long-term monthly evapotranspiration modeling by several data-driven methods without climatic data, Comput. Electron. Agric. 115 (2015) 66–77, https://doi.org/10.1016/j.compag.2015.04.015.

[31] M.A. Yassin, A.A. Alazba, M.A. Mattar, Artificial neural networks versus gene expression programming for estimating reference evapotranspiration in arid climate, Agric. Water Manag. 163 (2016) 110–124, https://doi.org/10.1016/j.agwat.2015.09.009.

[32] A.P. Patil, P.C. Deka, An extreme learning machine approach for modeling evapotranspiration using extrinsic inputs, Comput. Electron. Agric. 121 (2016) 385–392, https://doi.org/10.1016/j.compag.2016.01.016.

[33] Y. Feng, N. Cui, D. Gong, Q. Zhang, L. Zhao, Evaluation of random forests and generalized regression neural networks for daily reference evapotranspiration modeling, Agric. Water Manag. 193 (2017) 163–173, https://doi.org/10.1016/j.agwat.2017.08.003.

[34] S. Mehdizadeh, Estimation of daily reference evapotranspiration (ET0) using artificial intelligence methods: offering a new approach for lagged ET0 data-based modeling, J. Hydrol. 559 (2018) 794–812, https://doi.org/10.1016/j.jhydrol.2018.02.060.

[35] F. Granata, Evapotranspiration evaluation models based on machine learning algorithms – a comparative study, Agric. Water Manag. 217 (2019) 303–315, https://doi.org/10.1016/j.agwat.2019.03.015.

[36] V. Nourani, G. Elkiran, J. Abdullahi, Multi-station artificial intelligence based ensemble modeling of reference evapotranspiration using pan evaporation measurements, J. Hydrol. 577 (2019) 123958, https://doi.org/10.1016/j.jhydrol.2019.123958.

[37] B. Mohammadi, S. Mehdizadeh, Modeling daily reference evapotranspiration via a novel approach based on support vector regression coupled with whale optimization algorithm, Agric. Water Manag. 237 (2020) 106145, https://doi.org/10.1016/j.agwat.2020.106145.

[38] Y. Tikhamarine, A. Malik, A. Kumar, D. Souag-Gamane, O. Kisi, Estimation of monthly r3eference evapotranspiration using novel hybrid machine learning approaches, Hydrol. Sci. J. 64 (15) (2019) 1824–1842, https://doi.org/10.1080/02626667.2019.1678750.

[39] Z. Wu, X. Chen, N. Cui, B. Zhu, D. Gong, L. Han, L. Xing, S. Zhen, Q. Li, Q. Liu, P. Fang, Optimized empirical model based on whale optimization algorithm for simulate daily reference crop evapotranspiration in different climatic regions of China, J. Hydrol. 612 (part A) (2022) 128084, https://doi.org/10.1016/j.jhydrol.2022.128084.

[40] L. Gao, D. Gong, N. Cui, M. Lv, Y. Feng, Evaluation of bio-inspired optimization algorithms hybrid with artificial neural network for reference crop evapotranspiration estimation, Comput. Electron. Agric. 190 (2021) 106466, https://doi.org/10.1016/j.compag.2021.106466.

[41] L. Zhao, X. Zhao, X. Pan, Y. Shi, Z. Qiu, X. Li, X. Xing, J. Bai, Prediction of daily reference crop evapotranspiration in different Chinese climate zones: combined application of key meteorological factors and Elman algorithm, J. Hydrol. (2022), https://doi.org/10.1016/j.jhydrol.2022.127822.

[42] L. Ye, M.M. Abdul Zahra, N.K. Al-Bedyry, Z.M. Yaseen, Daily scale evapotranspiration prediction over the coastal region of southwest Bangladesh: new development of artificial intelligence model, Stoch. Environ. Res. Risk Assess. 36 (2022) pages451–471, https://doi.org/10.1007/s00477-021-02055-4.

[43] P. Zhang, S.T. Ariaratnam, Life cycle cost savings analysis on traditional drainage systems from low impact development strategies, Front. Eng. Manag. 8 (2021) 88–97, https://doi.org/10.1007/s42524-020-0063-y.

[44] B. Hadid, E. Duviella, S. Lecoeuche, Data-driven modeling for river flood forecasting based on a piecewise linear ARX system identification, J. Process Control 86 (2020) 44–56.

[45] S. Cramer, M. Kampouridis, A.A. Freitas, Decomposition genetic programming: an extensive evaluation on rainfall prediction in the context of weather derivatives, Appl. Soft Comput. 70 (2018) 208–224, https://doi.org/10.1016/j.asoc.2018.05.016.

[46] H. Fan, M. Jiang, L. Xu, H. Zhu, J. Cheng, J. Jiang, Comparison of long short term memory networks and the hydrological model in runoff simulation, Water 12 (1) (2020) 175.

[47] S. Kumar, T. Roshni, D. Himayoun, A comparison of emotional neural network (ENN) and artificial neural network (ANN) approach for rainfall-runoff modelling, Civil Eng. J. 5 (10) (2019) 2120–2130.

[48] R.S. Vilanova, S.S. Zanetti, R.A. Cecflio, Assessing combinations of artificial neural networks input/output parameters to better simulate daily streamflow: case of Brazilian Atlantic Rainforest watersheds, Comput. Electron. Agric. 167 (2019) 105080.

[49] L. Diop, S. Samadianfard, A. Bodian, Z.M. Yaseen, M.A. Ghorbani, H. Salimi, Annual rainfall forecasting using hybrid artificial intelligence model: integration of multilayer perceptron with whale optimization algorithm, Water Resour. Manag. 34 (2020) 733–746, https://doi.org/10.1007/s11269-019-02473-8.

[50] O. Kisi, M. Cimen, Precipitation forecasting by using wavelet-support vector machine conjunction model, Eng. Appl. Artif. Intell. 25 (2012) 783–792, https://doi.org/10.1016/j.engappai.2011.11.003.

[51] A. Solgi, V. Nourani, A. Pourhaghi, Forecasting daily precipitation using hybrid model of wavelet-artificial neural network and comparison with adaptive neurofuzzy inference system (case study: Verayneh Station, Nahavand), Adv. Civ. Eng. 2014 (2014) 1–12, https://doi.org/10.1155/2014/279368.

[52] M. Valikhan Anaraki, S. Farzin, S.-F. Mousavi, H. Karami, Application of hybrid least square support vector machine-whale optimization algorithm (LSSVM-WOA) for downscaling and prediction of precipitation under climate change (case study: Karun3 basin), J. Irrig. Water Eng. 11 (3) (2021) 252–271, https://doi.org/10.22125/IWE.2021.128204.

[53] A. Farpour, H. KhozeymeNezhad, Improving the performance of the Hymod Model using the Whale Optimization Algorithm, Watershed Eng. Manag. 14 (3) (2021) 376–386, https://doi.org/10.22092/IJWMSE.2021.352418.1850.

[54] X. Yuan, C. Chen, Y. Yuan, B. Zhang, Runoff prediction based on hybrid clustering with WOA intervals mapping model, J. Hydrol. Eng. 26 (6) (2021) 04021019, https://doi.org/10.1061/(ASCE)HE.1943-5584.0002087.

[55] Z. Lv, J. Zou, D. Rodriguez, Predicting of runoff using an optimized SWAT-ANN: A case study, J. Hydrol. Reg. Stud. 26 (2020) 100688, https://doi.org/10.1016/j.ejrh.2020.100688.

[56] M. Azmat, M.U. Qamar, S. Ahmed, M.A. Shahid, E. Hussain, S. Ahmad, R.A. Khushnood, Ensembling downscaling techniques and multiple GCMs to improve climate change predictions in cryosphere scarcely-gauged catchment, Water Resour. Manag. 32 (2018) 3155–3174, https://doi.org/10.1007/s11269-018-1982-9.

[57] S. Tripathi, V.V. Srinivas, R.S. Nanjundiah, Downscaling of precipitation for climate change scenarios: a support vector machine approach, J. Hydrol. 330 (2006) 621–640, https://doi.org/10.1016/j.jhydrol.2006.04.030.

[58] S. Ghosh, SVM-PGSL coupled approach for statistical downscaling to predict rainfall from GCM output, J. Geophys. Res. 115 (2010) D22102, https://doi.org/10.1029/2009JD013548.

[59] H. Chen, J. Guo, W. Xiong, S. Guo, C.Y. Xu, Downscaling GCMs using the smooth support vector machine method to predict daily precipitation in the Hanjiang Basin, Adv. Atmos. Sci. 27 (2010) 274–284, https://doi.org/10.1007/s00376-009-8071-1.

[60] A. Ahmadi, A. Moridi, E.K. Lafdani, G. Kianpisheh, Assessment of climate change impacts on rainfall using large scale climate variables and downscaling models – a case study, J. Earth Syst. Sci. 123 (2014) 1603–1618, https://doi.org/10.1007/s12040-014-0497-x.

[61] S. Kundu, D. Khare, A. Mondal, Future changes in rainfall, temperature and reference evapotranspiration in the Central India by least square support vector machine, Geosci. Front. 8 (2017) 583–596, https://doi.org/10.1016/j.gsf.2016.06.002.

[62] P. Sarzaeim, O. Bozorg-Haddad, A. Bozorgi, H.A. Loáiciga, Runoff projection under climate change conditions with data-mining methods, J. Irrig. Drain. Eng. 143 (2017) 04017026, https://doi.org/10.1061/(ASCE)IR.1943-4774.0001205.

[63] P. Parisouj, H. Mohebzadeh, T. Lee, Employing machine learning algorithms for streamflow prediction: a case study of four river basins with different climatic zones in the United States, Water Resour. Manag. 34 (2020) 4113–4131, https://doi.org/10.1007/s11269-020-02659-5.

[64] A. Tharwat, A.E. Hassanien, Chaotic antlion algorithm for parameter optimization of support vector machine, Appl. Intell. 48 (2018) 670–686, https://doi.org/10.1007/s10489-017-0994-0.

[65] P. Du, J. Wang, W. Yang, T. Niu, Multi-step ahead forecasting in electrical power system using a hybrid forecasting system, Renew. Energy 122 (2018) 533–550, https://doi.org/10.1016/j.renene.2018.01.113.

[66] Q.T. Bui, M. Van Pham, Q.H. Nguyen, L.X. Nguyen, H.M. Pham, Whale optimization algorithm and adaptive neuro-fuzzy inference system: a hybrid method for feature selection and land pattern classification, Int. J. Remote Sens. 40 (2019) 5078–5093, https://doi.org/10.1080/01431161.2019.1578000.

[67] M. Valikhan Anaraki, S. Farzin, S-F. Mousavi, H. Karami, Uncertainty analysis of climate change impacts on flood frequency by using hybrid machine learning methods, Water Resour. Manag. 35 (2021) 199–233, https://doi.org/10.1007/s11269-020-02719-w.

[68] S.G. Milan, A. Roozbahani, M.E. Banihabib, Fuzzy optimization model and fuzzy inference system for conjunctive use of surface and groundwater resources, J. Hydrol. 566 (2018) 421–434.

[69] H. Kardan Moghaddam, H. Kardan Moghaddam, Z.R. Kivi, M. Bahreinimotlagh, M.J. Alizadeh, Developing comparative mathematic models, BN and ANN for forecasting of groundwater levels, Groundwater Sustain. Dev. 9 (2019) 100237.

[70] V.H. Nhu, A. Shirzadi, H. Shahabi, S.K. Singh, N. Al-Ansari, J.J. Clague, A. Jaafari, W. Chen, S. Miraki, J. Dou, C. Luu, Shallow landslide susceptibility mapping: a comparison between logistic model tree, logistic regression, Naïve Bayes tree, artificial neural network, and support vector machine algorithms, Int. J. Environ. Res. Public Health 17 (8) (2020) 2749.

[71] A. Mirarabi, H.R. Nassery, M. Nakhaei, J. Adamowski, A.H. Akbarzadeh, F. Alijani, Evaluation of data-driven models (SVR and ANN) for groundwater-level prediction in confined and unconfined systems, Environ. Earth Sci. 78 (15) (2019) 489.

[72] P.T. Nguyen, D.H. Ha, A. Jaafari, H.D. Nguyen, T. Van Phong, N. Al-Ansari, I. Prakash, H.V. Le, B.T. Pham, Groundwater potential mapping combining artificial neural network and real AdaBoost ensemble technique: the DakNong province case-study, Vietnam, Int. J. Environ. Res. Public Health 17 (7) (2020) 2473.

[73] F.B. Banadkooki, M. Ehteram, A.N. Ahmed, F.Y. Teo, C.M. Fai, H.A. Afan, M. Sapitang, A. El-Shafie, Enhancement of groundwater-level prediction using an integrated machine learning model optimized by whale algorithm, Nat. Resour. Res. 29 (5) (2020) 3233–3252.

[74] O.H. Kombo, S. Kumaran, Y.H. Sheikh, A. Bovim, K. Jayavel, Long-term groundwater level prediction model based on hybrid KNN-RF technique, Hydrology 7 (3) (2020) 59.

[75] S. Maroufpoor, O. Bozorg-Haddad, E. Maroufpoor, Reference evapotranspiration estimating based on optimal input combination and hybrid artificial intelligent model: hybridization of artificial neural network with grey wolf optimizer algorithm, J. Hydrol. 588 (2020) 125060.

[76] K. Asefpour Vakilian, Machine learning improves our knowledge about miRNA functions towards plant abiotic stresses, Sci. Rep. 10 (2020) 3041.

[77] E. Sarlaki, A. Sharif Paghaleh, M.H. Kianmehr, K. Asefpour Vakilian, Valorization of lignite wastes into humic acids: process optimization, energy efficiency and structural features analysis, Renew. Energy 163 (2021) 105–122.

[78] M. Abd El Aziz, A.A. Ewees, A.E. Hassanien, Whale optimization algorithm and moth-flame optimization for multilevel thresholding image segmentation, Expert Syst. Appl. 83 (2017) 242–256.

[79] Y. Ling, Y. Zhou, Q. Luo, Lévy flight trajectory-based whale optimization algorithm for global optimization, IEEE Access 5 (2017) 6168–6186.

[80] B. Mohammadi, S. Mehdizadeh, Modeling daily reference evapotranspiration via a novel approach based on support vector regression coupled with whale optimization algorithm, Agric. Water Manag. 237 (2020) 106145.

[81] B. Vaheddoost, Y. Guan, B. Mohammadi, Application of hybrid ANN-whale optimization model in evaluation of the field capacity and the permanent wilting point of the soils, Environ. Sci. Pollut. Res. Int. 27 (12) (2020) 13131–13141.

[82] A. Seifi, F. Soroush, Pan evaporation estimation and derivation of explicit optimized equations by novel hybrid meta-heuristic ANN based methods in different climates of Iran, Comput. Electron. Agric. 173 (2020) 105418.

[83] Z. Kayhomayoon, S.G. Milan, N.A. Azar, H. Kardan Moghaddam, A new approach for regional groundwater level simulation: clustring, simulation and optimization, Nat. Resour. Res. 30 (2021) 4165–4185, https://doi.org/10.1007/s11053-021-09913-6.

[84] J. Radosavljević, A solution to the combined economic and emission dispatch using hybrid PSOGSA algorithm, Appl. Artif. Intell. 30 (5) (2016) 445–474, https://doi.org/10.1080/08839514.2016.1185860.

[85] A. Ben Alaya, A. Souissi, J. Tarhouni, K. Ncib, Optimization of Nebhana reservoir water allocation by stochastic dynamic programming, Water Resour. Manag. 17 (4) (2003) 259–272, https://doi.org/10.1023/A:1024721507339.

[86] K.L. Chong, S.H. Lai, A.N. Ahmed, W.Z. Wan Zurina, R.V. Rao, M. Sherif, A. Sefelnasr, A. El-Shafie, Review on dam and reservoir optimal operation for irrigation and hydropower energy generation utilizing meta-heuristic algorithms, IEEE Access 9 (2021) 19488–19505, https://doi.org/10.1109/ACCESS.2021.3054424.

[87] K.L. Chong, S.H. Lai, A.N. Ahmed, W.Z. Wan Zurina, A. El-Shafie, Optimization of hydropower reservoir operation based on hedging policy using Jaya algorithm, Appl. Soft Comput. 106 (2021) 107325, https://doi.org/10.1016/j.asoc.2021.107325.

[88] P.-S. Ashofteh, O.B. Haddad, H. Akbari-Alashti, M.A. Mariño, Determination of irrigation allocation policy under climate change by genetic programming, J. Irrig. Drain. Eng. 141 (4) (2015) 04014059, https://doi.org/10.1061/(ASCE)IR.1943-4774.0000807.

[89] M. Amirkhani, O. Bozorg-Haddad, E. Fallah-Mehdipour, H.A. Loáiciga, Multiobjective reservoir operation for water quality optimization, J. Irrig. Drain. Eng. 142 (12) (2016) 04016065, https://doi.org/10.1061/(ASCE)IR.1943-4774.0001105.

[90] V. Lai, Y.F. Huang, C.H. Koo, A.N. Ahmad, A. El-Shafie, Optimization of reservoir operation at Klang Gate Dam utilizing a whale optimization algorithm and a Lévy flight and distribution enhancement technique, Eng. Appl. Comput. Fluid Mech. 15 (1) (2021) 1682–1702, https://doi.org/10.1080/19942060.2021.1982777.

[91] M. Mohammadi, S.-F. Mousavi, S. Farzin, H. Karami, Optimal operation of dam reservoir using whale optimization algorithm and its hybrid with genetic algorithm based on multi-criteria decision making, Iran. J. ECO Hydrol. 6 (2) (2019) 281–293, https://doi.org/10.22059/IJE.2019.270039.990.

[92] A. Donyaii, A. Sarraf, H. Ahmadi, Optimization of reservoir dam operation using gray wolf, crow search and whale algorithms based on the solution of the nonlinear programming model, J. Water Soil Sci. 24 (4) (2021) 159–175, https://doi.org/10.47176/jwss.24.4.42751.

Chapter 45

A MTIS method using a combined of whale and moth-flame optimization algorithms

Taybeh Salehnia[a], Farid MiarNaeimi[b], Saadat Izadi[a], Mahmood Ahmadi[a], Ahmadreza Montazerolghaem[c], Seyedali Mirjalili[d,e], and Laith Abualigah[f,g,h,i,j,k]

[a]*Department of Computer Engineering and Information Technology, Razi University, Kermanshah, Iran,* [b]*Faculty of Engineering, Department of Civil Engineering, University of Sistan and Baluchestan, Zahedan, Iran,* [c]*Faculty of Computer Engineering, University of Isfahan, Isfahan, Iran,* [d]*Centre for Artificial Intelligence Research and Optimisation, Torrens University Australia, Brisbane, QLD, Australia,* [e]*Yonsei Frontier Lab, Yonsei University, Seoul, South Korea,* [f]*Computer Science Department, Al al-Bayt University, Mafraq, Jordan,* [g]*Hourani Center for Applied Scientific Research, Al-Ahliyya Amman University, Amman, Jordan,* [h]*MEU Research Unit, Middle East University, Amman, Jordan,* [i]*Department of Electrical and Computer Engineering, Lebanese American University, Byblos, Lebanon,* [j]*School of Computer Sciences, Universiti Sains Malaysia, Pulau Pinang, Malaysia,* [k]*School of Engineering and Technology, Sunway University Malaysia, Petaling Jaya, Malaysia*

45.1 Introduction

Segmentation is considered as an essential step in image processing. This process divides different parts of the image into several categories. Multi-level Thresholding is a method that facilitates this process. The problem is to correctly segment each image to find the best set of thresholds [1]. Thresholding usually uses image processing methods due to its consistency and low Computational Complexity (CC). Two main methods are Otsu's method [2–4] and Kapur's method [5,6]. However, such approaches have high CC for Multi-level Thresholding [7]. Thresholds help each other to separate interesting objects from their background. The higher splitting quality depends on the selected thresholds [8]. Recently, Meta-Heuristic (MH) algorithms like Particle Swarm Optimization (PSO) [9], Whale Optimization Algorithm (WOA) [10], Moth-Flame Optimization (MFO) [11] have been successfully applied for Thresholding problems [3,8,12], and ABC [13,14] and, Harris Hawks Optimizer (HHO) [15] are used in other problems.

MH algorithms have attracted the attention of researchers due to their excellent performance in finding threshold vectors in Multi-level Thresholding Image Segmentation (MTIS) systems. MH algorithms are either used separately in these problems, or been used in a combined version to solve the MTIS. Most MH algorithms are population-based and initially find a plausible answer by randomly moving through the search space. Such algorithms also include two phases of exploration and exploitation to search for the desired solution on the search space, through which the two phases search globally and locally, respectively. Therefore, several attempts have been made in the literature to achieve a better balance between exploration and exploitation phases to ensure maximum performance on a given optimization problem. In this chapter, our contribution is the design and implementation of an MTIS system using a combination of WOA, MFO, and the Inverse Otsu (IO) Function. This modification is developed using the operators of the MFO algorithm in an attempt to enhance the exploitation phase of WOA during the process of finding the optimal solution for a given optimization problem. It is used to increase the system's performance so that the combined MFWOA algorithm performs better than WOA and MFO and provides better solutions. Therefore, the optimal exploration and exploitation properties of MFO and WOA are used in the search space to find the best thresholds. The rest of our chapter is organized as follows: Section 45.2 presents an overview of related work. In Section 45.3, we describe the prerequisites used in the proposed method. Section 45.4 offers the proposed method. Section 45.5 describes the performance analysis and test results. Finally, Section 45.6 presents the conclusions.

45.2 Related work

The works that have been done so far in the field of MTIS using MH algorithms are single MH and Hybrid MH, which are briefly described in the following.

Handbook of Whale Optimization Algorithm. https://doi.org/10.1016/B978-0-32-395365-8.00051-8

45.2.1 Image segmentation using single meta-heuristics

This section briefly reviews the latest related work on image segmentation. Rodríguez-Esparza et al. proposed the HHO-based solver for image segmentation based on K-means and the Fuzzy IterAg machine learning algorithms [16]. The experimental results show that the proposed method improves accuracy, consistency, and quality compared to the other methods. Anitha et al. [17] presented a Modified WOA (MWOA) to optimize the Multi-level color image Thresholding. The experiments show that the proposed method using MWOA performs better with less CPU computing time, image quality, and feature protection than other state-of-the-art algorithms. Abd El Aziz et al. [3] tested the ability of each of the WOA and MFO algorithms separately. During the optimization, they used Otsu's Fitness Function. Their proposed method was performed on different benchmark images compared with five algorithms. Also, the proposed method is provided a good balance between exploration and exploitation and works better than other algorithms.

Doun et al. [18] provided an Improved Cuckoo Search (ICS) for the optimal Multi-level Thresholding. Two modifications were used to improve the cuckoo search algorithm. In the experiments, six benchmark test images and a series of measures were performed, including Fitness Function value and standard deviation, Peak Signal to Noise Ratio (PSNR), FSIM, and Structure Similarity Index (SSIM). The result shows that the ICS algorithm is superior to other MH algorithms. Salehnia et al. [19] performed three MFO, WOA, and Grasshopper Optimization Algorithm (GOA) for utilizing Multi-level Thresholds, which use a mathematical equation using the corresponding image features as a Fitness Function. The results show that these algorithms are better than other algorithms for the Fitness Function, and GOA achieves a higher performance.

45.2.2 Hybrid meta-heuristics

Abd Elaziz et al. [12] developed a method for determining the optimal threshold for image segmentation. Their proposed method is an enhanced HHO by considering the Salp Swarm Algorithm (SSA), which is called HHOSSA, to improve HHO. The evaluation results show that the proposed method compared to HHO, SSA, and other methods obtained excellent results and performance. Samantaray et al. [20] present a new algorithm, the Harris Hawks-Cuckoo Search (HH-CS) algorithm, based on Multi-level Thresholding. This paper uses eight different images for the breast cancer thermogram image analysis, and some metrics such as PSNR, Feature Similarity Index (FSIM), SSIM are used. HHO-CS algorithm is beneficial for analysis of image and Function optimization. Hosseinzade and Mozafari [5] provided a hybrid algorithm based on Genetic Algorithm (GA) and Simulated Annealing (SA) algorithm for MTIS. The advantage of GA is that it is precise, and the disadvantage is that it is time consuming. The advantage of the SA is that it is fast and has a simple search space, and the disadvantage is that it may stuck in local minima. They used Otsu and Kapur methods as Fitness Functions and obtained their results based on four benchmarks. Their results showed that their proposed algorithm outperforms other algorithms.

45.2.3 Weakness of single and combined algorithms used to solve MTIS problem

In all the papers reviewed in the literature section, MH algorithms have been used individually, improved, and combined to solve the MTIS problem. These algorithms have been trying to obtain relatively optimal thresholds or solutions to the MTIS problem. But according to the results of the algorithms seen in the papers and according to their evaluation, not all of them are able to find the best global answer. In other words, according to the numbers observed for PSNR, SSIM and processing time in these papers, the accuracy of the segmented image using the thresholds obtained from these methods is low and most of them have high computation time. This indicates that the algorithms used in the existing papers have not been able to obtain the optimal global answer and the thresholds obtained by the respective algorithms can not segment the image pixels more accurately. Therefore, in this chapter, in order to improve MFO, a combined MFWOA algorithm is proposed in which the operators used in WOA help to increase the power of MFO in finding the optimal answer.

45.3 Preliminaries

This section will discuss the Otsu method, WOA, and MFO.

45.3.1 Fitness function

In this chapter, the Otsu Thresholding method (IO Function) is used as a Fitness Function in the corresponding MH algorithms to determine the optimal threshold vector for classifying and boundaring image pixels (Eq. (45.4)). Otsu Thresholding method [21] is a popular method used as a Fitness Function in most MTIS methods that use MH algorithms. The Otsu method is an automatic Thresholding method obtained according to the image histogram. Using the Otsu method, the boundaries of objects in the desired image can be specified. The Otsu function is computed using Eq. (45.1).

$$F = \sum_{i=0}^{k} SUM_i (\mu_i - \mu_1)^2 \tag{45.1}$$

$$SUM_i = \sum_{j=T_i}^{T_{i+1}-1} P_j \tag{45.2}$$

$$\mu_i = \sum_{j=T_i}^{T_{i+1}-1} i \frac{P_j}{SUM_i} \quad \text{where } P_j = f(j)/NUM_p \tag{45.3}$$

$$Fit = 1/F \tag{45.4}$$

In Eq. (45.1), μ_1 is the image density average for $T_1 = 0$ and $T_2 = I$ (where I is the maximum pixel density of the image, which is 255 for gray images), μ_i is the density average of the C_i class for T_i and $T_{i+1} - 1$, k is the number of searched thresholds, and SUM_i is the sum of probabilities. In Eq. (45.2) and Eq. (45.3), P_j indicates the probability of the gray level j, $f(j)$ is the frequency of the gray level j, and NUM_p is the total number of pixels in the image. Eq. (45.4) is the same Fitness Function used in the algorithms in this chapter.

45.3.2 Whale optimization algorithm

In WOA [10], as with most optimization algorithms, the optimization process begins with a randomly generated set of candidate solutions ($\overrightarrow{X_i}$ vector) [10]. It should be noted that for the MTIS problem in this chapter, the position of each threshold value or the position of each solution ($\overrightarrow{X_i}$) is between the minimum pixel brightness and the maximum pixel brightness in the image. Each solution is represented as a vector according to Eq. (45.5). The solutions produced using Eq. (45.6) and evaluated using Eq. (45.4).

$$\overrightarrow{X_i} = (x_{i,1}, x_{i,2}, \ldots, x_{i,k}) \quad \text{where } 0 \leq x_{i,1}, x_{i,2}, \ldots, x_{i,k} \leq H \tag{45.5}$$

$$x_{i,j} = lb + rand(0,1) \times (ub - lb), \quad x_{i,j} \in \overrightarrow{X_i}, \quad j = 1, 2, \ldots, k \tag{45.6}$$

where in MTIS problem, lb and ub are the lower bound and the upper bound, respectively, $x_{i,k}$ represents each threshold of the threshold vector, $rand(0, 1)$ is a random number between 0 and 1, and H represents the maximum brightness of the pixels in the image. The input and output of WOA are the image histogram and the threshold vector, respectively. This algorithm is inspired by humpback whales' bubble-net hunting method. The WOA is performed in three phases as follows [10]:

- Siege hunting phase.
- Exploitation phase: The bubble net attacking method.
- Exploration phase: Hunting search.

Once the best search agent is identified, other search agents try to update their location to the best search agent. As:

$$\overrightarrow{D} = \left| \overrightarrow{C} . \overrightarrow{X}^*(t) - \overrightarrow{X}(t) \right| \tag{45.7}$$

$$\overrightarrow{X}(t+1) = \overrightarrow{X}^*(t) - \overrightarrow{A} . \overrightarrow{D} \tag{45.8}$$

where t denotes the current iteration, $\overrightarrow{A}$ and $\overrightarrow{C}$ are the coefficient vectors, $\overrightarrow{D}$ is the distance between the position of $\overrightarrow{X}^*(t)$ and $\overrightarrow{X}(t)$, $X^*(t)$ the location vector is the best solution obtained at present, and $\overrightarrow{X}(t)$ is the location vector. Vectors $\overrightarrow{A}$

and $\vec{C}$ are calculated as follows:

$$\vec{A} = 2\vec{a}.\vec{r} - \vec{a} \tag{45.9}$$

$$\vec{C} = 2\vec{r} \tag{45.10}$$

where $\vec{a}$ decreases linearly from 2 to 0 during iterations (in both exploration and exploitation phases), and $\vec{r}$ is considered a random vector between 0 and 1. Two methods have been designed to model the bubble net behavior of whales mathematically:

a. *Contractile blocking mechanism*

This behavior is achieved by increasing a value in Eq. (45.9). The oscillation range of $\vec{A}$ is reduced by a. In other words, $\vec{A}$ is a random value in the distance $[-a, a]$, and is decrease from 2 to 0 during iterations. The new location of the search agent can be defined by selecting random values of a in the range -1 to 1 anywhere between the primary area of the agent and the location of the current best agent.

b. *Spiral Updating Location*

This method first calculates the distance between the whale located in the bait's $\vec{X}$ and $\vec{Y}$ coordinates in $\vec{X}^*(t)$ and $\vec{Y}^*(t)$. A spiral equation is created between the whale's position and the bait to mimic the spiral-shaped movement of the humpback whale:

$$\vec{X}_i(t+1) = \begin{cases} \vec{X}^*(t) - \vec{A}.\vec{D}, & p < 0.5 \\ \vec{D}'(t).e^{bl}.cos(2\pi l) + \vec{X}^*(t), & p \geq 0.5 \end{cases} \tag{45.11}$$

where $\vec{D}'$ refers to the distance from the 1st whale to the bait (the best solution obtained so far), b is a constant for defining the shape of the logarithmic spiral, and l is a random number between -1 and 1. It is assumed that the whale to model this simultaneous behavior with a 50% probability chooses one of the contractile siege mechanism or spiral models to update the whales' position during optimization. Also:

$$\vec{D} = \left| \vec{C}.\overrightarrow{X_{rand}} - \vec{X} \right| \tag{45.12}$$

$$\vec{X}(t+1) = \overrightarrow{X_{rand}} - \vec{A}.\vec{D} \tag{45.13}$$

where $\overrightarrow{X_{rand}}$ is the current population's randomly selected position vector (random whale). A random search agent is selected in $|\vec{A}| > 1$ mode, while the best solution is selected when $|\vec{A}| < 1$ to update the position of the search agents. Finally, WOA stops by reaching the stop condition, and the best solution in the MTIS is the same threshold vector as the final answer or output of the algorithm. Fig. 45.1 shows the process of producing optimal thresholds in the MTIS using WOA.

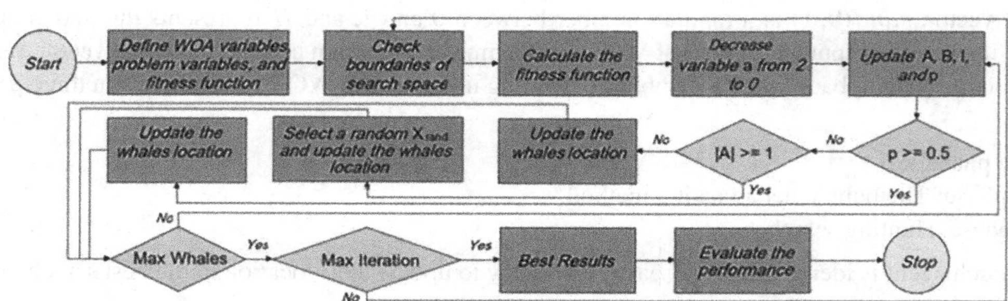

FIGURE 45.1 The structure of the WOA in the MTIS problem.

45.3.3 Moth-flame optimization algorithm

MFO Algorithm is another nature-inspired MH for solving optimization problems designed in the year 2016 [11]. Like other MH algorithms, the MFO starts the optimization process with an initial population $\vec{X}_i$ $(i = 1, 2, \ldots, N)$ of N moths that are randomly located in different locations. It should be noted here that moths and flames are both solutions. The difference between them is the way we treat and update them in each iteration. The moths are actual search agents that move around the search space, whereas flames are the best position of moths that obtains so far. In other words, flames can be considered as flags or pins that are dropped by moths when searching the search space. Therefore, each moth searches around a flag (flame) and updates it in case of finding a better solution. With this mechanism, a moth never lose its best solution. Each moth or solution ($\overrightarrow{X_i}$) is shown as Eq. (45.5). The position of each moth is initialized using Eq. (45.6) and evaluated using Eq. (45.4).

$$\overrightarrow{X_i} = \overrightarrow{D_l}.e^{bl}.cos(2\pi l) + \overrightarrow{F_u} \tag{45.14}$$

where $\overrightarrow{F_u}$ is the uth flame, b is a constant for defining the shape of the logarithmic spiral, $\overrightarrow{D_l}$ defines the distance between the ith moth $\overrightarrow{X_l}$ and the uth flame $\overrightarrow{F_u}$ ($\overrightarrow{D_l} = |\overrightarrow{F_u} - \overrightarrow{X_l}|$), and $l \in [-1, 1]$ is a random number. The Fitness Function is then calculated for each search agent [11]. Here, the locations update is repeated until the stop conditions are met. In MFO, the exploitation of the best solutions may degrade because of the updating of moths' position regarding to N different locations in the search space. So, a technique is used using Eq. (45.15) [11].

$$F_{num} = round\left(N - z \times \frac{N-1}{iter}\right) \tag{45.15}$$

where F_{num} is number of flames, z is the current number of iterations, and iter indicates the maximum number of iterations. The location and Fitness of the best target are ultimately given to the output as the best approximation of the global optimum. Fig. 45.2 shows the process of generating optimal thresholds in the MTIS problem using the MFO algorithm [11].

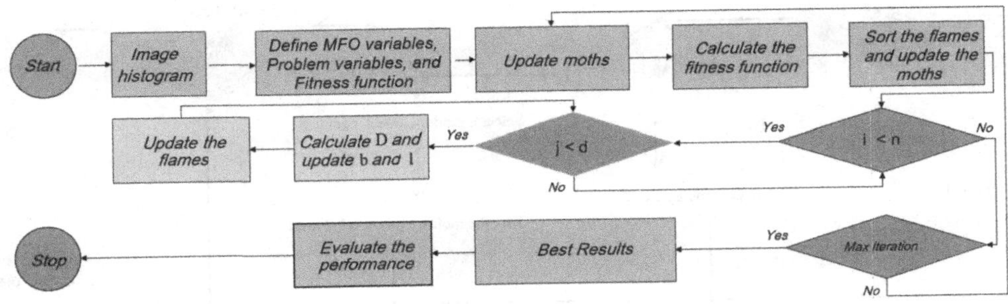

FIGURE 45.2 The structure of the MFO.

45.4 Proposed method

This chapter, uses a combination of two MH algorithms, i.e., WOA and MFO, to improvise the MFO and solve the MTIS problem. In most optimization algorithms, the process consists of two main stages: exploration and exploitation.

Exploration refers to the ability of the algorithm to search the search space globally, in which case the algorithm does not get stuck in the local optimization. Exploitation refers to the ability to discover solutions to improve their quality locally. The better the balance between these two phases of exploration and exploitation, the better the algorithm's performance. WOA is more concentrated in the exploration phase, and MFO is more concentrated in the exploitation phase. If WOA is combined with MFO, it can achieve much better performance. Therefore, in this chapter, we combined the exploitation phase of WOA with exploration phase of MFO and solve the MTIS problem. In MFWOA, the solutions during the exploitation phase are updated using the operators of WOA, and in the exploration phase, only the operators of MFO are used. Then, it computes the quality of each solution according to its Fitness Function value (Eq. (45.4)). Finally, MFWOA stops by reaching the stop condition, and the best solution in the MTIS is the same threshold vector as the final answer or output of the MFWOA. In this chapter, to determine the best threshold vector, the MFWOA algorithm is repeated 100 times on the search space (image histogram). Therefore, at the beginning of optimization, an initial population of solutions is randomly generated using Eq. (45.6). Solutions are distributed over the search space, and then the Fitness Function for the all solution is calculated according to Eq. (45.4). Then, in the search space exploration step, the position of the other solutions is updated based on the metric solution and according to Eq. (45.14) and Eq. (45.15). In the exploitation phase, the solutions are updated using high-powered WOA algorithm operators. In this case, the MFWOA does not get stuck in the local optimization at the beginning of the optimization and achieves a high improvement with the help of WOA algorithm operators in the exploitation phase. Therefore, in this stage, the solutions are updated using Eq. (45.11). The MFWOA output is the same as the optimal threshold vector. Fig. 45.3 shows the flowchart of the MFWOA algorithm for determining the threshold vector in the MTIS process.

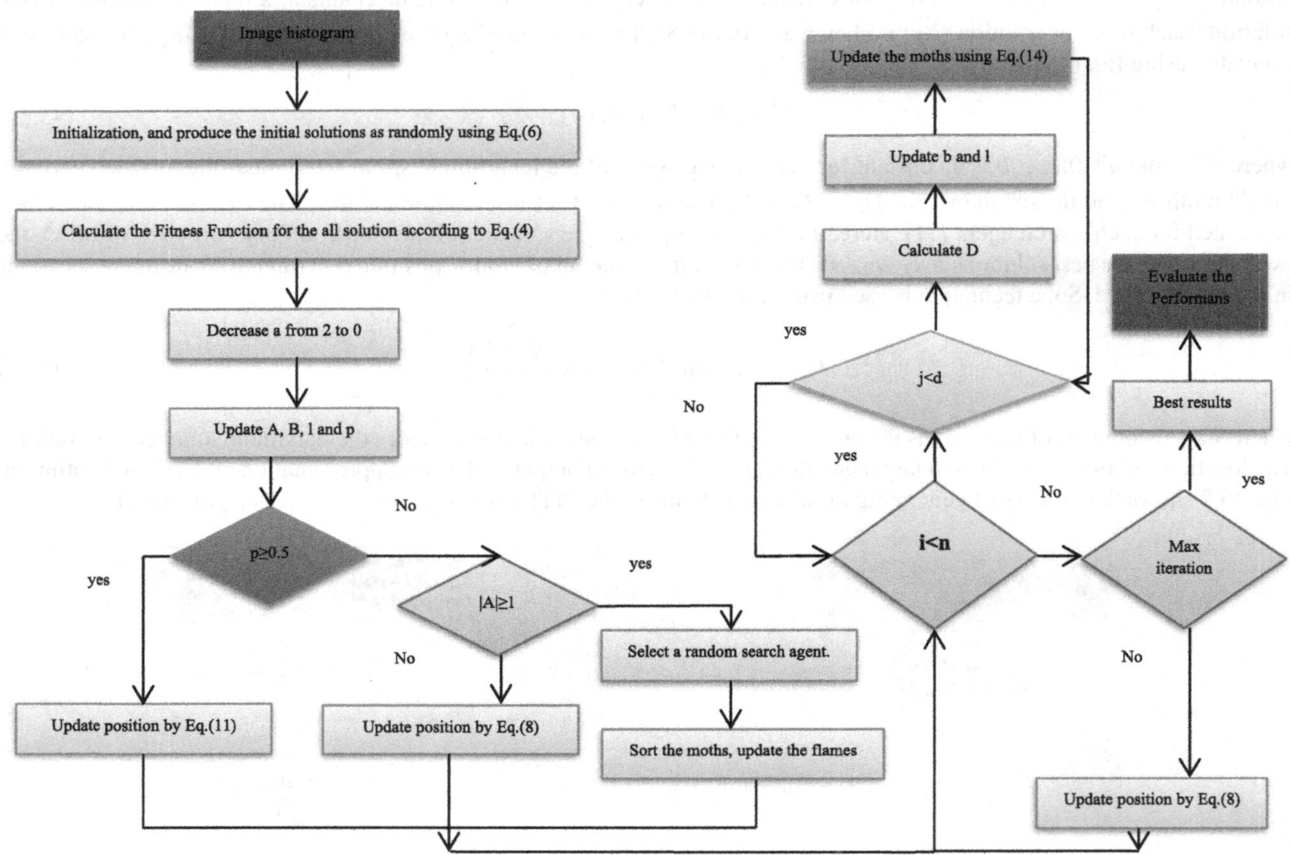

FIGURE 45.3 The proposed MFWOA structure.

45.4.1 Computational complexity of MFWOA

The CC is a field of computational theory that examines the cost of problem-solving process. The CC of MH algorithms is estimated based on the number of search agents, number of problem dimensions, and the maximum number of iterations [22]. The CC of the sorting process for N search agents at the best and worst state is equal to $CC(N \times logN)$ and $CC(N^2)$, respectively [10,11]. The CC of the position updating process in a D-dimensional space is also equal to $CC(N \times D)$. Assuming $It_{max}^{WOA} = It_{max}^{MFO} = T$, and applying an equal number of search agents for the WOA and MFO $\left(N^{WOA} = N^{MFO} = N\right)$. Therefore, the CC of the MFWOA during the first phase $\left(CC^{WOA}\right)$ and the second phase $\left(CC^{MFO}\right)$ optimization process can be defined as [11]:

$$\begin{aligned} O^{WOA} &= CC\,(T \times [CC\,(Sorting) + CC\,(position\ update)]) \\ &= O\left(T \times \left[N^2 + N \times D\right]\right) = O\left(T \times N^2 + T \times N \times D\right), \\ O^{MFO} &= O\left(T \times \left[N^2 + N \times D\right]\right) = O\left(T \times N^2 + T \times N \times D\right) \end{aligned} \tag{45.16}$$

The overall CC of the proposed MFWOA is obtained as;

$$O^{WOA-MFO} = O^{WOA} + O^{MFO} = O\left(T \times N^2 + T \times N \times D\right) \tag{45.17}$$

The CC of all three algorithms (WOA, MFO, MFWOA) is the same. Because all three algorithms (WOA, MFO, MFWOA) have almost the same structure.

45.5 Performance analysis and test results

In this section, various experiments that have been performed on the test images. We took these images from the Berkeley Segmentation Dataset and Benchmark (Fig. 45.4(a)-(d)), and **Ali Daei** images which is Iranian sports legend in the field of football (Fig. 45.4(e), (f)). For the corresponding MH algorithms, 100 search agents look up for the best threshold vector (the best solution) on the search space during 100 iterations. The reason why we have chosen 100 search agents and 100 iterations for the MH algorithms used in this chapter is that the higher the number of population members (search agents) and the number of iterations in the algorithms, these algorithms will achieve more accurate and better answers. The stop condition for any algorithm is to reach the iteration 100th (same conditions for all algorithms: the same Fitness Function, 100 search agents, and 100 iterations). In this case, it is better to calculate the statistical results using ANalysis Of VAriance (ANOVA) or P-Value for the corresponding algorithms and identify the best algorithm in terms of performance.

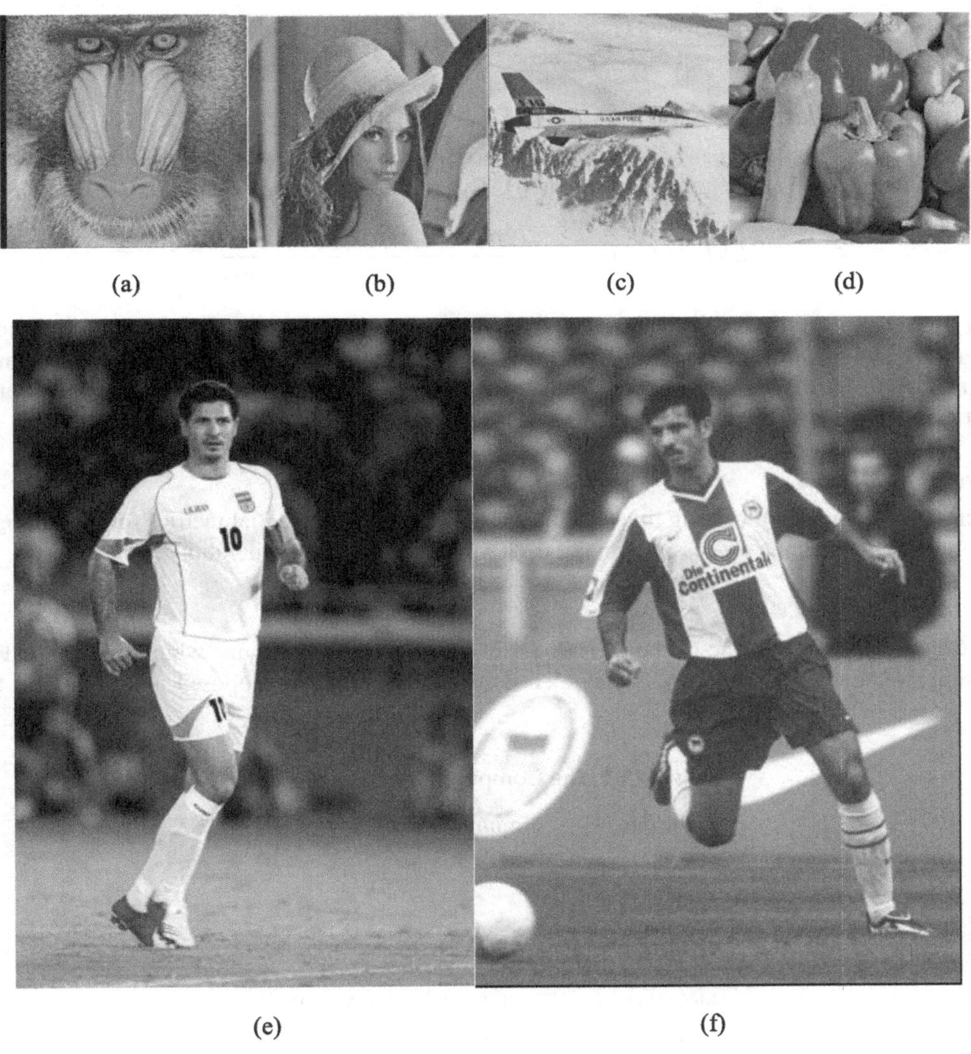

FIGURE 45.4 Test images in proposed methods. (a) Test 1 (b) Test 2 (c) Test 3 (d) Test 4 (e) Test 5 (f) Test 6.

45.5.1 Evaluation metrics

Like other researches in this field, this chapter uses SSIM, PSNR, processing time, CC, Fitness Function value, threshold values, and statistical test evaluation metrics to evaluate the algorithms and compare the proposed method (MFWOA technique) to similar algorithms. The proposed algorithm and the comparative mechanisms are programmed in "MATLAB® 2018b" and run in a "Windows 7-64bit" environment on a laptop with an Intel Core i4 GHz processor and "6 GB" memory.

45.5.1.1 Peak signal-to-noise ratio (PSNR)

The PSNR evaluation metric is a famous metric used to measure the similarity between the segmented and original images. The amount of PSNR for the image is obtained using Eq. (45.19) and depends on the Mean Squared Error (MSE) value [20].

$$MSE = \frac{1}{m \times n}\sum_{i=1}^{m}\sum_{j=1}^{n}(I_O(m,n) - I_S(m,n)) \tag{45.18}$$

$$PSNR(I_O, I_S) = 10Log_{10}(\frac{255^2}{MSE}) \tag{45.19}$$

For the image with a size of $m \times n$, $I_O(m,n)$ represents the original image pixels and $I_S(m,n)$ represents the segmented image pixels.

45.5.1.2 Structural similarity index measure (SSIM)

The SSIM is a famous metric in the image segmentation methods that is used to measure the amount of structural similarity between the original image (I_O) and segmented image (I_S). This metric is obtained using Eq. (45.20) [3].

$$SSIM(I_O, I_S) = \frac{(2\mu_1\mu_S + c_1)(2\sigma_{1,S} + c_2)}{(\mu_1^2 + \mu_S^2 + c_1)(\sigma_1^1 + \sigma_S^2 + c_2)} \tag{45.20}$$

Here, μ_1 and μ_S are the mean brightness intensity of the I_O and I_S, respectively. The σ_1 and σ_s, represent the standard deviation of images I_O and I_S images, respectively. The $\sigma_{1,S}$ represents covariance between I_O and I_S images. c_1 and c_2 are two constant values that are 6.50 and 58.52, respectively [3]. The higher the SSIM value in image segmentation methods and the closer it is to 1, the corresponding method is more effective.

45.5.1.3 Processing time

In image segmentation methods, processing time (second) is also one of the essential metrics for evaluating algorithms. In this chapter, we have calculated the processing time of each MH algorithm in 100 iterations.

45.5.1.4 Computational complexity

Computational Complexity (CC) is one of the metrics for evaluating MH algorithms, which can be used to compare algorithms with more accuracy and certainty. Table 45.1 shows the CC for the proposed algorithm and the comparable algorithms.

TABLE 45.1 The Computational complexity.

Algorithm	Computational complexity
HHO [24]	$O(N + T \times N \times D + T \times N)$
EO [25]	$O(T \times N \times D + T \times C \times N)$
MPA [26]	$O(T \times N + T \times N \times D + T \times C)$
WOA [10]	$O(T \times N^2 + T \times N \times D)$
MFO [11]	$O(T \times N^2 + T \times N \times D)$
MFWOA	$O(T \times N^2 + T \times N \times D)$

According to Table 45.1, T, N, C, and D represent the number of iterations, the number of population members, cost of Fitness Function, and the Dimensions size of each population member, respectively. As can be seen from Table 45.2, in this chapter, the CC of the WOA, MFO, EO and MFWOA is the same. MPA and HHO algorithms are also more CC. But in general, the order of CC in all of them is $T \times N^2$.

45.5.1.5 Fitness function value

The Fitness Function, which is essential in all MH algorithms, and its selection is a fundamental principle, is necessary for MTIS methods performed using MH algorithms. In this chapter, the optimal thresholds are obtained using the minimization of the Fitness Function that the exact inverse Otsu Function (Eq. (45.4)).

45.5.1.6 Threshold values

The values of the threshold vector are an essential metric in MTIS. The ultimate image segmentation is done in MTIS methods using the threshold vector. In this chapter, the obtained threshold vector by each algorithm is presented in Table 45.7.

45.5.1.7 Statistical test (P-Value)

As with previous papers in the MTIS field [4,19], in this chapter, to compare the proposed MFWOA method with other MH algorithms, we use the ANOVA or P-value with a significant level of 0.05 [23]. The P-Value for PSNR, SSIM, processing time and Fitness Function is calculated, and then the proposed method is compared with other algorithms. Similar to previous papers [4,19], there are two hypotheses of zero and alternatives. According to the hypothesis of zero, there should be no significant difference between the mean values of the compared algorithms and the MFWOA algorithm (P-Value should be more than 0.05). However, according to the alternative hypothesis, there should be a significant difference between the proposed MFWOA method and comparative algorithms (i.e., P-Value should be less than 0.05).

45.5.2 The results and discussions

In this section, the results of the proposed MFWOA method and comparable algorithms are thoroughly examined using evaluation metrics. In Table 45.2, the list of constant parameters used in each MH algorithm with their numerical values is recorded.

TABLE 45.2 The constant parameters of each algorithm and their values.

Algorithm	Parameters	Value
WOA [10]	A	[0, 2]
	B	1
	L	[−1, 1]
MFO [11]	B	1
	L	[−1, 1]
EO [25]	V	1
	a_1	2
	a_2	1
	GP	0.5
HHO [24] MPA [26]	E_0	[−1, 1]
	FADs	0.2
	P	0.5

We compare our proposed algorithm with WOA [10], MFO [11], HHO [24], Equilibrium Optimizer (EO) [25], and Marine Predators Algorithm (MPA) [26] algorithms. The reason for choosing the EO, MPA, and HHO algorithms to compare with our work is that these three algorithms are strong and new algorithms. So we chose them to compare with our work to show the superiority of our proposed algorithm over them. We tested our proposed MFWOA algorithm and others for different threshold levels of k ($k = 2, 3, 4, 5, 6, 7, 8, 9, 10, 16, 32$) on the eight images, as it can be seen in Fig. 45.4, to be able to make more accurate evaluations and comparisons using the relevant evaluation metrics. Table 45.3 shows the value of the Fitness Function obtained from the proposed MFWOA algorithm and other algorithms, for different thresholds for all test images, during 100 iterations of the algorithms. We introduce the maximum and minimum values of the Fitness Function by each algorithm at each threshold level for some images, which can be seen in Table 45.3. This chapter obtains

TABLE 45.3 Value of Fitness Function for different threshold levels during 100 runs for different images.

	k	Optimization Algorithms							k	Optimization Algorithms					
		HHO	EO	WOA	MPA	MFO	MFWOA			HHO	EO	WOA	MPA	MFO	MFWOA
Test 1	2	1559.8	1559.8	1559.9	1559.8	1559.8	1559.8	Test 4	2	2399.9	2399.9	2399.1	2399.9	2399.9	2297.8
	3	1990.8	2021	2014.3	2021	2021	**1990.5**		3	3154.6	3211	3209.9	3211	3211	3010.9
	4	2313.6	2443.3	2315	2455.5	2455.5	**2313**		4	3569.2	4008.9	3565	4044.1	4044.1	3443.3
	5	2784	2794.5	2771.5	2810	2784.1	2773.4		5	4295.5	4406.7	4422.1	4426.1	4406.4	4113.1
	6	3268.1	3239.1	3207	3244.5	3244.5	3053.7		6	4362.1	5190.6	4722.7	5185.2	5217.4	4119.1
	7	3515.7	3624.7	3531.6	3677	3599	3488.1		7	4526.3	5768.5	6040	5543.1	6050.7	4489.7
	8	3947.8	3992	3938.6	4033.5	4005.6	3930.1		8	5616.9	6392.3	6339.1	6326.3	6390.7	5545.4
	9	4373.6	4396.7	4339	4465.9	4433.7	4264.8		9	6179	7031	6355	7191.8	7224.1	6372.3
	10	3669.3	4813.7	4279.5	4898.4	4334.1	3576.6		10	6384.5	7569.8	6848.3	7443.4	6507.8	6355.1
	16	5915.4	7155.2	5609.6	7265.4	7263.4	5630.1		16	9344	11339	8746.1	11533	10343	8565.9
	32	10770	13326	10324	12680	12036	8482.3		32	15213	21019	13420	20175	19148	12483
Test 2	2	1952.3	1952.3	1951.8	1952.3	1952.3	1950.4	Test 5	2	2510	2510	2510	2510	2510	2510
	3	2600.6	2600.6	2520.7	2600.6	2600.6	2522.5		3	3518.9	3518.3	3518.3	3518.3	3518.3	3518.3
	4	2947.2	3158.1	3173	3174	3174	2929.7		4	3538.4	4060.8	3678.5	4102.3	3552.3	3578.1
	5	3565.1	3579.8	3563.4	3600.7	3600.7	3431.2		5	4910.8	4895.8	4967	4987.1	4555.6	4533.4
	6	3792.8	4122.4	3742.3	4174.2	4170.1	3749.2		6	4199.8	5547	5398.4	5561.5	5571.2	40118
	7	4413.1	4609	4061	4600.8	4519.8	4431.4		7	5254.1	6196.1	6456	6145.6	6024.5	5079.4
	8	4583.4	5094.2	4605.6	5174.3	5091	4454.8		8	4648.8	6841.5	6763	7030.4	7030.4	4111.3
	9	4926	5603.4	5596.9	5601	5358.9	4871		9	7560.4	7527.5	7331.9	7924.9	7604.7	6111.3
	10	4811	6048.9	5349.1	6093.4	6091.2	4450.7		10	8176	8285.6	8024.5	8387	8066.1	8050.4
	16	6669.4	9040.4	7514.6	9170.8	8924.3	6385.9		16	7306.4	12145	10091	10353	11900	6716
	32	11729	16658	12333	15563	14832	10257		32	16636	21332	17154	20705	17171	16237
Test 3	2	1962.1	1962.1	1961.7	1962.1	1962.1	1961.1	Test 6	2	5686.4	5686.4	5686.2	5686.4	5686.4	5685.8
	3	3176.1	3176.1	3176.1	3176.1	3176.1	3174.6		3	8068.4	8117.4	8120	8121	8121	8120.1
	4	3331.2	3549.2	3330.6	3607.8	3607.8	3207.8		4	8522.9	10507	8522.4	10507	85507	8522.2
	5	3758.2	4537.6	3757.4	4537.6	4537.6	3600.3		5	10859	10936	10884	10961	10902	10907
	6	4936.6	4940.3	4953.4	4969.3	4969.2	4944.8		6	11263	13251	11350	13347	13347	10328
	7	5897.3	5851.6	5104	5899	5400.9	5023.1		7	13655	15006	13725	15734	13708	12784
	8	5897.4	6264.3	6244.9	6055.6	6330.7	5706.9		8	15567	16145	15087	16187	15187	15167
	9	6091.4	7005	6403.1	7260.5	7260.5	5818.6		9	16435	17983	16540	18573	16552	16487
	10	7338.1	7654.3	7247.4	7692.2	7692.1	6521		10	14284	15385	14442	16960	14974	13875
	16	10138	11471	9951.5	11003	11003	8578.7		16	25440	28772	25925	31412	25174	25191
	32	16967	21469	16748	19655	21819	13493		32	39047	53566	36612	50999	39423	36698

the proposed MFWOA and other algorithms by minimizing the Fitness Function. Therefore, according to Table 45.3, if each algorithm's obtained Fitness Function value is lower, the corresponding algorithm performs better.

Also, as shown in Table 45.3, by increasing the value of k, the value of the obtained Fitness Function by all algorithms for Test3 and Test5 increases. For the Test1 image, the value of the obtained Fitness Function by HHO, WOA, and MFWOA decreases at $k = 10$ and increases with an increasing value of k. For the Test2 image, the value of the obtained Fitness Function by MFWOA decreases at $k = 10$ and increases with the value of k. For the Test4 image, the value of the obtained Fitness Function by the MFO decreases at $k = 10$ and then increases as the value of k increases. For the Test6 image, the value of the obtained Fitness Function by all algorithms decreases at $k = 10$ and then increases as the value of k increases. As per the results in Table 45.3, in most cases, the proposed MFWOA algorithm has a lower Fitness Function value than other algorithms. If it is higher, it does not differ much from different algorithms. It does not reduce the PSNR and SSIM values. Therefore, the proposed MFWOA algorithm has the necessary efficiency. It can be said that the proposed algorithm has better performance than other algorithms and can achieve the most suitable thresholds.

For Test1 image:

At the level $k = 2$, all algorithms have the same value. When $k = 3$, the lowest Fitness Function values are related to MFWOA, HHO, WOA, and other algorithms with the same value. Considering $k = 4$, the lowest Fitness Function values are related to HHO, WOA, MFWOA, EO, and MPA=MFO, respectively. At the $k = 5$, the lowest values of the Fitness Function are related to WOA, WOA=MFWOA, HHO=MFO, MPA, and EO, respectively. At the $k = 6$, the lowest values of the Fitness Function are associated with MFWOA, WOA, EO, MFO=MPA, and HHO, respectively. In the case of $k = 7$, the lowest values of the Fitness Function are related to MFWOA, WOA, HHO, MFO, EO, MFO, and MPA, respectively. Assuming $k = 8$, the lowest values of the Fitness Function are related to MFWOA, WOA, HHO, EO, MFO, and MPA, respectively. At the $k = 9$, the lowest values of the Fitness Function are related to MFWOA, WOA, HHO, EO, and MPA=MFO, respectively. At the $k = 10$, the lowest values of the Fitness Function are related to MFWOA, WOA, HHO, EO, MFO, and MPA, respectively. At the $k = 16$, the lowest values of the Fitness Function are related to MFWOA, WOA, HHO, EO, MFO, and MPA, respectively. At the $k = 32$, the lowest values of the Fitness Function are related to MFWOA, WOA, HHO, MFO, MPA, and EO, respectively.

For Test2 image:

When $k = 2$, the lowest values of the Fitness Function are related to MFWOA, WOA, and other algorithms that have the same value. At the $k = 3$, the lowest Fitness Function values are associated with MFWOA, WOA, and other algorithms with the same value. At the $k = 4$, the lowest Fitness Function values are related to MFWOA, HHO, EO, WOA, MPA=MFO, respectively. When $k = 5$, the lowest values of the Fitness Function are associated with MFWOA, WOA, HHO, EO, and MPA=MFO, respectively. At the $k = 6$, the lowest values of the Fitness Function are related to MFWOA, WOA=HHO, EO, MPA, and MFO, respectively. At the $k = 7$, the lowest values of the Fitness Function are associated with MFWOA, WOA, HHO, EO, MFO, and MPA, respectively. At the $k = 8$, the lowest values of the Fitness Function are related to MFWOA, HHO, WOA, MFO, EO, and MPA, respectively. At the $k = 9$, the lowest values of the Fitness Function are related to MFWOA, HHO, MFO, WOA, MPA, and EO, respectively. At the $k = 10$, the lowest values of the Fitness Function are related to MFWOA, HHO, WOA, EO, and MPA=MFO, respectively. At the $k = 16$, the lowest values of the Fitness Function are related to MFWOA, HHO, WOA, EO, MFO, and MPA, respectively. At the $k = 32$, the lowest values of the Fitness Function are related to MFWOA, HHO, WOA, MFO, MPA, and EO, respectively.

For Test3 image:

When $k = 2$, the lowest values of the Fitness Function are related to MFWOA, WOA, and other algorithms that have the same value. At the $k = 3$, all algorithms have the same value. At the $k = 4$, the lowest Fitness Function values are related to MFWOA, WOA, HHO, EO, MPA=MFO, respectively. When $k = 5$, the lowest values of the Fitness Function are associated with MFWOA, WOA, HHO, and MFO=MPA=EO, respectively. At the $k = 6$, the lowest values of the Fitness Function are related to HHO, EO, WOA, MFWOA, and MPA=MFO, respectively. At the $k = 7$, the lowest values of the Fitness Function are associated with MFWOA, WOA, MFO, EO, HHO, and MPA, respectively. At the $k = 8$, the lowest values of the Fitness Function are related to MFWOA, HHO, MPA, WOA, EO, and MFO, respectively. At the $k = 9$, the lowest values of the Fitness Function are related to MFWOA, HHO, WOA, EO, and MPA=MFO, respectively. At the $k = 10$, the lowest values of the Fitness Function are related to MFWOA, WOA, HHO, EO, and MPA=MFO, respectively. At the $k = 16$, the lowest values of the Fitness Function are related to MFWOA, WOA, HHO, MFO=MPA, and EO, respectively. At the $k = 32$, the lowest values of the Fitness Function are associated with MFWOA, WOA, HHO, MPA, EO, and MFO, respectively.

For Test4 image:

When $k = 2$, the lowest values of the Fitness Function are related to MFWOA, and other algorithms have the same value. At the $k = 3$, the lowest Fitness Function values are associated with MFWOA, HHO, WOA, and other algorithms with the same value. At the $k = 4$, the lowest Fitness Function values are related to MFWOA, WOA, HHO, EO, MPA=MFO, respectively. At the $k = 5$, the lowest values of the Fitness Function are associated with MFWOA, HHO, EO=MFO, WOA, and MPA, respectively. At the $k = 6$, the lowest values of the Fitness Function are related to MFWOA, WOA=HHO, EO, MPA, and MFO, respectively. At the $k = 7$, the lowest values of the Fitness Function are associated with MFWOA, HHO, MPA, EO, WOA, and MFO, respectively. At the $k = 8$, the lowest values of the Fitness Function are related to MFWOA, HHO, MPA, WOA, MFO, and EO, respectively. At the $k = 9$, the lowest values of the Fitness Function are related to HHO, WOA, MFWOA, EO, MPA, and MFO, respectively. At the $k = 10$, the lowest values of the Fitness Function are related to MFWOA, HHO, MFO, WOA, MPA, and EO, respectively. At the $k = 16$, the lowest values of the Fitness Function

TABLE 45.4 Value of PSNR for different threshold levels during 100 runs for different images.

	k	Optimization Algorithms							k	Optimization Algorithms					
		HHO	EO	WOA	MPA	MFO	MFWOA			HHO	EO	WOA	MPA	MFO	MFWOA
Test 1	2	24.3	24.3	24.298	24.3	24.3	24.304	Test 4	2	22.824	22.824	22.814	22.824	22.824	22.838
	3	21.773	22.9	22.401	22.9	22.9	22.948		3	22.049	20.55	20.317	20.55	20.55	22.582
	4	22.947	24.512	24.712	20.658	20.658	24.966		4	23.304	19.604	23.259	19.604	19.604	23.655
	5	24.339	22.9	25.988	22.9	25.423	25.989		5	22.096	23.578	23.337	23.579	23.524	23.972
	6	26.711	23.036	22.524	23.036	23.036	26.989		6	20.332	21.946	24.364	21.946	20.557	24.365
	7	27.306	20.658	26.151	20.658	22.9	27.932		7	23.99	19.604	19.981	23.887	19.605	23.998
	8	27.523	25.471	22.624	23.036	25.463	27.705		8	23.374	23.906	20.847	21.947	20.558	24.447
	9	22.253	23.036	24.003	23.036	21.897	24.585		9	22.289	20.55	21.128	21.947	20.558	24.925
	10	28.636	22.901	26.547	20.658	25.501	28.66		10	21.129	24.712	25.983	22.087	23.906	25.991
	16	22.267	26.662	25.341	23.036	22.901	26.677		16	24.034	23.955	21.198	23.398	21.168	26.176
	32	30.528	25.834	20.282	26.114	30.213	31.533		32	26.663	25.203	26.978	25.516	27.964	27.973
Test 2	2	22.973	22.973	22.966	22.973	22.973	22.986	Test 5	2	21.47	21.47	21.47	21.466	21.47	21.476
	3	21.229	21.229	22.373	21.229	21.229	22.483		3	19.323	19.636	19.636	19.636	19.756	19.836
	4	23.323	19.917	19.93	19.917	19.917	22.937		4	21.894	18.648	22.189	18.648	18.648	22.264
	5	20.996	21.233	21.816	21.254	21.233	22.663		5	18.668	19.636	20.313	19.636	22.753	22.805
	6	21.277	23.259	24.825	21.244	21.234	24.833		6	20.549	19.87	22.506	19.853	19.853	22.589
	7	23.423	24.632	22.37	21.233	24.632	25.097		7	22.186	22.791	19.636	18.899	22.791	22.861
	8	21.343	22.11	23.158	21.244	24.373	25.383		8	22.919	22.753	21.798	19.853	19.87	24.326
	9	22.848	24.678	23.35	21.233	24.802	24.883		9	22.445	22.791	22.477	19.636	19.979	22.847
	10	23.278	24.538	27.019	24.64	25.241	27.934		10	21.395	21.705	21.717	21.892	22.667	22.728
	16	20.895	23.298	20.019	21.26	25.94	25.984		16	20.27	23.339	25.026	25.163	22.574	25.288
	32	21.653	25.983	20.547	27.303	25.748	27.759		32	19.132	23.339	18.182	26.08	26.105	27.856
Test 3	2	24.985	24.985	24.972	24.985	24.985	24.985	Test 6	2	22.16	22.17	22.167	22.16	22.16	22.167
	3	23.304	23.304	23.304	23.304	23.304	23.304		3	20.436	20.426	20.265	20.426	20.426	20.45
	4	25.851	22.544	25.799	22.544	22.544	25.801		4	22.278	19.17	22.218	19.17	19.17	22.343
	5	25.332	23.304	26.124	23.304	23.304	26.235		5	21.02	23.072	23.564	20.426	23.123	23.583
	6	25.033	23.304	23.271	23.304	23.304	24.143		6	23.053	20.894	22.107	20.508	20.508	23.543
	7	23.33	23.304	25.431	23.304	22.544	25.485		7	20.933	19.17	23.478	19.17	23.276	23.695
	8	24.459	25.831	24.406	25.831	23.304	25.989		8	23.61	23.178	21.927	20.508	20.508	23.705
	9	25.536	23.304	25.388	23.304	23.304	25.706		9	21.055	20.508	20.06	20.508	23.322	23.399
	10	24.659	23.304	25.513	23.304	23.304	25.64		10	24.453	20.508	23.306	19.17	23.072	24.799
	16	25.468	23.304	23.254	25.978	25.975	28.053		16	22.073	20.932	22.785	19.17	22.615	22.712
	32	25.194	25.975	26.725	27.197	25.011	30.676		32	21.579	23.136	24.572	25.315	25.112	25.337

are related to MFWOA, WOA, HHO, MFO, EO, and MPA, respectively. At the $k = 32$, the lowest values of the Fitness Function are related to MFWOA, WOA, HHO, MFO, MPA, and EO, respectively.

For Test5 image:

When $k = 2$ and 3, all algorithms have the same value. At the $k = 4$, the lowest Fitness Function values are related to HHO, MFO, MFWOA, WOA, EO, and MPA, respectively. At the $k = 5$, the lowest values of the Fitness Function are related to MFWOA, MFO, EO, HHO, MPA, and WOA, respectively. At the $k = 6$, the lowest values of the Fitness Function are related to MFWOA, HHO, WOA, EO, MPA, and MFO, respectively. At the $k = 7$, the lowest values of the Fitness Function are related to MFWOA, HHO, MFO, MPA, WOA, and EO, respectively. At the $k = 8$, the lowest values of the Fitness Function are related to MFWOA, HHO, WOA, EO, and MFO=MPA, respectively. At the $k = 9$, the lowest values of the Fitness Function are associated with MFWOA, WOA, EO, HHO, MFO, and MPA, respectively. At the $k = 10$, the lowest values of the Fitness Function are related to WOA, MFWOA, MFO, HHO, EO, and MPA, respectively. At the $k = 16$,

TABLE 45.5 Value of SSIM for different threshold levels during 100 runs for different images.

	k	Optimization Algorithms							k	Optimization Algorithms					
		HHO	EO	WOA	MPA	MFO	MFWOA			HHO	EO	WOA	MPA	MFO	MFWOA
Test 1	2	0.75532	0.75532	0.75529	0.75532	0.75532	0.75602	Test 4	2	0.70335	0.70335	0.70209	0.70335	0.70335	0.70713
	3	0.6952	0.7155	0.69704	0.7155	0.7155	0.71644		3	0.6951	0.6794	0.67302	0.6794	0.6794	0.69895
	4	0.71683	0.76391	0.77424	0.62461	0.62461	0.7752		4	0.71472	0.64733	0.71068	0.64733	0.64733	0.72004
	5	0.75717	0.7155	0.81393	0.7155	0.80591	0.8173		5	0.70014	0.73655	0.72717	0.73654	0.73519	0.73579
	6	0.83815	0.72179	0.7233	0.72179	0.72179	0.83815		6	0.68101	0.69013	0.75031	0.69013	0.68094	0.75339
	7	0.83708	0.62461	0.83132	0.62461	0.7155	0.83986		7	0.77418	0.64733	0.66918	0.7516	0.64733	0.77696
	8	0.85276	0.81067	0.73891	0.72179	0.80813	0.85832		8	0.75212	0.7353	0.7021	0.69013	0.68095	0.75525
	9	0.69315	0.72179	0.77298	0.72179	0.70013	0.7746		9	0.71801	0.6794	0.70808	0.69013	0.68095	0.7769
	10	0.87498	0.7155	0.84179	0.62461	0.81031	0.87866		10	0.72621	0.76047	0.79503	0.69895	0.7353	0.79891
	16	0.69532	0.85805	0.82013	0.72179	0.7155	0.85611		16	0.79158	0.75352	0.72806	0.73423	0.72726	0.79323
	32	0.90707	0.82056	0.60426	0.82678	0.90384	0.93129		32	0.85374	0.77731	0.83593	0.79718	0.82723	0.85744
Test 2	2	0.68017	0.68017	0.67086	0.68017	0.68017	0.68151	Test 5	2	0.63566	0.63566	0.63566	0.6317	0.63566	0.6357
	3	0.69931	0.69931	0.68173	0.69931	0.69931	0.69942		3	0.51512	0.5289	0.5289	0.5289	0.53344	0.5389
	4	0.72248	0.64058	0.64053	0.64058	0.64058	0.72825		4	0.65709	0.48206	0.66658	0.48206	0.48206	0.66762
	5	0.70998	0.70009	0.71525	0.70132	0.70009	0.71595		5	0.4673	0.5289	0.56448	0.5289	0.7079	0.70855
	6	0.70653	0.70328	0.73991	0.70207	0.69942	0.73993		6	0.70239	0.53843	0.67887	0.53565	0.53565	0.67204
	7	0.70042	0.75025	0.69395	0.70009	0.75025	0.77269		7	0.67159	0.71211	0.5289	0.49342	0.71211	0.71629
	8	0.72172	0.67525	0.72049	0.70207	0.72391	0.75438		8	0.76008	0.7079	0.64822	0.53565	0.53843	0.76131
	9	0.74742	0.75157	0.70676	0.70009	0.74343	0.75839		9	0.71149	0.71211	0.66499	0.5289	0.54019	0.71369
	10	0.76407	0.72984	0.79361	0.75254	0.75629	0.79414		10	0.72801	0.64803	0.63847	0.65468	0.70859	0.7287
	16	0.70873	0.70452	0.67308	0.70143	0.78095	0.78137		16	0.7162	0.7357	0.78329	0.78231	0.70222	0.78444
	32	0.76002	0.82033	0.69726	0.80164	0.80385	0.82164		32	0.49859	0.7357	0.43112	0.82113	0.81574	0.8539
Test 3	2	0.82627	0.82627	0.82462	0.82627	0.82627	0.82628	Test 6	2	0.6727	0.6727	0.6723	0.6727	0.6727	0.67288
	3	0.80202	0.80202	0.80202	0.80202	0.80202	0.80206		3	0.65881	0.60842	0.60297	0.60842	0.60842	0.65907
	4	0.85073	0.78395	0.85124	0.78395	0.78395	0.85575		4	0.67564	0.57719	0.67188	0.57719	0.57719	0.67628
	5	0.83639	0.80202	0.85251	0.80202	0.80202	0.85577		5	0.66585	0.70198	0.71132	0.60842	0.70633	0.71231
	6	0.83252	0.80202	0.80241	0.80202	0.80202	0.83961		6	0.71637	0.6588	0.67332	0.61308	0.61308	0.71783
	7	0.80306	0.80202	0.83654	0.80202	0.78395	0.83781		7	0.66091	0.57719	0.71563	0.57719	0.71419	0.71651
	8	0.81083	0.85136	0.82185	0.85136	0.82202	0.85823		8	0.71285	0.71004	0.66783	0.61308	0.61308	0.71508
	9	0.84288	0.80202	0.82262	0.80202	0.80202	0.84371		9	0.67035	0.61308	0.60412	0.61308	0.71404	0.71547
	10	0.81873	0.80202	0.85379	0.80202	0.80202	0.85471		10	0.78293	0.61308	0.6754	0.57719	0.70198	0.78997
	16	0.86899	0.80202	0.80049	0.85863	0.85837	0.8727		16	0.74956	0.6608	0.69741	0.57719	0.76925	0.76984
	32	0.83703	0.85837	0.83454	0.87201	0.83475	0.9134		32	0.69714	0.70413	0.72049	0.72909	0.72309	0.75979

the lowest values of the Fitness Function are related to MFWOA, HHO, WOA, MPA, MFO, and EO, respectively. At the $k = 32$, the lowest values of the Fitness Function are related to MFWOA, HHO, WOA, MFO, MPA, and EO, respectively.

For Test6 image:

At the level $k = 2$, the lowest values of the Fitness Function are related to MFWOA, and other algorithms have the same value. At the $k = 3$, the lowest Fitness Function values are associated with HHO, WOA=MFWOA, EO, and other algorithms have the same value. At the $k = 4$, the lowest Fitness Function values are related to HHO=WOA=MFWOA and MPA=EO=MFO, respectively. At the $k = 5$, the lowest values of the Fitness Function are associated with HHO, WOA, MFO, MFWOA, EO, and MPA, respectively. At the $k = 6$, the lowest values of the Fitness Function are related to HHO, WOA, MFWOA, EO, and MFO=MPA, respectively. At the $k = 7$, the lowest values of the Fitness Function are associated with MFWOA, HHO, MFO, WOA, EO, and MPA, respectively. When $k = 8$, the lowest values of the Fitness Function are related to WOA, MFWOA, MFO, HHO, EO, and MPA, respectively. At the $k = 9$, the lowest values of the Fitness Function

TABLE 45.6 Value of execution time for different threshold levels during 100 runs for different image.

	k	Optimization Algorithm							k	Optimization Algorithm					
		HHO	EO	WOA	MPA	MFO	MFWOA			HHO	EO	WOA	MPA	MFO	MFWOA
Test 1	2	2.5885	35.545	1.482	2.3924	3.0818	2.3628	Test 4	2	2.5304	35.087	1.3939	2.5094	2.986	2.2661
	3	5.8835	49.898	4.497	5.9809	6.4076	5.3508		3	12.838	151.02	8.1752	8.4191	14.121	9.54
	4	6.7822	66.293	5.2668	7.0125	7.3707	6.2322		4	14.488	189.58	12.999	14.736	19.804	13.94
	5	7.4725	65.063	5.9615	7.9977	8.1565	6.9915		5	12.578	224.55	10.74	12.444	12.82	11.756
	6	8.3069	73.484	6.5406	15.31	16.713	7.8798		6	8.5955	71.098	6.4763	8.5368	8.9628	7.6614
	7	17.237	246.06	15.169	26.738	28.324	16.516		7	9.5911	76.536	7.4286	20.116	13.905	8.7317
	8	20.89	209.33	18.593	33.58	36.424	27.621		8	14.203	88.523	11.925	14.429	14.888	13.436
	9	26.945	113.04	23.851	26.723	27.196	25.491		9	11.537	91.224	8.93	11.604	11.987	10.517
	10	12.246	102.65	9.697	12.676	13.144	11.475		10	12.783	97.781	9.7056	12.727	13.228	11.422
	16	16.357	143.99	11.514	15.888	16.942	14.11		16	15.583	142.23	12.028	16.086	16.847	14.634
	32	26.21	323.04	19.098	26.871	28.344	23.974		32	25.36	251.9	19.156	27.068	27.055	24.158
Test 2	2	2.549	128.42	1.451	2.4985	6.6195	2.1516	Test 5	2	2.6203	57.697	1.391	2.4971	3.2826	2.228
	3	9.6909	186.56	8.2486	9.5985	9.9906	9.0861		3	6.2835	50.61	4.8264	6.2362	6.6442	5.7335
	4	6.543	57.122	5.0665	6.7625	7.3037	6.0152		4	6.7514	56.324	5.1596	13.529	17.062	6.1574
	5	7.9833	66.109	6.0976	7.9914	8.3577	7.1817		5	17.838	207.19	15.94	17.85	18.19	17.125
	6	8.8229	72.188	6.6915	8.6627	9.2596	7.903		6	8.9051	72.104	6.6849	8.8714	9.3535	7.9926
	7	9.8139	79.02	7.5362	19.877	14.386	8.8675		7	9.9042	101.6	7.671	10.287	10.415	9.1771
	8	14.869	88.98	12.273	14.719	15.003	13.77		8	10.791	127.49	8.2182	10.931	11.497	9.8845
	9	26.28	92.145	17.342	27.478	28.636	22.641		9	12.317	95.167	9.2849	12.138	12.741	11.044
	10	29.133	267.31	26.483	29.127	29.663	28.062		10	13.167	259.31	10.039	12.981	13.848	11.913
	16	15.299	136.88	11.539	15.586	15.958	13.82		16	17.257	347.67	12.222	16.544	17.589	14.839
	32	25.04	246.95	18.619	25.772	75.463	22.633		32	27.963	261.04	19.505	30.369	28.826	25.996
Test 3	2	3.6559	71.727	3.3928	8.0961	9.1527	6.1958	Test 6	2	2.6428	62.506	1.4912	2.3882	3.1865	2.2949
	3	12.349	122.93	10.851	18.88	15.85	14.015		3	6.3289	49.805	4.6819	6.0926	6.6972	5.5972
	4	13.928	60.724	8.8811	10.274	13.509	9.4412		4	7.1452	59.161	5.6421	7.4866	8.4836	6.6062
	5	10.871	89.487	9.2188	11.257	11.626	10.337		5	21.886	67.944	9.7285	11.672	17.889	11.621
	6	8.8765	75.884	6.9928	9.0694	9.3399	8.1975		6	28.705	98.079	22.445	27.226	29.218	26.46
	7	9.5359	80.774	7.3131	9.7526	9.8271	8.5936		7	21.654	134.25	18.968	21.65	21.815	20.477
	8	10.41	85.279	8.0442	10.447	10.864	9.361		8	11.129	118.45	8.424	11.156	11.659	10.145
	9	11.26	94.081	8.8589	11.642	25.777	10.467		9	12.908	101.33	9.5723	12.67	13.173	11.612
	10	26.129	268.21	23.428	26.423	37.202	25.088		10	13.316	126.21	10.485	13.549	14.324	12.261
	16	26.046	322.93	22.119	26.311	26.775	24.621		16	23.844	164.18	12.349	21.124	37.755	15.169
	32	26.014	477.14	18.658	26.333	27.581	23.482		32	48.453	316.88	39.833	48.312	49.638	45.08

are related to HHO, MFWOA, WOA, MFO, EO, and MPA, respectively. At the $k = 10$, the lowest values of the Fitness Function are related to MFWOA, HHO, WOA, MFO, EO, and MPA, respectively. Considering $k = 16$, the lowest values of the Fitness Function are related to MFO, MFWOA, HHO, WOA, EO, and MPA, respectively. At the $k = 32$, the lowest values of the Fitness Function are related to WOA, MFO, MFWOA, HHO, MPA, and EO, respectively.

Table 45.4 shows the PSNR values of the segmented image from the obtained optimal thresholds by each MH algorithm and the proposed MFWOA algorithm for different photos at different threshold levels. Inspecting the results in Table 45.4, the proposed MFWOA method has achieved the desired quality for the segmented images at different threshold levels. The proposed MFWOA method has a higher PSNR value than the MFO, WOA, HHO, EO, and MPA algorithms for all threshold levels and relevant images. According to Table 45.4, the value of the PSNR decreases by each algorithm, with increasing k and increasing at other levels.

Table 45.5 shows the value of SSIM for different images and different threshold levels after 100 runs. According to Table 45.5, the proposed method has a higher SSIM value for all images than the existing algorithms compared in

TABLE 45.7 The value of thresholds during 100 runs for different images.

Image	k	EO	WOA	HHO	MPA	MFO	MFWOA
Test 1	2	99, 159	86, 149	99, 156	98, 159	98, 159	94, 155
	3	29, 90, 117	29, 57, 117	29, 90, 116	29, 90, 117	29, 90, 116	29, 79, 117
	4	29, 116, 117, 236	29, 85, 119, 236	29, 90, 102, 159	29, 29, 90, 117	29, 109, 119, 218	29, 68, 103, 170
	5	29, 29, 90, 92, 117	29, 29, 59, 59, 133	29, 29, 92, 107, 109	29, 90, 111, 151, 231	29, 29, 69, 81, 104	29, 49, 90, 108, 150
Test 2	2	76, 140	76, 140	76, 140	76, 140	76, 140	76, 140
	3	125, 153, 253	118, 153, 254	4, 85, 127	4, 88, 126	127, 153, 254	42, 108, 169
	4	4, 125, 126, 254	24, 132, 149, 226	4, 125, 127, 252	4, 84, 85, 142	85, 147, 160, 254	37, 121, 131, 207
	5	4, 75, 125, 140, 253	4, 46, 116, 128, 254	4, 116, 130, 167, 225	4, 4, 94, 100, 118	109, 142, 144, 207, 222	4, 79, 123, 145, 244
Test 3	2	93, 158	93, 159	92, 157	91, 159	92, 156	92, 157
	3	129, 153, 255	95, 147, 249	1, 89, 125	1, 88, 128	127, 154, 255	32, 108, 167
	4	1, 129, 130, 255	1, 112, 138, 255	1, 51, 98, 128	1, 87, 94, 153	90, 137, 141, 254	30, 91, 111, 178
	5	128, 153, 155, 255, 255	129, 165, 166, 255, 255	1, 51, 87, 109, 167	1, 13, 87, 87, 130	138, 156, 170, 255, 255	43, 76, 113, 150, 184
Test 4	2	48, 125	46, 126	48, 125	48, 125	48, 125	47, 125
	3	104, 132, 255	74, 138, 255	1, 56, 100	1, 56, 100	107, 135, 255	25, 83, 151
	4	1, 100, 104, 255	1, 59, 108, 253	98, 141, 255, 255	107, 112, 194, 255	51, 105, 118, 255	53, 92, 140, 254
	5	104, 132, 137, 255, 255	1, 17, 107, 128, 255	1, 1, 53, 57, 100	1, 4, 7, 61, 99	82, 103, 130, 251, 254	1, 7, 55, 82, 151
Test 5	2	90, 168	82, 164	90, 167	89, 169	90, 168	87, 166
	3	1, 68, 97	1, 43, 99	1, 68, 98	1, 72, 94	90, 128, 255	30, 79, 150
	4	1, 97, 98, 255	1, 96, 98, 255	1, 38, 79, 97	1, 43, 82, 97	1, 91, 94, 196	1, 57, 85, 130
	5	1, 1, 68, 68, 97	1, 5, 97, 100, 253	1, 64, 101, 116, 255	1, 1, 70, 70, 95	1, 54, 105, 107, 232	1, 39, 91, 97, 194
Test 6	2	0.59534	3.7494	1.3487	0.80151	1.247	119, 177
	3	16, 139, 154	16, 141, 156	16, 139, 154	16, 139, 154	16, 139, 154	16, 139, 154
	4	16, 154, 154, 231	16, 133, 136, 174	16, 150, 160, 230	16, 16, 139, 155	17, 114, 139, 173	16, 87, 138, 167
	5	16, 17, 139, 139, 154	16, 16, 90, 141, 143	16, 16, 141, 141, 153	16, 16, 138, 141, 155	22, 22, 115, 120, 129	16, 16, 122, 140, 150

TABLE 45.8 The P-Value and Mean difference of the PSNR, SSIM, execution time, and Fitness values for the proposed method.

Mean difference	P-Value	Algorithms	Proposed Method	Metric	Mean difference	P-Value	Algorithms	Proposed Method	Metric
-0.2620	0.39 (*)	HHO	MFWOA	execution time	-117.3716	0.032 (*)	HHO	MFWOA	Fitness values
-181.8310	0.023 (*)	EO			-149.1045	0.048 (*)	EO		
11.6823	0.031 (*)	WOA			-135.0909	0.045 (*)	WOA		
0.7902	0.039 (*)	MPA			-194.0841	0.05	MPA		
2.5068	0.046 (*)	MFO			-193.6614	0.05	MFO		
0.0020	0.045 (*)	HHO	MFWOA	SSIM	0.0585	0.044 (*)	HHO	MFWOA	PSNR
0.0009	0.029 (*)	EO			0.0306	0.038 (*)	EO		
0.0075	0.030 (*)	WOA			0.3249	0.035 (*)	WOA		
0.0033	0.036 (*)	MPA			0.2165	0.034 (*)	MPA		
0.0049	0.028 (*)	MFO			0.3168	0.031 (*)	MFO		

Table 45.5. As the value of k increases, the obtained SSIM for all photos by all algorithms has an ascending/descending trend. The value of SSIM does not increase with the increasing value of k. Instead, at some levels, the threshold decreases and then rises again. In general, the value of SSIM and PSNR at higher threshold levels is much higher than at lower threshold levels, but as the threshold levels increase, the values of PSNR and SSIM often fluctuate. For example, in this chapter, the value of SSIM at the threshold level $k = 32$ is higher than the lower threshold levels. However, it can be seen from Table 45.4 and Table 45.5 that the proposed MFWOA algorithm has better results than all other algorithms at all levels of the lower, middle, and upper levels, and this is because according to the combination of the WOA and MFO algorithms.

This chapter considers each image separately as an optimization problem and a search space for each MH algorithm and the proposed MFWOA method. The results in some cases may be different for each parameter, depending on the structure of the respective MH algorithm. Table 45.6 shows the execution time for different algorithms over 100 runs. For all images, WOA and EO have the minimum and maximum execution time at all threshold levels, respectively. Also, the execution time of MFWOA is longer than WOA and less than MFO because the combination of WOA and MFO is used in the proposed MFWOA method.

Inspecting the results in Table 45.6, EO is slower than other algorithms, and WOA is faster than other algorithms. MFWOA is then faster than HHO, EO, MFO, and MPA. HHO is also faster than EO, MFO, and MPA. MPA is also quicker than EO and MFO. MFO is also faster than EO. This difference is in the speed of operation of algorithms is the difference in their structure and the use of special operators that each algorithm has used to achieve the final answer. For example, the HHO has fewer parameters, low complexity, and high speed. MFO is also very accurate and is one of the efficient algorithms. The parameters and operators in any MH algorithm will determine the degree of convergence and efficiency. Early convergence may occur if these parameters and operators are not appropriately selected. WOA has a straightforward structure, and fewer parameters are used in its design. WOA uses more straightforward operators. Therefore, it has less CC and is faster. The EO algorithm is slow because it examines the various conditions to increase its performance and search the search space locally and globally. MPA also has a relatively high CC due to its structure.

HHO, MFO, EO, MPA, and WOA have also made the best use of parameters and operators to achieve the best answer. In any case, the proposed MFWOA algorithm has better performance in terms of SSIM and PSNR than the other five compared algorithms, and this is because it uses a combination, which in this case also has the high capability. Some algorithms take advantage of the exploration phase and the capacity of algorithms in the exploitation phase. Therefore, using the combination of features of WOA and MFO algorithms in the proposed MFWOA method is the main factor of MFWOA superiority over the compared algorithms. Table 45.8 shows the values of the obtained thresholds for the various images obtained by each algorithm after 100 runs. Figs. 45.5 to 45.10 show the Thresholding images of "Test1", "Test2", "Test3", "Test4", "Test5", and "Test6" that obtained by all algorithms, respectively.

Table 45.8 shows the P-Value and mean value for the metrics of the Fitness Function, SSIM, PSNR, and execution time. In particular, the P-Value indicates the probability of error in accepting the validity of the observed results, valid in the sense that the experimental result well represents the community. For example, a P-Value of 0.05 indicates a 5% probability that the relationship we observed in the sample is "accidental". The lower the P-Value, the higher the accuracy of our work and the lower the error rate. In this chapter, Hypothesis Zero assumes that there is no significant difference between the mean values of the algorithms. However, the alternative hypothesis considers a considerable difference between them. The

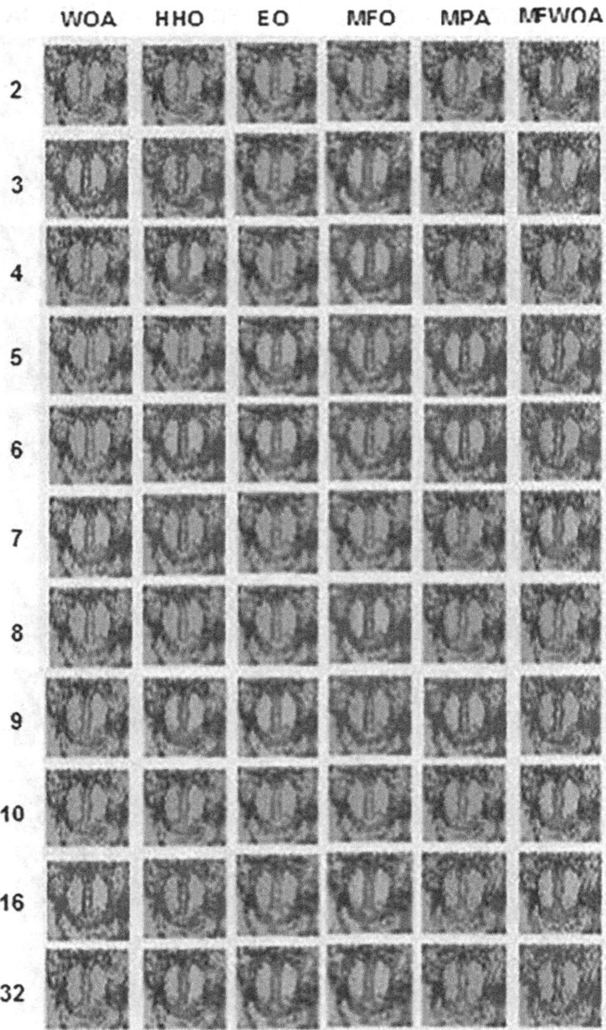

FIGURE 45.5 The Thresholding images of the "Test1" image obtained by all algorithms.

negative value difference in Table 45.8 indicates that the proposed algorithm performs worse than the compared algorithms in terms of the relevant evaluation metric. However, considering that in our proposed method, the threshold values are obtained by minimizing the Fitness Function, so the lower the value of the Fitness Function, the better the corresponding algorithm. So, if the value difference value for the Fitness Function metric is negative, the relevant algorithm has better performance than other algorithms. Also, the higher the mean difference per execution time for a negative algorithm, the faster the corresponding algorithm and the shorter the execution time than other algorithms.

Considering the results in Table 45.8, considering that the proposed algorithm has a negative value for Fitness Function values compared to different algorithms, the proposed algorithm is better than these algorithms in terms of evaluation metric values of the Fitness Function. Wherever the P-Value difference of the proposed algorithm with other algorithms is less than 0.05, it means that the performance of the proposed algorithm is generally better than the corresponding algorithms. In 18 cases, the P-Value difference of the proposed method with other algorithms is less than 0.05 and is indicated by (*). It shows a significant difference between the presented and compared algorithms, and the null hypothesis is incorrect.

Therefore, according to Table 45.8, the null hypothesis is rejected for 18 cases, and there is a significant difference between the proposed algorithm and other algorithms. In all three cases, the P-Value is 0.05, meaning there is a 5% chance that the relationship we observed in the sample is "accidental." In this chapter, based on Table 45.8, the P-Value difference of the proposed algorithm with other algorithms for SSIM is less than 0.05. Therefore, the null hypothesis for this metric is not accepted, and this shows a significant difference between the proposed algorithm and other comparative algorithms

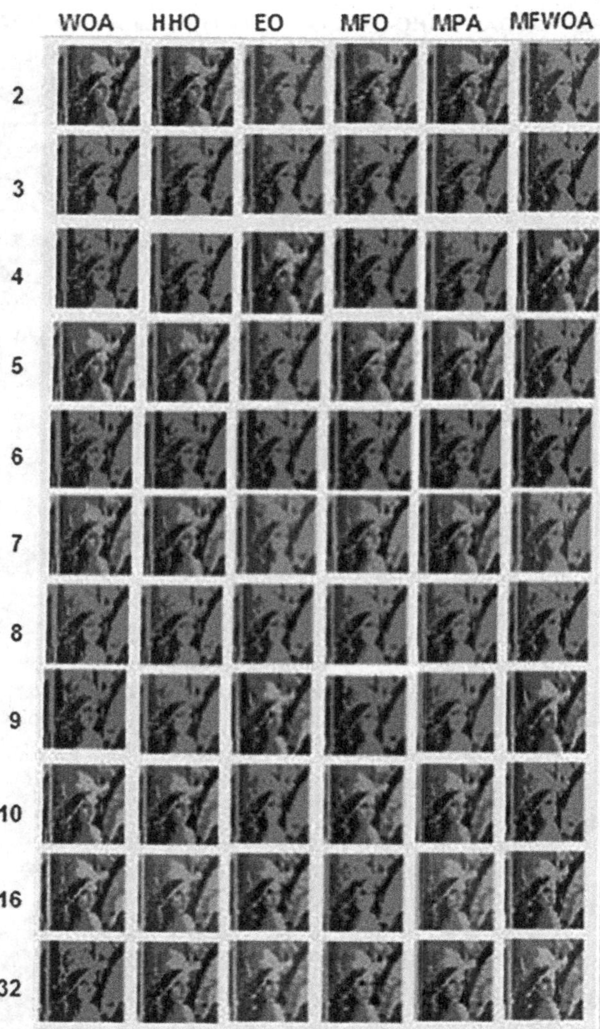

FIGURE 45.6 The Thresholding images of the "Test2" image obtained by all algorithms.

for this metric (SSIM). Also, the difference between the P-Value of the proposed algorithm and the HHO, EO, and WOA algorithms per Fitness values is less than 0.05. Therefore, hypothesis zero for this metric is not accepted. It shows a significant difference between the proposed algorithm and the corresponding algorithms for this metric (Fitness values). Also, for this metric, the value of the difference between the proposed method, MPA, and MFO algorithms is equal to 0.05. It shows that 5% may be "accidental" in the sample for the proposed algorithm and MPA and MFO algorithms.

Table 45.8 also revealed the difference between the P-Value of the proposed algorithm and the EO, WOA, MFO, and MPA algorithms per execution time is less than 0.05. Therefore, the zero hypotheses are rejected. It shows a significant difference between the proposed algorithm and these algorithms for this metric. Also, the difference between the P-Value of the proposed algorithm and the HHO algorithm per execution time metric is more than 0.05. Therefore, the alternative hypothesis is rejected, which means a significant difference between the proposed algorithm and this algorithm. The difference in the P-Value for the proposed algorithm and other algorithms for PSNR is less than 0.05. Therefore, the zero hypotheses are rejected. It shows a significant difference between the proposed algorithm and these algorithms for this metric. It should be noted that the alternative hypothesis is accepted in this chapter, and it is argued that there is a significant difference between the proposed algorithm and other algorithms. Because in most cases, the P-Value is less than 0.05. According to Table 45.9 and Table 45.10, the Mean value of the proposed MFWOA algorithm is positive with the other algorithms against the PSNR, Fitness Function, execution time and SSIM evaluation metric. The MFWOA algorithm performs better than the different algorithms. The mean difference value for both PSNR and SSIM evaluation metrics is

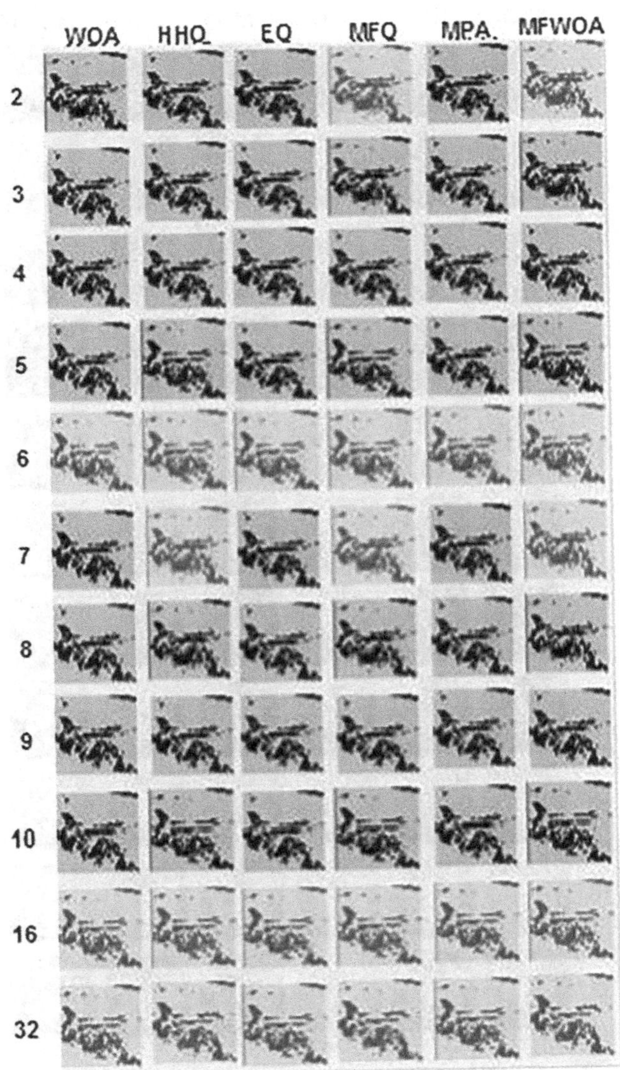

FIGURE 45.7 The Thresholding images of the "Test3" image obtained by all algorithms.

negative, meaning that the MFWOA algorithm performs worse than other algorithms. For the Fitness Function evaluation metric, given that we have used the minimization of this metric to obtain solutions, the lower the Fitness Function value for each algorithm, it can be said that the relevant algorithm is more efficient than other algorithms. Of course, it should be noted that in some cases, the value of the Fitness Function for MFWOA is slightly higher than different algorithms, but the results of PSNR and SSIM for the proposed MFWOA algorithm are better than other algorithms. So, anywhere in Table 45.9, the Mean difference is negative for the Fitness Function evaluation metric, meaning that the MFWOA algorithm performs better. Also, considering that the lower the value of execution time, the higher the speed, so for this metric (execution time), the difference between the proposed algorithm and other algorithms is negative, i.e., the proposed algorithm is faster than the algorithm has the desired.

As shown from Table 45.9, for the Fitness Function, for different images at different threshold levels, it has a positive Mean difference value in 35 cases, indicating that the Fitness Function value for MFWOA is higher than other algorithms. In 19 points, the value difference is zero, meaning that the value of the Fitness Function is the same for MFWOA and the corresponding algorithm. In other cases, the value difference is negative. As shown from Table 45.9, for the PSNR evaluation metric, the Mean difference value is negative in 5 cases, indicating that the PSNR value from the MFWOA is lower than the other algorithms. In 16 points, the Mean difference value is equal to zero, indicating that the PSNR value of the MFWOA is the same as the PSNR value of the other algorithms. In other cases, the mean difference value is positive,

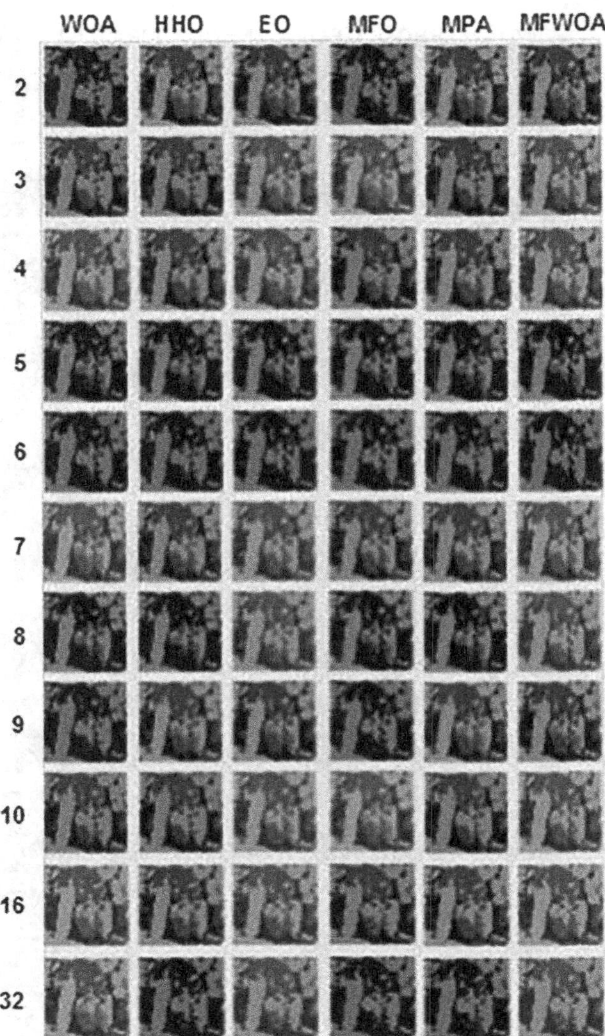

FIGURE 45.8 The Thresholding images of the "Test4" image obtained by all algorithms.

indicating the superiority of the proposed MFWOA algorithm over this metric (PSNR) is different from other algorithms. In the SSIM evaluation metric, Table 45.10 evidently shows that the Mean difference value is negative in one case only, indicating that the SSIM value obtained from MFWOA is less than the other algorithms. In one case, the Mean difference value is equal to zero, indicating that the SSIM value from the MFWOA is the same as the SSIM value from the other algorithms.

In other cases, the value difference is positive, indicating the superiority of the proposed MFWOA algorithm over this metric (SSIM) is different from other algorithms. For the execution time evaluation metric, for all cases, the Mean difference value of the two algorithms MFWOA and HHO for the Test1 and Test6 images is negative, indicating that the execution time value obtained from the MFWOA is lower than the HHO. The speedup in the MFWOA is higher than the HHO for the two corresponding images. The MFO is lower than other algorithms at those levels. In other cases, the mean difference is positive, indicating that the proposed MFWOA algorithm for this metric (execution time) is slower than different algorithms. In summary, the results of our experiments show that the use of MFWOA for MTIS is more efficient than other algorithms. However, WOA, compared to MFO, showed promising results for a few thresholds, while its performance is in most cases weaker than MFO (according to SSIM and PSNR results). It could be because the MFO can switch between the exploration and operation phases, which are the two main phases in any MH algorithm. MFOs better escape local optimization and early convergence and find more accurate answers to the problem. At the same time, WOA is trapped in the local optimization in the early stages of optimization and cannot find optimal global solutions in the

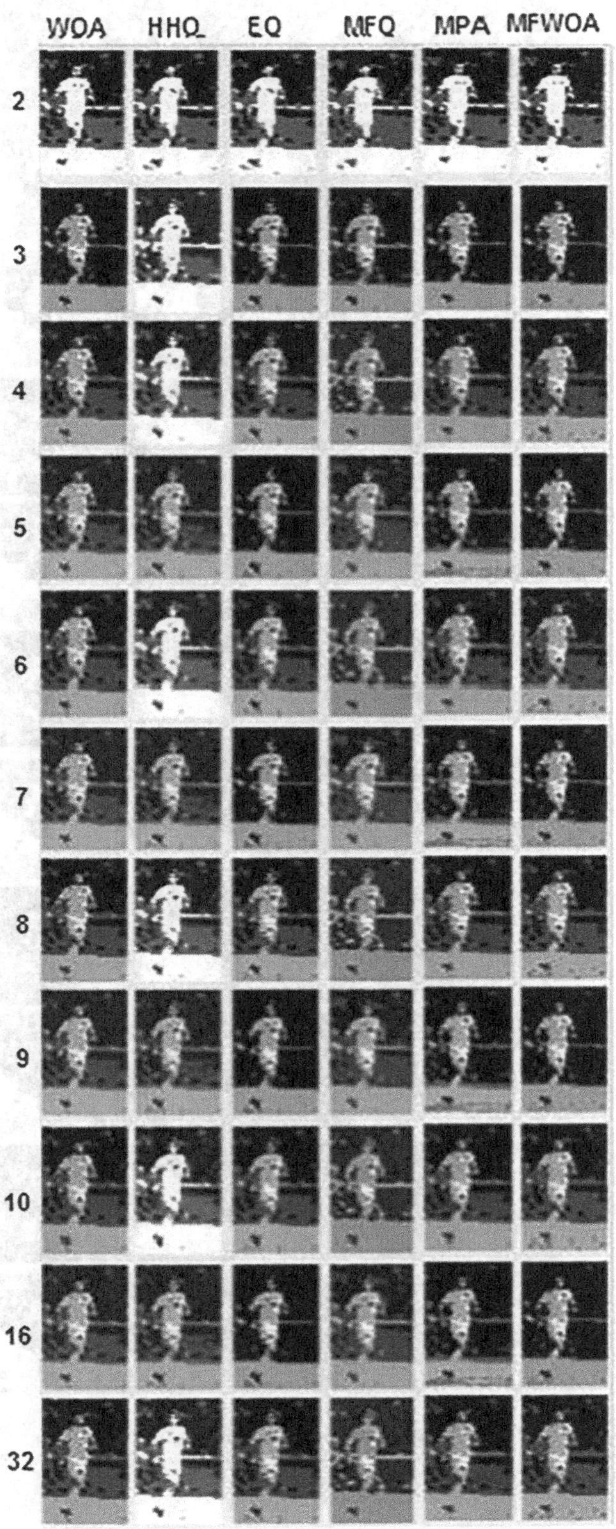

FIGURE 45.9 The Thresholding images of the "Test5" image obtained by all algorithms.

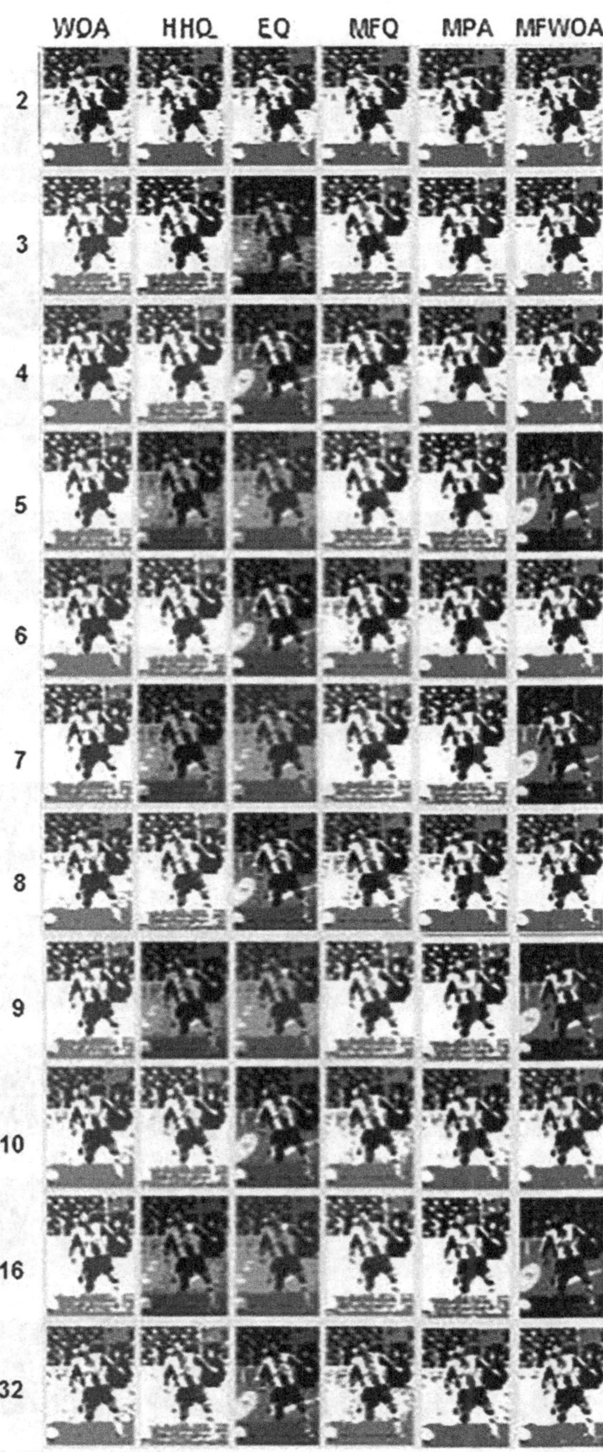

FIGURE 45.10 The Thresholding images of the "Test6" image obtained by all algorithms.

search space. Thus, combining MFO with WOA helps improve WOA and, after merging with MFO, makes WOA more capable of switching between exploration and operation phases and achieves better outputs. It should be noted that for all values obtained in Tables 45.9 and 45.10, significant difference at level P-Value <0.05.

TABLE 45.9 The Mean difference of the Fitness Function and PSNR values for the proposed method.

	k	PSNR					Fitness Function				
		HHO	EO	WOA	MPA	MFO	HHO	EO	WOA	MPA	MFO
Test 1	2	0.0040	0.0040	0.0060	0.0040	0.0040	0	0	-0.0001	0	0
	3	1.1750	0.0480	0.5470	0.0480	0.0480	-0.0003	-0.0305	-0.0238	-0.0305	-0.0305
	4	2.0190	0.4540	0.2540	4.3080	4.3080	0.0164	-0.1133	0.0150	-0.1255	-0.1255
	5	1.6500	3.0890	0.0010	3.0890	0.5660	-0.0106	-0.0211	0.0019	-0.0366	-0.0107
	6	0.2780	3.9530	4.4650	3.9530	3.9530	-0.2144	-0.1854	-0.1533	-0.1908	-0.1908
	7	0.6260	7.2740	1.7810	7.2740	5.0320	-0.0276	-0.1366	-0.0435	-0.1889	-0.1109
	8	0.1820	2.2340	5.0810	4.6690	2.2420	-0.0177	-0.0619	-0.0085	-0.1034	-0.0755
	9	2.3320	1.5490	0.5820	1.5490	2.6880	-0.1088	-0.1319	-0.0742	-0.2011	-0.1689
	10	0.0240	5.7590	2.1130	8.0020	3.1590	-0.0927	-1.2371	-0.7029	-1.3218	-0.7575
	16	4.4100	0.0150	1.3360	3.6410	3.7760	-0.2853	-1.5251	0.0205	-1.6353	-1.6333
	32	1.005	5.6990	11.2510	5.4190	1.3200	-2.2877	-4.8427	-1.8417	-4.1977	-3.5537
Test 2	2	0.0130	0.0130	0.0200	0.0130	0.0130	-0.0019	-0.0019	-0.0014	-0.0019	-0.0019
	3	1.2540	1.2540	0.1100	1.2540	1.2540	-0.0781	-0.0781	0.0018	-0.0781	-0.0781
	4	-0.3860	3.0200	3.0070	3.0200	3.0200	-0.0175	-0.2284	-0.2433	-0.2443	-0.2443
	5	1.6670	1.4300	0.8470	1.4090	1.4300	-0.1339	-0.1486	-0.1322	-0.1695	-0.1695
	6	3.5560	1.5740	0.0080	3.5890	3.5990	-0.0436	-0.3732	0.0069	-0.4250	-0.4209
	7	1.6740	0.4650	2.7270	3.8640	0.4650	0.0183	-0.1776	0.3704	-0.1694	-0.0884
	8	4.0400	3.2730	2.2250	4.1390	1.0100	-0.1286	-0.6394	-0.1508	-0.7195	-0.6362
	9	2.0350	0.2050	1.5330	3.6500	0.0810	-0.0550	-0.7324	-0.7259	-0.7300	-0.4879
	10	4.6560	3.3960	0.9150	3.2940	2.6930	-0.3603	-1.5982	-0.8984	-1.6427	-1.6405
	16	5.0890	2.6860	5.9650	4.7240	0.0440	-0.2835	-2.6545	-1.1287	-2.7849	-2.5384
	32	6.1060	1.7760	7.2120	0.4560	2.0110	-1.4720	-6.401	-2.076	-5.306	-4.475
Test 3	2	0	0	0.0130	0	0	-0.0010	-0.0010	-0.0006	-0.0010	-0.0010
	3	0	0	0	0	0	-0.0015	-0.0015	-0.0015	-0.0015	-0.0015
	4	-0.0500	3.2570	0.0020	3.2570	3.2570	-0.1234	-0.3414	-0.1228	-0.4000	-0.4000
	5	0.9030	2.9310	0.1110	2.9310	2.9310	-0.1579	-0.9373	-0.1571	-0.9373	-0.9373
	6	-0.8900	0.8390	0.8720	0.8390	0.8390	0.0082	0.0045	-0.0086	-0.0245	-0.0244
	7	2.1550	2.1810	0.0540	2.1810	2.9410	-0.8742	-0.8285	-0.0809	-0.8759	-0.3778
	8	1.5300	0.1580	1.5830	0.1580	2.6850	-0.1905	-0.5574	-0.5380	-0.3487	-0.6238
	9	0.1700	2.4020	0.3180	2.4020	2.4020	-0.2728	-1.1864	-0.5845	-1.4419	-1.4419
	10	0.9810	2.3360	0.1270	2.3360	2.3360	-0.8171	-1.1333	-0.7264	-1.1712	-1.1711
	16	1.5850	4.7490	4.7990	2.0750	2.0780	-1.5593	-2.8923	-1.3728	-2.4243	-2.4243
	32	5.4820	4.7010	3.9510	3.4790	5.6650	-3.474	-7.9760	-3.255	-6.162	-8.326
Test 4	2	0.0140	0.0140	0.0240	0.0140	0.0140	-0.1021	-0.1021	-0.1013	-0.1021	-0.1021
	3	0.5330	2.0320	2.2650	2.0320	2.0320	-0.1437	-0.2001	-0.1990	-0.2001	-0.2001
	4	0.3510	4.0510	0.3960	4.0510	4.0510	-0.1259	-0.5656	-0.1217	-0.6008	-0.6008
	5	1.8760	0.3940	0.6350	0.3930	0.4480	-0.1824	-0.2936	-0.3090	-0.3130	-0.2933
	6	4.0330	2.4190	0.0010	2.4190	3.8080	-0.2430	-1.0715	-0.6036	-1.0661	-1.0983
	7	0.0080	4.3940	4.0170	0.1110	4.3930	-0.0366	-1.2788	-1.5503	-1.0534	-1.5610
	8	1.0730	0.5410	3.6000	2.5000	3.8890	-0.0715	-0.8469	-0.7937	-0.7809	-0.8453
	9	2.6360	4.3750	3.7970	2.9780	4.3670	0.1933	-0.6587	0.0173	-0.8195	-0.8518
	10	4.8620	1.2790	0.0080	3.9040	2.0850	-0.0294	-1.2147	-0.4932	-1.0883	-0.1527
	16	2.1420	2.2210	4.9780	2.7780	5.0080	-0.7781	-2.7731	-0.1802	-2.9671	-1.7771
	32	1.3100	2.7700	0.9950	2.4570	0.0090	-2.7300	-8.5360	-0.937	-7.692	-6.665

continued on next page

TABLE 45.9 (*continued*)

	k	PSNR					Fitness Function				
		HHO	EO	WOA	MPA	MFO	HHO	EO	WOA	MPA	MFO
Test 5	2	0.0060	0.0060	0.0060	0.0100	0.0060	0	0	0	0	0
	3	0.5130	0.2000	0.2000	0.2000	0.0800	-0.0001	0	0	0	0
	4	0.3700	3.6160	0.0750	3.6160	3.6160	0.0040	-0.0483	-0.0100	-0.0524	0.0026
	5	4.1370	3.1690	2.4920	3.1690	0.0520	-0.0377	-0.0362	-0.0434	-0.0454	-0.0022
	6	2.0400	2.7190	0.0830	2.7360	2.7360	3.5918	3.4571	3.4720	3.4556	3.4547
	7	0.6750	0.0700	3.2250	3.9620	0.0700	-0.0175	-0.1117	-0.1377	-0.1066	-0.0945
	8	1.4070	1.5730	2.5280	4.4730	4.4560	-0.0537	-0.2730	-0.2652	-0.2919	-0.2919
	9	0.4020	0.0560	0.3700	3.2110	2.8680	-0.1449	-0.1416	-0.1221	-0.1814	-0.1493
	10	1.3330	1.0230	1.0110	0.8360	0.0610	-0.0126	-0.0235	0.0026	-0.0337	-0.0016
	16	5.0180	1.9490	0.2620	0.1250	2.7140	-0.0590	-0.5429	-0.3375	-0.3637	-0.5184
	32	8.7240	4.5170	9.6740	1.7760	1.7510	-0.0399	-0.5095	-0.0917	-0.4468	-0.0934
Test 6	2	0.0070	-0.0030	0	0.0070	0.0070	-0.0001	-0.0001	-0.0000	-0.0001	-0.0001
	3	0.0140	0.0240	0.1850	0.0240	0.0240	0.0052	0.0003	0.0000	-0.0001	-0.0001
	4	0.0650	3.1730	0.1250	3.1730	3.1730	-0.0001	-0.1985	-0.0000	-0.1985	-7.6985
	5	2.5630	0.5110	0.0190	3.1570	0.4600	0.0048	-0.0029	0.0023	-0.0054	0.0005
	6	0.4900	2.6490	1.4360	3.0350	3.0350	-0.0935	-0.2923	-0.1022	-0.3019	-0.3019
	7	2.7620	4.5250	0.2170	4.5250	0.4190	-0.0871	-0.2222	-0.0941	-0.2950	-0.0924
	8	0.0950	0.5270	1.7780	3.1970	3.1970	-0.0400	-0.0978	0.0080	-0.1020	-0.0020
	9	2.3440	2.8910	3.3390	2.8910	0.0770	0.0052	-0.1496	-0.0053	-0.2086	-0.0065
	10	0.3460	4.2910	1.4930	5.6290	1.7270	-0.0409	-0.1510	-0.0567	-0.3085	-0.1099
	16	0.6390	1.7800	-0.0730	3.5420	0.0970	-0.0249	-0.3581	-0.0734	-0.6221	0.0017
	32	3.758	2.2010	0.7650	0.0220	0.2250	-0.2349	-1.6868	0.0086	-1.4301	-0.2725

TABLE 45.10 The Mean difference of the SSIM and execution time for the proposed method.

	k	SSIM					Fitness Function				
		HHO	EO	WOA	MPA	MFO	HHO	EO	WOA	MPA	MFO
Test 3	2	0.0007	0.0007	0.0007	0.0007	0.0007	-1.8325	0.0007	0.0007	0.0007	0.0007
	3	0.0212	0.0009	0.0194	0.0009	0.0009	-5.1671	0.0009	0.0194	0.0009	0.0009
	4	0.0584	0.0113	0.0010	0.1506	0.1506	-6.0070	0.0113	0.0010	0.1506	0.1506
	5	0.0601	0.1018	0.0034	0.1018	0.0114	-6.6552	0.1018	0.0034	0.1018	0.0114
	6	0	0.1164	0.1148	0.1164	0.1164	-7.4688	0.1164	0.1148	0.1164	0.1164
	7	0.0028	0.2153	0.0085	0.2153	0.1244	-16.3971	0.2153	0.0085	0.2153	0.1244
	8	0.0056	0.0476	0.1194	0.1365	0.0502	-20.0317	0.0476	0.1194	0.1365	0.0502
	9	0.0814	0.0528	0.0016	0.0528	0.0745	-26.1704	0.0528	0.0016	0.0528	0.0745
	10	0.0037	0.1632	0.0369	0.2540	0.0684	-11.3673	0.1632	0.0369	0.2540	0.0684
	16	0.1608	-0.0019	0.0360	0.1343	0.1406	-15.5009	-0.0019	0.0360	0.1343	0.1406
	32	0.0242	0.1107	0.3270	0.1045	0.0274	-25.2787	0.1107	0.3270	0.1045	0.0274
Test 4	2	0.0013	0.0013	0.0106	0.0013	0.0013	0.0013	0.0013	0.0106	0.0013	0.0013
	3	0.0001	0.0001	0.0177	0.0001	0.0001	0.0001	0.0001	0.0177	0.0001	0.0001
	4	0.0058	0.0877	0.0877	0.0877	0.0877	0.0058	0.0877	0.0877	0.0877	0.0877
	5	0.0060	0.0159	0.0007	0.0146	0.0159	0.0060	0.0159	0.0007	0.0146	0.0159
	6	0.0334	0.0366	0.0000	0.0379	0.0405	0.0334	0.0366	0.0000	0.0379	0.0405
	7	0.0723	0.0224	0.0787	0.0726	0.0224	0.0723	0.0224	0.0787	0.0726	0.0224
	8	0.0327	0.0791	0.0339	0.0523	0.0305	0.0327	0.0791	0.0339	0.0523	0.0305
	9	0.0110	0.0068	0.0516	0.0583	0.0150	0.0110	0.0068	0.0516	0.0583	0.0150
	10	0.0301	0.0643	0.0005	0.0416	0.0378	0.0301	0.0643	0.0005	0.0416	0.0378
	16	0.0726	0.0768	0.1083	0.0799	0.0004	0.0726	0.0768	0.1083	0.0799	0.0004
	32	0.0616	0.0014	0.1244	0.0200	0.0178	0.0616	0.0013	0.1244	0.0200	0.0178

continued on next page

TABLE 45.10 *(continued)*

	k	SSIM					Fitness Function				
		HHO	EO	WOA	MPA	MFO	HHO	EO	WOA	MPA	MFO
Test 5	2	0.0000	0.0000	0.0017	0.0000	0.0000	0.0000	0.0000	0.0017	0.0000	0.0000
	3	0.0000	0.0000	0.0000	0.0000	0.0000	0.0000	0.0000	0.0000	0.0000	0.0000
	4	0.0050	0.0718	0.0045	0.0718	0.0718	0.0050	0.0718	0.0045	0.0718	0.0718
	5	0.0194	0.0538	0.0033	0.0538	0.0538	0.0194	0.0538	0.0033	0.0538	0.0538
	6	0.0071	0.0376	0.0372	0.0376	0.0376	0.0071	0.0376	0.0372	0.0376	0.0376
	7	0.0348	0.0358	0.0013	0.0358	0.0539	0.0348	0.0358	0.0013	0.0358	0.0539
	8	0.0474	0.0069	0.0364	0.0069	0.0362	0.0474	0.0069	0.0364	0.0069	0.0362
	9	0.0008	0.0417	0.0211	0.0417	0.0417	0.0008	0.0417	0.0211	0.0417	0.0417
	10	0.0360	0.0527	0.0009	0.0527	0.0527	0.0360	0.0527	0.0009	0.0527	0.0527
	16	0.0037	0.0707	0.0722	0.0141	0.0143	0.0037	0.0707	0.0722	0.0141	0.0143
	32	0.0764	0.0550	0.0789	0.0414	0.0786	0.0764	0.0550	0.0789	0.0414	0.0786
Test 6	2	0.0038	0.0038	0.0050	0.0038	0.0038	0.0038	0.0038	0.0050	0.0038	0.0038
	3	0.0038	0.0195	0.0259	0.0195	0.0195	0.0038	0.0195	0.0259	0.0195	0.0195
	4	0.0053	0.0727	0.0094	0.0727	0.0727	0.0053	0.0727	0.0094	0.0727	0.0727
	5	0.0357	-0.0008	0.0086	-0.0007	0.0006	0.0357	-0.0008	0.0086	-0.0007	0.0006
	6	0.0724	0.0633	0.0031	0.0633	0.0725	0.0724	0.0633	0.0031	0.0633	0.0725
	7	0.0028	0.1296	0.1078	0.0254	0.1296	0.0028	0.1296	0.1078	0.0254	0.1296
	8	0.0031	0.0200	0.0532	0.0651	0.0743	0.0031	0.0200	0.0532	0.0651	0.0743
	9	0.0589	0.0975	0.0688	0.0868	0.0959	0.0589	0.0975	0.0688	0.0868	0.0959
	10	0.0727	0.0384	0.0039	0.1000	0.0636	0.0727	0.0384	0.0039	0.1000	0.0636
	16	0.0017	0.0397	0.0652	0.0590	0.0660	0.0017	0.0397	0.0652	0.0590	0.0660
	32	0.0037	0.0801	0.0215	0.0603	0.0302	0.0027	0.0801	0.0215	0.0603	0.0302
Test 7	2	0.0000	0.0000	0.0000	0.0040	0.0000	0.0000	0.0000	0.0000	0.0040	0.0000
	3	0.0238	0.0100	0.0100	0.0100	0.0055	0.0238	0.0100	0.0100	0.0100	0.0055
	4	0.0105	0.1856	0.0010	0.1856	0.1856	0.0105	0.1856	0.0010	0.1856	0.1856
	5	0.2413	0.1796	0.1441	0.1796	0.0007	0.2413	0.1796	0.1441	0.1796	0.0007
	6	-0.0303	0.1336	-0.0068	0.1364	0.1364	-0.0303	0.1336	-0.0068	0.1364	0.1364
	7	0.0447	0.0042	0.1874	0.2229	0.0042	0.0447	0.0042	0.1874	0.2229	0.0042
	8	0.0012	0.0534	0.1131	0.2257	0.2229	0.0012	0.0534	0.1131	0.2257	0.2229
	9	0.0022	0.0016	0.0487	0.1848	0.1735	0.0022	0.0016	0.0487	0.1848	0.1735
	10	0.0007	0.0807	0.0902	0.0740	0.0201	0.0007	0.0807	0.0902	0.0740	0.0201
	16	0.0682	0.0487	0.0011	0.0021	0.0822	0.0682	0.0487	0.0011	0.0021	0.0822
	32	0.3552	0.1182	0.4228	0.0328	0.0382	0.3553	0.1182	0.4228	0.0328	0.0382
Test 8	2	0.0002	0.0002	0.0006	0.0002	0.0002	-1.9699	0.0002	0.0006	0.0002	0.0002
	3	0.0003	0.0507	0.0561	0.0507	0.0507	-5.6698	0.0507	0.0561	0.0507	0.0507
	4	0.0006	0.0991	0.0044	0.0991	0.0991	-6.4689	0.0991	0.0044	0.0991	0.0991
	5	0.0465	0.0103	0.0010	0.1039	0.0060	-21.1737	0.0103	0.0010	0.1039	0.0060
	6	0.0015	0.0590	0.0445	0.1048	0.1048	-27.9872	0.0590	0.0445	0.1048	0.1048
	7	0.0556	0.1393	0.0009	0.1393	0.0023	-20.9375	0.1393	0.0009	0.1393	0.0023
	8	0.0022	0.0050	0.0473	0.1020	0.1020	-10.4139	0.0050	0.0473	0.1020	0.1020
	9	0.0451	0.1024	0.1114	0.1024	0.0014	-12.1925	0.1024	0.1114	0.1024	0.0014
	10	0.0070	0.1769	0.1146	0.2128	0.0880	-12.5260	0.1769	0.1146	0.2128	0.0880
	16	0.0203	0.1090	0.0724	0.1926	0.0006	-23.0742	0.1090	0.0724	0.1926	0.0006
	32	0.0626	0.0557	0.0393	0.0307	0.0367	-47.6932	0.0557	0.0393	0.0307	0.0267

45.6 Conclusions

In this chapter, the problem of determining the optimal thresholds in a MTIS was considered as an optimization problem. So, a combination of WOA and MFO was used to improve the performance of WOA to solve the problem of MTIS that uses the Fitness Function minimization. Inverse Otsu was also employed as a Fitness Function in the MFWOA algorithm and other MH algorithms. The experimental results of the proposed MFWOA algorithm were compared with MPA, WOA, HHO, MFO, and EO algorithm on the eight different images using PSNR, SSIM, execution time, and Fitness Function evaluation metric. The results demonstrated that the MFWOA algorithm is better for all images regarding PSNR and SSIM than other algorithms. However, in terms of execution time, MFWOA seemed a little slower. Therefore, our proposed MFWOA algorithm performed better than other algorithms regarding PSNR, SSIM, segmentation time, and segmentation accuracy on the tested images. But in terms of execution time evaluation metric, the MFWOA algorithm is faster than WOA and slower than MFO. In some cases, the proposed algorithm was faster than other algorithms. Our next work is to use a combination of the WOA and the Artificial Neural Network (ANN) to improve the WOA as well as the MTIS problem as an optimization problem.

References

[1] E.H. Houssein, et al., A novel Black Widow Optimization algorithm for multilevel thresholding image segmentation, Expert Systems with Applications 167 (2021) 114159, https://doi.org/10.1016/j.eswa.2020.114159.

[2] R. Srikanth, K. Bikshalu, Multilevel thresholding image segmentation based on energy curve with harmony Search Algorithm, Ain Shams Engineering Journal 12 (1) (2021) 1–20, https://doi.org/10.1016/j.asej.2020.09.003.

[3] M. Abd El Aziz, et al., Whale optimization algorithm and moth-flame optimization for multilevel thresholding image segmentation, Expert Systems with Applications 83 (2017) 242–256, https://doi.org/10.1016/j.eswa.2017.04.023.

[4] L. Li, et al., Fuzzy multilevel image thresholding based on improved coyote optimization algorithm, IEEE Access 9 (2021) 33595–33607, https://doi.org/10.1109/ACCESS.2021.3060749.

[5] S. Mozafari, Provide a hybrid method to improve the performance of multilevel thresholding for image segmentation using GA and SA algorithms, in: IEEE-7th Conference on Information and Knowledge Technology (IKT), 2015, pp. 1–6, https://doi.org/10.1109/IKT.2015.7288751.

[6] G. Sun, et al., A novel hybrid algorithm of gravitational search algorithm with genetic algorithm for multi-level thresholding, Applied Soft Computing 46 (2016) 703–730, https://doi.org/10.1016/j.asoc.2016.01.054.

[7] B. Küçükuğurlu, E. Gedikli, Symbiotic organisms search algorithm for multilevel thresholding of images, Expert Systems with Applications 147 (2020) 113210, https://doi.org/10.1016/j.eswa.2020.113210.

[8] V.K. Bohat, K. Arya, A new heuristic for multilevel thresholding of images, Expert Systems with Applications 117 (2019) 176–203, https://doi.org/10.1016/j.eswa.2018.08.045.

[9] J. Kennedy, R. Eberhart, Particle swarm optimization, in: Proceedings of ICNN'95-International Conference on Neural Networks, vol. 4, IEEE, 1995, pp. 1942–1948, https://doi.org/10.1109/ICNN.1995.488968.

[10] S. Mirjalili, A. Lewis, The whale optimization algorithm, Advances in Engineering Software 95 (2016) 51–67, https://doi.org/10.1016/j.advengsoft.2016.01.008.

[11] S. Mirjalili, Moth-flame optimization algorithm: a novel nature-inspired heuristic paradigm, Knowledge-Based Systems 89 (2015) 228–249, https://doi.org/10.1016/j.knosys.2015.07.006.

[12] M. Abd Elaziz, et al., A competitive chain-based Harris Hawks Optimizer for global optimization and multi-level image thresholding problems, Applied Soft Computing 95 (2020) 106347, https://doi.org/10.1016/j.asoc.2020.106347.

[13] T. Salehnia, A. Fathi, Fault tolerance in LWT-SVD based image watermarking systems using three module redundancy technique, Expert Systems with Applications 179 (2021) 115058, https://doi.org/10.1016/j.eswa.2021.115058.

[14] H. Liu, Research on cloud computing adaptive task scheduling based on ant colony algorithm, Optik 258 (2022) 168677, https://doi.org/10.1016/j.ijleo.2022.168677.

[15] S. Raziani, et al., Selecting of the best features for the knn classification method by Harris Hawk algorithm, in: Conference: 8th International Conference on New Solutions in Engineering, Information Science and Technology of the Century ahead, 2021, https://civilica.com/doc/1196573/.

[16] E. Rodríguez-Esparza, et al., An efficient Harris hawks-inspired image segmentation method, Expert Systems with Applications 155 (2020) 113428, https://doi.org/10.1016/j.eswa.2020.113428.

[17] J. Anitha, et al., An efficient multilevel color image thresholding based on modified whale optimization algorithm, Expert Systems with Applications 178 (2021) 115003, https://doi.org/10.1016/j.eswa.2021.115003.

[18] L. Duan, et al., Multilevel thresholding using an improved cuckoo search algorithm for image segmentation, Journal of Supercomputing 77 (2021) 6734–6753, https://doi.org/10.1007/s11227-020-03566-7.

[19] T. Salehnia, et al., Multi-level image thresholding using GOA, WOA and MFO for image segmentation, in: 8th International Conference on New Strategies in Engineering, Information Science and Technology in the Next Century, Dubai, United Arab Emirates (UAE), Civilica, 2021, https://civilica.com/doc/1196572/.

[20] L. Samantaray, et al., A new Harris hawks-Cuckoo search optimizer for multilevel thresholding of thermogram images, Revue d'Intelligence Artificielle 34 (5) (2020) 541–551, https://doi.org/10.18280/ria.340503.

[21] Nobuyuki Otsu, A threshold selection method from gray-level histograms, IEEE Transactions on Systems, Man and Cybernetics 9 (1) (1979) 62–66, https://doi.org/10.1109/TSMC.1979.4310076.

[22] X.-S. Yang, Nature-Inspired Metaheuristic Algorithms, Luniver Press, 2008.

[23] Z. Xing, An improved emperor penguin optimization based multilevel thresholding for color image segmentation, Knowledge-Based Systems 194 (2020) 105570, https://doi.org/10.1016/j.knosys.2020.105570.

[24] A.A. Heidari, et al., Harris hawks optimization: algorithm and applications, Future Generation Computer Systems 97 (2019) 849–872, https://doi.org/10.1016/j.future.2019.02.028.

[25] A. Faramarzi, et al., Equilibrium optimizer: a novel optimization algorithm, Knowledge-Based Systems 191 (2020) 105190, https://doi.org/10.1016/j.knosys.2019.105190.

[26] A. Faramarzi, et al., Marine predators algorithm: a nature-inspired metaheuristic, Expert Systems with Applications 152 (2020) 113377, https://doi.org/10.1016/j.eswa.2020.113377.

Index

I

J

K

N

O

P

T

Printed in the United States
by Baker & Taylor Publisher Services